Process Automation Handbook

Jonathan Love

Process Automation Handbook

A Guide to Theory and Practice

Jonathan Love, MSc
School of Chemical Engineering and Advanced Materials
Newcastle University

Merz Court
Claremont Road
Newcastle upon Tyne
NE1 7RU
UK

http://www.ncl.ac.uk/pact/

British Library Cataloguing in Publication Data
Love, Jonathan
　Process automation handbook : a guide to theory and
　practice
　1. Process control 2. Automation
　I. Title
　670.4'27
ISBN-13: 9781846282812
ISBN-10: 1846282810

Library of Congress Control Number: 2007927990

ISBN 978-1-84628-281-2　　　e-ISBN 978-1-84628-282-9　　　Printed on acid-free paper

© Springer-Verlag London Limited 2007

Apart from any fair dealing for the purposes of research or private study, or criticism or review, as permitted under the Copyright, Designs and Patents Act 1988, this publication may only be reproduced, stored or transmitted, in any form or by any means, with the prior permission in writing of the publishers, or in the case of reprographic reproduction in accordance with the terms of licences issued by the Copyright Licensing Agency. Enquiries concerning reproduction outside those terms should be sent to the publishers.

The use of registered names, trademarks, etc. in this publication does not imply, even in the absence of a specific statement, that such names are exempt from the relevant laws and regulations and therefore free for general use.

The publisher makes no representation, express or implied, with regard to the accuracy of the information contained in this book and cannot accept any legal responsibility or liability for any errors or omissions that may be made.

9 8 7 6 5 4 3 2 1

Springer Science+Business Media
springer.com

This text is dedicated:

to my parents, David and Shiela, for their values,

to my wife, Astrid, for her trust,

to my daughters, Helen, Margaret and Katherine, for their hopes, and

to my grandchildren, Tiana and Stanley, for their laughter.

Overview

Foreword ix
Preface xi
Acknowledgements xv
Biography xix
Contents xxi

Part I: Technology and Practice

1 Introduction 3
2 Instrumentation 55
3 Final Control Elements 123
4 Conventional Control Strategies 147
5 Process Control Schemes 207
6 Digital Control Systems 277
7 Control Technology 377
8 Management of Automation Projects 463

Part II: Theory and Technique

9   Maths and Control Theory 565
10 Plant and Process Dynamics 699
11 Simulation 771
12 Advanced Process Automation 803
13 Advanced Process Control 963

Bibliography 1047
Abbreviations and Acronyms 1055
Index 1065

Foreword

The process industries are subject to increasing changes. These include globalization, increased earning per share expectation, increasing safety and environmental legislation, staff retirement, shortage of engineers and increasingly rapid evolution of technology.

To be successful, process manufacturing companies must respond by:

- Consistently improving safety and environmental performance
- Becoming more responsive to customers
- Reducing manufacturing costs
- Reducing working capital
- Improving equipment life

There are many approaches to achieving these improvements such as improvement of process equipment and flowsheets, staff training, work process improvement, organizational restructuring and, in particular, process automation.

Process automation covers the breadth of technologies that can be deployed to improve the performance of a plant including measurement, control, sequence automation, optimisation and decision support. These technologies have been deployed increasingly since the 1970s and the benefits are widely documented. Typical improvements include:

- Capacity increases of 2–5%
- Reduction of variable energy use by up to 15% or more
- Increases in yield of more valuable product of 2–10%

along with a breadth of other improvements such as reduced equipment wear and better manufacturing responsiveness. Payback times are often measured in weeks and months rather than years.

Many control texts focus on the theory and mathematics associated with control. This handbook, and the associated MSc course, whilst covering relevant theory, addresses the practical issues of applying these technologies in the process industries.

The text covers a wide range of topics from instrumentation and control systems to advanced control technologies as well as management aspects. It therefore provides an important new reference to those seeking to apply automation in the drive to improve business performance.

Andrew Ogden-Swift
Director
Honeywell Process Solutions, Southampton

Preface

Unlike the chapters of this text, which hopefully are entirely factual and objective, this preface is unashamedly coloured with the author's opinions and prejudices.

Scope of Subject

The terms "process" and "automation" mean different things to different people. In order to define the scope of this text, it is useful to attempt to articulate them in relation to the industry sector and to the nature of the operations carried out within it.

Process sector. This consists of those industries in which raw materials are processed. It is wide ranging and embraces oil, petrochemicals, agrochemicals, fine and speciality chemicals, dyestuffs, drugs and pharmaceuticals, biochemicals, food and drink, pulp and paper, plastics and films, synthetic fibres, iron and steel, minerals, cement, fuel and gas, power and water, nuclear processing, *etc.*

Economy. The importance of this sector to prosperity is not generally appreciated. The Chemical Industries Association reported (in January 2006) that the UK chemical industry, which is just *part* of this sector, had an annual turnover of some £32 billion (*i.e.* £3.2 × 10^{10}) with a trade surplus (exports over imports) of £4.5 billion and made capital investments of £1.9 billion. To put this into an international context, the sales of chemicals for the UK, EU and US were £32, £314 and £253 billion respectively.

Structure. The sector is essentially comprised of three categories of company: operating companies (end-users) who own the plant and undertake the manufacturing activity; contractors who design, build and commission the plant and provide project management capability; and suppliers (vendors) and integrators who produce the equipment from which the plant is built and who provide technical support and services.

Operations. Raw materials are processed in specialist plant and equipment using operations such as blending, distillation, filtration, reaction, *etc.* They invariably involve activities such as conveying, cooling, heating, mixing, pumping, storage, and so on.

Control. Process control concerns the design and specification of systems for the automatic control of process plant and equipment. It involves the measurement and manipulation of process variables such as flow, level, weight, temperature and composition. The objective is to drive measurements towards, and then maintain them at, a set of desired operating conditions. Modern process control is largely implemented by means of digital control systems and involves the application of a variety of techniques.

Automation. Process automation includes the immediate objectives of process control but also addresses the wider issues of enterprise management: operability, quality, reliability, safety and viability. This requires an integrated approach to plant operations, fundamental to which is real time data, the tools to extract information from that data, and the means for utilising the information.

Process automation is distributed throughout the sector. All but the smallest operating companies and contractors have whole groups dedicated to the automation aspects of projects and production. A large proportion of the supplier companies' sole business is instrumentation and control.

Control Viewpoint

It is interesting to take stock of the position of process automation in relation to the discipline as a whole.

From a design point of view, it has a major impact. Gone are the days when control was something of an afterthought, bolted onto the steady state plant design rather like some decoration on a Christmas tree. Gone also is the catalogue approach to control, a euphemism for the minimalist and cheapest solution. Nowadays, process automation is an early and integral part of plant design. Advanced control systems are seen as the primary means of reducing capital costs, increasing flexibility, maximising throughput, and so on.

From an operational point of view, control is central. It is inconceivable that any process plant of consequence would be built today without a comprehensive control system of some sort. It is through such control systems that plant is commissioned and through which all subsequent operations take place. Control systems provide the facility for making changes and for monitoring, controlling and optimising performance. In short, it is process automation that makes a plant or process come alive.

Through process monitoring, alarm and shut down systems, automation has a major part to play on the environment, protection and safety fronts. Indeed, for an existing plant, it is often the case that the only effective way to reduce its environmental impact or to enhance its safety is through the installation of additional measurements, typically of an analytical nature, linked in to alarms, trips and interlocks.

From a business point of view, automation is becoming the lead function. Control systems provide a platform of real time data for management information systems. Such systems, and the tools they support, enable decision making across a variety of applications such as production scheduling, plant utilisation, energy consumption, inventory control, performance monitoring, asset management and environmental audit.

On the technology front, the advances in electronics, digital devices and software are inexorable. These mostly find their way into instrumentation and control systems which make for a rapidly changing field. Functionality is forever increasing, leading to improvements across the board in reliability, flexibility and speed of response. This is in stark contrast to the evolution of pipes, pumps, exchangers, vessels and so on, the nuts and bolts from which the plant itself is built.

From a research point of view, whilst chemical engineering as a discipline is relatively mature, and all the obvious research has been done, there is much activity at its interfaces with other disciplines. Process automation is no exception. There is major interest and investment in the application of a range of advanced techniques, such as predictive control and optimisation, many of which are covered in the latter sections of the text.

Process automation can justly claim to be right at the heart of chemical engineering. The capital cost of instrumentation and control systems, as a percentage of the total cost of new plant, has steadily risen over the years. Typically from as little as 6% in the 1960s to some 12% at the millennium. Retrofits to existing plant invariably involve much higher percentage investments in automation. The continual striving for increased productivity, combined with the pressures due to legislation on the environmental and safety fronts, will ensure the trend continues. This is basically good news: there will never be a shortage of employment for control engineers. It is an exciting, rich and varied area to work in.

It is a simple fact that, in terms of the technology involved and the depth of understanding required, the design of a plant's control systems is usually more complex than the design of the plant itself. To function effectively, a control engineer requires a tremendous variety of technical expertise and practical experience. This not only embraces the technology of instrumentation and control systems, aspects of electrical and software engineering, and a host of underlying control theories and techniques, but also a good working knowledge of process dynamics and plant design. You can't design a process control system unless you understand how the plant works. Developing that understanding is both challenging and satisfying.

Approach

It is this variety that attracts personnel from a range of backgrounds into control engineering. The most common backgrounds are chemical and electrical engineering, but there are many chemists, physicists, mathematicians and computer scientists too. They are found throughout the sector working for suppliers in the development of systems, for contractors in their design, installation and commissioning, and for-end users in their operation and support. This text is aimed at all of them.

The philosophy has been to attempt to provide a sufficient coverage of most topics rather than an in-depth treatment of a few. There is a law of diminishing returns: something like 20% of the total knowledge available on any particular subject is probably sufficient to cope with 80% of all situations. A primary objective, therefore, has been to distil into a single volume the essentials of process automation in sufficient depth for most purposes. The challenge has been to strike a number of sensible balances: between breadth and depth, technique and technology, conventional and modern, theory and practice, information and understanding.

Fundamental to getting across that essential 20% is the presentation of key principles in as simple and concise a way as is practicable. An overly academic approach creates barriers: for that reason much of the usual mumbo-jumbo, red herrings and minutia have been deliberately filtered out. The keeping of feet on the ground and relating theory to practice is the paradigm.

The text is aimed at both advanced students and practising engineers. The 20/80 balance should enable students to get up the learning curve fairly quickly on quite a broad front. At the same time, it is intended to serve as a useful working reference for practising engineers. If the coverage of a particular topic isn't comprehensive or deep enough for the reader's purposes, there are many excellent specialist texts that are referenced to which the reader should refer. No apology is made for this: it is simply a consequence of striving for the 20/80 balance.

Getting that balance right is also important for another reason. There is a gulf between the perceptions of theoreticians and practitioners in control engineering, a gulf that is wider than in all other branches of engineering. It is all too easy for academics to get lost amongst the curly brackets and to dismiss implementation issues as being trivial: the simple truth is that theory is useless unless it can be related to practice in a meaningful way. Likewise, many practitioners are contemptuous of advanced control techniques on the basis of ignorance and perceptions that they don't work and cause too much grief. A primary objective of this text, therefore, is to help bridge that gulf.

The emphasis on practice stems from a strongly held conviction of the importance of doing the job properly. For that reason, aspects of what is generally considered to be good practice are woven into the text throughout. However, it must be recognised that this is not a substitute for experience: that can only be acquired through designing, building and operating real systems. It does nevertheless establish a basis for asking the right questions and provides a yardstick for interpreting the answers.

As far as changing technology is concerned, the text has been deliberately written in such a way that it stands the test of time. Wherever possible, the emphasis is on generic principles, common approaches and standards. In particular, it focuses on functionality and is not locked into any particular technology. The stance taken has been that, in general, issues such as processor power, display resolution, transmission rates, *etc.*, are all of little relevance provided they are sufficient for the purpose.

Acknowledgements

Throughout my career of over 30 years I have worked with many splendid colleagues, both academic and industrial, and some complete clowns. They, even the clowns, have subjected me to a rich variety of views and opinions which have significantly influenced the development of this text.

I am particularly indebted to:

- Several former Heads of Department, each of whom encouraged my appointment and then had the dubious privilege of having to manage me: Gordon Barwick at Huddersfield University, Colin McGreavy at Leeds University, Peter Fleming at Sheffield University and Julian Morris at Newcastle University.
  Through the space that they provided, I acquired the experience and expertise necessary for the task of writing this text. My dogged pursuit of this objective, coupled with an indifference to publishing papers and attending conferences, must have been a constant source of frustration.
- John Backhurst, a gentleman and former Head of Department at Newcastle University, whose friendship and support saw me through a particularly difficult period of my career.
- The School of Chemical Engineering and Advanced Materials (CEAM) at Newcastle University, of which I am part, contains the largest academic grouping of process control expertise in the UK. I am fortunate to be surrounded by such able and committed colleagues as Jarka Glassey, Elaine Martin, Gary Montague, Julian Morris, Ming Tham, Mark Willis and Jie Zhang.
  Without their support and that of Steve Bull, the current head of CEAM, much of what I have attempted to achieve would not have been possible. I hope that they can take some pride in the outcome: it should be something of real value to the industry to which we are all committed.
- The Partnership in Automation and Control Training (PACT) which is comprised of Newcastle University and a number of companies in the process sector. I am indebted to both Phil Males of Honeywell and David Witt, formerly of British Nuclear Fuels, for their leadership and to the many individuals within the PACT who have personally contributed source materials.
- The Integrated Graduate Development Scheme (IGDS) in Process Automation, which is offered through the PACT, has provided the framework for the development of this text. Interaction with the delegates has immeasurably improved my understanding of the subject.
  Paul Oram and Zaid Rawi of BP, Michael Phillipson of Foster Wheeler and Gerard Stephens of Corus have been especially diligent in proof reading and providing

comment: I appreciate the tact with which they pointed out mistakes in some of the drafts.
- The Process Control Subject Group of the IChemE, with which I have been associated since its formation in the early 1980s. The Group represents a significant proportion of the UK's process control community through which, to a large extent, I have established myself professionally and kept in touch with the real world. I have known many members of the Committee for a long time and value their loyalty and friendship. The current chair, Andrew Ogden-Swift of Honeywell, has been particularly supportive.
- Peter Burton, faithful friend, formerly Technical Director of ABB Kent Process Control and recently Manager of the PACT, from whom I have learnt most of what I know about the projects side of process automation. He has taken a personal interest in the development of the text and has been a source of independent and objective comment throughout.
- Brian Hoyle of Leeds University and Richard Wynne, formerly of Bournmouth University, whose friendship and interest goes back much further than the writing of this text.
- Last, but not least, my wife Astrid, who has loyally and devotedly "kept the ship afloat" at home over the decade or so that it has taken to write this text. Many evenings and weekends have been spent hidden away with my papers and laptop. I have consistently evaded domestic, family and social duties and am clearly going to struggle to come up with a better excuse in the future.

Contributors

Many colleagues and friends have contributed source materials that I have made use of in writing the chapters, as listed in the table below. To a large extent the materials have been used simply to help me better understand the topics concerned and to provide ideas on how best to present them. Where they have been adapted and edited by myself, I take full responsibility for any mistakes introduced.

To all of these people I say thank you. Not only have they contributed source materials but most have also willingly given their time and effort in discussing the content and reviewing the chapters. I trust that the list is complete and does justice to their contribution.

Contributor	Affiliation	Chapters
Jim Anderson	ICI (formerly)	35, 117
Roger Ashley	Zeneca (formerly)	63, 64
Steve Barber	BP	51, 54
Matthew Bransby		58
Richard Burkett	BP (formerly)	35, 59
Peter Burton	ABB (formerly)	18, 37, 49, 50, 60–67, 100
Andy Chipperfield	Southampton University	105, 110
Tony Cornah	Sheffield University (formerly)	99

Contributor	Affiliation	Chapters
Peter Fleming	Sheffield University	103, 110
Jarka Glassey	Newcastle University	107
Mike Hadfield	British Nuclear Group	65
Chris Hawkins	MDC Technology	106
Bruce Jerrit	BP (formerly)	106
Myke King	Whitehouse Consulting	35
Elaine Martin	Newcastle University	83, 101, 102
Frank McKenna	Catena Associates	53, 55–57
Gary Montague	Newcastle University	109, 112, 113, 117
Julian Morris	Newcastle University	101, 102
Andrew Ogden-Swift	Honeywell Process Solutions	100, 106
Martin Pitt	Sheffield University	17
Barrie Reynolds	Honeywell Control Systems	53, 55, 56
Ming Tham	Newcastle University	114–116
Ed Thorn	ABB (formerly)	40
Andrew Trenchard	Honeywell Process Solutions	59
Mark Willis	Newcastle University	104, 109, 111, 117, 118
David Wright	IChemE	67

Biography

Jonathan Love has MSc degrees in both chemical engineering from Newcastle University and automatic control from Manchester University.

A nomadic academic career has seen him working his way around the Universities in the north of England with lectureships at Huddersfield, Leeds, Sheffield and Newcastle. This has been interspersed with periods of recalibration in industry, initially as a commissioning engineer with Akzo Nobel, later as a control engineer with Honeywell and currently as a consultant with BP.

He is essentially an industrialist in academic's clothing and has consistently taken an interest in the application side of process control and automation: in the design and development of systems, in the management of projects, in the technology and its implementation, and in practice. This is all underpinned by a belief in the importance of being able to practice what you preach.

Over the last decade or so he has been responsible for establishing and managing the UK's national Integrated Graduate Development Scheme (IGDS) in process automation. This is based at Newcastle University and is offered through the Partnership in Automation and Control Training (PACT) which involves the University and a consortium of major companies in the process sector.

Professionally he is active within the UK process control community through the Process Control Subject Group of the IChemE of which he is a founder member and former chair. He is also member of the IET.

At a personal level, he subscribes to the protestant work ethic and has liberal principles. He is interested in current affairs, appreciates culture and enjoys travelling, especially in Europe. His principal recreations are sailing on the Mediterranean and skiing in the Alps.

June 2007

Contents

Part I: Technology and Practice

Section 1: Introduction

1 **Summary** 5

2 **P&I Diagrams** 9
    2.1   Bubbles and Symbols 9
    2.2   Tag Numbers 11

3 **Block Diagrams** 15
    3.1   Signals 15
    3.2   Structure 16
    3.3   Sub-systems 17
    3.4   Modes of Operation 17

4 **Signals** 19
    4.1   Signal Categories 19
    4.2   Standard Ranges 20
    4.3   Linearity 21

5 **Pneumatics** 23
    5.1   Air Supply 23
    5.2   Pilot Valves 23
    5.3   Flapper Nozzles 24
    5.4   Pneumatic Relays 24
    5.5   Force Balance 25

6 **Electronics** 27
    6.1   Semiconductors 27
    6.2   Diodes 28
    6.3   Transistors 28
    6.4   Logic Gates 30
    6.5   Flip-Flops 31
    6.6   Counters, Registers and Memory 32
    6.7   Operational Amplifiers 33

7	**Data Typing**		35
	7.1	Binary Representation	35
	7.2	Two's Complement	36
	7.3	Characters	37
	7.4	Identifiers	37
	7.5	Integer Literals	37
	7.6	Real Literals	38
	7.7	Time Literals	38
	7.8	String Literals	39
	7.9	Boolean Literals	39
	7.10	Derived Data Construct	40
	7.11	Array Data Construct	40
	7.12	Enumerated Data Construct	40
	7.13	Structured Data Construct	40
	7.14	Comments	41
8	**Structured Text**		43
	8.1	IEC Software Model	43
	8.2	Global and External Variables	44
	8.3	Input and Output Variables	44
	8.4	Memory Variables	45
	8.5	Miscellaneous	45
	8.6	Program Structure	45
	8.7	Specimen Program	46
	8.8	Comments	47
9	**Microprocessors**		49
	9.1	Overview	49
	9.2	The Control Unit	50
	9.3	The Arithmetic/Logic Unit	51
	9.4	The Memory System	51
	9.5	The Processor Cycle	52

Section 2: Instrumentation

10	**Characteristics**		57
11	**DP Cells**		61
	11.1	Diaphragm Capsules	61
	11.2	Pneumatic DP Cell	61
	11.3	Electronic DP Cell	62
	11.4	Calibration	63
	11.5	Installation	65
	11.6	Commissioning	66

12	**Flow: Orifices**	67
	12.1 Construction	67
	12.2 Operation	68
	12.3 Design Considerations	69
	12.4 Flow Equations	69
	12.5 Worked Example	70
	12.6 Specification	71
	12.7 Installation	72
	12.8 Nomenclature	75
13	**Flow Measurement**	77
	13.1 Overview	77
	13.2 Sight Glasses	78
	13.3 Rotameters	79
	13.4 Gap Meter	80
	13.5 Turbine Flowmeters	81
	13.6 Electromagnetic Flowmeters	82
	13.7 Ultrasonic Flowmeters	83
	13.8 Vortex Shedding Meters	84
	13.9 Mass Flowmeters	85
14	**Level Measurement**	87
	14.1 Gauge Glass	87
	14.2 Direct Use of DP Cells	87
	14.3 Use of DP Cells for Density Measurement	89
	14.4 Use of DP Cells for Interface Measurement	89
	14.5 Pneumercators	90
	14.6 Capacitance	91
	14.7 Ultrasonics	91
	14.8 Nucleonics	92
	14.9 Nomenclature	92
15	**Weight Measurement**	93
	15.1 Resistance Effects	93
	15.2 Strain Gauges	94
	15.3 Load Cells	95
	15.4 Installation	96
16	**Temperature Measurement**	99
	16.1 Thermocouple Principles	99
	16.2 Thermocouple Types	100
	16.3 Thermocouple Installation	100
	16.4 Resistance Temperature Devices	102
	16.5 RTD Construction	103
	16.6 RTD Installation	103

	16.7	Thermowells	104
	16.8	Comment	105
17	**pH Measurement**		107
	17.1	Nature of pH	107
	17.2	Strong Acids and Alkalis	108
	17.3	Weak Acids and Bases	109
	17.4	Mixing Effects	109
	17.5	The Glass Electrode	110
	17.6	Practical Issues	111
	17.7	Comments	111
18	**Chromatography**		113
	18.1	Principle of Chromatography	113
	18.2	Column Design	114
	18.3	Katharometer	115
	18.4	Flame Ionisation Detector	116
	18.5	Sliding Plate Valve	117
	18.6	Sampling and Conditioning	118
	18.7	Column Operation	118
	18.8	Multiple Columns	119
	18.9	Calibration and Signal Processing	121
	18.10	Comments	121

Section 3: Final Control Elements

19	**Valves and Actuators**		125
	19.1	Valve Types	125
	19.2	Globe Valves	126
	19.3	Butterfly Valves	129
	19.4	Ball Valves	129
	19.5	Diaphragm Valves	130
	19.6	Pneumatic Actuators	130
	19.7	Valve Selection	132
20	**Valve Sizing**		133
	20.1	Inherent Characteristic	133
	20.2	Installed Characteristic	134
	20.3	Comparison of Installed Characteristics	135
	20.4	Worked Example No 1	135
	20.5	Trim Selection	136
	20.6	Valve Sizing	136
	20.7	Worked Example No 2	137
	20.8	Critical Flow	139
	20.9	Nomenclature	141

21	**Valve Positioners**	143
21.1	Operation of Positioner	143
21.2	Uses of Positioner	143
21.3	Bias Example	144
21.4	Split Range Example	144

Section 4: Conventional Control Strategies

22	**Feedback Control**	149
22.1	Feedback Principles	149
22.2	Deviation Variables	150
22.3	Steady State Analysis	151
22.4	Worked Example	152
22.5	Nomenclature	153

23	**PID Control**	155
23.1	Single Loop Controller	155
23.2	Proportional Action	156
23.3	Integral Action	158
23.4	Derivative Action	159
23.5	Bumpless Transfer	160
23.6	Derivative Feedback	161
23.7	Integral Windup	161
23.8	Worked Example	162
23.9	Other Analogue Forms of PID Controller	162
23.10	Discretised Form of PID	162
23.11	Incremental Form of PID	163
23.12	Mnemonics and Nomenclature	163
23.13	Summary of Control Actions	163

24	**Controller Tuning**	165
24.1	Stability	165
24.2	Marginal Stability	167
24.3	Continuous Cycling Method	168
24.4	Reaction Curve Method	169

25	**Cascade Control**	173
25.1	Master and Slave Loops	173
25.2	Cascade Control of Heat Exchanger	174
25.3	Cascade Control of Jacketed Reactor	175
25.4	Implementation	177

26	**Ratio Control**	179
26.1	Scaling Approach	179
26.2	Direct Approach	180

	26.3	Indirect Approach	181
	26.4	Comments	182
27	**Feedforward Control**		183
	27.1	Feedforward Compensation	183
	27.2	Dynamic Compensation	184
	27.3	Feedforward Control	186
	27.4	Feedforward Control of a Heat Exchanger	187
	27.5	Implementation Issues	188
	27.6	Comments	189
	27.7	Nomenclature	189
28	**On-Off Control**		191
	28.1	On-Off Cycling	191
	28.2	On and Off Curves	192
	28.3	Lag Effects	193
	28.4	Worked Example	194
	28.5	Comments	194
	28.6	Nomenclature	194
29	**Sequence Control**		195
	29.1	A Charging Operation	195
	29.2	Simple Sequence	198
	29.3	Decision Trees and Tables	199
	29.4	Parameter Lists	200
	29.5	Timing Considerations	201
	29.6	Sequential Function Chart	201
	29.7	Parallelism	204
	29.8	Top Down Approach	205
	29.9	Comments	206

Section 5: Process Control Schemes

30	**Determining Strategy**		209
	30.1	Process of Determination	209
	30.2	Mixing Example	210
	30.3	Crude Oil Separator Example	211
	30.4	Absorber Example	212
	30.5	Flash Drum Example	214
	30.6	Methodology	215
	30.7	Comments	216
31	**Evaporator Control**		217
	31.1	Operational Issues	217
	31.2	The Vacuum System	218

	31.3	Reboiler and Separator	219
	31.4	Condenser with Barometric Leg	220
	31.5	Control Scheme	221
	31.6	Comments	223
	31.7	Nomenclature	223
32	**Control of Heat Exchangers**	225
	32.1	Heat Transfer Considerations	225
	32.2	Temperature Difference	226
	32.3	Overall Coefficient	226
	32.4	By-Pass Control	227
	32.5	Surface Area	228
	32.6	Comments	229
	32.7	Nomenclature	229
33	**Boiler Control**	231
	33.1	Basis of Control	232
	33.2	Signal Selection	233
	33.3	Ratio Trimming	234
	33.4	Comments	235
34	**Neutralisation Control**	237
	34.1	Classical Approach	237
	34.2	In-line Neutralisation	239
	34.3	Anti-Logging Approach	240
35	**Distillation Control**	243
	35.1	Strategic Considerations	243
	35.2	Inventory Considerations	244
	35.3	Cut and Separation	245
	35.4	Worked Example	246
	35.5	Measured Variables	246
	35.6	Mass and Energy Balance Control	248
	35.7	Control of Overhead Composition	249
	35.8	Dual Control of Overhead and Bottoms Composition	250
	35.9	Control of Column Feed	252
	35.10	Control of Column Pressure	253
	35.11	Internal Reflux Control	255
	35.12	Override Control	255
	35.13	Multicomponent Distillation	256
	35.14	Multicolumn Separations	257
	35.15	Comments	259
	35.16	Nomenclature	260

36	**Reactor Control**		261
	36.1	Steady State Considerations	261
	36.2	Dynamic Considerations	263
	36.3	Temperature Control Schemes	264
	36.4	Comments	264
	36.5	Nomenclature	266
37	**Batch Process Control**		267
	37.1	Variety of Plant Design	267
	37.2	Plant Structure	268
	37.3	Physical Model	269
	37.4	Process Model	270
	37.5	Procedural Model	270
	37.6	Methodology	272
	37.7	Recipe Model	274
	37.8	Activity Model	274
	37.9	Comments	276

Section 6: Digital Control Systems

38	**Systems Architecture**		279
	38.1	Advisory Control	279
	38.2	Supervisory Control	280
	38.3	Direct Digital Control	280
	38.4	Integrated Control	281
	38.5	Distributed Control	282
	38.6	Programmable Logic Controllers	283
	38.7	Supervisory Control and Data Acquisition	283
	38.8	Management Information Systems	284
	38.9	Computer Integrated Manufacturing	284
	38.10	Open Systems	286
	38.11	Comments	286
39	**Systems Hardware**		287
	39.1	Operator Station	287
	39.2	Physical Structure	288
	39.3	Card Types	290
	39.4	Personal Computers	291
40	**Communications**		293
	40.1	Messages	293
	40.2	Local Area Networks	294
	40.3	Token Systems	295
	40.4	Protocols	296
	40.5	Network Access	297

	40.6	Transmission	298
	40.7	Telemetry	298
	40.8	Radio Communications	299
	40.9	Telemetry Units	300
	40.10	Comments	300
41	**Software Overview**		**301**
	41.1	Operating System	301
	41.2	Operator's Control Program	302
	41.3	Direct Digital Control	302
	41.4	Archive	302
	41.5	Sequence Executive	303
	41.6	Batch Process Control	303
	41.7	High Level Language	303
	41.8	Engineers Control Program	303
	41.9	Distribution of Packages	305
	41.10	Comments	305
42	**Operator Interface**		**307**
	42.1	Access	307
	42.2	Tag Numbers	307
	42.3	Reserved Areas	308
	42.4	Display Systems	308
	42.5	Mimic Diagrams	309
	42.6	Group Displays	311
	42.7	Trend Diagrams	314
	42.8	Status Displays	314
	42.9	Text Displays	315
	42.10	Diagnostic Displays	316
	42.11	Keyboard Functions	316
43	**Integrated Alarm Environment**		**317**
	43.1	Database	317
	43.2	Display Systems	318
	43.3	Alarm Lists	319
	43.4	Integrated Safety Environment	319
	43.5	Alarm Management Policy	319
	43.6	Comments	320
44	**Analogue Control Loop**		**321**
	44.1	Input Interface	322
	44.2	Input Scaling	323
	44.3	Filtering	323
	44.4	PID Control	323
	44.5	Output Scaling	323

	44.6	Output Interface	324
	44.7	Nomenclature	325

45 Database Operations and Structure 327
	45.1	Input Scaling	327
	45.2	Filtering	329
	45.3	PID Control	329
	45.4	Output Scaling	330
	45.5	Database Structure	330

46 Discrete I/O Devices 333
	46.1	Output Interface	333
	46.2	Field Instrumentation	334
	46.3	Input Interface	335
	46.4	Input Processing	335
	46.5	Output Processing	336

47 Programmable Logic Controllers 337
	47.1	Ladder Constructs	337
	47.2	Ladder Execution	340
	47.3	Functions and Function Blocks	341
	47.4	Timers and Counters	342
	47.5	Development Environments	343
	47.6	Comments	343

48 Configuration 345
	48.1	Functions	345
	48.2	Function Blocks	346
	48.3	User Defined Function Blocks	348
	48.4	Nesting of Function Blocks	348
	48.5	Function Block Diagrams	349
	48.6	Comments	350
	48.7	Compliance	350

49 Open Systems 351
	49.1	Integration of Control and Information Systems	351
	49.2	Standards and Enabling Technologies	352
	49.3	Architecture	352
	49.4	Data Objects	353
	49.5	Object Linking	353
	49.6	Open Process Control	354
	49.7	OPC and the Internet	355
	49.8	Openness and Security	355
	49.9	Information and Control Security	356
	49.10	Information Security Management	357
	49.11	Firewalls	359

	49.12	Demilitarised Zones	360
	49.13	Malware Summary	361
	49.14	Anti-Virus Software	362
	49.15	Comments	363
50	**Fieldbus**		365
	50.1	HART Protocol	365
	50.2	Objectives	366
	50.3	History	367
	50.4	Physical Layer	368
	50.5	Data Link Layer	370
	50.6	Application and User Layers	371
	50.7	Hardware Configuration	372
	50.8	Software Configuration	372
	50.9	Function Block Attributes	373
	50.10	Project Management	373
	50.11	Benefits	374
	50.12	Comments	375

Section 7: Control Technology

51	**System Layout**		379
	51.1	Conventional Layout	379
	51.2	Power Supply	382
	51.3	Segregation Policy	383
	51.4	Air Supply	383
	51.5	Comments	384
52	**Intrinsic Safety**		385
	52.1	Intrinsic Safety	385
	52.2	Hazardous Areas	386
	52.3	Gas Group	388
	52.4	Temperature Class	388
	52.5	Common Explosive Gas-Air Mixtures	389
	52.6	Types of Protection	389
	52.7	Certification	389
	52.8	Flameproof Enclosures	390
	52.9	Pressurisation and Purge	390
	52.10	Other Types of Protection	391
	52.11	Ingress Protection	391
	52.12	Barriers	392
	52.13	Certification of Barriers	393
	52.14	ATEX Directives	393

53	**Reliability Principles**	397
	53.1 Unrepairable Systems	397
	53.2 Repairable Systems	398
	53.3 Proof Testing	399
	53.4 Elements in Series and Parallel	400
	53.5 Common Mode Failure	401
	53.6 Voting Systems	401
	53.7 Standby Systems	402
	53.8 Protection Systems	403
	53.9 Worked Example	404
	53.10 Hazard and Demand Rates	405
	53.11 Comments	406
54	**Hazard Analysis**	407
	54.1 HAZOP Studies	408
	54.2 Limitations of HAZOP	409
	54.3 CHAZOP Studies	410
	54.4 The Need for COOP Studies	413
	54.5 Control and Operability Studies	413
	54.6 Failure Mode and Effect Analysis (FMEA)	416
	54.7 Fault Tree Analysis (FTA)	418
	54.8 Minimum Cut Sets	418
	54.9 Fault Tree Evaluation	421
	54.10 Evaluation of Demand Rate	422
55	**Layers of Safety**	425
	55.1 Concept of Layers	426
	55.2 Passive Systems Layer	426
	55.3 Active Systems Layer	427
	55.4 Control Systems Layer	428
	55.5 The HSE Guidelines	428
	55.6 The EEMUA Guidelines	429
	55.7 Comments	431
56	**Protection Systems**	433
	56.1 Risk	433
	56.2 ALARP	434
	56.3 Safety Methodology	435
	56.4 Safety Integrity Levels	437
	56.5 Worked Example on Quantitative Approach to SIL	438
	56.6 Qualitative Approach to SIL	441
	56.7 Worked Example on Qualitative Approach to SIL	442
	56.8 Architectural Constraints	443
	56.9 Categories of Failure	444
	56.10 Protection System Design	445

	56.11	Independence	446
	56.12	Comments	447

57	**Safety Equipment**	449
	57.1 Equipment Approvals	449
	57.2 Failure Modes	451
	57.3 Voting Systems	451
	57.4 Electromagnetic Relay Logic	452
	57.5 DC Coupled Logic	452
	57.6 Dynamic Logic	453
	57.7 DCS and PLC Systems	453
	57.8 Special Purpose PES	453

58	**Human Factors**	457
	58.1 Operator's Role	457
	58.2 Psychological Model	458
	58.3 Anthropometrics	459
	58.4 Display Systems	459
	58.5 Alarm Systems	460
	58.6 Comments	461

Section 8: Management of Automation Projects

59	**Costs and Benefits Analysis**	465
	59.1 Life Cycle	465
	59.2 Conceptual Design	466
	59.3 Methodology	467
	59.4 Sources of Benefit	468
	59.5 Benefits for Continuous Plant	469
	59.6 Worked Example	470
	59.7 Benefits for Batch Plant	470
	59.8 Estimating Benefits	471
	59.9 Categorisation of Costs	471
	59.10 Estimating Costs	472
	59.11 Payback and Cash Flow	475
	59.12 Comments	478

60	**User Requirements**	479
	60.1 Process Description	479
	60.2 The P&I Diagrams	480
	60.3 Control Philosophy	480
	60.4 Particular Requirements	483
	60.5 General Considerations	486
	60.6 Golden Rules	487

61	**Tendering and Vendor Selection**		489
	61.1	Tender Generation	489
	61.2	Compliance Commentary	490
	61.3	Quotation	490
	61.4	Hardware Estimates	491
	61.5	Software Estimates	491
	61.6	Worked Example	492
	61.7	Tender Analysis and Vendor Selection	493
62	**Functional Specifications**		495
	62.1	Methodology	495
	62.2	Participants	496
	62.3	Contents of DFS	496
	62.4	Acceptance	497
63	**Design, Development, Testing and Acceptance**		499
	63.1	Waterfall Model	499
	63.2	Structured Programming	501
	63.3	Software Design	502
	63.4	Module Design	503
	63.5	Walkthroughs	503
	63.6	Software Development	503
	63.7	Support Tools	504
	63.8	Test Specifications	504
	63.9	Test Methods	505
	63.10	Module Testing	507
	63.11	Software Integration	507
	63.12	System Integration	508
	63.13	Acceptance Testing	508
	63.14	Project Management	508
	63.15	Personnel Management	510
	63.16	Comments	511
64	**Installation and Commissioning**		513
	64.1	Time Scale	513
	64.2	Installation	514
	64.3	Calibration of Field Instrumentation	515
	64.4	Pre-Commissioning of Field Instrumentation	515
	64.5	Commissioning Application Software	516
	64.6	Software Change Control	517
	64.7	Access to the System	518
	64.8	Personnel	519
	64.9	Documentation	519
	64.10	Works Acceptance	520
	64.11	Comments	520

65 System Management 525
- 65.1 Maintenance 525
- 65.2 Support 526
- 65.3 Obsolescence 527
- 65.4 Replacement 528
- 65.5 Upgrades 528
- 65.6 Comments 529

66 Quality Assurance 531
- 66.1 ISO 9001: Approach 531
- 66.2 ISO 9001: Content 532
- 66.3 Industry Guides 533
- 66.4 The GAMP Guide 533
- 66.5 Validation 534
- 66.6 Documentation 534
- 66.7 Procedures 534
- 66.8 Additional Supplier's Procedures 540
- 66.9 Additional End-User's Procedures 542
- 66.10 Comments 542

67 Contracts 543
- 67.1 Purpose of Contract 543
- 67.2 Contract Law 544
- 67.3 Relationships 544
- 67.4 Standard and Model Contracts 545
- 67.5 IChemE Model Conditions 545
- 67.6 Lump Sum *vs* Reimbursable Contracts 546
- 67.7 The Red Book 547
- 67.8 Project/Contract Management 550
- 67.9 Testing and Acceptance 553
- 67.10 Changes and Variations 555
- 67.11 Delays and Lateness 556
- 67.12 Defects, Site Accidents, Insurance and Exclusions 558
- 67.13 Payments 560
- 67.14 The Green Book 560
- 67.15 The Yellow Book 560
- 67.16 Comments 561

Part II: Theory and Technique

Section 9: Maths and Control Theory

68 Series and Complex Numbers 567
    68.1   Power Series 567
    68.2   Taylor's Series 568
    68.3   Complex Numbers 568

69 First Order Systems 571
    69.1   Example of Thermal System 572
    69.2   Example of Residence Time 572
    69.3   Example of RC Network 573
    69.4   Example of Terminal Velocity 573
    69.5   Example of Simple Feedback 574
    69.6   Comments 574

70 Laplace Transforms 575
    70.1   Definition 575
    70.2   Use of Laplace Transforms 577
    70.3   Partial Fractions 578

71 Transfer Functions 581
    71.1   First Order System 581
    71.2   Lags, Leads, Integrators and Delays 582
    71.3   The 3-Term Controller 583
    71.4   Block Diagram Algebra 584
    71.5   Open and Closed Loop Transfer Functions 584
    71.6   Steady State Analysis 585
    71.7   Worked Example No 1 585
    71.8   Worked Example No 2 586
    71.9   Characteristic Equation 587
    71.10   Worked Example No 3 588
    71.11   The Routh Test 589
    71.12   Worked Example No 4 589

72 Second and Higher Order Systems 591
    72.1   Second Order Dynamics 591
    72.2   Overdamped, $\zeta > 1$ 592
    72.3   Critically Damped, $\zeta = 1$ 592
    72.4   Underdamped, $0 < \zeta < 1$ 592

	72.5	Undamped, $\zeta = 0$	594
	72.6	Higher Order Systems	594
73	**Frequency Response**		597
	73.1	Attenuation and Phase Shift	597
	73.2	Substitution Rule	598
	73.3	Bode Diagrams	599
	73.4	Nyquist Diagrams	600
	73.5	Lags, Leads, Integrators and Delays	600
	73.6	Second Order Systems	600
	73.7	Compound Systems	602
	73.8	Worked Example No 1	602
	73.9	Bode Stability Criteria	603
	73.10	Worked Example No 2	604
	73.11	Worked Example No 3	604
	73.12	Gain and Phase Margins	605
74	**Root Locus**		607
	74.1	Worked Example No 1	607
	74.2	Angle and Magnitude Criteria	609
	74.3	Worked Example No 2	609
	74.4	Evans' Rules	610
	74.5	Comments on Evan's Rules	611
	74.6	Worked Example No 3	611
	74.7	Worked Example No 4 (with Time Delay)	613
	74.8	Second Order Systems	615
	74.9	Dominant Roots	616
	74.10	Worked Example No 5 (Effect of I Action)	617
	74.11	Pole Placement	618
75	**Z Transforms**		621
	75.1	Samplers and Holds	621
	75.2	Equivalence of Pulses and Impulses	622
	75.3	Analysis of Pulse Trains	623
	75.4	Transforms of Variables	623
	75.5	Difference Equations	624
	75.6	Inverse Z Transformation	626
	75.7	Worked Example	626
76	**Sampled Data Systems**		629
	76.1	Impulse Response and Convolution	629
	76.2	Pulse Transfer Functions	630
	76.3	Worked Example No 1	631
	76.4	Limitations	631
	76.5	Cascaded Elements	632

	76.6	Worked Example No 2	633
	76.7	Closed Loop Systems	633
	76.8	Equivalent Representations	634
	76.9	Worked Example No 3	634
	76.10	Implicit Input Signals	635
77	**Z Plane Analysis**		637
	77.1	S to Z Mapping	637
	77.2	Stability Analysis	638
	77.3	Modified Routh Test	638
	77.4	Worked Example No 1	638
	77.5	Pole Positions and Root Locus	639
	77.6	Worked Example No 2	639
	77.7	Pole Zero Cancellation	643
	77.8	Worked Example No 3	644
	77.9	Comments	645
78	**Impulse Compensation**		647
	78.1	PID Compensation	647
	78.2	Generic Closed Loop Response	648
	78.3	Dahlin's Method	648
	78.4	Deadbeat Method	649
	78.5	Direct Programs	650
	78.6	Cascade Programs	651
	78.7	Parallel Programs	652
	78.8	Comments	653
79	**Matrices and Vectors**		655
	79.1	Definitions and Notation	655
	79.2	Determinants	656
	79.3	Matrix Operations	656
	79.4	Matrix Inversion	657
	79.5	Use of Matrices	658
	79.6	Differentiation of/by Vectors	658
	79.7	Transforms of Vectors	659
	79.8	Worked Example No 1	659
	79.9	Eigenvalues and Eigenvectors	660
	79.10	Worked Example No 2	660
80	**State Space Analysis**		663
	80.1	Second Order System	663
	80.2	n Dimensional State Space	664
	80.3	Conversion of Transfer Functions	665
	80.4	Worked Example No 1	665
	80.5	The Transition Matrix	666

	80.6	Worked Example No 2	667
	80.7	Single Input Single Output Systems	667
	80.8	Multiple Input Multiple Output Systems	668
	80.9	Multivariable Control Systems	668
	80.10	Similarity Transformation	669
	80.11	Diagonalisation	669
	80.12	Worked Example No 3	670
	80.13	System Modes	670
	80.14	Worked Example No 4	671

81 Multivariable Control ... 673
	81.1	Case Study	673
	81.2	Compensator Concept	675
	81.3	Control System Model	675
	81.4	Compensator Design	675
	81.5	Worked Example No 1	676
	81.6	Decouplers	676
	81.7	Sampled Data Model	677
	81.8	Impulse Compensator Design	678
	81.9	Worked Example No 2	679
	81.10	Sampled Data Decoupler	680
	81.11	Comments	680

82 Stochastics ... 681
	82.1	Summary Statistics	681
	82.2	Multivariate Statistics	682
	82.3	Probability Distribution	683
	82.4	The Normal Distribution	685
	82.5	Correlation	686
	82.6	Properties of Correlation Functions	688

83 Linear Regression Analysis ... 689
	83.1	Basic Concepts	689
	83.2	Pre-Processing of Data	690
	83.3	Method of Least Squares	690
	83.4	Model Validation	691
	83.5	Goodness of Fit	692
	83.6	Worked Example No 1	692
	83.7	Multiple Linear Regression	693
	83.8	Variable Selection	694
	83.9	Worked Example No 2	694
	83.10	Worked Example No 3	695
	83.11	Comments	697

Section 10: Plant and Process Dynamics

84 Linearisation 701
- 84.1 The Need for Linearisation 701
- 84.2 Deviation Variables 702
- 84.3 The Process of Linearisation 702
- 84.4 Unsteady State Balances 703
- 84.5 Transfer Function Model 704
- 84.6 Worked Example 705
- 84.7 Nomenclature 706

85 Lumped Parameter Systems 707
- 85.1 Lumped Parameter Models 707
- 85.2 Steam Heated Jacketed Vessel 707
- 85.3 Water Cooled Jacketed Vessel 709
- 85.4 Worked Example 712
- 85.5 Nomenclature 713

86 Zero Capacity Systems 715
- 86.1 Steam Injection System 715
- 86.2 Worked Example 716
- 86.3 Significance of Dynamics 717
- 86.4 Dead Time Compensation 717
- 86.5 Blending System 718
- 86.6 Nomenclature 719

87 Compressible Flow Systems 721
- 87.1 Resistance to Flow 721
- 87.2 Volumetric Capacitance 722
- 87.3 Pressure Control 723
- 87.4 Worked Example 724
- 87.5 Boiler Dynamics 724
- 87.6 Nomenclature 726

88 Hydrodynamics 727
- 88.1 Nature of the Process 727
- 88.2 Energy Considerations 727
- 88.3 Energy Balance 728
- 88.4 Nomenclature 730

89 Multivariable Systems 731
- 89.1 Semi-Batch Reactor: Temperature Control 731
- 89.2 Temperature and Flow Coefficients 732
- 89.3 Semi-Batch Reactor: Pressure Control 733
- 89.4 Multivariable Control 733
- 89.5 Nomenclature 734

90 Multistage Systems 735
- 90.1 Vapour Flow Lags 735
- 90.2 Liquid Flow Lags 736
- 90.3 Concentration Lags 737
- 90.4 Dual Composition Control 739
- 90.5 Worked Example 740
- 90.6 L-V Strategy 741
- 90.7 Comments 743
- 90.8 Nomenclature 744

91 Reacting Systems 745
- 91.1 Mass Balance 745
- 91.2 Heat Balance 746
- 91.3 State Space Model 748
- 91.4 Stability Considerations 748
- 91.5 Nomenclature 749

92 Distributed Parameter Systems 751
- 92.1 Heat Exchanger Dynamics 751
- 92.2 Exchanger Process Transfer Function 752
- 92.3 Exchanger Load Transfer Function 753
- 92.4 Cooling Coil 753
- 92.5 Absorption Column Dynamics 754
- 92.6 Nomenclature 756

93 Anti-Surge Systems 757
- 93.1 Dynamics of Anti-Surge 757
- 93.2 Anti-Surge Design 758
- 93.3 Worked Example 759
- 93.4 Anti-Surge Control: Pressure Context 759
- 93.5 Comments 760
- 93.6 Nomenclature 761

94 Psychrometric Systems 763
- 94.1 Description of Spray Drier 763
- 94.2 Volume Balance 763
- 94.3 Mass Balance 764
- 94.4 Heat Balance 764
- 94.5 Nomenclature 766

95 Electro-Mechanical Systems 767
- 95.1 Simple Feedback System 767
- 95.2 Cascade Control System 768
- 95.3 Nomenclature 770

Section 11: Simulation

96 Numerical Integration 773
    96.1   Euler's Explicit Method 773
    96.2   Predictor-Corrector Method 774
    96.3   Worked Example No 1 774
    96.4   Runge Kutta Method 775
    96.5   Euler's Implicit Method 775
    96.6   Worked Example No 2 776
    96.7   Step Length 777
    96.8   Nomenclature 777

97 Procedural Simulation 779
    97.1   The Matlab User Interface 780
    97.2   Array and Matrix Operations 781
    97.3   Curve Fitting 781
    97.4   Root Finding 782
    97.5   Multiplying and Dividing Polynomials 782
    97.6   Differentiating Polynomials 783
    97.7   Finding Partial Fractions 783
    97.8   Display Functions 783
    97.9   Statistics Functions 784
    97.10   Import and Export of Data 785
    97.11   Script M-Files 785
    97.12   Program Structure 786
    97.13   Control Loop Simulation 787
    97.14   Function M files 789
    97.15   Comments 790

98 Block Orientated Simulation 791
    98.1   The Simulink Environment 791
    98.2   First Order Step Response 792
    98.3   Control Loop Simulation 793
    98.4   State Space Models in Simulink 795
    98.5   State Feedback Regulator Example 797
    98.6   Dynamic Process Simulators 800

Section 12: Advanced Process Automation

99 Relational Databases 805
    99.1   Alarm System for RDB Purposes 806
    99.2   Structure and Terminology 806
    99.3   Mappings and Constraints 808
    99.4   Structure and Syntax of SQL 809
    99.5   Other Constructs 811

	99.6	Dependency Theory 812
	99.7	Entity Relationship Modelling 813
	99.8	Database Design 815

100 Management Information Systems 817
- 100.1 Information Requirements 817
- 100.2 Functionality of MIS 818
- 100.3 Materials Resource Planning 819
- 100.4 Process MRP 820
- 100.5 Manufacturing Execution Systems 821
- 100.6 Project Planning for the Process Industries 821
- 100.7 Integration of Control and Enterprise Systems 825
- 100.8 Comments 825

101 Principal Components Analysis 827
- 101.1 PCA for Two Variables 828
- 101.2 Summary of Bivariate PCA 829
- 101.3 PCA for Multiple Variables 830
- 101.4 Reduction of Dimensionality 831
- 101.5 Worked Example No 1 831
- 101.6 Interpretation of Principal Components 832
- 101.7 Worked Example No 2 833
- 101.8 Loadings Plots 834
- 101.9 Worked Example No 3 834

102 Statistical Process Control 837
- 102.1 Data Collection 837
- 102.2 Data Pre-Screening 838
- 102.3 Control Charts 839
- 102.4 Average Control Charts 840
- 102.5 Moving Average Control Charts 840
- 102.6 Spread Control Charts 841
- 102.7 Six Sigma 842
- 102.8 Capability Indices 843
- 102.9 Multivariate SPC 844
- 102.10 Geometric Modelling 846
- 102.11 Comments 848

103 Linear Programming 849
- 103.1 Concept of Linear Programming 849
- 103.2 Worked Example on LP 850
- 103.3 The Simplex Method 852
- 103.4 Tableau Representation 854
- 103.5 Mixed Integer Linear Programming 857
- 103.6 Implicit Enumeration 858

104 Unconstrained Optimisation ... 863
- 103.7 Branch and Bound ... 860
- 103.8 Comments ... 861

104 Unconstrained Optimisation ... 863
- 104.1 One Dimensional Functions ... 863
- 104.2 Two Dimensional Functions ... 864
- 104.3 Newton's Method ... 864
- 104.4 Worked Example No 1 ... 865
- 104.5 Search Procedures ... 866
- 104.6 Steepest Descent ... 866
- 104.7 Levenberg Marquardt Algorithm ... 867
- 104.8 Line Searching ... 868
- 104.9 Worked Example No 2 ... 868
- 104.10 Comments ... 869

105 Constrained Optimisation ... 871
- 105.1 The Lagrangian Function ... 871
- 105.2 Worked Example No 1 ... 872
- 105.3 Worked Example No 2 ... 873
- 105.4 Generalised Lagrangian Function ... 873
- 105.5 Worked Example No 3 ... 874
- 105.6 Sensitivity Analysis ... 875
- 105.7 Kuhn-Tucker Conditions ... 875
- 105.8 Worked Example No 4 ... 876
- 105.9 Quadratic Programming ... 877
- 105.10 Worked Example No 5 ... 878
- 105.11 Recursive Form of QP ... 879
- 105.12 Sequential Quadratic Programming ... 880
- 105.13 Reduced Gradient Methods ... 882
- 105.14 Penalty Functions ... 882
- 105.15 Nomenclature ... 883

106 Real Time Optimisers ... 885
- 106.1 Steady State Optimisers ... 886
- 106.2 Models for SS Optimisers ... 888
- 106.3 Methodology for SS Optimisers ... 889
- 106.4 Steady State Detection ... 889
- 106.5 Steady State Model Updating ... 890
- 106.6 Solution of SS Optimisation Problem ... 891
- 106.7 Dynamic Optimising Controllers ... 892
- 106.8 Constraint Handling by DOCs ... 893
- 106.9 QP Solution of DOC Problem ... 894
- 106.10 Formulation of Constraints ... 894
- 106.11 Application of DOCs ... 895
- 106.12 Comments ... 897

107 Knowledge Based Systems 899
- 107.1 Architecture and Terminology 900
- 107.2 Inferencing 901
- 107.3 Rule Based Expert Systems 903
- 107.4 The Expert System Control Cycle 904
- 107.5 Semantic Nets 906
- 107.6 Frame Based Systems 907
- 107.7 Object Oriented Programming 910
- 107.8 Expert System Shells 911
- 107.9 Knowledge Elicitation 912
- 107.10 Development Life Cycle 914
- 107.11 Comments 914

108 Fuzzy Logic Control 917
- 108.1 Controller Structure 917
- 108.2 Fuzzification 919
- 108.3 Rule Base 920
- 108.4 Decision Logic 921
- 108.5 Defuzzification 922
- 108.6 Worked Example 923
- 108.7 Real Time Operation 924
- 108.8 Example with Non-Linearity 924
- 108.9 Example with Interaction 925
- 108.10 Alternative Approaches 926
- 108.11 Self Adaptive Fuzzy Control 927
- 108.12 Observations 928

109 Artificial Neural Networks 929
- 109.1 Multi Layer Perceptrons 929
- 109.2 Operation of MLP 931
- 109.3 Back Propagation Training Algorithms 932
- 109.4 Network Size and Generalisation 933
- 109.5 Evaluation of Jacobians 934
- 109.6 Data Encoding 936
- 109.7 Pre-Processing of Data 937
- 109.8 Radial Basis Function Networks 937
- 109.9 Training of RBF Networks 938
- 109.10 Worked Example 939
- 109.11 Dynamic Modelling with Neural Nets 940
- 109.12 Neural Nets for Inferential Estimation 942
- 109.13 Neural Nets for Optimisation 942
- 109.14 Comments 943

110 Genetic Algorithms 945
110.1 Chromosomes and Genes 945
110.2 Cost and Fitness Functions 946
110.3 Selection 947
110.4 Crossover 949
110.5 Mutation 950
110.6 Reinsertion 950
110.7 Structure of a GA 951
110.8 Parallel GAs 952
110.9 Multi-Objective GAs 955
110.10 Pareto Ranking 956
110.11 Pareto Ranking MOGAs 956
110.12 Visualisation 958
110.13 Design of a Batch Scheduling MOGA 960
110.14 Comments 962

Section 13: Advanced Process Control

111 Multiloop Systems 965
111.1 Relative Gain Analysis 965
111.2 Interpretation of RGA Elements 966
111.3 Worked Example: L-V Scheme 967
111.4 Worked Example: L-B Scheme 968
111.5 Worked Example: D-V Scheme 969
111.6 Effective Gain 969
111.7 Singular Values 969
111.8 Application to Blending System 970
111.9 Singular Value Decomposition 970
111.10 Worked Example No 4 971
111.11 Comments 971

112 State Feedback Regulators 973
112.1 The Control Law 973
112.2 Worked Example No 1 974
112.3 Set Point for Regulo Control 975
112.4 Worked Example No 2 976
112.5 Observer Design 976
112.6 Full Order Observer 977
112.7 Worked Example No 3 978
112.8 Reduced Order Observer 978
112.9 Integration of Observer and Controller 980
112.10 Implementation 980
112.11 Comments 981

113 Kalman Filtering 983
- 113.1 The Luenberger Observer 983
- 113.2 Kalman Filter Design 984
- 113.3 Formation of Riccati Equation 986
- 113.4 Solution of Riccati Equation 987
- 113.5 Realisation of Kalman Filter 988
- 113.6 Implementation Issues 988
- 113.7 Use of Kalman Filters in Control Systems 990
- 113.8 Worked Example 991
- 113.9 Nomenclature 992

114 Least Squares Identification 993
- 114.1 The Plant Model 994
- 114.2 Least Squares Estimation 995
- 114.3 Recursive Least Squares 996
- 114.4 Least Squares Using Instrumental Variables 998
- 114.5 Generalised Least Squares 1000
- 114.6 Extended Least Squares 1001
- 114.7 Comparison of Least Squares Estimators 1001
- 114.8 Extension to Non-Linear Systems 1002
- 114.9 Extension to Multivariable Systems 1002
- 114.10 Nomenclature 1004

115 Recursive Estimation 1005
- 115.1 Setting the Order of the Model 1005
- 115.2 Initialisation of Parameter Values 1006
- 115.3 Initialisation of the Covariance Matrix 1006
- 115.4 Forgetting Factors 1007
- 115.5 Covariance Resetting 1007
- 115.6 Numerical Instability 1008
- 115.7 Covariance Windup 1008
- 115.8 Variable Forgetting Factors 1009
- 115.9 Convergence 1010
- 115.10 Comments 1011

116 Self Tuning Control 1013
- 116.1 Explicit and Implicit STC 1014
- 116.2 Notation 1014
- 116.3 Minimum Variance Control 1015
- 116.4 Implementation of MV Control 1017
- 116.5 Properties of MV Control 1018
- 116.6 Generalised Minimum Variance Control 1018
- 116.7 Implementation of GMV Control 1019
- 116.8 Properties of GMV Control 1020
- 116.9 Comments 1021
- 116.10 Nomenclature 1022

117 Model Predictive Control 1023
- 117.1 Prediction and Control Horizons 1024
- 117.2 The Basis of Forward Prediction 1025
- 117.3 Prediction of the Output 1026
- 117.4 Controller Output Sequence 1028
- 117.5 Worked Example 1029
- 117.6 Recursive Implementation 1031
- 117.7 QP Solution of MPC Problem 1032
- 117.8 Extension to Multivariable Systems 1034
- 117.9 Models for MPC 1035
- 117.10 Proprietary Packages 1036
- 117.11 Nomenclature 1038

118 Non-Linear Control 1039
- 118.1 L/A Control 1039
- 118.2 Control Affine Models 1040
- 118.3 Generic Model Control 1041
- 118.4 Application of GMC to CSTR 1042
- 118.5 Worked Example on GMC 1043
- 118.6 Lie Derivatives and Relative Order 1043
- 118.7 Globally Linearising Control 1044
- 118.8 Application of GLC to CSTR 1045
- 118.9 Comments on GLC 1046
- 118.10 Nomenclature 1046

Bibliography 1047

Abbreviations and Acronyms 1055

Index 1065

Part I

Technology and Practice

Section 1

Introduction

Summary

This text is organised in two parts. The first covers, to a large extent, the less theoretical aspects of process automation. As such, it focuses on the basic technology and practice of the discipline. It is surprising how much material comes into this category. The second part develops a range of techniques, many of which are inherently mathematical in nature, and focuses on more advanced aspects of control and automation.

The text has been carefully structured into relatively self contained sections and partitioned into chapters in a logical way. Extensive cross referencing enables the connections between the topics to be readily established. Whilst most of the topics are generic and relatively timeless, some will inevitably become dated: these have been isolated into single chapters to simplify updating in future editions. The structure also enables new chapters to be added as the technology evolves.

Part 1

1. Introduction. This section introduces a variety of concepts, technology and terminology as a basis for subsequent sections. It is, hopefully, self explanatory: if not, the reader has problems!

2. Instrumentation. For common process measurements such as temperature, pressure, level, flow, weight and even composition, there is good quality instrumentation available off the shelf. It isn't necessary to know exactly how an instrument works, although it helps to understand the principles. The key issues are specifying the right instrument for the job, installing it correctly, and knowing how to use it. Remember, you can't control what you don't measure properly in the first place. Many a control system has foundered on the rocks of measurement.

3. Final Control Elements. This section covers the various control elements other than those directly concerned with measurement: these are often loosely referred to as instrumentation too. Despite the variety of applications there are relatively few such elements to consider. Of particular interest are actuators and valves used for manipulating the process and for isolation purposes. It is important to appreciate that the quality of control achieved is a function of all the elements in the loop. There is no point in striving for perfection with the measurement if the valve is sized wrongly.

4. Conventional Control Strategies. The 3-term PID controller is the basis of feedback control. It accounts for something like 80% of all continuous control as used in the process industries, so a thorough treatment of the use and application of 3-term control is easily justified. PID control provides the basis for a variety of other strategies, *e.g.* cascade and ratio control, and is often used as the basis for feedforward control. These together account for another 15% or so and are discussed fully.

5. Process Control Schemes. For any given plant there are invariably many ways of controlling it. Seldom is any single way correct; it is just that some ways are better than others. So deciding what strategies to use and formulating a control scheme is a question of judgement. Fundamental to this is an understanding of how the process works and a feel for how the items of plant and equipment interact. An approach to "determining" strategies is outlined and develop in relation to schemes for the control of a selection of items of plant.

6. Digital Control Systems. These are the norm for new plants, both continuous and batch: the benefits are well understood. They support both conventional and advanced strategies, and a variety of other activities such as alarm handling and operator interaction. To be able to apply digital control effectively it is necessary to have a reasonable grasp of how the systems work. Their functionality is explained from an applications point of view. Focusing on their functionality is the best approach: this enables the specifics of any particular system to be mastered relatively easily, no matter how large or complex it may be.

7. Control Technology. The realisation of modern control systems has to be achieved within a rather mundane framework that is properly engineered and is both safe and operable. Signals have to be physically routed to and from instrumentation scattered around the plant. Different types of i/o channel are used to illustrate layout and related issues such as intrinsic safety. The concept of "layers of safety" is introduced as a basis for considering various aspects of good practice. This embraces reliability principles and protection system design. The inexorable drive for safety, and environmental constraints, place a major responsibility on those involved in process automation.

8. Management of Automation Projects. It does not matter a great deal how sophisticated a control system is if its not "fit for purpose" or commercially viable. Fundamental to this is knowing when and how to apply control technology. This section, therefore, concentrates on the formulation of specifications and good practice with regard to project management. All aspects of the project life cycle covering justification, design and development, testing and acceptance, installation, commissioning and maintenance are covered. The time and effort necessary to produce properly documented specifications, and their role as the only meaningful basis for quality assurance, especially in relation to application software, is emphasised.

Part 2

9. Maths and Control Theory. Advanced process control is inherently quantitative and has a theoretical basis for which no apology is given. This section summarises various aspects of both classical and modern control theory. Coverage is sufficient to provide an understanding of subsequent sections. The reader is spared the full-blooded academic rigour of most texts on control theory. Interspersed amongst the chapters on control theory are others on the essential mathematics. A knowledge of differentiation and integration is assumed.

10. Plant and Process Dynamics. Usually it is sufficient to have a qualitative feel for the way a plant behaves. However, sometimes the dynamics are not easy to understand, perhaps because the plant is highly integrated or the process is complex. In such circumstances it may be necessary to build a quantitative model. In this section, models are built of a selection of items of plant and operations. Several modelling techniques and a variety of classical assumptions are introduced. The golden rules are to never start model building unless it is essential, and to keep the model as simple as possible consistent with reality.

11. Simulation. Occasionally, the equations of a model may be solved analytically. More usually, because of non-linearities, discontinuities and the like, solutions have to be obtained by means of simulation. Various packages exist for simulating the behaviour of plant and their control systems. They enable "what-if" scenarios to be investigated as a basis for making design decisions. Use of such packages is relatively straightforward provided certain common sense rules are followed. The principles of simulation are introduced. Key issues such as validation of results are discussed.

12. Advanced Process Automation. This section covers a diverse range of modern techniques and technology. To a large extent these are concerned with information and its manipulation. Database structures, such as relational and object oriented, and client-server technology are fundamental to this. Major topics covered include management information systems (MIS), statistical process control (SPC), optimisation, expert systems, fuzzy logic and neural nets. The emphasis throughout is on the underlying technology and on its application.

13. Advanced Process Control. This section covers a diverse range of modern control techniques. They are non-trivial in a mathematical sense. Major topics covered include state feedback, identification and estimation, self-tuning, model predictive control (MPC) and non-linear control. The coverage of these topics concentrates on their underlying principles and on the known practical problems in applying them.

P&I Diagrams

Chapter 2

2.1 Bubbles and Symbols
2.2 Tag Numbers

Piping and instrumentation (P&I) diagrams are of fundamental importance in process automation and are discussed, for example, by Coulson (2004). Their formulation is a major stage in the design of a process plant. They are subsequently used as the basis for the detailed design of the plant's control system, and become reference documents for other purposes such as HAZOP studies. P&I diagrams depict, from a process point of view, measurements and control schemes in relation to items of plant and their interconnecting pipework. They are represented by means of symbols and tag numbers.

2.1 Bubbles and Symbols

Symbols are used to represent individual elements such as sensors and valves, or combinations of elements such as measurement channels or control loops. They are linked together by signal lines. Tag numbers are written inside circles, referred to as bubbles or balloons. They consist of letter codes and reference numbers. Letter codes indicate the function of the elements and are generic. Reference numbers are specific to particular elements and are used for identification purposes.

The amount of detail shown on a P&I diagram varies throughout a project according to the amount of information available. In the early stages it tends to be sketchy and functional only; later on the information becomes more comprehensive and specific. An important point to appreciate is that the same set of symbols, signals and letter codes is sufficient to enable elements to be represented whether the version of P&I diagram be outline or detailed.

Consider Figure 2.1 which is a section of an outline P&I diagram that depicts a flow control loop associated with a centrifugal pump.

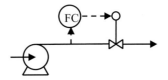

Fig. 2.1 P&I diagram of flow control loop

The diagram states that the function of the loop is flow control, a feedback strategy is implied and a regulatory type of control valve is indicated. No information is given about the type of measurement or signals involved. The elements of the loop are shown in their correct positions in a functional sense, *i.e.* the control valve is downstream of the pump and the flow measurement is between the valve and the pump. However, the symbols do not necessarily indicate their relative positions in a physical sense, *e.g.* the flow measurement and valve could be a long way from the pump.

Arrows are normally put on the signals to indicate the direction of flow of information. However, if this is obvious as in Figure 2.1, they may be omitted. Power and air supply lines to the various elements are not normally shown on P&I diagrams to avoid confusion with the signals.

Broken lines are used in Figure 2.1 to depict the signals. Strictly speaking, broken lines should only be used for electrical signals. However, on outline P&I diagrams it is common practice to use broken lines for most signals on the basis that they will probably be electrical anyway. Representations for the most common signal types are as shown in Table 2.1.

Table 2.1 Representation of signal types

Signal type	Representation
Capillary	—✗————✗—
Data link	—⊖————⊖—
Electrical	– – – – – –
Electromagnetic	—∿————∿—
Hydraulic	—L————L—
Mechanical	—●————●—
Pneumatic	—//————//—
Process	————————

Consider Figure 2.2 which is a section of a detailed P&I diagram that depicts the same flow control loop.

The process stream flows through a sight glass FG 47, an orifice plate FE 47 and a control valve FV 47. These are all in-line elements. Note the different use of the bubbles: as a symbol for the sight glass but as an identification for the orifice plate and control valve. Note also that hand operated valves are indicated on either side of all these elements so that they can be isolated prior to removal for maintenance or repair. Additionally, the control valve has a bypass line to enable flow to continue under manual control in the event of the control valve having to be removed.

The pressure drop across the orifice plate is measured using a dp cell FT 47 which transmits the flow rate as an electrical signal to the totaliser FQR 47, the low level switch FSL 47 and the controller FIC 47. If the flow drops below some pre-set lower limit the flow switch FSL 47 will activate a low level alarm FAL 47. The controller indicates the current value of the flow rate and generates a control signal which is transmitted, via an I/P converter FY 47, to the control valve. FV 47 is a pneumatically actuated diaphragm type of regulatory valve. It has a positioner attached, as indicated by the box on the stem of the valve symbol. The arrow on the stem

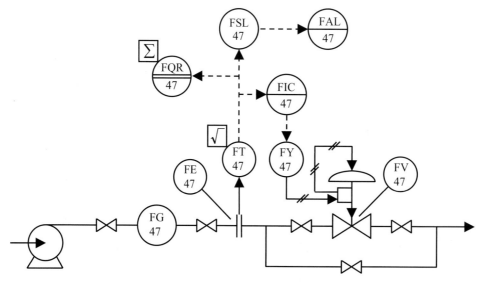

Fig. 2.2 Detailed P&I diagram of flow control loop

indicates that, in the event of a loss of air supply, the valve will fail shut.

Information about the location of an element is indicated by the presence or otherwise of a line through its bubble. No line means that the element is field mounted, *i.e.* it is installed on or adjacent to a pipe, vessel or some other item of plant. A single line through the bubble, as with FAL 47 and FIC 47, means that the element is located in a central control room. And a double line, as with FQR 47, means that it is panel mounted elsewhere, say on a field termination cabinet.

2.2 Tag Numbers

There are several national standards which relate to the representation of instrumentation and control schemes. The most important English language ones are BS 1646 and ISA S5.1. Also, most major companies have their own internal standards. Whilst there are many differences in detail between them, the basic symbols and letter code structure are essentially the same.

ISA S5.1 is used throughout this guide on the grounds that it is the most widely used standard in practice. Its letter code structure is as indicated in Table 2.2 which also lists the meaning of the more commonly used letters.

Letter codes are configured according to function. The first letter corresponds to the measured variable and, if necessary, may be qualified by a modifier. Succeeding letters describe an element's readout or control functions; these may themselves have modifiers. A certain amount of judgement may be necessary to establish the most appropriate letter code for an element.

It is important to appreciate that the letter code is determined by the function of the element and not by its design or installation. For example, the dp cell in Figure 2.2 is designated as a flow transmitter FT 47 rather than as PDT 47 even though it measures differential pressure across the orifice plate.

The first letter of all elements within a loop is that of the measured variable. Thus the I/P converter is designated FY 47 although its input is a current and its output a pressure.

Within the letter code, modifiers must immediately follow the letter that they qualify. For example, the modifier Q changes the measured variable F of element FQR 47 into a totalised flow. In addition, the modifier L states that the alarm FAL 47 will only be activated by low flows.

If a loop contains two or more elements with the same function they may be distinguished by means of suffixes. For example, if the flow loop had two control valves in a duplex arrangement, they could have been designated as FV 47A and FV 47B.

Sometimes the letter code is insufficient to give an adequate description of the function of an element. Additional information may be provided, either in a box attached to its bubble or as adjacent text. For example, FT 47 has a square root function to compensate for the square relationship inherent in the flow measurement by the orifice plate. The total flow computed by FQR 47 is obtained by means of an integral function. Some of the more common functional designations are given in Table 2.3.

Numbering of elements is in accordance with some plant based convention. There are two approaches, serial and parallel, of which serial is the more common. On a serial basis each channel, loop or scheme is allocated a unique number, 47 in the case of Figure 2.2. Regardless of the letter code, all its elements then assume the same number. On a parallel basis, blocks of numbers are allocated according to instrument type or function, depending on its letter code. This results in similar elements in different loops having contiguous numbers.

The bubbles used in Figure 2.2 indicate the function of the elements. It is implied that each element is a discrete item of instrumentation which may be either analogue or digital in nature. With computer-based control systems many of these individual functions are realised by means of configurable software in common hardware. For representation of such functions, ISA S5.1 recommends a different set of symbols. These are listed in Table 2.4: note that the symbols listed for input and output (I/O) channels are non-standard but are

Table 2.2 ISA letter codes for tag numbers

	First letter		Succeeding letters		
	Measured or initiating variable	Modifier	Readout or passive function	Output function	Modifier
A	Analysis		Alarm		
B	Burner or combustion		User's choice	User's choice	User's choice
C	User's choice, *e.g.* conductivity			Control	
D	User's choice, *e.g.* density	Differential			
E	Voltage or emf		Sensor or primary element		
F	Flow rate	Ratio or fraction			
G	User's choice		Sight glass or viewing device		
H	Hand				High
I	Current (electrical)		Indicate		
J	Power	Scan			
K	Time or schedule	Rate of change		Control station	
L	Level		Light (pilot)		Low
M	User's choice, *e.g.* moisture	Momentary			Middle or intermediate
N	User's choice		User's choice	User's choice	User's choice
O	User's choice		Orifice or restriction		
P	Pressure or vacuum		Point (test)		
Q	Quantity	Integrate or totalise			
R	Radiation		Record or print		
S	Speed or frequency	Safety		Switch	
T	Temperature			Transmit	
U	Multivariable		Multifunction	Multifunction	Multifunction
V	Vibration			Valve or damper	
W	Weight or force		Well or pocket		
X	Unclassified	X axis	Unclassified	Unclassified	Unclassified
Y	Event or state	Y axis		Relay, compute or convert	
Z	Position or dimension	Z axis		Driver, actuator or other control element	

2.2 Tag Numbers

Table 2.3 Additional functions for use with tag numbers

Function	Symbol
Average	AVG
Bias	+ or −
Boost	1:1
Characterise	$f(x)$
Derivative	$\frac{d}{dt}$
Difference	Δ
Divide	$\div$
High select	$>$
Integrate	$\int$
Low select	$<$
Multiply	$\times$
Raise to power	x^n
Ratio	$1:n$
Reverse	REV
Square root	$\sqrt{}$
Sum	Σ

Table 2.4 P&I diagram symbols for computer-based systems

Configurable functions	Symbol
Shared display and control	(circle in square)
Computer function	(hexagon)
Programmable logic controller	(square with diamond)
Undefined logic or sequence control	(diamond)
Input channel	(arrow in)
Output channel	(arrow out)

nevertheless commonly used. A solid line across the symbol means that the function has a shared, screen based display: the absence of a line means that the function is inaccessible to the operator.

Figure 2.3 shows an implementation of the previous flow control loop by means of a distributed control system (DCS). The square root extraction previously associated with FT 47, the shared dis-

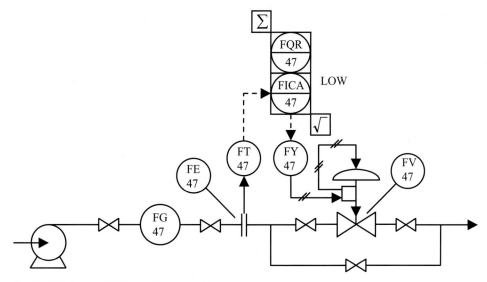

Fig. 2.3 P&I diagram of DCS-based flow control loop

play and control functions, and the low flow alarm previously handled by FSL 47 and FAL 47 are now all realised by FICA 47. The integration is realised by FQR 47 which logs, but does not display, the totalised flow.

The reader is encouraged to browse through the appendix to ISA S5.1 which contains a host of examples on the use of symbols, signals and tag numbers in a variety of contexts. The reader is also referred to ISA S5.3 on symbols for computer control systems.

Block Diagrams

3.1 Signals
3.2 Structure
3.3 Sub-systems
3.4 Modes of Operation

Consider the level control system shown in schematic form in Figure 3.1. The plant consists of a tank with inlet and outlet pipes, the outlet pipeline containing a centrifugal pump and a control valve.

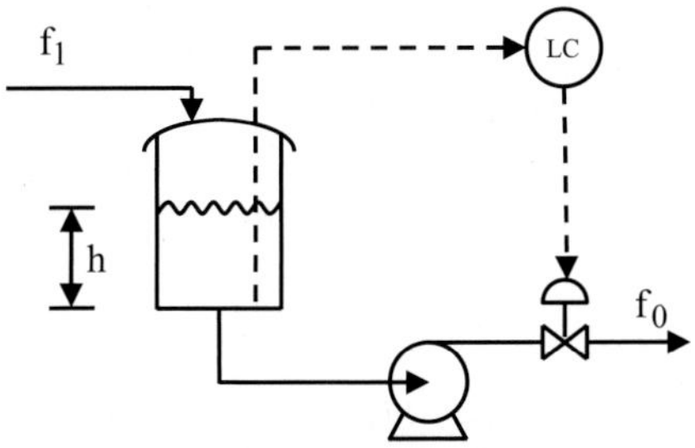

Fig. 3.1 Schematic of level control system

The flow rate f_1 of the inlet stream is determined by upstream conditions over which the level control system has no influence: f_1 is said to be wild. The level h is controlled by adjusting the flow rate f_0 of the outlet stream by means of a control valve. This is a typical feedback control system and is shown in block diagram form in Figure 3.2.

3.1 Signals

The various signals indicated on the block diagram all have generic names and most have acronyms too, as defined in Table 3.1.

Table 3.1 Generic names and acronyms of signals

Label	Signal	Acronym
h	Controlled (process) variable	CV (PV)
h_m	Measured variable (input)	IP
h_r	Reference (set point)	SP
e	Error	
u	Controller output	OP
f_0	Manipulated variable	MV
f_1	Disturbance variable	DV

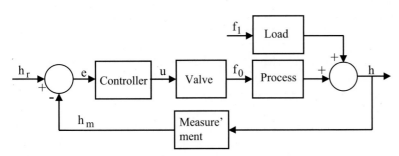

Fig. 3.2 Block diagram of level control system

The controlled variable, the level h, is also referred to as the process variable. It is measured and fed back to the controller where it is compared with the reference signal h_r, usually referred to as the set point, to produce the error signal e. The controller produces an output u, depending upon the error, which is applied to the valve to manipulate the outlet flow f_0 as appropriate. The resultant level in the tank depends upon the combined effect of its outlet flow and any disturbances in its inlet flow f_1.

The acronyms of CV, DV, MV, PV and SP are widely used in the process industries to refer to these particular variables. The acronyms IP and OP refer to the controller's input and output signals.

3.2 Structure

It is evident that a block diagram depicts the structure of a system and shows the functional relationship between its various elements. The blocks relate to the elements and represent functions. These functions invariably consist of combinations of conversion, or scaling, factors and dynamics. The lines between the blocks represent signals, the arrows indicating the direction of flow of information. Addition and subtraction of signals is represented by circles with signs as appropriate: the absence of signs implies addition.

Note the general layout of the block diagram. By convention, the controlled variable comes out on the right hand side and external inputs are normally shown entering from the left. Signals are represented by single arrows even though, for example, with electrical signals there may be two or more wires to establish the signal's circuit. Power and air supplies are not shown. The size of the blocks does not relate to the physical size of the elements!

It is appropriate to think in terms of the functions acting on the signals. Thus the input to one block is operated upon by its function to produce an output which in turn becomes the input to the next block, and so on. Elements are normally non-interacting. For example, the input to the valve u directly affects its output flow f_0, and not *vice versa*. The flow f_0 can only indirectly effect the control signal u by virtue of its affect on the level being fed back around the loop.

The block diagram of Figure 3.2 is somewhat simplistic: in reality there are many more elements and signals involved. Figure 3.3 is a more comprehensive block diagram of the same level control loop.

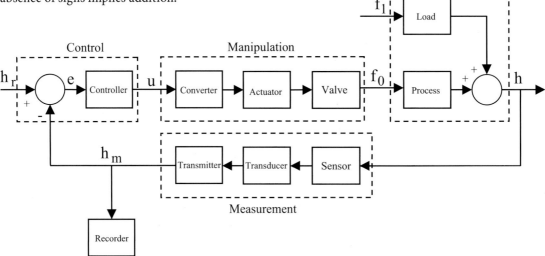

Fig. 3.3 Comprehensive block diagram of level control system

3.3 Sub-systems

There are essentially four sub-systems as indicated. The combined control, manipulation and plant sub-systems are often referred to as the feedforward path, and the measurement sub-system is referred to as the feedback path.

The plant sub-system consists of the process and its load. From a chemical engineering point of view, the process obviously relates to the vessel and its pump, the flow rates being the input and output as shown in Table 3.2. However, from a control engineering point of view, inspection of Figure 3.3 shows that the outlet flow rate is the input to the process and the level its output.

Table 3.2 Classification of input and output signals

Viewpoint	Input	Output
Chemical	f_1	f_0
Control	f_0	h

This different perspective is a source of much confusion and needs clarification. For control purposes, the term process is defined to be the way in which the manipulated variable (MV) affects the controlled variable (CV): it is a dynamic relationship. Similarly the load is defined as the relationship between a disturbance and the controlled variable. Given that there may be various sources of disturbance, it follows that there may be several different loads. According to the principle of superposition, the net change in controlled variable is the sum of the individual effects of the process on the manipulated variable and the loads on the disturbance variables.

The measurement sub-system normally consists of a sensor, transducer and a transmitter. In the level example, the sensor may be a back pressure device, referred to as a pneumercator, with the transducer and transmitter functions being realised by means of a dp cell. The sensor is the element directly in contact with the process which is the source of the measurement. Often the sensor is a low power device; invariably it produces a signal whose form is unsuitable for other purposes. The transducer therefore operates on the sensor output, amplifying, characterising and filtering it as appropriate. In addition, the transmitter provides the signal with the necessary power for transmission to the controller and recorder.

Functionally speaking, the controller sub-system consists of two parts, the comparator and the controller proper. In practice the comparator is considered to be part of the controller and, for single loop controllers, is an integral part of the unit. In essence, the comparator subtracts the measured value from the reference signal to produce an error. The controller, usually of the 3-term PID type, then operates on the error to produce the control signal. The objective of the controller is to vary u in such a way as to minimise e.

The manipulation sub-system normally consists of a valve, an actuator and a converter. Because manipulation invariably involves adjusting some flow, the most common final control element by far is the regulatory type of control valve. For various reasons, including cost and safety, control valves are almost always fitted with a pneumatic diaphragm actuator. Given that the controller output is usually electrical, and that the actuator requires a pneumatic input, an I/P converter is necessary.

It is essential that all the elements of a sub-system, and the sub-systems themselves, are compatible with each other with regard to form, range, response, power, reliability and so on. Many problems in process automation can be traced back to non-compatible specifications.

3.4 Modes of Operation

It is appropriate to introduce a number of other commonly used terms in relation to the generalised block diagram of Figure 3.3.

Most controllers have an auto/manual station. This is in effect a switch which enables the controller to be switched between automatic (AUTO) and manual (MAN) modes of operation. In automatic, the controller output varies according to how the PID terms operate on the error. In manual,

Table 3.3 Modes of operation of closed loop

	Set point	Disturbances	Objectives
Regulo	Constant	Intermittent (may be large)	Steady state
Servo	Varying	Varying (usually small)	Tracking

the output is not related to the error and can be adjusted by hand to any desired value.

The operation of a control loop may be described as being either closed or open. If all the elements of a loop are functioning (correctly or otherwise), interconnected and the controller is in auto such that automatic control occurs, the loop is said to be closed. If any element is disconnected or not powered up then the loop is said to be open. It is as if one of the signals has been cut. Whenever a controller is switched into manual the loop is opened.

An alternative way of categorising the mode of operation of a control loop is according to the type of input signals, as shown in Table 3.3.

Generally speaking, regulo control is the norm in process automation: the objective is to keep the plant operating under steady state conditions despite disturbances which occur. Servo control is more usually associated with electromechanical control systems, for example in the machine tool or aerospace sectors, where the objective is to track a moving set point. However, systems for the automatic start-up and shut-down of continuous plant, or for batch process control, inevitably have to handle both regulo and servo control.

Signals

4.1 Signal Categories
4.2 Standard Ranges
4.3 Linearity

Signals were described in the previous chapter as flows of information. Strictly speaking, a signal is a variable that conveys the value of a parameter between the elements of a system.

There are many types of signal in process automation and it is helpful to categorise them. This can be done in various ways. Table 4.1 lists common signal types by physical form with examples of each.

4.1 Signal Categories

More usefully, signals may be categorised by nature as being analogue, discrete or pulse.

Analogue signals are continuous in nature. They vary with time over some finite range and may have any value within that range. The limits of the range are normally determined by some physical phenomenon associated with the process, by the mechanical design of the plant, or by the specification of the instrumentation. Most process signals are inherently analogue.

Discrete signals are sometimes, confusingly, referred to as digitals. They have two or more states and, at any point in time, may only exist as one or other of those states. Discrete signals are used to convey status information such as auto/manual, on/off, open/closed, running/stopped, and so on.

Pulse signals consist of trains of pulses, each pulse being equivalent to a fleeting discrete signal. They are associated with rotary devices such as turbine meters and agitator shafts. A known number of electrical pulses are generated with each revolution. Counting of the pulses with respect to time yields an average shaft speed. Relative to analogues and discretes, pulse signals are not very common.

Table 4.1 Signal types by physical form

Form	Example
Audio	Sound from high level alarm
Capillary	Expansion in filled system
Electrical	Current from transmitter
	Power to motor windings
	Resistance of strain gauge
	Voltage across thermocouple junction
Electromagnetic	Radio link in telemetry system
	Magnetic coupling in proximity switch
Hydraulic	Flow of oil through spool valve
Manual	Change of set point by operator
	Contact of push button in relay circuit
Mechanical	Position of valve stem
Pneumatic	Pressure to valve actuator
Process	pH of process stream
	Speed of agitator shaft
	Weight of charge vessel
Visual	Colour of symbol on display
	Position of pointer on scale
	Status of controller on faceplate
	Number in register on counter

Another way of categorising signals is on a channel basis. In process automation it is normal to define channels with respect to the control function. Thus input channels relate to measurements and output channels to control signals. An input channel is essentially a circuit, consisting of various interconnected elements, wiring, *etc.*, that enables a signal to be routed from a sensor in the field through to the controller. Similarly, an output channel relates to the transmission of a signal from the controller through to a control valve.

The number of I/O channels broken down according to nature of signal and presented in the form of Table 4.2 is known as the point count. This is commonly used as a measure of the size of a computer control system.

Table 4.2 Template for point count

	Input	Output
Analogue		
Discrete		
Pulse		

Digital signals are associated with microprocessor-based instruments and control systems. They are handled internally as binary numbers with values being held as bit patterns in memory, as described in Chapter 7. For interfacing purposes, analogue signals have to be converted into binary form, the bit patterns occupying one or more bytes of memory depending upon the resolution required. Discrete signals are inherently binary and are stored as individual bits: they are usually grouped together bytewise. For pulse signals, the pulses are counted into registers on a binary basis, the number of bytes per signal depending on the range of numbers to be counted.

4.2 Standard Ranges

There are standard ranges for signals. Compatibility of input and output signals enables systems to be configured from off-the-shelf elements. It also provides for interchangeability of elements: this reduces stock requirements for spares and promotes competition between suppliers. The standard signal ranges for electrical and pneumatic analogue signals are as shown in Table 4.3. These are *de facto* industry standards and are used on a worldwide basis. There are no such standards for discrete or pulse signals: the figures given are simply consistent with common practice.

Table 4.3 Standard signal ranges

Form	Nature	Standard range
Electrical	Analogue	4–20 mA, 0–5 V
	Discrete	0/24 V
	Pulse	0/20 mA
Pneumatic	Analogue	0.2–1.0 barg (3–15 psig)
	Discrete	0/3.5 barg (0/50 psig)

A knowledge of the behaviour or calibration of the elements and of the nature of the signal is necessary to determine the value of the parameter being transmitted. Consider, for example, the temperature measurement shown in Figure 4.1.

Fig. 4.1 Temperature measurement block diagram

Suppose the instrument has a linear calibration with a direct current (d.c.) output range of 4–20 mA corresponding to an input range of 0–100°C. This data enables an output signal of, say, 10 mA to be established as equivalent to a temperature of 37.5°C.

Note that analogue signals have a non-zero lower limit to their range to enable zero valued signals to be distinguished from faulty ones. For example, in Figure 4.1 a signal of 4 mA corresponds to a temperature of 0°C whereas a signal of 0 mA indicates a faulty instrument.

4.3 Linearity

Linearity is an issue that pervades automation. Signals which are linear are easy to interpret intuitively. Elements that have linear responses are relatively easy to control. Unfortunately, non-linear input-output relationships occur frequently. They may be inherent in the process or arise from the measurement or control actions. A basic strategy of control system design is to eliminate non-linearities by utilising elements with opposite non-linear affects such that the overall relationship is roughly linear.

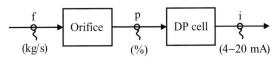

Fig. 4.2 Flow measurement block diagram

Consider, for example, Figure 4.2 in which the square relationship arising from the use of an orifice plate for flow measurement is offset by a square root characteristic applied in the dp cell. Suppose that the calibration of the orifice and dp cell is as shown in Table 4.4.

Table 4.4 Flow measurement calibration

	Input	Output
Orifice	0–f_m kg/s	0–100%
dp cell	0–100%	4–20 mA

Noting that each signal is expressed as a fraction of its range, the characteristic of the orifice is of the form of Equation 4.1. For overall linearity, the characteristic required of the dp cell must be of the form of Equation 4.2.

$$\frac{p}{100} = \left(\frac{f}{f_m}\right)^2 \tag{4.1}$$

$$\frac{(i-4)}{16} = \sqrt{\frac{p}{100}} \tag{4.2}$$

Effective signal processing is fundamental to process automation. Signals must be transmitted safely and reliably between elements in such a way that the inherent value of the signal is not changed. This theme is developed further in some of the following chapters.

Pneumatics

5.1 Air Supply
5.2 Pilot Valves
5.3 Flapper Nozzles
5.4 Pneumatic Relays
5.5 Force Balance

Historically, pneumatic instrumentation has made a major contribution to process automation. Some of the control loop elements produced were very sophisticated both in terms of mechanical design and precision engineering. However, to a large extent, electronics have superseded pneumatics. Only a limited number of pneumatic elements are still in widespread use, mostly associated with actuation. This chapter covers the devices from which those elements are built.

5.1 Air Supply

Pneumatic elements require a clean supply of air at constant pressure, usually supplied by a compressor. Various aspects of the supply and distribution of "instrument air" are covered in BS 6739. It is normal practice to withdraw air from the supply system on a local basis, delivering it to the elements through metal or nylon tubing via combined air filter/regulator units of the sort shown in Figure 5.1. The purpose of the filter is threefold: to remove any entrained solids, such as particles of rust from the pipework, to remove condensate, and to reduce the oil content, introduced during compression, to a fine mist. The regulator is set at an appropriate pressure P_s, typically of 1.4 bar (20 psi) for instruments operating in the 0.2–1.0 bar standard range. Distribution of air supply is discussed in more detail in Chapter 51.

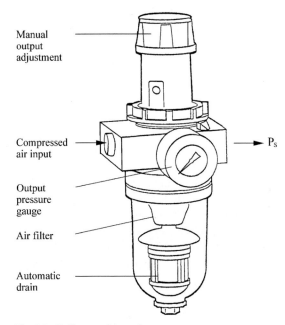

Fig. 5.1 Air filter regulator unit

5.2 Pilot Valves

Perhaps the simplest and most common pneumatic element, usually associated with discrete output channels, is the pilot valve, so called because of its use for switching on/off of the air supply to larger isolating valves. Pilot valves are ambiguously referred to in manufacturer's literature as control valves, solenoid valves or pneumatic relays. They are usually of a modular construction and consist of a port assembly and an actuator. The port assembly physically consists of a casting with supply, outlet and vent ports and, typically, a sprung diaphragm. There are various actuating mechanisms but it is the solenoid type that is of particular interest, as depicted in Figure 5.2.

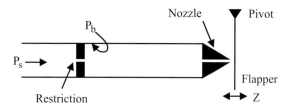

Fig. 5.3 Flapper nozzle assembly

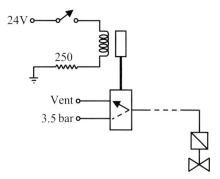

Fig. 5.2 Solenoid actuated pilot valve

the back pressure P_b, varies with the width of the gap between the nozzle and flapper as indicated in Figure 5.4.

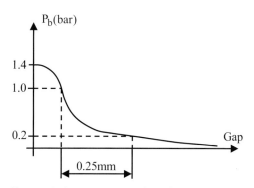

Fig. 5.4 Back pressure *vs* gap relationship

When the input to the solenoid is switched on, current flows through the coil creating a magnetic field. The force induced in the ferrous core is transmitted to the diaphragm, causing it to flex. This enables air to be routed through from the supply to the outlet, shown connected to the actuator of an isolating valve. When the current is switched off, the spring returns the diaphragm to its original position in which the outlet port is vented to atmosphere.

5.3 Flapper Nozzles

The flapper nozzle is the basic input detector mechanism used in pneumatic instruments. It is essentially a transducer which converts mechanical position z into pressure as shown in Figure 5.3.

An air supply P_s of approximately 1.4 bar is applied to a small tube containing a fine capillary bore restriction and a nozzle. The "flapper" is pivoted and is mechanically positioned across the nozzle by some sensing mechanism. The pressure between the restriction and the nozzle, known as

5.4 Pneumatic Relays

Unfortunately, the flapper nozzle relationship is non-linear and unsuitable for control purposes, so it has to be used in conjunction with a relay. Figure 5.5 depicts the continuous bleed type of pneumatic relay and Figure 5.6 is its equivalent symbol. Air is supplied at approximately 1.4 bar to the lower chamber and through the restriction to the upper chamber and nozzle. Each chamber has a plug and seat type of valve. The plugs are on a common stem, the upper plug being attached to a thin metallic diaphragm and the lower one being returned by a spring.

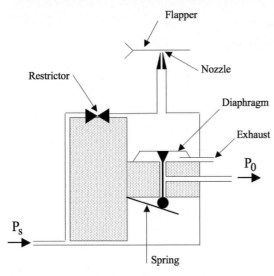

Fig. 5.5 Continuous bleed type of pneumatic relay

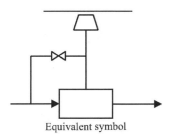

Fig. 5.6 Symbol for pneumatic relay

If the back pressure in the nozzle increases, the diaphragm flexes downwards slightly, causing the supply valve opening to increase and the exhaust valve opening to decrease. This results in a new equilibrium being established with a higher output pressure P_0 at which the air flow through the two valves is balanced. Conversely, if the back pressure reduces, the output system will vent itself through the exhaust and a reduced output pressure is established. Note that the continuous flow of air through the exhaust is by intent: it is often confused for a leak! Typical air consumption is 1–2 L/min.

This arrangement is very sensitive: movement of the flapper by only 0.025 mm is sufficient to give the relay an output range of 0.2–1.0 bar. By careful design, the operating range of the relay can be established on a linear section of the back pressure curve which, after amplification, gives an almost linear relationship over the whole of the output range. The relay also functions as a transmitter and, as such, amplifies the volume of the output. If the supply of air to the output system was dependant upon the tiny flow of air through the restriction, its operation would be too slow for practical use. However, by using the back pressure to manipulate the opening of the supply and exhaust valves, the full supply pressure is available for output purposes, as indeed is the exhaust for venting purposes.

5.5 Force Balance

Another very common pneumatic element, usually associated with analogue output channels, is the I/P converter which normally has an input i of 4–20 mA and produces an output P_0 of 0.2–1.0 bar. Most I/P converters are designed on the basis of a force balance, the construction of which is typically as depicted in Figure 5.7.

The force bar is a rigid rod, pivoted about a fulcrum, which acts as the flapper. The actuating force is generated by the dc input current flowing through the coil of a solenoid. Following an increase in current, a clockwise torque is applied to the bar. This positions the flapper closer to the nozzle resulting in an increase in relay output which is applied to the feedback bellows. As the bellows expand an anti-clockwise torque is developed which results in the bar taking up some equilibrium position. Clearly, by design, the various devices are positioned such that their relative mechanical advantages result in an input range of 16 mA corresponding to an output range of 0.8 bar. The zero spring is used to preload the bar such that with an input of 4 mA the output is 0.2 bar.

In a force balance type of device, such as the I/P converter described, the displacement of the force bar is almost negligible. This is because it is pivoted at a fixed fulcrum and the feedback bellows ensure that it is always positioned within the operating range of the flapper nozzle assembly. It is thus essential that no bending of the force bar should

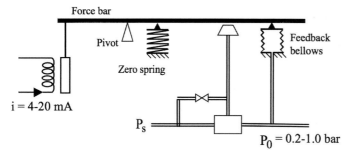

Fig. 5.7 Force balance type of I/P converter

occur. For this reason, force balances are of a robust construction and their housings are designed such that external stresses cannot be transmitted to the force bar. Because they have such little movement they suffer few wear and alignment problems.

The design of pneumatic dp cells is discussed in Chapter 11, actuators in Chapter 19 and positioners in Chapter 21.

Electronics

Chapter 6

6.1 Semiconductors
6.2 Diodes
6.3 Transistors
6.4 Logic Gates
6.5 Flip-Flops
6.6 Counters, Registers and Memory
6.7 Operational Amplifiers

Electronic circuits may be considered as being either passive or active. In essence, passive circuits consist of components associated with analogue signals such as resistors, capacitors and operational amplifiers. An example of a passive circuit is the simple filter used for smoothing noisy signals. Active circuits contain components associated with discrete signals such as diodes, transistors and gates. Active circuits invariably contain passive components too.

Simple circuits may be made up of components individually wired together as appropriate. However, it is normal practice, especially for complex circuits or those produced in quantity, for the components to be mounted on purpose made printed circuit boards (PCB). The connections between the components are established by dedicated tracks within the surface of the board with power supplies, earthing, *etc.* being distributed about the board on general purpose tracks. In computer control systems these PCBs are usually referred to as cards. Highly integrated circuits (IC) are commonly fabricated on silicon chips which are themselves mounted on PCBs referred to as mother boards.

This chapter introduces a selection of electronic components and circuit types that are of significance in the design of instrumentation and control systems. A good introduction to the subject is given by Duncan (1997): for a more comprehensive treatment the reader is referred, for example, to the classic text by Hughes (2005).

6.1 Semiconductors

Semiconductors are materials whose electrical conductivity lies between that of good conductors such as copper and good insulators such as polythene. Silicon is the most important semiconductor. Silicon has four valence electrons and forms a stable covalent lattice structure. At normal temperatures the atoms within the lattice are vibrating sufficiently for some of the bonds to break and release electrons. These enable a small current to flow if a potential difference is applied. The resistance of a semiconductor increases with temperature.

The use of semiconductors for diodes, transistors and the like depends on increasing their conductivity by the addition of small, but controlled, amounts of other elements, a process known as doping. The two elements most commonly used for doping purposes are phosphorus and boron. Doping with phosphorus, which has five valence electrons, results in a surplus of electrons which are available for conduction purposes: a so-called n-type semiconductor. Doping with boron, which has three valence electrons, results in a deficit of electrons which are analogous to positive charges: a p-type of semiconductor.

The operation of many semiconductor devices depends upon the effects which occur at the junction between p- and n-type materials formed within a continuous lattice structure.

6.2 Diodes

Diodes are semiconductor devices, the most common type being the so-called p-n junction diode. It is represented by the symbol shown in Figure 6.1 which also shows the characteristic for a silicon diode.

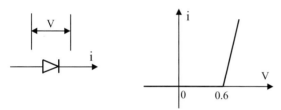

Fig. 6.1 Symbol and characteristic of a diode

In essence, the diode will permit current to flow through it in the direction of the symbol provided there is a voltage drop across it of approximately 0.6 V. This voltage drop is referred to as the forward bias. Normally a diode has to be protected by a resistor in series to limit the current through the diode to its rating. Negligible current flows if the polarity is reversed, unless the voltage difference is so large as to cause the junction to break down. Junction diodes are commonly used to prevent damage to circuits from reversed power supply. The symbol for a silicon junction zener diode and its characteristic is shown in Figure 6.2.

In forward bias the behaviour of the zener is the same as that of the p-n junction type of diode. However, in reverse bias there is a threshold beyond which the current changes at virtually constant voltage. Indeed, unless limited by a resistor in series, the current will damage the diode. The

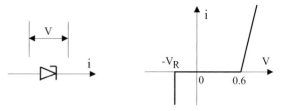

Fig. 6.2 Symbol and characteristic of zener diode

threshold value, known as the breakdown or reference voltage V_R, is determined by the composition of the semiconductor: a wide range of zeners with different nominal reference voltages is available. This property of the zener is exploited extensively for protection against over-voltages and for stabilising power supplies. Consider, for example, the circuit for the analogue input channel of a control system as depicted in Figure 6.3.

The dp cell, which has a 24-V d.c. supply, transmits a 4–20 mA signal. The objective is to produce a voltage drop V_0 of 1–5 V across the 250 Ω resistor R_0 which can be "read" by the controller. The output voltage of the dp cell V_1 varies according to the current being transmitted. For example, for a 20-mA current V_1 would be approximately 15.6 V allowing for the 0.6 V forward bias across the diode, 10 V across the 500 Ω line resistor R_1, and 5 V across R_0. The diode protects the dp cell because it prevents any reverse current flow, and the zener protects the controller from overvoltage. Since the maximum voltage under normal conditions across R_0 is 5 V, a zener with a reference voltage of approximately 5.5 V would be appropriate. In effect, V_0 could vary with current up to a maximum of 5.5 V beyond which the zener would stabilise V_0 at 5.5 V by "leaking" current to earth. The capacitor, in conjunction with the resistors, filters out high frequency noise.

6.3 Transistors

Transistors are semi-conductor devices manufactured either as discrete components or as parts of integrated circuits. There are two basic types of transistor: the junction transistor and the less

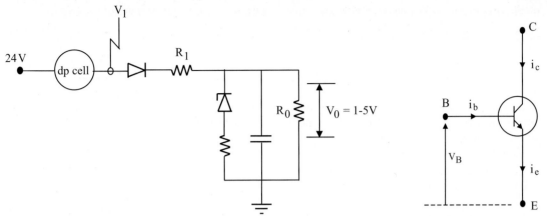

Fig. 6.3 Circuit for analogue input channel

Fig. 6.4 Symbol for a transistor

common field effect transistor. The most common type of junction transistor is the silicon n-p-n type, consisting of two p-n junctions in the same silicon crystal, and is represented by the symbol shown in Figure 6.4.

A transistor has three connections: base, collector and emitter, imaginatively denoted as B, C and E, which are used to establish two internal circuits. The input base circuit and the output collector circuit have a common connection, the emitter, for which reason the transistor is said to be in the common emitter mode. Note that for an n-p-n type of transistor both the base and collector must be positive relative to the emitter.

In essence, a transistor behaves both as a switch and as a current amplifier. If the base voltage $V_B < 0.6$ V then the base current i_b is zero. However, if $V_B > 0.6$ V then

$$i_c = k.i_b \quad \text{and} \quad i_e = i_b + i_c$$

The current gain k is typically a factor of 50–1000, depending on the type of transistor, and is approximately constant for any given transistor over a limited range of i_c values. However, k does vary significantly between transistors of the same type due to manufacturing tolerances.

Capable of being switched millions of times per second, transistors are used extensively for switching in integrated circuits. Also, having no moving parts, their life is almost indefinite, a distinct advantage over other electrically operated switches such as relays. Figure 6.5 shows a typical transistor switching circuit.

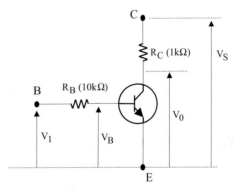

Fig. 6.5 Transistor switching circuit (for a NOT gate)

In effect the base circuit is a forward biased p-n diode and requires a large resistance R_B in series to protect the transistor from excessive currents and overheating. The collector circuit contains a load resistor R_C, the load and transistor forming a potential divider across the supply voltage V_S.

- For inputs less than 0.6 V the transistor is off and behaves like an open switch. If $V_1 < 0.6$ V then $i_b = i_c = 0$ and $V_0 = V_S$.
- For a range of inputs, greater than 0.6 V but less than some upper limit V_U, the transistor is partly

on and its output varies inversely with base current. If $V_U > V_1 > 0.6$ V then $i_c = k.i_b$ and $V_0 = V_S - i_C.R_C$.
Note that, because of the large value for the gain k, these relationships only hold true over a very narrow range of inputs.

- When the input reaches this upper limit, the base current will be such that the all of the supply voltage is dropped across the load. The transistor is fully on, behaving like a closed switch giving maximum output current and zero output voltage. If $V_1 = V_U$ then $i_C.R_C = V_S$ and $V_0 = 0$.
- Further increasing the input has no effect on the output: the transistor is said to be saturated. If $V_1 > V_U$ then $V_0 = 0$.

The behaviour of the switching circuit is summarised in Table 6.1.

Table 6.1 Behaviour of switching circuit

V_1	i_b	V_0	i_c
Low ($V_1 < 0.6$ V)	0	High (V_S)	0
$V_U > V_1 > 0.6$ V	i_b	$V_S - i_C.R_C$	$k.i_b$
High ($V_1 > V_U$)	V_1/R_B	Low (0)	High (V_S/R_C)

6.4 Logic Gates

Logic gates are switching circuits used in digital electronics for handling discrete signals. "High" is represented as logic level 1 and corresponds to a voltage close to the supply, e.g. +5 V, and "low" is represented as logic level 0 corresponding to a voltage close to 0 V. The output state of a gate depends on the states of its inputs. The switching circuit of Figure 6.5 is normally referred to as a NOT gate, or inverter, since if its input is high the output is low, and vice versa.

There are various common gate types, the behaviour of each of which is summarised in a truth table, as shown in Table 6.2.

In practice, digital electronic circuits consist of multiple gates using combinational logic.

Table 6.2 Symbols and truth tables for common gates

NOT

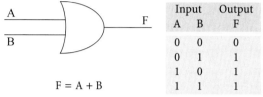

$F = \overline{A}$

Input A	Output F
0	1
1	0

OR

$F = A + B$

Input A	B	Output F
0	0	0
0	1	1
1	0	1
1	1	1

NOR

$F = \overline{(A + B)}$

Input A	B	Output F
0	0	1
0	1	0
1	0	0
1	1	0

AND

$F = A.B$

Input A	B	Output F
0	0	0
0	1	0
1	0	0
1	1	1

NAND

$F = \overline{A.B}$

Input A	B	Output F
0	0	1
0	1	1
1	0	1
1	1	0

Exclusive OR

Input A	B	Output F
0	0	0
0	1	1
1	0	1
1	1	0

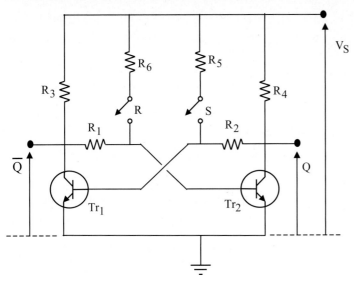

Fig. 6.6 Circuit for an SR bistable flip flop

6.5 Flip-Flops

The flip-flop is the basic building block of circuits used for counters, registers and memory. The circuit for a flip-flop based upon an SR type bistable is as depicted in Figure 6.6.

A flip-flop is said to be a bistable device because it has two stable states and is realised by a so called multivibrator circuit. In fact there are three types of multivibrator circuit: bistables, astables and monostables, which lend themselves to different applications. Any multivibrator circuit contains two transistors, the essential characteristic being that the collector of each is connected to the base of the other. In the case of the SR bistable, both of these connections are across resistances, shown as R_1 and R_2 in Figure 6.6.

The outputs of the bistable are the voltages at the two collectors, Q and $\overline{Q}$, which are complementary. There are also two switches which provide the input signals: these are denoted as S (set) and R (reset), hence the term SR bistable.

When the supply V_S is connected, both transistors draw base current but, because of slight differences, one has a larger base current and saturates quickly. Suppose that is transistor Tr_1. Thus Tr_1 draws base current through R_2 and R_4, virtually all of the supply voltage V_S is dropped across R_3 and the output of Tr_1 is zero. Thus the base current of Tr_2 drawn through R_1 is zero so the output of Tr_2 becomes the supply voltage V_S. This is a stable state.

Suppose the reset switch R is momentarily closed. This causes the base of Tr_2 to draw a pulse of current through R_6 resulting in the collector of Tr_2 drawing a pulse of current through R_4. The voltage drop across R_4 causes the output of Tr_2 to fall suddenly to zero. That in turn cuts off the base current through R_2 to Tr_1, the output of which jumps to the supply voltage V_S. The increase in output voltage of Tr_1 enables the base of Tr_2 to continue to draw current through R_1, even though the reset switch is no longer closed. Thus a new stable state has been established.

A similar analysis reveals that the effect of the set switch S being momentarily closed is to restore the original stable state. The logic of the SR bistable is depicted in truth table form in Table 6.3 in which the closure of either switch S or R and an output of V_S for either transistor is denoted by logic 1.

In summary, closing the set switch S, no matter how temporarily, causes the output of Tr_2 to flip to its high state and to remain there indefinitely. It

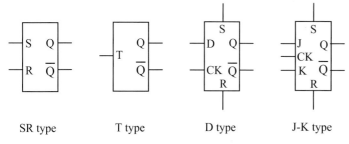

SR type   T type   D type   J-K type

Fig. 6.7 Symbols of SR, T, D and J-K types of bistable

Table 6.3 Truth table for an SR flip flop

State	S	R	Q	$\overline{Q}$
Set	1	0	1	0
Reset	0	1	0	1

will only flop to its low state when the reset switch R is closed, no matter how temporarily, where it will stay indefinitely, until the set switch is closed again. The output of Tr_1 does the opposite.

Besides the SR type of bistable there are also the so-called T type, D type and J-K types, all of whose symbols are as depicted in Figure 6.7.

- The T (triggered) type of bistable is a modified SR type with extra components that enable successive pulses, applied to an input referred to as a trigger, to toggle its output between its two output states. Two input pulses have to be applied to the trigger to cause one output pulse.
- The D (data latch) type of bistable is used for clocked logic. Most IC systems have very large numbers of bistables and it is essential that their operation is synchronised. For example, with memory, all the relevant bistables have to be changed simultaneously when new values are read in. The D type bistable has a clock (CK) input. Thus on, and only on, the rising edge of the clock pulse, the output Q is set to whatever is the state (0 or 1) of the data bit at the input D. The output Q latches onto that output until the occurrence of a rising edge with a different input. The S and R inputs are used for setting (Q = 1) or resetting (Q = 0) the output.
- The J-K type of bistable is the most versatile of all the bistables. It has two inputs, called J and K for no obvious reason. When clock pulses are applied to CK, the output Q:
  i. Retains its current value if J = K = 0
  ii. Acts as a D type flip-flop if J and K are different
  iii. Acts as a T type flip-flop if J = K = 1

6.6 Counters, Registers and Memory

A counter consists of a set of flip-flops interconnected such that the number of pulses in a series of pulses may be counted in binary code. The size of the counter, and hence the number of bits (0 or 1) in the binary word, is established by the number of flip-flops involved. Figure 6.8 depicts a simple 3-bit binary up-counter.

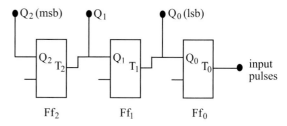

Fig. 6.8 A three-bit binary up-counter

The binary counter shown consists of T type bistables. The input train of pulses is applied to the trigger of the first (right hand) flip-flop. The output of the first flip-flop, which corresponds to the least significant bit (lsb) of the word, is used to toggle the trigger of the second, and so on. At any instant, the output of the counter is the set of values of the outputs of all three flip-flops. This counter is capable of counting up from 000 to 111 in binary, corresponding to 0 to 7 in decimal. Having reached 111, any further input pulse would progress the output to 000 and onwards. Clearly, in practice, counters that are much bigger than three bits are used. Note that binary notation is covered in detail in Chapter 7.

Binary counters can also be realised by D and J-K types of bistable. For D types the series of input pulses is applied to the CK input, with the not Q output being connected to the D input, and for J-K types the inputs are set with $J = K = 1$. The counter of Figure 6.8 is of the up-type: that is, starting with 000, pulse inputs cause the output to increment towards 111. Different circuits are available for down counters and decimal counters.

A shift register stores a binary number and transfers (shifts) it out onto a bus when required. The register consists of a number of D or J-K type of flip-flops, one for each bit of the word stored. Figures 6.9 and 6.10 depict serial and parallel arrangements respectively for a simple 4-bit shift register using D type flip-flops.

In the serial register, the Q output of each flip-flop is applied to the D input of the next. The bits are loaded one at a time, from the left as depicted, and move one flip-flop to the right every clock pulse. Four clock pulses are needed to enter a 4-bit word, and another four pulses to move it out serially. In the parallel register, all bits enter their D inputs simultaneously and are transferred together, by the same clock pulse, to their respective Q outputs from which they can then be read in parallel.

Registers are ubiquitous. They are used in calculators for storage of binary numbers prior to addition. The central processor unit of any microprocessor has multiple registers which are used for temporarily holding key values during processor operations, as described in Chapter 9. Registers provide the basis for computer memory where the IC equivalent of millions of flip-flops are etched onto silicon chips.

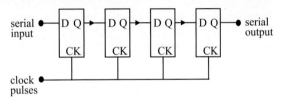

Fig. 6.9 A serial four-bit shift register

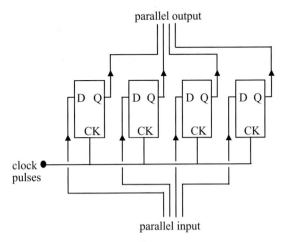

Fig. 6.10 A parallel four-bit shift register

6.7 Operational Amplifiers

Operational amplifiers (op-amps) are essentially high gain, high resistance, linear IC devices which, depending on the associated circuits, can be used for a variety of signal conditioning applications. Figure 6.11 shows an op-amp circuit used for scaling purposes.

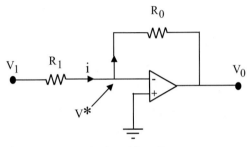

Fig. 6.11 Op-amp circuit used for scaling purposes

Functionally an op-amp has two inputs, one positive and the other negative, and one output although, in practice, it has other connections for power supplies, *etc*. The output voltage is given by

$$V_0 = k.V^* \qquad (6.1)$$

where V^* is the difference between the input voltages. The amplification factor k is typically of the order 10^8 which means that, in practice, V^* has to be virtually zero.

Noting that the op-amp has virtually infinite resistance and therefore draws negligible current, Ohm's law may be applied across the resistors:

$$i = \frac{V_1 - V^*}{R_1} = \frac{V^* - V_0}{R_0}$$

Substituting for V^* from Equation 6.1 and rearranging yields

$$V_0 = -\frac{kR_0}{(k-1)R_1 - R_0}.V_1$$

which, because of the value of k, results in

$$V_0 \approx -\frac{R_0}{R_1}.V_1 = -K.V_1$$

where K is the scaling factor applied. Clearly by specifying appropriate values of R_0 and R_1, any desired scaling factor, referred to as the gain, may be applied to the input voltage. There are many variations on the theme. For example, the op-amp circuit of Figure 6.12 is that of an integrator.

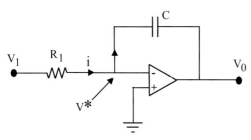

Fig. 6.12 Op-amp circuit for an integrator

Again, applying Ohm's law:

$$i = \frac{V_1 - V^*}{R_1} = C\frac{d}{dt}(V^* - V_0)$$

$$V_0 \approx \frac{1}{RC}.\int_0^t V_1 dt$$

By specifying appropriate values for R and C the input voltage may be integrated at any desired rate. A more complex op-amp circuit that is of particular interest for current transmission purposes is shown in Figure 6.13.

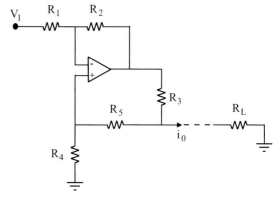

Fig. 6.13 Op-amp circuit used for current transmission

Detailed analysis of this circuit, which is not easy, shows that

$$i_0 = \frac{-R_2}{R_1.R_3}.V_1$$

provided that

$$R_2.R_4 = R_1.R_3 + R_1.R_5$$

The significance of this is that, within sensible limits, the output current i_0 is proportional to the input voltage V_1, irrespective of the line resistance R_L. Note that it is normal practice to transmit signals as currents rather than as voltages. This is because with voltage transmission the output voltage inevitably differs from the input due to line resistances, whereas with current transmission the current must be the same at either end of the line. Even if resistances are carefully matched to avoid loading effects, voltage transmission is vulnerable to distortion caused by changes in line resistance due to temperature effects or to components being replaced.

Data Typing

Chapter 7

7.1 Binary Representation
7.2 Two's Complement
7.3 Characters
7.4 Identifiers
7.5 Integer Literals
7.6 Real Literals
7.7 Time Literals
7.8 String Literals
7.9 Boolean Literals
7.10 Derived Data Construct
7.11 Array Data Construct
7.12 Enumerated Data Construct
7.13 Structured Data Construct
7.14 Comments

Representation of data, and its manipulation, is fundamental to the design and operation of digital devices. This chapter covers the common conventions for representing and storing numerical data and alphabetics. A range of terminology is introduced. These representations are used as a basis for data typing which underlies the IEC 61131 (part 3) standard for real-time programming.

Data typing involves defining the type of data associated with constants and variables used in programs. This is necessary for all types of program, whether of a procedural or configurable nature for control purposes, or of a relational or object oriented nature for other purposes. The standard supports a full range of "literals", often referred to in other languages as "primitives". The range includes integer, real, time and date, string and Boolean. Special constructs enable the more complex derived, enumerated and structured data types to be defined.

Data typing is explained in terms of structured text, which is one of the languages of the standard, and is itself covered in Chapter 8. For a more detailed coverage of data typing in relation to the IEC 61131 standard, the reader is referred to the text by Lewis (1998) or indeed to the standard itself.

7.1 Binary Representation

In digital systems, values are stored in bits. A bit is the smallest discrete entity of memory and has two possible states, as depicted in Table 7.1.

Table 7.1 Bit states

State	Binary	Boolean
On	1	True
Off	0	False

Single bits can only be of use for storing the values of variables that are inherently two state, such as logical variables and discrete I/O. To represent values greater than unity it is necessary to group bits together in multiples of 4, 8 and 16 bits, referred to respectively as nibbles, bytes and words. The term "word" is also used, ambiguously, to refer to groups of bits generically.

Table 7.2 shows the range of integer numbers 0 to 15 that can be represented by one nibble. The table depicts the binary, octal, decimal and hexadecimal equivalents of these numbers, which have bases of 2, 8, 10 and 16 respectively.

Table 7.2 Alternative representations of 0–15 decimal

Binary	Octal	Decimal	Hexadecimal
0000	0	0	0
0001	1	1	1
0010	2	2	2
0011	3	3	3
0100	4	4	4
0101	5	5	5
0110	6	6	6
0111	7	7	7
1000	10	8	8
1001	11	9	9
1010	12	10	A
1011	13	11	B
1100	14	12	C
1101	15	13	D
1110	16	14	E
1111	17	15	F

By inspection, it can be seen that a 1 in the nth significant bit, ie counting from the right, corresponds to 2**(n − 1) in decimal. Thus the decimal value N of a word consisting of m bits, the value b_n of the nth bit being 0 or 1, is given by

$$N_{10} = \sum_{n=1}^{m} b_n . 2^{(n-1)}$$

Also, by inspection, it can be seen that the range of values R that can be represented by a word consisting of m bits is given by

$$R_{10} = 0 \rightarrow (2^m - 1)$$

Table 7.3 shows the range of values that can be represented by different word lengths.

Table 7.3 Range of values *vs* word length

Number of bits	Octal	Decimal	Hexadecimal
1	0–1	0–1	0–1
4	0–17	0–15	0–F
8	0–377	0–255	0–FF
16	0–177777	0–65535	0–FFFF
32	–	$0–4.295 \times 10^9$	–
64	–	$0–1.845 \times 10^{19}$	–

7.2 Two's Complement

The values considered so far have all been positive and integer. In practice, process control systems have to be able to cope with negative and real numbers too. When a number could be either positive or negative, it is normal to use the first bit, *i.e.* the most significant bit, to indicate the sign. Thus, if the first bit is 0 the number is positive and if it is 1 the number is negative. The most common technique of representing signed numbers is referred to as two's complement. Table 7.4 shows, in descending order, the range of integer values +7 to −8 that can be represented by one nibble using two's complement.

By inspection, it can be seen that to change the sign of a number, each of the bits must be negated and 1 added. For example, to:

- Convert +7 into −7, binary 0111 is negated to become 1000 and adding 1 gives 1001
- Convert −5 into +5: binary 1011 is negated to become 0100 and adding 1 gives 0101

Note that when converting numbers in two's complement, care must be taken to negate any zeros in the most significant bits, referred to as leading zeros.

The concept of two's complement can perhaps be best appreciated by analogy with counting on the odometer of a car. Forward travel is self ex-

7.4 Identifiers

It is normal to assign names, referred to as identifiers, to constants and variables. This makes for easier understanding of programs. An identifier can be any string of letters, digits and underlines, provided that:

- The first character is not a digit
- Underline characters only occur singly
- The identifier contains no spaces
- The first six characters are unique
- It is not a keyword within the languages

TM_47 and COUNT are examples of valid identifiers. Note that the rules governing choice of identifier are consistent with conventions for tag numbers.

7.5 Integer Literals

Integers are typically used for counting purposes, and for identities such as batch numbers. There are eight integer data types as shown in Table 7.5. The type of integer used depends on the range of values concerned, and whether there are any negative values.

To minimise memory usage and, in particular, to reduce the processor loading, it is good practice to use the shortest data type consistent with the range of the variable concerned.

Table 7.5 Integer types *vs* range of values

Data type	Description	Bits	Range
SINT	Short integer	8	−128 to +127
USINT	Unsigned short integer	8	0 to +255
INT	Integer	16	−32768 to +32767
UINT	Unsigned integer	16	0 to $2^{16} - 1$
DINT	Double integer	32	-2^{31} to $+2^{31} - 1$
UDINT	Unsigned double integer	32	0 to $2^{32} - 1$
LINT	Long integer	64	-2^{63} to $2^{63} - 1$
ULINT	Unsigned long integer	64	0 to $2^{64} - 1$

Table 7.4 Binary *vs* decimal in 2's complement

Binary	Decimal
0111	7
0110	6
0101	5
0100	4
0011	3
0010	2
0001	1
0000	0
1111	−1
1110	−2
1101	−3
1100	−4
1011	−5
1010	−6
1001	−7
1000	−8

planatory. However, starting at 0000 and reversing the car for 1 km will cause the odometer to read 9999. Reversing another km will cause it to read 9998, and so on. Suppose the odometer counts in binary. Starting at 0000, if the car travels backwards, the odometer will decrement to 1111 after 1 km, to 1110 after 2 km, and so on.

7.3 Characters

In addition to numbers, systems also have to be capable of handling textual information. ISO 8859 (1987) defines 8 bit single byte coded graphic character sets based upon the former American Standard Code for Information Interchange (ASCII) code. Part 1 is a specification for the Latin alphabet which covers English and most other west European languages. The character set contains all the commonly used letters, digits, punctuation, printer control characters, etc, and allows for important international differences, such as between £ and $ symbols. In practice, textual information consists of strings or files of characters. Each character is stored and manipulated as a single byte of memory.

In structured text, both constants and variables are defined as being of a particular type by statements such as

```
COUNT:   usint;
```

in which the variable COUNT is declared as being of the integer type USINT. The colon means "is of the type" and the semicolon denotes the end of the statement. Note that throughout this chapter, and indeed the rest of the text, the convention used for all structured text is to write identifiers and other user defined terms in upper case and IEC 61131 keywords in lower case. It is good practice to use case to distinguish keywords in this way. It doesn't matter which case is used, as long as it is used consistently.

Variables may be initialised by having values assigned to them at their definition stage, otherwise they default to zero. For example:

```
COUNT:   usint:= 47;
```

in which case the variable COUNT is declared as being an integer and set to an initial value of 47 in the same statement. The assignment operator := means "becomes equal to".

All numbers in structured text are assumed to be to base 10 unless prefixed by 2#, 8# or 16# which denote binary, octal and hexadecimal respectively. For example:

```
2#101111 = 8#57 = 47 = 16#2F
```

7.6 Real Literals

Real numbers, *i.e.* those that contain a decimal point, are used extensively in calculations, for example in scaling analogue I/O signals or in evaluating control algorithms. There are two real data types, as shown in Table 7.6, the type used depending on the values concerned. The ranges cover both positive and negative numbers and enable both very large and small fractional values to be handled.

Real numbers are normally held in a floating point format consisting of a mantissa and expo-

Table 7.6 Real types *vs* range of values

Data type	Description	Bits	Range
REAL	Real numbers	32	$\pm 10^{\pm 38}$
LREAL	Long real numbers	64	$\pm 10^{\pm 308}$

nent. The values of both the mantissa and the exponent are stored in two's complement form, and an algorithm used to convert from one format to the other.

Note that exponential notation may be used with real variables. For example, a ramp rate of 72°C/h may be initialised in degrees Celsius per second as follows:

```
RAMP:   real:= 2.0E-2;
```

Several variables may be defined in the same statement, an example of which is as follows:

```
T1, T2, TAVG:   real;
```

7.7 Time Literals

Time and date literals are available for both elapsed and absolute time. An elapsed time is the duration of an activity since some instant in real time, such as the time taken since the start of a discrepancy check. Absolute time and/or date enables control activity to be synchronised relative to the calendar, such as starting a batch at the beginning of the next shift. Time and date types are as defined in Table 7.7.

Table 7.7 Time literals

Data type	Description	Short form
TIME	Time duration	T#
DATE	Calendar date	D#
TIME_OF_DAY	Time of day	TOD#
DATE_AND_TIME	Date and time of day	DT#

The following letters are used in defining times:

```
d = days, h = hours, m = minutes,
s = seconds, ms = milliseconds
```

A statement defining an elapsed time referred to as SOAK and initialising it with a value of 2.5 min is as follows:

```
SOAK:  time:=  T#2m30s;
```

Because the most significant field of a time literal may overflow and the last field may be given in decimal format, the following examples are permissible:

```
SOAK:=  T#61s;   SAMPLE:=  T#0.5s;
```

The formats for defining dates and time of day are shown in the following examples in which values are assigned to predefined variables as follows:

```
START:= TOD#12:00:00; (*midday*)
BIRTH:= D#1947-06-27; (*a broad hint*)
```

Note the use of the comment. Comments enable annotation of programs and may be inserted anywhere provided they are framed by brackets and asterisks as follows:

```
(*This is a comment.*)
```

7.8 String Literals

Strings are used for holding characters and are typically used for identity purposes or for passing messages. They consist of printable and non-printable characters and are framed by single quotes. Non-printable characters are inserted by prefixing the hexadecimal (ASCII) value of the character with a $ symbol. There are also a number of reserved letters, shown in Table 7.8, used with the dollar symbol to denote commonly used control characters.

Typical examples of string literals are:

```
'JUICE_47'       (*a batch identification*)
'CONFIRM LEVEL OK $N'
      (*a message with an embedded line feed*)
''               (*an empty or null string *)
```

Bit strings enable storage of binary data. Their most common usage is for storage of status information. There are five data types of different length, as shown in Table 7.9.

Table 7.9 Data types *vs* length/usage

Data type	Length (bits)	Usage
BOOL	1	Discrete states
BYTE	8	Binary information
WORD	16	Ditto
DWORD	32	Ditto
LWORD	64	Ditto

Table 7.8 String literals

Code	Interpretation
$$	Single dollar sign
$'	Single quote character
$L	A line feed character
$N	A new line character
$P	Form feed or new page
$R	Carriage return character
$T	Tabulation (tab) character

7.9 Boolean Literals

The Boolean data type is used to assign the state of a Boolean variable as being true or false. Its principal use is in testing and setting the status of discrete variables, such as discrete I/O signals. The Boolean data type is defined in Table 7.10: the analogy with Table 7.1 is obvious.

Table 7.10 Boolean literals

Data type	Binary	Boolean
BOOL	0 or 1	FALSE or TRUE

Example

A variable FLAG is declared as a Boolean variable and initialised by default as being FALSE.

```
FLAG: bool;
```

When the FLAG becomes true, the mode of PUMP_47 is set to RUNNING:

```
if
  FLAG = true
then
  PUMP_47 := RUNNING;
```

7.10 Derived Data Construct

Derived data types are used to enhance the readability of structured text. In essence they enable new data types to be generated from the previously defined literals. New data types are framed with TYPE and END_TYPE statements. For example:

```
type
  PRES,TEMP: real;
end_type
```

This allows variables corresponding to pressures and temperatures to be defined as being of the type PRES or TEMP and they are subsequently treated as if they are REAL variables.

There is often a requirement to restrict the range of values that can be assigned to certain types of variable. A sub-range can be specified within the derived data construct. For example, variables of the type PRES are limited to values within the range 0.2 to 1.0:

```
type
   PRES: real (+0.2..+1.0); (* bar.*)
end_type
```

7.11 Array Data Construct

Arrays are a compact means of storing large quantities of data and enable the data to be accessed efficiently. They too are framed by TYPE and END_TYPE statements. For example, an array for holding five real values could be as follows:

```
type VECTOR:
   array [1..5] of real;
end_type
```

One-dimensional arrays enable a technique known as indirect addressing, in which plant and data items are referenced via parameter lists. This technique is used extensively in batch process control and is explained in Chapter 29 and used in Chapter 37.

Arrays may be n dimensional. For example, an array of 10 × 50 elements, for holding 10 sets of 50 values may be defined as follows:

```
type MATRIX:
   array [1..10,1..50] of real;
end_type
```

7.12 Enumerated Data Construct

This construct enables different states of an integer variable to be named, such as the operational modes of a device. Enumerated data constructs are also framed by the statements TYPE and END_TYPE, with the named states being bracketed as in the following example:

```
type PUMP_MODE:
     (ENABLED, RUNNING, STOPPED,
      STANDBY, FAULTY);
end_type
```

Note that variables of the type PUMP_MODE can only be assigned values from the list of named states.

7.13 Structured Data Construct

Composite data types may be derived by defining a structure from existing data types. The composite is declared by framing the definition with STRUCT and END_STRUCT statements. In the following example, a structured data type called TEMP_TRANS is declared which contains all the data associated with a temperature transmitter; note that each element of the list can be given a meaningful name:

```
type TEMP_TRANS;
  struct
    STATUS:     bool;
    RANGE:      real;
```

```
    OFFSET:      real;
    HI_ALARM:    real;
    LO_ALARM:    real;
    NO_ALARMS:   int;
    DAMPING:     time;
    CALIBRATION: date;
  end_struct;
end_type
```

A generic list may thus be formulated, with default values, from which specific instances may be created. For example, the temperature transmitter T47 may be declared as an instance:

```
type
  T47:  TEMP_TRANS;
end_type
```

The elements of any such list may be accessed individually by means of a dot extension. For example, the status bit of T47 may be set by the statement

```
T47.STATUS := 1;
```

The structured data construct is one of the more powerful features of structured text. In effect, the construct enables objects to be handled, although it is important to appreciate that structured text is not an object oriented language, as described in Chapter 107.

7.14 Comments

It is good practice to define data and variable types in groups at the beginning of a program so that they are available throughout the program. The constructs used provide for consistency in software development and enable, through compilation, many errors to be detected very effectively. Although data typing has been explained in terms of structured text, it is generic to all five languages of the IEC 61131 standard. Strong data typing, as described in this chapter, is fundamental to the standard: it enables software developed in the different languages to be integrated.

Structured Text

8.1 IEC Software Model
8.2 Global and External Variables
8.3 Input and Output Variables
8.4 Memory Variables
8.5 Miscellaneous
8.6 Program Structure
8.7 Specimen Program
8.8 Comments

This chapter covers structured text, one of the five languages of the IEC 61131 (part 3) standard, the others being instruction lists, ladder diagrams, function block diagrams (FBD) and sequential function charts (SFC). Structured text was introduced in Chapter 7: it is a procedural language, designed for real-time programming, and is appropriate for process control. Although it is not a so-called high level language, it does support constructs associated with high level languages, such as nesting. Similarly, it is not an assembler language, although it supports low level constructs such as direct memory addressing. For this reason, structured text is often described as being like pseudo-code.

Many of the variables and constructs used in process control are different to those used in conventional programming. This is largely due to the real-time nature of programming, to the fact that programs invariably run simultaneously and interact, and that the database is often distributed between different devices. This chapter, therefore, focuses on these variables and constructs.

Structured text is used extensively throughout this text for explaining a variety of concepts. For a more detailed coverage of the language, the reader is referred to the text by Lewis (1998).

8.1 IEC Software Model

Figure 8.1 depicts the software model of the IEC 61131 standard. This allows for the partitioning of applications into entities referred to as program organisational units (POU). These entities include configurations, resources, programs and function blocks.

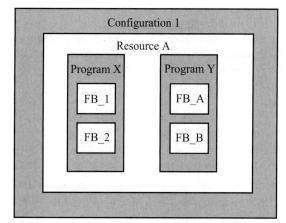

Fig. 8.1 Software model of IEC 61131 standard

A configuration is equivalent to all of the hardware and software of either a single programmable logic controller (PLC) system or of the node of a distributed control system (DCS): both PLCs and DCSs are considered in detail in Section 6. Distributed systems may be considered to consist of multiple configurations. Within each configuration there are one or more resources. A resource provides all the support necessary for executing programs and is, typically, associated with a processor on a mother board. Resources should be capable of operating independently of each other. Programs are run within resources under the control of tasks. Although Figure 8.1 depicts a program consisting of function blocks, a program may be written in any one of the five IEC 61131 languages, or a mixture thereof.

The declaration of variable type relates to these entities.

8.2 Global and External Variables

In general, variables are declared in lists within a software entity. If the variables are only to be used within that entity they are defined as follows:

```
var
   RAMP:          real:= 2.0E-2;
   T1, T2, TAVG:  real;
   SOAK:          time:= T#2m30s;
end_var
```

Global variables may be defined at the configuration, resource or program levels. By so doing, those variables are declared as being available for use by any software entity at a lower level within the parent entity. Such variables may be accessed from the lower level by declaring them to be external variables at that lower level. For example, the set point and mode of a temperature controller may be defined as a global variables within a program as follows:

```
var_global
   T47_SP:    real;
   T47_MODE:  CONT_MODE;
end_var
```

where the enumerated data type CONT_MODE has already been defined at the program level. Within the program values may be assigned to T47_SP and T47_MODE.

A temperature controller function block T47 within that program may then access the values of the set point and mode variables. To do so, within the definition of the function block, the variables must be defined as being external variables, as follows:

```
var_external
   T47_SP:    real;
   T47_MODE:  CONT_MODE;
end_var
```

Any variable defined within an entity that contains lower level entities is potentially accessible by any of those lower level entities. It is therefore capable of being inadvertently changed. By defining variables at the appropriate level, constraints are imposed as to what variables can be used within any particular entity. Use of the global and external construct makes that access explicit. Thus the scope for errors being introduced inadvertently is significantly reduced, making for enhanced software integrity. This is fundamental to the concept of encapsulation which is concerned with making software entities as self contained as is reasonably practicable.

8.3 Input and Output Variables

These relate to external sources of data and are used at the program, function block and function levels of POU. It is often the case that the value of a parameter used within one such entity is determined by another. Appropriate declaration of such parameters as input or output variables enables their values to be passed between entities. The most obvious example is in the use of function blocks, where the output of one block becomes the input of the next. Thus a slave controller function block may require declarations as follows:

```
var_input
   PC_47.OP:  real;
   TC_47.OP:  real;
end_var
```

```
var_output
  FC_47.OP: real;
end_var
```

With `PC_47`, `TI_47` and `FC_47` having been declared elsewhere as being of the structured data construct type, values for the set point and measured value of the slave loop may be imported and a value exported to the output channel.

8.4 Memory Variables

These are invariably used by input and output variables and enable memory locations to be referenced directly. The identity of such a variable starts with a % character and is followed by a two letter code, depending on the nature (input or output) of the memory location and the type (bit, byte, *etc.*) of the parameter stored there. The letter code is defined as in Table 8.1. If the second letter of the code is not given, the code is interpreted to be X.

The letter code is accompanied by a numerical reference. This may be to a specific memory location or to a hardware address and is clearly system specific. Some examples are as follows:

```
%I100        (* Input bit no 100 *)
%IX16#64         ditto
%IW47        (* Input word no 47 *)
%IW5.10.13  (* Rack 5, Card 10, Channel 13 *)
%QL27       (* Output long word no 27 *)
%MB8#377    (* Memory location byte no 255 *)
```

8.5 Miscellaneous

There are a number of attributes that may be assigned to variables at the declaration stage. `RETAIN` indicates that the value of a variable is held during loss of power. For example:

```
var_output retain
  PIDOUT: real;
end_var
```

Table 8.1 Memory variable codes

1st letter	2nd letter	Interpretation
I		Input memory location
Q		Output memory location
M		Internal memory
	X	Bit
	B	Byte (8 bits)
	W	Word (16 bits)
	D	Double word (32 bits)
	L	long word (64 bits)

`CONSTANT` indicates that the value of a variable cannot be changed. For example:

```
var constant
  MAXTEMP: real:= 78; (* degC *)
end_var
```

`AT` is used to attach an identifier, such as a tag number, to the memory location of a constant or variable. For example:

```
var
  TM19 at %ID47: real;
end_var
```

Variables may have initial values given to them at the declaration stage. These will override any default initial values defined for the data type.

Communication of data between configurations is realised by means of ACCESS variables, or by the use of dedicated send and receive type function blocks.

8.6 Program Structure

Inspection of the fragments of structured text used so far reveals a number of constructs for defining an input variable and a function, and for handling conditional statements. These, and similar, constructs are what provide the structure of structured text. Indeed, a program is itself created by means of a construct:

```
program NAME
  <constructs>
  <statements>
end_program
```

Within this construct a typical program consists of a set of constructs for declaring data and variable types, and constructs for user defined functions and function blocks. These are followed by the main body of the program consisting largely of conditional and iterative constructs and assignment statements. The conditional and iterative constructs used for program control provide much flexibility.

The conditional construct is of the general form

```
if <logic test> then
  <statements>
elsif <logic test> then
  <statements>
else
  <statements>
end_if
```

The logic test may be a simple check of a Boolean variable or the evaluation of a Boolean expression. An alternative CASE ... END_CASE construct enables selected statements to be executed according to the value of some integer variable.

There are several iterative constructs:

```
for <initial integer value>
  to <final integer value>
  by <increment> do
  <statements>
end_for

while <logic test> do
    <statements>
end_while

repeat
  <statements>
  until <logic test>
end_repeat
```

Note that there are no explicit goto or jump commands, which is consistent with the ethos of structured programming. Branching has to be realised implicitly by means of the conditional and iterative constructs, and the nesting thereof. An Exit command can be used within iterative constructs to terminate iteration.

Constructs consist of statements which are executed in program order. Most of the statements are assignments of type and/or value, and evaluations of arithmetic and/or Boolean expressions. Evaluation of expressions is conventional, expressions containing one or more operators, variables and functions. Operators are standard and executed in order of precedence, for example ( ), $*$, $/$, $+$, *etc*.

8.7 Specimen Program

A specimen program in structured text is presented in Program 8.1. It is presented here simply to demonstrate program structure and to illustrate how some of the various constructs may be used. The program is the structured text equivalent of Figure 29.2 which depicts a charging sequence:

Program 8.1. Specimen program in structured text

```
program CHARGE
var_input
  LT52 at %IW52: real;
  WT53 at %IW53: real;
end_var
var_output
  YS47 at %QX47: bool;
  YS48 at %QX48: bool;
  YS49 at %QX49: bool;
  YS50 at %QX50: bool;
end_var
var_external
  LMAX, TDES, WMIN, WREQ: real;
  TC51: PID;
end_var
var
  WBEG, WDIF: real;
  MESSAGE_1: string:= '$n LEVEL TOO HIGH $r'
  MESSAGE_2: string:= '$n WEIGHT TOO LOW $r'
end_var
```

```
if LT52 ge LMAX then
  MESSAGE_1
 elsif WT53 le WMIN then
  MESSAGE_2
 else
  YS47:= 0;
  YS48:= 1;
  WBEG:= WT53;
  YS50:= 1;
  repeat
   WDIF:= WBEG - WT53;
   until WDIF ge WREQ
  end_repeat
  YS50:= 0;
  YS48:= 0;
  YS49:= 1;
  TC51.SP:= TDES;
  TC51.ST:= 1;
 end_if
end_program
```

8.8 Comments

IEC 61131 is not prescriptive. Thus, for example, ladder diagrams may contain function blocks, function blocks in FBDs may be written in structured text, transitions in a SFC may be triggered by rungs of a ladder diagram, *etc*. Functional requirements may therefore be decomposed into parts, each of which may be developed in whichever language is most appropriate. The strength of IEC 61131, therefore, is that it provides a coherent framework for software development.

As stated, structured text does not support branching. This would be impractical if all control software, especially for complex sequencing, could only be written in the form of structured text programs. However, that is not the case. The primary purpose of structured text is as a vehicle for developing the component parts of the other IEC 61131 languages. Thus, for example, the functionality of steps and actions in SFCs (Chapter 29) and of function blocks in FBDs (Chapter 48) is normally articulated in structured text. Constructs within these other languages are then used to support branching, looping and parallelism as appropriate.

Microprocessors

9.1 Overview
9.2 The Control Unit
9.3 The Arithmetic/Logic Unit
9.4 The Memory System
9.5 The Processor Cycle

Many instruments and almost all control systems are microprocessor based. Whilst it is seldom necessary to understand exactly how any particular processor is designed, it is often helpful in their application to have some understanding of how microprocessors work. The emphasis in this chapter therefore is on the principal features and principles of operation of microprocessors. Detailed consideration of real-time operating systems is deferred until a subsequent edition. There are hundreds of texts to which the reader could be referred for a more comprehensive treatment of the subject: one of the better ones is by Kleitz (2003).

9.1 Overview

A microprocessor is essentially a very large scale integrated (VLSI) transistorised circuit on a single silicon chip. The so-called von Neumann machine, which is the classical microprocessor architecture, is depicted in Figure 9.1. This shows the functional relationship of the major hardware elements of a typical microprocessor, and how they are interconnected by the address, control and data buses.

The architecture can be divided functionally into three parts. First, the central processing unit (CPU), contained within the broken line, which consists of the control unit, the program counter (PC), the instruction register (IR), the arithmetic/

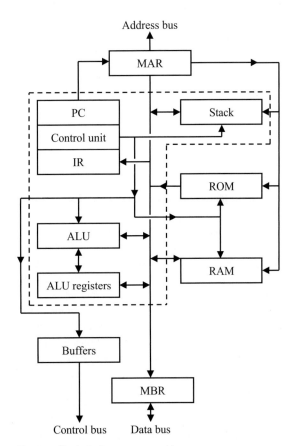

Fig. 9.1 Classical microprocessor architecture

logic unit (ALU) and its registers, and the register stack. Second, the main memory system consisting of random access memory (RAM) and read only memory (ROM). And third, the input/output system, consisting of the memory address register (MAR), the memory buffer register (MBR) and various other devices for communications purposes.

The register set consists of the IR, PC, ALU registers and the stack, the stack consisting of a number of special purpose and general purpose registers. An important point to appreciate is the distinction between the set's functions and its physical location. For example, the IR and PC are both functionally associated with the control unit but are physically part of the stack. Indeed, the stack may itself be distributed between the processor chip and main memory, *i.e.* on or off chip. This distinction is common to many aspects of microprocessor architecture.

Architectures other than that of Figure 9.1 exist. These enable, for example, parallel processing and/or reduced instruction set computing (RISC). However, they have yet to find their way into process automation to any significant extent.

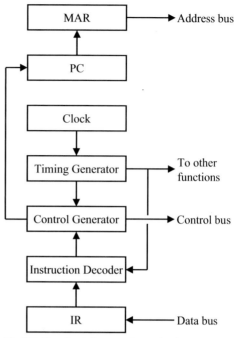

Fig. 9.2 Functional diagram of control unit

9.2 The Control Unit

A functional diagram of the control unit is shown in Figure 9.2.

The clock generates a continuous train of pulses at a rate determined by an external frequency source, *i.e.* by a quartz crystal. Current generation processors have pulse rates of the order of 1 GHz (1×10^9 pulses/s) with pulsewidths of 1 ns (1×10^{-9} s/pulse). These pulses determine the basic operating speed of the microprocessor. The timing generator, which consists of the integrated circuit equivalents of counters, registers and gates, is used to count the clock pulses and generate the timing signals, typically T_1–T_5 as shown in Figure 9.3. These establish the processor cycle and sub-cycle times.

The IR receives an instruction as a byte of information, typically 32 or 64 bits, from the data

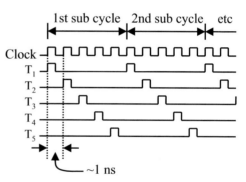

Fig. 9.3 Processor cycle and sub cycle times

bus and temporarily stores it whilst decoding takes place. This instruction is usually referred to as the operation code (op-code). The instruction decoder, in essence a complex integrated circuit, is enabled by the timing generator only when an op code is in the IR. The decoder interprets the op-code and, according to the nature of the operation to be carried out, determines how many sub-cycles are involved, which arithmetic/logic operations to execute within them, what data is to be put onto the control bus, and when.

The control generator enables the control bus with the decoder output at the appropriate instants, as determined by the timing generator. It also oversees the operation of the PC, updating it at each operation. The PC contents are transferred into the MAR at the start of an operation and held there throughout.

9.3 The Arithmetic/Logic Unit

A functional diagram of the ALU is shown in Figure 9.4.

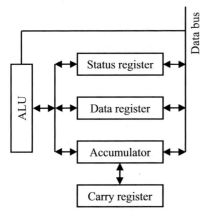

Fig. 9.4 Functional diagram of ALU

The ALU essentially consists of a gate network which enables arithmetic and logical operations to be carried out, and dedicated registers for temporarily storing operands. Movement of data between memory, registers and the ALU is bidirectional. This data flow, and manipulation of the data within the ALU, is under control of the instruction decoder and control generator.

The majority of ALU operations require two operands. Typically, one is stored in the data register and the other in the accumulator. In most microprocessors, the accumulator also stores the result of the operation, overwriting the operand that was originally stored there. Certain operations result in an overflow of the accumulator, *e.g.* a 33-bit answer to be stored in a 32-bit register. A carry register is provided to accommodate this.

All ALU operations result in flags being set. These flags are essentially independent single bit registers set to 0 or 1 according to the outcome of an operation or the state of the processor. The carry register is one such flag. These registers are individually referred to as status registers, and are collectively referred to as the condition code register.

The ALU itself is generally only capable of binary addition and subtraction, Boolean logic testing, and shift operations. More complex operations, *e.g.* multiplication, have to be realised by means of software.

9.4 The Memory System

The memory system consists of the register stack, cache, ROM, RAM, disc based memory, which may be hard or floppy, and other peripheral storage devices such as compact discs (CD), tapes, *etc*. The ROM and RAM are often referred to as main memory. Figure 9.5 shows how the stack, RAM and ROM are interconnected to the buses.

The general purpose registers are used principally for the manipulation and temporary storage of data, and are often referred to as scratch pad registers. The data in any one register may be moved to or exchanged with the contents of any other register. A feature of registers in microprocessors is that bytes may be treated in pairs to "trick" the system into thinking that longer words are available: this is particularly useful for addressing and arithmetic operations. For example, 32-bit words can be paired for 64-bit systems.

Selection of a general purpose register and the operation to be performed is accomplished by the control unit. When a byte in the IR is decoded, if a register operation is indicated, the register select is enabled by a signal on the control bus. This then directly decodes certain of the bits in the instruction byte, which is still on the data bus, and subsequently causes the contents of the registers to be placed on the data bus as appropriate.

Such register operations take only 1–2 timing periods and are much faster than ROM and RAM operations which involve loading data from mem-

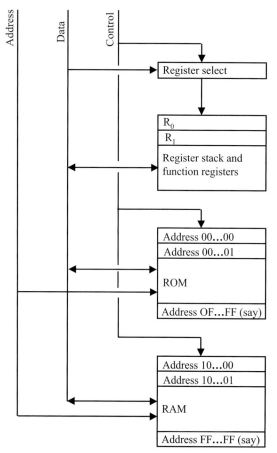

Fig. 9.5 Memory system connections to buses

ory onto the address bus, incrementing the PC, *etc.*, and take 1–2 processor sub-cycles. For this reason, processors always have an additional memory cache which is used for holding recently used information. On the grounds that recently used instructions and data are more likely to be used again than not, holding them in dedicated high speed registers local to the CPU saves time compared with retrieving them from main memory each time. This considerably speeds up program execution.

Sometimes the microprocessor has to suspend one operation in order to respond to another one of higher priority. When this happens the current values of the various registers are temporarily stored in RAM, the start address at which they are stored being entered into another special purpose register known as the stack pointer. Thus, when the lower priority task is resumed, the registers can be reloaded from RAM and the operation recommenced from where it was broken off.

Both RAM and ROM consist of 32- or 64-bit word storage locations, depending on the processor technology and memory architecture concerned. The primary difference between them is that ROM information is permanently stored and is not subject to change by actions that occur within the microprocessor, whereas the contents at RAM locations may be erased or changed under program control, and new information stored.

Each location has a unique address and is accessed by placing the desired address on the address bus. If the op-code is decoded as a read instruction, the data at that ROM or RAM location is put onto the data bus. However, if the op-code is decoded as a write instruction, the data on the data bus is written into memory at that RAM location.

A typical 32-bit microprocessor will have a 32-channel data bus to enable parallel transfer of data between memory and registers, *etc.*, and access to a maximum of some 4 Gbytes of memory, of which some 1 Gbyte would be main memory. Hard discs with a capacity of 120 Gbyte are not uncommon.

Main memory is often said to be interleaved which means that it is divided into two or more sections. The CPU can access alternate sections immediately without having to wait for memory to be updated due, say, to delays arising from execution of previous instructions. A disc cache is similar in principle to a memory cache, except that it is a reserved area of main memory used for temporarily storing blocks of information from disc that is in recent and/or regular use. This too can dramatically speed up CPU operations.

9.5 The Processor Cycle

A processor cycle is the sequence of operations required by the microprocessor to obtain information from memory and to perform the operations required by or on that information. A processor cycle is made up of a number of sub-cycles, typically two or three, depending on the complexity of the instruction to be carried out and the processor technology concerned. Each sub-cycle typically contains a maximum of five timing periods, as shown in Figure 9.3.

The basic strategy of a processor cycle may be described as fetch, decode and execute. The first sub-cycle of each processor cycle always treats the information obtained to be an op-code to be decoded. All of the remaining sub-cycles treat information obtained to be data to be operated on.

The first timing period of the first sub-cycle is used to move the contents of the PC into the MAR for the purpose of fetching the op-code from memory. During the second timing period the PC is incremented so that the next location to be addressed is readily available. The third timing period is used to move the op-code, at the address on the address bus, into the IR *via* the MBR. Depending on the op-code, the fourth and fifth timing periods are used to move information between other registers and memory *etc.*, or to execute simple arithmetic/logic operations.

The first timing period of any subsequent sub-cycle is used to place the contents of the PC into the MAR. The second timing period is used to increment the PC and the third to place the data, at the address on the address bus, into the data register of the ALU *via* the MBR. This is then operated upon during the fourth and fifth periods as appropriate. If the address in the PC is of a register in the stack, then data manipulation would begin with the second timing period.

It is important to appreciate that op-codes and data are stored contiguously in memory, and that the order in which they are stored is compatible with the microprocessor cycle. Thus incrementing the PC enables op-codes to be moved into the IR and data into the ALU sequentially, as appropriate.

Section 2

Instrumentation

Characteristics

Chapter 10

There are many excellent texts covering both instrumentation and measurement. Some are generic, one of the best being by Bentley (2004). Others, which are specific to the process sector, are referred to in the next few chapters.

As seen in Figure 3.3 there is usually some combination of sensor, transducer and transmitter associated with the measurement process. These devices may be characterised in many ways, *e.g.* according to principle of operation, performance, physical design or commercial considerations. For specification purposes the most important characteristics are as follows:

a. Accuracy. This is a measure of the closeness of the measurement of a variable to the true value of that variable. Consider, for example, an instrument for temperature measurement as shown in Figure 10.1. Its calibration is linear, with an input range of 75–100 °C (span of 25 °C) corresponding to an output range of 4–20 mA (span of 16 mA).

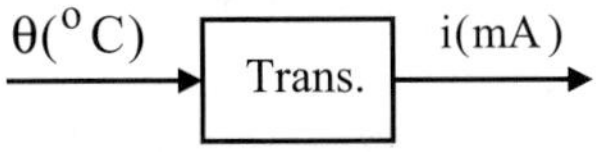

$\theta(°C) \rightarrow$ | Trans. | $\rightarrow i(mA)$

Fig. 10.1 Instrument for temperature measurement

Suppose a temperature of 85 °C is measured as 84.5 °C. In absolute terms the error is 0.5 °C and the instrument is said to have an accuracy of 0.5 °C. It is more usual to express accuracy on a percentage basis. The error may be expressed either as a percentage of the true value or, more commonly, as a percentage of the span of the instrument according to

$$\text{Accuracy} = \pm \frac{\text{Error}}{\text{Span}} \times 100\%$$

A temperature of 85 °C should give an output of 10.4 mA whereas the output corresponding to 84.5 °C is only 10.08 mA. Thus the accuracy of the instrument is ±2.0%. It does not particularly matter what basis is used for quoting accuracy, as long as it is properly understood.

b. Precision, which is often confused with accuracy, is associated with analogue signals and is a function of the scale used for measurement. Suppose the above temperature measurement was indicated on a gauge whose scale was calibrated from 75 to 100 °C in divisions of 0.5 °C between which it is possible to interpolate to within 0.1 °C. The precision is 0.1 °C or ±0.4% of scale. Thus the measurement which is only ±2.0% accurate can be read with a precision of ±0.4%.

c. Resolution is, strictly speaking, the largest change in input signal that can occur without any corresponding change in output. Resolution is associated in particular with digital signals and is a function of the number of bits used to represent the measurement. Again, using the 75–100 °C example, suppose the signal is stored in a 10-bit register. The register has a range of 1023_2, each bit corresponding to 25/1023 °C, giving a resolution of approximately 0.1% of range.

d. Repeatability is a measure of an instrument's ability to produce the same output signal for different instances of a given input signal. Lack of repeatability is usually due to random effects in the instrument or its environment. Repeatability is characterised by statistical metrics, such as average value and standard deviation. For process control purposes, where consistency of operation is of the essence, an instrument's repeatability is often more important than its accuracy.

e. Linearity, as discussed in Chapter 4, pervades automation. It is highly desirable that instruments have linear I/O characteristics. If any significant non-linearity exists, it is normal to quantify it in terms of the maximum difference between the output and what it would be if the characteristic were linear, expressed as a percentage of the instrument's span.

For the example of Figure 10.1, suppose the ideal (linear) and nonlinear outputs are given by

$$i_{id} = -44.0 + 0.64.\theta \quad i_{nl} = f(\theta)$$

and are as depicted in Figure 10.2

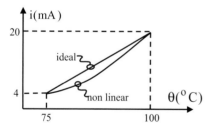

Fig. 10.2 Ideal and nonlinear characteristics

The percentage nonlinearity is thus

$$\frac{(-44.0 + 0.64.\theta - f(\theta))_{max}}{16} \times 100$$

f. Hysteresis is a particular form of non-linearity in which the output signal may be different for any given input signal according to whether the input is increasing or decreasing, as illustrated in Figure 10.3.

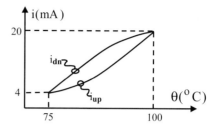

Fig. 10.3 Characteristic with hysteresis

Hysteresis is also quantified in terms of the maximum difference between the two outputs expressed as a percentage of the span, i.e.

$$\frac{(i_{dn} - i_{up})_{max}}{16} \times 100$$

g. Rangeability. Instruments are designed to have some specified I/O relationship, linear or otherwise. Normally this relationship is valid over the whole range. Sometimes, however, the relationship does not hold at one end or other of the range, usually the bottom end. Rangeability is expressed as the ratio of the proportion of the scale over which the relationship holds to the proportion over which it doesn't. For example, an instrument that is linear over the top 96% of its range has a rangeability of 24.

h. Drift. This is a time dependent characteristic, normally associated with analogue signals. It is the extent to which an output signal varies with time for a constant input. Drift is invariably due to heating effects. Many components take a while to heat up after being switched on so a device's output may initially drift prior to stabilising. Drift can also be an indication that a device is overheating.

i. Sensitivity, otherwise referred to as gain, is a steady state measure of an instrument's responsiveness. It is defined to be the change in output per unit change in input. Assuming that its I/O characteristic is linear across the whole range, sensitivity is given by

$$\text{Gain} = \frac{\text{Output span}}{\text{Input span}}$$

The gain of the example of Figure 10.1 is 0.64 mA/°C. Sensitivity is determined by the specification of the I/O ranges and, for control purposes, is arguably the most important of an instrument's metrics.

j. Speed of response. An instrument's dynamics are of particular interest. The types of instrumentation used in process automation are relatively slow. It is therefore normal practice to approximate their response to that of a first order system and to characterise it in terms of a time constant as explained in Chapter 69. If necessary, this can be found empirically by carrying out a step response test and measuring the time for a 63.2% change in output.

k. Reliability. Process instrumentation has to be designed to operate, with minimal maintenance, for 24 h/day throughout the life of the plant. High reliability is essential. For analogue devices, such as sensors and transmitters, the most useful means of quantifying reliability is either in terms of failure rates, *e.g.* number of failures/10^6 h, or in terms of the mean time between failures. MTBF is applicable to any type of instrument that can be repaired by the replacement of a faulty component or unit. For discrete devices used in protection systems, such as switches and solenoids, it is normal practice to express reliability in terms of the probability that the device will fail on demand. Reliability is discussed in detail in Chapter 53.

l. Intrinsic Safety. When an instrument has to be situated in a hazardous area, *i.e.* one in which flammable gas/air mixtures could exist, it must be intrinsically safe. The extent of the IS requirement depends upon the zone number, the gas group and the temperature class. If the IS specification for an instrument cannot be met, perhaps on the grounds of cost, then it is acceptable to install it in a flameproof enclosure. Depending on what an instrument is connected up to, it may be necessary to specify a barrier at the zone boundary. All of these safety related issues are explained in detail in Chapter 52.

m. Process Interface. Those parts of an instrument that come into direct contact with the process have to be capable of withstanding the temperatures and pressures involved. They also have to be chemically resistant to the process medium, whether it be acidic, caustic or otherwise. There is therefore a need to specify appropriate materials of construction. The process interface, surprisingly, is the most expensive part of an instrument. It usually consists of a casting of steel, or some other alloy, which houses the sensor. Its manufacture requires a significant amount of machining and assembly work, all of which is labour intensive.

n. Physical. Mundane, but of vital importance to the successful installation of an instrument, are some of its physical characteristics. For example, the correct fittings or flanges have to be specified for the process interface, appropriate brackets for panel or post mounting, suitable connectors for wiring purposes, *etc.* The intent with regard to power supply must be made clear. There are many options: the supply may be local to the instrument or provided *via* the signal lines, the voltage may be a.c. or d.c., the instrument may require special earthing arrangements, screening may be specified, *etc.*

o. Availability. There is much variety in the supply of process instrumentation: different techniques of measurement, alternative designs, choices in functionality, *etc.* There are many suppliers and the market is both competitive and price sensitive. The best way of identifying the principal suppliers of an instrument type is to consult a trade catalogue, such as the InstMC Yearbook. Looking at the technical press on a regular basis helps to keep abreast of current developments. Some journals publish surveys which are particularly useful.

p. Commercial. Most companies attempt to standardise on a limited range of instrumentation and to maximise on its interchangeability. This reduces the inventory of spares required, which could otherwise be extensive, and makes for more effective maintenance and repair. This standardisation inevitably leads to a preferred, *i.e.* restricted, supplier list. Factors taken into account include instrument performance, product support, company stability and pricing policy.

q. Cost. Whereas the purchase price of an instrument is relatively easy to establish, its true capital cost is not. This is because installation costs are specific to the application and can vary widely. It is not uncommon for the installed cost, *i.e.* the purchase price plus installation costs, to be more than double the purchase price. Strictly speaking, in specifying an instrument, its life cycle costs should be taken into account, *i.e.* its installed cost plus maintenance costs. This seldom happens in practice because capital costs and operating costs come from separate budgets.

Comments

Specifying an instrument inevitably involves making compromises between desirable technical characteristics and commercial criteria, subject to the constraints of company policy. To do this properly requires a good understanding of the application and its requirements. It also requires alot of detailed information. The extra time and effort spent in getting the detail right at the specification stage saves much money and inconvenience correcting mistakes later on.

DP Cells

Chapter 11

11.1 Diaphragm Capsules
11.2 Pneumatic DP Cell
11.3 Electronic DP Cell
11.4 Calibration
11.5 Installation
11.6 Commissioning

Differential pressure is normally measured with an instrument known as a dp cell. It consists of a sensor, a transducer and a transmitter combined in a single device. The dp cell is a very versatile instrument and, without doubt, is the most common of all process instruments. Depending on how it is installed, the dp cell can be used for measuring differential pressure, absolute pressure, vacuum, level, density and flow. Use of the dp cell for flow measurement is covered in Chapter 12 and for level, density and interface measurement in Chapter 14.

Following an overview of the design of dp cells, this chapter focuses on their installation for a variety of applications. Also covered are procedures for their calibration and commissioning. For more detailed information the reader is referred to BS 6739 in particular.

11.1 Diaphragm Capsules

Most dp cells have a diaphragm capsule as their sensor. Capsules consist of two circular metallic diaphragms formed into a disc, typically 5–15 cm in diameter by 1–2 cm in depth, and an internal backing plate, as shown in Figure 11.1. The diaphragms often have concentric corrugations to optimise their flexibility/rigidity. The backing plate has the same profile as the diaphragms and protects them from being over pressured. The space between the diaphragms and the backing plate is filled with silicone oil so that, by displacement, any force applied to one diaphragm is transmitted to the other. Thus, when a differential pressure is applied, both diaphragms flex by the same amount in the same direction.

The dimensions of the diaphragms and their material of construction vary widely according to application. In essence, the smaller the differential pressure the larger the diameter. Thickness is optimised to give sufficient flexing for high sensitivity subject to the constraint of linearity. The material has to be strong enough to withstand the forces involved, especially metal fatigue from continual flexing, and be resistant to corrosion and chemical attack. The two most common materials of construction are stainless steel and beryllium copper, the latter being particularly flexible.

11.2 Pneumatic DP Cell

The schematic of a classical pneumatic dp cell is shown in Figure 11.2.

The capsule is held between two flanged castings which form chambers on either side. These are designated as the high and low pressure sides of the dp cell as appropriate. The force bar is rigidly

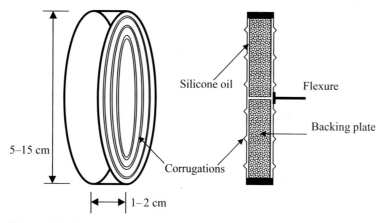

Fig. 11.1 Diaphragm capsule of dp cell

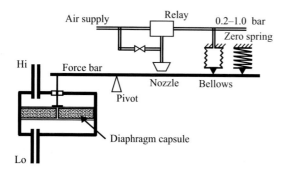

Fig. 11.2 Schematic of pneumatic dp cell

connected to the capsules' diaphragm by means of a flexure which passes through an alloy seal in the high pressure chamber. The force bar acts as a flapper with respect to the nozzle. The output of the pneumatic relay is connected to the feedback bellows, adjacent to which is the zero spring. These pneumatic components are protected by a characteristically shaped weather and dust proof housing.

Following an increase in differential pressure, the diaphragms are displaced downwards: the maximum displacement being some 0.25 mm. This causes an anti-clockwise moment to be exerted on the force bar which positions itself closer to the nozzle. The back pressure is amplified by the relay, whose output is within the range of 0.2–1.0 bar. As the feedback bellows expand, a clockwise moment is exerted on the force bar and a force balance is established.

Altering the position of the pivot varies the relative mechanical advantages of the forces exerted by the capsule and by the feedback bellows. For a fixed relay output, if the pivot is moved leftwards then the moment exerted by the feedback bellows is increased. To maintain balance a greater differential pressure is required; thus the range is increased. The externally adjusted zero spring creates a force which pre-loads the force bar to provide a relay output of 0.2 bar with zero differential across the capsule.

11.3 Electronic DP Cell

Externally, the general features of construction of electronic and pneumatic dp cells are much the same. Thus the diaphragm capsule is held between two flanged castings which form high and low pressure chambers on either side. And there is a weatherproof housing on top which contains the electronic circuits. There are many different transducer types in common usage. Cells using analogue electronics typically convert the deflection of the diaphragms into a change in resistance, inductance or capacitance which is then measured using an integrated circuit as appropriate. With digital electronics the force is typically applied directly to a sil-

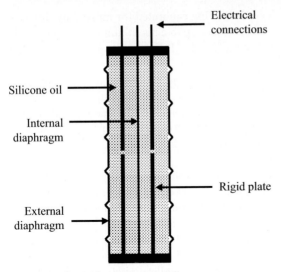

Fig. 11.3 Differential capacitance transducer

icon chip with embedded circuits. The principle of operation of a differential capacitance transducer is illustrated in Figure 11.3.

The diaphragm capsule contains two rigid plates and a stiff internal diaphragm which are electrically insulated from the capsule itself. These effectively form two separate parallel-plate capacitances, the internal diaphragm being common and the silicone oil acting as a dielectric. Suppose that, following an increase in differential pressure, the diaphragm is deflected slightly leftwards. Capacitance is inversely proportional to separation. Thus the right hand capacitance will decrease and the left hand one increase. The two rigid plates and the internal diaphragm are connected to a passive electronic circuit. This converts the difference in capacitance to a millivolt signal which is then amplified, filtered, characterised and transmitted as a 4–20 mA signal.

Alternatively, in a so-called microprocessor based "smart" or "intelligent" dp cell, the signal from the embedded circuits would be amplified, changed by an A/D converter into a bit pattern and stored in a RAM type database. Signal processing would take place by means of a ROM based programme operating on the database. Smart dp cells often have a temperature input as well to enable the differential pressure to be compensated for varia-

Picture 11.1 Typical smart dp cell (Emerson)

tions in temperature. The output would either be transmitted as a serial signal according to some protocol such as HART or Profibus (refer to Chapter 50) or else converted back into analogue form by a D/A converter and amplified for transmission as a 4–20 mA signal. A typical smart dp cell is as illustrated in Picture 11.1.

11.4 Calibration

Calibrating a dp cell is essentially a question of checking that the input and output ranges are as specified and, if not, making the necessary adjustments. There are normally two settings which can be adjusted: range and zero. Unfortunately, adjusting the range often upsets the zero setting, and *vice versa*. Calibration is therefore an iterative process. However, the iterations normally converge quickly on the correct calibration. The calibration is usually linear so it is sufficient to iterate between the top and bottom of the range only.

Clearly, for calibration purposes, it is necessary to have appropriate workshop test facilities. The differential pressure is normally generated pneumatically, for which an air supply, regulator and an accurate pressure gauge are required. For analogue electronic dp cells the output is measured by means of an ammeter in series with a load resistance, typically of 250 Ω, as shown in Figure 11.4. For digital electronic dp cells some form of interface is required for reading the output.

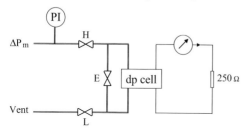

Fig. 11.4 Analogue electronic dp cell test facility

Note the isolating valves, labelled H and L, corresponding to the high and low pressure sides of the dp cell respectively and the equalising valve, labelled E, connected across them.

The procedure for calibrating the dp cell is shown in flow chart form in Figure 11.5. In this case it is being calibrated for measuring differential pressures up to a maximum of ΔP_m bar, *i.e.* an input range of $0-\Delta P_m$ bar corresponds to an output range of 4–20 mA.

This calibration procedure is appropriate for many instruments other than dp cells. When the procedure has been carried out, it is good practice to complete an instrument pre-installation calibration sheet, a specimen copy of which is given in BS6739.

A smart dp cell is essentially calibrated on the same basis. However, once calibrated, it may be reconfigured remotely, typically by means of some hand-held microprocessor based communications device connected to the dp cell *via* its 4–20 mA signal transmission lines. Analogue signal transmission is suspended whilst digital messages are transmitted in serial form by means of mA pulses. The communicator would have dedicated push-buttons and a simple LCD type of display for operator interaction. Thus, for example, by using pre-configured function blocks, a new range may be specified and down loaded into the dp cell's RAM.

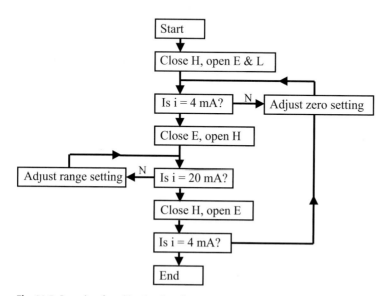

Fig. 11.5 Procedure for calibrating dp cell

11.5 Installation

The high and low pressure chambers of a dp cell have to be connected to the two process pressures whose difference ΔP is to be measured. The connecting pipes are referred to as impulse lines. It is essential that the differential pressure at the process end of the impulse lines is transmitted to the dp cell without distortion. There is much scope for distortion. For example:

- Sediment may accumulate in one or other of the impulse lines and block it.
- Air bubbles in liquid filled impulse lines may cause a difference in static head.
- Condensate may dribble into an impulse line causing a static head to build up.
- The orientation of the dp cell and impulse lines relative to the process may cause bias.

These problems are largely countered by proper design of the installation and sound commissioning of the dp cell. At the dp cell end of each of the impulse lines there should be an isolating valve, with an equalising valve connected across them as shown in Figure 11.6.

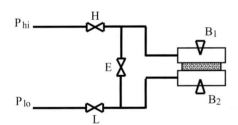

Fig. 11.6 Impulse lines, isolating and equalising valves

Picture 11.2 Installation showing dp cell, impulse lines, manifold and supports

These valves can either be installed individually or as a proprietary manifold which bolts directly onto the dp cell's flanges. The advantage of the manifold arrangement is that it is more compact, easier to install and reduces the scope for leakage.

Pneumatic dp cells require a clean supply of air at a constant pressure of approximately 1.4 bar. This is supplied by a local filter regulator as described in Chapter 5. Electronic dp cells normally obtain their power through the signal transmission lines, as discussed in Chapter 51, and do not need a local power supply. Whether pneumatic or electronic, the dp cell will also require trunking or conduit to carry the signal tube or cabling.

The dp cell is a heavy instrument and 5 kg is not untypical. Together with its impulse lines, valve manifold and any ancillary devices, there is an obvious need for support. It is usually mounted on a strategically situated post or pipe using proprietary brackets, as depicted in Picture 11.2. This is reflected in the installed cost for a dp cell which, including fitting and wiring, is roughly twice its capital cost.

11.6 Commissioning

Assuming the dp cell to be correctly calibrated and to be physically connected to the process by its impulse lines, the objective of commissioning is to successfully apply the differential pressure to be measured to the dp cell. Consider the scenario depicted in Figure 11.6. Suppose that the process is aqueous and the impulse lines are filled with water. Presume that P_{hi} and P_{lo} are both large and that ΔP is small:

$$\Delta P = P_{hi} - P_{lo}$$

It is necessary to flush out any bubbles of air in the impulse lines and in the high and low pressure chambers of the dp cell. Since air rises in water, the bleed nipples B_1 and B_2 at the top of the chambers would be used for this purpose. Note that bleed nipples at the bottom of the chambers would be used for bleeding out condensate if the impulse lines were air or gas filled. Opening valves H and B_1, for example, enables water to flow from the process through the impulse line and high pressure side of the dp cell and into the atmosphere. Bubbles of air cause a spitting effect as they vent through the bleed nipple so it is easy to tell when all the air has been bled out. Appropriate precautions must be taken for personnel protection during this bleeding process, especially if the pressures are high and the process medium is harmful.

The diaphragm capsule will have been specified on the basis of ΔP_m. If P_{hi} and P_{lo} are significantly greater than ΔP_m it is essential, to avoid damaging the diaphragms, that neither P_{hi} nor P_{lo} is applied to either side of the capsule in isolation. This can be achieved by appropriate use of the equalising valve. The procedure for commissioning the dp cell, which indicates the correct order for opening and closing the isolating and equalising valves, is shown in flow chart form in Figure 11.7, assuming all values are closed to begin with.

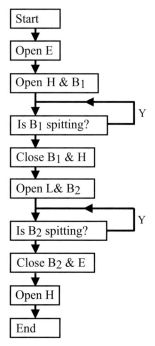

Fig. 11.7 Procedure for commissioning dp cell

Most manufacturers of dp cells make much of their profit from the sales of replacement capsules for those damaged during commissioning.

Flow: Orifices

12.1 Construction
12.2 Operation
12.3 Design Considerations
12.4 Flow Equations
12.5 Worked Example
12.6 Specification
12.7 Installation
12.8 Nomenclature

Orifice plates used in conjunction with dp cells are the most common form of flow measurement. The combination of orifice plate and dp cell is sometimes referred to as an orifice meter. Detailed guidance on design and specification is given in BS 1042, superseded by ISO 5167, and on installation in BS 6739. The standard ISO 5167 is in four parts as follows:

ISO 5167	Measurement of fluid flow by means of pressure differential devices inserted in circular cross-section conduits running full
Part 1	General principles and requirements
Part 2	Orifices
Part 3	Nozzles and venturi nozzles
Part 4	Venturi tubes

12.1 Construction

An orifice plate is essentially a thin circular plate, normally of stainless steel to resist both corrosion and erosion, with a sharp edged hole at its centre. The plate is sandwiched with gaskets between two flanges in a pipe. For relatively small diameter pipes, say up to 10 cm, there are proprietary orifice "carriers", as illustrated in Picture 12.1. The impulse lines are connected between "tappings" in the pipe or carrier, on either side of the orifice plate, and the dp cell.

For very small orifice diameters, say below 5 mm, proprietary carriers are available which bolt onto the dp cell, either directly or *via* a proprietary manifold, the impulse lines being realised by internal channels. These are referred to as integral

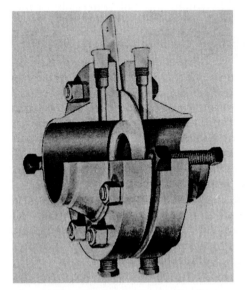

Picture 12.1 Proprietary carrier with orifice plate

orifices. A typical dp cell with manifold and integral orifice assembly is illustrated in Picture 12.2. Note that proprietary carriers and manifolds are relatively expensive.

Picture 12.2 Dp cell with proprietary manifold and integral orifice

It is normal for an orifice plate to have a handle which protrudes from the flanges of the pipe or carrier as shown in both Pictures 12.1 and 12.2. The handle has the orifice diameter stamped on it for reference purposes. There is also a hole drilled through the handle: the hole externally distinguishes between an orifice and a blank, blanks often being inserted during commissioning for isolating purposes.

12.2 Operation

Fluid flowing along the pipe is forced through the orifice resulting in a pressure drop across the orifice. This pressure drop is proportional to the square of the flow and is measured by means of a dp cell.

The fluid accelerates as it converges on the orifice, its pressure energy being converted into kinetic energy. A maximum velocity is reached just downstream of the orifice, at the *vena contracta*, after which the fluid expands to full pipe flow. This is an inefficient process because the kinetic energy is not all recovered by expansion: a signifi-

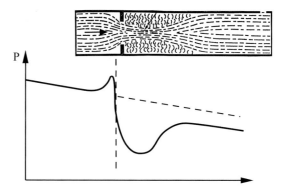

Fig. 12.1 Flow and pressure profile across orifice

cant proportion is dissipated as heat due to turbulence, referred to as friction losses, resulting in a net pressure drop across the orifice. Flow through the orifice and its pressure profile is depicted in Figure 12.1.

The flow nozzle and venturi are depicted in Figures 12.2 and 12.3 respectively. The flow nozzle is marginally more efficient than the orifice but much less so than the venturi. In the case of the flow nozzle the contraction is shaped by the nozzle's profile and is not sudden. However, most of the hydraulic inefficiency is attributable to the expansion, rather than the contraction, so flow nozzles are relatively ineffective. With the venturi, both of the contrac-

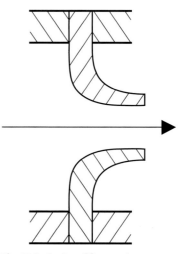

Fig. 12.2 Section of flow nozzle

Fig. 12.3 Section of venturi

tion and expansion are constrained by the conical inlet and outlet sections. The angle of the outlet cone is critical, typically approximately 7° relative to the axis of the venturi. This is a much more efficient arrangement because, in particular, there is no scope for back eddying downstream of the venturi's throat. A venturi is expensive compared to an orifice plate.

12.3 Design Considerations

The capital cost of an orifice meter is cheap relative to other methods of flow measurement. However, its operation is expensive in terms of pressure loss which can be equated to the energy costs for pumping or compression. The life cycle costs of flow measurement using orifice meters is undoubtedly much greater than other methods. The reason that orifice meters are so common is due to the simple fact that capital cost is normally the deciding factor on most projects.

Design is essentially a question of determining the right size of orifice for the application, establishing that the pressure drop is sensible and compatible with the range of the dp cell, specifying pipework layout and conditioners if necessary, deciding on the type of tappings, and considering the requirements for the impulse lines.

Readings of flow accurate to within 1–2% of full range can be achieved using an orifice meter, provided it is properly installed and carefully calibrated, although it is difficult to cover a wide range of flows with any single size of orifice because of the square relationship. However, orifice plates can be easily changed.

12.4 Flow Equations

Let subscripts 0, 1 and 2 refer to cross sections of the orifice, the pipe upstream of the orifice and to the *vena contracta* respectively, as shown in Figure 12.4.

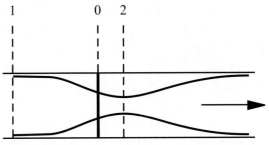

Fig. 12.4 Sections of orifice, pipe and *vena contracta*

Bernoulli's theorem may be applied across the pipe between the *vena contracta* and the upstream section:

$$0.5 \left(v_2^2 - v_1^2\right) + g\left(z_2 - z_1\right) + \int_1^2 v.dP = 0$$

If it is assumed that the fluid is incompressible, *i.e.* a liquid, and that any change in height is negligible, then

$$0.5 \left(v_2^2 - v_1^2\right) + \frac{1}{\rho}\left(P_2 - P_1\right) = 0$$

A volume balance gives

$$Q = A_1 v_1 = A_2 v_2$$

Substituting and rearranging gives

$$Q = A_2 \sqrt{\left(\frac{2\left(P_1 - P_2\right)}{\rho\left(1 - \left(A_2/A_1\right)^2\right)}\right)}$$

The cross sectional area of the *vena contracta* is related to that of the orifice by a coefficient of contraction: $c_c = A_2/A_0$. Hence:

$$Q = c_c A_0 \sqrt{\left(\frac{2\left(P_1 - P_2\right)}{\rho\left(1 - \left(c_c A_0/A_1\right)^2\right)}\right)}$$

In reality the flow rate will be significantly less than that predicted by this equation because of friction

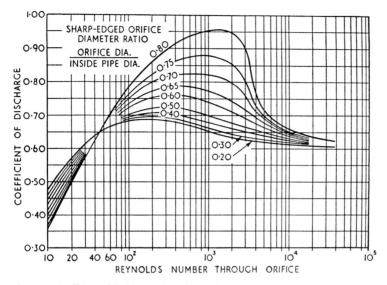

Fig. 12.5 Coefficient of discharge *vs* Reynolds number

losses. A coefficient of discharge c_d is introduced to account for these which incorporates c_c. Thus:

$$Q = c_d A_0 \sqrt{\left(\frac{2(P_1 - P_2)}{\rho(1 - \beta^4)}\right)} \qquad (12.1)$$

where $\beta = d_0/d_1$.

The coefficient of discharge is a complex function of β, the ratio of orifice to pipe diameters, Reynolds number Re_0 and the type of tappings. Accurate values of c_d are tabulated in ISO 5167. However, for most purposes, the data correlated in the form of Figure 12.5, which is independent of tapping type, is accurate enough.

Note that the Reynolds number is based on the velocity in and diameter of the orifice:

$$Re_0 = \frac{\rho d_0 v_0}{\mu}$$

12.5 Worked Example

An orifice assembly is to be used as the measuring element in a loop for controlling the flow of water at a rate Q of 0.5 kg/s through a horizontal pipe of internal diameter $d_1 = 5$ cm.

Data for water: viscosity $\mu = 1 \times 10^{-3}$ kg m^{-1} s^{-1}
density $\rho = 1000$ kg m^{-3}

The volumetric flow along the pipe is given by

$$Q = \frac{\pi d_1^2}{4}.v_1$$

$$0.5 \frac{\text{kg}}{\text{s}} \cdot \frac{1}{1000} \frac{\text{m}^3}{\text{kg}} = \frac{\pi}{4}\left(\frac{5}{100}\right)^2 \text{m}^2.v_1\frac{\text{m}}{\text{s}}$$

Hence $v_1 = 0.25$ m/s and $Re = \left(\frac{\rho d_1 v_1}{\mu}\right) = 12{,}500$.

This is the bulk Re; in the orifice it will be much higher. From Figure 12.5 it can be seen that $c_d \approx 0.61$.

Size the orifice such that under normal conditions, *i.e.* flow is at its set point, the pressure drop is at mid point in the range of the dp cell. Suppose that the dp cell is calibrated over the range 0–0.4 bar $\equiv$ 4–20 mA. At mid-range $\Delta P = 0.2$ bar $= 0.2 \times 10^5$ N m^{-2}. Since the pipe is horizontal, $\Delta z = 0$.
Substituting into Equation 12.1 gives

$$Q = c_d.A_0 \sqrt{\frac{2\Delta P}{\rho(1 - \beta^4)}}$$

$$\frac{0.5}{1000} = 0.61 \cdot \frac{\pi d_0^2}{4} \sqrt{\frac{2.0.2 \times 10^5}{1000 \cdot \left(1 - \left(\frac{d_0}{0.05}\right)^4\right)}}$$

where $d_0 = 0.01283$ m. An orifice plate of 13 mm diameter will be good enough. The assumption of $c_d = 0.61$ is now justified as follows:

$$F = \frac{\pi d_0^2}{4} \cdot v_0$$

$$\frac{0.5}{1000} = \frac{\pi \cdot 0.013^2}{4} \cdot v_0$$

giving $v_0 = 3.767$ m s^{-1}.

Hence Reynolds number in the orifice:

$$Re_0 = \frac{\rho \cdot d_0 \cdot v_0}{\mu} = \frac{1000 \times 0.013 \times 3.767}{1.0 \times 10^{-3}}$$
$$\approx 4.9 \times 10^4$$

Inspection of Figure 12.5 shows that Re_0 is well within the region for which c_d may be presumed to be 0.61.

12.6 Specification

Equation 12.1 can be used for both liquids and gases, despite the fact that incompressible flow was assumed in its derivation. This is because with gas flows the value of Re_0 is high and, as can be seen from Figure 12.5, for high values of Re_0 the value of c_d is independent of both β and Re_0. In fact, for $Re_0 > 10^5$ the value of c_d is approximately 0.61.

The design scenario: the procedure for specifying the diameter of an orifice for measuring flow is depicted in flow chart form in Figure 12.6.

The measurement scenario: the procedure for calculating the flow rate through an orifice from a measurement of the pressure drop across it is depicted in flow chart form in Figure 12.7.

Both procedures are iterative, lend themselves to a computer routine, and converge quickly on a solution.

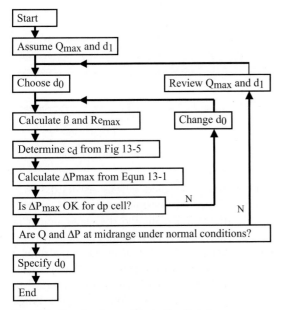

Fig. 12.6 Procedure for specifying orifice diameter

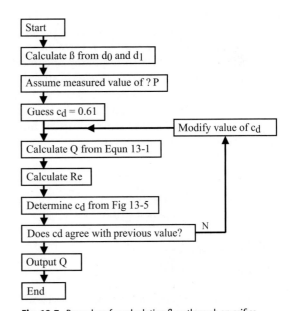

Fig. 12.7 Procedure for calculating flow through an orifice

12.7 Installation

An important consideration in specifying an orifice meter is the pipework layout. The orifice must be situated in a straight length of pipe and be far enough away from any bends, pipe fittings or other devices for the flow profile to be axially symmetrical. Of particular concern is swirl: Figure 12.8 depicts how two 90° bends in different planes cause swirl.

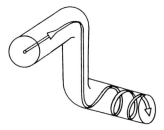

Fig. 12.8 Formation of swirl

ISO 5167 tabulates minimum up and down stream straight lengths as a function of the diameter ratio β for various types of bends, *etc*. As a rule of thumb, 25 d, *i.e.* 25 pipe diameters, upstream and 5 d downstream for values of $\beta < 0.5$ is good enough for most purposes.

If the minimum recommended up and down stream straight distances cannot be realised, especially if the orifice plate is to be located close to a swirl inducing device such as a pump or control valve, it is necessary to fit a straightener. A straightener essentially straightens out the flow. One type is the so called "etoile" straightener, which is depicted in Figure 12.9. Other types consist of bundles of tubes or perforated plates.

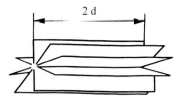

Fig. 12.9 Etoile type of straightener

Figure 12.10 indicates the minimum acceptable straight distances up and down stream of an orifice used with a straightener. If there is more pipe length available than the minimum, it is best to increase the distance between the inlet straightener and the orifice.

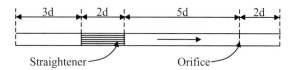

Fig. 12.10 Minimum distances from orifice with straightener

The three principal types of tappings are depicted in Figure 12.11. All are threaded for connection to the impulse lines. Bearing in mind the installation costs, the cheapest and most common type are flange tappings. These are perpendicular holes drilled through the flanges supporting the orifice plate. Corner tappings are mostly used for small bore pipes: the holes are drilled at an angle such that the pressure is measured on the orifice plate's surfaces. For large diameter pipes, flange tappings are expensive so d and d/2 tappings are used: the locations correspond to full pipe flow upstream of the orifice and to the *vena contracta* downstream. For venturis the tappings have to be of the d and d/2 type.

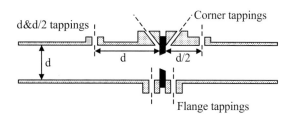

Fig. 12.11 Flange, corner, d and d/2 types of tappings

Other important considerations, as discussed in Chapter 11, are the positioning of the dp cell relative to the orifice plate and the arrangement of the impulse lines between them. The following diagrams depict the preferred arrangements for a variety of flow measurements. Note that, for simplicity, the dp cell's isolating and equalising valves are not shown: these are obviously required. When the freezing point of the process fluid is close to ambient temperature, the impulse lines should be heat traced and/or insulated.

Figure 12.12 is for liquid flow measurement. Under low or zero flow conditions, entrained air or gas bubbles tend to flow along the top of a pipe and sediment settles at the bottom, so the tapping points are at the side of the pipe. The impulse lines are full of process liquor which transmits the differential pressure. The cell is mounted below the tapping points so that any air bubbles rise into the pipe. The materials of construction of the dp cell and its diaphragm capsule must be specified carefully since they are in direct contact with the process liquor.

Figure 12.13 is for gas flow measurement. Note that the tappings are on top of the pipe. If there is any possibility of condensate forming in the impulse lines they should slope down towards the tappings and up towards the dp cell. Any condensate collects in the catch pots which can be drained off intermittently.

For measuring the flow of steam, or other condensing vapours, the best strategy is to accept that condensation is going to occur rather than try to prevent it. Thus the dp cell is operated with its impulse lines full of condensate which is used to transmit the differential pressure. Figures 12.14 and 12.15 show two arrangements, depending on

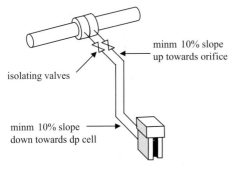

Fig. 12.12 Arrangement for liquid measurement: horizontal flow

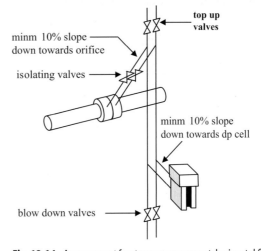

Fig. 12.14 Arrangement for steam measurement: horizontal flow

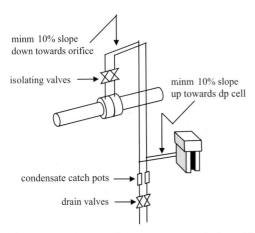

Fig. 12.13 Arrangement for gas measurement: horizontal flow

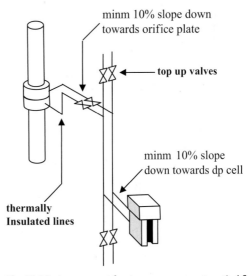

Fig. 12.15 Arrangement for steam measurement: vertical flow

whether the steam flow is in a horizontal or vertical pipe respectively. Because it could take a long time for the impulse lines to fill up by condensation, for commissioning purposes, filling tees are provided which enable them to be topped up to the same level with water.

When suspended solids are present as, for example, in measuring the flow of slurries or crystallising solutions, the best strategy is to recognise that deposits will occur and to make provision for removing them. Figures 12.16 and 12.17 show suitable rodding arrangements for both horizontal and vertical pipes respectively.

Figure 12.18 depicts a purged flow arrangement. Purge liquid flows *via* check (one way) and needle valves, through the impulse lines and into the process stream. The purge liquid supply pressure must obviously be greater than that of the process stream. The needle valves are used to regulate the flow of purge liquid. A small flow only is required. Downstream of the needle valves, the impulse lines are at the pressure of the process stream. Thus the purge liquid transmits the differential pressure to the dp cell. This arrangement is suitable for measuring the flow of slurries because the flow of purge liquid through the impulse lines prevents them from blocking up. The check valves guarantee that process liquor doesn't flow back into the impulse lines. Thus the purge liquid acts as a fluid barrier between the dp cell and the process stream. This arrangement is particularly suitable for measuring the flow of corrosive or reactive liquors where there are potential materials of construction problems. Because the purge liquid ends up in the process stream it is essential that this is operationally acceptable. When the process stream is aqueous it is normal to use water as the purge. For organic process streams an appropriate solvent has to be identified.

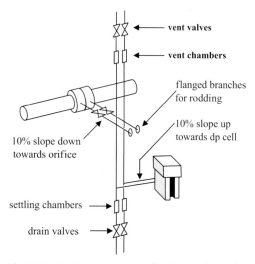

Fig. 12.16 Rodding arrangements for slurry: horizontal flow

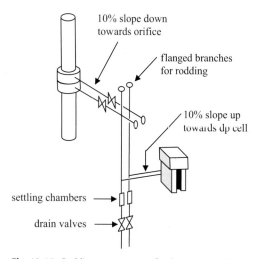

Fig. 12.17 Rodding arrangements for slurry: vertical flow

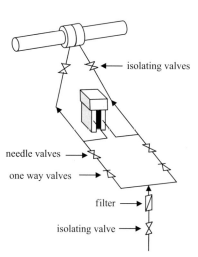

Fig. 12.18 Purge arrangement for corrosive liquid flow

12.8 Nomenclature

A	denotes	cross sectional area	m^2
d		diameter	m
g		gravity	ms^{-2}
P		pressure	Nm^{-2}
Q		volumetric flow rate	$m^3 s^{-1}$
v		velocity	ms^{-1}
z	denotes	height	m
β		ratio of orifice to pipe diameter	–
ρ		density	$kg\,m^{-3}$
υ		specific volume	$m^3\,kg^{-1}$
μ		viscosity	$kg\,m^{-1}s^{-1}$

Subscripts

0	orifice
1	upstream (full bore) flow
2	*vena contracta*

Acknowledgement. Note that the original of Figure 12.5 was published in Chemical Engineering, Volume 1, by Coulson, J.M. and Richardson, J.F. and published by Pergamon Press. It is reprinted here by permission of Butterworth Heinemann Publishers, a division of Elsevier.

Flow Measurement

Chapter 13

13.1 Overview
13.2 Sight Glasses
13.3 Rotameters
13.4 Gap Meter
13.5 Turbine Flowmeters
13.6 Electromagnetic Flowmeters
13.7 Ultrasonic Flowmeters
13.8 Vortex Shedding Meters
13.9 Mass Flowmeters

Starting with an overview, this chapter considers a variety of the more important methods of flow measurement other than the use of the orifice meters. For each instrument type there is an insight into the principle of operation, followed by a summary of the salient factors with regard to specification, installation and usage. There are many texts on flow measurement but the reader is referred, in particular, to the Shell handbook by Danen (1985).

13.1 Overview

A very effective summary of various flowmeter types is given in Table 13.1. The purpose of this table is to identify quickly, using simple criteria, those flowmeter types that are likely to be suitable for an application. Design effort can then be focused on establishing which of those identified is most suitable and developing a specification.

The figures quoted for accuracy in Table 13.1 are necessarily simplistic. In practice, accuracy depends on the appropriateness of the instrument specified, the effectiveness of the design of the installation, the efficiency of calibration and commissioning, and on the quality of maintenance. A useful feel for the range of accuracies that can be expected for any particular flowmeter type according to circumstances is given in Figure 13.1, reproduced from the Flomic report by Sproston (1987).

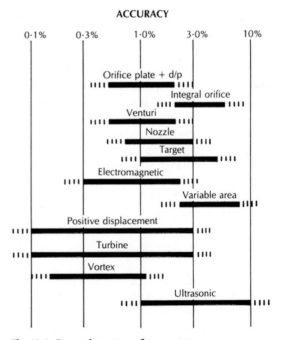

Fig. 13.1 Range of accuracy *vs* flow meter type

Table 13.1 Summary of flow meter types *vs* application

Flowmeter type	Accuracy %	Characteristic	Rangeability	Size (MAX) mm	Max. temp. C°	Max press. bar	Head loss	Gases	Liquids	Steam	Low flow/velocity	Slurries	Maintenance	Hygienic	Viscous fluids	Comment
Orifice plate	2	√	3	NL	800	1000	H	✓	✓	✓	✗	✗	M	✗	✓	Mostly universal
Venturi	1.5	√	3	NL	200	150	M	✓	✓	✗	✗	✗	M	✗	✗	Low head loss expensive
Flow nozzle	2	√	3	NL	200	150	M	✓	✓	✓	✗	✗	M	✗	✗	Best for steam
Pitot/annubar	2	√	3	NL	500	100	L	✓	✓	✓	✗	✗	M	✗	✗	Low cost low head loss
Variable area	2	L	10	100	250	100	M	✓	✓	✗	✓	✗	M	✓	✗	Small size
Turbine meter	0.5	L	10	500	200	500	H	✓	✓	✗	✓	✗	M	✗	✗	High accuracy pulse output
Positive displacement	0.5	L	20	150	100	50	H	✓	✓	✗	✓	✗	L	✗	✓	High accuracy high cost
Electromagnetic	1	L	10	3000	100	20	L	✗	✓	✗	✓	✓	L	✓	✓	Conductive fluids
Ultrasonic/doppler	2	L	10	1500	100	10	L	✗	✓	✗	✗	✗	M	✗	✓	Non-intrusive low head loss
Vortex shedding	1.5	L	10	200	200	100	H	✓	✓	✓	✗	✗	M	✓	✓	Pulse output
Target meter	2	√	3	200	200	100	H	✓	✓	✓	✗	✓	M	✗	✗	Not popular ok slurries
Mass/coriolis	0.75	L	10	150	150	100	H	✓	✓	✗	✓	✓	L	✓	✓	Costly limited size

✓ Suitable   ✗ Unsuitable   ✓ Normally suitable   ✗ Normally unsuitable   NL No limit

Similarly, the figures quoted in Table 13.1 for maximum size are somewhat simplistic. In effect they refer to the diameter of the largest pipe that the meter can be sensibly installed in. Not only is this a function of meter type but is very supplier dependant. Figure 13.2, also from Sproston (1987), gives a feel for the range of sizes, both maximum and minimum, of meter types that are available as off-the-shelf products.

13.2 Sight Glasses

The humble sight glass is a common form of flow indicator. Strictly speaking not a flowmeter, because it does not produce an analogue flow measurement, it does nevertheless give a discrete flow indication, *i.e.* either flow exists or it doesn't. Sight glasses are particularly useful when commissioning plant because they enable independent, visual checks on the state of flow. They are also useful for checking the condition of a process stream, *e.g.* its colour, whether clear or turbid, or whether it contains suspended solids or bubbles.

In essence a sight glass consists of a steel casting, with glass windows on opposite sides, that is installed between flanges in a pipe. The casting is designed to protect the windows from external pipework stresses. Flow is observed through the windows. Sight glasses are very simple and reliable devices. One common problem is that, depending on the nature of the flowing liquid, the windows become dirty on the inside. It is often necessary to install a lamp in line with the sight glass for effective observation.

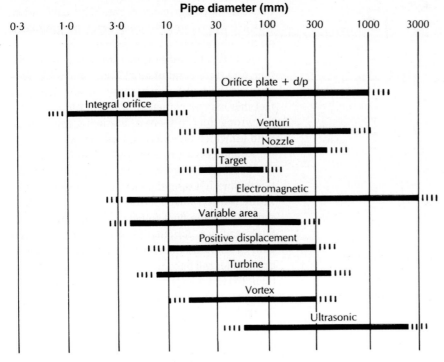

Fig. 13.2 Range of pipe sizes *vs* flow meter type

Sight glasses may be supplied with a hinged flap fitted across the inlet port. The opening of the flap, which is opposed by a spring, is indicated on a scale which can be calibrated against flow. Because, in essence, it is a mechanical device, there are significant operational problems. Both the hinge and spring are subject to metal fatigue due to being continually in motion, the inevitable consequence of which is failure. They can only be used for clean liquids, any solids could cause the flap to seize up. It is dubious practice to install such devices.

13.3 Rotameters

Another common form of flow indication, this time an analogue measurement, is the so-called rotameter. This consists of a vertically mounted, tapered, glass tube and "float" assembly, as shown in Figure 13.3. The tube is flange mounted and supported in a steel framework to protect it from pipework stresses.

Whereas with the orifice meter the orifice diameter is fixed and the pressure drop across it varies with flow, with the rotameter the pressure drop across the float is constant but the annular area varies with flow. At equilibrium, as depicted in Figure 13.3, the difference between the forces acting on the float due to gravity and to upthrust is balanced by the drag on the surface of the float, as follows:

$$V_f \left(\rho_f - \rho \right) g = \Delta P_f . A_x + R . A_s$$

where $R = fn(u_e)$, $u_e = fn(Q, A_a)$ and $A_a = fn(x)$.

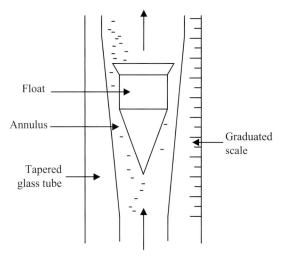

Fig. 13.3 Rotameter: showing glass tube and float assembly

A_a	denotes	cross sectional area of annulus	m²
A_s		effective surface area of float	m²
A_x		effective cross sectional area of float	m²
g		acceleration due to gravity	ms⁻²
ρ_f		density of float	kg m⁻³
ρ		density of liquid or gas	kg m⁻³
ΔP_f		pressure drop across float	N m⁻²
Q		flow rate	m³ s⁻¹
R		mean shear stress on float's surface	N m⁻²
u_e		equilibrium velocity in annulus	m s⁻¹
V_f		volume of float	m³
x		height of float in tube	m

For a given float and fluid, the balance of forces required for equilibrium corresponds to a particular velocity profile of the fluid around the float. The float therefore rises or falls to a position such that the volumetric flow rate through the annulus between the float and the tube wall corresponds to that velocity profile. The position of the float is thus calibrated against flow rate. This calibration, which is approximately linear, may be graduated on the tube wall or on an adjacent scale.

Note that the calibration is unique to that tube, float and fluid combination. If any one of these is changed the calibration becomes invalid. It is a sad fact that many rotameters in use in industry have the wrong float in them or are being used with a fluid for which they were not calibrated.

Rotameters are relatively cheap and quite effective but there are several important points of practice to note. It is essential that they are installed vertically. Non-symmetrical flow through the annulus, especially if the float touches the tube wall, gives false readings. Rotameters should only be used with clean fluids. As with sight glasses, the tube of a rotameter can become dirty on the inside and difficult to read. The weight due to any sediment that deposits on top of the float will cause false readings. Furthermore, they are easily blocked. It is important that the float is read at the correct level relative to the scale. As can be seen from Figure 13.4, there are many different float shapes and alternative positions to read from. There are no rules of thumb on this, the manufacturer's literature and/or calibration details should be adhered to.

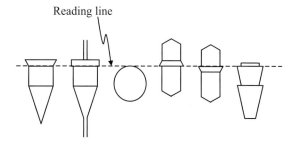

Fig. 13.4 Reading lines of various float shapes

13.4 Gap Meter

Another type of variable area meter is the gap meter. Whereas in the rotameter the annulus for flow is between a float and a tapered tube, in the gap meter the annulus is between a float and a fixed orifice. The cross sectional area of the annulus depends on the position of the float relative to the orifice. Under flow conditions the float establishes some equilibrium position, the height of the float indicating the flow rate. The float travel is small, typically up to 5 cm.

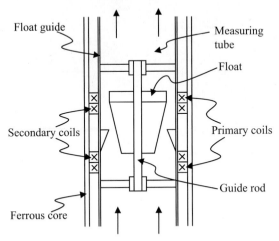

Fig. 13.5 Cross sectional view of a typical gap meter

As depicted in Figure 13.5, the float is guided, *i.e.* it slides along an axially mounted rod. Normally mounted vertically, the float returns under gravity, otherwise it is spring opposed. Embedded in the walls of the tube are the coils of a linear variable displacement transducer (LVDT). The principle of operation of an LVDT is as depicted in Figure 13.6.

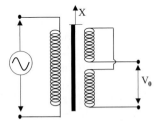

Fig. 13.6 Schematic of an LVDT

This consists of three fixed coils, one primary and two secondary, and a variable ferrous core. An a.c. input voltage V_1 is applied to the primary coil which induces voltages in the two secondary coils. The secondary coils are identical and reverse connected so the induced voltages tend to cancel. Indeed, when the ferrous core is symmetrical with respect to the coils, the output voltage V_0 is zero. Displacement x of the core from the mid position, along the axis of the coils, increases the voltage induced in one secondary coil and reduces that in the other. There is thus a linear amplification of the output:

$$V_0 = k.x$$

In the gap meter, the tube wall and float act as the core. The output is zeroed for no-flow conditions and the range adjusted as appropriate. Again the calibration is approximately linear. Electronically, the output voltage is conditioned, scaled, *etc.* and transmitted as an analogue signal, 4–20 mA or otherwise, proportional to the flow rate. Gap meters should only be used with clean fluids. As with all variable area meters, the gap meter is invasive. However, the pressure drop across a gap meter is less than that across an orifice meter of equivalent size.

13.5 Turbine Flowmeters

A turbine meter consists of a rotor axially mounted on a spindle and bearings, as shown in Figure 13.7. The speed of the rotor is proportional to the fluid flow rate, rotations normally being sensed magnet-

Fig. 13.7 Cut away view of a turbine meter

ically. The output of the sensing circuit is inherently a pulse signal which is counted and, if appropriate, converted into a 4–20 mA analogue signal.

Turbine meters are one of the most accurate types of flowmeter available. They are invariably used for custody transfer and/or fiscal purposes, for example where ownership changes hands, as in off-loading of petroleum products at terminals, or where tax or duty is payable, as in tanker loading of alcoholic beverages. Turbine meters are only suitable for clean, single phase flow, whether it be gas or liquid. In particular, they are unsuitable for slurries because solid particles damage the bearings, or for steam flow because the impact of the droplets of condensate damages the rotor blades. If accurate measurement of these or other types of flow is required, then some form of positive displacement flowmeter should be specified.

As with the orifice meter, the accuracy of a turbine meter is critically dependant upon straight lengths of pipe both upstream and downstream of the meter and, if necessary, flow straighteners. The pipework should be free of vibration and the meter mounted horizontally. Turbine meters are easily damaged. The meter should always have a strainer installed upstream to prevent foreign matter from damaging the rotor. For liquid flows care should be taken to ensure that the temperature and pressure of operation are such that cavitation cannot occur. If the liquid contains any bubbles of gas some form of disengagement device must be used upstream of the meter. During commissioning in particular, flow should only ever be introduced slowly to the meter to prevent damage of the rotor blades from hydraulic impact or overspeed. For further information on the specification, installation and calibration of turbine meters refer to ISA RP31.1.

13.6 Electromagnetic Flowmeters

The principle of operation of an electromagnetic flowmeter is as depicted in Figure 13.8. If an electrolyte flows axially through an electromagnetic

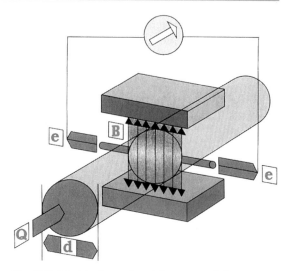

Fig. 13.8 Principle of operation of electromagnetic flowmeter

field, then an emf is generated radially. This emf, which can be measured using electrodes in contact with the electrolyte, is some average of the velocity profile and is directly proportional to the flow rate:

$$e = kBQd$$

where
e	denotes	emf generated	V
k		meter constant	
B		magnetic field strength	Wb
Q		volumetric flow rate	$m^3 \, s^{-1}$,
d		electrode spacing	m

The meter is effectively a section of pipe, often referred to as the primary head. It has to be electrically insulated to prevent the emf being short circuited by the pipe wall. The meter is therefore typically made from steel with a liner made from some insulating material. The liner is invariably plastic. The magnetic field is induced by a coil embedded in the wall of the meter to which is applied either an a.c. or a pulsed d.c. voltage. The electrodes are fitted flush with the inside surface of the liner, as depicted in Figure 13.9, and connected to electronic circuits in an externally mounted housing, the output normally being a 4–20 mA signal.

The orientation of an electromagnetic flowmeter is unimportant, provided that it is always full

Fig. 13.9 View through primary head showing electrode

of liquid. This normally means that for vertical installations the flow through the meter should be upwards. If it is horizontal the electrode axis should not be in the vertical plane. There is no requirement for straight lengths of pipe either side of the meter. It is essential that the process stream is at earth potential. For steel pipework this is realised by earthing the pipework directly. For plastic pipework metallic earthing rings or gaskets have to be installed and connected to earth.

Electromagnetic flowmeters are non-invasive. They therefore have negligible pressure drop across them and, subject to materials of construction constraints, are suitable for all types of liquid flow, including slurries and suspended solids. They are more than accurate enough for most purposes and are available for installation in a very wide range of pipe sizes. The output signal is a linear function of the flow profile and is virtually independent of pressure and temperature. Surprisingly, the output is also independent of the conductivity of the electrolyte. This is because the emf generated is measured using a high input impedance device which draws no current from the circuit. It follows that the meter is tolerant of electrode fouling. All that is required is enough clean electrode surface in contact with the process stream to establish the circuit.

The principle constraint on the use of electromagnetic flowmeters is that the process stream must be liquid and electrolytic. They cannot be used for gases or steam. They are unsuitable for solvents or other non-aqueous liquid flows in which there are no ionic species present. They are relatively expensive in terms of capital costs but, being non-invasive, their operating costs are negligible.

For further information on electromagnetic flowmeters refer to BS 5792.

13.7 Ultrasonic Flowmeters

Figure 13.10 depicts the more common "transit time" type of ultrasonic flowmeter. In essence, bursts of ultrasound are transmitted through the process stream at an angle to the axis of the pipe and detected by the receiver. The transmitter and receiver are typically ceramic piezo-electric elements. These may either be installed in the pipe wall, protected by some sort of "window", or clamped on the pipe externally, providing a non-invasive measurement.

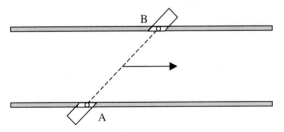

Fig. 13.10 Transit time type of ultrasonic flow meter

The transit time between the transmitter and receiver for each burst of ultrasound varies linearly with the mean velocity of the fluid. Clearly, if A is the transmitter and B the receiver, the transit time decreases with increasing flow rate, the maximum transit time corresponding to zero flow. Conversely, if B is the transmitter and A the receiver, the transit time increases with increasing flow rate, the minimum transit time corresponding to zero flow. The transit time can obviously be calibrated against flow rate for the liquid concerned.

In practice, the peizo electric crystals used can both transmit and receive. Thus pairs of signals in parallel but in opposite directions are produced with differing but corresponding flight times. This difference provides a basis for amplification and for cross checking. There are many configurations. In large diameter pipes it is not uncommon to have transmitter-receiver pairs in different planes and to compute the flow rate from average transit times. To minimise errors due to entrained gases or from solid deposits, the transmitter-receiver pairs should be installed in a horizontal plane across the pipe.

The cost of an ultrasonic flowmeter is largely independent of pipe size, so ultrasonics tends to be very cost effective for large diameter pipes. Ultrasonic flow measurement is only feasible for liquid flows, or for gases at pressures higher than typically 10 bar. The transit type of meter is only suitable for clean liquids: suspended solids cause a scattering of the bursts of ultrasound. For slurries the "Doppler" type of ultrasonic flowmeter is appropriate as its principle of operation is based upon scattering effects. Ultrasonic flowmeters are calibrated for a fully developed flow profile. For this reason there are minimum straight pipe length requirements both up and down stream of the meter. The speed of sound is a function of the liquid density which is itself dependant upon temperature and maybe pressure: both of these effects have to be compensated for in calibration. There are many sources of noise on a plant, all capable of being transmitted by the pipework and corrupting and/or distorting the flow measurement. Of particular importance are sources of cavitation, *e.g.* pumps and valves. Ultrasonic flowmeters should be located well away from such.

13.8 Vortex Shedding Meters

When a fluid flows past a blunt object, often referred to as a bluff body, vortices are formed. These vortices occur in an alternating pattern on either side of the bluff body, as depicted in Figure 13.11. The rate at which the vortices are formed is directly proportional to the flow rate. Measurement of flow is essentially a question of counting the rate at which the vortices are shed. Within any vortex, the increase in kinetic energy comes from a decrease in pressure energy. Thus, by measuring the pressure downstream of the bluff body, the presence of a vortex may be detected. Peizo-electric crystals are sensitive enough to measure the pulses in pressure.

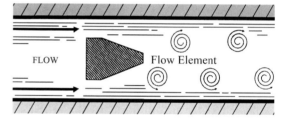

Fig. 13.11 Vortices being shed by bluff body

There are various designs of vortex shedding meter. However, all have a bluff body of some shape at the upstream end and pressure sensors downstream. The sensors may be on the trailing edge of the bluff body or mounted on a separate probe. It is normal to count the vortices on either side of the bluff body for cross checking purposes. The output of the counting circuit is inherently a pulse signal although it is often converted into a 4–20 mA analogue signal. Vortex shedding meters are best suited to liquid flow measurement, but are unsuitable for slurries and suspended solids. They have good accuracy for Reynolds numbers in the range of 10^4 to 10^6. Being an invasive meter, there is a significant loss of head.

13.9 Mass Flowmeters

Mass flowmeters operate on the basis of the Coriolis effect. This can be explained with reference to Figure 13.12. The pipework, in the form of a continuous tube, is rigidly fixed along the X axis. It is continuously vibrated, electro-mechanically, in the Y plane about the X axis. The flow of fluid through the tube causes it to twist slightly about the Z axis. The amount of twist, which is measured optically, is directly proportional to the mass flow rate of the fluid.

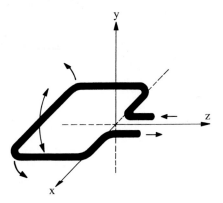

Fig. 13.12 Coriolis principle of operation of mass flow meter

Note that it is the mass flow rate that is measured and not the volumetric flow rate. The meter is therefore suitable for virtually all types of flow. It tends to be used mostly for difficult applications, especially two phase flow, such as liquids containing bubbles of gas, flashing liquids, liquids with suspended solids, gases with entrained solids, *etc*. Mass flowmeters are very accurate for single phase flow, liquids in particular, and are increasingly being used for custody transfer and fiscal purposes. They are also accurate for multiphase flow provided it is homogeneous: that is, the relative proportions of the phases are consistent. Mass flowmeters are non-invasive: whilst the tortuous nature of flow through the meter results in a significant pressure drop, there are designs available which minimise this. By suitable choice of material of construction for the tube they can be made corrosion resistant to virtually any process medium.

There are, however, a number of important constraints. The tube has to be thin enough to enable the flexing and twisting about its various axes, so there are metal fatigue considerations in the choice of material of construction. There is also an upper limit to the pressure of the process medium. It is essential that the X axis is firm and free from vibration itself. For small diameter pipework this can be achieved fairly easily by effective use of conventional pipe supports. However, for larger diameters, rigid mounting is a non-trivial issue and normally involves independent concrete foundations and steel support structures. This makes the installation very expensive.

Level Measurement

14.1 Gauge Glass
14.2 Direct Use of DP Cells
14.3 Use of DP Cells for Density Measurement
14.4 Use of DP Cells for Interface Measurement
14.5 Pneumercators
14.6 Capacitance
14.7 Ultrasonics
14.8 Nucleonics
14.9 Nomenclature

Dp cells are used for the majority of liquid level measurements in the process industries. The design of an installation varies according to context. For example, whether the liquid is under pressure, contains suspended solids, and so on. This chapter outlines good practice in relation to several common applications. It also show how DP cell, can be used for the related measurements of density and interface position. There are many other methods of level measurement, both for liquids and solids. Three of the more important ones, capacitance, ultrasonics and nucleonics, are considered here too. Again the reader is referred to BS6739 and to Bentley (2004).

14.1 Gauge Glass

The simplest form of level indication is the humble gauge glass, equivalent to the sight glass used for detecting liquid flow. The gauge glass typically consists of an externally mounted vertical glass tube with isolating and drain valves as depicted in Figure 14.1. The gauge glass is cheap and reliable with the major advantage that the level can be seen. The glass tube must be physically protected from impact and pipework stresses by a steel

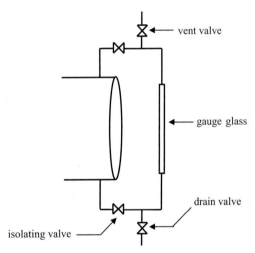

Fig. 14.1 Gauge glass with isolating and vent valves

framework. Because of the possibility of breakage, gauge glasses should not be used with flammable or toxic materials.

14.2 Direct Use of DP Cells

Figure 14.2 shows a dp cell being used for measuring level in a vented tank. The high pressure side of the dp cell is connected to a branch as close to the

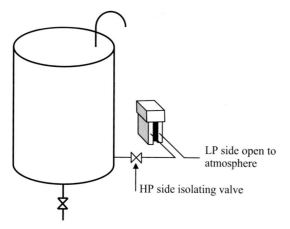

Fig. 14.2 Use of dp cell for level measurement in vented tank

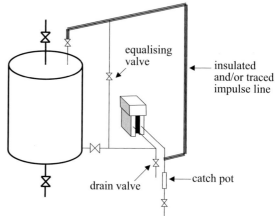

Fig. 14.3 Use of dp cell under pressure or vacuum conditions

bottom of the tank as is practicable. The low pressure side of the dp cell is open to the atmosphere.

This arrangement can only be used for clean liquids: any suspended solids could settle out and block the connecting pipe. Also, because the process liquor is in direct contact with the diaphragm of the dp cell, there are potential materials of construction problems.

Note that it is a pressure difference that is being measured, *i.e.* the static pressure at the bottom of the tank relative to atmospheric pressure. It is directly related to the head of liquid by the equation

$$\Delta P = H\rho g \qquad (14.1)$$

Many tanks and vessels are designed to be operated under pressure or vacuum. The measurement of static pressure at the bottom of such items of plant is therefore relative to the gas or vapour pressure in the space above the liquid. This is achieved by connecting the low pressure side of the dp cell to the top of the tank, as shown in Figure 14.3.

This arrangement is suitable for non-volatile liquids. Note the slope of at least 1 in 20 on the upper part of the low pressure side impulse line. This enables any condensate formed to drain back into the tank rather than down to the dp cell. If slight condensation does occur then it may be necessary to install heat tracing to prevent its formation. Alternatively, a catch pot may be fitted to collect any condensate that dribbles down towards the dp cell,

to be drained off on an intermittent basis. Also note the isolating, equalising and drain valves for commissioning purposes.

If condensation is significant, as for example with liquids close to their boiling point, it is necessary to use a "wet leg" as shown in Figure 14.4.

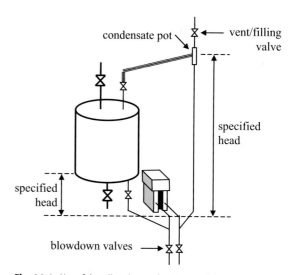

Fig. 14.4 Use of dp cell under condensing conditions

The strategy is the same as in the use of the dp-cell for measuring the flow of condensing vapours as shown in Figures 12.14 and 12.15. Thus the vapour is allowed to condense and fill the impulse line on the low pressure side of the dp cell, the condensate being used to transmit the pressure. Clearly in calibrating the dp cell it is necessary to bias the zero setting to offset the head of condensate. A filling tee is provided to save time during commissioning. Thus the wet leg may be pre-filled with liquid rather than having to wait for it to fill up by condensation.

14.3 Use of DP Cells for Density Measurement

Clearly, by rearrangement of Equation 14.1, density can be articulated as a function of differential pressure:

$$\rho = \frac{1}{Hg} \Delta P \qquad (14.2)$$

Thus, for a fixed height of liquid, a dp cell can be calibrated to provide a measure of density, as depicted in Figure 14.5 in which it is assumed that the tank is vented and the fixed height of liquid is established by an internal weir.

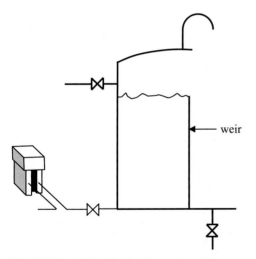

Fig. 14.5 Use of dp cell for density measurement

This is not a particularly accurate means of measuring density: the approach is only effective if the change in density being measured is significant relative to the accuracy of the dp cell. However, it does provide an average value of density across the depth of liquid which is useful if there are variations in density with depth due, for example, to layering effects.

14.4 Use of DP Cells for Interface Measurement

Dp cells can be used to determine the position of the interface between two immiscible liquids. Consider the vented tank depicted in Figure 14.6.

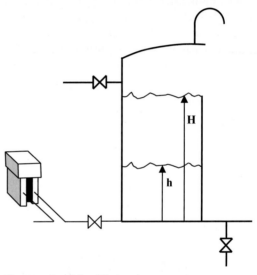

Fig. 14.6 Use of dp cell for interface measurement

The differential pressure is given by

$$\Delta P = h\rho_L g + (H - h)\rho_U g$$

Rearranging gives

$$h = \frac{1}{(\rho_L - \rho_U)g}\Delta P - \frac{H\rho_U}{(\rho_L - \rho_U)} \qquad (14.3)$$

There is thus a simple linear relationship between height of the interface and overall differential pressure which enables the dp cell to be calibrated. Again, this is not a particularly accurate measurement. Noting that the sensitivity of the instrument, $\Delta P/h$, is proportional to $(\rho_L - \rho_U)$, it follows that the greater the difference in density between the two layers, the more accurate the measurement becomes. However, with interface measurement, absolute accuracy is often not critical: what matters is that the interface is known to be somewhere between two given levels, for which this arrangement is very effective.

14.5 Pneumercators

A pneumercator enables the indirect use of a dp cell when it may be inappropriate to measure level through a branch at the bottom of a tank. It utilises a "dip-leg" which is a rigid tube, typically installed through a branch in the top of the tank and long enough to reach down to the bottom. The dip-leg is used in conjunction with a regulated air supply, needle valve and bubbler in an arrangement generally referred to as a pneumercator. The dip-leg is connected to the high pressure side of a dp cell as shown in Figure 14.7.

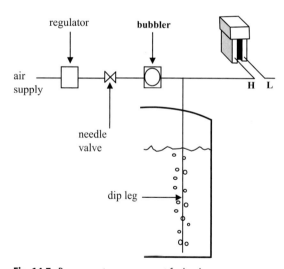

Fig. 14.7 Pneumercator arrangement for level measurement

The regulator is set to some pressure higher than the static pressure at the bottom of the tank. The needle valve is adjusted to give a small flow of air which flows down the dip-leg. When air is bubbling out of the dip-leg and into the liquid the pressure in the dip-leg, referred to as the back pressure, must be equal to the static pressure at the bottom of the tank. This back pressure is measured by the dp-cell and corresponds directly to the head of liquid.

The flow rate of air should be low enough for the pressure drop due to frictional losses in the dip-leg to be insignificant, but high enough to observe that the flow exists. Because it is not normally possible to see the bottom of the dip-leg, there should be an external bubbler with a window through which the bubbles may be observed. The regulator, needle valve and bubbler are normally located on top of, or adjacent to, the tank for access.

Pneumercators are cheap, easy to install and very reliable. They are particularly suitable for use with slurries and dirty liquids because a dip-leg is inherently self cleaning. If a blockage occurs due to accumulation of solids, which could only happen at the bottom of the dip-leg, the back pressure builds up towards the regulator output pressure. Eventually this forces the solids out and the air vents itself into the liquid. Pneumercators also enable dp-cells to be used for level measurement in corrosive environments, the air acting as a fluid barrier between the dp cell and process liquid. Sometimes, for example with biodegradable products, blanketing with nitrogen is necessary in which case the nitrogen can be bubbled in through the dip-leg.

If a dip-leg is to be used in a tank which is operated under pressure or vacuum, then a separate

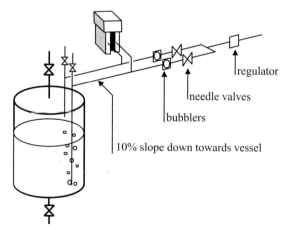

Fig. 14.8 Use of pneumercator under pressure or vacuum conditions

pneumercator may be necessary for measuring the pressure in the space over the liquid: this is connected to the low pressure side of the dp-cell, as shown in Figure 14.8.

Again note the slope of at least 1 in 20 for the impulse lines to enable any condensate formed to drain back into the tank. Also note the overhead space requirement for removal of the dip-legs.

14.6 Capacitance

The use of a capacitance probe for level measurement is illustrated in Figure 14.9. In principle, the capacitance between the probe and the vessel wall is a function of their geometry and on the dielectric of the medium between them. Thus, as the level changes there is a proportionate change in capacitance which can be measured electronically and converted into a signal for transmission.

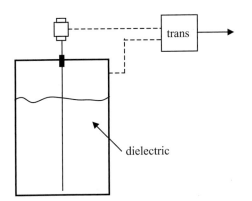

Fig. 14.9 Use of capacitance probe for level measurement

Capacitance can be used for measurement of the level of both liquids and solids, for depths of up to 3 m, over a wide range of temperatures and pressures. For liquids there are no major problems. For solids a useful feature is that the measurement is fairly insensitive to uneven surfaces. The principal sources of error are due to build-up of solids on the probe, variation of bulk density, and unrepresentative signals due to poor location of the probe.

14.7 Ultrasonics

The principle of operation of an ultrasonic type of level measurement is as depicted in Figure 14.10. The sensor consists of a combined transmitter and receiver. In essence, sonic pulses are emitted by the transmitter which are reflected off the surface and detected by the receiver. The transmission time is obviously a linear function of the level of the surface and is converted by an active circuit into an electrical signal.

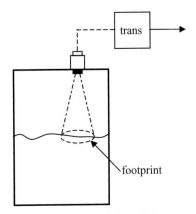

Fig. 14.10 Principle of ultrasonic level measurement

Ultrasonic sensors can be used for both liquid and solids level measurement and have a wide rangeability. They are particularly useful for measuring the level of solids in hoppers, silos and the like, where the surface can be very uneven, as the receiver converts the reflected sound into an average for the level within the footprint of the transmitter. Although the sensors are non-contact devices they are, nevertheless, exposed to the process environment so materials of construction need to be considered carefully. Ultrasonics should not be used with liquids where there is foam on the surface as they give false readings. Neither should they be used in dusty environments where the dust absorbs and/or scatters the sound pulses resulting in weak signals. Another potential problem is noisy environments because of the susceptibility of the receiver to pick-up spurious acoustics.

14.8 Nucleonics

Nucleonic level measurement is based upon the absorption of gamma rays. The amount of absorption for a given material is a function of the density of the material and the path length. The use of nucleonics is depicted in Figure 14.11. The source is a pellet of radioactive material which "shines" a beam of gamma rays through the vessel. The radiation is detected on the other side of the vessel by a Geiger Muller tube. In essence, the pellet is a point source whereas the detector is longitudinal. The intensity of the radiation detected is attenuated linearly in proportion to the level. Calibration is straightforward, with the level measurement being scaled on the basis of maximum and minimum intensity detected. The electronics automatically compensate for the declining strength of the source according to its half life.

The principal advantage of nucleonics is that it is non-invasive and can be used in "difficult" situations. For example, with high temperatures and pressures, in foam, spray or dusty environments, or with media that are corrosive or abrasive. Nucleonic devices are reliable and have very low maintenance requirements. Obviously, because of the radioactivity, there are health physics issues which require proper shielding but, if installed and operated properly, nucleonics is perfectly safe.

14.9 Nomenclature

ΔP	is the	pressure difference	Nm^{-2}
h		height of the interface	m
H		height of liquid/weir	m
ρ		density of liquid	$kg\,m^{-3}$
g		acceleration due to gravity	$m\,s^{-2}$

Subscripts

L	is the lower, denser layer
U	upper, lighter layer

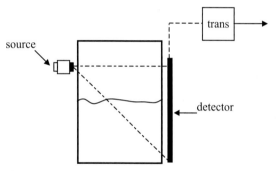

Fig. 14.11 Principle of nucleonic level measurement

Weight Measurement

Chapter 15

15.1 Resistance Effects
15.2 Strain Gauges
15.3 Load Cells
15.4 Installation

Weight is generally used as a measurement in situations where other measures of quantity, such as level, are not accurate enough. In particular, it is commonly used for batching out quantities of reagents for use in reactions. Weight has traditionally been measured by means of mechanical weighbridges in which, for example, a vessel and its contents are balanced by levered counterweights and springs. However, weight measurement by means of load cells, based upon strain gauges, is now the norm. This chapter focuses on the principle of operation of strain gauges, their interfacing, and on the use of load cells.

15.1 Resistance Effects

The electrical resistance of a length of wire is given by

$$R_0 = \rho L_0 / A_0$$

where the subscript 0 denotes some initial condition.

If the wire is stretched its length increases and its cross sectional area decreases. Its new resistance is thus

$$R = \rho (L_0 + \delta L)/(A_0 - \delta A)$$

but its volume remains approximately constant:

$$L_0 \cdot A_0 = (L_0 + \delta L)(A_0 - \delta A)$$

Substituting, expanding, ignoring second-order terms and rearranging gives

$$R \approx R_0 (1 + 2\delta L / L_0)$$

i.e.

$$\delta R / R_0 \approx 2 \delta L / L_0 \qquad (15.1)$$

in which the factor of 2 is referred to as the gauge factor. Thus the fractional change in resistance is directly proportional to the fractional change in length, otherwise known as the strain. However, according to Hooke's law, strain is itself directly proportional to stress, stress being the force applied per unit cross sectional area:

$$F/A = E \delta L / L_0 \qquad (15.2)$$

It follows that change in resistance can be used to measure force. The device used for doing this is known as a strain gauge. Note that electrical resistance is a function, not only of strain, but also of temperature:

$$R = R_0 (1 + a(T - T_0)) \qquad (15.3)$$

where

a is the	temperature coefficient	°C^{-1}
A	cross sectional area	m^2
E	Young's modulus of elasticity	N m^{-2}
F	force applied to wire	N
L	length of wire	m
ρ	resistivity	m Ω
R	resistance	Ω
T	temperature	°C

15.2 Strain Gauges

The most common type of strain gauge consists of a thin foil of metal etched onto a film of plastic. Terminals are formed by soldering wire onto the pads at either end of the foil. A typical gauge, as shown in Figure 15.1, is about the size of a small postage stamp.

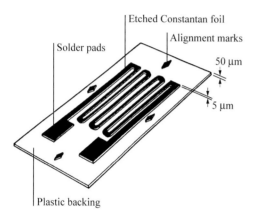

Fig. 15.1 Foil type of strain gauge

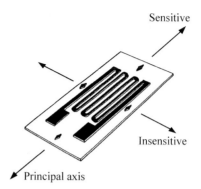

Fig. 15.2 Strain *vs* principal axis

Equation 15.1 reveals that for maximum sensitivity a large value of R_0 and a small value of L_0 are required. Thus design is a compromise between making A_0 small, the foil is typically 5 μm thick, and L_0 relatively large. To accommodate its length the foil is etched in folds which makes the gauge unidirectional. Thus, as depicted in Figure 15.2, it is sensitive to strain along its length, the principal axis, but tends to open out under cross strain.

Sensitivity is dependant on the gauge factor too. The metal foil is normally of constantan (55% copper and 45% nickel) whose gauge factor of approximately 2.0 is consistent with the "ideal" of Equation 15.1. Other alloys, with gauge factors ranging from 2 to 10, may be used but they suffer from being more temperature dependant. Semiconductor strain gauges are available with factors from 50 to 200. However, their response is non-linear. Whenever a strain gauge is supplied the manufacturer should state what its gauge factor is. The accuracy of the factor quoted is a measure of the quality of the gauge.

To measure force the strain gauge is bonded onto the surface of a rigid member. It is bonded such that its principal axis is aligned with the direction of the force being measured. Any stretching of the member's surface causes an equal stretching of the gauge. Thus, as tensile force is applied to the member, its strain is transmitted to the strain gauge. By measuring its change in resistance, the strain in the gauge and hence the force in the member can be calculated from Equations 15.1 and 15.2. This is the basis for calibration.

Effective bonding of the gauge to the surface is clearly critical to the measurement. This is usually realised by means of a proprietary cement or adhesive. If the metal foil is etched onto an acrylic film then methyl methacrylate based adhesive (superglue) should be used as this dissolves acrylic and forms a seamless bond. Clearly the bonding of the gauge to the member must not cause any distortion of the force being measured. In practice such loading effects are normally negligible because of the relative proportions of the gauge and member.

Essential to the accurate measurement of the gauge's resistance is the prevention of short circuiting. There are two aspects to this. First, the gauge must be effectively insulated from the surface to which it is bonded, the member invariably being metallic. Normally the plastic film and the adhesive provide adequate insulation. And second, the gauge must be kept completely dry: any trace of moisture will cause short circuiting between the folds of the foil. This problem is addressed by hermetic sealing.

Strain gauge resistance is almost universally measured by means of an integrated circuit equivalent to a Wheatstone bridge, as depicted in Figure 15.3. It is both accurate and sensitive enough for measuring the very small fractional changes in resistance involved.

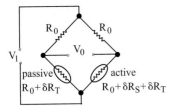

Fig. 15.3 Wheatstone bridge circuit: active and dummy gauges

To compensate for temperature effects, due to Equation 15.3, it is necessary to bond two strain gauges to the surface of the member. They are bonded close together so that they are subject to the same variations in temperature. The principal axis of the active gauge is aligned with the direction of the force being measured whereas that of the dummy gauge is at right angles. The active gauge thus responds to changes in both strain and temperature whereas the dummy gauge responds to temperature only. Since the two gauges are on opposite sides of the bridge, the changes in resistance due to temperature effectively cancel out.

If both the resistors are chosen to have the same resistance R_0 as the initial resistance of the strain gauges, δR_S is the change in resistance due to strain effects on the active gauge and δR_T the change in resistance due to temperature effects on both the active and dummy gauges, then the output voltage V_0 is given by:

$$V_0 \approx \frac{0.5 \cdot V_1}{2 \cdot R_0 + \delta R_S + 2\delta R_T} \cdot \delta R_S$$

The output voltage is thus approximately proportional to the strain effects and the bridge can be calibrated as such. The excitation voltage V_1 is typically a chopped 10V dc supply resulting in an output of some 0–20 mV. Signal conditioning would then result in either a 4–20 mA analogue signal proportional to force or a digital serial signal for transmission.

15.3 Load Cells

Strain gauges are delicate devices. For measuring weight in a plant environment they have to be mounted in special purpose structures referred to as load cells. These provide the necessary protection and physical robustness. The cell is then bolted into the load path.

Load cells are available in a variety of shapes and sizes, capable of individually weighing down to 25 g and, in combination, up to 5000 tonnes. Properly calibrated, they can be accurate to within 0.03% at full range, with linearity to within 0.02% and repeatability to within 0.01%. However, the zero setting on load cells is prone to drift so these figures are only meaningful for differences in weight rather absolute values. For specification purposes, the tare (empty) weight of the vessel should be some 5–50% of full range and the live weight some 20–75%. For batching purposes, the minimum resolution should be not less than 0.02% of full range.

There are various types of load cells. The canister type is used extensively for loads up to 250 tonnes. This has a central load bearing column with four strain gauges bonded to it, one pair mounted horizontally and the other vertically, as depicted in Figure 15.4.

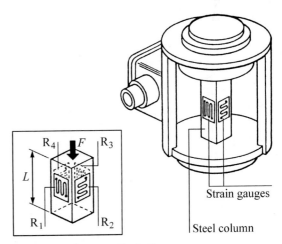

Fig. 15.4 Canister type of load cell

When load is applied, the central column distorts with vertical compressive strain in proportion to lateral tensile strain according to Poisson's law. The four gauges form a bridge, with various additional resistors for balancing and calibration purposes, as depicted in Figure 15.5. The bridge arrangement is inherently temperature compensating.

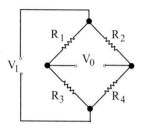

Fig. 15.5 Bridge circuit for four gauge arrangement

The shear beam type of load cell is used extensively for loads of up to 25 tonnes. It is in effect a cantilever arm, the load producing shear forces. The cell is formed by machining a central web in a block of steel. The web is an area of high, almost uniform, shear strain which is at a maximum at 45° to the vertical. The strain is thus measured by a pair of strain gauges mounted on opposite sides of the web, at 45° and 135° to the vertical, as depicted in Figure 15.6, which ensures that the bridge arrangement is inherently temperature compensating. The nature of shear forces in cantilever arms is such that shear cells are tolerant of variations in the position of the vertical load. They are also fairly insensitive to bending stresses, so sideways loads and torques can be tolerated. This makes shear cells particularly suitable for weighing agitated vessels.

15.4 Installation

For weighing process vessels load cells are invariably used under compression. It is normal practice to sit the vessel on the cells, as depicted in Figures 15.7 and 15.8, and to sum their outputs.

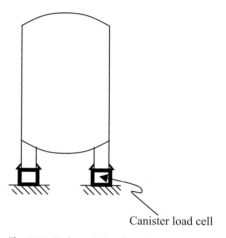

Fig. 15.7 Under tank three legged load cell arrangement

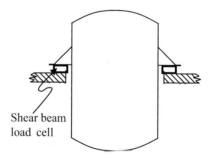

Fig. 15.8 Suspended mid floor three load cell arrangement

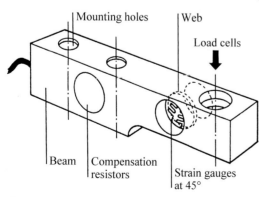

Fig. 15.6 Shear beam type of load cell

For vertical cylindrical vessels, three load cells with 120°C axial spacing is optimum: it guarantees a stable platform with equal sharing of load. Four or more load cells require careful alignment. With some loss of accuracy, some of the load cells in an installation may be substituted with dummies, or pivots, as depicted in Figure 15.9, with scaling

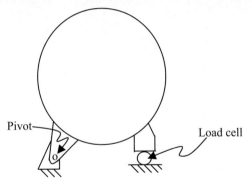

Fig. 15.9 Use of dummy load cells as pivots

factors being applied to the active cells' outputs as appropriate.

Load cells should be handled carefully, they can be damaged by sudden mechanical shock. They are particularly vulnerable when the vessel is being lowered into position. It is best to install the vessel without the load cells and then to jack it up for them to be carefully inserted underneath. The load being weighed and the cells must be aligned vertically. To minimise sideways loads it is common practice to design the mountings to allow a limited amount of sideways movement. The mountings must incorporate constraints to stop the vessel from falling off the cells, perhaps due to vibration, or lifting up under abnormal conditions. The mounting bolts for shear beam load cells must be of high tensile steel and correctly torqued, otherwise the bolts will stretch and distort the readings.

Forces exerted by pipework connected to a vessel are a major source of inaccuracy. Pipelines entering the vessel should be horizontal, rather than vertical, and should be distributed around the vessel rather than clustered on one side. Longer pipelines with bends are more flexible and less problematic. If possible, use short lengths of flexible pipe to connect up pipelines to the vessel, these effectively eliminate all external pipework stresses. And finally, although load cells are hermetically sealed, moisture can seep in around the signal cable: particular attention should be paid to sealing the cable gland.

Temperature Measurement

Chapter 16

16.1 Thermocouple Principles
16.2 Thermocouple Types
16.3 Thermocouple Installation
16.4 Resistance Temperature Devices
16.5 RTD Construction
16.6 RTD Installation
16.7 Thermowells
16.8 Comment

Temperature, obviously, is a common and important measurement in the process industries. Virtually all measurements from −250 to +650°C can be made using either thermocouples or resistance temperature devices (RTD). Thermocouples are cheaper than RTDs but not as accurate. Thermistors, which are semiconductor devices, may also be used within this range. Although they are much more sensitive than either thermocouples or RTDs, their response is very non-linear: for this reason they are seldom used. Above 650°C thermocouples, subject to materials of construction considerations, and radiation pyrometers are used.

This chapter considers thermocouples and RTDs only. Their principles of operation, construction and interfacing requirements are described in sufficient detail to enable meaningful specification of requirements. Finally, temperature probe assemblies and their installation are discussed. A good overview of the operation of thermocouples is given in BS1041 shown in Table 16.1.

16.1 Thermocouple Principles

When wires of two dissimilar metals are joined together to form a circuit, as shown in Figure 16.1, and one of the junctions is heated to a higher temperature than the other, an emf is generated and a current flows. This is known as the Seebeck effect. The emf generated is small (mV) and has to be measured on open circuit: any current drawn through the thermocouple will cause loading effects and distort the measurement. For this reason a very high input impedance transducer must be used, as shown in Figure 16.2.

Table 16.1 Standards on temperature measurement

BS 1041	Temperature measurement	Published
Part 3	Guide to selection and use of industrial resistance thermometers	1989
Part 4	Guide to the selection and use of thermocouples	1992
Part 5	Guide to the use and selection of radiation pyrometers	1989

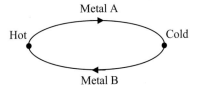

Fig. 16.1 Depiction of the Seebeck effect

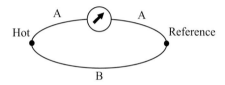

Fig. 16.2 Open circuit measurement of emf

For a given pair of dissimilar metals, the emf generated is dependant on the difference in temperature between the two junctions. For temperature measurement it is usual to use the hot junction as the sensor and to measure relative to the cold junction. Thus the temperature of the cold junction, referred to as the reference junction, must be known. In the laboratory a reference of $0\,°C$ is commonly established with melting ice. However, this is not normally practicable for plant purposes, so the reference is invariably taken as ambient temperature which is measured independently by means of an RTD.

Not only is emf a function of temperature difference – it also depends on the metals concerned. In principle, a thermocouple can be made from any pair of dissimilar metals. However, in practice, only a limited number of combinations of metals and alloys are used because of their high sensitivity, in terms of $mV/°C$, and chemical stability. The standard IEC 60584 provides values of emf tabulated against temperature for all commonly used types of thermocouple, together with their tolerances on accuracy, as shown in Table 16.2.

Table 16.2 Thermocouple reference data

IEC 60584		Published
Part 1	Thermocouples: reference tables	1995
Part 2	Thermocouples: tolerances	1982

16.2 Thermocouple Types

They are normally referred to by letter type, the more common ones being listed in Table 16.3. The tolerances quoted, either as a percentage of the true value or on an absolute basis, give an indication of the potential accuracy of thermocouple measurements. In practice, the accuracy achieved also depends upon errors due to faulty installation and signal processing.

16.3 Thermocouple Installation

A third dissimilar metal may be introduced at a junction without affecting the emf generated. Thus, for example, the two dissimilar metals at the hot junction may be brazed or soldered together. Introducing copper at the reference junction enables a more practical arrangement for measuring the emf, as depicted in Figure 16.3, the copper leads being the connections to the transducer. The reference junction is effectively split into two parts which must be kept close together to ensure that they are at the same temperature.

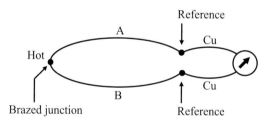

Fig. 16.3 Use of third metal for split reference junction

It is good practice for the reference junction to be situated away from any potential source of heat. In a typical installation involving multiple thermocouples it is normal for all the reference junctions to be located in a common area, such as an input signal termination cabinet, and to have a single shared RTD measurement of cabinet temperature for reference purposes.

Because thermocouple wire is of a very light gauge, it is common practice to use extension leads to extend back to the cabinet. These are of a gauge heavy enough for wiring purposes. Ideally, extension leads are of the same materials as the thermocouple itself. However, on the grounds of cost,

Table 16.3 Characteristics of different types of thermocouple

Type	Metals	Range and accuracy		Comments
		Continuous usage	Short term usage	
E	Chromel (10%Cr, 90%Ni)/Constantan (55%Cu, 45%Ni)	0 to 1100°C	−270 to 1300°C	Most sensitive type. Resistant to oxidation and to moisture corrosion. Must use compensating cable
J	Iron/Constantan	20 to 700°C 1% or 3°C	−180 to 750°C	Can be used in vacuum, reducing and inert atmospheres. Oxidises above 540°C. Unsuitable for moist or sulphurous conditions. Must use compensating cable
K	Chromel/Alumel (5%Al, 95%Ni)	0 to 1100°C 0.75% or 3°C	−180 to 1350°C	Resistant to oxidation, especially above 500°C. Unsuitable for reducing or sulphurous conditions. Must use compensating cable
R	Platinum/Platinum-Rhodium (87%Pt, 13%Rh)	0 to 1600°C 0.15% or 1°C	−50 to 1750°C	Suitable for oxidising and inert conditions. Rapidly poisoned by reducing atmosphere. Contaminated by metal vapours so non-metallic sheaths required, *e.g.* alumina. Must use compensating cable
T	Copper/Constantan	−185 to 300°C 1% or 1°C	−250 to 400°C	Can be used in vacuum, oxidising, reducing, moist and inert atmospheres. Not resistant to acid fumes. Only type to have defined limits of error below 0°C. Can use copper extension leads

it is usually necessary to use extension leads of appropriate proprietary "compensating cable", as depicted in Figure 16.4. Such cable has thermoelectric properties similar to thermocouple wire so the emf measured is not distorted. Also, the use of compensating cable minimises corrosion effects due to the formation of electrochemical cells between the dissimilar metals. Provided the split reference junctions are at the same temperature, variations along the length of the extension leads and/or differences at the other split junction do not matter.

To provide both chemical and physical protection for thermocouples, it is standard practice to use them in sheaths as depicted in Figure 16.5. The sheath is normally of stainless steel and up to 6 mm in diameter. The mineral packing is typically of

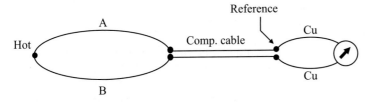

Fig. 16.4 Use of compensating cable for remote measurement

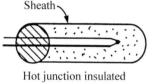

Hot junction insulated from sheath

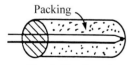

Bonded hot junction

Fig. 16.5 Standard thermocouple sheath

magnesia (MgO) and provides good heat transfer and electrical insulation. Practices differ, the hot junction may be bonded to the inside end of the sheath or else insulated. Clearly a bonded junction gives a faster response, but the bond is difficult to make and can become detached. A plastic or, for high temperatures, a glazed ceramic terminal block is used at the external end of the sheath. The polarity should be clearly marked for wiring purposes.

Thermocouples can be used for measuring average temperatures. Suppose an average of three measurements is required. Each may be measured independently and the results averaged. Alternatively, the three thermocouples may be wired in series, as depicted in Figure 16.6. The resultant emf, scaled by a factor of 3 prior to signal processing, yields a temperature averaged at source which is more accurate.

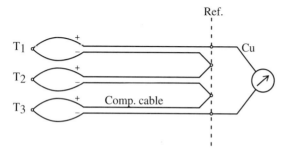

Fig. 16.6 Thermocouples wired in series for average measurement

Similarly, thermocouples may be used to measure temperature difference directly by wiring them up "back to back" as depicted in Figure 16.7.

Fig. 16.7 Thermocouples wired in parallel for difference measurement

16.4 Resistance Temperature Devices

Electrical resistance is a function of temperature. Thus, by measuring the change in resistance of a given resistor, temperature may be determined. The only RTD of any consequence in process automation is the platinum resistance thermometer as described in BS1904:

Above $0°C$
$$R = R_0 \left(1 + aT - bT^2\right) \quad (16.1)$$

Below $0°C$
$$R = R_0 \left(1 + aT - bT^2 - c(T - 100)T^3\right) \quad (16.2)$$

where R is the resistance Ω
  T   temperature $°C$

and subscript 0 denotes the zero condition. The values of the coefficients are as follows:

$a = 3.9083 \times 10^{-3} \quad °C^{-1}$
$b = 5.775 \times 10^{-7} \quad °C^{-2}$
$c = 4.183 \times 10^{-12} \quad °C^{-4}$

It is standard practice for R_0 to be 100Ω giving a value for R of 138.5 Ω at $100°C$. This is referred to as the Pt100 sensor. To satisfy BS1904, the output of a Pt100 sensor must fall within a defined range of accuracy about Equations 16.1 and 16.2. There are two classes of tolerance, A and B, of which Class A is the more stringent:

For Class A

dT = 0.15 + 0.002. |T|

For Class B

dT = 0.15 + 0.005. |T|

It is normal practice to use the Pt100 sensor for measurements up to 600°C. From 600 to 850°C it is usual to use the Pt10 sensor. This is of thicker gauge, for more reliable service, with an Ro value of 10 Ω.

16.5 RTD Construction

There are two types of platinum RTD: the film type and the wire wound. The film type is formed by the deposition of platinum on a ceramic substrate. Because of their flat structure film type RTDs are suitable for surface temperature measurements. However, they are not as reliable as wire wound RTDs and are unsuitable for use in a conventional sheath. The wire wound RTD, as depicted in Figure 16.8, is used almost universally for accurate temperature measurement.

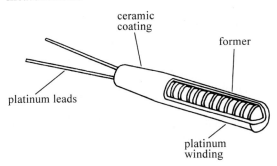

Fig. 16.8 Wire wound type of RTD

It consists of the platinum resistance wire wound around a glass or, for high temperatures, ceramic former and sealed with a coating of glass or ceramic. As with thermocouples, it is normal practice to use RTDs in mineral packed sheaths of stainless steel up to 6 mm in diameter. The platinum leads are insulated with silica tubing, to prevent short circuiting, terminating in a glazed ceramic block at the external end of the sheath.

16.6 RTD Installation

RTD resistance is almost universally measured by means of an integrated circuit equivalent to a Wheatstone bridge, the resistors being specified such that the bridge balances at 0°C. The two wire system of Figure 16.9 is adequate if the length of the leads from the RTD to the bridge is short. However, if the leads are long, as is invariably the case with process plant, their resistance would distort the measurement since both leads are on the same side of the bridge, in series with the RTD. It is therefore common practice to use a three wire system, as depicted in Figure 16.10.

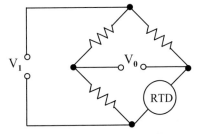

Fig. 16.9 Bridge circuit for 2-wire RTD arrangement

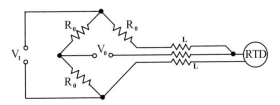

Fig. 16.10 Bridge circuit for 3-wire RTD arrangement

In a three wire system the leads to the RTD are on adjacent sides of the bridge so their resistances L, and any changes in such, effectively cancel out. Note that since the output V_o is measured by a high impedance transducer, which draws negligible current, the resistance L due to the third lead across the bridge is of no consequence. Any imbalance between the resistances of the leads will nevertheless lead to some inaccuracy.

For high accuracy measurements a four wire system is required utilising either null balance or constant current source techniques:

- Null balance involves switching between two field wire configurations and null balancing the bridge. One configuration involves wires A, B and D, the other involves wires A, C and D as depicted in Figure 16.11. The resulting null balance equations are used to derive the RTD resistance value. There is still scope for some minor inaccuracy due to differences introduced by the switching circuit and its contacts.

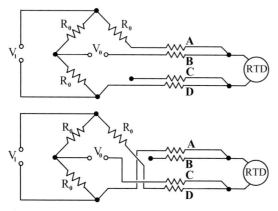

Fig. 16.11 Bridge circuit for 4-wire RTD arrangement

- Constant current source involves passing a small constant current through two connecting wires and the RTD. Another two wires are then used to measure the voltage across, and hence the resistance of, the RTD using a high impedance transducer. The voltage measured is compensated for offset due to thermocouple junction effects at the RTD. The amount of offset is established by switching off the constant current source for a short period and measuring the offset voltage directly. There is still scope for some minor inaccuracy due to the self heating effects of the constant current within the RTD.

RTDs may be used to measure temperature difference directly by wiring them into adjacent sides of a bridge, as depicted in Figure 16.12.

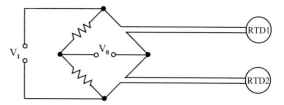

Fig. 16.12 Bridge circuit for RTDs used for difference measurement

In using a bridge circuit to measure the resistance of an RTD, it is very important that the other resistances are such that the voltage drop across the RTD is very low. Otherwise the current through the RTD will cause self heating of the RTD which will significantly distort the signal.

16.7 Thermowells

A temperature probe assembly consists of a thermowell, insert and head cap as illustrated in Figure 16.13.

Often referred to as a thermopocket, the thermowell is effectively part of the plant in which it is installed. It is in direct contact with the process medium and has to be capable of withstanding whatever process conditions exist in the plant. The thermowell has several functions. It enables thermal contact between the process and the sheath containing the thermocouple or RTD. The thermowell protects the sheath from the process. It enables the sheath to be withdrawn for maintenance without having to shut the plant down. In addition, during maintenance, it prevents contamination of the process by entry of air or dirt.

The thermowell may be either screw fitted, as in Figure 16.13, or flanged. The dimensions of a thermowell and its fittings generally conform to BS2765. Of particular importance is the internal diameter, normally 7 mm, which is consistent with the 6 mm maximum outside diameter of thermocouple and RTD sheaths. The length of a thermowell is determined by the application. The choice materials of construction is important, a comprehensive listing of appropriate materials is given by Pitt (1990).

The insert consists of the sheath, containing the thermocouple or RTD, and the terminal block assembly, as shown in Figure 16.14. The head cap is typically a steel casting with a screw cap that protects the terminals from dust and moisture and provides access for maintenance.

Although a thermocouple or an RTD has an inherently fast response, by the time it is packed into a sheath and inserted into a thermowell, there is an appreciable thermal capacity. Also, the air film between the sheath and thermowell is a significant resistance to heat transfer. The net effect is a sluggish response. If the sheath and thermowell are lumped together as a first order system, the time constant is typically 0.5–1.0 min but, in extreme cases, can be up to 5 min. Better thermal contact may be achieved by spring loading the sheath, such that its tip is in contact with the bottom of the thermowell. Heat transfer may also be enhanced by pouring a small amount of proprietary heat transfer oil into the annulus between the sheath and thermowell.

16.8 Comment

For effective temperature measurement it is essential that the probe assembly is properly located. It does not matter how accurate the sensor is if, for example, the probe does not reach down into the liquid whose temperature is being measured. It is much cheaper, and less embarrassing, to address these issues at the design and specification stage than when the plant is operational.

It is increasingly common practice to find head mounted transmitters in use with both thermocouples and RTDs. In essence, an intrinsically safe integrated circuit is mounted in the head cap which enables the temperature to be transmitted directly as a 1–5-V or 4–20-mA signal, or otherwise. For thermocouples the circuit provides amplification, filtering, linearisation and scaling, but the temperature measurement is only relative to the local ambient temperature conditions. Likewise for RTDs for which the circuit provides the bridge type of measurement.

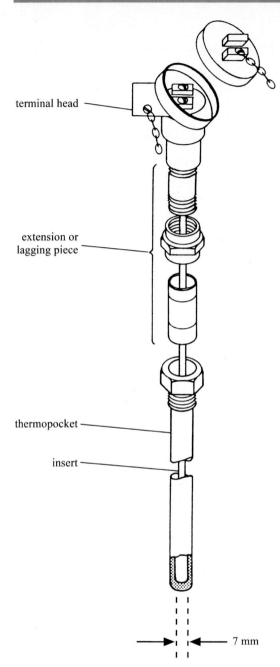

Fig. 16.13 Thermowell, insert and head cap assembly

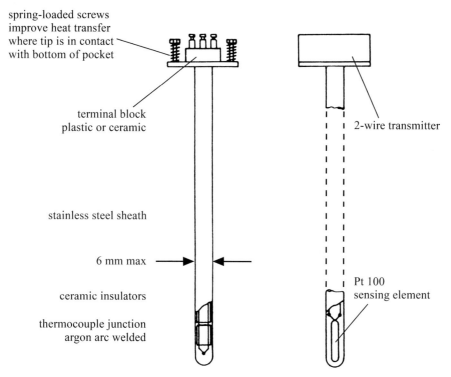

Fig. 16.14 Sheath and terminal block assembly

pH Measurement

Chapter 17

17.1 Nature of pH
17.2 Strong Acids and Alkalis
17.3 Weak Acids and Bases
17.4 Mixing Effects
17.5 The Glass Electrode
17.6 Practical Issues
17.7 Comments

pH is a measure of acidity and is a surprisingly common measurement. For example, in the chemical industry, the acidity of the reagents in many types of reactor has to be controlled to enable optimum reaction conditions. In addition, in the water industry, the acidity of fresh water for consumption and of effluent for discharge have to be controlled carefully to satisfy legislative requirements.

pH is an electro-chemical measurement, invariably made by means of the so-called glass electrode. It is a notoriously difficult measurement to make because of factors such as drift and fouling. Understanding the significance of measurements requires an appreciation of electro-chemical equilibria. And using pH for control purposes is problematic because of the inherent non-linearities and time delays, as described in Chapter 34.

Most textbooks on physical chemistry give detailed explanations of pH and some of the related issues. However, for a more control oriented approach the reader is referred to the rather dated but classic text by Shinsky (1973).

17.1 Nature of pH

There is a formal definition of pH:

$$pH = -\log_{10}[H^+] \quad (17.1)$$

where [ ] denotes concentration of ions in aqueous solution with units of g ions/L. In the case of hydrogen, whose atomic and ionic weights are the same, $[H^+]$ has units of g/L or kg m^{-3}. The logarithmic scale means that pH increases by one unit for each decrease by a factor of 10 in $[H^+]$.

Pure water dissociates very weakly to produce hydrogen and hydroxyl ions according to

$$H_2O \Leftarrow H^+ + OH^- \quad (17.2)$$

At equilibrium at approximately 25°C their concentrations are such that

$$[H^+].[OH^-] = 10^{-14} \quad (17.3)$$

The dissociation must produce equal concentrations of H^+ and OH^- ions, so

$$[H^+] = [OH^-] = 10^{-7}$$

Since pure water is neutral, by definition, it follows that for neutrality:

$$pH = -\log_{10}[10^{-7}] = 7$$

This gives rise to the familiar pH scale of 0–14, symmetrical about pH 7, of which 0–7 corresponds to acidic solutions and 7–14 to alkaline solutions.

To evaluate the pH of alkaline solutions it is usual to substitute for H^+ in Equation 17.1 from Equation 17.3:

$$pH = -\log_{10}\left[\frac{10^{-14}}{[OH^-]}\right] = 14 + \log_{10}[OH^-]$$

17.2 Strong Acids and Alkalis

Acids and alkalis dissociate upon solution in water. Strong acids and alkalis, such as hydrochloric acid and caustic soda, dissociate completely. This means that all their H^+ and OH^- ions exist as such in solution and are measurable by a pH electrode:

$$HCl \Rightarrow H^+ + Cl^-$$

$$NaOH \Rightarrow Na^+ + OH^-$$

Note the use of the word "strong" to describe HCl and NaOH. This is a measure of the fact that they dissociate completely, rather than an indication of their concentration. It is, of course, possible to have a dilute solution of a strong-acid and a concentrated solution of a weak-acid.

In principle, if equal quantities of solutions of strong acids and alkalis of the same concentration are mixed, then the resultant solution will be neutral with pH 7. For example:

$$HCl + NaOH \Rightarrow NaCl + H_2O$$

The salt NaCl is itself completely dissociated and neutral:

$$NaCl \Rightarrow Na^+ + Cl^-$$

and the water weakly dissociated according to Equation 17.2. In practice it is very difficult to get the concentrations and quantities exactly the same, so the resultant solution is likely to be somewhere between pH 6 and pH 8.

Figure 17.1 is a sketch of the titration curve for the neutralisation of HCl by the addition of NaOH, *i.e.* of a strong-acid by a strong-alkali. The plot is of pH *vs* volume of alkali added.

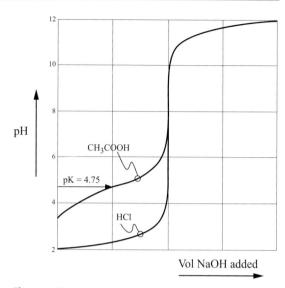

Fig. 17.1 Titration curve: HCl by NaOH

The dominant feature of the titration curve is the almost vertical slope in the vicinity of pH 7. This vividly depicts the logarithmic non-linearity and illustrates the nature of the control problem. Consider 1 m³ of HCl solution of pH 2 being neutralised with NaOH solution of pH 12. The amount of caustic soda required to raise its pH in unit steps from 2 to 7 is as shown in Table 17.1.

Table 17.1 Caustic required per unit pH change

pH change	L NaOH	Total L
2–3	818.2	818.2
3–4	162.0	980.2
4–5	17.8	998.0
5–6	1.8	999.8
6–7	0.2	1000.0

It is evident that, starting with a fairly coarse adjustment, the closer the pH is to 7 the finer the adjustment required becomes. Thus, to increase the pH from 2 to 7 requires a total of 1000 L of caustic soda, of which only 1.8 L is required to take the pH from 5 to 6, and a further 0.2 L to take the pH from 6 to 7. The addition of 1000 L to within 1.8 L, let alone 0.2 L, is beyond the accuracy of normal process flow instrumentation.

17.3 Weak Acids and Bases

Dissociation of weak acids and bases is incomplete and results in some equilibrium between the ions and the non-dissociated solute. For example:

$$HA \Leftrightarrow H^+ + A^- \qquad BOH \Leftrightarrow B^+ + OH^-$$

for which the equilibrium constants are defined to be

$$K = \frac{[H^+].[A^-]}{[HA]} \qquad K = \frac{[B^+].[OH^-]}{[BOH]}$$

The equilibrium constant is itself often articulated on a logarithmic basis as follows:

$$pK = -\log_{10} K$$

A knowledge of the pK value, and of the initial concentration of acid or base, enables the concentration of the various species at equilibrium to be determined, and hence the titration curve to be drawn. Some common weak acid and base dissociations are summarised in Table 17.2; others are listed in the text by Shinsky (1973).

Table 17.2 Common weak acid and base dissociations

Acid/base	Equilibrium	pK
Acetic acid	$CH_3COOH \Leftrightarrow CH_3COO^- + H^+$	4.75
Carbonic acid	$H_2O + CO_2 \Leftrightarrow H^+ + HCO_3^-$	6.35
	$HCO_3^- \Leftrightarrow H^+ + CO_3^{2-}$	10.25
Ammonia	$NH_3 + H_2O \Leftrightarrow NH_4^+ + OH^-$	4.75
Slaked lime	$Ca(OH)_2 \Leftrightarrow CaOH^+ + OH^-$	1.4
	$CaOH^+ \Leftrightarrow Ca^{2+} + OH^-$	2.43

Note that ammonia, which by most yardsticks would be considered to be chemically reactive, is a weak-base. Also note that some dissociations involve parallel equilibria, each of which has its own equilibrium constant.

Figure 17.1 also depicts the titration curve for the neutralisation of CH_3COOH by the addition of NaOH, i.e. of a weak-acid by a strong-alkali. Note that for low pH values the slope is much less severe than in the strong-acid strong-base case. This is due to a phenomenon known as buffering:

$$CH_3COOH \Leftrightarrow CH_3COO^- + H^+$$

$$NaOH \Rightarrow Na^+ + OH^-$$

$$H^+ + OH^- \Rightarrow H_2O$$

As caustic soda is added to the acetic acid, the OH^- ions combine with the H^+ ions to produce water. In effect, the H^+ ions are removed from solution, so the acetic acid dissociates further to maintain the equilibrium. As more caustic soda is added, the acetic acid continues to dissociate until most of it is used up. Thus, the change in $[H^+]$ within the buffer zone, and hence in pH value, is less than it would have been otherwise. This reduced sensitivity makes for easier control of pH in the buffer zone, but for more difficult control outside.

Similarly, if a weak-base is neutralised by a strong-acid, buffering occurs somewhere between pH 7 and 14. In the case of ammonia, the titration curve is the mirror image of that for acetic acid: this is because their pK values are the same. If the dissociation involves parallel equilibria, or if there are several weak acids and/or bases present, then there will be various buffer zones.

17.4 Mixing Effects

It is perhaps useful to consider the effects of mixing equal volumes of acid (say pH 3) and alkali (say pH 11). The resultant pH will be anywhere between pH 3 and 11 depending on strength as indicated in Table 17.3.

If either the acid or alkali was a solid, the pH of the resultant mixture would depend on how the solid reacts. For example, chalk reacts slowly with acid, the rate of reaction being determined by the rate at which the chalk dissolves:

$$CaCO_3 + 2HCl \rightarrow CaCl_2 + H_2O + CO_2$$

Thus the pH of the resultant mixture is only apparent once dissolution is complete.

Table 17.3 Effects of mixing strong/weak and concentrated/dilute solutions

Acid (pH 3)	Alkali (pH 11)	Mixture	Comment
Dilute strong	Dilute strong	About pH 6–8	Only pH 7.0 by chance
Concentrated weak	Concentrated weak	Between pH 5 and 9	Approximately neutral
Concentrated weak	Dilute strong	Perhaps pH 3.5	Buffering by acid, large pK
Dilute strong	Concentrated weak	Perhaps pH 10.5	Buffering by alkali, lower pK

17.5 The Glass Electrode

The pH sensor consists of a glass electrode and a reference electrode. The two electrodes are often combined into a single unit as depicted in Figure 17.2.

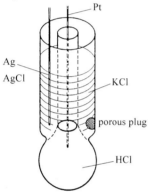

Fig. 17.2 pH sensor comprising glass and reference electrodes

The glass electrode has a thin glass bulb which is permeable to H⁺ ions. Inside the bulb is a solution of HCl acid, typically of pH 0 to 1. The electrode has a platinum wire inside to make an electrical connection. The reference electrode consists of a silver wire with a coating of silver chloride immersed in a standard solution of KCl, *i.e.* a solution of known strength. The porous plug establishes electrical contact between the KCl solution and the process solution, external to the electrode unit, whose pH is being measured.

Each electrode can be thought of as a half cell which generates an emf, the two electrodes together forming an electrochemical circuit. When the glass bulb is placed in the process, the difference in concentration of H⁺ ions across the bulb causes ions to diffuse through it. A tiny emf is generated directly proportional to the pH being measured. The purpose of the reference electrode is to complete the circuit and to generate a known emf against which that generated by the glass electrode can be measured. The difference between the two emfs is measured by a high impedance device so there is negligible current drawn from the circuit: this maintains the strength of the solutions inside the electrodes.

The output, *i.e.* the difference between the two emfs, is given by an equation of the form of the Nernst equation:

$$e = e_0 + k.T.\log_{10}[H^+]$$

where e is normally in mV and T is in deg K. The constants e_0 and k depend on the particular combination of electrodes and solutions used. This yields the pH as being directly proportional to the change in voltage measured, which gives some rationale to the definition of the pH scale:

$$pH = -\log_{10}[H^+] = \frac{e_0 - e}{k.T}$$

Classically, the KCl solution leached out through the porous plug, the leaching process helping to keep the plug clean. The slight outflow required the KCl solution to be topped up intermittently. Lack of KCl solution and/or the plug becoming blocked were the major causes of faults: that is, failure of the electrochemical circuit. Manufacturers have striven to minimise the amount of attention needed, the outcome being electrodes that are either disposable or only require periodic cleaning.

Some systems have the glass and reference electrodes as separate entities. Others have two reference and/or glass electrodes for redundancy and fault detection purposes. In case the Ag/AgCl/KCl cell is not suitable for particular process conditions, other reference cells are available. The most common other combination is $Hg/HgCl_2/KCl$, the so-called calomel electrode.

17.6 Practical Issues

pH electrodes have a reputation for being temperamental. Because glass electrodes are delicate they can easily be damaged. They are very susceptible to fouling which causes false readings. Being of an electrochemical nature they require constant attention: topping up of electrolyte, re-calibration, *etc*. However, modern pH sensor and transmitters are quite reliable supporting self-diagnostics, automatic re-calibration, *etc*.

The electrodes are normally installed in special purpose holders designed to enable contact with the process liquor whilst providing protection from mechanical damage. Holders are available for immersing the electrodes in a tank, typically inserted vertically through the roof or suspended from above the liquid surface. Also very common are in-line electrode holders. Perhaps the best arrangement, but most expensive, is to insert the electrodes in a sample stream in some separate analyser housing where the electrodes can be regularly inspected, cleaned and calibrated.

The most common problem with pH electrodes is fouling. The formation of an impervious film on the surface of the glass bulb, whether due to the deposition of grime or accumulation of residues, causes drift, hysteresis and sluggish response. Various self cleaning devices have been developed for use in dirty fluids: none of them are entirely satisfactory. If fouling cannot be prevented, say by means of a filter, then it is probably best to accept that regular replacement of the electrodes is necessary.

Variations in temperature affect pH, as indicated by the Nernst equation. The most common means of addressing this problem is by means of temperature compensation. Thus alongside the glass and reference electrodes is installed a temperature probe, wired into the pH transmitter, which automatically corrects the pH value. Alternately, if a sample stream is being used, the analyser housing can provide some means of thermostatic control.

There are also potential materials of construction problems. Whilst glass is generally regarded as chemically inert, it is attacked by strong caustic soda and hydrofluoric acid and, since the bulb is thin, can easily be damaged. Also the bulb normally has a plastic cap and/or sheath for mounting which clearly have temperature limitations. Thermal shock should be avoided with the glass electrode being brought up to operating temperature gradually. It is particularly important to specify if steam sterilisation or ultrasonic cleaning is to be used.

17.7 Comments

Neutrality, or a pH of 7, is based on the concept of pure water which is not to be confused with distilled water, de-ionised water or any other form of water. Very small amounts of contamination, including dissolved CO_2 from the air, will cause water to deviate from pH 7, so values from 6 to 8 may be considered to be neutral for industrial purposes. For this reason water should never be used for calibration purposes: buffer solutions of known pH are used.

Whilst the 0–14 scale for pH covers most situations, it is possible to obtain values of pH < 0 and pH > 14 for very concentrated solutions. However, for such solutions, it is unlikely that the dissociation is complete so there is uncertainty over the values for both $[H^+]$ and $[H_2O]$, so the pH values are not very meaningful.

The pH scale comes from the dissociation of water into H^+ and OH^- ions. It follows that pH measurements are only relevant for aqueous solutions. Whereas a measure of the pH of solvents, resins, *etc*. can be made using the glass electrode, or otherwise, the values obtained are meaningless relative to the pH scale.

The pH sensor is the best known specific ion electrode. It responds to changes in H^+ ion concentration and is little influenced by other ions. Manufacturers increasingly offer a range of selective electrodes, such as for nitrates and redox potential. In general, selective ion electrodes respond primarily to the named ion but may also give some response to high concentrations of other ions. For example, a typical chloride electrode will respond to bromide ions, but less strongly. The use of specific ion electrodes is commonplace in, for example, boiler water treatment, electro plating baths and effluent treatment.

Chromatography

Chapter 18

18.1 Principle of Chromatography
18.2 Column Design
18.3 Katharometer
18.4 Flame Ionisation Detector
18.5 Sliding Plate Valve
18.6 Sampling and Conditioning
18.7 Column Operation
18.8 Multiple Columns
18.9 Calibration and Signal Processing
18.10 Comments

Having covered the instrumentation used for conventional measurements in the previous chapters, the emphasis now shifts to analytical measurements. This chapter concerns chromatography: that on spectroscopy is deferred to a subsequent edition. These are undoubtedly the most common of the on-line analytical techniques used in the process industry.

Chromatography is a field in which there are hundreds of texts. An excellent industrial perspective is given by Liptak (1995) and a more in depth treatment by, for example, Braithwaite (1995). The reader is also referred to the useful EEMUA (1995) guide on calibration and EEMUA (2000) guide on maintenance: whilst these refer to analysers in general, much of the content is directly applicable to chromatographs.

18.1 Principle of Chromatography

Chromatography separates the mixture to be analysed into its components by interaction between two immiscible phases. A sample of the mixture is introduced into the so-called mobile phase and is "carried" through a thin column containing the stationary phase. Gas chromatography (GC) refers to the situation when the mobile phase is a gas, the stationary phase being either solid or liquid. Liquid chromatography is when the mobile phase is liquid, the stationary phase usually being solid. The vast majority of industrial chromatographs are of the GC type for which reason GC is the main thrust of this chapter.

In GC the mobile phase consists of the sample being analysed in a carrier gas which is typically hydrogen, helium or nitrogen. As the plug of sample flows through the column, a dynamic equilibrium is established with components of the sample adsorbing onto the surface of the stationary phase and then desorbing back into the carrier stream, the underlying mechanism being diffusion. For a given component, provided the carrier gas flow rate and its temperature are constant, the rates of adsorption and desorption are determined by its concentrations in the carrier gas and at the surface of the stationary phase:

$$N_A = k_G A \left(P_g - P_s\right)$$

Note that $k_G = f\left(Re, \theta_g\right)$ and $Re = g\left(u_g, \theta_g\right)$ where

N_A is the rate of mass transfer mol s^{-1}
k_G mass transfer coefficient   mol m^{-2} bar^{-1} s^{-1}
A   surface area   m^2
P   partial pressure   bar
Re   Reynolds number
θ temperature   °C
u   velocity   m s^{-1}
Subscripts
g   carrier gas stream
s   surface of stationary phase

These rates of mass transfer vary along the length of the column and result in the various components being held on the stationary surface for different amounts of time. This produces a separation of the components into bands within the carrier gas with components emerging at the end of the column in order of increasing affinity with the stationary phase. Generally speaking, this affinity relates to the volatility of the components, *i.e.* those with the highest volatility emerging first. A detector is used to detect the presence of components in the exit stream resulting in a plot of the amount of component *vs* time, referred to as a chromatogram, as depicted in Figure 18.1.

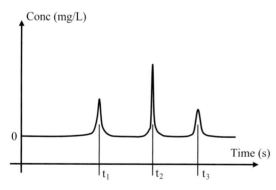

Fig. 18.1 A chromatogram: concentration *vs* time

The ordinate is the output signal of the sensor (typically mV) which, in practice, is scaled in terms of concentration (mg/L) of component in the carrier gas: the abscissa is obviously lapsed time.

Retention time is the time it takes a component to travel through the column. For the ith component:

$$T_i = T_{mi} + T_{si}$$

where T_{mi} is the time spent by the ith component in the mobile phase and T_{si} is the time it spends in the stationary phase. Components are often characterised for chromatography purposes in terms of their distribution ratio which is defined as

$$K_i = \frac{T_{si}}{T_{mi}}$$

For a given sample, the separation achieved depends on many factors including the carrier gas flow rate and temperature, column design, packing density, choice of stationary phase, *etc*. Careful specification of these parameters enables separation of the mixture into relatively distinct bands as opposed to overlapping ones. For streams with few components it is common to use the chromatograph to establish the fraction of each component in the sample, but for streams with many components the fraction of a few key ones is usually sufficient.

The layout of a typical GC is as depicted in Figure 18.2, the essential features of the key components being as described in the following sections.

18.2 Column Design

Traditionally, packed columns were nearly always used for GC. Typically the column would be of stainless steel, with an internal diameter of 2 mm and a maximum length, determined by pressure drop considerations, of some 3–4 m. The stationary phase may be either solid or liquid. If solid it is in the form of particles of an active packing material such as alumina, charcoal or silica gel. However, the activity of such packings is variable and leads to inconsistent results: generally speaking, liquid stationary phases are preferred.

The liquids used for the stationary phase are active in the sense that they promote adsorption. Typical liquids are mixtures of methyl silicone, phenyl and cyanopropyl or polyethylene glycol depending upon the application. The liquid is coated onto some passive packing material such as di-

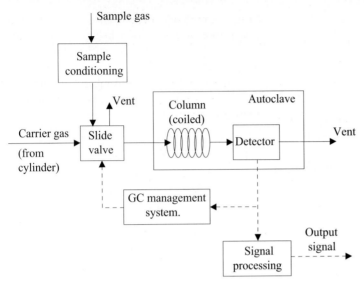

Fig. 18.2 Layout of typical gas chromatograph

atomaceous earth which has a high surface area to volume ratio.

It is now common practice to use capillary columns rather than packed columns. The column is usually of a fused silica with an external polyamide coating, typically with an internal diameter of 0.25 mm and length of up to 30 m. Being not dissimilar in construction to fibre optic cables, such capillary columns are both tough and flexible and, for installation purposes, can be coiled up into a convenient space. The active liquid stationary phase is coated onto the internal wall of the column and chemically bonded to the silica surface which makes for a stable level of activity.

Capillary columns are much more efficient than packed columns in terms of separation of components and speed of response. However, the volumes handled are much smaller so effectiveness is critically dependent on the precision of the injection system and the sensitivity of the detector.

There are many detector devices available but the two in most common usage are the thermal conductivity detector, generally referred to as a katharometer, and the flame ionisation detector (FID). Traditionally the katharometer was the GC detector of choice: it is sensitive enough for most purposes and very reliable. However, despite higher costs and increased complexity, use of the FID has become universal because of its greater sensitivity and speed of response.

18.3 Katharometer

A katharometer determines the change in composition of a gas stream indirectly. Central to its operation is an electrically heated element, either a tungsten or alloy filament or a thermistor, across which the gas stream flows. The temperature to which the element rises is determined by a thermal equilibrium between the electrical power dissipated and the rate of heat transfer into the gas stream by convection:

$$Q = i^2 R_e = hA(\theta_e - \theta_g)$$

Note that $R_e = f(\theta_e)$ and $h = g(F_g, k_g, \theta_g)$ where

Q	is the rate of heat transfer	kW
i	current	A
R	resistance	Ω
h	film coefficient	kW m^{-2}°C
A	surface area	m^2

θ	temperature	°C
F	flow rate	m³ s⁻¹
k	thermal conductivity	kW m⁻¹ °C

with subscripts

g	gas stream
e	element

For a given current, provided the gas stream flow rate and its temperature are constant, the temperature of the element depends solely on the conductivity of the gas stream. A rise in conductivity causes an increase in the film coefficient: the improved heat transfer results in a decrease in the temperature of the element and hence of its electrical resistance.

Within a katharometer, the element protrudes from a cavity into a channel in a metal block through which the carrier gas stream flows. There are three physical arrangements, as depicted in Figure 18.3:

a. All of the carrier gas flows across the element. This has a fast response to changes in conductivity but is sensitive to changes in the gas flow rate for which tight control is required.
b. There is limited circulation across the element. This arrangement is insensitive to changes in carrier gas flow rate but has a slow response to conductivity changes.
c. This is a compromise arrangement which provides adequate speed of response to conductivity changes and is sufficiently insensitive to carrier gas stream flow changes for most purposes. It is the most commonly used design.

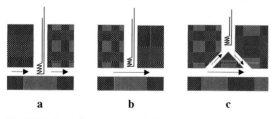

Fig. 18.3 Physical arrangements for katharometer

In practice there are two separate channels within the block with a pair of elements installed in each channel. The carrier gas stream flows through one channel and a reference gas flows through the other channel. Any change in resistance of the elements is measured by means of a bridge circuit as depicted in Figure 18.4. An alternative approach involves applying a constant current, irrespective of the element's resistance, and measuring the voltage required with a circuit not dissimilar to that of Figure 6.13.

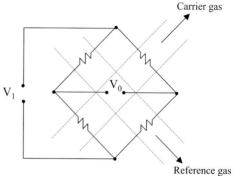

Fig. 18.4 Bridge circuit for katharometer

An important characteristic of katharometer design is that the volume of the cavities and channels within the block are as small as possible, for two reasons:

- The large thermal mass of the block relative to that of the carrier gas stream has the effect of filtering out any slight variations in temperature.
- The high sensitivity necessary for use with capillary columns requires that the detector volume be minimal.

18.4 Flame Ionisation Detector

Within a flame ionisation detector (FID), the exit stream from the column is burned in a minute oxygen rich hydrogen flame as depicted in Figure 18.5.

The exit stream is mixed with hydrogen and flows through a jet where it burns. A voltage of some 300 V d.c. is applied between the jet and an electrode, referred to as the collector, which is positioned above the flame. When a pure hydrogen-air mixture is burned there are no ions produced and the current between the jet and collector is virtually zero. However, as soon as organic material

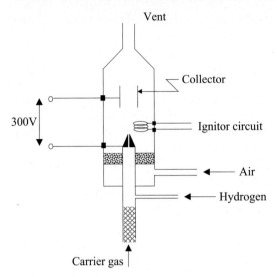

Fig. 18.5 Flame ionisation detector

18.5 Sliding Plate Valve

Valves are used for injection of the sample into the carrier gas and/or for switching the carrier gas stream between columns. There are various types of valve used in chromatography but the sliding plate valve, often referred to as a slide valve, is the most common. It comes in both linear and rotary forms but the linear is more usually used because its actuation is simpler. The slide valve consists of a PTFE plate sandwiched between two stainless steel blocks. Both PTFE and stainless can be machined which makes for precise internal volumes, tight sealing and smooth action.

The slide valve is designed to deliver a precise volume of sample. For the very small samples (µL) used with capillary columns, the volume is defined by the dimensions of the channels internal to the PTFE plate. For the larger samples (mL) used with packed columns, the volume is defined by the dimensions of an external sample loop as depicted in Figure 18.6.

from the sample enters the flame, ionisation occurs and a current due to the ions and free electrons flows. The speed of response of an FID depends upon the flow rates of carrier gas, hydrogen and oxygen.

The magnitude of the FID's response depends upon the concentration of sample material in the carrier gas and on the nature of the ionisation process. In essence, ionisation depends upon the structure of the molecule and the type of atoms involved: it is characterised by an effective carbon number (ECN). For any given structure, *e.g.* aliphatic, aromatic, *etc.*, the ECN is the sum of the contributions to the FID response due to the number of carbon atoms (per molecule) and to their combinations with non carbon atoms such as in –OH and –NH$_2$ groups. Values of ECN are tabulated: they are used for calibrating the FID and enable the component parts of its response to be distinguished.

A number of common gases including O_2, N_2, H_2O, CO, CO_2, H_2S, SO_2 and NH_3 cannot normally be detected by FID. However, they can be detected by doping the carrier gas with an ionisable gas such as CH_4 resulting in a negative response due to the dilution of the ionising gas as the sample passes through the FID.

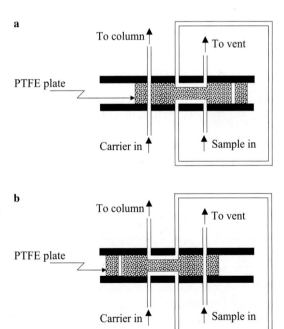

Fig. 18.6 Slide valve with external sample loop: **a** in flush mode, **b** in inject mode

Operation of the slide valve is simple. In stand-by mode, the carrier gas flows through the valve and into the column, flushing out any traces of the previous sample. The sample stream from the process passes around the external loop and is vented, as depicted in Figure 18.6A. For inject mode, the slider is moved across such that the sample of gas trapped within the slider and the external loop is pushed by the carrier gas into the column, the sample stream passing through the valve to vent, as depicted in Figure 18.6B. When the sample has passed through the column, the slider is returned to its original position with the valve in stand-by mode.

The six-port arrangement of Figure 18.6 is commonplace. However, for use with multiple columns, it is usual to have slide valves with ten ports or more. Four of the ports are assigned to the basic carrier, sample, column and vent streams. Design of the slide valve is such that alignment of internal channels with the other ports enables streams to be switched between columns according to the application and between modes depending upon the operations required.

18.6 Sampling and Conditioning

The importance of effective sampling and sample conditioning prior to injection cannot be overemphasised. There are a number of aspects to this:

- The sampling system must result in a sample stream that is representative of the process stream being sampled. The positioning of the sample points and orientation of the probe is critical. Thought must also be given to the temperature and pressure of the process stream and its general condition in terms of condensables, solids content, *etc*.
- The sample stream invariably requires some form of conditioning to get it into a form suitable for injection into the column. For example, for gas streams this may involve condensing out water vapour, adjusting the pressure and/or flow rate, *etc*. For liquid streams the filtering out of solids is usually required prior to vaporisation for GC.
- There must be a sufficient and constant flow of sample stream. Generally speaking the process pressure is relied upon to drive the sample stream through the conditioning system and slide valve but, if that is insufficient, pumping may be required. Typically sample stream flow rates are ≥ 2 L/min.
- The sample stream is usually either returned to the process or, depending upon its nature, vented to atmosphere along with the carrier gas.
- The time delay due to the sampling and conditioning process, together with that due to operation of the chromatograph, must be sufficiently small for control purposes. In practice this means situating the chromatograph close to the sampling point. Typically, the column and detector, in their autoclave, together with the slide valve and ancillary equipment are installed in a field mounted cabinet adjacent to the plant.

18.7 Column Operation

If the chromatograph is to give reproducible results in terms of retention times, and hence the profile of the chromatogram, then control of the carrier gas flow rate is fundamental to operation of the column. Whilst it is obviously possible to control the flow directly, it is more usual to control the gas flow indirectly by means of pressure regulators. That is because pressure regulation responds quickly to downstream disturbances due, for example, to switching the mode of the slide valve. Typically, regulation is a two stage process. Firstly the carrier gas is regulated at source, for example at the outlet of the cylinder in which it is stored. And secondly, it is regulated in-line immediately upstream of the slide valve.

Also of fundamental importance is control of the carrier gas temperature as this has a direct bearing on the rates of mass transfer within the column and hence on the retention times. It is common practice for the column and some of the ancillary equipment, especially the katharometer if used, to be mounted in an autoclave. In essence this is an oven, electrically heated with fan assisted hot air circulation, which enables a constant and uniform temperature throughout. Isothermal operation is the norm, but modern GC equipment provides the option for temperature profiling. Typically the autoclave is maintained at a constant temperature whilst the more volatile components elute from the column: the temperature is then ramped up at a fixed rate to accelerate the separation of the remaining components.

18.8 Multiple Columns

In practice, GC is more subtle than simply switching samples through a column as implied in the previous sections. Use of multiple columns and slide valves enables chromatographic techniques to be targeted according to the nature of the sample being analysed. There are several common configurations of columns and/or switching arrangements as follows:

- Parallel columns are used when the time delay associated with sampling and analysis is too long for control purposes. By having duplex or triplex columns with a common sampling system and conditioning system, and operating the columns out of phase with each other, the frequency of results can be doubled or trebled.
- Backflushing is used when only the lighter (more volatile) components in the sample are of interest: it enables them to be separated from the heavier ones. The arrangement consists of two columns in series, the first is referred to as the stripping or pre-cutting column and the second as the analysis column, as depicted in Figure 18.7. Functionally the arrangement may be thought of as having a six-port slide valve prior to each column although, physically, there would only be a single ten-port slide valve used. In essence:
  a. Initially both slide valves are in their standby modes with carrier gas flowing through both columns to vent.

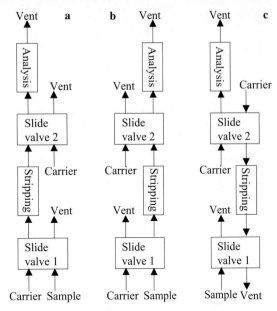

Fig. 18.7 Backflushing: **a** standby mode; **b** inject mode; **c** backflush mode

  b. The first slide valve is switched to inject mode and the sample is injected into the primary column in the normal way.
  c. Once the lighter components have passed through the stripping column, the intermediate slide valve is switched into backflush mode. The lighter components continue to be carried through the analysis column but the carrier gas flushes the heavier ones back out through the stripping column to vent.

- Heartcutting is used when the concentration of the components of interest are low relative to the bulk of the sample. Figure 18.8 depicts a chromatogram of a sample containing small quantities of benzene, ethyl benzene and paraxylene in a mixture which is mainly toluene. The ethylbenzene and paraxylene peaks are not clearly resolved and sit on the tail of the toluene peak. The arrangement for heartcutting also involves two columns in series, the primary and secondary columns, as depicted in Figure 18.9. Again, functionally, the arrangement may be thought of as having a simple slide valve prior

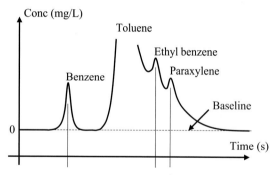

Fig. 18.8 Chromatogram for mixture with dominant component (toluene)

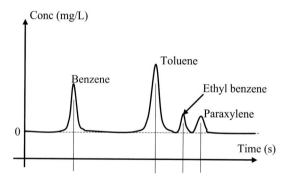

Fig. 18.10 Chromatogram after heartcutting

to each column although, physically, there would only be a single more complex slide valve used.

a. Initially both slide valves are in their standby modes with carrier gas flowing through both columns to vent.
b. The first slide valve is switched to inject mode and the sample is injected into the primary column in the normal way.
c. When the benzene has passed through the intermediate slide valve, it is switched into heartcut mode: the benzene continues to be carried through the secondary column whilst the toluene is vented from the first column.
d. Once most of the toluene has been vented, the slide valve is switched back into inject mode and the remainder of the toluene together with the ethyl benzene and paraxylene are carried into the secondary column.

Figure 18.10 depicts the resultant chromatogram.

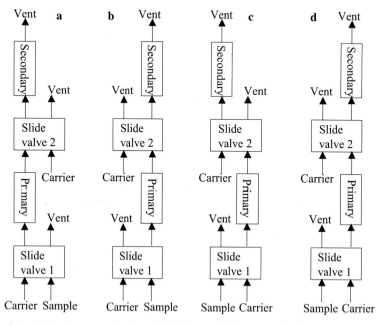

Fig. 18.9 Heartcutting: **a** standby mode; **b** inject mode; **c** heartcut mode; **d** inject mode

18.9 Calibration and Signal Processing

The distribution of peaks on the chromatogram, and their shape, is the underlying information about the composition of the sample being analysed. In essence, the position of a peak indicates which component is present and the height of, or area underneath, the peak is a measure of its concentration. However, to determine the value of the concentration of a particular component requires that the chromatograph be calibrated.

In the ideal world a chromatograph would be calibrated using a mixture containing the same components as the sample being analysed. However, it is not always possible to prepare such a mixture, perhaps due to chemical instability of the components or to problems with condensation. Fortunately, that approach is not necessary. Since the relative positions of the peaks for all volatile substances are well known, it is sufficient to use a reference mixture comprising known quantities of a few specified stable components to establish and calibrate a baseline. The chromatogram of any sample is then analysed in relation to that baseline. There are many variations on the theme but the essentials are as follows:

- Peak positions are determined by monitoring the slope of the chromatogram, with sufficient tolerance to allow for sample noise.
- Thinking of a distinct peak as being roughly triangular in shape, the start and end of a peak may be established by extrapolation of tangents on either side of the peak back to the baseline.
- Peak area is established by numerical integration of the area under a peak and above the baseline. For distinct peaks the integration is between the start and end of the peak as determined above. For unresolved peaks integration is started at the minimum immediately prior to the peak and continued to the minimum immediately afterwards.
- *A priori* knowledge of the mixture enables the use of gates: these are pre-determined time periods during which peaks are searched for and the area integrated.

Concentrations of individual components are then established by applying scaling factors to the peak areas, the scaling factors being derived from the chromatogram of the reference mixture.

18.10 Comments

Whilst they have been used off-line for years in the laboratory, on-line chromatographs have hitherto been seen as being both costly and unreliable: extensive maintenance was invariably required to enable meaningful results. With an installed cost of typically £50K per measurement, chromatographs are still expensive devices but their reliability and functionality has improved dramatically. A modern GC has automated sampling systems, microprocessor based analyser management, software for interpreting the data, displaying and recording the results, together with interfaces to control systems. It is the evolution of these systems and software, rather than developments in chromatographs *per se*, that have led to their increased effectiveness.

Section 3

Final Control Elements

Valves and Actuators

19.1 Valve Types
19.2 Globe Valves
19.3 Butterfly Valves
19.4 Ball Valves
19.5 Diaphragm Valves
19.6 Pneumatic Actuators
19.7 Valve Selection

It is invariably the case in process automation that the manipulated variable is a flow rate, either of a reagent or of a utility such as steam, heating oil or cooling water. At least 90% of all analogue output channels terminate in an automatic control valve, and some 50% of all discrete output channels terminate in an automatic isolating valve. It follows that the manufacture and maintenance of valves, actuators and associated devices is a substantial industry in its own right.

For P&I diagram purposes, a control valve, its actuator and I/P converter are depicted as shown in Figure 19.1, and an isolating valve, its actuator and pilot valve are depicted as shown in Figure 19.2.

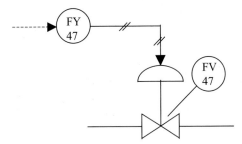

Fig. 19.1 Symbols for control valve, actuator and I/P converter

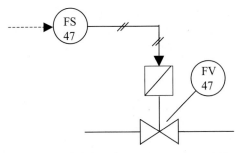

Fig. 19.2 Symbols for isolating valve, actuator and pilot valve

19.1 Valve Types

There is much variety in valve design. The more important types are ball, butterfly, diaphragm and globe. These may be categorised on the basis of actuation and function as in Table 19.1.

Table 19.1 Valve types categorised by function and actuation

		Function	
		Control	Isolating
Actuation	Linear	Globe	Diaphragm
	Rotary	Butterfly	Ball

The categorisation in Table 19.1 is consistent with normal industrial practice, but is not exclusive. For example, globe and butterfly valves are often used for dual control and isolating purposes, although a tight shut-off may not be achieved. Diaphragm and ball valves are sometimes used for control purposes, but are unlikely to provide good quality control.

This chapter outlines the principal features of construction of valves and actuators. For a more comprehensive treatment the reader is referred in particular to the text by Driskell (1983). Valve sizing is covered in Chapter 20 and valve positioners are considered in Chapter 21.

19.2 Globe Valves

These are described in detail, on the basis that they are the most common type of control valve. A typical pneumatically actuated globe valve is shown in Picture 19.1. The actuator, often referred to as the motor, and the valve body are functionally distinct although in practice they are normally supplied as a single unit. This unit is itself often loosely referred to as the valve. Most manufacturers have a standardised range of actuators and valve bodies which are bolted together as appropriate. The yoke provides rigidity.

The essential feature of a globe valve is its plug and seat assembly. Upwards movement of the stem causes the plug to lift out of its seat. This varies the cross sectional area of the annulus between the plug and its seat resulting in a change in flow. The relationship between the stem position and the flow through the annulus is known as the valve's characteristic. The plug shown in Picture 19.1 is of the solid needle type. The characteristic of this type of plug is determined by its profile, and by the size of the plug relative to its seat. Most manufacturers provide a range of plug profiles and sizes to enable selection of the characteristic. Needle plugs may be used for either liquid or gas flows. They are invariably used for "difficult fluids", *e.g.* slurries, dirty and sticky liquids.

There is a force, acting upwards along the stem, due to flow through the valve. The magnitude of this force depends on the pressure drop across the plug and its effective cross sectional area. If this pressure drop is large, the forces acting on the stem will result in the actuator not opening the valve to its correct position. In such circumstances it is good practice to use a double seated valve, as depicted in Picture 19.2.

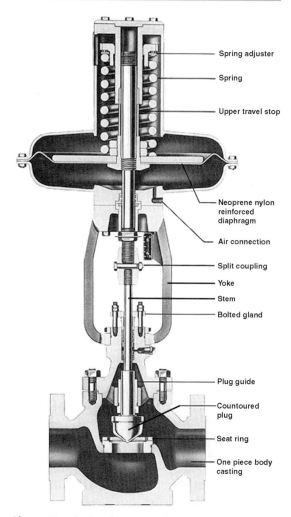

Picture 19.1 Section of pneumatically actuated globe valve

It is evident that the forces acting on the stem due to flow through each plug are in opposite directions and roughly cancel out. This enables the use of standard actuators. Because double seated valves are used in highly turbulent situations, it is necessary for the stem to be guided at both top and bottom of the body. This prevents the stem from vibrating and avoids fracture due to metal fatigue. The principal disadvantage of a double seated valve is that it is difficult to achieve-shut off in both seats. This is especially true at high temperatures due to differential expansion effects. For this reason dou-

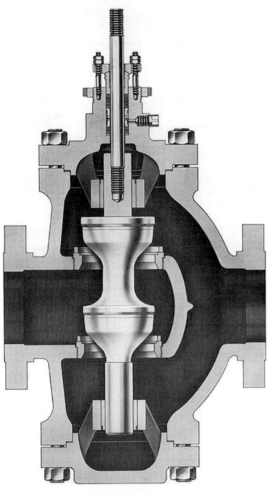

Picture 19.2 Section of body of double seated valve

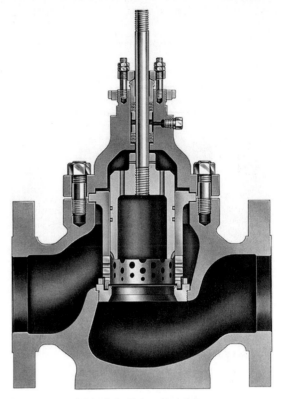

Picture 19.3 Globe valve with cage type of plug

ble seated valves should not be used for isolating purposes.

An alternative to the solid needle type of plug is the hollow cage type, as depicted in Picture 19.3. The skirt of the plug has a number of holes drilled in it. These align with holes in the cage. Fluid flows from the inlet port, through the cage and skirt, and down through the plug to the outlet port. As the stem lowers the plug within the cage, the holes move out of alignment and the area available for flow is reduced. There is much variety in design to realise different characteristics. The size of the holes and their distribution may vary axially. The holes may be non-circular or slotted. The skirt may be fluted. Cage type plugs are primarily used for gas, steam and vapour flows. They cannot be used if there are any solids present since build-up of particles in the annulus between the cage and plug would restrict the plug motion.

Acoustic noise is a potential problem for valves. Noise is caused, if the pressure drop across the valve is large, by virtue of the high velocities involved. A basic strategy in valve design is to counter noise at source by minimising velocities within the plug and seat assembly, and by avoiding sudden expansions and changes in direction of the flow path. Noise can be reduced by the use of concentric sleeves around the cage, by fitting a diffuser to the seat exit or, as shown in Picture 19.4, by installing baffles in the valve outlet, all of which effectively drop the pressure in stages.

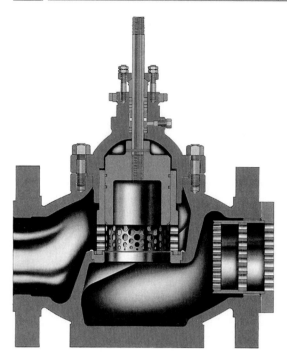

Picture 19.4 Cage type plug with baffles

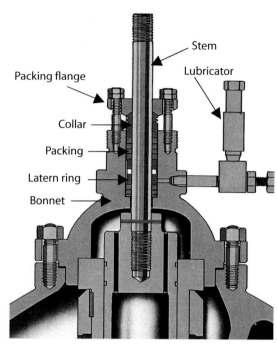

Picture 19.5 Stem packing assembly

Another important issue in the design of valves is leakage into the atmosphere of process fluid along the stem. Clearly the higher the pressure inside the body the greater is the potential for leakage. This is addressed by the use of packing around the stem. The packing assembly is shown in more detail in Picture 19.5. The stem passes through a cylindrical cavity, known as the stuffing box, drilled into the bonnet of the valve. This contains the lantern ring, packing and a steel collar. By tightening the packing flange the collar compresses the packing against the lantern ring: the packing fills the space between the stem and stuffing box wall and stops the leakage.

The packing is usually in the form of tight fitting chevron or split rings. Below 250°C the packing material is invariably PTFE, because it has a very low coefficient of friction and is inert to attack by most chemicals. At higher temperatures packings of laminated graphite are used. Unfortunately, graphite and steel react electrochemically. To prevent galvanic corrosion of the stem it is necessary to use zinc washers, the zinc being a sacrificial anode, with the washers being replaced periodically.

Older valves, many of which still exist in the field, have a variety of packing materials. These invariably suffer from stiction, *i.e.* the stem is unable to slide through the packing smoothly. Indeed, even for valves with PTFE or graphite packings, stiction is a problem for high pressure applications for which the packing must be tight. In such cases an external lubricator may be fitted, the lubricant being applied to the stem as it moves through the lantern ring.

Clearly, both the body and trim must be resistant to chemical attack by the process medium, especially at normal operating temperatures. The most common type of valve body is of mild steel. The casting has to be strong enough to withstand process pressures and temperatures, and external pipework stresses. For corrosive fluids the body would be lined with PTFE, or else a stainless steel or alloy casting would be used. Chemical attack can render a valve useless after only a few hours of operation. For cryogenic applications the casting would probably be of bronze.

In addition to resisting chemical attack, the valve trim, *i.e.* stem, plug, cage and seat, must also be hard enough to resist abrasion and erosion. The trim is usually made from stainless steel, the grade of stainless depending upon the extent to which the process stream contains abrasives. Great care must be taken in specifying the correct materials of construction.

19.3 Butterfly Valves

There is much variety in the design of butterfly valves, the one shown in Picture 19.6 is typical.

Picture 19.6 Wafer type of butterfly valve

The valve consists of a circular disc which is rotated in a body. The maximum rotation is about 90°. The body has integral rims against which the disc seats. The rims often have some elastomeric ring to provide a tight seal. The rotation of the disc may be symmetrical with respect to the axis of the pipe, or offset. The cross section of the disc is usually contoured to enhance the valve's characteristic. The most common type of body is referred to as a wafer. This is flangeless and is installed between the flanges of a pipe using long bolts. To provide corrosion resistance the body is typically fitted with a PTFE lining and the disc encapsulated. General rotation of the disc does not require much torque but tight shut off usually requires a high torque. The torque is applied by a stem and is normally provided by a pneumatic piston, either by means of a lever or a rack and pinion.

19.4 Ball Valves

A typical manually operated ball valve is as shown in Picture 19.7.

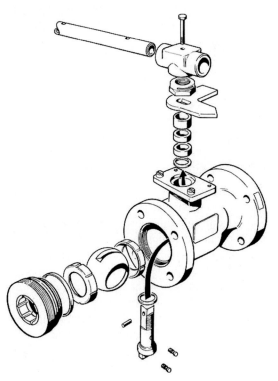

Picture 19.7 Hand actuated ball valve

It consists of a ball that rotates in a body. The ball is normally of stainless steel and the body of mild steel. Rotation of the ball varies the alignment of a cylindrical port through the ball with the flow passage through the body and pipework. The standard ball valve has a port diameter that is about 80% of the pipe diameter, which makes for a compact valve. The alternative full-port ball valve has a port that is the same diameter as the pipeline which enables full bore flow. The full-port ball valve is the only type of valve that is wholly non-invasive. However, it is more bulky and requires a more powerful actuator.

The ball rotates against seat rings which provide a seal between the ball and body. The pressure drop across the valve forces the upstream seat against the ball. Rotation is realised by applying an external torque to the ball by means of the stem. The ball is guided by the stem shaft. For larger ball valves there is often a stub, on the opposite side of the ball from the stem, for guiding purposes. An alternative design is for the ball to be floating inside the body, in which case it is supported by the seat rings. In this case both seats form a seal, *i.e.* the pressure drop across the valve forces the upstream seat against the ball and the ball against the downstream seat.

19.5 Diaphragm Valves

The most common diaphragm valve is the weir type as shown Picture 19.8.

The body provides a weir between its inlet and outlet ports which provides a contoured surface for seating the diaphragm. The actuator applies force to the diaphragm's backing plate which compresses the diaphragm against the weir. Diaphragm valves may be operated either manually or by means of an actuator. That depicted in Picture 19.8 is pneumatically actuated with a manual override.

19.6 Pneumatic Actuators

The actuator depicted in Picture 19.1 produces a linear motion. It consists of a nylon reinforced diaphragm, with a backing plate to provide rigidity, held in the flanges of the housing. The 0.2–1.0 bar signal from the I/P converter is applied through the

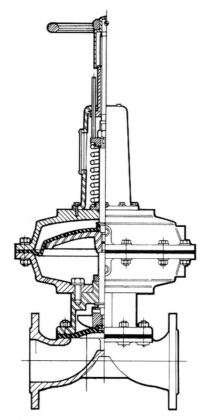

Picture 19.8 Weir type of diaphragm valve

air connection to the underside of the diaphragm, the force exerted on the diaphragm being opposed by the compression of the return spring. The stem, which is in two parts with a split coupling to simplify assembly and maintenance, is fixed at its upper end to the diaphragm and has the plug fixed to its lower end.

It is evident that movement of the plug relative to its seat is dependant upon a balance of forces. As discussed, there are forces due to the 0.2–1.0 bar control signal acting on the diaphragm, to compression of the spring, to friction in the packing and to the pressure drop across the plug.

There are many alternative arrangements used by the various manufacturers. The pressure may be applied to either side of the diaphragm, the return spring may be either above or below the diaphragm, the spring may be either compressed or

Table 19.2 Essential characteristics of valve types

	Globe	Butterfly	Ball	Diaphragm
Size	0.5–40 cm	2–500 cm	1.5–90 cm	0.5–50 cm
Maximum temperature[a]	650°C	1200°C	750°C	175°C
Maximum pressure drop[a]	10–500 bar	5–400 bar	5–100 bar	2–15 bar
Materials of construction	Body and trim in most metals. Small sizes in plastic and PTFE	Body and disc in most metals. Wide range of seals	Body and ball in wide range of metals. Also reinforced plastic bodies. Ball may be PTFE coated. Various seals	Bodies of various metals, plastic and glass lined. Several types of elastomeric diaphragm
Applications	Good for extreme conditions	Large size. Good with viscous fluids	OK for slurries but not gritty materials	Good for difficult fluids, *e.g.* corrosive, suspended solids, sticky liquids, *etc.*
Advantages	Wide variety of types, options and characteristics. Superior noise and cavitation handling. Good accuracy	Cheaper than globe valves for sizes above 5 cm. High capacity. Wide rangeability.	Moderate cost. High capacity. Good rangeability. Non-invasive. Low pressure drop. Low leakage	Low cost. Simple construction. Low leakage
Disadvantages	High cost. Moderate rangeability	High torque for shut-off. Disposed to noise and cavitation. Sensitive to specification errors	High friction. High torque required. Vulnerable to seal wear. Dead space in body cavity. Disposed to noise and cavitation	Limited operating conditions. Poor rangeability. Maintenance costs high due to diaphragm wear

[a] Note that the maximum temperature is highly dependent on the materials of construction and, in particular, the packing. Also, the maximum pressure drop across the valve depends on its diameter, the larger the diameter the smaller the maximum pressure drop.

extended, and the plug may be either lifted out of or pulled into its seat. A major consideration in the specification of a valve is its fail-safe requirement. In the event of a failure of the air supply, such that the input 0.2–1.0 bar signal becomes zero, should the return spring force the valve fully shut or wide open? The valve can be either of the air-to-open type in which case it fails-closed or of the air-to-close type which fails-open. The choice is a process design decision based solely upon safety considerations.

Spring opposed pneumatic actuators are the norm for all automatic control valves. There are several good reasons for this which are perhaps best appreciated by consideration of the electrical alternative. This would consist of an electric motor and gearbox, with switchgear for reversing polarity to enable both opening and closing. Pneumatic actuators are simpler to manufacture and much cheaper. They are accurate enough for most purposes and are very reliable. Pneumatically actuated valves are powerful and have a faster response. They cannot cause sparks so are safer for use in hazardous areas. The spring will force them to a fail-safe state in the event of air failure. The arguments for pneumatics are much the same in the case of automatic isolating valves where the alternative is a spring opposed solenoid actuator.

19.7 Valve Selection

Key features of the various valves are summarised in Table 19.2.

Acknowledgement. The diagrams of the valves depicted in Pictures 19.1 to 19.5 have been copied from originals provided by Kent Introl Valves Ltd whose permission to copy is gratefully acknowledged.

Valve Sizing

20.1 Inherent Characteristic
20.2 Installed Characteristic
20.3 Comparison of Installed Characteristics
20.4 Worked Example No 1
20.5 Trim Selection
20.6 Valve Sizing
20.7 Worked Example No 2
20.8 Critical Flow
20.9 Nomenclature

Effectiveness is critically dependant on selecting the right sort of valve for the job, specifying an appropriate characteristic and sizing it properly. This chapter provides an introduction to the complexities of sizing. For a more detailed treatment the reader is referred in particular to the texts by Driskell (1983) and Baumann (2003) and to IEC 60534 which supersedes various BS and ISA standards.

20.1 Inherent Characteristic

The inherent characteristic of a valve is theoretical to the extent that it is the relationship between the flow through the valve and its opening, assuming a constant pressure drop. With reference to Figure 20.1:

$$F = f(X)|_{\Delta P_V = const} \quad (20.1)$$

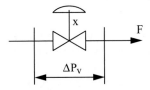

Fig. 20.1 Valve: symbols for flow, pressure drop and opening

As discussed in Chapter 19, the characteristic is determined by the plugs' shape and/or profile, by its size and by the position of the plug relative to its seat. In principle, there are many possible characteristics. However, in practice, most manufacturers have standardised on three trims which are readily available and satisfy most applications. These are referred to as square root, linear and equal percentage, and are depicted in Figure 20.2.

The inherent characteristics and their slopes are quantified in Table 20.1.

Table 20.1 Characteristics and slopes *vs* trim types

	$F = f(X)$	$\dfrac{dF}{dX}$
Square root	$F = k\sqrt{X}$	$F' = \dfrac{k^2}{2F}$
Linear	$F = kX$	$F' = k$
Equal %	$F = F_0 e^{kX}$	$F' = kF$

Valves with a square root characteristic are sometimes referred to as having a quick opening trim, but it cannot be assumed that any valve with a quick opening trim has a square root characteristic. Some manufacturers produce trims with a so called modified equal percentage characteristic

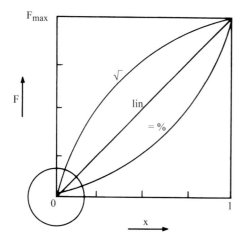

Inherent characteristics

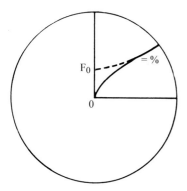

Fig. 20.2 Inherent characteristics for standard trims

whose profile lies somewhere between linear and equal percentage.

The equal percentage characteristic needs some explanation. First, if the equation for its slope is rearranged as follows:

$$\frac{\Delta F}{F} = k \Delta X$$

it can be seen that the fractional change in flow for a given change in valve opening is constant. That is, an equal percentage change in flow occurs for a given percentage change in valve opening, irrespective of whereabouts in the range the change occurs. Hence the name.

Second, note that if $X = 0$ then $F = F_0$, which implies that there is a flow of F_0 when the valve is shut! There is a reason for this. Over most of the valve's range of opening, the inherent characteristic holds true: it is determined by the geometry of the plug relative to its seat. However, in the vicinity of the origin, as depicted by the inset to Figure 20.2, the characteristic is determined by the annulus between the plug's rim and the top edge of the seat. When the rim touches the seat shut-off is obtained.

Note that the slopes of the characteristics are proportional to F^{-1}, F^0 and F^{+1} respectively. This enables the valves' characteristics to be matched against the plant and/or pipework. The inherent characteristic chosen is that which gives rise to the most linear installed characteristic, as discussed below.

20.2 Installed Characteristic

In practice, the pressure drop across the valve varies. A typical scenario is as depicted in Figure 20.3.

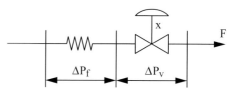

Fig. 20.3 Control valve in series with fixed resistance

Cooling water flows from a supply main, through some pipework and a control valve, and is discharged into a return main. If the valve is opened, the increase in flow will cause the pressure drop due to friction in the pipework to increase. Assuming that the overall pressure drop is approximately constant, if ΔP_F increases then ΔP_V must decrease. Thus the increase in flow will be less than that predicted by the inherent characteristic which assumes a constant ΔP_V.

The installed characteristic of a valve is the relationship between the flow and valve opening that would be obtained if the valve was calibrated empirically, and takes into account the variation of pressure drop.

$$F = f(X, \Delta P_V) \qquad (20.2)$$

Returning to Figure 20.3, the pipework may be treated as if it is a single *fixed* resistance to flow. Any pipe fittings, strainers, bends, contractions and expansions, orifices, open isolating valves, heat exchangers, jackets, *etc.* may be lumped in with the pipework. The control valve is treated as a single *variable* resistance to flow. The overall pressure drop may be considered to be approximately constant.

$$\Delta P_F + \Delta P_V = a \qquad (20.3)$$

The pressure drop across the fixed resistances due to frictional losses is of the form

$$F = b\sqrt{\Delta P_F} \qquad (20.4)$$

Assume, for example, that the valve has an equal percentage inherent characteristic:

$$F = F_0 e^{kX} \qquad (20.5)$$

For a given opening, the valve may be considered to behave like an orifice, in which case the flow through it depends on the pressure drop across the valve:

$$F_0 = c\sqrt{\Delta P_V} \qquad (20.6)$$

This set of four equations is the model of the system of Figure 20.3. These may be solved:

$$F = F_0 e^{kX} = c\sqrt{\Delta P_V} \, e^{kX}$$
$$= c\sqrt{a - \Delta P_F} \, e^{kX} = c\sqrt{a - \frac{F^2}{b^2}} \, e^{kX}$$

This gives an implicit relationship between F and X. The installed characteristic requires that it be explicit. Squaring both sides and rearranging gives

$$F = bc\sqrt{\frac{ae^{2kX}}{b^2 + c^2 e^{2kX}}} \qquad (20.7)$$

20.3 Comparison of Installed Characteristics

The characteristics of square root, linear and equal percentage valves are summarised in Table 20.2.

20.4 Worked Example No 1

A control valve and its pipework is as depicted in Figure 20.3. The valve's inherent characteristic is of the form

$$F = F_0 \, e^{4.0X}$$

What is the valve's installed characteristic if normal conditions are as follows?

$\overline{F} = 60.0 \, \text{kg/min} \quad \overline{X} = 0.5 \quad \overline{\Delta P_F} = \overline{\Delta P_V} = 1.0 \, \text{bar}.$

From Equation 20.3, if $\overline{\Delta P_F} = \overline{\Delta P_V} = 1.0$ bar then $a = 2.0$ bar.

From Equation 20.4, if $\overline{F} = 60$ kg/min and $\overline{\Delta P_F} = 1.0$ bar then $b = 60$ kg min^{-1} bar$^{-0.5}$

Comparison with Equation 20.5 reveals that $k = 4.0$.

Table 20.2 Installed characteristics *vs* trim types

	Square root	Linear	Equal percentage
Overall	$\Delta P_F + \Delta P_V = a$	$\Delta P_F + \Delta P_V = a$	$\Delta P_F + \Delta P_V = a$
Fixed resistance	$F = b\sqrt{\Delta P_F}$	$F = b\sqrt{\Delta P_F}$	$F = b\sqrt{\Delta P_F}$
Inherent characteristic	$F = k\sqrt{X}$	$F = kX$	$F = F_0 e^{kX}$
Variable resistance	$k = c\sqrt{\Delta P_V}$	$k = c\sqrt{\Delta P_V}$	$F_0 = c\sqrt{\Delta P_V}$
Installed characteristic	$F = bc\sqrt{\dfrac{aX}{b^2 + c^2 X}}$	$F = bc\sqrt{\dfrac{aX^2}{b^2 + c^2 X^2}}$	$F = bc\sqrt{\dfrac{ae^{2kX}}{b^2 + c^2 e^{2kX}}}$

Table 20.3 Installed characteristic for valve of Worked Example No 1

X	–	0	0.2	0.4	0.5	0.6	0.8	1.0
F	kg/min	11.4	24.6	47.4	60.0	70.8	81.0	84.0

Combining Equations 20.5 and 20.6 gives

$$F = c\sqrt{\Delta P_V}\, e^{kX}$$

If $\overline{F} = 60$ kg/min, $\overline{\Delta P_V} = 1.0$ bar and $\overline{X} = 0.5$ then $c = 60e^{-2}$.

Substituting for a, b, c and k into Equation 20.7 yields the installed characteristic:

$$F = 60\sqrt{\frac{2e^{8X-4}}{1 + e^{8X-4}}}$$

from which the points in Table 20.3 are obtained.

Hence the installed characteristic may be plotted. Inspection reveals a linear enough characteristic for most control purposes.

20.5 Trim Selection

For various reasons, as explained in Chapter 4, it is highly desirable for the output channel to have as linear an installed characteristic as possible. The shape of the installed characteristic is very sensitive to the relative values of ΔP_V and ΔP_F:

$\Delta P_V > \Delta P_F$. If most of the pressure drop is across the valve, then ΔP_V will not vary much and a valve with linear inherent characteristic should be used.

$\Delta P_V \approx \Delta P_F$. If the pressure drop across the valve and pipework are of the same order of magnitude, then an equal percentage valve should be used.

$\Delta P_V < \Delta P_F$. If the pressure drop across the valve is less than about 30% of the total available, effective control will not be possible. To achieve any change in flow the valve will have to be fully opened or closed, *i.e.* the valve functions in an on/off mode.

Note that the square root valve is used for special applications, such as when a square process characteristic is to be offset, or where a relatively "quick opening" is required.

20.6 Valve Sizing

Sizing is essentially a question of specifying the required characteristic and calculating the capacity required. The latter is articulated in terms of the hydraulic valve coefficient K_V, as defined by IEC 60534-1. The maximum flow through a control valve, *i.e.* when the valve is wide open and $X = 1$, is given by

$$Q_M = K_V\sqrt{\frac{\Delta P_{VM}}{G}} \quad (20.8)$$

where Q_M and K_V have units of m³/hr, ΔP_{VM} is in bar and G is the specific gravity of the liquid relative to water. Clearly, if $X = 1$, $\Delta P_{VM} = 1$ bar and $G = 1$, *i.e.* the liquid is water, then K_V is the volumetric flow rate in m³/hr.

Equation 20.8 appears to be dimensionally inconsistent due to a scaling factor of unity that is not dimensionless.

There is a widely used alternative basis of valve sizing, **not** used in this text, based upon the hydraulic valve coefficient C_V in which:

$$Q_M = C_V\sqrt{\frac{\Delta P_{VM}}{G}}$$

where Q_M and C_V have units of US gal/min, ΔP_{VM} is in psi and G is the specific gravity of the liquid relative to water. Clearly, if $X = 1$, $\Delta P_{VM} = 1$ psi and $G = 1$, *i.e.* the liquid is water, then C_V is the volumetric flow rate in US gal/min.

Note that 1 US gal = 3.785 litres and 1 psi = 0.0687 bar, whence $C_V = 1.16 K_V$.

The procedure for sizing a valve is depicted in Figure 20.4. It essentially consists of determining the

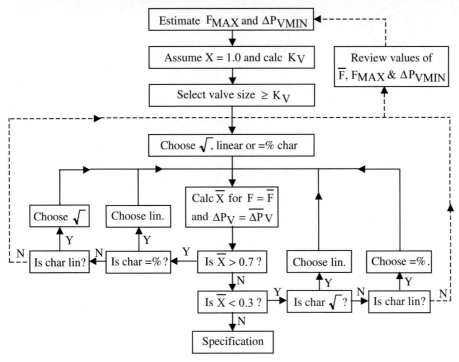

Fig. 20.4 Procedure for valve sizing

desired K_V value from a formula, based upon maximum flow and minimum pressure drop conditions. A valve is then selected from a manufacturer's catalogue. Invariably, none of the valves available will have the exact K_V required, so the next biggest is normally chosen. It is then checked to see that it is roughly half way open under normal conditions. The formulae to use are as follows:

For liquids

$$K_V = 0.06 F_M \sqrt{\frac{1}{\Delta P_{VM} G}} \qquad (20.9)$$

For gases and vapours

$$K_V = 2.44 F_M \sqrt{\frac{Z}{\Delta P_{VM}(P_{1M} + P_{0M})G}} \qquad (20.10)$$

Note that

- The specific gravity G is relative to water for liquids and to air for gases and vapours.
- For most gases and vapours, at upstream pressures less than approximately 6 bar, a compressibility factor of Z = 1 can be assumed.
- These formulae are based on the mass flow rate of fluid through the valve, for volumetric flow rates there are alternative formulae available.

These two formulae are sufficient for something like 80% of all applications. However, if critical flow exists then different formulae are necessary.

20.7 Worked Example No 2

The temperature of the contents of a reactor is to be controlled by regulating the flow of cooling water through an internal coil. The water is supplied from a supply main at approximately 4.0 bar g and flows through to a return main at approximately 0.4 bar g. Under design conditions the water flow rate required is 90 kg/min. It is estimated that, at

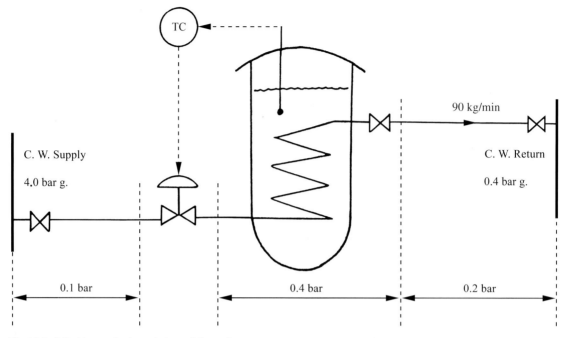

Fig. 20.5 Coil with control valve and pipework for cooling water

this flow rate, the pressure drop across the inlet and outlet pipework, isolating valves and coil will be as shown in Figure 20.5. It is anticipated that the temperature controller will need to vary the flow rate by as much as ±100%. Specify a suitable valve.

This arrangement reduces to that depicted in Figure 20.3 in which

$$\overline{\Delta P_F} = 0.7 \text{ bar} \quad \overline{\Delta P_V} = 2.9 \text{ bar} \quad \overline{F} = 90 \text{ kg/min}$$

Equation 20.3 is thus

$$\Delta P_F + \Delta P_V = a$$

where a = 3.6 bar.

Equation 20.4 applies to the pipework:

$$F = b\sqrt{\Delta P_F}$$

Substitute for normal conditions gives

$$b = 107.6 \text{ kg min}^{-1} \text{ bar}^{-1/2}.$$

Establish maximum flow conditions:

If $F_M = 180$ kg/min then $\Delta P_{FM} \approx 2.8$ bar hence $\Delta P_{VM} \approx 0.8$ bar.

Calculate theoretical K_V from Equation 20.9, assuming G = 1.0:

$$K_V = 0.06 F_M \sqrt{\frac{1}{\Delta P_{VM} G}} \approx 12.1$$

Suppose that inspection of a manufacturers catalogue reveals that a valve whose $K_V = 12.9$ is the next biggest available.

Choose a linear characteristic:

$$F = kX$$

where $k = c\sqrt{\Delta P_V}$.

From the definition of K_V:

If $\Delta P_V = 1$ bar and if $X = 1.0$ then
$F = 12.9$ m^3/hr $\equiv 215$ kg/min.
Hence $c = 215$ kg min^{-1} bar$^{-1/2}$.

Table 20.4 Inherent characteristic for valve of Worked Example No 2

X	%	0	10	20	30	40	50	60	70	80	90	100
F	US gpm	0	0.34	0.50	0.74	1.12	1.72	2.59	3.88	5.95	9.05	12.9

Check that the valve is half open under normal conditions.

If $\overline{\Delta P_V} = 2.9$ bar
then $k = 215\sqrt{2.9} = 366.1$ kg/min.

So if $\overline{F} = 90$ kg/min then $90 = 366.1\overline{X}$
Hence $\overline{X} = 0.246$.

The valve is only 25% open under normal conditions so try an equal percentage characteristic:

$$F = F_0 e^{kX}$$

where $F_0 = c\sqrt{\Delta P_V}$.

The inherent characteristic for a valve whose $K_V = 12.9$, as tabulated in the manufacturer's literature, is typically as shown in Table 20.4.

Thus, from the inherent characteristic, assuming a constant $\Delta P_V = 1$ bar:

If $X = 1$ then $F = 12.9$ m³/hr $\equiv 215$ kg/min
whence $215 = F_0.e^k$

If $X = 0.5$, say, then
$F = 1.72$ m³/hr $\equiv 28.67$ kg/min
whence $28.67 = F_0.e^{0.5k}$

Eliminating F_0 gives $k = 4.03$.

From the definition of K_V:

If $\Delta P_V = 1$ bar and if $X = 1.0$ then
$F = 12.9$ m³/hr $\equiv 215$ kg/min.

Substituting gives: $215 = c\sqrt{1.0}.e^{4.03}$
Hence $c = 3.82$ kg min⁻¹ bar$^{-1/2}$.

Check that the valve is half open under normal conditions.

If $\overline{\Delta P_V} = 2.9$ bar then
$\overline{F_0} = 3.82\sqrt{2.9} = 6.505$ kg/min.

So if $\overline{F} = 90$ kg/min then $90 = 6.505.e^{4.03\overline{X}}$
Hence $\overline{X} = 0.652$

The valve is 65% open under normal conditions. This is within the 30–70% range so is good enough. Therefore the valve required has a $K_V = 12.9$ and an equal percentage trim.

20.8 Critical Flow

In the context of valve sizing, critical flow is a term which relates to the onset of choked flow, cavitation or flashing. It is also, ambiguously, used in relation to sonic flow and is often mistakenly linked to critical pressure.

Choked flow describes the situation when, with the upstream conditions remaining constant, the flow rate through a valve cannot be further increased by reducing the down stream pressure. It is associated, in particular, with flashing. The onset of flashing is often described as critical flow. Thus sub-critical flow means that flashing is not occurring.

Critical flow is also the term used to describe the situation, with gas flow, when the velocity in the *vena contracta* approaches the speed of sound. This occurs when the upstream pressure is twice the downstream pressure, *i.e.* if $P_1 \geq 2P_0$. Note that when sonic flow occurs, the flow rate is a function of the gas density. Thus, although the velocity has reached a maximum, the mass flow rate increases with increasing upstream pressure. This is *not* related to choked flow, flashing or cavitation.

Note that the critical pressure P_C is that pressure above which a gas cannot be condensed by cooling alone. In the case of the water/steam system, for example, $P_C = 221$ bar. Critical pressure is a thermodynamic property of a substance and is *not* related to choked flow, flashing or cavitation.

The effects of cavitation and flashing are best explained by consideration of the saturated vapour pressure P_{SV} in relation to the pressure profile of

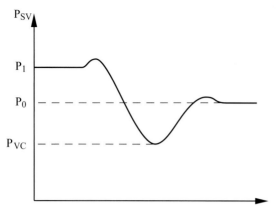

Fig. 20.6 Typical pressure profile for flow through a control valve

the fluid as it flows through the valve. The saturated vapour pressure of a pure liquid at a given temperature is the pressure at which the liquid boils at that temperature. For example, for water at 100°C its saturated vapour pressure is 1 atmosphere. A typical pressure profile is as depicted in Figure 20.6 which is essentially the same as Figure 12.1 for flow through an orifice. P_1 and P_0 are the upstream and downstream pressures respectively, and P_{VC} is the pressure at the vena contracta.

Four scenarios are considered:

1. $P_{SV} < P_{VC}$. This is the normal case and neither cavitation nor flashing occurs.
2. $P_{SV} = P_{VC}$. This case corresponds to the onset of cavitation, often referred to as incipient cavitation. As the liquid flows through the valve its pressure just drops to, but not below, the pressure P_{SV}. At this pressure the liquid is at its boiling point but vaporisation does not occur.
3. $P_{VC} < P_{SV} < P_0$. Cavitation occurs. As the liquid flows through the plug its pressure drops below P_{SV}. Some of the liquid vaporises and bubbles are formed, the amount of vaporisation depending upon the difference between P_{SV} and P_{VC}. As the two-phase mixture approaches the exit port its pressure rises above P_{SV} and the bubbles collapse. The mechanical shock causes permanent damage to the valve body and trim. In extreme cases extensive damage can be done to downstream pipework.

The onset of cavitation is predicted by a valve's critical flow factor which is defined to be

$$C_F = \sqrt{\frac{P_1 - P_0}{P_1 - P_{VC}}} \quad (20.11)$$

Values of C_F are tabulated against type and size of a valve by its manufacturer. A value of $C_F = 0.9$ is typical. Cavitation is avoided by preventing P_{VC} from dropping down to P_{SV}. This is normally achieved by specifying a valve whose C_F factor is such that $P_{SV} < P_{VC}$. If this is not possible, then it is best to change the process, *e.g.* by reducing the temperature of the process stream.

4. $P_0 < P_{SV} < P_1$. This case corresponds to flashing. A proportion of the liquid vaporises as it flows through the valve and, because $P_{SV} > P_0$, exits the valve as vapour. This results in increased velocities within the valve and leads to potential noise problems. These can be addressed by the use of diffusers and baffles as discussed in Chapter 19. Also, the increased volumetric flowrate may require a larger pipe diameter downstream of the valve.

For liquids, the criterion for the onset of choked flow, cavitation or flashing is that $\Delta P_V \geq C_F^2.(P_1 - P_{VP})$. When this occurs, provided that $P_{VP} < 0.5P_1$, where P_{VP} is the vapour pressure at the temperature of the flowing fluid, the basis for valve sizing is:

$$K_V = \frac{0.06F_M}{C_F}\sqrt{\frac{1}{(P_1 - P_{VP})G}} \quad (20.12)$$

For gases and vapours the criterion for the onset of choked flow is $\Delta P_V \geq 0.5C_F^2 P_1$. When this occurs, the basis for valve sizing is that:

$$K_V = \frac{2.82F_M}{C_F P_1}\sqrt{\frac{Z}{G}} \quad (20.13)$$

Note that C_F occurs in the denominator of both Equations 20.12 and 20.13. Given that $C_F < 1.0$, it follows that the effect of C_F is to increase the K_V value: a larger valve is specified to accommodate the increase in volume due to choked flow, cavitation or flashing.

When a valve is installed between pipe reducers, because the pipeline diameter is greater than that of the valve's body, there is a decrease in valve capacity. This is because the reducers create an additional pressure drop in the system by acting as a contraction and an enlargement in series with the valve. There is a correction factor that can be applied to take this into account.

Most valve manufacturers have programs for valve sizing and specification. The sizing procedure is essentially as depicted in Figure 20.4, the various sizing formulae being invoked according to the conditions. The programs normally provide for specification of all the mechanical details and materials of construction.

20.9 Nomenclature

C_V	hydraulic valve coefficient	gal(US) min^{-1}
C_F	critical flow factor	
F	mass flow rate through the valve	kg min^{-1}
G	specific gravity	–
K_V	hydraulic valve coefficient	m^3 hr^{-1}
P	pressure	bar (abs)
ΔP	pressure drop	bar
Q	volumetric flow rate	m^3 hr^{-1}
X	fractional opening of the valve	–
Z	compressibility factor	–

Subscripts

C	critical
F	fixed resistance
M	maximum flow and minimum pressure drop
SV	saturated vapour
V	variable resistance
VC	vena contracta
VP	vapour at temperature of flowing liquid
0	downstream
1	upstream

Valve Positioners

21.1 Operation of Positioner
21.2 Uses of Positioner
21.3 Bias Example
21.4 Split Range Example

A valve positioner is a device used for manipulating the opening of a control valve more accurately than it would be otherwise. For P&I diagram purposes it is represented symbolically as shown in Figure 21.1.

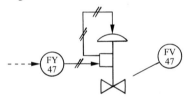

Fig. 21.1 P&I diagram symbol for valve with positioner

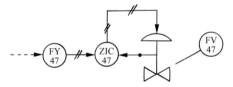

Fig. 21.2 Functionality of valve with positioner

21.1 Operation of Positioner

Functionally, a positioner is a high gain feedback controller dedicated to controlling the position of a valve stem, as depicted in Figure 21.2. In effect the 0.2–1.0 bar output of the I/P converter FY47 is the set point of the positioner ZIC47. This is compared with the measured value of the stem position to produce an error signal. The positioner output pressure, which varies in direct proportion to the error, is applied to the actuator. Thus the change in force produced by the actuator is related to the stem error.

Physically, a positioner is bolted onto the yoke of the valve. The feedback signal is established by means of a mechanical linkage between the stem and positioner. Often, the positioner itself is a pneumatic force balance type of device, as discussed in Chapter 5, and has settings for adjusting its range, zero and direction. There are normally small pressure gauges fitted to the positioner to indicate the values of its input and output signals and of the supply pressure. Alternatively, the I/P converter is incorporated in the positioner and the mechanism is electronic with a pneumatic output to the actuator.

An important point to appreciate is that a positioner usually has its own independent air supply. This local air supply is often regulated at a pressure significantly higher than the normal 1.4 bar supply for pneumatic instruments. This enables a wide range of pressures to be applied to the actuator diaphragm and, hence, a wide range of forces applied to the valve stem.

21.2 Uses of Positioner

A positioner should normally only be fitted to a valve to achieve one or more of the following ends. Otherwise, provided the valve and its actuator are properly sized, installed and maintained, they should function quite effectively without a positioner.

1. Accuracy. A control valve can only be positioned to an accuracy of some 1–2%. Given the feedback nature of a control loop, and its inherent ability to correct for errors, this is normally quite sufficient. Use of a positioner can increase the accuracy to about ±0.5%.
2. Power amplification. To prevent leakage from a valve it is necessary to tighten its packing. Despite the use of PTFE as a packing material, and lubrication, there can be much stiction and friction around the stem. This makes it difficult to position, leading to a lumpy motion and loss of resolution. Use of a positioner enables force to be applied by the actuator in a controlled way. Thus initially, following a change in I/P converter output, when the error is large and stiction is dominant, a large change of force is applied. However, once the stem is moving, less force is required to overcome the residual friction. So, as the stem approaches its correct position and the error reduces to zero, the change in positioner output reduces and a new balance of forces is established. This results in a smoother motion and much increased accuracy.
3. Shut-off. If the process stream through a valve is a slurry, or contains suspended solids, it is possible for deposits to form within the body of the valve. This may lead to the plug not seating properly resulting in leakage through the valve. Use of a positioner can overcome poor shut-off by virtue of the power amplification which forces the plug against its seat.
4. Speed of response. A control valve may be situated a long way from the I/P converter. This makes its response slow since all the air to operate the actuator has to flow through the connecting pneumatic tubing. Use of a positioner can increase the speed of response quite dramatically. This is because the 0.2–1.0 bar signal only has to pressurise a small bellows inside the positioner, the local air supply providing the power to actuate the valve.
5. Characterisation. Sometimes the action of a valve needs to be modified, perhaps because the valve being used was originally specified for a different purpose. For example, a bias may be applied to restrict the opening of the valve, *i.e.* to prevent it from fully opening and/or closing. Alternatively, for non-critical applications, the direction of action, *i.e.* air to open/close, may be reversed. Use of a positioner enables this by calibration as appropriate. This is illustrated by the two following examples.

21.3 Bias Example

Consider the arrangement depicted in Figure 21.2. Suppose that a bias must be applied to the control signal such that valve FV47 has a minimum opening of 10%. Assume standard signal ranges and that the valve is of the air-to-open type. The calibration of the positioner must be as indicated in Table 21.1.

Table 21.1 Calibration of positioner with bias on output

Element	Output values					Units
FY47	0.2	0.4	0.6	0.8	1.0	bar
ZIC47	0.28	0.46	0.64	0.82	1.0	bar
FV47	10	32.5	55	77.5	100	%

21.4 Split Range Example

An important use of positioners is in split range action. Sometimes referred to as duplex action, this enables two valves to be operated off one control signal. A typical scenario is of a temperature control system in which the flows of cooling water and steam are both manipulated, as depicted in Figure 21.3. The output of controller TC47 and I/P converter TY47 is simultaneously applied to both positioners ZC47A & ZC47B which open either TV47A or TV47B as appropriate.

Suppose that, from fail-safe considerations, the water valve must be air-to-close and the steam valve air-to-open. This dictates that the water valve

21.4 Split Range Example

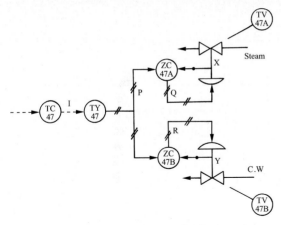

Fig. 21.3 Temperature control system with split range valves

used. The calibration of the two positioners must be as indicated in Table 21.2.

Table 21.2 Calibration of positioners for split range action

Signal	Values					Units
I	4	8	12	16	20	mA
P	0.2	0.4	0.6	0.8	1.0	bar
Q	0.2	0.2	0.2	0.6	1.0	bar
X	0	0	0	50	100	%
R	0.2	0.6	1.0	1.0	1.0	bar
Y	100	50	0	0	0	%

must work off the bottom end of the range of the I/P converter output signal and the steam valve off the top end. Assume that both valves are shut at mid range, and that standard signal ranges are

It should be appreciated that for computer controlled plant, for the cost of an extra I/P converter and an additional analogue output channel, it is normal to realise the duplex arrangement by means of software. This provides more flexibility and is less likely to produce a dead zone at mid-range.

Section 4

Conventional Control Strategies

Feedback Control

22.1 Feedback Principles
22.2 Deviation Variables
22.3 Steady State Analysis
22.4 Worked Example
22.5 Nomenclature

Feedback control was introduced in Chapter 2 in which the representation of a feedback loop for flow control was discussed. The functionality of a feedback loop for level control was considered in Chapter 3 leading to the generalised block diagram of Figure 3.3. This chapter develops the concept of proportional control from a steady state point of view. Chapter 23 deals with the 3-term PID controller and introduces aspects of control loop dynamics. Chapter 24 addresses the issue of tuning a PID controller for optimum performance.

22.1 Feedback Principles

Consider the level control system depicted in Figure 22.1 for which the corresponding block diagram is as shown in Figure 22.2.

The comparator generates the error signal by subtraction of the measured value from the set point. It is this subtraction that gives rise to the term negative feedback:

$$e = h_R - h_M \qquad (22.1)$$

If the measured value is a 4–20-mA signal, then both the set point and the error may be considered to have units of mA too. Assuming that the controller has proportional action only, its behaviour is as described by the following equation:

$$u = u_B \pm K_C \cdot e \qquad (22.2)$$

The controller output u is directly proportional to the error. The bigger the error the greater the change in output signal. The controller gain K_C is user definable and enables the sensitivity of the controller to be adjusted. Increasing K_C makes the controller more sensitive to changes in error. Note that if the measured value and the controller output have the same units, say they are both 4–20-mA signals, then K_C is dimensionless.

The proportional action $K_C e$ is superimposed on a constant term u_B, known as the bias. This is the value of the controller output when there is zero error. If the error is zero the level must be at its set point. If the level is at its set point, and steady, the flowrates f_1 of the inlet stream and f_0 of the outlet stream must be equal. If the outlet flowrate is at its normal value, the valve opening x must be at

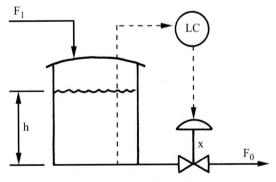

Fig. 22.1 Schematic of level control system

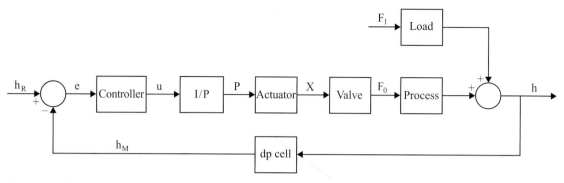

Fig. 22.2 Block diagram of level control system

its normal value too. The bias is the corresponding value of the controller output necessary to establish this equilibrium.

The sign of K_C depends on the direction of action. With modern controllers, if the sign is positive the controller is said to be forward acting, and if it is negative the controller is reverse acting. Suppose that the set point is constant but there is a sustained decrease in inlet flow. This will cause the level h to fall resulting in a decrease in measured value and an increase in error. To stop the level falling the controller must reduce the opening of the valve. Assuming that the valve is of the air-to-open type the output P of the I/P converter must decrease. This requires that the controller output decreases. A decrease in controller output, following an increase in error, can only be achieved if K_C has a negative sign, *i.e.* the controller is reverse acting.

Whether a controller needs to be forward or reverse acting depends on the context and can only be determined by analysis as above. If, in the example of Figure 22.1, the control valve manipulated the inlet stream *or* if the valve had been of the air-to-close type, then forward action would be required. However, if the valve was in the inlet stream *and* of the air-to-close type then reverse action would be required. Another term in common usage is direct action. This is associated with older controllers. If a controller is direct acting its output signal moves in the same direction as the measured value. Thus direct action is synonymous with reverse action as described above.

22.2 Deviation Variables

Any signal may be considered to be made up of its normal value plus some deviation about that normal value, usually denoted by a bar and a Δ symbol respectively. For example:

$$h = \bar{h} + \Delta h, \quad F_1 = \bar{F_1} + \Delta F_1, \quad X = \bar{X} + \Delta X, \ etc.$$

For design purposes, the normal value of a variable is its specified steady state value. In operation it is the average value of that variable. It is common practice in developing models of control systems to consider the deviations of the variables only. Thus, for example:

$$\Delta h_M = K_M \Delta h, \quad \Delta P = K_I \Delta u,$$
$$\Delta X = K_A \Delta P, \ etc. \quad (22.3)$$

where K_M, K_{IP} and K_A are the steady state gains of the dp cell, I/P converter and actuator respectively.

The use of deviation variables is a source of much confusion because, in practice, with the exception of the error, all the signals have non-zero normal values. The use of deviation variables for modelling purposes is discussed in more detail in Chapter 84.

For the comparator, Equation 22.1 may be put into deviation form:

$$e = h_R + \Delta h_R - (h_M + \Delta h_M)$$

Since, by definition, the normal values of h_R and h_M must be equal, this reduces to:

$$e = \Delta h_R - \Delta h_M \quad (22.4)$$

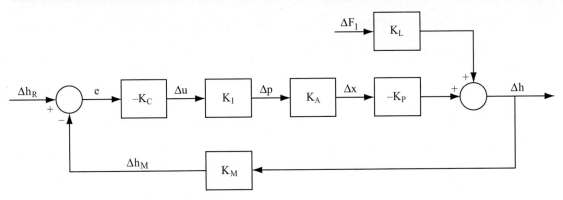

Fig. 22.3 Block diagram showing steady state gain values

As discussed, the controller bias u_B corresponds to zero error conditions and is the normal output. It follows that the deviation model of a reverse acting controller is simply

$$\Delta u = u - u_B = -K_C e$$

The head h of liquid in the tank clearly depends on the inlet and outlet flowrates. However, the outlet flow depends, not only on the valve opening, but also on the head of liquid in the tank. To establish the steady state gains for the process and load it is necessary to develop a model of the process taking these interactions into account. This is done in Chapter 84 in which the following relationship is established:

$$\Delta h = K_L \Delta F_1 - K_P \Delta X \quad (22.5)$$

A consequence of taking the valve characteristic into account is that the process gain relates the level to the valve opening rather than to the outlet flow. Also, note the negative sign of the process gain. This is negative because an increase in valve opening causes a decrease in level, and *vice versa*.

These various gains are incorporated in the block diagram shown in Figure 22.3. Note that the signals are all in deviation form. This implies that if there are no deviations then the signals are all zero.

22.3 Steady State Analysis

In the following analysis it is assumed that the system is initially at equilibrium with each signal at its normal value. It is also assumed that the system is inherently stable such that, following a disturbance, the transients decay away. Suppose that some sustained change in either F_1 and/or h_R occurs, that the control system responds to the changes and that some new steady state is reached. The net change in head is given by Equation 22.5:

$$\Delta h = K_L \Delta F_1 - K_P \Delta X$$

However, the change in valve opening is given by:

$$\Delta X = K_A \Delta P$$

Substituting for ΔX gives:

$$\Delta h = K_L \Delta F_1 - K_P K_A \Delta P$$

Similarly, ΔP can be substituted in terms of Δu, and Δu in terms of e according to:

$$\Delta P = K_I \Delta u \quad \text{and} \quad \Delta u = -K_C e$$

Whence: $\Delta h = K_L \Delta F_1 + K_P K_A K_I K_C e$

However, from Equation 22.4:

$$e = \Delta h_R - \Delta h_M$$

So $\quad \Delta h = K_L \Delta F_1$
$\qquad + K_P K_A K_I K_C (\Delta h_R - \Delta h_M)$

But $\quad \Delta h_M = K_M \Delta h$

Thus $\quad \Delta h = K_L \Delta F_1$
$\qquad + K_P K_A K_I K_C (\Delta h_R - K_M \Delta h)$

This is an implicit relationship between the change in output h of the control loop and its two inputs F_1 and h_R. The relationship can be made explicit by rearrangement to yield:

$$\Delta h = \frac{K_L}{1 + K_C K_I K_A K_P K_M} \cdot \Delta F_1$$
$$+ \frac{K_C K_I K_A K_P}{1 + K_C K_I K_A K_P K_M} \cdot \Delta h_R \quad (22.6)$$

The product in the denominator,

$$K = K_C K_I K_A K_P K_M,$$

known as the open loop gain, is of particular significance. It is the product of the steady state gains of all the elements in the loop. The open loop gain K should be dimensionless and, for negative feedback systems, should be a positive quantity. The numerator product $K_C K_I K_A K_P$ is known as the forward path gain.

Equation 22.6 is of the form

$$\Delta h = K_1 \Delta F_1 + K_2 \Delta h_R \quad (22.7)$$

where K_1 and K_2 are known as the closed loop gains for changes in F_1 and h_R respectively. This may be represented in block diagram form as depicted in Figure 22.4.

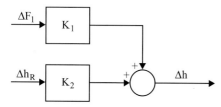

Fig. 22.4 Block diagram showing steady state closed loop gains

The significance of Equation 22.7 is that it reveals that there is an offset in level, *i.e.* a residual change in h. The controller has compensated for the changes in F_1 and/or h_R but has been unable to restore the level to its original value. Offset is explained by consideration of Equation 22.2. There needs to be a steady state error for the controller to produce a change in output to compensate for the changes in input. Offset is an inherent feature of proportional control. It can be reduced by increasing the controller gain. Inspection of Equation 22.6 shows that as K_C is increased K_1 tends towards zero and K_2 tends towards $1/K_M$. Thus, for disturbances in F_1 the change in level becomes small, and for changes in h_R the level tracks the set point closely. Unfortunately, it is not possible to eliminate offset by simply turning up the controller gain to some maximum value because there is normally a constraint on K_C determined by stability criteria.

22.4 Worked Example

The various concepts introduced are perhaps best illustrated by means of a numerical example.

Assume that the tank is 2 m deep and that the dp cell, I/P converter and actuator have linear calibrations as summarised in Table 22.1.

Suppose that under normal conditions the value of the inlet flow is 0.3 m³/min, that the tank is half full, that the valve is half open and that there is no offset. On this basis the normal value for each of the signals is as indicated in Table 22.2.

From the model developed in Chapter 84 the following values for the process and load gains may be assumed:

$K_P = 6.443$ m $\quad K_L = 6.667 \times 10^{-3}$ m min kg^{-1}

Table 22.1 Calibration of instrumentation for Worked Example

	Input	Output	Gain
Dp cell	0–2 m	4–20 mA	$K_M = 8$ mA m^{-1}
I/P converter	4–20 mA	0.2–1.0 bar	$K_I = 0.05$ bar mA^{-1}
Actuator	0.2–1.0 bar	0–1.0	$K_A = 1.25$ bar^{-1}

Table 22.2 Normal and steady state values of signals

Signal	Normal	Deviation	Absolute	Units
F_0, F_1	300.0	+60.0	360.0	kg min^{-1}
h	1.0	+0.05374	1.054	m
h_M	12.0	0.4299	12.43	mA
h_R	12.0	0	12.0	'mA'
e	0	−0.4299	−0.4299	'mA'
u	12.0	+0.8598	12.86	mA
P	0.6	0.04299	0.6430	bar
X	0.5	0.05374	0.5537	–

Substitution of the values for the various gains of the level control system into Equation 22.6 gives

$$\Delta h = \frac{6.667 \times 10^{-3}}{1 + 3.222 K_C} \Delta F_1 + \frac{0.4027 K_C}{1 + 3.222 K_C} \Delta h_R$$

Assume that the controller has a gain $K_C = 2.0$. This gives

$$\Delta h = 0.8956 \times 10^{-3} \Delta F_1 + 0.1082 \Delta h_R \quad (22.8)$$

Suppose, for example, the inlet flow increases by 20% from its normal value of 300 kg/min but the set point is constant. Thus the changes are $\Delta F_1 = 60$ kg/min and $\Delta h_R = 0$ m. Substituting gives $\Delta h = 0.05374$ m, *i.e.* an offset of some 5.4 cm.

Knowing the value of Δh, the steady state changes in all the other variables can be calculated using Equations 22.2–22.4. These are listed as deviations in Table 22.2. Addition of the normal values and the deviations gives the absolute values.

As a final check, the deviations so determined may be substituted into Equation 22.5:

$$\begin{aligned} \Delta h &= K_L \Delta F_1 - K_P \Delta X \\ &= 6.667 \times 10^{-3} \times 60.0 - 6.443 \times 0.05374 \\ &= 0.05377 \text{ m} \end{aligned}$$

which agrees well enough with the offset predicted by Equation 22.8.

22.5 Nomenclature

e	error signal	"mA"
h	level of liquid	m
F	mass flow rate	kg min^{-1}
K	gain	–
P	pressure	bar
u	controller output	mA
X	fractional opening	–

Subscripts

B	bias
C	controller
M	measured value
R	reference (set point)
0	outlet stream
1	inlet stream

PID Control

Chapter 23

23.1 Single Loop Controller
23.2 Proportional Action
23.3 Integral Action
23.4 Derivative Action
23.5 Bumpless Transfer
23.6 Derivative Feedback
23.7 Integral Windup
23.8 Worked Example
23.9 Other Analogue Forms of PID Controller
23.10 Discretised Form of PID
23.11 Incremental Form of PID
23.12 Mnemonics and Nomenclature
23.13 Summary of Control Actions

Proportional, integral and derivative (PID) control is often referred to as 3-term control. P action was introduced in Chapter 22. It provides the basis for PID control, any I and/or D action is always superimposed on the P action. This chapter is concerned with the functionality of PID control and its open and closed loop behaviour. An equation for PID control is first developed in analogue form, as used in pneumatic and electronic controllers. This is then translated into a discrete form for implementation as an algorithm in a digital controller.

The principles discussed in this chapter and the equations developed are essentially the same whether the PID controller is a dedicated single loop controller, a function provided within some other control loop element such as a dp cell, or is a configurable function within a distributed control system capable of supporting multiple loops simultaneously.

23.1 Single Loop Controller

The 3-term controller, often referred to as a single loop controller (SLC), is a standard sized unit for panel mounting alongside recorders and other control room displays and switches. Multiple SLCs are typically rack mounted. It is a single unit which contains both the comparator and the controller proper. Thus it has two input signals, the measured value and the set point, and one output signal. It would also have its own power and/or air supply.

The facia of a typical SLC is depicted in Figure 23.1. Often referred to as a faceplate, it has a scale for reading values of the measured value and set point. The range of the scale normally corresponds to the calibration of the measuring element, or else is 0–100% by default. With pneumatic controllers the signals would be indicated by pointers. With electronic controllers liquid crystal displays (LCD) or light emitting diode (LED) bar displays are used whereas with digital controllers a VDU representation of the faceplate is common practice. Faceplates also show relevant alarm limits. A

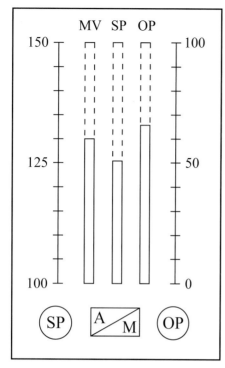

Fig. 23.1 Facia of a typical single loop controller

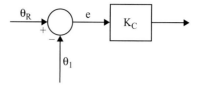

Fig. 23.2 Symbols for signals of PID controller

23.2 Proportional Action

With reference to Figure 23.2 the equations of the comparator and of a proportional controller are:

$$e = \theta_R - \theta_1 \qquad (23.1)$$

$$\theta_0 = \theta_B \pm K_C.e \qquad (23.2)$$

The gain K_C is the sensitivity of the controller, *i.e.* the change in output per unit change in error. Gain is traditionally expressed in terms of bandwidth, often denoted as %bw, although there is no obvious advantage from doing so. Provided the various signals all have the same range and units, gain and bandwidth are simply related as follows:

$$K_C = \frac{100}{\%bw} \qquad (23.3)$$

Thus a high bandwidth corresponds to a low gain and *vice versa*. However, if the units are mixed, as is often the case with digital controllers, it is necessary to take the ranges of the signals into account. Strictly speaking, bandwidth is defined to be the range of the error signal, expressed as a percentage of the range of the measured value signal, that causes the output signal to vary over its full range.

The issue is best illustrated by means of a numerical example. Consider Figure 23.3 in which θ_1 and θ_0 have been scaled by software into engineering units of 100–150 °C and 0–100% respectively. Suppose that the controller is forward acting, $K_C = 5$, θ_R is 125 °C and θ_B is 50%. What this means in practice is that signals within the range $115 < \theta_1 < 135$ °C, *i.e.* an error of ±10 °C, will cause the output signal to vary from 0–100%. According to the definition of bandwidth, $bw = 20/50 \times 100 = 40\%$, which is different from the value that would have been obtained by substituting $K_C = 5$ into

---

dial and/or dedicated pushbuttons enables the set point to be changed locally, *i.e.* by hand. Otherwise the set point is changed on a remote basis. There is often a separate scale to indicate the value of the output signal, range 0–100%, with provision for varying the output by hand when in MAN mode. A switch enables the controller to be switched between AUTO and MAN modes.

All of the above functions are operational and intended for use by plant personnel. Other functions of the controller are technical and are normally inaccessible from the faceplate. Thus, for example, the PID settings and the forward/reverse switch are typically adjusted either by dials internal to the controller or by restricted access pushbuttons. Despite its name, a modern digital SLC will provide two or more 3-term controllers, handle several discrete I/O signals, and support extensive continuous control and logic functions. Access to this functionality is restricted to engineering personnel, typically by means of a serial link.

Equation 23.3. This is clearly a source of confusion: it is best to avoid using the term bandwidth and to consistently work with gain.

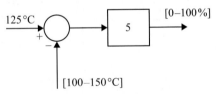

Fig. 23.3 Temperature controller for explaining bandwidth

Note that outside the range of $115 < \theta_1 < 135°C$ the controller output is constrained by the limits of the output signal range. Thus θ_0 would be either 100% for all values of $\theta_1 < 115°C$ or 0% for $\theta_1 > 135°C$. When a signal is so constrained it is said to be saturated.

In Chapter 22 the steady state response of a closed loop system was considered. In particular, the way in which increasing K_C reduces offset was analysed. Also of importance is an understanding of how varying K_C affects the dynamic response. It is convenient to consider this in relation to the same level control system of Figure 22.1. The closed loop response to step increases in set point h_R and inlet flowrate F_1 are as shown in Figures 23.4 and 23.5, respectively.

Two traces are shown on Figure 23.4. The exponential one corresponds to a low gain and depicts a so-called over-damped response. The change in set point causes a step increase in error and, because the controller is reverse acting, produces a sudden closing of the valve. Thereafter the valve slowly opens as the error reduces and the level gently rises towards the new set point with a steady state offset. Increasing K_C produces a faster asymptotic approach towards a smaller offset.

However, there is some critical value beyond which increasing K_C causes the response to become oscillatory, as depicted in the second trace. In effect, the controller is so sensitive that it has over-compensated for the increase in set point by closing the valve too much. This causes the level to rise quickly and overshoot the set point. As the level crosses the set point the error becomes negative so the controller increases its output. The resultant valve opening is more than that required to compensate for the overshoot so the level falls below the set point. And so on. Provided the value of K_C is not too high these oscillations decay away fairly quickly and the under-damped response settles out with a reduced steady state offset. Increasing K_C further causes the response to become even more oscillatory and, eventually, causes the system to become unstable.

A similar sort of analysis can be applied to Figure 23.5. The principal effects of P action are summarised in Table 23.1 on page 163.

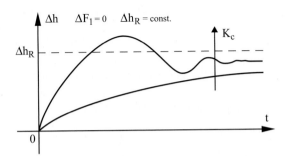

Fig. 23.4 Closed loop response to step change in set point

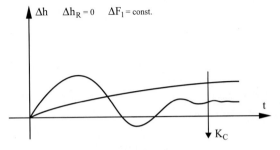

Fig. 23.5 Closed loop response to step change in inlet flow

23.3 Integral Action

The purpose of I action is to eliminate offset. This is realised by the addition of an integral term to Equation 23.2 as follows:

$$\theta_0 = \theta_B \pm K_C \left(e + \frac{1}{T_R} \int_0^t e \, dt \right) \quad (23.4)$$

T_R is known as the reset time and characterises the I action. Adjusting T_R varies the amount of I action. Note that, because T_R is in the denominator, to increase the effect of the I action T_R has to be reduced and *vice versa*. The I action can be turned off by setting T_R to a very large value. T_R has the dimensions of time: for process control purposes it is normal for T_R to have units of minutes.

The open loop response of a PI controller to a step change in error is depicted in Figure 23.6. Assume that the controller is forward acting, is in its AUTO mode and has a 4–20 mA output range. Suppose that the error is zero until some point in time, t = 0, when a step change in error of magnitude e′ occurs.

Substituting e = e′ into Equation 23.4 and integrating gives:

$$\theta_0 = \theta_B + K_C e' + \frac{K_C}{T_R} e' t$$

The response shows an initial step change in output of magnitude $K_C e'$ due to the P action. This is followed by a ramp of slope $K_C/T_R \cdot e'$ which is due to the I action. When t = T_R then:

$$\theta_0 = \theta_B + 2 K_C e'$$

This enables the definition of reset time. T_R is the time taken, in response to a step change in error, for the I action to produce the same change in output as the P action. For this reason integral action is often articulated in terms of repeats per minute:

$$\text{Repeats/min} = \frac{1}{T_R}$$

Note that the output eventually ramps up to its maximum value, 20 mA, and becomes saturated.

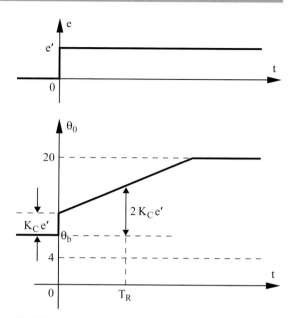

Fig. 23.6 Open loop response of PI controller to a step change in error

The closed loop response of a PI controller with appropriate settings to a step change in set point is depicted in Figure 23.7.

Again it is convenient to consider the level control system of Figure 30.1. Initially the P action dominates and the response is much as described for the under-damped case of Figure 23.4. However as the oscillations decay away, leaving an offset, the I action becomes dominant. Whilst an error persists the integral of the error increases. Thus the controller output slowly closes the valve and nudges the level towards its setpoint. As the error reduces, the contribution of the P action to the controller output decreases. Eventually, when there is zero error and the offset has been eliminated, the controller output consists of the bias term and the I action only. Note that although the error becomes zero the integral of the error is non-zero:

$$\theta_0 = \theta_B - \frac{K_C}{T_R} \int_0^t e \, dt$$

Thus the P action is short term in effect whereas the I action is long term. It can also be seen from Figure 23.7 that the response is more oscillatory

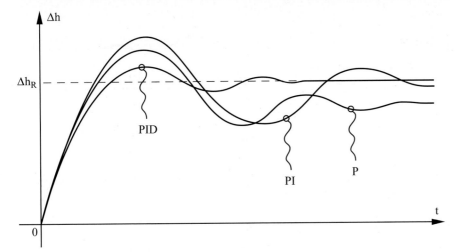

Fig. 23.7 Closed loop P, PI and PID response to a step change in set point

than for P action alone. The oscillations are larger in magnitude and have a lower frequency. In effect the I action, which is working in the same direction as the P action, accentuates any overshooting that occurs. It is evident that I action has a destabilising effect which is obviously undesirable. The principal effects of I action are also summarised in Table 23.1.

Equation 23.4 is the classical form of PI controller. This is historic, due to the feedback inherent in the design of pneumatic and electronic controllers. Note the interaction between the P and I terms: varying K_C affects the amount of integral action because K_C lies outside the bracket. An alternative non-interacting form is as follows:

$$\theta_0 = \theta_B \pm \left(K_C e + \frac{1}{T_I} \int_0^t e \, dt \right) \quad (23.5)$$

T_I is known as the integral time. It can be seen by inspection that $T_I = T_R/K_C$.

23.4 Derivative Action

The purpose of D action is to stabilise and speed up the response of a PI controller. This is realised by the addition of a derivative term to Equation 23.4 as follows:

$$\theta_0 = \theta_B \pm K_C \left(e + \frac{1}{T_R} \int_0^t e \, dt + T_D \frac{de}{dt} \right) \quad (23.6)$$

T_D is known as the rate time and characterises the D action. Adjusting T_D varies the amount of D action, setting it to zero turns off the D action altogether. T_D has the dimensions of time: for process control purposes it is normal for T_D to have units of minutes.

The open loop response of a PID controller to a sawtooth change in error is depicted in Figure 23.8. Again assume that the controller is forward acting, is in its AUTO mode and has a 4–20-mA output range. Suppose that the error is zero for $t < 0$ and for $t > t'$, and that the error is a ramp of slope m for $0 < t < t'$.

Substituting $e = mt$ into Equation 23.6 gives:

$$\theta_0 = \theta_B + K_C mt + \frac{K_C}{T_R} \frac{mt^2}{2} + K_C T_D m$$

The response shows an initial step change in output of magnitude $K_C T_D m$ due to the D action. This is followed by a quadratic which is due to the P and I actions. At $t = t'$ there is another step change as the contribution to the output of the P and D

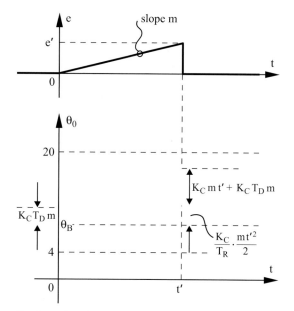

Fig. 23.8 Open loop response of PID controller to a sawtooth change in error

actions disappears. The residual constant output is due to the I action that occurred before $t = t'$.

The closed loop response of a PID controller with appropriate settings to a step change in set point is as depicted in Figure 23.7. It can be seen that the effect of the D action is to reduce the amount of overshoot and to dampen the oscillations. This particular response is reproduced in Figure 23.9 with the corresponding plots of e, Int(edt) and de/dt *vs* time.

Again, referring to the level control system of Figure 22.1, it is evident that as the level crosses the set point and rises towards the first overshoot, the sign of de/dt is negative. This boosts the controller output so that the valve opening is more than it would be due the P and I actions alone, with the obvious effect of reducing the amount of overshoot. As the level passes the first overshoot and starts to fall the sign of de/dt becomes positive and the valve is closed more than it would be otherwise. And so on. In effect the D action is anticipating overshoot and countering it.

D action depends on the slope of the error and, unlike P and I action, is independent of the mag-

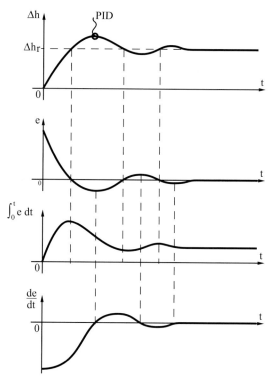

Fig. 23.9 Closed loop PID response to a step change in set point

nitude of the error. If the error is constant, D action has no effect. However, if the error is changing, the D action boosts the controller output. The faster the error is changing the more the output is boosted. An important constraint on the use of D action is noise on the error signal. The spikes of a noisy signal have large slopes of alternating sign. If the rate time is too large, D action amplifies the spikes forcing the output signal to swing wildly. The principal effects of D action are summarised in Table 23.1.

In addition to the effects of P, I and D actions, there are a number of other operational characteristics of 3-term controllers to be considered.

23.5 Bumpless Transfer

If changes in controller output are required, it is good practice to move in a regulated rather than a sudden manner from one state to another. Process plant can suffer damage from sudden changes. For example, suddenly turning off the flow of cooling water to a jacket may cause a vessel's glass linings to crack. Suddenly applying a vacuum to a packed column will cause flashing which leads to the packings breaking up. Such sudden changes in output are invariably caused by the controller being forced to respond to step changes in error.

One source of step change in error occurs when a controller is transferred from its MAN mode of operation into AUTO. If the measured value is not at its set point, depending on the settings, the controller output may well jump to saturation. A simple way round this problem is to adjust the output in MAN mode until the measured value and set point coincide prior to switching into AUTO.

Alternatively, a bumpless transfer function may be specified. In effect, on transfer into its AUTO mode, the set point is adjusted to coincide with the measured value and hence zero error. This results in the controller being started off in AUTO mode with the wrong set point, but that may then be ramped up or down to its correct value at an appropriate rate. The integral action is also initialised by setting it to zero and whatever was the output signal in MAN mode at the time of transfer normally becomes the value of the bias in AUTO. Thus,

$$\theta_R = \theta_1|_{t=0}$$
$$\int_{-\infty}^{0} e.dt = 0$$
$$\theta_B = \theta_0|_{t=0}$$

Set point tracking is an alternative means of realising bumpless transfer. Thus, whilst the controller is in MAN mode, the set point is continuously adjusted to whatever is the value of the measured value. This means that when the controller is switched into AUTO mode, there is zero error and transfer is bumpless. Again, the integral action is initialised by setting it to zero.

For transfers from AUTO into MAN mode the output is normally frozen at whatever was its value in AUTO at the time of transfer.

23.6 Derivative Feedback

Another source of step change in error occurs when the controller is in its AUTO mode and the set point is suddenly changed. Of particular concern is the D action which will initially respond to a large value of de/dt and may cause the output to jump to saturation. To counter this derivative feedback may be specified, as follows:

$$\theta_0 = \theta_B \pm K_C \left(e + \frac{1}{T_R} \int_0^t e.dt - T_D \frac{d\theta_1}{dt} \right) \quad (23.7)$$

Note that the D action responds to changes in the measured value rather than to the error. Also note the minus sign which allows for the fact that the measured value moves in the opposite direction to the error signal. Whilst the set point is constant the controller behaves in exactly the same way as the classical controller of Equation 23.6. However, when the set point changes, only the P & I actions respond. This form of 3-term control is common in modern digital controllers.

23.7 Integral Windup

Another commonly encountered problem is integral windup. If the controller output saturates whilst an error exists, the I action will continue to integrate the error and, potentially, can become a very large quantity. When eventually the error reduces to zero, the controller output should be able to respond to the new situation. However, it will be unable to do so until the error has changed sign and existed long enough for the effect of the integration prior to the change of sign to be cancelled out. The output remains saturated throughout this period and the controller is effectively inoperative. A simple way round this problem is to switch the controller into its MAN mode when the saturation occurs. Switching it back into AUTO when the situation permits causes the I action to be initialised at zero.

A more satisfactory way of addressing the issue is to specify an integral desaturation facility. In effect, whilst saturation of the output occurs, the I action is suspended. This may be achieved by the I action considering the error to be zero during saturation. Alternatively, in digital controllers, the instructions used for calculating the I action may be by-passed during periods of saturation, as illustrated later on.

23.8 Worked Example

A classical 3-term controller which is forward acting has the following settings:

40% bandwidth, 5 min reset time, 1 min rate time.

Also it is known that an output of 10 mA in MANUAL is required for bumpless transfer to AUTO. This information is sufficient to deduce the parameters of Equation 23.6:

$$\theta_0 = 10 + 2.5 \left(e + \frac{1}{5} \int_0^t e.dt + \frac{de}{dt} \right)$$

Suppose the controller is at steady state in AUTO when the measured value starts to decrease at the rate of 0.1 mA/min.

Thus, if $t < 0$ then $e = 0$ and $\theta_0 = 10$ mA, and if $t \geq 0$ then:

$$\frac{d\theta_1}{dt} = -0.1.$$

However, $e = \theta_R - \theta_1$. Assuming θ_R is constant then:

$$\frac{de}{dt} = -\frac{d\theta_1}{dt} = +0.1,$$

whence $e = +0.1\,t$.

Substituting for e into the PID equation gives:

$$\theta_0 = 10 + 2.5 \left(0.1\,t + \frac{1}{5} \int_0^t (0.1\,t)\,dt + \frac{d(0.1\,t)}{dt} \right)$$
$$= 10.25 + 0.25\,t + 0.025\,t^2$$

Inspection reveals a small jump in output at $t = 0$ of 0.25 mA followed by a quadratic increase in output with time. The controller output saturates when it reaches 20 mA, i.e. when:

$$20 = 10.25 + 0.25\,t + 0.025\,t^2$$

$$t^2 + 10\,t - 390 = 0$$

i.e. when $t \approx 15.4$ min. Note that this must be an open loop test: the input signal is ramping down at a constant rate and appears to be independent of the controller output.

23.9 Other Analogue Forms of PID Controller

Equation 23.6 is the classical form of PID controller and Equation 23.7 is the most common form used in digital controllers. There are, however, many variations on the theme which are supported by most modern controllers. For example, to make the controller more sensitive to large errors the P action may operate on the square of the error:

$$\theta_0 = \theta_B \pm K_C \left(\text{Sign}(e).e^2 + \frac{1}{T_R} \int_0^t e.dt - T_D \frac{d\theta_1}{dt} \right)$$

To make the controller penalise errors that persist the I action may be time weighted, with some facility for reinitialising the I action:

$$\theta_0 = \theta_B \pm K_C \left(e + \frac{1}{T_R} \int_0^t e.t.dt - T_D \frac{d\theta_1}{dt} \right)$$

It is important to be aware of exactly what form of 3-term control has been specified and what is being implemented.

23.10 Discretised Form of PID

Equation 23.7 is an analogue form of PID control and has to be translated into a discretised form for implementation in a digital controller. It may be discretised as follows:

$$\theta_{0,j} \approx \theta_B \pm K_C \left(e_j + \frac{1}{T_R} \sum_{k=1}^{j} e_k.\Delta t - T_D \frac{(\theta_{1,j} - \theta_{1,j-1})}{\Delta t} \right)$$

$$= \theta_B \pm \left(K_C.e_j + \frac{K_C.\Delta t}{T_R} \sum_{k=1}^{j} e_k - \frac{K_C.T_D}{\Delta t} (\theta_{1,j} - \theta_{1,j-1}) \right)$$

$$= \theta_B \pm \left(K_C.e_j + K_I. \sum_{k=1}^{j} e_k - K_D. (\theta_{1,j} - \theta_{1,j-1}) \right) \quad (23.8)$$

where j represents the current instant in time and Δt is the step length for numerical integration. This is the non-interacting discretised form of PID controller in which:

$$K_I = \frac{K_C . \Delta t}{T_R} \qquad K_D = \frac{K_C . T_D}{\Delta t}$$

Equation 23.8 may be implemented by the following algorithms written in structured text. These algorithms would be executed in order, at a frequency known as the sampling period which corresponds to the step length Δt. Note that Line 2 establishes whether the controller output is saturated and, if so, by-passes the instruction for calculating the I action on Line 3. Lines 5 and 6 constrain the controller output to its specified range:

```
    E= SP-IP
    if (OP=0 or OP=100) then goto L
    IA=IA+KI*E
L   OP=BI+(KC*E+IA-KD*(IP-PIP))
    if OP<0 then OP=0
    if OP>100 then OP=100
```

Sometimes the bias is deemed to be equivalent to some notional integral action prior to the loop being switched into its AUTO mode. It is then included in the algorithm implicitly as the initial value of the integral action:

$$\theta_{0,j} = \pm \left(K_C.e_j + K_I . \sum_{k=-\infty}^{j} e_k - K_D.(\theta_{1,j} - \theta_{1,j-1}) \right) \quad (23.9)$$

23.11 Incremental Form of PID

Equation 23.8 determines the absolute value of the output signal at the jth instant. There is an alternative, commonly used, incremental form which determines the change in output signal. At the j-1th instant the absolute output is given by

$$\theta_{0,j-1} = \theta_B \pm \left(K_C.e_{j-1} + K_I . \sum_{k=1}^{j-1} e_k - K_D.(\theta_{1,j-1} - \theta_{1,j-2}) \right)$$

Subtracting this from Equation 23.8 yields the incremental form:

$$\Delta\theta_{0,j} = \pm \left(K_C.(e_j - e_{j-1}) + K_I.e_j - K_D.(\theta_{1,j} - 2\theta_{1,j-1} + \theta_{1,j-2}) \right) \quad (23.10)$$

23.12 Mnemonics and Nomenclature

BI	θ_B	output bias
E	e	error
IA	–	integral action
IP	θ_1	measured value
KC	K_C	proportional gain
KD	K_D	derivative action
	T_D	rate time
KI	K_I	integral gain
	T_R	reset time
	T_I	integral time
OP	θ_0	controller output
PIP		previous measured value
SP	θ_R	set point

23.13 Summary of Control Actions

The affect of changing the controller settings on the closed loop response of a PID controller is summarised in Table 23.1.

Table 23.1 Summary of control actions

Action	Change	Effect
P	Increase K_C	Increases sensitivity
		Reduces offset
		Makes response more oscillatory
		System becomes less stable
I	Reduce T_R	Eliminates offset faster
		Increases amplitude of oscillations
		Settling time becomes longer
		Response becomes more sluggish
		System becomes more unstable
D	Increase T_D	Stabilises system
		Reduces settling time
		Speeds up response
		Amplifies noise

Controller Tuning

24.1 Stability
24.2 Marginal Stability
24.3 Continuous Cycling Method
24.4 Reaction Curve Method

The sort of closed loop response that can be obtained from a PID controller with appropriate settings was seen in Chapter 23. This gives rise to two issues. First, what is the best sort of response? This is normally characterised in terms of the response to a step input, the ideal being a fast response with no overshoot or offset. This is not physically possible because of a plant's dynamics. There is therefore a need to compromise between the speed of response and the amount of overshoot. For certain critical applications an overdamped response with no overshoot is essential, the penalty being a slow response. However, for most purposes, an underdamped response with a decay ratio of 1/4 and no offset is good enough, as depicted in Figure 24.1. Often referred to as the optimal response, this has no mathematical justification: it is just accepted good practice.

Second, what are the appropriate settings and how do you find them? The process of finding the optimum settings is generally referred to as loop tuning. They may be found by trial and error. However, given that each of the settings for K_C, T_R and T_D can typically be varied from 0.01 to 100, and that the various lags and delays associated with process plant are often large, this could be rather tedious. The settings may be predicted theoretically. However, this requires a model of the system that is reasonably accurate, and any realistic model usually needs simulation to provide the settings. The cost and time involved in modelling and simulation can seldom be justified. There is therefore a need for a practical approach.

This chapter outlines two practical methodologies; one is empirically based and the other theoretical, for establishing the so called optimum settings. An informed account of these methods is given by Coughanowr (1991) and a comprehensive coverage of both these and many other methods of tuning is given by Astrom (1995). However, first, an insight into the nature of stability is required.

24.1 Stability

Stability is a fundamental consideration in the design of control systems. Many open loop systems are stable. For example, the level control system of Figure 22.1 is open loop stable. Suppose the controller is in its manual mode and the valve opening is fixed at its normal value. Following a step increase in inlet flow, the level will rise until a new equilibrium is established at which there is sufficient head for the flow out to balance the flow in. Such a system is said to be self regulating and is relatively easy to control. However, with inappropriate controller settings, a system which is open loop stable can become closed loop unstable.

Some systems are open loop unstable. The classic example of an item of plant that is inherently unstable is the exothermic reactor. A slight increase in temperature will make the reaction go faster.

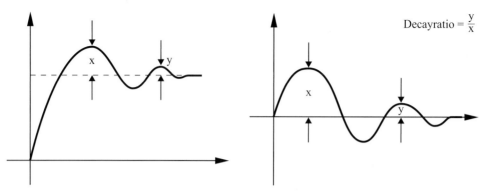

Fig. 24.1 The optimal response

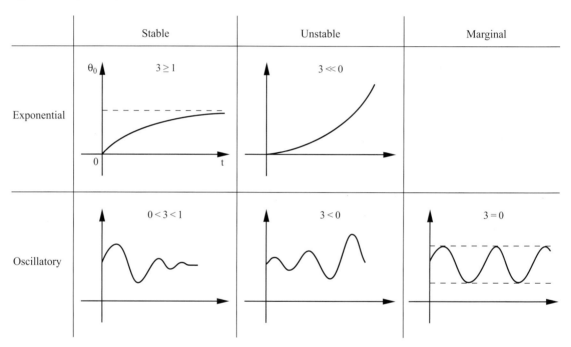

Fig. 24.2 Categorisation of signals on the basis of stability

This produces more heat which increases the temperature further, and so on. However, by applying feedback, such an open loop unstable system can be made closed loop stable.

From a control engineering point of view, stability manifests itself in the form of a system's signals. These may be categorised, depending on whether the system is stable or unstable, as being exponential or oscillatory as depicted in Figure 24.2. An important point to appreciate is that it is the system that is stable or otherwise, not its signals. Also, stability is a function of the system as a whole, not just of some parts of it. It follows that all the signals of a system must be of the same form. It is not possible, for example, for the controller output of the level control system of Figure 22.1 to be stable but for the level to be unstable. Similarly, it is not possible for the response of the level to be exponential whilst that of the valve opening is oscillatory.

As will be seen in Chapters 71 and 72, stability may be characterised by a so-called damping factor ζ in relation to a second order system. Table 24.1 relates the categories of stability to values of the damping factor.

Table 24.1 Stability as a function of damping factor

Factor	Damping	Nature of stability
$\zeta > 1.0$	Overdamped	Stable exponential
$\zeta = 1.0$	Critically damped	Limiting case
$0 < \zeta < 1$	Under-damped	Stable oscillatory
$\zeta = 0$	Undamped	Marginally stable
$\zeta < 0$	Self excited	Unstable oscillatory
$\zeta \ll 0$	Over excited	Unstable exponential

24.2 Marginal Stability

The case of marginal stability is of particular importance for design purposes. The most common design philosophy is to establish the conditions under which a system is marginally stable. An appropriate safety factor is then specified which ensures that the system's operation is stable for all foreseeable circumstances.

Marginal stability corresponds to the situation when all the signals in a system are sinusoidal with constant amplitude, *i.e.* the oscillations are neither growing nor decaying. Consider the feedback system of Figure 24.3 in which all the elements, other than the controller, have been lumped in with the process. The controller has P action only.

Suppose that the error signal is a sine wave of constant amplitude:

$$e = A.\sin(\omega_C t)$$

Now suppose that the frequency w_C, known as the critical frequency, is such that the effect of the process on the error is to produce a measured value whose phase is shifted by 180°. Also suppose that

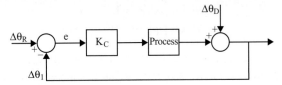

Fig. 24.3 Feedback system with P controller and lumped process

the controller gain has a value K_{CM} such that the controller and the process have no net effect on the amplitude of the signal. Thus, assuming that the signals are in deviation form:

$$\Delta\theta_1 = A.\sin(\omega_C t - 180)$$

So $\Delta\theta_1$ is the mirror image of e as shown in Figure 24.4. But

$$e = \Delta\theta_R - \Delta\theta_1$$

If the set point is constant, *i.e.* $\Delta\theta_R = 0$, then

$$e = -\Delta\theta_1 = -A.\sin(\omega_C t - 180) = A.\sin(\omega_C t)$$

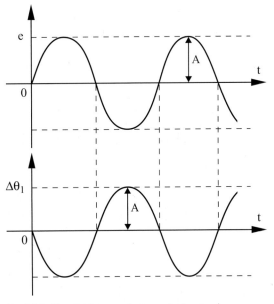

Fig. 24.4 Sinusoidal error and measured value signals

The comparator introduces a further 180° phase shift which converts the measured value back into the error. This is a self sustaining sine wave of constant amplitude. Thus the twin criteria necessary for marginal stability are

Open loop gain = 1.0
Open loop phase shift = −180°

These are known as the Bode stability criteria and will be considered in more detail in Chapter 73.

24.3 Continuous Cycling Method

The continuous cycling method is a simple practical method that tunes a control loop as installed, rather than as designed. The method is carried out at the controller, in its automatic mode, and takes advantage of the faceplate for observing the signals. The procedure for carrying out the method is shown in flow chart form in Figure 24.5.

In essence the process consists of changing the controller's gain K_C incrementally and observing the loop's response to small step changes in set point. If the oscillations decay then K_C is too low and if the oscillations grow then K_C is too high. The value of the marginal gain K_{CM} that forces the loop into self sustained oscillation of constant amplitude is noted. So too is the period P_U of that oscillation, sometimes referred to as the ultimate period. The critical frequency is given by

$$\omega_C = \frac{2\pi}{P_U}$$

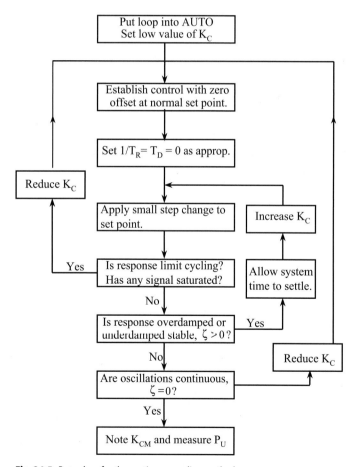

Fig. 24.5 Procedure for the continuous cycling method

Knowing the values of K_{CM} and ω_C, the optimum settings can be determined from the Zeigler and Nichols formulae given in Table 24.2.

Table 24.2 The Zeigler and Nichols formulae

	K_C	T_R	T_D
P	$K_{CM}/2.0$	–	–
PI	$K_{CM}/2.2$	$P_U/1.2$	–
PID	$K_{CM}/1.7$	$P_U/2.0$	$P_U/8.0$

The amplitude of the oscillations under conditions of marginal stability is a function of the system and cannot be controlled. Therefore, before carrying out the continuous cycling method on a plant, it should be established whether it is acceptable to do so. Whilst it may well be acceptable to force the plant into oscillation during commissioning or periods of shut-down, approval to do so is unlikely to be forthcoming during production!

Care should be taken to protect the system from external disturbances whilst the tests are being carried out so as not to distort the results. The most common source of disturbance is due to changes in the supply pressure of utilities such as steam and cooling water. Also, the control loop being tuned may interact with other loops as, for example, in cascade control. In such cases it is usually necessary to put the other loops into their manual mode to prevent them from trying to compensate for the oscillations in the loop being tuned.

Whether the controller has P, P&I or P, I&D actions, the continuous cycling method must be carried out with the controller set for P action only. The I and D actions are switched off by setting T_R = max and T_D = min respectively. Once the procedure has been completed, the optimum settings for the P, I and D actions can then be set as appropriate.

It is important to establish oscillations of constant amplitude that are sinusoidal. These can easily be confused with limit cycles which are of constant amplitude but non-sinusoidal. Limit cycles occur when a system is in oscillation and at least one signal is saturated. The most likely signal to saturate, and the easiest to observe, is the controller output. This may saturate at either the top and/or the bottom of its range, resulting in the valve opening being more square than sinusoidal in form. The process will filter out this squareness resulting in a measured value that may appear to be sinusoidal. Reducing the controller gain should stop the saturation and prevent limit cycles from occurring.

The optimum settings for a control loop will vary across the range of its measured value if there are any non-linearities present. It follows that a loop must be tuned for its normal operating conditions. The oscillations should therefore be established about the normal value of each signal. In particular, the normal set point should be used, and the step changes applied in alternate directions about it to ensure that the system stays close to normal.

The nature of the Zeigler and Nichols formulae needs some explanation. First published in 1941, they are used extensively in industry and have stood the test of time. The formulae are empirical, although they do have a rational theoretical explanation. They predict settings that are optimum on the basis of a decay ratio of 1/4. However, because the formulae are empirical, they do not predict the optimum settings precisely, and further tuning of a trial and error nature may be required. This might not seem to be very satisfactory but, noting that each of the settings typically has a range of 0.01 to 100, *i.e.* a rangeability of some 10^4, a method that predicts settings to within even 50% of the optimum as a first estimate is extremely useful. In practice the predictions are often to within 10% of the optimum.

24.4 Reaction Curve Method

Whereas forcing a plant into oscillation with no control over amplitude may be unacceptable, introducing a small step change of known size, which is the basis of the reaction curve method of predicting optimum settings, is another matter altogether.

This is an open loop method of controller tuning and is depicted in Figure 24.6. With the con-

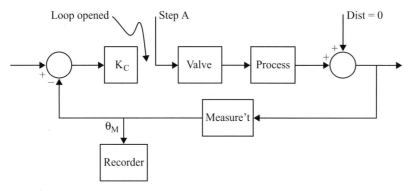

Fig. 24.6 The open loop reaction curve method

troller in its manual mode, the output is adjusted to its normal value and the system allowed to reach equilibrium. Then a small step change of known magnitude A is applied to the controller output. The system is allowed to respond and the measured value recorded for an appropriate period, as shown in deviation form in Figure 24.7.

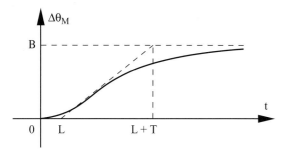

Fig. 24.7 Characteristic "S-shaped" reaction curve

The reaction curve shown, often referred to as being "S shaped" for some unknown reason, is characteristic of most process control systems. Indeed, as will be seen in Chapter 72 in relation to higher order systems, many systems can be approximated by a combination of a steady state gain K, a first-order system with a time constant of T min and a time delay of L min. The values of K, T and L can be estimated from the reaction curve.

First, the steady state asymptote B of θ_M is established. Since B is simply the steady state effect of the open loop elements operating on the step A, the gain can be obtained from the ratio $K = B/A$.

Second, by drawing a tangent to the reaction curve at the point of inflexion, and finding its intersection with the asymptote and the time axis as shown, the values of T and L are found. Knowing the values of K, T and L, the optimum settings can be determined from the Cohen and Coon formulae given in Table 24.3.

These formulae are theoretically derived on the assumption that the plant consists solely of a gain K, time constant T and time delay L. They predict settings that are optimum on the basis of a decay ratio of 1/4.

There are a number of important precautions. Care should be taken to protect the system from external disturbances whilst the tests are being carried out so as not to distort the results. The reaction curve method is much more susceptible to distortion by disturbances than the continuous cycling method. The step input A applied should be small enough for the response to stay within the bounds of linearity. The response of the recording system must be fast enough not to distort the reaction curve. It is normal to apply the step change at the controller output. However, it may be applied anywhere in the loop, provided that the reaction curve is the open loop response of all the elements of the loop. If the controller is included then it must have no effect, i.e. set $K_C = 1$, $T_R = \max$ and $T_D = \min$.

There are two major problems in using the reaction curve method. First, it is often difficult to insulate the plant from disturbances long enough to obtain a true reaction curve. And second, given

Table 24.3 The Cohen and Coon formulae

	K_C	T_R	T_D
P	$\dfrac{1}{K}\dfrac{T}{L}\left(1+\dfrac{L}{3T}\right)$		
PI	$\dfrac{1}{K}\dfrac{T}{L}\left(\dfrac{9}{10}+\dfrac{L}{12T}\right)$	$L\left(\dfrac{30+3L/T}{9+20L/T}\right)$	
PD	$\dfrac{1}{K}\dfrac{T}{L}\left(\dfrac{5}{4}+\dfrac{L}{6T}\right)$		$L\left(\dfrac{6-2L/T}{22+3L/T}\right)$
PID	$\dfrac{1}{K}\dfrac{T}{L}\left(\dfrac{4}{3}+\dfrac{L}{4T}\right)$	$L\left(\dfrac{32+6L/T}{13+8L/T}\right)$	$L\left(\dfrac{4}{11+2L/T}\right)$

the constraints on linearity, it is difficult to obtain a large enough change in output to predict with confidence the value of the asymptote B and the position of the point of inflexion.

Cascade Control

25.1 Master and Slave Loops
25.2 Cascade Control of Heat Exchanger
25.3 Cascade Control of Jacketed Reactor
25.4 Implementation

This chapter summarises the essential principles and practice of cascade control: it is discussed more fully by Murrill (1988). Cascade control is a powerful extension of conventional 3-term feedback control. It is a strategy which compensates for specific disturbances at source and largely prevents them from affecting the process being controlled.

25.1 Master and Slave Loops

A cascade control scheme has two controllers, the output of the master controller being used to adjust the set point of the slave controller. This is best illustrated by means of an example. Consider the boiler level control system of Figure 25.1.

Boiler drum level is notoriously difficult to control. One reason for this is that the steam pressure in the drum can vary significantly. Suppose the pressure in the vapour space above the liquid in the drum suddenly increases. This will cause the pressure drop across the control valve to fall. Thus the water flow will be reduced, irrespective of the level, even if the level is below its set point. The effect of steam pressure disturbances can be compensated for using a slave flow control loop as depicted in Figure 25.2.

The set point of the flow control loop u_M is manipulated by the level controller output. If the level h is too low the set point of the flow loop will be increased, and vice versa. The flow loop controls the flow of water F_W against this set point in the normal way. Again, suppose that the pressure in the drum P_S suddenly increases, perhaps due to a transient drop in steam load F_S, causing the flow of water to decrease. The flow loop will respond quickly by opening the valve to maintain the water flow F_W at the rate demanded by the level controller. In effect, the flow loop is insulating the level loop from, or re-

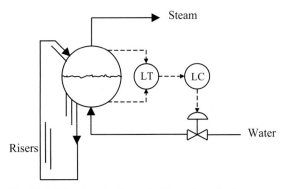

Fig. 25.1 Boiler drum level control with simple feedback

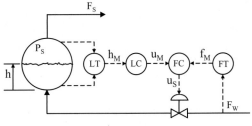

Fig. 25.2 Boiler drum level control with cascade system

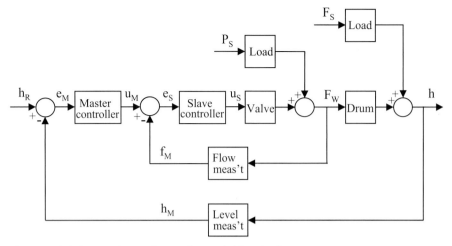

Fig. 25.3 Block diagram of system for cascade control of drum level

jecting disturbances due to, changes in steam pressure. The corresponding block diagram is given in Figure 25.3.

Its structure consists of two feedback loops, one nested inside the other. The outer loop is the master loop, sometimes referred to as the primary loop, and controls the level. The inner loop is the slave loop, sometimes referred to as the secondary loop, and controls the flow. The flow loop has a much faster response than the level loop, the dynamics of the later being dominated by the lags due to the capacity of the drum. It is often the case that the slave loop is a flow control loop. The need for the inner loop to have a faster response than the outer loop is characteristic of all cascade control schemes.

Note in particular the location of the disturbances. Changes in P_S affect the slave loop which compensates for them before they have any signif icant effect on the drum level. Changes in F_S affect the level and are compensated for by the master loop. Strictly speaking, there is a linkage between changes in F_S and P_S: the details of this are omitted here for simplicity.

25.2 Cascade Control of Heat Exchanger

Another example of the application of cascade control is in the control of heat exchangers. Consider Figure 25.4 in which a process liquor is heated on the tube side of an exchanger by condensing steam on the shell side. A conventional feedback control scheme uses the outlet temperature T_0 to manipulate the flow rate F_S of steam.

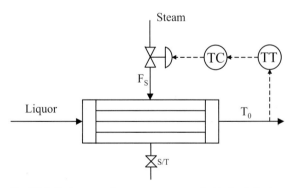

Fig. 25.4 Exchanger outlet temperature control with simple feedback

This scheme works well, but can be improved by cascade control. In particular, changes in the steam supply pressure affect the flow through the control valve which in turn affects the pressure inside the shell. Use of steam flow as a slave control loop would work, but steam temperature inside the shell would be much more effective as a slave loop since this relates directly to the rate of heat transfer: ultimately this is the manipulated variable that matters. Noting that the steam pressure is directly related to its temperature in condensing systems, it is logical to use a pressure loop instead of a temperature loop as the slave, as depicted in Figure 25.5. That is for two reasons: first, the rate of heat transfer is significantly more sensitive to the steam's pressure than its temperature. And second, given any uncertainty over the nature of the steam quality and its measurement, pressure is the more reliable metric. Note that the slave loop rejects disturbances due to steam supply pressure only, disturbances due to changes in the supply pressure of the process liquor and its temperature are handled by the master loop.

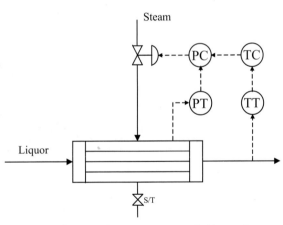

Fig. 25.5 Exchanger outlet temperature control with cascade system

25.3 Cascade Control of Jacketed Reactor

The choice of slave variable is not always straightforward. Consider the jacketed stirred tank reactor of Figure 25.6. Reagents flow into the reactor and displace products through the overflow at the same rate. The reaction is exothermic, heat being removed by cooling water being circulated through the jacket. Reactor temperature is controlled by a conventional feedback loop which manipulates the flow of water.

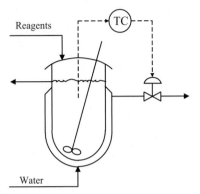

Fig. 25.6 Simple feedback of temperature in jacketed reactor

This scheme has a poor response because of the sluggish dynamics of both the reactor and its jacket. In particular, any disturbance in the cooling water supply pressure will cause a change in jacket temperature which, eventually, will affect the reactor temperature. Only when the reactor temperature moves away from its set point can the controller start to compensate for the disturbance. Significant errors occur before compensation is complete. This is a classical application for cascade control. There is a choice of three slave loops using either the water flowrate, pressure or temperature as the controlled variable, as depicted in Figures 25.7, 25.8 and 25.9 respectively.

- Flowrate: use of water flowrate as the slave variable is essentially the same as in the example of the boiler drum level control scheme of Figure 25.2. The slave loop specifically rejects disturbances in water flow rate due to changes in its supply pressure. The slave loop has a fast response since it is dependant only upon the hydrodynamics of the water system and the dynamics of the instrumentation and valve.
- Pressure: use of the jacket pressure as the slave variable also specifically rejects disturbances in water flow rate and has a fast response. How-

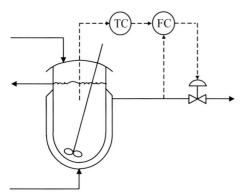

Fig. 25.7 Cascade control of reactor: jacket water flowrate as slave loop

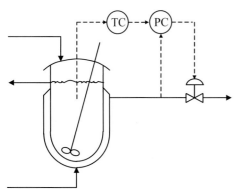

Fig. 25.8 Cascade control of reactor: jacket water pressure as slave loop

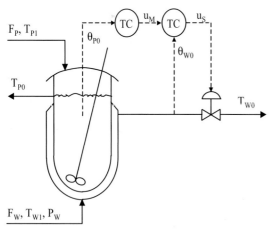

Fig. 25.9 Cascade control of reactor: jacket water temperature as slave loop

ever, the scheme is counter-intuitive and could confuse the operators:

i. The slave controller must be reverse acting. For example, following an increase in supply pressure the jacket pressure rises and the controller opens the valve. Assuming most of the resistance to flow is due to the jacket and the pipework upstream of the valve, opening the valve increases the flow and the pressure drop across the fixed resistances. Hence the jacket pressure falls. One would intuitively expect to close the valve to counter the effect of an increase in supply pressure.

ii. The master loop must be forward acting. Thus, following an increase in reactor temperature, the master controller reduces the set point of the slave controller. This results in the valve opening thereby thereby giving the increase in cooling water necessary to counter the temperature.

It would be bad practice to use this scheme given that there are viable alternatives.

- Temperature: as can be seen from the corresponding block diagram of Figure 25.10, use of jacket temperature T_{W0} as the slave variable rejects disturbances in water temperature T_{W1} as well as disturbances in water flowrate F_W due to changes in supply pressure P_W. Disturbances in the temperature T_{P1} and flowrate F_P of the reagents feed stream are handled by the master loop.

It is evident that the dynamics of the plant have been split, the reactor is still in the master loop but the jacket has been shifted into the slave loop. The response of this slave loop is relatively slow since it is dominated by the thermodynamics of the jacket. However, this is not necessarily a disadvantage, provided the slave loop's response is still faster than that of the master. Indeed, shifting the jacket's dynamics into the slave loop makes for better control of the master loop. Note that there are important interactions between the reactor and jacket so, strictly speaking, they cannot be split as described. The details of this are omitted here for simplicity, but are considered quantitatively in Chapter 85.

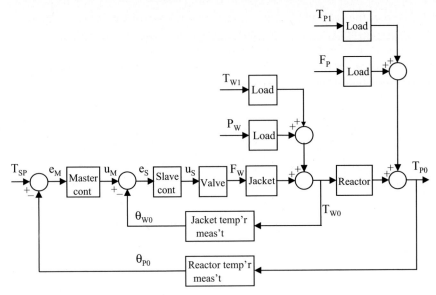

Fig. 25.10 Block diagram of cascade system with jacket temperature as slave loop

25.4 Implementation

Without doubt cascade control can bring about substantial improvements in the quality of control. However, the benefits are critically dependant upon proper implementation. The control scheme should be designed to target specific disturbances. If possible choose slave loops that reject disturbances from more than one source. It is essential that the slave loop's dynamics are significantly faster the master loop's to minimise the effects of interactions between the loops. As a rule of thumb, the dominant time constant in the slave loop must be less than one third of the dominant time constant in the master loop.

Because of the interactions between the loops, any non-linearity introduced by the slave loop will have an adverse effect on the behaviour of the master loop. The slave loop should not introduce non-linearity. It is particularly important when the slave loop is controlling flow that square root extraction is used.

When tuning the controllers, the basic strategy is to tune the inner loop first. Switching the master controller into its manual mode effectively disconnects the two loops. By applying step changes to the master controller output, the slave loop can be tuned using the continuous cycling method. Alternatively, by switching the slave loop into its manual mode, it can be tuned by the reaction curve method. Once the slave loop has been tuned, it can be switched into its automatic mode and the master loop tuned in the conventional way. In effect, the slave loop is treated as if it were any other element whilst tuning the master loop. It may be necessary to slightly detune the slave loop if its dynamics propagate into the master loop to the extent that the interactions adversely affect the master loop's performance.

If analogue single loop controllers are being used, remember that the slave controller is physically different to the master controller. The slave controller requires an input for a remote, or external, set point signal whereas the master controller's set point is local and adjusted manually.

It would appear that cascade control is more expensive than conventional feedback. It certainly requires an additional slave measurement, although it is often the case that the measurement is probably being made anyway, or ought to be if the disturbances are significant. Otherwise, the only additional costs are for an extra analogue input channel and additional function blocks. If a DCS, PLC or modern digital SLC is being used, it will support all of the functionality necessary for handling both the master and slave controllers. As far as the output channel is concerned, the same I/P converter, valve and actuator are used for cascade control as would be otherwise. So there truly is relatively little extra cost.

Ratio Control

26.1 Scaling Approach
26.2 Direct Approach
26.3 Indirect Approach
26.4 Comments

Ratio control is another control strategy commonly used in the process industries. It is used when the flow rates of two or more streams must be held in proportion to each other. Typical applications are in blending, combustion and reactor feed control systems. There are essentially three approaches, one is based upon the simple scaling of signals and the other two are based upon the PID controller. In the latter cases, the ratio may be controlled either directly using a proprietary ratio controller or else indirectly by means of a ratio station with a conventional controller. All three approaches are discussed. A good treatment of ratio control is given by Shinsky (1996).

26.1 Scaling Approach

Consider Figure 26.1 in which stream A is wild and stream B is manipulated to keep it in proportion to stream A. Assume that the flow transmitter is calibrated for the full range of the manipulated flow. Also assume that the valve is carefully sized such that its full range of flow corresponds to that of the manipulated flow. The flow transmitter output may be applied *via* an I/P converter directly to the valve to achieve the desired ratio control. This is a simple and effective means of ratio control, but is critically dependant upon the linearity of the elements. The flow transmitter may require square root extraction, and the control valve must have a linear installed characteristic.

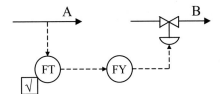

Fig. 26.1 Simple scaling approach to ratio control

In practice it is unlikely that the valve could be sized such that the ranges of the wild and manipulated flows are exactly in the desired ratio. A scaling factor is therefore necessary. This can be realised by changing the calibration of the flow transmitter, or by fitting a positioner to the valve and adjusting its range. An alternative is to use a so-called ratio station, as depicted in Figure 26.2.

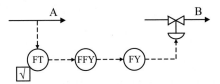

Fig. 26.2 Scaling approach with a ratio station

The ratio station, denoted by the code FFY, is simply a device for applying a user definable scaling factor K to a signal. For example, if the station's I/O are 4–20 mA signals, then its operation is described by:

$$\theta_0 = 4 + K(\theta_1 - 4)$$

Care has to be taken in deciding what scaling factor to apply. It depends on the calibrations of the transmitter, valve and/or positioner. Also, if the signals are within a digital system, they may well have been scaled into engineering units too.

26.2 Direct Approach

An example of the direct approach is given in Figure 26.3. Again stream A is assumed to be wild. Both flow rates are measured. The ratio controller FFC manipulates the flowrate of stream B to produce the desired ratio of B to A. Note that the ratio control loop rejects disturbances in stream B due to changes in supply pressure P_S.

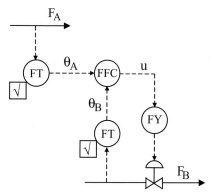

Fig. 26.3 Direct approach to ratio control

Assuming 4–20 mA signals, the ratio is calculated according to:

$$R_M = \frac{(\theta_B - 4)}{(\theta_A - 4)}$$

This measured ratio is then compared with the desired (reference) ratio R_R and an error signal e is generated. The ratio controller operates on the error to produce an output signal u. The block diagram is given in Figure 26.4.

Most proprietary ratio controllers physically combine the division, comparison and control functions into a single unit. Thus a ratio controller has two measured value inputs and one output signal. Typically the ratio of the measured values is displayed on the faceplate alongside the manually set desired ratio. In most other respects a ratio controller is much the same as a conventional 3-term controller. For example, it provides P, I and D actions, has a forward/reverse action switch and supports both automatic and manual modes of operation.

To the extent that each involves two measured values and one output signal, it is easy to confuse the P&I diagrams of ratio and cascade control schemes. For example, notwithstanding the fact that the processes are different, the ratio scheme of Figure 26.3 looks similar to the cascade scheme of Figure 25.2. However, comparison of the corresponding block diagrams reveals that their structures are fundamentally different.

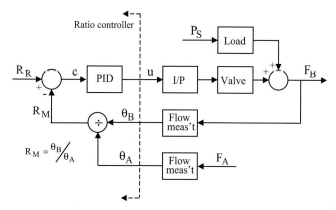

Fig. 26.4 Block diagram of direct approach to ratio control

26.3 Indirect Approach

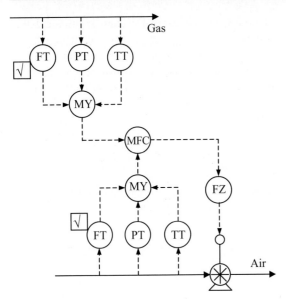

Fig. 26.5 Direct approach with temperature and pressure compensation

Ratio control of gas flows subject to changes in operating conditions may well require pressure and/or temperature correction, as depicted in Figure 26.5. In these circumstances it is best to calculate the mass flow rate of each stream, and to control the ratio of the mass flow rates. However, bearing in mind the scope for calibration and measurement errors, and their potential impact on the ratio calculated, the flow changes must be fairly significant to justify the increased complexity.

26.3 Indirect Approach

The indirect approach is both simple and effective, an example is given in Figure 26.6.

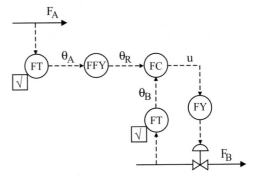

Fig. 26.6 Indirect approach to ratio control

The measured value θ_A of the flowrate of wild stream A is operated on by the ratio station. Assuming 4–20 mA signals it calculates the desired value θ_R for the flowrate of stream B according to the equation:

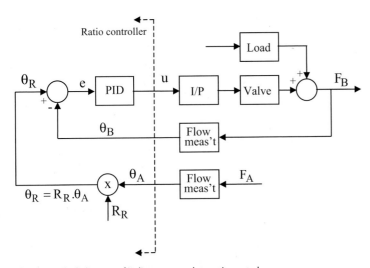

Fig. 26.7 Block diagram of indirect approach to ratio control

$$\theta_R = 4 + R_R (\theta_A - 4)$$

where R_R is the desired (reference) ratio. A conventional feedback loop is then used to control the flowrate of stream B against this set point, as depicted in the block diagram in Figure 26.7. Note again the rejection of disturbances in stream B.

26.4 Comments

The direct approach is commonly used throughout industry although, on two different counts, the indirect method is superior. First, with the direct ratio scheme there is the potential for zero division by θ_A giving rise to indeterminate ratios. This is particularly problematic with low flows: errors in the flow measurement become disproportionate and lead to very poor quality ratio control. This can also be a problem if ratio control is used as part of a model based control strategy involving deviation variables. The problem does not arise with the indirect ratio scheme because θ_A is multiplied.

Second, there is the question of sensitivity of the error to changes in the wild flow. Inspection of Figure 26.4 reveals that the sensitivity of the direct scheme is:

$$\left.\frac{de}{d\theta_A}\right|_{\bar{\theta}_B} = \left.\frac{d}{d\theta_A}\left(R_R - \frac{\theta_B}{\theta_A}\right)\right|_{\bar{\theta}_B} = \frac{\theta_B}{\theta_A^2} = \frac{R_m}{\theta_A}$$

whereas inspection of Figure 26.7 reveals that the sensitivity of the indirect scheme is:

$$\left.\frac{de}{d\theta_A}\right|_{\bar{\theta}_B} = \left.\frac{d}{d\theta_A}(R_R.\theta_A - \theta_B)\right|_{\bar{\theta}_B} = R_R$$

Clearly there is a linearity issue. The lower the flow the greater the sensitivity of the direct ratio scheme, whereas the indirect scheme's sensitivity is constant.

Feedforward Control

27.1 Feedforward Compensation
27.2 Dynamic Compensation
27.3 Feedforward Control
27.4 Feedforward Control of a Heat Exchanger
27.5 Implementation Issues
27.6 Comments
27.7 Nomenclature

The inherent limitation of feedback control is that it is retrospective. A feedback controller can only respond to disturbances once they have affected the controlled variable. For many processes this does not matter unduly. However, when the disturbances are large, or where the process dynamics are sluggish, feedback control results in significant and sustained errors. Using cascade control to reject specific disturbances can produce substantial improvements in performance. However, control is still retrospective. Ratio control is different. It responds to changes in one variable by adjusting another to keep them in proportion. In a sense it is anticipating the process needs and is a particular case of feedforward control.

What feedforward control offers is the prospect of control action which anticipates the effect of disturbances on the process and compensates for them in advance. This chapter develops the concept of feedforward control, considers some of its limitations, and introduces its implementation.

27.1 Feedforward Compensation

Consider a process and its load, with steady state gains of K_1 and K_2 respectively, as depicted in Figure 27.1. Suppose that the dynamics are negligible. Let θ_0, θ_1 and θ_2 be the controlled, disturbance and manipulated variables in deviation form.

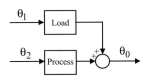

Fig. 27.1 Block diagram of process and its load

To compensate for changes in θ_1 a feedforward element of gain K_F may be introduced, as depicted in Figure 27.2.

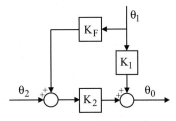

Fig. 27.2 Process and load with feedforward compensation

Steady state analysis yields:

$$\theta_0 = K_1.\theta_1 + K_2.(K_F.\theta_1 + \theta_2)$$
$$= (K_1 + K_2.K_F)\,\theta_1 + K_2.\theta_2$$

For ideal disturbance rejection, changes in θ_1 have no effect on θ_0 such that

$$\theta_0 = K_2.\theta_2$$

To satisfy this criterion it is necessary for:

$$(K_1 + K_2.K_F)\,\theta_1 = 0$$

Since θ_1 cannot be assumed to be zero, it follows that:

$$K_F = -K_1/K_2$$

In effect, the feedforward path is creating an inverse signal which will cancel out the effect of the load operating on the disturbance. Any practical implementation of this feedforward compensation requires a measurement of θ_1 and some means of applying the compensation. Let K_T, K_I and K_V be the steady state gains of the measuring element, I/P converter and control valve as depicted in Figure 27.3. It is now appropriate to consider θ_2 as being the output of a conventional controller and F_2 as the manipulated variable.

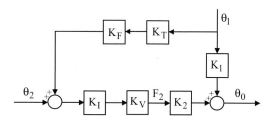

Fig. 27.3 Practical implementation of feedforward compensation

A similar steady state analysis to the above yields:

$$K_F = -K_1/(K_T.K_I.K_V.K_2) \qquad (27.1)$$

It is necessary that the values of these steady state gains are known. K_1 and K_2 may be determined either empirically or theoretically. K_T, K_I and K_V are known by specification and/or calibration. However, any inaccuracies will lead to a steady state offset in θ_0 in the event of a change in θ_1. The significance of this should not be underestimated.

27.2 Dynamic Compensation

It is unrealistic to ignore the dynamics of the process and load. Indeed, it is largely because of their dynamics that feedforward compensation is being considered. Whereas the gains of the various elements are relatively easy to establish, their dynamics are not. This is especially true of the process and load. Whilst the structure of their dynamic models may be determined from first principles, it is often difficult to predict the values of the parameters involved with any confidence. Therefore, in practice, it is usual to separate out the steady state gain from the dynamics, and to have two feedforward compensation terms as depicted in Figure 27.4.

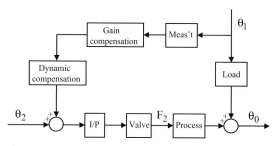

Fig. 27.4 Distinction between gain and dynamic compensation

The steady state feedforward gain is as defined by Equation 27.1. By a similar argument, assuming that the dynamics of the instrumentation is insignificant, the dynamic compensation term C(s) is the ratio of the dynamics of the load L(s) to those of the process P(s).

$$C(s) = \frac{L(s)}{P(s)}$$

The Laplace notation used here, necessary for articulating the dynamics, is explained in detail in Chapters 70 and 71. If the process and load dynamics are the same, which is often the case, they cancel and there is no need for dynamic compensation. Otherwise, the dynamic compensation term consists of time lags, leads and delay terms, such that its structure corresponds to the required dynamic ratio C(s). The parameters of the dynamic compensation term are then tuned empirically.

Because of the scope for offset due to errors in the values of the various steady state gains, and the approximate nature of the dynamic term, feedforward compensation is seldom used in isolation.

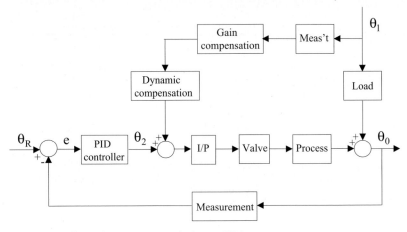

Fig. 27.5 Feedforward compensation grafted onto a PID loop

The most common strategy is to use it in conjunction with a conventional 3-term feedback control loop as depicted in Figure 27.5. The feedback loop will eliminate offset due to inaccuracies in the feedforward compensation, handle residual dynamic errors, and correct for other disturbances.

A practical example of the use of feedforward compensation in the control of a distillation column is depicted in Figure 27.6. It is used in conjunction with a cascade system which controls the composition in the bottom of the column by manipulating the flow of steam into the reboiler, steam pressure

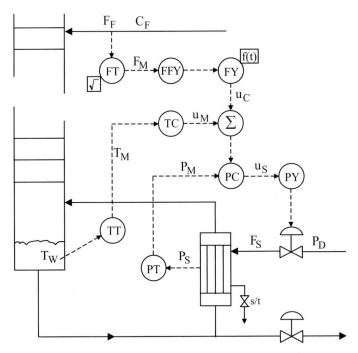

Fig. 27.6 P&I diagram of feedforward compensation for distillation column

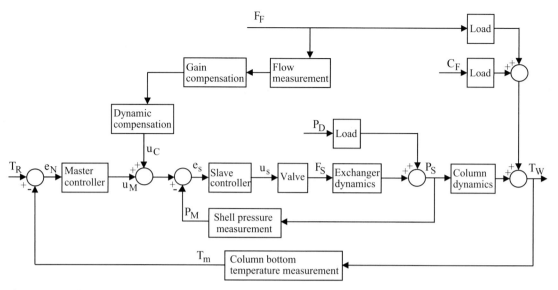

Fig. 27.7 Block diagram of feedforward compensation for distillation column

being the slave variable. The feedforward compensation varies the boil-up rate in anticipation of the effects of changes in the column feed rate by applying a bias to the set point of the slave loop. The corresponding block diagram is given in Figure 27.7.

27.3 Feedforward Control

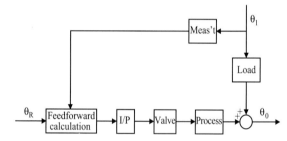

Fig. 27.8 Process and load with feedforward calculation

Alternatively, the feedforward control can incorporate the set point function, as depicted in Figure 27.8. The use of the set point as an input to the feedforward calculation is what distinguishes feedforward control from feedforward compensation. Indeed, it is the litmus test.

In practice, feedforward control as depicted in Figure 27.8 is seldom used in isolation because of the problems of offset and the need to handle other disturbances. The most common strategy is to use it in conjunction with a conventional 3-term feedback control loop as depicted in Figure 27.9. This feedback loop is analogous to the slave loop used in cascade control. Note that the steady state and dynamic compensation have again been separated out.

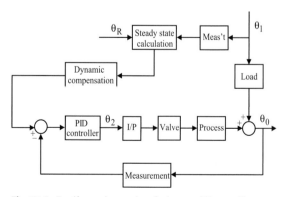

Fig. 27.9 Feedforward control grafted onto a PID controller

27.4 Feedforward Control of a Heat Exchanger

The practicalities of feedforward control are perhaps best illustrated by means of an example. Consider the heating up of a process stream on the tube side of an exchanger by the condensation of steam on its shell side, as depicted in Figure 27.10.

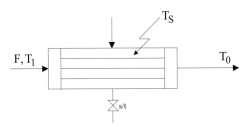

Fig. 27.10 Steam heated shell and tube exchanger

A steady state heat balance across the exchanger gives

$$Q = U.A.T_m = F.\rho.c_p(T_0 - T_1)$$

However, the log mean temperature difference is given by

$$T_m = \frac{T_0 - T_1}{\ln\left(\dfrac{T_S - T_1}{T_S - T_0}\right)}$$

Hence

$$\frac{T_S - T_1}{T_S - T_0} = \exp\left(\frac{UA}{F\rho c_p}\right)$$

Assume that all the resistance to heat transfer is due to the tube side film coefficient. From the Dittus Boelter correlation, the overall coefficient may be approximated by

$$U \approx kF^{0.8}$$

Also, for saturated steam and water,

$$T_S \approx mP_S$$

Hence

$$P_S \approx \frac{T_1 - T_0 \cdot \exp\left(\dfrac{kA}{\rho c_p F^{0.2}}\right)}{m\left[1 - \exp\left(\dfrac{kA}{\rho c_p F^{0.2}}\right)\right]}$$

This equation is, in effect, the steady state model of the process. For implementation as a feedforward controller, the outlet temperature T_0, which is arbitrary, may be replaced by its desired value T_R. The inlet temperature T_1 and flow rate F may be replaced by their measured values T_M and F_M respectively. The equation explicitly calculates the

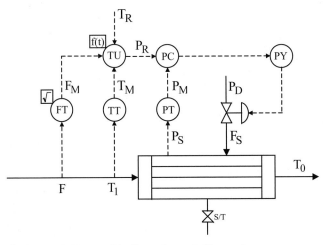

Fig. 27.11 P&I diagram of feedforward control of heat exchanger

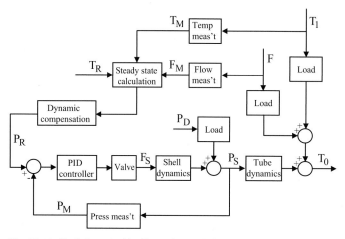

Fig. 27.12 Block diagram of feedforward control of heat exchanger

necessary steam pressure, which becomes the set point P_R for a conventional feedback control loop:

$$P_R \approx \frac{T_M - T_R \exp\left(\dfrac{kA}{\rho c_p F_M^{0.2}}\right)}{m\left[1 - \exp\left(\dfrac{kA}{\rho c_p F_M^{0.2}}\right)\right]} \quad (27.2)$$

This control scheme is shown in P&I diagram form in Figure 27.11 and the corresponding block diagram is shown in Figure 27.12.

27.5 Implementation Issues

Inaccuracy is the prime source of difficulty in implementing feedforward control. Any errors in the temperature and flow measurements will be propagated into the derived steam pressure set point through Equation 27.2. More fundamental though is the accuracy of Equation 27.2 itself. A model of the process has been developed. Various assumptions and approximations have been made. Even if

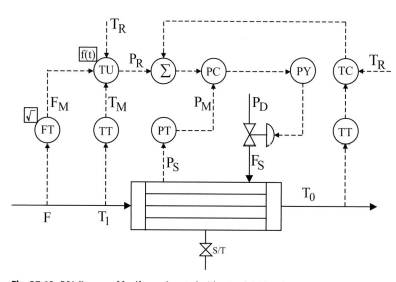

Fig. 27.13 P&I diagram of feedforward control with set point trimming

these are all correct, it is unlikely that accurate values are available for the parameters of the model. Furthermore, the model is steady state and ignores the dynamics of the process. Dynamic compensation by means of time lags, leads and delay terms is, at best, approximate.

There is likely, therefore, to be significant offset in the outlet temperature. This is best handled by another controller which trims the system. Trimming can be achieved by various means: in this case a bias is applied to the set point of the steam pressure control loop, as shown in Figure 27.13.

27.6 Comments

Feedforward control is not an easy option. Developing the model requires both experience and understanding of the process. There are major problems due to inaccuracy. Only specific disturbances are rejected. This results in feedforward control having to be used in conjunction with other loops. The outcome is that the control schemes are complex, some would say unnecessarily so. Finding the optimum form of dynamic compensation and tuning the loops is not easy. Nevertheless, it does work and does produce benefits. There are many feedforward control schemes in operation throughout the process industries. Modern control systems support all the functionality necessary for their implementation.

27.7 Nomenclature

A	mean surface area of tubes	m^2
c_p	specific heat	$kJ\ kg^{-1}\ K^{-1}$
F	flow rate	$m^3\ s^{-1}$
k	coefficient	$kJ\ s^{-0.2}\ m^{-4.4}\ K^{-1}$
m	coefficient	$°C\ bar^{-1}$
ρ	density	$kg\ m^{-3}$
P	pressure	bar
Q	rate of heat transfer	kW
T	temperature	$°C$
U	overall heat transfer coefficient	$kW\ m^{-2}\ K^{-1}$

Subscripts

M	measured
m	logarithmic mean
R	reference
S	shell side steam
1	tube side inlet
0	tube side outlet

On-Off Control

28.1 On-Off Cycling
28.2 On and Off Curves
28.3 Lag Effects
28.4 Worked Example
28.5 Comments
28.6 Nomenclature

On-off control, sometimes referred to as bang-bang control, is conceptually the same thing as proportional control, as described in Chapter 22, with a high controller gain. It is characterised by very small, but finite, errors causing the controller output to switch between maximum and minimum output according to the sign of the error. The response of a forward acting on-off controller to a sawtooth error is depicted in Figure 28.1.

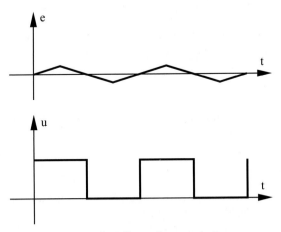

Fig. 28.1 Response of on-off controller to sawtooth error

28.1 On-Off Cycling

It is seldom that a proportional controller *per se* is used for on-off control. Much more typical is the use of amplifiers and relays in simple thermostats. On-off control is surprisingly common for simple, non-critical applications. There is more to it than meets the eye though.

Consider a tank containing a liquid whose temperature is thermostatically controlled as depicted in Figure 28.2.

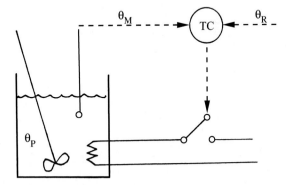

Fig. 28.2 Tank with thermostat

Depending on the temperature of the liquid in the tank, the power supply is either connected to or disconnected from the heating element. This results in the temperature cycling about the set point θ_R within a narrow band $\Delta\theta$, as depicted in Fig-

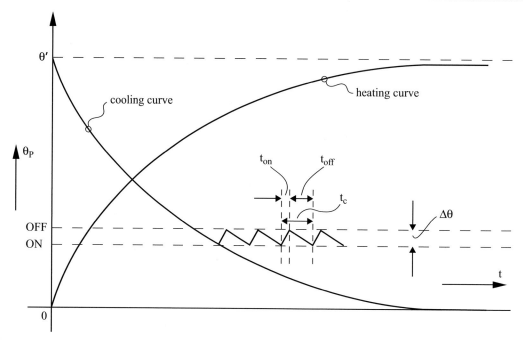

Fig. 28.3 Heating and cooling curves for tank contents

ure 28.3. Since $\Delta\theta$ is small, the sections of the heating and cooling curves may be approximated by straight lines.

28.2 On and Off Curves

Consider the heating curve. If the heater is switched on, assuming that the liquid is well mixed and that there are no heat losses due to evaporation, then an unsteady state heat balance for the contents of the vessel gives:

$$M.c_P \frac{d\theta_P}{dt} = W - U.A.\theta_P$$

where θ is measured relative to ambient temperature. This is a first order system, of the type described in Chapter 69, whose response is of the form:

$$\theta_P = \frac{W}{UA}\left(1 - e^{\frac{-UA}{Mc_P}t}\right)$$

If $t \to \infty$ then $\theta_P \to \theta' = \frac{W}{U.A}$. Defining the time constant $T_P = \frac{M.c_P}{U.A}$ yields:

$$\theta_P = \theta'.\left(1 - e^{-t/T_P}\right)$$

The slope of the heating curve is thus:

$$\frac{d\theta_P}{dt} = \frac{\theta'}{T_P}.e^{-t/T_P} = \frac{\theta' - \theta_P}{T_P}$$

Let $\overline{\theta}_P$ be the mean value of θ_P within the band $\Delta\theta$. Thus, within $\Delta\theta$:

$$\left.\frac{d\theta_P}{dt}\right|_{on} \approx \frac{\theta' - \overline{\theta}_P}{T_P} \qquad (28.1)$$

Now consider the cooling curve. If the heater is switched off, then an unsteady state heat balance for the contents of the vessel gives:

$$M.c_P \frac{d\theta_P}{dt} = -U.A.\theta_P$$

This too is a first-order system, whose response is of the form:

$$\theta_P = \theta'.e^{-t/T_P}$$

The slope of the cooling curve is thus:

$$\frac{d\theta_P}{dt} = \frac{-\theta'}{T_P} \cdot e^{-t/T_P} = \frac{-\theta_P}{T_P}$$

So, for values of θ_P within the band $\Delta\theta$:

$$\left.\frac{d\theta_P}{dt}\right|_{off} \approx \frac{-\overline{\theta}_P}{T_P} \qquad (28.2)$$

28.3 Lag Effects

A section of the sawtooth of Figure 28.3 is reproduced in Figure 28.4.

The controller period t_C is the sum of the t_{on} and t_{off} times. Thus:

$$t_C = \frac{1}{\left.\frac{d\theta_P}{dt}\right|_{on}} \cdot \Delta\theta + \frac{1}{\left.\frac{d\theta_P}{dt}\right|_{off}} \cdot \Delta\theta$$

Substituting from Equations 28.1 and 28.2 gives:

$$t_C = \frac{\theta' \cdot T_P \cdot \Delta\theta}{\overline{\theta}_P \cdot (\theta' - \overline{\theta}_P)} \qquad (28.3)$$

The measured temperature θ_M lags behind the tank temperature by the time constant T_M of the measuring element. Because of mechanical backlash in the switches, hysteresis, stiction, etc., the heater state will not change from on to off and vice versa until there exists a finite error about the set point. The corresponding band of measured temperature is called the differential gap $\Delta\theta_M$.

From Figure 28.4 it can be seen that:

$$\Delta\theta = \Delta\theta_M + \left.\frac{d\theta_P}{dt}\right|_{on} \cdot T_M + \left.\frac{d\theta_P}{dt}\right|_{off} \cdot T_M$$

Substituting from Equations 28.1 and 28.2 gives:

$$\Delta\theta = \Delta\theta_M + \frac{T_M}{T_P} \cdot \theta' \qquad (28.4)$$

The mean tank temperature is half way across $\Delta\theta$:

$$\overline{\theta}_P = \theta_B + \frac{\Delta\theta}{2} = \theta_B + \frac{\Delta\theta_M}{2} + \frac{T_M}{2 \cdot T_P} \cdot \theta'$$

Assume that the backlash, etc., is symmetrical about the set point θ_R:

$$\theta_R = \theta_B + \left.\frac{d\theta_P}{dt}\right|_{off} \cdot T_M + \frac{\Delta\theta_M}{2} = \theta_B + \frac{T_M}{T_P} \overline{\theta}_P + \frac{\Delta\theta_M}{2}$$

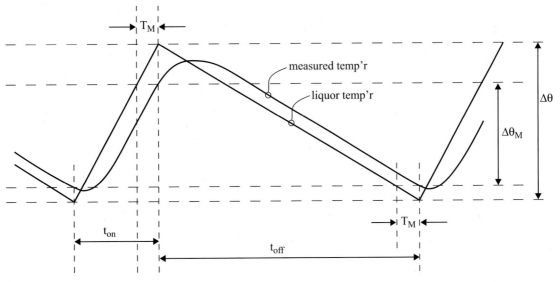

Fig. 28.4 On and off sections of sawtooth

For on-off control the offset is defined to be the difference $\bar{\theta}_P - \theta_R$, whence:

$$\text{Offset} = \bar{\theta}_P - \theta_R \equiv \frac{T_M}{T_P} \cdot \left(\frac{\theta'}{2} - \bar{\theta}_P\right) \quad (28.5)$$

28.4 Worked Example

The temperature of the liquid in a tank is controlled by an on-off controller. The trend diagram indicates that if the set point is 65°C then the on-time is 249 s, the off-time is 150 s and the differential gap is 1.5°C. If the heater is left switched on the temperature reaches 95°C eventually. What is the temperature variation in the tank?

$\Delta\theta_M = 1.5°C$, $t_{on} = 249$ s and $t_{off} = 150$ s.

Assume an ambient temperature of 15°C:

$\theta_R = 65 - 15 = 50°C$ and $\theta' = 95 - 15 = 80°C$.

From the slopes of the heating and cooling curves, Equations 28.1 and 28.2:

$$\Delta\theta = \frac{\theta' - \bar{\theta}_P}{T_P} \cdot t_{on} = \frac{\bar{\theta}_P}{T_P} \cdot t_{off}$$

Substituting values:

$$\left(80 - \bar{\theta}_P\right).249 = \bar{\theta}_P.150$$

whence $\bar{\theta}_P = 49.9248$

From the definition of offset, Equation 28.5:

$$\bar{\theta}_P - \theta_R \equiv \frac{T_M}{T_P} \cdot \left(\frac{\theta'}{2} - \bar{\theta}_P\right)$$

$$-0.0752 \equiv \frac{T_M}{T_P} \cdot -9.925$$

whence

$$\frac{T_M}{T_P} = 0.007577$$

Substituting into Equation 28.4 gives:

$$\Delta\theta = \Delta\theta_M + \frac{T_M}{T_P}.\theta' = 1.5 + 0.007577 \times 80 = 2.106$$

Hence temperature variation in the tank is approximately 2°C.

28.5 Comments

Most on-off systems operate with a relatively large cycle time and deviation. A long cycle time gives rise to large deviations in the controlled variable, whereas a short cycle time may cause excessive wear on the relays, actuators, *etc.* Thus there is a trade off between cycle time and deviation.

Equations 28.3–28.5 give the relationships between the cycle time, width of the band and its mean, the differential gap, set point, offset and time lags. To achieve a narrow band requires minimal backlash, *etc.*, and a fast measurement, *i.e.* $T_M \ll T_P$.

There will be no offset when $\theta' = 2\bar{\theta}_P$ in which case:

$$\left.\frac{d\theta_P}{dt}\right|_{on} = \left.\frac{d\theta_P}{dt}\right|_{off} = \frac{\theta'}{2\,T_P}$$

The on and off times will be equal when the slopes of the on and off curves are the same.

28.6 Nomenclature

A	effective surface area	m²
c_P	specific heat	kJ kg⁻¹ K⁻¹
M	mass of contents of vessel	kg
t	time	s
T	time constant	s
U	overall heat transfer coefficient	kW m⁻² K⁻¹
W	heater power	kW
θ	temperature	°C

Subscripts

B	ase position
C	controller
M	measurement
P	process
R	set point

Sequence Control

29.1 A Charging Operation
29.2 Simple Sequence
29.3 Decision Trees and Tables
29.4 Parameter Lists
29.5 Timing Considerations
29.6 Sequential Function Chart
29.7 Parallelism
29.8 Top Down Approach
29.9 Comments

Sequence control is used extensively throughout the process industries. It is concerned with operations that are sequential, or discontinuous, in nature. Some examples of situations where sequence control would be used are as follows:

- Transferring reagents from one vessel to another
- Progressing operations in a batch reactor
- Automatic start-up of plant
- Emergency shut-down of plant

These operations are typically characterised by:

- Manipulation of discrete I/O signals
- Time-ordered sets of events and actions
- Logic which is conditional upon plant and/or process status
- Integration of sequential and continuous control functions

Automatic sequence control is realised by means of sequences which are, in effect, programs. In terms of implementation, sequence control was historically realised by means of hard wired logic elements, such as pneumatic relays and gates, but nowadays it is invariably software based and realised by means of a PLC or DCS.

This chapter introduces the concepts of sequence control and the issues involved in its implementation, the emphasis being on sequence design and using the IEC-61131 (part 3) sequential function chart (SFC) language. This provides the basis for Chapter 37 on batch process control.

29.1 A Charging Operation

A charging operation is used as a vehicle to introduce the concepts. Consider the plant depicted in Figure 29.1, the objective being to transfer reagent under gravity from the charge vessel into the agitated process vessel and to start heating it.

A simple sequence to realise this may be as depicted in flow chart form, referred to as a sequence flow diagram (SFD), in Figure 29.2.

In essence, the operation consists of switching the isolating valves, agitator motor and temperature control loop in an appropriate order. Analogue input signals for level and weight provide information about plant conditions.

- First, the level in the process vessel LT52 is checked against some maximum value LMAX to ensure that there is sufficient space for the charge. If not, a message that the level is too high is displayed for the operator and the sequence aborted.

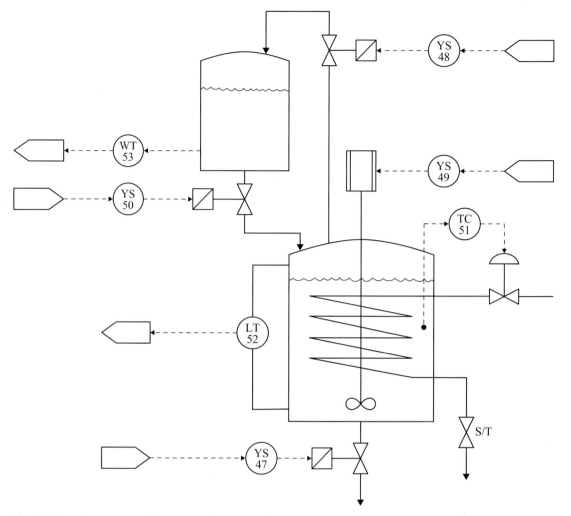

Fig. 29.1 Charging system comprising charge and process vessels

- Next the weight of the charge vessel WT53 is checked against some minimum quantity WMIN to ensure there is sufficient reagent for a full charge. If not, the operator is alerted to the situation and the sequence aborted.
- Then the process vessel drain valve YS47 is closed prior to charging: there is no point in charging to drain! It is good practice in sequence design to close the valve, even if it should already be closed, because some other previous operation could have left it open.
- Provided the level and weight are both acceptable, the vent valve YS48 is opened to equalise pressure during transfer, otherwise the process vessel could become pressurised and the charge vessel evacuated, both of which would inhibit the transfer.
- The initial weight WBEG of the charge vessel is noted, and the transfer valve YS50 opened to enable reagent to flow under gravity.
- Charging proceeds until the decrease in weight of the charge vessel reaches the required charge

WREQ, at which point both the transfer and vent valves are returned to their initial closed state.

- The agitator YS49 is then started, the set point of the temperature control loop TC51 is set to its desired value TDES and the loop switched into its automatic mode.

Further thought will reveal that there is much scope for making the sequence more robust. For example:

- Presumably there is an inlet pipe and isolating valve to the charge vessel. It would be sensible to make sure that this stays shut during the charge operation.
- The transfer pipe and valve could well be part of an inlet manifold interconnecting with other charge and process vessels, in which case it would be necessary to isolate the required transfer route during the charge operation.
- When an isolating valve is closed or the agitator started, it would be sensible to check that the action has indeed occurred. This may be achieved by means of a technique, known as discrepancy checking, which is explained in detail in Chapter 46.
- The reagent could be charged in increments, decreasing in size as WREQ is reached. Allowing for the "in flight" reagent potentially makes for more accurate charging.
- If the process vessel overfills, a level alarm could be activated.
- If the weight in the charge vessel is too low then, rather than defaulting to the end of the sequence, it might be better to provide the operator with some options such as filling up the charge vessel or making a batch of product with a smaller charge.
- Given that the increase in level should correspond to the weight charged, there is scope for cross checking the weight and level measurements and stopping the sequence if there is an inconsistency.

Nevertheless, the sequence as it stands has sufficient complexity to demonstrate the essential principles.

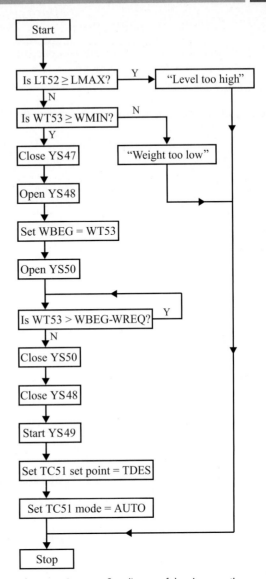

Fig. 29.2 Sequence flow diagram of charging operation

29.2 Simple Sequence

The structured text equivalent of the sequence of Figure 29.2 is depicted in Program 29.1 and is largely self explanatory.

Program 29.1. Structured text equivalent of sequence flow diagram

```
program CHARGE
  var_input
    LT52 at %IW52: real;
    WT53 at %IW53: real;
  end_var
  var_output
    YS47 at %QX47: bool;
    YS48 at %QX48: bool;
    YS49 at %QX49: bool;
    YS50 at %QX50: bool;
  end_var
  var_external
    LMAX, WMIN, WREQ, TDES: real;
    TC51: PID;
  end_var
  var
    WBEG, WDIF: real;
    MESSAGE_1: string:='$n LEVEL TOO HIGH $r'
    MESSAGE_2: string:='$n WEIGHT TOO LOW $r'
  end_var

  if LT52 ge LMAX then
    MESSAGE_1
  elsif WT53 le WMIN then
    MESSAGE_2
  else
    YS47:= 0;
    YS48:= 1;
    WBEG:= WT53;
    YS50:= 1;
    repeat
      WDIF:= WBEG - WT53;
    until WDIF ge WREQ
    end_repeat
    YS50:= 0;
    YS48:= 0;
    YS49:= 1;
    TC51.SP:= TDES;
    TC51.ST:= 1;
  end_if
end_program
```

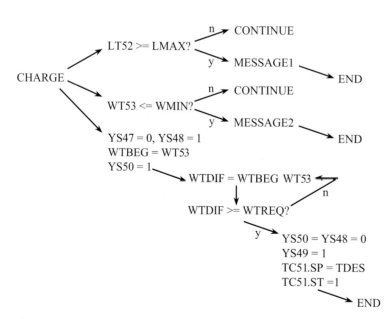

Fig. 29.3 Decision tree for sequence of charging operation

29.3 Decision Trees and Tables

Note that the analogue measurements are declared as input variables and the discrete signals are declared as outputs, and that the tag numbers are used to uniquely relate the variables to their database locations and I/O channels.

Also note that the temperature measurement and the output to the steam control valve have been configured into the loop TC51. This is assumed to have been declared globally as being of the function block type PID so, within this program, has to be accessed as an external variable. The loop set point TC51.SP and its status TC51.ST, being parameters of TC51, do not need to be declared separately. The realisation of an analogue control loop by means of configuration is described in detail in Chapters 44 and 48.

The SFD of Figure 29.2 is depicted in the form of a decision tree in Figure 29.3 on page 198 and, notwithstanding initialisation, is equivalent to Program 29.1. Decision trees are a fairly compact way of representing logic and are relatively easy to understand.

The corresponding decision table is depicted in Table 29.1. In essence, each column corresponds to a production rule. If the conditions specified in the upper part of a column are satisfied, then the actions specified in the lower part are realised. At each sampling instant, the system attempts to fire all the rules. By design, sequence progression starts with the left hand column and moves across

Table 29.1 Decision table for sequence of charging operation

Conditions	Rules							
	1	2	3	4	5	6	7	8
BEGIN=1	Y							
END=1		N	N	N	N	N	N	N
TRANSFER=1				N	Y	Y		
DONE=1				N			Y	
LT52 ≥ LMAX		Y						
WT53 ≤ WMIN			Y					
WDIF ≥ WREQ					N	Y		
TC51.ST=1								Y
Actions								
BEGIN=0	Y							
END=0	Y	N	N				N	
TRANSFER=0	Y			N		Y		
DONE=0	Y					N		
MESSAGE 1		Y						
MESSAGE 2			Y					
YS47=0	Y			Y				
YS48=0	Y			N		Y		
YS49=0	Y						N	
YS50=0	Y				N	Y		
WBEG=WT53				Y				
WDIF=WBEG-WT53				Y	Y			
TC51.ST=0	Y						N	
TC51.SP=TDES							Y	

the table rightwards. Note that additional flags have been introduced to handle the branching and looping. Decision tables enable logic to be documented in a very compact form but are not very understandable.

Whilst decision trees and tables are relatively common in the manufacturing industries, they are seldom used in the process sector.

29.4 Parameter Lists

In the above sequence, the tag numbers and the charging data are embedded within the sequence. Whilst this makes for clarity and simplicity, it is somewhat inflexible. If a different quantity was to be charged, or a different pair of charge and process vessels was to be used, all the tag numbers and data in the sequence would have to be changed. In effect, another identical sequence is required, but with different tags and data.

An alternative approach uses parameter lists and indirect addressing. All the plant and process specific data is separated out from the sequence and included in parameter lists. The sequence is designed to be generic. Then, according to the requirements for a particular charging operation, the sequence is interpreted in relation to the relevant parameter lists.

Suppose there are two parameter lists, an A list for **addresses** (tag numbers) and a B list for **batch** (operational) data, created with the array data type construct as follows:

```
type A:
  array [1..5] of bool;
  array [6..8] of real;
end_type

type B:
  array [1..4] of real;
end_type
```

and variables assigned to the elements as follows:

```
A[1]:= YS47;      B[1]:= LMAX;
A[2]:= YS48;      B[2]:= WMIN;
A[3]:= YS49;      B[3]:= WREQ;
A[4]:= YS50;      B[4]:= TDES;
A[5]:= TC51.ST;
A[6]:= TC51.SP;
A[7]:= LT52;
A[8]:= WT53;
```

Assuming that the parameter lists are declared and variables assigned at a higher level (of POU) as appropriate, the corresponding generic sequence is given in Program 29.2.

Program 29.2. Charging operation with parameter lists

```
program CHARGE
  var_external
    A: array;
    B: array;
  end_var
  var
    WBEG, WDIF: real;
    MESSAGE_1: string:= '$n LEVEL TOO HIGH $r'
    MESSAGE_2: string:= '$n WEIGHT TOO LOW $r'
  end_var

  if A[7] ge B[1] then
    MESSAGE_1
  elsif A[8] le B[2] then
    MESSAGE_2
  else
    A[1]:= 0;
    A[2]:= 1;
    WBEG:= A[8];
    A[4]:= 1;
    repeat
      WDIF:= WBEG - A[8];
    until WDIF ge B[3]
    end_repeat
    A[4]:= 0;
    A[2]:= 0;
    A[3]:= 1;
    A[6]:= B[4];
    A[5]:= 1;
  end_if
end_program
```

Note that the parameter lists themselves would be made generic and instances created and parameterised for particular charges. Also note that separate C and D lists could have been created too: typically the C list is of **common** variables (flags, integer and string variables) for cross referencing between sequences and the D list is of **data** (string variables) for handling messages.

29.5 Timing Considerations

The statements of a sequence are executed in order, progression being determined by the sequence logic according to timing constraints. These constraints may be explicit, for example in the form of an instruction to wait for a specified period. More usually, though, they are implicit and conditional upon plant status and process conditions.

It is good practice to separate out statements that are time dependent into distinct "steps". Also, when there are branches and/or loops in a sequence, it is normal for sequence flow to be forced to the start of a step. This results in plant and/or process oriented groupings of statements. They arise naturally, are relatively self contained, and are executed as composite groups.

Sequence progression is suspended when a step is encountered. Typically, the sequence waits for an elapsed time of one second since the start of the current step before beginning the next step. Thus, as a sequence is interpreted, its execution is typically at a maximum rate of one step *per* second.

The decomposition of a sequence into steps is depicted in Program 29.3.

Program 29.3. Charging operation decomposed into steps

```
STEP0 (* Start *)
  program CHARGE
STEP1 (* Null *)
  if LT52 ge LMAX then
    MESSAGE_1
    goto STEP5
STEP2 (* Check *)
  if WT53 le WMIN then
    MESSAGE_2
    goto STEP5
STEP3 (* Transfer *)
  YS47:= 0;
  YS48:= 1;
  WBEG:= WT53;
  YS50:= 1;
  repeat
    WDIF:= WBEG - WT53;
  until WDIF ge WREQ
  end_repeat
  YS50:= 0;
  YS48:= 0;
STEP4 (* Heat-up *)
  YS49:= 1;
  TC51.SP:= TDES;
  TC51.ST:= 1;
STEP5 (* Stop *)
end_program
```

Inspection of Program 29.3 reveals that the nested nature of the original sequence of Programs 29.1 and 29.2 has been replaced with a block structure. In practice, especially with more complex sequences, it is very difficult, and indeed unrealistic, to design them using the so-called nested approach that is encouraged by advocates of structured programming. Inevitably, sequences contain branches, jumps, loops, *etc.* They are a reflection of the process logic and plant constraints which inherently dictate the sequence structure. SFCs, whose structure comes through the decomposition of sequences into steps, are a recognition of this fact.

29.6 Sequential Function Chart

SFCs have evolved via a French standard known as Grafcet, which was itself based upon a representation known as the Petri-net. Grafcet describes the behaviour of a system in terms of states and transitions. SFCs, as defined in IEC 61131, are a graphical means of representing sequences in terms of steps and the transitions between them. The essential features of an SFC are as depicted in Figure 29.4 which, apart from the branching, corresponds to the charge sequence of Program 29.3.

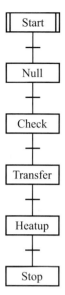

Fig. 29.4 Sequential function chart for charging operation

The steps are shown as rectangular boxes connected by vertical lines. Each connecting line has a horizontal bar representing a transition. Every SFC must have one start and one end step, the box of the start step having parallel sides. Two consecutive steps or transitions are not permitted. Each step in the sequence must have a unique name. A transition is triggered by a specific condition which, when true, causes the step before the transition to be deactivated and the step that follows to become active. Sequence control generally progresses down the page, but branches can force control back to previous steps.

Partitioning of a sequence into steps is very important, not only from the point of view of the design, development and testing of the sequence, but also in relation to its execution. There are two variables associated with the execution of each step. The X flag is a Boolean variable set true when the step becomes active and false when deactivated. It is used by steps in one sequence to establish the status of another. The elapsed time variable T is of the data type time. When a step is activated T is reset to zero and, whilst active, T is incremented in real-time. When the step is deactivated T holds the elapsed time of the step. The T variable is used for co-ordinating the activities of steps in different sequences. Both variables are accessed by dot extension, *i.e.* step_name.X or step_name.T.

Transitions may be triggered by logic tests in any of the IEC 61131 languages, as depicted in Figure 29.5.

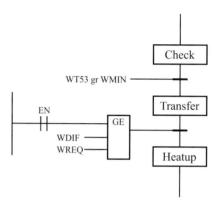

Fig. 29.5 SFC indicating transitions being triggered

Steps consist of one or more actions. An action is depicted as a rectangular box attached to a step. Each action has a name that is unique to the program or function block that contains the SFC. The action may be described in any of the five IEC 61131-3 languages. Note that this enables one SFC to contain another. Program 29.4 is the detailed SFC corresponding to the charge sequence of Figure 29.4.

Each action has a qualifier that determines when the action is executed. In the above example the N qualifier indicates that the outcome of the action continues for as long as the step is active, the S qualifier sets the action until it is explicitly reset, and the P qualifier causes the action to occur once only. The full set of qualifiers is listed in Table 29.2. Note that if no qualifier is specified it defaults to N. Also attached to an action is an optional "indicator" variable used for annotation purposes. The indicator is a variable changed by the action and is used to indicate that the action has completed its execution.

Program 29.4. SFC with steps being realised by actions

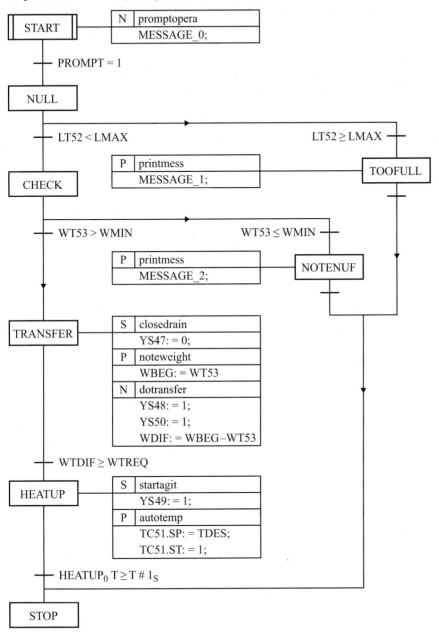

Table 29.2 Outcomes of qualifiers for actions

Qualifier	Outcome of action
N	Held whilst step active
S	Held (stored) beyond end of step until reset
R	Resets a stored action
L	Held for limited period, or until end of step if sooner
D	Held after a delay until end of step, unless step finishes first
P	Executed once only at start of step
SD	Stored and delayed
DS	Delayed and stored
SL	Stored and limited

29.7 Parallelism

The ability of SFCs to support parallelism is a prerequisite for their use in batch process control. It is possible for there to be two or more transitions from one step, thereby enabling the sequence to diverge to one of a selected number of steps, as has already been seen in Program 29.4. The branch that the sequence takes depends on which transition becomes true first. In the event of two transitions becoming true simultaneously, left to right evaluation is normal, unless the transitions are numbered otherwise. To avoid ambiguity, it is good practice to ensure that transition conditions in divergent paths are mutually exclusive.

Divergent paths are used to skip a section of a sequence, or to branch back to an earlier section, as depicted in Figures 29.6 and 29.7 respectively. If necessary, for clarity, arrows can be used to indicate program flow.

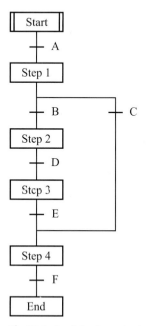

Fig. 29.6 Parallel path construct used to skip section of SFC

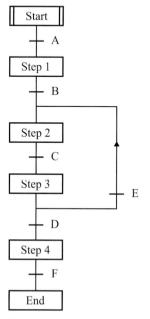

Fig. 29.7 Parallel path construct used to branch back within SFC

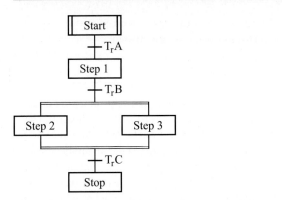

Fig. 29.8 Construct for simultaneous divergence and convergence

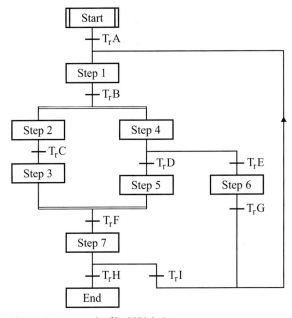

Fig. 29.9 An example of bad SFC design

Parallel lines are used to depict simultaneous divergence and convergence, as shown in Figure 29.8. Notice the position of the transitions. For simultaneous divergence there is only one transition above the parallel. Once Transition B becomes true, both Steps 2 and 3 become active. For simultaneous convergence, both Steps 2 and 3 must be complete before Transition C can be evaluated.

Certain sequence designs are potentially unsafe. For example, in the SFC of Figure 29.9, if the

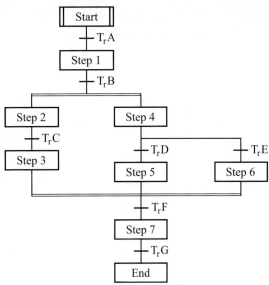

Fig. 29.10 Another example of bad SFC design

divergent path to Step 6 is taken, sequence control may progress *via* Transition G back to Step 1 whilst, say, Step 3 is still active. As execution of the SFC continues, it is possible for both Steps 2 and 3 in the same sub-sequence to become active, leading to unpredictable behaviour.

Another example of poor design is depicted in Figure 29.10. Convergence could not be achieved if Transitions D and E are mutually exclusive, resulting in the sequence getting stuck.

In general, the number of sequences that simultaneously converge should be the same as the number that diverged originally. Some systems are capable of identifying unsafe constructs such as this. However, great care should be taken in sequence design, especially when using parallel sequences, to avoid unpredictable behaviour.

29.8 Top Down Approach

SFCs are hierarchical in that one SFC may contain another. This enables a top down approach to sequence design. They may be written at any level of abstraction. For example, at a high level they may be used to capture user requirements or, at a low level, to articulate the detailed workings of a sequence. It is good practice to:

- Use meaningful names for steps, transitions and actions.
- Partition applications into high level sequences and low level sub-sequences.
- Use only SFCs at higher levels of design.
- Only use action blocks at lowest levels of design.
- Minimise the interaction between simultaneous sequences.
- Avoid different sequences manipulating the same variables.
- Ensure sub-sequences are suspended in an orderly manner, *i.e.* no loose ends.

29.9 Comments

Sequence control is used extensively, both in its own right and as a basis for batch process control. The fairly simple example of charging a vessel has been used to illustrate a number of principles, and the use of structured text and SFCs for sequence control has been demonstrated.

An important point to recognise is that the charging example has focused on the normal operational intent. It is relatively easy to think through how things should work if everything goes according to plan. However, there are many potential ways of deviating from the norm. What happens if the charge valve sticks open? What if the decrease in weight doesn't correspond to the increase in level? And so on. For the sequence to be robust enough for operation in practice, these "what if" questions must be asked within the sequence, and safe hold-states and recovery options provided to retrieve the situation by means of branches off into other sub-sequences. A sobering thought is to appreciate that the sequencing for handling normal situations is typically only 20–40% of the total; the other 60–80% is in providing the recovery options.

The issues of handling contentions, applications diagnostics and recovery options are introduced in Chapter 37.

Section 5

Process Control Schemes

Determining Strategy

30.1 Process of Determination
30.2 Mixing Example
30.3 Crude Oil Separator Example
30.4 Absorber Example
30.5 Flash Drum Example
30.6 Methodology
30.7 Comments

Imagine a plant designed and built without any controls. Suppose it is then started up by opening all the valves, turning on the pumps and agitators, and so on. Would the flows, temperatures and other process variables settle out as *per* design? Of course not. No doubt some sort of equilibrium would be established, but it is most unlikely that this would be as intended. Clearly the plant cannot achieve the desired set of operating conditions on its own. It has to be driven towards them, and then held there. That is the function of the control system.

For any process plant there are invariably many ways of controlling it. Seldom is any single way correct; it is just that some ways are better than others. So, deciding what strategies to use and formulating a control scheme is a question of judgement. Fundamental to this is an understanding of how the process works and a feel for how the items of plant and equipment interact. In this chapter a rational approach to determining control strategy is introduced and a methodology for its implementation developed. This approach is used in subsequent chapters for formulating control schemes for a selection of items of plant.

30.1 Process of Determination

The steady-state operation of a plant may be described by a set of independent steady-state mass and energy balance equations. The variables involved in these equations are said to be strategic and may be categorised as being:

- Wild, *i.e.* determined by upstream conditions.
- Controlled, either directly as specified by the operating conditions for the process or indirectly according to the design of the plant.
- Floating, *i.e.* free to vary subject to the constraints of plant design.
- Determined, by the mass and energy balances being simultaneously satisfied.

In determining the control strategy for a plant, the objective is to ensure that there are no floating strategic variables. Other, non-strategic variables, may float. These are generally of an inventory nature, typically related to operability of the plant, or variables manipulated to optimise process performance.

The difference between the number of strategic variables and equations is known as the number of degrees of freedom. The process of determination involves identifying and categorising the strategic variables and reconciling them with the number of degrees of freedom. This is best illustrated by means of an example.

30.2 Mixing Example

Consider the merging of two streams as depicted in Figure 30.1.

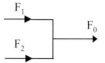

Fig. 30.1 Blending of two streams

The steady state mass balance is:

$$F_1 + F_2 = F_0 \qquad (30.1)$$

There is one equation, three strategic variables and two degrees of freedom. Consider some alternative scenarios. Suppose that F_1 and F_2 are both controlled, as depicted in Figure 30.2. In this case both degrees of freedom have been used up and there are none left to float. Thus F_0 is "determined" by the mass balance and there is no point in trying to control it.

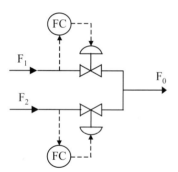

Fig. 30.2 Blending system that is determined

Suppose F_0 is controlled as well, as depicted in Figure 30.3. This scheme is doomed to failure because F_0 is already determined. It would not even work if the set point of the flow controller is set at the value for F_0 as determined by the mass balance. This is because of the inaccuracies in measuring the flows: the set point for F_0 would always be wrong! The scheme is said to be "over-determined". This means that too many things are being controlled simultaneously and the controllers would "fight" each other.

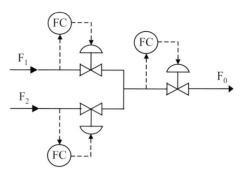

Fig. 30.3 Blending system that is over-determined

Next suppose that F_1 is wild and F_0 is controlled at some specified flowrate, as depicted in Figure 30.4. In this case both degrees of freedom have been used up and F_2 is determined by the mass balance.

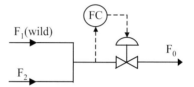

Fig. 30.4 Another blending system that is determined

The merging process of Figure 30.4 consists of pipework only and, for liquid flows, the inlet and outlet flow rates inevitably balance at all times. In reality, plant consists of reactors, columns and so on, all of which have capacity for hold-up. Consider the vessel of Figure 30.5.

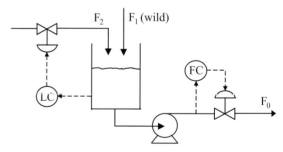

Fig. 30.5 Blending system with capacity

In the long term, the average values of F_1, F_2 and F_0 must balance out and Equation 30.1 still applies. So if F_1 is wild and F_0 is controlled then the average value of F_2 is determined by the mass balance as before. However, in the short term, the flows are unlikely to balance because of the capacity of the vessel. Subject to operability constraints, such as the vessel not overflowing or keeping the pump primed, it doesn't particularly matter what the level is. It is evident that the level is not strategic but inventory in nature and free to float.

Remembering that F_1 is wild, it is sensible to use a level controller to manipulate F_2. Note that F_2 is being manipulated but not controlled. The distinction is important. The average, long term value of F_2 is still determined by F_1 and F_0. Indeed, by de-tuning the level controller, *i.e.* operating it with a low gain, and allowing the level to rise and fall, the capacity of the vessel may be used to average out fluctuations in flow.

30.3 Crude Oil Separator Example

A crude oil consisting of insoluble gases and vapours, hydrocarbon liquid and water is separated into three streams as depicted in Figure 30.6.

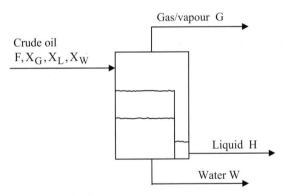

Fig. 30.6 Crude oil separator

The gases and vapours disengage from the liquid in the separator. The liquid separates into immiscible hydrocarbon and water layers. The hydrocarbon layer, being the less dense, is displaced over a weir and the water is withdrawn from the bottom of the separator.

An overall steady state mass balance gives:

$$F = G + H + W \quad (30.2)$$

and mass balances for the various components give:

gas/vapour: $\quad F.X_G = G \quad (30.3)$

hydrocarbon liquid: $\quad F.X_H = H \quad (30.4)$

water: $\quad F.X_W = W \quad (30.5)$

subject to the constraint: $\quad X_G + X_H + X_W = 1.0 \quad (30.6)$

where F, G, H and W denote flow rates (kg/s) and X denotes weight fraction in the crude oil.

In determining control strategy it is essential that the equations used are independent. However, the above five equations that describe the separator are not independent. That is because the sum of the component balances is the overall mass balance. This can be seen by the addition of Equations 30.3 to 30.5 and substitution from Equation 30.6 which yields Equation 30.2. In fact any one of the five equations can be eliminated so choose to eliminate Equation 30.5. This results in four equations, the minimum set necessary to determine the control strategy, with seven strategic variables and three degrees of freedom. This view is represented by the first column of Table 30.1.

Table 30.1 Analysis of degrees of freedom for separator

	Four equations with seven strategic variables	
Variables	First view	Second view
Wild	–	F, X_G, X_H
Controlled	–	–
Floating	$F, G, H, W, X_G, X_H, X_W$	–
Determined	–	G, H, W, X_W

In practice both the flow rate of crude oil and its composition may be considered to be wild. There are thus three wild variables: F and any two of the weight fractions, say X_G and X_H. Note that of the three weight fractions for the crude oil, only two can be deemed to be wild: the third X_W is determined by Equation 30.6. The three wild variables take up all three of the degrees of freedom available. That leaves four equations in four unknowns, G, H, W and X_W which can be solved. Thus G, H, W and X_W are determined. This view is represented by the second column of Table 30.1.

What this means is that none of the strategic variables can be controlled. Any attempt to do so would result in the system being over-determined. The only scope for control therefore is of an inventory nature. A sensible strategy is depicted in Figure 30.7. It is essential that effective control of the interface between the liquid layers be maintained, otherwise water could end up in the hydrocarbon stream or hydrocarbons in the water stream. This is realised by using the height of the interface, the measurement of which is explained in Chapter 14, to manipulate the water offtake. The level in the downcomer is controlled by manipulating the hydrocarbon flow rate: this maintains a seal and prevents gases and vapours escaping with the hydrocarbon liquid. Finally, a pressure controller manipulates the gas stream flow rate: this enables the capacity of the separator to be used to average out fluctuations in pressure to downstream units.

30.4 Absorber Example

A continuous ammonia recovery plant is depicted in Figure 30.8.

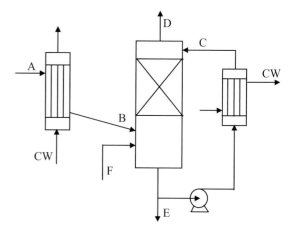

Fig. 30.8 Continuous ammonia recovery plant

Stream A contains a mixture of steam and air that is rich in ammonia, its flow rate and composition being determined by the plant upstream. It is fed into a condenser, in which the steam condenses and absorbs much of the ammonia. Stream B consists of a mixture of air, containing water vapour and undissolved ammonia, and the condensate. It is passed into the base of a scrubber where it disengages into gas and liquid phases.

The gas phase flows up through the packing of the scrubber, counter current to a flow of strong aqueous ammonia, stream C. The gas leaving the scrubber, stream D, is discharged into the atmosphere, its ammonia and water content being negligible. The flow of gas through the system is due to the pressure of stream A.

The liquid phase of stream B mixes with the aqueous ammonia dropping off the packing. Stream C is drawn from the base of the scrubber. It is pumped through a cooler, the absorption process being exothermic, and into the top of the scrubber in an external circulation loop. A fraction of the circulating liquor, stream E, is drawn off as product. Steam F is fresh water makeup.

Assuming the concentrations of NH_3 and H_2O in stream D are negligible, the minimum set of

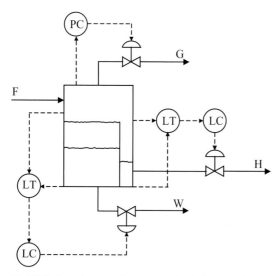

Fig. 30.7 Control strategy for separator that is determined

30.4 Absorber Example

Table 30.2 Analysis of degrees of freedom for recovery plant

	Three equations with seven strategic variables		
Variables	First view	Second view	Third view
Wild	–	A, Y_{NH_3}, Y_{H_2O}	A, Y_{NH_3}, Y_{H_2O}
Controlled	–	–	X_{NH_3}
Floating	$A, E, F, X_{NH_3}, X_{H_2O}, Y_{NH_3}, Y_{H_2O}$	E, F, X_{NH_3}, X_{H_2O}	–
Determined	–	–	E, F, X_{H_2O}

equations and strategic variables that completely define the process is found from mass balances for any two of the components as follows:

$$A.Y_{NH_3} = E.X_{NH_3} \quad (30.6)$$

$$A.Y_{H_2O} + F = E.X_{H_2O} \quad (30.7)$$

$$X_{NH_3} + X_{H_2O} = 1.0 \quad (30.8)$$

where X and Y denote the weight fractions in the liquid and gaseous streams respectively. Therefore, there are three equations with seven strategic variables and four degrees of freedom. The strategic variables may, initially, all be considered to be floating. This view is represented by the first column of Table 30.2.

Two further equations could have been included for completeness:

$$A.Y_{AIR} = D$$

$$Y_{NH_3} + Y_{H_2O} + Y_{AIR} = 1.0$$

Including these would yield five equations with nine strategic variables. There is nothing wrong with including the extra equations, other than the increased scope for confusion, since there are still only four degrees of freedom.

Also, an overall mass balance could have been included:

$$A + F = D + E$$

However, this can be produced by manipulation of the other equations and so is not independent. It is now evident that the original three equations are indeed the minimum set necessary to determine the control strategy.

Next consider the feed stream. Three of the degrees of freedom are taken up by A, Y_{NH_3} and Y_{H_2O} because the feed stream is wild, which leaves four floating variables. This view is represented by the second column of Table 30.2.

Finally, choose to control one of the remaining floating variables. Controlling X_{NH_3} at a high value would be sensible because it would maximise the efficiency of the process and yield a product that is of value. If X_{NH_3} is controlled then all four degrees of freedom are taken up, leaving three equations and three unknowns. Thus E, F and X_{H_2O}, the remaining strategic variables, are determined. This view is represented by the third column of Table 30.2.

A scheme for the control of this recovery plant is depicted in Figure 30.9. Note that there is a density control loop which controls the strategic variable X_{NH_3} directly, the density of the circulating

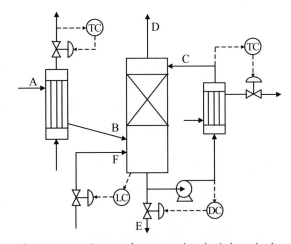

Fig. 30.9 Control strategy for recovery plant that is determined

liquor being used as an indirect measure of the ammonia concentration. A density rising above its set point would cause the product stream flow rate E to increase. The resultant decrease in level in the base of the scrubber would cause the makeup flow rate F to increase thereby diluting the circulating liquor and returning its concentration to the set point. It is evident that the level control loop is manipulating the makeup rate F for inventory purposes. Both of the temperature control loops are for inventory purposes too, the flow rates being manipulated to make effective use of the cooling water.

30.5 Flash Drum Example

This is a more complex example and involves simultaneous mass and heat balances. Consider the flash drum at the base of a stripping column as depicted in Figure 30.10. The flash drum acts as a reservoir at the base of the column. It receives fresh feed at a rate F_1 and column bottoms at a rate F_B. A fraction of the liquor in the drum is vaporised by condensing steam at a rate F_S in an external heat exchanger. The vapours produced are fed to the base of the column at a rate F_V, the balance being taken off as product at a rate F_0.

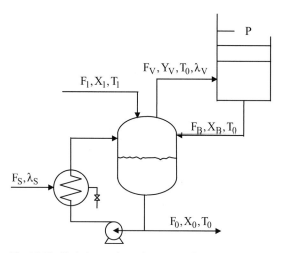

Fig. 30.10 Flash drum at base of stripping column

An overall steady state mass balance across the drum gives:

$$F_1 + F_B = F_V + F_0 \qquad (30.9)$$

where F denotes flow rate (kmol/s).

Assuming that the feed is a miscible binary mixture, a steady state mass balance for the more volatile component gives:

$$F_1.X_1 + F_B.X_B = F_V.Y_V + F_0.X_0 \qquad (30.10)$$

where X and Y denote mol fractions in the liquid and vapour phases respectively.

The following assumptions are made in relation to the heat balance:

- The feed is liquid and below its boiling point.
- The specific heats of all the liquid streams are approximately the same.
- The specific and latent heats are approximately constant.
- The steam supply is dry saturated.

On that basis, a steady state heat balance on the drum contents gives:

$$F_S.\lambda_S = F_1.c_P.(T_0 - T_1) + F_V.\lambda_V \qquad (30.11)$$

where λ denotes the latent heat (kJ/kmol) and c_P the specific heat (kJ/kmol °C). In effect, the heat input goes towards raising the drum feed to its boiling point and thereafter vaporising some of the drum's contents. Strictly speaking, the column's vapour feed stream and bottoms stream are unlikely to both be at the same temperature T_0 but, for the purposes of this analysis, the assumption is good enough.

In addition to the above three mass and heat balance equations, there are two important constraints. The first concerns volatility of the mixture: for any given pressure there is a unique relationship between the mol fractions of the more volatile component in the vapour and liquid phases in the drum:

$$Y_V = f(X_0, P) \qquad (30.12)$$

where P is the pressure (bar) in the drum which is essentially the same as in the column. And second, again for a given pressure, the boiling point of

Table 30.3 Analysis of degrees of freedom for flash drum

Variables	Five equations with 12 strategic variables		
	First view	Second view	Third view
Wild	–	$F_1, F_B, P, T_1, X_1, X_B$	$F_1, F_B, P, T_1, X_1, X_B$
Controlled	–	–	T_0
Floating	$F_0, F_1, F_B, F_V, F_S, P, T_0, T_1, X_0, X_1, X_B, Y_V$	$F_0, F_V, F_S, T_0, X_0, Y_V$	
Determined	–	–	F_0, F_V, F_S, X_0, Y_V

the mixture in the drum is uniquely related to its composition:

$$T_0 = g(X_0, P) \qquad (30.13)$$

There is thus a set of five equations, the minimum necessary to determine the control strategy, involving twelve strategic variables with seven degrees of freedom. This view is represented by the first column of Table 30.3.

The following assumptions are made regarding variables being wild:

- The feed rate to the drum, its composition and temperature are all wild.
- The column bottoms flow rate and composition are both wild.
- The pressure P, being determined by the column's overhead system, is wild.

Thus six of the variables are wild leaving six floating variables as indicated by the second column of Table 30.3. Therefore, of the seven degrees of freedom, six are spoken for by the wild variables leaving one floating variable which can be controlled directly. The obvious one to choose to control, as explained in Chapter 35, is the temperature T_0 of the product stream. That leaves five equations in five unknowns: the remaining strategic variables are determined, the view represented by the third column of Table 30.3. A scheme for the control of the flash drum is depicted in Figure 30.11.

It can be seen that the product stream temperature is controlled by manipulating the steam flow to the reboiler. A temperature T_0 falling below its set point would cause the steam flow F_S to increase. The resultant increase in the rate of vaporisation F_V would lead to the remaining liquor becoming

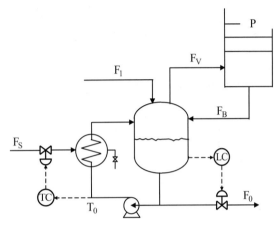

Fig. 30.11 Control strategy for flash drum that is determined

richer in the less volatile component, that is X_0 decreases, and its temperature T_0 is driven back towards the set point. The only other control loop necessary is of an inventory nature: a level loop manipulating the bottom product flow rate F_0 to ensure sufficient head in the drum to maintain circulation through the reboiler.

30.6 Methodology

The above process of determination is simple but effective. With experience, it can be applied to large and complex plant. However, a methodology for doing so is necessary.

First, for the plant as a whole, establish the overall mass and energy balances and identify the strategic variables. Categorise them as wild, controlled, determined and/or floating. There is no need, at this stage, to reconcile them with the degrees of freedom.

Second, carry out the process of determination on a unit (or sub-system) basis. In this context a unit may be considered to be a major equipment item such as a reactor, or an operation such as separation. For each unit:

- Establish the steady state mass and energy balances and identify the strategic variables.
- Identify the wild variables and, if appropriate, specify ratio control and/or feedforward control strategies.
- Identify the controlled variables and allocate feedback and/or cascade control loops as appropriate.
- Identify any inventory controls required and formulate additional loops as appropriate.
- Identify major sources of disturbance, especially with utilities such as cooling water and steam, and target slave loops on them as appropriate.
- Identify strong coupling between control loops and choose input-output combinations to minimise interactions.
- Reconcile the number of strategic variables and degrees of freedom. If there are any floating variables then use up the spare degrees of freedom by specifying more control loops.
- Check that the scheme proposed is not over-determined and, if so, remove control loops as necessary.

Third, consider the plant as a whole again. The plant is made up of units which should fit together. If each unit is properly determined then the plant as a whole should be determined too. In particular, when two units are connected, check that controlled or determined outputs from one unit appear as wild inputs to the other. At an overall level the number of strategic variables and degrees of freedom should now be reconciled. Any spare degrees of freedom identified earlier should have been used up at the unit level: there should be no floating variables left.

There are two categories of commonly made errors in applying the process of determination. One is errors arising from ill-defined mass and energy balances, such as:

- Equations used are not independent: check to see if any can be eliminated.
- Equations do not adequately represent the process: confirm that any assumptions and/or approximations made are valid.
- Some constraint equation has been missed out.
- Some strategic variable may have been overlooked.
- A variable used may be insignificant or approximately constant.

The other category is over-determination, typically:

- Single pipelines having two control valves in them.
- Pipelines connected up to a branch or junction all having valves in them.
- Trying to control variables that are indirectly controlled by other means.
- Trying to control wild variables.

30.7 Comments

This approach to determining strategy, and the methodology for implementing it, is fundamental to the successful design of process control systems. To do so quickly and effectively requires a significant amount of experience and expertise. It also essential that the process of determining strategy is carried out alongside the conceptual design of the plant. With modern flexible and responsive processing plant, the control systems are an integral part of the plant design. They cannot be bolted on afterwards, as if they are decorations on a Christmas tree.

Evaporator Control

Chapter 31

31.1 Operational Issues
31.2 The Vacuum System
31.3 Reboiler and Separator
31.4 Condenser with Barometric Leg
31.5 Control Scheme
31.6 Comments
31.7 Nomenclature

This chapter concerns the control of a thermosyphon type of evaporator. The issues are much the same for calandria and other single effect types of evaporator.

Figure 31.1 depicts the evaporator in which an aqueous stream containing suspended particulates (or dissolved salts) is concentrated by evaporating off water. The dilute feed stream mixes with concentrated liquor circulating at its boiling point. This is partially evaporated on the tube-side of the reboiler by condensing steam on its shell-side. The tube-side exit stream consists of water vapour with entrained droplets of liquor. The separator provides the space for the droplets to disengage from the vapour. The droplets fall into the liquor which is recirculated, a fraction being drawn off as concentrated product. The reboiler is operated under vacuum provided by a pump which sucks the vapour out of the separator. The vapour condenses on the shell side of a water cooled condenser, the condensate being discharged *via* a barometric leg and pumped away by a centrifugal pump.

31.1 Operational Issues

The separator, reboiler and interconnecting pipework form a manometer. Between the separator and the inlet of the reboiler, the pipework is full of liquor. However, the tubes of the reboiler and its outlet pipework contain a mixture of vapour and liquor. This density difference provides the driving force for the circulation of the liquor, known as a thermosyphon. It is essential that the head of liquor in the separator be maintained constant because, if circulation were to cease, the reboiler tubes could become blocked with aggregated particulates (or crystallised salts). However, the head must remain below the level of the separator inlet as otherwise the liquor would impede the vapour flow.

The reboiler is operated under vacuum to reduce the boiling point of the liquor. This increases the temperature difference between the steam and liquor which is the driving force for heat transfer. It is essential that a steady vacuum is maintained. Any sudden increase in pressure, and hence in boiling point, would cause the liquor to go off the boil and circulation would cease. Any sudden decrease in pressure would cause the liquor to flash off. The surge in vapour flow may be beyond the capacity of the condenser, resulting in carry over of condensate into the vacuum pump. Depending on the type of pump, this could cause damage.

The condensate formed in the condenser cannot be discharged using a steam trap because the condenser is operated under vacuum. A barometric leg is used, the principle of operation of which is depicted in Figure 31.2. A head of condensate is

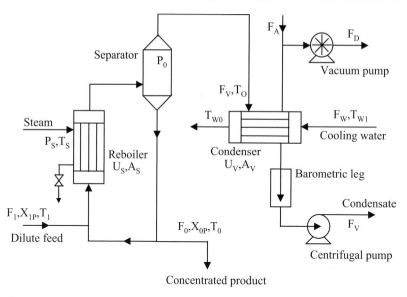

Fig. 31.1 Thermosyphon type of evaporator

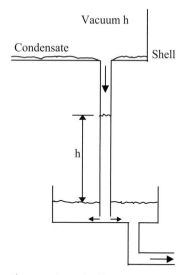

Fig. 31.2 Principle of barometric leg

established within the leg which balances the vacuum in the condenser. Any condensate entering the top of the leg displaces an equivalent amount into the reservoir at the bottom. It is essential that the level in the reservoir be maintained above the bottom of the leg, otherwise the vacuum seal would be broken and air sucked up the leg into the condenser.

For the purpose of determining control strategy, the evaporator is considered to consist of three sub systems: the vacuum system, the reboiler and separator, and the condenser with the barometric leg.

31.2 The Vacuum System

The pressure in the evaporator system is fundamental to its operation so consideration of the vacuum system is a convenient starting point for applying the process of determination. The vacuum system consists of the vacuum pump together with the vapour space upon which the vacuum pump operates. That space includes the shell side of the condenser, the disengagement space within the separator and the overhead pipework. In essence, that space is the system's capacity for vapour holdup.

It is evident that the pressure P_0 in the system depends upon four factors: the rate at which vapour is produced in the reboiler F_V, the rate at

Table 31.1 Analysis of degrees of freedom for vacuum system

	One equation and five strategic variables			
Variables	First view	Second view	Third view	Fourth view
Wild	–	F_V, Q_V	F_V, Q_V	F_V, Q_V
Controlled	–	–	–	P_0
Floating	F_A, F_D, F_V, P_0, Q_V	F_A, F_D, P_0	F_A, P_0	–
Determined	–	–	F_D	F_A, F_D

which vapour is condensed which is determined by the rate at which heat is removed in the condenser Q_V, the rate at which air leaks (or is bled) into the system F_A and the rate at which air/vapour is discharged by the vacuum pump F_D.

Thus, at equilibrium:

$$P_0 = f(F_V, Q_V, F_A, F_D) \quad (31.1)$$

There is thus one equation involving five strategic variables with four degrees of freedom. This view is represented by the first column of Table 31.1.

The rates at which the vapour is produced and condensed are determined by the design and operation of the reboiler and condenser so, as far as the vacuum system is concerned, these rates must be considered to be wild variables. Thus two of the degrees of freedom have been taken up leaving three floating variables as depicted in the second column of Table 31.1.

It is common practice in the design of evaporator systems to specify a vacuum pump that operates at a constant speed: the efficiency of a vacuum pump is relatively high only over a narrow range of operating conditions. Thus the discharge rate is determined by plant design considerations, as depicted in the third column of Table 31.1.

Given the constant speed vacuum pump, the pressure in an evaporator system is typically controlled by manipulating the air bleed rate. If the pressure is controlled then the remaining degree of freedom is spoken for. There is thus one equation with one unknown and the air flow rate is determined as depicted by the fourth column of Table 31.1.

31.3 Reboiler and Separator

An overall steady state mass balance for the contents of the thermosyphon gives:

$$F_1 = F_0 + F_V \quad (31.2)$$

and a steady state mass balance for the solids content gives:

$$F_1 . X_{1P} = F_0 . X_{0P} \quad (31.3)$$

where X denotes weight fraction and subscripts 1, 0 and V refer to the inlet, outlet and vapour streams respectively. As pointed out in Chapter 30, there is no need to use every possible equation. By definition:

$$X_{1P} + X_{1W} = 1$$
$$X_{0P} + X_{0W} = 1$$

where subscripts P and W refer to particulates and water content respectively. These two extra constraints introduce another two variables, X_{1W} and X_{0W}, but have no net effect on the number of degrees of freedom. There is no obvious benefit from including them and, since they potentially confuse the issue, are best discarded. Likewise a mass balance for the water content yields:

$$F_1 . X_{1W} = F_0 . X_{0W} + F_V$$

However, this equation is not independent: it can be formed by subtraction of Equation 31.3 from Equation 31.2 and substitution of the constraints, so it too is discarded.

The heat input through the reboiler goes partly into raising the temperature of the feed stream to that of the circulating liquor but mostly into partially evaporating the liquor. A steady state heat

Table 31.2 Analysis of degrees of freedom for reboiler and separator

Variables	Six equations and 11 strategic variables First view	Second view	Third view
Wild	–	X_{1P}, T_1, P_0	X_{1P}, T_1, P_0
Controlled	–	–	P_S, X_{0P}
Floating	$F_0, F_1, F_V, Q_S, T_0, T_1, T_S, P_0, P_S, X_{0P}, X_{1P}$	$F_0, F_1, F_V, Q_S, T_0, T_S, P_S, X_{0P}$	–
Determined	–	–	$F_0, F_1, F_V, Q_S, T_0, T_S$

balance for the contents on the tube side of the reboiler thus gives:

$$Q_S = F_1.c_P.(T_0 - T_1) + F_V.\lambda_V \qquad (31.4)$$

where Q_S is the heat duty of the reboiler, T denotes temperature and c_P and λ_V denote specific and latent heats respectively. However the rate of heat transfer is itself determined by the temperature difference between the shell and tube side of the reboiler:

$$Q_S = U_S A_S.(T_S - T_0) \qquad (31.5)$$

Because the circulation through the reboiler is roughly constant, its overall heat transfer coefficient U_S may be presumed to be constant too. Note that the steam on both sides of the reboiler is saturated, so both temperatures are uniquely related to their pressures. Thus, assuming that any elevation of the boiling point on the tube side is negligible, two further equations must be introduced:

$$T_0 = g(P_0) \qquad (31.6)$$

$$T_S = g(P_S) \qquad (31.7)$$

Equations 31.2–30.7 are the minimum set of six equations necessary for determining the control strategy for the reboiler and separator sub-system. They involve eleven strategic variables giving five degrees of freedom. This view is represented by the first column of Table 31.2.

Assume that the composition X_{1P} and temperature T_1 of the feed stream are both wild. The pressure P_0 is controlled by the vacuum system so, as far as the reboiler and separator are concerned, it is deemed to be wild. Thus three of the degrees of freedom have been taken up leaving eight floating variables: the view depicted by the second column of Table 31.2.

There are two remaining degrees of freedom which can be used for controlling any two of the floating variables although, in this case, it is fairly obvious which two to choose. Noting that P_0 is controlled by the vacuum system, assigning one degree of freedom to the shell side pressure P_S enables the temperature difference across the reboiler and hence the rate of evaporation to be controlled. It is sensible to assign the other degree of freedom to the control of product quality X_{P0}. This results in six equations in six unknowns: thus all the remaining strategic variables are determined. This view is represented by the third column of Table 31.2.

31.4 Condenser with Barometric Leg

The latent heat of condensation is removed by the cooling water. A steady state heat balance yields the duty of the condenser:

$$Q_V = F_W.c_P.(T_{W0} - T_{W1}) \qquad (31.8)$$

Noting that the temperature difference between the shell and tube sides is not uniform, a steady state heat balance across the tubes of the condenser yields:

$$Q_V = U_V.A_V.\frac{(T_{W0} - T_{W1})}{\ln\frac{(T_0 - T_{W1})}{(T_0 - T_{W0})}} \qquad (31.9)$$

Table 31.3 Analysis of degrees of freedom for condenser and barometric leg

Variables	Three equations and six strategic variables		
	First view	Second view	Third view
Wild	–	T_0, T_{W1}	T_0, T_{W1}
Controlled	–		T_{W0}
Floating	$F_W, U_V, Q_V, T_0, T_{W0}, T_{W1}$	F_W, U_V, Q_V, T_{W0}	
Determined	–	–	F_W, U_V, Q_V

The overall heat transfer coefficient cannot be presumed to be constant since the cooling water flow rate is a variable. Thus:

$$U_V = h(F_W) \tag{31.10}$$

Note that this particular relationship is explained in more detail in Chapter 32.

Note also that there are another two equations, for the condensation of steam, that could be considered as follows:

$$T_0 = g(P_0) \tag{31.6}$$
$$Q_V = F_V . \lambda_V$$

However, these simply introduce another two strategic variables: they do not affect the no of degrees of freedom and can be discarded.

Equations 31.8–31.10 are the minimum set of three equations necessary for determining the control strategy for the condenser sub-system. They involve six strategic variables giving three degrees of freedom. This view is represented by the first column of Table 31.3.

The cooling water inlet temperature T_{W1} is wild. As far as the condenser is concerned, the operating temperature T_0 must also be considered wild since it is determined by the pressure P_0 which is controlled independently. These take up two of the degrees of freedom as depicted in the second column of Table 31.3.

To make effective use of the cooling water it is common practice to control the condenser's outlet temperature T_{W0}. That uses up the remaining degree of freedom, so the other three variables are determined. This view is represented by the third column of Table 31.3.

31.5 Control Scheme

A scheme for the control of the evaporator system is depicted in Figure 31.3.

The evaporator is operated under vacuum. Its pressure P_0 is controlled by the controller PAC which manipulates the flowrate of an air bleed into the inlet of the vacuum pump. It doesn't particularly matter whether the pressure is measured in the separator, overhead pipework or condenser shell because the vacuum system has a relatively open structure. Manipulating a bleed valve is preferable to varying the pump speed because vacuum pumps are designed to run over a narrow range of speeds and their efficiency decreases rapidly outside this range.

The concentration of the circulating liquor X_{0P} is controlled by the controller CC which manipulates the product flow rate F_0. The set point for this control loop is the concentration specified for the product. The solids content would almost certainly be measured indirectly, perhaps optically or by means of density depending on the nature of the particulates (or conductivity for dissolved salts). It is important that the measurement is made in the circulating liquor. It should not be made in the product line because, in the event of the valve being shut by the controller, its content would be stagnant and independent of the concentration in the evaporator.

The level in the separator is controlled by the controller LC which manipulates the dilute feed flowrate F_1. It can be seen that these two control loops are highly interactive. If the concentration X_{0P} rises above its set point, the concentration controller would increase the product flow F_0. This

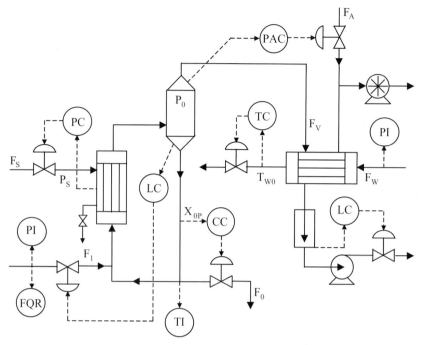

Fig. 31.3 Control strategy for evaporator that is determined

would cause the level in the separator to fall and force the level controller to increase the feed flow F_1. This would dilute the circulating liquor resulting in X_{0P} returning to its setpoint.

Note that both the inlet and outlet flows are being manipulated, but not controlled, which is consistent with the flows F_0 and F_1 being determined by the mass balances. F_0 is being manipulated to control the strategic variable X_{0P} and F_1 is being manipulated to control an inventory variable, the level in the separator.

Note also that these two loops can be interchanged, *i.e.* CC to manipulate F_1 and LC to manipulate F_0. It is often the case in process control that loops can be interchanged like this. The question arises as to what are the most appropriate couplings of controlled and manipulated variables. There is a technique, known as relative gain analysis, which enables the best arrangement to be identified. This is discussed in Chapter 111.

The steam pressure P_S on the shell side of the reboiler is controlled by the controller PC which manipulates the flow of steam to the shell. The combination of the two pressure controllers ensures that the temperature difference across the tubes of the reboiler is regulated. The rate of circulation, and hence the overall heat transfer coefficient, is roughly constant because the level controller LC controls the head of liquor in the separator. Given that the reboiler's temperature difference is controlled and that the overall coefficient is constant, the rate of heat transfer Q_S, and hence the rate of evaporation F_V, is determined.

The controller TC controls the condenser outlet temperature T_{W0} by manipulating the cooling water flow rate F_W. If the temperature increases then the flow rate is increased, and *vice versa*. The control valve would require a bias on its opening to guarantee a minimum flow for start-up purposes. This is an inventory control and ensures that effective use is made of the cooling water. The alternative, which is to have a flow control loop on the cooling water flowrate, would need a set point high enough to cope with all eventualities and would be wasteful of cooling water.

Condensate is discharged into the barometric leg's reservoir, the controller LC controlling the head in the leg by manipulating the product flowrate. If the level rises the rate at which the product is pumped away is increased. This is an inventory control which maintains the vacuum seal on the leg. It is important in situations such as this that the pump is of the centrifugal type. In the event of the valve being shut by the controller, a constant head is developed and the pump is not damaged. The impeller simply rotates the trapped liquor, which warms up due to the energy being dissipated. Closing the exit valve on the alternative positive displacement types of pump would result in a sudden and massive increase in pressure leading to damage to the pump and/or rupture of the pipework.

The totaliser FQR provides a record of the amount of dilute feed processed and is included for management purposes. The remaining pressure and temperature indicators are included for commissioning and diagnostic purposes.

31.6 Comments

The sub-systems of this evaporator plant are typical of the scope of the unit referred to in the methodology of Chapter 30. A careful review of the methodology will reveal that each of the issues raised has been explicitly addressed in the control scheme proposed. In particular:

- The only strategic variables that are controlled are those four which were deemed to be controlled by the process of determination: P_0, P_S, X_{OP} and T_{W0}.
- For the purpose of controlling these four strategic variables, another three strategic variables and one utility are manipulated: F_0, F_1, F_A and F_S.
- For inventory purposes, level loops manipulate a further two strategic variables: F_V and F_W.
- None of the strategic variables deemed to be determined are controlled (as opposed to manipulated).
- All strategic variables controlled in, or determined by, one unit are deemed to be wild variables in other units.
- No attempt is made to control any of the wild variables: X_{1P}, T_1 and T_{W1}.

31.7 Nomenclature

A	is the	surface area	m^2
c_P		specific heat	kJ kg^{-1} K^{-1}
F		flow rate	kg s^{-1}
P		pressure	bar
Q		rate of heat transfer	kW
T		temperature	K
U		overall heat transfer coefficient	kW m^{-2} K^{-1}
X		weight fraction	–
λ		latent heat	kJ kg^{-1}

Subscripts

0	outlet
1	inlet
P	particulates
S	steam/reboiler
V	vapour/condenser
W	cooling water

Control of Heat Exchangers

Chapter 32

32.1 Heat Transfer Considerations
32.2 Temperature Difference
32.3 Overall Coefficient
32.4 By-Pass Control
32.5 Surface Area
32.6 Comments
32.7 Nomenclature

Heat transfer within a heat exchanger is arguably the most common unit operation carried out in the chemical and process industries. From a process perspective it results in the heating or cooling of a fluid and may involve evaporation or condensation. In terms of equipment, exchangers may be of the plate type or of the shell and tube type. For operational purposes, there are many alternative designs and configurations such as with reboilers and thermosyphons. For control purposes, all this variety is accommodated by a relatively small number of generically similar control schemes. This chapter focuses on those schemes.

The control of heat exchangers in a variety of applications is covered, for example, in Chapters 31 and 35. The reader is referred to the text by Shinsky (1996) for an authoritative and more comprehensive treatment of schemes and strategies for the control of exchangers. A model of the dynamics of a shell and tube heat exchanger is developed in Chapter 92.

32.1 Heat Transfer Considerations

A single pass shell and tube exchanger in which a liquid process stream on the tube side is heated up by condensing steam on the shell side, as depicted in Figure 32.1, is used as a vehicle for comparison of the various control schemes.

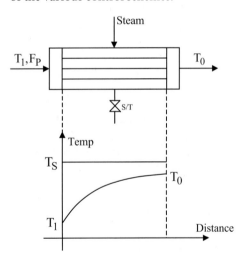

Fig. 32.1 Steam heated shell and tube exchanger

Also shown are the temperature profiles. Note that there is a uniform, but not necessarily constant, temperature throughout the shell. That is because the shell, being a fairly open structure, must have a uniform pressure throughout and, for condensing steam, the pressure and temperature are directly related. On the tube side, the process stream's temperature rises exponentially towards the steam temperature. This, of course, assumes that there is no phase change on the tube side, *i.e.* that the process stream does not reach its boiling point.

As stated in Chapter 27, the steady state heat balance is given by:

$$Q = U.A.T_M = F_P.\rho_P.c_P.(T_0 - T_1) \quad (32.1)$$

where the log mean temperature difference is given by:

$$T_M = \frac{T_0 - T_1}{\ln\left(\dfrac{T_S - T_1}{T_S - T_0}\right)} \quad (32.2)$$

The objective is to control the tube side stream outlet temperature. In essence, the rate of heat transfer Q depends on the overall coefficient, the surface area and the temperature difference, so varying any one of U, A and T_M enables Q to be manipulated and hence T_0 to be controlled. Each of the schemes described in this chapter involves varying one or other of these parameters.

Strictly speaking, U, A and T_M cannot be varied independently. For example, increasing F_P results in an increase in U which causes Q to increase which, for given values of T_S and T_1, causes T_0 to increase and hence T_M to decrease. Some of these interactions are negligible. However, in the case of U as a function of F_P, the interaction is substantial as described in Section 3.

32.2 Temperature Difference

Figure 32.2 depicts a control scheme in which T_M is varied.

Suppose that a decrease in outlet temperature T_0 occurs, perhaps because of changes in the process stream flow rate F_P and/or inlet temperature T_1. The controller action is such that the control valve opens. The increase in steam flow F_S causes the steam pressure P_S and hence its temperature T_S to rise. The resultant increase in T_M causes an increase in the rate of heat transfer Q which restores the outlet temperature to its original value.

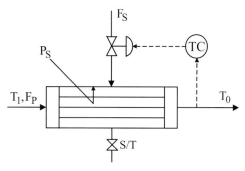

Fig. 32.2 Control scheme in which log mean temperature difference is varied

This is the most common scheme for controlling an exchanger used for heating purposes. It has a relatively high process gain, *i.e.* change in T_0 per unit change in P_S or T_S, and a fast response, both of which make for a good quality of control. The dynamics of this scheme are considered in detail in Chapter 92.

If the process stream was heated by cooling some other liquid stream (as opposed to steam) on the shell side, heat transfer would be less effective and the dynamics more sluggish.

32.3 Overall Coefficient

Figure 32.3 depicts a control scheme in which U is varied.

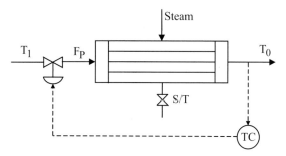

Fig. 32.3 Control scheme in which overall coefficient is varied

Following an increase in T_0 the controller action is such that it opens the control valve. The increase in F_P has two opposite effects. First, for a given rate of heat transfer the increase in sensible heat is fixed, so an increase in flow leads to a decrease in T_0. However, second, the increase in F_P leads to an increase in U and hence to a higher rate of heat transfer Q which tends to increase T_0. The first effect is the more dominant, the net effect being a decrease in T_0.

Increasing F_P causes more turbulence which reduces the tube side film thickness and hence its resistance to heat transfer. The film coefficient for heat transfer is related to the process stream velocity by the dimensionless Dittus Boelter equation:

$$Nu = 0.023 \, Re^{0.8} . Pr^{0.4}$$

$$\left(\frac{h_P d_w}{k_P}\right) = 0.023 \left(\frac{\rho_P . d_w . u}{\mu}\right)^{0.8} \left(\frac{c_P . \mu}{k_P}\right)^{0.4}$$

Assuming all the physical properties of the process stream to be approximately constant over the range concerned, this may be simplified to:

$$h_P = b.F_P^{0.8}$$

The resistances to heat transfer due to the tube and shell side films and to the wall itself are in series. For a thin walled tube:

$$\frac{1}{U} = \frac{1}{h_P} + \frac{x_w}{k_w} + \frac{1}{h_S}$$

Assume that the shell side film coefficient is large, which it will be for condensing steam, and allowing for the resistance to heat transfer due to the tube wall:

$$\frac{1}{U} = \frac{1}{h_P} + c$$

Substituting and rearranging gives:

$$U = \frac{bF_P^{0.8}}{1 + bcF_P^{0.8}} \quad (32.3)$$

Substituting Equations 32.2 and 32.3 into Equation 32.1 gives

$$\frac{bF_P^{0.8}}{(1 + bcF_P^{0.8})} A \frac{1}{\ln\left(\frac{T_S - T_1}{T_S - T_0}\right)} = F_P . \rho_P . c_P$$

Further rearrangement yields

$$T_0 = T_S - (T_S - T_1) . \exp\left(\frac{-bA}{\rho_P c_P F_P^{0.2} (1 + bcF_P^{0.8})}\right) \quad (32.4)$$

Inspection of Equation 32.4 reveals that an increase in F_P does indeed lead to a net decrease in T_0.

This scheme has various drawbacks. It has a relatively low process gain, *i.e.* a small change in T_0 per unit change in F_P, and does not have a particularly fast response. Also, its response is non-linear. All of these make for a poor quality of control.

Note also the difficulty in start-up of the control loop. Suppose that the valve is initially closed: there is no flow and the outlet temperature T_0 is cold. With the controller in automatic mode, its action is such that it tries to close the valve further! The outlet temperature therefore needs to be brought into the vicinity of the set point under manual control before switching into automatic mode. To avoid such problems in operation there is a minimum flow constraint.

Nevertheless, this scheme is commonly used in condensing applications, especially when the shell side pressure cannot be manipulated, to minimise the consumption of cooling water on the tube side.

32.4 By-Pass Control

Figure 32.4 depicts a control scheme in which U and T_M are varied.

This is essentially a blending system in which hot and cold streams are mixed. The controller manipulates the proportion of process stream flowing

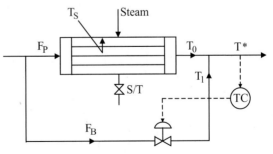

Fig. 32.4 By pass control scheme

Table 32.1 Analysis of degrees of freedom for by pass system

Variables	Three equations with five strategic variables		
	First view	Second view	Third view
Wild	–	T_1	T_1
Controlled	–	–	T^*
Floating	F_P, F_B, T_1, T_M, T^*	F_P, F_B, T_M, T^*	–
Determined	–	–	F_P, F_B, T_M

through the exchanger F_P and that which by-passes it F_B. Following an increase in the controlled variable T^*, the controller action is such that it opens the by-pass control valve. This reduces the flow F_P which results in a decrease in U and an increase in T_0. The combination of the reduction in F_P and the increases in T_0 and F_B at the mixing junction restores T^* to its initial value.

Such a by-pass control scheme has a high process gain, *i.e.* change in T^* per unit change in F_B, and is fairly linear. It also has a very fast response because its dynamics are a function of the blending process rather than of the exchanger itself. Indeed, the dynamics are largely determined by those of the temperature sensor and the control valve. This scheme is particularly effective when the exchanger is over sized, in which case the outlet temperature T_0 is almost independent of F_P provided there is no phase change on the tube side.

It is interesting to apply the process of determination to this control scheme. The minimum set of steady state heat and mass balance equations that is sufficient to fully describe the process is as follows:

$$U.A.T_M = F_P.\rho_P.c_P.(T_0 - T_1) \qquad (32.1)$$

$$T_M = \frac{T_0 - T_1}{\ln\left(\frac{T_S - T_1}{T_S - T_0}\right)} \qquad (32.2)$$

$$F_P.T_0 + F_B.T_1 = (F_P + F_B).T^* \qquad (32.5)$$

Suppose that T_S is approximately constant, maybe because it is controlled by some slave loop. Assuming that the exchanger is oversized it may be presumed that both U and T_0 are effectively constant. Also assume that the parameters ρ_P, c_P and A are constant. Thus there are three equations, five strategic variables and two degrees of freedom. This view is represented by the first column of Table 32.1.

Suppose that T_1 is wild. One of the degrees of freedom has been taken up leaving four floating variables. This view is represented by the second column of Table 32.1.

There is one remaining degree of freedom. Suppose that T^* is specified and realised by means of a control loop. This results in three equations in three unknowns. Hence F_P, F_B and T_M are all determined. This view is represented by the third column of Table 32.1.

32.5 Surface Area

Figure 32.5 depicts a control scheme in which A is varied.

Following an increase in T_0 the controller action is such that it closes the control valve. This causes condensate to accumulate in the shell, the rising level submerging the lower section of the tubes. In effect, the area exposed to the steam and available for heat transfer is reduced. This leads to a lower rate of heat transfer which restores the outlet temperature to its original value.

This scheme is not very common. It has a fairly high process gain, *i.e.* change in T_0 per unit change in F_S, which makes for good steady state performance. However, its dynamic response is poor because the level can only rise slowly, as determined

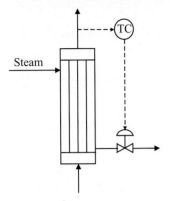

Fig. 32.5 Control scheme in which surface area is varied

by the rate of condensation. Preferred scenarios are vapours with a low latent heat (*i.e.* organics) and hence a relatively volumes of condensate per unit rate of heat transfer, and exchangers with a high ratio of surface area to shell volume. A further complication is non-symmetric dynamics. Whereas the time taken for the level to rise may be long, it can fall quickly if there is a large pressure drop across the valve. Careful sizing of the exchanger, valves and pipework is critical.

32.6 Comments

The exchanger schemes described are of a simple feedback nature. Whilst there is scope for applying more sophistication in the form of cascade control, ratio control and feedforward control, as in the example in Chapter 27, the underlying principles are unchanged. Also, the only scenario considered is that of condensing steam to heat up a process stream. Again, the principles developed are of a general nature and applicable in all situations.

32.7 Nomenclature

A	mean surface area of tubes	m^2
c_P	specific heat	kJ kg^{-1} K^{-1}
d	tube diameter	m
F	flow rate	m^3 s^{-1}
h	tube side film coefficient	kW m^{-2} K^{-1}
k	thermal conductivity	kW m^{-1} K^{-1}
Q	rate of heat transfer	kW
T	temperature	°C
u	velocity	m s^{-1}
U	overall heat transfer coefficient	kW m^{-2} K^{-1}
x	wall thickness	m
ρ	density	kg m^{-3}
μ	viscosity	kg m^{-1} s^{-1}

Subscripts

B	by-pass
M	logarithmic mean
P	process stream
S	steam
W	wall
1	inlet
0	outlet

Boiler Control

33.1 Basis of Control
33.2 Signal Selection
33.3 Ratio Trimming
33.4 Comments

The objective of most boiler plant used in the process industries is to raise steam for heating purposes, electricity production usually being of a secondary consideration. It is normal practice to produce and distribute the steam throughout the works at specific pressures, reducing it locally as required. Mains pressures are typically designated as being high (> 20 bar), intermediate (approximately 10 bar) and low pressure (< 5 bar gauge).

Most boilers used for raising steam are of the water tube type. In essence, they consist of many tubes, usually hundreds but maybe thousands, through which boiling water is circulated, as depicted in Figure 25.1. The boiler tubes, referred to as risers, are exposed to high temperatures in the combustion chamber where heat transfer takes place by a combination of radiation in the flame zone and convection in the exhaust gases. Most boilers have a steam drum. This acts as a header for the water entering the risers and as a disengagement space for the steam/water mixture leaving them. Fresh water is fed into the drum to compensate for the steam being withdrawn. It is not uncommon for the boiler feed water to be preheated in an economiser, by heat exchange with the flue gases, and for the steam produced to be further heated in a superheater.

There is much variety in both design and capacity of boilers. The nature of the fuel, whether it be gas, oil or pulverised coal, is fundamental to burner design and to the provision for fuel injection and air flow. Capacity ranges from a few hundred kW up to a thousand MW. From a control point of view the issues are much the same irrespective of feed type or capacity, although the complexity of the schemes used tends to increase with size. This is because with large boiler plant there is more scope for making significant savings through marginal increases in efficiency.

Coping with variable steam demand and the consequences thereof is the dominant design consideration. Demand can be subject to frequent, sudden and large changes as major steam users come on and off stream, although the capacity of the steam mains goes some way to averaging out the fluctuations. The steam flow is nevertheless wild. Boiler control essentially reduces to five main activities:

- Controlling steam supply pressure by manipulating the fuel flow rate
- Maintaining the correct fuel to air ratio
- Maintaining the water level in the boiler's drum
- Satisfying various environmental and safety criteria
- Maximising efficiency of combustion

Schemes for the control of a boiler taking these various factors into account are developed in this chapter. For a more comprehensive treatment the reader is referred to the text by Dukelow (1991).

33.1 Basis of Control

The most common strategy is to supply steam to the main at a constant pressure. A feedback loop may be used to manipulate the gas/oil flow, as shown in Figure 33.1. Thus, if the steam pressure drops the fuel flow is increased, and *vice versa*.

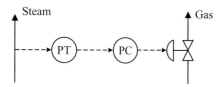

Fig. 33.1 Simple feedback control of steam pressure

To reject disturbances in the gas/oil supply system it is normal to use a slave loop in a cascade strategy, as depicted in Figure 33.2.

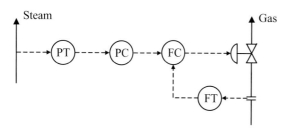

Fig. 33.2 Cascade control with slave loop for gas flow

Clearly, for combustion purposes, it is necessary to maintain the air flow in proportion to the fuel flow. This may be done directly using a ratio controller as shown in Figure 33.3.

Alternatively, it may be done indirectly using a ratio station and a flow controller, as depicted in Figure 33.4.

Note that disturbances in the air supply system are rejected in both direct and indirect approaches.

It is essential that the water level be maintained in the drum to ensure circulation through the boiler tubes and hence their physical protection. This is normally achieved by means of a cascade strategy, with the slave loop manipulating the boiler feed rate, and is discussed fully in Chapter 25.

A further complication with drum level control is the "swell" effect. Following a sudden decrease in steam pressure, the water in the drum flashes off and the bubbles formed temporarily increase its volume. This leads to an apparent increase in drum level even though the mass of water in the drum is decreasing. This is best countered by measuring drum level by means of pressure difference, as depicted in Figure 25.1, and by ensuring that the drum is large enough to accommodate any swell effects.

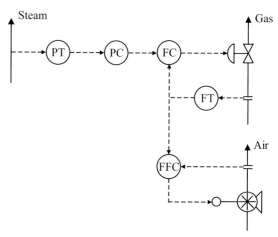

Fig. 33.3 Cascade control with ratio control between gas and air slave loops

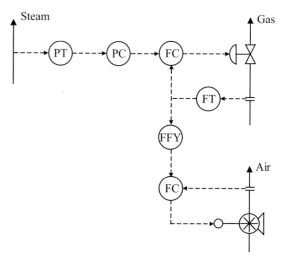

Fig. 33.4 Cascade control with indirect ratio control between slave loops

33.2 Signal Selection

There are important health and safety constraints in the operation of boiler plant. Incomplete combustion of fuel oil and pulverised coal leads to the emission of smoke, which essentially consists of particulates, and deposition of soot. The particulates are harmful to health and the soot deposits are a potential fire hazard. Also, incomplete combustion of gas, oil or coal produces carbon monoxide (CO) which is toxic, as opposed to carbon dioxide (CO_2). The CO can cause explosions when the hot flue gases come into contact with fresh air in the stack.

These effects are best countered by ensuring that there is always an excess of air present in relation to the stoichiometric requirements for combustion. This is particularly important when the fuel flow is being manipulated to meet changes in steam demand. An elegant approach to this problem is depicted in Figure 33.5.

Assume the system is at steady state and consider the effect of an increase in steam demand. The resultant decrease in pressure in the steam main would cause the output of PC47, which is the desired value of the gas flow, to increase. The low signal selector routes the lower of its two inputs to the set point of FC47A, the gas flow controller. Since the output of ratio station FFY47A is unchanged, it is this signal that would be routed through to FC47A, the gas flow remaining constant.

However, the high signal select routes the increase in desired gas flow through to ratio station FFY47B which multiplies it by a scaling factor R, the required air to gas ratio, to determine the corresponding desired air flow. This would then be applied to the set point of FC47B, the air flow controller. As the air flow increases, the ratio station FFY47A multiplies it by the reciprocal of R to determine the allowable gas flow. Since this signal would be lower than the output of PC47 it is routed through to the set point of FC47A which increases the gas flow. A new equilibrium would be established at which, provided the various elements have been calibrated correctly, the outputs of PC47 and FFY47A would be equal.

This scheme guarantees that, following an increase in steam demand, the increase in air flow leads the increase in fuel. A similar analysis reveals that, following a decrease in steam demand, the decrease in air flow lags behind the decrease in gas flow.

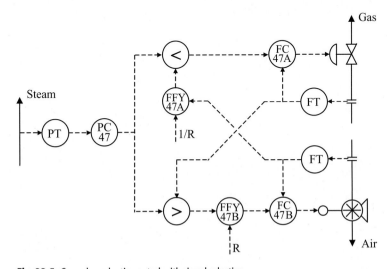

Fig. 33.5 Cascade and ratio control with signal selection

33.3 Ratio Trimming

The efficiency of boilers and their environmental impact are closely related. High flame temperatures are essential for efficient combustion. As stated, it is also necessary to have an excess of oxygen to prevent incomplete combustion. Given that air is 79% nitrogen, which is inert, it can be seen that only small amounts of excess oxygen will lead to significant dilution of the combustion process. This dilution reduces the flame temperature, leading to cooler combustion gases and less effective heat transfer to the water in the tubes. The excess must therefore be as small as is practicable.

However, low flame temperatures are desirable from an environmental perspective. The lower the temperature the less the scope for production of the NO_X gases (NO, N_2O_2 and NO_2). Also, most organic fuels contain sulphurous compounds, not necessarily in trace quantities. Again, low flame temperatures tend to mitigate against production of the SO_X gases (SO_2 and SO_3). All three forms of gaseous oxides, CO_X, NO_X and SO_X, contribute to global warming (greenhouse) effects and acid rain.

There is, therefore, a trade off between high efficiency and low oxide emissions. In practice, boiler plant is normally operated with some 1.0–1.5% excess air in relation to the stoichiometric requirements for complete combustion. The amount of excess air is normally monitored by measuring the amount of oxygen in the flue gas using a zirconium oxide based sensor. The output from the analyser is used to trim the air flow by adjusting the scaling factors R and 1/R applied by the ratio stations to the set points of the flow controllers. There are various ways of realising this.

- R may be adjusted by a ratio station in direct proportion to the O_2 concentration.
- R is the output of a PID controller AC47 whose set point is the desired oxygen concentration in the flue gas, as depicted in Figure 33.6.
- As above, except that the reference ratio R_r is externally preset and multiplied by a scaling factor which is the controller output as depicted in Figure 33.7. This output would normally be about unity and limited to a narrow band of, say, 0.98 to 1.02.

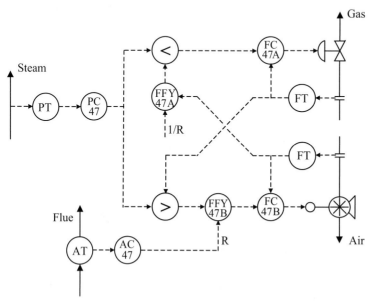

Fig. 33.6 Cascade control with signal selection and ratio trimming

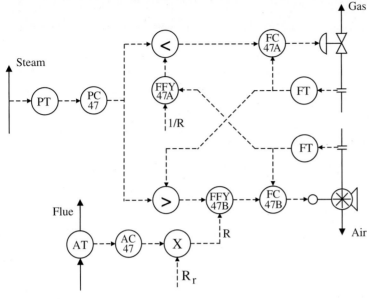

Fig. 33.7 Cascade control with alternative approach to ratio trimming

- As above, except that the reference ratio R_r is incremented by addition of the controller output. This output would normally be about zero and limited to a narrow band according to the calibration of the instrumentation.

An important bonus of the technique of using the oxygen content of the flue gas to trim the ratio setting is that it will respond to changes in the composition of the fuel. Suppose the composition changes such that the stoichiometric requirement increases, *i.e.* more oxygen is required per unit amount of gas. This will manifest itself in a reduction of the oxygen content of the flue gas resulting in the ratio being trimmed upwards to increase the air flow.

Also note that, for environmental purposes, it is common practice to monitor the carbon monoxide content of the flue gas and to use that as an input to a trip system which, in extreme circumstances, will shut down the boiler plant. CO monitoring also provides a useful means of cross checking the effectiveness of the O_2 based ratio trimming.

33.4 Comments

The boiler plant described is typical of that used in the process industries for steam raising and, as such, its control system is relatively simple. In the power industry, where the objective is to produce steam for driving turbines and generators, the boilers have very much greater capacity and steam pressures of 50 bar are not uncommon. Boiler plant is more highly integrated, involving economisers and superheaters, whilst ancillaries and back-up plant are more extensive. The control systems used on these boilers are more sophisticated with feedforward control, load scheduling, stand-by arrangements, *etc.*

Neutralisation Control

34.1 Classical Approach
34.2 In-line Neutralisation
34.3 Anti-Logging Approach

Neutralisation is the process whereby acid and base reagents are mixed to produce a product of specified pH. In the context of waste water and effluent treatment the objective is to adjust the pH to a value of 7 although, in practice, any value in the range 6–8 is good enough. In many chemical reactions the pH has to be controlled at a value other than 7, which could be anywhere in the range of 0–14. Neutralisation is always carried out in aqueous solutions, pH is a meaningless quantity otherwise. Note that a base that is soluble in water is usually referred to as an alkali.

pH is without doubt the most difficult of common process variables to control. For example, the measurement is electrochemical, made with a glass and reference electrode pair as described in Chapter 17, and is prone to contamination, hysteresis and drift. The signal produced, being logarithmic, is highly non-linear. The process being controlled invariably has a wide range of both concentration and flow rate. The rangeability of flow gives rise to variable residence times. To achieve satisfactory control, all of these issues have to be addressed.

This chapter focuses on the conventional approaches to pH control. For a more comprehensive treatment of such the reader is referred to the text by Shinsky (1973). Because of its inherent difficulty, pH control lends itself to more sophisticated techniques such as self-tuning and generic model control. Whilst the technical literature abounds with references to these methods, they have hardly been taken up by industry. This is largely due to wariness of advanced control methods, due in itself to a general lack of understanding, and to the difficulty in setting up and providing long term support for such.

34.1 Classical Approach

The classical approach is to change the pH stagewise, in a series of agitated vessels, with the pH being controlled at each stage by means of a feedback loop, as depicted in Figure 34.1 in which an acidic influent is treated with alkaline reagent.

There are certain ground rules that it is essential to comply with in neutralisation systems if good pH control is to be achieved. These are as follows:

- Average out fluctuations in flow and composition of the influent, if feasible, prior to neutralisation. Otherwise use level control for anti-surge purposes as shown.
- Good mixing of influent with neutralising reagents is essential. This also prevents sediment and solids from settling out. It doesn't particularly matter how the mixing is realised as long as it is good. Remember that mixing in cylindrical vessels is only effective if there are baffles fitted. Centrifugal pumps are very good mixers: mixing can be significantly enhanced by use of a centrifugal pump in an external loop with a high circulation rate.

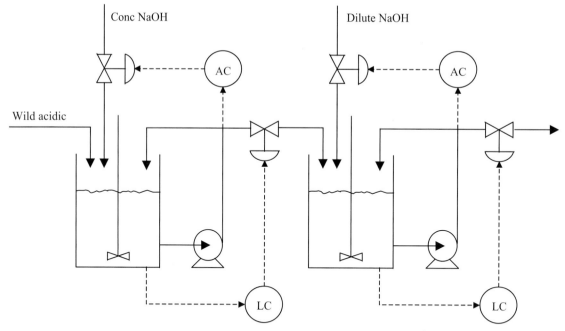

Fig. 34.1 Stagewise approach to pH control with agitated vessels

- Residence time should be sufficient for neutralisation, the issue being the time it takes for dissociation to occur. If a weak-acid-or weak-base is involved, provided the reagents are already in solution, a residence time of some 5 min is necessary. For neutralisations involving slurries, such as slaked lime, a residence time of some 20 min is required for complete solution of the particles. Strong-acid strong-base neutralisations take place almost instantaneously.
- The electrodes for pH measurement should be located such that they are in turbulent flow, but shielded from excessive turbulence and potential mechanical damage, to minimise their response time. Particular care should be taken when dealing with dirty effluents to prevent an impermeable film forming on the electrodes.
- Time delays should be minimised throughout the system.
- Use concentrated acid/alkali solutions for neutralisation at low/high pH values and dilute solutions in between.

- A maximum change of 2 pH per mixing vessel, corresponding to $[H^+]$ changing by a factor of 100, is the best that can normally be achieved with simple feedback. This is because the rangeability of a control valve, assuming it to be fitted with a positioner, is some 100:1 at best.
- Rangeability, and hence quality of control, can be improved if control valves with overlapping ranges are used. Equal percentage characteristics are presumed, as depicted in Figure 34.2. The valves are used in pairs, of different size, on a split range basis. Such a duplex arrangement is realised by means of appropriately calibrated positioners, as described in Chapter 21.
- If the influent can be either acidic or alkaline then a logical decision, based upon the pH value, is used to decide which reagent has to be added. Alternatively, both acid and alkali valves may be manipulated on a split range basis using positioners as depicted in Figure 34.3.
- Quality of control can also be improved if the controller gain is characterised according to the

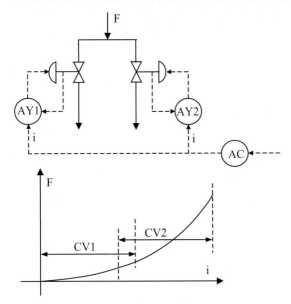

Fig. 34.2 Use of split range valves with overlapping ranges

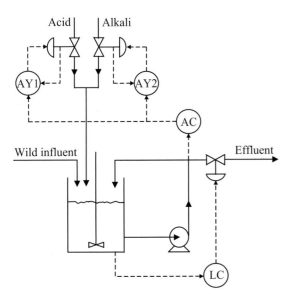

Fig. 34.3 Use of split range valves for mixed acid/alkaline neutralisations

measurement is offset by a variable controller gain: it is linearised.
- When the influent contains various species of acid and/or base, there will be multiple buffering effects and the titration curve is likely to be variable. In such circumstances, an estimate has to be made of the likely average titration curve for characterisation purposes. Self tuning control can also be considered.
- Storage tanks are desirable for environmentally sensitive effluent treatment schemes. In the event of failure, effluent can be diverted to storage whilst repairs are carried out. The size of the storage tanks, and hence the feasibility of such, is clearly a function of the throughput.

34.2 In-line Neutralisation

This is essentially an end-of-pipe approach which is growing in acceptance. The mixing vessels have been replaced with in-line mixers involving a variety of control strategies, as depicted in Figure 34.4. Again the influent is assumed to be a wild acidic stream and the neutralising reagent to be alkaline.

Typically the neutralisation is realised in two stages. Centrifugal pumps in dedicated recirculation loops provide the mixing for each stage:

- The first stage is a coarse adjustment, for which the set point is approximately pH 5. There is a feed forward strategy, based upon the flow rate and pH of the influent stream, assuming the pH of the neutralising reagent is known. A slave flow control loop is used to control the alkali addition. Down-stream of the in line mixer is a further pH measurement which is used to trim the feed forward calculation.
- The second stage is a fine adjustment for which the set point is nominally pH 7. This essentially involves a feedback loop, with variable gain according to the titration curve.

Clearly an in-line approach is infeasible for neutralisations that require significant residence time. Because of the lack of capacity within the system,

known titration curve, an example of which is depicted in Figure 17.1. This is achieved by means of gain scheduling. Thus the controller has a low gain when the slope of the titration curve is steep and vice versa. In effect, the pH

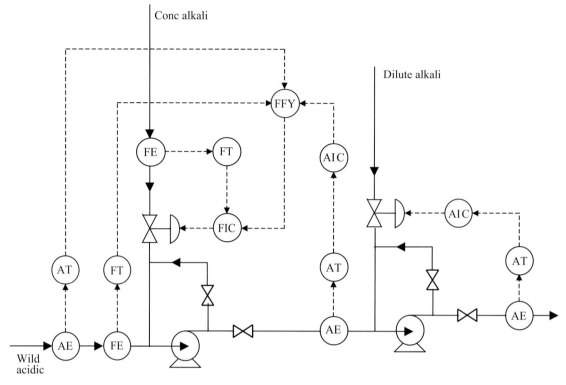

Fig. 34.4 Stagewise in-line approach to pH control

and the sophistication of the control scheme, the dynamics of the in-line system are fast. This makes it suitable for effluents with high rangeability of both flow rates and concentration. At the same time, it is very susceptible to errors in calibration, drift in measurement, offsets and bias in initialisation, *etc*. For these reasons a jacket of operability constraints is necessary to provide robustness.

34.3 Anti-Logging Approach

This is an interesting alternative approach. It recognises that the neutralisation process, which is the mixing of quantities of reagents in proportion to their concentrations, is inherently linear and that the non-linearity in pH control is introduced at the point of pH measurement. Therefore, it is logical to linearise the measurement by taking the anti-log of the pH signal. This is analogous to square root extraction in flow measurement in which the non-linearity is removed at source.

Anti-logging may be achieved by converting the pH measurement into an hydrogen ion concentration [H$^+$] on the following basis:

If pH ≤ 7 then [H$^+$] = $10^{(6-\text{pH})}$

If pH ≥ 7 then [H$^+$] = $-10^{(\text{pH}-8)}$

Figure 34.5 depicts the resultant quasi pH scale. Note that:

- A factor of 10^6 has been introduced such that the units of [H$^+$] are mg/m^3. This makes for more manageable values of [H$^+$] for control and display purposes and, in the vicinity of neutrality, is more intuitive.
- Alkaline solutions are treated as if they are negative acid solutions.
- There is a hole in the middle of the [H$^+$] scale, between $-0.1 \leq$ [H$^+$] $\leq +0.1$.

This hole is so small that it is of no practical consequence. As the measurement signal passes through pH 7 and jumps across the hole in the [H^+] scale, it is not possible to observe any kick in the controller output.

Anti-logging transforms the control problem into the linear [H^+] domain. The controlled variable becomes [H^+] which is often of more importance than pH *per se*. The set point for neutralisation is a zero value for [H^+]. Working in the [H^+] domain enables classical linear control system design techniques to be applied.

This approach has been demonstrated to work with strong-acid strong-base neutralisations but the extent to which it can accommodate buffering effects is not proven. However, ultimately, it too is critically dependent upon the accuracy of pH measurement and the rangeability of control valves.

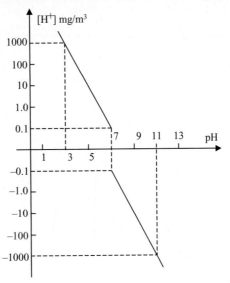

Fig. 34.5 Quasi pH scale for use with anti-logging approach

Distillation Control

Chapter 35

35.1 Strategic Considerations
35.2 Inventory Considerations
35.3 Cut and Separation
35.4 Worked Example
35.5 Measured Variables
35.6 Mass and Energy Balance Control
35.7 Control of Overhead Composition
35.8 Dual Control of Overhead and Bottoms Composition
35.9 Control of Column Feed
35.10 Control of Column Pressure
35.11 Internal Reflux Control
35.12 Override Control
35.13 Multicomponent Distillation
35.14 Multicolumn Separations
35.15 Comments
35.16 Nomenclature

Although distillation is a complex separation process, the control of a distillation column is relatively straightforward in principle. From an operational point of view, the objectives are to:

- Achieve safe and stable operation.
- Maintain mass and energy balances across the column.
- Maintain product qualities within specifications.
- Maximise recovery of the more valuable product.
- Minimise disturbances to downstream units.
- Optimise product yield subject to energy constraints.

Most, or even all, of these objectives can be simultaneously achieved by effective design of a column's control scheme with one crucial proviso: that the process design of the column is appropriate for its duty in terms of number of plates/trays, column diameter and capacity of reboiler and condenser. There is much variety: this chapter outlines some of the more common, conventional control schemes. The dynamics of one such scheme is considered in detail in Chapter 90.

Because distillation is of fundamental importance to the chemical and process industries, its dynamics and control have been studied extensively. There is a wealth of experience and many excellent texts: the reader is referred in particular to those by Buckley (1985), Deshpande (1985), Luyben (1992), Marlin (2000) and Shinskey (1977).

35.1 Strategic Considerations

Consider a continuous column as depicted in Figure 35.1.

An overall steady state mass balance across the column gives

$$F = D + B \qquad (35.1)$$

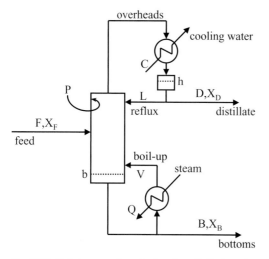

Fig. 35.1 Basic layout and terminology for a distillation column

Suppose that its feed is a binary mixture. It is also presumed throughout this chapter to be liquid and at its boiling point. A steady state mass balance for the more volatile component (MVC) gives

$$F.X_F = D.X_D + B.X_B \qquad (35.2)$$

Applying the process of determination, it is evident that there are two equations, six strategic variables and four degrees of freedom. Normally F is either wild or controlled by a flow control loop and X_F is wild, resulting in two equations with four floating variables. There are thus two degrees of freedom remaining. If any two of D, X_D, B and X_B are controlled then the other two will be determined.

At face value, the above analysis appears to be simplistic having ignored the underlying stagewise nature of distillation and not taking into account any of the vapour liquid equilibria, composition and flow constraints or heat balances. There are indeed many other relationships but, at steady state, none of the extra equations and variables have any net effect on the number of degrees of freedom: they are either not independent or cancel out.

Note, in particular, that constant molal overflow is presumed: that is, the heat released by unit amount condensing at each stage is sufficient to vaporise an equivalent amount at the same stage.

Thus the analysis reduces to a mass balance scenario only: the heat balances can be ignored.

35.2 Inventory Considerations

Notwithstanding the holdup on the plates of a column, which is fixed by the plate design, there are invariably three key inventory variables. With reference to Figure 35.1, these are:

- The column pressure P. This is a measure of the amount of vapour in the column and its overhead system: clearly the higher the pressure the greater the vapour inventory. It is normal practice to operate a column at constant pressure. Unless the column is operating at atmospheric pressure or under vacuum, its pressure is usually controlled by manipulating the flow of cooling water C to the condenser.
- The reflux drum level h. The reflux drum has several functions. It provides a reservoir of distillate which enables the reflux rate to be manipulated. Otherwise, if the reflux rate increased then the top product rate would decrease, and vice versa. The capacity of the drum also has an anti-surge function, averaging out sudden changes in the top product flow rate and composition. The drum also provides a head for flow through the reflux and top product valves. The drum level must be controlled at all times to prevent it from emptying or flooding.
- The still level b. Assuming the column does indeed have a still, it's principal function is to provide a reservoir of liquid for the reboiler and the bottom product pump. Otherwise, the capacity is provided by the reboiler itself. Like the reflux drum, it also has an anti-surge function. The still level must be controlled at all times to keep the thermosyphon functioning.

Thus, potentially, there is a maximum of seven controlled variables: that is four strategic and three inventory, D, X_D, B, X_B, P, h and b, and a maximum of five manipulated variables: that is D, L, B, Q and C. Establishing a control scheme for a column in-

35.3 Cut and Separation

volves specifying loops as appropriate by assigning MVs to CVs subject to satisfying the mass balances, ensuring that the column is not over determined and that all of the inventories are controlled.

There are many possible combinations of MVs and CVs. The process of specifying loops therefore requires some understanding of the underlying relationships. Table 35.1 provides an overview of the principal interactions. The entries in the table indicate whether an MV directly (d) or indirectly (i) affects a CV and provide a basis for deciding whether a proposed loop is sensible or not.

Table 35.1

Controlled variables (CV)		Manipulated variables (MV)				
		D	L	B	Q	C
Strategic variables	D	d	i	i	i	
	X_D	d	d	i	i	
	B	i	i	d	i	
	X_B	i	i	d	d	
Inventory variables	P				d	d
	h		d	d		i
	b				d	d

The fact that several manipulated variables can each affect a single controlled variable is a measure of the scope for interactions. Assigning MVs to CVs is normally done on the basis of experience. However, there is an analytical technique for establishing the most appropriate control loops, known as relative gain analysis (RGA), which is covered in Chapter 111.

It is normal practice with column control to use cascade control systems, as described in Chapter 25, to reject supply disturbances. There is invariably a slave flow control loop on the top and bottom product streams and on the reflux stream. For the reboiler, the slave may be either a flow or a pressure control loop on the steam supply. These are not shown explicitly in subsequent diagrams, but should be taken as read.

From an operational point of view the required "cut" is often specified. This is defined to be the ratio D/F and can be obtained by manipulation of Equations 35.1 and 35.2:

$$\text{cut} = \frac{D}{F} = \frac{X_F - X_B}{X_D - X_B} \quad (35.3)$$

If a particular cut is specified then, since it is articulated in terms of the strategic variables, the cut clearly has an impact on the process of determination. Thus, including Equations 35.1 and 35.2, there would be three equations, six strategic variables and three degrees of freedom. Again, assuming F and X_F to be controlled and/or wild, this reduces to three equations and four unknowns. If any one of the three floating variables is controlled, then the other three will be determined. In effect, specifying the cut has removed a degree of freedom.

Specifying the cut does not in itself fix the product compositions. This can be demonstrated by manipulating Equation 35.3 into the form

$$X_B = \frac{X_F - \frac{D}{F} \cdot X_D}{1 - \frac{D}{F}}$$

Thus, for a given feed composition, the cut determines the relative values of the top and bottom compositions rather than their absolute values.

Another performance parameter that is often specified is "separation" which is a function of the relative volatility and defined to be

$$\text{separation} = S = \frac{\frac{X_D}{1 - X_D}}{\frac{X_B}{1 - X_B}} \quad (35.4)$$

Now, assuming F and X_F to be controlled and/or wild, if both the cut and separation are specified, then there are four equations and four unknowns, thus both the product stream flowrates and compositions are determined. The separation has used up the final degree of freedom.

Inspection of Equations 35.3 and 35.4 reveals that increasing the cut is consistent with worsening the distillate quality (reduces X_D) and/or improving the bottoms quality (reduces X_B), whereas increasing separation is consistent with improving both qualities.

For a given feed and column design, the only way in which the separation can be changed is by manipulating the reflux and boil-up rates. However, there is an upper limit to the achievable separation which depends upon the cut specified. No amount of processing will enable separations to be achieved beyond that limit. In terms of control strategy, it follows that cut overrides separation. If the cut is wrong then it may well be impossible to achieve both desired product quantities and maybe neither.

Cut and separation may be thought of as operational constraints. In practice, there are also physical constraints on both cut and separation. These constraints are invariably due to:

- Flooding considerations, *i.e.* the maximum flow of liquid or (more usually) vapour through the trays/plates which is a function of the column's sizing.
- Capacity limitations on the column's ancillaries such as the thermal duty of its reboiler and/or overhead condenser, the sizing of its reflux drum/pump, *etc*.

35.4 Worked Example

The feed to a column may be considered to consist of 60 mol% of the light key. The distillate is required to contain not less than 99.5 mol% of the light key and the bottom product not less than 95 mol% of the heavy key. Thus, at the operational limit:

$$X_F = 0.6 \quad X_D = 0.995 \quad X_B = 0.05$$

From Equations 35.3 and 35.4 the cut and separation are given by

$$\frac{D}{F} = \frac{0.6 - 0.05}{0.995 - 0.05} = 0.582 \quad S = \frac{\frac{0.995}{0.005}}{\frac{0.05}{0.95}} = 3781$$

The scope for manipulating the cut and separation is now explored. Suppose the top product composition is held at its specified value. Then, as the cut is increased, the value of X_B reduces towards zero. The (theoretical) maximum value of cut is thus

$$\frac{D}{F} = \frac{0.6 - 0.0}{0.995 - 0.0} = 0.603$$

and, as X_B tends to zero, the separation becomes infinite. Now suppose the bottom product composition is held at its specified value. As the cut is reduced the value of X_D increases towards unity. The (theoretical) minimum value of cut is thus

$$\frac{D}{F} = \frac{0.6 - 0.05}{1.0 - 0.05} = 0.579$$

Again, as X_D tends to unity, the separation becomes infinite. It follows that the only feasible range of operation for the column is for values of $0.579 < D/F < 0.603$. Over the whole of this range the desired value of either X_D or X_B can be achieved, but a cut of 0.582 is necessary to achieve both desired values simultaneously.

35.5 Measured Variables

Composition measurement is clearly necessary for column control purposes. Ideally, the values of X_D and/or X_B will be measured directly by some analytical means. Analytical devices are much more reliable and accurate than hitherto. The most common form of on-line analyser used in column control is the chromatograph which, as explained in Chapter 18, has the advantage of being able to measure multiple components. Gas phase chromatographs are preferred to liquid phase because the higher vapour velocity reduces the sampling delay time. However, they are expensive: the installed cost of an on-line chromatograph, including its sampling system, is typically £100K.

It is much more usual to use temperature as an indirect measure of composition. Assuming pressure to be constant, the temperature of a binary mixture is uniquely related to its composition by the vapour liquid equilibria. Figure 35.2 depicts a

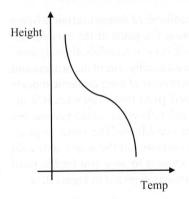

Fig. 35.2 Ideal profile of temperature *vs* height

typical steady state profile of temperature plotted against distance from the column base for an ideal separation.

It is evident that little separation occurs at the top or bottom of the column, because there are only small changes in temperature from plate to plate, with a region of high separation somewhere in the middle. That middle part of the profile with relatively low slope and high separation is often referred to as the "knee". It is helpful to think in terms of the function of the control system being to maintain the temperature profile, keeping the knee positioned over the same few plates in the middle.

As a means of indirectly measuring composition, the best place to measure temperature is in the middle of the column, where temperature is most sensitive to composition changes, and the worst place is at the top and bottom. However, for control purposes, the requirement is the exact opposite. The nearer the point of measurement is to the ends of the column, the closer it approximates to the product stream composition which, ultimately, is the controlled variable. Also, the closer the point of measurement is to the ends of the column, and thence to the means of manipulating reflux and boil-up rates, the faster is the response of the control loops. This is because of the lags associated with the holdup of the plates within the loops.

There obviously has to be a compromise. It is therefore normal practice to measure the temperature several plates down from the top of the column, or up from the bottom, where the temperature is sufficiently sensitive to variations in composition for changes to be meaningfully detected, whilst minimising the effect of the lags.

A further complication is the effect of pressure. Even when the column pressure is controlled, there will be small fluctuations in pressure which shift the vapour-liquid equilibria and cause changes in temperature. This can obviously obscure the effect of composition changes. It is important, therefore, that the temperature at the point of measurement is more sensitive to changes in composition than to the variations in pressure. These sensitivity considerations are of fundamental importance to the effectiveness of using temperature as an indirect measure of composition.

Assuming temperature is used as a measure of composition, measurements normally available for the purposes of controlling a column are as depicted in Figure 35.3. Not all of these measurements are necessarily available on any particular column, and there could be more available.

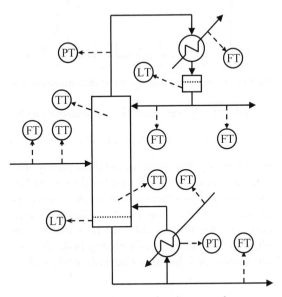

Fig. 35.3 Measurements available for column control

35.6 Mass and Energy Balance Control

Mass balance control is essentially a question of satisfying Equation 35.1. Any imbalance will manifest itself as a changing level in either the reflux drum or the reboiler. Two mass balance control schemes are depicted in Figures 35.4 and 35.5.

Assuming that F is wild, if either D or B is controlled then the other is determined by the mass balance. In the scheme of Figure 35.4 it is D that is controlled, B is the manipulated variable for the loop controlling the reboiler level which is simply of an inventory nature. In the scheme of Figure 35.5 it is the other way round: B is controlled and D is manipulated by the reflux drum level controller.

If a mass balance scheme is chosen, it is normal to put the flow controller on whichever is the smaller of D and B, and to use the level controller to manipulate the other.

The so-called "energy balance" control scheme is depicted in Figure 35.6. Neither D nor B is controlled directly, both being manipulated by level control loops. The inventory nature of the two level loops ensures that the mass balance is always satisfied, irrespective of the reflux and boil-up rates.

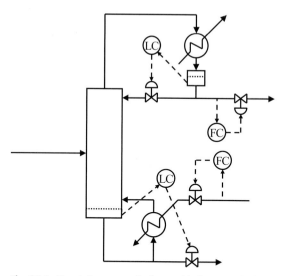

Fig. 35.4 Mass balance control scheme with flow control of distillate

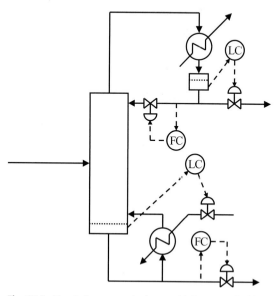

Fig. 35.5 Mass balance control scheme with flow control of bottoms

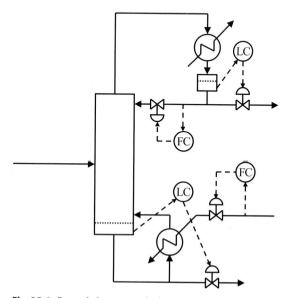

Fig. 35.6 Energy balance control scheme

The mass balance scheme is preferable for high reflux ratios (*i.e.* R = L/D > 5) and when frequent disturbances to the energy inputs occur, *i.e.* feed enthalpy, steam and cooling water supplies. The energy balance scheme is preferable for low reflux ratios (*i.e.* R < 1), when frequent disturbances to feed rate and composition occur, and when there is a large reflux drum. For many columns there is not a clear choice, but the following more sophisticated strategies deal with the weaknesses of both approaches and so make the choice less important.

Note that in all three of the mass and energy balance schemes there is no attempt to control composition: it is assumed that if the flows are set correctly then the desired compositions, cut and separation will be achieved.

35.7 Control of Overhead Composition

Control of composition is essentially a question of satisfying both Equation 35.1 and 35.2 simultaneously. Figure 35.7 depicts the most commonly used scheme for controlling the overhead composition X_D which is of the energy balance type.

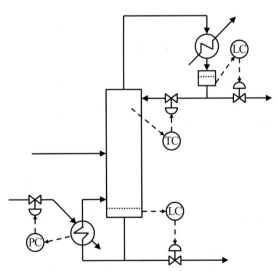

Fig. 35.7 Control of overhead composition by manipulating reflux rate

The signal from the column top temperature controller is used to manipulate the reflux rate L. Suppose that the composition of the top product stream drops off, *i.e.* X_D decreases. This would be revealed by the measured temperature increasing relative to its set point. The temperature controller would respond by opening the reflux valve, increasing L and hence increasing the reflux ratio R. In effect, this makes the column more efficient so a greater separation of the components is achieved, X_D rises towards its desired value and the temperature returns to its set point.

The boil-up rate V is maintained constant by the steam pressure control loop on the reboiler. However, at steady state the boil-up rate equates to the sum of the reflux and top product rates:

$$V = L + D$$

Since the reflux ratio L/D is indirectly controlled by the temperature control loop, the top product rate is itself indirectly controlled.

Assume that the feed rate and composition are both wild. Since the top product flow rate and composition are both controlled indirectly, then the bottom product flow rate and composition are both determined. It follows that any other control loop must be of an inventory nature.

A common variant of the scheme of Figure 35.7 is that depicted in Figure 35.8 in which the temperature control and the reflux drum level control loops having been interchanged.

In the scheme of Figure 35.9, following a decrease in X_D the temperature controller decreases V. Since L is controlled by the flow loop, this leads to a decrease in D and to an increase in R. As discussed, increasing R drives X_D back to its desired value. The wide separation of the temperature sensor and the steam control valve is bad practice from a dynamic point of view. However, if the sensor has to be several plates down from the top of the column, for sensitivity purposes, then this arrangement may be better because changes in vapour flow rate are transmitted through the column much faster than changes in reflux flow.

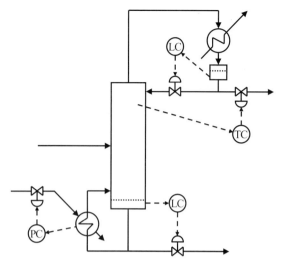

Fig. 35.8 Control of overhead composition by manipulating distillate flow

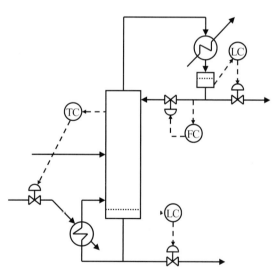

Fig. 35.9 Control of overhead composition by manipulating boil-up rate

35.8 Dual Control of Overhead and Bottoms Composition

In the so-called L-V scheme of Figure 35.10, X_D and X_B are both controlled indirectly by temperature control loops which manipulate L and V respectively. Since F and X_F are wild then D and B are determined by the mass balances. Clearly the two level loops are for inventory purposes only and the scheme is of the energy balance type. This is the same scheme as that used in Chapter 90 as a basis for explaining the dynamics of distillation.

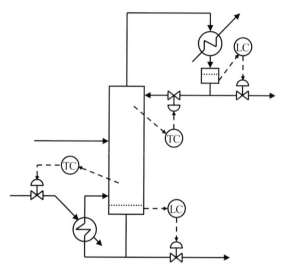

Fig. 35.10 Dual composition control: the L-V scheme

The top temperature and the reflux drum level control loops have been interchanged in the D-V scheme of Figure 35.11, and the bottom temperature and the reboiler level control loops have been interchanged in the L-B scheme of Figure 35.12. All three schemes are of the energy balance type: neither of the product streams is being controlled (as opposed to being manipulated) either directly or indirectly.

It is evident that, potentially, there are major interactions between the two composition control loops. The Ryskamp scheme addresses this problem. In particular, it keeps the reflux ratio, and hence the overheads quality, constant despite changes in boil-up rate due to bottom tempera-

35.8 Dual Control of Overhead and Bottoms Composition

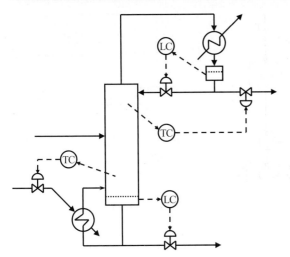

Fig. 35.11 Dual composition control: the D-V scheme

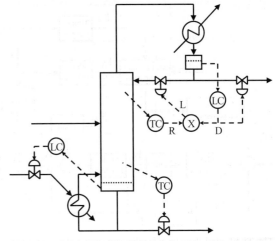

Fig. 35.13 Dual composition control: the basic Ryskamp scheme

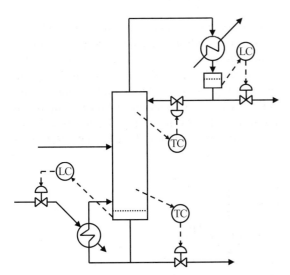

Fig. 35.12 Dual composition control: the L-B scheme

ture control. In effect, it decouples the interactions between the two composition control loops. This is achieved at the expense of worsened bottoms quality control so should only be used when overheads quality is more important than bottoms. Also, it does not handle feed composition changes well.

The basic Ryskamp scheme, as depicted in Figure 35.13, is an important and effective variant of the L-B scheme. The basic strategy is to maintain the reflux ratio constant. Since the reflux drum level controller varies the distillate flow, its output is a measure of the flow D, albeit subject to some scaling factor not shown in Figure 35.13. In the basic Ryskamp scheme this signal is multiplied by the desired reflux ratio R=L/D. This yields the required reflux rate L which becomes the set point for the reflux rate slave flow control loop. Thus, as the reflux drum level varies the distillate flow, the reflux flow is adjusted in proportion to the level to maintain a constant reflux ratio. A rising level will cause the reflux rate to be increased.

There are several variants on the Ryskamp scheme, of which one is depicted in Figure 35.14. The top temperature controller manipulates the slope of the upper operating line L/(L + D). The reflux drum level controller manipulates the total flow out of the reflux drum (L + D). Multiplication and subtraction of these signals as indicated in Figure 35.14 yields set points for the reflux and distillate flow slave loops. Again, relevant scaling factors are omitted.

The Achilles heel of dual composition control schemes is the potential for interaction between the control loops. The multivariable form of model predictive control (MPC), as explained in Chapter 117, is particularly effective at countering these interactions.

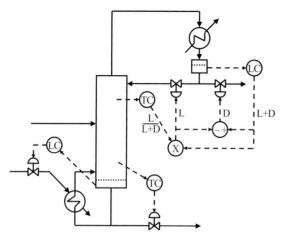

Fig. 35.14 Dual composition control: variant on the Ryskamp scheme

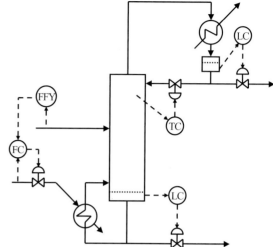

Fig. 35.15 Control of column feed: ratio control on boil-up rate

35.9 Control of Column Feed

The cost of buffer storage to average out fluctuations in F and/or X_F cannot usually be justified. It is normal therefore, when large fluctuations in feed rate may be predicted, to change V in proportion to the change in F. In effect, this anticipates the change in V that would eventually occur if any of the previous schemes were used.

The scheme depicted in Figure 35.15 uses a ratio control strategy to change V in proportion to F. Either the direct or the indirect approach may be used, as discussed in Chapter 26. Any increase in F will cause a proportionate increase in the set point of the slave steam flow control loop set point. This is a somewhat coarse approach and makes for sudden changes in bottoms flow rate.

In terms of strategic variables, F and X_F are wild, D and X_D are controlled indirectly and B and X_B are determined by mass balances. Both the overhead and bottoms level control loops are of an inventory nature.

The scheme depicted in Figure 35.16 uses a combination of feedforward and cascade control strategies to change V according to changes in F. The feed rate, bottoms temperature and flow rate, etc., are all inputs to the feedforward controller,

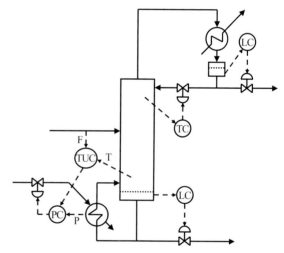

Fig. 35.16 Control of column feed: feedforward control on boil-up rate

which is described in detail in Chapter 27. This is more effective than simple ratio control because the dynamics of the column below the feed plate are taken into account.

In terms of strategic variables, F and X_F are wild, X_D and X_B are controlled indirectly and D and B are determined by the mass balances. Again, both the level control loops are of an inventory nature.

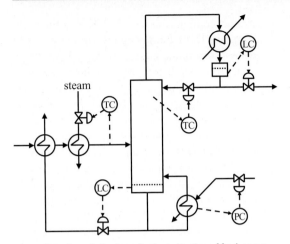

Fig. 35.17 Control of column feed: pre-heating of feed stream

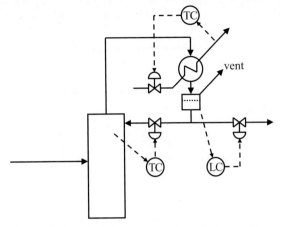

Fig. 35.18 Control of column pressure: atmospheric operation

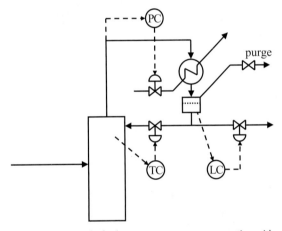

Fig. 35.19 Control of column pressure: pressure operation with low/no inerts

A cold feed causes more vapour to condense on the feed plate than would otherwise be the case, leading to increased liquid and vapour flows in the lower half of the column. It is normal practice to pre-heat the feed to its boiling point. This is realised by heat exchange with the bottom product stream, which makes for more efficient use of energy, and additional steam heating if necessary, as depicted in Figure 35.17. This scheme is otherwise the same as that depicted in Figure 35.7.

35.10 Control of Column Pressure

Controlling a column's pressure is essentially a question of balancing the energy put in through the reboiler with the energy taken out by the cooling system. Any imbalance manifests itself in the form of a change in pressure. There is a variety of schemes, depending on whether the column is operated at, above or below atmospheric pressure.

For distillation at atmospheric pressure, the reflux drum is vented to atmosphere, as indicated in Figure 35.18. Venting is subject to safety precautions regarding the potential release of toxic or flammable vapours, such as venting into a stack/flare system, and the reflux drum may need to be continuously purged with nitrogen to prevent the ingress of air. The cooling water outlet temperature is controlled to minimise water usage. Note the minimum flow constraint for this type of loop as explained in Chapter 32.

For distillations other than at atmospheric pressure, the pressure must be closely controlled if temperature is used as a measure of composition. The pressure throughout the overhead system, comprising the overhead pipework, condenser and reflux drum, is relatively uniform so it does not particularly matter where it is measured.

The type of control scheme chosen for pressure operation depends primarily on the amount of in-

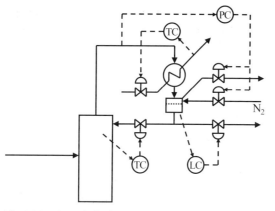

Fig. 35.20 Control of column pressure: pressure operation with high inerts

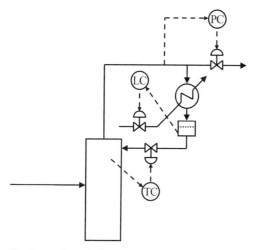

Fig. 35.21 Control of column pressure: pressure operation with vapour product

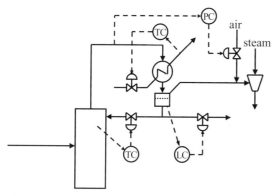

Fig. 35.22 Control of column pressure: vacuum operation

erts or non-condensables, typically air, in the overhead vapour. The scheme depicted in Figure 35.19 is the most common pressure control scheme. Suppose that the overhead pressure increases. The pressure controller would increase the cooling water flow rate until a new heat balance is established in the condenser with a lower exit water temperature and a higher rate of condensation. If there are low concentrations of inert gases present, it is necessary to have a small continuous purge to prevent the inerts from blanketing the condenser. It is normal to purge from the reflux drum to minimise the venting of vapours.

With a large amount of inerts, a pressure controller operating a valve in the line venting the drum gives satisfactory control, as depicted in Figure 35.20. Closing the control valve enables build up of inerts in the overheads system: the rate of heat transfer in the condenser is reduced by a blanketing effect which causes the pressure to rise. A potential problem with this arrangement is a non-symmetrical dynamic response. Opening the valve will vent the inerts fairly quickly if the column pressure is high, whereas the build up of inerts caused by closing the valve may be relatively slow. It may well be that injection of air or nitrogen is required to boost the inerts content, using a pair of valves in a duplex arrangement as described in Chapter 21. Valve sizing is critical.

If the overhead product is vapour and all the condensate formed is refluxed, the control scheme depicted in Figure 35.21 can be used. The pressure controller manipulates the valve in the top product line and the cooling water flow rate is manipulated to maintain the level in the reflux drum. If the drum level falls then the cooling water flow rate is increased, and *vice versa*.

In vacuum distillation, steam ejectors are commonly used to pull the vacuum. Ejectors operate efficiently only over a narrow range of steam supply pressure, especially if several are connected in series to pull a hard vacuum. So, rather than try to vary the steam supply, the overheads pressure is usually controlled by manipulating a bleed of air into the ejector inlet as depicted in Figure 35.22. Sometimes vapour from the ejector discharge system is used for the bleed stream. If the column pressure increases then the bleed rate is reduced and *vice versa*.

35.11 Internal Reflux Control

A fairly common problem associated with overhead condensation systems is that of sub-cooling. If the temperature of the liquor in the reflux drum is lower than its boiling point, then additional condensation of vapour will be required on the top plate of the column to restore boiling conditions. This results in the internal reflux being greater than its externally measured value. Based upon the scheme of Figure 35.19, that shown in Figure 35.23 depicts a simple on-line calculation to compensate for the effect of increased reflux.

If the amount of sub cooling is ΔT, then the heat required to restore the reflux to boiling is

$$q = L.c_p.\Delta T$$

Assuming constant molal overflow and a latent heat of ΔH, the rate of condensation must be

$$\frac{L.c_p.\Delta T}{\Delta H}$$

The internal reflux rate is thus given by

$$J = L + \frac{L.c_p.\Delta T}{\Delta H} = L\left(1 + \frac{c_p.\Delta T}{\Delta H}\right)$$

The calculated internal reflux rate becomes the measured value of the slave flow control loop, the external reflux rate being its manipulated variable. The master loop therefore manipulates the internal reflux rate, its set point typically being the temperature used for controlling overheads composition.

35.12 Override Control

An excessive boil-up rate in a column can cause flooding: the vapour velocity is such that the pressure drop across individual plates causes the liquid flow to back up in the downcomers. This can be detected by measuring the differential pressure across the column using the plates as orifices. Normally the differential pressure is low but rises rapidly at the onset of flooding. This can be exploited for override control, as depicted in Figure 35.24 which is based upon the L-V scheme of Figure 35.10. The TC is forward acting and manipulates V in the normal way. However, if the out-

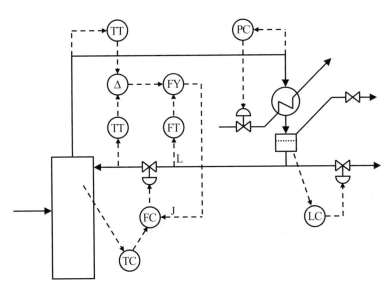

Fig. 35.23 Pressure operation with internal reflux flow compensation

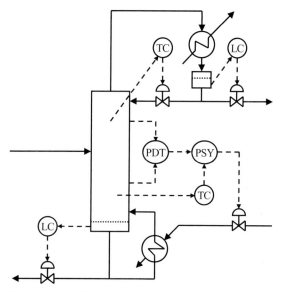

Fig. 35.24 L-V scheme with override controls

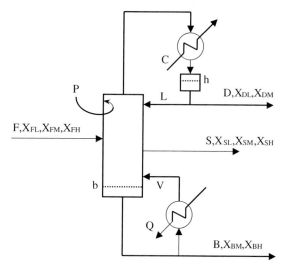

Fig. 35.25 Multicomponent distillation

put of PDT rises above some threshold value, then PSY reduces the steam flow to a set value until the flooding constraint is satisfied. Provision must obviously be made for bumpless transfer and integral desaturation when override occurs.

35.13 Multicomponent Distillation

Only binary mixtures have been considered in this chapter so far. Industrially, many columns distil multi-component mixtures and, strictly speaking, for the purpose of the process of determination, separate balances should be set up for each component. However, it is often the case that two key components, referred to as the light and heavy keys, can be identified as being representative of the separation of the mixture as a whole. Thus the various schemes based upon a binary mixture may be used as a basis for considering the control of multi-component separations too. In practice, a degree of over purification of the key components is always necessary in the steady state to ensure acceptable performance during transient conditions.

However, it is often the case with multicomponent distillations that there are sidestreams too which means that it may be necessary to consider three or more key components. A column with a single sidestream for which there are three key components is depicted in Figure 35.25.

An overall mass balance gives:

$$F = D + S + B$$

Component balances for the light, middle and heavy keys gives:

$$F \cdot X_{FL} = D \cdot X_{DL} + S \cdot X_{SL}$$
$$F \cdot X_{FM} = D \cdot X_{DM} + S \cdot X_{SM} + B \cdot X_{BM}$$
$$F \cdot X_{FH} = S \cdot X_{SH} + B \cdot X_{BH}$$

The composition constraints are:

$$X_{FL} + X_{FM} + X_{FH} = 1.0$$
$$X_{DL} + X_{DM} = 1.0$$
$$X_{SL} + X_{SM} + X_{SH} = 1.0$$
$$X_{BM} + X_{BH} = 1.0$$

Inspection reveals that, of these eight equations, seven are independent with 14 strategic variables, that is four flow rates and ten compositions, giving seven degrees of freedom. Presuming every-

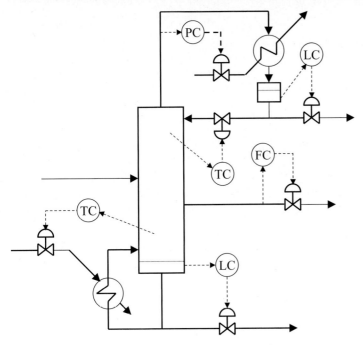

Fig. 35.26 Control of column with sidestream

thing about the feed to be wild accounts for three of the degrees of freedom. Thus, in principle, if four strategic variables are controlled, then the rest are determined. However, there are the three inventory variables b, h and P to be controlled too. Assume all six potential manipulated variables B, C, D, Q, V and S are available of which, say, B, C and D are assigned for controlling the inventories. That only leaves L, Q and S for controlling the four remaining strategic variables: suppose these are assigned to controlling X_{DM}, X_{BM} and S respectively as depicted in Figure 35.26.

Thus of the 14 strategic variables, F, X_{FL} and X_{FM} (say) are wild, X_{DM}, X_{BM} and S are controlled, X_{FH}, X_{DL} and X_{BH} are determined but X_{SL}, X_{SM} and X_{SH} are floating. This scheme would work, and work well, provided the quality of the sidestream was not of consequence. Alternative schemes are feasible, by assigning MVs to different CVs as appropriate, but a fully determined solution is not possible without more MVs.

35.14 Multicolumn Separations

In practice, separations are often affected using two or more columns, typically operating in a train at different pressures. In terms of operability and controllability, for a given duty, a pair or more of distillation columns are generally considered to provide more flexibility than operating a single larger column. Consider the pair of columns depicted in Figure 35.27.

It is certainly true to say that the number of potential manipulated variables has increased. Whereas there were six MVs with the single column, there are ten potential MVs in the case of the pair of columns: that is B, C_1, C_2, D, L_1, L_2, Q_1, Q_2, S and W. However, the number of inventory and intermediate variables has increased too. It is not clear without some analysis whether the net number of MVs available for controlling strategic variables has increased.

Presume the pair of columns to have the same duty as the single column in Figure 35.25. Thus the

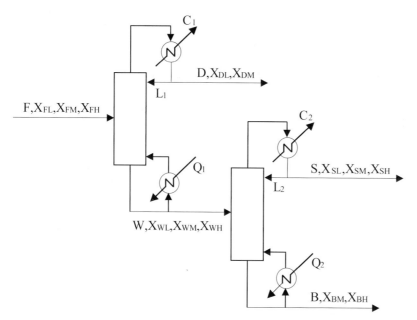

Fig. 35.27 Equivalent two column configuration

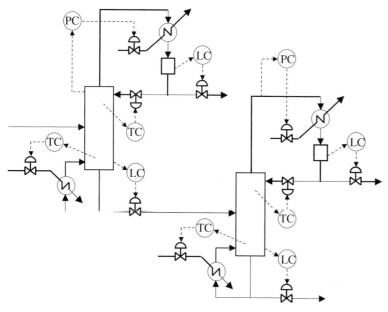

Fig. 35.28 Control of two column configuration

feed F and product D from the first column and products S and B from the second column are all as per the single column with sidestream of Figure 35.25. Let W be the intermediate flow from the bottom of the first column.

For the first column, an overall mass balance gives:

$$F = D + W$$

Component balances:

$$F.X_{FL} = D.X_{DL} + W.X_{WL}$$
$$F.X_{FM} = D.X_{DM} + W.X_{WM}$$
$$F.X_{FH} = W.X_{WH}$$

Constraints:

$$X_{FL} + X_{FM} + X_{FH} = 1.0$$
$$X_{DL} + X_{DM} = 1.0$$
$$X_{WL} + X_{WM} + X_{WH} = 1.0$$

Inspection reveals there to be six independent equations with 11 strategic variables giving five degrees of freedom. Again, presuming everything about the feed is wild accounts for three of the degrees of freedom. Thus, in principle, if two strategic variables are controlled, then the rest are determined. Of the five manipulated variables available, assume that C_1, D and W are assigned to controlling the three inventory variables P_1, h_1 and b_1 leaving L_1 and Q_1 available to be assigned to controlling two strategic variables, say X_{DM} and X_{WL}.

Thus of the 11 strategic variables, F, X_{FL} and X_{FM} (say) are wild, X_{DM} and X_{WL} are controlled and X_{FH}, D, X_{DL}, W, X_{WM} and X_{WH} are all determined. The energy balance scheme for the first column, as depicted in Figure 35.28, is thus fully determined.

For the second column a similar analysis gives:

$$W = S + B$$

Component balances:

$$W.X_{WL} = S.X_{SL}$$
$$W.X_{WM} = S.X_{SM} + B.X_{BM}$$
$$W.X_{WH} = S.X_{SH} + B.X_{BH}$$

Constraints:

$$X_{WL} + X_{WM} + X_{WH} = 1.0$$
$$X_{SL} + X_{SM} + X_{SH} = 1.0$$
$$X_{BM} + X_{BH} = 1.0$$

Again there are six independent equations with 11 strategic variables giving five degrees of freedom. As far as the second column is concerned, everything about its feed must be presumed wild which accounts for three of the degrees of freedom. Thus, again, if two strategic variables are controlled, then the rest are determined. Of the five manipulated variables available, assume that C_2, S and B are assigned to controlling the three inventory variables P_2, h_2 and b_2 leaving L_2 and Q_2 available to be assigned to controlling two strategic variables, say X_{SH} and X_{BM}.

Thus of the 11 strategic variables, W, X_{WL} and X_{WM} (say) are wild, X_{SH} and X_{BM} are controlled and X_{WH}, S, X_{SL}, X_{SM}, B and X_{BH} are all determined. The energy balance scheme for the second column, also depicted in Figure 35.28, is thus fully determined.

Since control of the single column arrangement resulted in floating variables and the control scheme for the pair of columns is fully determined, it is evident that splitting the duty between the two columns has indeed provided more control over the product streams.

35.15 Comments

This chapter has provided an overview of the many schemes for the control of distillation columns involving a variety of strategies. However, to understand fully the functioning of these schemes, it is necessary to have an appreciation of the basis of the process design of distillation columns. Of particular importance is the effect of varying the reflux and boil-up rates on the slope of the operating lines, and how this affects the separation achieved per stage. The reader is referred to any of the standard texts on distillation column design, such as that by Coulson (2004).

35.16 Nomenclature

b	still/reboiler level	m
B	bottoms (bottom product) flow rate	kmol s^{-1}
cp	specific heat	kJ kmol^{-1} °C^{-1}
C	cooling water flow rate	kg s^{-1}
D	distillate (top product) flow rate	kmol s^{-1}
F	feed flow rate	kmol s^{-1}
h	reflux drum level	m
ΔH	latent heat	kJ kmol^{-1}
J	internal reflux flow rate	kmol s^{-1}
L	external reflux flow rate	kmol s^{-1}
P	pressure	bar
q	rate of heat transfer	kW
Q	reboiler heat load	kW
R	reflux ratio: L/D	–
S	sidestream flow rate	kmol s^{-1}
T	temperature	°C
V	boil up flow rate	kmol s^{-1}
W	intermediate stream flow rate	kmol s^{-1}
X	mole fraction of more volatile component	–

Subscripts

B	MVC in the bottoms
D	MVC in the distillate
F	MVC in the feed
H	heavy key
L	light key
M	middle key
1	first column
2	second column

Reactor Control

36.1 Steady State Considerations
36.2 Dynamic Considerations
36.3 Temperature Control Schemes
36.4 Comments
36.5 Nomenclature

Many industrial processes such as dilution, mixing and reaction are carried out in agitated vessels. Invariably there is heat transfer associated with the process, for which heating or cooling is provided by means of a coil or jacket. In general, heating is more efficient than cooling: it is easier to get heat into a process mixture than to get it out. This is due to the size of the heat transfer coefficients involved. Typically, heating is realised by means of condensing steam for which the coefficients are high, whereas cooling is realised by means of circulating cooling water for which the coefficients are relatively low. Thus endothermic processes, *i.e.* those in which heat is absorbed, are fairly easy to control whereas exothermic processes, *i.e.* those in which heat is released, are more difficult to control.

Furthermore, endothermic processes are inherently stable: the process slows down in the event of insufficient heat transfer, and may even stop. In stark contrast to this, exothermic processes are inherently unstable: in the event of insufficient heat transfer they have the potential to overheat and, in extreme circumstances, to explode.

For a reactor with an exothermic reaction, a typical scenario is that an increase in temperature of 1°C may increase the rate of reaction by say 10%. This will lead to an increase in the rate of heat generation of 10%. If this is accompanied by an increase in rate of heat removal of 15% then the reactor is inherently stable, but if there is only a 5% increase in rate of heat removal, then the reactor is unstable. Fortunately, it is possible to achieve stable control of a reactor that is inherently unstable.

This chapter focuses on the control of exothermic reactions in continuous flow stirred tank reactors (CSTR). Control is considered from both steady state and dynamic points of view, followed by a review of conventional reactor control schemes. A detailed model of a semi-batch reactor is developed in Chapter 89 and of a CSTR in Chapter 91.

36.1 Steady State Considerations

Consider the reactor depicted in Figure 36.1.

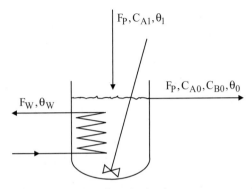

Fig. 36.1 Continuous flow stirred tank reactor

Suppose that the reaction taking place is of the first-order irreversible type. This is the simplest of all reactions but is quite adequate to illustrate the principle:

$$A \Rightarrow B \qquad (36.1)$$

where A is the reagent and B is the product. Consider unit volume of reaction mixture at some temperature θ_0. The rate at which A is consumed by the reaction, which is the same as the rate at which B is produced, is given by:

$$\frac{dc_{A0}}{dt} = -\frac{dc_{B0}}{dt} = -k.c_{A0} \quad \text{kmol m}^{-3}\text{ s}^{-1}$$

where k is known as the rate constant. Let V be the volume of the reactor's contents. Thus the overall rate of reaction is given by:

$$\Phi = -V.k.c_{A0} \quad \text{kmol s}^{-1}$$

If ΔH is the amount of heat released per unit mass of A reacted, then the rate of heat generation by the reaction at temperature θ_0 is given by:

$$Q_G = \Phi.\Delta H = -V.k.c_{A0}.\Delta H \quad \text{kW} \quad (36.2)$$

Note that, by convention, ΔH is negative for exothermic reactions so that Q_G is a positive quantity.

The rate constant is temperature dependant and varies according to Arrhenius' equation:

$$k = B.e^{\frac{-E}{R.\theta_0}} \quad \text{s}^{-1}$$

A rise in θ_0 thus causes an increase in k and hence in Q_G.

Figure 36.2 depicts a plot of Q_G against θ_0. Each point on the graph corresponds to an equilibrium. *Steady state* conversion is a parameter which varies along the curve. At low temperatures the rate of reaction is low so the rate of heat generation is low too. At high temperatures both the rate of reaction and the rate of heat generation are high.

Inspection of Equation 36.2 reveals that the slope of the graph initially increases because of the exponential effect of temperature on the rate constant. The slope subsequently decreases because

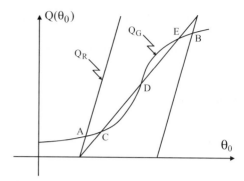

Fig. 36.2 Rate of heat generation/removal *vs* reactor temperature

of the resultant lower reagent concentration. The graph becomes asymptotic to the rate of heat release corresponding to complete conversion.

Some of the heat of reaction is removed as sensible heat by the product stream, but most is removed by the cooling water circulating through the coil. If the flow rate of cooling water is such that its rise in temperature is small, the rate of heat removal is given by:

$$Q_R = U.A.(\theta_0 - \theta_W) + F_P.\rho.c_p.(\theta_0 - \theta_1) \quad \text{kW} \quad (36.3)$$

If all the other parameters in this equation are constant, then Q_R is a linear function of θ_0. Graphs of Q_R vs θ_0 for three different combinations of θ_W and U are also plotted on Figure 36.2:

1. There is only one point of intersection at A. This is a stable operating point, since a slight increase in θ_0 leads to a greater increase in rate of heat removal than generation, which forces the temperature back to A.
2. θ_W is higher and the only intersection is at B, a stable operating point at almost complete conversion.
3. U is lower and there are three intersections. D is an unstable operating point, even though the heat balances are satisfied. A slight increase in θ_0 leads to a greater increase in rate of heat generation than removal, which results in a rapid rise in temperature to E. Similarly, a slight decrease in θ_0 results in a rapid fall in temperature to C.

Thus a necessary, but insufficient, condition for stability at the operating point is that:

$$\frac{dQ_R}{d\theta_0} > \frac{dQ_G}{d\theta_0} \quad kW\,°C^{-1}$$

36.2 Dynamic Considerations

For complete stability the dynamic response of the reactor must be considered for the worst case scenario. This corresponds to the maximum rate of increase in Q_G and the minimum rate of increase in Q_R.

First consider the maximum rate of increase in Q_G. If θ_0 increases then k will increase according to Arrhenius' equation:

$$k = B.e^{\frac{-E}{R.\theta_0}}$$

However, the increase in k causes a decrease in C_{A0} since:

$$\frac{dc_{A0}}{dt} = -k.c_{A0}$$

The dynamic response is due to the simultaneous changes in both k and C_{A0} since:

$$Q_G = -V.k.c_{A0}.\Delta H \qquad (36.2)$$

Whereas the effect on Q_G of the increase in k is immediate, the effect due to the decrease in C_{A0} is exponential and is negligible initially. The maximum rate of change of Q_G is thus found by assuming C_{A0} to be constant. Differentiating Equation 36.2 gives:

$$\begin{aligned}\left.\frac{dQ_G}{d\theta_0}\right|_{max} &= \left.\frac{\partial Q_G}{\partial \theta_0}\right|_{C_{A0}} \\ &= \frac{\partial}{\partial\theta_0}(-V.k.c_{A0}.\Delta H) \\ &= -V.c_{A0}.\Delta H.\frac{dk}{d\theta_0} \\ &= -V.c_{A0}.\Delta H.\frac{d}{d\theta_0}\left(B.e^{\frac{-E}{R.\theta_0}}\right) \\ &= -V.c_{A0}.\Delta H.B.e^{\frac{-E}{R.\theta_0}}.\frac{E}{R.\theta_0^2}\end{aligned}$$

Thus, under normal operating conditions:

$$\left.\frac{dQ_G}{d\theta_0}\right|_{max} = -V.\overline{c}_{A0}.\Delta H.B.e^{\frac{-E}{R.\overline{\theta}_0}}.\frac{E}{R.\overline{\theta}_0^2} = \frac{\overline{Q}_G.E}{R.\overline{\theta}_0^2}$$

Second, consider the minimum rate of increase in Q_R which occurs when the cooling water flow rate F_W is constant. This corresponds to F_W, the manipulated variable, not responding to the increase in θ_0 which would initially be the case. The cooling water temperature θ_W may be assumed to be initially constant too. The variation of Q_R with θ_0 is found by differentiating Equation 36.3:

$$\begin{aligned}\left.\frac{dQ_R}{d\theta_0}\right|_{min} &= \left.\frac{\partial}{\partial\theta_0}\left(U.A.(\theta_0-\theta_W)\right.\right.\\ &\qquad\left.\left. + F_P.\rho.c_P.(\theta_0-\theta_1)\right)\right|_{F_W} \quad (36.4)\\ &= U.A + A.(\theta_0-\theta_W).\left.\frac{\partial U}{\partial\theta_0}\right|_{F_W} \\ &\quad + F_P.\rho.c_P\end{aligned}$$

Note that the dependency of U upon θ_0 is largely due to changes in viscosity. Under normal operating conditions:

$$\left.\frac{dQ_R}{d\theta_0}\right|_{min} = \overline{U}.A + A.(\overline{\theta}_0-\overline{\theta}_W).\left.\frac{\partial U}{\partial\theta_0}\right|_{\overline{F}_W,\overline{\theta}_0} + \overline{F}_P.\rho.c_P$$

Thus, for complete stability:

$$\overline{U}.A + A.(\overline{\theta}_0-\overline{\theta}_W).\left.\frac{\partial U}{\partial\theta_0}\right|_{\overline{F}_W,\overline{\theta}_0} + \overline{F}_P.\rho.c_P > \frac{\overline{Q}_G.E}{R.\overline{\theta}_0^2}$$

Usually, most of the heat generated is removed through the coil, i.e. $\overline{U}.A \gg \overline{F}_P.\rho.c_P$.

If $\left.\frac{\partial U}{\partial\theta_0}\right|_{\overline{F}_W,\overline{\theta}_0}$ is small then, for stability,

$$\overline{Q}_G < \frac{\overline{U}.A.R.\overline{\theta}_0^2}{E}.$$

Unless these conditions are satisfied, a runaway reaction is possible. Note that this is the criterion for the first order reaction of Equation 36.1 only. Different criteria apply for other reactions although the underlying issues are the same.

36.3 Temperature Control Schemes

Temperature control is normally realised by manipulating the cooling water flow rate, as depicted in Figure 36.3.

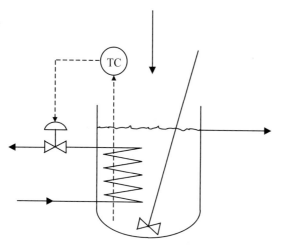

Fig. 36.3 Simple feedback control of reactor temperature

In practice, for a given agitator speed and reaction mixture, the overall heat transfer coefficient $U = f(F_W, \theta_W, \theta_0)$.

Thus an increase in θ_0 would cause an increase in F_W which leads to a greater U. Inspection of Equation 36-4 shows that increasing U has the effect of making the slope of the heat removal graph steeper, thereby improving the stability of the reactor. Indeed, this approach enables operation of a reactor under what would otherwise be unstable conditions. The dynamics of this particular scheme are analysed in detail in Chapter 91.

More effective control is realised by means of a conventional cascade control system as depicted in Figure 36.4, the slave loop's function being to reject cooling water supply pressure disturbances. However, in practice, the overall response is sluggish.

An alternative cascade arrangement, which has a fast response, is depicted in Figure 36.5. Refrigerant is pumped by a centrifugal pump from a reservoir into the jacket. The temperature inside the jacket is the boiling point of the coolant at the

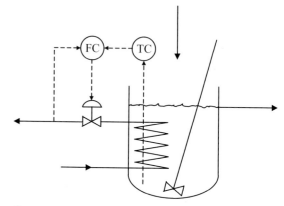

Fig. 36.4 Cascade control with cooling flow rate as slave loop

prevailing jacket pressure. As a result of heat transfer from the reactor, coolant is vaporised inside the jacket. The vapours flow into a heat exchanger where they are condensed, the condensate draining back into the reservoir.

The slave loop's function is to control the pressure inside the jacket. Following an increase in reactor temperature, the master controller reduces the set point of the slave controller. This opens the control valve, thereby increasing the flow of refrigerant vapour to the condenser and reducing the pressure in the jacket. The resultant decrease in boiling point of the refrigerant causes it to flash off. Heat is rapidly removed from the jacket, in the form of latent heat. The sudden increase in temperature difference between the reactor and jacket contents increases the rate of heat transfer from the reactor, producing the desired decrease in reactor temperature.

36.4 Comments

For most reactors the product concentration is not controlled directly but is held nearly constant by having a controlled reactor temperature, controlled feed composition and flowrates, and a constant hold up in the reactor. Often there is no point in controlling the composition because the maximum possible conversion is required.

The success of composition control schemes is critically dependant upon the composition mea-

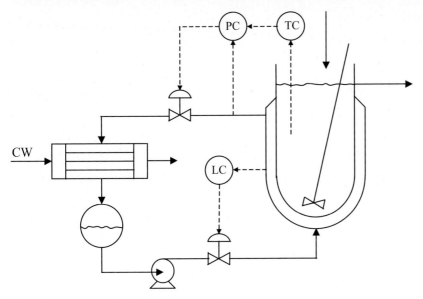

Fig. 36.5 Cascade control with refrigerant pressure in jacket as slave loop

surement. Direct measurement of composition is ideal but rapid and accurate measurement is not always feasible. Analytical instrumentation, such as chromatographs and spectrometers, cannot be used for many measurements and is expensive anyway. Indirect measurement, using parameters such as refractive index, viscosity, dissolved oxygen, pH, *etc.*, as a rough measure of composition is much more common. Composition is then controlled by means of manipulating variables such as reagent feed rate, catalyst concentration, gas/liquid holdup, *etc.*

Whilst the dynamics of reacting systems may be complex, as seen in Chapters 89 and 91, it should be remembered that very effective control can be achieved using fairly conventional control schemes and strategies.

36.5 Nomenclature

A	effective surface area of coil	m^2
B	frequency factor	s^{-1}
c	concentration	$kmol\ m^{-3}$
c_p	specific heat	$kJ\ kmol^{-1}\ K^{-1}$
E	activation energy	$kJ\ kmol^{-1}$
F	reactor feed rate	$m^3\ s^{-1}$
ΔH	heat of reaction	$kJ\ kmol^{-1}$
k	rate constant	s^{-1}
Q	rate of heat generation/removal	kW
Φ	rate of reaction	$kmol\ s^{-1}$
R	universal gas constant (8.314)	$kJ\ kmol^{-1}\ K^{-1}$
t	time	s
U	overall heat transfer coefficient	$kW\ m^{-2}\ K^{-1}$
V	volume of reactor contents	m^3
ρ	density	$kmol\ m^{-3}$
θ	temperature	K

Subscripts

A	reagent
B	product
D	*temperature*
G	generation
P	process
R	removal
W	cooling water
0	outlet
1	inlet

Batch Process Control

37.1 Variety of Plant Design
37.2 Plant Structure
37.3 Physical Model
37.4 Process Model
37.5 Procedural Model
37.6 Methodology
37.7 Recipe Model
37.8 Activity Model
37.9 Comments

When raw materials are processed in discrete quantities, the processes are said to be batch or semi-batch. These processes are normally carried out in the liquid phase although solids, usually in the form of a powder, may be fed into the mixture or gases bubbled through it. Batch operations are ones in which all the reagents are charged prior to processing whereas in semi-batch operations the bulk of the reagents are pre-charged but one or more reagents are subsequently trickle fed. Processes may be as simple as a mixing operation or involve complex multi-stage reactions. Such processes are used extensively in the manufacture of bio-chemicals, fine chemicals, pharmaceuticals, and so on. Typically, throughput is of a low volume and high value relative to the continuous processes associated with bulk and commodities chemicals manufacture.

Batch process control is concerned with the automatic control of batch processes. Control may consist of a few simple sequences or involve complex recipe handling and batch scheduling. It is invariably carried out by some form of digital control system. For an introduction to computer controlled batch processing, the reader is referred to the IChemE guide by Sawyer (1993). A more comprehensive treatment is provided in the text by Fisher (1990).

This chapter is based upon the IEC 61512 standard for batch process control. It was developed by the ISA and is generally referred to as S88. Part 1 of the standard focuses on the models and terminology used. Part 2 focuses on the underlying data structures. There are five key models within Part 1: physical, process, procedural, recipe and activity. These establish a framework for the specification of requirements for the control of batch processes and for the subsequent design and development of application software. For a more detailed understanding of S88 the reader is referred to the texts by Fleming (1998) and Parshall (2000) and to the standard itself.

37.1 Variety of Plant Design

A batch plant typically consists of a limited number of major equipment items (MEI), such as charge vessels, reactors, filters, *etc.*, each with its own ancillary equipment. Operations on a batch of materials are normally carried out to completion, or near completion, in a single MEI referred to as a unit.

The batch may then remain in that unit for further processing or be transferred to another unit. There is much variety in batch plant design, depending on the:

- Number of products, *i.e.* single product or multi-products.
- Throughput, in terms of the frequency and size of batches.
- Number of units, *i.e.* single or multiple units.
- Grouping of units into cells.
- Organisation of units within cells into streams/trains.
- Number of streams, *i.e.* single or multi-stream.
- Capacity of different streams.
- Number of batches being processed simultaneously in a single stream.
- Dedication of a stream to "campaigns" of batches of a single product.
- Materials of construction of particular units.
- Extent of ancillary equipment associated with the units.
- Sharing of ancillary equipment between units.
- Sharing of units between streams.
- Ability to reconfigure units within a cell, *i.e.* multipurpose plant.

An interesting perspective on this is depicted in Figure 37.1 in which type of batch plant is categorised on the basis of variety (number of products) against volume (frequency and size of batches).

37.2 Plant Structure

A process cell contains one or more units. If the units of a cell are not configured in any particular order, they are said to be clustered. In that case batches tend to move backwards and forwards between units according to what operations are required to be carried out and what ancillary equipment is associated with particular units.

Alternatively, units may be configured into streams. A stream is the order of units used in the production of a specific batch. Put another way, a stream is the route taken through the plant by a batch. Typically, each unit is dedicated to a certain type of operation, such as reaction or filtration, ordered according to the processing requirements, such that batches progress linearly from unit to unit along the stream. A train consists of the units and equipment modules used to realise a particular stream.

Analysis of the structure of a process cell gives a good insight into the nature of both the operations and the control requirements. The key to this analysis is identification of parallel and sequential operations. Consider, for example, the cell depicted in Figure 37.2 which consists of four processing units and two storage units. This clearly is a multi-stream plant. It is said to be networked because batches can be switched from one stream to another. As shown, there are potentially at least ten different streams. In practice, it is likely that only a limited number of streams and trains would exist.

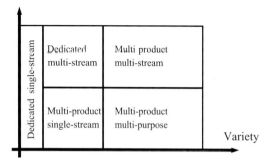

Fig. 37.1 Batch plant type categorised on basis of variety and volume

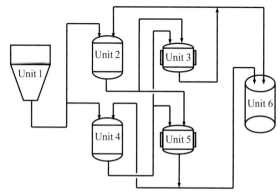

Fig. 37.2 Process cell with processing and storage units

37.3 Physical Model

The physical (equipment) model of the standard is shown in Figure 37.3. The unit is the key layer of the physical model. As stated, it is usually centred on an MEI, such as a mixing vessel or a reactor. A unit includes all of the logically related equipment necessary to perform the processing required in that unit. One unit operates relatively independently of other units.

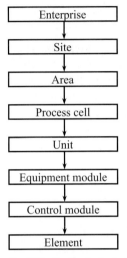

Fig. 37.3 The batch physical (equipment) model

Units normally, but not necessarily, consist of equipment modules which may themselves contain other equipment modules. Similarly, equipment modules consist of control modules. Although not formally included as a layer in the physical model, control modules obviously have to consist of elements.

Consider, for example, the mixing vessel shown in Figure 37.4. The mixing vessel and all of its ancillary equipment may be considered as a unit. Typically, there would be separate equipment modules for blanketing, charging, heating and pumping. For example, the equipment module for blanketing would consist of the all the pipework and control modules required for admitting the inert gas, venting and applying the vacuum. There would be at least two control modules, one for sequencing the isolating valves and the other for flow control. The sequence control module would consist of discrete i/o channels and the flow control module would be an analogue feedback loop. The various valves, sensors and so on would be the elements of the control modules.

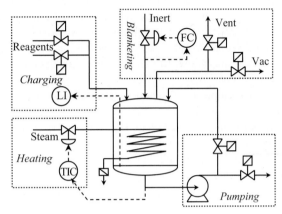

Fig. 37.4 Mixing vessel: one unit comprising four equipment modules

Although a unit frequently operates on, or contains, the complete batch of material at some point in the processing cycle, it may operate on only part of a batch, *i.e.* a partial batch. However, a unit may not operate on more than one specific batch at any point in time. Put another way, different parts of one unit cannot simultaneously contain parts of different batches. It follows that defining the boundary of a unit is a key design decision.

It is inevitable that some equipment modules and/or control modules will have to be shared between units. Weigh vessels, valve manifolds and receivers are obvious candidates. Such modules have to be deemed either as exclusive use resources, in which case each resource can only be used by one unit at a time, or as shared use resources which can be used by more than one unit at a time.

In drawing the boundaries around units questions inevitably arise about ownership of the pumps and valves between them. A pump is best deemed to be part of the unit containing the source vessel, *i.e.* it is a discharge pump, if there is one source vessel and one or more target vessels. However, it is deemed to be part of the unit containing the target vessel, *i.e.* a charge pump, if there are several source vessels and only one target.

Much the same reasoning can be applied to valves. A valve is deemed to be part of the unit containing the target vessel, *i.e.* it is an inlet valve, if there is only one valve between the source vessel and target vessel. An obvious exception to this is the last unit in a train which will also have an outlet valve. When there are two valves between any two vessels, one is deemed to be an outlet valve and the other as an inlet.

The upper three layers of the physical model, *i.e.* enterprise, site and area, whilst of interest from a corporate perspective, do not normally impinge directly on process automation and are not considered further in this chapter.

37.4 Process Model

A batch process may be subdivided hierarchically as depicted in Figure 37.5.

The process may be considered to be an ordered set of process stages, organised in series and/or in parallel. As such, any one process stage usually operates independently of the others. It results in a planned sequence of physical and/or chemical changes to the materials being processed. Typical stages are reaction and separation. Process stages may be decomposed further into ordered sets of process operations such as charge, and process actions such as heat or hold.

In principle, the process model may be decomposed without reference to the physical model of the target plant. For example, process stages are not constrained by unit boundaries. Indeed, process stages that are not associated with a single unit are of particular significance for decomposition purposes. For example, a single stage reaction could take place in a pair of agitated vessels, each of which is defined to be a unit. Also, if a unit is multifunctional, say a reactor is subsequently used for batch distillation, the process stage may need to be defined to include all of the functions because a unit is not permitted to simultaneously exist within different stages.

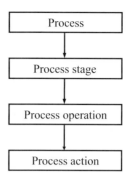

Fig. 37.5 The batch process model

The significance of the process model is apparent in relation to the recipe model as explained later in the chapter. In practice, the process model features at higher levels of the recipe model but, rather surprisingly, is relatively unimportant compared to the physical and procedural models at the lower levels.

37.5 Procedural Model

The procedural model is shown in Figure 37.6. Procedures address the operational requirements and contain all the sequencing necessary to enable batches of materials to be processed. A procedure is the highest level in the procedural hierarchy and is normally associated with a train or, at least, with

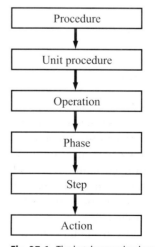

Fig. 37.6 The batch procedural model

a process stage. Procedures consist of ordered sets of unit procedures, operations and phases. Note that the operations and phases of the procedural model relate to the process operations and process actions of the process model respectively.

The operation is the key layer of the procedural model and consists of an independent processing activity such as transfer, charge, circulate, and so on. Operations are associated with and carried to completion within units. It is helpful in defining the scope of an operation to identify key process states of the unit and/or batch such as empty, full, ready and finished. An operation may be thought of as being the procedural entity that causes progression from one such state to the next. Operation boundaries should only occur where it is safe to put the unit into a hold-state.

Parallel operations on a single unit are not permitted within S88, operations may only occur sequentially. Thus, for example, two different operations could not be operating simultaneously on the same batch in one unit. Also, operations are not permitted to operate across unit boundaries. For example, an operation cannot simultaneously operate on parts of the same batch in different units, as may be required during a transfer. In practice, this would be dealt with by two separate operations working in parallel on different units, each handling parts of the same batch. It follows that defining the scope of an operation is a key design decision. These constraints are considered by some to be an unnecessary restriction: it remains to be seen to what extent suppliers will conform with S88 in this respect.

Whereas operations are only permitted to occur sequentially on a single unit, phases may occur both sequentially and/or in parallel within any single operation, as depicted in Figure 37.7. This, together with the constructs used for branching, looping, *etc.* within the implementation language, should provide sufficient flexibility to satisfy most requirements.

Operations are created by the configuration of phases drawn from a library of proven phases, somewhat analogous to the implementation of continuous control functions by configuration of

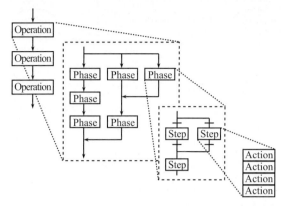

Fig. 37.7 Parallel and sequential phases within an operation

function blocks. Indeed, the drive towards configuration for batch has been one of the primary forces behind S88. A phase is the smallest practicable group of control functions that can be sufficiently decoupled from the rest of the operation to exist in isolation. Phases are associated with equipment and control modules. The objective, from a design point of view, is to establish the library of operations and phases.

To enable the development of phase logic, *i.e.* the software for implementing phases, further decomposition of phases into steps and actions is essential. Although not formally included as layers in the procedural model, steps and actions are defined in the IEC 61131-3 standard and are described in detail by Lewis (1998).

- Steps consist of actions grouped together for composite execution and offer natural breaks for timing purposes. Thus, by design, subject to the constraints of program flow, plant status and software logic, the rate at which a phase is executed is typically one step per second.
- Actions are typically implemented by means of individual instructions such as lines of code or rungs of ladder logic. Note that the actions of the procedural model do not relate to the process actions of the process model.

For any given batch or part batch of raw materials being processed within a unit, it is through the operations and phases of the procedural model that the operator interacts with the batch. That in-

Complete
Restarting
restart
Held
Holding
reset
hold
Idle
start
Running
pause
Pausing
abort
stop
resume
Aborting
Stopping
Paused
Aborted
Stopped
reset
reset

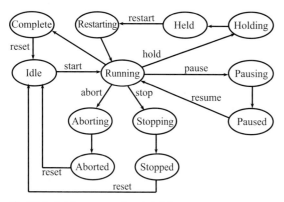

Fig. 37.8 Batch process and control states

teraction typically entails starting, holding, stopping and aborting batches as appropriate. Such intervention in the progression of a batch may be thought of in terms of changes to key control states of the unit and/or batch as depicted in Figure 37.8. The operator interface of the control system clearly needs to provide the functionality necessary to support the interaction required. That is system dependant and S88 is not prescriptive about this.

Thus, for example, all of the functionality depicted in Figure 37.8 could be supported at the level of the operation but only a subset of it supported at the level of the phase. There are many non trivial issues here, especially concerning how changes in state relate to each other. For example:

- if an operation is put into hold, is the currently running phase put into hold or allowed to progress to its end point?
- if a phase is stopped, is the parent operation stopped, paused or put into hold?
- if one phase of an operation is stopped, are other parallel phases within the same operation stopped too? And so on.

37.6 Methodology

The relationships between the layers of the models are as depicted in Figure 37.9. This provides a framework for decomposition of requirements for batch purposes.

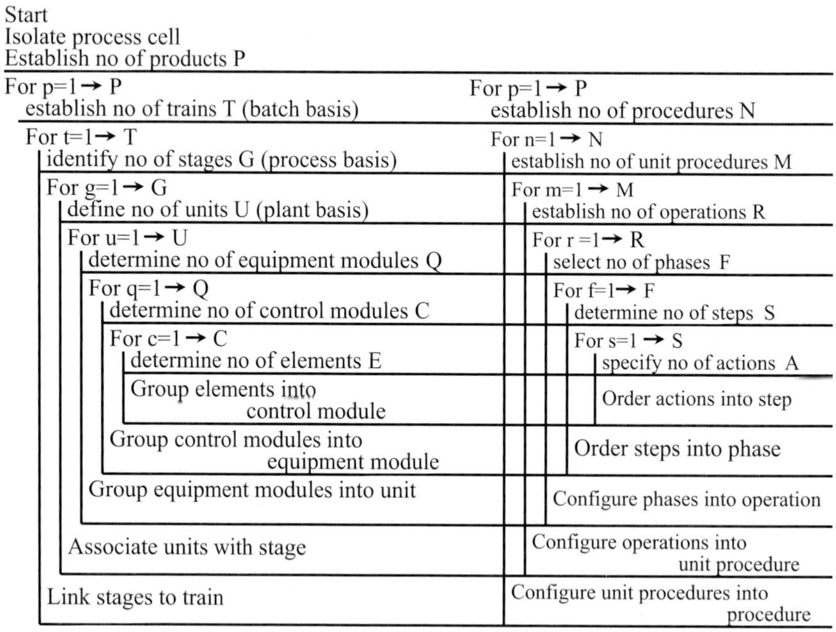

Start
Isolate process cell
Establish no of products P
For p=1→ P
establish no of trains T (batch basis)
For t=1→ T
identify no of stages G (process basis)
For g=1→ G
define no of units U (plant basis)
For u=1→ U
determine no of equipment modules Q
For q=1→ Q
determine no of control modules C
For c=1→ C
determine no of elements E
Group elements into control module
Group control modules into equipment module
Group equipment modules into unit
Associate units with stage
Link stages to train
For p=1→ P
establish no of procedures N
For n=1→ N
establish no of unit procedures M
For m=1→ M
establish no of operations R
For r =1→ R
select no of phases F
For f=1→ F
determine no of steps S
For s=1 → S
specify no of actions A
Order actions into step
Order steps into phase
Configure phases into operation
Configure operations into unit procedure
Configure unit procedures into procedure

Fig. 37.9 Relationship between layers of physical and procedural models

The flow chart has two halves, the left half focuses on the physical model and the right half on the procedural model. Note, in particular, the alignment of the layers of the two halves. One of the layers, that between units and operations, is quite explicit within the standard. Some layers, for example that between process stages and unit procedures, are implied. Other layers, such as that between elements and actions, are logical but outside the scope of the standard. Note also the nested nature of the constructs. For each entity established at one level, all the relevant entities have to be determined at the next.

The first (top) layer assumes that the boundary between the cell of interest and the rest of the site/area is known. Within that cell each of the P products for which batches are to be made is identified. This information should be available from production records and marketing forecasts.

The second layer establishes the number of trains T and the number of procedures N. This is done on a batch basis, for each product, by following the route through the plant structure that would be taken by a batch. Each such stream yields a train. If the plant is multi-stream and networked, as in Figure 37.2, then a likely outcome is several streams per train. There is no need to consider potential streams that will not occur in practice. As a rule of thumb, for each stream there will be at least one procedure. However, there will not necessarily be a different procedure for each product. Products which are generically similar should be produced by the same procedure using different parameter lists. Additional procedures may be required for handling inter-batch cleaning requirements.

The third layer involves identifying the number of process stages G and the number of unit procedures M. This is done on a process basis by considering the nature of the processing operations carried out on a batch. It is logical to assume that for each stage there will be a corresponding unit procedure, although it is quite permissible for a procedure to consist of operations and phases only. However, unless a process stage embraces more than one unit, this layer is academic. Process stages are invariably associated with single units, in which case this layer collapses into the next.

The fourth layer is, arguably, the most important because it is the most tangible within the standard. It defines the number of units U and establishes the corresponding number of operations R. This is done on a plant basis by considering all the MEIs, such as process vessels, as candidate units and all transitions that take a batch from one key state to another as candidate operations. Remember that a unit can contain only one batch (or partial batch) at a time and that only one operation can be carried out on a unit at a time. However, since operations may be carried out serially, it can be expected that $R > U$.

The fifth layer establishes the number of equipment modules Q and the number of phases F. In general, each of the equipment modules within a unit will be manipulated by a separate phase within an operation. Thus, for each equipment module, there will be at least one phase. There could be more than one phase per module because the equipment may need to be manipulated differently according to the process operations, say, so it can be expected that $F > Q$. If a library of appropriate phases exists, there is no need for further decomposition: the sixth and seventh layers are embedded as far as the user is concerned.

The sixth layer relates the number of control modules C to the number of steps S. Although equipment modules do not have to be decomposed into control modules, and control modules may be manipulated by phases directly, it is logical to think in terms of control modules being manipulated by steps. It is unlikely, in practice, that steps can be configured in the same way as phases. To a large extent steps will be bespoke, depending on context, so one can expect $S > C$.

The seventh layer is essentially a question of developing actions to enable implementation of the steps. Although there is an alignment between the elements and the actions, this is only notional. The actions will, in general, operate on the control modules rather than on elements directly. There is therefore little direct relationship between A and E, except that each is potentially large.

The lower half of the flow chart is essentially a question of ordering, grouping and configuring multiple entities at one level into single entities at the next.

37.7 Recipe Model

The recipe model is depicted in Figure 37.10. The standard defines four levels of recipe: general, site, master and control. These different levels enable a range of representations varying from the general, which are generic process descriptions typically focusing on the chemistry with normalised values, through to the control which are detailed and unambiguous. Although translation of recipes from the generic to the specific is essentially a question of detail, the process is quite complex. The site recipe is specific to plant type, eg batch reactor with jacket for cooling water and inlet manifold, whereas the master recipe is equipment specific in that it relates to the units and equipment modules in a particular cell and/or train. The control recipe is batch specific and fully parameterised. It is only the master and control types of recipe that are of immediate interest from a control point of view.

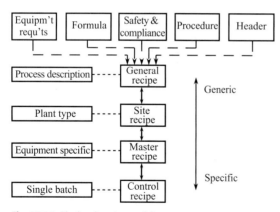

Fig. 37.10 The batch recipe model

A recipe is defined to be the complete set of information that specifies the control requirements for manufacturing a batch of a particular product. Recipes contain five categories of information: header, formulation, equipment, procedures, safety and compliance information. The information contained in each of the five categories has to be appropriate to the level of recipe.

- The header contains the batch name, lot no, product information, *etc*.
- The formulation covers the amounts of raw materials, operating conditions *etc*. In the control recipe this is invariably in the form of data in a parameter list (B-list).
- The equipment relates to the physical model and, for example, identifies equipment and control modules. In the control recipe this information invariably consists of channel addresses in a parameter list (A-list).
- Procedures enable the progression of batches, as described. At the master recipe level the phases are referred to by name only, but at the control level all the steps and actions are included.
- Safety and compliance information provides for miscellaneous information and comment as appropriate.

37.8 Activity Model

The batch control activity model is depicted in Figure 37.11. In a sense, this diagram is an overview of

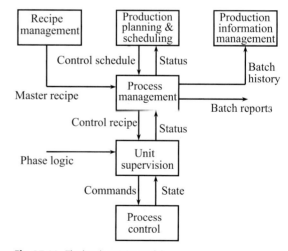

Fig. 37.11 The batch activity model

the S88 architecture. It shows the relationship between the various control activities, each of which is summarised below. The basis of this relationship is the flow of information between activities.

- Recipe management concerns the storage, development and manipulation of recipes. Master recipes are downloaded to process management from which control recipes are downloaded to unit supervision.
- Production planning and scheduling. Production planning receives customers' orders and generates a master schedule for a given time frame. This consists of a queue of recipes in the order they are to be run, each recipe uniquely relating a batch of a particular product to both a unit type and a procedure. Then, knowing what master recipes are available for each unit type, and plant status information from process management, the dynamic scheduler produces the control schedule. This queue relates batches to specific units and procedures, in the same time frame, taking into account delays in processing, availability of equipment, stock of raw materials, *etc*. In effect, the dynamic scheduler has a queue management function.
- Production information management concerns the logging of batch data, its storage and its transformation into batch reports. The standard details the requirements for journal structure, querying systems, historical data, batch tracking, batch end reports, *etc*.
- Process management functions at the level of the process cell. Master recipes are selected, edited as appropriate and translated into control recipes. Operational requirements are cast into batch parameter lists (B lists). Individual batches are initiated. Cell resources are managed according to the requirements of the control schedule, subject to the constraints of shared equipment. Cell and batch data is gathered.
- Unit supervision relates the equipment of a unit to the operations and phases of its control recipe. In particular, the phases, which are referred to by name only in recipes at higher levels of the recipe model, acquire their phase logic in terms of steps and actions. They are further parameterised with equipment address lists (A lists). Unit supervision includes the requesting and release of shared resources. Commands are received from process management and process control status information is returned.

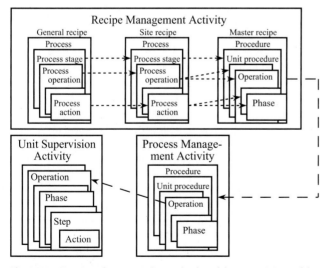

Fig. 37.12 Mapping of process and procedural models onto activity model

- Process control executes the phase logic. This is realised by means of conventional regulatory and discrete control functions. It is also responsible for the gathering and display of data. Note that plant and process protection lies outside the scope of S88.

Some of the relationships between the activity model and the recipe, process and procedural models are depicted in Figure 37.12.

Whilst not stated in the standard itself, it is appropriate to think in terms of the higher level activities of recipe management, production planning and scheduling and production information management being realised by some off-line management information system (MIS), with the lower level activities being handled by nodes within a DCS as appropriate.

37.9 Comments

The standard establishes a framework that addresses a variety of complex issues that are not recognised by many, let alone understood. By defining terminology and relating it to meaningful models much of the ambiguity which pervades batch process control is removed. More importantly, the framework of S88 should force the development of more structured specifications and lead to better quality application software.

Most major control system suppliers have committed themselves to evolving the batch control functionality of their systems towards conformance with the S88 framework. This will be enabled by Part2 of S88 which focuses on underlying issues such as data models, data exchange methodology, phase interfaces, general recipe definition and user depiction of a recipe procedure. It is easy to visualise tools being developed to enable users to articulate their requirements in terms of S88 objects and constructs, and for those requirements to be translated automatically into application software. To what extent it results in portability of S88 objects and code between systems remains to be seen.

Section 6

Digital Control Systems

Systems Architecture

Chapter 38

- 38.1 Advisory Control
- 38.2 Supervisory Control
- 38.3 Direct Digital Control
- 38.4 Integrated Control
- 38.5 Distributed Control
- 38.6 Programmable Logic Controllers
- 38.7 Supervisory Control and Data Acquisition
- 38.8 Management Information Systems
- 38.9 Computer Integrated Manufacturing
- 38.10 Open Systems
- 38.11 Comments

Architecture is the term used to describe the structure of a computer control system. It is directly related to the organisation of the system's I/O channels. New architectures have evolved over the years as the technology of computer control has advanced. This chapter surveys systems architecture and provides a historical perspective. Common terminology used to describe different architectures is explained. Detailed descriptions of systems architecture and terminology are provided by most of the major suppliers in their technical literature.

38.1 Advisory Control

Early computer systems were so unreliable that they could not be used for control, so the first systems used were in an off-line advisory mode, as depicted in Figure 38.1. The computer system operated alongside the conventional analogue instrumentation which controlled the plant. It was connected up to the input signals by means of an input interface (IIF) and was used for monitoring and data logging purposes only. From a control point

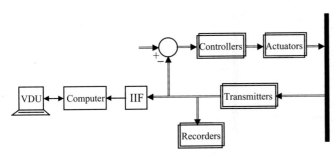

Fig. 38.1 Advisory control system

of view, the computer system achieved very little and the expenditure could only be justified under some research budget. The first advisory system was installed by Esso at their Baton Rouge refinery in Louisiana in 1958.

Note that double lines with arrows are used to indicate multiple signals.

38.2 Supervisory Control

Given that the inputs were being monitored, it made sense to put some optimisation programs into the computer and use them to generate set points. These were output to the analogue controllers by means of an output interface (OIF). Hence the so-called supervisory system, as depicted in Figure 38.2. Note that when the computer system failed the set points would stay put and the analogue loops would continue controlling the plant. In practice, the optimisation programs were often too complex for the computing power available and their benefits proved to be elusive. Whilst there were some marginal hardware savings, mainly in terms of reduced numbers of analogue recorders, the principal benefits realised were experience and confidence. The first supervisory system was installed by Texaco at their Port Arthur refinery in Texas in 1959.

38.3 Direct Digital Control

The next stage of development was to incorporate the computer system within the control loops. This is known as direct digital control (DDC) and is depicted in Figure 38.3. DDC enabled the control, display and recording functions to be realised by means of software. All the analogue controllers and recorders, apart from a few retained on critical loops, were replaced by visual display units (VDU), keyboards and printers. Because these were shared between the various loops, this enabled substantial economic benefits in terms of hardware. Given the

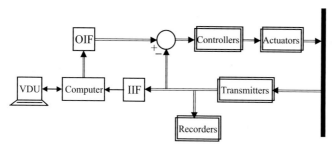

Fig. 38.2 Supervisory control system

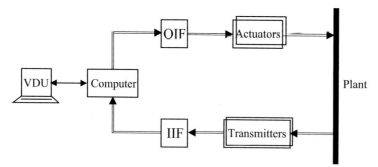

Fig. 38.3 Direct digital control (DDC) system

reliability constraints and the lack of an analogue fall-back position, the early applications had to be on non-critical plant. The first implementation of DDC was on an ICI soda ash plant at Fleetwood in Lancashire in 1962 using a Ferranti Argus system, as reported by Thompson (1964).

Note that the term DDC initially referred to an architecture which was centralised, or monolithic, in nature. However, over the years, it has become a more generic term and is now synonymous with digital devices being an integral part of the loop.

38.4 Integrated Control

This centralised type of architecture was used in particular for the control of complex batch plant. The systems evolved with advances in technology. In particular, from the early 1970s onwards, they became microprocessor based. Initially they were based upon single processors but, with the ever increasing demands for power and functionality, multiple processors became commonplace. Such systems were referred to as integrated control systems (ICS), as depicted in Figure 38.4. An ICS essentially consisted of three parts with dedicated links between: a plant interface unit (PIU), a process control unit (PCU) and an operator control station (OCS).

The PIU handled all the field I/O signals: it was sometimes referred to as a plant multiplexer assembly. The PCU supported all the control software. Note that, although functionally different, the PIU and the PCU were usually housed in the same cabinet. The OCS, often referred to simply as the operator station, was the human interface. An important point to appreciate is that the OCS was more than just a VDU and keyboard and, typically, had a processor which pre-processed signals from the PCU for display purposes.

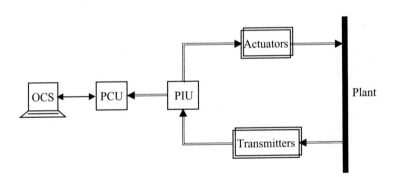

Fig. 38.4 Integrated control system (ICS)

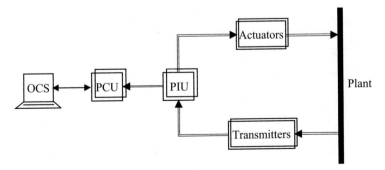

Fig. 38.5 Dual integrated control system

The obvious drawback to the ICS was that "all its eggs were in one basket" and, when the system failed, control of the plant was lost. Thus, for critical applications, it was necessary to have a dual system with common I/O signals, as shown in Figure 38.5, so that the stand-by system could take over when the controlling system failed. Fortunately, it was not usually necessary to have to go to the lengths, not to mention cost, of putting in dual systems.

ICS, as an architecture, fell into disuse as recently as the 1990s.

38.5 Distributed Control

The year of 1975 saw a step change in architecture with the launch by Honeywell of their microprocessor based TDC 2000 system. This was the first distributed control system (DCS) on the market and, in many respects, it became the de-facto standard for the next decade or so. The principal feature of the architecture of a DCS is its decentralised structure, consisting of PIUs, PCUs and OCSs interconnected by means of a proprietary highway. Figure 38.6 illustrates a so-called multi-drop type of DCS architecture. Note that the acronyms used are not universal.

The highway provides for communications between the PCUs and the OCSs, with current values and status information being passed from PCU to OCS and control commands in the opposite direction. It is usual for the highway to be dual to provide for redundancy, each PCU and OCS being connected to both highways. In the event of one highway failing, units can continue to communicate over the other. Indeed, this facility enables units to be connected to and removed from the system, one highway at a time, without disrupting operations.

Other modules are invariably connected to the highway, such as history modules (HM) and application modules (AM). Typically, an HM provides bulk memory for archiving purposes which enables much larger quantities of historical data to be stored and processed than is feasible in the memory of a PCU alone. Similarly, an AM enables advanced control packages to be run that require more processor power than is normally available in a single PCU. Such packages would be for optimisation or statistical process control or an expert system for decision support.

The PIUs and PCUs are normally organised on a local basis, handling the I/O signals and taking the control actions for a relatively self contained

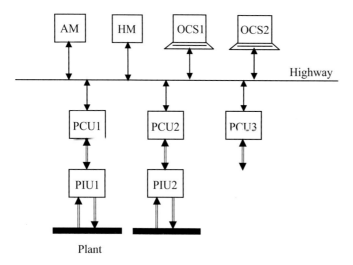

Fig. 38.6 Multi-drop type of distributed control system (DCS)

area of plant. This is a principal advantage of DCS: it enables processor power to be targeted very effectively. DCS lends itself to the control of continuous plant consisting of relatively autonomous production units. With modern DCSs, a single node consisting of a PIU, PCU and OCS is powerful enough to be considered as an ICS in its own right.

38.6 Programmable Logic Controllers

Programmable logic controller (PLC) systems have a different pedigree to PCUs. They emerged within the manufacturing industries as a microprocessor based alternative to hard wired relay logic circuits. Their architecture is not dissimilar to that of DCS systems, to the extent that a PLC has to have a PIU to handle plant I/O and that the PLCs and OCSs communicate over a highway, as depicted in Figure 38.7. However, from a software point of view, there are some very important differences. PLCs are discussed in more detail in Chapter 47.

38.7 Supervisory Control and Data Acquisition

Supervisory control and data acquisition (SCADA) is a term which is ambiguously used to describe a type of application rather than an architecture. Historically, SCADA was mainly associated with utilities and offshore applications but nowadays it is much more commonplace. In general, SCADA systems are used for monitoring and data logging purposes. Their control capability tends to be restricted to adjusting the set points of controllers: supervisory control rather than DDC. Thus SCADA systems have large numbers of inputs and relatively few outputs, as depicted in Figure 38.8.

In terms of hardware, SCADA systems are diverse. Hierarchically speaking, a typical system consists of a local area network (LAN) of personal computer (PC) type operator stations sitting on top of other systems. These other systems may be any combination of DCS nodes, PLCs, single loop controllers (SLC) and "packaged" instrumentation such as analysers. Connections to the network are by means of gateways (GW). These are microprocessor based devices which convert data from the protocol of one highway or network to that of an-

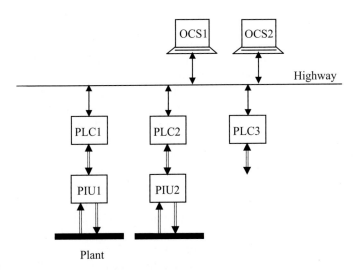

Fig. 38.7 Programmable logic controller (PLC) system

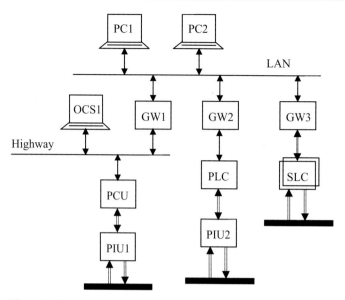

Fig. 38.8 Supervisory control and data acquisition (SCADA) system

other. Gateways also provide buffering to accommodate the different speeds of communication.

38.8 Management Information Systems

A management information system (MIS) typically consists of one or more "host" computers, typically file servers, connected up to the highway of a DCS by means of a gateway to provide access to the field I/O, as depicted in Figure 38.9. All the information within the DCS, and other systems connected to the highway, is thus available for storage and manipulation within the host and is typically available at PC type terminals or workstations. Connection of other systems to the highway is by means of other gateways.

The information stored in the database of an MIS is extensive and accessible to personnel beyond those directly concerned with the operation of the DCS. An MIS enables plant wide calculations, on-line, typically using model based methods that would be too complex or too extensive to be carried out within a PCU. Examples of such calculations are of process efficiencies, plant utilisation, materials inventory and utilities consumption. The database is invariably of a relational nature, as discussed in Chapter 99.

38.9 Computer Integrated Manufacturing

Computer integrated manufacturing (CIM) is in many respects an extension of MIS. The essential difference is that in CIM management information is used for controlling the plant. Thus information flows in both directions between the host and PCUs. For example, production may be scheduled in order to maximise throughput or performance optimised to minimise costs. CIM calculations are in real-time and have to take into account production requirements, availability of raw materials, plant utilisation, *etc.* It is normal within CIM to adopt a clustered approach, as depicted in Figure 38.10. Within a cluster, PCUs and OCSs send data to each other over the highway, such that each cluster can function independently of the others. Thus only data that needs to be transferred from

38.9 Computer Integrated Manufacturing

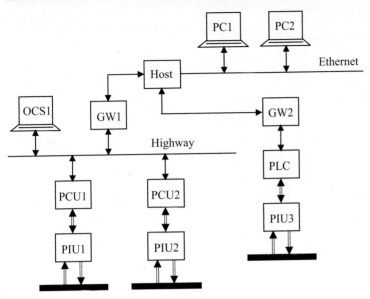

Fig. 38.9 Management information system (MIS)

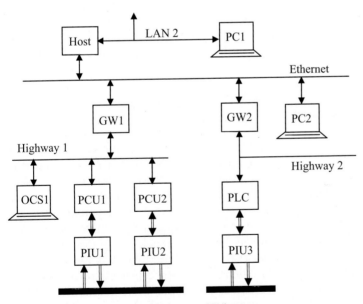

Fig. 38.10 Computer integrated manufacturing (CIM) system

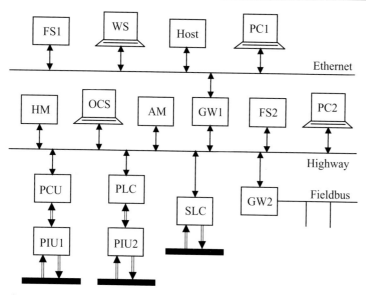

Fig. 38.11 Open system architecture

one cluster to another is transmitted over the network. MIS and CIM are covered in more detail in Chapter 100.

38.10 Open Systems

The advent of open systems is leading to flatter architectures as depicted in Figure 38.11.

In effect, evolution of systems is resolving itself into two distinct domains. The information domain is based upon a network, normally Ethernet, for management information and the control domain consists of devices connected to a highway for control purposes. Central to the concept of an open system is the use of industry standards for communications, databases, displays, *etc.*, which enable integration of non-proprietary equipment. Note the use of workstations (WS) and file servers (FS) on the networks for handling the very large databases associated with management information, communications, *etc.* It is not inconceivable with the advent of real-time Ethernet that evolution will result in just a single domain.

Aspects of communications and open systems are covered more fully in Chapters 40 and 49 respectively. Field bus is covered in Chapter 50.

38.11 Comments

There are many varieties of architecture and the distinctions between some of them, such as SCADA, MIS and CIM, are rather fuzzy. No particular architecture is "correct" for any given plant, although there are many examples of systems that have been installed that are inappropriate for the application. The golden rule is that the architecture of the system should match that of the plant. For example, as indicated, DCS are appropriate for continuous plant consisting of relatively autonomous production units, SCADA for applications with diverse control systems which need to be co-ordinated, PLCs when there are localised sequencing requirements, and so on.

Systems Hardware

Chapter 39

- 39.1 Operator Station
- 39.2 Physical Structure
- 39.3 Card Types
- 39.4 Personal Computers

There is much similarity in the hardware of different suppliers' systems, whether they are DCS, PLC or otherwise. This chapter provides an overview of various generic aspects of hardware. The emphasis is on developing an understanding of how a system's hardware fits together and of how it relates to the system's functionality. This should be sufficient to enable meaningful dialogue about a systems hardware requirements.

39.1 Operator Station

The operator's control station (OCS) is a term which refers to the hardware used by an operator. It typically consists of a colour graphics VDU, a keyboard, and some processor capacity for handling the operators control program (OCP). In essence the OCP is the software with which the operator interacts. It enables access to the system's database and handles the various displays.

The VDU is the primary means of displaying information about the plant being controlled in all but the most rudimentary of systems. Most modern OCSs require high resolution colour VDUs to enable the graphics supported by the OCP to be displayed. Even for the smallest of plants it is usual practice to provide at least two VDUs: on larger plants there are multiple VDUs. First, this provides a measure of redundancy in the event of a VDU failing. And second, it enables different displays to be displayed simultaneously which is often very convenient.

Keyboards are the primary means of manually entering data into a control system. They need to be in some ruggedised form for use in control rooms. Flat panel membrane keyboards are commonly used because they are coffee proof. There has been much effort invested in the design of keyboards. Most have both the standard qwerty alphabetic and numeric keys, and a combination of dedicated and user definable keys. There is a trade off between the number of dedicated keys which make for user friendliness and the number of user definable keys which provide flexibility.

Because process control systems make extensive use of displays, it is necessary to have some form of cursor control. This is typically provided by means of dedicated keys, mouse or trackball. An expensive alternative is use of a touch screen. Note that provision of a touch screen does not obviate the need for a keyboard.

The VDU and keyboard, or alternatives, are often engineered into proprietary desk-type furniture, in which case they are referred to collectively as a console. Again, much effort has been invested by some suppliers in trying to determine the optimum ergonomic layout of consoles in terms presentation of information, access to functions and operator comfort.

For engineering purposes, it is usual to have separate VDUs and keyboards, often located in an office away from the control room. This enables software development to be done off-line. In general, the engineers control program (ECP), through which software development is realised, does not have the same high resolution colour graphics requirements as the OCP and, typically, a PC is used.

The principal peripherals associated with control systems are compact disc (CD) drives and printers. The disc drives are used for system loading and long term archiving of historic data. Normally there will be two printers, one dedicated to event logging and the other for printing out reports, database listings, *etc*. This also provides a minimum level of redundancy. Some printers have dedicated keyboards which may provide emergency backup for the OCS keyboards.

The functionality of an OCP is described in much more detail in Chapter 42.

39.2 Physical Structure

PIUs, PCUs and PLCs were described in Chapter 38 as being functionally distinct. However, they are not necessarily physically distinct. In practice, systems are built from cards which slot into racks, the combination of cards being determined by the application. The racks, often referred to as bins, cages or files, are mounted in frames which are usually installed in a cabinet. The cabinets typically have self sealing doors at front and back for access. Large systems have several cabinets. Figure 39.1 depicts the typical physical structure.

The dimensions of cards, racks, *etc*. are covered by the DIN 41494 (Part 1) and IEC 60297 (Part 1) standards. Construction is modular and is sized on a unit (U) basis, 1 U = 1.75 in = 44.5 mm A standard 19-inch rack has an effective width of 10 U and is capable of housing 5 cards of width 2 U, 10 cards of width 1 U, 20 cards of width U/2, or combinations thereof. The nominal height is determined by that of the front panel and is a multiple of the U value. For example, a 3-U panel is 13.35 cm high. This modular basis enables suppliers to configure systems as appropriate. A commercially available

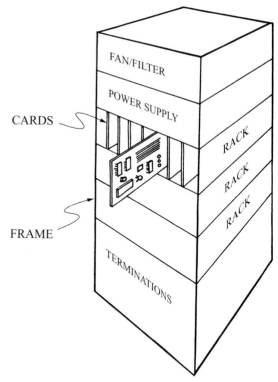

Fig. 39.1 Typical card, rack and frame structure

frame will typically house 12 racks of height 3 U, or whatever. Cards have guides along their top and bottom edges which enable them to be slid into grooves in the rack, which makes for easy removal. The location of a card/groove is often referred to as a slot.

Across the back of each rack is a so-called backplane. This is a printed circuit board (PCB) which has dedicated tracks for the system's buses, *i.e.* the address, control and data buses, and general purpose tracks for power supply and earthing. Each card has a multichannel connector on its back edge which engages in a socket, either directly or by means of a ribbon cable, on the back plane. This extends the buses from the back-plane through to the card, as depicted in Figure 39.2.

Slots may be dedicated to a certain type of card, for example a DIN slot for a DIN card, in which case each card will have a keyway, maybe built into the edge connector, to ensure that cards are plugged

39.2 Physical Structure   289

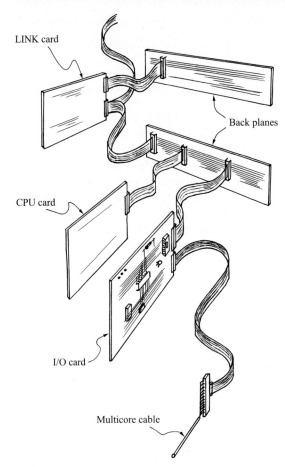

Fig. 39.2 Connections between cards and back plane

into the correct type of slot. Otherwise slots may be multipurpose, in which case the type of card in a given slot is established by means of software. Whether slots are dedicated or multipurpose is a function of the design of the backplane.

The back planes are themselves connected to each other to extend the buses from rack to rack and from frame to frame. This is achieved either directly, by means of ribbon cables in a daisy chain, or indirectly by means of LINK cards, also depicted in Figure 39.2.

Note the termination rack, power supply and fan shown in Figure 39.1.

The termination rack is for handling the I/O signals from the field. These normally arrive at the cabinet *via* multicore cables, having been pre-grouped on a like-for-like basis in some separate marshalling cabinets. For example, one multicore cable may carry 4–20 mA current inputs destined for a particular AIN card. Each multicore is either wired into a termination assembly, or is connected *via* a panel mounted multicore socket and plug, to connect up with a ribbon cable. This ribbon cable is connected to the appropriate I/O card by means of an edge connector, again depicted in Figure 39.2. System layout is described in more detail in Chapter 51.

Computer control systems are essentially comprised of light current circuits and devices. Their components become hot under normal operation and require cooling. Cabinets are therefore provided with a fan to draw air into the cabinet, through a filter and blow it across the cards.

The power supply to a system is normally either 110 or 240 V a.c. This is rectified to d.c. in a power supply unit (PSU) and regulated at levels appropriate to the device types used on the cards. Typically the PSU supplies power at +24 V, +10 V and ±5 V d.c. Power is distributed from the PSU throughout the cabinet by means of d.c. rails, connections to general purpose tracks on the back-planes and, *via* the edge connectors, to the cards. If the system has several cabinets, it is usual to have a single PSU in one cabinet connected up to the d.c. rails in all the others.

An important point to appreciate is that the PSU is usually sized to provide the power for the computer control system and for all the field instrumentation too. Thus, for example, all the 4–20-mA signal transmission is typically driven by the 24-V system supply. This is a reliability issue. There is no point in providing a high integrity power supply for the computer control system if the power supply for its instrumentation is unreliable. Chapter 51 covers power supply, back-up and distribution in more detail.

39.3 Card Types

Cards are printed circuit boards with dedicated tracks upon which are mounted a variety of integrated circuit chips, passive and active electronic components. There are essentially three categories of card: I/O signals, systems and communications:

I/O cards:

AIN: analogue input cards. These typically handle 8, 16 or 32 analogue input channels per card. There are different types of AIN cards for 4–20 mA, 0–10 V, mV thermocouple signals, *etc.*

AOT: analogue output cards. They handle 4, 8 or 16 analogue output channels per card, the channels normally being of 4–20 mA.

PIN: pulse input cards. For handling pulse input signals from rotary devices such as turbine meters.

DIN: discrete input cards. Typically handle 16 or 32 discrete input signals. The discrete signals are usually in the form of 0/10 or 0/5 V.

DOT: discrete output cards. Normally enable 16 or 32 relays to be operated. There are different types of DOT card according to the power to be handled by the relays.

Note that all I/O cards generally have both active and passive circuits to provide protection against electrical damage. They also provide for significant pre and post processing of I/O signals. The functionality of some of these cards is described in more detail in Chapters 44 and 46.

System cards:

CPU card. The central processor unit card is the heart of any control system. The functionality of the CPU is described in detail in Chapter 9. The CPU card has the real time clock (RTC) mounted on it. The function of all the other cards is to support CPU activity. Note that in modern control systems there are often several CPUs operating in a multi-processing environment, each CPU being targeted on specific areas of activity.

ROM cards. These cards contain read only memory (ROM) which is inherently permanent. ROM memory is always in the form of chips. Although generically referred to as ROM, there are a number of variations on the theme such as PROM (programmable), EPROM (erasable and programmable) and EAROM (electrically alterable), *etc.*

RAM cards. These cards contain random access memory (RAM) which is not permanent and is often described as being volatile. If the power supply fails the information stored in RAM is lost. For this reason most systems have battery back-up for the RAM which will provide protection for a few hours.

CHECK cards. These essentially check the correct functioning of other cards. They have a diagnostics capability which either intercepts/interprets the transactions between other cards and/or communicates with them directly. It is sensible to have one check card per frame. Not every system has them.

DISC drivers. It is common practice to mount the system's discs, compact or hard, and their drivers in slot based modules and to connect them up to the back-plane as if they were cards. The discs provide bulk memory. There should be at least two separate bulk memory systems to provide for redundancy.

It is normal practice, to make for efficient processing, for the CPU, ROM, and RAM cards to all be within the same rack.

Coms cards:

LINK cards. Typically each rack, apart from that housing the CPU, will have one. Link cards may be either active or passive. An active link card receives requests to send/receive data from the CPU and, if the request relates to a card within its rack, will direct the request towards the appropriate card. A passive link card essentially consists of a bus driver and enables extension of the back-plane. In effect, any request to send/receive data is repeated to every card connected to the buses.

PORT cards. These are for serial communication with devices, such as the OCS, connected to the highway. They convert the data from the bus system into the protocol of the highway, and *vice versa*. They also provide buffering to accommodate the different communications speeds.

GATE cards, referred to as gateways. They convert data from the protocol of the DCS highway to that of other networks associated with, for example, MIS or CIM. Gateways also provide buffering to accommodate the different communications speeds of the highway and network.

COMS cards. These are for either serial (UART) or parallel communication with peripheral devices connected into the bus structure of the system.

Whereas the functionality of these various cards is reasonably common from one system to another, there is little agreement on the terminology of card types other than for I/O cards. The terminology used in this chapter is representative only.

39.4 Personal Computers

Current generation systems use PC technology for both the OCS and PCU functions. This was generally considered to be unacceptable practice, even for small non-critical applications, because PCs did not have real-time operating systems and could not support multi-tasking. However, with the advent of NT, which enables both real-time operations and multi-tasking, there has been a massive expansion in the use of PCs for display and monitoring purposes, if not for the I/O processing and control functions.

Another important consideration is memory type because PCs are normally disc based. Disc drives used to be unreliable but that is no longer an issue. Also, disc access is not deterministic in a real-time sense, although that is not necessarily a problem in many applications. However, what is a very real problem is potential memory corruption. For this reason it is normal for the operating system and control routines to be ROM based and for the database to be loaded into RAM from hard disc on start up. PCs can be fitted with both ROM and RAM boards to meet these criteria. Back up copies of the database are ideally kept on redundant hard discs although use of autonomous partitions on a common disc is not uncommon on non-critical applications. The hard disk also provides bulk memory for archiving of data.

Communications

Chapter 40

40.1 Messages
40.2 Local Area Networks
40.3 Token Systems
40.4 Protocols
40.5 Network Access
40.6 Transmission
40.7 Telemetry
40.8 Radio Communications
40.9 Telemetry Units
40.10 Comments

Communications, in particular of a digital nature, are fundamental to modern process control systems and some of the concepts have already been encountered in previous chapters. The significance of highways and networks were established in Chapter 38 on systems architecture and various types of communications cards were considered in Chapter 39 on systems hardware.

The technology of communications is vast, embracing different types of networks and transmission media. This chapter, therefore, attempts to summarise some of the more important aspects in relation to process control systems, with an emphasis on the relevant standards and protocols. Fieldbus is deferred to Chapter 50. For a more comprehensive coverage the reader is referred to standard texts such as that by Thompson (1997) and Willis (1993).

40.1 Messages

A message is the term loosely used to describe data transmitted over a network. Messages can be transmitted between any two devices, usually referred to as nodes, connected to the network. In practice, depending on the nature of the network, a message consists of a structure known as a frame which contains the data being transmitted. Figure 40.1 depicts an IEEE 802 type of frame.

As can be seen, the frame consists of a header, the data being transmitted, and a trailer. The header consists of

- A preamble which announces the start of transmission
- A start delimiter which enables the start of the frame to be detected
- The address of the node to which the message is being transmitted: the destination
- The address of the node which generated the message: the source
- The frame type and/or length which enables the destination node to determine how to handle the message
- Status information, such as flags and functional options

The trailer consists of:

- A frame check sequence for checking whether the data transmitted has been received error free
- An end delimiter which enables the end of the frame to be detected

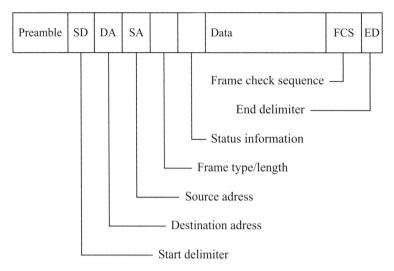

Fig. 40.1 IEEE 802 type of frame

Any device connected to the network must have the necessary hardware and software to be capable of both reading and/or generating such a frame.

40.2 Local Area Networks

The most commonly used type of communications system is the local area network (LAN). The IEEE has developed standards for LANs which are generally adhered to by systems suppliers. The three LANs of particular interest for process control purposes are Ethernet, token bus and token ring. The essential difference between them is the method by which messages on the LAN, known as traffic, are controlled.

The IEEE 802.3 standard applies to a LAN type known as the carrier sense multiple access with collision detection (CSMA/CD). The most widespread implementation of CSMA/CD is Ethernet, developed jointly by Xerox, Intel and DEC during the 1980s, although it should be appreciated that Ethernet is not the standard itself. Industrial Ethernet is the technology compatible with the IEEE 802.3 standard that has been designed and packaged to meet the requirements and rigor of industrial applications.

Within an Ethernet system, no one node has priority over another. Thus, before transmitting a message, a node must establish that no other node is using the network. If the network is free then the node can start to send data. If not, it must wait until the other node has completed its transmission: normally there will be minimal delay before an idle condition is detected. When a node transmits, all the other nodes receive the frame and compare the destination address with their own. That node to which the message is directed receives it, the others ignore it.

However, with such a non-priority system, it is possible for two or mode nodes to attempt to start transmitting simultaneously. This is referred to as a collision. If collisions were allowed to occur then the data on the network would become corrupted. Thus the nodes' hardware contain circuits which recognise the occurrence of collisions and abort transmission. Each node then enters a wait mode before trying again. To reduce the scope of a further collision, each node has a randomised wait period built into its transmission logic. This uncertainty about the delay associated with collisions means that Ethernet transactions are not deterministic with respect to time: the LAN is said to be probabilistic.

Clearly the more nodes there are on a network, the greater the chance of collisions. Also, the longer the network becomes, the more often collisions occur due to the propagation delay between nodes. Both of these factors affect the throughput of the network and thus the response time between nodes. Provided the loading on the network is less than some 30% of the system's capacity, CSMA/CD is the most efficient LAN type. It is easy to see that CSMA/CD will cope well with intermittent operations, such as downloading of programs, file transfer and so on. For this reason Ethernet is commonly used at higher levels in MIS and CIM type applications.

However, for process control purposes, the requirement is to handle many short messages at a relatively high frequency. CSMA/CD would probably cope with normal operating conditions within the 30% of its capacity. However, under abnormal conditions, even if there are only a few nodes, the volume of traffic due to alarms, diagnostics and so on escalates rapidly and leads to a massive increase in the number of collisions. Thus a significant amount of LAN time is spent "timed out" and its performance deteriorates. It is possible, but statistically improbable, that some messages may not get through at all. For this reason token systems are used for control purposes.

40.3 Token Systems

Token systems use a mechanism, known as a token, which avoids collisions on the network. It is conceptually equivalent to the baton used in relay races: the nodes take it in turn to hold the token, and only one node can hold the token at a time. The node which currently holds the token may transmit to any other node on the network. It is appropriate to think of the token as being a flag held within the data frame that indicates whether the frame is "free" or otherwise.

The token bus is a LAN, often referred to as a highway, to which the IEEE 802.4 standard applies. Token bus topology, which is essentially the same

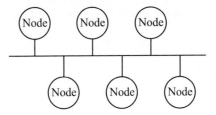

Fig. 40.2 Open structure of token bus topology

as for Ethernet, has an open structure as depicted in Figure 40.2.

An empty frame containing a "free" token is passed from node to node in a fixed order. If the node receiving the frame does not have any data to transmit, it is passed on to the next node, referred to as its successor, and so on. In essence, the address of the successor node is loaded into the frame and transmitted. All the other nodes receive the frame and compare the successor address with their own: the successor receives it and the others ignore it. The order of succession is predefined and includes all the nodes on the highway: it is not necessarily the order in which they are physically connected. Although the token bus has an open structure, it can be seen that it is controlled logically as a ring in the sense that the token circulates around all the nodes in a fixed order.

Suppose that an empty frame arrives at a node waiting to transmit data. The token is marked as "busy", the frame is loaded with the addresses of the source and destination nodes and the data itself, and the frame is transmitted. All the other nodes receive the frame and compare the destination address with their own: that node to which the frame is directed receives it and the others ignore it. The destination node then retransmits the frame to the original source node: in effect, the addresses having been swapped over. The source node then either transmits another frame, if it has time, or else marks the token as "free" and passes it on to its successor.

It is important to distinguish between the two modes of operation. If there are no messages, the empty frame with the "free" token passes around the bus in successor order looking for a message. The time taken for an empty frame to pass around all the nodes is known as the token rotation time. However, when data is being transmitted, it goes directly from source to destination node and back again. The time allowed for this is known as the slot time, and is different from the token rotation time.

Token passing is most efficient when the network is heavily loaded because there are no collisions: it is inefficient at low loadings since the frame is empty most of the time, but that doesn't matter. Token bus is entirely deterministic because the token rotation time is fixed by the number of nodes and each node's slot time is specified. Also, no single node can monopolise the network. For these reasons token bus is used for real-time control in industry.

Token ring is another type of LAN to which the IEEE 802.5 standard applies. It is essentially the same as token bus, except that the ends of the bus are physically connected to form a ring, as depicted in Figure 40.3. One obvious advantage of a ring network is that transmission occurs around the ring in both directions which provides for redundancy in the event of it being physically severed.

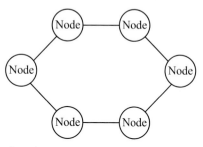

Fig. 40.3 Loop structure of token ring topology

In some systems, when the token is passed from a node to its successor, the node listens for the successor to either transmit data or pass the token on. If there is no evidence of this, the node assesses the status of the successor and, if a failure is detected, subsequently by passes it.

The process of receiving signals and retransmitting at a node is referred to as repeating. It has the effect of boosting the strength of a signal. Because signals degrade over long distances, i.e. they become distorted and pick up noise, repeaters are used to extend the length of a network.

40.4 Protocols

The frames transmitted on a network must conform to some form of communications protocol. At its simplest this defines the format of the frames and any control signals that are required to ensure correct data transfer. More complex protocols contain multiple layers which control not only the transmission of the data itself, but also the presentation of that data to and from the applications. The open systems interconnection (OSI) protocol, known as the OSI model, published in ISO 7498-1 (1995) for reference purposes, has seven layers whose functionality is as summarised in Table 40.1.

Table 40.1 Functionality of OSI seven layer model

Layer		Description
7	Application	User interaction
6	Presentation	Translates data
5	Session	Controls dialogue
4	Transport	Provides transparency
3	Network	Routes transmission
2	Data link	Detects errors
1	Physical	Connects device

The OSI model is generic in that it describes the functionality of an "ideal" protocol and is independent of any physical medium or type of connection. It is designed to cover the most complex communications networks involving world-wide communication of all types and volumes of data transmitted *via* multiple computer systems. Not all seven layers would necessarily exist in any specific protocol. Imagine data being transmitted from a user program in a source computer over a network to a user program in a destination computer. The data travels down the various levels in the source until it reaches the physical layer, is transmitted *via* the physical medium to the destination, and then up through the various layers of the destination to its user program. The functions of each layer are as follows:

- Physical layer: transmits and receives bit patterns over the network. It is concerned with the electronic aspects of the transfer, such as the voltage levels representing binary 0 and 1 and the control signals used for synchronisation, and with the mechanical aspects of plugs and pin assignments.
- Data link layer: makes communications at the physical layer appear to higher layers as though they are error free. It embraces error checking mechanisms and organises for the retransmission of corrupted data.
- Network layer: concerned with routing messages across the network. It takes a message from the transport layer, splits it into "packets", identifies the packets' destinations and organises their transmission.
- Transport layer: hides all the network dependent characteristics from the layers above it, essentially providing transparent data transfer. Thus a user on one computer system can communicate with a user on another without concern for the underlying network being used to transfer the data.
- Session layer: establishes and maintains the communication path between two users of the network during the period of time that they are logically connected.
- Presentation layer: concerns the format of data and handles its encryption and decryption. For example, text may be translated into/from ASCII code, and data may be compressed by using codes for commonly used phrases.
- Application layer: this is the highest level in the model and is the environment in which the user's programs operate.

Many different protocols exist: some are specific to given suppliers and/or systems and others have been defined by bodies such as the International Telegraph and Telephone Consultative Committee (CCITT) and ISO. Protocols can be categorised as being either:

- Master/slave (primary/secondary) communication. The master node is in overall control of the network. Typical of this type of protocol is the ISO high-level data-link control (HDLC).
- Peer to peer communication, as used on most LANs, in which no node has overall control. Perhaps the two best known protocols in this category are General Motors' manufacturing automation protocol (MAP) and the transmission control protocol / internet protocol (TCP/IP) standard.

40.5 Network Access

Access concerns gaining control of the network in order to transmit messages. The three layered access model of IEEE 802 is depicted in Figure 40.4 and relates to both Ethernet and token bus types of network. The most commonly used access protocol is TCP/IP which has three layers. It is consistent with the two lower layers of the OSI model but doesn't have its full functionality.

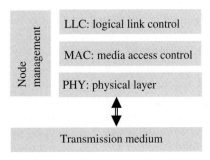

Fig. 40.4 Three layer access model of IEEE 802

Node management provides the functionality that fragments and/or reconstructs messages into/from frames of data which are of the correct length for transmission. The three layers are:

- Logic link control (LLC) layer: this puts the source and destination addresses, the data to be transmitted and the checksum into the frame.
- Media access control (MAC) layer: this handles collisions and imposes random waits for Ethernet type LANs, or handles the token/flags and enables repeats for token bus type LANs.

- Physical layer: this handles the encoding and decoding of data, if necessary, and the transmission and/or receipt of messages between nodes.

The hardware interface and software drivers necessary to support these layers of access must be provided by any PORT card, as described in Chapter 39, to enable its node's access to the network.

Because of the variety of protocols in use, it may not be possible to connect a particular node directly to a network or highway. Incompatibilities are handled by protocol converters, normally referred to as gateways, referred to in Chapter 38.

40.6 Transmission

LANs are available either as broadband or as base band systems:

- A broadband system uses analogue technology, where a modem is used to introduce a frequency carrier signal onto the transmission medium. The carrier signal is modulated by the digital data supplied by the node. Broadband systems are so named because the analogue carrier signals are high frequency: some 10–400 MHz.
- A baseband system uses digital technology. A line driver introduces voltage shifts onto the transmission medium. The voltage changes directly represent the digital data. For example, the RS 232 standard specifies a voltage between +3 and +12 V to represent a "0" and between −3 and −12 V to represent a "1".

The cabling used for carrying LANs is typically screened copper cable. This may be armoured if it is to be passed through areas where there is a possibility of damage. In many systems dual highways are provided, mainly to increase the security of communication, although some systems use the second highway to increase throughput.

Increasingly, fibre optic cables are being used with networks for communications. This involves converting electrical signals into light pulses before transmission. The advantages of fibre optics include:

- Fibre optics enable longer networks because power loss is very low. Typical distances are 120 km without repeaters.
- Because there is no electrical current being carried by the cable, the cable is unaffected by electro-magnetic interference, so it is possible to run fibre optic cables alongside power cables with no adverse effects. Also, there is no induction between different fibre optic channels.
- The absence of electrical energy in the cable makes them particularly useful in hazardous areas, where short circuits are a potential source of ignition.

The principal disadvantages are:

- Limited curvature on bends.
- High skill levels are required for installation and maintenance.
- Cost: the hardware necessary for electrical/optical conversion is expensive, although the advantages gained often outweigh the additional cost.

For all these reasons, fibre optics is an important alternative to conventional copper cabling.

40.7 Telemetry

Some process industries, such as oil, gas and water, have pipeline networks in which the distances between, say, pumping stations and central control rooms are measured in hundreds of kilometres. In such situations LANs are inappropriate and wide area networks (WAN) may be used. Historically WANs provided their long distance capability by means of modulated frequency techniques through the use of modems. However, with the advent of services such as BT's integrated services digital network (ISDN), digital transmission has become more common.

The transmission medium used in a WAN may be a dedicated private line or one leased from a communications company. Alternatively, the link can be set up temporarily by dialling a connection *via* the public switched telephone network

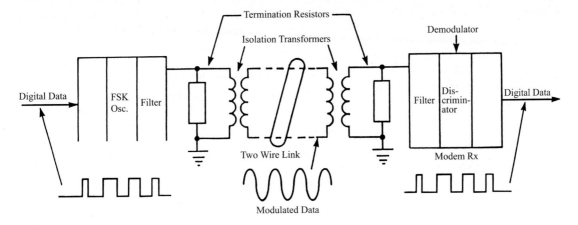

Fig. 40.5 Frequency shift keying for telemetry

(PSTN). Such transmission is modulated using a carrier frequency in the voice frequency band of 300–3400 Hz.

Often the lines used carry several channels of communication. This is achieved by using a technique known as frequency division multiplexing (FDM) for which the carrier frequency required is determined by the rate at which data is to be transferred. Typically this would be 1700 Hz at 1200 baud as specified in CCITT publications.

One modulation technique used in telemetry systems is called frequency shift keying (FSK). In this technique the carrier frequency ω is shifted by the digital data being transmitted. For example, a binary "1" might be represented by $\omega + \Delta\omega$ and a binary "0" by $\omega - \Delta\omega$ where, for a 1200 baud system, $\Delta\omega$ is typically 50 Hz. This technique is depicted in Figure 40.5. Other modulation techniques include pulse code modulation (PCM) and phase modulation.

40.8 Radio Communications

As an alternative to land lines a telemetry system may use radio communications, or some hybrid of telephone lines and radio channels. The use of radio is similar to using land lines except that the power levels required for signals are higher and transmission delays can be longer.

Radio systems normally use frequency modulated (FM) transceivers operating in the range 450–470 MHz ultra high frequency (UHF) band. At UHF and higher frequencies, radio signals can be made very directional with antenna systems using parabolic reflectors producing very narrow beams. In such systems a line of sight transmission path is required, which may be difficult to achieve in certain terrain.

To overcome obstacles in the transmission path a number of different methods may be used:

- Repeaters. A repeater may be placed on top of a hill or tall building, which is acting as a barrier to communication, to pick up the signal, amplify and retransmit it.
- Troposcopic scatter. Radio energy is directed towards part of the Earth's atmosphere called the troposphere, situated at about 10 km above ground level. This forward scatters the signal to the receiver and thus overcomes any obstacles.
- Satellites. Geo-stationary satellites are increasingly being used for all types of communication and can be used with UHF radio.

40.9 Telemetry Units

Telemetry systems require a controlling master station at the control centre which will communicate with one or more slaves, known as remote telemetry units (RTU), *via* the selected transmission medium. Because the master station may have to deal with many RTUs using a common link, it scans them on a time division multiplexing (TDM) basis. An RTU may be completely passive, in that it only performs actions when commanded to do so by the master station. Alternatively, an RTU may have a level of intelligence which enables it to perform some functions unsolicited, such as calling up the master station in the event of an alarm.

The messages transmitted between RTUs and master station must adhere to some protocol such as HDLC. A typical message from the master station will indicate the RTU address and the function it is to perform. If data is also to be sent this will be included in the message. The final part of the message will contain checksum information to allow the RTU to check the incoming message for errors.

The addressed RTU responds to the command by either reading from or writing to the local instrumentation and/or control systems and returning a message to the master station.

Because of the inherent delays associated with communications used in telemetry systems, real-time functions such as closed loop control are difficult to achieve.

40.10 Comments

As can be seen, communications is an extensive and complex field. In general it is not necessary for control engineers to have a detailed knowledge of the subject. Control systems are normally supplied with a communications capability adequate for the application, so it is essentially a question of being able to make a sensible specification in the first place. However, when interconnecting two dissimilar system, a surprisingly common problem, a grasp of the principles is essential.

Software Overview

41.1 Operating System
41.2 Operator's Control Program
41.3 Direct Digital Control
41.4 Archive
41.5 Sequence Executive
41.6 Batch Process Control
41.7 High Level Language
41.8 Engineers Control Program
41.9 Distribution of Packages
41.10 Comments

Within any computer control system there are two main categories of software: system and application. System software is that upon which the system depends to function, irrespective of the type of plant to be controlled. It is purchased as part of the system and can be looked upon as if it is hardware. System software is standard to the system and should never need to be altered. It operates upon the application software for controlling the plant. Application software is plant specific. When a control system is purchased for a plant much effort has to be invested in specifying, developing and testing the application software. If the application software is not correct the system will be unable to control the plant as intended. System software includes a variety of tools used for developing application software.

The distinction is well illustrated by way of analogy. A word-processor may typically consist of an environment such as Windows within which exists a word-processing package such as Word, which itself supports a variety of tools such as a spell-checker. These may be used to create a letter which is subsequently printed. Windows, Word, the spell checker and the printer routines are all systems software. They are part of the system and can be assumed to function correctly. The user needs to know how to use them to the extent that it is necessary to edit text, save files and print copies. The Word file is equivalent to application software. It is just a file of text. It may well be free of spelling mistakes and grammatically correct but unless the right words are used in the correct order the letter will be meaningless. The printer will print it out irrespective of whether it makes sense or not.

This chapter describes the major system software packages within a typical computer control system and identifies the associated application software. They are depicted in Figure 41.1. A brief description is given of each.

41.1 Operating System

The real-time operating system (RTOS) is, hierarchically speaking, at the top of the system. It is the RTOS which co-ordinates and synchronises all the other activities, referred to as tasks, within the system. The RTOS controls both hardware and software activity. It handles interrupts and determines priorities.

Application software associated with the RTOS includes:

- Task processing files

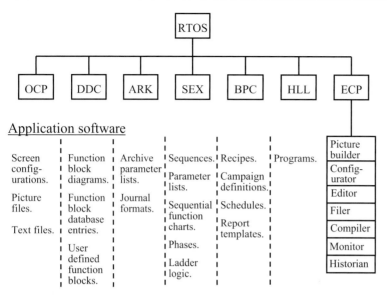

Fig. 41.1 System *vs* application software

41.2 Operator's Control Program

This is the system software with which the operator interacts. It handles both data entry *via* the keyboard, or otherwise, and all the VDU displays. The OCP is described in detail in Chapter 42.

Application software associated with the OCP package includes:

- Screen configurations for group and trend displays
- Picture files for mimic diagrams
- Text files for messages and help displays

41.3 Direct Digital Control

It is this package that provides the real-time control capability of any computer control system. It handles the I/O signals, and applies the control functions to them. These functions are largely configurable, *i.e.* they consist of pre-defined software routines which operate on the database. Often referred to as "function blocks", each one relates to a block of data in the database. Thus configuring a control scheme consists of selecting appropriate function blocks from a library and linking them in the appropriate order according to the application. Most systems support a wide variety of continuous and logic blocks for handling both analogue and discrete signals. The DDC package is discussed in detail in Chapters 44 and 45.

Application software associated with the DDC package includes:

- Function block diagrams
- Function block database entries
- User-defined function blocks

41.4 Archive

The archive package (ARK) concerns long term storage of data in bulk memory. Typical contents of the archive include values of input signals, event logs and batch records. The functionality of the archive enables data to be compressed by, for example, selecting the frequency of sampling and/or the basis for averaging samples. It also enables historic data to be retrieved and displayed or printed as appropriate.

Application software associated with the ARK package includes:
- Archive parameter lists
- Journal formats

41.5 Sequence Executive

Most process control systems provide some form of sequence executive (SEX) package. These vary in complexity from some simple sequencing facility through to handling multiple sequences running in parallel with parameter lists. Simple and short sequences may be realised by means of configurable logic blocks, the use of two such blocks for the realisation of a discrete device are described in Chapter 46. More usually, sequencing is realised by means of a procedural language or a sequential function chart, as explained in Chapter 29, or in the case of PLCs by means of ladder logic as described in Chapter 47. An important aspect of SEX is its interface with DDC because, in reality, sequence control involves a combination of both continuous and sequential activities.

Application software associated with the SEX package includes:
- Sequences
- Phases
- Ladder logic
- Sequential function charts
- Parameter lists

41.6 Batch Process Control

A batch process control (BPC) package is necessary when the batch processing required is other than simple sequence handling. Typically BPC supports recipe handling, campaign definition, real-time scheduling and batch reporting. The functionality of BPC is described more fully in Chapter 37. It may well also support a proprietary batch language, with process oriented instructions and constructs for complex batch control. An important aspect of BPC is its interface with SEX because, in reality, batch process control is realised by means of sequences.

Application software associated with the BPC package includes:
- Recipes
- Campaign definitions
- Schedules
- Batch report templates

41.7 High Level Language

The ability to perform calculations in a high level language (HLL) is extremely useful for modelling purposes. For example, with real time optimisers (RTO), a model of the plant is used to predict controller set-points. Some systems provide an HLL package as part of the system software supported within the PCU. However, recognising that the HLL capacity is of most use in the plant-wide type of calculations associated with MIS and CIM type of applications, many suppliers prefer to provide the HLL package within a separate application module (AM) or in some host mini-computer. Opportunities for the use of an HLL for advanced process automation and control are given in Sections 12 and 13.

Application software associated with the HLL package includes:
- Programs

41.8 Engineers Control Program

The ECP is an integrated suite of tools used for creating new, or editing existing, application software. In general the ECP is used in an off-line mode. Existing application software has to be disabled before it can be edited.

The principal ECP tools are as follows:

Picture Builder

This is the tool used for developing OCP displays, as follows:

1. Configurable displays, such as group displays and trend diagrams, which use pre-determined

templates. Thus the task of picture building consists of identifying the correct templates to use, configuring them on the screen, and attaching parameters to them as appropriate.
2. Free format graphics displays, such as mimic diagrams. These typically consist of a background display, showing pipelines and the outline of plant objects selected from a shape library, upon which is superimposed a foreground display consisting of live data.

Configurator

This is the tool used for database development. Modern configurators are block oriented. In operation, functions are selected from a library and their blocks are dragged across the screen. Connectivity, *i.e.* connections between the blocks, is established by means of drawing lines between them as appropriate. Some of the database entries associated with each block are inherited according to block type, established by virtue of their connectivity, others are entered manually in a template. The database is fundamental to the activity of all the major system software packages, especially DDC, and is discussed in detail in Chapter 45.

Recipe Builder

This tool is used for creating and editing recipes, defining campaigns, creating batch reports, *etc.* Typically the recipe builder is of a configurable nature, with extensive use made of menus, predefined templates and fill in boxes. However, for more extensive and/or detailed data entry, an interface with the editor is normally provided.

Editor

This tool tends to be general purpose and is used for creating and editing text for all the other packages supported by the system. For example, text files for OCP, database for DDC, parameter lists for ARK, sequences for SEX and programs for HLL. It is not uncommon for a word-processing package to be provided for editing purposes.

When there is an MIS capability, the word-processor is invariably accompanied by a spreadsheet. Data may be imported, typically from the archive, which enables the production of management reports.

Compiler

When a program is created, whether it be a high level language program, in the form of a sequence, or whatever, it will need to be compiled into the assembly language of the processor used for implementation: the target system. Cross compilers may be required to translate the programs into an intermediate form. For example, a sequential function chart may need to be cross compiled into structured text prior to being compiled into assembly language.

Filer

Closely associated with the editor is a file handling facility. This enables template files to be copied and edited as well as new files to be created, deleted, listed, printed, *etc.*

Monitor

This tool is primarily associated with SEX and HLL and is used for debugging sequences and programs. It allows the software being tested to be stepped through in a one-step-at-a-time mode which enables its logic to be checked. The software under test normally drives either dummy I/O, *i.e.* a hardware model, or a software model of the plant. Software testing is covered in Chapter 63.

Historian

This is the name given to the tool which is used for interacting with the archive. It allows variables for archiving to be specified and associated parameters to be defined. The historian also enables templates for journals, used for event logs, *etc.*, to be created.

Access to the ECP has to be strictly controlled and procedures for software management imposed. This is discussed further in Chapters 63 and 64.

41.9 Distribution of Packages

Although the functionality of the packages is much the same from one system to another, the detail varies tremendously. In particular, the extent to which the packages are distributed within the hardware makes a big difference to their implementation.

Sometimes hardware is provided specific to a package, provided the package is self-contained. A good example of this is the ARK package which, within a typical DCS, would reside exclusively within a dedicated history module (HM), with bulk memory, and supported by a local historian.

Some PCUs have to support several packages simultaneously because they need to interact. The SEX, for example, may reside within a single PCU but needs to interact with DDC. A typical scenario is that, under sequence control, SEX will initiate a change in loop status, say from manual to auto, which has to be implemented by DDC. Given that they reside in the same PCU they may well have a common configurator and editor.

Other packages, in particular DDC and OCP, must be distributed across several PCUs and supported locally by ECP functions as appropriate. This is realised by replicating in each PCU those parts of the packages that support local functions, using the highway to enable sharing of global functions.

41.10 Comments

Note that there is no general agreement on terminology for these packages and tools, the names and acronyms used in this chapter are representative only.

Operator Interface

Chapter 42

42.1 Access
42.2 Tag Numbers
42.3 Reserved Areas
42.4 Display Systems
42.5 Mimic Diagrams
42.6 Group Displays
42.7 Trend Diagrams
42.8 Status Displays
42.9 Text Displays
42.10 Diagnostic Displays
42.11 Keyboard Functions

Various aspects of the hardware of operator stations (OCS) were described in Chapter 39, and the operators control program (OCP) was introduced in Chapter 41 as the system software package through which the operator interacts with the system. This chapter describes in more detail the functionality of typical OCP type packages. It covers both the different types of screen displays and the keyboard functions.

42.1 Access

Fundamental to the use of an OCP is the level of access to the system that it provides. In general there will need to be at least four levels:

1. No access. Provides very restricted access to the OCP. For example, it may enable personnel to call up and observe overview displays only, but not permit any interaction with the database. This level of access is that in which the system should be left when unattended.
2. Operator. Provides full access to the functionality of the OCP. For example, it enables most displays to be called up, and permits routine interaction with control functions. For example, changing set points, switching loops between auto and manual modes, acknowledging alarms, starting sequences, *etc.*, would all be permitted.
3. Supervisor. This is the same as for the operator but, in addition, provides access to facilities for editing schedules, batch logs, *etc.* It enables access to restricted management information displays.
4. Engineer. Provides full access to all OCP and ECP functions. It also provides full access to all displays, including system diagnostic displays.

Access is normally restricted by means of keys or passwords. The control of access is an important management issue that is covered in Chapter 64.

42.2 Tag Numbers

For reference purposes, it is necessary to have some numbering system to relate the data displayed to both the database and the plant. Most systems have a fairly flexible numbering system that supports combinations of letters and numbers. It therefore makes sense, and is common practice, to use the same tag numbers used on the P&I diagrams of the plant as a basis for referencing within the control system. Thus, for example, the tag number of a control loop would be used to refer both to its database block and to its display templates. Extensions often have to be attached to the tag number to discriminate, for example, between an input or an output signal of the same loop, or between an alarm or status display of the same signal.

42.3 Reserved Areas

The OCP divides the screen into two or more areas. The main area is reserved for the standard displays described below. However, there are also one or more so-called reserved areas, normally above, beside and/or below the main area. The reserved areas are themselves normally divided up into sub areas used for displaying specific process orientated functions such as the date/time, access level, current alarm areas, data entry, status display, messages, *etc*. This information is displayed on the screen at all times, irrespective of what displays are in the main area. When the reserved area consists of dedicated lines across the screen they are commonly referred to as banner lines. Other reserved areas are used for displaying generic toolbars and icons.

42.4 Display Systems

The main types of screen display are menu, mimic, group, trend, status and text. For each of these there will be many instances. It is not uncommon for systems to have several hundred displays. Given that these displays are the principal means of access to data within the system, it is essential that they are organised within a framework with some structure. This enables the operator to find particular displays quickly. Figure 42.1 shows a typical display system structure.

It is usual to have menu displays at the top of the display hierarchy. Menu displays are normally organised on an area basis with, typically, one menu for a number of functionally related major equipment items (units). The function of menu displays is simply to provide access to other display types. These may be organised hierarchically, in which case mimic diagrams would be the next most significant. Otherwise, they would be organised laterally as indicated for the remaining display types. Every display within the system must have a unique reference name/number.

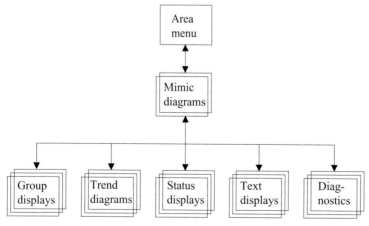

Fig. 42.1 Display system structure

There are various means of selecting displays. Some, such as the use of dedicated keys, require a knowledge of the display reference. Cursor addressing enables the display to be selected from a list. More flexibly, on-screen buttons enable the user to toggle back and forth, up and down, within the display hierarchy. There is a trade-off. For displays whose reference is known the use of dedicated keys is undoubtedly fastest. Otherwise, it is best to logically work through the hierarchy until the required display is found. Most systems support at least two of these approaches.

Two important alternatives are soft keys and touch screens. These are context specific and provide means of selecting one display from another. References for relevant displays that can be called up are built into the application software of the current display. In the case of soft keys, these references are displayed in boxes on the screen which are aligned with user defined keys on the keyboard, one of which is selected as appropriate. For touch screens the references are positioned on the display in pre-defined locations, one of which is selected by the operator touching the screen.

42.5 Mimic Diagrams

Mimic diagrams are screen versions of P&I diagrams. In general, any P&I diagram contains too much information to be contained in a single mimic and has to be broken down into sections. Given that most plants have several P&I diagrams there are inevitably many mimics.

A typical mimic consists of a background display of static information onto which is superimposed a foreground display of dynamic data. The static display is stored as a file on disc and the dynamic data is extracted from the database. When the mimic is selected, the OCP superimposes the dynamic data in positions on the static display that were specified during the design of the static display.

The picture builder tool is used to create mimic diagrams. The background display consists of the outline of plant objects, equipment items, essential pipelines, *etc*. The objects are typically selected from a library of shapes, dragged across the screen to their desired positions, enlarged or reduced if necessary, and connected by drawing lines. Other items, such as valves and controllers, may be selected from a menu of symbols and attached to the shapes in appropriate positions. A typical menu will have dozens of symbols. Some of the more commonly used ones include:

- Agitators
- Centrifugal pump
- Control valve
- Controller bubble
- Isolating valve
- Motors
- Orifice assembly
- Switch

These symbols typically have associated faceplates which enable the display of live data adjacent to the symbol, as illustrated in Picture 42.1. For example, the faceplate for an analogue input may have its tag number, current value and units. That for a discrete output may have its tag number and current status. Some systems support embedded characters in the mimic diagram. These are cursor addressable and enable faceplates to be opened up on the mimic. This is a very effective way of integrating mimics, faceplates and trends.

The use of colours, of which there is often a full palate of 256 choices available, and special effects can considerably enhance the effectiveness of mimics. For example:

- Coloured lines to distinguish between process streams and utilities.
- Different width lines to emphasise the importance of different streams.
- Vertical bars to fill vessel shapes to convey a sense of level.
- Shading switched on and off to create the effect of motion.
- Colour changed to indicate valve status, such as black for closed.
- Symbol colour changed to red to indicate alarm conditions.

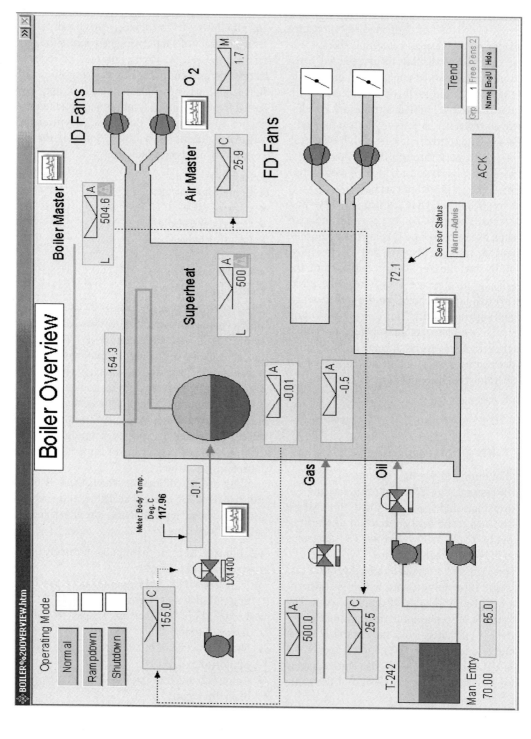

Picture 42.1 Typical mimic display with faceplates (Honeywell)

- Flashing of symbols or vessel outlines to indicate a particular status.
- Pop-up message boxes to provide context sensitive information.
- Strategically positioned soft keys to select mimics of adjacent items of plant.
- Integration of templates for bar displays and trend diagrams.

The design and style of mimic diagrams is subjective: there are no standards *per se* but there are, nevertheless, well recognised industry conventions. The most common, and widely adhered to, convention is the use red to denote alarm conditions and yellow for warnings. It is important that colour coding and special effects are used consistently across the whole set of mimics, to avoid confusion. For example, it would be bad practice to use red for alarm conditions on one mimic and red for stop/green for go on another, unless stop was an alarm condition. It should also be remembered that the objective of mimic diagrams is to present information effectively. The most common mistake is to create displays that are too cluttered and/or colourful. In general, mimics should only relate to one major item of plant each and display useful information only. There is a consensus that subdued colours, especially shades of grey, together with 3-D effects such as shadows, are most effective for backgrounds and major items on the static display.

A costly but very effective alternative to mimic diagrams is the rolling map. In effect, rather than break down the P&I diagrams into separate mimics, a single large mimic of the whole P&I diagram is created. The screen is then treated as a window through which any part of the P&I diagram may be observed. Using a trackball, joystick or mouse, the screen is scrolled across the mimic until the relevant section is found. The only obvious disadvantage to this approach is that, for very large mimics, unless the user is familiar with the topology of the plant, it could take some time to find the relevant section. Put differently, the user can get lost!

42.6 Group Displays

Group displays provide an overview of the state of a section of plant: indeed, they are often referred to as overviews. Typically a group display is configured from a number of faceplates as depicted in Figure 42.2 and illustrated in Picture 42.2.

Using the picture builder tool, standard templates are selected from a template library and allocated to a space in a designated group display. Typically, templates exist for:

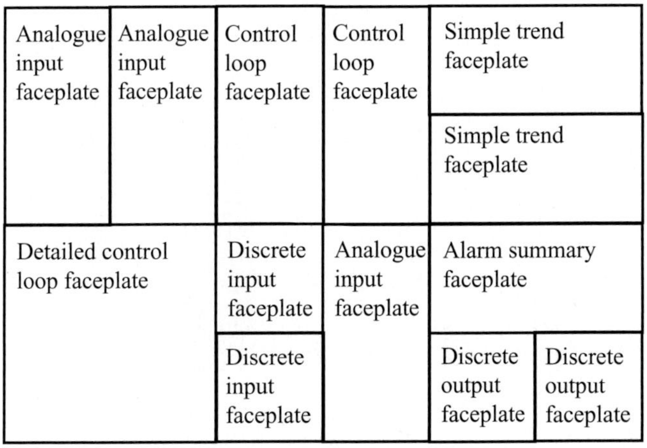

Analogue input faceplate	Analogue input faceplate	Control loop faceplate	Control loop faceplate	Simple trend faceplate	
				Simple trend faceplate	
Detailed control loop faceplate		Discrete input faceplate	Analogue input faceplate	Alarm summary faceplate	
		Discrete input faceplate		Discrete output faceplate	Discrete output faceplate

Fig. 42.2 Group display configured from faceplates

42 Operator Interface

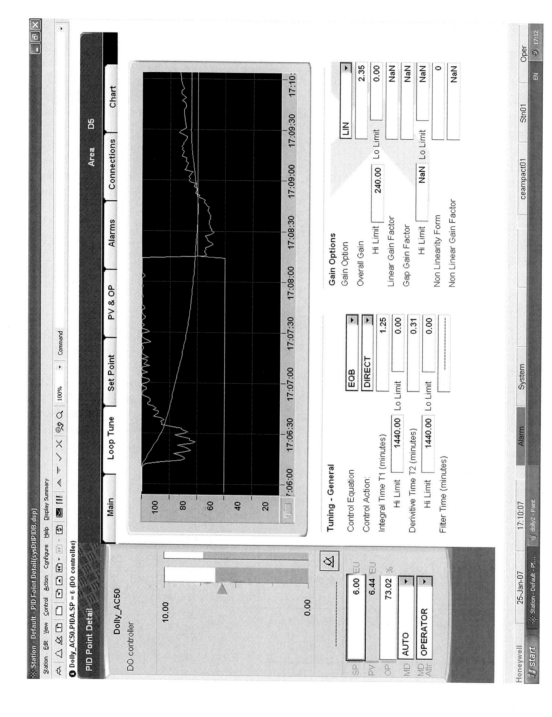

Picture 42.2 Typical group display (Honeywell)

- Analogue inputs.
- Analogue loops, both standard and full.
- Simple trend diagrams.
- Discrete inputs.
- Discrete I/O pairs, which are often referred to as entities.
- Logical variables, *etc.*

Faceplates come in rectangular shapes of varying size and proportions, according to type and the amount of information to be displayed, which provides for flexibility in designing group displays. For example, a standard control loop faceplate, which is quite large, will typically display the same information that is on the facia of a conventional single loop controller, as illustrated in Figure 23.1:

- Its tag number.
- The units of its measured value.
- Numerical values of input, set point and output.
- Bar displays of input, set point and output.
- Alarm limits on the bar displays.
- Its auto/manual status.

Sometimes it is necessary to display more information about a control loop than this so typically a full control loop faceplate would additionally show:

- Controller tuning parameters.
- Choice of algorithm.
- Forward/reverse acting.
- Default settings.
- Status on power-up, *etc.*

In contrast to this, a discrete input template, which is small, may simply show its tag number, current status and desired status. There is a trade-off between the size and number of templates in a group. The smaller the template the less the detail about any particular function but the greater the sense of an overview.

Once a tag number is assigned to a template, current values are extracted from the database and included in the template whenever that group is displayed. As the data changes the OCP updates the display in real-time.

A simple but very effective form of faceplate for use with group displays uses single and/or double lines of text with an embedded bar display, as depicted in Figure 42.3, for an analogue measurement. On the left of the upper line is the signal reference number and a field to display the engineering units. On the right is a standard length bar, scaled in proportion to the range of the measured value, with the current value of the measurement indicated by a pointer or otherwise. The bar would typically also indicate the upper and lower alarm limits and the set point. The lower line contains fields for a description of the measurement and for displaying numerical values in engineering units for the top and bottom of the range and for the current value of the measurement.

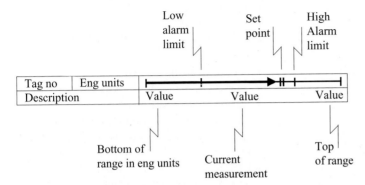

Fig. 42.3 Bar display type of template

42.7 Trend Diagrams

Trend diagrams are essentially graphs of variables plotted against real-time. Each trend is just a series of values of some variable extracted from the database. Typically several related variables are plotted on a normalised vertical scale against time, as illustrated in the trend faceplate of Picture 42.2.

Trend diagrams are configured using standard trending templates. The configuration process entails specifying:

- Which trend template to use.
- What reference number will be used for the trend diagram.
- Which variables to trend, up to the limit for that template type.
- Choice of points, lines or bars.
- What colours/shading to use for which trend.
- What scale to use for the vertical axis, given that the default will be 0–100 %.
- The scan rate.
- The scale required for the time axis.
- The basis for data compression, if necessary.

Note that the scan rate refers to the frequency at which the database is sampled to update the trend. There is no point in this being greater than the rate at which the database itself is updated by sampling the I/O signals. Note that the scan rate must be the same for all variables on the same trend diagram. The scan rate establishes the step length, *i.e.* the time between values displayed. It should be consistent with the resolution of the time scale, *i.e.* there should be as many values to display as there are points along the time axis. Sometimes, for example if there is a noisy signal, a higher scan rate may be specified, in which case some form of data compression will be required. This normally involves the grouping of consecutive values and trending the average value of the groups.

From an operational point of view there are two categories of trend diagram, live and historic. In live operation, the OCP adds new values to the display area, as time progresses, and the trends advance across the screen from left to right. There are two approaches to handling the situation when they reach the right hand edge of the display area, the trends may be reset or shuffled. With reset the left hand half, say, of each trend is deleted, the right hand half moved across to the left hand side of the display area, and trending continued. Alternatively, the first set of values drops off the left hand edge, all the remaining values are shuffled leftwards by one step length and the next set of values enters on the right hand side. This creates the impression that the trends are moving from right to left.

Historic trends utilise the bulk data archived by the ARK package. This may be retrieved and displayed using the trend facilities as described. The only major additional factor to be considered concerns the span of time scale to the extent that the start and stop times for the display must be specified as absolute dates and/or times.

42.8 Status Displays

These are essentially structured lists which are kept up to date in real-time. There are normally three such lists used for operational purposes, alarm lists, event logs and batch status lists, although there are also systems lists indicating task status, *etc.* for engineering purposes:

1. Alarm lists. Illustrated in Picture 42.3, these are discussed in detail in Chapter 43.
2. Event logs. These are lists which record, in chronological order, pre specified events such as:
   - Alarms being activated.
   - Changes in discrete output status.
   - Loops being switched between auto & manual modes.
   - Set points being changed.
   - Sequences being stopped and started.
   - Operator interventions.

   Whenever any such event occurs it will be logged, as a matter of course, with a date and time stamp and the relevant tag number for subsequent analysis. The most recent events appear at the top of the list, which may run to several pages. Since the number of events logged

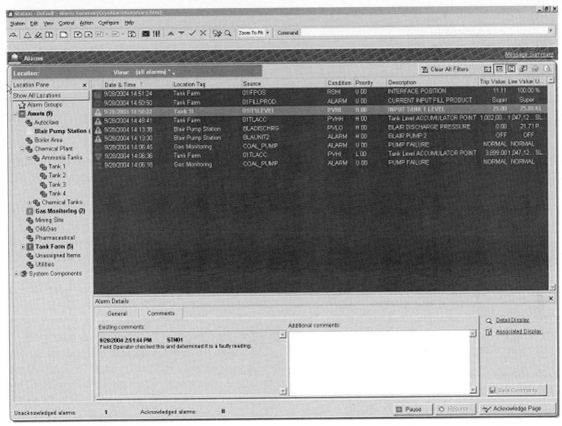

Picture 42.3 Typical alarm list (Honeywell)

can be extensive there has to be some constraint on the amount of memory used up. This may simply be the number of entries permitted in a fixed length list. Once the list is full, any new entry at the top displaces one at the bottom. Alternatively, the list may be of variable length but time related, *eg.* events are deleted automatically from the log after 24 h. The basis for constraining the log length is normally user-definable. It is also normal practice to have a printer dedicated to producing an on-line paper event log which is usually kept for posterity.

3. Batch status. This list essentially logs and displays the current status of various batches being made, if any, on the plant. The following items are typically listed:
   - Batch and lot numbers.
   - Time when batch started.
   - Train/unit being used.
   - Recipe being used.
   - Current batch status.
   - Sequence being used.
   - Current phase/step number.

42.9 Text Displays

These are basically files of text that can be displayed. The two most commonly used categories of text displays are for messages and providing help.

1. Messages. These may be established and stored in a library file. They are then triggered by application software when particular events or combinations of circumstances occur. Messages are normally aimed at the operator. For

example, at the start of a batch, the operator may be requested to make a visual check of a particular item of plant.

Alternatively, messages may be left on the system by operators, supervisors and engineers to remind each other of work to be done, faulty equipment, *etc*.

2. Help displays. These are text files that may be called up from disc which explain the functionality of the system. They are related to both hardware and software and are often extensive. The purpose of help displays is to provide on-screen technical support to personnel.

42.10 Diagnostic Displays

These are special purpose displays, typically in group or list form, which provide information about the health and/or status of the system. They relate to system diagnostics functions supported by the operating system, and include:

1. Hardware. An overview of the status of system hardware indicating which, if any cards, are faulty and the nature of the fault.
2. Communications. A summary of the status of all devices connected up to the system's highway and/or network, indicating which are on/off line and which are disconnected.
3. Tasks. A listing of the status of all current tasks, *e.g.* ready, queuing, active, suspended, failed, aborted, *etc*.

42.11 Keyboard Functions

Keyboards were discussed in general terms in Chapter 39. When a key is pressed, it is the OCP that responds . To an extent, the scope of the keyboard is a measure of the functionality of the OCP. In general, the keys on a keyboard are grouped according to function, as follows:

1. Alphabetic/numeric. These are used for entering data. The most commonly entered data are tag numbers as arguments for the dedicated keys and values for parameters.
2. Display. Various dedicated keys for accessing the standard displays of area, group, trend, mimic, event and message. There will usually be associated keys for clearing, recalling, freezing, printing and saving displays.
3. Alarm. Dedicated keys for acknowledging and accepting alarms, calling up the alarm list and locating faceplates that are in alarm.
4. Loop control. This has keys dedicated to switching loops into automatic or manual modes, for changing set points, for raising or lowering analogue outputs, for switching discrete outputs, *etc*.
5. Sequence control. Dedicated keys for assigning sequences to units, for starting, holding, continuing, single stepping and stopping sequences, for calling up the sequence status display, *etc*.
6. Change parameters. Access to these dedicated keys is usually strictly limited. Parameters for which there are dedicated keys normally include controller PID settings, deadbands, ramp rates, display scales, alarm limits, scan rates, *etc*.
7. General. User definable keys, cursor addressing keys, system access keys, test keys, printer control keys, *etc*.

In general, there is a procedure for interacting with the OCP. Typically a dedicated function key is pressed. This is then followed by a reference number which will cause the current value/status of the function to be displayed at the cursor. A new value is then entered using the alphanumeric keys and, to confirm the intent, an accept key has to be pressed.

Whilst there is no doubt that modern OCPs are very sophisticated, they are relatively easy to use. Given a grasp of their functionality, and an understanding of the plant being controlled, they can be mastered with a few hours effort.

Integrated Alarm Environment

43.1 Database
43.2 Display Systems
43.3 Alarm Lists
43.4 Integrated Safety Environment
43.5 Alarm Management Policy
43.6 Comments

Computer control makes an important contribution to safety. This is discussed in more detail in Chapter 55. Of particular significance is the capacity for the systematic handling of alarms which pervades system design. Thus, for example, alarm features are built into the database for processing by both the DDC and OCP packages. This chapter summarises those features that are typical. It is also used as a vehicle to introduce various aspects of database design.

For a more comprehensive treatment of the subject the reader is referred to the guide on the design, management and procurement of alarm systems published by EEMUA (1999).

Table 43.1 Analogue input function and data block

Function block Slot	Description	Datablock Value
1	Block no.	B1005
2	Block type	AIN
3	Tag no.	LRC 47
4	Description	Buffer tank level
5	Block status	On
6	Sampling frequency	5
7	Frame/rack/card/channel no.	1/2/10/06
8	Characterisation	LIN
9	Bias	−0.1
10	Span	2.1
11	Engineering units	m
12	Display area	06
13	Alarm priority	0
14	High alarm limit	1.8
15	Low alarm limit	−0.1
16	Deadband	0.05
17	Message code	0
18	Result	

43.1 Database

Function blocks are configurable and consist of a routine, *i.e.* a program or set of algorithms, which operate upon associated blocks of data in the database. The function blocks exist within a DDC package and are under control of the RTOS. Their associated blocks of data are invariably referred to as data blocks. The use of function blocks is elaborated upon in the following chapters and their configuration is explained in Chapter 48.

Consider an analogue input signal. This may, for example, relate to some level control loop LRC 47. The signal is processed by a function block of the type AIN whose purpose is to scale analogue input signals. The AIN function block handles the scaling and associated alarm status. A typical data block for an AIN function block is as shown on the right hand side of Table 43.1.

The values in slots 1–11 will be discussed in Chapter 45. It is sufficient to say that the measurement of the level in a tank varies between 0 and 2 m, and its calculated value is stored in slot no 18. The contents of slots 12–17 all relate to alarm handling as follows:

12. This defines the area menu of the group display which contains the template for LRC47. Whenever the level goes into alarm, the OCP will prompt/guide the operator to the relevant area/alarm. This is typically realised by a dedicated banner line across the top of the display system which immediately identifies areas with new alarms.
13. When an alarm occurs, it is automatically entered into the alarm list and event log. The alarm priority determines where it enters on the list, as discussed below. The value of 0 in slot 13 implies it has no priority.
14. The value in this slot specifies the upper alarm limit. Whenever value in slot 18 goes above 1.8 m the alarm will be activated. This is obviously a warning that the tank is nearly full.
15. The value in this slot specifies the lower alarm limit. The fact that a value of −0.1 m is specified, which can never be reached because of the offset on the level measurement, implies that a low level does not matter. It is common practice to set alarm limits at the end of, or outside, the measurement range when there is no need for an alarm.
16. A deadband of 0.05 m is specified. The purpose of the deadband is to suppress the spurious alarms that would occur due to noise on the level measurement when it is close to the alarm limit. An obvious source of such noise is waves on the surface of the liquid. Thus, once an alarm has been triggered by the value for the level rising above 1.8 m, the level will stay in alarm until it has fallen below 1.75 m.
17. A message code is specified. Whenever an alarm occurs, a specified message can be printed in a reserved area of the OCP display. Messages are typically stored in library files, each message having a unique code. The code of 0 in this case implies that there is no message attached.

These alarm features are established when the database is built. All other analogue inputs would have similar data blocks. Similarly, all other signal and variable types will have alarm features built into their data blocks. Thus alarm handling criteria will be dispersed throughout the database. The DDC package will systematically apply the appropriate routines to these data blocks and, if necessary trigger alarms. The OCP will then systematically display and log any alarms that are triggered.

43.2 Display Systems

The basic alarm handling functionality of most computer control systems is broadly similar. Whenever an alarm occurs, the following takes place automatically and simultaneously:

- An audible alarm will be annunciated, typically a buzzer, which the operator may acknowledge and accept as appropriate.
- A message may appear in the reserved alarm message area of the OCP display.
- The alarm will be entered into the alarm list and thereafter its status monitored and the alarm list entry kept up to date.
- A printout of the alarm will occur on the event log.
- The area within which the alarm has occurred will be highlighted on the banner line to assist the operator in locating the relevant group display.
- Some feature of the relevant group display faceplate will change. For example, the colour of a bar will change to red, or some text characters may start to flash.
- Similar colour changes and/or special effects will occur on all relevant faceplates integrated into the mimic diagrams.

43.3 Alarm Lists

These are structured lists which are kept up to date in real-time. The essential features are as follows:

- Grouping of alarms on separate lists according to plant areas. Whether this is necessary depends on the size of the plant, the extent of the areas and the number of alarms.
- Alarms within an area are grouped together according to priority, the highest priority grouping being at the top of the list.
- Within any grouping, alarms are listed in chronological order, the most recent being at the top of the group.
- The list entry consists of the tag no, description, date/time of occurrence, type of alarm and current alarm status.
- The length of the list will be constrained by the no of entries or the time lapsed since occurrence, as described for event logs in Chapter 42.

The above alarm handling functions form an integrated alarm environment. They are built into the system and happen as a matter of course. Fundamental to this are the alarm related values built into the database.

43.4 Integrated Safety Environment

An integrated alarm environment is concerned with alarm handling only. It is not the same thing as an integrated safety environment in which a variety of additional functions may be supported. Some of these functions may be more sophisticated means of alarm handling. Others will enable corrective action to be taken in the event of an alarm occurring:

- Warnings. The data block for LRC 47 above supported a single pair of high and low alarm limits. This is normal. However, some systems support multiple limits on a single signal. For example, there can be a warning band set within the alarm band. These limits are referred to as Lolo, Lo, Hi and Hihi. The Lo and Hi limits trigger a warning but not an alarm.
- Alarm filtering. When large numbers of alarms occur over a short period of time, it can be difficult to interpret them. Some systems provide a filtering function. This effectively identifies critical alarms and suppresses others: suppression being either a reduction in priority, a delay or even deletion. Clearly the filtering criteria have to be specified very carefully.
- Trips and interlocks. These are closely allied to alarms, are defined in Chapter 55 and, typically, are realised by means of configurable logic blocks. They take pre-defined corrective or preventative action in the event of an alarm occurring. Trips and interlocks are used extensively and, after alarms, are by far the most common safety function supported by any system.
- Sequences. These can be used to carry out cross checking, application diagnostics and automatic shut-down. Sequence control was introduced in Chapter 29.
- Expert systems. These have been used in an online mode to provide decision support in relation to alarm handling. In essence, by analysing the pattern of alarm occurrences, the expert system attempts to deduce the underlying causes and recommend appropriate actions. An introduction to expert systems is given in Chapter 107.

43.5 Alarm Management Policy

The purpose of an alarm management policy is to ensure that the system generates alarms, warnings and/or messages in as effective a way as possible with regard to the operators. It is fairly obvious, but often forgotten, that:

- Alarms, warnings and messages provided to an operator must be sufficient to enable correct analysis and /or diagnosis. Information that is too sparse cannot be interpreted reliably.
- Alarms and warnings must be provided early enough to enable timely intervention, if neces-

sary. Alarms that occur too late are next to useless.
- The volume of information presented must be such that the operator is not overloaded. The term "deluge" is used to describe alarms that are too many or too frequent for an operator to handle meaningfully.

There are various strategies that can be used, the most important being:

- Categorise alarms as being either global or local in relation to the architecture of the control system. For example, global alarms will be displayed throughout the system whenever they occur, whereas local alarms will only be displayed, listed, *etc.* on the operator station used for the area to which the alarm relates.
- Group and prioritise alarms, according to area, priority, chronology, *etc.* as described previously.
- Minimise standing alarms. These are variables that persist in an alarm state, albeit having been acknowledged. This is often because the alarm limits are inappropriate. It is good management practice to review standing alarms on a regular basis, to decide if the alarm is necessary and, if so, whether its limits are sensible.
- Avoid duplication of alarms. There is much scope for duplication, for example:
  - An AIN block may generate alarms according to predefined limits.
  - Further alarms may be generated if the same input signal goes out of range or, with smart devices, if the transmitter goes out of calibration.
  - There may be alarms in associated function blocks, such as deviation alarms attached to error signals in associated PID blocks and range limits in analogue output (AOT) blocks, as seen in Chapter 44.
- Prevent spurious alarms. Use of deadband to prevent alarms due to noise, as described above for slot 16. Such spurious alarms are bad practice because they cause the alarm to fall into disrepute.
- Suppress known, predictable alarms such as during start-up and shut-down.
- Provide operator support by means of alarm filtering, acknowledgement records, application diagnostics, *etc.*

The combination of strategies that is most appropriate for a given context depends on factors such as the nature of the process and/or plant, the expertise of the operators, and the functionality of the control system. It is essential that the alarm requirements are thought through, properly specified, and reviewed regularly.

43.6 Comments

Many of the major control system vendors have developed proprietary alarm management system (AMS) packages, as described in Chapter 99, which can be installed retrospectively on DCS and SCADA systems. They provide much of the functionality outlined in Section 43.4 and enable alarm management policy as outlined in Section 43.5. More sophisticated AMS packages are used for abnormal situation management (ASM). Whilst these packages can make a major contribution to safety, and it is certainly inappropriate to argue against usage of such, their installation is tantamount to an admission of failure. There is sufficient functionality in any DCS or SCADA system to cope with any sensible alarm handling requirements without the need for bolt-on packages. The problem is that those requirements have to be properly thought through at the specification stage and incorporated in the detailed functional specification (DFS) as appropriate. Unfortunately, that is a corner which is cut all too easily and far too commonly in the interest of expediency.

Analogue Control Loop

44.1 Input Interface
44.2 Input Scaling
44.3 Filtering
44.4 PID Control
44.5 Output Scaling
44.6 Output Interface
44.7 Nomenclature

This chapter explains how a typical analogue control loop is realised by means of computer control. A simple level control loop with a tag no LRC 47 is considered. The signal is followed from the level transmitter, through the system and back out to the control valve. Both hardware and software aspects are covered. An explanation of the structure of the database is deferred to Chapter 45.

Consider again the level control loop depicted in Figure 3.1, the block diagram for which is shown in Figure 3.2, and whose implementation by computer control is depicted in Figure 44.1.

This may be considered to consist of four sub systems:
1. Process and load (F/H), and instrumentation comprising current to pressure (I/P) converter, valve (P/F) and dp cell (H/I).
2. Input interface consisting of current to voltage (I/V) converter, sampler and analogue to digital converter (A/D).

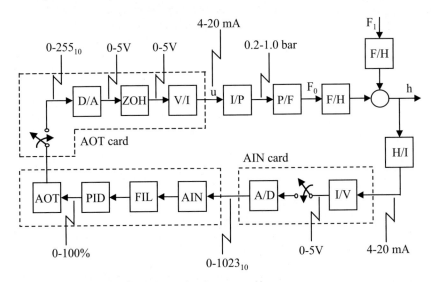

Fig. 44.1 Block diagram for realisation of analogue control loop

3. Function blocks for input scaling (AIN), filtering (FIL), control (PID) and output scaling (AOT).
4. Output interface consisting of sampler, digital to analogue converter (D/A), zero order hold (ZOH) and transmitter (V/I).

44.1 Input Interface

The three elements of I/V, sampler and A/D are all realised by the circuits of an AIN card. Assume that the dp cell's output is one of a number of 4–20 mA channels handled by the card, the circuit for which is as depicted in Figure 44.2.

The I/V conversion essentially consists of converting the 4–20 mA signal into a 0–5 V signal compatible with the input range of the A/D converter. The mA signal is dropped to earth across a 250 Ω resistor and converted into a 1–5 V signal. This is subsequently scaled into a 0–5 V signal by an op-amp circuit, as described in Chapter 6. Note the RC network for filtering high frequency noise, such as mains frequency and harmonics thereof, as described in Chapter 69. Also note the barrier for electrical protection which is described fully in Chapter 52. The barrier depicted is external but on some systems is realised by circuits on the AIN card.

The sampling process, sometimes referred to as scanning, is necessary because the A/D converter is usually shared between all the input channels on the card. The sampler is often referred to as a multiplexer and sampled signals are said to be multiplexed. In operation, under control of the RTOS, the 0–5 V signal is switched through to the input of the A/D converter, and held there long enough for the A/D conversion to take place. The analogue signal must be sampled frequently enough for the samples to be a meaningful representation, but not so often as to cause unnecessary loading on the system. At the same time the sampling frequency must not be too low as to cause aliasing effects. The various input channels would be sampled at different frequencies, as appropriate.

The A/D converter is an integrated circuit chip. It converts the sampled 0–5 V signal into a bit pattern. Given that the original level measurement is from a dp cell, whose accuracy is ±1% at best, one would expect a 10-bit word with a range of $0–1023_{10}$ and a resolution of approximately ±0.1% to be adequate. However, this is insufficient resolution for some routines. For example, in PID control, the derivative action operates on the difference between successive input samples and, to avoid numerical instability, higher resolution is required. Therefore, it is not uncommon for A/D conversion to use at least 14 bit words with a range of $0–16383_{10}$ and a resolution of approximately ±0.006%.

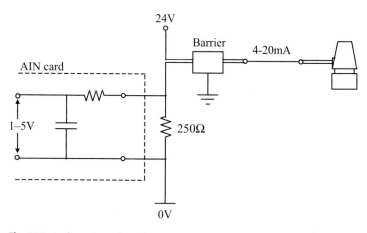

Fig. 44.2 Analogue input channel

44.2 Input Scaling

The AIN function block of Figure 44.1 represents an analogue input scaling routine. It operates on the output of the A/D converter. The following equation is used universally for linear scaling. For brevity, a 10-bit A/D converter is assumed:

$$\theta_0 = \text{bias} + \frac{\theta_1}{1023}.\text{span}$$

where θ_1 is the decimal value of the binary output of the A/D converter.

A typical algorithm, written in structured text, for implementing it would be:

$$\text{OP} := \text{BI} + \text{SN}*\text{IP}/1023 \quad (44.1)$$

For example, a bit pattern of 1000011001_2 corresponds to $\theta_1 = 537_{10}$ which, given a bias of -0.1 and a span of 2.1, yields a value for θ_0 of approximately 1.0 with a resolution of approximately 0.1%. The resultant value of 1.0012 is stored in slot 18 of Table 43.1 which obviously corresponds to the tank being half full.

44.3 Filtering

The FIL function block of Figure 44.1 represents a filter routine. It operates on the output of the AIN function block. Strictly speaking, given the high frequency filtering of the AIN card, this is only necessary if there is lower frequency noise, which is unlikely on the level in a buffer storage tank. The filter is nevertheless included as a typical example of signal processing.

The most common type of filter is the simple first order lag, as explained in Chapter 69, which is of the form:

$$T\frac{d\theta_0}{dt} + \theta_0 = \theta_1$$

which may be rearranged:

$$\frac{d\theta_0}{dt} = \frac{\theta_1 - \theta_0}{T}$$

Using Euler's first order explicit method of numerical integration:

$$\theta_{0,j+1} = \theta_{0,j} + \left.\frac{d\theta_0}{dt}\right|_j .\Delta t = \theta_{0,j} + \frac{\theta_{1,j} - \theta_{0,j}}{T}.\Delta t$$

$$= \theta_{0,j} + k.(\theta_{1,j} - \theta_{0,j})$$

A typical algorithm for implementing the filter would be

$$\text{OP} := \text{OP} + \text{FC}*(\text{IP} - \text{OP}) \quad (44.2)$$

44.4 PID Control

The PID function block of Figure 44.1 represents the routine for a 3-term controller. It operates on the output of the FIL function block. The functioning of PID control was discussed in detail in Chapter 23. In particular, a discretised version of the absolute form of the classical PID controller with derivative feedback was developed in Equation 23.8. Typical algorithms, written in structured text, for implementing this controller would be

$$\begin{aligned}
& \text{E} = \text{SP} - \text{IP} \\
& \text{if (OP} = 0 \text{ or OP} = 100) \text{ then goto L} \\
& \text{IA} = \text{IA} + \text{KI}^*\text{E} \\
\text{L} \quad & \text{OP} = \text{BI} + (\text{KC}^*\text{E} + \text{IA} - \text{KD}^*(\text{IP} - \text{PIP})) \\
& \text{if OP} < 0 \text{ then OP} = 0 \\
& \text{if OP} > 100 \text{ then OP} = 100 \quad (44.3)
\end{aligned}$$

44.5 Output Scaling

The AOT function block of Figure 44.1 represents an analogue output scaling routine. It operates on the output of the PID function block. The output of the AOT function block is a decimal number whose binary equivalent is consistent with the input range of the D/A converter.

Given that the output signal is eventually used to position a valve, which can be done to within $\pm 1\%$ at best, one would expect an 8-bit word with a range of $0-255_{10}$ and a resolution of approximately $\pm 0.25\%$ to be adequate. However, in practice, most D/A converters use at least 10-bit words with a range of $0-1023_{10}$ and a resolution of approximately $\pm 0.1\%$.

The following equation is used extensively for output scaling. The range of the controller output is normally 0–100% and, for brevity, an 8-bit A/D converter is assumed:

$$\theta_0 = 255 \cdot \frac{(\theta_1 - \text{bias})}{\text{span}}$$

The output θ_0 is stored in the database as a bit pattern. The algorithm used for the output scaling is typically

$$\text{OP} := 255 * (\text{IP} - \text{BI})/\text{SN} \qquad (44.4)$$

44.6 Output Interface

The four elements of sampler, D/A, ZOH and V/I are all realised by the circuits of an AOT card. Assume that the output is one of a number of 4–20 mA channels handled by the card, the circuit for which is as depicted in Figure 44.3. Again note the barrier for electrical protection.

The sampler is virtual. On a regular basis, under control of the RTOS, values of the AOT output are extracted from the database and routed through to the D/A converter. The sampler is held open long enough for the conversion to take place.

The D/A converter is an integrated circuit chip. It converts the sampled bit pattern into a 0–5 V signal. Since D/A converters are relatively cheap, it is normal practice for each output channel to have its own dedicated D/A converter.

The ZOH holds, or latches onto, the output of the D/A converter in between sampling periods. This effectively converts the pulse output from the D/A converter into a quasi, or piecewise linear, analogue signal that can be output to the I/P converter and thence to the control valve. Figure 44.4 depicts the construction of the quasi analogue signal.

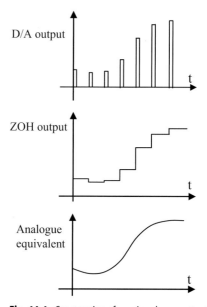

Fig. 44.4 Construction of quasi analogue output signal

The transmitter V/I provides the scaling and power for signal transmission. The 0–5 V latched signal from the D/A is a very low power, TTL or otherwise, signal. After scaling into a 1–5 V signal it is converted into a 4–20 mA signal. The op-amp circuit shown in Figure 6.13 is typical of the type used for such power amplification.

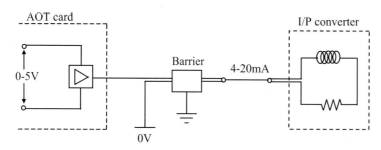

Fig. 44.3 Analogue output channel

44.7 Nomenclature

Symbols

θ	some variable
T	filter time constant
Δt	step length which is, in effect, the sampling period
k	filter constant, $\Delta t/T$

Subscripts

0	output
1	input
j	current value

Pseudocode

BI	bias
E	error
FC	filter constant
IA	integral action
IP	input
KC	proportional gain
KD	derivative gain
KI	integral gain
OP	output
PIP	previous input
P	set point
SN	span

Database Operations and Structure

45.1 Input Scaling
45.2 Filtering
45.3 PID Control
45.4 Output Scaling
45.5 Database Structure

The operation of the AIN, FIL, PID and AOT function blocks, as depicted in Figure 44.1, conveys the impression that a signal is progressed around the loop, from block to block. In effect this is true but, in reality, the function blocks interact with the database, as depicted in Figure 45.1.

As stated previously, function blocks consist of routines which operate upon associated data blocks. Each routine is interpreted, independently of the others, under control of the RTOS, using its own data block. An input value, specified in the data block, is operated upon by the algorithm, using data taken from the data block, and a result is calculated which is returned to the data block for storage. The routines are executed at the same frequency. Progression of the signal around the loop is by virtue of the result stored in one block being specified as the input to another.

This is best appreciated by consideration of the algorithms in relation to their own data blocks, as follows.

45.1 Input Scaling

The AIN block of Figure 45.1 represents an analogue input scaling routine. Its algorithm operates on the generic data block shown in Table 45.1.

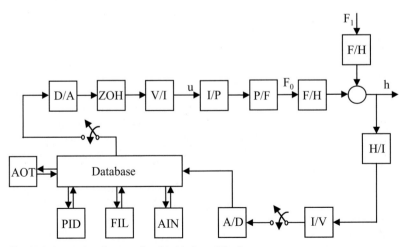

Fig. 45.1 Interactions between function blocks and database

Table 45.1 Analogue input function and data block

Function block		Datablock
Slot	Description	Value
1	Block no.	B1005
2	Block type	AIN
3	Tag no.	LRC47
4	Description	Buffer tank level
5	Block status	On
6	Sampling frequency	5
7	Frame/rack/card/channel no.	1/2/10/06
8	Characterisation	LIN
9	Bias	−0.1
10	Span	2.1
11	Engineering units	m
12	Display area	06
13	Alarm priority	0
14	High alarm limit	1.8
15	Low alarm limit	−0.1
16	Deadband	0.05
17	Message code	0
18	Result	

This is identical to the data block considered in Chapter 43, which was written with this example in mind, and is reproduced for convenience:

$$OP := BI + SN*IP/1023 \quad (44.1)$$

1. Slot 1 is the reference to that part of the database which contains this data block.
2. The block type establishes which routine is to be used for processing the signal. In this case it is an analogue input scaling routine for 4–20-mA signals. Specifying the block type determines the number of slots in the block, what type of data has to be entered in which slot, and the number of bytes of memory per slot. If, for example, the analogue input was a millivolt signal from a thermocouple, a different block type would have been specified which would have a different number of slots and data requirements.
3. The tag number relates the function block and its data block to the control scheme, as defined in the P&I diagram. This tag number is used extensively by the OCP for display and logging purposes.
4. The description is supplementary to the tag number.
5. The block status is typically on or off. If set to on, the routine operates on the data block. If set to off, execution of the routine is suspended. This would be appropriate, for example, during configuration, or if there is a hardware fault that causes the input signal to be in a permanent state of alarm.
6. The 0–5-V signal corresponding to the level is sampled at a frequency of 5 s.
7. The entry in slot 7 uniquely relates the data block B1005 and the AIN routine to the hardware channel from which the level is sampled.
8. Most AIN routines support several characterisations. Their function is to compensate for any non-linearities introduced in the measurement process. In this case, since there is a linear relationship between the level in the tank and the dp cell output, no characterisation is required and the linear scaling function LIN is specified. If, for example, the input had been from a dp cell associated with an orifice plate for flow measurement, square root extraction would have been specified.
9. Because the dp cell is situated 10 cm below the bottom of the tank, it is necessary to offset the calculated levels by 0.1 m.
10. The tank is 2 m deep and the dp cell has a calibrated range of 0–2.1 m.
11. The units associated with the measurement are specified for display purposes. They are often referred to as engineering units because the original signal has to be reconstructed from the A/D bit pattern. A common default is percentage.

12–17. These slots relate to alarm handling and were covered in Chapter 43.

45.2 Filtering

The FIL function block of Figure 45.1 represents a filter routine. Its algorithm operates on the generic data block shown in Table 45.2.

Table 45.2 Filter function and data block

Function block Slot	Description	Datablock Value
1	Block no.	B1947
2	Block type	FIL
3	Tag no.	LRC47
4	Block status	On
5	Input block	B1005
6	Sampling frequency	5
7	Filter constant	0.1
8	Result	

To a large extent the slots are self explanatory. Note that:

5. When the algorithm is executed, the value used for the input IP is the output of the AIN scaling block.
8. During execution of the algorithm, the value of the output OP used for subtraction from IP is the current output stored in slot 8. This will obviously be overwritten with the next value of OP.

$$OP := OP + FC*(IP - OP) \quad (44.2)$$

45.3 PID Control

The PID function block of Figure 45.1 represents a 3-term controller routine. Its algorithms operate on the generic data block shown in Table 45.3:

$$E := SP - IP$$
$$IA := IA + KI*E$$
$$OP := BI + (KP*E + IA - KD*(IP - PIP)) \quad (44.3)$$

This data block makes that for the FIL routine look positively trivial. Again, many of the slots are self explanatory. Note that:

Table 45.3 PID control function and data block

Function block Slot	Description	Datablock Value
1	Block no.	B2706
2	Block type	PIDABS
3	Tag no.	LRC47
4	Block status	Auto
5	Sampling frequency	5
6	Input block	B1947
7	Set point	1.0
8	Engineering units	m
9	Status on initialisation	Auto
10	Set point tracking on switch	Yes
11	Deadband on error	0.0
12	Bias	50.0
13	Reverse action	Yes
14	Proportional gain	1.0
15	Integral gain	0.1
16	Initial value of integral action	0.0
17	Integral action	
18	Integral desaturation	Yes
19	Derivative gain	3.0
20	Derivative feedback	Yes
21	Previous value of input	
22	Alarm priority	0
23	High output alarm limit	100
24	Low output alarm limit	0
25	Result	

2. The absolute form of PID controller is chosen as opposed to the incremental.
6. When the algorithms are executed the value used for the input IP is the output of the FIL filter block.
7. The set point is given a numerical value. Some systems have separate function blocks for handling the set point. If the controller had been a slave controller in a cascade system, then a block number would have been given for a remote set point.

9. This enables the mode of the loop to be specified as being either MAN or AUTO on start up of the system.
10. Set point tracking may be specified. This enables bumpless transfer when the loop's mode is switched from MAN to AUTO or on start up.
11. A deadband may be set about the error within which the error is considered to be zero. This minimises valve wear.
17. Slot 17 is used to store the current value of the integral action IA from one execution of the algorithm to the next.
21. The previous value of the input PIP, or of the error if derivative feedback is not chosen, is stored in slot 21. This is then overwritten with the current value of the input IP.

45.4 Output Scaling

The AOT function block of Figure 45.1 represents an analogue output scaling routine. Its algorithm operates on the generic data block shown in Table 45.4:

$$OP := 255 * (IP - BI)/SN \quad (44.4)$$

Table 45.4 Analogue output function and data block

| Function block | | Datablock |
Slot	Description	Value
1	Block no.	B2907
2	Block type	AOT
3	Tag no.	LRC47
4	Block status	On
5	Sampling frequency	5
6	Input block	B2706
7	Frame/rack/card/channel no.	1/2/12/06
8	Characterisation	LIN
9	Bias	0.0
10	Span	100.0
11	Engineering units	%
12	Result	

45.5 Database Structure

Databases are organised differently according to application. Three types of database are of relevance in process automation. This chapter is concerned with block structured databases, as used by table driven algorithms for real-time control. Relational databases, as used in management information systems, are covered in Chapter 99 and object oriented databases, as used in knowledge based systems, are covered in Chapter 107.

Control is realised by means of the RTOS. The routines are executed in relation to their data blocks. The algorithms listed in Equations 44.1–44.4 are in structured text but, in practice, would consist of assembly language instructions. As the algorithm is interpreted by the CPU, data from the data block is routed through the bus system to the ALU, as explained in Chapter 9.

The data arriving at the ALU must be in the order required by the algorithm. The most effective way of organising this is in the tabular form of a data block, hence the term "table-driven". For this reason it is essential that values are entered into the slots of the data block in the correct order. They must also be in the correct format: integer, decimal, alphabetic, *etc.* When a block number is entered as an input, the value in the last slot of that data block is returned.

Data is accessed from the database with reference to a so-called block pointer. When an algorithm is to be interpreted, the RTOS sets the pointer to the relevant data block number. The data block is located by virtue of the number in its first slot. Subsequent values are identified by specifying the slot number relative to the block pointer. The length of a data block is determined by the block type.

Note that the second slot in each data block is the block type. This enables the RTOS to determine which routine to apply to what data block. The sampling frequency establishes how often the routine is applied. It is helpful to think in terms of the RTOS systematically switching routines from one data block to another.

Like data blocks are grouped contiguously within the database: this simplifies the design of the database and makes it easier to understand. It also maximises processing efficiency. Thus, the RTOS can quickly switch a routine from one data block to the next, until it runs out of like blocks. It can then move onto applying a different routine to another group of like data blocks.

Remember that the database is used by all the main systems packages. Thus, not all the data in the slots of a data block are used by the DDC routine. For example, the tag number is used by the OCP for display purposes. This is taken into account in the process of interpreting the algorithms which, in effect, skip the redundant slots.

An overview of the process of configuration is given in Chapter 48, together with a more detailed explanation of the functionality of function blocks and function block diagrams.

Discrete I/O Devices

Chapter 46

46.1 Output Interface
46.2 Field Instrumentation
46.3 Input Interface
46.4 Input Processing
46.5 Output Processing

This chapter explains how a typical discrete device is realised by means of computer control. Discrete devices, sometimes referred to as entities, relate to a system's discrete I/O channels. A device consists of some two to four discrete input and/or output channels that are logically connected. The logic is normally handled by function blocks. Typical applications are in enabling trips and interlocks, discrepancy checking and polling redundant signals. Larger numbers of discrete I/O are normally handled by sequences.

The discrepancy checking of a valve position is used as a vehicle for explaining the operation of devices. The device consists of a discrete output channel which is used for opening an isolating valve, and a discrete input channel connected to a proximity switch used for checking whether the valve is open or not. This is depicted in P&I diagram form in Figure 46.1.

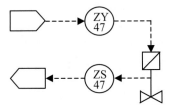

Fig. 46.1 Representation of an isolating valve as a discrete device

The corresponding block diagram is shown in Figure 46.2.

This may be considered to consist of four sub systems:

- An output interface consisting of a sampler, discrete to voltage converter D/V, zero order hold ZOH and relay V/V
- Field instrumentation comprising a solenoid actuated pilot valve V/P, pneumatically actuated isolating valve P/Z and a proximity switch Z/V
- An input interface consisting of a voltage converter V/V, sampler and voltage to discrete reader V/D
- Function blocks for handling discrete input DIN and discrete output DOT signals

46.1 Output Interface

The four elements of sampler, D/V, ZOH and V/V are all realised by the circuits of a DOT card. Assume that the output to the isolating valve is one of a number of 0/24 V channels handled by the card, the circuit for which is as depicted in Figure 46.3. Note the barrier for electrical protection.

The sampler is virtual. On a regular basis, under control of the RTOS, the status of the DOT block output is sampled, *i.e.* it is copied from the database into a register. It is usual for all the DOT outputs associated with a card to be stored in the bits of the register and to be "scanned" simultaneously.

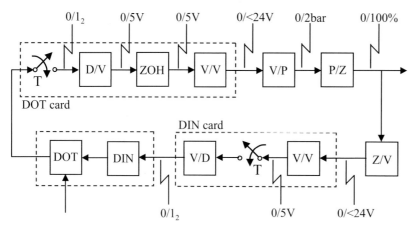

Fig. 46.2 Block diagram for realisation of discrete device

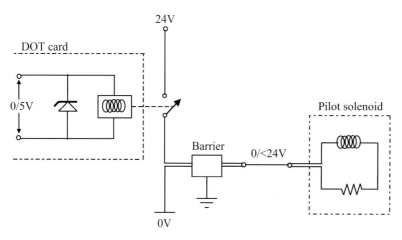

Fig. 46.3 Discrete output channel

The D/V conversion of the logical 0/1 in the register into a 0/5 V signal is handled by an op-amp type of circuit. This is then latched by the ZOH in between samples as described in Chapter 44. The relay depicted is electro-mechanical in nature. This, and the transistor based type of relay, are the most common forms of relay used for discrete outputs signals.

46.2 Field Instrumentation

When the relay is closed, the 24 V power supply is routed through the barrier to the solenoid of a pilot valve, as described in Chapter 5. This enables compressed air to be applied to the diaphragm actuator of the isolating valve.

Suppose that the valve is of the air-to-open type and its closed position is of interest. The proximity switch is attached to the yoke of the valve and the magnet to its stem, both being positioned at the closed end of stem travel. Also suppose that the switch is normally open, but is closed by the magnet. Thus, when the valve is shut the output of the proximity switch is approximately 24 V. If the valve is open, fully or otherwise, or if the proximity switch fails, the output is 0 mA.

46.3 Input Interface

The three elements of V/V, sampler and V/D are all realised by the circuits of a DIN card. Assume that the input from the proximity switch is one of a number of 0/24 V channels handled by the card, the circuit for which is as depicted in Figure 46.4.

The two zener diodes would typically have threshold values of $z_1 = 20$ V and $z_2 = 25$ V. These zeners establish a window through which only voltages of approximately 24 V can pass. This guards against false inputs due to spurious induction effects in the field circuits. Note the RC network for filtering out noise due to bounce on the proximity switch. Again note the barrier for electrical protection.

The V/V conversion concerns scaling the 0/24 V signal into a 0/5 V signal using an op-amp type of circuit. On a regular basis, under control of the RTOS, the 0/5 V signal is sampled, converted into a logical 0/1, and written into a register from which it is read into the database. It is usual for all the inputs associated with a DIN card to be stored in the bits of the register and to be scanned simultaneously.

46.4 Input Processing

The DIN function block of Figure 46.2 represents a discrete input processing routine. Its algorithm operates on the generic data block shown in Table 46.1.

Table 46.1 Discrete input function and data block

Function block		Datablock
Slot	Description	Value
1	Block no.	B0027
2	Block type	DIN
3	Tag no.	ZS47
4	Description	Proximity switch
5	Block status	On
6	Sampling frequency	1
7	Frame/rack/card/channel no.	1/2/4/16
8	Display area	06
9	Alarm priority	0
10	Alarm on high	No
11	Alarm on low	No
12	Message code	0
13	Current status	

The content of these slots is largely self explanatory. Note the following slots in particular:

5. If set to off, this would suspend execution of the DIN routine, despite the fact that the input

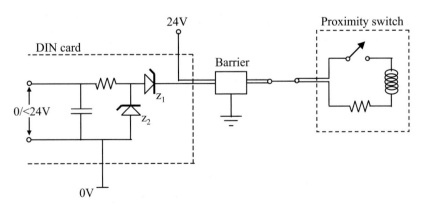

Fig. 46.4 Discrete input channel

is being sampled. Such a course of action would be appropriate, for example, if there is a hardware fault that causes the input signal to be in a permanent state of alarm.

10 and 11. These enable alarms to be attached to either state of the discrete input. Neither is used with the proximity switch because of the discrepancy checking on the discrete output to the solenoid valve.

12. This slot permits the time delay allowed for in the discrepancy checking to be specified.

13–15. If a discrepancy occurs, it is treated as an alarm for which an area has to be specified, a priority attached and an option to generate messages provided.

These DIN and DOT function blocks have been explained on the basis of their use for discrepancy checking. However, it should be appreciated that the routines are general purpose and are used extensively for handling conventional discrete I/O that are not logically connected.

46.5 Output Processing

The DOT function block of Figure 46.2 represents a discrete output processing routine. Its algorithm operates on the generic data block shown in Table 46.2.

The content of many of these slots is self explanatory. Note the following slots in particular:

7. This identifies the source of the required status of the discrete output corresponding to the valve opening, typically either 0 for shut or 1 for open. The status is normally determined by some other logic block, B0470 as indicated, by a sequence or by an operator decision.
9. An option for discrepancy checking is provided. For example, following a decision to close the solenoid valve, and allowing a time delay for the valve to close, the input from the proximity switch is checked.
10. This identifies the discrete input signal to be used for discrepancy checking.
11. A discrete output of 0 should correspond to an input of 1, and *vice versa*. Hence the binary codes of 01 and 10 are specified as acceptable combinations.

Table 46.2 Discrete output function and data block

Function block		Datablock
Slot	Description	Value
1	Block no.	B0270
2	Block type	DOT
3	Tag no.	ZY47
4	Description	Isolating valve
5	Block status	On
6	Sampling frequency	1
7	Input block	B0470
8	Frame/rack/card/channel no.	1/2/7/16
9	Discrepancy checking	Yes
10	Discrepancy input block	B0027
11	Discrepancy criteria	01,10
12	Discrepancy delay	5
13	Display area	06
14	Alarm priority	0
15	Message code	0
16	Current status	

Programmable Logic Controllers

47.1 Ladder Constructs
47.2 Ladder Execution
47.3 Functions and Function Blocks
47.4 Timers and Counters
47.5 Development Environments
47.6 Comments

PLCs were described in Chapter 38 as a microprocessor based alternative to hard wired relay logic circuits. Historically their strength has been in handling discrete I/O signals, sometimes in very large numbers, and processing the associated logic. For this reason they are used extensively throughout the manufacturing industries. However, modern PLCs have the capacity to handle analogue I/O signals as well, and PLCs are now commonplace in the process industries too. They are ideal for sequencing applications and for simple batch process control.

In terms of architecture, as depicted in Figure 38.7, a typical PLC has an associated PIU for handling the I/O and is linked to an OCS for operator interaction. However, from a hardware point of view, the PIU and PLC are often separate IC boards in a single unit. This unit is designed either for rack mounting or is ruggedised for installation on the plant. Typically the facia of a PLC has a set of LEDs to indicate I/O status. Two very important characteristics of PLCs are that they are both cheap and reliable. For this reason PLCs are often used as front ends to DCS and SCADA systems.

This chapter focuses on the software aspects of the use of PLCs and, in particular, on the use of ladder logic. For a more comprehensive treatment of PLCs, the user is referred to the texts by Warnock (1988) and Lewis (1998).

47.1 Ladder Constructs

Ladders are made up of rungs. Each rung notionally represents the flow of power from a left hand power rail to a right hand one. Every rung consists of one or more contacts and one coil. These contacts and coils represent the state of boolean variables. They may be real or virtual, examples of which are listed in Table 47.1:

Table 47.1 Contact and coil types

Real contact	A switch connected *via* a discrete input channel
Virtual contact	The status of an enumerated data type of integer variable, as explained in Chapter 7
Real coil	A coil associated with a solenoid or, say, the pilot valve on an isolating valve
Virtual coil	A Boolean variable used as a flag

The symbols in common usage are listed in Table 47.2. This variety enables PLCs to utilise standard discrete I/O channel hardware irrespective of application.

A typical simple rung is as depicted in Figure 47.1. The contacts are indicated by parallels and the coils by brackets. In practice there are many types of field device used for which a variety of contacts and coils are required.

Table 47.2 Symbols in common usage for contacts and coils

Contact/coil	Symbol	Status/action
Contact	-----| |-----	Open=0, Closed=1 Normally open
Negated contact	-----|/|-----	Closed=0, Open=1 Normally closed
Positive transition contact	-----|P|-----	Open=0, Closed=1 Normally open Closes on rising edge True for one cycle only
Negative transition contact	-----|N|-----	Open=0, Closed=1 Normally open Closes on falling edge True for one cycle only
Coil	-----( )-----	Off=0, On=1 Normally off Goes on when rung true
Retentive memory coil	-----(M)-----	Off=0, On=1 Normally off Goes on when rung true Stays on if power fails
Negated coil	-----(/)-----	Off=0, On=1 Normally on Goes off when rung true
Positive transition coil	-----(P)-----	Off=0, On=1 Normally off Goes on with rising edge True for one cycle only
Negative transition coil	-----(N)-----	Off=0, On=1 Normally off Goes on with falling edge True for one cycle only
Set coil	-----(S)-----	Off=0, Set=1 Normally off Set on when rung true Remains set until reset
Set retentive memory coil	-----(SM)-----	As for set coil Stays set on if power fails
Reset coil	-----(R)-----	Reset=0, On=1 Normally on Set off when rung true Remains reset until set
Reset retentive memory coil	-----(RM)-----	As for reset coil Stays set off if power fails

47.1 Ladder Constructs

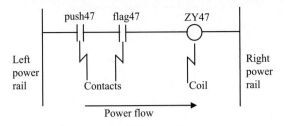

Fig. 47.1 Typical rung of a ladder

The convention is that contacts are considered to be normally open (false) and coils to be normally de-energised (false). As the contacts become closed (true) power flows across the rung and energises the coil (true). It is normal practice to label the contacts and coils as appropriate, and for their status to be animated during operation.

The ladder logic of the rung is that coil 47 becomes energised when both push button 47 and flag 47 are true. This may be expressed in structured text as follows:

```
ZY47:=PUSH47 and FLAG47
```

In effect, contacts across the rung represent AND gating.

A more complicated rung is depicted in Figure 47.2. Note the slash across the parallels of a contact which reverses the logic. Such a contact is normally closed (false) and power flows across the rung when it opens (true). Vertical connections between parallel rungs represent OR gating.

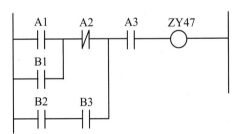

Fig. 47.2 A more complicated rung

The logic of this rung represents a combination of AND and OR gates as follows:

```
X1:=(((A1 or B1) and not A2) or (B2 and B3))
     and A3
```

Just as multiple contacts may be considered using AND and OR gates to drive a single coil, so too may a single rung drive multiple coils as depicted in Figure 47.3 for which

```
C1=C2:=not A2 and (A1 or B1)
```

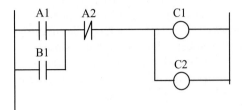

Fig. 47.3 A rung with multiple coils

Parts of a ladder may be inhibited from execution by setting up a dummy variable as a coil, and subsequently using the dummy variable as a contact, as depicted in Figure 47.4. In this example, the second rung cannot execute until the boolean variable DUM is set by the first rung. Use of dummy variables in this way enables ladder diagrams to be partitioned into logical sections.

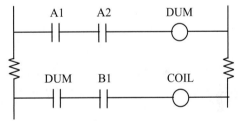

Fig. 47.4 Use of a dummy variable to inhibit part of a ladder

Labels may be used to force jumps in program flow, as depicted in Figure 47.5.

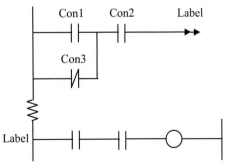

Fig. 47.5 Label used to force jump in program flow

Interlocks are commonly realised by means of ladder logic. Consider the starter motor circuit of Figure 47.6. The start contact is in series with an enable flag and a field stop button so the motor cannot be started unless the enable flag is set and the stop button has been released, *i.e.* when START, FLAG and not STOP are all true. Note that the stop button would be wired to fail safe, *i.e.* it has to be energised to enable the motor to run. An auxiliary motor contact AUX in parallel with the start contact keeps the motor running until either the enable flag is reset or the stop button is pushed.

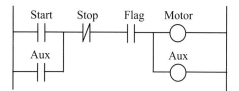

Fig. 47.6 Starter motor circuit with interlock

47.2 Ladder Execution

The execution of a ladder program is depicted in Figure 47.7. It is helpful to think in terms of the ladder being continuously executed by the PLC in a cycle, with system tasks such as updating the real-time clock and diagnostics being carried out in the background. There are three parts to the cycle:

1. At the start of a cycle, the PLC reads the value of the signals in all its input channels. It is said to scan the inputs into memory, or to refresh them. It then moves onto executing the ladder.
2. Starting at the top, it works its way down the ladder, one rung at a time. At each rung, the PLC reads the current status of the contacts, which may be either input signals or variables. According to the logic, the coil will be set to true or false as appropriate. These new values of the outputs are saved in memory. The PLC progresses down the rungs until it reaches the bottom of the ladder.
3. The PLC writes the values of all its output signals from memory to their respective channels,

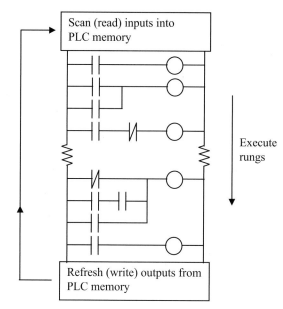

Fig. 47.7 Execution of a PLC ladder program

i.e. the outputs are refreshed. The cycle then begins again.

It is important to understand this cyclic nature of execution. Not doing so can lead to unexpected results. For example, consider a pump started by a push button, as depicted in Figure 47.8.

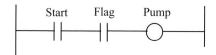

Fig. 47.8 Rung for push button pump start (faulty design)

Suppose that the start button is of the fleeting contact type, *i.e.* its closure is not latched and is transient in nature. Assume that the PLC's execution cycle is fast enough to detect the closure within at least one cycle, which is normally the case. Thus when a closure occurs, and presuming the enable flag is set, the rung becomes true and the pump starts. However, as soon as the closure has passed, the contact would revert to its open state and the pump stops. To get round this problem, a positive transition sensing contact and set type of coil are used as depicted in Figure 47.9. The rising edge

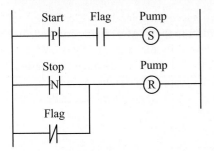

Fig. 47.9 Push button pump start with set/reset coil

contact senses the closure and latches on for one execution cycle: the rung becomes true and the set type of coil prevents the pump from stopping, until it is reset by the stop button on another rung.

As explained in Chapter 29, it is often necessary with sequencing to enable parallel activities such as the simultaneous operation of different phases on the same unit of plant. Given that a PLC normally can only support a single ladder, parallelism has to be handled by means of partitioning. The logic for each parallel activity takes up different parts of the ladder, the activities being switched on/off by dummy variables as described above. Thus, although there is only a single ladder and execution cycle, the sequences appear to be running in parallel.

For more complex sequencing, such as the simultaneous running of separate operations on different units of plant, it may be necessary to establish parallelism by assigning operations to different tasks and to exercise control over the tasks independently which, of course, presumes that the PLC system supports multi-tasking.

47.3 Functions and Function Blocks

The nature of ladder logic does not lend itself to calculations and data manipulation. Furthermore, a structured approach to software development is necessary. For example, similar operations, perhaps using different parameters, can be realised by replicating parts of the ladder, but this leads to very long programs. Both of these problems are addressed by incorporating functions and function blocks within ladders. This provides a measure of encapsulation which makes programs easier to understand and simpler to test, resulting in better quality software.

Figure 47.10 depicts a rung of a ladder which contains two functions, ADD and SQRT. Their functionality is explained in detail in Chapter 48.

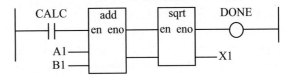

Fig. 47.10 Rung with functions

The logic of this rung is as follows:

```
if CALC then
    X1:= sqrt (A1 + B1);
    DONE:= 1.0;
end_if
```

Note that, when a function is used in a ladder, it is possible to control when the function executes by means of a special input EN and output ENO. If EN is set true the function executes and when successfully completed ENO is set true. By chaining the ENO of one function to the EN of the next, it is possible to control the order of execution of the functions and to ensure that the result of a chain of functions is valid. Any modern PLC supports a wide range of standard maths and logic functions.

Similarly Figure 47.11 depicts a rung of a ladder which contains a function block INTEGRAL. Its functionality is also explained in detail in Chapter 48.

This function block totalises the flow F47 whilst the GO button is pressed. If the RESET button is pressed, the TOTAL is set to zero and the READY coil energised.

It is important to note that whilst changes in the status of the GO and RESET contacts are subject to execution of the ladder by the PLC cycle, the INTEGRAL block functions in real time as a background task. In particular, execution of the ladder continues whilst the function block executes.

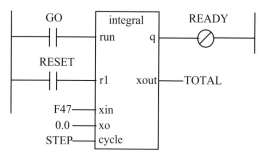

Fig. 47.11 Rung with function block

Most PLCs provide function blocks for supporting a wide range of signal processing and continuous control functions such as filtering and PID control.

47.4 Timers and Counters

The model of Figure 47.7 explains the essence of PLC operations. However, it is evident that the cycle time depends on the number of I/O signals being refreshed, the number of rungs in the ladder and the logic involved. For process control purposes timing needs to be deterministic: it is not adequate to be based upon something as variable as execution cycle time. For this reason a set of timer function blocks are defined. These are used for a wide range of purposes such as allowing time for transfers to take place, timing of interlocks, *etc*. The most common type of timer function block is the so-called "on-timer" as depicted in Figure 47.12. Other timer function blocks include off-timers and pulse timers, and options with either the input or output negated.

The timer is initiated when its input IN becomes true. The elapsed time ET then increments from zero to its user defined preset value PT, shown as 2 min and 30 s in this case, when the output Q becomes true. The timer's output COIL47 is then used to drive succeeding ladder logic. It can be seen that, subject to the FLAG47 being enabled, the timer is controlled by the contact CON47. Whenever either CON47 or FLAG47 are opened, whether the timer has finished incrementing or not, the elapsed time is reset to zero. This logic is depicted in Figure 47.13.

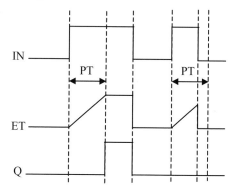

Fig. 47.13 Example of the logic of the "on-timer" rung

Various types of special purpose counter are also provided as standard on all PLCs. These enable events to be counted and the count used to determine subsequent progression of the ladder logic. An "up-counter" type of function block is depicted in Figure 47.14. Provided the reset R is not true, the CTU block counts the number of rising edges at the input CU and increments the output CV for each edge detected. When the integer variable INT becomes equal to (or greater than) the user defined target PV, which has a value of 47_{10} in this example, the output variable Q goes true and sets the COIL47. Once set, the output Q remains set until the reset R goes true whereupon the output Q goes false and the count CV is cleared.

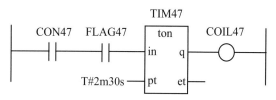

Fig. 47.12 Rung with "on-timer" type of function block

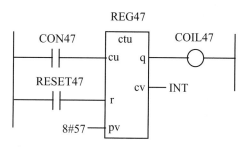

Fig. 47.14 Rung with "up-counter" type of function block

47.5 Development Environments

The tools provided for software development on most PLC systems are not dissimilar to those outlined under ECP in Chapter 41. For example:

- The configurator is used to develop ladder logic. Typically it depicts the power rails and rungs: contacts and coils are selected from a menu and dragged to and/or positioned on the rungs as appropriate. The number of elements on a rung is constrained by the size and resolution of the VDU: seven is typical. Contacts and coils may be annotated and colour coded. Selection from a menu of junction types and jumps enables the branching of the ladder structure necessary for realising complex logic. Functions and function blocks may be chosen from a library and built into the ladder. At a higher level, ladders may be partitioned, treated as function blocks, and replicated which makes for efficient software engineering. Once created, the ladder is compiled and the database generated. Syntax errors and structural faults would be detected at this stage.
- The monitor allows the logic to be tested by stepping through the ladder in a one-rung-at-a-time mode: progression through the ladder is invariably animated. The logic is tested against a "model" of the process. The model consists of a sequence of specified contact and/or coil states which the ladder responds to and/or sets. The model may be in either paper form or simulated: in the latter case the user can interact with the model to test alternate scenarios.

The advent of IEC 61131, underpinned by common data typing, has forced suppliers into providing development environments in which PLCs can be programmed in languages other than ladder logic. Thus cross compilers enable programs in structured text and sequential flow charts, or mixtures thereof, to be translated into ladder logic for downloading onto the PLC.

47.6 Comments

This chapter has concentrated on the use of ladder diagrams *per se* and on the incorporation of standard functions such as timers and function blocks. It should, however, be appreciated that ladder logic can itself be used within other languages. For example, ladder logic may be used both to trigger transitions and to define the content of the steps and actions within sequential function charts, as described in Chapter 37.

Configuration

48.1 Functions
48.2 Function Blocks
48.3 User Defined Function Blocks
48.4 Nesting of Function Blocks
48.5 Function Block Diagrams
48.6 Comments
48.7 Compliance

Traditionally, configuration is associated with continuous control functions, as described in Chapter 44 and depicted in Figure 44.1 for an analogue control loop. However, configuration is not restricted to continuous functions: function blocks are commonly used for handling the logic associated with discrete signals, as described in Chapter 46, and are increasingly being used for simple sequencing applications.

Configuration is the process by which predefined routines, usually referred to as function blocks, are strung together to perform some overall function using some configuration tool. Thus, as explained in Chapter 41, configuring a control scheme consists of creating instances of function blocks from generic types in a library, and linking them together in the order required by the application. Each function block relates to a block of data in the database which requires values to be specified for a variety of parameters.

This chapter explains functions and function blocks, as per the IEC 61131 (Part 3) standard, and the creation of function block diagrams. In particular, it gives an insight into their underlying functionality which is all too easy to lose sight of with the use of configurators. For a more detailed explanation, the reader is referred to the text by Lewis (1998). There is a further standard, IEC 61499, which concerns the use of function blocks for industrial process measurement and control systems. A more detailed explanation of this is given in another text by Lewis (2001).

48.1 Functions

A function is a software element which, when used with a given set of inputs, produces a single value as a result. Functions can be used in any of the five languages of IEC 61131. Figure 48.1 depicts the function for finding the average of two real variables.

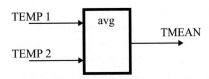

Fig. 48.1 Function for averaging two real variables

Its functionality is defined in structured text as follows:

```
function AVG: real
   var_input
      INPUT1, INPUT2: real;
   end_var
      AVG:=(INPUT1+INPUT2)/2;
end_function
```

This function may then be invoked to calculate the mean, say of two temperatures, by a statement of the form:

```
TMEAN:=AVG(TEMP1, TEMP2)
```

Functions may be strung together to realise a more complex function. For example, the same average could be computed from add and divide functions, as depicted in Figure 48.2.

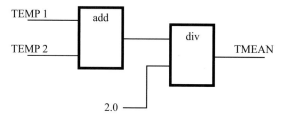

Fig. 48.2 Alternative functions for averaging

There are many standard functions defined in IEC 61131 for converting data types, arithmetic, trigonometric and boolean operations, variable selection and comparison, bit and character string manipulation, and for handling the date and time. The intent is that the functions are overloaded which means that, as far as is practicable, each function can handle a variety of data types.

When a function is used in either a ladder diagram or a function block diagram, it is possible to control when the function executes by means of a special input EN and output ENO. If EN is set true the function executes and when successfully completed ENO is set true. By chaining the ENO of one function to the EN of the next, it is possible to control the order of execution of the functions and to ensure that the result of a chain of functions is valid. An example of this is depicted in Figure 48.3 which shows two functions in the rung of a ladder diagram.

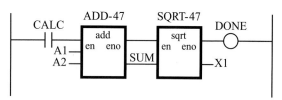

Fig. 48.3 Use of EN and ENO for controlling order of execution

Its structured text equivalent is as follows in Program 48.1:

Program 48.1. Structured text equivalent of Figure 48.3

```
type
   A1, A2, SUM, X1: real;
   CALC, DONE: bool;
   ADD_47: add;
   SQRT_47: sqrt;
end_type

if CALC then
   ADD_47.EN:=1;
   SUM:=A1+A2;
   ADD_47.ENO:=1;
end_if;
if ADD_47.ENO then
   SQRT_47.EN:=1;
   X1:=SQRT(SUM);
   SQRT_47.ENO:=1;
end_if;
if SQRT_47.ENO then
   DONE:=1;
end_if;
```

48.2 Function Blocks

Function blocks are different from functions in that they are not restricted to a single output and can retain values between executions. The integral function block, for example, is depicted in Figure 48.4.

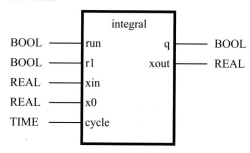

Fig. 48.4 The integral function block

Its structured text equivalent is as follows in Program 48.2:

Program 48.2. Structured text equivalent of the integral function block

```
function_block INTEGRAL
  var_input
    RUN: bool; (* Integrate=1, Hold=0 *)
    R1: bool; (* Reset*)
    XIN: real; (* Input signal *)
    X0: real; (* Initial value *)
    CYCLE: time; (* Sampling time equiv to
                    integration step length *)
  end_var
  var_output
    Q: bool; (* Not reset=1*)
    XOUT: real; (* Integrator output *)
  end_var
  Q := not R1;
  if R1 then
    XOUT := X0;
  elsif RUN then
    XOUT := XOUT+XIN * time_to_real(CYCLE);
  end_if
end_function_block
```

The functionality of this integral function block is indicated by the annotated comments. In essence, the following algorithm is used for numerical integration:

$$x_0(j) = x_0(0) + \sum_{i=0}^{j} x_1(i).\Delta t$$

The first half of the function block is taken up with type statements for input and output variables. The reset flag R1 is used to initialise integration. If R1 is on (true) then the output flag Q is off (false) indicating that integration is not occurring and XOUT is set to X0, the initial value for integration. Otherwise, R1 is off and Q is on, indicating that integration is occurring and, provided that RUN is on too, summation of the input XIN with respect to time CYCLE occurs.

If R1 and RUN are both off (false), as determined by the external logic, then the integrator is idle and its status is ambiguous:

- It could have been reset and waiting to start integration, or
- Integration could be suspended and waiting to be continued, or
- It could have finished integrating and be waiting to be reset.

The accuracy of the numerical integration is critically dependent upon the value of the input CYCLE time being equal to the sampling period corresponding to the frequency at which the function block is executed. This is an implementation issue, normally resolved at task level. It is implicit in function block diagrams and sequential function charts, but is not necessarily so with ladder diagrams whose cycle time is usually variable.

An instance of this function block can then be declared, or instantiated, within any program where an integration is required, as follows:

```
var
    FIT47_TOT: INTEGRAL
end_var
```

The instance is thereafter referred to within the program by name. However, to be executed, the inputs and outputs of the instance need to be related to those of the function block. An example of a statement for doing such is as follows:

```
FIT47_TOT (RUN:=FLG47_1, R1:=FLG47_2,
          XIN:=FIT47,   X0:=ZERO,
          FLG47_3:=Q,   F47TOT:=XOUT);
```

When an instance of a function block is executed it is said to have been invoked. Those variables explicitly declared as parameters of the function block when it is invoked must have been defined elsewhere, and the names of the remaining variables are assumed to be the same in the function block as in the calling program.

The integral function block's functionality was articulated in structured text for convenience. It could have been done in terms of functions, or indeed any of the other IEC 61131 languages and mixtures thereof, but the explanation would have been more cumbersome.

The standard currently defines a small number of fairly basic function blocks such as bistables, edge detectors, counters, timers, real-time clock, ramp, hysteresis, I and D actions, PID control and ratio. It can be anticipated that a variety of other function blocks, such as cascade and lead/lag, will be incorporated through extensions to the standard and that system suppliers will provide libraries of function blocks.

48.3 User Defined Function Blocks

Function blocks can be user defined. For example, a function block for charging based upon the sequence of Figure 29.2 could be defined, as depicted in Figure 48.5.

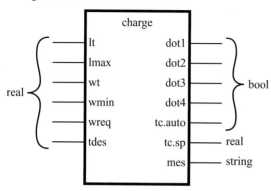

Fig. 48.5 User defined function block for charging sequence

The structural text equivalent, adapted from Program 29.1, is as follows in Program 48.3:

Program 48.3. Structured text equivalent of charging sequence

```
function_block CHARGE
    var_input
        LT, WT: real;
        LMAX, TDES, WMIN, WREQ: real;
    end_var
    var_output
        DOT1, DOT2, DOT3, DOT4: bool;
        TC: pid;
        MES: string;
    end_var
    var
        WTBEG, WTDIF: real;
    end_var
```

```
    DOT1:=0;
    if LT ge LMAX then
        MES:='$n LEVEL TOO HIGH $r'
    elsif WT le WMIN then
        MES:='$n WEIGHT TOO LOW $r'
    else
        DOT2:=1;
        DOT3:=1;
        WTBEG:=WT;
        DOT4:=1;
        repeat
            WTDIF:=WTBEG-WT
            until WTDIF ge WTREQ
        end_repeat
        DOT4:=0;
        DOT2:=0;
        TC.SP:=TDES;
        TC.AUTO:=1;
    end_if
end_function_block
```

An instance of this function block could then be declared within a program, for each identical arrangement of charge and process vessel.

48.4 Nesting of Function Blocks

Function blocks may be nested in the sense that one function block may be defined in terms of others. The PID function block, for example, is depicted in Figure 48.6.

Assume that this function block is equivalent to the PID controller of Equation 23.6:

$$\theta_0 = \theta_b \pm K_C \left(e + \frac{1}{T_R} \int_0^t e.dt + T_D \frac{de}{dt} \right)$$

$$= \pm K_C \left(e + \frac{1}{T_R} \left(\frac{T_R}{K_C} . \theta_b + \int_0^t e.dt \right) + T_D \frac{de}{dt} \right)$$

$$\equiv \pm K_C \left(e + \frac{1}{T_R} \left(\int_{-\infty}^0 e.dt + \int_0^t e.dt \right) + T_D \frac{de}{dt} \right)$$

48.5 Function Block Diagrams

Function block diagrams (FBD) are used to express the behaviour of functions and function blocks in terms of the flow of signals between them. They depict the multiple connections between the various elements involved in realising control schemes. An insight into the scope of these interconnections is provided by Figure 48.7 which is a relatively simple FBD. It is usual for control schemes to consist of several function blocks as was seen, for example, in Chapter 44 in which AIN, FIL, PID and AOT function blocks were configured. Clearly an FBD depicting in detail all the interconnections for this would be extensive.

The IEC 61131 standard defines graphical formats for depicting the elements and signals, and recommends conventions for layout. For example, inputs should occur on the left and outputs on the right hand side of a FBD and, in general, signal flow should be from left to right. Signals may be

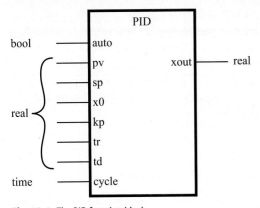

Fig. 48.6 The PID function block

Its functionality may be explained utilising integral and derivative function blocks as depicted in function block diagram form in Figure 48.7.

The standard, IEC 61499, currently being developed, addresses the issues relating to hierarchical models of nested function blocks.

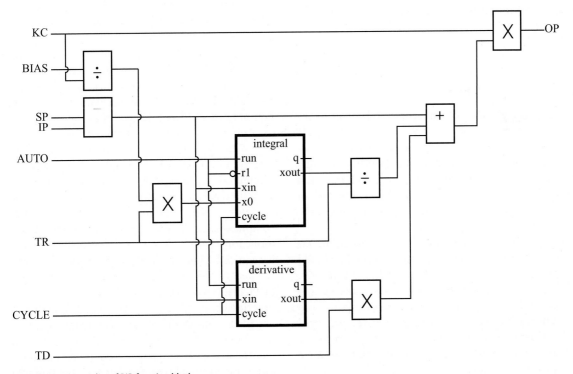

Fig. 48.7 Functionality of PID function block

fed back, right to left, from outputs of one block to inputs of another. This feed back may be either explicit or by means of connectors.

The elements on FBDs should be arranged so as to enable the principal signals to be traced readily. A good configurator enables the elements and their connections to be dragged across the FBD and positioned such that the crossing of signals and changes of direction are minimised. There is no limit on the size or complexity of a FBD although, in practice, there are physical constraints imposed by the resolution of the screen and system dependent constraints on the number of function blocks *per* configuration.

The order of execution of functions and function blocks is implicit from the their positions in the FBD. When a function block is executed, its outputs are updated: it follows that, in general, signals propagate from left to right. However, the order can be made explicit by chaining of the EN and ENO signals, as discussed. The order of execution is not always obvious in FBDs with feedback paths and can lead to subtle variations in behaviour between systems.

48.6 Comments

From a users point of view, configuration of a control scheme is fairly straightforward. Invariably there is a configuration tool which guides the user with prompts, options and default settings through the processes of instantiating function blocks and then parameterising them. However, the feel of the tool will vary substantially from one system to another.

Typically, the configurator of a DCS will provide, for each function block instance, a template based display of the corresponding datablock type. The user is then prompted for values to be entered into the slots, as depicted in Chapter 45. Apart from type testing and memory allocation, the principal outcome of the subsequent compilation process is to automatically establish the connections between the function blocks.

In contrast to this, the configurator of a SCADA or PLC based system is more likely to be graphical. It will enable instances of function blocks to be created and "dragged" into place on a function block diagram. Connections have to be established explicitly by "wiring" them together. The principal outcome of the subsequent compilation process is to automatically establish the underlying datablocks.

48.7 Compliance

One of the principal driving forces behind development of the IEC 61131 standard was the portability of application software from one system to another. There is a huge potential benefit to end-users in terms of re-usability through the modularity of design encouraged by the standard. To this end an independent compliance testing institute, PLCopen, has been established. PLCopen's purpose is to test system's languages, tools, *etc.* against IEC 61131-3 and to establish the extent to which they are compliant with the constructs, functionality, *etc.* of the standard. There are different levels of compliance and certification identifies those aspects of a system that are not compliant. Note the deliberate use of the word "compliance" and the distinction between compliance and accordance. Systems which are claimed to be in accordance with the standard do not necessarily comply with it.

Application software, which may be created in any of the five IEC 61131 languages, or mixtures thereof, ultimately results in files. These files are created, or subsequently operated upon, by tools such as configurators, graphical editors, compilers, *etc.* A fairly fundamental constraint on full compliance are the interfaces between the various tools, for which a common file exchange format is required. Thus, for portability of programs developed in the graphical languages, *i.e.*, function block diagrams (FBD), ladder logic and sequential function charts (SFC), the source code must be in one of the textual languages. Noting that application software must eventually be cross compiled into the assembly language of the target machine, rather than into instruction lists (IL), the de-facto source code for file exchange is structured text.

Open Systems

49.1 Integration of Control and Information Systems
49.2 Standards and Enabling Technologies
49.3 Architecture
49.4 Data Objects
49.5 Object Linking
49.6 Open Process Control
49.7 OPC and the Internet
49.8 Openness and Security
49.9 Information and Control Security
49.10 Information Security Management
49.11 Firewalls
49.12 Demilitarised Zones
49.13 Malware Summary
49.14 Anti-Virus Software
49.15 Comments

An open system is one that is based on widely used industrial standards. This covers programming languages, operating systems, display technologies and communications networks. There are three principal issues involved in systems being open: access, inter-operability and security. An open system provides unrestricted access, from anywhere in the system, to data for information and display purposes. Inter-operability features include:

- Portability of application software, *i.e.* software developed for one system capable of being run on another
- Scalability of hardware, *i.e.* the ability to migrate to a different, more powerful, processor without having to rewrite any software
- Ability to mix and match hardware and software from different suppliers

Open systems are the antithesis of the monolithic ICS and early DCS systems, as described in Chapter 38, which were based on proprietary operating systems and networks. It was difficult enough to connect one system up to a another from the same supplier, let alone from a different supplier. The porting of software and implantation of new processors was simply out of the question.

49.1 Integration of Control and Information Systems

The ability to merge and manipulate real-time data from control systems with cost, quality and scheduling data from information systems enables operators and managers to make timely decisions about processes and production. There is an inexorable blurring of the boundaries between traditional so-called islands of automation, leading to integration of control and information systems. A variety of commercial pressures on both end-users and control system suppliers are reinforcing this move towards open systems. For example:

- Access to on-line control data enables performance against increasingly stringent environmental and safety regulations to be monitored more effectively.
- Advanced control and optimisation techniques are commonly used to maximise profits. These techniques are information rich and require access to data about costs and production requirements.
- Responsiveness to market demands requires flexibility in processing to meet rapidly changing production requirements. Thus control schedules need to take operational constraints into account.
- Flatter management structures mean that responsibility for decision making is pushed down to the lowest levels practicable. This is only possible if appropriate information is readily accessible at all levels.
- The advent of intelligent instrumentation provides much more control data than hitherto. Also, data related to instrument performance is available, for which there is a requirement for integration with maintenance planning.
- Control systems' hardware is increasingly becomes a commodity product. To retain profitability, system suppliers are shifting towards the provision of advanced control techniques and the integration of control functions with MIS and CIM capabilities.
- The end-user's investment in application software may well be the most valuable part of a control system as it represents the intellectual output of some its best engineers. The use of standard operating systems, languages and protocols enables portability, if not yet between manufacturers but at least between systems from the same supplier.
- Industrial quality control system hardware has to be rugged for plant and control room use. Office equipment, such as PCs and workstations, is relatively cheap. There are real cost benefits from being able to mix industrial and commercial equipment in the same system. This is only possible if they work to the same communications standards and protocols.
- The cost of developing quality system software is such that suppliers are entering into third party agreements rather than develop it themselves. Thus graphics packages, relational databases, *etc.* are supplied as part of a system. This is only feasible with a common platform.

49.2 Standards and Enabling Technologies

It is obvious that a necessary prerequisite for open systems is the existence of common standards, whether they are of an industry *de facto* nature or produced by a standards authority such as IEC or IEEE. Standards take a long time to develop which, in part, explains why open systems have taken so long to come about. This is largely because the issues are complex and it is often difficult to reach genuine agreement between end-users and suppliers, and between different countries. Choices have to be made between differing practices and technologies and invariably there are conflicting interests. Nevertheless, a set of common standards has emerged and others are coming to fruition. The enabling technologies of particular note are:

- Operating systems: Windows 2003/XP, UNIX, Linux
- Languages: C++, Visual Basic, Java, SQL, IEC 61131, DDL, *etc.*
- Networks: Ethernet (IEEE 802.3), token bus (IEEE802.4), fieldbus (IEC 61158), *etc.*
- Data access: OLE (ActiveX), OPC, *etc.*
- Data structures: ISA S95 and IEC 61512

49.3 Architecture

The impact of open systems is to flatten the architecture of control systems into two domains, control and information, as depicted in Figure 38.11.

Elements at the control level are connected to a highway which is typically token bus. If the highway is proprietary and/or non-standard, gateways are required to provide compatibility between them. Gateways also enable serial links to intelligent instruments and to HART, Profibus and/or fieldbus devices.

The information domain is invariably based on Ethernet using TCP/IP, as described in Chapter 40, which is the *de facto* multivendor network. This enables any required combination of NT or UNIX based workstations, X-Windows terminals, computers and file servers to be used.

When all the elements of a system have a common standard real-time operating system and common communications and display standards, data can be accessed and displayed at all levels: operator displays at management levels and management displays at operator level, subject to access control as appropriate. Likewise, data held in relational databases can be accessed by client-server techniques at all levels.

An important development is the evolution of operator control stations (OCS) based on a ruggedised form of workstation which provides operators and engineers with a single point of access to both control and plant-wide information. Thus a single workstation doubles up as a human interface for system development and as an operators control program (OCP) for control purposes.

It is essential in handling information in the control domain that the systems are:

- Robust in the sense that the operating systems do not crash
- Deterministic in that events are guaranteed to take place on time
- Permanent in the sense that critical software cannot be corrupted
- Reliable and able to support redundancy if necessary
- Secure from faults in the information system
- Tolerant of operator errors
- Resistant to unauthorised external access

49.4 Data Objects

Data objects are items such as process variables, function blocks, parameter lists, programs, *etc*. They exist as discrete entities within a data base.

The use of data objects is only meaningful in the context of distributed databases within an open system in which the objects are globally accessible. There are various advantages in using data objects:

- Applications can access data in the form of objects by name only, typically a tag or block number, without reference to precise data addresses.
- The data object exists as a single point within the system, thereby ensuring that the data accessed is up-to-date, complete and consistent.
- A standard application program interface (API) can be readily defined and provided to third parties so that their applications and devices can be seamlessly integrated.
- Structured query language (SQL) extensions can be defined to access data objects within a client-server architecture which can connect smoothly and reside securely within larger, enterprise wide, proprietary relational databases.

49.5 Object Linking

Object linking and embedding (OLE) is an open standard for connecting applications in a Windows or UNIX environment. It provides dynamic real-time links with databases and networks for process control using standard interfaces. Using OLE concepts, applications can be developed with a standard interface to other applications developed to the same standard, thereby offering true software inter-operability. This is depicted in Figures 49.1 and 49.2 which contrast bespoke customised interfaces with OLE based ones.

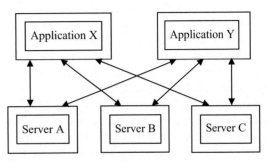

Fig. 49.1 Customised interfaces

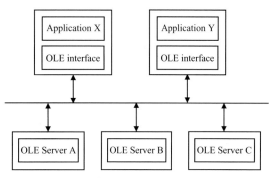

Fig. 49.2 OLE based interfaces

OLE allows programs to be broken down into objects, or components, which can be considered as building blocks. They are built around standard protocols and hence can communicate with each other, are platform independent and are reusable. Components are combined into an application by a configuration type of process referred to as aggregation. There is no limit to aggregation and aggregated components may themselves aggregate.

OLE permits parallel transfer of data between applications, with practically no limit to the size of data block, which is very efficient. This is in stark contrast with dynamic data exchange (DDE), another Windows standard which relates to computer aided design (CAD) environments, in which data transfer is serial and block size is limited. DDE is unsuitable for most control purposes, although "*fast* DDE" does permit multiple block transfers.

OLE-DB incorporates the open database connectivity (ODBC) access interface. It is being developed to enable corporate databases, either in computers or of a relational nature in file servers, to communicate with PC based data. The programmer will then be able to use OLE components to access data anywhere in the organisation.

The OLE standard is now referred to as ActiveX and its components are referred to as controls. The Microsoft.Net development environment, using the Visual Basic language, provides facilities for accessing and linking ActiveX components. This enables fully functional systems to be created from libraries of open components provided by the suppliers of I/O hardware, display systems, *etc.*, and interfaced with standard packages such as Excel and Access for calculation and logging.

Developments on the OLE front have caused a revolution in productivity in the design of systems for automation, especially in enabling the integration of control hardware and intelligent devices with information systems. The cost of putting the components together in building such systems is but a fraction of what would be the cost of writing them in the first place.

49.6 Open Process Control

OLE for process control (OPC) is an open standard for process control. It focuses on the definition of the components required for real-time transfer of data between an OPC server and an OLE compliant application. An OPC server is simply a device that collects data that is consistent with the OPC protocol and makes that data available simultaneously to any number of OPC clients. The standard consists of an evolving set of plug-and-play OLE interfaces intended for linking DCS, SCADA and PLC type systems with each other and for integration with applications such as control and data acquisition.

The objective of OPC is to enable interoperability between different automation applications written in different languages and running on different platforms anywhere within an open system. Fundamental to this objective is acceptance of information system standards such as ISA S95. The benefits of OPC are:

- Instrument and system hardware suppliers will only have to produce one OPC compliant interface for each device.
- Users will be able to choose the best mix of hardware for a given application,
- Suppliers' costs will be contained because the same standard components will interface to different hardware.
- Efficient data transfer from devices to applications.
- Standard techniques for addressing information in process control systems and devices.

An important extension to OPC is the development of OPC-DX which acts as a software gateway. Each device with an OPC-DX interface has both a client and a server part. The client part consumes data from other devices for use by its application (read) and the server part produces data from its device for other applications (write). Connection of devices using DX interfaces is done by means of a configuration tool from a library of device types using drag and drop techniques.

OPC-DX is of particular significance in the context of intelligent instrumentation. Networking of field devices has resulted in a number of widely accepted fieldbus protocols such as HART and Profibus, as explained in Chapter 50, resulting in the need for multiple bespoke gateways to establish effective communication with third party devices. DX offers the prospect of becoming the *de facto* API for communication with and between fieldbus function blocks.

49.7 OPC and the Internet

Increasingly process control systems are being connected to the internet. This might seem to be a bizarre thing to do, given the security issues described in the following section, but there are good reasons for doing so. For example, under certain safety or operability related circumstances, it may be appropriate for a control system to alert key personnel to a situation by automatically generating an e-mail. Sometimes, suppliers have access *via* the internet to control systems that they have supplied for diagnostic and maintenance purposes. Internet connections are increasingly used to abstract information from a control system into other applications, such as SPC packages. Access is often provided over the internet to data within the control for management purposes.

For internet applications, the hypertext markup language (HTML) is used to write small programs (scripts). HTML scripts contain commands which are executed when the program is run. However, HTML is a scripting language largely concerned with display formats and does not support re-usability of data. For example, if the same data is to be displayed and printed, two separate script files must be created with different format instructions for video and printer.

An important extension to OPC is the extensible markup language (XML) which allows instructions to be distinguished from data. OPC XML scripts enable identifiable packets of data to be exchanged over the internet. XML is a web service technology and control systems suppliers are starting to incorporate programs written in XML to distribute process and other data across TCP/IP networks.

49.8 Openness and Security

The Achilles heel of openness is security. The essential problem is that the more open a system becomes the more vulnerable it is to unauthorised access. This is especially so when MIS and (even) control systems are connected up to intranets and the internet. There are various dangers which may be categorised as follows:

- Inadvertent. For example, persons using the internet to legitimately access data within an MIS, or functions within a control system, may inadvertently corrupt data or introduce a virus. A summary of virus types is provided in Section 13.
- Mischief. Hacking is a well known phenomenon for which tools and training is readily available. Hackers are often motivated by the challenge and satisfaction of breaking into a system rather than for the damage they can cause. However, a breach in security is damaging in itself because, when revealed, usually by a message left by the hacker, there is uncertainty as to what damage may have been caused.
- Spying. Use of the internet, by hacking or otherwise, provides access to data in a company's MIS. Data such as recipes and production schedules can potentially give a competitor significant commercial advantage.

- Malicious intent or sabotage. For somebody with a grudge against a company, say a current or former employee, unauthorised access would provide the opportunity to wreak havoc. This may be by the simple but effective means of deleting files, programs or data within the MIS. Or it may be more subtle by, for example, changing key parameters that lurk in a control system and cause it to function incorrectly or even dangerously in the future.

Restriction of access to authorised users depends upon how they are trying to access a system. For internal personnel there is an obvious need for access to the MIS and/or control system. The common approach is to have a security policy which determines level of access according to job function and expertise. This is typically realised by means of a password and, for greater security, can involve keys, swipe cards or even signatures such as finger, retina or voice prints. The essential thing is that the policy is policed with named individuals, levels of access, dates of issue and periods of validity all being monitored, reviewed regularly and changed if/as appropriate. Systems are at their most vulnerable following personnel changes, or when third parties are given temporary access for some specific project. See also Chapter 64.

For external personnel, access is usually *via* the internet and is restricted by means of a so-called firewall. Firewalls are discussed in more detail in Section 49.11. In essence, a firewall is a device through which messages *via* internet connections are forced to pass, as depicted by the box labelled FW in Figure 49.3. The messages are examined by a program within the firewall and either accepted or rejected according to some rule base. A firewall will prevent unauthorised access and search attachments to messages for known viruses, but cannot prevent unknown (to the firewall) viruses from getting through to the control system.

It should be remembered that there are other means of remote access to a control system such as dial-up connections, wireless systems and third party connections. In focussing on internet access and firewall solutions, these other forms of intrusion shouldn't be overlooked.

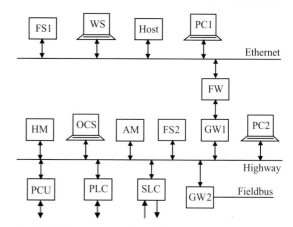

Fig. 49.3 Open system with simple firewall

49.9 Information and Control Security

There are two generic standards, IEC 17799 (Part 1) and BS 7799 (Part 2), concerning information security management systems. These have emerged from the commercial world rather than control and automation, and do not explicitly address the problems of IT security in the context of real-time systems. However, many of the principles involved are common and are certainly applicable to any on-line MIS connected to a control system, DCS or otherwise, and to any off-line systems used for design purposes.

There are another two US standards which are more specific to the process industries – AGA 12 and API 1164. These concern the security from cyber attack of SCADA systems used in the gas and oil industries respectively in relation to the national critical infrastructure.

The body concerned with cyber attack in the UK is the National Infrastructure Security Coordination Centre which has produced a guide to principles of good practice for process control security through the development of a security framework. Although this NISCC (2005) guide specifically refers to SCADA systems, it is just a applicable to DCS, PLC and hybrid systems. Three guiding principles are employed:

1. Protect, detect and respond:
   - Put in place protection measures to prevent and discourage electronic attacks.
   - Deploy means to rapidly identify actual or suspected attacks.
   - Take appropriate action in response to confirmed security incidents.
2. Defend in depth:
   - Implement multiple protection measures to avoid single points of contact.
3. Manage protection:
   - Recognise the contribution of procedural and managerial measures such as change control, firewall monitoring and assurance, and training.

49.10 Information Security Management

The NISCC (2005) guide focuses on seven key themes, summarised below, for each of which further more detailed guidance is being developed:

1. Understand the business risks.
   The objective is to gain a thorough understanding of the risks to the business from threats to its control systems in order to identify and implement the appropriate level of security protection. Good practice entails understanding:
   - What systems are involved: their role, where they are located, who owns, manages and supports them, and how they interact.
   - The threat: identify and evaluate the threats facing the process control systems.
   - The impact: identify potential consequences of a breach of security.
   - How the systems are vulnerable: this includes networks, applications, remote access connectivity, *etc*.
2. Implement secure architecture.
   The objective is to implement appropriate security protection measures to provide a secure operating environment for the control system. Good practice requires that:
   - The number of external connections be identified and reduced to an absolute minimum. Segregate and/or isolate process control systems from other networks, with dedicated infrastructure for safety related systems.
   - Connections between the control system and network connections be protected with firewalls. Implement effective management of firewall configuration and change control.
   - Control systems be hardened to prevent network based attacks. Thus all unused services and ports in the operating system and applications should be removed or disabled, and all inbuilt system security functions enabled.
   - E-mail and internet access from the control system be disabled.
   - Wireless networking be avoided wherever possible.
   - Control systems be protected with anti-virus software on servers and workstations. Accreditation and configuration guidance from the control system supplier is strongly recommended.
   - Any new device connected to the control system is proven to be virus free.
   - Security systems, that is firewalls and anti-virus software, be kept up to date. Procedures should be based upon supplier certification of patches, testing of patches, and staged deployment to minimise the risk of disruption.
   - Remote access be logged and managed effectively. This involves maintaining an inventory of connections, restricting access to specified machines, users and times, auditing and systematically reviewing access.
   - A variety of user orientated measures be put in place. This includes personnel background security checks, procedures for issue and change of passwords, training, authorisation of new users and removal of former users.
   - There be proper documentation for access control.
   - Effective back-up and recovery procedures are in place.

- A variety of physical security measures are put in place. Typical examples are drive locks, tamper proof casings, intruder alarms, access control systems, CCTV, *etc.*
- Control system activity be monitored to indicate health of the system. This involves monitoring network activity and time taken for specific tasks against baselines for normal operation.

3. Establish response capabilities.
   The objective is to establish procedures for monitoring, evaluating and taking appropriate action in response to security events. Good practice entails:
   - Forming a computer emergency response team (CERT) to respond to suspected security incidents. Responses may include increased vigilance, isolation of control systems and application of patches.
   - Ensuring that appropriate security response and business continuity plans are in place for the control systems, and that the plans are maintained, rehearsed and tested.
   - Establishing an early warning system to notify appropriate personnel of security alerts and incidents, and ensuring that all such warnings are formally recorded and reviewed.

4. Improve awareness and skills.
   The objective is to increase process control security awareness throughout the organisation and to ensure that all personnel have the appropriate knowledge and skills to fulfil their role. Good practice requires that:
   - Management understands the business implications of the security risk to control systems and therefore commits to the management of the risk and provision of tools and training.
   - IT personnel understand the differences between the security of control systems and IT security in general. For example, control system operations cannot be suspended whilst patches are installed or reboots made to enable upgrades.
   - Control engineers develop IT security skills and commit to system security procedures.
   - Links be established between IT personnel and control teams to build working relations, share skills and facilitate knowledge transfer.

5. Manage third party risks.
   The objective is to ensure that all security risks from suppliers, contractors and other third parties are managed. Good practice requires that:
   - All third parties that have legitimate access to the system should be identified.
   - The basis of third party access be detailed contractually at the procurement stage, defining the terms of the connection and prompt notification of vulnerabilities.
   - Third parties agree to be bound by the end-users security regime.
   - System suppliers agree to provide anti-virus protection, security support and patches, and agree to system hardening procedures.
   - Other third parties be prevented from having access to the control systems until their equipment, systems and software is proven to not be a security risk.
   - There be regular security audits and reviews of third party access.

6. Engage projects.
   The objective is to ensure that any project that may impact on control system security is identified and that, early in the project's life cycle, appropriate security measures are included in its design and specification. Good practice involves:
   - Identifying all projects and developments that could potentially have an impact on the control system's security.
   - Ensuring that a named individual has responsibility for security risk management throughout the project life cycle.
   - Addressing security issues in the URS and DFS documentation and subsequent contracts, and ensuring that security policies are adhered to.
   - Carrying out security reviews and testing security at key points in the control system's development cycle.

7. Establish on-going governance.
   The objective is to provide clear direction for the management of risks to control system security and to ensure on-going compliance and review of policy and standards. Good practice requires that:
   - The roles and responsibilities of all concerned with the control system's security are defined.
   - Security policy and standards be defined, documented and disseminated.
   - Procedures be put in place for the management of the policy and standards, which includes provision for their occasional review.

This may all seem to be rather tedious: yes, but unfortunately necessary. Gone are the days when IT security of control systems could be taken for granted.

49.11 Firewalls

It is evident that effective firewalls are fundamental to information security management. The NISCC commissioned a guide to good practice on the deployment of firewalls in SCADA and process control networks which has subsequently been published.

This NISCC (2005) guide makes the point strongly that there is more to deploying a firewall at the interface between an IT network and a control system than meets the eye. The two central issues are that:

- The goals of IT and process control personnel can be fundamentally different: the IT world sees performance and data integrity as being paramount whereas, ultimately, the control community's commitment is to issues such as operability, quality, reliability, safety and viability. These differences in perspective, which can potentially lead to conflicts in security practice, have been noted and addressed in the previous section.
- Whilst it is recognised as good practice that IT and control networks be separated by firewalls, that separation is unlikely to be effective without careful design, configuration and management. The rest of this section deals with those issues.

A firewall can be a separate hardware device physically connected to a network, a combined hardware and software function unit, or even a completely software based solution installed on the host machine to be protected. The first two categories, separate hardware or combined hardware and software, are referred to as network firewalls and typically provide the most secure solution for the separation of control and IT networks. They can be hardened to resist all but the most ingenious assaults and offer the best firewall management options.

As stated, messages (network traffic), comprised of packets of information, are routed through the network firewall. Each packet consists of a header (source and destination addresses, status information, *etc.*), data and a trailer, as described in Chapter 40. A firewall, upon receiving such a packet, analyses its characteristics and determines what action to take. The firewall decides whether to allow it through immediately, buffer it temporarily, redirect it elsewhere or to block it. The decision is based upon a set of rules, referred to as access control lists (ACL).

According to their ACL sophistication, firewalls can be classified as follows:

1. Packet filter firewalls. This is the simplest type of firewall. It uses so-called static rules to check the internet protocol (IP) addresses and port numbers of the packets on an individual basis. This approach, known as static filtering, lacks the ability to understand the relationships between a series of packets and is the most readily "hacked into" type of firewall. It is the cheapest type of firewall but, in its favour, happens to have the least impact on performance.
2. Stateful firewalls. This is a more sophisticated type of firewall. It tracks the interrelationships between packets allowed through it. By keeping a history of accepted packets and being aware of the state of current connections, it can accept only anticipated messages. The power of

this approach, known as dynamic packet filtering, is due to the fact that the rules can be made conditional. For example: a packet will only be accepted for a particular destination if it was received in response to a specific request for a certain type of data sent to a particular source address.

Stateful firewalls offer a high level of security, good performance and transparency to users. However, they are more expensive and, because of their complexity, can be less secure than packet filters if not administered by competent personnel.

3. Proxy firewalls. These work at the application layer of the OSI model as described in Chapter 40. The packets are opened, processed according to ACL rules, reassembled, and forwarded to the intended target device. The firewall is typically designed to handle a variety of protocols, through the one device, and then forward the messages to individual host computers for servicing. Thus, instead of connecting directly to an external server, the client connects directly to the proxy firewall which in turn initiates a connection to the requested external server.

Proxy firewalls can provide significant additional security functionality. For example, ACL can be used to require users or systems to provide additional levels of authentication before access is granted. Also rules can be created that are protocol specific. For example, a proxy firewall can be used to block all inward bound HTTP messages that contain scripts whereas, by contrast, a filter based firewall could block all HTTP messages, or none, but not a subset.

Whilst proxy firewalls offer a high level of security, they do have a significant impact on network performance. Furthermore, most proxy firewalls only support common internet protocols, such as file (FTP), hypertext (HTTP) and simple mail (SMTP) transfer protocols. Thus messages based upon control protocols such as common industrial protocol (CIP) and Modbus/TCP, *etc*., will still require the firewall to process the messages by filter or stateful means.

The current position is that most firewalls on the market use a combined stateful and application proxy approach.

4. Deep packet inspection (DPI). A DPI firewall offers deeper filtering into the application layer than the traditional proxy firewall, without performing a full proxy on the TCP connection. For example, DPI firewalls can inspect simple object access protocol (SOAP) objects in XML on web connections and enforce policy on what objects are allowed through the firewall. This firewall market is still in development.

Firewall suppliers usually offer other services besides their basic message handling functionality. Such services may comprise intrusion detection, deployment of anti-virus software, authentication services, secure encrypted "tunnels" and network address translation. These additional services obviously increase the cost and complexity and reduce the performance of the system. However, making good use of them can significantly improve the overall security of the control system.

49.12 Demilitarised Zones

There are various architectures available for deploying a firewall between an IT network and a control system. These configurations involve routers and switches, multiple ports, single and multiple firewalls, and demilitarised zones (DMZ). Of these, on the grounds of security, manageability and scalability, the DMZ configurations are overwhelmingly the most effective. One common DMZ configuration is depicted in Figure 49.4.

This configuration requires that the firewall has three or more ports. Thus one port is used to interface with the control system, a second with the IT network, a third to an intermediate local area network referred to as a process information network (PIN) and possibly a fourth to either a wireless local area network (WLAN) or remote and third party access systems.

Common servers such as history (HM) and application (AM) modules would typically be located on the PIN and remote users would have le-

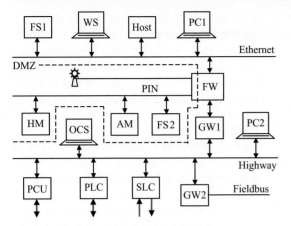

Fig. 49.4 Demilitarised zone with single firewall

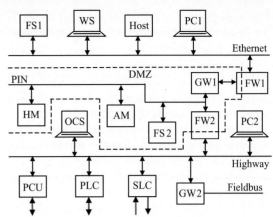

Fig. 49.5 Demilitarised zone with twin firewalls

gitimate access to them *via* the internet and the firewall. Likewise, for updating and/or application purposes, the control system has access to these modules *via* the gateway and firewall. By locating the HM and AM modules inside the DMZ, no direct communication channels are required between the IT network and the control system: each effectively ends in the DMZ. The objective of ACL design is to maintain a clear separation between the IT network and control system.

The primary security risk with this particular architecture is that if one of the modules inside the DMZ is compromised, it can be used to launch an attack on the control network. This risk is obviously reduced by hardening the devices on the PIN and by actively keeping the firewall itself up to date.

An alternative DMZ configuration, which uses a pair of firewalls between the IT network and the control system, is as depicted in Figure 49.5.

Again, common servers such as the HM and AM are located on a PIN within the DMZ. The first firewall FW1 provides security at the IT network interface and the second firewall FW2 secures the interface with the process control network. An attractive feature of the twin firewall approach to DMZ is that it allows both IT and control groups to have clearly separated device responsibility since each can independently manage its own firewall. The principal disadvantage of the twin firewall approach is increased cost and management complexity.

If firewalls from different suppliers are used then the diversity of design provides enhanced security. Indeed, the obvious direction for the control industry to evolve is for gateway design and firewall functionality to become integrated within the same physical unit.

An extension to the concept of diverse firewalls is the use of disjoint protocols across the DMZ. Thus, for example, if Ethernet is used for the IT network it is explicitly not allowed between the DMZ and the control network. Likewise, if token bus is used for the highway, then it is explicitly not allowed between the DMZ and the IT network.

Once the firewall architecture is fixed, the main focus of effort is in determining what traffic to allow through the firewalls. The NISCC (2005) guide provides detailed guidance on recommended practice for the design of ACL rule sets and on the management of firewalls.

49.13 Malware Summary

Malware is the generic term that refers to viruses, trojans, worms, *etc.* A virus is a program, or macro, that is capable of attaching itself to files and replicating itself repeatedly, typically without the user's knowledge or permission. Some viruses attach to files such that when the file is executed so too is the virus. Other viruses sit in a computer's memory and infect files as they are created, opened, edited, copied, *etc*. They may display symptoms and/or cause damage, but neither symptoms nor damage are essential in the definition of a virus: a non-damaging virus is still a virus.

Viruses spread from computer to computer *via* network connections, *via* shared storage media such as file servers and *via* portable storage media such as compact discs and USB sticks. By far the most common form of infection is by viruses attached to e-mails. The effects of virus infection are various and include:

- Just leaving a simple screen message
- Accessing e-mail address books, replicating messages to every address, and overloading the network
- Deleting and/or corrupting files

There are many virus variants:

1. Boot Sector Infector. This virus infects the boot sector of a disc which contains the boot record which is used by the computer to boot itself up. When the computer reads and executes the program in its boot sector, the virus is transferred into memory and gains control over basic computer operations.
2. Encrypted. The virus' code begins with a decryption algorithm: the rest of the virus is in encrypted code. Each time it infects, it automatically encodes itself differently and thus attempts to avoid detection by anti-virus software.
3. File. This type of virus usually replaces or attaches itself to .com and .exe files. Most are resident.
4. Macro. A macro is a series of instructions designed to simplify repetitive steps. A macro virus is typically attached to a document file, such as Word or Excel. When the document is opened, the macro runs, does its damage and copies itself into other documents.
5. Resident. A resident virus loads into memory and remains inactive until triggered, say by a date such as Friday 13th. When the event occurs, the virus activates and does its damage.
6. Stealth. This type hides itself by feeding anti-virus software a clean image of infected files or boot sectors. For example, the size of an infected file would be returned as the size of the file without its virus.
7. Trojan. This is a malicious program that pretends to be benign and causes an application to behave in an unpredictable way. Strictly speaking, a Trojan is not a virus because it does not replicate.
8. Worm. This is a parasitic program which replicates but, unlike a virus, does not infect other programs or files. It can replicate itself on the same computer or send copies to other computers *via* a network, typically as an e-mail attachment.

49.14 Anti-Virus Software

Anti-virus software scans a computer's memory and disc drives for viruses. If a virus is found, the software informs the user and may wipe clean, disinfect or quarantine any files, directories or discs affected. Detection is by means of signature scanning using a so-called signature file. The file holds a search pattern, often simple strings of characters and bytes, for every known virus. Signature files are huge and growing. Generally speaking, anti-virus programs will detect 100% of known viruses, provided the signature file is up to date: it is their ability to detect unknown ones that is critical.

Disinfection reverses the effects of viruses. This requires a complete understanding of what the virus does when it infects. Once detected, the virus is removed from the system and, whenever possible, the affected data is recovered. Disinfection alone is considered to be inadequate because the user may inadvertently re-introduce a virus.

Anti-virus software therefore includes live activity checking (LAC) in the operating system kernel which monitors all operating system activity for viruses. Modern systems are invariably supplied with LAC already installed in the operating system. An alternative approach is to use a background file pattern scanner which intermittently checks a system's hard disc for virus signatures. If the background scanner is effective there should be no need for LAC too. However, in practice, LAC is invariably used because of the potential for introducing a virus during the installation or updating of anti-virus software.

49.15 Comments

Open systems are concerned with the unrestricted access to data and interoperability. They enable the integration of control and information systems. It is evident that most of the prerequisites for open systems in the form of industry standards for operating systems, communications, programming, display and so on already exist. The move towards open systems is being driven by strong commercial pressures. However, protection against unauthorised access and malicious intent is a fundamental constraint. The scope for damage is huge. The cost of protection will rise inexorably and, in years to come, the notion that system's used to have open access will be seen to be quaint.

Fieldbus

50.1 HART Protocol
50.2 Objectives
50.3 History
50.4 Physical Layer
50.5 Data Link Layer
50.6 Application and User Layers
50.7 Hardware Configuration
50.8 Software Configuration
50.9 Function Block Attributes
50.10 Project Management
50.11 Benefits
50.12 Comments

The 4–20-mA current loop, used to transmit signals to and from process instrumentation, became an industry *de facto* standard during the latter part of the twentieth century. It was an open standard to the extent that instruments from different manufacturers could be interconnected to form control loops or interfaced with control systems to form I/O channels. The advent of smart, or intelligent, field devices has led to the need to transmit further information, such as range and bias settings for remote calibration and status signals for condition monitoring. Such data has to be transmitted serially. Initially this was realised by means of modulation, with the data being superimposed on the 4–20-mA signals on-line, or else with the instrument in an off-line mode. Of the various proprietary protocols and national standards that were established to enable this, the HART protocol became dominant. Latterly transmission has become bus based, with dedicated serial buses, and given rise to the so-called fieldbus technologies.

This chapter first considers HART as a vehicle for introducing fieldbus. That is followed by a potted history of fieldbus. Then comes consideration of the various layers of fieldbus, its configuration, application and management. Finally, the benefits of fieldbus are summarised. Much of the literature about fieldbus is in the form of journal articles and technical publications. However, an important guide to the application of fieldbus in the process industry was produced by EEMUA (1998).

50.1 HART Protocol

The highway addressable remote transducer (HART) protocol was developed by the Rosemount company which placed its specification in the public domain in order to encourage other suppliers to add its functionality to their instruments. As a consequence it became the most commonly used protocol based on 4–20-mA signals. It is a *de-facto* industry standard, reliable, field proven and in use world-wide. Commitments have been made to its long term support and for its connection into fieldbus. Aspects of the protocol are explained to provide insight to some of the issues involved in fieldbus.

HART is based upon the current transmission of serial messages in the form of a series of 8-

bit bytes generated by means of a universal asynchronous receiver and transmitter (UART) type of device. To distinguish between analogue and serial transmission, the logical 0 is deemed to be a current less than 3.8 mA and the logical 1 to be a current greater than 20.5 mA. Remember that to each byte transmitted is added a start bit, a parity bit and a stop bit. The protocol is of a master/slave nature: that is, the transmitting device is the master and the receiving device the slave. Its message structure, or frame, is as depicted in Figure 50.1:

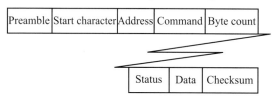

Fig. 50.1 Message structure of HART protocol

- Preamble bytes. Three or more bytes of 1s (FF characters in hex), to allow the receiving device to synchronise the signal after any pause in transmission.
- Start character. Specifies the frame format as being short or long depending on whether the slave address is unique, and whether the master or slave is the message source.
- Address field. Contains the master and slave addresses, and a bit to indicate whether the slave is sending data repeatedly in burst mode.
- Command byte. An integer in the range 0–255 (0–FF hex) which represents one of the HART commands. This byte determines the nature of the data bytes.
- Byte count. The number of bytes in the status and data sections, *i.e.* in the remainder of the message excluding the checksum byte. This allows the receiving device to identify the checksum byte and determine the end of the message.
- Status byte. Only used for messages from slaves to convey error codes.
- Data bytes. Up to 24 data bytes for unsigned integers, floating point numbers or text strings. The first bytes specify the types and quantity of data.
- Checksum byte. The result of an exclusive OR on all preceding bytes beginning with the start character. This provides a further check on transmission integrity in addition to the parity bit at UART level.

Commands within the HART protocol are concerned with reading and writing data to/from the slave, or instructions to the slave to perform actions. There are three categories of command: universal, common practice and transmitter specific:

1. Universal commands provide functions which are implemented in field devices from all suppliers. Typical commands are:
   - Read input and units
   - Read output
   - Read up to four predefined variables
   - Read/write 8 character tag number
   - Read/write 16 character device descriptor
   - Read sensor serial number
2. Common practice commands provide functions which are implemented in many field devices, but not all. Obviously these commands can only be used if the device is capable of realising them. Examples are:
   - Write transmitter range
   - Calibrate (set zero and range)
   - Perform self test
   - Reset
   - Write MV units
   - Write sensor serial number
3. Transmitter specific commands provide functions which are more or less unique to a specific supplier or type of sensor: For example:
   - Read/write low flow cut-off
   - Read/write density correction factor
   - Choose primary value (*e.g.* mass flow or density)

50.2 Objectives

It is now well recognised that the 4–20-mA standard is too restrictive a medium for intelligent instrumentation. The 4–20 mA wiring does not provide a multidrop capability. The low bus speeds do not permit deterministic sampling for control purposes, although some devices became available with local PID control functionality. Also, the low bus speeds limit transfer of the increasing amounts of data available in field devices.

These constraints have led to the development of fieldbus which is an all digital, two-way, multidrop, high speed communications system for instrumentation. The objective is to provide a single, standard, open, interoperable network protocol for enabling:

- Interconnection of field devices such as transmitters, valves, controllers and recorders.
- Connection of such field devices to PLCs, PCs, I/O multiplexers, *etc*.
- Integration of field devices and control systems within plant wide networks.
- Abstraction of real-time data in a seamless manner.

irrespective of the manufacturer of the devices.

50.3 History

The evolution of fieldbus has been tortuous involving various national standards bodies. Several protocols of a proprietary nature have been competing to be adopted as the fieldbus standard but, in effect, they can all be considered to be intermediate stages in its development. The most significant ones are as follows:

- IEC Fieldbus. Two committees were set up by the IEC and ISA in 1985 to work jointly towards a fieldbus standard. The protocol is based upon the physical, data link and application layers of the OSI seven layer model, described in Chapter 40, together with a user layer. The intent was that the physical layer would embrace all three transmission media: wire, telemetry and fibre optics. The user layer is based on function blocks for process control.
- FIP (factory instrumentation protocol). This is a French national standard and has been adopted in Europe by CENELEC as an EN-50170 standard. It is an arbitration (master/slave) based protocol. Chip sets which incorporate the FIP protocol and interface cards are supplied by a number of PLC and PC manufacturers. Development tools are also available. FIP was later expanded into WorldFIP through the involvement of mostly American suppliers. Although much of FIP has since been adopted as the fieldbus standard and been implemented by the Fieldbus Foundation, FIP itself is still being developed by a consortium of mostly French and Italian companies.
- Profibus (process fieldbus). This is a German standard (DIN 19245) which is another of the EN-50170 standards and has the largest installed base of all fieldbus technologies for manufacturing automation worldwide. It is a token passing based protocol. There are two versions: Profibus DP which is a high speed general purpose communications medium, and Profibus PA which was developed specifically for process automation. Both DP and PA use the same protocols, but PA enables devices to be powered over the bus and can be used for intrinsically safe (IS) applications. Profisafe is a version of PA dedicated to safety and protection systems.

Interface cards for PCs, PLCs and intelligent remote I/O supporting the Profibus protocol are available from a number of manufacturers. Together with a complete toolset, these enable field device development, network configuration and testing.

- ISP (interoperable systems project). Supported by a consortium of major control system suppliers, this project's intent was to accelerate the establishment of a single international fieldbus standard. It adopted the IEC/ISA physical layer standard and the user layer specification. The protocol was largely Profibus but incorporated aspects of FIP's synchronisation and the HART device definition language.
- Fieldbus Foundation. This was formed by the merger of WorldFIP and ISP in 1994 and Foundation Fieldbus (FF) now has the largest installed base of all fieldbus technologies for process automation worldwide. FF technology is based on:

1. The IEC/ISA physical layer standard
2. The IEC/ISA data link layer specification, which supports both FIP bus arbitration and Profibus token passing
3. The ISP Rev 3.0 application and user layers
4. The ISP/WorldFIP network and system management
5. The HART device description language

Chip sets incorporating the FF protocol are now available, as are technical specifications, device descriptions and training for FF products. Of particular importance has been the formation of two independent compliance testing organisations, the Frauenhofer Institute and the Institute for Automation and Control. Their brief is to establish interoperability test environments to enable manufacturers to have fieldbus devices tested and certified against the FF protocol.

The FF model is as depicted in Figure 50.2. It is based on the physical, data link and application layers of the OSI model, plus a user layer. The combination of the data link and application layers are often referred to as the communications stack.

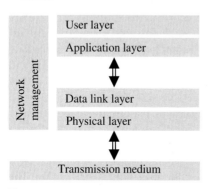

Fig. 50.2 Model underpinning the FF protocol

- Fieldbus International. This is a consortium of companies with the Norwegian Government who are developing an alternative transmission system for fieldbus based on FF for sub-sea use. The approach, which uses inductive coupling for connecting devices rather than conventional push fit plugs and sockets, substantially reduces electrical losses. A further bonus is that the maximum number of IS devices per bus is increased from 6 to 30.

Many other protocols have been developed during the period in which fieldbus has evolved. To a large extent these are aimed at other sectors and/or niche markets and many will continue to evolve in their own right. Some of the better known ones are Lonworks, P-NET (Danish origins), CAN (controller area network), DeviceNet, SDS (smart distributed system) and ASI net.

The IEC published its fieldbus standard in 2000 which is now into its third edition, IEC61158 (2004). This specifies physical, data link and application layers of ten different systems for a broad range of applications and performance requirements. The ten systems are, listed in the order of the standard: i) Foundation Fieldbus, ii) ControlNet, iii) Profibus, iv) P-Net, v) FF high speed ethernet, vi) SwiftNet, vii). WorldFIP, viii) Interbus, ix) FF FMS and x) Profinet.

The different fieldbus types include a range of selectable and configurable options within their detailed specifications and only certain combinations of options will work correctly on the basis of the layers specified in IEC 61158. These combination of options, referred to as profiles, are collected in Part 1 of IEC 61784 (2003). Part 2, which concerns real-time industrial Ethernet, is currently being developed.

50.4 Physical Layer

There are three bus speeds specified at the fieldbus physical layer. Focusing on the slowest of these, the 31.25 kbaud rate, this represents the fieldbus replacement for 4–20-mA signals between sensors and actuators. There are three topologies for connecting field devices to the bus, as depicted in Figure 50.3. These are trunk with spurs, trunk with splices which is usually referred to as "daisy chain", and tree and branch which is often referred to as "chicken's foot".

By optimising the cable route, the trunk with spurs and/or splices topology will produce the

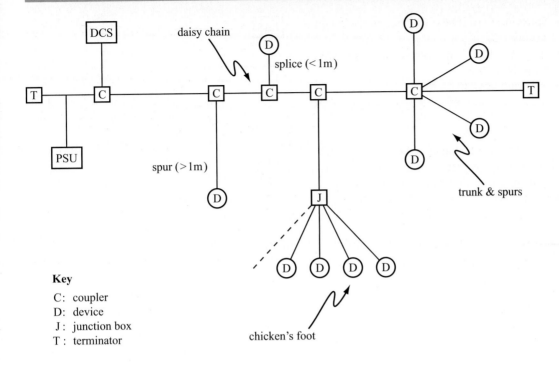

Fig. 50.3 Topologies for connecting field devices to the bus

minimum cabling cost solution to an application, albeit with a large number of couplers. However, the potential problems of having to connect in extra instruments if no spare couplers exist mean that this topology will only be attractive to suppliers of packaged equipment, such as compressor sets, where the design is fixed *a priori*.

For most applications the tree and branch approach will become the normal topology. This is because of its inherent flexibility: provision can be made in the junction box of the chicken's foot for connecting up extra instruments. Terminations in the junction box should be fitted with isolators to enable individual field devices to be isolated. Each spur from the junction box should be fused to protect the bus from faults in the spur or its attached device.

These topologies may be combined on any single bus, sometimes referred to as a segment, subject to various constraints. The cable length, which is the total length of the trunk and any spurs, depends on the quality of the cabling. The maximum length is 1900 m provided shielded, twisted pair (Type A) cabling of 18 gauge (AWG) is used: this maximum is reduced for lesser quality cables. Other constraints are:

1. Maximum number of devices on the bus:
   - 32 devices if self powered and non-IS
   - 12 devices if bus powered and non-IS
   - 6 devices if IS
2. Devices are connected to the bus with couplers:
   - Maximum of 15 couplers per 250 m of the bus
   - Terminators of 10 μF and 50 Ω are required at each end of the bus
   - All device interfaces to the bus must be galvanically isolated
3. Spurs: maximum spur length depends on total number of devices:
   - 120 m for one device, 30 m for four devices, etc.

- Spurs not permitted if number of devices on the bus >25
4. Spurs of less than 1 m are referred to as splices. Total splice length is:
   - Maximum of 8 m for bus < 400 m
   - Maximum of 2% of bus length if bus > 400 m

The bus power supply must be between 9 and 32 V d.c. Any device connected to the bus must be capable of withstanding 35 V of either polarity. For IS systems the barriers determine the supply voltage.

The slower bus speed does not support redundancy. Redundant bus/host interface cards should be considered for multiple buses since failure of this card is a common mode failure. The most likely cause of bus failure is through power supply problems. Any loss, or partial loss, of power which draws the bus voltage below the 9 V threshold will result in a loss of communications.

The two faster bus speeds of 1 Mbaud and 2.5 Mbaud specified within fieldbus are intended for higher level communications such as with PLCs, PCs or DCS. These communications will inevitably be superseded by the use of Ethernet at 100 Mbaud within fieldbus applications.

50.5 Data Link Layer

This is responsible for the transmission and receipt of data over the bus between devices. The data has to be converted into the correct format for transmission and then correctly interpreted upon receipt. This activity has to be co-ordinated by, for example, the link active scheduler (LAS) with FF or by token passing with Profibus. Figure 50.4 depicts the operation of the LAS, the following explanation of which is based upon the FIP bus arbitrator.

Each variable in the system is uniquely owned by one device which may be considered to be its producer (or source). However, there may be several devices that require the value of that variable, the consumers (or receivers). Thus the LAS broadcasts a request for a variable and the producer responds with its value. This is then simultaneously

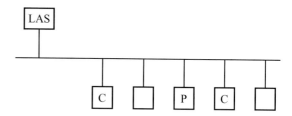

Key
C: consumer
P: producer

Fig. 50.4 Operation of the link active scheduler (LAS)

captured by all the consumers and the LAS requests the next variable.

Each bus has its own LAS, the software for which would normally be installed on some segment interface card in a junction box. Failure of that card would cause failure of the segment. A second, redundant, LAS may be installed in another designated device on the segment, typically a smart positioner, to which bus control would default.

Because the bus is intended for real-time control purposes, transactions must be both synchronous and deterministic. Synchronous means that variables are sampled at a fixed frequency. Deterministic means that the time taken for the sampling is fixed too. Together, these enable the bus loading to be calculated, thereby making it possible to guarantee that the bus capacity is adequate.

The schedule is essentially a table containing:

- A list of the variables to be sampled, or scanned
- The sampling frequency
- The variable types, *e.g.* integer, floating point, *etc.*
- The scan period, *i.e.* the time taken per sample, for each variable type

The schedule is constructed so as to spread the loading as evenly as possible. It consists of a macrocycle comprising a number of elementary cycles of equal duration. The macrocycle is based upon the lowest sampling frequency and the elementary cycle on the fastest. The LAS will infinitely perform the same macrocycle.

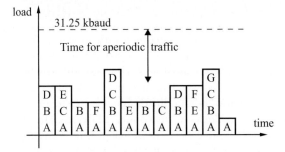

Fig. 50.5 LAS macrocycle for reading variables A–G

Figure 50.5 is an example of a macrocycle for a bus reading seven variables A–G. The time axis clearly indicates which devices are read and in what order. The vertical axis corresponds to the sum of the scan periods for each transaction in the elementary cycle. The loading is clearly within the 31.25 kbaud capacity of the bus.

Only variables to be scanned periodically are included in the schedule: this should not occupy more than some 30% of the bus capacity. Once each elementary cycle has been completed, the remaining time can be used by the scheduler for aperiodic transfers. A producer can, during the course of its normal scan, set a flag to indicate that it wants to make an intermittent write transaction. The scheduler then asks for and receives a list of variables which is queued and scanned at the next suitable aperiodic transfer period using the standard mechanisms.

50.6 Application and User Layers

The application layer is responsible for encoding and decoding user layer messages depending, respectively, on whether they are being transmitted or received. The layer includes:

- Read/write capability to both local and remote buses
- Bus arbitration services comprising scan list definition and balancing
- Interface to network management
- Support for device description and function block definition

The user layer provides device descriptions and function block definitions, and performs function block scheduling.

Device description is based upon the HART categories of universal, common practice and transmitter specific data. It provides the interoperability of fieldbus in that it enables any device from any supplier to be connected to the bus. The description of a device can be considered to be the device's driver: it is the code loaded into the device that enables communication between the rest of the system and the device. The description is programmed using a standard template type of device description language (DDL), provided by FF, which is then compiled into code. The code contains the standard access routines for universal and common practice data together with special routines for transmitter specific parameters. FF also provides device description services (DDS): a standard library of programs which can read the device code. Any control system or host with DDS can communicate with the device.

FF have extended DDL to include the definition of function blocks. These are essentially the same as IEC 61131 function blocks as described in Chapter 48. They permit the distribution of signal processing and control algorithms to any device on the bus. The emphasis is on the environment of the block: the input and output signals, their data types and the functionality. Thus the manner in which a function block type performs will be common from one supplier to another, although the algorithms used within the block may vary. Subtle differences in functionality, together with the provision of user defined blocks, will lead to loss of interoperability but provide suppliers with the means to differentiate their products. An important and evolving alternative to DDL is the field device tool (FDT) which has similar functionality to DDL but is generic as opposed to being FF specific.

50.7 Hardware Configuration

This is essentially concerned with deciding on how many fieldbus segments to use and on what devices should be connected to which bus, resulting in so-called "segment diagrams". Connection on the basis of a single loop to a bus is advised for the majority of process control applications for the following reasons:

- Multiple loops sharing the same segment are vulnerable to common mode failure.
- Single loops straddling two or more segments require gateways between the buses for data transfer.
- Scheduling of function blocks for loops shared between buses is more complicated.

The recommended procedure is as follows:

- Identify all the devices such as transmitters and valves associated with each control loop and assign them a segment to maintain single loop integrity.
- Treat simple control schemes, such as cascade or ratio control, as if they are single loops.
- Identify which devices remain, *i.e.* those not associated with any control loop, and assign them to segments according to their physical locations, taking into account the constraints on fieldbus topology.
- Check the segments assigned against the devices' location: there will be situations when it is preferable to trade off the inconvenience of cross bus loops against reduced cabling costs.

It may not be economic to provide simple devices such as contacts and thermocouples with fieldbus interfaces, in which case they may be connected to the control system as I/O channels directly, connected to another fieldbus device that has auxiliary channels, connected to a fieldbus multiplexer, or connected to a different bus/protocol with a fieldbus gateway.

50.8 Software Configuration

This is essentially concerned with allocating function blocks to devices, and specifying values for function block attributes. The segment to which a loop is assigned can be considered to be a "virtual controller" and the designer is free to allocate function blocks to any connected device. The sort of factors that need to be taken into account in allocating function blocks are the impact of the device failing and the impact on loop scheduling. In general, the advice is to:

- Place the function blocks for signal conditioning, *e.g.* linearisation, in a smart transmitter so that the characterising is done at source. This will minimise the number of function blocks required for characterisation if the signal is used by several other devices. Also, the closer to the signal's source that the characterising is done, the more accurate the associated time stamping becomes.
- Place the function blocks for control as close as possible to the valve, typically in a smart positioner, although this may not be feasible with split range arrangements. Given that many control schemes have more than one input signal but most have only one output, to a valve, putting the control blocks at the "end of the loop" normally maximises the efficiency of function block scheduling.

Note that a consistent approach, such as putting function blocks for signal conditioning in the transmitter and control blocks in the positioner, will make it easier for the operator to appreciate the consequences of device failure.

Remember that there is a constraint that every device must have at least an input (AIN or DIN) or an output (AOT or DOT) function block.

The use of fieldbus will lead to the need to represent functionality on P&I diagrams in a slightly different way. The basic concept of representing a control loop element by means of a bubble containing a tag number, *i.e.* letter code plus reference number, according to the rules explained in Chapter 2, stays the same. But, attached to each bubble, is a box indicating the function blocks used, the tag numbers for which conform with the same rules. The function block type should be obvious from the tag number and context without having to indicate the type explicitly. Figure 50.6 is the equivalent P&I diagram of the cascade scheme for boiler level control as depicted in Figure 25.2 in which both the flow and level transmitters and the valve positioner are smart devices.

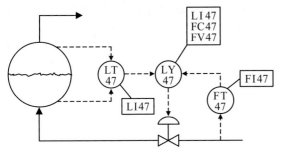

Fig. 50.6 Representation of fieldbus devices on P&I diagram

The outcome of this approach is fewer bubbles but with more detail. For multivariable instruments, such as an intelligent valve, the device should be named according to its primary measurement (*e.g.* FV 47) and secondary measurements should be named according to their type but with the same number (*e.g.* PT 47, TI 47, FX 47, *etc.*).

50.9 Function Block Attributes

Specifying values for function block attributes is a time consuming activity and requires careful attention to detail. Many design decisions have to be made about functionality. Also to be identified are those attributes to be made available for display to, and access by, the operator. The basic approach is to make available those attributes which affect the operability of the control system: it should be recognised that the operator has limited interest in the function blocks *per se* and zero interest in the underlying algorithms. The following defaults are recommended since they can be easily understood by an operator:

- AIN and AOT blocks to be restricted to two modes: out-of-service, to be used during configuration, and in-service. The in-service mode should enable blocks to support both local set points for automatic (AUTO) and remote set points for cascade (CASC). Control.
- On device failure, AOT block attributes to be specified such that the output channel is driven to a failsafe condition.
- PID (or other) control blocks: operator to be able to change mode between AUTO and MAN, to be able to change set point in AUTO and to be able to change output in MAN.
- On loss of communications, PID control blocks should fail to user defined AUTO or MAN mode. If failure forces into AUTO mode, then the set point should also be pre-defined.
- Smart positioners should display both the actual stem position and its desired value.
- On start-up, AIN and AOT blocks should be set to in-service, and controller blocks set to MAN.

One aspect of specifying function block attributes that needs particularly careful attention is alarm handling. The basic principles in relation to fieldbus are much the same as explained under alarm management policy in Chapter 43. The essential difference with smart devices is deciding how to handle situations in which the device detects a fault with itself. There are three categories: suspect and bad measured values, and a failed device. It is prudent to alert the operator to the situation in each case, but whether to assume pre-determined values for signals or take other actions depends on the context. Also to be thought through is how to handle the situation when a device recovers from a fault condition.

50.10 Project Management

With conventional control systems, it is possible to size the system on the basis of an estimate of the I/O point count early on in the project life cycle. However, fieldbus systems cannot be accurately sized until the segment diagrams are available. This will require more front end effort and as a consequence the sizing will not be available until later in the cycle. This will squeeze the subsequent time scale.

With fieldbus, functionality is distributed to the field devices and, potentially, responsibility for their configuration could be delegated to the suppliers. However, in the interests of consistency of configuration, change control and project coordination, it is preferable for the design and configuration activity to be single sourced.

Although smart devices are capable of being configured individually and independent of the rest of the system, there is clearly scope for the whole system to be configured centrally using a single integrated configuration tool. This will enable effective management of the complexity introduced by bus scheduling and cross bus control schemes. The functionality required of such a configurator is as follows:

- Ability to read and store device descriptions in DDL or FDT
- Seamless configuration between the control system and field instruments
- Support of network management, especially allocation of function blocks, scheduling, programming of the LAS, and handling bus loading calculations
- Ability to check the instrument database with the version held in the configurator
- Support version control, authorisation and change management, and be self documenting

Indeed, as the engineering and configuration costs of control systems increases relative to their hardware costs, and the timescales for projects are squeezed, the quality and scope of the configurator could well become a determining factor between suppliers.

Note that current work on data standardisation by the PISTEP consortium and standards for open systems such as OPC will, hopefully, enable a significant amount of system design that is independent of the supplier to be carried out earlier in the project life cycle. Thus function blocks would be able to automatically assimilate information from P&I diagrams and their associated datasheets.

Normal practice for control system testing is to undertake stringent factory acceptance testing (FAT) to ensure that the both the system hardware and software perform to specification. Migration of functionality to the instrumentation will reduce FAT to little more than a check on the operator interface. The first time that the full system will come together will be on site and only then can full testing begin, a scenario of organised chaos. Otherwise, most of the principles of testing and commissioning as outlined in Chapters 63 and 64 apply to fieldbus based systems.

50.11 Benefits

There are three driving forces behind fieldbus: commercial, technological and economic. The commercial is the most obvious and least significant: all the major suppliers have committed themselves to fieldbus because they would be vulnerable otherwise.

From a technological point of view, fieldbus is the vehicle that will enable the widespread implementation of smart field devices. The benefits to the end user of smart instrumentation are relatively easy to identify, some of the more obvious ones being:

- Interoperability, the freedom to choose the best device and system for each application
- Increased accuracy through digital signal processing for characterisation of signals (*e.g.* linearisation) and compensation (*e.g.* for hysteresis)
- Transmission of values in engineering units such that the full accuracy of the field device is propagated
- Improved reliability through self checking and diagnostics
- Quicker installation due to reduced cabling and faster commissioning due to remote calibration
- Digital signals less prone to interference in electrically noisy environments
- Service records for preventive maintenance and diagnostic status for predictive maintenance available in devices' databases for remote access

- Asset management in terms of an inventory of devices, serial numbers, locations, service history, specifications, configuration parameters, etc.
- Increased device functionality means fewer variants and hence reduced spares holdings
- Technology allows the development of innovative instruments, e.g. multi variable transmitters
- Smart devices enable techniques such as neural nets for parameter estimation or pattern recognition, and fuzzy logic for nonlinear control, etc.

From an economic point of view, the benefits to the end user of using fieldbus are very tangible. Bearing in mind that the cost of smart devices is unlikely to be significantly different from that of conventional instrumentation, it has been estimated that:

- Some 40% savings on field installation costs will be realised through reductions in cabling costs, trunking, etc. due to many devices being connected to a single bus *via* multi-drop serial connections.
- There will be a reduction of about two-thirds in the volume (or footprint) of I/O equipment, such as marshalling cabinets, terminations, I/O cards, barriers, power supplies, etc., through the use of fieldbus interface cards.
- Savings of about £200/year per device will result from reduced maintenance costs due to smart devices' remote re-calibration facilities.
- An average of 60% per valve will be saved on repair costs by being able to determine which valves can be fixed in situ and which must be removed for repair through on-line analysis of valve performance.

These benefits should, however, be tempered with a knowledge of the potential drawbacks. Some of the more obvious ones are:

- Due to the scope for providing user defined function blocks, not all devices will be fully interoperable.
- The process of allocating function blocks to devices and programming the LAS is non-trivial.
- The handling of control schemes that straddle more than one segment requires the use of gateways. It may be easier to realise complex schemes within the DCS and simply use the fieldbus for handling the I/O signals.
- The reduced scope for doing FAT means that little testing and commissioning of a fieldbus based system can be commenced until it is fully installed.

50.12 Comments

Without doubt, fieldbus and open system technologies are changing the nature of process automation. Fieldbus certainly offers the prospect of more sophisticated instrumentation with lower installation and running costs, and the potential drawbacks are well understood. However, the process industry is notoriously conservative and end users are looking for hard evidence that the benefits can be realised in practice. It is not just the technology that is different but also working practices in terms of maintenance, support and training. Typically, experience is sought on smaller projects, such as retrofits, before moving onto larger projects.

It is mostly on new plant that the opportunity it taken to go down the fieldbus route. A policy decision can be made to commit to the technology across the plant. For existing plant, it is difficult to justify wholesale replacement of existing 4–20-mA devices that are functioning effectively simply for technological advantage. Even when a conventional analogue instrument fails, it probably makes sense to replace it on a like for like basis because of factors such as spares policy, maintenance, and so on. There remains a massive installed base of analogue 4–20-mA type devices which will not disappear overnight.

Section 7

Control Technology

System Layout

51.1 Conventional Layout
51.2 Power Supply
51.3 Segregation Policy
51.4 Air Supply
51.5 Comments

This chapter explains the physical organisation of the hardware of a control system in relation to the plant. Given, as stated in Chapter 38, that the architecture of a system should match that of the plant, it follows that its layout will too. However, there is much variety depending on the nature of the process, the size of the plant, the I/O point count as defined in Chapter 4, and the technology used for control. No two layouts are the same. This chapter, therefore, focuses on typical systems covering the location of major hardware items, the wiring and cabling, marshalling of signals, power supply and air distribution. Whilst much of this will eventually be overtaken by fieldbus, as described in Chapter 50, there is nevertheless a massive installed base of systems with conventional layouts with which familiarity will be necessary for years to come.

There is a plethora of standards that relate to this area. Perhaps the most important, despite being somewhat dated, is BS 6739 on the installation of instrumentation in process control systems: this is the dominant UK industry standard and has yet to be superseded by an IEC standard. It is well complemented by the EEMUA (1994) design guide on electrical safety. The equivalent American documents are the API recommended practices (RP) 551, 552 and 554. For wiring purposes BS 7671 (the IET Wiring Regulations) is the relevant standard as is the IEC 61000 standard on the electromagnetic compatibility front. The text by Atkey (2005) is well regarded too.

51.1 Conventional Layout

The conventional layout of a DCS based control system is depicted in Figure 51.1.

The field instrumentation is hard-wired into field termination panels (FTP), sometimes referred to as marshalling junction boxes. These are located around the plant, either on an area or on a major equipment item (MEI) basis, according to the amount of instrumentation involved. Typically there would be some 50–100 I/O channels per FTP. Channels, mostly of a 4–20 mA nature, are individually wired up to one side of a termination strip in the FTP. Multicore cables, carrying some 16, 32 or 64 cores, referred to as 8, 16 or 32 pairs, capable of handling one channel per pair, are wired up to the other side of the FTP. Multicores are typically armoured for protection with individual cores screened to minimise interference.

Segregation is a key issue, as explained in Section 51.3 below. Intrinsically safe (IS) signals are always segregated, from source, throughout the system. Low level signals, such as from thermocouples, are normally segregated from source too. It is good practice to segregate other signals, on a like for like basis, at the FTP as far as is practicable or economic to do so. Otherwise, and inevitably, many of the multicores carry a mixture of analogue and discrete, input and output signals which have to be segregated in the marshalling cabinets (MC) situated in the local equipment centre

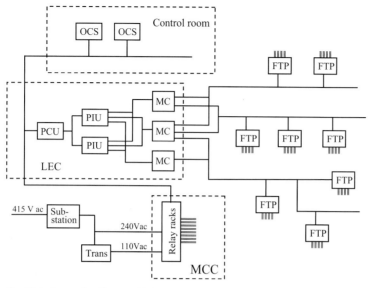

Fig. 51.1 Conventional layout of DCS type system

(LEC), sometimes referred to as the interface room. Clearly the more segregation that can be done at the FTP level, the less there is to do in the LEC.

The multicores from the FTPs are terminated, typically on DIN rails, in the marshalling cabinets. The significance of the DIN rails is that they are of a standard size for which there are many proprietary types of terminal block to enable terminations that are screwed or push fitted (pins or spades), fused and/or isolated, individual or grouped. It is not uncommon for the multicores to terminate in preformed multipin plugs which engage in dedicated multipin sockets. There is inevitably a trade off here between cost and flexibility.

Within the marshalling cabinets the I/O channels are cross connected to the terminations of other multicores which are wired through to the plant interface units (PIU), also in the LEC. It is inside the marshalling cabinets that the signals from/to the field are grouped together on a like-with-like basis and, notwithstanding segregation done at the FTP level, the segregation policy is applied. The marshalling cabinet is the equivalent of an old-fashioned telephone exchange and requires careful design to prevent it becoming like spaghetti junction.

Also designed for DIN rail compatibility are a host of I/O devices which can be mounted amongst the terminations and cross connections as appropriate. Such devices include barriers (shunt diode, galvanic, or otherwise as discussed in Chapter 52), isolators, pilot valves (see Chapter 5) and relays. When large numbers of these devices are involved, it is not uncommon for them to be grouped together in separate marshalling cabinets, again on a like with like basis.

The multicores from the marshalling cabinets are connected to termination assemblies local to each PIU and, typically *via* ribbon cables, to the I/O cards within the PIUs. The layout of a PIU comprising cards, racks, frames and cabinets is described in Chapter 39. The process control unit (PCU), only one is shown but there may be several, is depicted as being a functionally distinct entity although, in practice, it is physically located in the PIU cabinet. Data is transmitted by their PORT cards *via* a highway realised by means of co-axial cable to the operator control stations (OCS) in the control room and *via* gateways to higher level management information systems (MIS).

Figure 51.2 shows a typical "channel diagram" in which the route of two analogue signals, one in-

51.1 Conventional Layout 381

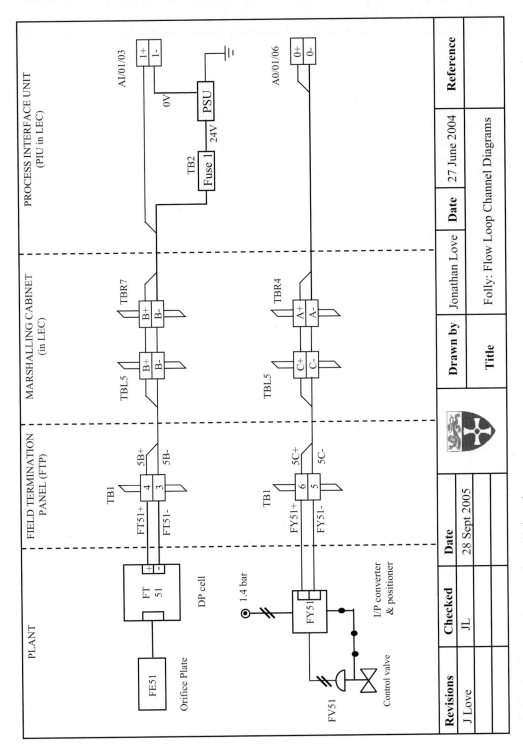

Fig. 51.2 Channel diagram for two analogue I/O channels

put and one output, through the system layout is depicted. The input channel is from an orifice plate FE51 and the output channel is to a control valve FV51. The diagram shows the devices, cables, terminal numbers, cabinets, power supplies, *etc*. The input channel shows how the flow transmitter draws its power from the power supply unit (PSU): the specifics vary from one system to another. Whilst this channel diagram is more detailed than those shown in Chapters 44 and 46, it is nevertheless incomplete in that, for example, there are no barriers shown.

It is common practice for the LEC to be situated fairly close to the control room although this is by no means necessary. For plants with large numbers of I/O signals, it is normal to have several LECs spread geographically, typically on a major unit or train basis. Indeed, the highway nature of a DCS enables plants that are widespread to have several LECs and satellite control rooms. There is therefore much scope for optimising the cost of cabling by strategically situating the FTPs relative to the LECs and the LECs relative to each other.

51.2 Power Supply

The power supply for light and heavy current devices is handled differently. For light current devices, such as transmitters and converters, the power typically comes from the dc supply within the PIU and is routed *via* the signals to the field through the LEC directly. This is as depicted in Figures 44.2, 44.3, 46.3 and 46.4. However, for heavy current equipment, such as pump and agitator motors, power is distributed through the motor control centre (MCC), also shown in Figure 51.1. Every sub-system, circuit, *etc*. should be individually protected by its own fuse or miniature circuit breaker. Elements that may be used independently should be capable of being isolated and separately protected.

Three phase 415 V alternating current (a.c.) mains supply is normally brought into the MCC *via* a substation, from which single phase 240 V supplies are taken off. It is common practice to drive plant items with single phase 110 V a.c. which is generated locally from the mains supply by means of a transformer. In total, it is normal for the power supply system to have some 25% excess capacity with spare circuits to allow for future modifications. Diverse power supplies must be provided for critical applications such as ESD systems. This is typically realised by means of stand-by generators which cut in on mains power failure and/or battery back-up.

The MCC usually contains racks of relays used for switching the power on and off. A typical modern relay is solid state and has a dedicated I/O card with four channels configured as appropriate. Two are DINs used for Auto/Manual and Run/Stop and two are DOTs for Start and Stop. These are manipulated by discrete I/O signals from the DCS *via* the highway and a serial interface. If there are not many relays, then the discrete I/O can be routed between the LEC and the MCC directly using multicores from a marshalling cabinet.

It is fairly obvious that the cabling of even a moderately sized control system is a major electrical engineering commitment. The LEC and MCC are normally adjacent to each other. By optimising the positions of the FTPs relative to the LEC and MCC significant reductions in cable length may be realised. For plants with large point counts it is normal to have several LECs and MCCs providing scope for further optimisation. Cabling costs can be reduced further by the use of field multiplexors.

It is not just the wiring and terminations that are expensive but also the supporting infrastructure of racks, frames, cabinets, conduit, trunking, cable trays, *etc*. For plants with point counts in the thousands it is often necessary, because of the sheer volume of cabling, to have false floors under the LEC to provide the space to enable the cables to be brought into the marshalling cabinets and PIUs.

There are well developed standards and codes of practice on wiring and cabling, covering issues such as size/types to use, colour coding, earthing, *etc*., with which all competent electrical contractors should be familiar. In the UK the standard is BS 7671, otherwise known as the IET Wiring Regulations (2001).

51.3 Segregation Policy

This is a major factor in the design of a plant and must be planned for carefully as opposed to being allowed to evolve by default. There are two aspects to this:

1. Segregation of electrical power lines from instrument signal cables. These must be kept apart otherwise electrical noise, mains frequency and power spikes will be induced from the power lines onto the instrument signals. As a rule of thumb, for a.c. power cables up to 10 A rating run in parallel with instrument cables carrying 4–20 mA or 5–50 V d.c. signals, a minimum separation of 25 cm is required. For instrument cables containing low level signals, say 0–5 V d.c., a minimum separation of 30 cm is required from other instrument cables.

    The extent to which induction occurs, not just between cables but also between devices, enclosures, cabinets, *etc.*, depends upon the quality of screening as well as the distances between them. Indeed, screening, bonding and earthing is fundamental to noise free signals. A variety of techniques and practices exist for which the IEC 61000 standard on electromagnetic compatibility (EMC) is the yardstick.

2. Segregation of one type of instrument signal from another. Instrument channels may be segregated on the basis of:
    - Signal type
    - Plant item/type
    - Process function

    as determined by the requirements for:

    - Card/rack organisation
    - Redundancy
    - Intrinsic safety
    - Electrical isolation
    - Emergency shut down
    - Manual operation
    - Local/remote power supply
    - Power amplification

One common category of segregation is between IS and non-IS signals which is typically realised by means of using different multicores and designated IS marshalling cabinets. Indeed, whereas power cables are normally coloured black and instrument multicores are grey, the multicores for IS circuits are often colour coded blue. For the purposes of isolation and emergency shut-down, it is common practice to have the I/O signals for specified MEIs routed through designated marshalling cabinets. Otherwise, segregation is largely realised by means of the cross connections within the marshalling cabinets.

51.4 Air Supply

The air supply for pneumatic equipment is always provided by means of some central compressor and distributed throughout the plant by pipework. Ideally the compressor should be of the oil-free type but, for larger installations, the lubricated type is often necessary. A typical compressor configuration is as depicted in Figure 51.3 in which the mains powered rotary compressor is backed up by a diesel powered reciprocating compressor, each of which is capable of being isolated for repair. The compressors discharge into a receiver vessel.

Typically air is supplied at 8.0 bar gauge pressure. The compressor switches on when the receiver pressure drops to, say, 7.8 bar and switches off when it reaches, say, 8.2 bar. Clearly a simple on-off control system of the type described in Chapter 28 is required. The compressor should have some 50% spare capacity in relation to the total instrument air consumption and, for reliability purposes, sized such that it is on load for some 50% of its time. For small systems the receiver is integral with the compressor but for larger installations it is normally a separate vessel. The receiver's capacity should be such as to continue to provide air for up to 30 min after failure of a single compressor but obviously, for a dual compressor system, a much smaller receiver is permissible.

It is normal for a compressor to be fitted with an after cooler/separator to remove most of the free water from the compressed air. The most unreliable part of a compressor system, and the most likely cause of total loss of air supply, is its cooling

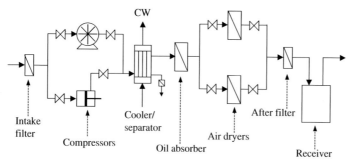

Fig. 51.3 Typical compressor configuration

system which invariably uses cooling water. Flow and temperature sensors, linked into alarms and trips as appropriate, are essential for detecting abnormal conditions.

The quality of the air is important to prevent condensation in the pipework and deposition of dust, oil and water in the working parts of pneumatic devices. A typical specification is as follows:

- Dew point of 10°C less than the minimum ambient temperature. To meet this criterion, it is often necessary to install a pair of air dryers downstream of the cooler/separator. These are usually of the adsorption type, the adsorbent being a desiccant which is automatically regenerated. The dryers are installed in parallel which enables one to be on duty with the other on regeneration, their operations being intermittently switched over.
- Amount of oil should be <1 ppm. If the compressor used is of the lubricated type, usually necessary for larger installations, then an oil absorber is also required. This is normally installed upstream of the dryer to protect the desiccant from contamination.
- No dust particles >3 μm in size. This is principally realised by means of filters on the air intake, but filters can also be installed downstream of the dryers.

Distribution of compressed air about the plant is by means of pipework. The pipework should be installed so that it is self draining with blow down valves at the lowest points. The main air headers are pipes of up to 50 mm diameter, typically of stainless steel, copper or aluminium. On smaller headers blow down is normally realised by means of the filters associated with individual elements, as described in Chapter 5.

Down at the level of individual elements, it is becoming increasingly common in manufacturing industries to use flexible nylon tubing, typically of 6 mm nominal bore. However, in the process industries, because of the potential for tubing to be exposed to heat, perhaps because of proximity to hot pipes and/or vessels, it is normal to use metal tubing. When a number of tubes are run together on the same route, multicore tubing is usually a cheaper option. Multicores typically contain 12 or 19 tubes laid-up and sheathed overall with plastic. The principal constraint on the use of multi-core tubing is their lack of flexibility and the minimum radius of bending.

51.5 Comments

As articulated in Chapter 59, the control and instrumentation budget is likely to be some 5–10% of the capital cost of a plant and the electrical budget a further 5–10%. Within these, the distribution of signals and the provision of power and air supplies, both in terms of equipment and installation costs are substantial. Whilst these topics are of a Cinderella nature, and would never arouse any enthusiasm in academia, the control and electrical infrastructure is nevertheless of fundamental importance to the effective design and operation of process plant.

Intrinsic Safety

52.1 Intrinsic Safety
52.2 Hazardous Areas
52.3 Gas Group
52.4 Temperature Class
52.5 Common Explosive Gas-Air Mixtures
52.6 Types of Protection
52.7 Certification
52.8 Flameproof Enclosures
52.9 Pressurisation and Purge
52.10 Other Types of Protection
52.11 Ingress Protection
52.12 Barriers
52.13 Certification of Barriers
52.14 ATEX Directives

If a flammable material is mixed with air, the mixture produced is normally explosive over a wide range of concentrations. This applies not just to gases, but to mixtures (with air) of flammable vapours, mists and dusts too. When using electrical equipment, such as instrumentation, in explosive atmospheres it is essential that the equipment cannot ignite the mixture. This can be caused by electronic components which become very hot during normal operation. Ignition can also be caused by electrical sparks produced under fault conditions. Other devices, such as switches and relays, are sources of sparking too. Instrumentation that is specifically designed for use in such hazardous areas is said to be intrinsically safe (IS).

IS certified instruments may not be available for a given application, or may be prohibitively expensive, in which case installation of non-IS instrumentation in some form of enclosure will have to be considered. Furthermore, it is not sufficient just to have IS instrumentation. The circuits between instrumentation in hazardous areas and the control and/or protection systems in non-hazardous areas are themselves sources of potential overvoltage and excessive current. These have to be protected against by means of barriers.

This chapter provides a working knowledge of intrinsic safety and related issues. It is a surprisingly complex area and the reader is referred for more detailed information to the texts by Garside (1991, 1995). There are many standards that relate to IS, in particular IEC 60079, together with the explosive atmosphere (ATEX) 100A directive that became mandatory with effect from 2003.

52.1 Intrinsic Safety

The European standard which covers the general requirements for the selection, installation and maintenance of electrical apparatus for use in potentially explosive atmospheres is EN 50014 (1998) and the standard which specifically relate to intrinsic safety is EN 50020 (1995). The dates of these standards are rather misleading: most are well established national standards which have been re-released as EU standards. Note that these relate to low power electrical equipment in general and not specifically to instrumentation.

Intrinsic safety is concerned with the probability of a fault occurring in an instrument (a fault that could ignite an explosive mixture if it were present) and of the simultaneous presence of an explosive mixture. An IS device is one whose circuits have been designed such that the amount of energy available from them is restricted. Thus the energy available must be low enough to ensure any sparks or thermal effects produced cannot ignite the specified explosive mixture for which the device is certified. The testing carried out includes both normal operation and specified fault conditions.

There are two standards of intrinsic safety:

ia is the higher standard for which safety is maintained by equipment design that provides protection for tolerance of two independent faults.

ib has a lower safety margin since it is only maintained with a single fault.

Intrinsic safety prevents explosions being caused by hot surfaces and electrical sparks only. It does not protect against ignition by other causes such as static discharge and frictional sparking.

52.2 Hazardous Areas

A hazardous area is one in which explosive gas-air mixtures are, or may be expected to be, present in quantities such as to require special precautions for the construction and use of electrical equipment.

There are three categories of hazardous area, referred to as zones as defined in Table 52.1, depending on the probability of an explosive mixture being present. The emphasis is largely on gases and vapours although, for reference, the equivalent standards for dust explosions are listed. America is moving from its traditional division based categorisation towards zones.

Table 52.1 Categories of hazardous zone

IEC 60079 (part10) Gases	IEC 6124 (part 3) Dusts	NEC 505 Gases	NEC 500 Gases & dusts
For UK and EU		For US	
Zone 0	An explosive gas-air mixture is continuously present, or present for long periods, typically ≥ 1000 h/year		
	Zone 20	Zone 0	Division 1
Zone 1	An explosive gas-air mixture is likely to occur, either occasionally or intermittently, in normal operation, typically 10–1000 h/year		
	Zone 21	Zone 1	Division 1
Zone 2	An explosive gas-air mixture is not likely to occur in normal operation and, if it does, will exist for only a short time, typically ≤10 h/year		
	Zone 22	Zone 2	Division 2

An area which is not classified as being Zone 0, 1 or 2 is assumed to be a non-hazardous or "safe" area by default. Only instrumentation certified to ia standard is suitable for use in Zone 0 but can be used in Zones 1 and 2. Instrumentation certified to ib standard is generally acceptable for Zones 1 and 2 only.

The zoning of a plant, *i.e.* the designation of areas as being Zone 0, 1 or 2, or as being safe, is essentially a process design decision. Typical examples are:

Zone 0: Close to the vent of a vessel containing flammable volatile liquids.
Close to the gland of a control valve in a high pressure pipeline containing flammable materials.
Close to a pump with a leaking seal.

Zone 1: Close to a sample point where samples of flammable materials are withdrawn from a vessel for analysis.
Close to a filter whose cloth/medium is changed intermittently.
Close to a drain which handles occasional spillages.

Zone 2: Close to all items of plant and pipework in which flammable materials are handled.

Safe: Control rooms, interface rooms, motor control centres, *etc*.

Zoning essentially involves identifying potential sources of release of flammable materials and deciding upon the "closeness". A zone generally consists of a region around the source of release, the extent of the zone being the distance from the source to the locus of the lower explosive limit (LEL). The shape and size of a zone is determined by a number of factors such as rate of release of flammable material, rate of evaporation, rate of dispersion, wind speed and direction, ambient temperature, density, position of walls, *etc*. Typically a spherical zone is specified about the source of release, the radius of which is equivalent to the LEL with some margin for error built in. A conical zone is sometimes specified for gases and vapours whose density is substantially different to that of air.

The basic approach is to consider each source of release, or potential source of release, in turn. First, consider a source under normal operating conditions. Then consider the same source under abnormal or fault conditions. This typically leads to a relatively small Zone 0 or 1 with a larger surrounding Zone 2 area. Note that it is not necessary to consider catastrophic failure conditions: that is a function of plant design with which zoning is not intended to cope.

It is inevitable that plants contain a mixture of zones and overlapping zones. Often, to avoid confusion, whole areas of plant will be designated as the "worst case" zone. There is a cost penalty for this in terms of more expensive instrumentation than would otherwise have been the case, not to mention the increased maintenance and support costs throughout the life of the plant. With multi-products and multi-purpose plant, where the gases and zones are not necessarily known *a priori*, worst case zoning is the only option.

Table 52.2 US and EU classification of gas groups

Representative gas	NEC 500 (gases and vapours) group	IEC 60079 (part 0) group	Comment
Methane	–	I	Relates to mining applications
Propane	D	IIA	
Ethylene	C	IIB	
Hydrogen	B	IIC	Referred to as either the H_2 or C_2H_2 group
Acetylene	A		

There is much scope for variability and much need for common sense because little help is available in determining the shape and size (radius) of a zone. Many companies have developed in-house standards for zoning based upon previous experience. The most useful document in the public domain is the Institute of Petroleum (IoP) Code of Practice Part 15 (1990). The widely quoted American Petroleum Institute (API) recommended practice RP 14C (1998) provides generic guidance on system layout for offshore structures but nothing on zone sizing.

52.3 Gas Group

The gas group, sometimes referred to as the apparatus group, defines the gases and vapours in which an instrument is certified to operate. This is done by considering the electrical energy, in the form of an arc or spark, which is needed to ignite the gas. The gas group is therefore of particular importance when specifying enclosures. All common industrial gases are allotted to one of four groups. These groups are often referred to by the name of the gas indicated in Table 52.2, the gas being representative of that group.

Group I is the least and group IIC the most explosive. It follows that instrumentation certified for use in a particular gas group can be used in gas groups above it. For example, a Group IIB instrument may be used in Group IIA.

52.4 Temperature Class

This defines the maximum surface temperature of the components in an instrument. The surface temperature must not exceed the gas ignition temperature under specified conditions. There are six maximum surface temperature bands, classes T1 to T6, as defined as in Table 52.3. The essential difference between the UK/EU and the US classes is that the US classes provide for a finer categorisation.

Table 52.3 US and EU definition of temperature classes

Maximum surface temperature deg C	US (NEC 500) temperature class	UK/EU and US (NEC 505) temperature class
450	T1	T1
300	T2	T2
280	T2A	
260	T2B	
230	T2C	
215	T2D	
200	T3	T3
180	T3A	
165	T3B	
150	T3C	
135	T4	T4
120	T4A	
100	T5	T5
85	T6	T6

All common industrial gases are allocated to one of the six classes, according to ignition temperature. For example, pentane, which has an ignition temperature of 285°C, cannot be used with instrumentation of UK/EU temperature class T2 because, potentially, some components could be as hot as 300°C. Thus pentane must have a temperature class of T3, and any instrument of class T4, T5 or T6 could also be used. On the US scale a temperature class of at least T2A is required. Note that the temperature class and the gas group are not related. For example, hydrogen requires very little spark energy to ignite it, but the surface temperature for ignition is very high.

Also note that the surface temperature of a component depends on its rate of heat loss to the ambient, which in turn depends upon the temperature of the ambient. Therefore, the temperature classification is stated with reference to an arbitrary maximum ambient temperature of 40°C. This is high enough for most purposes. However, if an instrument is operating in conditions hotter than this, a higher temperature classification may be necessary.

52.5 Common Explosive Gas-Air Mixtures

The gas (apparatus) group and temperature classes for a number of common explosive gas-air mixtures are given in Table 52.4.

52.6 Types of Protection

To identify the technique used to ensure safety in a potentially explosive atmosphere, it has been agreed internationally that the letters "E Ex" shall be followed by a letter code, as indicated in Table 52.5. The identification is attached to the instrument by means of a permanent label.

52.7 Certification

Certification is issued on the basis of type testing by a notified body: various such bodies exist within the UK, EU and US. Thus, a representative instrument is subjected to testing and, if approved, all other instruments made to the same design and manufactured to the same quality are certified.

The two certifying authorities relevant to the UK and EU are the British Approvals Service for Electrical Equipment in Flammable Atmo-

Table 52.4 Groups and classes for common explosive mixtures

Gas/vapour	Gas group required	UK/EU temperature class required
Acetic acid	IIA	T1
Acetone	IIA	T1
Acetylene	IIC	T2
Ammonia	IIA	T1
Butane	IIA	T2
Carbon disulphide	IIC	T6
Cyclohexane	IIA	T3
Di-ethyl ether	IIB	T4
Ethanol	IIA	T2
Ethylene	IIB	T2
Hydrogen	IIC	T1
Kerosene	IIA	T3
Methane non-mining applications	IIA	T1
Methanol	IIA	T1
Methyl ethyl ketone	IIA	T1
Natural Gas	IIA	T1
Propane	IIA	T1
n-Propyl alcohol	IIB	T2
iso-Propyl alcohol	IIA	T2
Toluene	IIA	T1
Xylene	IIA	T1

Table 52.5 E Ex types of protection

Type of protection	Code	Zone	IEC 60079	CENELEC Standard
General requirements	–	–	Part 0	EN 50014
Combined methods (category 1G)	–	–	Part 26	EN 50284
Intrinsic safety	ia	0	Part 11	EN 50020
	ib	1		
Combinations of IS	–	–	Part 25	EN 50039
Flameproof enclosures	d	1	Part 1	EN 50018
Powder filling	q	1	Part 5	EN 50017
Pressurised apparatus	p	1	Part 2	EN 50016
Encapsulation	m	1	Part 18	EN 50028
Oil immersion	o	1	Part 6	EN 50015
Increased safety	e	1	Part 7	EN 50019
Special	s	0 or 1		None
Non-incendive	n	2	Part 15	EN 50021

spheres (BASEEFA), which is part of the Health and Safety Executive (HSE), and the European Electrical Standards Co-ordination Committee (CENELEC). Equipment certified by BASEEFA is certified to the latest CENELEC regulations and is recognised as such in all its member countries.

Identification of a CENELEC certified instrument is as follows:

E Ex ia IIC T4 (BASEEFA No. Ex 537687 J)

the component parts of which indicate respectively that:

1. The instrument is certified to the appropriate CENELEC standard.
2. The instrument provides protection against explosion.
3. Its intended use is for Zone 0 but may be used in all other zones.
4. It may be used with gases in the hydrogen or any other group.
5. It should not be used with any gas whose ignition temperature is below 135°C.
6. The instrument has been type tested for which the certificate numbered has been issued by the certifying authority.

It is evident that intrinsic safety is very specific. The IS requirements of an instrument have to be determined according to the application in terms of zone, gas group and temperature classification. IS certification is not blanket: an instrument being "IS" doesn't mean it can be used in any hazardous situation.

52.8 Flameproof Enclosures

The use of flameproof enclosures provides "type d" protection and is often referred to as explosion proofing. For instrumentation it is an expensive alternative to intrinsic safety, but enclosures are used extensively with heavy current electrical equipment such as motors and switch gear. The approach is to recognise that non-IS equipment is a potential source of ignition and that contact with an explosive mixture is inevitable. The principle of design is to allow ignition but to remove heat at source to prevent propagation of an explosion. The principal design criteria for enclosures are that they must:

- Completely surround the item of electrical equipment and be of sufficient strength to withstand internal explosions.
- Suppress any internal flame such that it cannot ignite external explosive mixtures through joints or structural openings in the enclosure.
- Restrict the maximum external surface temperature to below the ignition temperature of any external explosive mixture.

A classic example is in the use of d.c. motors. These are heavy current devices so IS is not an option. They are often cooled internally by a draught of air. In a potentially explosive gas-air mixture this is problematic. A purpose designed enclosure is the solution. The enclosure is a strong steel casing with deep narrow openings to enable air to be sucked in/blown out. The openings are heavily flanged such that, in the event of an internal explosion, heat is absorbed by the flanges and the flame suppressed before the combustion gases exit the enclosure.

Enclosures are specific to:

- The rating of the equipment for which they are designed, both under normal operation, overload and defined fault conditions
- The surrounding explosive atmosphere

As with IS equipment, flameproof certification is not "blanket" and enclosures approved for one application cannot necessarily be used for another.

The powder (or sand) filled "type q" of protection is conceptually not dissimilar to that of flameproof enclosures in that the objective is to contain the explosion and/or quench any subsequent flame. Type q protection is not common in the UK/EU.

52.9 Pressurisation and Purge

Pressurisation provides "type p" protection. If IS instrumentation is not available, as is often the case with analytical instrumentation, and flameproof enclosures are not viable, this alternative approach to protection should be considered. The strategy is simple. Install the instrumentation inside a pressurised cabinet: if the pressure inside the cabinet is higher than outside, any leakage will be outwards and potentially explosive gases will be excluded. There are a number of important issues regarding pressurised cabinets:

1. Pressurisation is normally realised by means of compressed air. The source of air must not itself potentially contain explosive mixtures. For example, a small compressor local to the cabinet would simply draw in air from the same atmosphere against which protection is being provided! The works compressed air supply is usually adequate. Otherwise air or nitrogen cylinders may be considered.
2. The cabinet pressure does not need to be large; 0.02 bar gauge is quite sufficient, but it must exist. A pressure alarm will therefore be needed to indicate a drop in cabinet pressure.
3. The main source of air leakage from a pressurised cabinet is through the conduits *etc.* through which the wires and cables enter. These normally have glands fitted to reduce the losses but, in practice, it is almost impossible to stop the leakage altogether. Leakage through the cabinet provides for purging.
4. Access to the cabinet for maintenance *etc.* may need to be strictly controlled: clearly it cannot be pressurised if the door is wide open.

The encapsulation "type m" and oil immersion "type o" protection is conceptually not dissimilar to that of pressurisation and purge in that the objective is to keep the explosive mixture away from potential sources of ignition. Type m protection is becoming commonplace for many components and sub assemblies, solenoid valves being a good example. Type o is often used with transformers.

52.10 Other Types of Protection

"Type e" protection means increased safety that is realised by means of design. A typical example is a junction box for which the number of terminals, their type and spacing is restricted to minimise the scope for short circuiting. Induction motors are often of type e.

"Type s" means special protection which is a miscellaneous category for any proven method of protection for which there is no standard. Devices of "type s" may be used in whatever zone they are certified for: this includes Zone 0.

"Type n" protection is commonplace. It means that electrical equipment has been designed and constructed to be non-incendive. As such it is not capable, under normal operating conditions, of igniting a surrounding explosive atmosphere, and any fault capable of causing ignition is not likely to occur. Good examples of this are switches, junction boxes and light fittings used in hazardous areas.

Note that mixed technology devices are becoming increasingly common with, for example, a "type m" device having "type e" terminations.

52.11 Ingress Protection

There is no point in carefully specifying devices and/or systems to provide protection against particular requirements if they are not mounted or installed in such a way as to guard against ingress of extraneous materials. Thus EN 60529 (1992) is a standard for certifying cases and housings for ingress protection (IP). A two lettered IP code is used as defined in Table 52.6.

Thus, for example, a casing rated as IP 54 will provide protection against both dust and splashing water and is suitable for most indoor applications. IP 55 is typically specified for outdoor use and IP 66 for offshore applications.

Table 52.6 Ingress protection codes

First numeral	Second numeral
0 No protection	0 No protection
1 Objects greater than 50 mm	1 Falling water drops (vertical)
2 Objects greater than 12 mm	2 Falling water drops (to 15° from vertical)
3 Objects greater than 2.5 mm	3 Spraying water
4 Objects greater than 1.0 mm	4 Splashing water
5 Dust protected	5 Water jets
6 Dust tight	6 Powerful water jets
	7 Temporary immersion
	8 Continuous immersion

52.12 Barriers

As mentioned, the circuits between instrumentation in hazardous areas and the control and/or safety systems in non-hazardous areas are themselves sources of potential ignition. This is because of the scope for excessive current and/or overvoltage caused by short circuiting, cross wiring, use of the wrong power supply, *etc.* Barriers provide protection against such events.

Circuit diagrams indicating the use of barriers with both analogue and discrete I/O channels are shown in Figures 44.2, 44.3, 46.3 and 46.4. In these diagrams the barriers and the I/O cards would all be in safe areas, the field devices being in some hazardous area. There are different types of barrier. The most common type is the so called shunt diode barrier, as depicted in Figure 52.1, which is based upon part of Figure 44.2.

The basic principle of a barrier is to restrict, to a safe level, the amount of energy that can be transmitted through a channel under fault conditions, without significant attenuation of the control

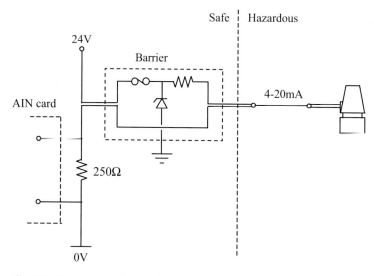

Fig. 52.1 Circuit of shunt diode barrier

signal under normal conditions. A simple barrier consists of a resistor, fuse and zener diode. The resistor limits the flow of current through the circuit in normal operation. It also establishes a voltage across the zener. If the current becomes excessive the fuse blows. If the voltage across the zener rises to its breakdown value, current is leaked to earth. This may also blow the fuse. The values of the fuse capacity, resistance and breakdown voltage are carefully specified according to the maximum permissible current and voltage in the circuit as a whole.

Barriers that have been activated need to be checked carefully to ensure that the zener has not been damaged. Modern barriers normally have duality of components to provide redundant protection, as indicated in Figure 52.2. Note also the diode in the signal path which provides protection for the AIN card and power supply.

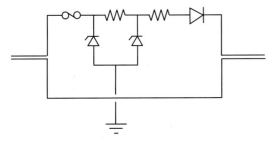

Fig. 52.2 Circuit of dual shunt diode barrier

Shunt diode barriers are self contained devices which are of a modular construction, typically designed for mounting on an earthing busbar in a termination rack. The importance of good earthing cannot be over emphasised.

The constraints imposed by the shunt diode barrier can be reduced, at an increased cost, by the use of isolating IS interface devices that do not tie the hazardous and safe area circuits to earth or to each other. The most common types are galvanic isolators, solid state relays and optical couplers. The galvanic type is based upon a transformer using high frequency modulation to handle dc currents. The high input impedance of solid state devices forms a natural barrier. Optical coupling is of particular use with discrete signals.

There is no point in using IS instrumentation and barriers if the cables between them contain sufficient electrical energy to cause ignition in the event of a discharge. This problem is addressed by means of certification.

52.13 Certification of Barriers

A modular approach is used. The barrier certificate will specify:

- Maximum power, *i.e.* current (amps) and voltage (volts), under fault conditions for which it is rated
- Maximum capacitance (Farads) and inductance (Henrys) for the circuit protected, *i.e.* for the IS device and its cabling

The IS device certificate will specify:

- Maximum current and voltage that the instrument can receive
- The instrument's internal capacitance and inductance

The instrument can then be used safely with any barrier whose maximum current and voltage are less than or equal to the maxima for the instrument.

The maximum capacitance and inductance permitted for the cable are found by subtracting the internal values for the instrument from the maximum values specified for the barrier.

52.14 ATEX Directives

The ATEX 100A directive is aimed at manufacturers of equipment intended for use in hazardous areas, compliance with which became a legal requirement in 2003. It details how manufacturers are to meet essential safety requirements in relation to all equipment and protective systems intended for use in potentially explosive atmospheres. The directive applies to all systems made, imported or sold in the EU. In essence, equipment is categorised as being either Group I for the mining industry or Group II for other industries, with sub categories according to zones and sub divisions for gases, vapours and mists (G) and dusts (D) as depicted in Table 52.7.

Table 52.7 Equipment categories of the ATEX directive

ATEX Equipment Category	Area classification	Protection type
II 1 G or II 1 D	Zone 0	ia
II 2 G or II 2 D	Zone 1	ib
II 3 G or II 3 D	Zone 2	n

The conventions for ATEX certification and CENELEC identification contain different information and will co-exist. For example, CENELEC contains gas group and temperature classification data which ATEX doesn't. An example of the ATEX convention is as follows:

$$CE\ 0600\ Ex\ II\ 1\ G$$

the component parts of which indicate respectively that:

1. The instrument conforms with the electromagnetic compatibility directive 89/336/EC.
2. Reference to the notified body. Codes: 0123 for TUV, 0518 for SIRA, 0600 for BASEEFA, etc.
3. Protection against explosion.
4. Equipment: Group II for chemical and process industry applications.
5. Equipment: Category 1, 2 or 3 according to zone and type of protection.
6. Type of atmosphere: G for gases, vapours and mists or D for dust.

There are different requirements on the implementation of ATEX 100A for the suppliers of equipment for use in hazardous areas and for the end-users. These are summarised in Table 52.8 which is self explanatory. Note that for Group II.2 a distinction is made between electrical and mechanical equipment and/or subsystems. Thus different criteria will apply to, say, the actuator and body of a solenoid operated isolating valve.

ATEX 137, another directive which also became effective from 2003, addresses personnel related issues in hazardous areas. This requires production of an explosion protection document which quantifies the basis on which equipment is specified according to area classification, equipment cate-

Table 52.8 Implementation requirements of the ATEX directive

Equipment category	II 1	II 2 elec	II 2 mech	II 3	100A annex
Supplier's design					
Certification by notified body	Yes	Yes			III
Self certification			Yes	Yes	VIII
Surveillance of manufacture					
QA of production by notified body	Yes				IV
QA of product by notified body		Yes			VII
QA by supplier			Yes	Yes	VIII
End-user's installation					
Unit verification by notified body	Yes				
Inspection by notified body		Yes			
Results of risk analysis lodged with notified body			Yes		
Self certification				Yes	

gories, gas groups, temperature classes, *etc.* Organisational measures such as training requirements, maintenance instructions, issue of permits, provision of warning signs, *etc.*, must also be articulated.

A likely consequence of ATEX 137, because of the extra costs involved in conformance, is more careful targeting of area classification as opposed to the current practice of blanket zoning.

Reliability Principles

53.1 Unrepairable Systems
53.2 Repairable Systems
53.3 Proof Testing
53.4 Elements in Series and Parallel
53.5 Common Mode Failure
53.6 Voting Systems
53.7 Standby Systems
53.8 Protection Systems
53.9 Worked Example
53.10 Hazard and Demand Rates
53.11 Comments

Underpinning the design of safety and protection systems are calculations about reliability. This chapter therefore explains the concepts of reliability and defines the terminology. The term "reliability" is itself confusing in that it has two meanings. In a general sense it is used to loosely describe the failure characteristics of any device or system. However, for systems that are unrepairable it has a very precisely defined probabilistic meaning. Various commonly used relationships of a simple mathematical nature are developed. The approach here is to establish the principles and to defer their application to subsequent chapters. For a more comprehensive treatment of the subject the reader is referred to the texts by Andrews (1993) and Goble (1998).

53.1 Unrepairable Systems

Much of the reliability theory has been developed in relation to unrepairable systems. These are systems which are dispensed with once a fault occurs, such as missiles and sub-sea valves. The life of such an element or system is characterised by its reliability (R) which is the probability that it will operate to an agreed level of performance. An example of the performance criterion for an instrument could simply be that it does/doesn't work to an accuracy of ±2%.

Reliability varies with time: an instrument that has just been checked and calibrated should have a reliability of 1.0 when first brought into use but may only have a reliability of 0.99 six months later. Unreliability (uR) is the complement of reliability. Since an element or system can only be failed or not failed, the sum of reliability and unreliability must be unity:

$$R(t) + uR(t) = 1.0 \qquad (53.1)$$

The failure rate (λ) is the average number of faults, *per* device, *per* unit time. Failure rates are calculated from mean time to failure (MTTF) data. Suppose that a large number of new devices are put into service, the devices and conditions being identical, and allowed to operate until they fail. The time (hours) taken for each device to fail is noted and that device is then taken out of service. The average of these times, when all of the devices have failed, is the MTTF:

$$\lambda = \frac{1}{\text{MTTF}} \quad (53.2)$$

If the units of MTTF are hours then those of λ are failures per hour. However, λ is more commonly articulated in terms of failures per year (8760 h) or failures per million hours (FPMH).

Reliability R(t) and failure rate (λ) are related. Consider n_o identical devices set in operation at time t = 0. Suppose that n_f have failed after a time t. Assuming that there are no repairs, the number surviving is

$$n_s = n_o - n_f \quad (53.3)$$

Let an additional Δn_f fail during the following time interval Δt. The number of faults, per device, per unit time is thus given by

$$\lambda \approx \frac{1}{n_s} \cdot \frac{\Delta n_f}{\Delta t}$$

which gives in the limit:

$$\lambda = \frac{1}{n_s} \cdot \frac{dn_f}{dt}$$

Differentiating Equation 53.3, noting that n_o is constant, and substituting gives

$$\frac{dn_s}{dt} = -\frac{dn_f}{dt} = -\lambda . n_s$$

Assuming λ is constant, integration yields

$$\int_{n_o}^{n_s} \frac{dn_s}{n_s} = -\lambda \int_0^t dt$$

whence:

$$n_s = n_o . e^{-\lambda t}$$

From its definition, reliability may be articulated as the ratio of n_s to n_o which gives the exponential relationship

$$R(t) = e^{-\lambda t} = e^{-t/\text{MTTF}} \quad (53.4)$$

The two assumptions underlying this exponential relationship are emphasised: that there are zero repairs and that the failure rate is constant.

53.2 Repairable Systems

Repairable systems are the norm in the process industry sector. If a system fails, the faulty parts are repaired and/or replaced, and the system put back into service. The failure rate of a typical component, device or system varies throughout its life as depicted in Figure 53.1, the so-called "bath-tub" curve.

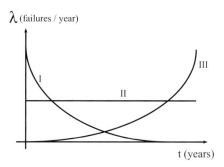

Fig. 53.1 The bath tub curve

Inspection of the bath-tub curve reveals three distinct phases, each with different characteristics which are somewhat exaggerated in Figure 53.1:

- The first phase is that of infant mortality in which failures due to faulty design and manufacture dominate. Modifications are made and faulty components are repaired or replaced with good ones resulting in a decreasing failure rate. The majority of infant mortality failures can be expected during testing by the manufacturer or during commissioning.
- The second phase is that of normal working life, often referred to as service life, and characterised by a constant failure rate (λ) due to random component failure. All the metrics for repairable systems relate to this phase for which there are two principal sub-divisions, namely continuous mode and demand mode of operation.
- The third phase is that of longevity: increasing failure rates due to wear and ageing of components.

A particularly useful metric is the mean time between failures (MTBF) of an element or system. Suppose that a large number n of identical devices are tested over a period of time T. Each fault is recorded, the device repaired and put back into service, and the total number of faults n_f during the period T found. The observed MTBF is

$$\text{MTBF} \approx \frac{n.T}{n_f} \quad (53.5)$$

For example, if 150 faults are recorded for 200 dp cells over 1.5 years the MTBF is 2.0 years.

It is evident that MTBF must include the mean time to repair (MTTR) of the devices. The balance is known as the mean time to fail (MTTF):

$$\text{MTTF} + \text{MTTR} = \text{MTBF} \quad (53.6)$$

Note that MTTR requires careful definition. An equipment vendor's estimate of MTTR will always assume that a fully trained technician, complete with appropriate spares and test equipment, is on hand for 24 h per day, and that the failed equipment is immediately available for repair. Vendor's estimates of 1 h for MTTR are common. In the real world, staffing levels (8- or 24-h days, 5- or 7-day weeks, *etc.*) and working procedures such as isolation of plant, work permits, post repair checkout and re-commissioning have to be taken into account. This means adding a standard figure of say, at least, 8 h to vendors' MTTR data. The sum of a vendor's MTTR plus other delays is sometimes referred to as the mean time to reinstate or as the mean down time (MDT) but these definitions are by no means universal.

There is significant scope for ambiguity in the use of these "mean times" since replacement may be part of the repair of a more complex system, such as a PLC or DCS. Strictly speaking MTBF can only be applied to repairable systems whereas MTTF relates to both repairable and unrepairable devices. Fortunately, in practice, it doesn't matter much since normally MTTF $\gg$ MTTR whence MTTF and MTBF are approximately equal.

Availability (A) is the probability that a system will be functioning correctly when needed. Put another way, it is the fraction of the total time that a device or system is able to perform its required function:

$$A = \frac{\text{MTTF}}{\text{MTBF}} = \frac{\text{MTTF}}{\text{MTTF} + \text{MTTR}} \approx \frac{\text{MTBF}}{\text{MTBF} + \text{MTTR}} \quad (53.7)$$

Note that although the latter approximation is commonly used for estimating availability, it is not valid to do so unless MTTF $\gg$ MTTR.

Availability is particularly useful for assessing the suitability of control systems for continuous processes as it provides an estimate of the "downtime" of the system and enables repair and maintenance requirements to be costed. Unavailability (U) is the complement of availability:

$$U = 1 - A = 1 - \frac{\text{MTTF}}{\text{MTBF}} = \frac{\text{MTTR}}{\text{MTTF} + \text{MTTR}} \quad (53.8)$$

Dividing by MTTF and substituting from Equation 53.2 gives

$$U = \frac{\text{MTTR}^*\lambda}{1 + \text{MTTR}^*\lambda} \approx \text{MTTR}^*\lambda \quad (53.9)$$

Again, the latter approximation assumes that MTTF $\gg$ MTTR.

53.3 Proof Testing

From Equation 53.4 it can be seen that reliability starts with a value of unity and decays. Suppose that after a period of time known as the proof test interval (PTI) the system is tested, any faults repaired and the system put back into service. Since the system is known to be working correctly at that point in time, its reliability is 1.0 and the clock is effectively reset to zero. The minimum reliability, which is the worst case, is thus given by

$$R_{min} = e^{-\lambda.\text{PTI}} \quad (53.10)$$

If this process of testing is repeated systematically, at regular intervals, the outcome is a significant increase in the reliability of the system, as depicted in Figure 53.2. Provided there are alternate means of maintaining operations whilst the system is being tested and/or repaired, such proof testing is a legitimate means of increasing reliability.

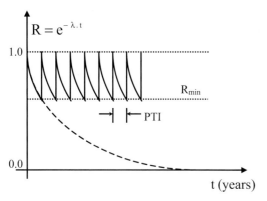

Fig. 53.2 Effect of proof testing on reliability

Since it is not known when unrevealed faults occur, they are presumed to occur half way through the proof test interval which, on average, must be true. Thus the proof test and repair time is given by

$$\text{PTRT} = (\text{PTI}/2) + \text{MTTR} \qquad (53.11)$$

The proof test and repair time is an effective MTTR for use in reliability calculations when faults are revealed by proof testing. Substituting for MTTR in Equations 53.7 and 53.8 gives

$$A \equiv \frac{\text{MTTF}}{\text{MTTF} + \text{PTRT}} \quad \text{and} \quad U \equiv \frac{\text{PTRT}}{\text{MTTF} + \text{PTRT}}$$

Provided that MTTF $\gg$ PTRT, which is usually the case, substituting from Equation 53.2 gives

$$U \equiv \frac{\text{PTRT}}{\text{MTTF} + \text{PTRT}} \approx \frac{\text{PTRT}}{\text{MTTF}} = \text{PTRT} \cdot \lambda \quad (53.12)$$

This may be further simplified, provided that PTI $\gg$ MTTR, which is not always the case, by substituting from Equation 53.11 to give

$$U \approx \frac{\text{PTI}}{2} \cdot \lambda$$

53.4 Elements in Series and Parallel

From a hardware point of view, control and protection systems are largely comprised of I/O channels which consist of elements such as sensors, logic solvers, actuators and valves that are connected in series. Thus evaluating the reliability of a series of elements is of fundamental importance. Consider a system of n devices in *series* with failure rates of $\lambda_1, \lambda_2, \ldots \lambda_n$ as depicted in Figure 53.3.

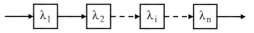

Fig. 53.3 System of devices in series

The system will only survive if every device survives: if any one device fails then the system fails. The system reliability is the product of the individual device reliabilities:

$$R_{\text{ser}} = R_1 . R_2 \ldots R_i \ldots R_n = \prod_{i=1}^{n} R_i \qquad (53.13)$$

Substituting from Equation 53.4:

$$e^{-\lambda_{\text{ser}} . t} = e^{-\lambda_1 . t} . e^{-\lambda_2 . t} \ldots . e^{-\lambda_n . t} = e^{-(\lambda_1 + \lambda_2 + \ldots + \lambda_n) . t}$$

whence the failure rate of a system of devices in series is the sum of the individual failure rates:

$$\lambda_{\text{ser}} = \lambda_1 + \lambda_2 + \ldots \lambda_n \qquad (53.14)$$

For repairable systems, it is the availability rather than the reliability that counts. Thus:

$$A_{\text{ser}} = A_1 . A_2 \ldots A_i \ldots A_n = \prod_{i=1}^{n} A_i \qquad (53.15)$$

Substitute $A = 1 - U$, expand, assume that $0 < U \ll 1.0$ such that second (and higher) order terms may be ignored, to give

$$U_{\text{ser}} \approx U_1 + U_2 + \cdots + U_n = \sum_{i=1}^{n} U_i \qquad (53.16)$$

So the overall series unavailability is the sum of the individual device unavailabilities.

Whilst *parallel* elements occur naturally in some systems, it is often the case that parallelism is deliberately introduced. For example, the use of parallel I/O channels to provide redundancy is one of the most common means of increasing the reliability of a system. Thus, for repairable systems, evaluating the availability of elements in parallel is also of fundamental importance. Consider a system of n devices in parallel with unavailabilities of $U_1, U_2, \ldots U_n$ as depicted in Figure 53.4.

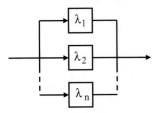

Fig. 53.4 System of devices in parallel

The overall system will only fail if every single device fails, whence:

$$U_{par} = U_1.U_2 \ldots U_i \ldots U_n = \prod_{i=1}^{n} U_i \quad (53.17)$$

So the overall parallel unavailability is the product of the individual device unavailabilities.

53.5 Common Mode Failure

A fundamental assumption underlying all of Equations 53.13–53.17 is that the reliabilities of the devices are independent of each other. Thus, say, for a system of two elements in parallel that are truly independent of each other:

$$U_{sys} = U_1.U_2$$

However, in practice, it is quite common for a single fault to affect many elements. Such causes and effects are referred to as common mode failures. These dependencies have the effect of reducing availability:

$$U_{sys} > U_1.U_2$$

For the purposes of reliability analysis, it is necessary to allow for such dependencies. For obvious common mode failures, such as failure of air or power supply, they are allowed for explicitly by, for example, dedicated branches in fault trees. But, in practice many of the dependencies are more subtle, such as the effects of operator error, and are allowed for by introducing a dependency factor, β. Thus:

$$\lambda_{sys} = \lambda_{ind} + \lambda_{dep}$$

where λ_{IND} is the failure rate of the system assuming that all its elements are independent and λ_{DEP} is the contribution of the dependencies to the net failure rate λ_{SYS}. This contribution is evaluated from

$$\lambda_{dep} = \beta.\lambda_{max}$$

where λ_{MAX} is the failure rate of the most unreliable element in the system under analysis. Values of β are tabulated: they are context dependent and vary according to the amount of redundancy in the system and the diagnostic coverage (DC). Whence:

$$\lambda_{sys} = \lambda_{ind} + \beta.\lambda_{max} \quad (53.18)$$

53.6 Voting Systems

As stated, parallel channels are deliberately introduced to enable redundancy. There are various redundancy and voting strategies that can be applied, the choice being context and cost dependent as explained in Chapters 56 and 57. For example, for continuous mode operations:

- The average value of two analogue inputs is used, unless one is out of range in which case the other is used.
- The middle value from three analogue input channels is used.
- The average of the two closest of three analogue input channels is used.

For shutdown purposes, typical scenarios are:

- 1oo2: a trip occurs if either one out of two channels detects a trip condition.
- 2oo2: a trip occurs if both channels agree that a trip condition exists.
- 1oo3: a trip occurs if any one out of three channels detects a trip condition.
- 2oo3: a trip occurs if any two channels out of three agree that a trip condition exists.

The voting strategy used affects the reliability of the overall system. So, in addition to evaluating the reliability of individual devices, channels and redundancy structures, the reliability of the strategy itself must be evaluated.

Figure 53.5 depicts a 2oo3 voting system which comprises three identical channels and a voting device. Suppose that the objective is to establish agreement between two channels that a shutdown situation exists. To establish this, any two out of the three channels must be functioning correctly. In other words, any two channels must have not failed.

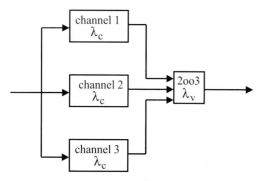

Fig. 53.5 2oo3 voting system with identical channels

Consider any two channels: since these are in parallel, and being identical have the same unavailability U_c, the probability of both channels failing is given by Equation 53.17:

$$U_{2\times 1ch} = U_c^2$$

Next consider the third channel. In fact there are three pairs of channels, 1-2, 2-3 and 1-3, so the chances of any one pair failing is three times greater:

$$U_{3\times 2ch} = 3U_c^2$$

Strictly speaking, there is also the possibility of all three channels failing simultaneously:

$$U_{3\times 1ch} = U_c^3$$

To avoid double counting this latter scenario, the overall unavailability of the voting system is

$$U_{2oo3} = 3U_c^2 - U_c^3 \approx 3U_c^2$$

provided that $0 < U \ll 1.0$, which should always be the case. For identical channels, this latter result may be generalised to more complex cases of k out of n voting for which the overall availability is

$$U_{koon} \approx \frac{n!}{k!.(n-k)!}.U_c^k \qquad (53.19)$$

Returning to the 2oo3 system of Figure 53.5, its overall availability is given by

$$A_{2oo3} \approx 1 - 3U_c^2$$

There is also the availability A_V of the voting device itself to be taken into account. Since this is in series with the three channels, Equation 53.15 is used to find the availability of the voting system as a whole:

$$A_{sys} \approx (1 - 3U_c^2).A_V$$

53.7 Standby Systems

Consider the scenario depicted in Figure 53.6 which consists of system A which is normally operational and system B which is on standby. When system A fails then system B is brought into use by some switching system S.

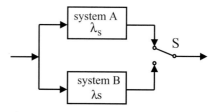

Fig. 53.6 Standby system with switching system

The nature of the switching is critical to the analysis. Suppose that the switch is in its de-energised position when system A is working and that all of the unreliability of the switch is associated with its energised state. Thus, for a complete failure there must be a simultaneous failure of system A and either system B or the switch. Strictly speaking, system B can only fail its standby duty when called upon to function, *i.e.* after system A has failed.

However, to all practical intents and purposes, the simultaneous failure of A and B may be considered.

Assume that both systems A and B have the same unavailability U_S and that of the switching system is U_{SW}. From Equation 53.16, the unavailability of the standby system is

$$U_{s/b} = U_s + U_{sw}$$

From Equation 53.17, the unavailability of the combined system is

$$U_{sys} = U_s.(U_s + U_{sw})$$
$$A_{sys} = 1 - U_s.(U_s + U_{sw})$$

If the unreliability of the switch is assumed to be equally distributed between its energised and de-energised states, i.e. it is just as likely to fail open as closed, then the unavailability of each channel is the same and

$$A_{sys} = 1 - U_{sys} = 1 - \left(U_s + \frac{1}{2}U_{sw}\right)^2 \quad (53.20)$$

As ever, the devil is in the detail. The above assumption that both systems A and B have the same unavailability U_s is, strictly speaking, only likely to be true if B is a hot standby system. Such a system is running and ready for use when called upon, whereas a cold standby has to be started up when called upon. In practice, if system B is a cold standby, its probability of failure on demand (PFD) is likely to be higher than that of system A.

53.8 Protection Systems

As stated, availability is particularly useful for assessing the suitability of control systems for continuous mode operation, say in terms of their percentage down-time or their maintenance and repair costs. In evaluating availability, overall failure rate (λ) values are used and no distinction is made between the failure modes of devices and/or systems. That is not unreasonable: something is either working or it isn't, and if it isn't then it's not available.

However, protection systems, which are always of a repairable nature, have a demand mode of operation, i.e. when called upon they are expected to function correctly, and it is necessary to distinguish between dangerous and safe failures. For example, a pressure switch which is closed under normal conditions, has two failure modes: dangerous, which is when it won't open on demand, and safe when it opens without a demand. When a dangerous failure occurs the protection system is normally expected to trip something: a valve, a shut down system, or whatever. If a trip occurs as a result of a safe failure it is referred to as a spurious trip.

For design purposes, it is essential that the failure rate λ values and the MTBF data is for dangerous failures only. That being so, the unavailability (U) is referred to as the probability of failure on demand (PFD) and its reciprocal as the hazard reduction factor (HRF). For example, if A = 0.999 then PFD = 0.001 and HRF = 1000 times.

Another metric used in the design of protection systems is the safe failure fraction (SFF). This concerns the use of diagnostics to detect failures, either for individual devices or at a system level. The number of undetected dangerous failures is expressed as a proportion of the total number of failures, whence the proportion of safe and detected dangerous failures:

$$\text{SFF} = 1 - \frac{\lambda_{UD}}{\lambda_S + \lambda_D} = \frac{\lambda_S + \lambda_{DD}}{\lambda_S + \lambda_D} \quad (53.21)$$

where
- λ_S is the failure rate for all safe (spurious) failures
- λ_D all dangerous failures
- λ_{DD} detected (overt) dangerous failures
- λ_{UD} undetected (covert) dangerous failures

Clearly the SFF increases as the number of undetected dangerous failures decreases. Note that if there are no on-line diagnostics, all dangerous failures must be considered to be undetected since they cannot be detected until proof testing occurs.

Yet another metric used is the diagnostic coverage (DC) which is the fraction of the dangerous failures that are detected by means of the diagnostics:

$$DC = \frac{\lambda_{DD}}{\lambda_D} \qquad (53.22)$$

53.9 Worked Example

Consider the protection system depicted in Figure 53.7 which comprises:

- Three parallel input channels, each consisting of a sensor and transmitter in series
- A 2oo3 voting system to decide whether a shutdown is required
- Two parallel output channels, each consisting of a pilot valve and an isolating valve

The dangerous failure rate for each device is stated on Figure 53.7. The figures quoted, in terms of number of failures per year (fpy), are unrealistically high to exaggerate the principles involved, as also are the proof testing interval of one month and the mean time to repair of one day.

The objective of the worked example is to analyse the reliability of the system, assuming that there are no common mode failures.

1. Single input channel. From Equation 53.14 the combined failure rate for a single input channel is

$$\lambda_{1 \times i/p} = \lambda_{sens} + \lambda_{trans} = 0.6 \text{ fpy}$$

PTI = 1 month = 0.08333 years and MTTR = 1 day. So, from Equation 53.11:

$$PTRT = PTI/2 + MTTR = 0.08333/2 + 1/365$$
$$= 0.0444 \text{ year}$$

Thus, from Equation 53.12 the unavailability of one input channel is

$$U_{1 \times i/p} \approx PTRT \cdot \lambda_{1 \times i/p} = 0.0444 \times 0.6$$
$$= 0.02664$$

2. Voting system. There are two aspects to this, the voting process which relates to the three input channels, and the reliability of the voting device itself.

   From Equation 53.19 for a 2oo3 voting system, the input sub-system's unavailability is

$$U_{3 \times i/p} \approx 3 \cdot U_{1 \times i/p}^2 = 3 \times 0.02665^2 = 0.00213$$

   Thus, from Equation 53.8 the availability of the input sub-system is

$$A_{2oo3} = 1 - U_{3 \times i/p} \approx 0.9979$$

   For the voting device itself, its availability comes from Equations 53.8 and 53.12:

$$A_{vote} \approx 1 - PTRT \cdot \lambda_{vote} = 1 - 0.04441 \times 0.1$$
$$= 0.9956$$

3. Output channel. From Equation 53.14 the combined failure rate for a single output channel is

$$\lambda_{1 \times o/p} = \lambda_{pilot} + \lambda_{isolat} = 0.5 \text{ fpy}.$$

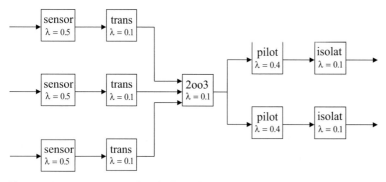

Fig. 53.7 Protection system with multiple I/O channels

So, from Equation 53.12, the unavailability of one output channel is

$$U_{1 \times o/p} \approx PTRT \cdot \lambda_{1 \times o/p} = 0.0444 \times 0.5 = 0.0222$$

But the output channels are in parallel so, from Equation 53.17:

$$U_{2 \times o/p} \approx U_{1 \times o/p}^2 = 0.0222^2 = 0.0004929$$

And from Equation 53.8 the availability of the output sub-system is given by

$$A_{2 \times o/p} = 1 - U_{2 \times o/p} \approx 0.9995$$

4. System characteristics. The protection system may be thought of as the input subsystem in series with the voting system and the output sub-system, whose availability is given by Equation 53.15:

$$\begin{aligned} A_{overall} &= A_{2oo3} \cdot A_{vote} \cdot A_{2 \times o/p} \\ &\approx 0.9979 \times 0.9956 \times 0.9995 = 0.9930 \end{aligned}$$

What this means in practice is that the protection system will function correctly 993 times out of every 1000 demands upon it. Put another way, its probability of failure on demand is 7 times in a 1000 giving a hazard reduction factor of 143. From Equation 53.8:

$$PFD = U_{overall} = 1 - A_{overall} \approx 0.007$$

The system failure rate is found from Equations 53.12:

$$U_{overall} \approx PTRT \cdot \lambda_{overall}$$
$$0.007 \approx 0.04441 \cdot \lambda_{overall}$$

whence $\lambda_{overall} \approx 0.1576$ failures per year.

From Equation 53.2,

$$MTTF_{overall} = \frac{1}{\lambda_{overall}} \approx 6.344 \text{ years.}$$

Noting that the PTRT is an effective MTTR, the MTBF is found from Equation 53.6:

$$\begin{aligned} MTBF_{overall} &= MTTF_{overall} + PTRT \\ &\approx 6.344 + 0.04441 \\ &= 6.389 \text{ years} \end{aligned}$$

53.10 Hazard and Demand Rates

Reliability metrics provide the basis for the design of protection systems. Central to that design is application of the following equation:

$$HR = DR \times PFD \tag{53.23}$$

- The demand rate (DR) is the unmitigated frequency (incidents/year) of the hazardous event, *i.e.* the frequency at which it would occur without any protection system. The DR has to be evaluated for a given process and/or plant design, including its normal control and safety systems, a non-trivial task explained in detail in Chapter 54.
- The PFD, as defined by Equation 53.8, is that of the protection system designed to guard against hazardous events. For design purposes the maximum PFD has to be specified. Values of PFD are banded into so-called safety integrity levels (SIL). These are defined in detail in Chapter 56.
- The hazard rate (HR) is the frequency (incidents/year) to which the hazardous event is reduced by the protection system. This is the residual frequency due to the simple fact that the protection system cannot provide 100% protection.

The objective of design, for a given DR, is to determine the SIL required to reduce the HR to an acceptable level:

$$Risk = HR \times C(E) \tag{53.24}$$

C(E) is the consequence of the hazardous event (deaths/incident) which, combined with the hazard rate yields the risk (deaths/year) to which personnel (or the public) is exposed. What constitutes an acceptable HR is a fundamental question of judgement on risk. It is subjective and contentious, but zero risk is not an option, and is discussed in more detail in Chapter 56.

53.11 Comments

The most fundamental problem for reliability analysis, and indeed for protection system design, is access to meaningful reliability data. Bodies such as the UK Health and Safety Executive (HSE) have consistently declined to commission and publish data, let alone target values, so the only official data that exists tends to have military origins. Much data of a proprietary nature is available but is commercially valuable and only available on a restricted access basis. An important exception is the Norwegian OREDA (2002) database which, whilst it has been produced on behalf of a consortium of oil companies operating in the North Sea, is not necessarily restricted to consortia members only. Most major control system suppliers and equipment vendors have their own data which is confidential and used in-house. Otherwise, there is a dearth of such data in the public domain. In the absence of access to proprietary data, the reader is referred to the texts by Lees (1996) and Smith (2005) which are well established and recognised sources of useful data.

Hazard Analysis

Chapter 54

54.1 HAZOP Studies
54.2 Limitations of HAZOP
54.3 CHAZOP Studies
54.4 The Need for COOP Studies
54.5 Control and Operability Studies
54.6 Failure Mode and Effect Analysis (FMEA)
54.7 Fault Tree Analysis (FTA)
54.8 Minimum Cut Sets
54.9 Fault Tree Evaluation
54.10 Evaluation of Demand Rate

There are many types of hazard. Some, such as those due to moving parts and rotating machinery, are no different to what are encountered in general manufacturing. Others, such as flammability and toxicity, are more specific to the process industries. Legislation in the UK covering hazardous installations is provided by the Control of Substances Hazardous to Health (COSHH) Regulations (1988) and the Control of Major Accident Hazards (COMAH) Regulations (1999).

Suppose that a plant is being designed to carry out some process that is inherently hazardous and for which some protection system is likely to be required. Clearly, as a basis for design of the protection system, the cause of any potential hazard must be identified and, if significant, quantified. That is the essence of hazard analysis for which various techniques are available.

The qualitative technique of a hazard and operability (HAZOP) study is used extensively to identify potential hazards that are of significance. The techniques of computer hazard and operability (CHAZOP) study and control and operability (COOP) study focus respectively on the hardware and software of the plant's control system. An alternative approach is failure mode and effect analysis (FMEA). Those hazards that are deemed to be significant, and which cannot be addressed by means of changes to the design of the process, the plant or its operation, normally result in the provision of one or more protection systems.

Having identified the need for a protection system, the quantitative method of fault tree analysis (FTA) may be used to understand the cause and effect relationships more thoroughly. In particular, FTA provides a basis for establishing the probability of a hazardous occurrence. This is the unmitigated demand rate (DR) which is required for use in Equation 53.23 for the design of protection systems. The other factor to be quantified is the consequence of a hazardous event, if it occurs. This is essentially a process engineering consideration. The consequence of the event C(E) is related to the hazard rate (HR) by Equation 53.24 for design purposes.

This chapter outlines some of the above-mentioned techniques: there are others such as event tree analysis and Markov modelling which are not covered here. For a more comprehensive coverage, the reader is referred to the texts by Kletz (1999), Wells (1996) and Andrews (1993): the latter includes a useful glossary of terms.

Table 54.1 Interpretation of basic guidewords for HAZOP

Guideword	Meaning	Comments
NO or NOT	The complete negation of the intention	No part of the intention is achieved and nothing else happens
MORE and LESS	Quantitative increase or decrease	Refers to quantities and properties (such as flow rates and temperatures) as well as to operations (such as charge, heat, react and transfer) related to the intention
AS WELL	Qualitative increase, something extra	All the design and operating intentions are achieved, together with some additional activity
PART OF	Qualitative decrease, the intention is not completed	Only some of the intentions are completed, others aren't
REVERSE	The logical opposite of the intention	The reverse of the intended action or the opposite of some effect occurs
OTHER THAN	Complete substitution	No part of the original intention is achieved. Something quite different happens. WHERE ELSE may be more useful in relation to position

54.1 HAZOP Studies

An authoritative introduction to hazard and operability (HAZOP) studies is provided by the CIA Guide (1992) and guidelines to best practice have been published by the IChemE (1999). The standard IEC 61882 is an application guide to HAZOP.

HAZOP is a systematic and critical examination, based upon the design of a process and/or plant, of the potential hazards due to malfunctions and/or mal-operations of individual items of equipment and the consequential effects of such. Whilst normally carried out at the design stage, when the cost of making changes is relatively small, there is no reason why a HAZOP study cannot be carried out retrospectively on the design of an existing plant. Note that the cost of carrying out a HAZOP study is not insignificant but, if done properly, should be recouped through having identified problems in advance that would otherwise be costly to address at the commissioning stage.

HAZOP studies involve systematic appraisal of conceivable departures from design intent, involving searches backwards for possible causes and forwards for possible consequences. The design, normally in P&I diagram form, is scrutinised vessel by vessel, pipe by pipe, to ensure that all potentially hazardous situations have been taken into account. In essence, using a framework of *guidewords*, the *causes* and *consequences* of all possible significant *deviations* in process operations are examined against the *intention* of the plant design to reveal potential *hazards*. The basic guidewords and their interpretation are listed in Table 54.1.

The methodology for carrying out a HAZOP study is depicted in Figure 54.1.

An important variation on HAZOP, which is used extensively for continuous plant, is the parameter based approach as described by Kletz (1999). In essence this focuses on deviations of process parameters such as flow and temperature, some examples of which are given in Table 54.2.

HAZOP studies are normally carried out by small multi-disciplinary teams with understanding of the plant design and operational intent, supplemented with specialist knowledge as appropriate. Membership of a team typically includes project, process, mechanical and instrument engineers, supplemented with chemists, control and safety specialists.

It is very important that the study sessions are accurately recorded, particularly with respect to

54.2 Limitations of HAZOP

the hazards identified, the solutions proposed and those responsible for follow up actions. This is normally realised by means of tables. Typically, the first column contains a guideword and the second column an abbreviated description of the deviation. Three other columns are used for possible causes, possible consequences and the action required to prevent the occurrence of the hazardous situations identified. There are computer packages available for aiding this task. A commonly used technique for visualising and keeping track of progress is to colour in the vessels and pipes as they are dealt with on hard-copy of the P&I diagrams.

54.2 Limitations of HAZOP

It should be evident that the HAZOP study is process and/or plant oriented and does not focus on the functionality of the control systems per se. Common practices for handling control systems are to treat them:

- As if they don't exist. This is simply unacceptable because the effects of failure have the potential to cause hazardous situations.
- As though they were realised simply by means of hardwired elements, whether they be DCS, SCADA or PLC based or otherwise. This is potentially unsafe because it overlooks the potential for common mode failures of both the hardware and software: see Chapter 53.
- As black boxes, *i.e.* perfection in the control systems and their operations is assumed. This is ridiculous: the reality is that control systems' hardware is subject to failure and there is much scope for making mistakes in the design and development of the application software.
- Separately from the plant. This is a pragmatic and valid approach because the issues to be addressed, whilst still related to the process/plant, are nevertheless different and in many respects far more complex. Also, the expertise required and the time scales involved are different.

When a control system fails, or partially fails, there is a whole spectrum of potential failure modes,

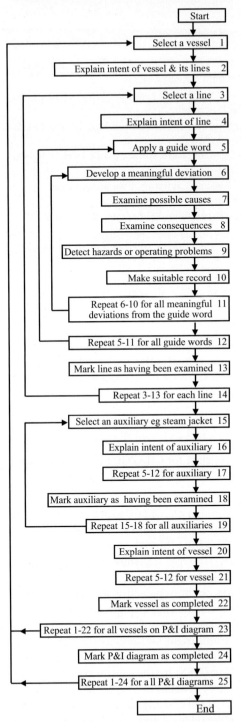

Fig. 54.1 Methodology for carrying out a HAZOP study

Table 54.2 Interpretation of deviation approach guidewords for HAZOP

Guideword	Type of deviation	Typical problems
None of	No flow / No pressure	Blockage. Pump failure. Valve closed or failed shut. Leak. Suction vessel empty. Delivery pressure too high. Vapour lock.
Reverse of	Reverse flow	Pump failure. Pump reversed. Non-return valve failed. Non-return valve reversed. Wrong routing. Back siphoning.
More of	More flow	Reduced delivery head. Surging. Supply pressure too high. Controller saturated. Valve stuck open. Faulty flow measurement.
	High temperature	Blockage. Hot spots. Cooling water valve stuck. Loss of level in heater. Fouling of tubes. Blanketing by inerts.
Less of	Lower flow / Low vacuum	Pump failure. Leak. Partial blockage. Cavitation. Poor suction head. Ejector supply low. Leakage. Barometric leg level low.
Part of	Change in composition	High or low concentration. Side reactions. Feed composition change.
As well as	Impurities / Extra phase	Ingress of air, water, *etc*. Corrosion products. Internal leakage. Wrong feed.
Other than	Abnormal operations	Start-up and shut-down. Testing and inspection. Sampling. Maintenance. Removal of blockage. Failure of power, air, water, steam, inert gas, *etc*. Emissions.

many of which may be unexpected. A good example is the failure of an analogue output card. In general, segregation policy should be such that all the output channels from one card are associated with a particular item of plant or processing function. Thus, if the card fails, or is removed for isolation purposes, that item of plant only is directly affected. However, if one of the outputs goes to a valve on another item of plant in a different area, perhaps because it was wired in at a later stage, that will fail too. Such failures would appear to be sporadic.

It is evident that there are various aspects of the control system that need to be subject to some form of HAZOP study. Ideally, these should be considered as part of the HAZOP study of the process and/or plant but, in practice, the design of the control system is seldom sufficiently complete at that stage of the design cycle for an integrated HAZOP study. Therefore it is necessary to carry out a seperate computer hazard and operability (CHAZOP) study on the control system. Recognising that the design of the application software always lags behind the design of the rest of the system, it is appropriate that the CHAZOP study concentrates on the control system's hardware design, its I/O organisation and the system software. Consideration of the application software is deferred to a later control and operability (COOP) study.

54.3 CHAZOP Studies

It is appropriate for CHAZOP to be carried out in two stages:

1. A preliminary CHAZOP which may well be integrated with the full HAZOP study.
   The thinking behind the preliminary CHAZOP is that strategic decisions about segregation policy, system layout and the handling of gross failures ought to be addressed early in the design cycle. Making changes later is expensive and causes delays to the project.
   The results of the preliminary CHAZOP should be incorporated in the user requirements specification (URS) for the control system, which is described in detail in Chapter 60.
2. A full CHAZOP study carried out at a later stage when the relevant detailed information is available.

Table 54.3 Interpretation of guidewords for CHAZOP

Guideword	Meaning	Comments
LOSS	The complete or partial absence of a function	The impact on the design intent of the loss of a function. May apply to either power supply, processor capability, memory, communications channels or to I/O signals.
RANGE	The distortion of an I/O signal	The impact on the design intent of a signal either being distorted (such as non-linearity or hysteresis) or going out of range.
MIXTURE	The combination of I/O channels	The failure pattern of inappropriate combinations of I/O channels in relation to the hardware organisation of the system.
VERSION	Incompatibility of and/or changes to functionality of the system software.	The potential consequences of either changes to the hardware platform or upgrades (new versions of or changes) to the system software on the integrity of the application software, either in relation to its development or to its subsequent support and maintenance.
SECURITY	The integrity of the system	The potential consequences of unauthorised access to the system, malicious or otherwise.

Results of the full CHAZOP should be incorporated in the hardware aspects of the detailed functional specification (DFS) described in Chapter 62.

When applying CHAZOP the same terms are used as in conventional HAZOP but they take on more specific meanings. Thus:

- *Intention* relates to the transfer of information (signals, commands, actions) between external elements of the control system and its internal software functions (both application and system) *via* either the system's I/O and/or communications *channels* or by means of operator interaction.
- *Guidewords* that are more appropriate for CHAZOP are listed in Table 54.3.
- *Deviations* are partial or total failures of either communications channels or processing functions.
- *Causes* are those combinations of events that result in deviations, the *consequences* being outcomes that could lead to operability difficulties and/or hazardous situations.
- *Recommendations* are additional requirements for inclusion in either the URS and/or DFS.
- *Actions* are reports of errors identified in the design upon which action is to be taken.

The basic methodology for CHAZOP is similar to that outlined in Figure 54.1, except that rather than being based on the vessels and pipes, the focus is on the hardware design of the control system, DCS or otherwise. In turn, consider each I/O signal to establish whether it is used for any safety related function. Clearly any signal identified in the HAZOP study for the process and/or plant as being safety related is a candidate for CHAZOP (these signals are normally identified as consequences in the HAZOP). Then identify all the communications channels used by any such signal. For every such channel:

- Review the functionality of the channel.
- Follow the physical route of the channel in terms of cabling, termination and marshalling cabinets, I/O cards, racks, *etc*.
- Note the arrangements for segregation (see Chapter 51) of the channel and the provision for redundancy, if any.
- Apply the guidewords as appropriate. In doing so, particular attention should be paid to the segregation and redundancy requirements.

Table 54.4 Examples of the use of CHAZOP guidewords

Guideword	Type of deviation	Typical problems
Loss of	No/frozen signal	I/O signal failed low/high. I/O channel disconnected (off scan). Polarity reversed. Wrong channel used. Component failure on I/O board. Board removed. Poor contact with back plane. Intermittent fault.
	No power	Local 24V d.c. supply failed. Mains 110V a.c. supply failed. Watchdog timed out. OCS dead. Operating system crashed.
	No communication	Highway disconnected or damaged. Highway overloaded. PORT and/or GATE cards disabled.
	Memory corrupted	Can't access data. Disc scratched. Read/write heads damaged. RAM chip failed.
Range of	Distortion of signal	Signal ranged incorrectly. Signal in wrong direction. Non-linearity. Sluggish dynamics. Damping effects. Noise and interference suppression. Drifting signals. Hysteresis.
	Signal out of range	Device disconnected for calibration or repair. Signal detected as bad. Impact of automatic ranging.
Mixture of	Channel segregation	Sporadic I/O failures. Segregation not on plant item/ function basis. Insufficient redundancy. Signals routed through two channels/cards. Schemes spread across cards/racks.
	Failure modes	Isolation and shut down. Manual operation. Local/remote power supply. Different failure modes across cards/racks.
Version of	Operating system	May require more processor power and/or memory than existing hardware platform with implications for timings and loadings.
	System software packages	Application software functions (such as alarm handling) may not be fully supported by new version of operating system. Different versions of system packages may not be fully compatible with each other.
Security of	Abnormal behaviour and/or loss of functionality	Operating system and/or system software vulnerable to hacking and/or viruses over the internet or intranet.
	Breach of confidentiality	Commercial information in application software unlawfully abstracted.
	Breach of trust	Application software deliberately changed without authorisation.

It is worth emphasising that it is not the intent that the above methodology be applied to every I/O channel but only to those used for safety related signals. It clearly makes for effective use of time if generic channel types can be identified and considered on a type basis. Some typical examples of the use of these CHAZOP guidewords are given in Table 54.4.

The design upon which the CHAZOP is based is articulated by various documents, the key ones being as listed below. In practice, they will be in various states of completeness according to the stage in the design cycle at which the HAZOP is done: the more detailed the better:

- All of the documentation for the HAZOP of the process and/or plant.
- Copies of the URS and DFS (see Chapters 60 and 62).
- P&I diagrams.
- Architecture of control system depicting networks, highways, links, bridges, gates, *etc*.

- Configuration of control system hardware: operator stations, cards/racks/cabinets, *etc.*
- Physical layout and siting of system cabinets, I/O racks, terminals, *etc.*
- Organisation of infrastructure: marshalling cabinets, field termination racks, *etc.*
- Power and wiring arrangements: types of multicore cables, connectors, colour coding, *etc.*
- Channel/loop diagrams.
- Details of system malfunctions: failure modes, fail-safe states, *etc.*
- Description of functionality of any non-standard equipment.

It is likely that the CHAZOP team will be more specialist, comprising instrument, control and electrical engineers, but there should always be some cross membership of the HAZOP and CHAZOP teams to provide continuity of thought.

It should be stated there is little published information about CHAZOP in the public domain, although the technique is carried out within many of the larger and more responsible companies on an in-house basis. A survey of industrial practice was commissioned by HSE and published by HMSO (1991) and an overview is provided by Kletz (1995). The text by Redmill (1999) provides a thorough overview of CHAZOP but advocates a different approach.

54.4 The Need for COOP Studies

Whilst all modern control systems support a wide variety of safety functions and clearly contribute to safe operations, their purpose is to control plant. Control systems are not protection systems. They do, nevertheless, contribute to plant safety to the extent that effective control reduces the demand on the protection systems. The converse is also true. If a control system leads to an increase in demand rate on a protection system, that contribution must be taken into account in the design of the protection system. This is recognised by the IEC 61508 standard on safety systems.

Central to the argument is the design of the application software and human factors. Quite simply, people interact with control systems to operate plant. It is very easy for operators to make mistakes: by making wrong decisions, by making changes at the wrong time, by not making decisions, and so on. Thus the application software needs to permit access by the operator to certain functions necessary for carrying out the operator's role but, at the same time, it needs to override inappropriate decisions and not permit certain interventions.

It follows that, to prevent the protection systems being exercised unnecessarily, there should be some systematic check on the design and functionality of the application software. This is best deferred to a separate control and operability (COOP) study since, as stated, the design of the application software lags behind the design of the rest of the system, usually to a considerable extent. In essence, a COOP study is used to check that the design of the application software has properly taken into account all conceivable and relevant human factors. It should also check that the logic is sound for the decisions being made by the system. Unless this is done systematically, by means of COOP or otherwise, it is not possible to argue that the control system is not contributing to the demand on the protection system.

54.5 Control and Operability Studies

Given that a COOP study is based upon the design of the application software for a particular system, it is obvious that the study cannot take place until its design is available. Referring to Figure 63.1, it can be seen that this is later in the design cycle than agreement on the DFS. As with CHAZOP studies, it makes sense to think in terms of carrying out COOP studies in two stages:

1. A preliminary COOP study. This is most appropriate at the software design stage, as described in Chapter 63, when strategic design

Table 54.5 Interpretation of guidewords for COOP

Guideword	Meaning	Comments
ACCESS	The scope for operator intervention	The impact on the design intent of an operator making inappropriate interventions. Applies to interaction of any nature, planned or otherwise.
TIMING	Frequency of recurrent events and/or order of logical events	The potential consequences of actions (by the operator or otherwise) being (or not being) carried out before, during or after specific events, or being done in the wrong order. SOONER/LATER may be more useful in relation to absolute time or logical events and LONGER/SHORTER may be more useful in relation to elapsed time.
STRUCTURE	The parallelism and/or sequential nature of control schemes and procedures	The potential consequences of components (such as function blocks and phases) not being selected and/or configured correctly. Also concerns the impact on progression of the interfaces between components being disjointed.
CONFLICT	The scope for high level interactions of an adverse nature	The potential for conflicting decisions that are not addressed by the previous guidewords. Applies in particular to outcomes from one application package being overriden by another.

decisions are being made. That is particularly so when there is any complex sequencing as, for example, with automatic start-up or shutdown, or when there is batch processing involving recipes and procedures. For simple designs, the preliminary COOP could be combined with the full COOP study. Any changes required to the DFS as a result of the preliminary COOP must be subject to formal change control procedures.

2. A full COOP study based upon the detailed module designs and prior to software development. For hazardous plant it is appropriate that this be done instead of the module design walkthrough as advocated in Chapter 63. Any changes required at this stage should feed back into software and/or module design and be subject to formal change control procedures.

When applying COOP the same terms are used as in CHAZOP but they take on different meanings. Thus:

- *Intention* relates to the execution of some *function* (configurable or procedural) on relevant signals (input, output, internal, status, *etc.*) in response to events and/or actions that are either systematic (sample updates, time ordered, logic based, *etc.*) or operator initiated.
- *Guidewords* that are more appropriate for COOP are listed in Table 54.5.
- *Deviations* are outcomes of the execution that are inconsistent with (or inappropriate to) the requirements of the DFS and/or good practice.
- *Causes* are those combinations of events and/or actions that result in deviations, the *consequences* being outcomes that could lead to operability difficulties and/or hazardous situations.
- *Recommendations* are changes to the design of the application software and/or operating procedures upon which *actions* are to be taken.

The basic methodology for COOP is similar to that outlined in Figure 54.1, except that rather than being based on the vessels and pipes, the focus is on the design of the application software. In turn, consider each control scheme and procedure to establish whether it uses and/or generates any safety related signal. Clearly any signal identified in the CHAZOP study for the system hardware as being safety related is a candidate for COOP. Then:

- Review the functionality of every such control scheme and procedure.

Table 54.6 Examples of the use of COOP guidewords

Keyword	Type of deviation	Typical problems
Access to	Change of mode	Function put in wrong mode of operation: DISABLED, START-UP, STAND-BY, OFF-LINE, OFF-SCAN, *etc.* Applies to system as whole as well as to components such as function blocks and phases.
	Loop status	Loop's AUTO/MAN status changed by mistake. Output in MAN too high/low. No limits on set point in AUTO. Functions not disabled.
	Sequence status	Can change status anywhere in sequence: START, STOP, READY, HOLD, *etc.* Operator access not inhibited.
	Declaration level	Variables declared globally that only need local access, and vice versa.
	Decision logic	No constraints on operator overrides on sequence logic. Need to override operator decisions.
	Application software	Authorisation mechanisms for access and/or change control not in place.
Timing of	Events (before/after)	Loops not initialised before switched into AUTO. No bumpless transfer or set point tracking. Wrong status/output on power-up and/or shut-down.
	Actions (during)	Loop settings, constraints and status not same throughout sequence. Operator interventions not safe at all stages. Alarm settings fixed.
Structure of	Function type	Components used having the wrong functionality for the application. Embraces both wrong choice of functions as well as incorrect attributes.
	Control schemes	Function blocks configured incorrectly. Switching between configurations not bumpless. Criteria ambiguous. Loop status not consistent with sequence logic.
	Procedures (sequences)	Phases in wrong order. Phases don't end in safe hold-states. Sequence restart not operable. Branching by-passes safety functions. Parallel phases not synchronised. Recipe parameters in sequence rather than lists.
	Commentary	Lack of comment. Difficult to follow. Intent not clear.
Conflict	Interactions	Different packages simultaneously changing the same variable. Optimiser relaxing constraints. Predictive controllers changing outputs. Statistics packages adjusting input values. High level programs overriding operator interventions.

- Identify what software components (*e.g.* function blocks, phases, *etc.*) operate upon which safety related signals.
- Note the order in which the components are arranged, especially the criteria for initiating and ending parallel activities.
- Apply the guidewords as appropriate. Particular attention should be paid to the role of the operator and the scope for inappropriate intervention. Searching questions should be asked, such as:
  i. What should happen?
  ii. Does it matter?
  iii. How will the operator know?
  iv. What is the operator expected to do?
  v. Can the failure propagate?
  vi. What is the operator not permitted to do?
  vii. Are any changes needed?

Some typical examples of the use of these COOP guidewords are given in Table 54.6. Further inspiration can be sought from Lists 64.1 and 64.2 at the end of Chapter 64 on commissioning.

It is the design of the application software that is subject to COOP rather than the software itself.

Design of application software, both configurable and procedural, is discussed in detail in Chapter 63. The design upon which the COOP is based is articulated by various documents, the key ones being as listed below. As with CHAZOP they will be in various states of completeness according to the stage in the design cycle at which the COOP is done: the more detailed the better:

- All of the documentation from the HAZOP and CHAZOP studies
- Copies of the URS and DFS (see Chapters 60 and 62)
- P&I diagrams
- Description of configurable software design, *e.g.* configuration charts, function block diagrams (FBD), database listings, *etc.*
- Description of procedural software design, *e.g.* sequential function charts (SFC) and sequence flow diagrams (SFD), recipe parameter lists, *etc.*

The COOP team is likely to be comprised of process and control engineers, with support from specialists such as software engineers as appropriate. Again, it is important that there be some cross membership with the HAZOP and CHAZOP teams to provide continuity of thought. For turnkey contracts, the COOP team members must be drawn from both the end-user and supplier.

The approach to COOP is much the same as for HAZOP, but the methodology of Figure 54.1 needs to be adapted. An obvious adaptation for visualising and keeping track of progress, in the absence of packages and tools for the task, is to colour in the signals, function blocks, logic, *etc.* on hard copy of the FBDs, SFCs and SFDs as each function is dealt with.

Conventional wisdom is that operability issues related to application software are best handled by means of walkthroughs at the design stage and by subsequent testing procedures. However, this is critically dependent on the quality of the detailed functional specifications (DFS) and, in any case, is not structured. The place for COOP studies is as an integral part of the development of the DFS, the outcomes of the COOP study being formally incorporated in the DFS itself.

Many companies simply don't bother with any form of COOP or equivalent study. However, given the contribution to safety through the reduction in demand rate (DR), there is growing acceptance of the need for COOP studies. Much work needs to be done here, both in developing the methodology and the framework of keywords.

There is virtually nothing in the public domain about COOP studies. MOD 00-58 is a defence standard on the use of HAZOP studies for systems which include programmable electronics, although the emphasis here is on information systems rather than control systems.

54.6 Failure Mode and Effect Analysis (FMEA)

FMEA is a bottom up search. It has an inductive approach, the objective being to determine the effects of each failure mode of each component on the rest of the system. The methodology for performing an FMEA is depicted in Figure 54.2.

The starting point in FMEA is the system description. Block diagrams of the system configuration are one such representation. Analysis of the system enables preparation of a table of components, their modes of failure, and the effects of each component's modes of failure. To assure a systematic and thorough coverage of all failure modes, the information is usually arranged in a tabular format. The table consists of at least three columns, one for the component name and its reference number, a second for the failure modes and the third for the effects. Other columns might include failure detection method, corrective action and criticality.

The effects of each failure mode are usually categorised according to the following classification:

1. Safe: no major degradation of performance, no equipment damage or personnel injury.
2. Marginal: degradation of performance which can be countered or controlled without major damage or injury to personnel.
3. Critical: degradation of performance, damage to equipment, or hazard requiring immediate

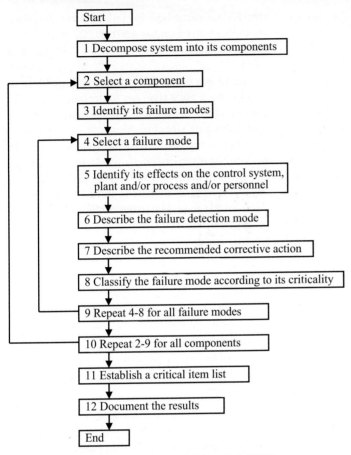

Fig. 54.2 Methodology for failure mode and effect analysis (FMEA)

corrective action for personnel or equipment survival.

4. Catastrophic: severe degradation of performance with subsequent equipment loss and/or death or multiple injuries to personnel.

The penultimate step in the performance of FMEA is the preparation of the critical item list. This generally includes those failure modes categorised as being either critical or catastrophic.

FMEA can be performed using either a hardware or functionally orientated approach. The hardware approach considers the failure modes of identifiable items of hardware such as sensors, actuators, interface devices, I/O cards, serial links, *etc*. As an example of the hardware approach, the analysis of a solenoid actuated pilot valve would consider the effects of the valve being stuck in the open and closed states.

By way of contrast, the functional approach considers the failure modes of sub-systems such as a control loop, trip or interlock. For example, if the function of a trip is to close an isolating valve when the temperature exceeds a certain value, then some of its failure modes are:

- Trip operates at a higher than specified temperature.
- Trip operates at a lower than specified temperature.
- Pilot valve sticks permanently open.
- Pilot valve sticks permanently shut.

- Isolating valve fails open.
- Isolating valve fails shut.

54.7 Fault Tree Analysis (FTA)

Fault tree analysis is a top down approach. Various failure modes will have been identified by HAZOP, or otherwise, as being critical and a fault tree is generated for each. Thus one of the failure modes is deemed to be the "top event" under which the branches of the tree are developed. By a process of deduction, the possible causes of the top event are identified. Each of these causes is then decomposed into lower level causes and so the tree is developed until a set of "basic events" is established. The combination of causes is handled by means of AND and OR gates.

Figure 54.3 depicts a tank with level control and trip systems. The control system consists of a level transmitter, controller and an air to open valve. The trip system consists of a normally closed level switch, a trip amplifier and a heavy duty relay. The corresponding fault tree is depicted in Figure 54.4, the top event being an overflow from the tank.

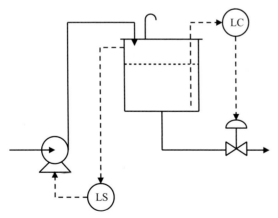

Fig. 54.3 Tank with level control and trip systems

The structure of the tree is self explanatory. The failure mode and intermediate conditions are depicted by rectangles and basic events by diamonds and circles. Events depicted by diamonds are characterised by probabilities. Events depicted by circles are characterised by failure rates. Standard symbols are used for the AND and OR gates. There are seven gates and twelve events numbered from E1 to E12.

There is a certain amount of skill involved in developing a fault tree. An understanding of the cause and effect relationships is fundamental to establishing its structure. Pragmatism is required in determining the size of the tree: for example, in deciding upon external boundaries such as failure of power, air and water supplies. Judgement about what events to include and when to stop decomposing is essential in developing the branches. Note that in Figure 54.4:

- Both events E3 (d.c. power supply failure) and E9 (control valve fails shut) are included twice.
- The manual actions of switching the controller between AUTO and MAN are undeveloped.
- All failures are worst case scenarios, *e.g.* there is no distinction between the control valve failing stuck or closed.
- Loss of mains supply is not included as an event because the pump would stop and no overflow can occur.

It is worth noting that, for analysis purposes, it is the basic events for which probability data is required. There is no point in decomposing branches beyond the level for which there is data available.

54.8 Minimum Cut Sets

A cut set is a combination of basic events which, if they all occur simultaneously, will cause the top event to occur. The minimum cut set (MCS) are those cut sets that are sufficient to account for every possible cause of the top event occurring. The basic approach is to establish all the possible cut sets: OR gates give rise to additional cut sets, AND gates make the cut sets more complex. Then the number of cut sets is reduced to the minimum by using the absorption and idempotence rules of Boolean algebra. Once the MCS are evaluated the fault tree may be quantified. This is best illustrated by means of an example such as the fault tree of Figure 54.5.

54.8 Minimum Cut Sets

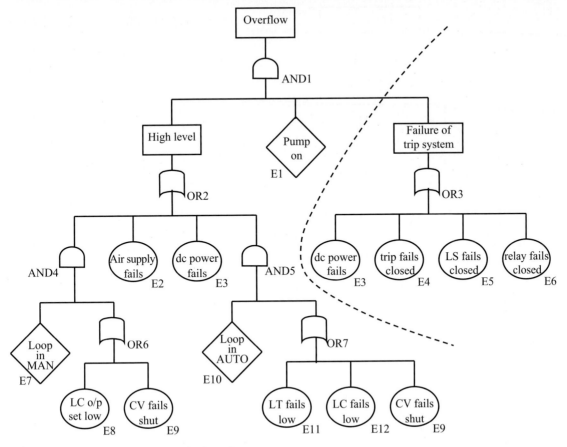

Fig. 54.4 Fault tree analysis for top event of tank overflow

The analysis begins with the OR1 gate immediately below the top event whose three inputs may be listed as separate rows:

AND1
OR2
AND2

Since the gate AND1=OR3 and AND3, the first row may be expanded:

OR3, AND3
OR2
AND2

Expansion of OR3=E1 or E2 leads to

E1, AND3
E2, AND3

OR2
AND2

Substituting for AND3=E2 and E3 gives

E1, E2, E3
E2, E2, E3
OR2
AND2

Continuing in this fashion produces a table of five cut sets:

E1, E2, E3
E2, E2, E3
E1, E3
E2, E4, E5
E1, E2, E4, E5, E5

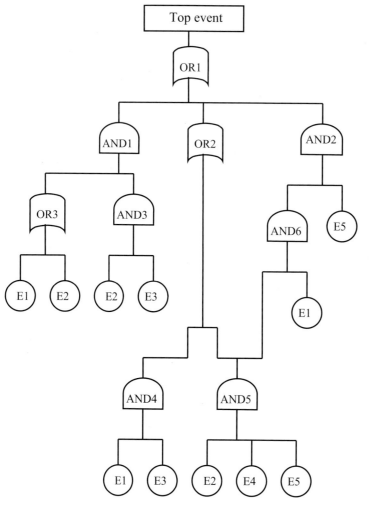

Fig. 54.5 Fault tree for minimum cut set (MCS) analysis

The second cut set may be reduced to E2, E3 by deleting the redundant E2 by means of the so-called idempotence rule. It is sufficient that these two basic events occur to cause an occurrence of the top event. Similarly an E5 is redundant in the fifth cut set:

E1, E2, E3
E2, E3
E1, E3
E2, E4, E5
E1, E2, E4, E5

Next the first cut set may be eliminated by the so-called absorption rule. That is because the first set contains the second set and, provided events E2 and E3 occur, the top event will occur irrespective of event E1. By the same argument the fifth cut set contains the fourth and may also be eliminated. The remaining three cut sets cannot be reduced further so the minimum cut set is

List 54.1 Minimum cut set for fault tree

E2, E3
E1, E3
E2, E4, E5

Thus the MCS consists of three combinations of basic events. Note that this approach is restricted to trees consisting of AND and OR gates only. There is no guarantee that the approach will produce the MCS for trees which contain the NOT operator, or in which NOT is implied by the exclusive OR operator. However, it is usually possible to structure the design of a fault tree to avoid the use of NOT gates. The same MCS as List 54.1 could have been established by a bottom up approach. Thus basic events are combined by lower level gates whose outputs are combined by intermediate level gates, and so on.

The system of Figure 54.5 was relatively simple to analyse, as indeed is that of Figure 54.4 whose MCS has the nineteen cut sets listed in List 54.2:

List 54.2 Minimum cut set for fault tree for tank overflow

E1, E2, E4,
E1, E2, E5,
E1, E2, E6,
E1, E3,
E1, E4, E7, E8,
E1, E4, E7, E9,
E1, E4, E9, E10,
E1, E4, E10, E11,
E1, E4, E10, E12,
E1, E5, E7, E8,
E1, E5, E7, E9,
E1, E5, E9, E10,
E1, E5, E10, E11,
E1, E5, E10, E12,
E1, E6, E7, E8,
E1, E6, E7, E9,
E1, E6, E9, E10,
E1, E6, E10, E11,
E1, E6, E10, E12.

However, systems are often more complex and/or extensive resulting in much larger fault trees: it is not uncommon to have several thousand cut sets with individual sets containing dozens of basic events. Truncation is used to keep the task manageable. This involves eliminating cut sets that are believed to contribute negligibly to the top event. Truncation may be based on size: cut sets whose length exceeds a specified value are discarded. Alternatively, it may be on the basis of probability: cut sets with a probability of occurrence less than some specified cut-off value are discarded.

There are packages available for handling fault tree synthesis and analysis. In essence, an environment of constructs and entities with attributes enables synthesis of the fault tree. Whereas synthesis is essentially a manual process, tools are provided for analysis which automatically generate the cut sets, handle the truncation, identify redundant events and eliminate cut sets as appropriate. These tools also enable quantitative evaluation of fault trees.

54.9 Fault Tree Evaluation

Of particular interest in the analysis of fault trees is the probability of some hazardous top event occurring based upon the probabilities of the bottom events. Because fault trees of real systems invariably contain repeated basic events, fault tree evaluation is always based upon the MCS. Consider the MCS of List 54.1. The top event can be articulated as a Boolean expression, the + signs corresponding to OR gates and the multiplications to AND gates:

$$\text{Top} = E1.E3 + E2.E3 + E2.E4.E5 \quad (54.1)$$

the probability (P) of which may be articulated thus:

$$P(\text{Top}) = P(E1.E3 + E2.E3 + E2.E4.E5) \quad (54.2)$$

The underlying logic is straightforward. The cut sets are each capable of causing the top event, so their contribution to the probability of the top event occurring is cumulative. However, for each cut set to cause the top event, all of its basic events must occur simultaneously. The probability of each combination thus relates to the product of the probability of the individual basic events occurring:

$$P(\text{Top}) \approx P(E1.E3) + P(E2.E3) + P(E2.E4.E5) \quad (54.3)$$

Each cut set contributes a term to the value of P(Top) which can be thought of as first term combinations. However, they are not mutually exclusive. For example, E1, E2 and E3 could occur simultaneously which would satisfy both of the first cut sets. This is referred to as a second term combination. In fact there are three second term combinations and, strictly speaking, their contribution to P(Top) must be taken into account. Similarly, a third term combination of all five basic events exists. For accuracy, Equation 54.3 is modified thus:

$$\begin{aligned} P(Top) = {} & P(E1.E3) + P(E2.E3) + P(E2.E4.E5) \\ & - \bigl(P(E1.E2.E3) + P(E2.E3.E4.E5) \\ & + P(E1.E2.E3.E4.E5)\bigr) \\ & + P(E1.E2.E3.E4.E5) \end{aligned} \quad (54.4)$$

Note that in Equation 54.4, by chance, the last of the second terms cancels out with the third term. The length and complexity of the expansion escalates exponentially with the number of cut sets. If, for example, there had been four first terms, then there would have been six second terms, fifteen third terms and one fourth term.

Inspection of Equation 54.4 reveals that the contribution of the first terms to the top event occurring are numerically more significant than the second, and the second are more significant than the third. Not only do the terms become less significant as the expansion increases in length, they also alternate in sign. For long expansions, evaluation of the series adds the contribution of successive odd numbered terms and subtracts that of even numbered terms. Thus, the series always converges. Truncating the series after an odd numbered term will give an upper bound on the exact value of P(Top), truncation after an even numbered term gives a lower bound.

It is evident that if the probability of the basic events is small, the first terms will dominate the value of P(Top) and the approximation of Equation 54.3 will be good enough. This is known as the "rare event approximation" and is always a conservative estimate of the full expansion.

54.10 Evaluation of Demand Rate

The probability of the top event occurring may be articulated in different ways depending upon how the basic events are quantified. For example, for repairable systems, basic events may be in terms of unavailability (U) or probability of failure on demand (PFD). In the design of protection systems it is common practice to use a fault tree to establish the unmitigated demand rate (DR) for use in Equation 53.23. Thus the top event is a demand on the protection system and is articulated as a frequency (events/year).

Consider again the tank with level control and trip systems of Figure 54.3. Since it is the unmitigated demand rate that is required, the fault tree must be based upon the plant and its control system only. Thus the right hand branch of Figure 54.4, corresponding to the trip system, must be excluded from the fault tree as indicated by the broken line. Fault tree analysis of the remainder of the tree yields the following MCS (List 54.3):

List 54.3 Fault tree analysis for tank overflow excluding trip system

```
E1, E2
E1, E3
E1, E7, E8
E1, E7, E9
E1, E9, E10
E1, E10, E11
E1, E10, E12
```

Assuming the rare event approximation, the probability of the top event occurring is given by

$$\begin{aligned} P(Top) \approx {} & P(E1.E2) + P(E1.E3) + P(E1.E7.E8) \\ & + P(E1.E7.E9) + P(E1.E9.E10) \\ & + P(E1.E10.E11) + P(E1.E10.E12) \end{aligned} \quad (54.5)$$

Remember that the + signs correspond to OR gates: the contributions of the cut sets on the right hand side of Equation 54.5 to the top event is cumulative. In effect, the cut sets are in series. Thus, in terms of failure rates, the number of top events per year relates to the sum of the failure rates of the cut sets, as *per* Equation 53.14.

Inspection of the fault tree reveals that events E1, E7 and E10 may all be articulated as probabilities: they are dimensionless with values in the range $0 \leq P \leq 1.0$. They are established, typically, by analysis of historical data: some, such as E1 and E10, may have values close to 1.0 and others, such as E7, will be close to zero. All the other basic events may be quantified as failure rates (failures/year). Thus:

$$\begin{aligned}\lambda_{Top} \approx\ & P(E1).\lambda_{E2} + P(E1).\lambda_{E3} \\ & + P(E1).P(E7).\lambda_{E8} \\ & + P(E1).P(E7).\lambda_{E9} \\ & + P(E1).P(E10).\lambda_{E9} \\ & + P(E1).P(E10).\lambda_{E11} \\ & + P(E1).P(E10).\lambda_{E12}\end{aligned} \quad (54.6)$$

All of the terms on the right hand side of Equation 54.6 have units of failures/year, their combination resulting in the top event frequency with units of events/year. The fact that each of these terms results from the product of one or more probabilities and a failure rate is dimensionally consistent. It also provides confidence in the correctness of the fault tree:

- A cut set resulting in the product of probabilities only suggests that an AND gate is missing.
- A cut set resulting in the product of two or more failure rates suggests that one or more OR gates are missing.

Layers of Safety

Chapter 55

55.1 Concept of Layers
55.2 Passive Systems Layer
55.3 Active Systems Layer
55.4 Control Systems Layer
55.5 The HSE Guidelines
55.6 The EEMUA Guidelines
55.7 Comments

Companies have a general responsibility to their employees and others for safety. In the UK this is covered by the Health and Safety at Work Act (1974) and responsibility for monitoring its implementation is vested in the Health and Safety Executive (HSE). In particular, there is a duty for

> "the provision and maintenance of plant and systems of work that are, so far as is reasonably practicable, safe and without risks to health."

Safety, therefore, is fundamental to the design and operation of process plant and pervades all operability and viability considerations. Designs must ensure that all plant is as safe as is reasonably practicable under all normal and most abnormal conditions. In the event of a hazardous incident occurring, for which it can be proven that not all reasonable safety measures were taken, any company involved in the design, manufacture, installation or operation of the plant and/or its safety systems is potentially liable to prosecution. So too are the individuals involved: they cannot hide behind the company.

Plant designs which are inherently safe are best, they do not need additional protection. However, in practice, few plants are inherently safe and it is necessary to enhance their safety by the installation of protection systems. Indeed, for existing plants whose design cannot be changed, it is often the case that additional protection systems are the only practicable way of enhancing safety.

Protection systems are designed on the basis of reducing, to an acceptable level, the probability of some hazardous event occurring. Normally the design is driven by safety considerations: that is, the consequences of the hazardous event in terms of risk to life and limb. However, depending upon the nature of the process, such a system may well also provide protection against environmental damage, say in terms of toxic release or pollutant discharge, protection against plant damage, or indeed protection against lost production. It is also true to say that protection systems designed on the basis of these other criteria will probably reduce the risk to life and limb too.

This chapter considers the contribution to safety made by automation and the relationship between plant design, protection systems and process control issues. Reference is made to industrial codes and standards as appropriate. A good introduction to the subject is given by Kletz (1995) but for a more comprehensive treatment the reader is referred to the text by Lees (2005).

55.1 Concept of Layers

Hazard assessment, as described in Chapter 54, provides a basis for deciding upon the appropriate preventive and protective measures to be incorporated in the design of the plant. There are many means of protection which may be categorised, on a broad brush basis, into three concentric layers of passive, active and control systems as depicted in Figure 55.1.

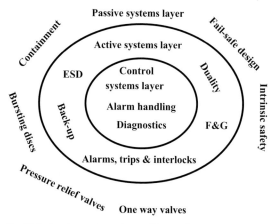

Fig. 55.1 Passive, active and control layers of protection

The concept of layers of safety is a proven approach to protection system design. The outer layer comprises a combination of safe plant design measures and civil/mechanical means. Protection devices are passive in the sense that they are mechanical, fail-safe and require no power supply. Ultimately, if everything else fails, it is this outer layer that provides the final protection to personnel, plant and the environment. Clearly, from a design point of view, the objective is to never have to exercise this layer.

Protection defaults to the outer layer only if the active systems of the middle layer fail. These protection systems are of an active nature in the sense that, typically, they require a power supply to function. Similarly, protection defaults to the middle layer only if the basic process control system (BPCS) of the inner layer fails, or fails to cope. Under all normal and most abnormal circumstances, plant operation should be maintained by the control system within the inner layer.

As stated in Chapter 54, basic process control systems are not protection systems. In the event of the failure of a control system, it cannot itself be relied upon to provide any means of protection. For this reason HAZOP applies to the outer and middle layers of protection only. If the inner layer is being relied upon to provide protection, the design is faulty.

Nevertheless, control systems make a significant contribution to plant safety to the extent that effective control, alarm management, and so on, reduce the demand on the protection systems. To prevent the protection systems being exercised unnecessarily, CHAZOP and COOP studies should be carried out on the design of the control system and the functionality of its application software. It follows that CHAZOP and COOP studies only apply to the inner layer.

55.2 Passive Systems Layer

Safe plant design is essentially a chemical engineering issue. The contribution of control engineering personnel to the outer layer of safety is through participation in the HAZOP study team, in the specification of any passive devices used and in their subsequent testing and/or maintenance. Protection at this layer is realised by means of:

1. Inherent safety. Whenever options exist, select items of plant which are inherently safe or choose process routes which minimise the extent of hazard. For example, stationary plant is safer than rotating equipment so use filters rather than centrifuges. Avoid solvent based reactions if aqueous routes are viable. Use non-toxic chemicals if there alternatives available.
2. Minimum inventory. Choose operations that minimise the inventory of hazardous materials. In general, continuous operations have lower inventories than batch. Avoid the use of buffer tanks between continuous operations.
3. Containment systems. Provide facilities such as quench tanks, bund walls and segregated drains to contain reagents in the event of unscheduled

releases. Use flame traps to prevent the propagation of fire.
4. Passive devices. Use bursting discs, pressure relief valves and vacuum breakers to protect plant against over and under pressures. Install one-way valves to prevent reverse flows. Fit over-speed bolts on rotating machinery. These are passive in the sense that they are all mechanical and not dependant upon a power supply.
5. Fail-safe design. For example, specify the action of pneumatically actuated valves such that in the event of air failure they fail-safe. Typically, a cooling water valve will be required to fail open whereas a steam valve will be required to fail closed.
6. Intrinsic safety. Specify instrumentation that does not in itself constitute a hazard. If this is not possible then use enclosures as appropriate. This is discussed in detail in Chapter 52.

55.3 Active Systems Layer

The middle layer is comprised of independent active protection systems and augments the passive features of the outer layer. These systems are active in the sense that they require a power supply to function. Clearly there is scope for either the protection system or its power supply to fail, in which case safety provision defaults to the outer layer. In safety critical applications, active protection systems must be designed to guarantee some minimum reliability criteria: this is discussed in detail in Chapter 56.

The middle layer contains the following means of protection:

7. Alarms, trips and interlocks. These are closely related and it is convenient to treat them together. They are by far the most common type of active system and it makes sense to define them formally:

An alarm indicates to an operator that some abnormal condition or event has occurred and that some action may be required. It has the following syntax:

"If *some condition becomes true or false* then *annunciate or display the change in status*".

There are normally two ways in which an alarm can be activated. Either some analogue input signal, such as a level measurement, has reached a threshold value or because a discrete signal, such as that from a level switch, has changed its status. The outcome is usually an audio and/or visual signal to the operator.

A trip automatically takes some action due to the occurrence of an alarm condition. It has the following syntax:

"If *some alarm condition occurs* then *change the status of some output signal*".

The change in output is normally associated with a discrete signal. Typically, if a tank is full then an isolating valve in the inlet will be closed or a pump in the outlet started.

An interlock is generally used for prevention purposes and is, in effect, the converse of a trip. The syntax of an interlock is of the following form:

"Unless (*or while*) *some condition occurs (exists)* do not *change the status of some output signal*".

A typical example with machinery would be: unless the guard is in place, do not enable the start button. A more typical process example would be in charging a vessel: while the inlet valve is open, do not open the drain valve or close the vent valve.

8. Emergency shut-down (ESD) systems. Under certain circumstances, it may be necessary to automatically shut down a plant. This is invariably triggered by some prescribed critical condition or combination of conditions. The syntax is typically:

"If *some critical condition occurs* then *force an emergency shut down*".

There are various shut-down strategies. One simple but effective strategy is to switch off a common power supply to a number of output channels grouped together for that purpose. Thus, in effect, all those channels are forced into a fail-safe condition consistent with the shut-down mode of the plant. Another strategy is to shut down the plant by manipulating the systems' outputs in a time ordered sequence. Sequence control was introduced in Chapter 29.

9. Fire and gas detection systems (F&G). These are commonly used on off-shore gas and oil installations. Typically, an F&G system consists of sensors to detect gas leaks or the presence of fire. F&G systems do not necessarily have shut-down capability in themselves but are invariably connected up to ESD systems. The syntax is thus:

> "If *a gas leak or a fire is detected* then *force an emergency shut down*".

10. Dual systems. A common approach to increasing the reliability of active safety systems is to provide duality. Typically, critical instruments may be duplicated, or even triplicated, with some means of cross-checking or polling of the measurement. In extreme cases whole systems may be duplicated.

11. Back-up systems. An active protection system cannot function in the event of failure of its own power supply or of any of the utilities that it is required to manipulate. Sometimes it may be necessary to provide back-ups. For example, mains power supply may be backed up by diesel generators which switch on automatically when the power fails. A head tank of cooling water may be installed as a back-up for failure of the pumps on the works cooling water supply main.

It is good practice to separate the active protection systems from the control systems. Thus each uses different sensors and actuators, and their signals are carefully segregated with colour coded cabling. Active protection systems usually utilise discrete signals, typically inputs from switches, logic elements, and outputs to relays. It is normal practice for these elements and signals to be individually hard wired.

55.4 Control Systems Layer

The inner layer consists of the plant's control systems. These contribute to safety in various ways:

12. Control schemes. Effective control schemes and strategies mean that processes are normally under control. Thus the active protection systems are only called upon to function in the event of an incident such as a failure of the control system or a mistake in operation. Maintaining control at the inner layer, and not exercising the middle and outer layers unnecessarily, represents an enhancement in safety.
13. Integrated alarm environment. Supported as standard by all modern control systems, DCS or otherwise, integrated alarm environments make a major contribution to the safe operation of process plants. These are discussed in detail in Chapter 43.
14. Application diagnostics. These are sequences which monitor complex situations and, depending upon circumstances, initiate either some recovery option or an emergency shut down.
15. Recovery options. These are normally associated with complex batch processes. They are, in effect, sequences that are activated under prescribed circumstances which are designed to retrieve abnormal process situations and return the plant into a safe hold state.

55.5 The HSE Guidelines

Traditionally, all active protection systems of an electrical nature were comprised of distinct elements such as sensors, switches, relays and actuators using analogue and/or discrete signals. In particular they were characterised as being hard-wired. The advent of microprocessor based devices and systems led to pressure to permit software based protection systems. To address the vacuum in standards and codes of practice, the HSE Guidelines (1987) on the use of programmable electronic systems (PES) in safety related applications were produced.

What the Guidelines said, in essence, was that under certain circumstances, and subject to various important constraints, it is acceptable for active safety functions to be realised by means of software based systems such as PLCs instead of hard-wired systems. The criterion used was to provide levels of protection that were at least as good as what would be provided by conventional hard-wired systems. In that sense, the Guidelines were both constructive and forward looking and were a landmark in plant protection.

The Guidelines were largely qualitative, their emphasis being on the approach to the provision of protection and on sound engineering practice. Although they were prescriptive with regard to methodology, focussing on the design process, they nevertheless provided a good degree of flexibility with regard to implementation.

The fundamental question to be asked was whether a protection system fell within the scope of the Guidelines or not. Any active protection system that was hard-wired, using conventional analogue and/or discrete elements, and was separate from the control systems, fell outside the scope of the Guidelines. Many companies had, and still have, a policy that active protection is only ever provided by such hard-wired systems.

An important precedent was set by the guidelines in relation to the use of software in instrumentation. If any part of a protection system was user programmable, by configuration or otherwise, then the HSE Guidelines applied. Thus, for example, most single loop controllers and any PLC fell within the scope of the Guidelines. However, if the electronics were embedded within a device such that, to all practical intents and purposes, the software was inaccessible to the user, then the device was not covered by the Guidelines. An example of this is the intelligent dp cell in which the embedded ROM based software is treated as if it was hardware.

The status of the Guidelines was slightly ambiguous. They were neither a formal standard nor a code of practice. They may as well have been though given that they were published by the HSE, the regulatory body, and represented what it considered to be good practice.

55.6 The EEMUA Guidelines

The HSE Guidelines were generic in that they were applicable to all sectors of industry. Second tier guidance specific to the process industry was developed and published by EEMUA (1989). It should be noted that the principles defined were not peculiar to PES, they were just as valid for all classes of protection systems. The concept of layers of safety is the basis of the EEMUA Guidelines, as depicted in Table 55.1.

The distinction between the different categories is essentially on the basis of the consequences of failure.

Category 0 systems are the passive devices referred to in Section 55.2 above. Failure of a Category 0 system results in risk to life and limb.

Table 55.1 EEMUA vs IEC 61508/61511 categories of protection

Figure 55.1 layers	EEMUA category	EEMUA consequences	EEMUA divisions	Comment	IEC 61508
Passive	0	Risk to life and limb			ERRF
Active	1	Risk to life and limb		ESD, HIPS	
	2	Damage to plant	2 (a)	PSD, HIPS	SRS (SIS)
		Lost production	2 (b)	PSD	
Control	3	Off spec product, loss of efficiency, *etc.*		DCS, SCADA	BPCS

Category 1 systems are instrument based systems equivalent to the passive devices of Category 0. These systems are only required when Category 0 protection cannot meet the safety requirements, such as when dynamics are involved. For example, it is better to anticipate shut-down of an exothermic reaction by monitoring the rate of change of temperature than waiting to vent the reactor by means of a pressure relief valve. Category 1 systems are also referred to as emergency shut-down (ESD) systems.

Category 1 protection is best applied on a per potential hazard basis, such as over heating of a reactor, over-pressure of a vessel or over-speed of a compressor. Keeping protection systems entirely separate from each other, as well as from the control systems, has the advantage of simplifying system design and facilitates the use of FMEA and FTA. It also minimises the scope for common mode failure. Category 1 protection systems should react by removing the primary cause such as heat or power. ESDs of this type represent a "narrow" view of the plant: secondary effects are ignored.

Category 1 protection systems must have a lower probability of failure to meet a demand than Category 2 or 3 systems. This lower PFD is obtained by using special purpose equipment, described in Chapter 57, and regular manual verification (proof testing), both of which make Category 1 systems expensive to purchase and operate. For this reason, instances of Category 1 protection should be kept to a minimum.

Category 2 systems are also referred to as process shut-down (PSD) systems. If an abnormal situation occurs which the control system cannot handle, the PSD shuts down the plant area automatically, *i.e.* without operator intervention. It may also shut down associated plant in other areas to affect an orderly shut-down rather than shut-down by the domino effect. This minimises the risk of damage to plant and simplifies subsequent start-up. PSDs represent a "plant-wide" view of protection. The worst case scenario for failure by a PSD is of a financial nature through damage to plant and/or lost production. Failure of a PSD results in a demand on the Category 1 systems or Category 0 devices.

In many processes there will be protection systems which are critical from a production point of view but, not being responsible for life and limb protection, are Category 2. For such systems the end-user may wish to provide protection using the same technologies and procedures as for Category 1. These applications are sometimes referred to as Category 2(a) and the remainder as Category 2(b), the split being application dependant. When the same technology is employed, some end-users lump Categories 1 and 2(a) together and refer to them as high integrity protection systems (HIPS).

Category 3 systems are typically DCS, SCADA or PLC systems used for maintaining the plant under normal operating conditions, for handling many abnormal conditions, and for supporting an integrated alarm environment. The worst case scenario for failure of a basic process control system (BPCS) is of a process nature such as off-spec product, loss of efficiency, *etc.* Failure of a BPCS results in a demand on a Category 2 system.

Whilst Category 2 and 3 systems are not mandatory, if they are used then care must be taken in giving them credit for any reduction in demand on the Category 0/1 devices and/or systems. Such credit could lead to an expensive and time consuming third party validation of the Category 2/3 systems, plus additional in-service costs associated with maintenance and modification to ensure that the validation is not compromised.

In the context of the IEC 61508 and 61511 standards, Category 0 systems are equivalent to the so-called external risk reduction facilities (ERRF) and Category 1 and 2 systems are equivalent to safety related systems (SRS). The EEMUA hierarchy of categories is consistent with the safety layer model of the ISA S84 standard in which Category 1 and 2 systems are referred to as safety instrumented systems (SIS).

Also, the EEMUA layers fit well with API RP 14C (1998) which requires that there should be two levels of protection "independent of and in addition to the control devices used in normal process operation".

55.7 Comments

The HSE and EEMUA Guidelines have been superseded by the IEC 61508 standard. It is nevertheless important to understand them because:

1. Much of the thinking behind the HSE and EEMUA Guidelines underpins the concepts and terminology of IEC 61508.
2. There is a large installed base of protection systems that have been designed, implemented, validated and documented on the HSE and/or EEMUA basis which will be in use for decades to come.

IEC 61508 is a generic standard in that it applies to all aspects of safety for all sectors of industry. As such it is extensive: there are seven sections of which Parts 1 to 3 are normative (mandatory) and Parts 4 to 7 are advisory. Second tier guidance is provided on a sector basis. That for the process sector, IEC 61511, specifically concerns instrumentation aspects of safety systems for the process industries. It is in three parts of which Part1 is normative. Second tier guidance for the nuclear sector is IEC 61513.

In the US the standard which is equivalent to IEC 61511 is ISA S84 which is itself complemented by ISA TR84.

A fundamental difference between the HSE and EEMUA Guidelines and the IEC/ISA standards is that targets have been set in the latter for the PFD of the various grades of protection system. These targets, known as safety integrity levels (SIL), are quantitative and represent a major step forward in system design. In effect, for the first time, acceptable levels of unreliability have been articulated on an authoritative basis. SILs are discussed in detail in Chapter 56.

Protection Systems

Chapter 56

56.1 Risk
56.2 ALARP
56.3 Safety Methodology
56.4 Safety Integrity Levels
56.5 Worked Example on Quantitative Approach to SIL
56.6 Qualitative Approach to SIL
56.7 Worked Example on Qualitative Approach to SIL
56.8 Architectural Constraints
56.9 Categories of Failure
56.10 Protection System Design
56.11 Independence
56.12 Comments

This chapter is concerned with the design of protection systems for plant in which hazardous processes are carried out and where the potential consequences of specified incidents are serious. It starts with consideration of the nature of risk and introduces the concept of ALARP. Then comes the ISA S84 life cycle model of a protection system. A design methodology based upon this model is explained. Next follows consideration of target levels for unreliability, the latter as quantified in both IEC 61508 and ISA S84, and discussion of their specification. These are used in two worked examples on the design of a protection system, one quantitative and the other qualitative. Next comes a categorisation of the different types of failure. This provides a basis for handling the various quality aspects of protection systems design.

56.1 Risk

Risk is an emotive and irrational thing. Consider the number of people killed on their way to/from work in a major city such as Manchester, and assume the numbers travelling by car and train to be the same. If the number of people killed in road accidents was one per week there would be little public reaction. However, if there was one rail crash per year killing 52 people per crash there would be outrage.

Also, there is the problem that some industries have traditionally accepted higher risks to life and limb than others. Deep sea fishing is one of the most dangerous industries. Even a minor accident out in the North Sea is likely to lead to serious non-recoverable injuries due to the lack of immediate access to emergency services. In addition to this, the sinking of vessels with the loss of several lives is not uncommon. If London office workers were exposed to the same levels of risk there would be more outrage.

Another factor is payment to take risks. Many people will accept higher pay than average in return for taking risks. Scaffolding workers are a common example and deep sea divers an extreme case.

These observations lead to the conclusion that there is a nonlinear relationship between acceptable risk (deaths/year), hazard rate (incidents/year) and consequences (deaths/incident). One model that has been proposed is an adaptation to Equation 53.24:

$$\text{Risk} = \text{HR} \times C(E)^n \quad (56.1)$$

where n is referred to as the aversion factor. Suggested values for the aversion factor are

If $C(E) \geq 10$ deaths/incident, then $n = 2$

If $C(E) < 10$ deaths/incident, then $n = 2/3$

This is an area that the standards and regulatory bodies have yet to address and, in the meantime, it is the ALARP principle that applies.

56.2 ALARP

The as low as reasonably practical (ALARP) concept was formalised by the HSE and is as depicted in Figure 56.1.

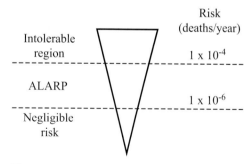

Fig. 56.1 Levels of risk and ALARP

The idea is that there exists some upper level of risk beyond which no operation should be allowed. Beneath this there is some lower level of risk which is insignificant. Between these two levels is the ALARP region. If a process has a potentially unacceptable risk, then means of reducing the risk into the ALARP region must be applied. Zero risk is not an option.

Upper and lower limits of 1×10^{-4} and 1×10^{-6} deaths per year respectively are accepted values within the process industries, although these are neither specified in IEC 61508 nor necessarily recognised by HSE.

Designers of protection systems must consider the consequences to employees and the public at large when setting ALARP limits. The concept of doubling the risk to employees paid to take a risk is acceptable in some areas. On the other hand, the concept of subjecting those not involved in the activity to a risk not greater than five times less than those employed in the activity is widely accepted.

So, if a process has the potential to cause more than 1×10^{-4} deaths per year, what are the options? First review the process design to see where safety can be increased by means of Category 0 (passive) systems, as described in Chapter 55, and referred to as external risk reduction facilities (ERRF) in IEC 61508. Good process design is the most effective route to safety. Second comes a review of operating procedures. This includes occupancy of the danger area. If there is nobody there when the hazardous event occurs then the number of fatalities and/or injuries is reduced to zero. This may seem obvious, but reducing occupancy can achieve an order of magnitude improvement in safety at little or no extra cost.

And third, if other means are insufficient or impracticable, comes the use of Category 1 (active) protection systems. These are referred to as safety related systems (SRS) within IEC 61508. The objective is to design the protection system such that the risk is brought into the ALARP region. This is essentially a question of applying the two fundamental design equations articulated in Chapter 53, in which the various terms were also defined:

$$\text{HR} = \text{DR} \times \text{PFD} \quad (53.23)$$
$$\text{Risk} = \text{HR} \times C(E) \quad (53.24)$$

The basic approach is to design a protection system such that its probability of failure on demand (PFD), when combined with the unmitigated demand rate of the plant (incidents/year) and the consequences of a hazardous event (deaths/incident) yields a residual risk (deaths/year) that is well within the ALARP region. Design invariably revolves around establishing a value for the hazard reduction factor (HRF), ambiguously referred to as risk reduction in IEC 61508:

$$\text{HRF} = \frac{1}{\text{PFD}} = \frac{\text{DR}}{\text{HR}}$$

Remember that the reliability data for plant and instrumentation is far from accurate and a significant safety margin must be built into any design.

56.3 Safety Methodology

The safety life cycle envisaged within IEC 61508, which is of a generic nature, is as depicted in Figure 56.2 and is self explanatory.

Box 9 of Figure 56.2 relates to the detailed design of electrical, electronic and programmable electronic systems (E/E/PES) for Category 1 protection. Another safety life cycle, from ISA S84 (1996), is as depicted in Figure 56.3 in which there is more of a focus upon the design of the protection system *per se*.

In many respects, all of the phases of the life cycle and their related issues are no different to those discussed in Chapters 60–66 on the project engineering and management of control systems. Note that the numbering of the phases in Figure 56.3 refers to the relevant parts of Section 4.2 of ISA S84. The essentials of the design process are as follows:

1. Hazard analysis and risk assessment (S84: 4.2.2).

    The techniques of HAZOP, CHAZOP and FMEA, as described in Chapter 54, are the most

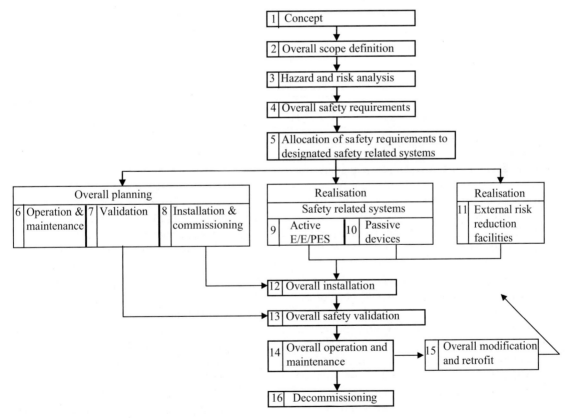

Fig. 56.2 The IEC 61508 protection system life cycle

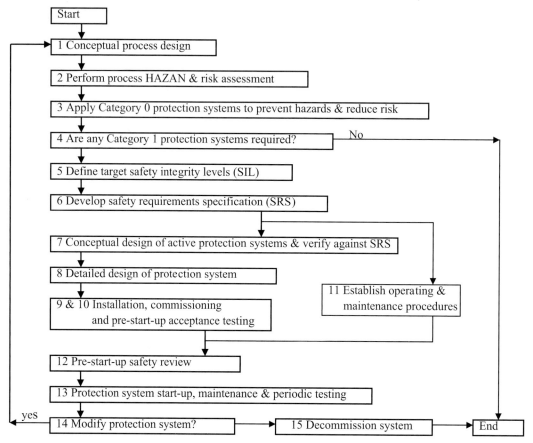

Fig. 56.3 The ISA S84 protection system life cycle

appropriate means of systematically analysing a proposed process design to identify potential hazards and to determine whether further measures are required.

Hazard analysis should be carried out early in the project life cycle to maximise its effectiveness. The later it is done the more expensive and the less satisfactory the solutions become. The unmitigated demand rate (DR) must be established for each potentially hazardous event. This is the probability of the occurrence of the hazardous event (incidents/year) in the absence of a protection system. It is normally quantified, if significant, by means of fault tree analysis (FTA), also described in Chapter 54.

2. Identify any Category 1 systems required (S84: 4.2.4).

Each potentially hazardous event that cannot be effectively handled by means of the application of Category 0 design methods and/or passive devices (ERRF) requires the installation of one or more Category 1 active systems (SRS). Remember that Category 1 protection is best applied on a per potential hazard basis.

3. Define target safety integrity levels (S84: 4.2.5). Safety integrity levels (SIL) are bands for the probability of failure on demand (PFD) of the protection system. In essence, the target SIL value for a protection system comes from an evaluation of its PFD which is calculated by working backwards from a specified acceptable level of risk (deaths/year), an estimate of the consequence $C(E)$ of the hazardous event occurring (deaths/incident) and a determina-

tion of the unmitigated demand rate (incidents/year). There are two approaches: qualitative and quantitative. Worked examples of both are provided in following sections.

4. Design protection systems against the targets (S84: 4.2.7).

   This is essentially a question of specifying structures with sufficient redundancy and equipment with high enough reliability such that the overall PFD satisfies the SIL requirement. A worked example on this is provided in Section 56.5 below.

5. Review constraints on architecture of protection system design.

   This requires an estimate of the safe failure fraction (SFF), as described in Chapter 53, for the system designed. The SFF is used to determine the SIL value that the architecture assumed is deemed to be capable of supporting in terms of fault tolerance.

   If SFF yields a lower SIL value to that used for design purposes then the SFF considerations override those of PFD. Generally speaking, SFF is a more stringent design criterion that PFD.

56.4 Safety Integrity Levels

Four grades of protection system, SIL 1 to SIL 4, are defined in both IEC 61508 and ISA S84 in terms of allowable bands for their PFD as shown in Table 56.1. A fifth band of SIL 0 is included for completeness.

There are several points about this table worth noting:

- The PFD, availability and failure rates are all for dangerous failures occurring as opposed to all (dangerous and spurious) failures.
- Normally the data for PFD and availability in demand mode of operation would be used. However, for high demand rates, when either DR > 1 or DR > 2/PTI incidents per year, then the data for continuous mode should be used.
- The SIL ranges apply to the whole protection system on an end-to-end basis, *i.e.* includes all elements from the sensor, switch, through any logic to the final relay or actuator.
- The upper bound on SIL 1 is a PFD of 10^{-1}. Thus any protection system with a hazard reduction factor (HRF) of less than 10 falls outside the scope of IEC 61508. This enables basic process control systems (BPCS), when evaluating the unmitigated demand rate (DR), to be categorised as SIL 0 and credited with an HRF of up to 10 in respect of the contribution that they make to safety.
- SIL 4 is defined in IEC 61508: the ISA S84 document only refers to SILs 1 to 3. It is true that the majority of applications in the process industries result in a SIL 2 or SIL 3 requirement but there is no reason to believe a-priori that SIL 4 requirements won't be encountered.

Clearly the SIL required for any protection system must reflect both the probability of the hazardous event occurring and the consequences of such. The

Table 56.1 Quantitative definitions of SIL levels

Safety integrity level SIL	Hazard reduction factor HRF	Demand mode of operation		Continuous mode
		PFD (fractional)	Availability A (fractional)	Failure rate λ (failures per hour)
0	> 10^0	1 to 10^{-1}	< 0.9	> 10^{-5}
1	> 10^1	10^{-1} to 10^{-2}	0.9 to 0.99	10^{-5} to 10^{-6}
2	> 10^2	10^{-2} to 10^{-3}	0.99 to 0.999	10^{-6} to 10^{-7}
3	> 10^3	10^{-3} to 10^{-4}	0.999 to 0.9999	10^{-7} to 10^{-8}
4	> 10^4	10^{-4} to 10^{-5}	0.9999 to 0.99999	10^{-8} to 10^{-9}

target SIL value, and the values of DR and C(E) upon which it is based, is a key component of the safety requirements specification (SRS).

IEC 61511 introduced the concept of vulnerability to allow for the fact that, in many cases, the operators may be able to avoid the consequences of a hazardous event if/when it occurs. For example, they could run away from it. Vulnerability is a factor (VF) used to reduce the consequence C(E) according to the nature of the hazard. Thus Equation 53.24 may be modified as follows:

$$\text{Risk} = \text{HR} \times \text{VF} \times \text{C(E)} \qquad (56.2)$$

The vulnerability factor enables SIL values to be reduced from those determined by PFD considerations alone. Suggested values for VF are given in Table 56.2.

Table 56.2 Interpretation of vulnerability factors

VF	Example	Reduction
0.01	Small release of flammable or toxic material	2 SIL levels
0.10	Large release of flammable or toxic material	1 SIL level
0.50	Large release with a high probability of ignition or else highly toxic	None
1.00	Rupture or explosion	None

In practice, companies are wary of using vulnerability factors. Not only is it a matter of chance as to whether or not an operator can take evasive action, the use of vulnerability factors can so distort the SIL values as to make a nonsense of their quantitative basis. They create the impression of a fix to retrospectively make a bad design seem OK.

Vulnerability should not be confused with the preventive parameters, P1 and P2, referred to in Chapter 57, which allow for operators taking manual action to prevent the hazardous event occurring in the first place.

56.5 Worked Example on Quantitative Approach to SIL

A protection system is proposed in which a high pressure gas supply is isolated in response to an emergency stop command as depicted in Figure 56.4a. It consists of a manually operated pushbutton switch which trips a solenoid valve. The push button switch has a "normally closed, open to trip" contact. The solenoid valve is normally open and is of the "normally energised, de-energise to trip" type. A spring-return mechanism closes the valve when power is removed from the solenoid.

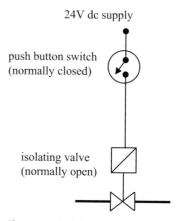

Fig. 56.4a Push button switch operated type of solenoid trip

The basic reliability data for this system is as listed in Table 56.3. It should be assumed that there will be a full inspection and test every two years at system shutdown.

1. **Establish Desired PFD and SIL**

Presume the objective of design is to reduce the risk to midway across the ALARP range.

Thus the acceptable risk = 1×10^{-5} deaths/year. Suppose the probability of somebody in the area being killed when the hazardous event occurs is 0.01. Assume that the probability of the area being occupied is 0.1.

56.5 Worked Example on Quantitative Approach to SIL

Table 56.3 Reliability data for Worked Example on SIL

Device	Failure rate (fpmh)	Mode/nature of failure	Probability of failure
Push button switch	0.2 (all faults)	Contact fails open (fails to close or stay closed)	80%
		Contact fails closed (fails to open and/or a false closure)	20%
Solenoid valve	1.0	Spring action fails to close valve when power is removed: valve fails open	
	8.0	Spurious closure: valve fails shut	
Operator	–	Operator fails to push button when required	5%
Pressure switch	1.1	Contact fails open	
	0.1	Contact fails closed	
Trip amplifier	0.4 (all faults)	No information available / No diagnostic coverage	

Hence the consequence of the event $C(E) = 1 \times 10^{-3}$ deaths/incident.

$$\text{Risk} = \text{HR} \times C(E) \quad (53.24)$$

Thus the required hazard rate $\text{HR} = 0.01$ incidents/year, i.e. once per hundred years.

Presume an *a priori* unmitigated demand rate of once per year, i.e. $\text{DR} = 1.0$:

$$\text{HR} = \text{DR} \times \text{PFD} \quad (53.23)$$

Whence, the required PFD is 0.01. Reference to Table 56.1 shows that a PFD of 0.01 is on the SIL 1/2 boundary. Prudence requires that SIL 2 be the target for design.

2. Human Factor

The push button switch is manually operated and there is obvious scope for human error. It is best to think of the operator as being in series with the protection system. Thus, assuming dangerous failure modes, from Equation 53.16:

$$\text{PFD} = U_{\text{overall}} = \sum U_j \quad (56.3)$$
$$= U_{\text{operator}} + U_{\text{switch}} + U_{\text{valve}}$$

It is interesting to note that if the operator is absent the protection system cannot be activated. In effect, Equation 56.3 defaults to

$$\text{PFD} = U_{\text{operator}} = 1.0$$

whence $\text{HR} = \text{DR}$.

However, if the operator is present but fails to push the button when required, Equation 56.3 applies. Consider all failures to push the button to be dangerous. The 5% chance of not pushing the button is equivalent to the operator not being available when required. Thus:

$$U_{\text{operator}} = 0.05$$

Substituting:

$$0.01 = 0.05 + U_{\text{switch}} + U_{\text{valve}}$$

The contribution of the operator alone exceeds the design criterion: clearly the manually operated system as proposed is not viable, irrespective of the quality of the push button switch and the isolating valve.

Vulnerability criteria could now be considered. It is not unreasonable to presume that the consequence of failure of the protection system is a large release of flammable material through a pressure relief valve. Inspection of Table 56.2 shows that this corresponds to a vulnerability factor of 0.1, depending upon the likelihood of ignition. Thus, in principle, the SIL level may be reduced by one, corresponding to an increase in the PFD criterion by a factor of 10.0.

Taking the design criterion into the range of $0.01 < \text{PFD} \leq 0.1$ clearly makes the push button design viable. However, nothing is known about the prospect of the operators taking evasive action. There is no basis for reducing the SIL level on the grounds of vulnerability, so SIL 2 has to remain as the target for design.

3. Estimate PFD of Alternative Design

One has to question whether it is sensible to rely upon a manually operated push button for a SIL 2 application. A pressure (or flow) switch hard wired *via* a trip amplifier (relay) into the solenoid would be more practicable, as depicted in Figure 56.4b.

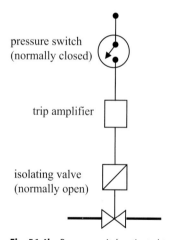

Fig. 56.4b Pressure switch activated type of solenoid trip

The pressure switch is of the "normally closed, open to trip" type. Depending upon the voltage output of the switch, the trip amplifier produces a current sufficient to power the solenoid. As before, the solenoid valve is normally open and is of the "normally energised, de-energise to trip" type.

The fastest way to estimate the PFD of the trip system is to recognise that both the switch and the valve are in series and to combine their failure rates into an overall figure. This approach is valid since both elements are subject to the same PTI and have the same MTTR.

The pressure switch is normally closed so its dangerous failure mode is closed, *i.e.* it fails to open on demand, for which the failure rate is 0.1 fpmh.

The trip amplifier has 0.4 fpmh. Since nothing is known about its distribution of failures, it must be assumed that these are all dangerous: it produces, or continues to produce, a current output when the input voltage is zero.

The solenoid valve is normally open so its dangerous failure mode is open, *i.e.* it fails to close on demand, for which the failure rate is 1.0 fpmh:

$$\lambda_{\text{ser}} = \lambda_1 + \lambda_2 + \ldots \lambda_n \quad (53.14)$$

whence the dangerous failure rate for the alternative isolating system is

$$\lambda_D = 0.1 + 0.4 + 1.0 = 1.50 \text{ fpmh}$$

Proof testing of the isolation system will be at two year intervals, *i.e.* PTI = 17520 h.

Assume MTTR is insignificant:

$$\text{PTRT} = (\text{PTI}/2) + \text{MTTR} \quad (53.11)$$

whence the effective proof test period PTRT $\approx$ 8760 h.

Noting that

$$U \approx \text{PTRT} * \lambda \quad (53.12)$$

and assuming dangerous failure modes:

$$\text{PFD}_D = U \approx \text{PTRT} * \lambda_D = 8760 * 1.50 \times 10^{-6}$$
$$\approx 0.01314$$

This corresponds to SIL 1 protection. Given that the original target was on the SIL 1/2 boundary, a higher specification design is required. Increasing the proof testing frequency to an annual basis, *i.e.* once per year, results in a PTRT of 4380 h.

$$\text{PFD}_D = U \approx \text{PTRT} * \lambda_D = 4380 * 1.50 \times 10^{-6}$$
$$\approx 0.00657$$

which corresponds to SIL 2 protection and clearly satisfies the requirement.

4. Spurious Trip Rate

Spurious trips are those which occur as a result of failure into a safe mode. For the alternative system of Figure 56.4b this is either when the pressure switch fails open or when the solenoid valve fails shut. The trip amplifier is not considered since all of its failures have already been deemed to be dangerous.

The safe failure rate for the pressure switch is 1.1 fpmh and the safe failure rate for the isolating valve is 8 fpmh:

$$\lambda_S = 1.1 + 8.0 = 9.1 \text{ fpmh}$$

whence:

$$\text{MTTF}_S = \frac{1}{9.1 \times 10^{-6}} \approx 110,000 \text{ h} \approx 12.5 \text{ years}$$

5. Safe Failure Fraction

This was defined in Chapter 53 as

$$\text{SFF} = 1 - \frac{\lambda_{UD}}{\lambda_S + \lambda_D} \quad (53.21)$$

Since there is no provision for testing either the pressure switch or the isolating valve, either automatically or manually, it must be presumed that all dangerous failures are undetected (covert), in which case:

$$\text{SFF} \equiv 1 - \frac{\lambda_D}{\lambda_S + \lambda_D} = 1 - \frac{1.5}{10.6} = 0.858$$

With reference to Table 56.6, it can be seen that an SFF of 85.8% for a type A simplex system provides SIL 2 protection, thus the architectural constraint is satisfied.

56.6 Qualitative Approach to SIL

The qualitative approach to selecting an appropriate SIL for a protection system utilises the so-called hazardous event severity matrix, as depicted in Figure 56.5. This has been absorbed into IEC 61508 but its roots are in the TUV (pronounced "toof") scheme for equipment testing in relation to the German DIN standards. The matrix is fairly simple to use but the results obtained require careful interpretation.

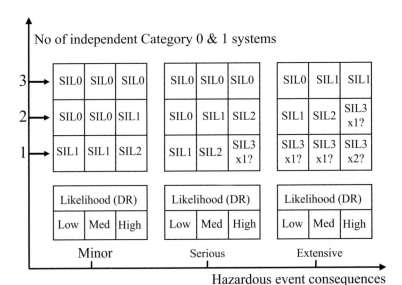

Fig. 56.5 Hazardous event severity matrix

Table 56.4 Categories and interpretation of consequences

Consequences	IEC 61508 category	TUV category – Figure 57.1	Fatalities (deaths/event)	Interpretation
Minor	A	C1	< 0.03	Serious but temporary injuries, short term loss of functions
Serious	B	C2	0.03–0.30	Non-recoverable health problems, permanent injuries, loss of limbs, may be single death
Extensive	C	C3	0.30–3.00	Single to several deaths
	D	C4	> 3.00	Catastrophic event, multiple deaths

Table 56.5 Categories and interpretation of likelihood

Likelihood	TUV category – Figure 57.1	Demand Rate (events/year)	Interpretation
Low	W1	< 0.01	Extremely unlikely, essentially zero
Medium	W2	0.01–0.10	Occasional occurrence
High	W3	0.10–1.00	Probable occurrence
	W3	> 1.00	Frequent occurrence

First, the potential consequences of a hazardous event have to be estimated and categorised as being minor, serious or extensive. This is subjective, but a sensible interpretation is as indicated in Table 56.4. To provide a feel for consequences, bands of average death rates are suggested for each category.

This establishes which of the three severity sub-matrices to use. Next, the likelihood of the hazardous event occurring has to be categorised as being low, medium or high. This is the unmitigated demand rate and is interpreted in Table 56.5. To provide a feel for likelihood, recognised bands of values for DR are stated for each category.

This establishes which column of the severity sub-matrix to use. Next the total number of independent protection systems is determined. This is essentially a question of counting the Category 0 (passive) and Category 1 (active) protection systems. This establishes which row of the severity sub-matrix to use. The required target SIL value is read off at the appropriate intersection.

The entries in the severity matrix need some explanation:

SIL0   The protection system is probably not required.
x1?   One SIL 3 protection system may not be sufficient.
x2?   One SIL 3 system is not sufficient. Further protection and/or SIL 4 is required.

Use of the hazardous event severity matrix is entirely subjective and, in itself, is an insufficient basis for design. The target SIL value derived from the matrix must be validated using safe failure fraction (SFF) criteria as explained in Section 56.8 below.

56.7 Worked Example on Qualitative Approach to SIL

Hydrocarbons are transferred from a high pressure vessel to a low pressure vessel as depicted in Figure 56.6.

The consequence of failure to protect the low pressure vessel against over-pressure will be rupture of the vessel and vapour release. Since ignited

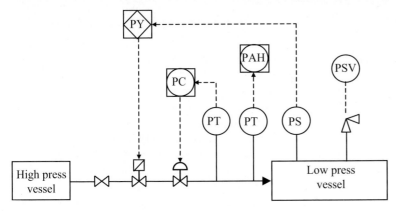

Fig. 56.6 Hydrocarbon transfer from high to low pressure vessels

vapour clouds may cause several deaths, Table 56.4 indicates that the "extensive" sub-matrix be selected.

Next comes consideration of the likelihood. Pressure in the low pressure vessel is controlled by the control loop PC, failure of which is the main cause of over-pressure. Although the control function is realised by means of a proven DCS, which is known from experience to only fail very occasionally, the other elements in the loop are less reliable. Thus the pressure sensor/transmitter is likely to have impulse lines which may be subject to intermittent fouling and/or distortion due to air bubbles. Also, the control valve is handling a high pressure stream and will be prone to leakage and sealing problems. Being prudent, Table 56.5 indicates that the "medium" column be selected.

Then comes the counting of the number of independent Category 0/1 systems:

- The DCS is the potential cause of failure for which protection is being provided, so it cannot count. In any case, a DCS is at best Category 2 and normally a Category 3 system.
- Manual response to the hardwired alarm PAH is unlikely to be fast enough to prevent over-pressure so it provides no protection and cannot be counted.
- The pressure relief valve PSV, which discharges into a vent system, can handle the worst case of over-pressure without risk to life or limb. This is a passive device and counts as Category 0 protection.
- The PLC based trip system clearly counts as a Category 1 protection system. Indeed, the whole point of this analysis is to establish the SIL required for the trip system.

Thus there are two independent protection systems and "row 2" of Figure 56.5 applies. The intersection of the medium column of the extensive sub-matrix with row 2 indicates that SIL 2 is the target value for the trip system.

56.8 Architectural Constraints

As stated previously, once the protection system design has been established on the basis of PFD considerations, it is necessary to review the architecture of that design for hardware fault tolerance. This requires an estimate of the safe failure fraction (SFF), as described in Chapter 53. The SFF is used to determine the SIL value for the protection that the architecture is deemed to be capable of providing. The SIL value is tabulated against SFF and the number of parallel channels of which there are three categories: simplex, duplex and triplex.

Type A safety related sub-systems are "simple" ones. This may be interpreted as those systems for which the failure modes are known and defined, and for which the failure rate data is known. For

Type A systems the architectural constraints are given by Table 56.6.

Table 56.6 Safe failure fractions (SFF) for Type A fault tolerance

Safe failure fraction SFF	Type A hardware fault tolerance		
	0 (Simplex)	1 (Duplex)	2 (Triplex)
< 60%	1	2	3
60–90%	2	3	4
90–99%	3	4	4
99–100%	3	4	4

SFF is a more stringent criterion that PFD and if SFF yields a lower SIL value to that used for design purposes then the SFF considerations are overriding. For example, consider a protection system for which the design requirement is SIL 3. Suppose that its design is of a simple nature with no redundancy, i.e. simplex, and its SFF<90%. According to Table 56.6 the design can only be considered to be of SIL 2 quality and other measures, such as introduction of redundancy, must be introduced.

For the more "complex" Type B systems the architectural constraints are given by Table 56.7.

Table 56.7 Safe failure fractions (SFF) for Type B fault tolerance

Safe failure fraction SFF	Type B hardware fault tolerance		
	0 (Simplex)	1 (Duplex)	2 (Triplex)
< 60%	Not allowed	1	2
60–90%	1	2	3
90–99%	2	3	4
99–100%	3	4	4

56.9 Categories of Failure

If a system is to be used to provide protection, then the factors most likely to cause the system itself to fail must be properly taken into account. There are different types of failure which may be categorised as being random hardware, systematic and common cause.

1. **Random Hardware Failures**

These correspond to the flat part of the bathtub curve described in Chapter 53 and are characterised by a constant failure rate. It is impossible to predict exactly when a system will break down because of the random nature of failure of its components. The reliability of a system has to be calculated from that of its components and sub-systems, as explained in the worked example of Chapter 53.

The most effective precaution against random hardware failures is redundancy. This means providing additional components or systems in parallel. If one fails the other will continue working. When redundancy is necessary, it is usually sufficient to provide dual redundant systems but, for very high integrity applications, triple redundancy may be required.

2. **Systematic Failures**

These are due to mistakes made at some stage in the process of specification, development, construction or operation of a system. The system will fail to function as intended every time a particular set of conditions occurs. Because it is impossible to test for every possible set of conditions, systematic failures may lie hidden in a system. There are three main types of systematic error:

- Specification: includes mistakes and omissions in the original design of the system.
- Hardware: may occur at any stage in the design, manufacture, installation or operation of the equipment.
- Software: made when the software was initially developed or can be introduced subsequently.

Redundancy through the provision of identical protection systems in parallel clearly gives no protection against systematic errors because the working and redundant systems are both likely to fail in the same set of conditions. Systematic errors are normally best countered by redundant diverse systems in parallel involving different designs and components, but these are obviously more expensive than identical systems.

Table 56.8 Classification of protection against types of failure

	Random hardware failures	Systematic failures	Common cause failures
Always occur under the same conditions	No	Yes	Yes
Prevented by identical redundancy	Yes	No	No
Prevented by diverse redundancy	Yes	Yes	Yes

3. Common Mode Failures

As stated in Chapter 53, it is quite common for a single fault to affect many elements. Such cause and effects are referred to as common mode failures. The original cause may be the result of either random hardware or systematic failures, or due to circumstances for which the protection system was never intended to cope. Some typical examples of causes are:

- Loss of electrical power
- Loss of air supply
- Radio frequency interference
- Corruption of database
- Failure of communications device
- Failure of an I/O board
- Malicious intent or sabotage.

Common mode failures can only be protected against by the provision of redundant systems targeted on specific potential common cause failures, such as the use of a standby generator for protection against mains power supply.

The protection afforded by different configurations against these categories of failure type is classified in Table 56.8.

56.10 Protection System Design

There is more to the design of a protection system than the technical issues and much of the emphasis in IEC 61508 and 61511 is on the design process itself. This is of particular importance if the protection system contains elements that are programmable. In general, the design strategy should focus on the three aspects of reliability, configuration and quality, as depicted in Figure 56.7.

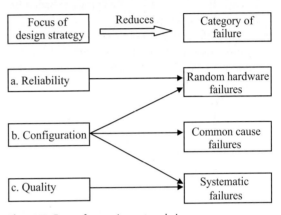

Fig. 56.7 Focus of protection system design

1. Reliability

This relates to protection against the consequences of random hardware failures. If the consequence of failure C(E) is not serious such that Category 2 protection only is required, see Table 55.1, then a qualitative assessment of reliability based on sound engineering judgement is a sufficient basis for design.

If the consequence of failure is serious then Category 1 protection is required at least. Thus a protection system is designed to meet a specification in the sense that it satisfies some SIL which, for a given demand rate, will yield a hazard rate that is low enough to ensure an acceptable risk. As seen, the SIL required may be determined either quantitatively or qualitatively. Depending on the reliability characteristics of the equipment used, the SIL value is achieved by means of introducing parallelism, such as redundancy and/or voting, and appropriate proof testing.

Hardware design should be subjected to searching questions about function, detection of failures, testing frequency, failure modes, *etc.*

2. Configuration

This refers to the organisation of the Category 0/1 protection system(s). Three issues need to be considered in particular:

- How many independent protection systems are needed?
- What is the extent of parallelism established by the reliability design?
- Can the failure of any single sub system, such as a power supply, I/O board, *etc.*, give rise to common mode failures not taken into account in the reliability design? If so, the configuration needs changing.

Recognising that common mode failures are best countered by means of redundant and/or diverse systems, the options are:

- Additional passive devices, *i.e.* Category 0 protection.
- Additional I/O channels. If these are of a different design, they provide effective protection against both common cause and systematic failures.
- Diverse software, *i.e.* different sets of software running in parallel systems, which also protects against systematic failures.

3. Overall Quality

Systematic errors can be greatly reduced by the combination of good design practice and quality engineering. Of fundamental importance to the quality of manufacture and implementation of protection systems are:

- Understanding of the safety principles involved by the personnel responsible for the specification and design work.
- Use of an established quality assurance system, such as ISO 9001, for producing the software.
- The experience and expertise of the personnel who are engineering the protection system.

56.11 Independence

Protection systems are disproportionately expensive compared with control systems of similar functionality, size, *etc.* That is because of the effort involved in quantifying the demand rate, failure rate, risk, *etc.* In a commercial environment there is much scope for reducing these costs by cutting corners. Recognising this, IEC 61508 lays down criteria for independence of the personnel involved in the design and, in particular, the validation of the design of protection systems. There are two aspects to this:

1. General Involvement

Involvement in any aspect of the life cycle of a protection system, other than Box 9 of Figure 56.2 which corresponds to detailed design, requires a degree of independence as indicated in Table 56.9, where HR denotes highly recommended and NR is not recommended.

Table 56.9 Independence required *vs* consequence

Minimum level of independence	Consequence (see Table 56.4)			
	A	B	C	D
Independent person	HR	HR	NR	NR
Independent department		HR*	HR	NR
Independent organisation			HR*	HR

The asterisk indicates that the requirement for an independent department may have to be met by using an external organisation. Companies may use an internal department skilled in risk assessment and in the application of protection systems instead of using an external organisation. However, the department must be managerially independent of those departments that have ownership of the plant and/or project.

2. Detailed Design

Involvement in Box 9, which roughly corresponds to Figure 56.3, requires a strong degree of independence, as indicated in Table 56.10.

Table 56.10 Independence required *vs* SIL

Minimum level of independence	Safety integrity level (SIL) (see Table 56.1)			
	1	2	3	4
Independent person	HR	HR	NR	NR
Independent department		HR*	HR	NR
Independent organisation			HR*	HR

The definition of HR, HR* and NR are the same as for Table 56.9.

56.12 Comments

IEC 61508 came into effect in 1998 since when the design of all protection systems have become subject to its provisions. It is not retrospective, thus any plant's protection systems previously designed and installed lie outside of its provision. However, if at any stage those systems are changed or extended, then all of that plant's protection systems becomes subject to the standard.

Whilst IEC 61508 specifically refers to E/E/PES systems, which may well include other elements, it is difficult to argue that it is not applicable to systems that are wholly hydraulic, mechanical or pneumatic. Ultimately, protection systems are designed on the basis of the hazards associated with the plant or process and the measures necessary to reduce the risk to an acceptable level. The issues, methodology and desired outcome are largely independent of the technology used.

Safety Equipment

57.1 Equipment Approvals
57.2 Failure Modes
57.3 Voting Systems
57.4 Electromagnetic Relay Logic
57.5 DC Coupled Logic
57.6 Dynamic Logic
57.7 DCS and PLC Systems
57.8 Special Purpose PES

In the previous chapter it was seen that central to the design of protection systems is the issue of determining the required safety integrity level (SIL). For a given system design, the question then is one of specifying equipment whose reliability and quality characteristics satisfy the SIL criterion. Whilst many suppliers of instrumentation and control systems will provide reliability data, and all will claim to deliver on quality, some form of independent testing is required to provide a basis for confidence. This chapter therefore starts with an introduction to equipment approvals. That is followed by consideration of failure modes and voting strategies, both of which provide a basis for an overview of the different technologies available for safety equipment. Finally, the pros and cons of these technologies are summarised.

57.1 Equipment Approvals

Technischer Überwachungs-Verein (TÜV) is the longest established and most credible organisation in Europe for the approval of equipment for protection system duties. The service is offered to suppliers whose equipment is type tested to German (DIN) standards and certified accordingly. The certification process is exhaustive and covers everything from the formulation and documentation of the original design concepts to the manufactured product and its suitability for the defined application. The TUV testing procedures result in equipment being deemed to belong to one of the four SIL categories.

Thus, for any protection system design, equipment selection is essentially a question of deciding which SIL category applies and ensuring that any equipment purchased has a matching SIL classification, or higher. In practice, most chemical and process industry applications result in SIL categories of 2 or 3. The selection process is qualitative, as depicted in the form of a so-called risk graph in Figure 57.1.

Starting from the left, select the most appropriate route through the risk graph according to the criteria listed below. (Note that the categories denoted by C, F, P and W in IEC 61508 are referred to as S, A, G and W respectively by TÜV).

- C = Extent of consequences of failure to meet a demand (see also Table 56.4):
  i. C1 = Recoverable injury
  ii. C2 = Single death or non-recoverable injury (*e.g.* loss of limb)
  iii. C3 = Death of more than one person
  iv. C4 = Catastrophic event, many deaths

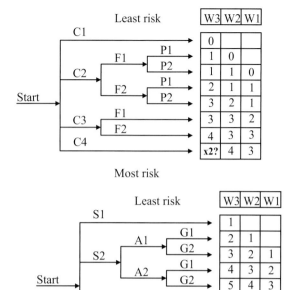

Fig. 57.1 Risk graph: IEC (upper) and TÜV (lower) versions

- F = Occupancy of the danger zone:
  i. F1 = Rare to frequent (occupied for say < 10% of the time)
  ii. F2 = Frequent to continuous (for say > 10% of the time)
- P = Preventive action (manual):
  i. P1 = Possible under most conditions (say > 80% chance)
  ii. P2 = Unlikely (say chance of < 20% of taking action)

The entries in the W matrix are the SIL values. Once the row of the W matrix has been determined, a column is selected according to the following criteria:

- W = Probability of undesired event, *i.e.* unmitigated demand rate DR (see also Table 56.5):
  i. W1 = Unlikely to happen
  ii. W2 = Small probability that event will happen
  iii. W3 = Relatively high probability of event

It is important to appreciate that TÜV certification is device oriented rather than end to end. The certification is not blanket but specific to a particular device, system configuration and/or certain types of application. Furthermore, a TÜV certificate may contain provisos, such as minimum proof testing intervals (PTI), which have to be considered carefully to ensure that they do not affect the suitability of the equipment for the intended application. Nevertheless, TÜV approval is of significant help to designers of protection systems as it greatly simplifies equipment selection.

Note that there are eight original TÜV classes, of which the eighth is for use in the most critical of applications. Because the TÜV testing procedures pre-date the work of IEC 61508 and ISA S84, the TÜV classes and SIL values do not align. The mapping between them is as shown in Table 57.1, the SIL values being the entries in the table.

Table 57.1 Mapping between TV classes and SIL values

	TÜV Class no							
	1	2	3	4	5	6	7	8
W1	0	1	1	2	3	3	–	–
W2	0	1	1	2	3	3	4	–
W3	0	1	1	2	3	3	4	4

The risk graph of Figure 57.1 is said to be risk averse in the sense that no allowance is made for P1/P2 at the higher levels of C3 and C4. The implication is that for those events with serious consequences it doesn't matter how much effort or capital is expended in making provision for manual preventive action. This seems perverse and so, within IEC 61508, an alternative form of risk graph is presented as depicted in Figure 57.2.

In this respect the standard appears to be inconsistent in that for a given set of parameters, according to which risk graph is used, different SIL values may be deduced. The key point here is that the risk graph needs to be calibrated according to the context and application. The onus is on the designer to agree on the calibration with the end user and for both to be able to justify its interpretation. Guidance on calibration is given in IEC 61511.

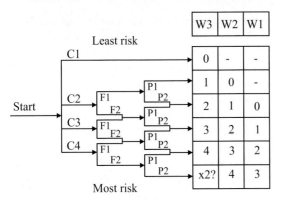

Fig. 57.2 Alternative risk graph for equipment selection

The non-SIL valued entries in the risk graph are as follows:

SIL0   The protection system is probably not required.
x2?   A single protection system is not sufficient.

57.2 Failure Modes

Failure modes are various and depend upon the technology used. For example, switch based sensors have the advantage of simplicity, but they also suffer from covert (hidden) failure modes, such as stuck open or shut, which are potentially dangerous. Frequent manual proof testing is the principal defence against this type of failure mode.

Use of analogue transmitters has the advantage of being able to distinguish between normal, out of range, and zero signals, the latter being consistent with failure of either the power supply or of the input channel itself. Set against this though is the increased complexity of the signal processing and its own scope for covert failures.

The advent of smart sensors and fieldbus has led to growth in both local and remote diagnostic testing. This has the ability to identify covert failures as well as overt ones. Despite the increase in complexity and the scope for failure, the reliability performance to date of this technology is impressive and looks set to become the norm for the future.

Any failure of a protection system may be categorised as being safe/dangerous and overt/covert. These are defined as follows:

- Dangerous failure: one which prevents the protection system performing its safety function.
- Safe failure: one which does not inhibit the safety function (spurious).
- Overt failure: one which is immediately obvious to the operators, or else is detected by means of diagnostics.
- Covert failure: one which is neither obvious to the operators nor is detected by means of diagnostics. Covert failures are only revealed by subsequent events, such as failure to meet a demand or by testing of the system.

These failure modes are depicted in the form of a grid in Figure 57.3. Note that metrics for articulating them have been defined in Chapter 53.

	Overt	Covert
Safe	Unimportant	Less important
Dangerous	Important	Important and very dangerous

Fig. 57.3 Overt/covert *vs* safe/dangerous failure modes

Testing of protection systems may be carried out either manually or by means of system diagnostics. The objective of testing is to convert covert failures, especially dangerous ones, into overt failures, preferably safe ones.

57.3 Voting Systems

Protection systems are essentially comprised of sensors, logic elements and actuators arranged in series and/or parallel. The logic element may be a simple hardwired relay or some software based combined diagnostic/voting system. The purpose of the diagnostics is to reveal covert failures, and that of the voting system is to determine whether to initiate a shut down or to continue operation with the residual protection.

A variety of equipment redundancy and voting strategies is available, as previously referred to in Chapter 53. These strategies may be optimised for meeting a demand or maintaining process availability. The more common ones are as follows:

- 1oo2. The most popular 1 out of n strategy in the process industries is 1oo2. This is based on the philosophy of "if in doubt, shut down". If two measurements, for example, do not agree to within some tolerance, then the system shuts down irrespective of whether the discrepancy is due to a genuine difference in the signals, an inherent fault in one of the measurements, or to failure of a measurement channel. The spurious shut-down rate in a simple system such as this may be unacceptably high.
- 2oo2. This strategy minimises the spurious shut-downs by requiring both sub-systems to agree that some condition has occurred before a shut-down is forced. However, if one sub-system fails there can never be agreement. In effect, the failed sub-system is unable to vote for a shut-down, and the whole of the protection system is disabled. Theses systems have a lower spurious trip rate than 1oo2 but a higher probability of failure on demand.
- 2oo3. Both capital and operating costs increase with redundancy, so the maximum m out of n strategy encountered on the vast majority of process plants is 2oo3. Put simply, this combines the qualities of both 1oo2 and 2oo2 strategies by maximising the probability of shutting down and minimising spurious trips. In addition, many 2oo3 systems permit on-line maintenance by reverting to 1oo2 or 2oo2 during maintenance, depending on the application.

Next comes an overview of the various technologies available for handling the logic and/or voting necessary for realising protection systems:

57.4 Electromagnetic Relay Logic

This is the oldest technology available. It has a proven history and a highly predictable failure mode when used in "de-energise to trip" systems. Most experts would agree that over 95% of failures result in the loss of the magnetic field and release of the ferrous core to the de-energised state, *i.e.* 95% of failures lead to spurious trips and only 5% to covert failures.

Note that if an electronic device, such as a timer or transistor based switch, is incorporated in the relay logic circuit, the predictability of its failure mode is lost. This is because the failure mode of electronic devices is 50/50 dangerous *vs* safe.

The logic of a relay system is in the hardwired connections of the relays. It is steady state d.c. coupled logic in that the circuit conditions are static, mainly energised, until a demand occurs whereupon they change to the opposite state. It is not practical to incorporate dynamic testing into a relay system so the principal defence against covert failures is regular manual proof testing done off-line.

Manual on-line proving requires a dual system, so that one half can provide protection whilst the other half is being tested. Triplicated systems are used occasionally: these optimise the probability of meeting a demand, minimise spurious trips and facilitate manual on-line proof testing.

Systems can always be constructed from electromagnetic relay logic elements to meet the requirements of Category 1 protection. The pros and cons of relay logic are summarised in Table 57.2.

57.5 DC Coupled Logic

This technology is based on logic gates such as AND, OR, *etc*. Modern systems use integrated circuit chips surrounded by protective circuitry and ruggedised for use in industrial environments.

The failure modes are as for any other electronic device: some 50% of failures will result in spurious trips and the other 50% in covert failures. However, the reliability of the technology is such that the MTBF is so high that the frequency of covert failures is very low.

It is possible to incorporate automatic dynamic on-line testing to convert covert dangerous failures into overt failures such as spurious trips or annunciation. This involves injecting a short pulse into the system's inputs and checking the associated outputs. The pulse width is chosen to be too short for the system's outputs, typically solenoid valves which are relatively sluggish, to be able to respond. The provision of such a facility obviously makes the system more complex and expensive.

As with electromagnetic relays, the logic is established by means of hard wired connections. Again, multiple systems are required to permit on-line maintenance and proof testing. The pros and cons of using d.c. coupled logic for protection are summarised in Table 57.2.

57.6 Dynamic Logic

The logic functions provided by dynamic logic are similar to those provided by d.c. coupled logic with the fundamental difference that the logic 1 level is represented by means of an a.c. signal. All other signals are treated as logic 0. The final output of the system is coupled *via* a transformer to a simple diode capacitor circuit for rectification. This provides a d.c. output suitable for driving solenoids.

By careful design, failure of any component in the circuitry will result in loss of the a.c. input signal to the transformer, thereby causing the d.c. output to decay to zero. Any failure of the rectifier circuit will have the same effect. This technology has the advantage that there are virtually no covert dangerous failure modes: all known failures are overt, a useful bonus being that analysis (FMEA and FTA) is greatly simplified. However, the conversion of covert failures into overt ones means that the spurious trip rate is higher than with d.c. coupled logic, and about the same as pulse tested d.c. coupled logic.

To facilitate on-line maintenance, dual protection systems are used where high process availability is essential. This duality can also be employed to reduce spurious trips by using 2oo2 logic. Because of the way that dynamic logic works, this would automatically revert to 1oo1 on a single failure and thereby annunciate the failure without tripping the system. This is a very cost effective solution in many applications.

Dynamic logic protection equipment has been around for many years and has an excellent track record. Devices are available "off the shelf" with TUV approval up to Class 7, making them suitable for virtually all Category 1 applications in the chemical and process industries. Note however, that whilst the devices may have TUV approval, the application logic itself will not be certified. The pros and cons of using dynamic logic for protection purposes are summarised in Table 57.2.

57.7 DCS and PLC Systems

General purpose DCS- and PLC-based systems are more than able to perform the logic functions required of any protection system. Their reliability has been established through millions of system-hours of operation. However, because of their complexity, the probabilities of safe/dangerous failure modes are approximately equal. Given that their self-diagnostics only cover power supply, CPU, memory and communications and generally do not cover the I/O subsystems, they are not acceptable for protection system duty.

It would be possible, by means of additional interconnecting inputs and outputs and the development of special purpose application software, to extend the coverage of the diagnostics to the I/O subsystems and thence realise Category 2 protection. At one time this approach was cost effective. However, it is no longer practical because of the high cost relative to the other technologies, not to mention the expense of third party validation and approval. The pros and cons of using conventional DCS and PLC systems for protection are summarised in Table 57.2.

57.8 Special Purpose PES

Special purpose protection systems are designed, through formal methods of handling and correcting faults, to minimise the probability of design or manufacturing error. PESs of this type have the look and feel of conventional PLCs and support the same application functionality. Indeed, in addition to their logic functions, some support PID control and other continuous functions. The essential difference is an integrated environment of self-diagnostics in the background that is invisible

Table 57.2 Pros and cons of protection system technologies

	Electro-magnetic relays	D.C. coupled logic	Dynamic logic	DCS- and PLC-based systems	Special purpose PES
Simplicity	Simple	Simple to design	Simple to design	Complex to design	Simple to use
Cost	Low	Relatively low	Not cost effective for large applications	Hardware relatively cheap, but approach expensive	Not cost effective for small applications
Pedigree	Proven	Proven	Proven	Not proven	Proven
Flexibility	Hardwired logic, so difficult to change. For some applications this is a bonus			Flexible	Very flexible
Failure mode	De-energise to trip. Highly predictable, 95% of failures lead to spurious trips	50/50 spurious/covert unless pulse tested. Greater MTBF than relays	Virtually 100% predictable. High trip rate may be reduced by a 2oo2 system	50/50 spurious/covert unless additional I/O used plus extra diagnostics	Virtually 100% predictable, but high spurious trip rate without redundancy
Approval	Designs are bespoke so type approvals are not available	Products with pulse testing and credible track records are available	TÜV approval up to Class 7 exists for devices but not applications	Designs bespoke so type approval not available	TÜV approval up to Class 5
Proof testing[a]	Manual, so cost of ownership high	Not required on products with automatic pulse tests	Not required	Depends on diagnostics developed	Not required
Size	Bulky, heavy	Smaller than relay equivalent	Similar to d.c. coupled logic, but depends on quantity of I/O signals	Depends on quantity of I/O signals	Complexity of logic has no effect on size for given I/O
Power consumption	Heavy	Low	Low	Moderate	Low

[a] The comments in the proof testing row of Table 57.2 refer to the logic solver function only. Scheduled end-to-end (sensor to actuator) testing is still required for the protection system as a whole.

to the user. These diagnostics cover all of the hardware and potential software errors. Thus the PES may be programmed without consideration of the effects of the program, or the I/O and their failure modes, on the diagnostics.

The primary aim of the self-diagnostics is to convert covert, potentially dangerous failures into overt failures. If these overt failures always lead to a spurious trip, the trip rate would probably be unacceptable. To overcome this, multiple systems are employed using voting strategies according to the application. An important difference between these systems and dynamic logic systems is that a vote for a trip is only accepted from a system which believes itself to be healthy, thereby minimising the spurious trip rate. For example, a duplicated system using 1oo2 may revert to 1oo1 on detection of a failure rather than tripping out. Similarly a 2oo3 may revert to 1oo2 or 2oo2.

The above voting strategies assume that the PESs are totally independent with some form of common "voter logic", using inter-wiring of solid state relays at the output. If done with great care to avoid the introduction of potential common mode failure, it is possible to increase the security of multiple systems. This is done by letting each monitor the performance of the other(s) and taking appropriate action. Unfortunately, with a dual system, situations will arise in which neither system knows which is right. This is avoided with triplicated modular redundant (TMR) systems in which it is assumed that the odd one out is wrong.

With TMR, the systems share their view of the inputs and outputs to detect discrepancies. They also check that the current version of the software in all three systems is the same. In this way, rouge I/O signals are detected and voted out without a spurious trip or loss of protection: in addition protection against software discrepancies is provided.

Some such special purpose PESs have TUV approval, typically up to Class 5 making them suitable for all Category 2 duties and some Category 1 applications. The pros and cons of using special purpose PESs for protection are summarised in Table 57.2.

Human Factors

58.1 Operator's Role
58.2 Psychological Model
58.3 Anthropometrics
58.4 Display Systems
58.5 Alarm Systems
58.6 Comments

Human factors concerns the important and often overlooked aspects of system design that relate to the role of the users of a control system, primarily its operators, although it is recognised that engineers and management are users too. As such it focuses on the design of the human interface, referred to as the man-machine interface (MMI), and on the interactions between the system and the operators. The design objective is to enable operators to carry out their role effectively, by which means they can contribute positively to the profitability of the end-user. There is also a particular requirement to minimise the scope for making mistakes. It follows that the starting point is consideration of the role of the operator. This leads on to the design of the interface, especially the layout of the hardware and presentation of information. The latter is of fundamental importance, the majority of operator errors relate to the quantity and/or quality of information being presented. A salutary overview of the subject is provided by Kletz (1995).

Human factors have been considered in several other chapters as appropriate. This chapter, therefore, for the sake of coherence, pulls together and further develops some of the thinking about human factors. For a more comprehensive treatment of the subject, the reader is referred to the text by Lees (1996), to the EEMUA Guide (1999) on the design, management and procurement of alarm systems and to another EEMUA Guide (2002) on process plant control desks utilising human-computer interfaces.

58.1 Operator's Role

Humans are a potential source of error. One obvious option is to automate out the role of the operator. However, that is not normally feasible because some operations, such as the addition of small quantities, taking of samples and visual checks are too impractical and/or expensive to automate. Also, this results in operators losing their "feel" for the plant which could be crucial in an emergency. So, current thinking is to give the operator a constructive role involving decision making at an appropriate level. This requires the interface to be designed to reduce the scope for making mistakes, and the application software to be designed to accommodate any potentially serious mistakes.

Defining the operator's role is a two-stage process. First are the questions of principle. What type of tasks should the operator be doing? What type of decisions should the operator be making? These questions are properly addressed as part of the control philosophy at the user requirements specification (URS) stage of a project as considered in Chapter 60. It is particularly important to think through the role of the operator in maintaining the safety of the plant. Even on plant with extensive automatic protection systems there are often some safety functions carried out by the operators.

Second comes the questions of detail. For each task some sort of task analysis is required. How does the operator know whether the task needs doing, and when? What information is required and is it accessible? What options in terms of decisions are there? What are the consequences of a wrong decision? How long does it take the operator to make a decision? Is the decision simply entered *via* the keyboard, or are other actions required? What happens in the event of no decision? And so on. These questions of detail are properly addressed by the application study in development of the detailed functional specification (DFS), as described in Chapter 62.

The outcome of such considerations influences the design of both the operator interface and the application software, the latter being what ultimately determines the role of the operator. The design itself is then subject to checking, either informally by means of walkthroughs as described in Chapter 63 or formally by means of a COOP study as described in Chapter 54.

58.2 Psychological Model

For the purpose of designing the operator's tasks, it is helpful to think in terms of each being comprised of the three distinct psychological stages of perception, cognition and action as follows:

Perception

This embraces the processes of attention and recognition. First the operator's attention has to be drawn to a situation, such as an abnormal condition or to some decision pending. Normally this is achieved by means of changing colours, flashing lights, sounding buzzer, *etc*. And second, the operator has to recognise that something needs to be done.

Whereas most process control is realised by means of proprietary DCS, PLC and SCADA systems, and there is little choice about the functionality of their operator control programs (OCP), there is nevertheless much scope for influencing the layout and style of their display systems and alarm environments at an application level. There are three very important principles here:

Text is difficult to read if the character size is too small. For a person with normal eyesight the "two hundred" rule of thumb applies:

$$\text{Character height} = \text{Viewing distance}/200 \quad (58.1)$$

For example, viewed from a distance of 2 m any character needs to be 1 cm high. If the contrast is poor then bigger characters are required. Ignoring this rule results in the density of information being too high which can lead to alarms/prompts being overlooked.

There needs to be consistency across all displays: for example, red corresponds to an alarm condition, flashing is a warning, *etc*. Failure to have consistency results in confusion. Visual effects can be very effective, but remember that 8% of the male population has some form of colour vision defect, so don't rely upon visual effects alone.

The limits that are applied to alarms must be sensible to prevent spurious occurrences. The use of deadband on alarm limits and their combination with warning limits is discussed in Chapter 43. When the frequency at which alarms and events are drawn to the attention of the operator is too high, they tend not to be taken seriously and fall into disrepute.

Cognition

This embraces the processes of interpretation, evaluation and synthesis. Given that a situation has been recognised by the operator, the relevant information has to be interpreted. All the relevant information must be readily available. Ideally this should be on a single display but, failing that, on a minimum of logically connected displays. The importance of having a simple, easy to navigate, intuitive display hierarchy is emphasised in Chapter 42.

Interpretation of the information available is analogous to identifying symptoms and is of a diagnostic nature. Evaluation concerns relating the symptoms to the various potential underlying cause and effect relationships, the essence of which is abstraction. Interpretation and evaluation provide a platform for synthesis which, in the context of control systems operation, is decision making. What actions are needed to make the plant safe? What needs to be changed to make the process more efficient?

Irrespective of whether a situation is complex or simple, the operators must have sufficient time to make the decisions, even if the actions are trivial. Also fundamental to the avoidance of mistakes is a sufficient understanding of the process, plant and its control system. A key function of display system design, therefore, is to enable the operator to develop an appropriate mental model of the plant, and to reinforce that model through usage. There is no short cut here: it is simply a question of training and experience.

Action

In general, once decisions have been made, the actions required to realise them should be fairly straightforward if the user interface has been well designed and is appropriate to the task. Depending on the context, the actions typically comprise combinations of switching loops between auto and manual, changing outputs, stopping sequences, acknowledging alarms, *etc*.

As stated in Chapter 54, the design of the application software must accommodate any potentially serious mistakes that could be made by the operator. This includes overriding inappropriate operator decisions and not permitting certain interventions.

58.3 Anthropometrics

Anthropometrics is concerned with the shape and size of people. Of particular importance in relation to the design of operator control stations (OCS) is the sitting position for which manikin diagrams of the 95 percentile man and the 5 percentile woman are as shown in Figure 58.1.

The hardware of an OCS is described in Chapter 39. In designing the layout, factors to be taken into account are the relative heights of chairs and work surfaces, distance from the VDU, comfortable VDU viewing angles, need for VDUs to be mounted on top of each other, *etc*. Generally speaking, the hardware of the OCS of a proprietary DCS is engineered into a console: these factors will have been taken into account by the supplier and the end-user has little choice. However, for PLC and SCADA systems, the hardware is normally loose and the end-user has to decide upon sensible layouts.

Other related factors, such as access to and redundancy of OCPs, are considered as part of the URS in Chapter 60.

58.4 Display Systems

The whole of Chapter 42 is concerned with the functionality of typical displays supported by proprietary OCPs. Whilst at the system level the functionality of display systems is generally fixed, there is much choice at the application level. That choice invariably involves making compromises, the key issues being:

- Number of displays *vs* the density of material per display.
- Number of trends, templates, *etc*. per display *vs* detail/resolution per entity.
- Use of standard templates/displays *vs* mimic based displays.
- Simplicity of displays *vs* complexity of display system hierarchy.
- Tabulated data *vs* pictorial representations.
- Use of text *vs* symbols.
- Use of dedicated *vs* user definable and/or soft keys for access to displays.
- Restrictions *vs* flexibility regarding access to data.

In most respects, design of a display system is no different to that of any other part of the control system and falls within the scope of the URS, DFS and system design as described in Chapter 62. The golden rule is to identify the end-user's needs at an

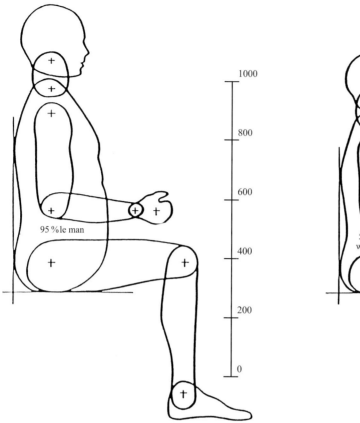

Fig. 58.1 Ninety-five percentile man and five percentile woman

early stage and to involve the end-user in all subsequent stages. The design process involves creativity, logical thinking, recognition of good ergonomic principles and common sense. It should always be borne in mind that operators must be able to find their way around a display system intuitively, and that they should be able to locate specific information within two to three actions at most. Remember that this will, sooner or later, have to be done under emergency conditions.

58.5 Alarm Systems

The whole of Chapter 43 is concerned with alarm systems and alarm management. Surveys consistently reveal that, in practice, their use and/or design leaves much to be desired. Most alarms are of low operational importance. Many channels or functions are alarmed unnecessarily. Alarms are often acknowledged and left standing. This all leads to alarms not being taken seriously or even ignored.

These issues can be addressed by systematically reviewing the most commonly occurring alarms on a regular basis, say weekly, especially in the early life of a system. Consider, for example, the ten most commonly occurring alarms, or the ten alarms that persist for the longest time. The review should focus on the causes of the alarms. Given the existence of alarm lists and event logs this is a relatively easy task. Systematically determine if changes can be made to alarm limits and/or deadband to prevent/reduce occurrences. Also consider whether an alarm's status can be reduced. Subject to software change control procedures, see Chapter 64, make whatever changes can be justified to reduce the occurrence of alarms.

It is a fact that alarms come in floods after upsets to the plant and/or process, and especially during emergency conditions. Retrospective alarm analyses, after major incidents, typically reveal that some 50% of all alarms are either repeats, *i.e.* multiple occurrences of the same alarm, or are being generated by plant items out of service. To reduce the scope for floods, systematically remove alarms that are duplicated, for example alarms on both measured variable and error signals in the same loop. Again, subject to change control procedure, change the status of alarms for plant items out of service or known to be faulty.

Of fundamental importance in designing and managing alarm systems is acceptance of the fact that if an alarm is to be recognised, the symptoms diagnosed and decisions made about what action to take, there is a human limitation on the rate at which this can be done. It is unrealistic to expect an operator not to make mistakes when alarms are flooding in. A useful rule of thumb for evaluating the operator workload from alarms is:

$$W = R \times T \quad (58.2)$$

where

R	is the	average alarm rate	min^{-1}
T		average time to deal with each alarm	min
W		fractional operator workload	–

For example, if one alarm occurs every two minutes (R = 0.5) and it takes 30 s to deal with each alarm (T = 0.5) then the operator workload is 25% (W = 0.25). This means that only 75% of the operator time is available for dealing with other matters.

The EEMUA guidelines recommend sensible upper limits on the number of alarms that an operator can reasonably be expected to cope with as follows:

Standing alarms:	≤ 10 alarms per operator station
Background alarms:	≤ 10 alarms per hour
Alarm floods:	≤ 10 alarms in first 10 min after an upset

In the latter case of alarm floods, the maximum alarm rate of 1.0 alarms/min, assuming the operator is able to concentrate fully on the emergency, that is W = 1.0, corresponds to a response time of T = 1 min/alarm.

58.6 Comments

Alarm handling is right at the heart of thinking about human factors. Display and alarm systems should be designed such that, under emergency conditions, the operators are aided in the process of making correct decisions. It is a simple fact of life that shortcomings are exacerbated by urgency. The alarm system must steer the operator's attention towards those particular plant/process conditions that require timely intervention. This inevitably means that alarms need prioritising and, maybe, filtering. All alarms should be properly defined and justified. If it isn't possible to define the operator's response to an alarm, it shouldn't be an alarm in the first place! And when an alarm does occur, the relevant information for making the necessary decisions must be readily accessible.

In a sense, this is all common sense and ought to be obvious. Unfortunately, the evidence is that human factors tend to be a relatively low priority and alarm handling something of an afterthought in the design of many systems. The reasons are fairly obvious. First, human factors are inherently of a qualitative "touchy-feely" nature and there is no "correct answer". Second, thinking through the issues at the detailed functional specification stage, which is when they ought to be addressed, takes time and effort and requires understanding. This costs money which is difficult to justify in terms of costs and benefits, not to mention project time scales, other than in the negative sense of the damage and/or cost of lost production in the event of a major incident. Difficult to justify maybe, but not justifiable.

Section 8

Management of Automation Projects

Costs and Benefits Analysis

59.1 Life Cycle
59.2 Conceptual Design
59.3 Methodology
59.4 Sources of Benefit
59.5 Benefits for Continuous Plant
59.6 Worked Example
59.7 Benefits for Batch Plant
59.8 Estimating Benefits
59.9 Categorisation of Costs
59.10 Estimating Costs
59.11 Payback and Cash Flow
59.12 Comments

Any automation system has to be properly engineered if it is to work effectively. There is much scope for things going wrong. When this happens it is invariably due to misunderstandings and lack of communication, poor planning, or to cutting corners. Key operational requirements may be overlooked. The complexity of some operations may be underestimated. The functionality of the control system may not be fully appreciated. The software design may not be flexible enough. And so on. The inevitable result of mistakes is systems being more expensive than budgeted for and delays in start-up.

This section of the Handbook concerns the project engineering of control systems. It is based on the life cycle of a typical project starting, in this chapter, with costs and benefits analysis. Subsequent chapters follow the project life cycle to its completion. Aspects of good practice are emphasised throughout. Whilst this section does not focus on safety and protection systems, it is interesting to note that the IEC 61508 standard uses the same project life cycle as its basis. A good introduction to costs and benefits analysis is given in the texts by Fisher (1990) and Lee (1998).

59.1 Life Cycle

For a system to be successfully engineered, all the relevant factors have to be taken into account, in sufficient detail, at every stage of the project's life cycle. These stages are listed in Table 59.1 which indicates, for a typical turnkey project, the involvement of the system supplier, end-user, or both. Supplier is the term used to describe either the manufacturer of a system or a systems integrator (a specialist control and/or electrical contractor) utilising hardware and/or software to provide an engineered working system. End-user is the term used to describe the purchaser of the system, *i.e.* the customer, which will be either an operating company or a process contractor acting on its behalf.

Note the role of the contractor. Major process contractors usually have a systems integration capability and so are able to buy hardware and system software from manufacturers and function as if they are suppliers. It is often the case that end-users have only limited technical capacity and are unable to handle large turnkey control projects effectively, so contractors commonly act on behalf of end-users. Sometimes they act both as supplier and end-user: whether that is desirable or not is a moot point. For the purposes of this chapter, and indeed the following ones, it is presumed that there is an arms length relationship between supplier and end-user: the role of the contractor has to be interpreted as appropriate.

Table 59.1 Project stage *vs* end-user supplier involvement

Stage	Involvement
Conceptual design	End-user
Costs and benefits analysis	End-user
User requirements specification	End-user
Tender/quotation	Supplier
System/vendor selection	End-user
Detailed functional specification	Both
Design, development and testing	Supplier
Acceptance	Both
Installation and commissioning	Both
Operation and maintenance	End-user
Upgrading and replacement	Both

Note the pattern of involvement. This is inherently different to that for, say, the purchase of major plant items in which, once the order is placed, there is normally little meaningful end-user involvement until commissioning. For a turnkey control system it is important that the end-user's engineers have a strong involvement in the early stages because it is not practicable for the supplier to work in isolation.

This life cycle applies to all projects, irrespective of whether:

- The project is large or small in terms of size of system
- The plant is new or being extended
- The system is a retrofit or an upgrade
- The system is DCS-, PLC- or SCADA-based

Note that retrofit is the term used to describe the installation of a control system on an existing plant that previously had either no control system at all or only a limited control capability. The term upgrade is used to describe the refurbishment or replacement of an existing control system.

It is also true to say that the life cycle applies to all aspects of a project, whether it be the instrumentation and control system *per se*, a management information system (MIS), an emergency shut-down system, or whatever.

The amount of work involved in the various phases of the life cycle will obviously reflect the nature of the project. For example, the system selection phase is trivial when there is a preferred vendor policy. Complex batch projects will require much more design effort than continuous control projects with a similar point count. Protection systems will require more rigorous design and testing than others.

59.2 Conceptual Design

Fundamental to the costs and benefits analysis is the conceptual design which essentially consists of an outline control philosophy document and an outline P&I diagram. Whilst the formulation of these is largely a process engineering activity, it is perhaps worthwhile reviewing their content and origins.

The control philosophy document covers, for example, the extent and scope of automation, the key control schemes and strategies, policy regarding operator involvement, safety and shut-down strategy, *etc*. The philosophy is developed in outline form alongside the preliminary design of the plant. This is appropriate because the design of the plant is influenced by the way in which it is to be operated, and *vice versa*. It must be emphasised that the demands of automation may require changes to the plant design. The document is further developed at the user requirements specification stage, as described in Chapter 60.

The P&I diagram is based upon and reflects the control philosophy. It is a statement of the:

- Operational intent with regard to the plant
- Conceptual design of the control schemes and strategies
- Extent and scope of automation required
- Functionality necessary of the control system

Formulation of the P&I diagram comes from an understanding of the way in which it is intended that the plant be operated. Control schemes and strategies are proposed on the basis of the process of determination, as explained in Chapter 30. Key control schemes and strategies are described in Chapters 22 to 37.

As the conceptual design evolves, decisions are made about the extent and scope of automation. It is, of course, possible to automate everything in sight, so some judgement about the extent of automation is required. There are some operations, such as visual inspection, sampling or addition of small quantities of reagents, which are either too expensive or impractical to automate. Judgement is also required about the scope of automation. This essentially concerns "ring fencing" the project such that only those plant items lying inside the fence are to be automated.

The P&I diagram that emerges from conceptual design is only in outline form: the detail is filled in at later stages in the life cycle. It must, nevertheless, be in sufficient detail to adequately represent the operational intent of the plant and extent of automation. If that information is not available, it is unlikely that a costs and benefits analysis can be carried out accurately enough to be meaningful.

There is an element of "chicken and egg" about the costs and benefits analysis and the conceptual design. Costs and benefits cannot be estimated fully until the P&I diagram is complete. However, some of the more marginal aspects of the control system cannot be justified until the viability of the project as a whole has been estimated. It is normal for there to be some iteration between design and estimation.

As the extent and scope of automation is established, the point count for the control system is determined. This is a key metric in determining both the functionality and size, and hence the cost, of the control system. Costing is fairly sensitive to the point count so a reasonably accurate figure is required.

59.3 Methodology

In the early days of computer control, it was not uncommon to carry out a feasibility study to determine whether or not it was technically feasible to automate a plant. It is now taken for granted that the technology is available to be able to implement as advanced or as extensive a control strategy as is required. The issue today is to decide on what is required, whether it can be justified and, if so, how best to go about it.

Justification is established by means of a costs and benefits analysis. Sometimes this is not necessary. For example, a similar process may have been successfully automated. Or perhaps the plant cannot be operated without computer control. In some companies the benefits of computer control are accepted to the extent that its use is expected irrespective of cost. However, in general, a costs and benefits analysis is required to determine the extent of automation that can be justified.

The analysis is best carried out by a small team within an agreed framework. Team members should be drawn from within the company according to the expertise they can contribute, as indicated in Table 59.2.

Table 59.2 Composition of costs and benefits team

Department	Expertise/experience
Business unit	Production forecasts, planning, costs, *etc.*
Process engineering	Process design, operational intent, *etc.*
Plant management	Operability, safety, human factors, *etc.*
Control engineering	Control, electrical, instrumentation, software
Miscellaneous	Independence

Clearly the size of the team depends on the scale of the project and the effort involved. The exact size doesn't matter: what is of fundamental importance is that collectively there is an understanding of the operational issues and of the control system functionality required.

Note the requirement for an independent person, ideally the chairperson of the team, who is somebody from a different plant/site, with no particular axe to grind. Costs and benefits studies inevitably challenge perceptions about how a plant ought to be operated, especially if the project concerns a retrofit or an upgrade, and the chair will be required to act as honest broker and maybe arbiter.

Another important person, unlikely to be a member of the team but crucial to its success, is a senior person in the company who is the 'champion' for the project. The champion may well have commissioned the costs and benefits study in the first place and will almost certainly use its outcome as a basis for arguing the case with 'management' for funding the project. It clearly makes sense that the champion's views are taken into account during the study and that s/he be kept abreast of progress.

As with any other project involving a team, the project needs to be broken down into separate tasks, assigned to individuals as appropriate, with the chairperson having oversight and responsibility for pulling it all together.

The outcome of the analysis should be a report identifying principal costs and benefits with an estimate of payback time. There should be some explanation of the basis of the various calculations, supplemented with tables of data as appropriate. Whatever assumptions have been made must be stated unambiguously. Recommendations about the operability of the plant and/or functionality of the control system must be made clear.

59.4 Sources of Benefit

There are two categories of potential benefits: tangible and intangible, some of the more obvious ones being listed in Table 59.3. Tangible benefits are those that can be quantified and provide the basis for estimating payback. Note that most of them have a converse: increased yield is the same as reduced losses, better consistency results in less off-spec product, *etc.*, and must not be double counted! The intangible benefits are those that cannot be quantified and have to be left out of the analysis. In effect, their realisation is a bonus of automation.

The potential benefits from automation depend on the extent of automation in the first place:

- For a new system/plant, the basis of comparison has to be the level of automation on comparable plants elsewhere. It is unrealistic to compare with the scenario of having either no control at all or very limited control.
- For a retrofit, where there is limited automation to begin with, the benefits have to be costed on the basis of improvements that can be achieved.
- For an upgrade, where the extent of automation is high to begin with, the benefits have to be costed on the basis of any increase in scope of automation and/or functionality.

The potential benefits also depend upon the nature and scale of the process. It is convenient to distinguish between continuous and batch processes. In general, most of the benefits that can be achieved on continuous plant can also be achieved on batch plant, but the converse is not true. There are additional benefits that can be realised on batch plant that do not apply to continuous plant.

Table 59.3 Tangible and intangible sources of benefit

Tangible	Intangible
Increased efficiency	More flexibility
Tighter control	Enhanced safety
Lower energy consumption	Better/more information
	More modern image
Higher throughput	Environmental impact
Reduced give-away	Improved job satisfaction
Improved yield	Better regulatory compliance
Better consistency	*etc.*
More responsive	
Improved plant utilisation	
Faster start-up	
Reduced manning levels	
Greater reliability	
Lower maintenance/ support	

59.5 Benefits for Continuous Plant

Efficiency, in the last analysis, measured in terms of the cost per unit amount of product, is the primary objective of automation. Increased efficiency cannot be achieved in itself, it is realised by means of other improvements such as:

- Tighter control. More extensive automation enables tighter control. This means that there is less variation in key process parameters. Thus plant can be "driven harder" with operating conditions set closer to the margins in terms of safety limits, quality and environmental constraints. There are substantial potential benefits. For example, reactors may be operated at higher temperatures resulting in increased rates of reaction and higher throughput.
- Reduced energy consumption. Tighter control also enables significant reduction in energy costs. Three examples. First: poor temperature control on a reactor leads to a cycle of overheating and undercooling, as depicted in Figure 59.1.

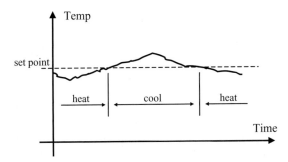

Fig. 59.1 Cycle of excessive heating and cooling

Reduced variation leads to a lower consumption of both steam and cooling water. Second, in overhead condensation: power consumption for pumping cooling water through an overhead condenser can be reduced by using temperature control to manipulate flow rate. Third, in agitation of viscous mixtures: much energy can be saved by linking agitator speed to changes in viscosity as, for example, in polymerisation.

- Reduced give-away. The probability distribution of some product composition for which there is a quality specification is as depicted in Figure 59.2. The objective is to produce product which is as close as possible to the specification. Consider the flatter of the two profiles. This has a wide spread which suggests that there is scope for much tighter control. Experience suggests that, by making more effective use of existing controls and/or introducing advanced controls, a 50% reduction in standard deviation can be achieved. However, there is no point in tightening control if the amount of product beyond the specification increases as, usually, over processing of product has a cost penalty. The strategy, therefore, is to shift the mean towards the specification whilst not increasing the size of the tail. This is discussed further in the Worked Example below.
- Better responsiveness. Modern controls enable more responsive processing. For example, feedforward arrangements enable major disturbances to be accommodated without the need for buffer storage, an obvious capital benefit. Operating strategies can be changed quickly to meet changed production requirements in response to market demand.
- Improved yield. Many items of plant, especially reactors and separation plant, operate at maximum efficiency over a fairly narrow range of conditions. Clearly operating such plant as close as possible to their design criteria leads to increased feedstock conversion and to an improved overall process yield.
- Faster start-up and shut-down. Automating both start-up and shut-down enables plant to be brought on-stream and taken off-stream faster than would otherwise be the case. This translates directly into a reduction in lost production time. It may also result in less off-spec product being produced during start-up and shut-down with savings on both raw materials and processing costs (especially energy consumption).
- Lower manning levels. In general, increased automation leads to a reduction in the number of operators required. Careful estimates need to be

made of the manning levels necessary to operate the system and to handle whatever operator interventions are required. Relatively small reductions in operator requirements can lead to substantial savings because of the need for 24-h coverage. Note that automation requires additional instrumentation and system support: this comes into the costs side of the equation.

- Increased reliability. Modern instrumentation and control systems, properly engineered, have much higher levels of reliability than the plant being controlled. In practice, it is possible to plan for continuous production, subject to the vagaries of the plant itself. Any reduction in lost production due to unplanned shut-down may be translated directly into benefits. Modern instrumentation also needs less maintenance.
- Effective maintenance and support. The typical life of a system is 8–10 years compared with that of a plant which may be up to 30 years. As a system becomes obsolete, its cost of support increases. For example, spare parts become unavailable and maintenance contracts escalate. There comes a point when there are significant savings to be made by going for an upgrade.

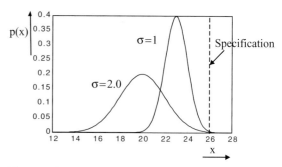

Fig. 59.2 Composition spread *vs* tightness of control (Matlab)

Going for a 50% reduction in standard deviation, that is setting

$$u = 26.0, \quad \sigma_a = 1.0, \quad \sigma_b = 2.0 \quad \text{and}$$
$$\overline{x}_b = 20.0 \text{ results in } \overline{x}_a = 23.0$$

Thus the mean has been shifted towards the specification by 3 units. Note that the size of the tail is still only 0.13%.

59.6 Worked Example

The two probability density distributions of Figure 59.2 are of the normal (Gaussian) type. The flatter distribution has a mean of 20 and a standard deviation σ of 2.0. The quality specification has a value of 26 below which, for this particular distribution, 99.87% of the data lies. In other words, there is a tail accounting for 0.13% of the product.

There is a useful rule of thumb which can be applied for determining the impact of tighter control:

$$\overline{x}_a = u \cdot \left(1 - \frac{\sigma_a}{\sigma_b}\right) + \overline{x}_b \cdot \frac{\sigma_a}{\sigma_b}$$

where $\overline{x}$ denotes mean, u denotes the (upper) specification and subscripts a and b denote the conditions after and before tightening control respectively.

59.7 Benefits for Batch Plant

- Improved product consistency. Computer control leads to greater repeatability. For batch plant this translates into improved product consistency. Given raw materials of the right quality, there is no reason for producing off-spec batches. Reducing the number of rejected batches is a major source of benefit. It is not just the cost of the lost raw materials and wasted energy, but also the lost production and the reworking costs.
- Increased throughput. By automating sequential operations, batch cycle times can be reduced substantially. Reduced cycle times translate into potential increased throughput. The benefit of this can be costed on the basis of the interest charges on the capital that would be required to de-bottleneck or expand the plant for the same increase in production.
- Better plant utilisation. This comes about, partly through the increased throughput, but also by the ability to implement control schemes that involve the effective sharing of common equipment items between different streams and units.

Again, this translates into a reduction in capital expenditure.

59.8 Estimating Benefits

The starting point is to assume that the control schemes and strategies of the P&I diagram are realised by some form of digital control system, DCS, PLC, SCADA or otherwise. Next, realistic assumptions are made about the tightness of control, consistency and reliability of operation, *etc*.

Then, from an operational point of view, the potential benefit of every control and safety function is analysed on a stage by stage, unit by unit and stream by stream basis. Each and every control function has to be justified against the key performance indicators (KPI) such as throughput, efficiency, yield, *etc*. Analysis will involve consideration of factors such as:

- Previous best performance
- Component and overall mass balances
- Losses of feed and products
- Variations in key process parameters
- Energy balances and excess energy
- Measurement of key variables
- Sources of disturbance
- Known controllability issues
- Time taken for operator interventions
- Likely incidents and frequency of occurrence
- Potential future changes: feed, catalyst, products, *etc*.

The benefits analysis is critically dependant upon the availability of good quality data and information. Typical requirements are:

- Price of raw materials, value of products, cost of utilities
- Different economic scenarios
- Quality constraints and margins
- Production plans and schedules
- Proposed operating conditions
- Control room log sheets: available for retrofits and upgrades only
- Historic data: strip charts and trend diagrams
- Experience from similar plants, *etc*.

As the benefits study progresses and alternative scenarios are considered, a profile of potential benefits emerges, each of which can be quantified.

59.9 Categorisation of Costs

The costs to be estimated may be categorised as shown in Table 59.4. Whilst the breakdown of costs will vary considerably from one project to another, it is useful to have a feel for typical cost scenarios.

- The control and instrumentation (C&I) budget for a new plant/system can be expressed as a percentage of the overall project cost. For a continuous plant this is likely to be some 5–10% and for a batch plant some 8–15% of the total.
- The electrical budget is typically some 5–10% of the overall project cost. Demarcation is normally on the basis of power: light current signals (mA) and devices come under the C&I budget and heavy current (amps) are in the electrical budget.
- Field instrumentation, *i.e.* all the elements associated with I/O channels, at list price is likely to be some 25% of the C&I budget but, allowing for installation, will cost approximately 50% of the budget.
- System hardware, plant and operator interface are likely to be 10–15% of the C&I budget. This also includes the system software which, to all practical intents and purposes, is treated as if it were hardware.
- Application software is approximately 10–20% of the C&I budget. For simple continuous plant this is at the 10% end of the range and for complex batch plant at the 20% end.
- Infrastructure costs are some 5–10% of the C&I budget. Note that power supply for motor drives *etc*., and the heavy current side of the motor control centres are provided for in the electrical budget.

An alternative perspective is to consider the cost *per* loop. The following typical metrics are for a new continuous plant with a DCS and some 500 analogue loops:

Table 59.4 Categorisation of costs

Main category	Sub-category	Example
Field instrumentation	Input channels	Sensors, transmitters
	Output channels	Actuators, valves, *etc.*
Hardware	System hardware	Memory, processors, communications cards, low power supply, *etc.*
	Plant interface	I/O cards, racks, cabinets, *etc.*
	Operator interface	OCPs, peripherals, *etc.*
Software	System	Operating system, communications, tools and packages
	Application	Configurable
		Graphical
		Procedural
Infrastructure	Electrical	Trays, trunking, termination cabinets, power and air supply
	Pneumatic	
Project management		
Miscellaneous	Documentation	
	Training	

- Overall project cost is approximately £730K *per* loop.
- C&I budget is some £60K *per* loop.
- Field instrumentation and infrastructure is £40K *per* loop.
- System hardware, plant and operator interface costs £7K *per* loop.

There is little published information about the detailed breakdown of costs. Sawyer (1993) provides such information for a multi-purpose, multi-products batch plant.

59.10 Estimating Costs

a. **Field instrumentation.** The starting point is to establish the point count as accurately as possible. This requires that the project be "ring fenced". Instrumentation and controls associated with items of plant inside the fence have to be costed and those outside don't. Getting the fence in the right place has a significant impact on the overall costs. It is usually fairly obvious whether a plant item is inside or outside the fence. What tends to be problematic is deciding how to handle streams that go through the fence. This is particularly true of utilities like steam and cooling water. In general, it is best to distinguish between the supply of a utility and its use.

Consider, for example, a boiler plant for raising steam and an exchanger in which steam is condensed, as depicted in Figure 59.3. Unless the boiler plant is itself to be automated, it is best to assume that both the boiler and its steam main are outside the fence. Suppose that the exchanger is inside the fence. Since the process stream temperature is to

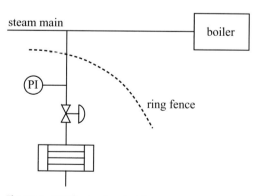

Fig. 59.3 Ring fencing the plant items

59.10 Estimating Costs

Table 59.5 Instrumentation costs on a per channel basis

Type	Variable	Elements	List price £	Installed cost £	Comments
Analogue input	Level, pressure, differential pressure	Dp cell, pneumerstat, air set[b], barrier	850 100 175 100	2500[a]	[a] Including impulse lines, equalising valves, *etc.* [b] Filter/regulator
	Flow	Orifice assembly	350	1750	+ dp cell *etc.*, excluding sraighteners
		Electromagnetic	1800	3000	IS or including barrier
		Coriolis	3500	5500	ditto
	Temperature	T/c or RTD, well/pocket	150 200	700	ditto
	Weight	3 load cells, amplifier/transmitter	2500 500	5500	ditto
	Conductivity	In-line housing, cell, amplifier/transmitter	100 200 550	1700	ditto
	pH	In-line housing, sensor/transmitter	250 550	1500	ditto
Analogue output	Stem position, flow	Barrier, I/P converter, positioner[c], air set, globe valve	100 300 250 100 1500	3050	[c] Intelligent positioner ≈ £850
Discrete input	Proximity	Detector, barrier	250 50	450	Assumes barrier is 2 channel
	Limit	Switch, barrier	275 50	500	ditto
Discrete output	Valve status	Relay, isolating valve, barrier	50 750 50	1500	ditto
	Motor status	Relay and/or contactor, barrier	0[d] 50	150	[d] Covered by electrical budget
Pulse input	Flow	Turbine meter barrier	1000 100	1500	Excluding straighteners

be controlled, the valve for manipulating the steam flow should be within the fence. Similarly, if the steam supply pressure is to be used for diagnostics purposes, say, its measurement should be deemed to be inside the fence.

The cost of the field instrumentation is determined directly from the point count. Typical figures are given in Table 59.5. Note that the following assumptions apply:

- List price costs cover all the elements of the I/O channels excluding the I/O cards.
- List prices are exclusive of VAT and carriage.
- Installed costs allow for branches and/or flanges, mechanical support, connections for hook up and local wiring.
- Local wiring is to a field termination panel or equivalent. Wiring and multi-core cabling through to the I/O cards is covered by the electrical and/or infrastructure budgets.
- Flow measurement and valve costs assume a line bore of 5 cm or less. Larger sizes will obviously cost more.
- Pipe fittings, flanges, impulse lines, isolating valves, etc, assumed to be in mild steel: other materials of construction will be more expensive.
- Castings for instrumentation such as dp cells, valves, etc in mild steel but internals such as diaphragm capsules, valve trims, etc in stainless steel.
- Costs are approximate at 2007 prices.
- Currency equivalents (2007):
  £2.00 ≈ €3.00 ≈ $4.00.

b. Hardware. Estimates for the system hardware and plant interface costs can only be meaningfully provided by systems' suppliers in the form of written estimates. The key information to be provided here is the point count and any anticipated needs for segregation and hardware redundancy.

Estimates for the operator interface also can only be meaningfully provided by systems' suppliers. This is largely determined by the numbers of operators, supervisors and engineers requiring access to the system, and the amount of redundancy required.

c. Software. System software is best treated as if it were hardware and costed by the supplier.

Application software has to be estimated on the basis of the effort required to produce it. Metrics for estimating the cost of developing, testing and documenting it are provided in Section 61.5.

d. Infrastructure. This is best costed by an electrical contractor using industry standard metrics.

e. Project Management. Some of the management costs are spread across the other categories. For example, management of the installation of the field I/O will invariably be covered by sub-contractors as part of the instrumentation costs. Management of the software development is usually provided for in the application software costs.

However, a significant proportion of the management costs have to be budgeted for directly. This includes, for example, effort in producing specifications, liasing between suppliers and contractors, carrying out acceptance tests, commissioning, *etc*. There is no easy way to cost this other than to make realistic estimates of the effort involved. It is particularly important not to underestimate the effort required of the end-user to produce the detailed specifications from which the application software can be developed. This is covered in Chapters 60–62.

f. Documentation. The cost of documentation is largely spread across those activities where it is generated. There are certain documents which are essential, for example:

- Specifications
- P&I diagrams
- I/O channel diagrams
- Termination listings
- Database design/listings
- Function block diagrams
- Sequential function charts, *etc*.
- Operating instructions
- System manuals
- Maintenance guides

The objective is to reduce the amount of documentation to the minimum necessary, consistent with maintaining quality standards and/or satisfying validation criteria. This is enabled by using generic documentation as far as is practicable, with application specific information being held separately. The approach is enabled by the use of CAD packages in which use is made of template diagrams, charts, *etc.*, and document management systems. When a document is required, the template is populated with application specifics from a database and printed off.

System and maintenance manuals are invariably software based and can be accessed on the system itself using prompts, keywords, *etc*. This should be provided for under the system software costs.

g. Training. Presuming appropriate levels of education and experience, this is relatively easy to cost given that system suppliers provide product specific courses at a variety of levels. It is essentially a question of assessing who needs what training at which level. Account must be taken of needs from operators through to system support, and is clearly context dependant.

h. Contingency. It isn't necessary to budget for a contingency at the costs and benefits stage. It should be noted that a contingency will nevertheless be required for the project proper.

i. Profit. Any contractor or supplier involved will require to make a profit. If the profit margins are too low then there is a danger that corners will be cut on the quality front. Profit is usually allowed for in the cost of field instrumentation and system hardware and software, but has to be budgeted for explicitly in relation to application software and project management.

j. Running Costs. These are relatively straightforward to assess. The principal categories are:

- Power. The electrical power required to drive the system and all it I/O channels is not insignificant. Estimates of power consumption have to be based on the system's ratings and by assuming average currents for all the I/O channels.
- Maintenance contracts. These are normally placed with the supplier for maintenance of the system. The cost of the contract depends on the speed of response, level of support and inventory of spares required. For critical applications, it is normal to have an agreed call-out response time, measured in hours, and "hot standby" spares on site. For less critical applications, some form of spares sharing agreement with other system users is not uncommon.
- Technician support. Additional electrical and instrumentation technician support is required to support the system and, in particular, to maintain the field instrumentation and I/O channels. Sensible estimates of effort required have to be made.
- Field instrumentation spares. A stock of these will have to be provided to enable effective instrumentation maintenance and repair. Clearly use of preferred vendor and instrument type policies will reduce the scope of the inventory.
- Hardware replacement. It is essential to budget for on-going hardware replacement to enable the system to evolve as opposed to becoming locked into a particular generation of technology. This will counter obsolescence and avoid the whole system having to be replaced at once.
- Software maintenance. It is not usual to budget for software support which is invariably required for utilising the flexibility of the system to realise enhanced operability, additional safety functions, *etc*. Given that these are all intangible benefits that have not been allowed for in the benefits analysis, it is appropriate not to count the costs of realising them.

59.11 Payback and Cash Flow

Payback, which is the time taken to recover the capital investment, is often calculated according to the following formula:

$$\text{Simple payback(yr)} = \frac{\text{capital costs (£)}}{\text{benefits (£/yr) - running costs (£/yr)}} \quad (59.1)$$

This formula for determining viability is often good enough for small projects and/or when the payback time is short. Payback for instrumentation and control projects of two to three years is not uncommon: for advanced control projects, where the control infrastructure is already in place, payback of less than less than six months is the norm. More sophistication is probably required for longer paybacks.

The essential shortcoming of simple payback is that it does not take into account the phasing of expenditure and return. Thus costs and benefits are presumed to be immediate and capital is presumed to be unlimited. In practice, costs and benefits are concurrent and there are constraints on capital. Simple payback is unable to differentiate between different spend rates or to distinguish different phases of a project. In particular, cash flow (CF) needs to be taken into account:

$$CF(\pounds \text{ year}^{-1}) \quad (59.2)$$
$$= \text{benefits} - \text{costs (capital + running)}$$

There are different categories of costs:

- Fixed capital. The lumpsum costs incurred by the purchase of major system items, bulk purchase of instrumentation and the placement of fixed price contracts for infrastructure, *etc*.
- Working capital. The variable costs incurred by the placement of reimbursable contracts, together with the cash on hand necessary for paying wages, for spares and repairs, and for the purchase of consumables used during commissioning, *etc*.

- Running costs. These are the recurrent costs incurred once beneficial operation of the system has commenced. They are identified in j above.

Cumulative cash flow (CCF) is the summation (£) of the CF over a period of n years:

$$CCF(n) = \sum_{j=1}^{n} CF(j) \quad (59.3)$$

Figure 59.4 depicts the costs, benefits, CF and CCF profiles for a typical project, based upon the data in the first five columns of Table 59.6.

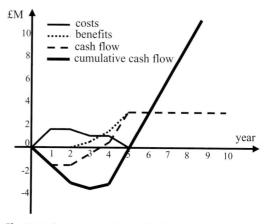

Fig. 59.4 Costs and cash flow profiles for a typical project

Table 59.6 Data for costs and cash flow profiles

Year (j)	Costs £K	Benefits £K	CF DR = 0.00	CCF	PV DR = 0.10	NPV	PV DR = 0.365	NPV
0	0	0	0	0	0	0	0	0
1	1500	0	−1500	−1500	−1364	−1364	−1099	−1099
2	1500	0	−1500	−3000	−1240	−2604	−805	−1904
3	1000	400	−600	−3600	−451	−3055	−236	−2140
4	1000	1500	500	−3100	342	−2713	144	−1996
5	0	3000	3000	−100	1863	−850	633	−1363
6	0	3000	3000	2900	1693	843	464	−899
7	0	3000	3000	5900	1539	2382	340	−559
8	0	3000	3000	8900	1400	3782	249	−310
9	0	3000	3000	11900	1272	5054	182	−128
10	0	3000	3000	14900	1157	6211	134	6

Aspects of the CCF profile are as depicted in Figure 59.5. Note that costs are incurred from the beginning of the project whereas benefits cannot occur until beneficial use of the system commences. The cash flow becomes positive when the benefits exceed the costs, and levels out only when the capital costs cease. Payback occurs when the CCF profile becomes positive. It is important to appreciate that by adjusting the phasing of the capital costs and hence the onset of benefits, significant differences can be made to the payback time.

The CF and CCF profiles of Figure 59.4 presume that money does not change in value. Unfortunately that is not the case. For example, interest charges paid on the capital raised cause it to lose value. Thus at 5% interest (say), each £1000 of capital borrowed today will only have a purchasing power of some £1000/1.05 ≈ £952 in a year's time, assuming zero inflation.

The rate at which future values of cash flow are depreciated is referred to as the discount rate (DR) expressed on a fractional basis. For example, a discount rate of 10% gives DR = 0.1. Thus the value today of the cash flow (£ year^{-1}) in j years time, known as its present value (PV), is evaluated using a discount factor (DF) as follows:

$$PV(j) = CF(j).DF(j)$$

where
$$DF(j) = \frac{1}{(1+DR)^j} \quad (59.4)$$

Note that the discount rate is not normally the same as the interest rate charged on the open market. The present value, for a given discount rate, is also referred to as the discounted cash flow (DCF) return.

Since capital for projects is invested over the first couple of years or so and benefits are realised sometime thereafter, the economics are worked out over a longer time scale taking the discount rate into account. The net present value (NPV) is calculated (£) as follows:

$$NPV(n) = \sum_{j=1}^{n} PV(j) = \sum_{j=1}^{n} \frac{CF(j)}{(1+DR)^j} \quad (59.5)$$

Also listed in Table 59.6 are values of PV and NPV for different values of DR, the profiles for NPV be-

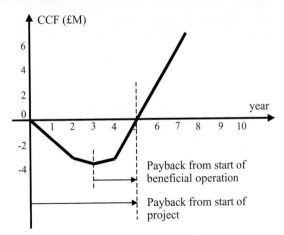

Fig. 59.5 The cumulative cash flow (CCF) profile

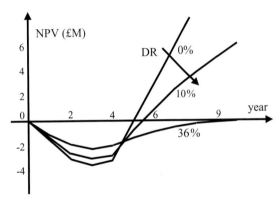

Fig. 59.6 Net present value (NPV) vs discount rate (DR)

ing as depicted in Figure 59.6. Note that the effect of discounting is to flatten off the profiles and delay payback. At higher values of DR the NPV struggles to become positive and the payback horizon becomes untenable. The golden rule of project finance is that the NPV must become positive.

The internal rate of return (IRR) is defined as that value of the discount rate DR which results in an NPV of zero. For example, for the project of Table 59.6, over a period of 10 years, IRR ≈ 0.36. Thus IRR is a measure of the benefits returned from investment in a project taking into account the time value of money. It is used for comparing and contrasting a variety of projects which are competing for investment capital. Companies often specify a minimum IRR return for projects to force the distinction between those that are clearly viable and those that are only marginally so.

Not only does discounting affect the current value of money but so too does inflation. Assuming a constant rate of inflation, a fractional inflation factor (IF) may be introduced:

$$\begin{aligned} NPV(n) &= \sum_{j=1}^{n} \frac{CF(j)}{(1+DR)^j \cdot (1+IF)^j} \\ &\approx \sum_{j=1}^{n} \frac{CF(j)}{(1+DR+IF)^j} \end{aligned} \quad (59.6)$$

Note that the cash flow is made up of many elements (hardware, software, services, *etc.*). The contracts/orders for these are all purchased/paid for at different stages of the project and probably subject to different inflation rate indexes, not to mention varying discount rates and exchange rates. For accuracy the cash flows should be calculated on a monthly basis rather than yearly. Clearly the calculation of detailed cash flow projections is a non-trivial task for which a spreadsheet or accounting package is required rather than a simple formula.

A further complication which can significantly affect the projections is the way in which benefits are taxed and the extent to which depreciation can be set off against taxation. In effect, Equation 59.2 for CF has to be modified as follows:

$$\begin{aligned} CF(\text{£ year}^{-1}) &= \text{benefits (after taxation)} \\ &\quad - \text{costs (after depreciation)} \end{aligned} \quad (59.7)$$

There is more to this than meets the eye. Taxation will depend upon whether the plant is a business in its own right or just part of a larger entity, and will vary annually according to profitability. Depreciation of assets is invariably used as a means of reducing tax liability and is clearly a function of company accounting policy. However, taxation and depreciation are often not taken into account in a costs and benefits study: it would be futile to attempt to do so without understanding the relevant law and/or policy.

There are several common mistakes made in calculating payback:

- Costs and benefits are calculated on a different economic basis. For example, costs for equipment are usually at market prices whereas the cost of labour may be at internal rates, which could be artificially high or low. Again, with benefits, there is scope for calculating these using either market or internal rates. The important thing is that a consistent basis is used.
- Marginal costings. This involves working out the incremental cost or benefit as opposed to the real value. For example, the cost of a few extra I/O channels is small if the rest of the control system is assumed to exist already or the project is going ahead anyway.
- The benefit of reduced manpower is lost if redeployment is not practicable.
- Costs and benefits not attributable to the project are included. This is most likely to occur with retrofits when the opportunity is taken to de-bottleneck the plant. For example, an external circulation loop and pump may be installed to improve process mixing but costed as a sampling system for additional instrumentation. Or the benefits from replacing a heat exchanger appear as improvements in temperature control.

59.12 Comments

A costs and benefits analysis is inevitably approximate. At best the costs can be estimated to within 10% and the benefits to within 20%. There is no point in looking for perfection where it does not exist. Detailed and more accurate costings are made at later stages in the project life cycle and, obviously, are subject to on-going review. The benefits tend not to be reworked, unless some major new factor emerges. It is good practice to maintain an audit of costs and benefits for future reference. Apart from anything else, it provides useful information for subsequent costs and benefits analyses.

User Requirements

60.1 Process Description
60.2 The P&I Diagrams
60.3 Control Philosophy
60.4 Particular Requirements
60.5 General Considerations
60.6 Golden Rules

Assuming that the outcome of the costs and benefits analysis has impressed senior management, the next step is to generate a user requirements specification (URS). The end-user writes the URS. Its function is to articulate what functionality is required of the control system. The purpose of the URS is to provide a clear basis for tendering and for making a meaningful quotation. It should therefore be correct, as complete as is practicable, and consistent. The URS must be in a form that can be understood by a system supplier or integrator. Also, it should enable the end-user to judge whether the supplier has understood the requirements.

The amount of detail required in an URS varies considerably depending on the extent of automation and the functionality required. In particular, it depends on who is going to develop the application software. Is the supplier expected to provide hardware and system software only or a turnkey solution? This chapter addresses the turnkey scenario.

There are essentially four parts to an URS: the process description, the P&I diagrams, the control philosophy and the particular requirements. This chapter outlines the factors to be taken into account in formulating an URS. There are various industry guides on the formulation of specifications and for more detailed information the reader is referred to the STARTS handbook (1989), the IEE guidelines (1990) and the GAMP guide (2002). The text by Lee (1998) also provides an introduction to the subject.

60.1 Process Description

This provides a basis for the supplier to understand and interpret the user's process objectives. The point of providing a process description is that it enables the supplier to make informed judgements about the requirements for control.

An overview of the process should be provided in sufficient detail for a process engineer to develop a feel for the plant and its automation needs. This should, for example, include:

- An outline process flow diagram
- An explanation of the function of each major unit and equipment item
- Information about the nature of the process carried out in each unit, *e.g.* exothermic or endothermic, under reflux, pressure or vacuum operation, precipitation of solids, evolution of gases, agitated, *etc.*
- A description of each stream, *e.g.* continuous or intermittent, liquid or gaseous, clear or slurry, viscous or otherwise
- For a multi-products/stream batch plant, typical production schedules indicating the number and frequency of products, number of stages

involved, permissible parallel operations, critical ordering of batches, inter-stage cleaning requirements, *etc.*
- An indication of any unusual features about the operating conditions, such as very slow ramp rates or small difference measurements, and any particularly stringent criteria such as very tight tolerances on error signals
- Nature of any constraints, *e.g.* maximum temperature differences, minimum flows, *etc.*

The more information that can be provided, the less scope there is for the supplier to claim a lack of understanding later on.

Remember, though, that this is only the URS. It is not necessary for the end-user to give away commercially sensitive information. For example, the detailed chemistry of the process, the capacity of the plant and precise operating conditions need not be divulged. Once the project has progressed beyond tendering and a contract for supply has been placed, it will be necessary to provide more detailed information. At that stage, if appropriate, confidentiality agreements can be signed.

60.2 The P&I Diagrams

The P&I diagrams are key working documents for the supplier. As stated in Chapter 59, they are based upon and reflect the control philosophy. Availability of the P&I diagrams is fundamental to an understanding of the control philosophy. Indeed, the P&I diagrams are a pictorial representation of much of the control philosophy.

They will probably be available in a more detailed form than was available at the costs and benefits analysis stage of project, but are still unlikely to be fully detailed. Again, the more information that can be provided, the less scope there is for the supplier to claim a lack of understanding later on.

60.3 Control Philosophy

The control philosophy document is essentially a collation of statements about design policy and the principles underlying key decisions relating to the control system. Those key decisions and the operational intent are themselves part of the control philosophy. The point is that the control philosophy enables the supplier to understand the basis on which the control system is being specified. As such, it provides an informed basis for interpreting the user's requirements.

Note that the principles established in the preliminary CHAZOP study, as discussed in Chapter 54, must feed into (and indeed become part of) the formulation of the control philosophy.

Note also the distinction between the principles which are the control philosophy and the details which form the particular requirements. A clear indication is required of:

a. Scope of project. The end-user needs to state unambiguously what is expected of the supplier in terms of:

- The supply of hardware and standard system software only, or provision of turnkey system including application software, or other variants.
- Requirements by end-user for involvement in software development, witnessing of tests and acceptance.
- Involvement of supplier in provision and/or integration of the control system with third party packages and systems as, for example, with MIS.
- Involvement of supplier during installation and commissioning.
- Documentation expected.
- Timescale for deliverables and deadlines.
- Any BS, IEC or ISA standards and guides such as GAMP (2002) which have to be complied with.
- Statutory requirements, such as FDA approval.
- Quality assurance, such as ISO 9000 approval for development procedures and documentation, for which a quality plan should be requested.

b. Scope of automation. The project needs to be ring fenced as indicated in Chapter 59. It is particularly important to identify parts of the plant:

- Shown on the P&I diagrams but which are not to be automated.
- Not shown on the P&I diagrams which are to be automated.

c. **Continuous operations.** For continuous plant, the key policy decisions concern continuity of operations:

- Is the plant to be operated continuously or intermittently?
- Is all the plant operated continuously, or are there are parts of it that are operated batch-wise? If so, are they semi-automatic or manual?
- How will any batch or semi-batch parts of the plant be brought on or taken off stream?
- Is automatic start-up required and, if so, what is the strategy? Units may be brought on stream one at a time, in process order, or all started up simultaneously, or some combination thereof. Start-up from empty may well require a different strategy to start-up following a shut-down.
- If automatic start-up is provided, it is usual to have automatic shut-down too. Again, what is the strategy? Remember that automatic shut down is very different from emergency shut-down (ESD).

d. **Control schemes and strategies.** The approach to tag numbering needs to be specified to ensure consistency throughout the system.

On a unit by unit basis, the principal control schemes and strategies such as cascade, ratio and feedforward control need to be identified, as outlined in Chapter 30.

Any non-standard requirements for control algorithms or associated features should be identified.

Any requirements for advanced process control (APC) techniques, such as model predictive control (MPC) or real-time optimisation (RTO) must be made explicit.

e. **Batch operations.** The general requirements for the control of batch operations need to be articulated. Key policy decisions here include:

- Whether all batches are to be produced under sequence control. If not, what are the criteria?
- Are inter batch/stage cleaning operations to be handled by sequencing?
- If a recipe management facility is required, to what extent are recipes likely to have to be created, modified and deleted?
- Will recipes have to be modified on-line for batches during processing?
- Will different products be made simultaneously in parallel streams?
- In the event of failed batches, is there a requirement to be able to go back in the sequence structure to repeat phases and/or steps, or to go forward and omit them?
- Is there a requirement to organise batches into campaigns? If so, how are campaigns defined and initiated, and how is production switched between campaigns?
- Is any form of dynamic scheduling of batches, units and recipes required?
- Does any provision need to be made for anticipated plant expansion?

f. **Operator involvement.** There needs to be some policy on operator involvement, covering the nature and scope of operator activity, as this determines the extent of automation. It is important that:

- Some indication of manning levels is given as this provides a feel for the extent of automation.
- Types of operation to be carried out manually, such as additions, sampling and visual checks, are identified.
- The range of manual actions for areas of plant that are to be semi-automated is clearly understood.
- Decisions to be made by operators are identified, such as when to start, continue, stop or abort a batch.
- Information to be entered into the system by operators, supervisors, *etc.* is categorised according to activity type.
- Policy exists regarding access to information and levels of authority for changing it.

g. Sequence design. The user needs to explain what approach to sequence development is acceptable. For example:

- Are sequences to be generic with parameter lists, as in Chapter 29, rather than being of a dedicated nature?
- Is the operations approach to sequence design, based upon plant units (MEIs), as explained in Chapter 37 to be adopted.
- Do operations need to be established by configuration of phases from a library of pre-defined phases?
- Is it a requirement that the plant be capable of being put into a safe hold-state at the end of each operation?
- In the event of an instrumentation, plant or process failure being detected by the sequence logic, is sequence flow forced into a safe hold-state?
- To what extent are recovery options to be built into sequences?
- Are parallel sequences to be used for application diagnostics?

h. Failure philosophy. The central issues here is consequence of failure, which may be of a hazardous or economic nature, or both. Policy decisions are therefore required about:

- The minimum levels of system reliability. This enables the requirements for duality to be determined, if at all.
- Power supply redundancy and standby power failure arrangements.
- Action on system hardware failure, *e.g.* are outputs forced to predefined values, activate the ESD, or what?
- On reinitialising after failure, should loops restart in manual, with their normal set points and zero outputs, or what?
- In the event of instrumentation or plant failure, should the outputs of control loops and schemes fail shut, open or stay put? Note that these options are not always available for configuration within DDC packages.
- Action to be taken in the event of failure of third party systems and/or packages to which the control system is interfaced.

i. Human factors. The user needs to explain what approach to the development of standard displays is acceptable, and to develop an alarm management policy as explained in Chapter 43. For example:

- The organisation and numbering of standard displays to maximise the operators ability to navigate the display system.
- Guidelines for the amount of detail to be included on standard displays. For example, one MEI per mimic display, or all loops associated with an MEI in a single group display.
- Conventions on colour codes, symbols and visual effects to ensure consistency throughout the display system.
- The basis for prioritising and grouping of alarms.
- The scope for suppression of alarms and provision of diagnostics.

j. Approach to layout. The general layout of the control system can affect its cost and functionality. Policy decisions need to be made with regard to:

- Relative positions of the rooms and/or buildings in which the operator stations, engineers terminals, I/O card racks, marshalling cabinets, motor control centre, field termination cabinets, *etc.* are to be situated.
- Locations of field stations for remote input of information by operators.
- Routing of instrumentation and power cables to a level of detail appropriate to the project scope.
- The clustering of units (MEIs) into cells such that units between which there is significant interaction can be handled within the same node of a DCS. This will allow the capacity of the nodes to be sized correctly.

k. Management information. Inevitably, there will be interfaces between the control system and other systems and/or packages for MIS purposes.

Whether the MIS is provided by the supplier or is of third party origin, the essential functionality required and the nature of the interface need to be specified. To a large extent, the approach to specifying an MIS is much the same as that for the control system itself. Chapter 100 provides more detail.

It is useful at the URS stage to identify any key performance indicators (KPI) that are likely to be required. KPIs can seldom be measured directly and invariably require calculation. That is typically done either in real-time by the DCS or off-line within the MIS. It is important to think through the nature of the calculations involved, to identify any necessary measurements, and to understand the data requirements. Typical KPIs are:

- Volumetric throughput
- Plant/unit efficiency
- Process yield
- Energy consumed per unit throughput
- Raw materials consumed per unit of product
- Emissions/discharges, *etc.*
- Variation in product quality
- Batch cycle times
- Plant utilisation
- Profitability
- Bonus rates, *etc.*

60.4 Particular Requirements

This part of the URS provides further, more detailed information to supplement the control philosophy document. Some of the more obvious examples of such information are categorised below.

Note that the particular requirements identified in the preliminary CHAZOP study, as discussed in Chapter 54, must feed into (and indeed become part of) the formulation of the control philosophy.

a. System hardware. The URS does not normally cover system hardware. However, whether the scope of supply includes installation or not, the environment within which the system is to be installed should be articulated. Of particular importance is the space available for the proposed system layout, and any constraints on physical access to that space. Likewise for the provision of power, normally a 110 V a.c. supply, and earthing arrangements.

Details and functionality of the interfaces to other systems and their peripherals should be provided. This includes serial links to other control systems, such as PLCs and packaged equipment, as well as links to other computers for MIS and APC purposes.

b. Plant interface. The point count must be specified as accurately as possible to enable the number of I/O cards to be determined. Remember to distinguish between static and fleeting contacts for discrete I/O.

Requirements for segregation of I/O signals must be specified, especially of IS and non-IS signals, since such constraints can have a significant effect on the numbers of I/O cards. Expansion and spares requirements in each I/O category should be identified.

Requirements for serial links to intelligent instruments must be specified to enable the number and type of gateways required to be determined.

c. Operator interface (hardware). Information needs to be provided about likely manning levels in each control room to enable decisions about appropriate numbers of operator and engineer stations, workstations and peripherals. The manning levels need breaking down on the basis of operators, supervisors and engineers, together with levels of access as appropriate.

Provision must be made for redundancy of OCPs. This enables simultaneous and/or multiple displays and allows for failure. Make allowance for both emergency and commissioning situations when access to OCPs becomes a bottleneck.

The requirements for any remote, field based, operator stations need to be detailed.

Any non-standard features, such as an interface to a hard wired mimic diagrams, need to be detailed.

d. System software. Any unusual requirements should be specified in full: if in doubt, specify it. The supplier's responsibility is to identify any additional system software that is necessary to provide the functionality required to meet the users' requirements. Areas for which system software may be required include:

- Special input conversion routines, control algorithms and device types.
- Implementation of the action on failure philosophy.
- Drivers for gateways to intelligent instrumentation to support, for example, HART or fieldbus protocols.
- Integration of I/O states in external devices with, for example, those in the DCS or PLC database.
- Special display and data entry requirements, *e.g.* to support any remote operator station functionality.
- Abnormal archiving requirements: the numbers, types of point and time scale for archiving should be defined. For example, time scales of a year, not uncommon in the water industry, may well not be within the scope of a typical DCS.
- Non-standard event recording: define what constitutes an event and what information must be recorded with it. The associated data is likely to be extensive to satisfy FDA requirements.

e. Configurable software. Simple control loop details can be determined from the P&I diagrams but any complex schemes should be specified in detail.

Typical I/O channel sampling frequencies should be identified, especially if high frequencies are necessary. This can have a significant effect on control processor loading and it is the supplier's responsibility to ensure that the equipment quoted has the necessary capacity.

Note that the contractor doing the plant design may be using a CAD package which allows instrument data to be ported into another system. If this is the case, include its details: the system supplier should interface to it if possible as this will save a great deal of database configuration and testing time. Publication of ISO 10303, a standard for the exchange of product model data (STEP), has focused attention on the potential for data storage and portability across the various project phases. An extension to this is being developed by the process industries STEP consortium (PISTEP).

f. Display system. If the turnkey effort includes configuration of standard displays of the type described in Chapter 42, the number and extent of each type involved must be defined.

g. Sequence software. Assessing the volume of turnkey sequence software is difficult: if the effort is initially underestimated, the chance is that it will lead to contractual problems during the project. The following guidelines apply to a multi product / stream plant which is the worst case scenario. The best approach is based on a two-stage analysis as follows:

First stage: quantify the total extent of the sequences:

- Identify all the products to be made and categorise them into generically similar groups on the basis that within any one group batches of each product could be made by the same procedure but with different parameter lists.
- Establish what products are to be made in which stream and hence identify the number of MEIs required for each product.
- Consider any inter batch cleaning operations on the same basis of sequences and parameter lists.
- Analyse each of the procedures and identify all the operations to be carried out on each of those MEIs for each of the products. Hence determine the commonality of operations between products.

Second stage: qualify the detail using representative sequences:

- Select three products, whose procedures are representative of all the others, that can be classified as being simple, typical or complex according to the complexity of sequencing involved.

- For each of these three products, based on a selection of the MEIs involved, break down their operations into phases and steps. Details for a mix of typical phases and steps should be articulated in the form of sequential function charts, or otherwise.

The cost of the procedures is estimated from the total amount of code and the effort involved. The amount of code is calculated from the number of procedures, their classification, the number of operations involved, the number of phases anticipated and the size of typical steps. The greater the commonality of operations and/or phases, the less the overall cost.

The above two-stage analysis is non-trivial and requires experience. If the end-user requires the supplier or some third party to carry it out, relevant information in sufficient detail must be provided. Of particular importance are the P&I diagrams for MEIs such as reactors, filters, dryers, *etc*. If some of the products were previously manufactured manually, the process instruction sheets are extremely useful, expurgated if necessary.

The supplier should be asked to program one of the operations and include it with the quotation as a worked example.

h. Batch software. In addition to the requirements for sequencing, further information for batch purposes needs to be provided:

- Estimates of the minimum, average and maximum number of products to be manufactured simultaneously
- The expected length of a typical campaign
- The maximum number of batches in a campaign

This information allows the supplier to establish that the functionality of the batch process control package (BPC) can accommodate the requirements for recipe handling and campaign definition.

Specify any complex or non-standard logging and/or reporting requirements in detail, including any requirements for access to RDB or MIS for such. Note that if FDA or GAMP conformance is required, a combination of messages and events will be required to be held for each batch of each campaign on a vessel by vessel basis. Some suppliers' reporting functions are more comprehensive than others so it is best to be cautious.

i. Higher level packages. These are generally required for calculations for which the amount of processing is beyond the scope of normal control processors. Typically, for management information type applications, the package will provide a front-end to a relational database in another computer, the package handling both access and communications. Or, for advanced control type applications, packages are often of a third party proprietary nature, usually running in a separate processor.

For MIS applications, it is of particular important to establish:

- Which variables are to be read into the database, RDB or otherwise, from the PLC, SCADA or DCS, and how often they are to be read. It is likely to be a large number of variables but only read occasionally.
- What levels of access to those variables exist within the MIS, given that the whole point of an MIS is to provide access to such information.
- That security exists to prevent values being written from the MIS into the control system database. If not, on what basis can writing occur?
- That the MIS and control system data types are consistent, both for tag numbers and numerical values.
- What protocols are in use and that the necessary drivers are available.
- That the MIS proposed has the required functionality.

The nature of RDB and MIS are covered in detail in Chapters 99 and 100. It is sufficient to say that establishing RDB and MIS links with control systems is not well standardised so the requirements should be specified in some detail.

For APC applications, such as SPC, MPC and RTO, key issues to establish are:

- Which variables are to be read into the database from the PLC, SCADA or DCS, and how often they are to be read. It is likely to be relatively few variables which are read often.
- Which variables are to be written from the APC package into the control system database. These are likely to be either set points for control loops or output values. If not, on what basis can writing occur?
- Any requirements for complex calculations or development of process models, *etc.*
- If the APC is model based, say for predictive control or optimisation purposes, how often the model runs and how it is initialised. Can operation of the model be suspended and, if so, how is it resumed? Are there additional variables required for any of these activities?
- The ways in which the APC package can fail. What happens if the packages loses its synchronisation with real time? How does the package handle out of range inputs? Are out of range outputs limited? What happens if there is a calculation failure?
- The extent to which the control system is robust to failure of the APC package. Does control revert to pre-determined set points in AUTO and/or outputs in MAN control? The action on failure philosophy may require significant design effort and must be carefully specified.
- The form of the operator interface to the control system. This clearly has to be consistent with other aspects of the operator interface.
- That the APC package and control system data types are consistent, both for tag numbers and numerical values.
- What protocols are in use and that the necessary drivers are available.
- That the APC package proposed has the required functionality.

If the supplier provides packages for APC, some negotiation over the model may well be required. In general, the process model will be developed by the end-user or some third party and will be highly confidential. However, there is no reason why an overview of its functionality cannot be provided.

Likewise, if the software already exists, the language and number of lines of code it occupies can be given. Further discussion with potential suppliers prior to their quotations can examine the possibility of porting existing programs.

60.5 General Considerations

As should be obvious, writing a good quality URS is a major task. So too is interpreting it. For this reason it is important that the document is well written. The following general considerations are worth noting:

- The user requirements should be as complete, concise, consistent and correct as possible. They should neither be duplicated nor contradicted. Take care to avoid vagueness and ambiguity. Avoid verbosity. Only use terminology and acronyms as defined in standards and texts. Avoid unnecessary jargon and technobabble.
- Only provide information that is relevant and reliable. Information whose quality is uncertain is best left out altogether: it is easier to provide supplementary information at a later stage than to correct misinformation.
- Where relevant, a distinction should be drawn between essential requirements and things that "would be nice to have".
- Each requirement statement should be uniquely numbered to enable cross referencing by the supplier.
- Each requirement must be testable in that there should be some means of deciding whether the solution proposed satisfies the requirement.
- A list of contact names for queries should be included in the URS.
- Specify a return date for quotations and give the suppliers enough time to respond in detail.

Reference should be made to contract terms and a copy of the user's Conditions of Contract appended. For large contracts, the best terms are ones that support a mixture of lump sum and reimbursable payments for different parts of the project. Of particular importance are arrangements for stage payments: if these are not available, that should be out in the open at the tendering stage. It should be stated when flexibility is allowable, rather than attempting to force the supplier to comply with fixed terms, often at additional cost.

60.6 Golden Rules

It is important for the end-user to specify requirements rather than solutions. The URS is a requirements specification. It is up to the supplier to interpret those requirements and tender a solution utilising the functionality of the system being offered.

The URS is invariably and inevitably incomplete. The supplier therefore has to interpret the requirements and interpolate between them. In general, the better the quality of information provided, and the better the understanding so obtained, the less likely it is that the project will go wrong.

Encourage the suppliers to visit the plant and discuss specification queries. Much can be learnt on both sides with such a meeting.

Different suppliers' systems and packages may well have different functionality, but still be able to meet the requirements specification. Care should be taken not to generate detailed specifications for packages which the supplier provides as standard.

Do not be tempted to develop the URS by working backwards from the detailed description of a particular supplier's system or package: it will be difficult for other suppliers to satisfy such an URS. Also, it is obvious when this has been done and suggests that the key decision about the choice of supplier, system and/or package has already been made. The tendering process is reduced to a charade.

A good quotation is one which enables the supplier to meet fully the user's specified requirements, subject to agreed quality procedures. It must be cheap enough to be accepted by the end-user but at the same time provide enough profit to support the viability of the supplier's business.

Tendering and Vendor Selection

61.1 Tender Generation
61.2 Compliance Commentary
61.3 Quotation
61.4 Hardware Estimates
61.5 Software Estimates
61.6 Worked Example
61.7 Tender Analysis and Vendor Selection

As with the previous chapter, this one addresses the turnkey scenario and covers tender generation, compliance commentary, tender analysis and vendor selection. In essence, on the basis of the user requirements specification (URS), suppliers are invited to tender for a contract to deliver a system. Having received the tenders, the end-user then has to analyse them to decide which best satisfies the requirements and meets the cost criteria.

Three tenders should be sufficient and four the maximum, provided they are from appropriate suppliers. Much effort is involved in tendering and the end-user is unlikely to be able to meaningfully analyse more than three or four tenders. Frivolous invitations to tender, perhaps to satisfy some corporate purchasing policy, are in nobody's interest.

61.1 Tender Generation

The objectives of the tender document are to:

- Give the end-user confidence in the supplier's capability to deliver a system.
- Demonstrate the supplier's experience and expertise relevant to the project.
- Commit the supplier to providing a system that satisfies the end-user's requirements.
- State unambiguously what the proposed system consists of.
- Explain exactly how the system proposed satisfies the requirements specification.
- Clearly identify those aspects of the specification that are either only partially satisfied or not satisfied at all.
- Provide a price for the contract, in the form of a quotation.
- Form a sound contractual base for the project.

Prior to generating the tender, the supplier's engineers should read the URS very carefully in order to understand the process and the intent with regard to automation. Any queries should be noted and a meeting with the end-user requested to clarify them. A site visit to the plant, or one like it, is useful to "get a feel" for the process.

As stated in Chapter 60, the URS should define the problem rather than formulate solutions. However, if solutions have crept in, then it is important to assess the end-users reaction to alternative solutions and what potential there is for flexibility. Lack of flexibility can lead to extensive "special" system software and unnecessary cost.

61.2 Compliance Commentary

This is a major part of the tender. It provides a commentary on the requirements specification, clause by clause, indicating compliance or otherwise. Every requirement should be addressed. The numbering system used for the compliance commentary should be consistent with that of the requirements specification to enable meaningful cross referencing.

At the beginning of the compliance commentary is a description of the system proposed in terms of system hardware, plant and operator interfaces, system software and application software. System manuals should be provided specifying standard functionality. Also, a recommended spares holding should be stated.

The proposed solution to each requirement should then be briefly outlined. It is to both parties advantage that this be achieved, as far as is practicable, using the standard functionality of the system software and hardware. It is cheaper, the software should be field proven and the system more reliable.

When aspects of the specification are not satisfied, give reasons and propose alternative solutions as/if appropriate. These alternatives should be specified in sufficient detail for the end-user to be able to assess them. Any "special" system software needed should be identified and an estimate of the additional cost given.

A provisional project plan should be included in the form of a Gantt chart which depicts projected effort and time scales. This can be used by the end-user to substantiate the application software estimates. These must include provision for management of the application software effort which is significant for large turnkey projects.

If the time scales are such that the deadlines for delivery cannot realistically be met, then the supplier should say so. Other suppliers will have the same problem, unless they haven't understood the requirements properly! Unrealistic expectations by end-users about time scales and self deception by suppliers are common. They are also inconsistent with delivery on quality. Corners cut at the tender stage invariably cause costly delay.

The supplier's policy on quality should be explained, together with a summary of its quality procedures. A quality plan should be provided for the project which identifies validation activities, responsibilities and procedures. This is covered in more detail in Chapter 66. Clearly aspects of the quality plan may be integrated with the project plan.

Fundamental to the ability to deliver on quality are the experience and expertise of the supplier's personnel. Typical curriculum vitae of key personnel such as the contract manager, application software development manager and engineers/programmers should be provided.

It is useful to include a summary of the company's history and stability, as well as a commentary on the company's experience of similar projects for the same or other end-users. The importance of such information for the confidence of the end-user should not be underestimated.

61.3 Quotation

Generating the quotation is essentially a question of systematically identifying everything that has to be supplied in terms of deliverables, and the work entailed in producing them. The principal categories of deliverables are system hardware, plant and operator interfaces, system software and application software. The quotation should also provide for services, such as training and support during commissioning.

The quotation should be on an itemised basis, otherwise there is scope for misinterpretation. It should clearly distinguish between those parts of the project being handled on a fixed-price lumpsum basis and those parts that are on a reimbursable basis.

Contractual aspects of the quotation to be addressed include deadlines for deliverables, penalty clauses, stage payments, provision for inflation, arrangements for handling variations, quality procedures, acceptance arrangements, software licence agreements, maintenance contracts, *etc.*

61.4 Hardware Estimates

The system hardware and plant interface costs are largely determined by the point count and the requirements for redundancy and/or segregation. Knowing these, the number of I/O cards, racks, frames, cabinets, processors, memory, *etc.* can be estimated. List prices for hardware take into account the cost of assembly, testing, documentation, *etc.*, and are based on experience of building previous systems. They obviously allow for profit too.

Estimates for the hardware costs can only be meaningfully provided by systems' suppliers in the form of written estimates. It is important to understand the pricing policy which varies significantly between suppliers, as depicted in Figure 61.1 which plots system cost against number of I/O channels.

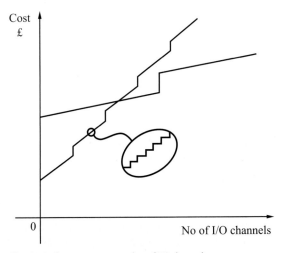

Fig. 61.1 System cost *vs* number of I/O channels

The intercept and slope depend upon the expandability of the system. Some suppliers offer very competitive prices for small "entry level" systems, but the incremental cost per additional loop is high making them unsuitable for large applications. Other systems have a high system cost but can be expanded relatively cheaply: these are usually unsuitable for small applications. Note that there are jumps in the plots due to constraints on the expandability. For example, the number of I/O channels can be increased until a new card is required, or the number of cards can be increased until a new rack is required, *etc.*

The operator interface consists of operator and engineer stations, workstations and peripherals. Numbers of such are largely determined by the requirements for access to the system and redundancy. Estimates for the operator interface are made on the basis of suppliers' list prices.

61.5 Software Estimates

System software is best treated as if it were hardware and costed by the supplier.

Application software has to be estimated on the basis of the effort required to produce it. Critical to the effectiveness of the estimating process are the supplier's metrics gained from experience of working on similar projects, and confidence in those metrics. Typical metrics are as follows:

Configurable software:

- Configurable functions and database blocks:
  i. Continuous: approximately ten analogue I/O or five loops per day.
  ii. Discrete: approximately ten discrete I/O devices with logic blocks per day.
- Configurable displays: some 20 bar displays, trend diagrams, *etc.* per day.
- Semi-configurable graphics: say 0.5 mimic diagrams per day.

Procedural software for sequencing and batch and high level language applications:

- Procedural code: average of 10–15 instructions, rungs or actions, *etc.* per day.

The above metrics all allow for the cost of design, development, testing and documentation of the software. The metrics assume that all the necessary tools are available for development and testing. Allowance must be made for the variety of expertise and experience of the personnel involved, and for managing this activity, as indicated in the example below.

It is also assumed that the detailed functional specification (DFS), from which the design of the application software is realised, will be fully available at the design stage. For the configurable software this comprises information about the I/O channels, derived variables, loops, control schemes, *etc*. For the procedural software it consists of procedures with parameter lists, the detail of the procedures being in the form of sequence flow diagrams, sequential function charts, or otherwise. The metrics do not allow for the effort involved in specifying these detailed requirements. This critical task is normally carried out by the end-user and should be provided for under the end-users project management costs.

It is particularly important to check that the estimates for the amount of sequence software required look realistic. These will have been determined at the URS stage by scaling up from estimates based on representative sequences, and will almost certainly have to be confirmed at the DFS stage. There is much scope for error. What is often overlooked is the extent of software required for error handling, safe hold-states and recovery options. As noted in Chapter 29, it is not unknown for this to account for 60–80% of the procedural software alone.

All application software estimates should be given in detail, with the metrics used provided to substantiate the estimates.

61.6 Worked Example

Suppose that an analysis of the sequencing requirements yields the following estimates:

Number of procedures = 6 and average number of operations per procedure = 5

Assume number of phases per operation = 12

Hence total number of phases = 360

But by inspection there is some 75% commonality.

Thus number of different phases = 90

Typical number of steps per phase = 8 and approximate number of actions per step = 10

Hence the total number of lines of code to be designed, developed and tested = 7,200

Use metric of 10–15 lines of code per person-day, say 70 lines per person-week.

Hence 103 person-weeks effort required assuming experienced personnel.

Allow an efficiency of 80% for holidays, *etc*. gives 129 person-weeks effort.

Assume a lapsed time of 25 weeks to meet the project deadline.

Allow 2 weeks for delivery gives 23 weeks per person.

Hence no of persons required

129/23 = approximately 5.6

Next to be taken into account is the programming experience of the engineers, for which some typical values are as indicated in Table 61.1:

Table 61.1 Programming experience *vs* efficiency

Experience (months)	Efficiency (%)
0–6	60
6–12	70
12–18	80
18–24	90
24–30	100

Assume that the pool of engineers likely to be available for the programming have the following mix of experience: 0, 5, 10, 10, 15, 20, 20, 25, 30 and 50 months. The average team efficiency is thus approximately

$(2 \times 60 + 2 \times 70 + 80 + 2 \times 90 + 3 \times 100)/10 = 82\%$.

Hence number of persons required

5.6/0.82 = approximately 6.8.

But this software team needs to be managed. The management effort required can be estimated from Table 61.2.

Hence a team of approximately 6.8 persons needs something like 90% of a full time manager.

It would clearly make sense in this case to allocate a team of eight persons to the project of which one would be a senior engineer dedicated to its management. The excess of 0.3 persons would provide some slack: it is better to have a little slack than be too tight.

Table 61.2 Management effort *vs* team size

Team size	% time
0	0
1	16
2	31
3	45
4	58
5	70
6	81
7	91
8	100

Such calculations of the software effort, showing the assumptions made and metrics used, should be included in the quotation so that, as the need for changes are revealed at the DFS stage, the estimates can be re-calculated on the same basis. Or, if an alternative basis has to be chosen, they can be used to justify the difference.

Overtime should ideally only be used for emergency purposes. A small amount, say 10–15%, can be sustained over a reasonable period of time but beyond this it results in a loss in efficiency, not to mention morale.

61.7 Tender Analysis and Vendor Selection

It is important that the tender analysis is thorough and the vendor selection is objective, otherwise there is no point in going through the process of tendering.

Also of fundamental importance is the need to determine the best system for the application on the basis of its functionality and life cycle costs. Note that the latter takes into account both the capital costs as well as the maintenance and support costs. Essentially there is a trade-off between functionality and costs, the best buy being some optimum, recognising that tenders rarely comply fully in all areas. The lowest priced quotation is a flawed approach because it implies that all tenders are fully compliant.

It is recommended that:

- The tenders are read carefully. Any query should be clarified with the supplier. If any misunderstandings are identified, the supplier should be asked to amend the tender and revise the quotation as appropriate. At the end of the day, it is vital to compare apples with apples.
- Visits be arranged to one or more of the suppliers' existing end-users, especially if they have similar automation requirements. A great deal of useful information, both technical and about the supplier, can be gained from such visits.
- The requirements be prioritised and costs balanced against proposed functionality. This is best done by means of a spreadsheet, as follows.

Create a separate sheet for each distinct set of requirements and allocate a weighting factor to every sheet. Typical sheets would be for costs, commercial considerations, project implementation capacity, user interface, sequence/batch capability, development environment, *etc*. For each sheet identify the key facets of interest. Then, for each facet for each system, a score can be entered, say on a scale of 0–10. By summing the scores for each system and taking a weighted average across the various sheets an aggregate score is achieved: the highest score is the best.

The principal advantage of such an approach is that the basis of vendor selection is transparent and objective: especially important when there are cultural and ethical differences between supplier and end-user and political considerations are brought to bear. The effects of changing weighting factors and scores can be explored, which gives a feel for the sensitivity of the criteria being used.

The end-user is then in a position to award the contract. In recognition of the effort involved in tendering, a common courtesy is to provide some feedback to the unsuccessful suppliers as to why their tenders were not accepted.

Functional Specifications

62.1 Methodology
62.2 Participants
62.3 Contents of DFS
62.4 Acceptance

Following completion of the system selection stage of a project's life cycle comes development of the detailed functional specification (DFS), as seen in Table 59.1. This is a complete and unambiguous statement of requirements which supersedes the user requirements specification (URS). The reader is referred, in particular, to the IEE guidelines (1990) for a comprehensive treatment of the content of the DFS. Note that, in this chapter, the DFS embraces both hardware and software.

The purpose of the DFS is to:

- Finalise the scope of supply, both hardware and software.
- Allow the supplier to finalise planning of resource requirements.
- Enable a delivery date to be fixed.
- Become the principal reference for both the supplier's and end-user's engineers.
- Help generate an atmosphere of mutual confidence and respect.

62.1 Methodology

The DFS is realised through an application study which is a joint study between the supplier and end-user with the supplier taking the lead role. The URS, compliance commentary and the quotation are used as the basis for the study. The study consists of a detailed and systematic discussion of all the requirements, general and particular, specifying the solutions in terms of the functionality and tools of the system selected.

Joint meetings should take place, typically once or twice per week over a period of one to four months depending on the quantity of application software required. Minutes should be produced by the study consultant immediately after each meeting and circulated to all members of the application study group. These minutes should include:

- A summary of the technical discussions.
- An action list identifying actions as completed or outstanding, persons responsible and completion dates.
- Outline topics for discussion at the next meeting together with any data to be gathered.
- The results of any post meeting analyses carried out by the study consultant on implementation issues.

It is useful to standardise the structure and reference number system for the minutes, such as meeting number, topic, sub-topic, *etc*. This makes it easier to collate discussions on the same topic which took place over a number of meetings. It also helps to prevent topics being forgotten, especially when there are many meetings.

It is important that the DFS is complete and agreed by the end-user before it is signed off. Clearly the more complete and detailed it is, the more precise are the estimates for software effort, resource requirements, project plans, delivery dates and costs. Once the DFS is agreed it becomes a contractual document, superseding the URS and subject to formal change control procedures. It contains revised estimates for effort required and a revised project plan for implementation and delivery. Provided there are no more changes, these estimates can be considered as final.

In revising the estimates for the effort involved, particular attention needs to be paid to the assumed values for the decomposition of procedures into operations, phases, steps, *etc.*, and to the estimate of the commonality. Inspection of the worked example of Chapter 61 reveals how sensitive the effort calculations are to these assumed values.

Implementation must not commence until the DFS is agreed. End-users invariably underestimate the amount of information they still have to provide at the DFS stage: acquiring that information often becomes the rate determining factor. The urge to proceed with implementation beforehand, because the DFS is taking longer than originally planned, should be firmly resisted. The inevitable consequence of late changes to the DFS is unplanned software modification. This causes chaos with the project plan leading to wasted effort and delay: even small changes can be pervasive and have a major impact. Changes increase cost, the later the change the bigger the increase in cost. In fact, the increase in cost is roughly proportional to the exponential of the lapsed project time.

62.2 Participants

The application study group should have a core of four or five members, drawn from both the supplier and end-user, augmented by specialists as appropriate. It is essential that the study consultant and the end-user's project manager have the authority to make the majority of technical decisions without having to refer back into their own companies.

The supplier's team comprises:

- Study consultant, the lead engineer. A senior employee with an in-depth understanding of the system's functionality and experience of applications similar to the end-users requirements.
- Application software manager, for turnkey projects only. This person will manage the team of application engineers who will design, develop and test the application software.
- Specialists, as and when required, in networking, system software, hardware, *etc.*
- Contract manager, as and when required, whose responsibility is for the whole contract embracing hardware and software, including commercial considerations.

The end-user's team comprises:

- Project manager, hopefully the control engineer responsible for installation and commissioning of the system.
- Process engineer, who has detailed knowledge of the process and the design of the plant.
- Instrument engineer, as and when required, whose responsibility is for the instrumentation and electrical side of the project.

62.3 Contents of DFS

The DFS is usually a substantial document running to several volumes depending on the quantity of application software. It should be self contained and provide a detailed understanding of the project, its automation requirements and implementation. As such the DFS is a reference document for both the supplier's and end-user's teams throughout the subsequent stages of the project. All the supplier's team members should be familiar with its content because, although each will only be involved in implementing part of it, it is important that they understand the overall requirements.

Typical contents are:

- Introduction.
- Standard system description:
    i. An overview only, with reference to standard system manuals as appropriate.

ii. Hardware details: final configuration, I/O point count, *etc.*
  iii. Physical layout: siting of system cabinets, I/O racks, terminals, *etc.*
  iv. Power and wiring requirements: lengths and type of multicore cables, plugs *etc.*
  v. Memory requirements, timing and loading estimates.
  vi. Performance metrics: accuracy, resolution, speed of response, *etc.*
  vii. System attributes: reliability, redundancy, maintainability, adaptability, *etc.*
  viii. System malfunctions: failure modes, fail-safe states, emergency operations, *etc.*

- Non-standard system requirements – detailed description of any non-standard requirements and includes:
  i. Links to other machines: user interface, drivers, protocols, connections *etc.*
  ii. Operator interface: non-standard displays.
  iii. Packages: database requirements, hardware platform, security, *etc.*

- Process description – description as per URS, amended by the end-user if necessary.

- Automation requirements:
  i. P&I diagrams.
  ii. Control philosophy as per URS, amended by the end user as appropriate.
  iii. Particular requirements as per the URS, but expanded upon as appropriate in the light of the application study.

- Functional specifications:
  i. Detailed explanation of the solution to every particular requirement, with cross references to minutes of the application study group meetings. This embraces control schemes, sequencing, operator interface, *etc.*
  ii. Operability issues: initialisation, status on start-up and shut-down, recovery from failure, provision for operator intervention (scheduled or otherwise), access security, on-line support, fail-safe states, *etc.*

- Resource estimates:
  i. Updated project plan and revised delivery date, if agreed.
  ii. Estimates for software effort, showing basis of calculations. Significant changes from those in the URS should be identified with "before" and "after" calculations.

- System acceptance specification for testing, as explained below.

- Appendices:
  i. Minutes of application study group.
  ii. Tables of outcomes of CHAZOP and COOP studies.
  iii. Configuration charts or otherwise for control loops and schemes.
  iv. Templates for mimic diagrams.
  v. Sequence flow diagrams for all sequencing requirements.
  vi. Software designs, *e.g.* function block diagrams and sequential function charts, if available.

The amount of time and effort required to produce a quality DFS cannot be over-emphasised. Whilst it is relatively easy to articulate requirements for situations when everything is going according to plan, what takes far more effort than is usually appreciated is the handling of abnormal situations. For continuous control this essentially concerns change of mode and operability issues. For sequencing and batch process control, the issue is safe hold-states and recovery options. These are the solutions to "what if" questions within the sequences and, as pointed out in Chapter 29, can account for some 60–80% of the sequence software.

62.4 Acceptance

It is important to distinguish between factory and works acceptance. Factory acceptance refers to factory acceptance tests (FAT) carried out at the supplier's factory, prior to delivery of the system to the end-users works. Works acceptance refers to acceptance tests carried out at the end-user's works after installation of the system has been completed. It is normal for the majority of acceptance testing to be done at factory acceptance where all the equipment and support systems are available.

Whereas the hardware, once it has been assembled, can be tested and accepted relatively quickly, the application software is potentially much more complex and takes far longer to test. It is not practicable, for example, to defer comprehensive witness tests until all the development has been completed. So, acceptance of the application software is not straightforward and the basis and time scale for its acceptance has to be agreed in advance.

Is it to be done by means of spot checks or by witnessing the functioning of every single line of code/function? When is it to be done and by whom? What notice is required? Is the testing to be model based or by using real hardware? What equipment is required? How are the results of testing going to be collected. What documentary evidence is required? Once it has been accepted, what provision is there to safeguard its integrity? What are the arrangements for handling mistakes found during acceptance. What testing is going to be left for works acceptance? And so on.

It is of fundamental importance that there be a clear understanding of the basis for acceptance because it is normal for a substantial part of the cost of the system be paid to the supplier upon successful completion of acceptance. Unless there have been previous stage payments, this can typically be some 80% of the contract value. For this reason, it is common practice to make provision for acceptance within the DFS which, as noted, is of a contractual nature. This normally leads on to a separate software acceptance specification and to an acceptance test schedule.

Design, Development, Testing and Acceptance

63.1 Waterfall Model
63.2 Structured Programming
63.3 Software Design
63.4 Module Design
63.5 Walkthroughs
63.6 Software Development
63.7 Support Tools
63.8 Test Specifications
63.9 Test Methods
63.10 Module Testing
63.11 Software Integration
63.12 System Integration
63.13 Acceptance Testing
63.14 Project Management
63.15 Personnel Management
63.16 Comments

As with previous chapters, this one addresses the turnkey scenario and covers the design, development and testing of the system. These activities are carried out by the supplier, as indicated in Table 59.1, the end user becoming involved again at the acceptance stage. This chapter concentrates on the application software since hardware design for process control, which comprises the configuration of the supplier's standard units, is invariably relatively straightforward. For a more thorough treatment of the issues the reader is again referred to the STARTS handbook (1989), the IEE guidelines (1990) and the GAMP guide (1995).

63.1 Waterfall Model

An understanding of the system life-cycle provides the basis for this chapter: it is detailed in the draft standard IEC 61713. In essence, a system development project is broken down into logical stages, as depicted in the form of the so-called waterfall model of Figure 63.1. There is a direct correspondence between the various design and development stages and the testing and integration stages with which they are aligned.

The waterfall model is generic in the sense that it applies to any system development of which there is a large software component. It is particularly appropriate for projects where the application software is to be developed and it can be presumed that standard system software (tools, operating system, *etc.*) exists. The model does largely apply to pro-

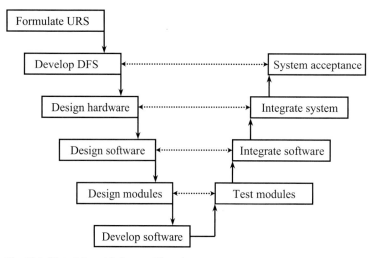

Fig. 63.1 Waterfall model of system life cycle

cess automation projects, although some stages are much more significant than others. Nevertheless, all the stages do exist and the model provides a useful framework for any project.

Starting with the DFS, it breaks the project down into a hardware design and a software design. It might seem odd that hardware design is included in what is essentially the software life cycle but this is necessary because, ultimately, the software has to be integrated with the hardware for testing and acceptance purposes.

Software design essentially consists of partitioning, or decomposing, the software requirements for the system into sub-systems or modules. The modules are autonomous, as far as is reasonably practicable, so that they can be developed independently of each other. It follows that a major consideration at the software design stage is the interactions between the modules and the interface between the modules and the rest of the system. Of particular interest here is the control of modules, such as their mode of operation, and the flow of data and status information between them.

For every module a detailed specification is developed in terms of its functionality and data requirements. The software for each module is developed and tested against that module's specification. Modules are then integrated, i.e. combined in groups as appropriate, and tested against those parts of the software design that applies. The software can then be installed on the system hardware and system integration tests carried out. Finally, the integrated system can be tested against the DFS for acceptance purposes.

The principal advantage of the waterfall model is that each stage is defined in terms of its deliverables, i.e. designs, test procedures, code, etc. and is not complete until the deliverables have fully materialised. The stages therefore can be considered as project milestones. If metrics exist for each milestone, they can be used to quantify progress with the project.

As a general rule, the next stage of a project should not start until its previous stage is complete, but it is important not to apply this rule too rigidly. If the specification for a standalone module has been completed, there is no reason why coding of the software for that module cannot start, even though other module specifications are incomplete.

Sometimes, software development requires a certain amount of experimentation to establish the general design concepts. The rigid staged approach of the waterfall model is inappropriate for this and prototyping should be allowed to proceed with the minimum of formal procedures until the concepts become established. Then, the discipline of developing design specifications and formal testing is imposed.

The waterfall model encourages a structured approach to software development. It should be evident that for each stage there has to be a careful analysis of the design requirements prior to the detailed software design itself.

63.2 Structured Programming

Structured programming is an approach to analysis and design that is strongly advocated by software scientists. The two most commonly referred to methods are due to Jackson and Yourdon: both are consistent with the waterfall model. Their approach comprises:

- Decomposition of complex systems into simpler sub-systems or modules.
- Analysing, designing and implementing the interfaces between the modules.
- Analysing, designing and implementing the processes that go on within the modules.

Structured programming relies heavily upon diagrammatic representations of structures, data flows and processes:

1. Structure charts. These depict the decomposition of the software requirements into modules, as shown in Figure 63.2, and show the overall software structure. In particular, they indicate the relationships between the modules, *i.e.*. whether they are in parallel or sequential. Each module is controlled by the next highest in the structure. The convention is that modules would be executed in the order depicted unless the logic within the modules determines otherwise.
The approach of decomposition first, followed by the detailed design of individual modules, is often referred to as being "top-down".

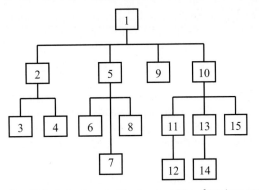

Fig. 63.2 A structure chart for decomposition of requirements

2. Data flow diagrams. These are an analysis tool, the basic components of which are as depicted in Figure 63.3. The circles indicate a process that adds to, modifies or deletes data within the module and the arrows indicate the direction of flow into or out of the processes. Boxes indicate an external entity: a person or another module outside the boundary of the module being studied. External entities originate or receive data being processed by the module. Rectangles indicate data storage, either of a temporary nature or a permanent data file.

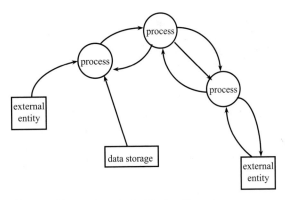

Fig. 63.3 The components of a data flow diagram

It is important to appreciate that data flow diagrams do not indicate how the a module is to be physically realised. They can be developed independent of any hardware considerations.

3. Process descriptions. Once a data flow diagram is complete, the operations that go on within

each process have to be specified in sufficient detail to provide a basis for complete and unambiguous coding to be developed. It doesn't matter how they are articulated: the purists advocate both decision tables and decision trees but, in practice, the most common means of representation are pseudo code, sequence flow diagrams (SFD) and sequential function charts (SFC). All of these are described in Chapter 29.

A sacred cow in structured programming is the avoidance of jumps in program flow, other than forwards. Thus designs in pseudocode should avoid the use of "goto" and "jump to" label types of construct. Branching has to be realised implicitly by means of the conditional and iterative constructs, and the nesting thereof. This prevents the development of spaghetti code and makes for easier understanding, testing, *etc*.

Another important principle is that each sub-module, *e.g.* phase, should have a single entry and exit back to its parent module which makes for good maintainability. This point is fundamental to the S88 concept of creating procedures and/or operations by configuring pre-defined phases from a library.

The concepts of structured programming are applicable to process automation, but only to a limited extent. That is because the majority of projects are based on an existing control system, such as a DCS, SCADA or PLC, chosen through the vendor selection process. Such systems have their own software environments and support extensive standard software, both for both configurable functions and procedural code. These impose their own structure on the application software design.

Structured programming is therefore most useful in relation to those parts of a project that are not structured a priori. The obvious example is in the use of higher level language programs used for realising more advanced aspects of process control, in optimisation for example.

63.3 Software Design

Starting with the DFS, and assuming the hardware design exists, the application software may be designed. Remembering that the objective is to decompose the software requirements into modules that are as autonomous as is practicable, the top down approach of Figure 63.2 prevails. First, the application software is categorised according to the software types supported by the system, typically:

- Configurable for analogue functions such as control schemes, loops and channels.
- Semi-configurable for displays, logs, *etc*.
- Procedural for sequencing, batch process control, *etc*.
- High level language for more esoteric applications.

For the configurable software, design is essentially a question of identifying every analogue control scheme, decomposing each into control loops, channels, *etc*. and allocating them to modules as appropriate. Similarly for discrete devices, logic functions and channels. The subsequent module design is straightforward.

For the semi-configurable software, module design is almost trivial. It is essentially a question of identifying the displays, trends, *etc*. required and grouping them as appropriate within the display hierarchy as described in Chapter 42.

For the procedural software, the top down approach of structured programming applies resulting in structure charts. Decomposition into modules is largely determined by the software structures supported by the system. For example, using S88 constructs, decomposition of procedures would typically be into operations and phases according to the methodology outlined in Chapter 37. The decomposition is a non-trivial task because, as emphasised in Chapter 37, great care has to be taken in defining the boundaries between entities such that the correct relationships between the procedural and physical models are established. Note that the phases are identified by name and function only at this software design stage.

Any non-trivial high level language applications could be decomposed into modules according to the principles of structured programming. However, these are few and far between: for most systems the emphasis is overwhelmingly on the configurable and procedural categories.

The software design provides the basis for developing specifications used for testing the modules at the software integration stage. It also provides the basis for any preliminary COOP studies, as described in Chapter 54.

63.4 Module Design

For configurable functions the design is relatively mechanical. For each module standard function blocks are identified which, when configured, will realise the functionality required. The detailed design comes in the specification of the corresponding data blocks. The design thus consists of configuration diagrams and database tables. Examples are provided in Chapters 44–46.

For sequencing applications the detailed design is much more involved. The basic approach of developing a data flow diagram for each module applies, as per Figure 63.3. This may not be necessary for every module: it all depends on the complexity of the module and the structure imposed by the system. Noting that a module typically equates to an S88 operation, the processes (circles) identified in the data flow diagrams may be related to phases and/or steps. These then have to be specified in detail, normally by means of pseudo code, SFDs and SFCs, all of which are described in Chapter 29:

- Pseudocode: typically simple structured English language. This is relatively easy to write and understand and converts readily into code such as structured text.
- Sequence flow diagrams: probably the most popular means which are also easy to write and understand. They give the impression of being structured but are in fact free format.
- Sequential function charts. These are the highest level of the IEC 61131 languages and are very structured, perhaps too much so for design purposes, but are directly executable.

For high level language applications the same basic approach of developing a data flow diagram for each module is appropriate, as per Figure 63.3, resulting in detailed designs for each of the sub-modules identified.

63.5 Walkthroughs

A design walkthrough helps to verify that a module's design fulfils the requirements of the DFS. A formal meeting, *i.e.* one that is minuted, is arranged between the designer of the module, the team leader, and one or two other team members whose own designs interface to the module under consideration. The relevant documentation is circulated prior to the meeting and the designer "walks" the team through the design. It is important that there be a constructive attitude during this process: designers are often overly sensitive to criticism! Such walkthroughs provide the opportunity to ensure that the module's functionality and interfaces are correct. It is not uncommon for such walkthroughs to lead to significant design changes.

If the process and/or plant is of a hazardous nature and malfunctions and/or mal-operation of the control system can potentially contribute to the occurrence of hazardous events, then the design of the application software should be subject to a formal COOP study rather than to a simple walkthrough. This is as described in Chapter 54.

63.6 Software Development

Configuration and coding can commence once the module designs and module test specifications are complete. These clearly have to use the standard functions and languages supported by the target system.

It is important to encourage good configuration and coding practices since these make the application software more understandable and hence easier to modify and maintain. To this end, suppliers will have documented guidelines for good practice in addition to their normal programming manuals and the quality procedures must ensure their adherence.

As a module is being configured or coded, the programmer should be encouraged to check it as carefully as possible. Upon completion, the module should be informally checked for consistency with the module design and conformance with the relevant guidelines and standards. This will generally involve the programmer and team leader only. It is not necessary at this stage, but does no harm, to refer to the test specification.

63.7 Support Tools

Both computer aided software engineering (CASE) tools and integrated programming support environments (IPSE) exist to support the development of software and can make a major contribution to the quality and cost effectiveness of the product. CASE tools alone are little used by suppliers involved in application software development: the principal benefits are realised from an IPSE.

CASE tools are computer based means of developing and maintaining all the documentation generated during the life cycle of a software project. This includes both data flow diagrams, so good quality graphics are essential, and pseudocode. A particularly powerful CASE feature is the ability to enable executable prototypes to be developed from user generated displays. Thus, for example, the structured text could be automatically generated from a sequential function chart.

Typically, an IPSE is a network of PCs with software for supporting both software development and project management. In addition to CASE tools, an IPSE provides a language sensitive editor (LSE) and a code management system (CMS).

The LSE is used for code development. It provides templates of standard language constructs for the programmer, based on the supplier's guidelines for good practice. These can be cut, pasted and edited as appropriate for the program. So, not only does this speed up software development but also encourages conformance with the guidelines.

CMS automatically provides a complete version control system by logging all changes between one version of a program and the next, recording parent-offspring relationships. Also, the application software comprises many modules, several thousand is not uncommon, most of which are realised by separate programs. So, another important feature of CMS is its system build facility which ensures that when the application software is compiled, no programs are inadvertently omitted and the latest version of each is used. Version control is fundamental to quality procedures.

Other benefits of IPSE include:

- Availability of all software on-line, both for product and project management.
- Ability to develop application software away from the target machine.
- Automatic file back up.
- Sophisticated configuration and text editing facilities.
- Logging of standard software faults and fixes.
- Facility to test software using simulated I/O signals and track the response.
- File ownership enabled by grouping of files on user/account/project basis.
- File protection based on ownership and access privileges for users and projects.

There is evidence that use of an IPSE for application software development increases efficiency by some 25% compared with conventional means.

63.8 Test Specifications

It is at the software design stage of the waterfall model, where the interactions between the modules are first considered, that the test specifications for each module are formulated. They will almost certainly have to be refined and expanded in the light of the detailed module designs. Many of the requirements for testing will not be identified, or cannot be resolved, until the detailed module design stage.

The key issues that must be addressed for each module, and for logical groupings of modules, are the:

- Functionality of the module.
- Expected outcomes for given input scenarios.
- Data to which the module requires access.
- Initialisation of the module in terms of data and status information.
- Basis for entering and exiting a module.
- Means of controlling the module's mode, *e.g.* run, hold, stop, *etc.*
- Data within the module that is accessed and/or changed by other modules.
- Scope for the module producing invalid results.
- Robustness of the module at handling/rejecting invalid inputs.

There should be a separate test specification for each module, and it must be possible to relate an individual test to the module to which it belongs. Each specification must specify the test environment, the tests themselves, test inputs and expected outputs. All the tests for a particular module should be held in a single document which bears the module name for traceability. Note that these are different from the test specifications required for software integration.

The module test specifications cannot be finalised until the detailed module design is complete, and software development should not commence until the test specifications have been formally approved. These test specifications are the basis for module testing and for the software integration specifications. As such, they become an integral part of the software quality procedures.

63.9 Test Methods

Testing is realised by means of a mixture of simulation and physical methods. The methods used vary according to availability of equipment, to the functionality of the system, to the time scale and size of the project, and in accordance with what was agreed by the end-user in the DFS. There is no single correct approach.

In relation to the waterfall model, at the lower levels of module testing and software integration, testing tends to be done by means of simulation, especially if the development is IPSE based in the first place. Physical means tend to be used at the higher levels of the model, especially where the end-user is involved for acceptance purposes.

However, it is worth noting that physical means require access to the system's hardware which can become a project bottleneck. Testing by means of simulation decouples testing from hardware availability. This means that simulation based testing of modules and software integration can start earlier and in parallel to hardware testing. There is much evidence that this approach can substantially reduce the time scale and cost of system development.

Testing by simulation means testing the functionality of the module with simulated I/O signals. The principle is simple: known simulated inputs are applied to the module, separately or in combination as appropriate, the module is executed, and the real outputs are tested for expected values.

Analogue I/O signals are created by setting values of signals in the database at known values and either holding them constant, applying step changes, or ramping them up or down at user defined rates. Feedback can be established by connecting outputs to inputs and incorporating scaling factors, lags and delays to provide some dynamics. Discrete I/O signals are simply set true or false in specified combinations and orders, often driven by a user defined table.

A powerful extension to this, supported by more sophisticated systems, is to use the application software to test itself. Thus, for example, the sequence executive (SEX) enables a sequence to be run in a test mode in which the sequence is disconnected from the real I/O. Analogue loops and discrete devices are automatically closed with pre configured lags, delays, ramps, *etc.* to emulate the plant. Alternatively, inputs can be configured not to respond to check faulty device operation. The sequence is then executed and a monitor tool, as described in Chapter 41, used to observe progress.

An alternative IPSE based means of testing modules is to write a separate test program which is designed to exercise the module thoroughly and to log the response. The test program consists of a string of commands written to vary the inputs, or order and combination of inputs, in a specified way. It can also include commands for testing the effect of fault conditions, change of mode, *etc.* The module being tested and the test program are run simultaneously: both interact with the same database values.

Testing by physical means requires that the hardware of the system being delivered is available. Or, if the application software only is being tested, then some equivalent development system must be available. The following test methods apply:

1. I/O channels: for each channel, input or output, analogue or discrete, the signal must be exercised over its full range and the expected outcome observed. For example, varying an analogue input enables the functioning of the channel itself to be tested, the database values and the I/O function blocks' operation to be checked, and the changes observed at the operator interface as appropriate.

   The equipment required for these tests is relatively simple. In general, signal generators are used for producing known inputs, *e.g.* 4–20 mA, 0/10 V, *etc.* which are then observed using the system's database editor and/or display system. The system itself is used for generating outputs which are observed using ammeters and/or voltmeters. Additionally, for testing discrete I/O, switches and lamps are used.

   Note that this procedure simultaneously tests the functioning of the I/O channel hardware, database entries, function block operation and the user interface. Clearly there is scope for combining aspects of hardware testing with software integration.

2. Configuration: given that the I/O channels have themselves been tested, the remainder of configuration testing focuses on the functionality of the analogue loops and the logic of the discrete devices. The time taken to perform this testing thoroughly is often underestimated.

The order in which the function blocks, or algorithms, are executed in a loop can be critical and should be tested carefully. Some systems have a loop testing facility which allows each algorithm to be run in turn and the result inspected. This is particularly helpful for checking calculations.

It is also normal to test out the non-control functionality of loops and/or schemes at this stage. This embraces initialisation, fail-safe features, mode changes, fallback options, *etc.*

For configuration tests, in addition to equipment listed above, extensive use is made of the configurator and editor tools as described in Chapter 41.

Loop dynamics may be tested using the system itself to simulate plant dynamics, using holds, steps and ramps as described above. However, the testing of loop dynamics is normally deferred until commissioning when the loop tuning itself is carried out.

3. Sequences. The basic approach is to exercise the sequence by simulating inputs, analogue or discrete, as appropriate and observing the sequence's response. It is important that the sequence be tested both for likely as well as for unlikely, but nevertheless possible, combinations of inputs.

   In terms of equipment, the easiest approach is to use that listed above together with the system's own group displays and mimic diagrams. An obvious proviso is that the test boxes contain enough switches and lamps to simulate the I/O of any single sequence.

   An alternative, but more expensive approach, is to construct a simple mimic out of plant P&I diagrams glued onto hardboard. Stuck onto the mimic in the correct functional places are switches for discrete inputs and lamps for discrete outputs, together with potentiometers and ammeters for analogue signals if appropriate. These are then hardwired to the I/O cards and can be used for checking I/O signals instead of test boxes. Sequences are checked by manually operating the mimic's switches and potentiometers in response to the sequence driven lamps, setpoint ramps, *etc.* This approach provides a very good overview of all the sequences, is tangible to the end-user during acceptance, and is a valuable operator training tool.

For sequence testing access to a monitor tool, also described in Chapter 41, is essential.

An SFD or SFC is a useful vehicle for recording test progress. Each part is highlighted once it has been successfully tested and any untested backwaters are easily identifiable.

63.10 Module Testing

This is best done in three stages. The first two stages involve the programmer who developed the software for the module working alone as the tester. The third stage involves both the tester and a witness. It is usual practice for the witness to be another of the supplier's employees:

1. The software is carefully inspected and thoroughly tested against that module's test specification in an informal sense. Errors are identified and corrected as appropriate.
2. The software is formally tested against that module's test specification. Formal recording of the results of the various tests in the specification is done at this stage and any re-work and re-test are also recorded.
   It is not inconceivable that faults in the test specification itself will be found. Any amendments to such must obviously be subject to formal change control (as opposed to version control) procedures and properly recorded.
3. The software is formally spot tested by the witness who verifies that the test results recorded are valid. This is done on a sample basis with more detailed checks carried out if inconsistencies are revealed.
   The witness should be sufficiently familiar with the project to be able to make judgements about the effectiveness of the testing and to suggest additional tests if necessary. Any faults found should be handed back to the programmer for correction and re-testing.

It is normal practice for module testing to be carried out independent of the system software or indeed the hardware. Typically it is a mixture of paper based testing and software based simulation. Paper based testing involves careful and systematic checking of all the detail of the module against its specification. Simulation means testing the functionality of the module with simulated I/O signals, almost certainly within the IPSE.

63.11 Software Integration

Once the modules have been tested individually, they must be integrated and tested to check that they communicate effectively. There are two aspects to this: the interactions between the various modules, both configurable and procedural, and the interfaces between the modules and the rest of the system.

The interfaces with the rest of the system can only meaningfully be established if the application software is tested under the operating system being supplied, with the major system software packages as described in Chapter 41. This requires that the hardware is available, or that some equivalent development system is available for testing purposes, or that the IPSE is capable of emulating the operating system. Any of these scenarios is acceptable: it depends on the time scales involved, availability of equipment and systems, and what has been agreed with the end-user.

The software integration test specification should be based upon the software design and held in a single integration test schedule. The testing is best done by introducing modules one by one. The specification should be sufficiently flexible to enable the order in which modules are introduced to be changed. Thus, if unexpected interactions are revealed, or an interface is found to be faulty, integration can be switched to a different sub-set of modules whilst corrections are made.

Both positive and negative testing should be provided for. Positive testing checks for desired module interface functionality and, since this is known, these tests can be specified. Negative testing tests for undesired module interactions. These cannot be predicted and hence negative tests cannot be specified, other than to be aware of unexpected behaviour during normal testing and system use. If this occurs, the cause should be identified, corrected, and tests added to the integration test schedule to prove that the fault has been corrected.

63.12 System Integration

Assuming that hardware acceptance has been completed, system integration is simply a question of installing the application software on the target system and checking that it works under the real-time operating system. Provided that the application software has passed the software integration stage, and the correct version of the operating system is used, there is little scope for things going wrong. The system is then ready for acceptance testing.

The above presumes that the project concerns development of application software only, and that it is being integrated in an environment of proven system software. Where there is much scope for things going wrong is if the project has required changes to the system software. This can give rise to unexplained and/or unexpected interactions which can be very difficult to identify. The golden rule is to avoid changes to system software at all costs.

63.13 Acceptance Testing

As explained in Chapter 62, acceptance is normally a two-stage process: factory acceptance testing (FAT) and works acceptance testing. At this stage of the waterfall model, it is factory acceptance that is of concern. Works acceptance is covered in Chapter 64.

The basis for acceptance should have been agreed and incorporated within the DFS. The various acceptance tests necessary to satisfy that agreement should be combined into a single acceptance test schedule and agreed with the end-user.

FAT is essentially a question of proving to the end-user that the application software running under the real-time operating system on the hardware of the target system with simulated I/O is correct and functioning in accordance with the DFS.

It is good practice for the supplier to run through the complete FAT schedule prior to the end-user's visit. These tests should be formally recorded too and any faults corrected and re-tested.

The end-user's role in FAT is to verify that the testing and integration has been carried out properly and thoroughly, *i.e.* the emphasis is on verification. The supplier will have recorded the results of all module tests, software and system integration, and pre-acceptance tests. The end-user should only need to do spot checks on a representative and random sample of the tests. Provided all the spot checks verify those already recorded, there is a basis for confidence that the system has been thoroughly tested and works to specification.

Another school of thought is that all software and system integration tests should be witnessed and verified by the end-user. This can be a very time consuming operation which is open to misuse by end-users who have not properly articulated their requirements in the DFS. Acceptance provides an opportunity to correct this functionality at the expense of the supplier. This is contrary to a quality approach and contractual fair play as, apart from other considerations, changes made at this stage invariably have major implications on the software design with consequent re-work.

63.14 Project Management

So far, this chapter has concentrated on technical aspects of the software life cycle. However, there is a commercial basis to the cycle too, and the design, development and testing of application software has to be managed carefully. The key issues are making sensible estimates of the time and effort required at each stage, planning the deployment of both human and equipment resources, monitoring progress against milestones, ensuring that deliverables materialise and that quality procedures are adhered to.

Estimates of the software effort involved will have been made as part of the tendering process. Metrics were provided in Chapter 61, together with an example calculation. They allow for the cost of design, development, testing and documentation of the software. However, they do not distinguish between these activities and, for project management purposes, a further breakdown is required. A typical split in relation to the waterfall model is given in Table 63.1. Note that the split may vary according to the relative amounts of configurable and procedural software. Nevertheless, it gives a good feel for the proportion of effort required to complete each stage, both for the overall project and for an average module.

Table 63.1 Application software costs *vs* project stage

Stage	%
Design software	20
Design modules	15
Develop software	30
Test modules	20
Integrate software	15

Given the overall estimates of the software effort predicted in the tendering process, and the above split, it is possible to assign reasonably accurate estimates for the effort required for each stage for each module. It follows that an important management task, at the software design stage, is to make estimates of the effort required for each milestone for each module.

A project plan is required. This essentially allocates human resources, and perhaps equipment too, to the project. Initially the resource estimates will be of a global nature but when the software design is complete it should be possible to identify resources allocated to individual modules. A detailed Gantt type of chart for software implementation should be developed showing the tasks for which each programmer is responsible, with milestones. It is most important that these are realistic and should, for example, include holidays and allow for management effort and programmer efficiency: refer to Tables 61.1 and 61.2. Depending on the timescales involved and the resources available, it may be necessary to do some critical path analysis.

With the project plan in place and resources allocated, progress must be measured. As the modules progress through the life cycle, time sheets are kept of the various activities against the milestones, to an accuracy of say one hour. Deliverables are used to measure completion of milestones. This data is then used to monitor progress against the budget on a daily/weekly basis. The monitoring process lends itself to spreadsheet analysis with deadlines, estimates, lapsed times, efficiencies, *etc*. broken down against individuals, activities, milestones, *etc*. Some useful metrics for this are listed in Table 63.2.

It is also useful to present progress to date graphically. The project plan represents the desired progress and can be used to draw a graph of target progress. Actual progress can be plotted on the same graph week by week. A complex project may have an implementation phase of a year or longer and several teams working in parallel. Morale is often a problem after several months into such a project and progress graphs can be used to advantage to demonstrate that progress is being made and to introduce an element of competition between the teams.

The above time sheet data can also be used, at the end of the project, to refine the metrics. Average values can be calculated for the different activities and the metrics achieved determined. These can be compared with the metrics used for prediction and fed back into the estimating process. Similarly, the split between the stages can be determined and refined for future use.

It is very important that this analysis be carried out: the feedback is invaluable in ensuring

Table 63.2 Metrics for progress *vs* activity

Activity	Rate per person
Configurable functions and database blocks:	
Develop and test control functions	15 I/O channels/day
Other configurable functions:	
Develop and test group displays, trends, *etc*.	30 tags/day (total)
Develop and test historic data.	50 tags/day
Semi-configurable graphics:	
Develop and test new mimics	0.75 mimics/day
Modify and test old mimics	2.0 mimics/day
Procedural code:	
Produce flowcharts: SFD and/or SFCs	4 pages/day
Modify and draw new SFDs and/or SFCs	8 pages/day
Walthrough sequence designs	15 pages/day
Review (COOP) of SFDs, SFCs, *etc*.	5 pages/day
Develop (code) new sequences	3 pages/day
Modify old sequences	5 pages/day
Compile, link and print sequences	10 pages/day
Test sequences	4 pages/day
Other semi-configurable functions:	
Develop and check recipe parameter lists	4 lists/day
Develop and test new batch reports	0.2 reports/day
Modify and test old batch reports	0.5 reports/day

Notes:

1 page of sequence code is approximately 30 instructions, rungs or actions.

Use in this table of the words design, develop and test is deliberate and consistent with the waterfall model of Figure 63.1.

The metrics in this table are different from those in Chapter 61 because the latter also allow for specification, management, acceptance and also commissioning.

that the metrics used for estimating remain useful, effective and relevant. Also, the software manager should produce an end of project report. This should contain an overview of the problems encountered during the project and how these may be avoided in future. The outcome of such deliberations may result in proposals to modify existing procedures, standards and guidelines.

63.15 Personnel Management

It should be obvious that a successful project is critically dependant upon the availability of sufficient human resource and its sensible deployment. The resource required is estimated along the lines of the worked example in Section 61.6, taking into account experience, management effort, and so on. The deployment of that resource in terms of planning and metrics is covered in the section above on project management. However, organisation of the human resource is a separate issue.

Some suppliers organise their personnel into relatively small teams on an industry sector basis: for example, pulp and paper, utilities, petro-chemicals, oil, pharmaceuticals, *etc*. The rationale is that the team builds up expertise in a sector of activity and the supplier is able to tender for projects in that sector confident of having the experience and expertise to follow it through. An alternative and proven approach is to organise personnel into larger groups as depicted in Figure 63.4.

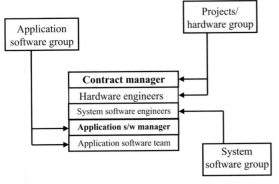

Fig. 63.4 Personnel groups for sourcing project teams

In essence, the expertise is held within three large groups irrespective of sector. The projects group handles the contracts and accounts for projects. It is responsible for managing all aspects of projects apart from the application software. The application software group is only involved in turnkey contracts. The system software group is largely concerned with the on-going development of the system and its functionality.

For any turnkey project, group members are seconded to project teams as appropriate as follows:

- Contract manager. A senior engineer with an understanding of the system's functionality, experience of applications similar to the end-users requirements, and proven project management experience. The contract manager's choice of team with the right mix of skills is critical.
- Hardware engineers. Assuming that standard hardware items only are required, little hardware effort is required. These typically come from the projects group too and tend to be electrical and/or electronic engineers. They report to the contract manager.
- System software engineers. Assuming that standard system software only is required, little system software effort is required. These come from the system software group and tend to be computer scientists. They report to the contract manager.
- Application software manager. This person will manage the team of application engineers who will design, develop and test the application software. The manager must have a good understanding of the project's requirements and be familiar with the various procedures and processes involved. Experience of managing similar projects is essential. This person also reports to the contract manager.
- Application software engineers. Likely to be the largest part of the project team, these are typically chemical or process engineers: they are best able to understand the application and appreciate the control requirements. Typically they will have a range of experience and expertise. They report to the application software manager.

There are several advantages of this approach as opposed to that based on sectors. Experience and expertise are spread more evenly within the groups, it is more flexible and less vulnerable to non-availability of key personnel, and it provides more variety to individuals. Also, being in larger groups enables a career structure that is independent of the vagaries of projects.

63.16 Comments

This chapter, and previous ones, have addressed the turnkey scenario in which the application software is developed by the supplier on behalf of the end-user. The basic issues are unchanged when the development is done by a contractor or by the end-user instead. The same documents, such as the URS and DFS, and the same procedures are required. There are, however, other considerations.

Often, when a new plant is to be built or a major refurbishment of an existing process undertaken, a main contractor is appointed to handle the whole project of which the control system is but a part. Given its strategic position, it is likely that the contractor will formulate the URS, handle the tendering process with the suppliers, and develop the DFS on behalf of the end-user. The contractor may well do the subsequent design and development work in-house, simply buying the hardware and system software from the supplier. Alternatively, the contractor may sub-contract the work to the supplier. In the latter case, there is scope for confusion over the contractor's role in managing the project. At the very least it is an extra layer of management between the end-user and supplier with potential for confusion and delay.

The end-user should think carefully about this relationship. It is not uncommon, and arguably better, for the end-user to deal with the supplier directly. Indeed, many end users have partnership agreements to that effect. In general, a supplier has a more in-depth understanding of its systems than any contractor, has a greater breadth of experience of applying those systems, and probably has better development facilities, tools and procedures. Also, a supplier has more of a vested interest in the longer term success of an implementation than the contractor. Set against that, an established contractor should have a breadth of experience of a variety of systems and be well positioned to advise the end-user on alternative solutions. Contractors tend to produce the cheapest solution, rather than the best, because they are maximising their own profit.

Irrespective of whether a system is developed by the contractor or supplier, in either case there will be a contractual relationship with the end-user. This focuses the mind on costs, deliverables, time scales, and so on. That is not normally the case when systems are developed by the end-user in-house. Whilst few end-users have the capability to engineer a system properly themselves, relying upon some supplier to provide hardware and system software, many attempt to do so. Such projects are notorious for lack of discipline, under-resourcing, inadequate procedures, incomplete documentation, moving goalposts, *etc*.

Installation and Commissioning

64.1 Time Scale
64.2 Installation
64.3 Calibration of Field Instrumentation
64.4 Pre-Commissioning of Field Instrumentation
64.5 Commissioning Application Software
64.6 Software Change Control
64.7 Access to the System
64.8 Personnel
64.9 Documentation
64.10 Works Acceptance
64.11 Comments

The emphasis in this chapter is on the commissioning of control systems that are digital in nature: no attempt has been made to distinction between DCS-, SCADA- or PLC-based systems. The principles are essentially generic and, to a large extent, apply to conventional analogue systems too. The chapter addresses commissioning activities that are post factory acceptance. Its content is based on the text by Horsley (1998) on process plant commissioning, to which the reader is referred for an appreciation of the broader context.

64.1 Time Scale

The sequential and parallel nature of the various testing, integration, installation and commissioning activities is depicted in Figure 64.1.

Note that the hardware testing, software testing and system integration are independent of the testing and pre-commissioning of the field instrumentation. However, it is only when the control system is connected up to the field instrumentation that the system as a whole can be commissioned. It is not feasible to commission them independently because of their functional integration.

Much of the instrumentation cannot be installed until construction of the plant itself is largely complete, and the control system is often one of the last things to be delivered to site. Consequently the installation and commissioning of the control system lags behind the rest of the plant: there is an overlap. Inevitably, if the construction of the plant gets behind schedule and the start up date is fixed, the time available for commissioning is reduced. Because of the overlap, the time scale for commissioning the control system is squeezed most of all. Often, the least time is available for commissioning the most complex part of the plant!

It is obvious that the overlap needs careful and strong management to ensure that corners are not cut in the supposed interests of the overall project. Remember, much effort has gone into testing, integrating and validating the application software: it doesn't take much in the way of undisciplined editing to completely prejudice its integrity.

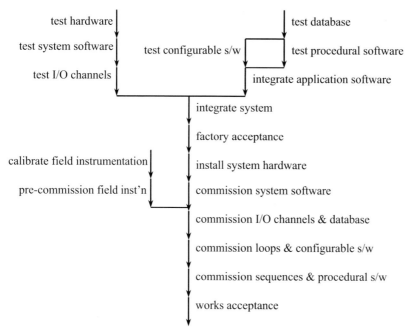

Fig. 64.1 Parallel and sequential nature of testing and commissioning

64.2 Installation

The key to successful installation is thinking through the electrical requirements of the system at the DFS stage. Thus the physical layout of the system should have been specified, from which the cabling requirements both for power and signals can be determined. The obvious things to check are that the cables are long enough, are of the correct duty, in the case of multicore cables have sufficient channels, have the correct plugs and sockets, and are properly labelled. This is normally handled by the site's main electrical contractor who will also, probably, been responsible for wiring up the instrumentation too.

The other point, often overlooked, is access. Will the system's cabinets fit through doorways? Are there any steps or stairs and, if so, are lifts and/or hoists available? If the system hardware is delivered before installation is due, where can it be stored safely?

Given the above, site installation of the control system should be relatively straightforward and often only takes a couple of days. It is usually carried out by the system supplier and essentially consists of:

- Installing the system cabinets and I/O racks/frames in an interface room.
- Connecting up the I/O cards to prewired termination cabinets, some of which may be in a separate motor control centre (MCC), usually by means of multi-channel cables and connectors.
- Positioning the operator stations and peripherals in the control room and connecting up their serial interfaces.

Following installation of the hardware of the control system, the system software may be commissioned. This apparently major stage of commissioning is almost trivial. It basically consists of powering up the system and carrying out standard diagnostics routines. Given that the system software is the ultimate test program, not only do the diagnostics prove the system software itself but also the system hardware.

64.3 Calibration of Field Instrumentation

Instrumentation is always calibrated carefully by the manufacturer and despatched to site with appropriate packaging. Upon receipt, instruments need to be carefully stored until time for installation. Anything whose packaging has obviously been damaged in transit or storage should be thoroughly checked. It is normal practice to check the calibration prior to installation. When calibrating remember to:

- Remove all packaging, especially plastic seals over sensors and caps on fittings.
- Check that the tag number and instrument specification agree.
- Calibrate over the whole range and set the bias correctly.
- Confirm that any additional functions are valid, *e.g.* square root extraction for dp cells.
- Use the correct static conditions when testing differential devices such as dp cells.
- Use the correct process medium when calibrating analysers.
- Check the local/remote switches function correctly.
- Ensure that any contacts for initiating alarms are set correctly.

Clearly it is necessary to have access to appropriate test rigs and equipment for calibration purposes. If the plant being commissioned is on an existing site then there will almost certainly be an instrument workshop available. Otherwise one will have to be provided.

64.4 Pre-Commissioning of Field Instrumentation

As implied in Figure 64.1, the instrumentation can only be fully commissioned as part of commissioning the database once the control system has been installed and the I/O channels connected. However, much checking can be done before that stage is reached. Ideally, the field instrumentation and process connections, such as impulse lines for dp cells, should have been properly installed, as should the infrastructure: power and air supplies, conduits, trunking, cable trays, wiring of termination cabinets, tags, *etc.* In practice this is rarely so, and the primary objective of the pre-commissioning is to identify faulty installation.

For detailed guidance on most relevant aspects of process, pneumatic and electrical installation practice, the reader is referred to BS 6739.

An essential first step in the pre-commissioning of instrumentation is a visual inspection of the installation: its mounting, associated pipework and wiring. Check that the workmanship is of an acceptable standard: would you pay for it if it was your own money?

It is then necessary to systematically check, throughout the system, that:

- The correct instrument has been installed in the right place, the right way round.
- The primary sensors installed are of the correct type and consistent with their transmitter ranges, *e.g.* orifice plates and dp cells.
- The process interface is sound, *i.e.* no leaks, impulse lines have correct orientation, right sealing fluids are used.
- The correct pneumatic or electrical connections have been made.
- The grade/quality of all tubing and cabling, single core or multicore, is appropriate to the duty, *e.g.* proximity to sources of heat.
- All electrical signal and power lines are carried in separate trays, *etc.*
- All control and solenoid valves fail-safe and, in the case of control valves, that their stroke length and direction is correct.
- All ancillary devices are properly installed, *e.g.* alarms, limit switches, air filters, positioners, *etc.*

It should not be necessary to test out every signal and supply line during pre-commissioning. However, in the event of any element having to be disconnected, maybe because of some fault that has been found, it is wise to check out those lines when the installation has been restored.

For pneumatic signals and supply lines, instrumentation must be disconnected as necessary to prevent over-ranging during pre-commissioning. Lines should be flow tested for continuity and pressure tested to approximately 2.0 bar for leakage. Remember that all air lines must be clean and dry before being brought into service.

For electrical signal channels and power lines, continuity, earthing and screening should be checked. If necessary, compliance with the requirements for maximum loop resistance should be confirmed, likewise the minimum requirements for insulation.

It is essential to check that the segregation policy, explained in Chapter 51, has not been compromised in the course of making modifications during commissioning. It is all too easy, for example, to connect up and configure an additional signal using an apparently free channel and to overlook the segregation implications.

The principal check that always needs to be made for each instrument, whether it be for measurement or actuation, is that its calibration, as installed, with regard to both range and bias, is consistent with its specification and duty. This is best done when the plant items to which the instruments are connected are themselves being pre-commissioned. Normally, during plant pre-commissioning, fluids such as water and compressed air are used: vessels are filled and emptied, pumps are switched on, *etc*. These simulated process conditions are quite adequate for checking the range and bias of the instrumentation.

Whilst in general the main electrical contractor will have overall responsibility for the pre-commissioning, there is much to be said, for familiarisation purposes, for involving the site instrument maintenance personnel in the process.

64.5 Commissioning Application Software

The principal objective of the commissioning process is to identify faulty installation and/or operation of the hardware and mistakes in the application software. Its scope therefore embraces inaccurate calibrations, wrongly wired I/O channels, faulty configuration, incorrect database values, incomplete operator displays, illogical sequence flow *etc*.

The essence of commissioning instrumentation and control systems is to test systematically the functioning of elements and sub-systems against the DFS and module specifications. The big difference between this and the testing done hitherto is that the control system has been connected up to the plant and the signals are real. Thus known inputs are applied to a sub-system or element and the output tested for expected values. For inputs, the test signals are simulated process signals, either generated during air and water tests on the plant, or applied manually. For outputs, control signals generated by the computer system are used.

The elements referred to may be either hardware such as transducers and valves, or configurable software functions such as data blocks or display faceplates. The sub-systems will be a mixture of hardware and software, varying in complexity from single I/O channels, interlocks, closed loops and simple sequences through to complex control schemes, recipes and optimisers.

Once the control system has been installed and the I/O channels connected, commissioning of the application software can commence. As depicted in Figure 64.1, there is a logical order for this, each providing a platform for the next:

1. I/O channels and all configurable software directly associated with I/O signal processing: function blocks, associated data blocks and faceplate or group displays.
2. Control loops and schemes and all configurable software associated with them: function blocks, data blocks and mimic displays. Likewise for all discrete devices.
3. General purpose displays: alarm lists, trend diagrams, *etc*.
4. Procedures: operations, phases and associated semi-configurable software, *e.g.* recipes, batch logs, sequence displays, *etc*.
5. Procedural code for high level language programs.

Checklists of factors to be considered during the functional testing of control loops and sequences are given in Lists 64.1 and 64.2 respectively at the end of the chapter.

An important point to appreciate is that it is not necessary to functionally test every single element. This is because the application software has already been subject to detailed software and integration testing. Thus, if a sub-system is found to function correctly, then it may be assumed that all of its elements are functioning too. For example, applying a known analogue output test signal and observing the correct opening of a valve establishes that all the elements in that analogue output channel are functioning effectively. In the event of the channel being found to be faulty, then functional testing is focused on the elements of that channel.

In addition to the control system, its displays and safety related functions, there are often separate systems dedicated to:

- Emergency shut down
- Management information
- "Packaged" items of plant, *e.g.* compressors, batch filters
- Specialised instruments, *e.g.* on-line chromatographs
- Back up generators and/or compressors, *etc.*

From a commissioning point of view, they will all consist of similar elements and sub-systems, and the approach to commissioning them is the same. However, they invariably have different functional specifications and particular care needs to be taken at the interfaces between them.

Each of the commissioning activities needs to be broken down into a number of manageable tasks. And for each task a schedule needs to be established with bench marks for monitoring purposes. Just as with the software testing and integration, the rate of commissioning is measurable in terms of the number of loops, displays or sequence steps commissioned per day. This enables progress to be reviewed regularly and presented graphically. The rate of progress should be at least two to four times faster than the metrics given in Table 63.3. Commissioning is faster than testing and integration because it is mostly carried out at sub-system level rather than at element level. It is perhaps worth noting that delays in commissioning the control system and its application software are common: they are invariably due to other factors such as process constraints or unavailability of plant.

For each sub-system tested, the engineer responsible should sign an appropriate test form, preferably accompanied by proof of testing, typically in the form of printout. Examples of various test forms for instrument calibration, alarm system checks, loop tests, *etc.* are given in BS 6739.

64.6 Software Change Control

It is during commissioning that the control system is first used for controlling the plant. At this stage, inevitably, mistakes and omissions in the DFS will be revealed: they are usually minor, but not always, and there may be many of them. Also, mistakes in the software from misinterpreting the DFS will be revealed. They will all probably necessitate modifications to both the configurable and procedural application software.

It is therefore necessary to have a procedure to enable software change. This is discussed further in Chapter 66. Noting the implications for software integrity, not to mention safety, this procedure must make authorisation of change difficult and its subsequent implementation easy. This approach ensures that software changes are treated just as seriously as modifications to the plant, but takes advantage of the editing and configuration tools to make the changes quickly.

The procedure typically involves the completion of an appropriate modification control form, an example of which is given in Form 64.1 at the end of the chapter.

Key features of the form are:

- Identification of the function to be modified.
- Description of the problem: an explanation of why the change is required and an outline of the proposed solution.

- A grid, for collecting dates and signatures as appropriate, to enable progression of the modification to the software to be monitored.
- Another grid, for collecting dates and signatures as appropriate, to enable monitoring of the progression of the revisions to the documentation resulting from the software change.

Underpinning use of the change control procedure must be a change control policy. Formulating the policy is a non-trivial issue but, nevertheless, it is essential that there is a clear understanding of the scope of the policy and the accompanying levels of authorisation:

1. The scope of the policy is context dependant. For example, changing a set point would normally lie outside the scope of the policy but, in the context of temperature control of an exothermic reactor, would almost certainly be subject to the change control procedure for safety reasons. Clearly, safety issues are important: however, for the majority of changes the issue is operability.
   A sensible approach is categorise the potential changes, *e.g.* controller settings, control loop modes, initialisation, sequence progression, alarm priorities, *etc.* These categories may be tabulated against major equipment items in a grid, with ticks or crosses depending on whether the policy applies. Every proposed change is then vetted against the grid by a designated person to see whether the procedure applies.
2. The levels of authorisation necessary vary according to the nature of the change proposed. Thus, for any proposed change that does fall within the scope of the policy, the designated person decides who should be involved in the change and what level of authorisation is required at each stage of progression. There should be agreed guidelines on this. Again, it is sensible for this guidance to be categorised on the same grid basis as above.

In general, the nature and scope of changes arising from mistakes and omissions in the DFS should be such that further HAZOP studies are unnecessary. However, if any change has involved hardware, such as extra I/O channels, it may well be necessary to carry out a further CHAZOP study. Any subsequent application software changes related to I/O channels which have a safety related function will almost certainly require a further COOP study (see Chapter 54).

However, mistakes revealed due to misinterpretation of the DFS can be corrected without further CHAZOP or COOP considerations, provided the DFS itself has not been changed. This is because the studies would have been based on the DFS in the first place. Some important considerations regarding the modification of safety related software are given in checklist form in the HSE (1987) guidelines.

Although implicit in the above, it is nevertheless worth emphasising that the discs, CD or otherwise, containing the system software, database and application software are themselves covered by the procedure for making software changes. The discs, including back-up discs, should be clearly labelled to indicate which version of software is stored on them. The master should be kept separately and not updated until the software tests have been fully completed.

The modification control forms become an important part of the system documentation.

64.7 Access to the System

Access to the system during commissioning is an important consideration. There are two types of access:

- Access to the application software by the control engineers for commissioning the system itself.
- Access through the system by process engineers for commissioning the plant.

These needs for access are not necessarily conflicting, but may be. Access to the system needs to be strictly controlled. This is normally accomplished by means of keyswitches and/or passwords, with the levels of access being determined on the basis of the nature of the work to be done and the individual's experience and expertise. Passwords and levels need to be recorded and reviewed regularly, with written consent for changes if necessary.

To accommodate the needs for access, which peak during commissioning, temporary extra OCPs and keyboards may be necessary.

Because of the distance between the various elements of the loops, interface racks, control room, *etc.* the use of portable phones for two-way communication during testing is essential. This, of course, is subject to intrinsic safety and interference requirements, especially with regard to corruption of memory.

64.8 Personnel

Successful commissioning of instrumentation and control systems needs to be considered within the context of the overall commissioning programme. Good planning, co-ordination, communications, leadership, teamwork and training are essential.

The commissioning team will need to consist of a mixture of computer specialists, control, instrument and process engineers. The size of the team and the balance of engineers to technicians obviously depends on the nature and scope of the system. Typical responsibilities are identified in Table 64.1.

In general, the application software should not be commissioned by the engineers who developed and tested it. It is best commissioned by the process engineers who will ultimately have responsibility for operating the plant: they have a vested interest in its correct functioning. Also, it is preferable that the instrument engineers should not be involved in commissioning the database: this can lead to software solutions to hardware problems. For example, if a signal's polarity is reversed, it is much easier to introduce a minus sign and change the range in the software than to sort out where the wiring is crossed.

It should be stated that, as the levels of automation and management information on plants increase, the control function is increasingly becoming the lead discipline upon which the progress of the other disciplines depend. It is important therefore to have adequate resourcing at the correct level to handle the inevitable peak demands.

There is much to be gained from involving the operators in the commissioning process. It familiarises them with the system and, as such, constitutes training. And they can contribute to the modifications, *e.g.* mimic diagram layout, warning and/or alarm settings, *etc.*

64.9 Documentation

The documentation associated with the instrumentation and control system is extensive. Quite apart from the DFS of Chapter 62 and the various software designs and test specifications of Chapter 63, there are also:

Table 64.1 Commissioning team: tasks *vs* disciplines

	Instrument	Control	Process	Computer
Instrumentation	X			
Field wiring	X			
I/O system	X	X		
Computer hardware		X		
System software		X		X
Database		X	X	
Configurable software		X	X	
Procedural software		X	X	

- P&I diagrams
- Channel/loop diagrams
- Termination rack layouts
- Wiring and circuit diagrams
- Database tables
- Tag number schedules
- System manuals
- Function block diagrams
- Sequential function charts
- Procedural coding
- Recipe parameter lists

It is imperative that the documentation is kept up to date and is accurate. As modifications are made to the hardware and changes to the software, the appropriate corrections must be made to the documentation. All involved in the commissioning of the system have a responsibility to identify any inconsistencies and to ensure that they are rectified. This issue is addressed by the lower part of the modification control form, Form 64.1.

64.10 Works Acceptance

It is a moot point as to when works acceptance occurs. There is a strong argument that it should be at the point when the system hardware has been installed and the system software successfully commissioned. The logic here is that whereas the hardware can be damaged in transit and during installation, the software cannot. So if the application software was acceptable at the factory acceptance stage, it must still be so.

However, it is more common for works acceptance to be after commissioning of the application software. It all depends on what was agreed in the DFS and the contract. Typically the costs and payments for the hardware and application software will have been staged over the development cycle with some 80% of the total system cost having been paid upon completion of factory acceptance. The remaining 20% is paid upon works acceptance. That 20% may be split between completion of commissioning the system software and the application software.

It is normal practice for the system supplier to provide specialist support during site commissioning on a *pro rata* basis: it would be contractually difficult to arrange otherwise, there being so many factors beyond the supplier's control that influence the time scale of commissioning and hence the cost involved.

64.11 Comments

The control system may be thought of and used as a sophisticated commissioning tool. For example, the availability of group displays makes the checking of I/O much easier. Similarly, the availability of both I/O and mimics enables the testing of plant items. That being so, there is much to be said for early installation of the system whereas, in fact, it tends to be one of the last things to be installed. This would enable a phased approach to commissioning. Thus, for example, major equipment items, together with their pipework and electrical infrastructure, could be deliberately built out of phase so that they do not all become available for commissioning at once.

Such a phased approach could have a significant impact on the time scale for bringing a new plant on stream, not to mention making the commissioning smoother. However, the requirements of a phased approach need to be considered carefully with regard to the DFS and software integration tests: likewise the planning of the installation and pre-commissioning of the field instrumentation.

List 64.1 Checklist for functional testing of control loop

Input channel:

Check channel no.
Correct input function block.
Correrct tag no?
Correct scaling: range/bias/engineering units.
Test input signal range: top/middle/bottom.
Is signal characterisation necessary?
    e.g. square root, linearisation, *etc.*
Does the input signal need filtering?
    If so, check the filter constant is sensible.
Check alarm and warning limits: hihi/hi/lo/lolo.
Check sampling frequency.

Group/faceplate display:

Is group in correct area/menu?
Right type of template used?
Correct tag number?
Check information is as per input channel.
Correct engineering units shown?
Is set point shown OK?
Is fail safe position of output shown correctly?

Trend diagram:

Is trend in correct area/menu?
Correct tag no?
Is the time scale appropriate?
Check display update frequency.
Are the range/bias/units of the ordinate OK?
Correct colour coding?
Does output signal need to be trended?

Archiving:

Correct tag number?
Correct logging frequency?
How long is data to be archived for?
Check scope/need for data compression.
    If so, what is basis of compression?
Are the engineering units OK?

Alarm handling:

Is alarm put into alarm list directly?
Is alarm put into the right area/group?
Check priority/colour coding is appropriate.
Is alarm logged on printer with date/time stamp?
Do tag number and alarm description correspond?
Can operator override the alarm?
Any special requirements for annunciation?
Are standard facilities for acknowledgement OK?
Are there any associated trips or interlocks.
    If so, check correct functioning.

Control functions:

Are correct function blocks used?
Is set point local or remote?
    If local, is its value set correctly?
    If remote, is its block reference correct?
Check set point ramp rate.
Is set point tracking required?
Is integral desaturation requird?
Does access to loop status functions need to be inhibited?
Can operator change set point in auto?
Can operator change output in manual?
Does loop need to be disabled during start-up?
Does the loop power up in auto or manual?
Are hi/lo alarms on error signal necessary/OK?.
Has correct control algorithm been selected?
    eg. derivative feedback.
Is proportional gain OK regarding oscillation?
Is reset time OK regarding offset?
Is rate time OK regarding speed of response?
Does loop satisfy other performance criteria?
    e.g. overshoot, noise, *etc.*?
Is the controller action in the correct direction?
Is the controller output bias sensible?
Is the output in absolute or incremental form?
Are there any alarms on the output signal?

Output channel:

Check channel number.
Correct output function block?
Correct tag no?
Correct output scaling: range/bias/engineering

units?
Are the output channel and controller outputs consistent?
Check valve opening at the maximum and minimum output signals.
Check that valve action is fail safe.
Is there any constraint on the output signal?
    e.g. lower limit on valve opening.
Check functioning of any limit switches.
Are time delays for testing change of status sensible?
Check output signal sampling frequency.

Documentation:

Have the function block diagrams been completed.
Are the function blocks in the correct order.
Are they consistent with the P&I diagram?
Are the channel/loop diagrams correct?
Are the database tables correct?
Is the loop tag number correct throughout?
Is the operator's manual complete?
Is the database listing up to date?
Have the backup discs been updated?

List 64.2 Checklist for functional testing of a sequence

Sequence Check List

Declarations:

Is sequence number and name correct?
Check variables correctly named.
    Integer/flag/block/signal/floating point, etc.
Are variable types correctly defined?
Are global & external variables declared correctly.
Check addresses of I/O variables correctly listed.
Have constant values been correctly assigned?
Have initial/default values been correctly stated?

Structure:

Are the procedures and operations correctly named?
Are there any parallel operations?
Are the operations configured from the correct phases?
Are the phases in the right order?

Are the criteria for progression from step to step OK?
Check the logic for all branching.
    Confirm there are no "loose ends".
Check for mutually exclusive decisions.
    Confirm no 'never ending' waits.
Are the correct parameter lists accessed?

Timing:

Are all timing constants sensible/correct?
    Check both absolute and lapsed times.
Are there any unnecessary waits/delays?
Have flight times been allowed for?
Is there sufficient time for discrepancy checking?
Check synchronisation with other sequences.

Contention handling:

Are any equipment modules shared or exclusive use resources?
Have criteria for handling contentions been specified?
    First come first served basis?
Check logic for handling contentions.
    Are all relevant signals considered?
    Are correct flags set/reset?
Are criteria the same throughout the sequence?

Operator access:

Is manual intervention required?
Does operator have full control over sequence?
    Check start/stop/hold/restart.
Can sequence be stopped anywhere?
Can the sequence be progressed manually?
Does it have to be restarted from the same step?
    Check on need to skip/repeat steps.
Does the sequence need to override operator action?

Recovery options:

What/where are the safe hold-states?
Are correct criteria used for initiating recovery?
Are criteria the same throughout the sequence?
Check on reserved/released variables.

Are recovery options correct?
> *e.g.* branch forward/back, hold, shutdown, *etc.*

Are actions the same throughout the sequence?
Branch into separate sequence or operation?
If so, return to same place?
Check need to activate/suspend other sequences?

Recipe handling:

Does the operator assign the recipe to the batch?
How are the parameter lists attached to the sequence?
Are there separate address (A), batch (B), common (C) and data (D) lists?
Are the addresses in the A list OK?
Is the A list in the order required by the sequence?
Are the formulations of the B list OK?
> Quantities and order?

Can the operator change the quantities formulation?
What provision is there for rejecting out of range values?
How are "as weighed" values handled?
Are the operating conditions of the B list OK?
> Check set points, ramp rates, time delays, etc.
> Have alarm limits been defined?

Is the B list in the order required by the sequence?
Are the values of the C list OK?
> Check status flags, initial conditions, message codes, etc.
> Check discrete outputs, flags and integer variables.

Is the C list in the order required by the sequence?
Are the values on the D list OK?
> Check message codes.
> Check messages in correct position on list.

Firmware interface:

Is loop status changed from within sequence?
> Confirm output OK if put into MAN.
> Confirm setpoint OK if put into AUTO.

Is configuration changed from within sequence?
> Check correct function block references used.
> Check correct parameters set up.
> Confirm alarm settings OK.

Check need to change mode of other loops and schemes?

Sequence display

Is sequence in correct display area/group?
Is sequence number/name correct?
Confirm messages fit into reserved area.
Check messages appropriate to steps.
Are operator prompts intelligible?
Is annunciation/acknowledgement of prompts OK?

Batch logging:

Is standard batch log in use?
> Check appropriate format?

What identification is required?
> Batch/lot number/sequence/recipe.

Is all recipe information included?
> Parameter lists?
> What about "as weighed" values.

Are all actions/events logged?
> If not, check that appropriate ones are?

Check provision for logging abnormal conditions.
> Recovery options.
> Manual interventions.
> Maximum/minimum values of signals.

Documentation:

Are the overview sequence flow diagrams correct?
> eg. procedures, operations, phases.

Have the sequential function charts been updated?
> eg. phases, steps and actions.

Are the parameter lists properly recorded?
Is the operator's manual complete?
Is the listings of sequence coding up to date?
Have the backup discs been updated?

Form 64.1 Software modification control form

Modification Control Form						**Request No:**	
Identification							
	System	Loop	Seq'nce	Display	Trend	Log	Other
Tag no / Ref							

Description of Problem

Progression (initials & date as appropriate)							
	Plant manager	Works chemist	Safety advisor	Project engin'r	Process engin'r	Control engin'r	Inst engin'r
Initiated							
Authorised							
Designed							
CHAZOP							
COOP							
Checked							
Tested							
Implemented							

Documentation							
	Opera's manual	P&I diags	Channel diags	Func block diags	SFCs	D'base listing	Discs
Page/diag no							
Updated							
Checked							

System Management

Chapter 65

65.1 Maintenance
65.2 Support
65.3 Obsolescence
65.4 Replacement
65.5 Upgrades
65.6 Comments

Having commissioned a control system, the emphasis shifts to its operation. This chapter therefore covers system maintenance and support, and concludes by consideration of aspects of upgrade and replacement.

65.1 Maintenance

This essentially concerns the field instrumentation and the hardware of the control system. Modern instrumentation is mostly electronic: self diagnostics and remote calibration, either by means of fieldbus or otherwise, are becoming the norm. With three obvious exceptions, the amount of maintenance required is fairly minimal. Those exceptions are instruments used in protection systems, for which proof testing at specified intervals is mandatory, analytical measurements and control valves.

From a reliability point of view, if an instrument survives its first few hours of operation successfully then its subsequent operation is likely to be only subject to random failure of its component parts. The key issue then is having available sufficient spares to enable replacement in the event of failure. It is normal practice for companies to maintain a stock of spares. Given the number of different types and/or size of instrument, careful thought needs to be given to the spares inventory. Otherwise a substantial amount of capital is tied up in spares which are doing nothing. Use of preferred suppliers for purchasing purposes has the very desirable effect of reducing the number of types of instrument for which spares have to be held. Standardising on certain models of instrumentation also reduces the stockholding.

Whereas most instruments themselves require little in the way of maintenance, or none, their process interfaces often require extensive maintenance. Particularly notorious are impulse lines and sampling systems. As discussed in Chapters 11 and 12, for example, there is much scope for distortion of signals due to unwanted bubbles of condensate and/or air in the impulse lines, blockages due to deposition of solids in sample lines, failure of air or nitrogen supplies for dip legs, and so on. These issues can only be addressed from a knowledge of which instruments are problematic and instigating preventive maintenance at appropriate intervals.

Analytical instrumentation is well known for requiring disproportionate amounts of routine maintenance. Many in-line measurements involve direct contact of an electrochemical sensor with the process fluid. There is clearly scope for the sensor becoming dirty or contaminated. Often the only feasible solution is to use disposable sensors and to replace them regularly. Analytical measurements usually involve use of a reference electrode: the standard solutions contained therein degrade with time and have to be refreshed on a regular basis. Also, the transmitters associated with analytical measurements are prone to drift, hysteresis, *etc.*, and are in constant need of recalibration, although modern transmitters are more robust. Sampling systems for off-line analysers, such as chromatographs, require even more maintenance: they usually have filters requiring regular replacement, carrier gases or liquids which need replenishing, and so on.

Control valves, with their moving parts and pneumatic positioners, require some form planned maintenance. The most common problem is loss of accuracy in the range and zero calibration due to wear, partly of the actuator but mostly of the positioner. Continual movement of the stem causes the packing to lose its tightness resulting in leakage. Also, due to vibration and shock, other devices such as limit switches and alarm contacts work loose. The plug and seat are subject to erosion under extreme flow conditions and, in time, will lose their characteristic and shut off capability. For all these reasons, control valves require regular and routine maintenance. The advent of smart valves will not obviate the need for maintenance.

As far as the hardware of the control system itself is concerned, it is normal practice to have a maintenance contract with the system supplier. There are different types of maintenance contract according to the level of support required. The two key issues are callout time and provision of spares. The callout time is essentially the supplier's guaranteed speed of response, *i.e.* the maximum time it takes the supplier to get an engineer onto site to repair faults. The callout time required depends on the financial impact of a disruption to production. The shorter the callout time the more expensive the maintenance contract.

Short callout times can be avoided if the end-user's engineers are capable of doing some hardware maintenance themselves. This typically involves them attending a few basic hardware maintenance courses provided by the supplier to gain the necessary skills, and keeping a sufficient stock of spares such as I/O cards. System diagnostics enable hardware faults to be readily identified and, in general, they can be quickly resolved by swapping cards, boards, *etc.* from stock. It is crucial that the faulty ones are returned to the supplier and replaced with new to ensure the stock of spares is functional.

65.2 Support

The final act of acceptance should be the handing over by the supplier to the end-user of a complete and up to date set of documentation, or else of those parts that haven't been handed over already. The end-user must keep them up to date so an effective change control policy must be instituted addressing the changes both to the application software itself and to the related documentation. In the interests of continuity and consistency, it makes sense to use the same policy and modification control form as used during commissioning and described in Chapter 64.

During the life of a control system there will inevitably be many changes made for process reasons such as improvements to operability and safety, extensions to the plant, addition of new products, *etc.* There will also be changes made for control reasons: enhancements to the control strategies, introduction of optimisation functions, *etc.* It is important that the plant management should recognise the value of on-going enhancement and accept the need for deploying control engineering resource to support these changes.

That resource essentially consists of one or more engineers, who have relevant applications expertise and experience of using the system, whose time is allocated wholly or in part to system enhancement. Obviously that resource includes the necessary development environment and test tools too.

65.3 Obsolescence

A key long-term decision is whether to allow the control system to become obsolete or to attempt to upgrade it incrementally over the years. Central to this is an understanding of the product cycle of a control system. This is depicted in Figure 65.1:

- Once a system has been developed and released for sale it enters an active sales phase. During this period, both for commercial purposes and the need to maintain a competitive edge, the supplier is likely to make various upgrades to the hardware and enhancements to the system software.
- At some stage, typically ten years after the system was first released, users are given six months notice that the system is to be withdrawn from production, whereupon it enters its classic phase of life. The classic phase lasts up to some seven years during which there is minimal investment in the system by the supplier.
- Towards the end of the classic phase, users are given a "last chance to buy" before the system enters its limited phase of life. During this phase it is not possible to buy new parts from the supplier for existing systems and, usually, a repairs only service is offered.
- A supplier will typically commit to supporting a system for up to about ten years from the end of the active sales phase. The system is then said to be obsolescent, as opposed to obsolete which is what it becomes at the end of the limited phase.
- Generally, as one system enters its classic phase of life, the supplier releases the next generation of system which enters its active sales phase. This is known as system evolution.

Whereas the production cycle of a system is some twenty years, the service life is more likely to be 10–12 years. That is because the hardware becomes unreliable, maintenance costs start to rise and the potential for loss of production increases. Also, there is the difficulty, as a system approaches the end of its service life, of finding the resource to support it. Nobody wants to work with old hardware platforms using historic languages and obsolescent operating systems.

Given that the average life of a process plant is typically 30 years or more, it can be anticipated that any plant's control system will be replaced at least twice in the life of a plant. With the excep-

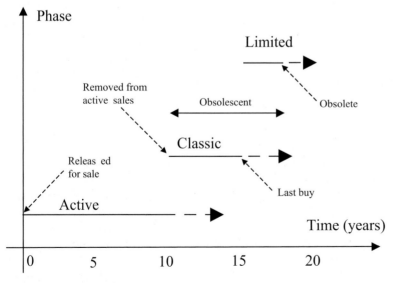

Fig. 65.1 Control system product cycle

tion of the field instrumentation, which tends to be replaced as and when necessary, such replacements are generally complete: new I/O and other cards, racks and frames, processors, operator stations and peripherals.

65.4 Replacement

Replacement is the lazy option and, in the short term, is clearly the cheapest option. This may be a conscious decision or happen by default. In essence, it means doing nothing, allowing the system to drift into obsolescence, and then to replacing the whole system lock stock and barrel at the end of its effective service life.

Replacement can be considered to be a repeat of the system life cycle, starting with an URS, progressing through a DFS, and so on. The same rigour should be applied. However, provided that the original DFS has been kept up to date, there is an existing documentary basis for articulating the requirements of the new system. The effort involved must be substantially less.

Particular care should be taken to ensure that the new system software can support the functionality of the existing system. In practice, the replacement system is likely to offer more but different functionality than the old. This should be carefully considered and taken advantage of as appropriate.

A crucial consideration is the portability of the application software. Many person-years of effort will have been invested in developing and refining it. In many respects the application software encapsulates all the existing knowledge about the operability and safety, if not the viability, of the plant. It is essential that some means exists to transfer this from the old system to the new. As far as the database and configurable software is concerned, transfer should be relatively straightforward provided the function blocks offer the same or similar functionality.

However, procedural software is a different matter. Suppliers must provide software tools which enable upgrade routes. The conversion may not be complete but should realise at least 80% automatic conversion of code from ancient to modern. The advent of the IEC 61131 and IEC 61512 standards, with their formalised physical and procedural models, should lead to automatic translation becoming more effective and widespread.

Another key decision is whether the replacement process should be cold or hot. Cold implies that the whole system is replaced at once, typically during a plant shut-down, whereas hot implies replacement on a loop by loop basis whilst the plant is still running. This clearly depends on how critical the process is from both production and safety points of view.

65.5 Upgrades

The alternative to replacement is a policy of continuous, incremental upgrading. This requires considerable effort on an on-going basis. It is certainly more expensive in the short term and may even be so in the long term, but does have the very obvious attraction of not having to find the capital to replace the whole system all at once. Also, because the system is up to date, finding the resource for its maintenance is not as problematic.

Throughout the active sales phase, the supplier will enhance and redesign the system's hardware and system software and issue new releases. These will:

- Involve processor upgrades
- Provide additional functionality which may not be required, but could be taken advantage of
- Require modifications to existing application software

These releases provide the end user with a migration path for an existing control system to evolve towards the next generation. That is the essence of the upgrade strategy as depicted in Figure 65.2.

The upgrade path is truly incremental. It is not necessary, or indeed desirable, to realise every new release because some are only minor and can be bundled into more significant upgrades. Each upgrade needs to be planned and treated as a project in its own right for which the original, but updated,

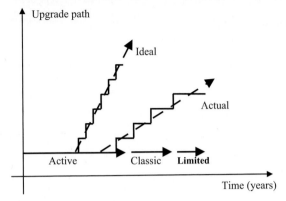

Fig. 65.2 Migration path for system upgrades

DFS documents are the basis. There may be a some re-design work but the project would normally consist largely of the usual stages of development, testing, commissioning and acceptance. For application software, the same issues about portability apply as were discussed under replacement.

Inspection of Figure 65.2 reveals two upgrade paths: the ideal and the actual. The ideal implies realisation of significant upgrades in a timely manner. The actual depicts the outcome of a lack of commitment: too little and too late. A gap emerges which, if not addressed, will result in partial obsolescence. That is worse than either of the replacement or upgrade strategies as the inevitable outcome is replacement but some of the costs of upgrading have been incurred too.

65.6 Comments

As stated, the replacement strategy is the easy option and cheapest in the short term at least. It is undoubtedly the preferred option of the accountants. The upgrade strategy is the preferred option of the suppliers as it reduces the number of previous releases the supplier has to maintain. Equally, it is in the end users interest to avoid obsolescence. There are clearly pros and cons. However, if an upgrade strategy is to be adopted, a policy decision is needed at the outset so that the cost of upgrading can be budgeted for. Also, there is a potential trap with the upgrade strategy: it can lead to an up to date version of an old system and, inevitably, to replacement. Confidence, by the end user, that the migration path does indeed enable evolution to the next generation is therefore essential.

Quality Assurance

66.1 ISO 9001: Approach
66.2 ISO 9001: Content
66.3 Industry Guides
66.4 The GAMP Guide
66.5 Validation
66.6 Documentation
66.7 Procedures
66.8 Additional Supplier's Procedures
66.9 Additional End-User's Procedures
66.10 Comments

The standard BS 5750 on quality systems was introduced in 1979 to address standards and specifications in the engineering industries. It rationalised the different quality assurance (QA) standards of the day, produced by various purchasing and third party organisations, which were themselves based upon Ministry of Defence standards. Consisting of four parts, BS 5750 has itself been superceded by the identical ISO 9000 (1987) standards, correspondence between the two being as listed in Table 66.1.

This chapter focuses on ISO 9001 which is the standard that applies to the production of a control system by a supplier for an end-user on a turnkey basis. It also draws upon ISO 12207, a generic standard for information technology regarding the software life cycle.

66.1 ISO 9001: Approach

ISO 9001 requires organisations to establish, document and maintain an effective and economical "quality system". ISO 9001 is not in itself a quality system but provides guidelines for constructing one. It seeks to set out the broad requirements of a quality system, leaving the supplier to fill in the details as appropriate. The essence of the standard is that it puts the onus onto the supplier to provide documentary evidence to demonstrate that all of its activities and functions are oriented towards the attainment of quality.

Thus a supplier is required to consider and define its purpose, management structure and individual's responsibilities, and to create a "quality policy". Oversight of that policy becomes the responsibility of a "quality manager" who reports directly to the chief executive. This ensures that the quality manager has direct access to policy making and enables reporting that is objective and independent.

Detailed plans and procedures are required for every process and activity, organised in accordance with the quality policy. There must be checks and balances to ensure that quality is achieved at all stages or, if not, that action will be taken quickly and reliably. Traceability is a key requirement: the ability to trace back from an error in order to identify and correct its cause. To this end, every stage in a system's development must be documented in such a way that this backwards traceability can be achieved. This requires records to prove who did

Table 66.1 Summary of quality standards

ISO 9000	BS 5750 Part 0 Section 1	A guide to selection and use of the appropriate parts of the standard
ISO 9001	BS 5750 Part 1	Relates to quality specifications for design/development, production, installation and servicing when the requirements are specified by the end-user in terms of how a system (or service) must perform and which is then provided by the supplier (or contractor)
ISO 9002	BS 5750 Part 2	Sets out requirements where a supplier (or contractor) is manufacturing systems (or offering a service) to a published specification or to the end-user's specification
ISO 9003	BS 5750 Part 3	Specifies the quality system to be used in final inspection and test procedures
ISO 9004	BS 5750 Part 0 Section 2	A guide to overall quality management and to the quality system elements within the standard

what, where, when, how, to what, with what and why.

A "quality manual" has to be created which consists of the various procedures, work instructions, *etc.* created as a consequence of the quality policy. This manual is to be accessible to all staff and kept up to date. Clearly staff have to be familiar with those parts of the manual that apply to their own job functions.

Most suppliers provide services as well as systems. For example, not only do they provide the hardware and system software but they also provide application software, after-sales support, maintenance contracts and upgrades. These require the quality system to address the customer interface and the less tangible considerations of communications, care, treatment, satisfaction, *etc.*

ISO 9001 certification can be applied for once the procedures have been developed, put in place and used for sufficient time, typically one to two years, to provide sufficient documentary evidence that they are functioning effectively. Certification provides contractors and end-users with an assurance that the supplier has procedures which do what it says they should do. However, this does not provide any guarantee that the system sold is what the end-user wants or that the application software developed will do what is required.

66.2 ISO 9001: Content

The standard provides brief guidance on the quality requirements for each of the following activities:

1. Management responsibility: policy, organisation and review
2. Quality system: documented procedures and effective implementation
3. Contract review: defined requirements, queries resolved, capability to meet requirements
4. Design control: planning, requirements, implementation, verification, change control
5. Document control: approvals, changes and modifications
6. Purchasing: assessment of sub-contractors, specification of requirements, verification
7. Purchaser supplied (free issue) product: storage and maintenance procedures
8. Product identification and traceability
9. Process control: work instructions, monitoring and approval
10. Inspection and testing
11. Inspection, measuring and test equipment
12. Inspection and test status
13. Control of non-conforming product
14. Corrective action
15. Handling, storage, packing and delivery
16. Quality records
17. Internal quality audits

18. Training
19. Servicing
20. Statistical techniques

66.3 Industry Guides

As stated, the standards simply set out the broad requirements of a quality system, leaving the supplier to fill in the details for the industry as appropriate. This has led to the formation of second tier guidance. Three guides have been produced relating to quality software: the STARTS (1989) guides concentrate mainly on the various techniques and tools used in software development, the TickIT (1990) guide concerns quality systems for software production and the IEE guidelines (1990) relate to the documentation of software for control (and other) systems.

There are two further guides specifically produced for the pharmaceuticals industry. The HMSO (1997) rules and guidance (the Orange Book) for pharmaceutical manufacturers and distributors brings together the various EC directives, the code of practice for qualified persons and the standard provisions for manufacturer's licences and includes an annexe on computerised systems. Its title is somewhat misleading given that the regulations contained are statutory, having roughly the same status as the US Codes of Federal Register, and go far beyond mere guidance.

Regulatory bodies in the pharmaceuticals industry in general, and the US Food and Drugs Administration (FDA) in particular, refuse to recognise ISO 9001 certification without validation. To address this the GAMP (2002) guide on good automated manufacturing practice was developed. This relates directly to quality systems for validation of automated systems. Although developed for the pharmaceutical industry, the GAMP guide is generic and may be applied throughout the process sector. The rest of this chapter is based on GAMP.

66.4 The GAMP Guide

The guide's purpose is to help suppliers of automation to the pharmaceutical industry ensure that systems are developed with quality built in and with documentary evidence that their systems meet the agreed specifications. It is intended primarily for suppliers when developing systems and not for the retrospective validation of existing systems.

It promotes a formal system for the development, supply and maintenance of automated systems by the supplier. Adherence to this management system by the supplier will provide sufficient documentary records for subsequent acceptance by the end-user and regulatory body. The formal acceptance of an automated system and its documentation by the end-user is an integral part of the validation system.

The scope of the guide is broad and covers a range of automated systems including automatic plant items, control systems, automated laboratory equipment, manufacturing execution systems (MES), and manufacturing and laboratory database systems. It addresses both the hardware and software components of automated systems together with the controlled functions and associated documentation.

The guide distinguishes between systems embedded in equipment, such as in complex filter presses and compressor sets, and standalone systems such as control systems, laboratory information systems, *etc.* The procedures for embedded and standalone systems are similar so the rest of this chapter focuses on standalone control systems of the DCS, SCADA and PLC type.

The benefits that the principles defined in GAMP bring to end-users and suppliers include:

- Increase mutual understanding.
- Eliminate the need for retrospective validation, which is costly.
- Clarify the division of responsibility.
- Provide a framework for defining and agreeing standards for quality.
- Enable delivery on time, within budget and to agreed quality standards.

- Reduce the cost and shorten the period of validation through better project visibility.

66.5 Validation

Validation is defined in GAMP as "establishing documented evidence which provides a high degree of assurance that a specific process will consistently produce a product meeting its pre-determined specifications and quality attributes".

It follows that the existence of a separate and complete description of the product, that is a control system, is a prerequisite to the process of producing validation documentation. In the context of a control system this is the detailed functional specification (DFS): this must be rigorously maintained and kept up to date.

The objective of validation therefore is to produce documented evidence which provides a high degree of assurance that all the parts of a system will consistently work correctly when brought into use. Traditionally, there are four stages to validation:

1. Demonstration that the system has been designed correctly.
2. Demonstration that the system has been installed as specified.
3. Demonstration that the system works as specified.
4. Demonstration that the system produces acceptable product quality when running correctly.

These stages can be related to the waterfall model, as depicted in Figure 63.1, in which stage 1 refers to the URS, DFS and software/module design, stage 2 is broadly equivalent to hardware acceptance, stage 3 aligns with system acceptance (FAT) and stage 4 relates to commissioning.

66.6 Documentation

The various documents involved in a quality management system are shown in Figure 66.1 which relates to the waterfall model. It should be recognised that these documents include or refer to most/all of those identified in Chapter 60–64.

The management system requires procedures to control the format, scope, contents, circulation, production and approval of each of these documents. A typical format is as follows:

Front page.
    Name of procedure.
    Reference no.
    Issue number, date, author, status (*e.g.* draft or final).
    Authorisation.
    Circulation.

Revision status.
    Changes since the last issue.

Procedure.
    Overview of procedure.
    Scope (*i.e.* context in which procedure is to be applied).
    General guidelines (*e.g.* limitations to be strictly defined, ambiguities avoided).
    Contents (*e.g.* functions, data, interfaces, responsibilities, *etc.*)

References.
    Cross references to other procedures as appropriate.

66.7 Procedures

GAMP identifies ten major procedures which are important for a quality management system. They are covered in the following sections. There are many other procedures necessary, both for the supplier and the end-user: these are listed afterwards.

66.7.1 Procedure for Project and Quality Plan

This is produced by the supplier from the end-user's validation plan which sets out the regulatory and project documentation requirements.

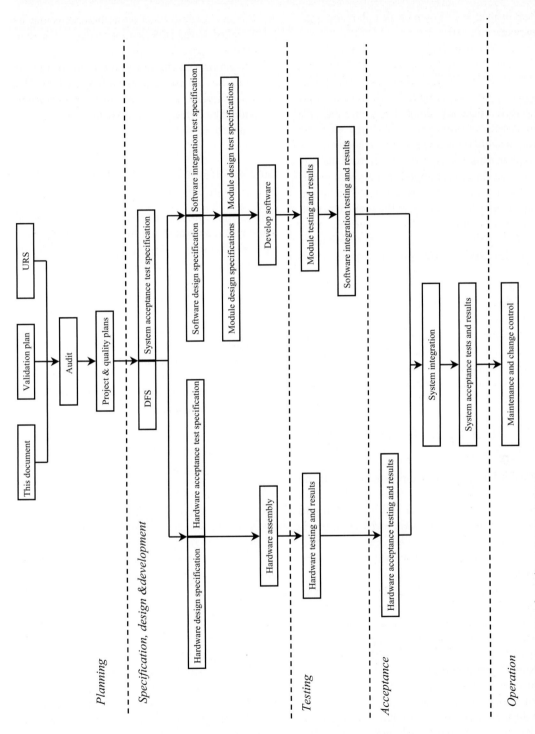

Fig. 66.1 Documentation for quality management system

1. Each project has its own project plan and quality plan.
2. The quality plan defines validation activities, responsibilities and procedures to be used. The end-user's quality requirements take precedence over the supplier's. However, where the differences are small it is preferable that the supplier's procedures stand: the end-user should avoid imposing pedantic changes on the supplier. There is less scope for error if the supplier's engineers are applying their own procedures with which they are familiar. The important thing is that they cover the requirements.
3. A project organisation chart is to be drawn:
   - Naming the project team and their job titles.
   - Identifying the single points of contact, normally the project manager for the end-user and the contract manager for the supplier.
   - Showing the interface to the supplier's QA department.
4. The project plan should be produced in sufficient detail to allow accurate progress monitoring. It should be regularly reviewed and updated and hence is best included as an appendix.

66.7.2 Procedure for Detailed Functional Specification

The methodology behind the production of the DFS is covered in detail in Chapter 62. It is written by the supplier with detailed end-user input and the final document must be approved by the end-user. It replaces the URS and becomes the principal reference document against which the system is tested.

1. The software effort should be re-estimated and compared with that in the quotation: refer to Chapter 61. Changes should be justified and variation orders generated as appropriate. Provided there is no further change to the DFS requested by the end-user, this estimate can be considered to be final.
2. Complex functions may be specified in outline in the main body of the document, with the detail of the functionality included as appendices.
3. For smaller systems, design documents may be attached to the DFS as appendices rather than exist as separate documents.
4. Project implementation should not commence until the end-user has formally agreed the DFS.
5. The system acceptance test specifications should be developed immediately after the DFS is approved. Refer to Chapters 63 and 64.

66.7.3 Procedure for Hardware Design Specification

This defines the system's hardware. For a system comprising standard hardware items in a standard configuration most of the detail will already be available on the supplier's specification sheets which should be cross referenced in this document.

Any special hardware will be cross referenced to its requirements in the DFS and specified further as appropriate in this document. It may well require a further detailed design specification which may be appended to this document.

The hardware acceptance test specification should be developed immediately after this document has been completed.

66.7.4 Procedure for Software Design Specification

This follows on from the DFS and defines the software sub-systems (modules) and the interfaces between them. The emphasis is on a top-down approach to design: structured programming techniques such as structure charts are appropriate. Refer to Chapter 63.

System Software

1. Diagrams providing an overview of the system depicting the software structure, system files and data structures are recommended.
2. For a standard system, the majority of the system software modules will already be specified in the standard system documentation and only need to be cross referenced.

3. Any special system software will be cross referenced to its requirements in the DFS and decomposed into modules as appropriate in this document. Non-trivial modules will require further detailed module design specifications.

Application Software

1. A separate specification should be produced for each substantial and distinct module, *e.g.* control scheme or procedure.
3. Control schemes should be decomposed into loops and channels as appropriate.
4. Procedures should be decomposed into operations and phases in the correct order. Remember that the phases are supposed to be configurable and are identified by name only at this level of design.
5. The higher level constructs of IEC 61131 and 61512 are appropriate for articulating the sequence software design.

Note that it is allowable to have several software design specifications provided that each refers back unambiguously to the DFS.

The software integration test specification should be developed once this document has been completed.

66.7.5 Procedure for Module Design Specification

This is essentially the same as for software design in Section 66.7.4 above except that is at a lower and/or more detailed level, the differences being:

1. Loops, I/O channels and other configurable functions are defined in terms of the parameters required by their function blocks whose functionality is defined by the target system. Refer to Chapter 48.
2. Structured programming techniques such as data flow diagrams may assist in the detailed design of phases and/or steps. Refer to Chapter 63 again.
3. It doesn't particularly matter how the detail of sequence design is depicted, provided it is unambiguous. Sequence flow diagrams with plain English statements or pseudo code is not uncommon. The SFC, ladder logic and structured text languages of IEC 61131 are all appropriate. Refer to Chapters 29 and 47.

The types of software, language, and relevant standards and guidelines used may be defined elsewhere and cross referenced.

Design walkthroughs should be conducted regularly during the design process. Once completed, each module design specification should be checked and formally approved before configuration and/or coding commences.

Likewise, the module design test specifications should be developed before configuration and/or coding begins.

66.7.6 Procedure for Development, Control and Issue of Software

The development of software, whether it be system or application, either by means of coding or configuration, is controlled by this procedure which sets out the software production and version management requirements. It applies to all the software produced by a supplier as part of a project.

1. GAMP states that there should be project specific sets of standards: languages, reference manuals, good practice guidelines, *etc.* which must be adhered to when developing software. However, it is desirable on the grounds of familiarity and efficiency for the supplier to have generic standards for use on all turnkey projects, irrespective of end-user. Also, from the point of view of long term support, project specific standards are not in the end-user's interest and it is better to adopt the supplier's without alteration whenever possible.
2. GAMP recommends that naming conventions for modules, programs, directories, files, *etc.* should be project specific. Again, it is better that they are set globally, provided they enable each

project to be identified, say by incorporating the project number in the reference.
3. Module headers should include name, version, date, author, cross references to any shared sub-modules, description and change history.
4. Modules should be inspected and the code or configuration approved before formal testing commences. Refer to Chapter 63.
5. Traceability. Once formal testing has commenced, any changes to a module should be identified in the source file. Additions and deletions should be identified by a commentary which includes a cross reference to the change control request. Deleted code may be commented out.
6. Version control. The GAMP view on this is restricted to handling updates consistently, by giving each module a unique name which allows traceability to its design specification, and by providing a version control procedure. The supplier should go further than this and provide an IPSE to support version control: it is much better handled automatically than manually.

66.7.7 Procedure for Production, Control and Issue of Documentation

This is a system for the production, review, approval, issue, withdrawal, change and archiving of documents. It applies in particular to those documents identified in Figure 66.1 and should be company wide.

Production

1. Documentation standards should be specified covering layout, style, contents, naming, approval and status.
2. Draft documents will have alphabetic issues starting with A, released documents will be numeric starting with 1.
3. A document is its author's responsibility prior to release, it belongs to the project afterwards.

Review and Approval

1. All documents are subject to formal review.
2. After the first review, a record is opened for that document and updated after each subsequent review. The GAMP guidance is that this history is part of a master document file but it is probably best kept with the document and recorded on a revisions page.
3. Any corrective actions consequent upon the review should be completed, and themselves reviewed before it is re-issued.
4. There should be at least two approval signatures, typically the author and a manager. This will vary according to the document type and this relationship should be defined here. Whereas the procedures for review and approval should be approved at Director level, responsibility for implementation should be devolved as far as is sensible.
5. The persons responsible for approving documents must accept that they have a responsibility to read the documents carefully, and make available the time for doing so. Document approval is not intended to be a rubber stamping exercise.
6. Any changes of substance to the content must be subject to the change control (as opposed to version control) procedure.

Document Issue

1. Approved documents must be issued to all copy holders specified on the document front page or elsewhere.
2. Each document must be added to the project document index when it is first issued.
3. A transmittal notice should be issued with each document sent between supplier and end-user, and *vice versa*.
4. The GAMP guidelines advocate the concept of controlled copies with each copy of a document numbered. This is overcomplicated for most project documents but some, such as the quality procedures themselves, may be handled so.

5. Any superseded documents should be clearly stamped as such and stored in the archive document file and the entry in the project document index amended accordingly.
6. Any changes of substance are subject to the change control procedure. Following approval of the changes, the issue number must be updated prior to release.
7. A master document file will hold the master of each project document. It will have an index listing each document's name, title and current issue status.

66.7.8 Procedure for Document and Software Reviews

This covers formal document and software reviews. The purpose of a review is to check that the functionality is correct and also that the relevant procedures and guidelines have been followed correctly. Reviews are formal and must be minuted with all actions clearly identified together with the name of the person responsible and the date for completion.

Document Reviews

Again, this applies to all the documentation identified in Figure 66.1. A review must be carried out prior to formal issue of any document. Circulation to the review team does not count as an issue!

1. Copies of a document should be circulated to the team before the review, giving enough time for the reviewers to read it properly.
2. The composition of a review team should be part of the quality plan. As a minimum it should include the author and the team leader. Other members of the team, especially those who are working on related modules, should also attend: this prevents misunderstandings and keeps module interfaces correct.
3. Each section of a document should be covered. "No comments" as well as comments should be recorded.
4. Corrective actions should be completed and the document updated by the dates agreed. For major updates it may be necessary to hold a further review prior to issue.
5. GAMP guidance is that each version of a document reviewed should be retained. This seems unnecessary provided that all changes are identified within the document and the revision page updated. As long as the changes are traceable, the simplest method may as well be used.

Software Reviews

The review will check that the software is following the documented design, adheres to configuration and coding standards and to good practice guidelines. This relates to the section on module testing in Chapter 63.

1. The review team must include the programmer.
2. Formal testing should not start until at least one formal review has been carried out.
3. Corrective actions must be completed by the dates specified in the review minutes.
4. Changes should be annotated within the files and/or code as in Clause 5 of Section 66.7.6 above.

66.7.9 Procedure for Change Control

Change control is fundamental to maintaining validated systems and software. It applies to changes in documentation, software, hardware, *etc*. Note that replacement, typically of hardware, on a like-for-like basis does not count as a change. The issues underlying change control policy were considered in Chapter 64: this section addresses its implementation.

Changes occur for three reasons:

1. Changes to correct faults in the software with respect to a correct DFS. Since fault correction has no effect on functionality, other than that the function now works, it lies outside the scope of the change control procedure. However, if module testing is into its formal stages, any such changes will still need to be recorded.
2. Changes to correct functionality due to misinterpretation of the DFS by the supplier. Each change should be requested using a separate

modification control form. Minor changes, *i.e.* those that only affect a single module, will require modifications to that module's design and test specification. Any such changes implemented must obviously be formally tested and recorded.

Major changes are likely to affect several modules. Depending on the complexity of the changes, various module design and test specifications will almost certainly have to be modified. If testing has already started, repeat testing will be required. Repeat testing of related modules may well also be required to ensure that existing functionality has not been inadvertently affected.

3. Changes in functionality, requested by the end-user, due to mistakes and omissions in the DFS. Changes in functionality will require modifications to the DFS itself, to the software design specification and to the software integration test specification, as well as to individual module design and test specifications. Such changes should be specified on the basis of a variation order to the contract and quoted for. Only upon acceptance by the end-user of the quotation should the change enter the change control procedure and be implemented. Clearly the later in a project that changes are initiated, the greater the impact and cost.

The GAMP line is that a modification control form has to be completed for every change requested. This is clearly not necessary for correcting faults or for changes made prior to formal testing, provided the functionality is not affected. Neither is it necessary for changes that lie outside the scope of the change control policy. Otherwise, for each change made, the modified software should cross-reference the number of the modification control form to provide traceability.

66.7.10 Procedure for Testing an Automated System

The various test specifications are:
- Module design testing.
- Software integration testing.
- Hardware acceptance testing.
- System acceptance testing (FAT: at supplier's factory).
- System acceptance testing (at end-user's works).

The procedure is similar for each type. The test specifications are written as soon as the implementation specification has been completed.

The following format is recommended for test specifications:

- Unique reference. Each test must be individually numbered and cross reference the relevant test specification.
- Test title. Brief overview in plain English of the purpose of the particular test.
- Pre-requisites. All items needed to run the test, such as simulated I/O, mimics, *etc.* and other software modules that must already be operational.
- Test description. Step by step details of how the test is to be carried out, including setting up initial conditions, and the expected results for a successful test.
- Data to be recorded. A list of the specific test data to be collected and appended to the test results sheet, such as equipment serial numbers, calibration certificates, loop responses, *etc.*

A test results sheet is used to record the results of each test together with any additional data and verifies, for example, that the initial conditions were set up as required. Also, all failed tests should be recorded, together with details of fault correction and re-test results. Any changes required should be handled through the change control procedure.

66.8 Additional Supplier's Procedures

The ten procedures outlined in Section 66.7 above are all very important as far as software development is concerned. However, they are themselves critically dependant upon underpinning by numerous other procedures in the supplier's quality management system. For example, the GAMP guidelines hardly touch upon either the estimation of software effort or the monitoring of project progress. These are arguably the two most important procedures of all since inadequate estimates and ineffective monitoring are guaranteed to lead to software chaos and project panic, attributes not normally associated with QA.

Main Procedures

1. General company organisation. The procedure outlines the function of each department and its interfaces to other departments with inputs and outputs clearly specified. This relates to the notion of internal customers with each department buying and selling products, services and resources to/from each other.
2. Application software manager's guidelines and project implementation procedures. These fully define the manager's responsibilities and cross references all relevant procedures, standards and working practices. Refer to personnel management in Chapter 63.
3. Software effort estimation procedure. This is the method with metrics used to determine the effort required for software development. It should allow for efficiency, management, *etc*. Refer to the worked example of Chapter 61.
4. Timesheet procedure. This defines how timesheet data is to be recorded and analysed for progress monitoring and metric evaluation.
5. Progress monitoring procedures. These are based upon metrics of an empirical nature gathered from work on previous projects. Progress is monitored in relation to these metrics. Refer to the section on project management in Chapter 63.
6. Monthly report procedure. There should normally be monthly progress meetings with a written monthly progress report. This covers progress, issues, corrective actions, *etc*.
7. Procedure for software report. Normally written by the application software manager at the end of the project, this consists of a detailed analysis of actual *vs* estimates of software effort. It should include detailed breakdowns into types of activity, based upon aggregated timesheet returns, for feedback into the metrics. The report should also provide a frank commentary on the problems encountered and how to avoid them in the future.
8. Procedure for project report. Normally written by the contract manager at the end of the project. Similar to the software report but providing an overview of the project as a whole, including a detailed analysis of costs. It is normal for a summary of this report to be produced for general circulation. Such project profiles are useful documents for sales support.

General Procedures

Procedures are required for handling all of the following:

1. Overtime forecasts and guidelines. The need for consistent overtime to recover project time scales should be identified and planned.
2. Holiday and illness forms. Staff holidays must be reflected in project planning.
3. Secondment of end-user personnel. Involvement of end-user staff in the supplier's project team is to be encouraged, but the seconded engineer must be competent and able to work as a team member.
4. Works visit reports. Visits to the end-user's works must be recorded, not only for invoicing purposes but also to check that documentation has been formally updated from the knowledge gained.
5. Recruitment. Job descriptions, training records and staff assessment all go towards ensuring that personnel are correctly qualified to carry out their allotted tasks and that documentary evidence exists to justify involvement in the project team.
6. Rules of professional conduct. These should be based upon, and staff need to be aware of, the guidelines of the professional institutions. For example, how to respond to end-user pressure for what they believe to be unsafe designs.
7. Cost and timesheet codes. Base data for metrication and progress monitoring.

66.9 Additional End-User's Procedures

There will be the equivalent of many of the GAMP procedures in the end-user's organisation, albeit in a form oriented towards the end-user's process sector. As far as the control and automation function is concerned, additional underpinning procedures are required to address the following:

- Analysis of costs and benefits. Refer to Chapter 59.
- Formulation of user requirements specification. Refer to Chapter 60.
- Process of vendor selection. Refer to Chapter 61.
- Auditing of suppliers' quality procedures. Supplier's quality systems need to be audited prior to vendor selection and thereafter on a regular basis.
- Monitoring of project progress. Attendance at monthly progress meetings is essential. The monthly reports from the supplier's contract manager should be received and carefully reviewed. Causes, and potential causes, of delay must be examined critically.
- Secondment of personnel to the supplier's project team.

66.10 Comments

It is perhaps useful to draw a distinction between quality assurance, quality control (QC) and total quality management (TQM). Quality assurance is effectively open-loop: you say what you are going to do, you do it, and you check that you've done it. This is underpinned by a raft of procedures and documentation to provide evidence. However, there is no feedback.

With QC the loop is closed. Thus defects in the processes of manufacturing a system and developing its application software are identified. Feedback then occurs to reduce the number of defects by improving the procedures upon which those processes are based. Such feedback and improvement is most readily realised in a manufacturing environment where a limited number of virtually identical products are being produced in bulk, say on a production line. It is not so easy in the context of the supplier of control systems where the products are complex and every one is different.

The environment within which quality is realised is said to be TQM if it satisfies the four absolutes:

- The definition of quality is conformance to requirements.
- The system for causing quality is prevention, *i.e.* QC, and not appraisal.
- The performance standard is zero defects.
- The measurement of quality is the price of non-conformance.

The characteristics of TQM are management commitment, employee involvement, non-adversarial management-employee relationships, recognition of success and determination.

Observation

Good quality systems are not cheap.
Cheap systems are not good quality.

Acknowledgement. Permission to quote from the GAMP guide and to include Figure 66.1, which is an adaptation of the GAMP diagram on documentation in the life-cycle, is gratefully acknowledged.

Contracts

Chapter 67

67.1 Purpose of Contract
67.2 Contract Law
67.3 Relationships
67.4 Standard and Model Contracts
67.5 IChemE Model Conditions
67.6 Lump Sum *vs* Reimbursable Contracts
67.7 The Red Book
67.8 Project/Contract Management
67.9 Testing and Acceptance
67.10 Changes and Variations
67.11 Delays and Lateness
67.12 Defects, Site Accidents, Insurance and Exclusions
67.13 Payments
67.14 The Green Book
67.15 The Yellow Book
67.16 Comments

Whenever a control system is ordered, there must be a contract covering its purchase. Given the complexity and cost of systems and the issues involved, it is useful for the more senior engineers managing such a contract to have some understanding of its basis.

This chapter starts by introducing various aspects of contract law and then goes on to consider the pros and cons of the two basic contract scenarios: standard and model contracts. However, the main thrust of the chapter focuses on the IChemE* Model Form of Conditions of Contract for Process Plant, subsequently referred to simply as the Model Conditions. These were specifically developed for contracts for the purchase/sale of process plant rather than control systems, so the emphasis here is on interpreting the Model Conditions in the context of control systems. The scenario addressed is that in which a supplier produces a control system for an end-user. It is assumed that the main contractor and other subcontractors are responsible for the field instrumentation and electrical infrastructure. The so-called Red Book is considered in particular since most contracts for control systems are on a lump sum basis.

This chapter is based upon the text by Wright (2004)* which is an engineer's guide to the Model Conditions, referred to as the purple book. For a more detailed understanding, the reader is encouraged to read both the text by Wright and the Model Conditions for themselves.

67.1 Purpose of Contract

The purpose of a contract is to set out a legal framework within which a supplier is responsible for designing and developing a working control system for an end-user. Every contract performs five functions:

- It defines the normal performance required of each party to the contract.
- It defines how risk is shared between the parties: technical, commercial, *etc.*
- It specifies how the contract can be varied which allows changes to be accommodated without having to re-negotiate the whole contract.
- It defines contract rules and procedures and, in effect, becomes the rule book for running the contract.
- It identifies most of the more predictable problems that might arise during the contract and describes how they should be handled.

67.2 Contract Law

All projects are affected by the requirements of the law. On issues such as planning, health and safety, and environmental protection the law imposes requirements that are both specific and detailed. However, as far as the law of contract is concerned, it only imposes requirements where the national interest is involved, such as prohibiting restraints on freedom of trading. Otherwise, it is largely concerned with the procedural aspects of a contract, setting out the basis for:

- Who can make contracts
- What is required of a contract
- How a contract can be created
- How a contract can be ended
- How legal disputes concerning contracts should be conducted
- How damages for breach of contract should be calculated

One of the project/contract manager's objectives is to ensure that no legal problems arise and, if they do so, that the rules have been followed. Most contracts complete with any disputes being settled without recourse to the law and without too much emphasis on what the contract says.

However, if a more serious dispute arises resulting in litigation, the precise meaning of the actual words becomes vitally important. The law assumes that a written contract is a complete and precise statement of exactly what terms the parties have agreed. The lawyer therefore asks "What did each of the parties promise to do, and did they do it?" The question is answered by reading the contract, concentrating on the precise meaning of the words used.

Engineers are not used to using words precisely – they use words in order to be understood by other people. Communication is largely by oral means; it certainly isn't done using the written word only. And when engineers do communicate in writing they use jargon and colloquialisms. However, a contract uses words precisely and it is clearly important for the managers to understand precisely what they mean: reference to a good dictionary is essential.

Whereas the law protects individuals against commercial exploitation, both as an employee and as a consumer, the law of contract provides companies with no such protection. Companies are deemed to be competent and free to make their own contracts, and mistakes. Every company is free to drive as hard a bargain as it is powerful or clever enough to achieve, even if the resultant contract is damaging to another company. The golden rule of commercial contracts is "if you don't like the terms, don't take it".

Many international contracts are subject to English law, largely by choice. This is partly because most main contractors progress many projects through their UK offices, and partly because many countries' legal systems are based upon English law. However, since the needs of commerce are the same the world over, most countries' laws of contract are similar, and conversion to another country's legal system is well within the competence of a contract specialist.

67.3 Relationships

Throughout contract law reference is made to the "purchaser" and to the "contractor". The parties to whom these terms apply depends upon the strategy for buying the control system:

Table 67.1 Relationships between end-user, contractor and supplier

Relationship	1	2	3
End-user	Purchaser		Purchaser
Main contractor	Contractor	Purchaser	
Supplier	Subcontractor	Contractor	Contractor

1. The end-user may place a single contract with a main contractor for the design and construction of a plant which includes the plant's control system, with the system supplier being handled as a subcontractor. This is the arrangement that most main contractors prefer.
2. The main contractor places a separate contract for the control system with the supplier, rather than handling the supplier as a subcontractor. This arrangement is unusual.
3. The end-user places one contract for the plant with the main contractor and a separate contract for the control system with the supplier directly. This is an increasingly common relationship for large and/or complex systems.

These three relationships are as depicted in Table 67.1.

It is the third of these scenarios that is considered throughout the rest of this chapter. The project manager is considered to be an employee of the end-user and the contract manager an employee of the supplier.

67.4 Standard and Model Contracts

Historically, contracts for control systems were always of the standard type. A standard contract is typically drawn up by one of the parties, so is based upon either the end-users' standard conditions of contract or the supplier's standard conditions of sale. There then follows some negotiation between the two parties, the outcome of which is a compromise depending upon the relative strengths of their negotiating positions. Such contracts are unique in the sense that they relate to a particular end-user/supplier pair and to the purchase/sale of a specific system.

A standard contract is invariably biased in favour of the company that drafted it, and the law recognises this. So, if a dispute arises and results in litigation, the law will generally interpret the conditions against that company's interest. However, this is not necessarily of benefit to the other company because, to realise the benefit, it has to embark on costly litigation and the outcome is not guaranteed.

Increasingly, contracts for control systems are of the model type. A model contact is based upon model conditions of contract, typically produced by a professional body or trade organisation. Such contracts are generic in the sense that negotiation between the end-user and supplier is simply a question of agreeing key issues such as cost, delivery dates, penalty clauses, *etc*. The model conditions are then accepted by both end-user and supplier for the specific project/control system.

A model contract describes the various stages of a project, allocates risks in a sensible way and provides appropriate solutions to the more predictable problems that may arise. Once accepted, its model conditions are deemed to have been agreed. The law accepts model conditions to be fair to both the end-user and supplier and consequently does not interpret the conditions against either party's interest.

67.5 IChemE Model Conditions

The IChemE developed its Model Conditions to address the turnkey scenario in which a main contractor is appointed by an end-user to build a plant on its behalf. They allow for plant design: this is inherently different to other model conditions. For example, in the building sector, the ICE and RIBA model conditions provide for building a structure on the basis of designs produced by independent architects. The main contractor's role is to ensure that the architect's design is correctly realised. In other sectors, the joint IMechE/IEE model forms of general conditions of contract MF/1 (2002) and MF/2 (1999) apply: they include design work and are result based, but are more oriented towards items of equipment rather than whole systems.

This emphasis on design is fundamental and is of particular importance in the context of control systems. The contract is not just to produce a working control system, but to produce a control system that enables a particular plant to produce specified products by some defined process. A major part of the contract, therefore, concerns proving that the control system functions exactly as required by the contract and enables effective operation of the plant/process. If it doesn't, for reasons which are not the responsibility of the end-user, then the supplier is in breach of contract.

There is a set of nine IChemE model forms of contract, each referred to by colour, of which the Red Book relating to lump sum contracts, Green Book relating to reimbursable contracts and the Yellow Book relating to sub-contracts are the most relevant. Both the Red and Green Books are suitable for the turnkey scenario in which the supplier is responsible for:

- Provision of all the resources and facilities necessary to realise the contract.
- Design of the control system, both hardware and software.
- Procurement of hardware and construction of the system.
- Development and testing of the software and integration of the system.
- Delivery of the system to the end-user's works and installation.
- Provision of support during commissioning of the plant/process.
- Issue of testing and performance guarantees.

This will typically involve:

- Provision of extensive information by the end-user for production of the DFS.
- Provision for verification of testing by the end-user.
- Liaison with electrical subcontractors regarding installation at the end-user's works.
- High quality project management by both the end-user and the supplier.
- Employment of a range of engineering skills: chemical, electrical, software, *etc*.

Control systems are complex and this is reflected in the relationship between the end-user and the supplier. It involves a number of different interfaces between the parties at the various stages of specification, design, testing, acceptance, installation and commissioning. All of these interfaces must be controlled throughout the contract. The IChemE's model conditions are the only ones available that are sufficiently sophisticated for handling this complexity.

67.6 Lump Sum *vs* Reimbursable Contracts

Contracts for control systems are normally on a lump sum basis, although there is a strong argument for saying that the hardware should be of a lump sum nature and the application software paid for on a reimbursable basis.

67.6.1 Lump Sum

A lump sum contract is one in which the price of developing a system is either fixed or else is subject to a cost/price escalation formula. It may include some items priced on a day-rate basis.

For the supplier to agree to a fixed price contract it must be possible to assess all the technical and commercial risks so that these can be allowed for in the contract price. This requires that the end-user's enquiry, *i.e.* the URS, be detailed. In practice, the URS is seldom in sufficient detail for suppliers to give a fixed price at that stage, as discussed in Chapter 60. There follows a competitive tendering process, as described in Chapter 61, in which each supplier is given reasonable time to make its best possible estimate of the costs involved. Some of these costs, typically the hardware and system software, may be fixed but the application software estimate will almost certainly be subject to later revision. Vendor selection invariably involves negotiation about costs and time scale, and results in a contract being placed. The DFS is then formulated jointly by the end-user and supplier, and the software estimates revised into a fixed price as explained in Chapter 62.

The contract should then be run "at armslength" by a project manager on behalf of the end-user. The project manager is given the powers and information necessary to determine the supplier's progress and hence exercise control over the project. However, the supplier's contract manager is in control of the work being done and the end-user may not interfere with that, although the end-user may instigate changes. Most fixed price contracts run smoothly but there is always a confrontational element in the project management relationship.

For the supplier a lump sum contract is comparatively high-risk and high-profit. Thus the Red Book is suitable for contracts where the supplier can identify and price for the risks involved, *i.e.* where the requirements of the control system are clear before the contract is signed and the timescale is not excessive.

67.6.2 Reimbursable

A reimbursable contract is one in which the end-user buys skilled resource from the supplier which is then paid for on a day-rate basis. It may include some items on a fixed price basis, such as purchase of hardware and sub-systems, third party software and subcontracted work.

Because it is basically effort only that is being provided, the supplier is much more certain of making a profit on a reimbursable contract: it is low risk and should only be low profit. However, for the end-user, a reimbursable contract is comparatively high risk. The cost is much less certain than for a fixed price contract: indeed, it is potentially open ended. The end-user carries the risk because the supplier cannot price for risk as with a lump sum contract.

Reimbursable contracts are more suitable than lump sum for projects that have a high degree of uncertainty in design, timescale or risk. They are particularly suitable for projects where:

- The project is small and the cost of setting up a fixed price contract is disproportionate.
- The project is fast track: setting up a reimbursable contract takes far less time than a lump sum contract. The end-user only has to roughly specify what is required to enable the supplier to estimate the resource required for the contract to be agreed.
- The end-user wishes to pool resources with the supplier in a joint development.

Since a collaborative, rather than confrontational, approach is the essence of a reimbursable contract, the project manager is given many more powers to obtain information from, and to control the day to day work of, the supplier and any subcontractors. The end-user therefore needs to provide a higher level of project management than for a lump sum contract.

The essential differences between lump sum and reimbursable contracts are summarised in Table 67.2.

67.7 The Red Book

The Red Book is of particular relevance to contracts for control systems. Its various clauses may be grouped into categories, as in the following Sections 7 to 13. A commentary is provided on those clauses that seem to be the most significant in the context of automation. To simplify cross reference, the paragraphs are indented with the clause numbers [in brackets] from the Red Book as appropriate.

Table 67.2 Differences between lump-sum and reimbursable contracts

	Lump sum	Reimbursable
Relationship	Arms length. Adversarial and potentially confrontational	Collaborative. Requires higher level of management
Discipline	Penalty clauses impose discipline on supplier. Specification to be complied with. Fixed price and timescale to achieve	No price constraint: more time spent the bigger the supplier's profit. Specification and time scale may not be fixed
Management	End-user's project manager observes supplier's progress and administers contract	Project manager needs to impose restraint through target cost arrangements, reporting procedures, *etc.*
Incentive	Supplier's incentive is to reduce costs by running contract efficiently and designing down to specification. End-user must ensure specification is adequate	Supplier encouraged to over-design. Contract must give end-user incentive to manage the supplier effectively
Risk	Supplier takes high risk but this is reflected in cost of the contract. Low risk to end-user	Supplier's risk is low. Profit is low but certain. High risk to end-user
Change	Rigid contract structure makes coping with change difficult. Late changes have major impact on cost and time scale. Strong management of change required	Designed to cope with change, but this still has to be managed

Procedure and General Clauses

The Red Book sets out the form of agreement to be used in setting up a contract. It lists the additional documents needed to make up the contract:

- The specification, *i.e.* the URS and/or DFS. This part of the contract cannot be written by a lawyer.
- The special conditions which adapt the Model Conditions to suit the particular contract.
- The schedules: description of work to be done, documentation for approval, times and stages for completion, *etc.*

Contract Interpretation

A list of defined terms start most contracts. These are a form of shorthand and will always have their defined meaning wherever they are used in the contract. For example:

- "Documentation" means all forms of documentation, technical or otherwise, including drawings, calculations and information held in computer readable form such as database listings.
- "Materials" means everything that the supplier provides that is intended to be part of the control system. This obviously includes the system hardware and both the system and application software.
- "Agreed rate" defines the basis upon which interest is to be paid on late payments. The Model Conditions take the view that the rate should be high enough to discourage deliberate late payment.

Responsibilities

[3.2] All work has to be carried out by the supplier safely and in accordance with good engineering practice but subject to the express conditions of the contract. This means that once the contract is agreed, of which the DFS is part, the system must be developed to the satisfaction of the project manager.

[3.6] The end-user can demand that the supplier retains the necessary resources within the project to enable satisfactory completion.

[3.8] Both the end-user and supplier are required to maintain a QA system as described in Schedule 13.

The provision of information by the end-user requires careful control since delays can cause major problems to the supplier for which extra cost and time can be claimed.

[4.1] The end-user must provide, through the project manager, all the agreed information required in a timely manner to enable the project to be completed on time.
[6.1] The supplier undertakes to bear the cost of carrying out the contract and takes the risk of correctly interpreting the information provided.
[6.2] The end-user takes responsibility for all the information provided through the project manager. Inaccurate (and incomplete) information can lead to the end-user paying the supplier additional costs and profit thereon.

The supplier will require information to be in writing, and to be formally handed over. When information provided by the end-user prior to the contract being awarded is to be used, that should be included in the contract as an additional schedule.

Intellectual Property

One concern is the licensing and use of third party packages as part of the control system.

[8.1] The supplier must enter into a licence agreement with the third party for use/supply of the package which covers the supplier's needs for integrating the package with the control system and testing it. The cost of the package is incorporated in the contract with the end-user.
[8.3] Each party (end-user and supplier) is responsible, and carries the cost of any dispute, if information it provided is found to infringe intellectual property rights. In practice, if infringement occurs, the third party is most likely to take the end-user to court: the end-user is more vulnerable having made the investment in the system which may not be viable if the claim is successful.
[8.4] Sets out the procedure to be followed when the supplier is responsible for the infringement and ensures that the problem is dealt with promptly and the end-user is kept fully informed.

A key issue is ownership and possible future use by the end-user of the supplier's design information.

[8.6] The supplier retains the copyright in the drawings and other documents provided under the contract and has the right to redress for any unauthorised use made of them.
[8.7] The end user has the right to use them in relation to the operation and maintenance of the plant/process for which the control system was supplied. After seven years, the end-user also has the right to use them for other purposes in relation to that control system, such as upgrading the system or replacing it.
[8.9] Within one month of signing the Acceptance Certificate, the supplier must hand over to the end-user all relevant code and documentation for the application software, and such documentation as is necessary for the end-user to maintain, modify and support the application software.

An area not specifically covered by the Red Book is ownership of the software, either system or application, provided by the supplier. However, the law is that copyright in software remains with the writer until transferred, and thus will normally remain with the supplier since there is nothing in the Red Book to transfer this to the end-user.

Termination

There are fairly standard clauses which enable the end-user to terminate the contract, either in case of breach of contract by the supplier or for the end-user's convenience.

[42] Allows termination at the end-user's convenience. Likely circumstances are that the end-user has decided to delay, or abandon,

building of the plant for which the control system was ordered. Such termination is at the end-users expense, which can be considerable.

[43.1] Covers termination if the supplier is in financial difficulty.
[43.2] Covers termination if the supplier ceases to continue with the contract or is in major breach.
[43.3] Permits the end-user to make alternative arrangements to provide the control system, *e.g.* appoint another supplier.

In the latter case, jumping ships in mid stream is very risky and is seen as the last resort. Furthermore, the complexity of contracts for control systems is such that both the end-user and supplier are regularly in breach. It is most unlikely that the end-user would be totally blameless and, as a consequence, the supplier will always strongly contest termination for breach of contract.

Decisions

This lays down rules governing communications between parties.

[5.1] All decisions must be made in writing. Any oral decision must be confirmed in writing within seven days.
[5.2] Any challenge to a decision must be made, in writing, within 14 days of receipt.

Note that there is no provision for communicating decisions by fax or email. Whilst these are generally considered to be equivalent to "made in writing", it is recommended that they be covered by a Special Condition.

Disputes Provision

If disputes do arise, the hope is that parties can settle by negotiation rather than litigation. The Model Conditions provide three methods of dispute resolution: mediation, use of an expert or by arbitration.

[45.7] Mediation is the quickest but requires both parties to be willing to admit to their own errors and hence amenable to compromise. The mediation process is normally completed in one day with the mediator first understanding each party's case and then encouraging compromise.
[47] The expert is used to resolve differences of opinion, rather than disputes, between the end-user and supplier on factual matters. This allows speedy resolution and prevents a disagreement from dragging on and souring relations. The expert's decision is binding and not open to appeal. The rate of pay for the expert is agreed in advance, but which party pays what proportion is decided by the expert as part of the resolution.
[48] Arbitration is more formal. Both parties present their cases to an arbitrator who is technically competent to understand the problem. The arbitrator will then give a ruling which the parties must abide by. This procedure is cheaper and often preferable to court action because the level of technical competence within the judiciary is extremely limited.

67.8 Project/Contract Management

Large tracts of the Red Book are concerned with the management of the contract and the project. These are pulled together in the following section.

Management Personnel

The essential structure of the contract involves a manager appointed by each party. These two persons, the end-user's project manager and the supplier's contract manager, must be individually identified by name in the contract and are responsible for its overall management. Each has full authority to control the running of the contract and to act on behalf of the company they represent.

[11.1] Sets out the standard of conduct expected of the project manager who, in particular, is required to take care to avoid any significant degree of bias when making any decision which has an effect on the technical or commercial interests of the supplier.

[12.4] Requires the supplier to make every effort not to move key personnel named in Schedule 8 until their contribution has been completed.

[12.5] Allows the project manager to take action to remove any member of the suppliers team at the end-users works for incompetence or serious misbehaviour. Realistically, this could only be an issue during the installation and commissioning phases of a project. Because removal of a team member is likely to be contentious, the clause specifically prevents the project manager from delegating this disciplinary function.

Program and Time Management

The Model Conditions deal with the supplier's obligations with respect to time in a number of different ways. The contract:

- Will include a Schedule 11 setting out the dates for various stages of completion. It is normal to include Gantt diagrams of the project's time scales.
- Will include a Schedule 12 setting out the liquidated damages (this is simply a legal term for agreed penalties) for non-compliance with the Schedule 11 dates.

It is important that both the end-user and supplier satisfy themselves as to each other's capability to comply with the intended program before the contract is signed.

[13] The supplier must submit a program of work to be done which contains adequate detail for project management. This has to be done within 14 days of commencement of the project, which is usually taken to be the contract date, the intention being not to waste time.

This may seem to be a short time scale for the detailed planning of a complex project, but most of the major issues of resource requirements and deployment should have been considered during the tendering process and an outline program already produced.

The supplier should indicate what documentation and other information is required of the end-user, and when it is needed, to enable the project manager to plan the end-users resource.

The project manager is given considerable powers in the event of the supplier being in serious delay. The supplier can be instructed to take proper action to retrieve the situation. The project manager can make revisions to the program if the supplier fails to put forward a proper proposal and can demand that the supplier's best endeavours be used to remedy the delay at the supplier's cost."Best endeavours" means that the supplier will only be excused for failure to succeed if it can be demonstrated that it was virtually impossible to succeed.

Documents and Approvals

[20] Sets out the provision for approval by the project manager of the system drawings and software design submitted by the supplier.

This seems a little bit odd because the end-user does not necessarily have the expertise to approve detailed designs. Nevertheless, this enables the project manager to:

- Identify information missing from the documentation that is the end-user's responsibility to provide.
- Raise comments and queries on any aspects of the designs.
- See, perhaps for the first time, the detailed design of the system in its entirety.

After approval, the supplier's obligations are increased to the extent that a system is to be supplied that complies not only with the contract but also in accordance with the approved documents.

The list of documents should be jointly agreed and included as Schedule 2 and their dates of submission included in Schedule 11. The DFS and the various test specifications must be included.

[21.3] Deals with the situation when the project manager is unwilling to approve the supplier's designs.

The Model Conditions assume that, since the supplier is taking the risk and responsibility to design the control system for the contract, the supplier has the right to decide what that design should be. The project manager may only refuse to approve if the design does not comply with some express provision of the contract or is contrary to good engineering practice. This prevents the project manager from influencing the design according to the end-user's opinion. Any disputed approval must be referred to an expert.

[21.6] Gives the project manager the right to inspect, but not to copy or keep, all system drawings and software designs (including those of any third parties) other than those which contain confidential information.
[21] Deals with the practical problems caused by mistakes in documentation. The supplier will charge for correcting errors caused by erroneous information supplied by the end-user. If the end-user has carried out abortive work due to incorrect documentation provided by the supplier, the latter must bear the cost, subject to a cash limit of 1% of the contract value.

Subcontracting

In any major project there is a potential conflict of interest over subcontractors although, as far as control systems are concerned, most suppliers tend to prefer to do the work in-house. Nevertheless, in as much as sub-systems and third party software involving subcontractors are used, the supplier will want freedom of choice to minimise costs. These may well be dominated by the effort associated with integration and testing rather than the direct costs. The end-user will want to standardise on certain preferred suppliers on commercial, spares and maintenance grounds.

[9] This gives the supplier maximum freedom to sub-contract. However, no sub-contracting reduces or changes the supplier's responsibilities to the end-user under the contract. The only exception to this is for so-called nominated subcontractors.

The options available to resolve this are to:

- Give the supplier complete freedom of choice.
- Give the supplier freedom of choice subject to approval by the contract manager for major third party packages or sub-systems.
- Include a list of approved suppliers in the contract.
- Use nominated subcontractors.

[10.7] Requires the end-user to accept the responsibility for failure to perform of any nominated subcontractor.

The nominated subcontractor concept, logical though it may seem, can be difficult to handle, especially if the end-user has no bargaining power. There are two precautions that can be taken to minimise the risk:

- To ensure that a potential subcontractors is unaware of its nominated status.
- To set up a separate agreement with the subcontractor in advance of nomination to set out the contractual arrangements should Clause 10.7 be activated.

Project Progress

[29] Provides for monthly progress meetings. These should be chaired by the project manager which allows for direct communication by both teams with the end-user on a regular basis.
[3.7] Sets out the requirement for monthly progress reports by the supplier.

Pre-Delivery Inspection

This only relates meaningfully to the system's hardware (see also hardware acceptance below). Repairing or replacing defective equipment is much more expensive in time and money once the system has left the suppliers factory. The contract will therefore include descriptions of pre-delivery tests to be carried out at the supplier's premises before being accepted for delivery to site.

- [22] Gives the project manager the right to visit the premises of the supplier and any subcontractors to observe these tests. There is also the right to visit to inspect equipment in the course of manufacture.
- [22.4] Allows the project manager to notify the supplier if tests are required by the end-user in addition to those specified in the contract.

Note that the testing requirements are not normally known at the time of placing the contract. That is because, in general, they cannot be specified until the design work is complete. The details are included in the DFS and test specifications approved under Clause 20.

Also note that any subcontractor will charge the supplier for additional testing not covered by its original purchase order. This could lead to a claim for a variation order. The project manager should indicate, as far as is practicable, any such tests required so that the supplier can make adequate provision when ordering.

Site Activities

The Red Book says little about the installation of the control system on the end-users plant. In essence it leaves the interface between the control system, the field instrumentation and the rest of the electrical infrastructure to the professionalism of the main contractor.

It also assumes that the approved test specifications and procedures are sufficient to prove that the control system performs exactly as required by the contract to produce the desired result in relation to the plant/process.

- [25] This is a standard "vesting" clause which minimises the disruption to the end-user in the event of the financial collapse of the supplier.

If a company goes into liquidation, the liquidator's duty is to convert the company's assets into money for distribution to its creditors. Those assets will therefore be sold to the highest bidder. The purpose of a vesting clause is to transfer ownership of materials to the end-user before acceptance of the system is complete. Thus when the end-user makes stage payments, the ownership of those materials is acquired. This applies to all materials, whether developed by the supplier or produced by a subcontractor.

67.9 Testing and Acceptance

This is the most critical part of a contract for any control system because it is the culmination of everything the supplier is committed to doing and, upon successful completion, the supplier is paid. The tests are also critical for the end-user who then takes full ownership of the system and assumes responsibility for the risk, *i.e.* the guarantee/defects liability period starts.

The purpose of testing is to demonstrate that the system meets its specification. The Model Conditions assume that an agreed set of tests is a fair examination of whether or not the system satisfies the contract. If the tests are too stringent or too lax, the result may be unfair to one party or the other. The three stages of testing articulated in the Model Conditions should be interpreted in the context of a control system as being hardware acceptance, factory acceptance (FAT) and works acceptance. The various test specifications should have all been included in Schedule 16 and approved as part of the contract under Clause 20, as described above.

Hardware Acceptance

This demonstrates that assembly of the system hardware is complete and that standard system software functions effectively on the hardware platform (see also pre-delivery inspection above).

- [32] Requires the supplier to generate a hardware assembly completion report for submission to the project manager when the system is ready for inspection.

The supplier should prove that the system meets its hardware acceptance test specification. This should be witnessed by the project manager who should check in particular that:

- The system has been assembled in accordance with the hardware design specification.
- No hardware items are in poor or damaged condition.
- That the system hardware functions effectively as per its hardware acceptance test specification.

System Acceptance (Factory)

Factory acceptance testing (FAT) is essentially a question of proving to the end-user that the application software, running under the real-time operating system on the hardware of the target system with simulated I/O, is correct and functioning in accordance with the DFS.

FAT was introduced in Chapter 62. Its requirements will have been developed as part of the design process, as outlined in Chapter 63.

[33] Permits a major stage payment to the supplier upon completion of factory acceptance.

This will depend on what stage payments have already been made, but some 80% of the total contract value should have been paid upon completion of FAT. This is not unreasonable since, by this stage, the system hardware has been built and all the effort in developing and testing the application software has been expended.

System Acceptance (Works)

The purpose of works acceptance is to demonstrate that the control system functions exactly as required by the contract and enables effective operation of the plant/process for which it was designed. Formal acceptance normally comes after the application software has been commissioned. Testing for commissioning purposes is covered in detail in Chapter 64. It should be recognised that during commissioning the plant is still at the teething stage and the control system will still need fine tuning.

[35] Referred to as performance tests in the Red Book but as commissioning tests in practice, these should:

- Relate to the DFS and the agreed acceptance test schedule.
- Be of a functional nature, concentrating on sub-systems.
- Each have an unambiguous expected outcome.
- Be selective: it is not possible to test everything.
- Not be unreasonably long since the supplier's support is paid for on a *pro rata* basis.

Works acceptance is of great interest to both parties. The end-user wants the control system functioning as soon as possible to enable the plant to be operated. However, at the same time, the end-user does not want to accept a control system that is not functioning effectively because, in doing so, the supplier is relieved of responsibility for correcting faults. Acceptance results in the balance of the contract value being paid, typically 20%, starts the guarantee period and frees the supplier from a major contractual hook.

Performance Guarantees and Liquidated Damages

Works acceptance testing is used to determine whether the control system performs effectively and, if not, to establish the extent to which its functionality is lacking. If the shortfall is minor and due to reasons attributable to the supplier, then the end-user may accept the system subject to payment by the supplier of liquidated damages. These are often related to a number of different parameters. Hardware parameters such as reliability, speed of response, *etc.* are relatively easy to quantify: software parameters such as functionality, correctness of logic, *etc.* are difficult. Guidelines for setting up liquidated damages are:

- Parameters which cannot be measured accurately should not carry liquidated damages.

- Only parameters that relate to the performance of the control system, as opposed to the plant, should carry liquidated damages.
- Always allow for measurement tolerance on the parameters.
- Three parameters at most should be subject to liquidated damages.
- One, perhaps two, parameters should be mandatory with no shortfall allowed.
- Performance tests and liquidated damages should be included in Schedule 16.

If the shortfall is major and due to the supplier then substantial damages will become payable. Also, the system will require to be modified and re-submitted for testing.

Acceptance Certificate

[36] To be issued by the project manager upon the successful completion of acceptance tests and allows the supplier to be paid.

67.10 Changes and Variations

This is the area most likely to cause management difficulty. The issue has already been considered in Chapters 64 and 66 from technical and quality points of view respectively. Change is very much a function of the complexity of a project, rather than its size, and of the timescale. Invariably, changes to the control system arise from modifying the plant to improve its performance, or changing the process to meet market conditions. These are largely within the control of the end-user. Occasionally changes occur outside the control of both parties, such as the requirements of regulatory bodies.

From a contractual point of view, managing change is problematic:

- On a lump sum contract, the supplier is in a monopoly situation and the project manager's position is weak.
- Variations orders are not the most cost effective way of getting work done by the supplier. Not only do they interfere with the existing project planning but also have to be done piecemeal and often with urgency. Variations therefore cost more than the same work would have done if they were part of the main contract.
- The project manager has difficulty in assessing the true cost of the work involved.
- The contract manager will demand an increase in cost to avoid a loss of profit to the supplier, and an increase in timescale to avoid being penalised for late delivery. The project manager is obliged to defend the end-user against both of these increases.
- Control systems are very vulnerable to change because, being highly integrated, changes in one sub system may well lead to changes in others.

The Model Conditions assume that some change is inevitable and provide the project manager with a range of options. They also assume that the project manager is sufficiently competent to properly negotiate variations orders with the contract manager. With regard to the control system, the only relevant categories of change are due to changes in legislation and variations orders.

Changes in Legislation

[7.3] If these occur after the contract is signed and increase the cost of the contract, the end-user has to reimburse the supplier's additional costs, without additional profit, and agree to an extension on delivery dates.

Variations Orders

Only the project manager can initiate a variation order, and this must be done in writing.

A variation is considered to be:

- Any change to the DFS that affects the work being carried out by the supplier, no matter how small. This excludes clarifications.
- Any change imposed on the supplier's methods of working not explicitly articulated in the DFS, such as the methods of testing or documenting software.

- Any change to the program of work: refer to Chapters 63 and 66.

The Model Conditions consider that it is unacceptable for the project manager to have the right to change a contract more than is reasonable since this could put excessive strain on the supplier's ability to deliver.

[16.7] Allows the contract manager to object to any variation that:
- Brings the total cost of all variations to more than 25% of cost of the original contract.
- Causes infringement of any third party agreement, such as a software licence.
- Requires expertise beyond the supplier's capability, unless a suitable subcontractor exists.

[16.4] Gives the project manager the right to call upon the supplier's expertise to explore potential variations. This must be paid for at reasonable profit.

Valuation of Variations

Many variations will be carried out on the basis of a fixed price agreed in advance. However, the project manager may agree the price of a variation at any time: either in advance, during or after the work has been done. The cost of the variation must allow the supplier a profit. The supplier's books must be opened to the project manager.

[19.1] When a variation order is issued, its cost must be a reasonable amount which includes a reasonable profit.

This clearly depends on circumstances. The end-user should expect to pay a higher pro-rata price for the variation than the contract because of the uncertainty. As far as the reasonable profit is concerned, this would normally be above, but not excessively, the profit level built into the contract which must have been acceptable to both parties in the first place, otherwise they wouldn't have signed it.

[16.9] Allows the project manager to use the open book approach to the pricing of complex variations. The supplier must preserve the records of all the costs of carrying out all the work on any variation whose price is yet to be agreed, and allow the project manager to inspect these.

Disputes

Disputes over the effects of a variation order on price, specification and delay are always a potential problem in large contracts. One solution is to leave resolution of the dispute, as opposed to the variation itself, until the end of the contract because everybody is too busy to handle the extra work. This causes difficulties for both parties and variations problems are best settled at the time they occur.

[16.7] Provides that either side can insist on the appointment of an expert who could resolve the issue in weeks, rather than the months which it may well take otherwise. This reduces the risk of acrimonious argument during or at the end of the contract.

Supplier's Proposals for Variations

[3.5] Requires the supplier to make the project manager aware of any possible improvements to the DFS for the control system which would enhance the operability and/or safety of the plant/process and thereby benefit the end-user.

[17.1] Requires the supplier to put forward a proposal to change the hardware and/or software for such an improvement.

[17.3] Covers the situation where the supplier has detected an error in the DFS.

67.11 Delays and Lateness

Liability for lateness is a serious risk for a supplier because, in practice, most projects are late for most of the time. It is impossible to predict at the start of a contract how late a system will be, why it will be late and what the effect of the lateness will be on the end-user. But, without doubt, serious delay of the delivery of a control system or its successful commissioning will prevent the plant from coming on stream and the consequential losses are potentially massive. The supplier must therefore take steps to limit that liability and any claims for losses.

Under the UK law of contract there are only three contractual situations regarding time: time of the essence, estimated delivery date and liquidated damages.

1. Time of the essence. The time dimension is a major factor in a commercial contract and a promise to complete by a certain date is of particular importance. This date is to be regarded as "of the essence" and a failure to meet that commitment is treated as a serious breach of contract. Note that it is not necessary to state that time is of the essence, although many do: the mere fact that a fixed date or period is stated is sufficient. Thus, if the supplier is late by just one day, the end-user may claim damages for breach of contract or treat the contract as cancelled and refuse delivery.
2. Estimated delivery date. In the event of there being no date for delivery or completion in the contract, the law requires the supplier to complete within a reasonable time. If the contract sets a date that is stated to be approximate or an estimate, the supplier must complete within a reasonable time of that date. In practice this is taken to be within 10–15% of the original period. If the supplier fails to deliver within that time, the end-user can claim as if it is a normal "time of the essence" situation.

The damages that may be claimed for are those that can be proved to have been suffered by the end-user and which the supplier could have reasonably foreseen, or was told about, at the time the contract was signed. This is a serious position for both parties:

- There is no right in law to claim an extension to the delivery date for reasons beyond the supplier's control.
- The supplier may be liable for damages but has no way of knowing the amount.
- The end-user has to prove that the losses have occurred: this is often harder that it may seem.

Whatever benefit the end-user may get from recovering damages, what is really required is a working control system as early as possible, even if it is a little bit late. The end-user does not want to go to law, risk losing the contract, and then have to start from scratch again with a new supplier.

The Model Conditions provide the means which allow the end-user to direct a supplier to minimise lateness.

[15] This is the liquidated damages clause (together with Schedule 12). It encourages the supplier to strive to minimise lateness and gives the end-user some redress without having to prove too much in the way of losses.

3. Liquidated damages. A sum agreed by both parties as a genuine pre-estimate of the damage caused to one party by a breach of contract by the other. The pre-estimate need not be accurate provided that both parties have agreed it and it bears some relation to reality. In practice, liquidated damages bear little relation to the potential maximum or minimum damage that might be suffered by the end-user in the event of serious lateness by the supplier.

Unlike normal contract law, the Model Conditions provide for extensions of time.

[14.1] Sets out the procedure for claiming extensions of time and the rights of the supplier to claim extensions in a range of situations apart from *force majeure*. One such situation is the late supply of information by the end-user.
[14.2] Uses the term *"force majeure"* to mean events outside a party's control, such as arson, and allows an extension to the contract period in the event that either the end-user or supplier are so delayed.
[14.6] Allows either party to bring the contract to an end if *force majeure* circumstances cause

work to be stopped for a continuous four month period.

Also the Model Conditions provide for suspension of contracts. There is always the risk with a long contract that some unexpected serious problem will occur. The supplier may have committed a serious design error, or some intractable system software fault may have emerged. The end-user's market conditions may have changed requiring time to investigate whether alternatives are viable.

[41] Allows the end-user to suspend the contract.

This clause should only be used when really necessary. It will reduce the supplier's costs, not stop them, and will result in additional cost to the end-user in the long term. The supplier's team has to be kept together and any subcontractors kept idle but ready to restart when requested in a situation of some uncertainty. Neither party can allow suspension to last indefinitely.

67.12 Defects, Site Accidents, Insurance and Exclusions

Human error is inevitable. So there will be faulty hardware and defective software design in any control system supplied to an end-user. There will also be mistakes made during commissioning. So, a very real risk exists that the system could be responsible for forcing the plant or process into a dangerous state. Any such incidents could result in a major claim for damages, direct or consequential, if they lead to accidents or significant loss of production. It is impossible to predict when such incidents occur and how serious they will be.

There are three different types of liability under which the supplier may have to pay damages to the end-user or third parties: liability under contract, liability in tort, and liability under statute.

1. Liability under Contract. This is covered by the Sale and Supply of Goods Act (1994). Amongst other things, this requires that goods must be of satisfactory quality and reasonably fit for purpose. Clearly if a control system fails to meet its specification (DFS) as agreed with the end-user, there may be a breach of the act. The supplier is also responsible for any subsystems or third party software provided by subcontractors. The end-user can claim for damages, reject the system, or both. The damages that may be claimed are for direct (as opposed to consequential) losses and damage which the supplier could, and should, reasonably have foreseen would have resulted from the breach. The damages are not limited in any way.

2. Liability in Tort. Tort is a private or civil wrong committed to a person. Thus the supplier is liable to anyone to whom there is a duty not to cause damage or injury through negligent actions or omissions. The law of negligence requires a duty to take care. Thus, if it can be foreseen that negligent or reckless actions could cause damage or injury to any person, the supplier owes a duty to that person to take reasonable care. When one party owes another a duty of care and then causes damage or injury to that person, the former is liable to compensate for all loss of any type that could be foreseen, however great the scope. The law allows the parties to a contract to agree to exclude claims in tort.

3. Liability under Statute. This includes, under the Employer's Liability Act, responsibility to provide insurance cover against industrial injury suffered by employees. Accidents at the end-users works, typically during commissioning, will be dealt with under the law of negligence. The arguments are similar to the above.

The supplier carries unlimited liability under these laws and could face enormous claims. However, subject to the agreement of the end-user, the law permits such liabilities to be excluded by clear wording of the contract.

The end-user is also in an unsatisfactory position because:

- The law of contract requires systems to be functioning effectively when supplied but does not allow for maintenance contracts and after sales support. This is not of much practical use. If a

system is defective, the end-user wants it sorting out as soon as possible and does not want to waste time and money going to court to claim damages.
- To be able to claim damages in the event of major loss, the end user not only has to prove damages, but also that the defect which caused the damage was in the system when supplied and not caused by misuse by the end-user. In the context of a control system, unless change control is enforced rigorously, it is not at all easy to prove that minor software changes by the end-user did not inadvertently cause other defects.
- The only way money can be guaranteed to cover losses is through insurance. However, small suppliers will only have limited insurance cover and even a large supplier will not have enough to cover a major loss. Furthermore, processing a complex claim against several suppliers can take several years to settle.

Insurance

The simplest solution to all three of these difficulties is for the end-user to insure the whole plant, including the control system, and to take out cover to the total value of those assets. Insurance cover is no longer uncertain. Blanket cover taken out by the end-user is cheaper than a multiplicity of policies taken out by the suppliers, all of whom have to pass on their premiums through the contract. By avoiding disputes between insurance companies, claims are resolved more quickly. Proving the cause of the problem is no longer an issue. The end-user can also set up an agreement with the supplier to provide free support during the guarantee period. Thus the end-user can release the supplier from liability. The end-user requires blanket insurance cover anyway to protect against claims by the end-user's own employees causing a major loss.

The Model Conditions advocate a coherent insurance scheme that provides adequate cover and minimises the risk of cross-claims. Its intent is to limit the supplier's liability to a fixed maximum amount, but below that limit the supplier is expected to be liable for claims.

[31.1] The supplier is responsible for arranging the insurance for the control system. This is to be in the joint names of the end-user, supplier and any third parties involved. Cover is provided for full replacement cost of the system for all insurable risks, unless the contract states otherwise. The period of cover should be from when the system is delivered to the end-user's works until works acceptance is complete, but should also provide cover for work done during the guarantee period.

[31.2] Requires the end-user to cover the whole works and adjacent property, other than the control system until acceptance is completed, for all normal risks subject to the supplier being liable for damages up to £5M per accident caused by and related to the system.

[31.3] Requires the supplier to provide cover for accidental damages to both the end-user and any third party on the end-user's works.

Guarantees

The defects repair clauses enable the end-user to ensure that the supplier repairs or replaces any equipment supplied which is found to be defective either before or during the guarantee period. The supplier pays for the repair when liable: otherwise the cost of repair is determined as though it were a variation.

[37.2] The basic defects liability period starts when the system is delivered to site (the end-user's works) and continues for one year after acceptance is complete.

[37.5] Provides for the repeat of tests to show that the system is functioning correctly after repair.

[37.6] Gives the end-user a new defects liability period of one year on individual items of equipment that have been repaired or replaced.

[37.7] Allows the end-user to back charge costs when the supplier has refused to honour defects repair obligations.

Final Certificate

This confirms that the defects liability period has been completed and that the supplier's obligations have ended. Once signed, it bars the end-user from taking any legal action for breach of contract.

67.13 Payments

The general approach is that the end-user and supplier are quite capable of agreeing the contract price and payment arrangements and concentrate on the procedure to be followed for the approval of the supplier's invoices and the time scale for payments.

[40] The lump sum may be fixed, or subject to adjustment by an agreed formula which takes account of any rise or fall in the supplier's costs during the life of the project. Prime cost sums may be included: an amount earmarked for a specific purpose such as purchase of some hardware or third party package. Not all of the contract needs to be fixed price. Provisional sums may be included for application software which cannot be accurately priced at the date of the contract. Day rates may be agreed for extra work that may be required.

[39.1] Establishes the program for payments by instalment on a milestone basis as agreed in Schedule 8. Refer to Chapter 64.

[39.3] Sets out the procedure and time scale for payment. The permitted timescale is three weeks from submission of the draft invoice: this allows one week for the project manager to approve it and two weeks for the end-user to pay it. It is recommended that variations are paid as the contract progresses to avoid build up of unexpectedly high claims at the end of the contract.

67.14 The Green Book

The legal interpretation of the Model Conditions of the Green Book are essentially the same as for the Red Book, except that they have been written to reflect the non-adversarial, collaborative approach. It is the only set of model conditions for reimbursable contracts in the UK and has been described by the International Bureau of Comparative Law in Switzerland as being the most fair in the western world.

The project and contract managers' tasks are more difficult with a reimbursable contract because:

- The relationship between end-user and supplier is less well defined.
- The supplier has the right to be reimbursed for reasonable expenditure, so the contract manager has to find the resource to oversee and control the supplier's work. Apart from variations work, that is not the case with lump sum contracts in which the end-user doesn't need to check the supplier's use of financial resource.
- The supplier's resource is paid for on a day rate basis with a relatively low profit margin: this does not include the contingencies present in a fixed price contract. Thus the end-user has to pay for more of the remedial costs relating to faulty work by the supplier, as if the resource was the end-user's own employees.
- The end-user has to provide resource to contribute to design and other work.

The net result is that there has to be mutual confidence and professional trust between the parties for the relationship to be effective. It works best when the two parties have complementary skills and respect each other's abilities.

67.15 The Yellow Book

The Model Conditions for subcontracts have been written to complement those for lump sum and reimbursable contracts. As such, the Yellow Book is normally used in the first of the relationships described in Table 67.1. However, this chapter largely concerns the third relationship, that in which the contract for a control system is placed with the supplier directly. Whilst there will be few (if any) subcontractors, typically for specialist hardware and third party packages, the Model Conditions are nevertheless applicable.

The potential conflicts that have to be resolved are:

- The supplier's desire to pass on contractual obligations to the subcontractor, and the latter's reluctance to accept such.
- The supplier's wish to buy third party equipment, packages or resource at as low a price as possible.
- The types of risk that subcontractors are prepared to accept. For example, they will only accept very limited liability for any software failure.
- The relative importance of the contract: every system is of importance to the supplier, but not every subcontract is of equal importance to the subcontractor.
- The different technical skills and understanding of the parties.

Back-to-back Subcontracts

If subcontractors have the same obligations in terms of work, risk and liability as the supplier, there is a basis for confidence that in meeting their own obligations the subcontractors will be satisfying those of the supplier. Back-to-back contracts, therefore, endeavour to pass on to the subcontractor conditions of contract which are as similar as possible to those of the main contract. They also pass onto the subcontractor as much risk as possible. However, a little thought will reveal that it is neither fair nor possible to pass on the same conditions and undue risk.

The IChemE has produced a Model Form of Subcontract, for use with Red and Green Book contracts. There is a presumption that each individual subcontract will be a stand-alone agreement. To avoid ambiguity, the end-user, main contractor (if necessary), supplier and subcontractor should be identified by name. The Model Form of Subcontract provides a series of schedules to allow the supplier to transfer the necessary technical and other information from the main contract to the subcontract.

On the Red Book basis, the Model Form will normally only be used for subcontracts which:

- Are fundamental to the design of the system
- Are large enough to form a significant proportion of the system, such as the provision of an MIS
- Form a complete subsystem or software package in themselves

On the Green Book basis, the supplier passes onto the subcontractor terms corresponding to the main contract as far as they are applicable. Employment of key personnel as consultants could be handled on this basis.

67.16 Comments

The world is becoming increasingly litigious with parties resorting to the law to resolve disputes sooner than hitherto. Disputes arise because one party feels to have been wronged. There are clearly many possible causes for this, but a common cause is that the original contract was not reasonable and the party concerned did not fully understand the obligations entered into. This is avoided by the use of model forms of contract as they are inherently fair to all parties.

Despite that, most control systems are purchased on a standard contract basis. However, the underlying issues are generic and independent of the form of contract. This chapter has used the IChemE Model Conditions as a vehicle for considering a wide range of contractual issues in relation to control systems. A good understanding of those issues and the related commentary provides a sound basis for interpretation and/or negotiation of contracts. It should also go a long way towards the avoidance of disputes.

Copyright and Liability

* It is acknowledged that David Wright is the author of the Engineer's Guide to the Model Forms of Conditions of Contract for Process Plant (the Purple Book) and that copyright of the Model Conditions themselves (Purple, Red, Yellow, Books, *etc.*) is vested in the IChemE.
* That any views expressed in this chapter are not necessarily those of the IChemE and the IChemE accepts no liability for them.

Part II

Theory and Technique

Section 9

Maths and Control Theory

Series and Complex Numbers

68.1 Power Series
68.2 Taylor's Series
68.3 Complex Numbers

Section 9 of the Handbook covers control theory which requires an understanding of various aspects of mathematics. These are introduced in separate chapters as appropriate. This chapter covers series and complex numbers, Chapter 70 covers Laplace transforms, Chapter 75 covers Z transforms, Chapter 79 covers matrices and vectors, Chapter 82 covers stochastics and Chapter 83 covers regression analysis. There are many texts on engineering mathematics which cover all these topics in depth, those by Jeffrey (2002) and Stroud (2003) being particularly good examples. The reader is also referred to the relevant sections of some of the more comprehensive texts on control theory such as Dutton et al. (1997) and Wilkie et al. (2002).

68.1 Power Series

These form the basis for many of the identities which are used in the analysis of signals. Consider some function y = f(x) as depicted in Figure 68.1.

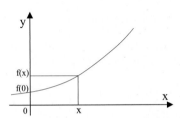

Fig. 68.1 An arbitrary function y = f(x)

Subject to certain constraints, for example on continuity and convergence, any such function may be represented by a power series expansion of the form:

$$y = f(x) = a_0 + a_1 x + a_2 x^2 + a_3 x^3 + a_4 x^4 + \ldots \quad (68.1)$$

If $x = 0$ then $a_0 = f(0)$.

Differentiating with respect to x gives:

$$\frac{dy}{dx} = f'(x) = a_1 + 2a_2 x + 3a_3 x^2 + 4a_4 x^3 + \ldots$$

and if $x = 0$ then $a_1 = f'(0)$.

Differentiating again gives:

$$\frac{d^2 y}{dx^2} = f''(x) = 2a_2 + 6a_3 x + 12a_4 x^2 + \ldots$$

and if $x = 0$ then $2a_2 = f''(0)$.

A process of successive differentiating and setting $x = 0$ gives $6a_3 = f'''(0)$, and so on.

Substituting back into Equation 68.1 for the coefficients yields:

$$y = f(x) = f(0) + f'(0)x + \frac{f''(0)}{2}x^2 \quad (68.2)$$
$$+ \frac{f'''(0)}{3!}x^3 + \frac{f^{iv}(0)}{4!}x^4 + \ldots$$

This is known as Maclaurin's series and can be used for finding a power series expansion for any f(x). For example:

$$\sin x = x - \frac{x^3}{3!} + \frac{x^5}{5!} - \frac{x^7}{7!} + \ldots \quad (68.3)$$

$$\cos x = 1 - \frac{x^2}{2} + \frac{x^4}{4!} - \frac{x^6}{6!} + \ldots \quad (68.4)$$

$$e^x = 1 + x + \frac{x^2}{2} + \frac{x^3}{3!} + \frac{x^4}{4!} + \ldots \quad (68.5)$$

$$(1+x)^n = 1 + nx + \frac{n(n-1)}{2}x^2 + \frac{n(n-1)(n-2)}{3!}x^3 + \ldots$$

This latter power series is known as the binomial expansion.

For convergence of a power series, the ratio of the magnitude of successive terms must be less than unity. Thus, with reference to Equation 68.1:

$$\left| \frac{a_{n+1} x^{n+1}}{a_n x^n} \right| < 1.0$$

The straight line brackets denote the "modulus" of the ratio which is its magnitude irrespective of sign. Clearly if the ratio is greater than unity the series diverges.

68.2 Taylor's Series

Suppose that the origin is shifted from $x = 0, y = 0$ to some arbitrary point $x = x, y = f(x)$, and define some new axes Δx and Δy that correspond to deviations in x and y about the new origin, as depicted in Figure 68.2.

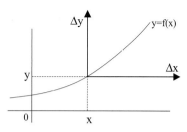

Fig. 68.2 The same arbitrary function with deviation variables as axes

Applying Maclaurin's series to the function relative to the new origin gives

$$y + \Delta y = f(x + \Delta x) = f(x) + f'(x)\Delta x + \frac{f''(x)}{2}\Delta x^2 + \frac{f'''(x)}{3!}\Delta x^3 + \ldots$$

Subtracting $y = f(x)$ yields Taylor's series:

$$\Delta y = f'(x)\Delta x + \frac{f''(x)}{2}\Delta x^2 + \frac{f'''(x)}{3!}\Delta x^3 + \ldots \quad (68.6)$$

Taylor's series provides the basis for approximating non-linear signals as deviations of variables. This is a commonly used technique in process modelling and enables linear control theory to be applied to non-linear systems. In particular, if Δx is small, then:

$$\Delta y \approx f'(x).\Delta x$$

In the above derivation some arbitrary point x, f(x) was used as the new origin. For modelling purposes it is common practice to use the normal values of the variables as the origin:

$$\Delta y \approx \left. \frac{df(x)}{dx} \right|_{\bar{y}} .\Delta x \quad (68.7)$$

where $\bar{y} = f(\bar{x})$. The accuracy of this approximation clearly depends upon the curvature of the function f(x) and the length of the step Δx.

Taylor's series can be extended to functions of two or more variables. For example, if $z = f(x, y)$ then:

$$\Delta z \approx \left. \frac{\partial f(x, y)}{\partial x} \right|_{\bar{z}} .\Delta x + \left. \frac{\partial f(x, y)}{\partial y} \right|_{\bar{z}} .\Delta y \quad (68.8)$$

where $\bar{z} = f(\bar{x}, \bar{y})$. Note that the derivatives are now partial.

68.3 Complex Numbers

Complex numbers arise from the roots of negative quantities. For example, consider the quadratic equation:

$$ax^2 + bx + c = 0$$

the solution to which is:

$$x = \frac{-b \pm \sqrt{b^2 - 4ac}}{2a}$$

If $4ac > b^2$ then:

$$x = \frac{-b \pm j\sqrt{4ac - b^2}}{2a}$$

where $j = \sqrt{-1}$. This is the notation normally used in engineering texts although it is common, especially in texts on mathematics, for the symbol $i = \sqrt{-1}$ to be used.

The above roots have a real part $\frac{-b}{2a}$ and imaginary parts $\pm \frac{\sqrt{4ac - b^2}}{2a}$.

Complex numbers arising from models of real systems always occur in conjugate pairs, for example:

$$x_1 = \sigma + j\omega$$
$$x_2 = \sigma - j\omega$$

The real and imaginary parts may be represented in cartesian form on an Argand diagram as shown in Figure 68.3.

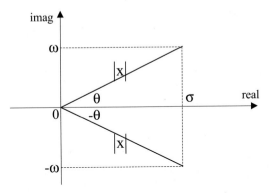

Fig. 68.3 Cartesian and polar forms of complex numbers

In polar co-ordinates the complex numbers may be represented thus:

$$x_1 = |x| \angle \theta$$
$$x_2 = |x| \angle -\theta$$

where:

$$|x| = \sqrt{\sigma^2 + \omega^2}$$
$$\theta = \tan^{-1}\left(\frac{\omega}{\sigma}\right)$$

$|x|$ is said to be the modulus of x, which is its magnitude, and θ is said to be the argument of x, which is its angle.

Complex numbers may be represented in exponential form. From Equation 68.5:

$$e^{j\theta} = 1 + j\theta - \frac{\theta^2}{2} - j\frac{\theta^3}{3!} + \frac{\theta^4}{4!} + j\frac{\theta^5}{5!} + \ldots$$
$$= \left(1 - \frac{\theta^2}{2} + \frac{\theta^4}{4!} + \ldots\right)$$
$$+ j.\left(\theta - \frac{\theta^3}{3!} + \frac{\theta^5}{5!} + \ldots\right)$$

Substituting from Equations 68.3 and 68.4 gives:

$$e^{j\theta} = \cos\theta + j.\sin\theta \qquad (68.9)$$

Similarly:

$$e^{-j\theta} = \cos\theta - j.\sin\theta$$

Addition and subtraction gives, respectively:

$$e^{j\theta} + e^{-j\theta} = 2\cos\theta$$
$$e^{j\theta} - e^{-j\theta} = 2j\sin\theta \qquad (68.10)$$

These relationships are particularly useful for transforming wave forms from one form to another.

It can be seen that the following representations of complex numbers are completely equivalent:

$$x = \sigma \pm j\omega \equiv |x| \angle \pm \theta \equiv |x|.(\cos\theta \pm j\sin\theta)$$
$$\equiv |x|.e^{\pm j\theta}$$

Complex numbers can be manipulated algebraically on the following basis.

Addition and subtraction:

$$(a + jb) + (c + jd) = a + c + j(b + d)$$

Multiplication:

$$(a + jb)(c + jd) = ac - bd + j(ad + bc) \qquad (68.11)$$

Division is achieved by multiplication of both the numerator and denominator by the complex conjugate of the denominator:

$$\begin{aligned}\frac{a+jb}{c+jd} &= \frac{(a+jb)}{(c+jd)} \cdot \frac{(c-jd)}{(c-jd)} \\ &= \frac{ac+bd+j(bc-ad)}{c^2+d^2} \\ &= \frac{ac+bd}{c^2+d^2} + j\frac{(bc-ad)}{c^2+d^2}\end{aligned}$$

By manipulation of Equation 68.11, it can be seen that:

$$|(a+jb)(c+jd)| = |a+jb| \cdot |c+jd|$$

and, in general, if:

$$x = \prod_{i=1}^{n}(\sigma_i + j\omega_i)$$

then:

$$|x| = \prod_{i=1}^{n}|\sigma_i + j\omega_i| \qquad (68.12)$$

Similarly:

$$\angle x = \sum_{i=1}^{n}\angle(\sigma_i + j\omega_i) \qquad (68.13)$$

where Π and Σ denote product and sum respectively.

First Order Systems

69.1 Example of Thermal System
69.2 Example of Residence Time
69.3 Example of RC Network
69.4 Example of Terminal Velocity
69.5 Example of Simple Feedback
69.6 Comments

A first order system is one whose behaviour may be described by a first order differential equation of the form:

$$T\frac{d\theta_0}{dt} + \theta_0 = \theta_1 \qquad (69.1)$$

This may be represented in block diagram form as shown in Figure 69.1 wherein θ_1 is the input signal and θ_0 the output signal.

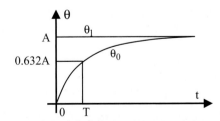

Fig. 69.1 First order system block diagram

The behaviour of first order systems is discussed in many texts, such as Coughanowr (1991). The response to a step input of magnitude A, assuming zero initial conditions, is as follows:

$$\theta_0(t) = A\left(1 - e^{-t/T}\right) \qquad (69.2)$$

which is depicted in Figure 69.2.

Fig. 69.2 Step response of first order system

The coefficient T is known as the time constant and can be seen to have units of time. Note that if $t = T$ then the output is 0.632A. Thus the time constant of a first order system may be found empirically by applying a step change and observing the time taken for its response to reach 63.2% of its steady state value.

The response of a first order system to a ramp input of slope m, assuming zero initial conditions, is given by Equation 69.3 and is depicted in Figure 69.3:

$$\theta_0 = mt - mT + mT.e^{-t/T} \qquad (69.3)$$

Note that, once the initial transients have decayed away, the output lags behind the input by the time constant T and has a dynamic error of mT.

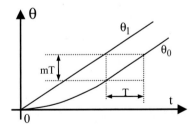

Fig. 69.3 Ramp response of first order system

The input θ_1 is often described as the forcing function. Step and ramp inputs are of particular interest in control systems design. They are typical of the sort of inputs that are commonly experienced in practice and, in their different ways, represent worst case scenarios. If a system can be designed to cope with step and ramp inputs then it ought to be capable of coping with anything else. Step and ramp inputs are also relatively easy to analyse.

First order systems are very common in process automation. Many items of plant and instrumentation have first order dynamics, or behave as multiple first order systems. Five simple but representative examples follow:

69.1 Example of Thermal System

First, consider a sensor monitoring the temperature of a process stream as shown in Figure 69.4. Suppose that the process stream temperature and its indicated value are θ_1 and θ_0 respectively.

Fig. 69.4 Temperature measurement of process stream

An unsteady state balance is of the general form:

Rate of Accumulation = Input − Output   (69.4)

This may be applied to the sensor. Assume heat transfer is by convection at the sensor surface and that there are no heat losses. Also, assume that the temperature throughout the sensor is uniform. With a datum of 0°C, an unsteady state heat balance gives:

$$\frac{d}{dt}(Mc_p\theta_0) = UA(\theta_1 - \theta_0) - 0$$

where:

θ	is the temperature	C
t	time	s
M	mass of the sensor	kg
c_p	specific heat	kJ kg^{-1} °C^{-1}
U	overall heat transfer coefficient	kW m^{-2} °C^{-1}
A	effective surface area	m^2

Rearranging gives:

$$\frac{Mc_p}{UA}\frac{d\theta_0}{dt} + \theta_0 = \theta_1 \qquad (69.5)$$

69.2 Example of Residence Time

Now consider an agitated vessel with an overflow arrangement. The level inside is roughly constant so the flow rate of the outlet stream may be assumed equal to that of the feed stream. Let C_1 be the concentration of some component in the feed and C_0 be its concentration in the outlet, as depicted in Figure 69.5.

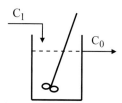

Fig. 69.5 Agitated vessel with overflow

Suppose that the vessel is well mixed such that the concentration throughout its liquid content is uniform. Since the outlet stream comes from within, the concentration inside the vessel must be C_0. Systems with such characteristics are said to be of a lumped parameter nature. An unsteady state mass balance for the component inside the vessel thus gives:

$$\frac{d}{dt}(VC_0) = F.C_1 - F.C_0$$

where:

C	is the concentration	kg m^{-3}
V	volume of liquid in vessel	m^3
F	flow rate	m^3 s^{-1}

Rearranging gives:

$$\frac{V}{F}\frac{dC_0}{dt} + C_0 = C_1 \qquad (69.6)$$

where the time constant V/F is often referred to as the residence time. If there is no residence time, *i.e.*

$V = 0$, then $C_0 = C_1$. It is evident that the capacity of the vessel averages out fluctuations in concentration. This process of averaging out fluctuations is often referred to as filtering or damping. Clearly the bigger the residence time the greater the damping.

69.3 Example of RC Network

Next consider an electrical filter: a circuit used for averaging out fluctuations in voltage as shown in Figure 69.6. The fluctuating input is V_1 and the damped output is V_0.

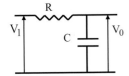

Fig. 69.6 RC network: a filter circuit

Suppose that the device for measuring V_0 has a very high input resistance so that it draws negligible current from the circuit. The only flow of current therefore is from V_1 to the capacitor. An unsteady state charge/current balance across the capacitor gives:

$$C \frac{dV_0}{dt} = \frac{V_1 - V_0}{R}$$

where:
V	is the	voltage	V
C		capacitance	A s V^{-1}
R		resistance	Ω

Rearranging gives:

$$RC \frac{dV_0}{dt} + V_0 = V_1 \qquad (69.7)$$

where the value of the time constant RC determines the amount of damping.

69.4 Example of Terminal Velocity

A steel ball is allowed to fall from rest through a viscous fluid. The drag on the ball is proportional to its velocity, as depicted in Figure 69.7.

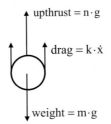

Fig. 69.7 Steel ball falling through viscous fluid

Noting that the upthrust is given by Archimede's principle and that Newton's law applies, an unsteady state force balance gives

$$m \cdot \frac{d^2x}{dt^2} = (m - n) \cdot g - k \cdot \frac{dx}{dt}$$

where:
k	is the drag coefficient	kg s^{-1}
g	gravitational acceleration	m s^{-2}
m	mass of the ball	kg
n	mass of water displaced	kg
v	velocity of ball	m s^{-1}
x	distance fallen	m

This may be articulated in terms of velocity rather than distance:

$$m \cdot \frac{dv}{dt} = (m - n) \cdot g - k \cdot v$$

which may be rearranged into the general form:

$$\frac{m}{k} \cdot \frac{dv}{dt} + v = \frac{(m - n)}{k} \cdot g \qquad (69.8)$$

Assume that the ball weighs 1 kg. If the SG of steel is 8.5 and that of the fluid is 0.8, then $n = 0.09412$ kg.

Suppose that the drag on the ball has a value of 5 N when the velocity is 1 m s^{-1}. Thus $k = 5$ kg s^{-1}. Assuming $g = 9.81$ m s^{-2} and substituting gives:

$$0.2 \frac{dv}{dt} + v = 1.777$$

whence, from Equation 68.2:

$$v = 1.777 \left(1 - e^{-5 \cdot t}\right)$$

The terminal velocity is that reached at steady state, the asymptote, $v = 1.777$ m s^{-1}.

69.5 Example of Simple Feedback

Figure 69.8 depicts a tank containing water whose temperature is controlled by a thermostat.

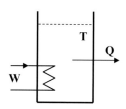

Fig. 69.8 Thermostatically controlled tank

An approximate unsteady state heat balance gives:

$$\frac{d}{dt}(Mc_p T) = W - Q = K(T_R - T) - U.A.T$$

where:

A	is the effective surface area	kg s^{-1}
c_p	specific heat	kJ kg^{-1} °C^{-1}
K	thermostat sensitivity	kW °C^{-1}
M	mass of water	kg
Q	rate of heat loss	kW
T	temperature relative to ambient	°C
U	overall heat transfer coefficient	kW m^{-2} °C^{-1}
W	heater power	kW

Suppose $K = 50$ kW °C^{-1}, $M = 10^4$ kg, $c_p = 4.18$ kJ kg^{-1} °C^{-1} and $UA = 2.0$ kW °C^{-1}.

Substituting and rearranging gives:

$$\frac{4.18 \times 10^4}{52} \frac{dT}{dt} + T = \frac{50}{52} T_R \qquad (69.9)$$

If the water is initially at ambient temperature and the thermostat switched on with a setting of $T_R = 50$°C, then this may be solved to give:

$$T = 48.08 \left(1 - e^{-0.001244\,t}\right)$$

By inspection it can be seen that the temperature would reach steady state at a temperature of approximately 48°C above ambient. The temperature reached after, say, 1 h is 47.53 which is to within about 0.5°C of the steady state value.

69.6 Comments

Although the five systems modelled are inherently different, Equations 69.5–69.9 are all of the form of Equation 69.1. Their responses to step and ramp inputs would be similar in every respect, except for speed of response as characterised by their time constants.

Laplace Transforms

70.1 Definition
70.2 Use of Laplace Transforms
70.3 Partial Fractions

Laplace transformation is used extensively in control engineering. It is a powerful technique for describing and solving sets of differential equations. When first encountered Laplace transforms may appear daunting although, in practice, they are fairly easy to use. It is best to learn how to use the technique first without fully understanding how it works. Having mastered the technique a deeper understanding will follow. This process is analogous to learning to drive. You don't need to know how an engine works to learn to drive, but being able to drive does give an insight into how it works. For a deeper understanding the reader is referred to classical control texts such as those by Dutton (1997) and Wilkie (2002).

70.1 Definition

The Laplace transform is defined by the following equation:

$$f(s) = L\{f(t)\} = \int_0^\infty f(t).e^{-st}.dt \qquad (70.1)$$

It is normal for the variable being transformed to be an analogue signal that is some function f(t) of time but it could, for example, be a function f(x) of distance. If the function f(t) is known explicitly, then its transform can be found by the integration defined in Equation 70.1. If the function f(t) is not known explicitly then its transform cannot be found. However, the variable may be considered to have been transformed in which case it is denoted as such by putting (s) after it. Thus f(s) is the Laplace transform of f(t).

The integration of Equation 70.1 involves the Laplace operator "s" which is a complex variable that is, by definition, independent of time. Because the integration is carried out over the range 0 to ∞ the resultant transform is an algebraic expression in s and is independent of time. This is best illustrated by some examples:

Example 1. Consider a step change in flow of magnitude A at time t = 0 as depicted by Figure 70.1. Substituting in Equation 70.1, assuming zero initial flow, gives:

$$f(s) = \int_0^\infty A.e^{-st}.dt = A.\left[\frac{e^{-st}}{-s}\right]_0^\infty = \frac{A}{s}$$

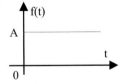

Fig. 70.1 A step change of magnitude A

Example 2. Consider an exponential decay $c = e^{-at}$ in concentration as depicted by Figure 70.2. Substituting in Equation 70.1, again assuming zero initial conditions, gives:

$$c(s) = \int_0^\infty e^{-at}.e^{-st}.dt = \left[\frac{e^{-(s+a)t}}{-(s+a)}\right]_0^\infty = \frac{1}{s+a}$$

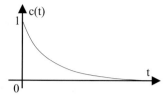

Fig. 70.2 The unit exponential decay

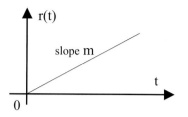

Fig. 70.3 A ramp signal of slope m

Example 3. Consider a ramping signal as depicted in Figure 70.3. Assume that the ramp has a slope of m. Integrating by parts gives:

$$r(s) = \int_0^\infty mt.e^{-st}.dt$$

$$= m.\left[t.\frac{e^{-st}}{-s}\right]_0^\infty - m.\int_0^\infty \frac{e^{-st}}{-s}dt = \frac{m}{s^2}$$

Fortunately, it is not normally necessary to have to work out the transform of a function from first principles. Extensive tables of Laplace transforms exist. Table 70.1 lists the more commonly used ones in process control. Table 70.2 lists the more important properties of Laplace transforms. These properties, which are all proved in texts such as Coughanowr (1991) and Ogata (1990), will be made use of in subsequent chapters.

Table 70.1 Table of common Laplace transforms

$f(t)_{t \geq 0}$ $(f(t)_{t<0} = 0)$	$f(s)$
A	$\dfrac{A}{s}$
t	$\dfrac{1}{s^2}$
t^n	$\dfrac{n!}{s^{n+1}}$
e^{-at}	$\dfrac{1}{s+a}$
$e^{-t/T}$	$\dfrac{T}{Ts+1}$
$1 - e^{-t/T}$	$\dfrac{1}{s.(Ts+1)}$
$t.e^{-t/T}$	$\dfrac{T^2}{(Ts+1)^2}$
$\sin(\omega t)$	$\dfrac{\omega}{s^2+\omega^2}$
$\cos(\omega t)$	$\dfrac{s}{s^2+\omega^2}$
$e^{-at}.\sin(\omega t)$	$\dfrac{\omega}{(s+a)^2+\omega^2}$
$e^{-at}.\cos(\omega t)$	$\dfrac{s+a}{(s+a)^2+\omega^2}$

Table 70.2 Important properties of Laplace transforms

$f(t)_{t \geq 0}$ $(f(t)_{t<0} = 0)$	$f(s)$
$f_1(t) + f_2(t)$	$f_1(s) + f_2(s)$
$k.f(t)$	$k.f(s)$
$\dfrac{d}{dt}f(t)$	$s.f(s) - f(0)$
$\dfrac{d^2}{dt^2}f(t)$	$s^2.f(s) - s.f(0) - \left[\dfrac{d}{dt}f(t)\right]_{t=0}$
$\int_0^t f(t)dt$	$\dfrac{f(s)}{s}$
$f(t - L)$	$e^{-Ls}.f(s)$
$\delta_0(t)$	1
$\delta_L(t)$	$\dfrac{1 - e^{-Ls}}{Ls}$
$f(\infty)$	$\lim_{s \to 0} s.f(s)$
$f(0)$	$\lim_{s \to \infty} s.f(s)$

Note that in the above examples, and for all the transforms given in Tables 70.1 and 70.2, the time $t = 0$ is both the instant from which the function exists and from which integration commences. Also, the function's prior values are specified as being zero. This is consistent with the practice of using deviation variables, as discussed in Chapter 22, and assuming that the variable was at its normal value prior to the period of interest.

70.2 Use of Laplace Transforms

Laplace transforms are used for solving ordinary linear differential (ODE) equations with constant coefficients of the general form:

$$a_n \frac{d^n f(t)}{dt^n} + a_{n-1} \frac{d^{n-1} f(t)}{dt^{n-1}} + \ldots$$
$$+ a_1 \frac{df(t)}{dt} + a_0 f(t) = u(t) \quad (70.2)$$

Thus Laplace transforms cannot normally be used for solving the following equations on the grounds that they are not ordinary, linear or with constant coefficients respectively:

$$a \frac{\partial f(t, x)}{\partial x} + f(t, x) = u(t)$$
$$a \frac{df(t)}{dt} + [f(t)]^2 = u(t)$$
$$a(t) \frac{df(t)}{dt} + f(t) = u(t)$$

There is essentially a four-step procedure to solving ODEs using Laplace transforms:

1. Transform both sides of the ODE using the transforms of Table 70.1. The first two properties listed in Table 70.2 are important here. First, the transform of the sum of two or more variables is the sum of the transforms of the individual variables. Thus the terms of the ODE may be transformed one at a time. And second, the transform of the product of a coefficient and a variable is the product of the coefficient and the transform of the variable. Thus the coefficient of any term in an ODE has no effect on the transformation process.
2. Solve the algebraic equation resulting from Step 1 for the transformed variable f(s).
3. Reduce the function f(s) resulting from Step 2 into partial fractions that occur on the right hand side of Table 70.1. For simple functions this step is trivial but for complex ones it can be very tedious.
4. Find the inverse transform of the equation resulting from Step 3 by using Table 70.1. Again, make use of the first two properties of Table 70.2 which enable the inverse transforms to be found one term at a time and independently of their coefficients.

Example 4. To illustrate this procedure reconsider the sensor used for monitoring the temperature of a process stream, as discussed in Chapter 69 and depicted in Figure 69.4. The sensor is a first order system and its dynamics are described by the first order ODE:

$$T \frac{d\theta_0}{dt} + \theta_0 = \theta_1 \quad (70.3)$$

where $T = Mc_p/UA$ as defined in Equation 69.5.

Suppose there is a step change of magnitude A in the process stream temperature. Thus, defining $t = 0$ to be the instant when the step change occurs, and assuming θ_0 to be in deviation form:

$$T \frac{d\theta_0}{dt} + \theta_0 = A$$

Step 1. Transform both sides of the equation:

$$T(s\theta_0(s) - \theta_0(0)) + \theta_0(s) = \frac{A}{s}$$

Since θ_0 is in deviation form, $\theta_0(0)$ is presumed to be zero:

$$Ts\theta_0(s) + \theta_0(s) = \frac{A}{s}$$

Step 2. Solve for $\theta_0(s)$:

$$\theta_0(s) = \frac{A}{s(Ts + 1)}$$

Step 3. There is no need to split this into partial fractions because $\theta_0(s)$ is on the right hand side of Table 70.1.

Step 4. Inverse transform:

$$\theta_0(t) = A(1 - e^{-t/T}) \qquad (69.2)$$

This step response is as depicted in Figure 69.2.

Example 5. Suppose that the process stream temperature is ramping upwards with a slope m. Define $t = 0$ to be the instant when the ramp starts and, again, assume θ_0 to be in deviation form:

$$T\frac{d\theta_0}{dt} + \theta_0 = mt$$

Step 1. Transform both sides of the equation:

$$T(s\theta_0(s) - \theta_0(0)) + \theta_0(s) = \frac{m}{s^2}$$

As before, θ_0 is in deviation form so $\theta_0(0)$ is zero:

$$Ts\theta_0(s) + \theta_0(s) = \frac{m}{s^2}$$

Step 2. Solve for $\theta_0(s)$:

$$\theta_0(s) = \frac{m}{s^2(Ts+1)}$$

Step 3. Split into partial fractions:

$$\theta_0(s) = \frac{m}{s^2} - \frac{mT}{s} + \frac{mT^2}{(Ts+1)}$$

Step 4. Inverse transform:

$$\theta_0(t) = mt - mT + mTe^{-t/T} \qquad (69.3)$$

This ramp response is as depicted in Figure 69.3.

70.3 Partial Fractions

There are two approaches to this. In both cases all possible partial fractions must be identified by inspection. Consider the function $\theta_0(s)$ in Example 5 which may be expanded as follows:

$$\frac{m}{s^2(Ts+1)} = \frac{a}{s^2} + \frac{b}{s} + \frac{c}{(Ts+1)} \qquad (70.4)$$

Note that the denominator of $\theta_0(s)$ has three roots, s^2, s and $(Ts+1)$, the s root being obscured by the s^2 root. Any such "hidden" roots must be allowed for in the expansion. If, in the above example, the hidden s root did not exist then its coefficient b would be found to be zero.

Approach 1. Equating coefficients.

Putting the right-hand-side of Equation 4 over a common denominator gives:

$$\frac{m}{s^2(Ts+1)} = \frac{(bT+c)s^2 + (aT+b)s + a}{s^2(Ts+1)}$$

The coefficients of the left and right hand numerators may be equated:

$$\begin{array}{ll} s^2 & 0 = bT + c \\ s^1 & 0 = aT + b \\ s^0 & m = a \end{array}$$

These three equations may be solved for the three unknowns, yielding: $a = m, b = -mT$ and $c = mT^2$. Hence:

$$\frac{m}{s^2(Ts+1)} = \frac{m}{s^2} - \frac{mT}{s} + \frac{mT^2}{(Ts+1)}$$

The equating coefficients approach is realistic for simple functions when the number of simultaneous equations to be solved is small. However, for more complex functions the method of residues is more appropriate.

Approach 2. The method of residues.

This method involves isolating the coefficients in turn by multiplying throughout by the roots as appropriate. The value of the coefficient is the residual when a value for s is substituted such that all the other numerator terms become zero. This is best illustrated by means of the same example.

Multiplying Equation 70.4 throughout by s^2 gives:

$$\frac{m}{(Ts+1)} = a + bs + \frac{cs^2}{(Ts+1)} \qquad (70.5)$$

Since this equation is true for all values of s, it must be true for any particular value. Choose $s = 0$ and substitute gives $a = m$.

Now multiply Equation 70.4 throughout by $(Ts + 1)$:

$$\frac{m}{s^2} = \frac{a(Ts + 1)}{s^2} + \frac{b(Ts + 1)}{s} + c$$

Choose $s = -1/T$ and substitute gives $c = mT^2$.

The value for b cannot be found in the same way. Multiplying Equation 70.4 throughout by s does not enable b to be isolated because of the s^2 in the denominator. In such circumstances there are two options.

Option 1. Substitute the values found for a and c into Equation 70.4 as follows:

$$\frac{m}{s^2(Ts + 1)} = \frac{m}{s^2} + \frac{b}{s} + \frac{mT^2}{(Ts + 1)}$$

Again this equation is true for all values of s so it must be true for any particular value. Choose $s = 1$ say and substitute gives $b = -mT$ whence:

$$\frac{m}{s^2(Ts + 1)} = \frac{m}{s^2} - \frac{mT}{s} + \frac{mT^2}{(Ts + 1)}$$

Option 2. Since s is a variable and all the other parameters are constants, Equation 70.5 may be differentiated with respect to s:

$$\frac{-mT}{(Ts + 1)^2} = b + \frac{2cs}{(Ts + 1)} - \frac{cTs^2}{(Ts + 1)^2}$$

This has now isolated b which can be found by substituting any convenient value for s, say $s = 0$, which gives $b = -mT$.

It is important to understand the processes involved and to be able to handle them manually for simple functions. However, for more complex functions, any modern maths or control systems design package will have routines for finding the roots and partial fractions of polynomial functions, which obviously simplifies the process of solving ODEs.

Transfer Functions

71.1 First Order System
71.2 Lags, Leads, Integrators and Delays
71.3 The 3-Term Controller
71.4 Block Diagram Algebra
71.5 Open and Closed Loop Transfer Functions
71.6 Steady State Analysis
71.7 Worked Example No 1
71.8 Worked Example No 2
71.9 Characteristic Equation
71.10 Worked Example No 3
71.11 The Routh Test
71.12 Worked Example No 4

In Chapter 3, in the context of block diagrams, the concept of an input signal being operated upon by some function to produce an output signal was developed. If the input and output signals are in transformed form, then the function is said to be a transfer function G(s) and the operation is that of multiplication, as depicted in Figure 71.1:

$$\theta_0(s) = G(s).\theta_1(s)$$

Fig. 71.1 Block diagram representation of a transfer function

Any system described as being a single-input single-output (SISO) system can be described by such a transfer function.

71.1 First Order System

Consider the first order system:

$$T\frac{d\theta_0}{dt} + \theta_0 = \theta_1$$

This may be transformed, assuming zero initial conditions, to give:

$$Ts\theta_0(s) + \theta_0(s) = \theta_1(s)$$

which may be rearranged into the form:

$$\theta_0(s) = \frac{1}{Ts+1}.\theta_1(s)$$

Thus the transfer function of a first order system is:

$$G(s) = \frac{1}{Ts+1} \qquad (71.1)$$

It is often the case, when the input and output signals are of a different nature and hence have different units, that the transfer function is of the form:

$$G(s) = \frac{K}{Ts+1}$$

where K is the steady state gain which is, in effect, a scaling factor.

Table 71.1 Characteristics of a lag, lead, integrator and delay

Lag	Lead	Integrator	Delay
(agitated vessel diagram with θ_1 in, θ_0 out)	Does not occur naturally	(agitated vessel diagram with θ_1 in, θ_0 out)	(pipe diagram with θ_1 in, θ_0 out)
$T \cdot \dfrac{d\theta_0}{dt} + \theta_0 = \theta_1$	$T \cdot \dfrac{d\theta_1}{dt} + \theta_1 = \theta_0$	$T \cdot \dfrac{d\theta_0}{dt} = \theta_1$	$\theta_0(t) = \theta_1(t - L)$
$\theta_1(s) \rightarrow \boxed{\dfrac{1}{Ts+1}} \rightarrow \theta_0(s)$	$\theta_1(s) \rightarrow \boxed{Ts+1} \rightarrow \theta_0(s)$	$\theta_1(s) \rightarrow \boxed{\dfrac{1}{Ts}} \rightarrow \theta_0(s)$	$\theta_1(s) \rightarrow \boxed{e^{-Ls}} \rightarrow \theta_0(s)$
(step response: θ_0 rising exponentially to θ_1)	Ill determined	(ramp response: θ_0 rising linearly, θ_1 constant)	(delayed step: θ_0 appears at $t=L$)

71.2 Lags, Leads, Integrators and Delays

The transfer functions of a lag, lead, integrator and delay are listed in Table 71.1. These are four of the simplest but most important transfer functions in process control. For each is given a physical example, the equation from which the transform function derives, and its response to a step input.

1 **Lag.** A lag is a first order system in which the $Ts + 1$ term is in the denominator of the transfer function, as in Equation 71.1. Lags occur naturally and are very common in process systems. Lags are associated with capacity. Numerous examples were given in Chapter 69, of which the temperature sensor was considered in detail in Chapter 70. The specific example given in Table 71.1 is of the agitated vessel described by Equation 69.6.

2 **Lead.** A lead is a first order system in which the $Ts + 1$ term is in the numerator of the transfer function. Leads do not occur naturally in isolation. Their response is indeterminate and cannot be visualised. Leads do however exist within the transfer functions of more complex systems. They are sometimes specified in the design of control systems, to cancel out the effects of undesired process lags, and are typically realised by means of software.

3 **Integrator.** The easiest way to visualise an integrator is as a vessel with an inlet but no outlet. The level is a measure of the integral of the flow in. A more realistic example is of a vessel in

which the inlet flow is wild but the outlet flow is constant, perhaps determined by the capacity of a pump. Steam boilers are characterised by integrators: following a step increase in combustion, but with a steady demand, the steam pressure will ramp upwards.

4 **Delay.** Most process control systems have a certain amount of delay, distance-velocity effects being the most common cause. For example, a change in concentration at the inlet to a pipe will, eventually, cause an identical change in concentration at the outlet. The word "identical" is significant: time delays cause a shift in the time scale of a signal but have no affect on its magnitude or form. Time delays are confusingly referred to as distance-velocity lags, transportation lags and dead-time. It is important to understand that delays and first order lags are inherently different, as is illustrated by their response to a step input.

For some analytical purposes the transfer function of a delay is difficult to handle, so a first order "Pade" approximation is often used. This is based upon the first two terms of a McClaurin series expansion as follows:

$$e^{-Ls} = \frac{e^{-Ls/2}}{e^{+Ls/2}} \approx \frac{1 - Ls/2}{1 + Ls/2} \quad (71.2)$$

71.3 The 3-Term Controller

Another important transfer function is that of a 3-term controller. The classical PID controller is given by Equation 23.6. If the variables are in deviation form, such that the controller bias may be ignored, this reduces to:

$$\theta_0 = \pm K_C \left(e + \frac{1}{T_R} \int_0^t e \, dt + T_D \frac{de}{dt} \right)$$

Considering the case of forward action, and assuming zero initial conditions, each of these terms may be transformed in turn to yield the transfer function:

$$\theta_0(s) = K_C \left(1 + \frac{1}{T_R s} + T_D s \right) . e(s) \quad (71.3)$$

which is represented in block diagram form as shown in Figure 71.2.

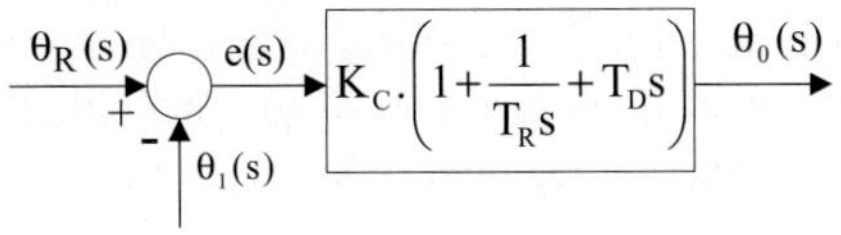

Fig. 71.2 Transfer function representation of a PID controller

A 3-term controller with derivative feedback, as defined by Equation 23.7 and assuming deviation variables, has the following transfer function:

$$\theta_0(s) = K_C \left(1 + \frac{1}{T_R s} \right) . e(s) - K_C T_D s . \theta_1(s) \quad (71.4)$$

There are many variations on the theme, as discussed in Chapter 23. Industrial controllers often have a modified form of derivative action, two of the more common forms being as follows:

$$\theta_0(s) = \pm K_C \left(\left(1 + \frac{1}{T_R s} \right) . e(s) \right. \quad (71.5)$$
$$\left. - \left(\frac{T_D s}{\alpha T_D . s + 1} \right) . \theta_1(s) \right)$$

$$\theta_0(s) = K_C \left(1 + \frac{1}{T_R s} \right) . \frac{(T_D s + 1)}{(\alpha T_D . s + 1)} . e(s) \quad (71.6)$$

In Equation 71.5, which has derivative feedback, the derivative action has an associated first order lag of value αT_D. Typically $\alpha = 0.1$. This lag effectively filters out high frequency transients, such as noise on the measurement signal, that might otherwise cause derivative kicks.

In Equation 71.6 filtering is realised by means of "lead compensation" consisting of single lead and lag terms, the derivative action being characterised by both T_D and the filter constant α. The numerator term is dominant because, by specification, $\alpha < 1.0$. Note that in this equation T_D is a time constant and is not the same as the rate time of Equations 71.3–71.5.

71.4 Block Diagram Algebra

Block diagram algebra is the name given to the process of manipulating the transfer functions of the individual elements of a system to establish some overall input-output relationship. There are two important properties of transfer functions that are made extensive use of in block diagram algebra. First, the overall transfer function of two or more elements in series is the product of the individual transfer functions, as depicted in Figure 71.3.

$$y(s) = G_2(s).x(s) = G_2(s)G_1(s).u(s)$$
$$= G_1(s)G_2(s).u(s)$$

Fig. 71.3 Transfer function of elements in series

Second is the principle of superposition. This states that the response of a system to two or more simultaneous inputs is the sum of the responses that would be obtained if each of the inputs were applied separately. This is depicted in Figure 71.4.

$$\theta_0(s) = G_1(s).\theta_1(s) + G_2(s).\theta_2(s)$$

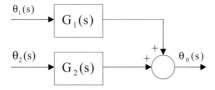

Fig. 71.4 Principle of superposition

An important constraint on the use of the principle of superposition is that the system must be linear, *i.e.* capable of being described by an ODE. It is also normal practice, but not essential, that the signals are in deviation form and to assume zero initial conditions.

71.5 Open and Closed Loop Transfer Functions

Consider the process control system depicted in Figure 71.5 in which $C(s), L(s), M(s), P(s)$ and $V(s)$ are the transfer functions of the controller, load, measurement, process and valve respectively. Note that, for simplicity, the converter and actuator are lumped in with the valve.

The techniques of block diagram algebra may be applied to the dynamic analysis of this system. The approach is much the same as the steady state analysis carried out in Chapter 22:

$$\begin{aligned}
\theta_0(s) &= L(s).\theta_1(s) + P(s).f(s) \\
&= L(s).\theta_1(s) + P(s)V(s).u(s) \\
&= L(s).\theta_1(s) + P(s)V(s)C(s).e(s) \\
&= L(s).\theta_1(s) \\
&\quad + P(s)V(s)C(s).(\theta_R(s) - \theta_M(s)) \\
&= L(s).\theta_1(s) \\
&\quad + P(s)V(s)C(s).(\theta_R(s) - M(s).\theta_0(s))
\end{aligned}$$

This is an implicit relationship between the controlled variable $\theta_0(s)$, the disturbance $\theta_1(s)$ and the

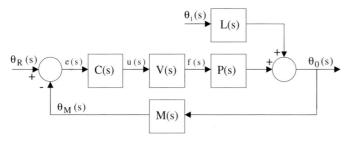

Fig. 71.5 Transfer function representation of a process control system

set point $\theta_R(s)$. The relationship can be made explicit by rearrangement to give:

$$\theta_0(s) = \frac{L(s)}{P(s)V(s)C(s)M(s)+1} \cdot \theta_1(s)$$
$$+ \frac{P(s)V(s)C(s)}{P(s)V(s)C(s)M(s)+1} \cdot \theta_R(s)$$

which is of the general form:

$$\theta_0(s) = G_1(s).\theta_1(s) + G_2(s).\theta_R(s) \quad (71.7)$$

where:

$$G_1(s) = \frac{L(s)}{P(s)V(s)C(s)M(s)+1} = \frac{L(s)}{OLTF+1}$$
$$G_2(s) = \frac{P(s)V(s)C(s)}{P(s)V(s)C(s)M(s)+1} = \frac{FPTF}{OLTF+1}$$

which is depicted in block diagram form in Figure 71.6.

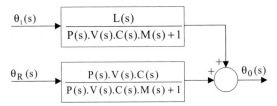

Fig. 71.6 Closed loop transfer functions of same system

The product $P(s)V(s)C(s)$ is known as the system's forward path transfer function (FPTF) and the product $P(s)V(s)C(s)M(s)$ is known as the open loop transfer function (OLTF). The terms $G_1(s)$ and $G_2(s)$ are known as the closed loop transfer functions (CLTF) for changes in disturbance and set point respectively.

71.6 Steady State Analysis

Although transfer functions are primarily used for describing dynamic behaviour, a system's steady state response can be determined from its overall transfer function by application of the final value theorem. In general terms:

$$\theta_0(\infty) = \underset{s \to 0}{\text{Lim}} \, s\theta_0(s)$$

For example, the closed loop response of the system in Figure 71.5 to a unit step change in set point is given by:

$$\theta_0(s) = \frac{1}{s} \cdot \frac{P(s)V(s)C(s)}{(P(s)V(s)C(s)M(s)+1)}$$

Thus, once the transients have decayed away, the steady state output is given by:

$$\theta_0(\infty) = \underset{s \to 0}{\text{Lim}} \, \frac{P(s)V(s)C(s)}{(P(s)V(s)C(s)M(s)+1)}$$

71.7 Worked Example No 1

The application of block diagram algebra techniques is best appreciated in relation to a specific example. Consider the process control system with a PI controller as depicted in Figure 71.7.

By inspection:

$$FPTF = K_C \cdot \left(1 + \frac{1}{T_R s}\right) \cdot \frac{1}{(0.2s+1)} \cdot \frac{5e^{-s}}{(10s+1)}$$

$$OLTF = K_C \cdot \left(1 + \frac{1}{T_R s}\right) \cdot \frac{1}{(0.2s+1)}$$
$$\times \frac{5e^{-s}}{(10s+1)} \cdot \frac{1}{(0.5s+1)}$$

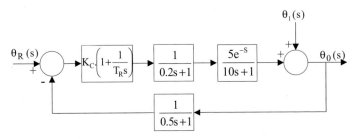

Fig. 71.7 Process control system with a PI controller

These may be substituted into the general form of Equation 71.7:

$$\theta_0(s) = \frac{1}{\text{OLTF}+1} \cdot \theta_1(s) + \frac{\text{FPTF}}{\text{OLTF}+1} \cdot \theta_R(s)$$

The CLTF for disturbances is thus:

$$G_1(s) = \frac{T_R s (10s+1)(0.5s+1)(0.2s+1)}{T_R s (10s+1)(0.5s+1)(0.2s+1) + 5K_C e^{-s}(T_R s+1)}$$

The closed loop response of the system to a step change of magnitude A in disturbance is given by:

$$\theta_0(s) = \frac{A}{s} \cdot \frac{T_R s (10s+1)(0.5s+1)(0.2s+1)}{T_R s (10s+1)(0.5s+1)(0.2s+1) + 5K_C e^{-s}(T_R s+1)}$$

Applying the final value theorem gives:

$$\theta_0(\infty) = \lim_{s \to 0} s\theta_0(s) = 0$$

There is no steady state change in output, which is what would be expected for such a system with PI control and subjected to step disturbances.

The CLTF for changes in set point is:

$$G_2(s) = \frac{5K_C e^{-s}(T_R s+1)(0.5s+1)}{T_R s (10s+1)(0.5s+1)(0.2s+1) + 5K_C e^{-s}(T_R s+1)}$$

The closed loop response of the system to a unit step change in set point is given by:

$$\theta_0(s) = \frac{1}{s} \cdot \frac{5K_C e^{-s}(T_R s+1)(0.5s+1)}{T_R s (10s+1)(0.5s+1)(0.2s+1) + 5K_C e^{-s}(T_R s+1)}$$

Applying the final value theorem gives:

$$\theta_0(\infty) = \lim_{s \to 0} s\theta_0(s) = 1$$

There is a unit steady state change in output. Thus, given the unit step change in set point, there is no offset which is what would be expected for such a system with PI control.

71.8 Worked Example No 2

The block diagram of a cascade control system is depicted in Figure 71.8.

Both the forward path and open loop transfer functions of the slave loop are found by block diagram algebra to be:

$$\text{FPTF} = \text{OLTF} = 2 \cdot \left(1 + \frac{1}{5s}\right) \cdot \frac{0.5}{5s+1} = \frac{1}{5s}$$

The closed loop transfer function of the slave loop for set point changes is thus:

$$\text{CLTF} = \frac{\text{FPTF}}{\text{OLTF}+1} = \frac{\frac{1}{5s}}{\frac{1}{5s}+1} = \frac{1}{5s+1}$$

The forward path transfer function of the master loop is thus:

$$\text{FPTF} = K_C \cdot \left(1 + \frac{1}{T_R s}\right) \cdot \frac{1}{(5s+1)} \cdot \frac{(s+1)}{(10s+1)}$$

and the open loop transfer function of the master loop is:

$$\text{OLTF} = K_C \cdot \left(1 + \frac{1}{T_R s}\right) \cdot \frac{1}{(5s+1)(10s+1)}$$

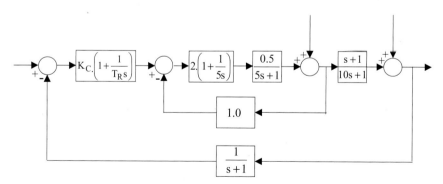

Fig. 71.8 Transfer function representation of a cascade control system

Hence the closed loop transfer function of the master loop for set point changes is:

$$\text{CLTF} = \frac{\text{FPTF}}{\text{OLTF} + 1}$$

$$= \frac{K_C \cdot \left(1 + \dfrac{1}{T_R s}\right) \cdot \dfrac{(s+1)}{(5s+1)(10s+1)}}{K_C \cdot \left(1 + \dfrac{1}{T_R s}\right) \cdot \dfrac{1}{(5s+1)(10s+1)} + 1}$$

$$= \frac{K_C \cdot (T_R s + 1)(s + 1)}{T_R s \cdot (10s+1)(5s+1) + K_C \cdot (T_R s + 1)}$$

Consider the responses of this system according to whether the master controller has P or P+I action.

First, suppose that the master controller is P only. Let $K_C = 1$ and $T_R = \infty$:

$$\text{CLTF} = \frac{\dfrac{(s+1)}{(5s+1)(10s+1)}}{\dfrac{1}{(5s+1)(10s+1)} + 1}$$

$$= \frac{(s+1)}{(10s+1)(5s+1) + 1}$$

Its closed loop response to a unit step input is:

$$\lim_{s \to 0} s \cdot \frac{1}{s} \cdot \text{CLTF} = \lim_{s \to 0} \frac{(s+1)}{(10s+1)(5s+1) + 1} = 0.5$$

Given that there was a unit step change in set point, there is an offset of 0.5 in the steady state response of the output. An offset is to be expected with P control only.

Second, suppose that the master controller is P+I with $K_C = 1$ and $T_R = 5$:

$$\text{CLTF} = \frac{(5s+1)(s+1)}{5s(10s+1)(5s+1) + (5s+1)}$$

Its closed loop response to a unit step input is:

$$\lim_{s \to 0} s \cdot \frac{1}{s} \cdot \text{CLTF}$$

$$= \lim_{s \to 0} \frac{(5s+1)(s+1)}{5s(10s+1)(5s+1) + (5s+1)} = 1.0$$

Given that there was a unit step change in set point, there is no offset in the steady state response of the output which, of course, is the objective of introducing integral action.

71.9 Characteristic Equation

Note that the denominators of both the CLTFs of Equation 71.7 are the same. This is usual. If the denominator of the CLTF is equated to zero it is referred to as the system's characteristic equation:

$$P(s).V(s).C(s).M(s) + 1 = 0 \qquad (71.8)$$

The roots of the characteristic equation are of particular significance as they determine the nature of the system's closed loop response and, in particular, its stability. To find the response, Equation 71.7 has to be solved. This entails use of the four step procedure for solving ODEs as described in Chapter 70. Consider the response to changes in set point only, *i.e.* assume $\theta_1(s)$ is zero. Suppose that Equation 71.7 is the outcome of Step 2:

$$\theta_0(s) = \frac{P(s)V(s)C(s)}{P(s)V(s)C(s)M(s) + 1} \theta_R(s) = \frac{N(s)}{D(s)}$$

where N(s) and D(s) are polynomial functions of s. This has to be reduced to partial fractions that occur on the right hand side of Table 70.1:

$$\theta_0(s) = \frac{N(s)}{D(s)} = \frac{c_1}{s + a_1} + \frac{c_2}{s + a_2} + \ldots + \frac{c_n}{s + a_n}$$

It is evident that the denominators of the partial fractions, which are the factors of D(s), come from the roots $s = -a_i$ of the characteristic equation. Taking the inverse transforms yields the response:

$$\theta_0(t) = c_1 e^{-a_1 t} + c_2 e^{-a_2 t} + \ldots + c_n e^{-a_n t} \qquad (71.9)$$

Note that the roots of the characteristic equation occur in the exponents of the exponential terms. The nature of the response depends upon the values of these roots. Clearly any negative real root will give rise to terms with an exponentially decaying response. Consider the case where the characteristic equation has two complex conjugate roots, giving rise to the partial fractions:

$$\frac{c}{s + \sigma + j\omega} + \frac{c}{s + \sigma - j\omega}$$

Table 71.2 Form and nature of response *vs* values of roots

Value of root	Form of response	Nature of response
$s_1 = -\sigma_1$	$c_1 e^{-\sigma_1 t}$	Exponential decay
$s_2 = +\sigma_2$	$c_2 e^{+\sigma_2 t}$	Exponential growth
$s_3 = 0$	c_3	Step
$s_4 = -\sigma_4 \pm j\omega_4$	$c_4 e^{-\sigma_4 t} \cos \omega_4 t$	Sinusoidal decay
$s_5 = +\sigma_5 \pm j\omega_5$	$c_5 e^{+\sigma_5 t} \cos \omega_5 t$	Sinusoidal growth
$s_6 = \pm j\omega_6$	$c_6 \cos \omega_6 t$	Continuous oscillations

Taking their inverse transforms, these will contribute the following terms to the response:

$$c.e^{-(\sigma+jw)t} + c.e^{-(\sigma-j\omega)t}$$
$$= c.e^{-\sigma t}(e^{j\omega t} + e^{-j\omega t})$$
$$= 2c.e^{-\sigma t} \cos \omega t \quad (71.10)$$

The form and nature of the closed loop response for different types of roots of the characteristic equation are summarised in Figure 71.9 and Table 71.2. These are essentially the same as the signal forms depicted in Figure 24.2.

By inspection of Equations 71.9 and 71.10 and of Figure 71.9 it is evident that roots that lie on the real axis correspond to an exponential response. In particular, those roots that lie on the real axis in the left hand half of the so-called "s plane" correspond to closed loop stable behaviour. They are said to be stable roots. The closer such roots are to the imaginary axis the slower the response. Real roots that lie in the right hand half are unstable. All complex roots, *i.e.* those that lie off the real axis, correspond to closed loop oscillatory behaviour. Those with negative real parts, *i.e.* in the left hand half, are stable. Those that lie in the right hand half are unstable and those on the imaginary axis are marginally stable. The further complex roots are from the real axis the higher the frequency of oscillation.

An important point to appreciate is that the characteristic equation is a function of the system and is independent of any external signals. It follows that the nature of the system's response does not depend on the source of input changes, such as set point or disturbance. It is the system that is stable or otherwise. However, both the form and magnitude of the response do depend upon the type of input, such as step or ramp, *etc.* and on its size.

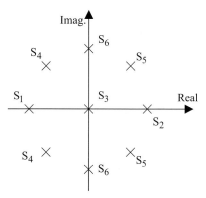

Fig. 71.9 Positions of roots of characteristic equation in the s-plane

71.10 Worked Example No 3

The significance of the roots of the characteristic equation is perhaps best illustrated by reference to the example system of Figure 71.7. The characteristic equation of this system is:

$$K_C \left(1 + \frac{1}{T_R s}\right) \frac{1}{(0.2s+1)} \frac{5e^{-S}}{(10s+1)} \frac{1}{(0.5s+1)} + 1 = 0 \quad (71.11)$$

To find the roots of this it is necessary to use the Pade approximation and to substitute known values for K_C and T_R. Supposing that $K_C = 1$ and $T_R = 5$ this gives:

$$\frac{(5s+1)(1-0.5s)}{s.(10s+1)(0.5s+1)^2(0.2s+1)} + 1 = 0$$

Following expansion and rearrangement this gives:

$$0.5s^5 + 4.55s^4 + 12.45s^3 + 8.7s^2 + 5.5s + 1 = 0$$

which may be factorised as follows:

$$(4.04s+1)(0.0538s^2 + 0.453s + 1)$$
$$\times (2.30s^2 + 1.00s + 1) = 0$$

from which the roots of the characteristic equation are found to be:

$$s = -0.247$$
$$s = -4.21 \pm j0.945$$
$$s = -0.218 \pm j0.622$$

The real root is negative, which contributes an exponential decay to the response. There are two pairs of complex conjugate roots which contribute sinusoidal components to the response. Since the real part of these complex roots is negative the oscillations will decay away. The system's response is therefore inherently stable for both disturbance and set point changes.

71.11 The Routh Test

This is a simple test that establishes whether or not any of the roots of a polynomial lie in the right half of a complex plane. It is used to establish the value of some parameter, typically the controller gain, that causes a closed loop system to become marginally stable. The test is based on the fact that for a stable system the coefficients of the characteristic equation should all be positive. Any root that is neither real and negative nor complex with negative real parts would cause negative coefficients.

Suppose that the characteristic equation of interest is fifth order of the form:

$$0.5s^5 + 4.55s^4 + 12.45s^3 + 8.7s^2 + 5.5s + 1 = 0 \quad (71.12)$$

The coefficients are written down in the first two rows of the Routh array, as depicted in Table 71.3.

The remaining elements are calculated across the rows according to a progressive formula until a value of zero is returned:

Table 71.3 Coefficients of the characteristic equation in Routh array form

s^5	a_5	a_3	a_1	0
s^4	a_4	a_2	a_0	0
s^3	b_1	b_2	0	
s^2	c_1	c_2	0	
s^1	d_1	0		
s^0	e_1			

$$b_1 = \frac{a_3.a_4 - a_2.a_5}{a_4} \qquad b_2 = \frac{a_1.a_4 - a_0.a_5}{a_4}$$

The process is repeated for each remaining row in turn:

$$c_1 = \frac{a_2.b_1 - b_2.a_4}{b_1} \qquad c_2 = \frac{a_0.b_1 - 0.a_4}{b_1}$$

$$d_1 = \frac{b_2.c_1 - c_2.b_1}{c_1}$$

$$e_1 = \frac{c_2.d_1 - 0.c_1}{d_1}$$

This might seem complicated but is fairly easy to do: the key thing is to recognise the pattern of the formulae. Routh's test for stability requires that all the coefficients of Equation 71.12 are positive and that all the terms in the first column of Table 71.1 have positive signs.

71.12 Worked Example No 4

Consider again the system described by Equation 71.11.

With $T_R = \infty$ and using the Pade approximation, the characteristic equation becomes:

$$\frac{5K_C.(1-0.5s)}{(10s+1)(0.5s+1)^2(0.2s+1)} + 1 = 0$$

Following expansion and rearrangement this gives:

$$0.5s^5 + 4.55s^4 + 12.45s^3 + (11.2 - 2.5K_C)s + 5K_C + 1 = 0$$

Table 71.4 Worked example of Routh array

s^4	0.5	12.45	$5K_C + 1$	0
s^3	4.55	$11.2 - 2.5K_C$	0	0
s^2	$11.22 + 0.2747K_C$	$5K_C + 1$	0	
s^1	$\dfrac{121.1 - 47.72K_C - 0.6868K_C^2}{11.22 + 0.2747K_C}$	0	0	
s^0	$5K_C + 1$	0		

The Routh array may thus be developed as in Table 71.4.

Assuming that $K_C > 0$, otherwise positive feedback will occur, only one sign change in the first column is possible, when

$$121.1 - 47.72K_C - 0.6868K_C^2 = 0$$

the solution to which gives:

$$K_C = -71.94 \quad \text{or} \quad K_C = +2.45.$$

The condition for stability is thus $K_C = \leq 2.45$.

Second and Higher Order Systems

72.1 Second Order Dynamics
72.2 Overdamped, $\zeta > 1$
72.3 Critically Damped, $\zeta = 1$
72.4 Underdamped, $0 < \zeta < 1$
72.5 Undamped, $\zeta = 0$
72.6 Higher Order Systems

An understanding of the dynamics of second order systems is justified on two counts. First, the dynamics of many items of plant and instrumentation are inherently second order. And second, the behaviour of many closed loop systems, whether they are second order or not, are often described or specified in terms of the characteristics of second order systems. For a more comprehensive treatment the reader is referred to the texts by Dutton (1997) and Wilkie (2002).

72.1 Second Order Dynamics

A second order system is one whose dynamic behaviour may be described by a second order ODE of the general form:

$$\frac{1}{\omega_n^2}\frac{d^2\theta_0}{dt^2} + \frac{2\zeta}{\omega_n}\frac{d\theta_0}{dt} + \theta_0 = K.\theta_1 \quad (72.1)$$

where the signals θ_0 and θ_1 are both in deviation form, and:

ω_n is the natural frequency
ζ damping factor
K steady state gain

Transforming both sides of Equation 72.1, assuming zero initial conditions, and rearranging yields the transfer function of a second order system. This is depicted in Figure 72.1.

$$\theta_0(s) = \frac{K.\omega_n^2}{s^2 + 2\omega_n\zeta s + \omega_n^2}\theta_1(s) \quad (72.2)$$

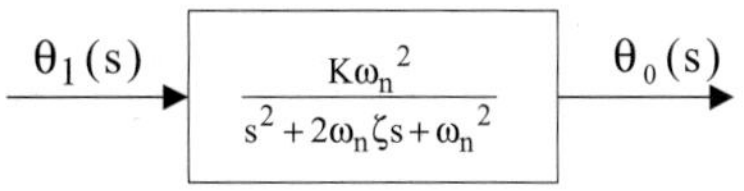

Fig. 72.1 Transfer function representation of a second order system

Suppose that the system is subjected to a step input of magnitude A. Substituting for $\theta_1(s) = A/s$ and factorising gives:

$$\theta_0(s) = \frac{A.K.\omega_n^2}{s.\left\{s + \omega_n\left(\zeta + \sqrt{\zeta^2-1}\right)\right\}.\left\{s + \omega_n\left(\zeta - \sqrt{\zeta^2-1}\right)\right\}} \quad (72.3)$$

The solution to Equation 72.3 is found by the usual procedure of reducing into partial fractions and finding their inverse transforms. The form of the solution is determined by the value of the damping factor ζ. The step response is usually categorised according to the value of ζ, as summarised in Table 24.1, and sketched in Figure 72.2.

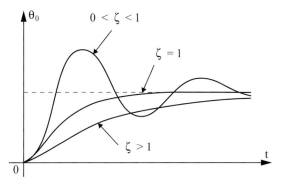

Fig. 72.2 Step response *vs* damping factor for second order system

72.2 Overdamped, $\zeta > 1$

Assume that $\zeta > 1$ such that the operands of the square roots in the denominator of Equation 72.3 are positive. The resulting negative real roots of s yield the step response:

$$\theta_0(t) = AK\left\{1 - e^{-\omega_n \zeta t}\left[\cosh\left(\omega_n\sqrt{\zeta^2 - 1}.t\right)\right.\right.$$
$$\left.\left. + \frac{\zeta}{\sqrt{\zeta^2 - 1}}\sinh\left(\omega_n\sqrt{\zeta^2 - 1}.t\right)\right]\right\} \quad (72.4)$$

A closed loop response of this overdamped form, in which the output asymptotically approaches its steady state value, is specified for control systems where it is essential that there is no oscillation.

72.3 Critically Damped, $\zeta = 1$

Assume that $\zeta = 1$ such that the operands of the square roots in the denominator of Equation 72.3 are zero. The resulting repeated roots of $s = -\omega_n\zeta$ yield the step response:

$$\theta_0 = AK\left\{1 - e^{-\omega_n t}(1 - \omega_n t)\right\} \quad (72.5)$$

This is the fastest response that can be obtained without any oscillation. However, closed loop systems should not be designed for critical damping. Any slight error in the design or measurements could result in a specified exponential response becoming oscillatory. Critical damping is nevertheless used as a basis for design. Typically, the criteria to establish critical damping are determined and a safety factor is then applied to guarantee overdamping.

72.4 Underdamped, $0 < \zeta < 1$

Assume that $0 < \zeta < 1$ such that the operands of the square roots in the denominator of Equation 72.3 are negative. The resulting complex conjugate roots of s yield the step response:

$$\theta_0(t) = AK\left(1 - \frac{1}{\sqrt{(1-\zeta^2)}}e^{-\omega_n \zeta t} \cdot \sin(\omega_d t + \phi)\right)$$
$$(72.6)$$

where ω_d is the damped frequency given by:

$$\omega_d = \omega_n\sqrt{1 - \zeta^2}$$

and ϕ is the phase shift given by:

$$\phi = \tan^{-1}\left(\frac{\sqrt{1-\zeta^2}}{\zeta}\right)$$

Both of these identities may be represented trigonometrically as shown in Figure 72.3.

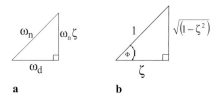

Fig. 72.3 a Identity for damped frequency, **b** Identity for phase shift

Inspection of Equation 72.6 reveals an exponentially decaying sinusoid which is sketched and labelled in Figure 72.4.

Various characteristics of this response are identified and defined in Table 72.1, which is largely self explanatory. Note that the formulae for all the characteristics are functions of ω_n and ζ only. Thus, specifying any two characteristics results in two equations which may be solved for ω_n and ζ, which determines the values of the other characteristics.

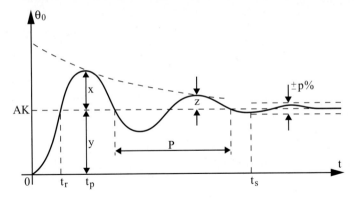

Fig. 72.4 Step response for underdamped second order system

Table 72.1 Characteristics of underdamped second order response

Characteristic	Symbol	Formula
Damped frequency	ω_d	$\omega_n\sqrt{1-\zeta^2}$
Period	P	$\dfrac{2\pi}{\omega_n\sqrt{1-\zeta^2}}$
Rise time	t_r	$\dfrac{1}{\omega_n\sqrt{1-\zeta^2}}\left\{\pi - \tan^{-1}\dfrac{\sqrt{1-\zeta^2}}{\zeta^2}\right\}$
Peak time	t_p	$\dfrac{\pi}{\omega_n\sqrt{1-\zeta^2}}$
Overshoot	$\dfrac{x}{y}$	$e^{-\dfrac{\pi\zeta}{\sqrt{1-\zeta^2}}}$
Decay ratio	$\dfrac{z}{x}$	$e^{-\dfrac{2\pi\zeta}{\sqrt{1-\zeta^2}}}$

The response limit is a band of arbitrary width $\pm p$ about the equilibrium value. The value of p is expressed as a percentage of the steady state change in output. In practice it is normally taken as either $\pm 5\%$ or $\pm 2\%$. The settling time t_s is the time taken for the output to fall within the response limit and to remain bounded by it.

The curve which is tangential to the peaks of the step response is known as its envelope, and is also sketched in Figure 72.4. The equation of the envelope is given by:

$$\theta_0 = AK.\left(1 + \dfrac{e^{-\omega_n\zeta t}}{\sqrt{1-\zeta^2}}\right)$$

Note that the time constant for the exponential decay of the envelope, $T = \dfrac{1}{\omega_n\zeta}$.

The response limit and settling time are related to this time constant by the following useful approximations:

If $p = \pm 5\%$ then $t_s \approx 3T$.
    $p = \pm 2\%$ $t_s \approx 4T$

72.5 Undamped, $\zeta = 0$

Substituting $\zeta = 0$ into Equation 72.3 results in imaginary roots of s which reduce into partial fractions as follows:

$$\theta_0(s) = AK \left(\frac{1}{s} - \frac{1}{2(s+j\omega_n)} - \frac{1}{2(s-j\omega_n)} \right)$$

Inverse transforming yields the step response:

$$\theta_0(t) = AK \left(1 - 0.5 \left(e^{-j\omega_n t} + e^{+j\omega_n t} \right) \right)$$

$$= AK \left(1 - \cos \omega_n t \right) \qquad (72.7)$$

This is a sine wave with oscillations of constant amplitude as depicted in Figure 72.5. For control system design and tuning purposes this clearly corresponds to the case of marginal stability as discussed in Chapter 24.

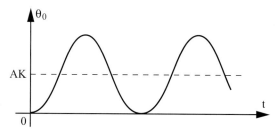

Fig. 72.5 Sinusoidal response of undamped second order system

72.6 Higher Order Systems

Consider a series of n identical first order systems, as depicted in Figure 72.6.

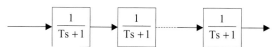

Fig. 72.6 A series of n identical first order systems

The series' overall transfer function is given by:

$$\frac{\theta_0(s)}{\theta_1(s)} = \left(\frac{1}{Ts+1} \right)^n$$

The system's response to a step input of magnitude A is sketched in Figure 72.7 as a set of curves with n as the parameter. If n=0 the output follows the input exactly. If n = 1 the response is that of a first order system. With n = 2 the response is that of a second order system with repeated roots and critical damping. For n ≥ 3 the response becomes more sluggish and "S shaped".

Process plant and control loop elements invariably consist of combinations of non-identical first order systems, overdamped second order systems and steady state gains, as depicted in Figure 72.8.

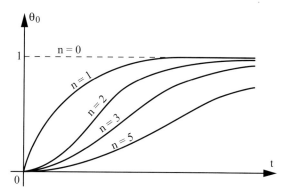

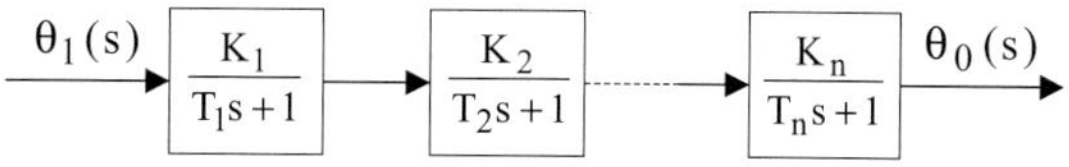

Fig. 72.7 Step response of n first order systems in series

Fig. 72.8 A series of n first order systems in series

Often it is impractical to determine the exact transfer function for each element. A common approach is to lump them together in an overall transfer function consisting of a single lag, delay and gain as follows:

$$\frac{\theta_0(s)}{\theta_1(s)} = \prod_{j=1}^{n} \frac{K_j}{T_j s + 1} \approx \frac{K e^{-Ls}}{Ts + 1} \qquad (72.8)$$

The effect of this approximation is to replace the "S shaped" response with that of a first order system subject to a time delay, as depicted in Figure 72.9.

The approximation is good enough for most purposes. For example, it provides the basis for the reaction curve method of controller tuning, described in Chapter 24, in which the control loop is assumed to consist of a 3-term controller in series with the lumped approximation, as depicted in Figure 72.10.

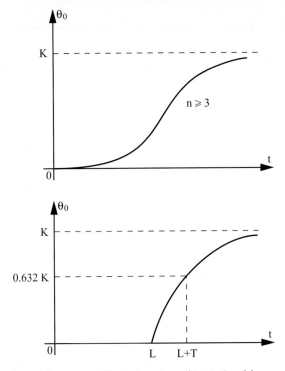

Fig. 72.9 Response of first order system subject to time delay

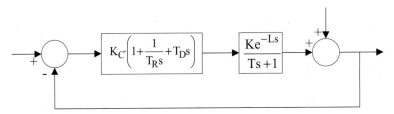

Fig. 72.10 PID controller with lumped approximation of plant

Frequency Response

73.1 Attenuation and Phase Shift
73.2 Substitution Rule
73.3 Bode Diagrams
73.4 Nyquist Diagrams
73.5 Lags, Leads, Integrators and Delays
73.6 Second Order Systems
73.7 Compound Systems
73.8 Worked Example No 1
73.9 Bode Stability Criteria
73.10 Worked Example No 2
73.11 Worked Example No 3
73.12 Gain and Phase Margins

Having seen in Chapters 71 and 72 how to represent the behaviour of systems in the Laplace domain, this chapter introduces frequency response. This is an important frequency domain means of analysis and design. In essence, it concerns the behaviour of systems that are forced by sinusoidal inputs. A system may be forced sinusoidally and, by measuring its output signal, the nature of the system may be deduced. Alternatively, a system may be designed such that the sinusoidal relationship between its input and output signals is as specified. Frequency response is of particular importance for design purposes from a stability point of view.

Being a standard technique, frequency response is covered extensively in classical texts such as Ogata (2002) and Wilkie (2002) to which the reader is referred for a more thorough treatment.

73.1 Attenuation and Phase Shift

Consider the first order system as shown in Figure 73.1.

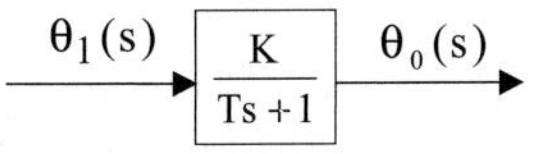

$$\theta_1(s) \longrightarrow \boxed{\frac{K}{Ts+1}} \longrightarrow \theta_0(s)$$

Fig. 73.1 First order system

If θ_1 is a sine wave of amplitude A and frequency ω rad s^{-1}:

$$\theta_1(t) = A \sin \omega t, \qquad \theta_1(s) = \frac{A\omega}{s^2 + \omega^2}$$

then the response is given by:

$$\theta_0(s) = \frac{K}{(Ts+1)} \cdot \frac{A\omega}{(s^2 + \omega^2)}$$

By splitting this into its partial fractions, inverse transforming and some non-trivial trigonometric manipulation, the solution may be found. Once

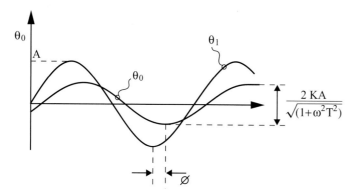

Fig. 73.2 Steady state response of first order system to sinusoidal input

the exponential transient has decayed away, the response becomes:

$$\theta_0(t) = \frac{KA}{\sqrt{(1+\omega^2 T^2)}} \cdot \sin(\omega t + \phi) \quad (73.1)$$

which is also a sine wave as depicted in Figure 73.2.

In Figure 73.2 the amplitude of θ_0 is $\frac{KA}{\sqrt{(1+\omega^2 T^2)}}$ which comprises the amplitude A of the input amplified by the steady state gain K and attenuated by the factor $\frac{1}{\sqrt{(1+\omega^2 T^2)}}$.

The product of gain and attenuation factor $\frac{K}{\sqrt{(1+\omega^2 T^2)}}$ is known as the amplitude ratio.

The frequency ω of θ_0 is the same as that of θ_1 but is shifted in phase by an amount ϕ where:

$$\phi = -\tan^{-1}(\omega T) \quad (73.2)$$

73.2 Substitution Rule

As was seen in Table 71.2, and will be further developed in Chapter 74, oscillatory behaviour is associated with roots of the form $s = \pm j\omega$. This fact may be exploited in determining the frequency response of a system. Consider the first order system:

$$G(s) = \frac{1}{Ts + 1}$$

Substitute $s = j\omega$ and form the complex conjugate giving:

$$G(j\omega) = \frac{1}{1 + j\omega T} = \frac{1}{1 + \omega^2 T^2} - j\frac{\omega T}{1 + \omega^2 T^2}$$

This may be depicted in Argand diagram form as shown in Figure 73.3.

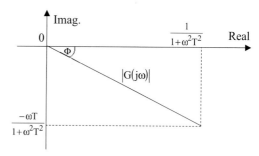

Fig. 73.3 Cartesian and polar forms of frequency response

It may also be expressed in polar co-ordinates as follows:

$$G(j\omega) = |G(j\omega)| \angle G(j\omega)$$

where the modulus, pronounced as "mod G jω", which is the length of the vector, is given by:

$$|G(j\omega)| = \sqrt{\left(\left(\frac{1}{1+\omega^2 T^2}\right)^2 + \left(\frac{\omega T}{1+\omega^2 T^2}\right)^2\right)}$$

$$= \frac{1}{\sqrt{1+\omega^2 T^2}} \quad (73.3)$$

and the argument, pronounced as "arg G jω", which is the angle of the vector, is given by:

$$\angle G(j\omega) = -\tan^{-1} \frac{\left(\frac{\omega T}{1+\omega^2 T^2}\right)}{\left(\frac{1}{1+\omega^2 T^2}\right)}$$

$$= -\tan^{-1}(\omega T) \qquad (73.4)$$

Comparison of Equations 73.1–73.4 shows that, for a first order system:

$$\text{Attenuation} = |G(j\omega)| = \frac{1}{\sqrt{(1+\omega^2 T^2)}}$$

and

$$\text{Phase shift}, \phi = \angle G(j\omega) = -\tan^{-1}(\omega T)$$

Although it has only been demonstrated here for a first order system, the frequency response of any system may be found by substituting $s = j\omega$ into $G(s)$ and rearranging into polar form.

Because of the complexity, it is convenient to present information about frequency response in graphical form. The two most common forms are Bode and Nyquist diagrams.

73.3 Bode Diagrams

These consist of graphs of attenuation and phase shift plotted against frequency, as shown for a first order system in Figure 73.4.

The attenuation graph is of $|G(j\omega)|$ along the ordinate (y-axis) vs ω along the abscissa (x-axis). $|G(j\omega)|$ is plotted using either a logarithmic scale or else in terms of decibels (dB) on a linear scale and frequency is plotted on a logarithmic scale:

$$\text{no of dB} = 20 \log_{10} |G(j\omega)|$$

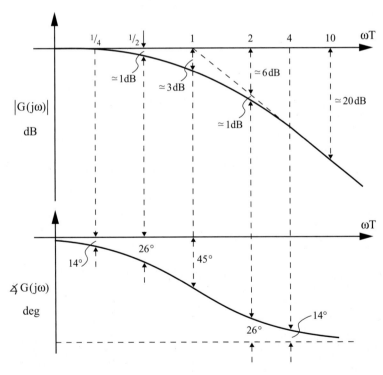

Fig. 73.4 Bode diagram for a first order system

Of particular interest are the asymptotes. For a first order system:

$$|G(j\omega)| = \frac{1}{\sqrt{(1 + \omega^2 T^2)}}$$

If $\omega \to 0$ then $|G(j\omega)| \to 1$ and $dB \to 0$, i.e. the graph is asymptotic to the 0 dB line.

If $\omega \to \infty$ then $|G(j\omega)| \to \frac{1}{\omega T}$ and $dB \to -20 \log_{10} \omega T$, i.e. the graph is asymptotic to a line of slope -20 dB/decade.

The two asymptotes intercept when

$$|G(j\omega)| = 1 = \frac{1}{\omega T},$$

i.e. when $\omega = \frac{1}{T}$ which is known as the corner frequency.

The phase shift graph is of $\angle G(j\omega)$ vs ω in which $\angle G(j\omega)$ is plotted on a linear scale. Again, a logarithmic scale is used for the frequency axis:

$$\angle G(j\omega) = -\tan^{-1}(\omega T)$$

If $\omega \to 0$ then $\angle G(j\omega) \to 0$.
If $\omega = \frac{1}{T}$ then $\angle G(j\omega) = -45^0$.
If $\omega \to \infty$ then $\angle G(j\omega) \to -90^0$.

Thus the graph of $\angle G(j\omega)$ vs ω is asymptotic to the 0 and $-90°$ lines.

For most purposes it is unnecessary to plot Bode diagrams accurately – a sketch is good enough. The data shown in Figure 73.4 is a sufficient basis for sketching.

73.4 Nyquist Diagrams

The Nyquist diagram for the system G(s) is the locus as ω varies from 0 to ∞ of the vector $|G(j\omega)| \angle G(j\omega)$, an example of which for a first order system is as shown in Figure 73.5.

Often referred to simply as a polar plot, a Nyquist diagram essentially consists of a graph of the modulus of G(jw) vs its argument in polar co-ordinates on an Argand diagram with ω as

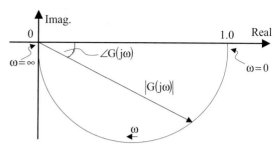

Fig. 73.5 Nyquist diagram for a first order system

a parameter. It combines into a single graph all the data contained in the two graphs of a Bode diagram. Note the convention that if the argument is zero the vector lies along the positive real axis from which negative phase angles are measured clockwise. Also, note that the value of the modulus is not plotted in terms of dB.

73.5 Lags, Leads, Integrators and Delays

The frequency response of a first order lag, a first order lead, an integrator and a time delay are as listed in Table 73.1. For each of these is given its mathematical form, transfer function, attenuation, phase shift and a sketch of its polar plot.

73.6 Second Order Systems

The transfer function of a second order system is:

$$G(s) = \frac{\omega_n^2}{s^2 + 2\zeta\omega_n s + \omega_n^2}$$

Substituting $s = j\omega$, finding the complex conjugates and putting into polar form yields:

$$|G(j\omega)| = \frac{1}{\sqrt{\left\{\left(1 - \left(\frac{\omega}{\omega_n}\right)^2\right)^2 + 4\zeta^2 \left(\frac{\omega}{\omega_n}\right)^2\right\}}}$$

(73.5)

73.6 Second Order Systems

Table 73.1 Frequency response of a lag, lead, integrator and delay

	Lag	Lead	Integrator	Delay
Mathematical form	$T \cdot \dfrac{d\theta_0}{dt} + \theta_0 = \theta_1$	$T \cdot \dfrac{d\theta_1}{dt} + \theta_1 = \theta_0$	$T \cdot \dfrac{d\theta_0}{dt} = \theta_1$	$\theta_0(t) = \theta_1(t - L)$
Transfer function	$\theta_1 \to \boxed{\dfrac{1}{Ts+1}} \to \theta_0$	$\theta_1 \to \boxed{Ts+1} \to \theta_0$	$\theta_1 \to \boxed{\dfrac{1}{Ts}} \to \theta_0$	$\theta_1 \to \boxed{e^{-Ls}} \to \theta_0$
Attenuation $\lvert G(j\omega) \rvert$	$\dfrac{1}{\sqrt{1+\omega^2 T^2}}$	$\sqrt{1+\omega^2 T^2}$	$\dfrac{1}{\omega}$	1
Argument $\angle G(j\omega)$	$-\tan^{-1}\omega T$	$+\tan^{-1}\omega T$	$-\dfrac{\pi}{2}$ rad $= -90$ deg	$-\omega L$ rad $= -57.3\omega L$ deg
Polar plot				

and

$$\angle G(j\omega) = -\tan^{-1}\left\{ \frac{2\zeta\left(\dfrac{\omega}{\omega_n}\right)}{1 - \left(\dfrac{\omega}{\omega_n}\right)^2} \right\} \quad (73.6)$$

The Bode and Nyquist diagrams of a second order system are as shown in Figures 73.6 and 73.7 respectively.

For the attenuation graph:

If $\omega \to 0$ then $\lvert G(j\omega) \rvert \to 1$ and dB $\to 0$, i.e. the graph is asymptotic to the 0 dB line.

If $\omega \to \infty$ then $\lvert G(j\omega) \rvert \to \dfrac{1}{\left(\dfrac{\omega}{\omega_n}\right)^2}$ and

dB $\to -40 \log_{10}\left(\dfrac{\omega}{\omega_n}\right)$, i.e. the graph is asymptotic to a line of slope -40 dB/decade.

The two asymptotes intercept when

$$\lvert G(j\omega) \rvert = 1 = \frac{1}{\left(\dfrac{\omega}{\omega_n}\right)^2}, \text{ i.e. when } \omega = \omega_n.$$

The natural frequency ω_n is thus the corner frequency too.

If $\omega = \omega_n$ then $\lvert G(j\omega) \rvert = \dfrac{1}{2\zeta}$

Note that if $\zeta < \sqrt{0.5} \approx 0.707$ then $\lvert G(j\omega) \rvert$ has a maximum:

$$\lvert G(j\omega) \rvert_{max} = \frac{1}{2\zeta\sqrt{(1-\zeta^2)}}$$

which occurs at the resonant frequency ω_r:

$$\omega_r = \omega_n \sqrt{(1 - 2\zeta^2)}$$

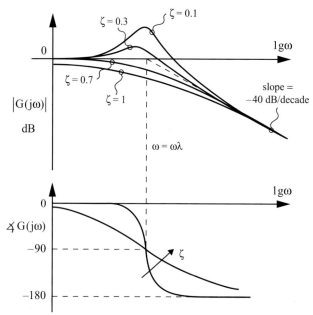

Fig. 73.6 Bode diagram for a second order system

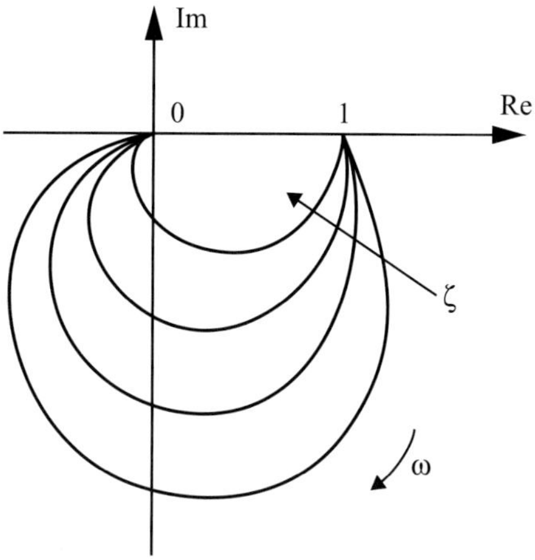

Fig. 73.7 Nyquist diagram for a second order system

For the phase shift graph:

If $\omega \to 0$ then $\angle G(j\omega) \to 0$.

If $\omega = \omega_n$ then $\angle G(j\omega) = -90°$.

If $\omega \to \infty$ then $\angle G(j\omega) \to -180°$.

Thus the graph of $\angle G(j\omega)$ vs ω is asymptotic to the 0 and $-180°$ lines.

73.7 Compound Systems

If $G(s)$ factorises into, say:

$$G(s) = G_1(s)G_2(s)G_3(s)\ldots G_n(s)$$

and for each $G_i(s)$:

$$G_i(j\omega) = |G_i(j\omega)| \angle G_i(j\omega)$$

then, because of its complex nature, the overall attenuation is given by:

$$|G(j\omega)| = \prod_{i=1}^{n} |G_i(j\omega)| \qquad (73.7)$$

or, in terms of decibels:

$$dB = \sum_{i=1}^{n} 20 \log_{10} |G_i(j\omega)|$$

and the overall phase shift is given by:

$$\angle G(j\omega) = \sum_{i=1}^{n} \angle G_i(j\omega) \qquad (73.8)$$

These rules considerably simplify the sketching of Bode and Nyquist diagrams for compound systems.

73.8 Worked Example No 1

Consider the system:

$$G(s) = \frac{0.1s + 1}{s(0.5s + 1)}$$

which consists of first order lag and lead terms, with time constants of 0.5 and 0.1 min respectively, and an integrator. Its Bode diagram is sketched in Figure 73.8.

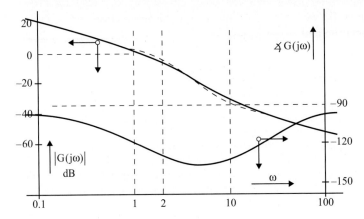

Fig. 73.8 Bode diagram for Worked Example No 1

The attenuation graph is given by:

$$|G(j\omega)| = \frac{\sqrt{(1 + 0.01\omega^2)}}{\omega\sqrt{(1 + 0.25\omega^2)}}$$

It was constructed as follows:

For $\omega < 2$ rad/min the asymptote is a line of slope -20 dB/decade passing through $\omega = 1$ due to $\frac{1}{s}$.

At $\omega = 2$ rad/min is the corner frequency of $\frac{1}{0.5s+1}$.

For $2 < \omega < 10$ the asymptote has a slope of -40 dB/decade of which -20 dB/decade is due to $\frac{1}{s}$ and another -20 dB/decade is due to $\frac{1}{0.5s+1}$.

$\omega = 10$ rad/min is the corner frequency of $(0.1s + 1)$. Note that the frequency response of $(0.1s + 1)$ is the mirror image of that due to $\frac{1}{0.1s+1}$.

For $\omega > 10$ the asymptote has a net slope of -20 dB/decade of which -40 dB/decade are due to $\frac{1}{s}$ and $\frac{1}{0.5s+1}$ and $+20$ dB/decade are due to $(0.1s + 1)$.

Having established the asymptotes, points may be located 3 dB below the corner frequency at $\omega = 2$ and 3 dB above the corner frequency at $\omega = 10$. These points are sufficient to enable the attenuation curve to be sketched in.

The phase shift graph is given by:

$$\angle G(j\omega) = +\tan^{-1}(0.1\omega) - 90 - \tan^{-1}(0.5\omega)$$

Clearly the asymptotes for this are $-90°$. The graph is drawn by selecting a few strategic values for ω.

73.9 Bode Stability Criteria

These were introduced qualitatively in Chapter 24 to explain the concept of marginal stability. Consider the feedback system depicted in Figure 73.9 in which all the elements, other than the controller, have been lumped in with the process. The controller has P action only.

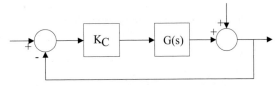

Fig. 73.9 Loop with P controller and single process transfer function

Let $K_C.G(s)$ be the open loop transfer function, where K_C is the controller gain and:

$$G(s) = \prod_{i=1}^{n} K_i.G_i(s)$$

where the K_i are steady state gains.

The Bode stability criteria may now be expressed quantitatively in terms of attenuation and phase shift as follows:

If:
$$K_{CM}.K.|G(j\omega_C)| = 1.0$$

where:
$$K = \prod_{i=1}^{n} K_i$$

and
$$|G(j\omega_C)| = \prod_{i=1}^{n} |G_i(j\omega_C)|$$

and if:
$$\angle G(j\omega_C) = -180°$$

where:
$$\angle G(j\omega_C) = \sum_{i=1}^{n} \angle G_i(j\omega_C)$$

then the closed loop system is marginally stable.

K_{CM} is known as the controller gain for marginal stability.

ω_C is referred to as the critical frequency at which the period of oscillation is P_U:

$$P_U = \frac{2\pi}{\omega_C}$$

73.10 Worked Example No 2

Consider the system whose open loop transfer function is as follows, the time constants being in minutes:

$$G(s) = \frac{6K_C}{(s+1)(s+2)(s+3)}$$
$$= \frac{K_C}{(s+1)(0.5s+1)(0.333s+1)}$$

Start with the phase shift which is given by:

$$\angle G(j\omega) = -\tan^{-1}\omega - \tan^{-1}(0.5\omega) - \tan^{-1}(0.333\omega)$$

If $\omega = \omega_C$ then $\angle G(j\omega_C) = -180°$ whence $\omega_C \approx$ 3.32 rad/min and $P_U \approx$ 1.894 min.

Now move onto the attenuation:
$$|G(j\omega)| = \frac{1}{\sqrt{(1+\omega^2)}\sqrt{(1+(0.5\omega)^2)}\sqrt{(1+(0.333\omega)^2)}}$$

Thus $|G(j\omega_c)| \approx 0.1$.

For marginal stability:
$$K_{CM}.K.|G(j\omega_C)| = 1.0$$

whence $K_{CM} \approx 10.0$

Using the Zeigler and Nichols formulae of Table 24.2, for a decay ratio of 0.25 and assuming P + I + D control action, the optimum settings are thus:

$$K_C \approx 5.9, \quad T_R \approx 0.95 \text{ min} \quad T_D \approx 0.24 \text{ min}.$$

Hint. A useful trigonometric identity for evaluating ω_c for third order systems with lags of T_1, T_2 and T_3 mins is as follows. If:

$$\tan^{-1}\omega_C T_1 + \tan^{-1}\omega_C T_2 + \tan^{-1}\omega_C T_3 = -180$$

then:
$$\omega_C = \sqrt{\frac{T_1 + T_2 + T_3}{T_1.T_2.T_3}}$$

73.11 Worked Example No 3

Consider the system shown in Figure 73.10 in which the time delay and time lags all have units of minutes. This is essentially the same system as that depicted in Figure 71.7.

The open loop phase shift is given by:

$$\angle G(j\omega) = -\tan^{-1}(10\omega) - \tan^{-1}(0.5\omega)$$
$$- \tan^{-1}(0.2\omega) - 57.3\omega$$

Note the factor of 57.3 to convert the contribution due to the time delay from radians into degrees. By trial and error it can be found that if $\omega = 1.005$ then $\angle G(j\omega) = -180°$.

Thus $\omega_C \approx 1.0$ rad/min.

Its open loop attenuation is given by:

$$|G(j\omega)| = \frac{1}{\sqrt{(1+100\omega^2)}\sqrt{(1+0.25\omega^2)}\sqrt{(1+0.04\omega^2)}}$$

Thus $|G(j\omega_C)| = 0.0873$

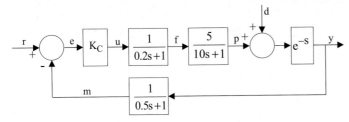

Fig. 73.10 Block diagram of control loop for Worked Example No 3

For marginal stability:

$$K_{CM}.K.|G(j\omega_C)| = 1.0$$

Substituting $K = 5$ and $|G(j\omega_C)| = 0.0873$ gives $K_{CM} = 2.29$.

Thus, for stability, $K_C < 2.29$.

Note that this value of K_{CM} is slightly smaller, but more accurate, than the value of 2.45 calculated in Chapter 71 in which a Pade approximation was used for the time delay.

Using the Zeigler and Nichols formulae of Table 24.2, the optimum settings for P + I + D action are thus:

$K_C \approx 1.34$, $T_R \approx 3.15$ min, $T_D \approx 0.78$ min.

73.12 Gain and Phase Margins

These arise from the Bode stability criteria and are a means of quantifying the stability of control systems. They are essentially safety factors and, as such, vary according to the application. Typical specifications are for a GM $\geq$ 1.7 and/or a PM $\geq 30°$.

Gain and phase margins are best explained in relation to Bode diagrams. With reference to the system of Figure 73.9, three cases are shown in Figure 73.11.

By definition:

$$GM = \frac{1}{K_C K |G(j\omega_C)|}$$

or, in terms of decibels:

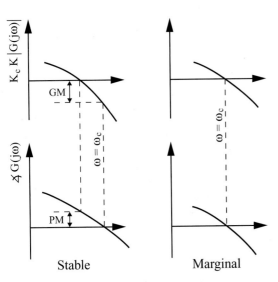
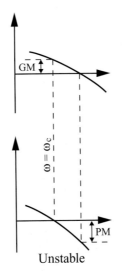

Fig. 73.11 Gain and phase margins from Bode diagrams

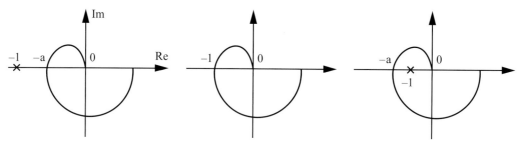

Fig. 73.12 Gain and phase margins from polar plots

$$GM = -20 \log_{10} K_C K \left|G(j\omega_C)\right|$$

Also, by definition:

$$PM = 180 + \phi$$

where ϕ is the value of $\angle G(j\omega)$ at the frequency for which $K_C K \left|G(j\omega)\right| = 1.0$.

For GM > 1 the system is closed loop stable. The GM is the factor by which K_c may be increased without causing instability.

If GM = 1 the closed loop system is marginally stable.

For GM < 1 the system is closed loop unstable. The GM is the factor by which K_c must be reduced to make the system marginally stable.

An alternative way to establish gain and phase margins is by examination of the open loop polar plots of $K_C K \left|G(j\omega)\right|$ vs $\angle G(j\omega)$ as shown in Figure 73.12 for a third order system.

Let $-a$ be the intersection of the polar plot with the negative real axis. By inspection:

$$GM = \frac{1}{oa}$$

If oa < 1, i.e. the polar plot passes between the origin and the point $(-1, 0)$, then the system is closed loop stable.

If oa = 1, i.e. the polar plot passes through the point $(-1, 0)$, then the closed loop is marginally stable.

If oa > 1, i.e. the polar plot encompasses the point $(-1, 0)$, then the system is closed loop unstable.

The PM is found from the intersection of the polar plot of $K_C K \left|G(j\omega)\right|$ vs $\angle G(j\omega)$ with the unit circle, as shown in Figure 73.13.R

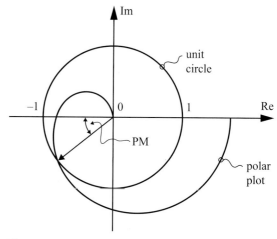

Fig. 73.13 Intersection of a polar plot with the unit circle

Root Locus

74.1 Worked Example No 1
74.2 Angle and Magnitude Criteria
74.3 Worked Example No 2
74.4 Evans' Rules
74.5 Comments on Evan's Rules
74.6 Worked Example No 3
74.7 Worked Example No 4 (with Time Delay)
74.8 Second Order Systems
74.9 Dominant Roots
74.10 Worked Example No 5 (Effect of I Action)
74.11 Pole Placement

It was seen in Chapter 71 that the nature of the roots of the characteristic equation determine the form of response and stability of a closed loop system. Indeed, the significance of the position of the roots in the s plane was summarised in Table 71.2 and depicted in Figure 71.8. In particular, the complex nature of $s = \sigma + j\omega$, in which the real part relates to an exponential component and the imaginary part to a frequency component, was introduced.

Root locus concerns plotting the roots of a characteristic equation in the s plane as some parameter, typically the controller gain, varies. The locus therefore indicates how that parameter affects the system's stability. This chapter demonstrates, by way of example, how to draw root loci, and how to interpret them.

74.1 Worked Example No 1

Consider the third order system depicted in Figure 74.1. In fact, this is the same system as that considered in Chapter 73.

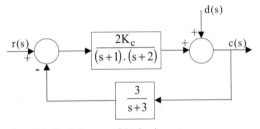

Fig. 74.1 Block diagram of third order system

Here the open loop transfer function is:

$$F(s) = \frac{6K_C}{(s+1)(s+2)(s+3)}$$

The open loop poles of $F(s)$ are the values of s in the denominator for which $F(s) = \infty$. In this case the poles are at $s = -1, -2$ and -3.

The open loop zeros of $F(s)$ are the values of s in the numerator for which $F(s) = 0$. In this case there are none.

The system's characteristic equation is thus:

$$\frac{6K_C}{(s+1)(s+2)(s+3)} + 1 = 0$$

i.e.

$$s^3 + 6s^2 + 11s + 6(K_C + 1) = 0$$

Table 74.1 Roots of characteristic equation vs controller gain

K_C	r_1	r_2	r_3
0.00	−3.00	−2.00	−1.00
0.038	−3.10	−1.76	−1.14
0.0642	−3.16	−1.42	−1.42
0.263	−3.45	−1.28 − j0.754	−1.28 + j0.754
1.00	−4.00	−1.00 − j1.41	−1.00 + j1.41
4.42	−5.09	−0.453 − j2.49	−0.453 + j2.49
10.0	−6.00	0.00 − j3.32	0.00 + j3.32
16.7	−6.72	+0.358 − j3.96	+0.358 + j3.96

The roots of this characteristic equation for various values of K_c are listed in Table 74.1 and plotted in Figure 74.2. These roots are the closed loop poles of the system. K_c is the parameter of interest.

Note that there are three loci, or branches, each emerging from one of the open loop poles. The value of K_c at the poles is zero. The direction of increasing K_c is indicated on the locus by an arrow. The root locus is symmetrical about the real axis. This is because the roots of real systems only occur in conjugate pairs.

Table 74.2 Damping as a function of controller gain

Value of K_c	Nature of damping
$0 < K_C < 0.0642$	Overdamped
$K_C = 0.0642$	Critically damped
$0.0642 < K_C < 10.0$	Underdamped
$K_C = 10.0$	Undamped
$10.0 < K_C$	Self excited

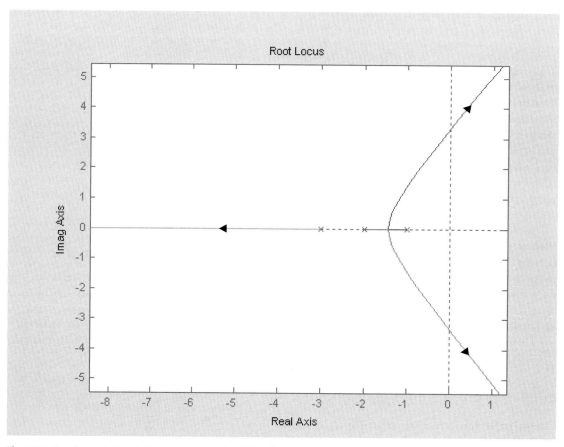

Fig. 74.2 Plot of roots of characteristic equation in s-plane (Matlab)

Following a step change in either r(t) or d(t), the form of the response of c(t) will be as indicated in Table 74.2.

Thus, at a glance, the root locus conveys the complete stability picture.

74.2 Angle and Magnitude Criteria

These provide the basis for Evans' rules for sketching root loci and for the software routines which generate them automatically.

Consider the characteristic equation:

$$F(s) + 1 = 0$$

Rearrange it into the form:

$$\frac{K(s+z_1)(s+z_2)\ldots(s+z_m)}{(s+p_1)(s+p_2)\ldots(s+p_n)} = -1 + j.0 \quad (74.1)$$

where K is the parameter of interest. This is typically, but not necessarily, the controller gain: it is shown later how to handle the reset time T_R as the parameter of interest.

Since the various poles and zeros are complex, in general, any point on the root locus must satisfy both the angle criterion:

$$\sum_{i=1}^{m} \angle(s+z_i) - \sum_{j=1}^{n} \angle(s+p_j) = \pm(2k+1)\pi \text{ rad}$$

(74.2)

where k = 0, 1, 2 ... and the magnitude criterion:

$$\frac{K \prod_{i=1}^{m} |s+z_i|}{\prod_{j=1}^{n} |s+p_j|} = 1.0 \quad (74.3)$$

The angle criterion establishes the loci of the roots and the magnitude criterion is used to determine the value of the parameter K at points along the loci.

74.3 Worked Example No 2

Consider the system for which the open loop transfer function is:

$$F(s) = \frac{K_C(s+5)}{s(s+1)(s+2)}$$

This has three open loop poles at s = 0, −1 and −2 and one open loop zero at s = −5.

From Equation 74.2:

$$\angle(s+5) - \angle s - \angle(s+1) - \angle(s+2) = (2k+1).180°$$

Consider the point s = −0.3 + j, as depicted in Figure 74.3:

$$\angle(4.7+j) - \angle(-0.3+j) - \angle(0.7+j)$$
$$- \angle(1.7+j)$$
$$= \tan^{-1}\frac{1}{4.7} - \left(180 - \tan^{-1}\frac{1}{0.3}\right) - \tan^{-1}\frac{1}{0.7}$$
$$- \tan^{-1}\frac{1}{1.7}$$
$$= 12.01 - 180 + 73.30 - 55.01 - 30.47$$
$$= -180.17$$

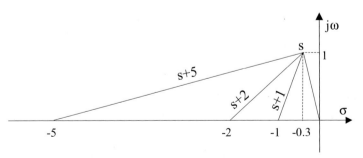

Fig. 74.3 Angle and magnitude criteria at s = 0.3 + j

which is close enough to $-180°$. Thus $s = -0.3 + j$ is a point on the root locus of $F(s) + 1 = 0$.

From Equation 74.3:

$$\frac{K_C |s+5|}{|s| |s+1| |s+2|} = 1$$

Again, if $s = -0.3 + j$ then:

$$K_C = \frac{|-0.3 + j| |0.7 + j| |1.7 + j|}{|4.7 + j|}$$

$$= \frac{\sqrt{(1+0.3^2)}\sqrt{(1+0.7^2)}\sqrt{(1+1.7^2)}}{\sqrt{(1+4.7^2)}}$$

$$= 0.523$$

That this point satisfies the criteria can be verified by measurement of Figure 74.3.

74.4 Evans' Rules

These rules, which are due to Evans, are generic and enable root loci to be sketched and interpreted. They are explained in detail in Ogata (2002). Accurate drawing of root loci is very tedious but, fortunately, due to the availability of computer aided design tools, is no longer necessary.

1. Rearrange the characteristic equation into the form:

$$F(s) + 1 = \frac{K(s+z_1)(s+z_2)\ldots(s+z_m)}{(s+p_1)(s+p_2)\ldots(s+p_n)} + 1 = 0 \quad (74.4)$$

   where $F(s)$ is the open loop transfer function and K is the parameter of interest. The system has m open loop zeros and n open loop poles. $n \geq m$ for all real systems.

2. Plot the open loop zeros (o) and poles (x) of $F(s)$ in the complex s plane:

$$s = -z_1, -z_2, \ldots, -z_m$$

$$s = -p_1, -p_2, \ldots, -p_n$$

3. The number of branches to the root locus is equal to the number of roots of the characteristic equation, usually n. Each branch starts at an open loop pole and ends at an open loop zero.

   If $n > m$ the remaining $(n-m)$ branches terminate at $(n-m)$ implicit zeros at infinity along some asymptotes. In the case of a qth order pole, q loci emerge from it. Likewise for a qth order zero, q loci terminate there.

4. Any point on the real axis is a part of the root locus if the total number of poles and zeros that lie on the real axis to the right of that point is odd.

   Note that in applying this rule, any qth order pole or zero must be counted q times. Also note that any complex conjugate poles and zeros need not be counted at all.

5. Root loci are always symmetrical with respect to the real axis. Thus it is only necessary to construct the root loci in the upper half of the s plane and copy their mirror image into the lower half.

6. The point at which two root loci emerging from adjacent poles (or moving towards adjacent zeros) on the real axis intersect and then leave (or enter) the real axis may be found by one of two methods:

   i. Rearrange the characteristic equation into the form:

   $$K = \frac{-\prod_{j=1}^{n}(s+p_j)}{\prod_{i=1}^{m}(s+z_i)} \quad (74.5)$$

   The breakaway points are then found by solving for the roots of: $\frac{dK}{ds} = 0$.

   ii. The breakaway points may also be found by solving for the roots of:

   $$\sum_{i=1}^{m} \frac{1}{s+z_i} = \sum_{j=1}^{n} \frac{1}{s+p_j} \quad (74.6)$$

   Note that the root loci break away from (or into) the real axis at angles of $\pm \pi/2$.

7. There are $(n-m)$ loci which asymptotically approach $(n-m)$ straight lines radiating from the centre of gravity of the open loop poles and

zeros. The centre of gravity, which always lies on the real axis, is given by:

$$s = \frac{\sum_{i=1}^{m} z_i - \sum_{j=1}^{n} p_j}{n - m} \quad (74.7)$$

Note that the z_i and p_j in this equation are not the zeroes and poles of the system but the values in the open loop transfer function $F(s)$.

8. The angles made by the asymptotes with the real axis are given by:

$$\phi = \frac{(2k + 1)\pi}{n - m} \text{ rad}$$
$$\forall k = 0, 1, 2 \ldots (n - m - 1) \quad (74.8)$$

Note that the symbol $\forall$ means "for all values of".

The asymptotes are equally spaced, the angles between them being $\frac{2\pi}{n-m}$ rad.

9. For single poles (or zeros) on the real axis, the angle of departure (or approach) of the root loci is either 0 or π.

10. For multiple poles (qth order) on the real axis, and for complex poles (single ($q = 1$) or multiple) at $s = s^*$, the angles of departure of the q loci are given by:

$$\theta = \frac{1}{q}\left((2k + 1)\pi + \sum_{i=1}^{m} \angle(s^* + z_i) - \sum_{j=1, j \neq *}^{n} \angle(s^* + p_j)\right) \text{ rad}$$
$$\forall k = 0, 1, 2 \ldots (q - 1) \quad (74.9)$$

Similarly, the root loci approach multiple zeros at $s = s^*$ at angles given by:

$$\theta = \frac{1}{q}\left((2k + 1)\pi + \sum_{j=1}^{n} \angle(s^* + p_j) - \sum_{i=1, i \neq *}^{m} \angle(s^* + z_i)\right) \text{ rad}$$
$$\forall k = 0, 1, 2 \ldots (q - 1) \quad (74.10)$$

11. The intersections of the root loci with the imaginary axis, and the values of K at the intersections, may be found by substituting $s = j\omega$ into the characteristic equation. The real and imaginary parts may then be set to zero, and the resulting equations solved simultaneously for ω and K.
Note that the values of K at the intersections may also be found by the Routh test.

12. Knowing the breakpoints, asymptotes, angles of departure, *etc.*, the root locus may be sketched, using the angle criterion as appropriate. The value of K at points of interest on the locus may then be found using the magnitude criterion.

74.5 Comments on Evan's Rules

Interpretation of the root locus is easier if the same scales are used for both the real and imaginary axes.

Usually it is unnecessary to have to use all the rules to construct a root locus.

A useful analogy is to consider the root loci to correspond to the paths taken by a positively charged particle in an electrostatic field established by positive poles and negative zeros. In general, a loci is repelled by a pole and attracted by a zero.

Another useful aid to plotting root loci is the fact that for $n - m \geq 2$ the sum of the roots $(r_1 + r_2 + \ldots + r_n)$ is constant, real and independent of K. Thus the motion of any branch in one direction must be counterbalanced by the motion of the other branches in their directions.

74.6 Worked Example No 3

Consider again the system of Example 2.

1.
$$F(s) = \frac{K_C(s + 5)}{s(s + 1)(s + 2)}$$

2. Zero at $s = -5$. Poles at $s = 0, -1$ and -2.

3. Three branches, two of which go to implicit zeros at ∞ asymptotically.
4. Real axis is part of root locus for:

$$-1 < \sigma < 0 \qquad -5 < \sigma < -2$$

6. $K_C = \dfrac{-s(s+1)(s+2)}{(s+5)} = \dfrac{-(s^3 + 3s^2 + 2s)}{(s+5)}.$

Thus:

$$\dfrac{dK_C}{ds} = \dfrac{(s^3 + 3s^2 + 2s)}{(s+5)^2} - \dfrac{(3s^2 + 6s + 2)}{(s+5)} = 0$$

so:

$$s^3 + 9s^2 + 15s + 5 = 0$$
$$(s + 6.943)(s + 1.611)(s + 0.447) = 0$$

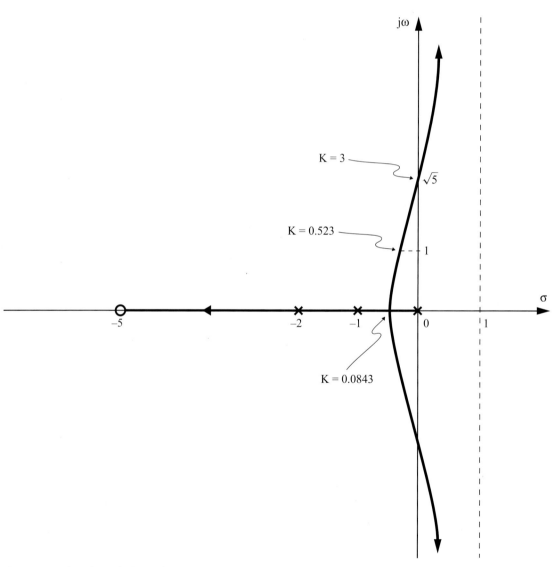

Fig. 74.4 Root locus for Worked Example No 3

By inspection, the breakaway point is:
$$s = -0.447$$

7. The centre of gravity is at $s = \frac{(5)-(0+1+2)}{2} = 1$
8. Angle of asymptotes, $\phi = \frac{(2k+1)}{2}.180°$. Since $k = 0$ and 1 then $\phi = 90$ and $270°$.
11. Characteristic equation is:
$$\frac{K_C(s+5)}{s(s+1)(s+2)} + 1 = 0$$
i.e.
$$s^3 + 3s^2 + (2 + K_C)s + 5K_C = 0$$
Substituting $s = j\omega$ gives:
$$(5K_C - 3\omega^2) + j\omega(2 + K_C - \omega^2) = 0$$
whence:
$$5K_C - 3\omega^2 = 0$$
$$2 + K_C - \omega^2 = 0 \quad \text{(or else } \omega = 0\text{)}$$
Solving gives $\omega = \pm\sqrt{5}$ and $K_C = 3$
12. The root locus is thus as depicted in Figure 74.4.
The gain at $s = -0.447$ is found from the magnitude criterion:
$$K_C = \frac{|-0.447||0.553||1.553|}{|4.553|} = 0.0843$$

74.7 Worked Example No 4 (with Time Delay)

Consider the system of Figure 74.5 consisting of a PI controller, a delay and a lag in which $T_R = 1$, $L = 2$ and $T = 25$ min.

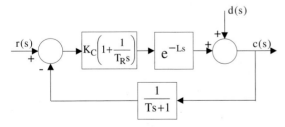

Fig. 74.5 Loop with PI controller and process with time delay

1. Approximation of the delay with a Pade function gives the open loop transfer function:
$$F(s) = K_C.\left(1 + \frac{1}{s}\right).\left(\frac{1-s}{1+s}\right).\frac{1}{25s+1}$$
$$= K_C.\left(\frac{1-s}{s}\right).\frac{1}{25s+1}$$

Rearrangement gives the characteristic equation in the required form:
$$F(s) + 1 = K_C.\frac{-0.04(s-1)}{s.(s+0.04)} + 1 = 0$$

2. Zero at $s = +1$. Poles at $s = 0$ and $s = -0.04$.
4. The angle criterion is of the general form:
$$F(s) = -1 + j.0$$
whence, for the system of Figure 74.5:
$$K_C.\frac{0.04(s-1)}{s.(s+0.04)} = +1 + j.0$$

It is evident that the angle criterion has been reversed, and the real axis is part of the root locus for the ranges $1 < s < \infty$ and $-0.04 < s < 0$.

6. Rearrange the characteristic equation:
$$K_C = \frac{25.s.(s+0.04)}{s-1}$$

The breakpoint occurs when:
$$\frac{dK_C}{dt} = \frac{25.(s^2 - 2s - 0.04)}{(s-1)^2} = 0$$

Solving gives $s = -0.02$ and $s = +2.02$.
Back substitution gives:
If $s = -0.02$ then $K_C = 0.0098$ corresponding to breakout.

If $s = +2.02$ then $K_C = 102$ corresponding to breakin.

11. Expand the characteristic equation:

$$s^2 + 0.04(1 - K_C).s + 0.04.K_C = 0$$

Substituting $s = j\omega$ gives:

$$0.04.K_C - \omega^2 + j.0.04(1 - K_C).\omega = 0 + j.0$$

Equating real and imaginary parts gives:

$$0.04.K_C - \omega^2 = 0$$
$$0.04(1 - K_C).\omega = 0$$

Hence the intersection with the imaginary axis is when $\omega = \pm 0.2$ and $K_C = 1.0$.

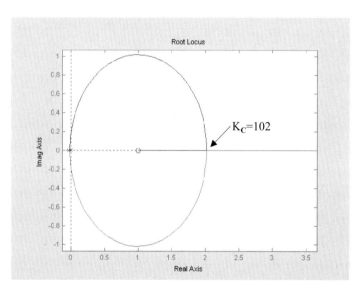

Fig. 74.6 Root locus for Worked Example No 4 (Matlab)

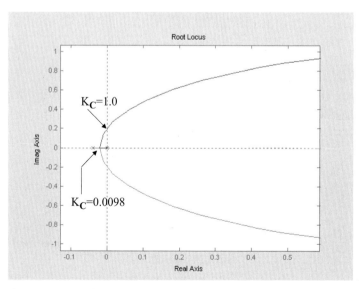

Fig. 74.7 Close up of root locus near to imaginary axis (Matlab)

12. The root locus is thus as depicted in Figure 74.6 and the section of interest, close to the imaginary axis, is expanded in Figure 74.7.

Note that for this system, with the Pade approximation for the delay, the optimum settings according to the Zeigler and Nichols criterion are $K_C \approx 0.45$ and $T_R \approx 26$ min.

74.8 Second Order Systems

Consider the system of Figure 8 in which the *open loop* transfer function is second order.

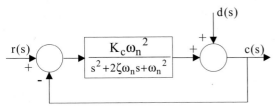

Fig. 74.8 System whose open loop transfer function is second order

Its open loop poles are the roots of:

$$s^2 + 2\zeta\omega_n s + \omega_n^2 = 0$$

As was seen in Chapter 72, this factorises into:

$$\left(s + \omega_n\zeta + \omega_n\sqrt{(\zeta^2 - 1)}\right) \times \left(s + \omega_n\zeta - \omega_n\sqrt{(\zeta^2 - 1)}\right) = 0$$

which gives the roots:

$$s = -\omega_n\zeta \pm \omega_n\sqrt{(\zeta^2 - 1)}$$

The characteristic equation is:

$$\frac{K_C\omega_n^2}{s^2 + 2\zeta\omega_n s + \omega_n^2} + 1 = 0$$

i.e.

$$s^2 + 2\zeta\omega_n s + (K_C + 1)\omega_n^2 = 0$$

The open loop pole positions and the root loci for varying K_C are shown in Figure 74.9 for different ranges of the damping factor.

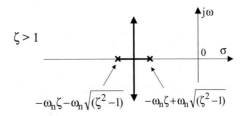

Fig. 74.9a Root locus for overdamped second order system

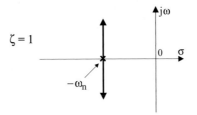

Fig. 74.9b Root locus for critically damped second order system

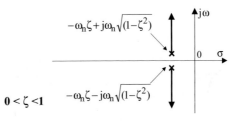

Fig. 74.9c Root locus for underdamped second order system

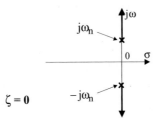

Fig. 74.9d Root locus for undamped second order system

Note that in the case for $\zeta > 1$ the breakpoint is exactly halfway between the two poles. This is necessary to satisfy the angle criterion.

Now consider the system of Figure 74.10 in which the *closed* loop transfer function is second order.

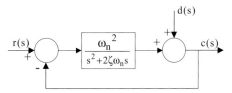

Fig. 74.10 System whose closed loop transfer function is second order

$$c(s) = d(s) + \frac{\omega_n^2}{s^2 + 2\zeta\omega_n s} \cdot (r(s) - c(s))$$

$$= \frac{s^2 + 2\zeta\omega_n s}{s^2 + 2\zeta\omega_n s + \omega_n^2} \cdot d(s)$$

$$+ \frac{\omega_n^2}{s^2 + 2\zeta\omega_n s + \omega_n^2} \cdot r(s)$$

Its characteristic equation is:

$$s^2 + 2\zeta\omega_n s + \omega_n^2 = 0$$

whose roots are:

$$s = -\omega_n\zeta \pm j\omega_n\sqrt{(1 - \zeta^2)}$$

The characteristic equation may be rearranged into the form:

$$\zeta \cdot \frac{2\omega_n s}{(s^2 + \omega_n^2)} + 1 = 0$$

where ζ is the parameter of interest. This has one open loop zero at $s = 0$ and two poles at $s = \pm j\omega_n$. The root locus for this system for varying ζ is shown in Figure 74.11.

Note that for $0 < \zeta < 1$ the root locus is a semicircle in the left half of the s plane. It follows from the geometry that:

1. The radius of s is ω_n.
2. The angle of s is $\pi - \cos^{-1}\zeta$.
3. The imaginary part of s is the frequency of the damped oscillation:

$$\omega_d = \omega_n\sqrt{(1 - \zeta^2)}$$

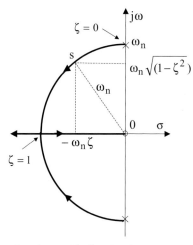

Fig. 74.11 Root locus with damping factor as parameter of interest

74.9 Dominant Roots

The further away, in a negative sense, that a root is from the imaginary axis, the less significant it becomes. This may be illustrated by considering again the system of Worked Examples No 2 and 3, for which part of its root locus is enlarged in Figure 74.12.

The characteristic equation is:

$$s^3 + 3s^2 + (2 + K_C)s + 5K_C = 0$$

If $K_C = 0.523$, say, then:

$$s^3 + 3s^2 + 2.523s + 2.615 = 0$$

i.e.

$$(s + 2.4)(s + 0.3 + j)(s + 0.3 - j) = 0$$

For regulo control, the closed loop response is given by:

$$c(s) = \frac{1}{1 + F(s)} \cdot d(s)$$

and the response to a step change in $d(s)$ is of the form:

$$c(s) = \frac{c_0}{s} + \frac{c_1}{s + 2.4} + \frac{c_2}{s + 0.3 + j} + \frac{c_2}{s + 0.3 - j}$$

whence $c(t) = c_0 + c_1 e^{-2.4t} + c_2 e^{-0.3t} \cos t$.

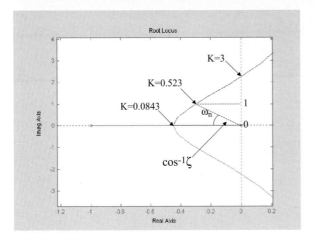

Fig. 74.12 Root locus close to origin for system with dominant roots (Matlab)

The exponential $e^{-2.4t}$ decays away very quickly relative to $e^{-0.3t}$. Thus, unless $c_1 \gg c_2$:

$$c(t) \approx c_0 + c_2 e^{-0.3t} \cos t$$

It is often the case, for systems with multiple roots, that the dominant roots are a pair of complex roots just to the left of the imaginary axis. In such cases it is sufficient to assume the closed loop response is approximately second order and to characterise it in terms of ω_n and ζ. These can be found by direct measurement, by assuming a radius of magnitude ω_n at an angle of $\pi \pm \cos^{-1} \zeta$ passes through the roots.

Thus for $K_C = 0.523$:

for which the dominant roots are $s = -0.3 \pm j$, the equivalent values are $\omega_n = 1.04$ and $\zeta = 0.287$.

In the previous worked examples the parameter of interest was the controller gain K_C. The effects of varying reset time T_R is now explored using the concept of dominant poles.

74.10 Worked Example No 5 (Effect of I Action)

Consider again the system of Worked Example No 1 for which the characteristic equation is:

$$\frac{6.K_C}{(s+1)(s+2)(s+3)} + 1 = 0$$

Note that $K_C = 10.0$ for marginal stability as seen in Table 74.1, and indeed in Chapter 73. Choosing a controller gain of $K_C = 5.0$, which gives a gain margin of a factor of 2, the dominant roots are:

$$s = -0.3928 \pm j2.598$$

for which the closed loop characteristics are $\omega_n = 2.63 \text{rad min}^{-1}$, $\zeta = 0.150$ and decay ratio = 0.387.

Consider the introduction of integral action, as depicted in Figure 74.13:

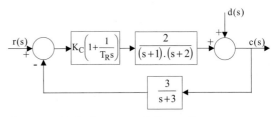

Fig. 74.13 Third order system with PI controller

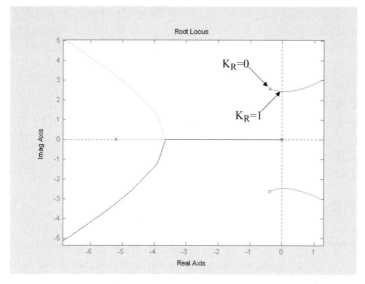

Fig. 74.14 Root locus for Worked Example No 5 with K_R as parameter (Matlab)

The characteristic equation thus becomes:

$$K_C \left(1 + \frac{1}{T_R s}\right) \frac{6}{(s+1)(s+2)(s+3)} + 1 = 0$$

Keeping the value of $K_C = 5$ and introducing a dummy variable $K_R = \frac{1}{T_R}$ results in:

$$5\left(1 + \frac{K_R}{s}\right) \frac{6}{(s+1)(s+2)(s+3)} + 1 = 0$$

Note that by introducing K_R rather than using T_R as the parameter of interest ensures that $n \geq m$ and that the root locus is well behaved. The characteristic equation may now be rearranged into the form required by the first of Evan's rules:

$$\frac{30.K_R}{s^4 + 6s^3 + 11s^2 + 36s} + 1 = 0$$

The root locus for this is as depicted in Figure 74.14 with K_R being the parameter of interest.

With a value of, say, $K_R = 0.2$ corresponding to $T_R = 5.0$ min, the dominant roots are:

$$s = -0.3247 \pm j2.552$$

for which the closed loop characteristics have changed moderately to $\omega_n = 2.57 \text{rad min}^{-1}$, $\zeta = 0.126$ and decay ratio $= 0.450$.

Increasing the amount of integral action makes for reduced stability. This can be seen by exploring the locus of Figure 74.14. Thus increasing the parameter K_R to 1.0 and hence reducing the reset time to a value of $T_R = 1.0$ min results in poles of

$$s = -1.0, \quad s = -5.0 \text{ and } s = 0.0 \pm j2.450$$

which clearly correspond to marginal stability. Inspection of the root locus demonstrates that it is very sensitive to small changes in position of the dominant poles.

The effect of derivative action can similarly be explored by varying the rate time T_D for a specified K_C and T_R.

74.11 Pole Placement

Pole placement is a technique used for design purposes. In essence it concerns choosing values for parameters such as K_C and T_R such that the root locus passes through dominant pole positions to give a specified closed loop performance.

Consider the PI control of a plant consisting of a single lag as depicted in Figure 74.15.

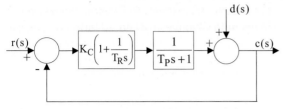

Fig. 74.15 Loop comprising PI controller and first order lag

The open loop transfer function is as follows:

$$F(s) = K_C \left(1 + \frac{1}{T_R s}\right) \frac{1}{T_P s + 1}$$

Introducing the dummy variable K_R as in the worked example previous, and assuming that both K_C and T_P are known quantities, the characteristic equation can be rearranged into the form required by the first of Evan's rules as follows:

$$K_R \cdot \frac{K_C}{T_P} \cdot \frac{1}{s \cdot \left(s + \frac{1+K_C}{T_P}\right)} + 1 = 0$$

There are two open loop poles at $s = 0$ and $s = -(1 + K_C)/T_P$.

The root locus is as depicted in Figure 74.16 with K_R being the parameter of interest.

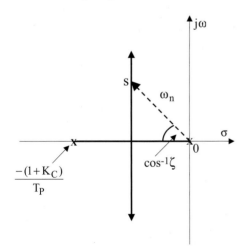

Fig. 74.16 Root locus for consideration of pole placement

It is evident from Evans' rules that the root locus is symmetrical: the breakpoint is halfway between the two poles and the branches are at right angles to the real axis. The two closed loop poles must be dominant and the intersection of the root locus with any vector from the origin, as depicted in Figure 74.16, will determine the values of both ω_n and ζ.

The breakpoint is at $s = -(1 + K_C)/2T_P$.

Suppose the damping factor ζ is specified. Given that the angle of the vector is $\cos^{-1} \zeta$, this determines the pole position on the root locus at which the closed loop performance has the specified values of ζ. The length of the vector is given by:

$$\omega_n = \frac{1 + K_C}{2 \cdot \zeta \cdot T_P}$$

Now apply the magnitude criterion, Equation 74.3, at the point of intersection:

$$K_R \cdot \frac{K_C}{T_P} \cdot \frac{1}{|s| \cdot \left|s + \frac{1 + K_C}{T_P}\right|} = 1$$

Whence:

$$K_R \cdot \frac{K_C}{T_P} \cdot \frac{1}{\left(\frac{1 + K_C}{2 \cdot \zeta \cdot T_P}\right)^2} = 1$$

Rearrange to give:

$$K_R = \frac{(1 + K_C)^2}{4 \cdot K_C \cdot T_P \cdot \zeta^2}$$

Thus for any plant of known lag T_P and arbitrary controller gain K_C, this determines the value of K_R required to establish the closed loop pole position on the root locus corresponding to the specified ζ. In effect, correct choice of the combination of values for K_C and T_R locates the pole positions on the root locus corresponding to the desired performance.

The above example is revealing but unrealistic for two reasons. First, the plant considered is trivial which enabled an analytical solution to be found: that is seldom possible for real plant. And second, the structure of a PID controller is fixed: it is not normally feasible to find values of K_C, T_R and T_D that satisfy arbitrarily specified dominant pole positions. Even if such values can be found, there is often only very limited scope for affecting the values of ω_n and ξ by adjusting the parameters of interest as can be seen, for example, with the reset time in Figure 74.14.

However, these limitations can be overcome by the use of compensators for the control function. A compensator is the generic term for a controller of arbitrary structure, of which a PID controller is but a particular instance. Pole placement is covered in detail in Chapter 77 where it underpins the use of pole-zero cancellation for compensator design, and in Chapter 112 in which it provides the basis for state feedback regulator design.

Copyright. Note that Matlab is a registered trademark of The MathWorks, Inc.

Z Transforms

75.1 Samplers and Holds
75.2 Equivalence of Pulses and Impulses
75.3 Analysis of Pulse Trains
75.4 Transforms of Variables
75.5 Difference Equations
75.6 Inverse Z Transformation
75.7 Worked Example

Z transforms are used in the analysis and design of systems in which the signals are sampled on a regular basis. Referred to as a sampled data signals, such a signal consists of a train of pulses which represent some continuous variable. Sampled data systems are common in practice. For example, many analytical instruments are intermittent in nature: they sample their inputs and produce pulse outputs. Digital controllers are inherently sampled data in nature: their algorithms are executed in phase with the sampling of the I/O signals.

This chapter introduces some important concepts in relation to the sampling process and pulse trains, defines the Z transform, introduces its properties, and shows how to use the Z transform for solving difference equations. These provide the basis for Chapter 76 on sampled data systems analysis and Chapter 77 on impulse compensator design. Many texts cover Z transforms and sampled data theory, the reader is referred in particular to Dutton (1997) and Ogata (2002).

75.1 Samplers and Holds

Consider the sampler and hold device of Figure 75.1.

The sampler is essentially a switch which closes for a finite time Δt every T s, the sampling period T normally being significantly shorter than the dominant time constants of the system. This has the effect of converting the continuous signal x(t) into a train of pulses x*(t) at the sampling instants, as depicted in Figure 75.2.

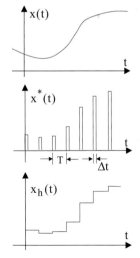

Fig. 75.2 The effect of a sample/hold device on a continuous signal

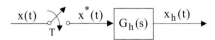

Fig. 75.1 Representation of sampler and hold devices

Note that the height of the pulse varies according to the value of x(t) at the sampling instant, and that the sampler transmits no information in between the sampling instants. Note also that two completely different signals that happen to have the same values at the sampling instants would produce the same sampled signal. It is therefore obvious that a sampled data signal contains less information than its original continuous signal.

The hold device converts the sampled signal $x^*(t)$ into a continuous one $x_h(t)$ which approximately reproduces the continuous signal x(t). The simplest, and most common, hold device is the so-called zero order hold (ZOH) which converts the sampled signal into one which is constant between consecutive sampling instants. The ZOH is often described as a latch because it latches onto the value of the last pulse. Its transfer function is as follows:

$$G_h(s) = \frac{1 - e^{-Ts}}{s} \tag{75.1}$$

which is two steps, in opposite directions, the second being delayed by the sampling period T.

75.2 Equivalence of Pulses and Impulses

Consider a pulse at time t = 0 of finite duration Δt and magnitude $1/\Delta t$, as depicted in Figure 75.3.

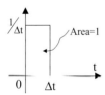

Fig. 75.3 A pulse of unit area

The Laplace transform for this is:

$$f(s) = \frac{1}{\Delta t.s}\left(1 - e^{-\Delta t.s}\right)$$

The impulse $\delta(t)$ may be considered to be the limiting case of the pulse as Δt tends to zero, *i.e.* it is of zero duration, infinite magnitude and unit area. A useful insight into the nature of an impulse is given by finding its Laplace transform:

$$\begin{aligned}L\{\delta(0)\} &= \frac{\text{Lim}}{\Delta t \to 0} f(s) \\ &= \frac{\text{Lim}}{\Delta t \to 0}\left(\frac{1}{\Delta t.s}\left(1 - e^{-\Delta t.s}\right)\right) \\ &= \frac{\text{Lim}}{\Delta t \to 0}\left(\frac{1}{\Delta t.s}\left(1 - \left(1 - \Delta t.s \right.\right.\right. \\ &\quad \left.\left.\left. + \frac{1}{2}(\Delta t.s)^2 - \frac{1}{3!}(\Delta t.s)^3 + \ldots\right)\right)\right) = 1\end{aligned}$$

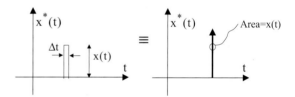

A classical assumption in sampled data theory is that a pulse function may be represented by an impulse function, where the magnitude of the pulse function is equal to the area under the impulse function. This is depicted in Figure 75.4.

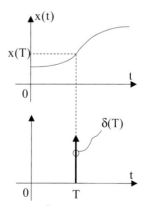

Fig. 75.4 The equivalence of a pulse and an impulse function

This assumption may be justified by considering a sampled data signal to be the product of a continuous and an impulse signal. Consider the two signals depicted in Figure 75.5.

Fig. 75.5 The product of a continuous and an impulse signal

Suppose that x(t) is a continuous signal and that $\delta(T)$ is a single impulse of unit area, referred to as the unit impulse, that occurs when t = T. If $x^*(t)$ is the product of the two signals, it can be seen by inspection that:

for t < T and t > T $x^*(t) = 0$
and for t = T $x^*(T) = \delta(T).x(T)$

where $x^*(T)$ is the unit impulse weighted by a factor of x(T), which is consistent with the classical assumption.

75.3 Analysis of Pulse Trains

Consider the sampler shown in Figure 75.6.

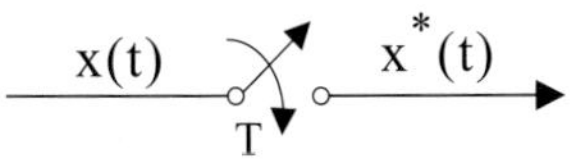

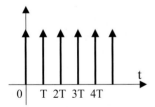

Fig. 75.6 A sampler with sampling period T

Suppose that x(t) is sampled at regular intervals of time, *i.e.* t = 0, T, 2T, 3T, 4T....

This results in a train of pulses of varying magnitude, each of which may be represented by a weighted impulse. Thus:

$$x^*(t) = \delta_T(t).x(t)$$
$$= \delta(0)x(0) + \delta(T)x(T) + \delta(2T)x(2T) + \ldots$$

where $\delta_T(t)$ is a train of unit impulses spaced a time T apart, as depicted in Figure 75.7.

Fig. 75.7 A train of unit impulses

However, each of the unit impulses is identical to the first, *i.e.* that at t = 0, except for being delayed by some multiple of the sampling period T. Thus:

$$x^*(t) = \delta(t)x(0) + \delta(t-T)x(T) + \delta(t-2T)x(2T) + \ldots$$
$$= \sum_{k=0}^{n} \delta(t-kT).x(kT)$$

Laplace transform, remembering that $L\{\delta(t)\} = 1$, this gives:

$$x^*(s) = \sum_{k=0}^{\infty} x(kT).e^{-kTs}$$

Now define z to be a complex variable according to $z = e^{Ts}$. It can be seen that z is equivalent to a time advance of duration T. Also, since both T and s are independent of time, so too is z. Whence:

$$x^*(s) = \sum_{k=0}^{\infty} x(kT).z^{-k} \quad (75.2)$$

This is the definition of the Z transform. By notation:

$$Z\{x(t)\} = x(z) = x^*(s) = L\{x^*(t)\} \quad (75.3)$$

75.4 Transforms of Variables

The formation of Z transforms is best illustrated by means of a couple of examples.

Example 1

The unit step function x(t) = 0 for t < 0, and x(t) = 1 for t ≥ 0.

$$Z\{1(t)\} = \sum_{k=0}^{\infty} 1(kT).z^{-k}$$
$$= +1 + z^{-1} + z^{-2} + z^{-3} + \cdots$$
$$\underline{z.Z\{1(t)\} = z + 1 + z^{-1} + z^{-2} + z^{-3} + \cdots}$$
$$(z-1).Z\{1(t)\} = z$$
$$Z\{1(t)\} = \frac{z}{(z-1)}$$

Example 2

The exponential function $x(t) = 0$ for $t < 0$, and $x(t) = e^{-at}$ for $t \geq 0$.

$$Z\{e^{-at}\} = \sum_{k=0}^{\infty} e^{-akT} \cdot z^{-k}$$

$$= +1 + e^{-aT}z^{-1} + e^{-2aT}z^{-2}$$
$$+ e^{-3aT}z^{-3} + \cdots$$

$$e^{aT} \cdot z \cdot Z\{e^{-at}\} = e^{+aT}z + 1 + e^{-aT}z^{-1}$$
$$+ e^{-2aT}z^{-2} + e^{-3aT}z^{-3} + \cdots$$

$$(e^{aT}z - 1) \cdot Z\{e^{-at}\} = e^{aT}z$$

$$Z\{e^{-at}\} = \frac{z}{(z - e^{-aT})}$$

Example 3

The function whose Laplace transform is:

$$x(s) = \frac{1}{s(s+1)}$$

Expanding into partial fractions gives:

$$x(s) = \frac{1}{s} - \frac{1}{s+1}$$

The Z transform is found from the inverse Laplace transform as follows:

$$x(z) = Z\{x(t)\} = Z\{L^{-1}\{x(s)\}\}$$
$$= Z\left\{L^{-1}\left\{\frac{1}{s} - \frac{1}{s+1}\right\}\right\} = Z\{1(t) - e^{-t}\}$$
$$= Z\{1(t)\} - Z\{e^{-t}\} = \frac{z}{z-1} - \frac{z}{z-e^{-T}}$$
$$= \frac{z(1 - e^{-T})}{(z-1)(z-e^{-T})}$$

Note that an important property of Z transforms, that the transform of the sum of two functions is equal to the sum of the transforms of the individual functions, has been used in this calculation.

Fortunately, it is not normally necessary to have to work out the transform of a function from first principles. Extensive tables of Z transforms exist. Table 75.1 lists the more commonly used ones in process control. Table 75.2 lists the more important properties of Z transforms.

Table 75.1 Table of commonly used Z transforms

$x(t)_{t \geq 0}$ $(x(t)_{t<0} = 0)$	$x(z)$
$\delta(t)$	1
$\delta(t - kT)$	z^{-k}
1	$\dfrac{z}{z-1}$
a^k	$\dfrac{z}{z-a}$
$a^k \cos(k\pi)$	$\dfrac{z}{z+a}$
t	$\dfrac{Tz}{(z-1)^2}$
t^2	$\dfrac{T^2 z(z+1)}{(z-1)^3}$
e^{-at}	$\dfrac{z}{z - e^{-aT}}$
$1 - e^{-at}$	$\dfrac{1 - e^{-aT}z}{(z-1)(z - e^{-aT})}$
$t \cdot e^{-at}$	$\dfrac{Tze^{-aT}}{(z - e^{-aT})^2}$
$\sin(\omega t)$	$\dfrac{z \cdot \sin(\omega T)}{z^2 - 2z \cdot \cos(\omega T) + 1}$
$\cos(\omega t)$	$\dfrac{z \cdot (z - \cos(\omega T))}{z^2 - 2z \cdot \cos(\omega T) + 1}$
$e^{-at} \sin(\omega t)$	$\dfrac{z \cdot e^{-aT} \cdot \sin(\omega T)}{z^2 - 2z \cdot e^{-aT} \cdot \cos(\omega T) + e^{-2aT}}$
$e^{-at} \cos(\omega t)$	$\dfrac{z^2 - z \cdot e^{-aT} \cdot \cos(\omega T)}{z^2 - 2z \cdot e^{-aT} \cdot \cos(\omega T) + e^{-2aT}}$

Note that the notation has been simplified in the tables by writing x(k) instead of x(kT), x(k + 1) instead of x((k + 1)T), *etc.*

75.5 Difference Equations

Just as Laplace transforms are used for solving differential equations, so Z transforms are used for solving difference equations. Again there is a four step procedure which exploits the first two proper-

75.5 Difference Equations

Table 75.2 Important properties of Z transforms

$x(k)_{k \geq 0}$ $(x(k)_{k<0} = 0)$	$x(z)$
$x_1(k) + x_2(k)$	$x_1(z) + x_2(z)$
$c.x(k)$	$c.x(z)$
$x(k+1)$	$z.x(z) - z.x(0)$
$x(k+2)$	$z^2.x(z) - z^2.x(0) - z.x(1)$
$x(k+m)$	$z^m.x(z) - z^m.x(0)$ $-z^{m-1}.x(1)$ $-\ldots - z.x(m-1)$
$k.x(k)$	$-z.\dfrac{d}{dz}x(z)$
$e^{-ak}.x(k)$	$x(z.e^a)$
$a^k.x(k)$	$x\left(\dfrac{z}{a}\right)$
$x(\infty)$	$\lim_{z \to 1}\{(z-1).x(z)\}$, provided that $\dfrac{-1}{z}.x(z)$ is analytic on and outside the unit circle
$x(0)$	$\lim_{z \to \infty} x(z)$, provided that the limit exists
$\sum_{k=0}^{\infty} x(k)$	$x(1)$
$\sum_{k=0}^{n} x(k).y(n-k)$	$x(z).y(z)$

ties of Table 75.2. These enable the transforms, and their inverses, of expressions to be found one term at a time and independently of their coefficients.

1. Transform both sides of the difference equation using the transforms of Table 75.1 as appropriate. This results in an algebraic equation with the complex variable z replacing kT as the independent variable.
2. Solve the algebraic equation resulting from Step 1 such that the transformed variable x(z) is expressed as a function of z.
3. Reduce the function x(z) resulting from Step 2 into partial fractions that occur on the right hand side of Table 75.1. For simple functions this step is trivial but for complex ones it can be very tedious.
4. Find the inverse transforms, *i.e.* the values of x(kT) in the left hand side of the table, corresponding to the partial fractions. This results in an equation in which the dependant variable x is expressed as a function of time kT.

This procedure is best illustrated by means of a couple of examples.

Example 4

Find the response $x(k)$ of the system whose behaviour is described by the difference equation:

$$x(k+2) + 3x(k+1) + 2x(k) = 0$$

given the initial conditions that $x(0) = 0$ and $x(1) = 1$.

Transform both sides using the third and fourth properties in Table 75.2, the forward difference properties:

$$z^2.x(z) - z + 3z.x(z) + 2x(z) = 0$$

Solve for x(z) and find partial fractions:

$$x(z) = \frac{z}{z^2 + 3z + 2} = \frac{z}{z+1} - \frac{z}{z+2}$$

Inverse transform both sides:

$$x(k) = (-1)^k - (-2)^k \quad \text{for } k = 0, 1, 2, 3, \ldots$$

Example 5

Solve the difference equation:

$$x(k+2) - 3x(k+1) + 2x(k) = u(k)$$

given the initial conditions that $x(k) = 0$ for $k \leq 0$, and that the forcing function $u(0) = 1$ and $u(k) = 0$ for $k < 0$ and $k > 0$.

The initial condition x(1) will also be required for the solution. This may be found by putting $k = -1$ in the difference equation, which gives $x(1) = 0$.

Transform both sides:

$$z^2.x(z) - 3z.x(z) + 2x(z) = u(z)$$

However, $u(z) = \sum_{k=0}^{\infty} u(kT).z^{-k} = 1$

Solve for x(z) and split into partial fractions to give:

$$x(z) = \frac{1}{z^2 - 3z + 2} = \frac{-1}{z-1} + \frac{1}{z-2}$$

To find the inverse transform from Table 75.1 it is necessary to have a z in the numerator of each partial fraction. The forward difference property may be exploited:

$$Z\{x(k+1)\} = z.(x(z) - x(0))$$
$$= z.x(z)$$
$$= \frac{-z}{z-1} + \frac{z}{z-2}$$

Inverse transform both sides gives:

$$x(k+1) = -1 + 2^k \quad \text{for } k = 0, 1, 2, 3 \ldots$$

Alternatively:

$$x(k) = -1 + 2^{k-1} \quad \text{for } k = 1, 2, 3, 4 \ldots$$

75.6 Inverse Z Transformation

The simplest method of inverse transformation is by expansion of x(z) into a time series in z^{-1}, which can be solved by inspection since:

$$x(z) = \sum_{k=0}^{\infty} x(kT).z^{-k}$$
$$= x(0) + x(T).z^{-1} + x(2T).z^{-2}$$
$$+ x(3T).z^{-3} + \ldots$$

Consider again the function x(z) of Example 4:

$$x(z) = \frac{z}{z^2 + 3z + 2}$$

This can be rewritten as:

$$x(z) = \frac{z^{-1}}{1 + 3z^{-1} + 2z^{-2}}$$

which gives by long division:

$$x(z) = z^{-1} - 3z^{-2} + 7z^{-3} - 15z^{-4} + \ldots$$

By inspection:

$$x(0) = 0, \quad x(T) = 1, \quad x(2T) = -3, \quad x(3T) = 7,$$

etc. which agrees with the analytical solution:

$$x(k) = (-1)^k - (-2)^k \quad \text{for } k = 0, 1, 2, 3, \ldots$$

Note that in general it is difficult to obtain the analytical solution x(k) from the set of values x(0), x(1), x(2), x(3), etc..

Finding the inverse by long division is often impracticable unless only the first few values of the time series are required, perhaps to get a feel for the form of the response. Otherwise the inverse has to be found by splitting x(z) into partial fractions.

Note the non-uniqueness of partial fractions. Again consider the x(z) of Example 4. Direct expansion yields the partial fractions:

$$x(z) = \frac{z}{z^2 + 3z + 2} = \frac{2}{z+2} - \frac{1}{z+1}$$

These are of no use for inverse transformation because there are no zs in the numerators. However, x(z)/z may be expanded into partial fractions as follows:

$$\frac{x(z)}{z} = \frac{1}{z^2 + 3z + 2} = \frac{1}{z+1} - \frac{1}{z+2}$$

This gives:

$$x(z) = \frac{z}{z+1} - \frac{z}{z+2}$$

whose partial fractions contain a z in their numerator, thereby enabling the transforms in Table 75.1 to be used for inverse transformation.

75.7 Worked Example

Find x(k) for the function:

$$x(z) = \frac{z(1 - e^{-T})}{(z-1)(z - e^{-T})}$$

Expanding x(z) into a convergent time series in z^{-1} gives:

$$x(z) = \frac{(1 - e^{-T}).z^{-1}}{1 - (1 + e^{-T}).z^{-1} + e^{-T}.z^{-2}}$$

Long division yields:

$$x(z) = 0 + \left(1 - e^{-T}\right).z^{-1} + \left(1 - e^{-2T}\right).z^{-2} + \left(1 - e^{-3T}\right).z^{-3} + \ldots$$

This may be compared with Equation 75.2:

$$x(z) = x(0) + x(1).z^{-1} + x(2).z^{-2} + x(3).z^{-3} + \ldots$$

which, by inspection gives:

$$x(k) = 1 - e^{-kT}$$

Sampled Data Systems

76.1 Impulse Response and Convolution
76.2 Pulse Transfer Functions
76.3 Worked Example No 1
76.4 Limitations
76.5 Cascaded Elements
76.6 Worked Example No 2
76.7 Closed Loop Systems
76.8 Equivalent Representations
76.9 Worked Example No 3
76.10 Implicit Input Signals

Z transforms provide the basis for representing and analysing the behaviour of sampled data systems. This chapter focuses on their representation, analysis being deferred to Chapter 77. Pulse transfer functions (PTF) are the principal means of representation, both for open and closed loop systems. PTFs in the Z domain are the discrete equivalent of transfer functions in the Laplace domain already encountered in Chapter 71 in relation to continuous systems. This chapter, therefore, starting from the basis of impulse response and convolution, develops the techniques for formulating and manipulating PTFs. However, whilst there are obvious analogies with continuous systems, there are also many subtle differences in their usage. These differences are emphasised.

76.1 Impulse Response and Convolution

Consider the continuous system depicted in Figure 76.1.

This may be represented in the Laplace domain by the equation:

Fig. 76.1 Transfer function of a continuous system

$$y(s) = G(s).x(s)$$

If $x(t)$ is the unit impulse $\delta(t)$ then $x(s) = 1$ and $y(s) = G(s)$, whence:

$$y(t) = g(t)$$

which is known as the impulse response. Now consider the sampled data system depicted in Figure 76.2.

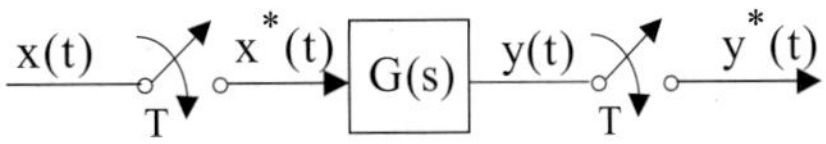

Fig. 76.2 Continuous system with sampled input and output signals

Let $x^*(t)$ be a train of k weighted impulses. By superposition, the response $y(t)$ is the sum of the individual impulse responses:

$$y(t) = x(0).g(t) + x(T).g(t - T) + x(2T).g(t - 2T) + \ldots$$

$$= \sum_{h=0}^{k} x(hT).g(t - hT)$$

If the output sampler is synchronous with the input sampler, then:

$$y(kT) = \sum_{h=0}^{k} x(hT).g(kT - hT) \qquad (76.1)$$

which is referred to as the convolution summation and is often written as:

$$y(kT) = g(kT) \bullet x(kT)$$

Convolution is the means of articulating, in the time domain, the process whereby an input signal is operated upon by an element to produce an output signal. The corresponding equation for continuous systems is:

$$y(t) = \int_{0}^{t} x(\tau).g(t - \tau).d\tau \qquad (76.2)$$

$$= \int_{0}^{t} g(\tau).x(t - \tau).d\tau$$

76.2 Pulse Transfer Functions

The concept of a transfer function for sampled data systems may be developed from the convolution summation by manipulating the limits of Equation 76.1 and taking its Z transform.

Since $g(kT - hT) = 0$ for $h > k$, because the input impulses haven't occurred, then:

$$y(kT) = \sum_{h=0}^{\infty} x(hT).g(kT - hT)$$

Taking the Z transform gives:

$$y(z) = Z\{y(kT)\} = \sum_{k=0}^{\infty} \sum_{h=0}^{\infty} x(hT).g(kT - hT).z^{-k}$$

Substituting $m = k - h$ gives:

$$y(z) = \sum_{m=-h}^{\infty} \sum_{h=0}^{\infty} x(hT).g(mT).z^{-(m+h)}$$

Again $g(mT) = 0$ for $m < 0$, i.e. for $h > k$. Thus:

$$y(z) = \sum_{m=0}^{\infty} \sum_{h=0}^{\infty} x(hT).g(mT).z^{-(m+h)}$$

$$= \sum_{m=0}^{\infty} g(mT).z^{-m} . \sum_{h=0}^{\infty} x(hT).z^{-h}$$

$$= G(z).x(z)$$

which may be represented as shown in Figure 76.3, where G(z) is known as the PTF.

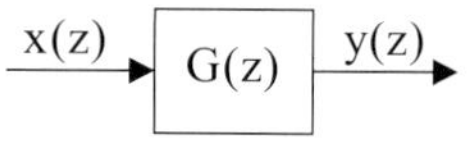

$$x(z) \rightarrow \boxed{G(z)} \rightarrow y(z)$$

Fig. 76.3 The pulse transfer function equivalent of Figure 76.2

Summary

If:
$$y(s) = G(s).x^*(s) \qquad (76.3)$$

then:
$$y(t) = L^{-1}\{G(s).x^*(s)\} \qquad (76.4)$$

and
$$y(z) = Z\{L^{-1}\{G(s).x^*(s)\}\} = G(z).x(z) \qquad (76.5)$$

where:
$$G(z) = Z\{g(t)\} = Z\{L^{-1}\{G(s)\}\} \qquad (76.6)$$

$$= G^*(s) = \sum_{k=0}^{\infty} g(kT).z^{-k}$$

The general 3-step procedure for finding PTFs is as follows:

1. Obtain the overall transfer function G(s) between the samplers.
2. Determine the impulse response from $g(t) = L^{-1}\{G(s)\}$.
3. Evaluate the PTF from $G(z) = \sum_{k=0}^{\infty} g(kT).z^{-k}$.

76.3 Worked Example No 1

Consider the system depicted in Figure 76.4.

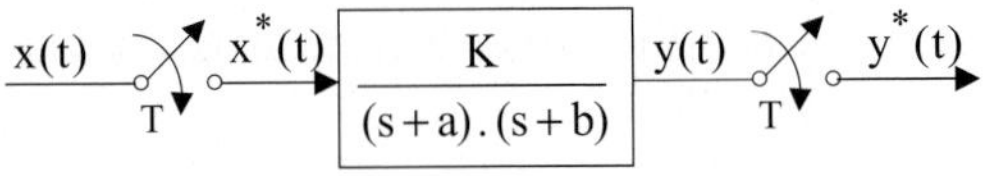

Fig. 76.4 Second order system with sampled input and output signals

Between the samplers:

$$G(s) = \frac{K}{(s+a)(s+b)}$$

$$= \frac{K}{(b-a)}\left(\frac{1}{(s+a)} - \frac{1}{(s+b)}\right)$$

Inverse Laplace transform:

$$g(t) = \frac{K}{(b-a)}\left(e^{-at} - e^{-bt}\right)$$

Z transform:

$$G(z) = \sum_{k=0}^{\infty} \frac{K}{(b-a)}\left(e^{-akT} - e^{-bkT}\right).z^{-k}$$

However:

$$\sum_{k=0}^{\infty} e^{-akT}.z^{-k} = 1 + e^{-aT}z^{-1} + e^{-2aT}z^{-2} + e^{-3aT}z^{-3} + \ldots$$

$$= 1 + e^{-aT}z^{-1}\left(1 + e^{-aT}z^{-1} + e^{-2aT}z^{-2} + \ldots\right)$$

$$= 1 + e^{-aT}z^{-1}\sum_{k=0}^{\infty} e^{-akT}.z^{-k}$$

Hence:

$$\sum_{k=0}^{\infty} e^{-akT}.z^{-k} = \frac{1}{1 - e^{-aT}.z^{-1}}$$

Substituting back into G(z) gives:

$$G(z) = \frac{K}{(b-a)}\left(\frac{1}{1 - e^{-aT}z^{-1}} - \frac{1}{1 - e^{-bT}z^{-1}}\right)$$

$$= \frac{K}{(b-a)} \cdot \frac{z.\left(e^{-aT} - e^{-bT}\right)}{\left(z - e^{-aT}\right)\left(z - e^{-bT}\right)}$$

76.4 Limitations

There are three important constraints:

1. The theory of PTFs is based upon the assumption that the train of sampled pulses can be adequately represented by a train of weighted impulses, the magnitude of each pulse being equal to the area under the corresponding impulse. This assumption is only valid if the pulse width Δt is small relative to the sampling period T which is itself small relative to the dominant time constants of the system.
2. The output $y(z) = G(z).x(z)$ only specifies the values of $y(t)$ at the sampling instants and does not contain any information about $y(t)$ in-between sampling instants.
3. The transfer function $G(s)$ of the continuous part of the system must have at least two more poles than zeros, otherwise the response obtained $y(t)$ may be incorrect.

The latter point may be demonstrated by considering the response of the continuous first order system depicted in Figure 76.5.

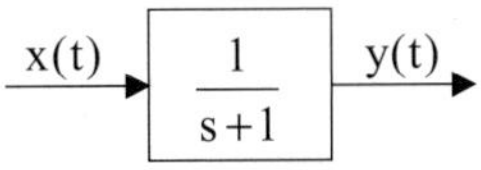

Fig. 76.5 A continuous first order system

If $x(t) = 1$, i.e. $x(s) = \frac{1}{s}$, then

$$y(s) = G(s).x(s) = \frac{1}{s(s+1)}$$

and $\quad y(t) = 1 - e^{-t}$.

The response $y(t)$ is the continuous curve of Figure 76.7.

Now consider the same first order system with samplers, as depicted in Figure 76.6.

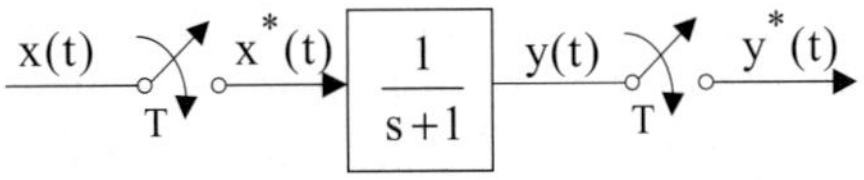

Fig. 76.6 First order system with sampled input and output signals

If $x(t) = 1$, i.e. $x(s) = \frac{1}{s}$, then $x(z) = \frac{z}{z-1}$.

If $G(s) = \frac{1}{s+1}$ then $G(z) = \frac{z}{z - e^{-T}}$.

Thus $y(z) = G(z).x(z) = \frac{z^2}{(z-1)(z-e^{-T})}$.

Suppose that $T = 1$, in which case:

$$y(z) = \frac{z^2}{z^2 - 1.368z + 0.368}$$
$$= \frac{1}{1 - 1.368z^{-1} + 0.368z^{-2}}$$
$$= 1 + 1.368z^{-1} + 1.504z^{-2} + 1.554z^{-3} + \ldots$$

However, by definition:

$$y(z) = y(0) + y(1)z^{-1} + y(2)z^{-2} + y(3)z^{-3} + \ldots$$

The corresponding response $y(k)$ is approximated by the broken curve of Figure 76.7.

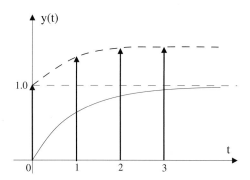

Fig. 76.7 Inconsistent responses to the same step input

The basic input $x(t)$ was the same in both cases but, in the case of the sampled data system, it was broken up by virtue of the sampling process. Thus the sampled data system was not forced by as strong an input signal as the continuous system, so its output signal should be weaker. The Z transform analysis suggests otherwise, which is clearly inconsistent.

76.5 Cascaded Elements

It is important to appreciate that, in general, the position of the samplers makes a big difference to the PTF. Consider the two systems depicted in Figures 76.8 and 76.9.

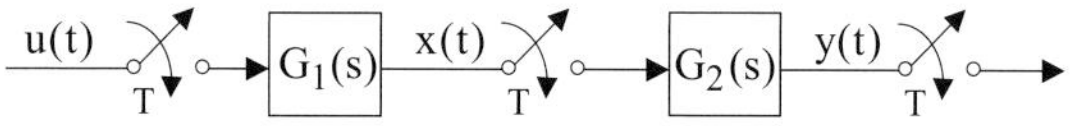

Fig. 76.8 Input and output sampling of both elements in series

By inspection:

$$y(z) = G_2(z).x(z) = G_2(z).G_1(z).u(z)$$

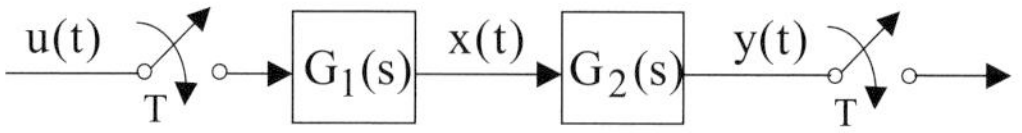

Fig. 76.9 Elements in series without intermediate sampling

Applying the 3-step procedure:

$$y(s) = G_2(s).G_1(s).u^*(s)$$

Thus:

$$y(z) = Z\{L^{-1}\{G_2(s).G_1(s).u^*(s)\}\}$$
$$= G_2G_1(z).u(z)$$

where:

$$G_2G_1(z) = Z\{L^{-1}\{G_2(s).G_1(s)\}\} \quad (76.7)$$
$$= G_2G_1^*(s)$$

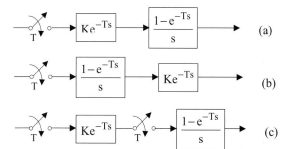

Fig. 76.10 Time delay and hold device synchronous with sampling period

The only obvious exception to this involves combinations of time delays and hold devices which are synchronous with the sampling period. The classic example of such is an on-line analyser, as depicted in Figure 76.10.

All three combinations have the identical PTF: $G(z) = K.z^{-1}$.

76.6 Worked Example No 2

Consider the system comprising two elements in series as depicted in Figure 76.11.

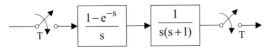

Fig. 76.11 Sampler, hold and second order system in series

Their combined transfer function is:

$$G(s) = \frac{1 - e^{-s}}{s^2.(s+1)}$$

whence:

$$G(z) = (1 - z^{-1}).Z\left\{L^{-1}\left\{\frac{1}{s^2.(s+1)}\right\}\right\}$$

$$= (1 - z^{-1}).Z\left\{L^{-1}\left\{\frac{1}{s^2} - \frac{1}{s} + \frac{1}{s+1}\right\}\right\}$$

Assuming a sampling period of 1 s:

$$G(z) = (1 - z^{-1}).\left(\frac{z}{(z-1)^2} - \frac{z}{(z-1)} + \frac{z}{(z-e^{-1})}\right)$$

$$= \frac{z + e - 2}{ez^2 - (e+1).z + 1}$$

Substituting for e gives:

$$G(z) = \frac{z + 0.718}{2.718.z^2 - 3.718.z + 1}$$

76.7 Closed Loop Systems

Consider the system depicted in Figure 76.12 in which the error signal only is sampled.

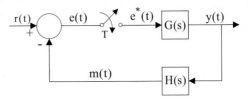

Fig. 76.12 Closed loop system with sampled error signal

For the comparator:

$$e(t) = r(t) - m(t)$$

Using the summation property of Z transforms:

$$e(z) = r(z) - m(z)$$

A useful trick in sampled data systems analysis is to introduce imaginary samplers as depicted in Figure 76.13.

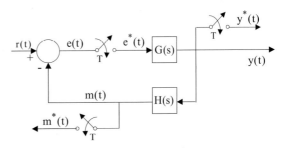

Fig. 76.13 Introduction of imaginary samplers

Thus, although the feedback signal m(t) is not physically sampled, the sampled signal m*(t) may be considered. Likewise for the output y(t). Using block diagram algebra techniques:

$$m(z) = HG(z).e(z)$$

whence:

$$e(z) = r(z) - HG(z).e(z)$$

$$= \frac{1}{1 + HG(z)}.r(z)$$

Thus:

$$y(z) = G(z).e(z)$$

$$= \frac{G(z)}{1 + HG(z)}.r(z) = F(z).r(z)$$

which may be depicted as in Figure 76.14.

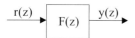

Fig. 76.14 Closed loop pulse transfer function

Here F(z) is the closed loop PTF. For a given F(z) and r(z) the value of y(z) may be determined, the inverse of which yields the response y(t) at the sampling instants only.

76.8 Equivalent Representations

The sampled data representation of digital controllers, whether they are single loop controllers or realised by means of a DCS, PLC or whatever, is a source of much confusion. They are correctly depicted as shown in Figure 76.15.

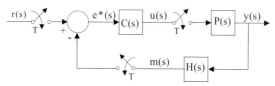

Fig. 76.15 Representation of a sampled data controller within a feedback loop

The comparator and controller are algorithms operating on values in the database. There are three samplers, each with the same sampling period. The samplers are synchronised with the execution of the algorithms, the sampling period being the same as the step length for numerical integration. The feedback sampler is real in the sense that the measurement signal m(t) is analogue and has to be physically sampled to provide an input value for the database. The set point sampler is virtual in the sense that the set point value has to be extracted from the database. The output sampler is also virtual in the sense that the algorithm output has to be extracted from the database and routed through to the D/A converter. Strictly speaking, the output channel hardware devices, such as the D/A converter and zero order hold, are part of the plant/process transfer function P(s). Note that the only sampler which has a hold device associated with it is that in the output channel. That is because, of the three samplers, it is the only one that produces a continuous analogue signal.

In many texts a digital controller is represented as depicted in Figure 76.16. This is functionally equivalent to Figure 76.15.

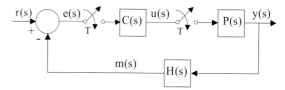

Fig. 76.16 Functionally equivalent representation of sampled data controller

76.9 Worked Example No 3

Find the closed loop PTF for the system depicted in Figures 76.16. Starting with the output of the digital controller:

$$u(z) = C(z).e(z)$$

However:

$$e(z) = r(z) - m(z)$$

whence:

$$u(z) = C(z).r(z) - C(z).m(z).$$

Assuming an imaginary sampler on the measurement m(s) gives

$$m(z) = HP(z).u(z)$$

whence:

$$u(z) = C(z).r(z) - C(z)HP(z).u(z)$$
$$= \frac{C(z)}{1 + C(z).HP(z)}.r(z)$$

Assuming another imaginary sampler on the output signal y(s) gives:

$$y(z) = P(z).u(z)$$
$$= \frac{P(z).C(z)}{1 + C(z).HP(z)}.r(z)$$

whence the closed loop PTF:

$$G(z) = \frac{y(z)}{r(z)} = \frac{P(z).C(z)}{1 + C(z).HP(z)} \quad (76.8)$$

76.10 Implicit Input Signals

Note that the closed loop PTF cannot always be isolated from the input signal as can be demonstrated for the system shown in Figure 76.17 which corresponds to a conventional analogue control loop with an analyser in the feedback path.

$$y(s) = G(s).e(s)$$
$$= G(s).r(s) - G(s).H(s).y^*(s)$$
$$y(t) = L^{-1}\{G(s).r(s)\} - L^{-1}\{G(s).H(s).y^*(s)\}$$
$$y(z) = Z\{L^{-1}\{G(s).r(s)\}\}$$
$$\quad - Z\{L^{-1}\{G(s).H(s).y^*(s)\}\}$$
$$= GR(z) - GH(z).y(z)$$
$$= \frac{GR(z)}{1 + GH(z)}$$

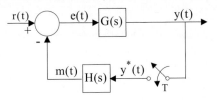

Fig. 76.17 System with implicit input in closed loop pulse transfer function

Because sampled data systems contain both sampled and continuous signals, the system equations often contain both sampled and continuous versions of the same signal. This means that analytical solutions can be difficult to obtain. There is no general procedure for finding closed loop PTFs, the most appropriate substitutions are learned with practice.

Z Plane Analysis

Chapter 77

77.1 S to Z Mapping
77.2 Stability Analysis
77.3 Modified Routh Test
77.4 Worked Example No 1
77.5 Pole Positions and Root Locus
77.6 Worked Example No 2
77.7 Pole Zero Cancellation
77.8 Worked Example No 3
77.9 Comments

Z plane analysis concerns the behaviour of sampled data systems. There are many analogies between s-plane analysis, as described in Chapters 70–74, and z-plane analysis. Therefore, rather than developing the techniques of z-plane analysis from first principles, they are established by means of analogy. Arguably, the two most important techniques are root locus for analytical purposes and pole-zero cancellation for design purposes. This chapter, therefore, focuses on these two techniques and the principles involved. Chapter 78 covers other means of designing impulse compensators and their realisation.

77.1 S to Z Mapping

From Equation 75.2:

$$z = e^{Ts}$$

However, as seen in Table 71.2, for example, s is itself a complex variable:

$$s = \sigma + j\omega$$

where both σ and ω may be either positive or negative. Substituting gives:

$$z = e^{(\sigma + j\omega).T}$$

which yields in polar form:

$$|z| = e^{\sigma T} \text{ and } \angle z = \omega T$$

Hence points in the s-plane may be mapped into the z-plane. In particular, consider the "primary strip" of the s-plane as shown in Figure 77.1.

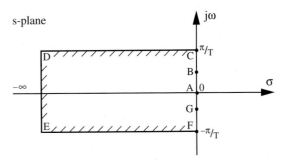

Fig. 77.1 The primary strip in the s-plane

By inspection this can be seen to map into the unit circle centred on the origin of the z plane, as shown in Figure 77.2.

The left hand half of the s plane may be considered to be made up of the primary strip and an infinite number of complementary strips, as shown in Figure 77.3.

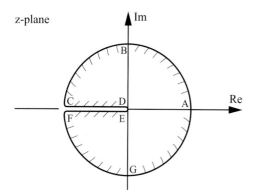

Fig. 77.2 The unit circle in the z-plane

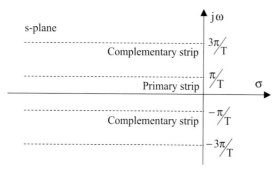

Fig. 77.3 The primary and complementary strips of the left half of the s-plane

Again, by inspection, each of these complementary strips can be seen to map into the unit circle. Thus the whole of the left half of the s-plane maps into the unit circle in the z-plane, the $j\omega$ axis mapping into the unit circle itself. Conversely, any point inside the unit circle in the z plane maps into a multiplicity of points in the left half of the s-plane. It is usual to consider the primary strip only.

77.2 Stability Analysis

Consider the sampled data system of Figure 77.4, or similar:
Its closed loop pulse transfer function is:

$$\frac{G(z)}{1 + GH(z)}$$

The stability of such a system may be determined from the location in the z-plane of the roots of the characteristic equation:

$$1 + GH(z) = 0$$

For stability, all the roots of the characteristic equation must lie inside the unit circle, i.e.

$$|z_i| < 1.0$$

The closed loop system is unstable if any closed loop pole lies outside the unit circle and/or any multiple poles lie on the unit circle. Note that a single pole at z = 1 does not imply marginal stability. The example later in this chapter demonstrates that such a pole arises from the combination of an integrator and a zero order hold device in the open loop.

77.3 Modified Routh Test

The Routh test, explained in Chapter 71, establishes whether or not any of the roots of a polynomial lie in the right half of a complex plane, and cannot be used directly to determine where roots lie relative to the unit circle. However, the bilinear transform:

$$z = \frac{r+1}{r-1}$$

maps the interior of the unit circle in the z-plane into the left half of the "r-plane", to which the Routh test may be applied in the usual way.

77.4 Worked Example No 1

Consider the system shown in Figure 77.4 for which $H(s) = 1$ and $G(s) = \dfrac{K}{s(s+1)}$.

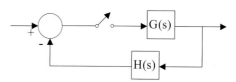

Fig. 77.4 Feedback system with sampled error signal

The open loop pulse transfer function is found as follows:

$$G(s) = \frac{K}{s(s+1)} = K\left(\frac{1}{s} - \frac{1}{s+1}\right)$$

$$G(z) = K\left(\frac{z}{z-1} - \frac{z}{z-e^{-T}}\right)$$

$$= \frac{K(1-e^{-T}).z}{z^2 - (1+e^{-T}).z + e^{-T}}$$

The characteristic equation is formed thus:

$$1 + G(z) = 1 + \frac{K(1-e^{-T}).z}{z^2 - (1+e^{-T}).z + e^{-T}} = 0$$

$$z^2 + \left(K(1-e^{-T}) - (1+e^{-T})\right).z + e^{-T} = 0$$

Make the bilinear transform:

$$\left(\frac{r+1}{r-1}\right)^2 +$$

$$\left(K(1-e^{-T}) - (1+e^{-T})\right)\left(\frac{r+1}{r-1}\right) + e^{-T} = 0$$

Expansion and rearrangement gives:

$$K.r^2 + 2r + \frac{2(1+e^{-T})}{(1-e^{-T})} - K = 0$$

from which the Routh array may be developed (Table 77.1).

Table 77.1 Routh array for Worked Example No 1

r^2	K	$\frac{2(1+e^{-T})}{(1-e^{-T})} - K$	0
r^1	2	0	–
r^0	$\frac{2(1+e^{-T})}{(1-e^{-T})} - K$	0	–

The criterion for stability is thus:

$$\frac{2(1+e^{-T})}{(1-e^{-T})} - K > 0, \text{ that is } K < \frac{2(1+e^{-T})}{(1-e^{-T})}.$$

For example, if the sampling period T = 1 s, the limiting value of K is 4.32.

77.5 Pole Positions and Root Locus

The impulse response of a sampled data system is determined by the position of its dominant closed loop poles in the z-plane: the response for various pole positions is as depicted in Table 77.2.

The techniques for plotting the root locus of continuous systems, explained in Chapter 74, may be applied directly to sampled data systems. Consider the system whose closed loop pulse transfer function is as follows:

$$\frac{G(z)}{1+GH(z)}$$

The behaviour of such a system, as its effective forward path gain increases from 0 to ∞, may be found from a plot of the roots of its characteristic equation:

$$1 + GH(z) = 0$$

according to the magnitude and angle criteria:

$$|GH(z)| = 1.0$$

$$\angle GH(z) = -(1 \pm 2n)\pi \quad \text{for } n = 0, 1, 2, \ldots$$

which simply requires a knowledge of its open loop pulse transfer function GH(z). The rules for construction of root locus in the z-plane are exactly the same as the Evans' rules detailed in Chapter 74. The root loci of some simple systems are shown in Table 77.3. Note that they all consist of straight lines and circular segments only.

77.6 Worked Example No 2

Consider the system shown in Figure 77.4 for which H(s) = 1 and the open loop pulse transfer function is:

$$G(z) = \frac{K(z+0.9)}{(z-1)(z-0.4)}$$

The root locus is constructed using Evan's rules as appropriate.

Table 77.2 Impulse response *vs* dominant closed loop pole positions in z-plane

Pole positions in s-plane	Pole positions in z-plane	Transient response
a)		
b)		
c)		
d)		
e)		
f)		
g)		

Table 77.2 Impulse response *vs* dominant closed loop pole positions in z-plane (Continued)

Pole positions in s-plane	Pole positions in z-plane	Transient response
h)		
i)		
j)		

1. G(z) is already in the correct form.
2. There is one open loop zero at z = −0.9 and two open loop poles at z = 1 and z = 0.4.
3. There are two branches.
4. The real axis is part of the root locus for 0.4 < z < 1 and −∞ < z < −0.9.
6. The breakpoints are found from the characteristic equation:

$$1 + \frac{K(z+0.9)}{(z-1)(z-0.4)} = 0$$

Rearrange and differentiate:

$$K = -\frac{(z-1)(z-0.4)}{(z+0.9)} = \frac{-z^2 + 1.4z - 0.4}{z+0.9}$$

$$\frac{dK}{dz} = \frac{-z^2 - 1.8z + 1.66}{z^2 + 1.8z + 0.81}$$

The breakpoints occur when:

$$-z^2 - 1.8z + 1.66 = 0.$$

Solution gives z = 0.6716 or z = −2.472.

11. Rearrange the characteristic equation to find the limiting value of K:

$$z^2 + (K - 1.4)z + 0.9K + 0.4 = 0$$

Make the bilinear transform and rearrange:

$$1.9K.r^2 + (1.2 - 1.8K)r + 2.8 - 0.1K = 0$$

from which the Routh array may be developed (Table 77.4).

Table 77.4 Routh array for Worked Example No 2

r^2	1.9K	2.8 − 0.1K	0
r^1	1.2 − 1.8K	0	−
r^0	2.8 − 0.1K	0	−

The system becomes unstable when K is increased to a value such that either:

$$1.2 - 1.8K = 0 \quad \text{or} \quad 2.8 - 0.1K = 0$$

for which the corresponding values are K = 0.667 and K = 28. Clearly K = 0.667 is the limiting value.

The root locus is of the form depicted in Table 77.3(VI).

77 Z Plane Analysis

Table 77.3 Root loci of some simple sampled data systems

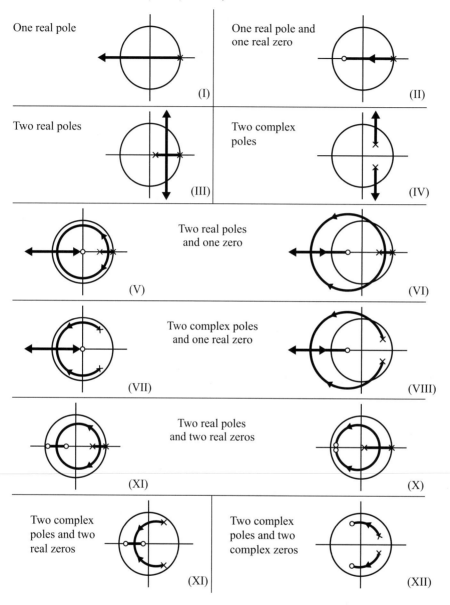

77.7 Pole Zero Cancellation

This is a design method. Consider the sampled data system depicted in Figure 77.5.

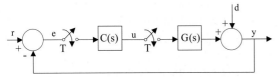

Fig. 77.5 Sampled data controller with plant in feedback loop

The objective is to find the impulse compensator C(z) for a given plant G(z) to produce a specified impulse response. The method normally assumes that a second order response is required of the closed loop system to a step input. The positions of the desired dominant closed loop poles are specified and the compensator designed to ensure that the root locus passes through them.

Step 1. Specify the form of response required in terms of damping factor and natural frequency. Remember that for a system that is second order, or may be considered to be so by virtue of a pair of dominant roots, ξ and ω_n are interrelated as shown in Figure 74.11.

If the desired response is articulated in terms of rise time, settling time, overshoot, *etc.*, these may be converted into ξ and ω_n using the formulae in Table 72.1.

Step 2. Map the desired closed loop pole positions into the z-plane using:

$$z = e^{Ts} = \alpha \pm j\beta$$

Note that a value for the sampling period T must be specified at this stage.

Step 3. Find the pulse transfer function of the plant G(z).

Step 4. Specify any compensator C(z) such that the root locus of:

$$1 + C(z)G(z) = 0$$

passes through $z = \alpha + j\beta$. The simplest design is to choose a C(z) such that cancellation of poles and zeros with G(z) results in a C(z)G(z) which has two open loop poles only and no zeros, *i.e.*

$$C(z)G(z) = \frac{K}{(z + p_1)(z + p_2)}$$

The root locus will thus be a vertical line whose breakaway point is halfway between the open loop poles $z = -p_1$ and $z = -p_2$. By appropriate choice of p_1 and p_2, that is:

$$p_1 + p_2 = 2\alpha$$

all the closed loop poles of the locus will have:

$$Re(z) = \alpha$$

thereby ensuring that the locus passes through the desired dominant pole positions.

Note that by specifying one of the open loop poles to be on the unit circle, say p_1 such that $z = -p_1 = 1.0$ which corresponds to an integrator, ensures that the closed loop response has integral action and eliminates offset. Also, placing one of the open loop poles on the unit circle maximises the distance between the two open loop poles: this increases the range of K values along the real axis which makes the design less sensitive to changes in controller gain.

Step 5. Find the forward gain K required to complete the compensator design. Since for any point on the root locus:

$$|C(z)G(z)| = 1.0$$

then at the desired dominant closed loop pole positions:

$$\frac{K}{|\alpha + j\beta + p_1||\alpha + j\beta + p_2|} = 1.0$$

whence K. Note that the pole zero cancellation method must be used with care due to the numerical sensitivity of z-plane analysis. Thus the effects of any inaccuracy in G(z), say due to modelling errors or due to changes in the system's parameters during operation, must be carefully considered at the design stage. There are many ways in which the compensator design may be refined and, indeed, other approaches to design, some of which are described in Chapter 78.

77.8 Worked Example No 3

With reference to Figure 77.5, consider a plant whose transfer function is:

$$G(s) = \frac{(1 - e^{-2s})}{s} \cdot \frac{1}{s(10s + 1)}$$

where the time delay and lag have units of seconds. Design an impulse compensator C(z), with a sampling period of 2 s, such that the closed loop response to a step input has a damping factor of 0.7 and a settling time of 20 s (±2% criterion).

Step 1. Specify the form of response required:

$$\xi = 0.7$$
$$t_s \approx \frac{4}{\omega_n \xi} = 20 \quad \text{Hence } \omega_n = 0.2857.$$

In Cartesian co-ordinates the desired dominant closed loop pole positions are thus:

$$s = -0.20 \pm j.0.204$$

Step 2. Map these pole positions into the z-plane:

$$z = e^{2s} = e^{2(-0.2 \pm j.0.204)}$$
$$= 0.6703 \angle \pm 23.38°$$
$$= 0.6153 \pm j.0.266$$

Step 3. Determine the pulse transfer function of the plant:

$$G(s) = \frac{(1 - e^{-2s})}{s} \cdot \frac{1}{s.(10s + 1)}$$
$$= (1 - e^{-2s}) \left(\frac{1}{s^2} - \frac{10}{s} + \frac{100}{(10s + 1)} \right)$$

Whence:

$$G(z) = (1 - z^{-1})$$
$$\cdot \left(\frac{2z}{(z-1)^2} - \frac{10z}{(z-1)} + \frac{10z}{(z - e^{-0.2})} \right)$$
$$= \frac{0.1873.(z + 0.9354)}{(z - 1)(z - 0.8187)}$$

Step 4. Design the compensator:

Choose one open loop pole such that $z = -p_1 = 1.0$. If $\alpha = 0.6153$ then the other open loop pole should be at $z = -p_2 = 0.2306$.

Specify:

$$C(z)G(z) = \frac{K}{(z + p_1)(z + p_2)}.$$

Thus:

$$\frac{C(z).0.1873.(z + 0.9354)}{(z - 1)(z - 0.8187)} = \frac{K}{(z - 1)(z - 0.2306)}$$

Hence:

$$C(z) = \frac{5.339.K.(z - 0.8187)}{(z + 0.9354)(z - 0.2306)}$$

Step 5. Find the controller gain required.
At the desired dominant closed loop pole positions:

$$\frac{K}{|0.6153 + j.0.266 - 1.0| \cdot |0.6153 + j.0.266 - 0.2306|} = 1.0$$

Hence the forward gain required is:

$$K = 0.2188$$

The impulse compensator designed is thus:

$$C(z) = \frac{1.168.(z - 0.8187)}{(z + 0.9354)(z - 0.2306)}$$

Check. That the response is of the correct form can be checked as follows.
Analysis of Figure 77.5 yields the closed loop pulse transfer function for disturbances:

$$y(z) = \frac{1}{1 + C(z)G(z)}.d(z)$$

Assume a unit step disturbance, for which $d(z) = \dfrac{z}{(z-1)}$.

Substitute for C(z) and G(z):

$$y(z) = \dfrac{\dfrac{z}{(z-1)}}{1 + \dfrac{1.168.(z-0.8187)}{(z+0.9354)(z-0.2306)} \cdot \dfrac{0.1873.(z+0.9354)}{(z-1)(z-0.8187)}}$$

$$= \dfrac{z.(z-0.2306)}{(z^2 - 1.231z + 0.4494)}$$

$$= \dfrac{1 - 0.2306z^{-1}}{1 - 1.231z^{-1} + 0.4494z^{-2}}$$

$$= 1 + z^{-1} + 0.7812z^{-2} + 0.5119z^{-3} + 0.2788z^{-4} + 0.1131z^{-5} + \ldots$$

the inverse transform of which is:

y(0)	y(T)	y(2T)	y(3T)	y(4T)	y(5T)	y(nT)
1.0	1.0	0.7812	0.5119	0.2788	0.1131	–

This step response is depicted in Figure 77.6. It is evident that the feedback cannot have any effect until $n \geq 2$ because of the inherent delay of T s in the zero order hold. Otherwise, the output y rapidly converges on the set point. If the power series expansion for y in terms of z^{-1} were continued further, it would be seen that the response is indeed that of an underdamped second order system.

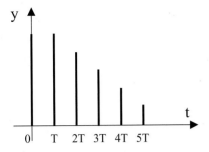

Fig. 77.6 Step response of system for Worked Example No 3

Once the transients have decayed away, the compensator should have rejected the disturbance and the steady state output reduced to zero. This can be confirmed by applying the final value theorem:

$$y(\infty) = \lim_{z \to 1} (z-1).y(z)$$

$$= \lim_{z \to 1} (z-1) . \dfrac{z.(z-0.2306)}{(z^2 - 1.231z - 0.4494)}$$

$$= 0$$

The fact that there is no offset is to be expected since an integrator was explicitly built into the compensator design of Step 4 by virtue of the $(z-1)$ in the denominator of the open loop pulse transfer function C(z)G(z).

77.9 Comments

Compensator design by means of root locus and pole cancellation can only be justified when the sampling period T is greater than say 0.1 to 0.2 times the system's dominant time constant. The two obvious scenarios where this is likely to occur are with digital controllers when the plant's response is fast and with analysers when the sampling period is long.

Impulse Compensation

78.1 PID Compensation
78.2 Generic Closed Loop Response
78.3 Dahlin's Method
78.4 Deadbeat Method
78.5 Direct Programs
78.6 Cascade Programs
78.7 Parallel Programs
78.8 Comments

Having covered the pole-zero cancellation method in Chapter 77, three other commonly used methods of compensator design are now introduced. One is the discrete form of the familiar 3-term PID controller, the others are Dahlin's and the deadbeat methods. The pros and cons of these different methods are discussed. There is of course no point in designing compensators unless they can be realised in practice. Being inherently discrete, they lend themselves to being implemented by means of software. An overview of three conventional approaches to such follow, the so-called direct, cascade and parallel programming methods. An understanding of these techniques is fundamental, not only to realising compensators, but to many other advanced control techniques such as filtering and identification, as covered in subsequent chapters.

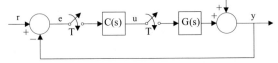

Fig. 78.1 Closed loop comprising impulse compensator and plant

Throughout this chapter it is presumed, as depicted in Figure 78.1, that $C(s)$ is the impulse compensator to be designed and $G(s)$ is the plant which is assumed to include the ZOH.

78.1 PID Compensation

The transfer function of a conventional 3-term controller was established in Chapter 71:

$$C(s) = K_C \left(1 + \frac{1}{T_R s} + T_D s\right) \quad (71.3)$$

Consider the integral action alone:

$$u(t) = \frac{K_C}{T_R} \int_0^t e(t).dt$$

$$\frac{du}{dt} = \frac{K_C}{T_R}.e$$

This may be discretised using a step length equivalent to the sampling period T:

$$\frac{u(k) - u(k-1)}{T} = \frac{K_C}{T_R}.e(k)$$

$$(1 - z^{-1}).u(z) = \frac{K_C.T}{T_R}.e(z)$$

$$u(z) = \frac{K_I}{(1 - z^{-1})}.e(z)$$

which clearly corresponds to a pole at $z = 1$ on the unit circle.

Similarly for the derivative action:

$$u(t) = K_C T_D . \frac{de(t)}{dt}$$

$$u(k) = K_C T_D . \frac{e(k) - e(k-1)}{T}$$

$$u(z) = K_D . (1 - z^{-1}) . e(z)$$

The three actions may thus be combined:

$$C(z) = K_C + \frac{K_I}{(1 - z^{-1})} + K_D . (1 - z^{-1})$$

$$= \frac{K_C . (1 - z^{-1}) + K_I + K_D . (1 - z^{-1})^2}{(1 - z^{-1})}$$

$$= \frac{K_C + K_I + K_D - (K_C + 2K_D) . z^{-1} + K_D . z^{-2}}{(1 - z^{-1})}$$

where K_C, K_I and K_D are known as the proportional, integral and derivative gains respectively.

Whence:

$$C(z) = \frac{a_0 + a_1 . z^{-1} + a_2 . z^{-2}}{1 - z^{-1}} \quad (78.1)$$

Relating $C(z)$ to the output and error signals gives:

$$(1 - z^{-1}) . u(z) = (a_0 + a_1 . z^{-1} + a_2 . z^{-2}) . e(z)$$

Inverse transforming gives a time series which can be realised by means of a simple algorithm:

$$u(k) = u(k-1) + a_0 . e(k) + a_1 . e(k-1) + a_2 . e(k-2) \quad (78.2)$$

Note that assigning an appropriate value to u for $k = -1$ establishes the bias on the controller output.

In Chapter 71, many variations to the classical form of the 3-term controller were introduced, such as derivative feedback. These can all be accommodated in alternative forms of $C(z)$. Whereas the compensator design being realised will have the form of Equation 78.1, it is important to appreciate that the user interface on any control system, DCS or otherwise, will present the gain parameters, K_C, K_I and K_D and sample period T explicitly to the user.

78.2 Generic Closed Loop Response

Both Dahlin's and the deadbeat methods of compensator design are based upon the closed loop pulse transfer function $M(z)$ for step changes. Applying block diagram algebra to the generic system of Figure 78.1 yields:

$$e(z) = r(z) - y(z) = r(z)\left(1 - \frac{y(z)}{r(z)}\right)$$

$$= r(z)(1 - M(z)) \quad (78.3)$$

where $M(z)$ is the closed loop transfer function.

Also:

$$e(z) = r(z) - C(z).G(z).e(z) = \frac{1}{1 + C(z).G(z)} . r(z)$$

Thus:

$$y(z) = \frac{C(z).G(z)}{1 + C(z).G(z)} . r(z)$$

Whence:

$$M(z) = \frac{C(z).G(z)}{1 + C(z).G(z)}$$

Hence:

$$C(z) = \frac{M(z)}{(1 - M(z)).G(z)} \quad (78.4)$$

78.3 Dahlin's Method

This method involves specifying the desired closed loop response to a step input. The response is articulated in terms of a time lag λ and a delay L, as depicted in Figure 78.2, and is inclined to be sluggish.

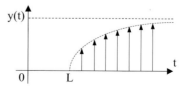

Fig. 78.2 Desired closed loop response to a step input

The approach has obvious parallels with the reaction curve method of tuning described in Chapter 24 and the approximation for higher order systems described in Chapter 72.

The designer may choose any sensible values for λ and L:

$$y(s) = \frac{e^{-Ls}}{s.(\lambda s + 1)} \quad (78.5)$$

If $L = n.T$ then:

$$y(z) = z^{-n}.Z\left\{L^{-1}\left\{\frac{1}{s.(\lambda s + 1)}\right\}\right\}$$
$$= z^{-n}.Z\left\{1 - e^{-t/\lambda}\right\}$$
$$= z^{-n}.\left(\frac{z}{z-1} - \frac{z}{z - e^{-T/\lambda}}\right)$$
$$= \frac{z^{-n+1}\left(1 - e^{-T/\lambda}\right)}{(z-1)\left(z - e^{-T/\lambda}\right)}$$

Assume a step input:

$$r(z) = \frac{z}{z-1}$$

Thus, for the chosen values of λ and L, the system of Figure 78.1 has a closed loop transfer function of:

$$M(z) = \frac{(z-1)}{z}.y(z) = \frac{z^{-n}\left(1 - e^{-T/\lambda}\right)}{\left(z - e^{-T/\lambda}\right)}$$

Equation 78.4 gives the compensator required:

$$C(z) = \frac{M(z)}{(1 - M(z)).G(z)}$$
$$= \frac{\dfrac{z^{-n}\left(1 - e^{-T/\lambda}\right)}{\left(z - e^{-T/\lambda}\right)}}{\left(1 - \dfrac{z^{-n}\left(1 - e^{-T/\lambda}\right)}{\left(z - e^{-T/\lambda}\right)}\right).G(z)}$$
$$= \frac{z^{-n-1}\left(1 - e^{-T/\lambda}\right)}{\left(1 - e^{-T/\lambda}z^{-1} - \left(1 - e^{-T/\lambda}\right)z^{-n-1}\right).G(z)}$$

which is always realisable for any sensible λ and L. The obvious disadvantage of Dahlin's method is that to calculate C(z) the value of G(z) is required.

If G(s) is not known, as is usually the case, it must be estimated. This may be achieved by developing a model of the plant from first principles, by empirical tests or by means of identification. Fortunately, the method is fairly tolerant of any estimating errors.

78.4 Deadbeat Method

Often referred to as minimal response design, the method enables a compensator to be designed such that the closed loop response exhibits a fast rise time and a short settling time with zero offset: the ideal response.

The zero offset is established using the final value theorem. From Equation 78.3:

$$e(z) = r(z).(1 - M(z))$$
$$e(\infty) = \lim_{k \to \infty} e(kT) = \lim_{z \to 1}(z-1).e(z)$$
$$= \lim_{z \to 1}(z-1).r(z).(1 - M(z))$$

Two cases will now be considered.

Case 1: Unit Step Input

If $r(z) = \dfrac{z}{z-1}$ then $e(\infty) = \lim_{z \to 1} z.(1 - M(z))$.

It is required that $e(\infty) = 0$. This will be achieved if, say:

$$z.(1 - M(z)) = z - 1$$
$$1 - M(z) = 1 - z^{-1}$$
$$M(z) = z^{-1}$$

The closed loop response is thus:

$$y(z) = M(z).r(z)$$
$$= z^{-1}.\frac{z}{z-1}$$
$$= z^{-1} + z^{-2} + z^{-3} + \ldots$$

Inverse transforming gives:

$$y(kT) = 0 \quad \text{if } k = 0$$
$$y(kT) = 1 \quad \text{if } k = 1, 2, 3, \ldots$$

which is depicted in Figure 78.3.

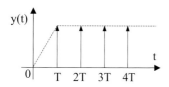

Fig. 78.3 The deadbeat method: closed loop response to a step input

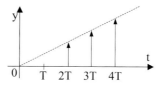

Fig. 78.4 The deadbeat method: closed loop response to a ramp input

The compensator required is given by Equation 78.4:

$$C(z) = \frac{M(z)}{(1 - M(z)) \cdot G(z)}$$

$$= \frac{z^{-1}}{(1 - z^{-1}) \cdot G(z)} = \frac{1}{(z - 1) \cdot G(z)}$$

Case 2: Unit Ramp Input

If $r(z) = \dfrac{Tz}{(z-1)^2}$ then

$$e(\infty) = \lim_{z \to 1} \frac{Tz}{(z-1)} (1 - M(z)).$$

A value of $e(\infty) = 0$ can be achieved if, say:

$$\frac{z}{(z-1)} \cdot (1 - M(z)) = \frac{z-1}{z}$$

$$1 - M(z) = \frac{(z-1)^2}{z^2} = 1 - 2z^{-1} + z^{-2}$$

$$M(z) = z^{-1}(2 - z^{-1})$$

The closed loop response is thus:

$$y(z) = M(z) \cdot r(z)$$

$$= \frac{z^{-1}(2 - z^{-1}) \cdot Tz}{(z-1)^2} = \frac{T \cdot (2 - z^{-1})}{z^2 - 2z + 1}$$

$$= T \cdot (2z^{-2} + 3z^{-3} + 4z^{-4} + \ldots)$$

Inverse transforming gives:

$$y(kT) = 0 \quad \text{if } k = 0, 1$$
$$y(kT) = kT \quad \text{if } k = 2, 3, 4, \ldots$$

which is depicted in Figure 78.4.

The compensator required is again given by Equation 78.4:

$$C(z) = \frac{M(z)}{(1 - M(z)) \cdot G(z)}$$

$$= \frac{z^{-1}(2 - z^{-1})}{(1 - z^{-1}(2 - z^{-1})) \cdot G(z)} = \frac{2z + 1}{(z-1)^2 \cdot G(z)}$$

As demonstrated in the two cases, the deadbeat design is optimal only for the specific inputs considered. For other inputs the response may well be unsatisfactory. Furthermore, the method gives minimal response at the sampling instants only. In practice there is often a lot of inter-sample ripple which may lead to violent overshoot, *etc.* Again, there is the need to know G(s). However, unlike Dahlin's method, this method is extremely sensitive to small errors in the plant model. It should also be noted that, for a given M(z), the only parameter that can be adjusted in the deadbeat method is the sampling period.

For these various reasons, the deadbeat method is seldom used in process control, despite its supposed popularity in other areas of application.

78.5 Direct Programs

Figure 78.5 depicts the pulse transfer function of an impulse compensator.

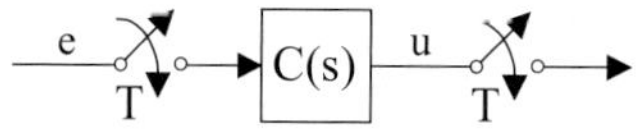

Fig. 78.5 Pulse transfer function of an impulse compensator

Any such compensator may be articulated in terms of a time series expansion for C(z):

$$C(z) = \frac{u(z)}{e(z)} = \frac{a_0 + a_1 z^{-1} + a_2 z^{-2} + \ldots + a_m z^{-m}}{b_0 + b_1 z^{-1} + b_2 z^{-2} + \ldots + b_n z^{-n}}$$

(78.6)

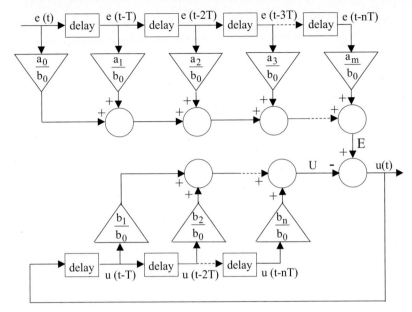

Fig. 78.6 Realisation of impulse compensator by direct programming

The power series must not contain any positive power in z, which would imply that an output signal could be produced before an error signal is applied. Thus $n \geq m$ for realisability. If $n = m$ then it is important that $b_0 \neq 0$.

Equation 78.6 may be rearranged:

$$\left(b_0 + b_1 z^{-1} + b_2 z^{-2} + \ldots + b_n z^{-n}\right).u(z)$$
$$= \left(a_0 + a_1 z^{-1} + a_2 z^{-2} + \ldots + a_m z^{-m}\right).e(z)$$

Inverse transform:

$$b_0.u(k) + b_1.u(k-1) + b_2.u(k-2) + \ldots$$
$$+ b_n.u(k-n)$$
$$= a_0.e(k) + a_1.e(k-1) + a_2.e(k-2) + \ldots$$
$$+ a_m.e(k-m)$$

Rearrange:

$$u(k) = \frac{1}{b_0} \sum_{i=0}^{m} a_i.e(k-i) - \frac{1}{b_0} \sum_{j=1}^{n} b_j.u(k-j)$$
$$= E - U$$

Thus the current value of the output depends on the current and previous values of the error, and on previous values of the output. The structure of the software for the direct programming technique is best illustrated in flow diagram form as depicted in Figure 78.6.

78.6 Cascade Programs

Any compensator of the form of Equation 78.6 may be factorised such that:

$$C(z) = \frac{a_0.\prod_{i=1}^{m}\left(1 + \alpha_i z^{-1}\right)}{b_0.\prod_{j=1}^{n}\left(1 + \beta_j z^{-1}\right)}$$

where the various $-\alpha_i$ and $-\beta_j$ are the zeros and poles respectively of C(z).

The compensator C(z) may be considered to be a series of simple pulse transfer functions cascaded as depicted in Figure 78.7.

$$e(z) \rightarrow \boxed{C_1(z)} \rightarrow \boxed{C_2(z)} \rightarrow \boxed{C_3(z)} \dashrightarrow \boxed{C_n(z)} \rightarrow u(z)$$

Fig. 78.7 Decomposition of compensator into product of factors

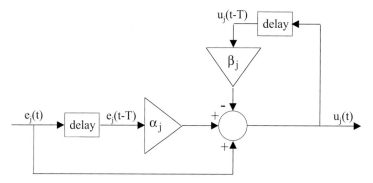

Fig. 78.8 Realisation of impulse compensator by cascade programming

Thus:

$$C(z) = \prod_{j=1}^{n} C_j(z)$$

where, for $n > m$:

$$C_1(z) = \frac{a_0 \cdot (1 + \alpha_1 z^{-1})}{b_0 \cdot (1 + \beta_1 z^{-1})} \quad \text{if } j = 1.$$

$$C_j(z) = \frac{(1 + \alpha_j z^{-1})}{(1 + \beta_j z^{-1})} \quad \text{for } j = 2, 3, \ldots m.$$

and:

$$C_j(z) = \frac{1}{(1 + \beta_j z^{-1})} \quad \text{for } j = m+1, m+2, \ldots n.$$

Each $C_j(z)$ may be implemented with software as, for example, in Figure 78.8.

Analysis yields:

$$u_j(z) = (1 + \alpha_j z^{-1}) \cdot e_j(z) - \beta_j z^{-1} \cdot u_j(z)$$

whence:

$$C_j(z) = \frac{u_j(z)}{e_j(z)} = \frac{1 + \alpha_j z^{-1}}{1 + \beta_j z^{-1}}$$

78.7 Parallel Programs

The same $C(z)$ of Equation 78.6 may be split into partial fractions such that:

$$C(z) = \frac{1}{b_0} \sum_{j=1}^{n} \frac{\alpha_j}{(1 + \beta_j z^{-1})}$$

where the various $-\beta_j$ are the poles of $C(z)$.

The compensator $C(z)$ may be considered to be a parallel set of simple pulse transfer functions as depicted in Figure 78.9.

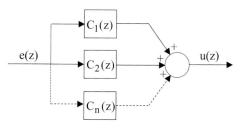

Fig. 78.9 Decomposition of compensator into sum of partial fractions

Each $C_j(z)$ may be implemented with software as depicted in Figure 78.10.

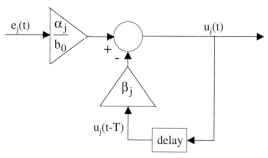

Fig. 78.10 Realisation of impulse compensator by parallel programming

Analysis yields:
$$u_j(z) = \frac{\alpha_j}{b_0}.e(z) - \beta_j z^{-1}.u_j(z)$$

whence:
$$C_j(z) = \frac{u_j(z)}{e(z)} = \frac{\alpha_j}{b_0\left(1 + \beta_j z^{-1}\right)}$$

78.8 Comments

Of the three methods, the direct method is the easiest in that no decomposition into factors or partial fractions is required prior to programming. Of the other two, the parallel method is marginally more efficient in computational effort than the cascade method. However, the principal attraction of the cascaded and parallel methods is that they lend themselves to a modular approach to software development. In particular, the provision of function blocks for realising cascaded and parallel elements within DCS systems leads itself to a more widespread application of impulse compensator designs.

Matrices and Vectors

79.1 Definitions and Notation
79.2 Determinants
79.3 Matrix Operations
79.4 Matrix Inversion
79.5 Use of Matrices
79.6 Differentiation of/by Vectors
79.7 Transforms of Vectors
79.8 Worked Example No 1
79.9 Eigenvalues and Eigenvectors
79.10 Worked Example No 2

This chapter introduces matrices and vectors. These are used extensively in the state-space descriptions of process plant and systems and in the design of multivariable controllers. A number of conventional matrix operations are covered and a variety of different types of matrix are defined. For a more comprehensive treatment the reader is referred to appropriate texts on mathematics, such as those by Jeffrey (2002) and Stroud (2003).

79.1 Definitions and Notation

A matrix is an array of subscripted elements organised in rows and columns as follows:

$$A = [a_{ij}] = \begin{bmatrix} a_{11} & a_{12} & a_{13} \\ a_{21} & a_{22} & a_{23} \\ a_{31} & a_{32} & a_{33} \end{bmatrix}$$

A matrix is normally denoted by an upper case letter and its elements contained within square brackets. Referring to the ijth element, the first subscript i always refers to the row number and the second subscript j to the column number.

The size and shape of a matrix are characterised by its dimensions, which indicate the number of rows and columns respectively. Thus A above is (3 × 3) square, it has the same number of rows as columns, and B below is (3 × 2) rectangular:

$$B = [b_{ij}] = \begin{bmatrix} b_{11} & b_{12} \\ b_{21} & b_{22} \\ b_{31} & b_{32} \end{bmatrix}$$

A matrix that has only one row or column is said to be a vector. C below is a row vector:

$$C = [c_j] = \begin{bmatrix} c_1 & c_2 & c_3 \end{bmatrix}$$

The elements of a matrix are usually the coefficients of an equation. These may be either numbers, real or complex, algebraic expressions or functions. In multivariable control the elements invariably are transfer functions.

In the special case of a vector whose elements are variables, it is normal practice to denote the vector with a lower case letter. Throughout this text such vectors will be emphasised by a bar underneath, as in the column vector below:

$$\underline{x} = [x_i] = \begin{bmatrix} x_1 \\ x_2 \\ x_3 \end{bmatrix}$$

A diagonal matrix is a square matrix in which all the elements other than those on the leading diagonal are zero:

$$A = \begin{bmatrix} a_{11} & 0 & 0 \\ 0 & a_{22} & 0 \\ 0 & 0 & a_{33} \end{bmatrix}$$

An objective in multivariable controller design is to produce systems with diagonal structures such that there is no interaction between the system's loops.

A diagonal matrix of particular significance in matrix manipulation is the identity matrix, normally denoted by I, in which all the non-zero elements have unity value:

$$I = \begin{bmatrix} 1 & 0 & 0 \\ 0 & 1 & 0 \\ 0 & 0 & 1 \end{bmatrix}$$

The zero matrix is one in which all the elements are zero.

The transpose of a matrix is one in which the rows and columns of the matrix have been interchanged and is normally denoted either by a prime or, as throughout this text, by a T superscript:

$$B^T = \begin{bmatrix} b_{11} & b_{21} & b_{31} \\ b_{12} & b_{22} & b_{32} \end{bmatrix}$$

$$\underline{x}^T = \begin{bmatrix} x_1 & x_2 & x_3 \end{bmatrix}$$

79.2 Determinants

For any square matrix there exists a determinant. The determinant of a matrix is usually denoted by the operator det and its elements contained within straight brackets:

$$\det(A) = |A| = \begin{vmatrix} a_{11} & a_{12} & a_{13} \\ a_{21} & a_{22} & a_{23} \\ a_{31} & a_{32} & a_{33} \end{vmatrix}$$

The value of det(A) is determined as follows:

$$\det(A) = a_{11} \begin{vmatrix} a_{22} & a_{23} \\ a_{32} & a_{33} \end{vmatrix} - a_{21} \begin{vmatrix} a_{12} & a_{13} \\ a_{32} & a_{33} \end{vmatrix}$$

$$+ a_{31} \begin{vmatrix} a_{12} & a_{13} \\ a_{22} & a_{23} \end{vmatrix}$$

$$= a_{11}(a_{22}a_{33} - a_{23}a_{32})$$
$$- a_{21}(a_{12}a_{32} - a_{13}a_{32})$$
$$+ a_{31}(a_{12}a_{23} - a_{13}a_{22})$$

For larger square matrices the determinant is evaluated by a similar process of reduction to determinants of size (2 × 2), with the same pattern of signs.

A matrix whose determinant is zero is said to be singular.

79.3 Matrix Operations

It is important to understand that whilst the elements of a matrix may each have a value, the matrix itself does not have a value as such.

For two matrices to be equal they must have the same number of rows and columns, *i.e.* the same dimensions, and all their corresponding elements must be identical. For example, if E is a (3 × 2) matrix then B = E if $b_{ij} = e_{ij}$ for all values of i and j.

Two matrices can be added if they have the same dimensions. Thus, for example:

$$B + E = \begin{bmatrix} b_{ij} + e_{ij} \end{bmatrix} = \begin{bmatrix} b_{11} + e_{11} & b_{12} + e_{12} \\ b_{21} + e_{21} & b_{22} + e_{22} \\ b_{31} + e_{31} & b_{32} + e_{32} \end{bmatrix}$$

Each element of B is added to the corresponding element of E. Similarly, for subtraction:

$$B - E = \begin{bmatrix} b_{ij} - e_{ij} \end{bmatrix}$$

Multiplication of a matrix by a scalar or a function gives a matrix in which each element is multiplied by that scalar or function. For example:

$$k.\underline{x}^T = \begin{bmatrix} k.x_1 & k.x_2 & k.x_3 \end{bmatrix}$$

$$f(s).I = \begin{bmatrix} f(s) & 0 & 0 \\ 0 & f(s) & 0 \\ 0 & 0 & f(s) \end{bmatrix}$$

Multiplication of one matrix by another is only possible between conformable matrices, *i.e.* the number of columns of the first matrix must be equal to the number of rows of the second. Otherwise multiplication of matrices is meaningless. By definition, if A is an n × m matrix and B is an m × p matrix, then the product AB is given by:

$$AB = \left[\sum_{k=1}^{m} a_{ik}.b_{kj}\right]$$

for values of i = 1, 2, ..., n and j = 1, 2, ..., p, where AB is an n × p matrix. For example,

if $D = \begin{bmatrix} d_1 & d_2 \end{bmatrix}$ and $\underline{u} = \begin{bmatrix} u_1 \\ u_2 \end{bmatrix}$

then $D.\underline{u} = [d_1 u_1 + d_2 u_2]$.

Similarly,

$$AB = \begin{bmatrix} a_{11}b_{11} + a_{12}b_{21} + a_{13}b_{31} & a_{11}b_{12} + a_{12}b_{22} + a_{13}b_{32} \\ a_{21}b_{11} + a_{22}b_{21} + a_{23}b_{31} & a_{21}b_{12} + a_{22}b_{22} + a_{23}b_{32} \\ a_{31}b_{11} + a_{32}b_{21} + a_{33}b_{31} & a_{31}b_{12} + a_{32}b_{22} + a_{33}b_{32} \end{bmatrix}$$

Note that even if A and B are conformable for AB they may well not be conformable for BA. For this reason it is essential that the correct order of matrices be preserved during matrix operations, hence the terms pre and post multiplication.

The following laws hold for matrix multiplication:

$$(AB).C = A.(BC)$$
$$(A + B).C = AC + BC$$
$$C.(A + B) = CA + CB$$

The transpose of a product is given by:

$$(AB)^T = B^T A^T$$

The inverse of a square matrix A is denoted either by inv(A) or by A^{-1} and is such that:

$$AA^{-1} = A^{-1}A = I$$

The following laws of inversion apply, provided that the matrices A and B are both square:

$$(AB)^{-1} = B^{-1} A^{-1}$$
$$(A^{-1})^{-1} = A$$
$$(A^{-1})^T = (A^T)^{-1}$$

Division of one matrix by another is meaningless. The equivalent operation is multiplication by the inverse of a matrix. For example, consider the system described by:

$$F.\underline{x} = B.\underline{u}$$

To solve this for the vector $\underline{x}$ both sides of the equation must be pre-multiplied by F^{-1}:

$$F^{-1}.F.\underline{x} = F^{-1}.B.\underline{u}$$
$$\underline{x} = F^{-1}.B.\underline{u}$$

Similarly, by post-multiplication, solving $\underline{x}F = B\underline{u}$ for $\underline{x}$ yields $\underline{x} = B.\underline{u}.F^{-1}$.

79.4 Matrix Inversion

Whilst using the inverse of a matrix is a fairly straightforward operation, the procedure for finding the inverse is both complicated and tedious. There are four steps involved:

1. Find the matrix of minors. If the ith row and jth column are deleted from an (n × n) matrix A the determinant of the resultant (n − 1 × n − 1) matrix is called the minor M_{ij} of A. The matrix of minors is thus given by:

$$\min(A) = [M_{ij}]$$

2. Find the matrix of cofactors. This involves changing the sign of alternate elements of the matrix of minors. Thus:

$$\text{cof}(A) = [(-1)^{i+j} M_{ij}]$$

3. Find the adjoint matrix. This is defined by:

$$\text{adj}(A) = [\text{cof}(A)]^T$$

4. Find the inverse matrix from:

$$\text{inv}(A) = \frac{\text{adj}(A)}{\det(A)}$$

which clearly requires that A be non-singular, i.e. that $\det(A) \neq 0$.

This is best illustrated by a numerical example.

If:
$$A = \begin{bmatrix} 1 & 2 & 0 \\ 3 & -1 & -2 \\ 1 & 0 & -3 \end{bmatrix}$$

then:
$$\min(A) = \begin{bmatrix} 3 & -7 & 1 \\ -6 & -3 & -2 \\ -4 & -2 & -7 \end{bmatrix}$$

and:
$$\text{cof}(A) = \begin{bmatrix} 3 & 7 & 1 \\ 6 & -3 & 2 \\ -4 & 2 & -7 \end{bmatrix}$$

so:
$$\text{adj}(A) = \begin{bmatrix} 3 & 6 & -4 \\ 7 & -3 & 2 \\ 1 & 2 & -7 \end{bmatrix}$$

hence:
$$\text{inv}(A) = \frac{1}{17}\begin{bmatrix} 3 & 6 & -4 \\ 7 & -3 & 2 \\ 1 & 2 & -7 \end{bmatrix}$$

A useful inversion to remember for convenience is that of a (2 × 2) matrix.

If:
$$A = \frac{1}{c}\begin{bmatrix} u & v \\ w & x \end{bmatrix}$$

then:
$$A^{-1} = \frac{c}{(ux - vw)}\begin{bmatrix} x & -v \\ -w & u \end{bmatrix}$$

Fortunately, it is not normally necessary to have to find the inverse of a matrix by hand, there are packages available for doing this.

79.5 Use of Matrices

Matrices provide a powerful means of representing and solving sets of simultaneous equations involving algebraic or transformed variables with real or complex numbers. The equations may be linear or differential, continuous or discrete. Consider the following simple example:

$$x + 2y = 4$$
$$3x - y - 2z = 7$$
$$x - 3z = 5$$

These equations may be presented in matrix form:
$$\begin{bmatrix} 1 & 2 & 0 \\ 3 & -1 & -2 \\ 1 & 0 & -3 \end{bmatrix}\begin{bmatrix} x \\ y \\ z \end{bmatrix} = \begin{bmatrix} 4 \\ 7 \\ 5 \end{bmatrix}$$

Premultiplying both sides of the equation by inv(A) yields:
$$\begin{bmatrix} x \\ y \\ z \end{bmatrix} = \frac{1}{17}\begin{bmatrix} 3 & 6 & -4 \\ 7 & -3 & 2 \\ 1 & 2 & -7 \end{bmatrix}\begin{bmatrix} 4 \\ 7 \\ 5 \end{bmatrix} = \begin{bmatrix} 2 \\ 1 \\ -1 \end{bmatrix}$$

The solution is thus $x = 2$, $y = 1$ and $z = -1$.

79.6 Differentiation of/by Vectors

More complex operations are carried out on matrices in much the same way as on scalar quantities, subject to both the rules of the operation and of matrix manipulation. For example, the derivative of a matrix is found by differentiating each element of the matrix. This operation is most commonly done on vectors of variables, as follows:

$$\underline{\dot{x}} = \frac{d}{dt}\underline{x} = \left[\frac{dx_i}{dt}\right] = \begin{bmatrix} \dot{x}_1 \\ \dot{x}_2 \\ \dot{x}_3 \end{bmatrix}$$

Less common, and certainly less obvious, is the differentiation of a scalar quantity with respect to a vector. Consider the scalar quantity S formed from the above (3 × 1) column vector $\underline{x}$:

$$S = \underline{x}^T.\underline{x} = x_1^2 + x_2^2 + x_3^2$$

Differentiating S with respect to each element of $\underline{x}$ in turn gives

$$\frac{dS}{dx_1} = 2x_1, \quad \frac{dS}{dx_2} = 2x_2, \quad \frac{dS}{dx_3} = 2x_3$$

Hence

$$\frac{dS}{d\underline{x}} = \begin{bmatrix} \frac{dS}{dx_1} \\ \frac{dS}{dx_2} \\ \frac{dS}{dx_3} \end{bmatrix} = 2 \begin{bmatrix} x_1 \\ x_2 \\ x_3 \end{bmatrix} = 2\underline{x}$$

The presence of matrices of constants can be treated in much the same way as in the differentiation of scalars involving constants. Suppose that K is a matrix of constants and both x and y are column vectors.

If $S = \underline{x}^T.K.\underline{y}$ then $\dfrac{dS}{d\underline{x}} = K.\underline{y}$

If $S = \underline{y}^T.K.\underline{x}$ then $\dfrac{dS}{d\underline{x}} = K^T.\underline{y}$

If $S = \underline{x}^T.K.\underline{x}$ then $\dfrac{dS}{d\underline{x}} = 2K.\underline{x}$ provided $K^T = K$

The latter may be demonstrated as follows:

$$S = \begin{bmatrix} x_1 & x_2 \end{bmatrix} \begin{bmatrix} k_{11} & k_{12} \\ k_{21} & k_{22} \end{bmatrix} \begin{bmatrix} x_1 \\ x_2 \end{bmatrix}$$

$$= k_{11}.x_1^2 + k_{12}.x_1.x_2 + k_{21}.x_1.x_2 + k_{22}.x_2^2$$

Differentiating:

$$\frac{dS}{dx_1} = 2k_{11}.x_1 + k_{12}.x_2 + k_{21}.x_2$$

$$\frac{dS}{dx_2} = k_{12}.x_1 + k_{21}.x_1 + 2k_{22}.x_2$$

Whence:

$$\frac{dS}{d\underline{x}} = \begin{bmatrix} \frac{dS}{dx_1} \\ \frac{dS}{dx_2} \end{bmatrix}$$

$$= \begin{bmatrix} k_{11} & k_{12} \\ k_{21} & k_{22} \end{bmatrix} \begin{bmatrix} x_1 \\ x_2 \end{bmatrix} + \begin{bmatrix} k_{11} & k_{21} \\ k_{12} & k_{22} \end{bmatrix} \begin{bmatrix} x_1 \\ x_2 \end{bmatrix}$$

$$= K.\underline{x} + K^T.\underline{x}$$

Provided that $K^T = K$ then $\dfrac{dS}{d\underline{x}} = 2K.\underline{x}$.

79.7 Transforms of Vectors

The Laplace transform of a vector is found by transforming each element of the vector, as follows:

$$L\{\underline{\dot{x}}(t)\} = L[\dot{x}_i(t)] = \begin{bmatrix} sx_1(s) - x_1(0) \\ sx_2(s) - x_2(0) \\ sx_3(s) - x_3(0) \end{bmatrix}$$

$$= s\underline{x}(s) - \underline{x}(0)$$

As will be seen in Chapter 80, the feedback control system of Figure 74.1, which was used as a vehicle for studying root locus, may be described in the generalised state-space form as follows:

$$\underline{\dot{x}} = A.\underline{x} + B.u \quad (79.1)$$
$$y = C.\underline{x}$$

Laplace transforming yields:

$$s\underline{x}(s) - \underline{x}(0) = A.\underline{x}(s) + B.u(s)$$
$$y(s) = C.\underline{x}(s)$$

Assuming zero initial conditions, this model may be solved by rearrangement and premultiplication as follows:

$$[sI - A].\underline{x}(s) = B.u(s) \quad (79.2)$$
$$\underline{x}(s) = [sI - A]^{-1}.B.u(s)$$
$$y(s) = C.[sI - A]^{-1}.Bu(s)$$
$$= C\frac{\mathrm{adj}[sI - A]}{\det[sI - A]}Bu(s)$$

provided that $[sI - A]$ is non-singular, i.e. $\det[sI - A] \neq 0$.

Note the care with which the correct order of the matrices has been preserved to ensure conformability.

79.8 Worked Example No 1

For the system of Figure 74.1:

$$[sI - A] = \begin{bmatrix} s & 0 & 0 \\ 0 & s & 0 \\ 0 & 0 & s \end{bmatrix} - \begin{bmatrix} 0 & 1 & 0 \\ -2 & -3 & -2K \\ 3 & 0 & -3 \end{bmatrix}$$

$$= \begin{bmatrix} s & -1 & 0 \\ 2 & s+3 & 2K \\ -3 & 0 & s+3 \end{bmatrix}$$

The four step procedure for finding the inverse of $[sI - A]$ yields:

$$\text{min}\,[sI - A] = \begin{bmatrix} s^2+6s+9 & 2s+6+6K & 3s+9 \\ -s-3 & s^2+3s & -3 \\ -2K & 2Ks & s^2+3s+2 \end{bmatrix}$$

$$\text{cof}\,[sI - A] = \begin{bmatrix} s^2+6s+9 & -2s-6-6K & 3s+9 \\ s+3 & s^2+3s & 3 \\ -2K & -2Ks & s^2+3s+2 \end{bmatrix}$$

$$\text{adj}\,[sI - A] = \begin{bmatrix} s^2+6s+9 & s+3 & -2K \\ -2s-6-6K & s^2+3s & -2Ks \\ 3s+9 & 3 & s^2+3s+2 \end{bmatrix}$$

$$\det[sI - A] = s(s^2+6s+9) - 2(-s-3) - 3(-2K)$$
$$= s^3 + 6s^2 + 11s + 6 + 6K$$

Substituting the values for $B^T = \begin{bmatrix} 0 & 2K & 0 \end{bmatrix}$ and $C = \begin{bmatrix} 1 & 0 & 0 \end{bmatrix}$ into Equation 79.2 gives:

$$y(s) = \frac{\begin{bmatrix} 1 & 0 & 0 \end{bmatrix}}{(s^3+6s^2+11s+6+6K)}$$
$$\times \begin{bmatrix} s^2+6s+9 & s+3 & -2K \\ -2s-6-6K & s^2+3s & -2Ks \\ 3s+9 & 3 & s^2+3s+2 \end{bmatrix}$$
$$\times \begin{bmatrix} 0 \\ 2K \\ 0 \end{bmatrix} u(s)$$

which yields the closed loop transfer function:

$$y(s) = \frac{\begin{bmatrix} 1 & 0 & 0 \end{bmatrix}}{(s^3+6s^2+11s+6+6K)}$$
$$\times \begin{bmatrix} 2K(s+3) \\ 2K(s^2+3s) \\ 6K \end{bmatrix} u(s)$$
$$= \frac{2K(s+3)}{(s^3+6s^2+11s+6+6K)} u(s)$$

Many other examples of the use of matrices for solving sets of differential equations, transformed or otherwise, are given in the following chapters.

79.9 Eigenvalues and Eigenvectors

The eigenvalues $\lambda_1, \lambda_2, \ldots \lambda_n$ of an n × n matrix A are the roots of the equation:

$$\det[\lambda I - A] = 0 \quad (79.3)$$

This is the form of the characteristic equation for multivariable control systems. The roots of this equation may be either real or complex, occurring in conjugate pairs. Just as with single-input single-output systems, these roots determine the nature of the system's stability.

The eigenvectors $\underline{r}_1, \underline{r}_2, \ldots \underline{r}_n$ of an n × n matrix A satisfy the equation:

$$[\lambda_j I - A]\,\underline{r}_j = 0 \quad (79.4)$$

where each $\underline{r}_j$ is an n × 1 vector.

79.10 Worked Example No 2

For the system of the previous worked example, if $K = 0.038$ then Equation 79.3 gives:

$$\det[\lambda I - A] = \lambda^3 + 6\lambda^2 + 11\lambda + 6.228 = 0$$

From Table 74.1 it can be seen that this has three real and distinct roots, its eigenvalues:

$$\lambda = -1.14, -1.76 \text{ and } -3.1$$

Consider the first of these. Substitution of $\lambda_1 = -1.14$ into Equation 79.4 gives:

$$\left(\begin{bmatrix} -1.14 & 0 & 0 \\ 0 & -1.14 & 0 \\ 0 & 0 & -1.14 \end{bmatrix} - \begin{bmatrix} 0 & 1 & 0 \\ -2 & -3 & -0.076 \\ 3 & 0 & -3 \end{bmatrix} \right)$$
$$\times \begin{bmatrix} r_{11} \\ r_{21} \\ r_{31} \end{bmatrix} = 0$$

Hence:

$$\begin{bmatrix} -1.14 & -1 & 0 \\ 2 & 1.86 & 0.076 \\ -3 & 0 & 1.86 \end{bmatrix} \begin{bmatrix} r_{11} \\ r_{21} \\ r_{31} \end{bmatrix} = 0$$

Expansion gives:

$$-1.14\, r_{11} - r_{21} = 0$$
$$2\, r_{11} + 1.86\, r_{21} + 0.076\, r_{31} = 0$$
$$-3\, r_{11} + 1.86\, r_{31} = 0$$

This set of equations is indeterminate: there is no unique solution but rather an infinity of solutions. However, with eigenvectors, it is the direction that counts rather than the magnitude. The direction is determined by the ratios of the values of the vector's elements. So any combination of r_{i1} that satisfies the set of equations will suffice. Arbitrarily choosing the value of one of the elements will enable the others to be found.

Choose $r_{11} = 1.0$ gives:

$$\underline{r}_1 = \begin{bmatrix} 1.0 \\ -1.14 \\ 1.62 \end{bmatrix}$$

Similarly for $\lambda_2 = -1.76$

$$\underline{r}_2 = \begin{bmatrix} 1.0 \\ -1.76 \\ 2.42 \end{bmatrix}$$

and for $\lambda_3 = -3.1$

$$\underline{r}_3 = \begin{bmatrix} 1.0 \\ -3.10 \\ -30.3 \end{bmatrix}$$

The eigenvectors are usually assembled into a matrix R as follows:

$$R = \begin{bmatrix} \underline{r}_1 & \underline{r}_2 & \underline{r}_3 \end{bmatrix} = \begin{bmatrix} 1.0 & 1.0 & 1.0 \\ -1.14 & -1.76 & -3.10 \\ 1.62 & 2.42 & -30.3 \end{bmatrix}$$

Note that a value of $r_{11} = 1$ was chosen arbitrarily: this enabled the values of the other elements of $\underline{r}_1$ to be determined. As stated, it is the ratio of the values of the elements of an eigenvector that matters: an eigenvector may be scaled by any factor.

State Space Analysis

80.1 Second Order System
80.2 n Dimensional State Space
80.3 Conversion of Transfer Functions
80.4 Worked Example No 1
80.5 The Transition Matrix
80.6 Worked Example No 2
80.7 Single Input Single Output Systems
80.8 Multiple Input Multiple Output Systems
80.9 Multivariable Control Systems
80.10 Similarity Transformation
80.11 Diagonalisation
80.12 Worked Example No 3
80.13 System Modes
80.14 Worked Example No 4

The techniques of control system analysis and design used in previous chapters, such as frequency response and root locus, are essentially concerned with single-input single-output (SISO) systems. However, for multiple-input multiple-output (MIMO) systems, generally referred to as multivariable systems, more powerful design techniques are necessary. These involve matrix manipulation for which state-space analysis provides the basis.

Although state-space techniques are primarily used with multivariable systems, they are fairly ubiquitous. They may be applied to SISO systems as well as to MIMO systems, used in either continous or discrete time, and with variables in real time or transformed into either Laplace or Z domains. They may also be applied in open and closed loop contexts, both to multivariable and multi-loop systems.

This chapter focuses on matrix techniques and state-space. However, these are useless unless models of the plant to be controlled are available in state-space form too. Several such models are developed in later chapters, in particular that of a blending system in Chapter 86, of a semi-batch reactor in Chapter 89 and of a distillation column in Chapter 90.

80.1 Second Order System

As an example, consider the second order system depicted in Figure 80.1.

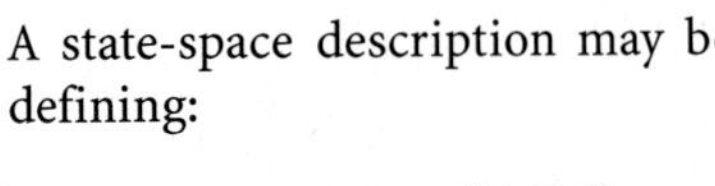

$$u \longrightarrow \boxed{\frac{\omega_n^2}{s^2 + 2\zeta\omega_n.s + \omega_n^2}} \longrightarrow y$$

Fig. 80.1 Transfer function of second order system

The dynamics are described by the equation:

$$\ddot{y} + 2\zeta\omega_n\dot{y} + \omega_n^2.y = \omega_n^2.u$$

A state-space description may be formulated by defining:

$$x_1 = y$$
$$x_2 = \dot{y}$$

Thus:

$$\dot{x}_1 = x_2$$
$$\dot{x}_2 = -\omega_n^2 . x_1 - 2\zeta\omega_n . x_2 + \omega_n^2 . u$$

which may be expressed in matrix form:

$$\begin{bmatrix} \dot{x}_1 \\ \dot{x}_2 \end{bmatrix} = \begin{bmatrix} 0 & 1 \\ -\omega_n^2 & -2\zeta\omega_n \end{bmatrix} \begin{bmatrix} x_1 \\ x_2 \end{bmatrix} + \begin{bmatrix} 0 \\ \omega_n^2 \end{bmatrix} u$$

This is the state equation of the system. The output equation is:

$$y = \begin{bmatrix} 1 & 0 \end{bmatrix} \begin{bmatrix} x_1 \\ x_2 \end{bmatrix}$$

These would normally be denoted in the generalised form:

$$\underline{\dot{x}} = A.\underline{x} + B.u \qquad (80.1)$$
$$y = C.\underline{x}$$

where:
- x_1 and x_2 are the state variables,
- $\underline{x}$ is the state vector,
- u is a scalar input,
- y is a scalar output, and
- A, B and C are matrices of constant coefficients.

The A matrix is often referred to as the system matrix.

For a given set of initial conditions, a unique plot of x_1 vs x_2 may be drawn, with time as a parameter. The locus of the point x_1, x_2 is known as a trajectory and the two dimensional state-space is referred to as the phase plane.

80.2 n Dimensional State Space

Consider an nth order SISO system whose transfer function is as depicted in Figure 80.2.

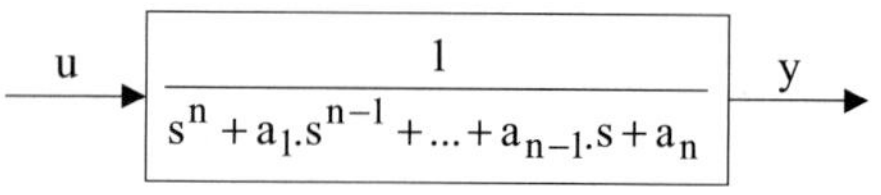

Fig. 80.2 Transfer function of nth order system

The dynamics are described by the equation:

$$\frac{d^n y}{dt^n} + a_1 \frac{d^{n-1} y}{dt^{n-1}} + a_2 \frac{d^{n-2} y}{dt^{n-2}} + \cdots + a_{n-2} \frac{d^2 y}{dt^2} + a_{n-1} \frac{dy}{dt} + a_n y = u$$

Any nth order differential equation of this type may be reduced to a set of n first order differential equations by defining:

$$x_1 = y$$
$$x_2 = \dot{y}$$
$$\vdots$$
$$x_n = y^{n-1}$$

Thus:

$$\dot{x}_1 = x_2$$
$$\dot{x}_2 = x_3$$
$$\vdots$$
$$\dot{x}_{n-1} = x_n$$
$$\dot{x}_n = -a_n x_1 - a_{n-1} x_2 - \ldots - a_1 x_n + u$$

Whence:

$$\begin{bmatrix} \dot{x}_1 \\ \dot{x}_2 \\ \vdots \\ \dot{x}_{n-1} \\ \dot{x}_n \end{bmatrix} = \begin{bmatrix} 0 & 1 & 0 & \cdots & 0 \\ 0 & 0 & 1 & \cdots & 0 \\ \vdots & \vdots & \vdots & \vdots & \vdots \\ 0 & 0 & 0 & \cdots & 1 \\ -a_n & -a_{n-1} & -a_{n-2} & \cdots & -a_1 \end{bmatrix} \times \begin{bmatrix} x_1 \\ x_2 \\ \vdots \\ x_{n-1} \\ x_n \end{bmatrix} + \begin{bmatrix} 0 \\ 0 \\ \vdots \\ 0 \\ 1 \end{bmatrix} u \qquad (80.2)$$

which yields the state equation:

$$\underline{\dot{x}} = A.\underline{x} + B.u$$

A system matrix of this particular form is known as the "companion" matrix.

The output equation is:

$$y = \begin{bmatrix} 1 & 0 & 0 & \cdots & 0 \end{bmatrix} \begin{bmatrix} x_1 \\ x_2 \\ \vdots \\ x_{n-1} \\ x_n \end{bmatrix}$$

which is again of the form:

$$y = C.\underline{x}$$

For a given set of initial conditions, the set of state equations describe a unique trajectory through an n dimensional space. Clearly this is difficult to visualise for $n > 3$.

80.3 Conversion of Transfer Functions

Particularly useful is the ability to convert a SISO model in transfer function form into state-space form. Consider the transfer function:

$$\frac{y(s)}{u(s)} = \frac{a_2.s^2 + a_1.s + a_0}{s^3 + b_2.s^2 + b_1.s + b_0}$$

This may be decomposed into numerator and denominator transfer functions in series as depicted in Figure 80.3, where $v(s)$ is some intermediate variable.

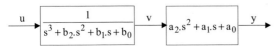

Fig. 80.3 Decomposition into denominator and numerator transfer functions

Inverse transforming the denominator transfer function yields:

$$\dddot{v} + b_2.\ddot{v} + b_1.\dot{v} + b_0.v = u$$

Define:

$$x_1 = v$$
$$x_2 = \dot{v}$$
$$x_3 = \ddot{v}$$

which yields the following equations:

$$\dot{x}_1 = x_2$$
$$\dot{x}_2 = x_3$$
$$\dot{x}_3 = -b_0.x_1 - b_1.x_2 - b_2.x_3 + u$$

These may be assembled into state-space form:

$$\underline{\dot{x}} = \begin{bmatrix} 0 & 1 & 0 \\ 0 & 0 & 1 \\ -b_0 & -b_1 & -b_2 \end{bmatrix} \underline{x} + \begin{bmatrix} 0 \\ 0 \\ 1 \end{bmatrix} u = A.\underline{x} + B.u$$

Inverse transforming the numerator transfer function yields:

$$a_2.\ddot{v} + a_1.\dot{v} + a_0.v = y$$

Assuming again that $x_1 = v$ *etc.* yields the output equation:

$$a_2.x_3 + a_1.x_2 + a_0.x_1 = y$$

which may be put into matrix form:

$$y = \begin{bmatrix} a_0 & a_1 & a_2 \end{bmatrix} \underline{x} = C.\underline{x}$$

80.4 Worked Example No 1

This example is of the application of state-space to the SISO feedback control system depicted in Figure 80.4. Note that this is the same system that was considered in detail in Example 1 of Chapter 74 on root locus.

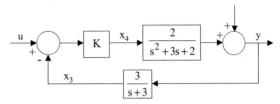

Fig. 80.4 Control system with third order open loop transfer function

For the second order element:

$$\ddot{y} + 3\dot{y} + 2y = 2x_4$$

Define:

$$x_1 = y$$
$$x_2 = \dot{y}$$

Note that the second state x_2 does not correspond to any signal within the control loop. It can be thought of as some internal variable related to the state of the process. Whence:

$$\dot{x}_1 = x_2$$
$$\dot{x}_2 = -2x_1 - 3x_2 + 2x_4$$

For the controller:

$$x_4 = K(u - x_3)$$

Thus

$$\dot{x}_2 = -2x_1 - 3x_2 - 2Kx_3 + 2Ku$$

For the feedback element:

$$\dot{x}_3 + 3x_3 = 3y$$
$$\dot{x}_3 = 3x_1 - 3x_3$$

These equations may be assembled in state-space form:

$$\begin{bmatrix} \dot{x}_1 \\ \dot{x}_2 \\ \dot{x}_3 \end{bmatrix} = \begin{bmatrix} 0 & 1 & 0 \\ -2 & -3 & -2K \\ 3 & 0 & -3 \end{bmatrix} \begin{bmatrix} x_1 \\ x_2 \\ x_3 \end{bmatrix} + \begin{bmatrix} 0 \\ 2K \\ 0 \end{bmatrix} u$$

$$y = \begin{bmatrix} 1 & 0 & 0 \end{bmatrix} \begin{bmatrix} x_1 \\ x_2 \\ x_3 \end{bmatrix}$$

Note that the system is third order, so it must have three states to describe it. However, the choice of states is arbitrary so great care must be taken in choosing appropriate ones. In general, there must be one state associated with each capacity for energy storage. The ability to measure certain states is another important consideration in multivariable design. In this example the following state vector could have been used but would clearly have resulted in a different model.

$$\underline{x} = \begin{bmatrix} x_1 & x_2 & x_4 \end{bmatrix}^T$$

80.5 The Transition Matrix

Consider the state equation

$$\underline{\dot{x}} = A\underline{x}$$

which is the simplest case, i.e. no inputs or outputs. The solution to this is, by inspection:

$$\underline{x} = e^{At} \cdot \underline{x}(0)$$

where $\underline{x}(0)$ is the initial condition vector and e^{At} is the matrix exponential, known as transition matrix Φ, defined as follows:

$$\Phi(t) = e^{At} = I + A.t + \frac{A^2 t^2}{2} + \frac{A^3 t^3}{3!} + \ldots \quad (80.3)$$

The transition matrix is so called because it "transfers" the state of the system at time 0 to time t. Of particular significance in control system design and analysis is the Laplace transform of Φ:

$$\Phi(s) = L\{e^{At}\}$$
$$= \frac{I}{s} + \frac{A}{s^2} + \frac{A^2}{s^3} + \frac{A^3}{s^4} + \ldots$$

$$[sI - A]\Phi(s) = [sI - A]$$
$$\times \left[\frac{I}{s} + \frac{A}{s^2} + \frac{A^2}{s^3} + \frac{A^3}{s^4} + \ldots\right]$$
$$= I^2 + \frac{A}{s} + \frac{A^2}{s^2} + \frac{A^3}{s^3} + \ldots$$
$$- \frac{A}{s} - \frac{A^2}{s^2} - \frac{A^3}{s^3} - \ldots$$
$$= I$$

Thus:

$$\Phi(s) = [sI - A]^{-1} \quad (80.4)$$

Whence:

$$\Phi(t) = e^{At} = L^{-1}\{[sI - A]^{-1}\}$$

80.6 Worked Example No 2

Find the transition matrix $\Phi(s)$ for the system of Worked Example No 1:

$$[sI - A] = \begin{bmatrix} s & 0 & 0 \\ 0 & s & 0 \\ 0 & 0 & s \end{bmatrix} - \begin{bmatrix} 0 & 1 & 0 \\ -2 & -3 & -2K \\ 3 & 0 & -3 \end{bmatrix}$$
$$= \begin{bmatrix} s & -1 & 0 \\ 2 & s+3 & 2K \\ -3 & 0 & s+3 \end{bmatrix}$$

$\Phi(s)$ is found from the inverse:

$$\Phi(s) = [sI - A]^{-1} = \frac{\text{adj}\,[sI - A]}{\det\,[sI - A]}$$

The adjoint and determinant for this system were found in Example 1 of Chapter 79:

$$\text{adj}\,[sI - A] = \begin{bmatrix} s^2 + 6s + 9 & s + 3 & -2K \\ -2s - 6 - 6K & s^2 + 3s & -2Ks \\ 3s + 9 & 3 & s^2 + 3s + 2 \end{bmatrix}$$

$$\det\,[sI - A] = s\left(s^2 + 6s + 9\right) - 2\,(-s - 3) - 3\,(-2K)$$
$$= s^3 + 6s^2 + 11s + 6 + 6K$$

Whence:

$$\Phi(s) = \frac{\begin{bmatrix} s^2 + 6s + 9 & s + 3 & -2K \\ -2s - 6 - 6K & s^2 + 3s & -2Ks \\ 3s + 9 & 3 & s^2 + 3s + 2 \end{bmatrix}}{s^3 + 6s^2 + 11s + 6 + 6K}$$

$\Phi(t)$ is found by assembling the various $\Phi_{ij}(t)$ into a matrix. The $\Phi_{ij}(t)$ are found individually. This entails finding the partial fractions of each:

$$\Phi_{ij}(s) = \frac{\text{adj}\,[sI - A]_{ij}}{\det\,[sI - A]}$$

and taking their inverse transforms. This is clearly a very tedious but, fortunately, seldom necessary process.

80.7 Single Input Single Output Systems

The state-space equations for a SISO system are as previously found:

$$\begin{aligned} \dot{\underline{x}} &= A.\underline{x} + B.u \\ y &= C.\underline{x} \end{aligned} \tag{80.1}$$

The solution to these was given in Equation 79.2 but is repeated for completeness. Assuming zero initial conditions:

$$\begin{aligned} [sI - A]\,.\underline{x}(s) &= B.u(s) \qquad (80.5) \\ \underline{x}(s) &= [sI - A]^{-1}.B.u(s) \\ &= \frac{\text{adj}\,[sI - A]}{\det\,[sI - A]}.B.u(s) \\ &= \Phi(s).B.u(s) \\ y(s) &= C.\Phi(s).B.u(s) \end{aligned}$$

provided that $[sI - A]$ is non-singular, *i.e.* $\det\,[sI - A] \neq 0$.

For the system of Worked Example No 1:

$$y(s) = \frac{\begin{bmatrix} 1 & 0 & 0 \end{bmatrix}}{s^3 + 6s^2 + 11s + 6 + 6K} \times \begin{bmatrix} s^2 + 6s + 9 & s + 3 & -2K \\ -2s - 6 - 6K & s^2 + 3s & -2Ks \\ 3s + 9 & 3 & s^2 + 3s + 2 \end{bmatrix} \times \begin{bmatrix} 0 \\ 2K \\ 0 \end{bmatrix} u(s)$$

When multiplied out this yields:

$$y(s) = \frac{2K\,(s + 3)}{s^3 + 6s^2 + 11s + 6 + 6K}\,u(s)$$

which is the closed loop transfer function as depicted in Figure 80.5.

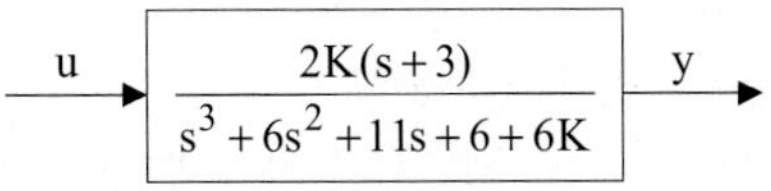

Fig. 80.5 Closed loop transfer function of the same system

This, of course, is the same closed loop transfer function that would have been obtained by applying conventional block diagram algebra to Figure 80.4. Note, in particular, that the denominator of the CLTF forms the characteristic equation:

$$\det [sI - A] = 0 \quad (80.6)$$

80.8 Multiple Input Multiple Output Systems

The usual state-space description of a MIMO system is of the form

$$\dot{\underline{x}} = A.\underline{x} + B.\underline{u} \quad (80.7)$$
$$\underline{y} = C.\underline{x} + D.\underline{u}$$

or, in discrete time domain:

$$\underline{x}(k+1) = A.\underline{x}(k) + B.\underline{u}(k) \quad (80.8)$$
$$\underline{y}(k) = C.\underline{x}(k) + D.\underline{u}(k)$$

where the input, output and state vectors, $\underline{u}$, $\underline{y}$ and $\underline{x}$ are of dimensions l, m and n, and the matrices A, B, C and D have dimensions $(n \times n), (n \times l), (m \times n)$ and $(m \times l)$ respectively.

The time domain block diagram representation of these state equations is depicted in Figure 80.6 in which all of the signals are vectors.

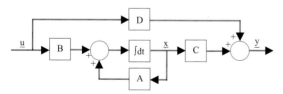

Fig. 80.6 Time domain representation of the state equations

The solution, as usual, is obtained by transformation:

$$s.\underline{x}(s) - \underline{x}(0) = A.\underline{x}(s) + B.\underline{u}(s)$$
$$[sI - A].\underline{x}(s) = \underline{x}(0) + B.\underline{u}(s)$$
$$\underline{x}(s) = [sI - A]^{-1}[\underline{x}(0) + B.\underline{u}(s)]$$
$$= \Phi(s).\underline{x}(0) + \Phi(s).B.\underline{u}(s)$$

However

$$\underline{y}(s) = C.\underline{x}(s) + D.\underline{u}(s).$$

Hence

$$\underline{y}(s) = C.\Phi(s).\underline{x}(0) + C.\Phi(s).B.\underline{u}(s) + D.\underline{u}(s) \quad (80.9)$$

Figure 80.7 is the equivalent block diagram representation, with $\underline{x}(0) = 0$ and $D = 0$.

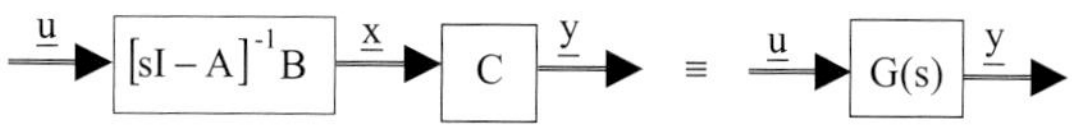

$$\underline{u} \rightarrow \boxed{[sI-A]^{-1}B} \xrightarrow{\underline{x}} \boxed{C} \xrightarrow{\underline{y}} \equiv \underline{u} \rightarrow \boxed{G(s)} \xrightarrow{\underline{y}}$$

Fig. 80.7 Matrix representation of the state equations

Again, note that the denominator of every term $\Phi_{i,j}(s)$ in the equation for $\underline{y}$ is the same.

These denominators form the characteristic equation

$$\det [sI - A] = 0 \quad (80.6)$$

In practice, all the useful information can be extracted from Equation 80.9 directly. However, if the inverse is required, it is obtained using the convolution integral, Equation 76.2, as follows:

$$\underline{y}(t) = C.e^{At}.\underline{x}(0) + C.L^{-1}\{\Phi(s).B.\underline{u}(s)\}$$
$$+ D.\underline{u}(t)$$
$$= C.e^{At}.\underline{x}(0) + C \int_0^t e^{A(t-\tau)}.B.\underline{u}(\tau)d\tau$$
$$+ D.\underline{u}(t) \quad (80.10)$$

80.9 Multivariable Control Systems

Block diagram algebra techniques may be applied to multivariable closed loop systems of the form of Figure 80.8 as follows

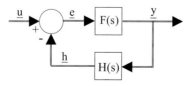

Fig. 80.8 Block diagram representation of multivariable control system

$$\underline{y}(s) = F(s).\underline{e}(s)$$
$$= F(s)[\underline{u}(s) - \underline{h}(s)]$$
$$= F(s)[\underline{u}(s) - H(s).\underline{y}(s)]$$
$$[I + F(s).H(s)].\underline{y}(s) = F(s).\underline{u}(s)$$
$$\underline{y}(s) = [I + F(s).H(s)]^{-1}.F(s).\underline{u}(s)$$
$$= G(s).\underline{u}(s) \quad (80.11)$$

The "transfer" matrix $G(s)$ of a MIMO system relates the output $\underline{y}(s)$ to the input $\underline{u}(s)$.

$G(s)$ is an $m \times l$ matrix. It contains ml transfer functions. The ith output is related to the jth input by the element $g_{ij}(s)$.

The design of multivariable control systems is considered in more detail in Chapter 81.

80.10 Similarity Transformation

In the analysis of multivariable systems, it is occasionally convenient to change the co-ordinates in state-space. Consider the system:

$$\underline{\dot{x}} = A.\underline{x} + B.\underline{u}$$
$$\underline{y} = C.\underline{x}$$

Let:
$$\underline{x} = R\underline{z} \quad (80.12)$$

Substituting:
$$R.\underline{\dot{z}} = A.R.\underline{z} + B.\underline{u}$$
$$\underline{\dot{z}} = R^{-1}.A.R.\underline{z} + R^{-1}.B.\underline{u}$$

Transform, assuming zero initial conditions:
$$s\underline{z}(s) = R^{-1}A.R.\underline{z}(s) + R^{-1}B.\underline{u}(s)$$
$$[s.I - R^{-1}A.R].\underline{z}(s) = R^{-1}B.\underline{u}(s)$$
$$\underline{z}(s) = [s.I - R^{-1}A.R]^{-1} R^{-1}B.\underline{u}(s)$$

Whence:
$$\underline{y}(s) = C.R.\underline{z}(s) \quad (80.13)$$
$$= C.R[s.I - R^{-1}A.R]^{-1} R^{-1}B.\underline{u}(s)$$

The characteristic equation of this system is thus:
$$\det[s.I - R^{-1}A.R] = 0 \quad (80.14)$$

It is important to appreciate that the eigenvalues of the system are unchanged by such a transformation. This is because

$$\det[\lambda.I - R^{-1}A.R] = |\lambda.R^{-1}R - R^{-1}A.R|$$
$$= |R^{-1}[\lambda.I - A].R|$$
$$= |R^{-1}|.|\lambda.I - A|.|R|$$
$$= |R^{-1}|.|R|.|\lambda.I - A|$$
$$= |R^{-1}R|.|\lambda.I - A|$$
$$= |\lambda.I - A|$$

80.11 Diagonalisation

Diagonalisation is a technique used in the analysis and design of multivariable systems. It is essentially concerned with transformation of the system matrix into the so-called "canonical" form in which the eigenvalues lie along the diagonal.

Consider the companion matrix of Equation 80.2. Suppose that it has real and distinct eigenvalues $\lambda_1, \lambda_2, \ldots \lambda_n$.

Provided R is of the form below, such a system matrix may be diagonalised using the similarity transformation $\underline{x} = R.\underline{z}$:

$$R = \begin{bmatrix} 1 & 1 & \cdots & 1 \\ \lambda_1 & \lambda_2 & \cdots & \lambda_n \\ \lambda_1^2 & \lambda_2^2 & \cdots & \lambda_n^2 \\ \vdots & \vdots & \vdots & \vdots \\ \lambda_1^{n-1} & \lambda_2^{n-1} & \cdots & \lambda_n^{n-1} \end{bmatrix} \quad (80.15)$$

This yields the so called modal decomposition:

$$\Lambda = R^{-1}A.R = \begin{bmatrix} \lambda_1 & 0 & \cdots & 0 \\ 0 & \lambda_2 & \cdots & 0 \\ \vdots & \vdots & \vdots & \vdots \\ 0 & 0 & \cdots & \lambda_n \end{bmatrix} \quad (80.16)$$

However, in general, the system matrix is not of the form of the companion matrix and diagonalisation has to be realised by means of a similarity transformation involving its eigenvectors, as follows.

80.12 Worked Example No 3

Real Eigenvalues

Consider again the system of Worked Example No 1. If K = 0.038 then the system matrix becomes

$$A = \begin{bmatrix} 0 & 1 & 0 \\ -2 & -3 & -0.076 \\ 3 & 0 & -3 \end{bmatrix}$$

From Table 74.1 it can be seen that if K = 0.038 there are three real and distinct eigenvalues for which the system has an overdamped stable response to step inputs:

$$\lambda = -1.14, \ -1.76 \text{ and } -3.1$$

It is shown in Example 2 of Chapter 79 that these eigenvalues correspond to the matrix of eigenvectors:

$$R = \begin{bmatrix} 1.0 & 1.0 & 1.0 \\ -1.14 & -1.76 & -3.10 \\ 1.62 & 2.42 & -30.3 \end{bmatrix}$$

Similarity transformation yields the modal decomposition:

$$\Lambda = R^{-1}.A.R = \begin{bmatrix} -1.14 & 0 & 0 \\ 0 & -1.76 & 0 \\ 0 & 0 & -3.10 \end{bmatrix}$$

Complex Eigenvalues

Consider yet again the system of Example 1. If K = 1.0 then the system matrix becomes:

$$A = \begin{bmatrix} 0 & 1 & 0 \\ -2 & -3 & -2 \\ 3 & 0 & -3 \end{bmatrix}$$

From Table 74.1 it can be seen that if K = 1.0 there are three distinct eigenvalues, one real and two complex, for which the system has an underdamped stable response to step inputs:

$$\lambda = -4.00 \text{ and } -1.00 \pm j1.414$$

The matrix of eigenvectors for this system is:

$$R = \begin{bmatrix} 1.00 & 1.00 & 1.00 \\ -4.00 & -1.00 + j1.414 & -1.00 - j1.414 \\ -3.00 & 1.00 - j0.707 & 1.00 + j0.707 \end{bmatrix}$$

Bearing in mind that eigenvectors are arbitrary, to the extent that it is the values of the elements relative to each other within the vector that counts, R may be rearranged as follows:

$$R = \begin{bmatrix} 1.00 & 1.00 & 0 \\ -4.00 & -1.00 & -1.414 \\ -3.00 & 1.00 & 0.707 \end{bmatrix}$$

Note that the real and imaginary parts of the complex eigenvectors have been grouped into separate vectors. In effect, it is the projections of the complex vectors onto the real and imaginary axes that form the second and third columns respectively of R. This respects the information held in the original vectors but simplifies R. Using this R matrix, similarity transformation yields the modal decomposition:

$$\Lambda = R^{-1}.A.R = \begin{bmatrix} -4.00 & 0 & 0 \\ 0 & -1.00 & -1.414 \\ 0 & 1.414 & -1.00 \end{bmatrix}$$

Note that the real parts of the complex eigenvalues lie on the diagonal of Λ and the imaginary parts are off diagonal.

80.13 System Modes

The physical significance of the eigenvalues is that they correspond to the possible forms of dynamic behaviour of a MIMO system, known as its modes. The roots of the characteristic equation are the only factors which can occur in the denominators of the elements of the transfer matrix. However, they do not necessarily all occur due to cancellation with common factors in their numerators. Therefore, to determine the full set of modes of the system, it is necessary to carry out the similarity transformation to establish the modal decomposition Λ.

For each real and distinct eigenvalue λ, which is the root $s = \lambda$ of the characteristic equation, there is an element of value λ which lies on the diagonal of the similarity transform Λ. For each distinct pair of complex eigenvalues $s = \sigma \pm j\omega$ there is a 2×2 block on the diagonal of Λ. For example:

$$\Lambda = R^{-1}A.R = \begin{bmatrix} \lambda & 0 & 0 \\ 0 & \sigma & \omega \\ 0 & -\omega & \sigma \end{bmatrix} \quad (80.17)$$

The matrix exponential $\exp(\Lambda.t)$ may be found from power series expansions:

$$e^{\Lambda.t} = I + \Lambda t + \frac{\Lambda^2 t^2}{2} + \frac{\Lambda^3 t^3}{3!} + \ldots \quad (80.3)$$

$$= \begin{bmatrix} e^{\lambda t} & 0 & 0 \\ 0 & e^{\sigma t}\cos\omega t & -e^{\sigma t}\sin\omega t \\ 0 & e^{\sigma t}\sin\omega t & e^{\sigma t}\cos\omega t \end{bmatrix} \quad (80.18)$$

Clearly negative real eigenvalues λ or negative real parts σ of complex eigenvalues correspond to decaying exponentials which are inherently stable. It only needs one such value to be positive for the system as a whole to be unstable. Only values of $\omega > 0$ are meaningful and correspond to oscillatory behaviour. If $\sigma = 0$ then the complex eigenvalue corresponds to a sustained oscillation of frequency ω.

Hence, by transforming each element:

$$[sI - \Lambda]^{-1} = L\{e^{\Lambda t}\} \quad (80.19)$$

$$= \begin{bmatrix} \frac{1}{s-\lambda} & 0 & 0 \\ 0 & \frac{s-\sigma}{(s-\sigma)^2+\omega^2} & \frac{-\omega}{(s-\sigma)^2+\omega^2} \\ 0 & \frac{\omega}{(s-\sigma)^2+\omega^2} & \frac{s-\sigma}{(s-\sigma)^2+\omega^2} \end{bmatrix}$$

In the above explanations it is assumed that the eigenvalues are distinct. If repeated eigenvalues exist, the standard form of Λ is more complicated, the Jordan form, and the effect on $\Phi(t)$ is that the exponential and sinusoidal terms may have coefficients which are polynomials in t.

Analysis of the modes of a system might, at first sight, appear something of an academic exercise. However, it is in design work that their principal benefits are realised. Being able to articulate the desired performance of a multivariable system by specifying values of λ, σ and ω in, say, Equation 80.17, enables the controller itself to be specified by means of block diagram algebra.

80.14 Worked Example No 4

Consider the following SISO system:

$$A = \begin{bmatrix} 0 & -1 & 0 \\ 0 & -1 & 2 \\ 1 & 0 & -1 \end{bmatrix} \quad B = \begin{bmatrix} 0 \\ 1 \\ 0 \end{bmatrix} \quad C = \begin{bmatrix} 0 & 0 & -1 \end{bmatrix}$$

Its eigenvalues are $\lambda = -2$ and $\lambda = \pm j$ for which the eigenvectors are $\begin{bmatrix} 1 \\ 2 \\ -1 \end{bmatrix}$ and $\begin{bmatrix} \pm 2j \\ 2 \\ 1 \pm j \end{bmatrix}$.

For similarity transformation,

$$R = \begin{bmatrix} 1 & 0 & 2 \\ 2 & 2 & 0 \\ -1 & 1 & 1 \end{bmatrix}$$

whence

$$\Lambda = R^{-1}AR = \begin{bmatrix} -2 & 0 & 0 \\ 0 & 0 & 1 \\ 0 & -1 & 0 \end{bmatrix}.$$

Note that the first column of R is the eigenvector corresponding to the real eigenvalue and the other two columns are the real and imaginary parts of the other eigenvectors.

The matrix exponential $\exp(\Lambda t)$ may be found from power series expansions:

$$e^{\Lambda.t} = I + \Lambda t + \frac{\Lambda^2 t^2}{2} + \frac{\Lambda^3 t^3}{3!} + \ldots \quad (80.3)$$

$$= \begin{bmatrix} 1 & 0 & 0 \\ 0 & 1 & 0 \\ 0 & 0 & 1 \end{bmatrix} + \begin{bmatrix} -2 & 0 & 0 \\ 0 & 0 & 1 \\ 0 & -1 & 0 \end{bmatrix} t$$

$$+ \frac{1}{2!} \begin{bmatrix} 4 & 0 & 0 \\ 0 & -1 & 0 \\ 0 & 0 & -1 \end{bmatrix} t^2$$

$$= \begin{aligned} &+ \frac{1}{3!} \begin{bmatrix} -8 & 0 & 0 \\ 0 & 0 & -1 \\ 0 & 1 & 0 \end{bmatrix} t^3 + \ldots \\ &\begin{bmatrix} 1 - 2t + \frac{(2t)^2}{2!} - \frac{(2t)^3}{3!} + \ldots & 0 & 0 \\ 0 & 1 - \frac{t^2}{2!} + \ldots & t - \frac{t^3}{3!} + \ldots \\ 0 & -t + \frac{t^3}{3!} - \ldots & 1 - \frac{t^2}{2!} + \ldots \end{bmatrix} \end{aligned}$$

Whence:

$$e^{At} = \begin{bmatrix} e^{-2t} & 0 & 0 \\ 0 & \cos(t) & \sin(t) \\ 0 & -\sin(t) & \cos(t) \end{bmatrix}$$

The modes of the system are thus a combination of decaying exponential and pure sinusoidal responses. This is consistent with its open loop transfer function which, since the system is only SISO, can be found readily:

$$G(s) = C[s.I - A]^{-1}.B = \frac{1}{(s+2)(s^2+1)}$$

Multivariable Control

Chapter 81

81.1 Case Study
81.2 Compensator Concept
81.3 Control System Model
81.4 Compensator Design
81.5 Worked Example No 1
81.6 Decouplers
81.7 Sampled Data Model
81.8 Impulse Compensator Design
81.9 Worked Example No 2
81.10 Sampled Data Decoupler
81.11 Comments

Some of the techniques of state-space were introduced in Chapter 80. The emphasis there was principally on the analysis of the behaviour of multivariable systems. This chapter introduces the use of matrix techniques, and diagonalisation in particular, for design purposes. A case study based upon a blending system is used as a realistic, but simple, example to demonstrate the principles. The approach used is known as internal model control because the inverse of the model of the plant to be controlled is embedded in the design of the controller. This gives an insight to the problems of the design of large multivariable control systems which provides a basis for considering related techniques such as sensitivity analysis in Chapter 111, state feedback in Chapter 112 and model predictive control in Chapter 117.

81.1 Case Study

An in-line blending system is depicted in Figure 81.1.

In summary, two streams f_1 and f_2 are blended to produce a third stream f_0 of concentration $c^\#$. Its

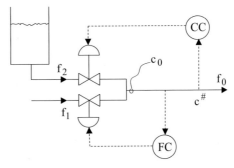

Fig. 81.1 An in line blending system

dynamics are fairly fast because the blending system is in-line. Also it is a highly interactive system. Any change in either flow f_1 or f_2 will affect both f_0 and $c^\#$. Note that changes in flow are propagated throughout the system instantaneously, whereas changes in concentration have an associated time delay.

A model of this system in matrix form is developed in Chapter 86 using deviation variables:

$$\begin{bmatrix} f_0(s) \\ c^\#(s) \end{bmatrix} = \begin{bmatrix} 1 & 1 \\ -K_1.e^{-Ls} & K_2.e^{-Ls} \end{bmatrix} \begin{bmatrix} f_1(s) \\ f_2(s) \end{bmatrix} \quad (81.1)$$

which may be denoted:

$$\underline{x}(s) = P(s).\underline{f}(s)$$

This is depicted in block diagram form in Figure 81.2.

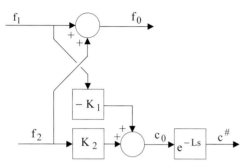

Fig. 81.2 Model of in line blending process

From a control point of view, this is a MIMO system with two inputs and two outputs. Suppose that both the flow rate and concentration of the dilute product stream are to be controlled simultaneously. A scheme consisting of two conventional feedback control loops as depicted in Figure 81.1 will work. The block diagram for this is depicted in Figure 81.3.

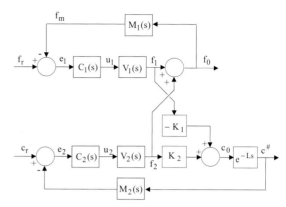

Fig. 81.3 Model of blending system with simple feedback loops

However, to minimise interaction, it is necessary to detune the controllers, *e.g.* by using low gains and large reset times. This makes for poor dynamic performance in terms of speed of response and poor disturbance rejection in terms of steady state offset. Much better control can be realised by means of a 2 × 2 multivariable controller, as depicted in block diagram form in Figure 81.4.

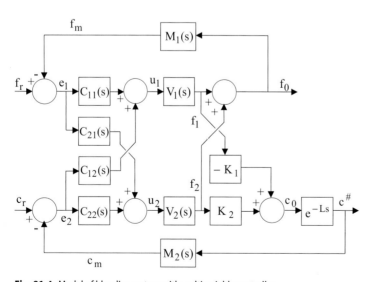

Fig. 81.4 Model of blending system with multivariable controller

81.2 Compensator Concept

The multivariable controller may be considered to consist of four compensators. In principle, compensators $C_{11}(s)$ and $C_{22}(s)$ handle the flow and concentration control loops, in a conventional way, and compensators $C_{21}(s)$ and $C_{12}(s)$ handle the interactions. Thus, for example, following an increase in flow set point f_r, compensator $C_{11}(s)$ will increase flow f_1 and hence f_0 to its desired value. However, in doing so, it will cause concentration c_0 to decrease via the gain K_1. This interaction is compensated for by $C_{21}(s)$ which increases f_2 to produce an equal but opposite effect on c_0 via K_2. If the compensators are designed correctly, other signals in the concentration control loop, such as e_2 and $c^{\#}$ should be unaffected by the change in f_0. Similarly, $C_{12}(s)$ protects f_0 against disturbances from the concentration control loop.

81.3 Control System Model

Two feedback loops are established such that both f_0 and $c^{\#}$ may be controlled against set points. The measurement functions may be articulated in matrix form:

$$\begin{bmatrix} f_m(s) \\ c_m(s) \end{bmatrix} = \begin{bmatrix} M_1(s) & 0 \\ 0 & M_2(s) \end{bmatrix} \begin{bmatrix} f_0(s) \\ c^{\#}(s) \end{bmatrix} \quad (81.2)$$

which may be denoted:

$$\underline{m}(s) = M(s).\underline{x}(s)$$

The comparators are handled simply by vector addition:

$$\begin{bmatrix} e_1(s) \\ e_2(s) \end{bmatrix} = \begin{bmatrix} f_r(s) \\ c_r(s) \end{bmatrix} - \begin{bmatrix} f_m(s) \\ c_m(s) \end{bmatrix} \quad (81.3)$$

which may be denoted:

$$\underline{e}(s) = \underline{r}(s) - \underline{m}(s)$$

Four compensators are required. In essence, C_{11} and C_{22} handle the feedback requirements for control and C_{21} and C_{12} counteract the process interactions:

$$\begin{bmatrix} u_1(s) \\ u_2(s) \end{bmatrix} = \begin{bmatrix} C_{11}(s) & C_{12}(s) \\ C_{21}(s) & C_{22}(s) \end{bmatrix} \begin{bmatrix} e_1(s) \\ e_2(s) \end{bmatrix} \quad (81.4)$$

which may be denoted:

$$\underline{u}(s) = C(s).\underline{e}(s)$$

The compensator outputs are applied to the control valves via I/P converters:

$$\begin{bmatrix} f_1(s) \\ f_2(s) \end{bmatrix} = \begin{bmatrix} V_1(s) & 0 \\ 0 & V_2(s) \end{bmatrix} \begin{bmatrix} u_1(s) \\ u_2(s) \end{bmatrix} \quad (81.5)$$

which may be denoted:

$$\underline{f}(s) = V(s).\underline{u}(s)$$

Equations 81.1–81.5 are a complete, but generic, description of the system.

81.4 Compensator Design

The starting point is to establish the transfer matrix of the closed loop system from Equations 81.1–81.5. Using block diagram algebra gives:

$$\underline{x}(s) = P(s).\underline{f}(s)$$
$$= P(s).V(s).C(s).(\underline{r}(s) - M(s).\underline{x}(s))$$

Matrix manipulation gives:

$$\underline{x}(s) = (I + P(s).V(s).C(s).M(s))^{-1}$$
$$. P(s).V(s).C(s).\underline{r}(s)$$

Let $G(s)$ be some user defined transfer matrix that specifies the desired closed loop response:

$$\underline{x}(s) = G(s).\underline{r}(s)$$

The objective is to minimise interaction between the loops. The ideal is zero interaction which corresponds to making $G(s)$ diagonal such that changing set point f_r only affects f_0 and changing set point c_r only affects $c^{\#}$:

$$\begin{bmatrix} f_0(s) \\ c^{\#}(s) \end{bmatrix} = \begin{bmatrix} G_{11}(s) & 0 \\ 0 & G_{22}(s) \end{bmatrix} \begin{bmatrix} f_r(s) \\ c_r(s) \end{bmatrix}$$

Whence:

$$G(s) = (I + P(s).V(s).C(s).M(s))^{-1}) \\ . P(s.V(s).C(s)$$

Solving for C(s) gives the desired multivariable controller design:

$$C(s) = V(s)^{-1}.P(s)^{-1}.G(s) \\ . (I - M(s).G(s))^{-1} \qquad (81.6)$$

C(s) is totally dependent upon the accuracy of the model and the sensible choice of G(s). If some G(s) is chosen that is unrealistic, perhaps looking for a faster response than the system is physically capable of achieving, then C(s) will be unrealisable.

81.5 Worked Example No 1

The dynamics have been chosen as typical of process systems. If the measurements of flow and concentration are made by means of an electromagnetic flow meter and conductivity cell respectively, time constants of 2 s are reasonable. Likewise, for the pneumatically actuated control valves, time constants of 3 s are assumed. The time delay for the blending process, which only affects the concentration measurement, is 2 s:

$$M(s) = \begin{bmatrix} \dfrac{1}{2s+1} & 0 \\ 0 & \dfrac{1}{2s+1} \end{bmatrix}$$

$$V(s) = \begin{bmatrix} \dfrac{1}{3s+1} & 0 \\ 0 & \dfrac{1}{3s+1} \end{bmatrix}$$

$$P(s) = \begin{bmatrix} 1 & 1 \\ -e^{-2s} & e^{-2s} \end{bmatrix}$$

Note that a value of unity is assumed for all the steady state gains. This implies that either all the signals have the same range, such as 4–20 mA, or that they have been scaled on a common basis, such as percentage of range.

Choose a sensible closed loop transfer matrix:

$$G(s) = \begin{bmatrix} \dfrac{1}{4s+1} & 0 \\ 0 & \dfrac{e^{-2s}}{4s+1} \end{bmatrix}$$

In the case of the flow loop, which has first order lags of 2 s and 3 s, a closed loop response equivalent to a first order lag of 4 s is reasonable. Likewise for the concentration loop except that a 2 s delay in the closed loop response has been allowed for the measurement delay.

Substitution of these values into Equation 81.6 and manipulation yields

$$C(s) = \begin{bmatrix} \dfrac{(3s+1)(2s+1)}{4s(4s+3)} & \dfrac{-(3s+1)(2s+1)}{2((4s+1)(2s+1)-e^{-2s})} \\ \dfrac{(3s+1)(2s+1)}{4s(4s+3)} & \dfrac{(3s+1)(2s+1)}{2((4s+1)(2s+1)-e^{-2s})} \end{bmatrix}$$

All four compensators consist of combinations of lead, lag, integrator and delay terms. This is quite normal for multivariable control, the compensators having been designed to give a specified response for a particular plant.

81.6 Decouplers

The principal disadvantage of multivariable controllers is that the compensators $C_{11}(s)$ to $C_{22}(s)$ do not have the structure of a PID controller. Their form is not intuitive which is undesirable for operational reasons. An alternative approach is to use decouplers, as depicted in Figure 81.5.

In this case the two compensators, $C_1(s)$ and $C_2(s)$, may be realised by conventional PID controllers:

$$C(s) = K_C \left(1 + \dfrac{1}{T_R s} + T_D s\right)$$

The decouplers, $D_1(s)$ and $D_2(s)$, and compensators are related according to:

$$\begin{bmatrix} u_1(s) \\ u_2(s) \end{bmatrix} = \begin{bmatrix} C_1(s) & C_2(s).D_2(s) \\ C_1(s).D_1(s) & C_2(s) \end{bmatrix} \begin{bmatrix} e_1(s) \\ e_2(s) \end{bmatrix} \qquad (81.7)$$

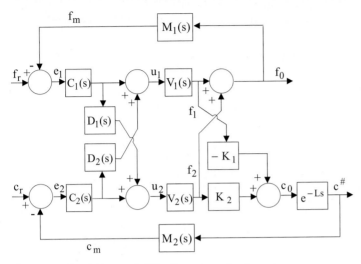

Fig. 81.5 Multivariable control of blending system using decouplers

The design technique essentially consists of choosing appropriate $C_1(s)$ and $C_2(s)$, empirically or otherwise, and then finding $D_1(s)$ and $D_2(s)$ for the known $P(s), M(s)$ and $V(s)$ to give a specified $G(s)$.

For the system of Figure 81.5, the decouplers can be determined by inspection. Firstly consider the decoupler $D_1(s)$. Following a change in error e_1, the affect on C_0 due to the interaction inherent in the process is given by:

$$c_0(s) = -K_1 V_1(s) C_1(s).e_1(s)$$

The corresponding affect on C_0 due to the decoupler is given by:

$$c_0(s) = K_2 V_2(s) D_1(s) C_1(s).e_1(s)$$

For the decoupler to be effective, these two affects must be equal and opposite: that is they cancel each other. Thus:

$$-K_1 V_1(s) C_1(s) e_1(s) = -K_2 V_2(s) D_1(s) C_1(s) e_1(s)$$

Hence:

$$D_1(s) = \frac{K_1 V_1(s)}{K_2 V_2(s)} \approx \frac{K_1}{K_2}$$

By a similar argument, for a change in error e_2 to have no net affect on f_0, the decoupler $D_2(s)$ required is simply:

$$D_2(s) = -1$$

81.7 Sampled Data Model

Multivariable controllers are realised by means of software. The dynamics of the sampling process can be ignored if they are fast compared with the process. However, if they are of the same order of magnitude, as with this blending system, they must be taken into account. Either way, it is convenient to do the design in the Z domain because the subsequent compensators are easily realisable.

The block diagram of the sampled data multivariable control system for the blending plant is as depicted in Figure 81.6.

Note that both the error and output signals are sampled, but only the outputs are held by zero order hold devices. Whereas previous values of the error signals may be stored in memory, the output signals must be held between sampling instants to create pseudo continuous signals. It is a common mistake to assume hold devices for the error signals.

Equations 81.1–81.3 still apply. Allowing for the samplers, the equation which describes the behaviour of the compensators becomes:

$$\begin{bmatrix} u_1(s) \\ u_2(s) \end{bmatrix} = \begin{bmatrix} C_{11}(s) & C_{12}(s) \\ C_{21}(s) & C_{22}(s) \end{bmatrix} \begin{bmatrix} e_1^*(s) \\ e_2^*(s) \end{bmatrix} \quad (81.8)$$

which may be denoted:

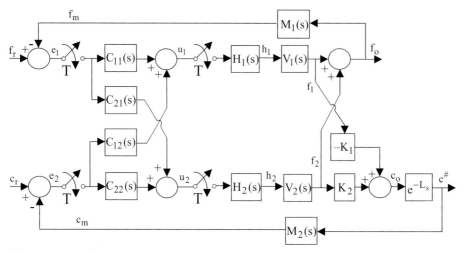

Fig. 81.6 Control of blending system using impulse compensators

$$\underline{u}(s) = C(s).\underline{e}^*(s)$$

An additional equation must be introduced to accommodate the hold devices:

$$\begin{bmatrix} h_1(s) \\ h_2(s) \end{bmatrix} = \begin{bmatrix} H_1(s) & 0 \\ 0 & H_2(s) \end{bmatrix} \begin{bmatrix} u_1^*(s) \\ u_2^*(s) \end{bmatrix} \quad (81.9)$$

which may be denoted:

$$\underline{h}(s) = H(s).\underline{u}^*(s)$$

The outputs of the holds are then applied *via* I/P converters to the two valves which manipulate the process inputs:

$$\begin{bmatrix} f_1(s) \\ f_2(s) \end{bmatrix} = \begin{bmatrix} V_1(s) & 0 \\ 0 & V_2(s) \end{bmatrix} \begin{bmatrix} h_1(s) \\ h_2(s) \end{bmatrix} \quad (81.10)$$

which may be denoted:

$$\underline{f}(s) = V(s).\underline{h}(s)$$

81.8 Impulse Compensator Design

Block diagram algebra yields:

$$\underline{e}(z) = \underline{r}(z) - \underline{m}(z)$$
$$= \underline{r}(z) - MPVH(z).\underline{u}(z)$$
$$= \underline{r}(z) - MPVH(z).C(z).\underline{e}(z)$$

Solving for $\underline{e}(z)$ gives:

$$(I + MPVH(z).C(z)).\underline{e}(z) = \underline{r}(z)$$
$$\underline{e}(z) = (I + MPVH(z).C(z))^{-1}.\underline{r}(z)$$

Assuming there are imaginary samplers on the controlled variables:

$$\underline{x}(z) = PVH(z).C(z).\underline{e}(z)$$
$$= PVH(z).C(z).(I + MPVH(z).C(z))^{-1}.\underline{r}(z)$$
$$= G(z).\underline{r}(z)$$

$G(z)$ is the closed loop transfer matrix which determines the overall system performance:

$$G(z) = PVH(z).C(z) \quad (81.11)$$
$$.(I + MPVH(z).C(z))^{-1}$$

If $G(s)$ is diagonal then any change in either set point will only affect one of the controlled variables, *i.e.* the system is decoupled. For a specified $G(s)$, and for given $M(s)$, $P(s)$, $V(s)$ and $H(s)$, there must be a unique $C(s)$:

$$G(z).(I + MPVH(z).C(z)) = PVH(z).C(z)$$
$$G(z) = (PVH(z) - G(z).MPVH(z)).C(z) \quad (81.12)$$
$$C(z) = (PVH(z) - G(z).MPVH(z))^{-1}.G(z)$$

81.9 Worked Example No 2

Assume that $M(s)$, $V(s)$ and $P(s)$ are as before, and that the zero order holds have a one second delay:

$$H(s) = \begin{bmatrix} \dfrac{1-e^{-s}}{s} & 0 \\ 0 & \dfrac{1-e^{-s}}{s} \end{bmatrix}$$

The 1 sec delay associated with the zero order holds implies a 1 sec sampling period.

Choose a closed loop transfer matrix, taking into account the dynamics of the sampling process:

$$G(s) = \begin{bmatrix} \dfrac{e^{-s}}{4s+1} & 0 \\ 0 & \dfrac{e^{-3s}}{4s+1} \end{bmatrix}$$

The rationale for choosing $G(s)$ is essentially the same as in the continuous case, except that a 1 sec delay has been included in the flow loop to allow for the hold. Likewise for the concentration loop where a 3 sec delay has been allowed for both the measurement delay and the hold.

Assume a sampling period of $T = 1$ s.

Extensive manipulation yields

$$C(z) = \begin{bmatrix} \dfrac{0.441\left(1 - 1.323z^{-1} + 0.4346z^{-2}\right)}{1 - 1.441z^{-1} + 0.430z^{-2}} & \dfrac{-0.441\left(1 - 1.323z^{-1} + 0.4346z^{-2}\right)}{1 - 1.385z^{-1} + 0.4724z^{-2} - 0.05597z^{-3} - 0.04239z^{-4}} \\ \dfrac{0.441\left(1 - 1.323z^{-1} + 0.4346z^{-2}\right)}{1 - 1.441z^{-1} + 0.430z^{-2}} & \dfrac{0.441\left(1 - 1.323z^{-1} + 0.4346z^{-2}\right)}{1 - 1.385z^{-1} + 0.4724z^{-2} - 0.05597z^{-3} - 0.04239z^{-4}} \end{bmatrix} \quad (81.13)$$

which is of the form:

$$C(z) = \begin{bmatrix} \alpha(z) & -\beta(z) \\ \alpha(z) & \beta(z) \end{bmatrix} \quad (81.14)$$

It can be seen by inspection that all four compensators are realisable, i.e. there are no terms in z^{+1}. The coefficients of z^{-1} in successive terms of both the numerator and denominator series are of reducing magnitude and, with the exception of the least significant term of the second denominator, are of alternate sign. This is necessary for stability. The compensators may be realised by means of software using any of the techniques described in Chapter 78. Remember that the variables in the

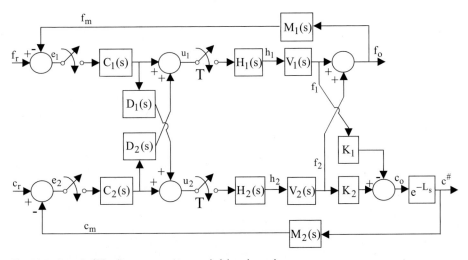

Fig. 81.7 Control of blending system using sampled data decouplers

model are all in deviation form so a steady state bias must be added to the controller outputs for implementation purposes.

81.10 Sampled Data Decoupler

The block diagram of a sampled data decoupler type of control system for the blending plant is as depicted in Figure 81.7.

Apart from the compensators and decouplers being of an impulse nature, the approach to design is essentially the same as for the continuous system of Figure 81.5. Control action is of the form

$$\begin{bmatrix} u_1(z) \\ u_2(z) \end{bmatrix} = \begin{bmatrix} C_1(z) & C_2 D_2(z) \\ C_1 D_1(z) & C_2(z) \end{bmatrix} \begin{bmatrix} e_1(z) \\ e_2(z) \end{bmatrix} \quad (81.15)$$

Comparison of Equations 81.14 and 81.15 reveals that, to meet the same closed loop performance criterion G(s) for the physical example of the blending system, the decouplers must be

$$D_1(s) = -D_2(s) = 1$$

Given the assumption that $K_1 = K_2 = 1$, this result is entirely consistent with that obtained in Section 81.6.

Note that it is often sufficient to realise the two compensators $C_1(z)$ and $C_2(z)$ by means of the discrete form of PID controllers developed in Chapter 78:

$$C(z) = \frac{a_0 + a_1 z^{-1} + a_2 z^{-2}}{1 - z^{-1}} \quad (78.3)$$

The approximation may simply be based on the first terms of $\alpha(z)$ and $\beta(z)$ which are the dominant terms. Otherwise, the PID compensator is taken as a factor out of $C_1(z)$ and $C_2(z)$. The parameters of the compensator would be presented to the operators in terms of the familiar gain, reset and rate times, whose values may be tuned empirically.

81.11 Comments

Although all these compensators have been designed for servo operation as opposed to regulo control, *i.e.* the effects of disturbances haven't been taken into account explicitly, they can nevertheless be just as effective at disturbance rejection.

Stochastics

82.1 Summary Statistics
82.2 Multivariate Statistics
82.3 Probability Distribution
82.4 The Normal Distribution
82.5 Correlation
82.6 Properties of Correlation Functions

Stochastics is the study of statistics in relation to signals. A variety of statistical techniques can be applied to signals which enables them to be characterised and relevant information extracted. This chapter essentially summarises standard stochastic terminology and functions. They are largely listed as formulae with minimal, but hopefully sufficient, explanation to enable cross referencing from other chapters.

The primary purpose of this chapter is to provide a foundation for Chapters 83, 101 and 102 on regression analysis, principal components analysis and statistical process control respectively. There are many excellent texts on applied statistics, such as that by Mongomerie (2006), to which the reader is referred for a more comprehensive treatment.

82.1 Summary Statistics

Summary statistics are single valued metrics that give a feel for the nature of a signal, the most obvious ones being mean and variance. If the value of a signal is available as a function of time, its mean or average value is:

$$\bar{x} \approx \frac{1}{2T} \int_{-T}^{T} x(t).dt \qquad (82.1)$$

If the signal is periodic, with a period of 2T, then this will be the true mean. Otherwise, the true mean is given by:

$$\bar{x} = \lim_{T \to \infty} \frac{1}{2T} \int_{-T}^{T} x(t).dt$$

If the signal is sampled, again over some period 2T, the mean value of a set of n samples is given by:

$$\bar{x} \approx \frac{x_1 + x_2 + \ldots + x_n}{n} = \frac{1}{n} \sum_{i=1}^{n} x_i \qquad (82.2)$$

This is necessarily an approximation to the true mean which is given by:

$$\bar{x} = \lim_{n,T \to \infty} \frac{1}{n} \sum_{i=1}^{n} x_i$$

Note that, for limiting purposes, both the sampling period and the number of samples per unit period increase.

Similarly, the mean square value for continuous signals is:

$$\overline{x^2} \approx \frac{1}{2T} \int_{-T}^{T} x^2(t).dt$$

And the mean square value for sampled signals is:

$$\overline{x^2} \approx \frac{1}{n} \sum_{i=1}^{n} x_i^2$$

Note that the mean and median values of a variable are not necessarily the same thing. The median is defined to be the middle value of x when the measurements of x are arranged in increasing order of magnitude.

Another important metric is variance. An essential distinction here is between samples of size (number) n and populations where n is either very large or infinite. For samples, variance is defined to be:

$$\sigma^2 \approx \frac{1}{n-1} \sum_{i=1}^{n} (x_i - \bar{x})^2 \qquad (82.3)$$

where σ, the positive square root of variance, is known as the standard deviation. In practice σ is not meaningful unless n > 5 and normally n > 30 is presumed for good estimates. Note that σ has the same units as the original signal x.

For hand calculation of variance it is quicker to use the following identity obtained by expansion:

$$\sigma^2 \approx \frac{1}{n-1} \sum_{i=1}^{n} \left(x_i^2 - 2x_i.\bar{x} + \bar{x}^2\right)$$

$$= \frac{1}{n-1} \left(\sum_{i=1}^{n} x_i^2 - 2\bar{x} \sum_{i=1}^{n} x_i + n\bar{x}^2 \right)$$

$$= \frac{1}{n-1} \left(\sum_{i=1}^{n} x_i^2 - 2n\bar{x}^2 + n\bar{x}^2 \right)$$

$$= \frac{1}{n-1} \left(\sum_{i=1}^{n} x_i^2 - n\bar{x}^2 \right)$$

For populations, the true value of variance is given by:

$$\sigma^2 = \lim_{n \to \infty} \frac{1}{n} \sum_{i=1}^{n} (x_i - \bar{x})^2$$

Again, for populations, variance, mean square and square mean values are related by:

$$\sigma^2 = \overline{x^2} - \bar{x}^2$$

82.2 Multivariate Statistics

The above summary statistics are all univariate in that only one signal was being considered. In practice, it is often the case that there are several interrelated signals. Suppose that there are p signals, each of which is sampled simultaneously to give n sets of data. The sample covariance between any two of the variables x_i and x_j is defined to be:

$$\sigma_{ij}^2 = \frac{1}{n-1} \sum_{k=1}^{n} (x_{ki} - \bar{x}_i)(x_{kj} - \bar{x}_j) \qquad (82.4)$$

The sample covariances between all of the p signals may then be assembled into a matrix of dimension (pxp), known as the covariance matrix, which is symmetrical:

$$S = \begin{bmatrix} \sigma_1^2 & \sigma_{12}^2 & \cdots & \sigma_{1p}^2 \\ \sigma_{21}^2 & \sigma_2^2 & \cdots & \sigma_{2p}^2 \\ \vdots & \vdots & \ddots & \vdots \\ \sigma_{p1}^2 & \sigma_{p2}^2 & \cdots & \sigma_p^2 \end{bmatrix} = \begin{bmatrix} \sigma_1^2 & \sigma_{12}^2 & \cdots & \sigma_{1p}^2 \\ \sigma_{12}^2 & \sigma_2^2 & \cdots & \sigma_{2p}^2 \\ \vdots & \vdots & \ddots & \vdots \\ \sigma_{1p}^2 & \sigma_{2p}^2 & \cdots & \sigma_p^2 \end{bmatrix}$$
(82.5)

However, for summary purposes, covariances are seldom used: they are difficult to interpret because they depend on the units in which the two variables are measured. Thus covariance is often standardised by dividing by the product of the standard deviations of the two variables to give a quantity called the correlation coefficient:

$$\rho_{ij} = \frac{\sigma_{ij}^2}{\sigma_i . \sigma_j} \qquad (82.6)$$

It can be shown that ρ_{ij} must lie between -1 and $+1$. If there is no correlation between the two signals then the product terms will be random in both size and sign and, when summed, will cancel out leading to a zero coefficient. The closer the coefficient is to unity the stronger the correlation, the maximum value of unity occurring when the two signals are identical. The sign of the coefficient depends simply on the slope, or direction, of the underlying relationship between the two variables.

If $0.75 < |\rho_{ij}| < 1.0$ then the correlation between x and y is strong.

If $0.25 < |\rho_{ij}| < 0.75$ then some correlation exists but is weak.
If $0.0 < |\rho_{ij}| < 0.25$ then the correlation is not significant.

Take care. A small correlation coefficient does not necessarily mean that two variables are independent. Similarly a large correlation coefficient does not necessarily indicate a cause and effect relationship. Engineering judgement is required in interpreting these coefficients.

Correlation coefficients may also be assembled into a matrix of dimension (pxp), known as the correlation matrix, which is also symmetrical:

$$P = \begin{bmatrix} 1 & \rho_{12} & \cdots & \rho_{1p} \\ \rho_{12} & 1 & \cdots & \rho_{2p} \\ \vdots & \vdots & \ddots & \vdots \\ \rho_{1p} & \rho_{2p} & \cdots & 1 \end{bmatrix} \qquad (82.7)$$

82.3 Probability Distribution

In the context of signals, probability distributions are concerned with the frequency of occurrence of the signal within a specified band. Consider Figure 82.1.

Suppose there are a total of n sampled values of which m lie within the jth band. The probability of the signal being within this band is thus:

$$P_j \approx \frac{m}{n}$$

Clearly probability is a positive quantity, its true value being given by:

$$P_j \approx \lim_{n \to \infty} \frac{m}{n}$$

An alternative approach, for continuous measurements, is depicted in Figure 82.2.

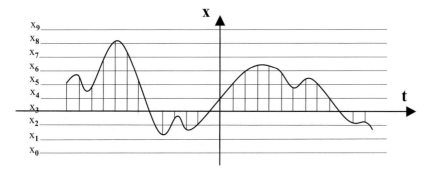

Fig. 82.1 Frequency of occurrence of signal within specified band

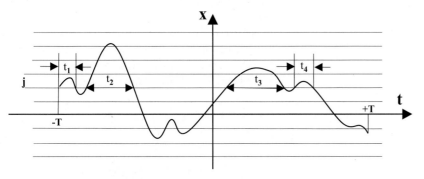

Fig. 82.2 Duration of signal within or above specified band

Probability is based upon the duration that the signal lies within or above a specified band, the jth band:

$$P_j \approx \frac{t_1 + t_2 + \ldots + t_n}{2T} = \frac{1}{2T} \sum_{i=1}^{n} t_i \qquad (82.8)$$

Again, the true value is found by taking a long record of the signal:

$$P_j = \lim_{T \to \infty} \frac{1}{2T} \sum_{i=1}^{n} t_i$$

The cumulative probability distribution (CPD) is depicted in Figure 82.3 which shows, for each band, what the probability is of the signal having a value that lies within or below that band.

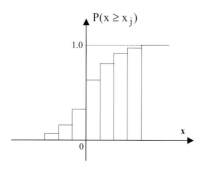

Fig. 82.3 Cumulative probability distribution (CPD) on a banded basis

As the width of the bands decreases, the CPD tends towards a continuous curve, as depicted in Figure 82.4. The ordinate (y axis) of the CPD is probability, which is dimensionless, and the maximum value of the curve is unity.

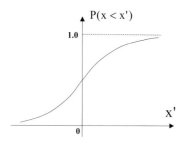

Fig. 82.4 CPD on a continuous basis

Figure 82.5 depicts the corresponding probability density distribution (PDD). The PDD is a graph of the slope of the CPD curve. The ordinate is known as the probability density function p(x) and is a measure of the probability of a signal having a value within a specified range. Its units are the reciprocal of those of the signal. The area under the PDD is unity.

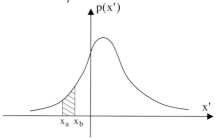

Fig. 82.5 Probability density distribution (PDD)

The CPD and PDD are defined by the following equations:

$$P(x \leq x') = \int_{-\infty}^{x'} p(x).dx \qquad (82.9)$$

$$P(x_a < x \leq x_b) = \int_{x_a}^{x_b} p(x).dx$$

$$= p(x \leq x_b) - P(x \leq x_a)$$

$$p(x) = \frac{d}{dx} P(x \leq x') \qquad (82.10)$$

$$\int_{-\infty}^{\infty} p(x).dx = 1$$

Many naturally occurring phenomena yield PDD curves that are of a known form. The three most common are sketched in Figure 82.6, for which the mean is assumed to be zero.

The equations for these are as in Table 82.1.

The mean and mean square values of a signal are related to its PDD by the integrals:

$$\bar{x} = \int_{-\infty}^{\infty} p(x).x dx \qquad (82.11)$$

$$\overline{x^2} = \int_{-\infty}^{\infty} p(x).x^2 dx$$

Table 82.1 Functions for Gaussian, Rayleigh and Maxwell forms of PDD

Gaussian (normal)	Range	$p(x) = \dfrac{1}{\sigma\sqrt{2\pi}} \exp\left(-\dfrac{(x-\bar{x})^2}{2\sigma^2}\right)$
Rayleigh	$x \geq 0$	$p(x) = \dfrac{x}{\sigma^2} \exp\left(\dfrac{-x^2}{2\sigma^2}\right)$
	$x < 0$	$p(x) = 0$
Maxwell	$x \geq 0$	$p(x) = \dfrac{x^2}{\sigma^2}\sqrt{\dfrac{2}{\pi}} \exp\left(\dfrac{-x^2}{2\sigma^2}\right)$
	$x < 0$	$p(x) = 0$

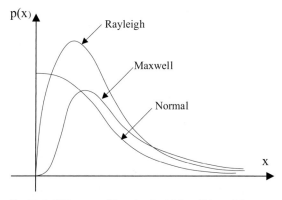

Fig. 82.6 PDD curves of Gaussian, Rayleigh and Maxwell forms

CPD is the probability of some signal being less than some specified value. This can be extrapolated to the simultaneous probability of two signals being less than a pair of specified values. This gives rise to the concept of joint CPD and PDD:

$$P(x \leq x', y \leq y') = \int_{-\infty}^{y'} \int_{-\infty}^{x'} p(x,y).dx.dy \quad (82.12)$$

$$p(x,y) = \frac{\partial^2}{\partial x.\partial y}.P(x \leq x', y \leq y') \quad (82.13)$$

$$\int_{-\infty}^{\infty} \int_{-\infty}^{\infty} p(x,y).dx.dy = 1$$

82.4 The Normal Distribution

The Gaussian distribution, usually referred to as the normal distribution, is of particular importance because it characterises random effects:

$$p(x) = \frac{1}{\sigma\sqrt{2\pi}} \exp\left(-\frac{(x-\bar{x})^2}{2\sigma^2}\right) \quad (82.14)$$

Random effects can be articulated in terms of the standard deviation. For example, as σ increases the distribution becomes flatter, as depicted in Figure 82.7.

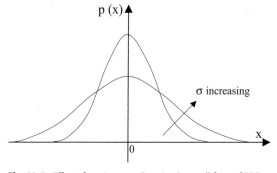

Fig. 82.7 Effect of varying σ on Gaussian (normal) form of PDD

The distribution is a measure of the closeness of sampled values to the mean. The proportion of samples lying within bands about the mean are depicted in Figure 82.8 and summarised in Table 82.2. Thus, for example, it can be expected that some 95% of samples lie within two standard de-

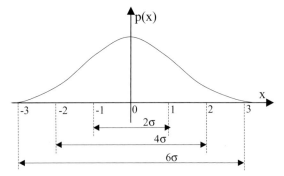

Fig. 82.8 Bands about the mean of the normal PDD

Table 82.2 Standard deviation *vs* percentage of samples for normal PDD

Number of standard deviations	% of samples inside limits	% of samples outside limits
$\pm 0.5\sigma$	38.30	61.70
$\pm 1.0\sigma$	68.26	31.74
$\pm 1.5\sigma$	86.64	13.36
$\pm 2.0\sigma$	95.44	4.56
$\pm 2.5\sigma$	98.76	1.24
$\pm 3.0\sigma$	99.74	0.26
$\pm 3.5\sigma$	99.95	0.05
$\pm 4.0\sigma$	99.994	0.006
$\pm 6.0\sigma$	–	$\sim 2 \times 10^{-7}$

viations of the mean. Put differently, if anything other than approximately 1 in 20 of the sampled values lie outside the 95% confidence limit (beyond two standard deviations of the mean), then the distribution probably isn't normal.

82.5 Correlation

Correlation concerns the extent to which one signal depends upon another. Consider a set of n values of each of two signals, x and y, which would normally correspond in time. Their dependency is articulated by the correlation coefficient defined to be:

$$\rho_{xy} = \frac{(x_1 - \bar{x})(y_1 - \bar{y}) + (x_2 - \bar{x})(y_2 - \bar{y}) + \ldots + (x_n - \bar{x})(y_n - \bar{y})}{(n-1).\sigma_x \sigma_y}$$

$$= \sum_{i=1}^{n} \frac{(x_i - \bar{x})(y_i - \bar{y})}{(n-1).\sigma_x \sigma_y} \qquad (82.15)$$

This is essentially the same as Equations 82.4 and 82.6.

However, what is often of particular interest in signal processing are the cross and auto correlation functions (not coefficients) because these more readily identify the effects of time delay (or time shift). Consider, for example, the two signals depicted in Figure 82.9.

For a large number of samples, the cross correlation function is given by:

$$R_{xy}[k] \approx \frac{1}{n}\left(x_1 y_{1+k} + x_2 y_{2+k} + x_3 y_{3+k} + \ldots + x_n y_{n+k}\right)$$

Clearly the function may be evaluated for different values of k, *i.e.* for integer multiples of the sampling period, and plotted as a function of k, as shown in Figure 82.10 which indicates a maximum correlation for a shift of approximately seven intervals.

In practice, only a finite number of samples is normally available, so for each value of k only $n-k$ products may be evaluated:

$$R_{xy}[k] \approx \frac{1}{n - |k|} \sum_{i=1}^{n-|k|} x_i y_{i+k} \qquad (82.16)$$

This is the so called "unbiased" form of the cross correlation function. Clearly the true value of the cross correlation function is given by:

$$R_{xy}[k] = \lim_{n \to \infty} \frac{1}{2n} \sum_{i=-n}^{n} x_i y_{i+k}$$

For a continuous signal the cross correlation function is given by:

$$R_{xy}(\tau) = \lim_{T \to \infty} \frac{1}{2T} \int_{-T}^{T} x(t).y(t+\tau).dt \qquad (82.17)$$

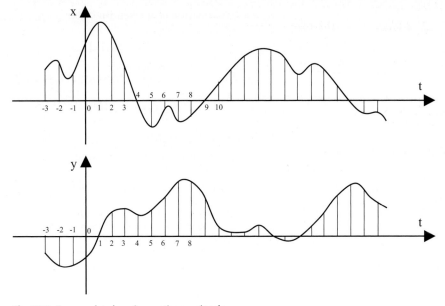

Fig. 82.9 Two correlated varying continuous signals

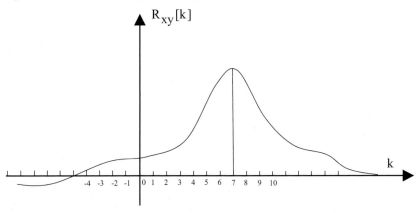

Fig. 82.10 Cross correlation function *vs* number of time shifts

Two important properties of cross correlation functions can be established by inspection, for example, of Figure 82.9:

$$R_{xy}[k] = R_{yx}[-k] \qquad R_{xy}(\tau) = -R_{yx}(-\tau)$$

The greatest value of the cross correlation function only occurs at k = 0 if there is no delay (or shift) between the two signals.

Whereas cross correlation functions relate to dependency between two signals in time, auto correlation functions relate to the dependency of one signal with itself. Auto correlation is particularly useful for detecting cyclic patterns in a signal.

For a large number of samples, the auto correlation function is given by:

$$R_{xx}[k] \approx \frac{1}{n}(x_1 x_{1+k} + x_2 x_{2+k} + x_3 x_{3+k} + \ldots + x_n x_{n+k})$$

For a finite number of samples:

$$R_{xx}[k] \approx \frac{1}{n-k} \sum_{i=1}^{n-k} x_i x_{i+k} \qquad (82.18)$$

The true value of the auto correlation function is given by:

$$R_{xx}[k] = \lim_{n \to \infty} \frac{1}{2n} \sum_{i=-n}^{n} x_i x_{i+k}$$

For a continuous signal the auto correlation function is given by:

$$R_{xx}(\tau) = \lim_{T \to \infty} \frac{1}{2T} \int_{-T}^{T} x(t).x(t+\tau).dt \qquad (82.19)$$

Auto correlation functions may be plotted as a function of k and interpreted in exactly the same way as cross correlation function plots.

82.6 Properties of Correlation Functions

Some of the more important properties of auto correlation functions are:

1. If there is zero shift, then

$$R_{xx}[0] = \overline{x^2}$$

2. If a signal consists of a normal value plus some deviation, i.e. $x = y + \overline{x}$, then

$$R_{xx}[k] = R_{yy}[k] + \overline{x}^2$$

3. Auto correlation is an even function:

$$R_{xx}[-k] = R_{xx}[k]$$

4. Maximum auto correlation occurs at zero shift:

$$R_{xx}[k] \leq R_{xx}[0]$$

5. If a signal contains sinusoidal components, then so too does its auto correlation function:

$$x(t) = y(t) + x_1 \sin(\omega_1 t + \phi_1) + \ldots$$

$$R_{xx}[k] = R_{yy}[k] + \frac{1}{2} x_1^2 \cos(\omega_1 t) + \ldots$$

6. The auto correlation of the sum of two signals is the sum of the auto and cross correlation functions:

$$z = x \pm y$$

$$R_{zz}[k] = R_{xx}[k] \pm R_{xy}[k] \pm R_{yx}[k] + R_{yy}[k]$$

If x and y are uncorrelated, for example y could be noise on a deterministic signal x, then the two cross correlations would be of zero value such that

$$R_{zz}[k] = R_{xx}[k] + R_{yy}[k]$$

If y is indeed noise, then its auto correlation function will be zero such that

$$R_{zz}[k] = R_{xx}[k]$$

Linear Regression Analysis

83.1 Basic Concepts
83.2 Pre-Processing of Data
83.3 Method of Least Squares
83.4 Model Validation
83.5 Goodness of Fit
83.6 Worked Example No 1
83.7 Multiple Linear Regression
83.8 Variable Selection
83.9 Worked Example No 2
83.10 Worked Example No 3
83.11 Comments

Regression analysis is concerned with curve fitting. Given a set of empirical data relating two or more variables, what is the best straight line or curve that fits the data? Whereas the data may be plotted and a line or curve fitted by inspection, regression analysis is a more rational means of doing so. The most commonly used method is the so called "least squares" approach which minimises the error involved in curve fitting. Regression analysis provides the basis for principal components analysis (PCA) and for statistical process control (SPC) as described in Chapters 101 and 102 respectively. In both cases the interest is in understanding the relationship between process variables.

The concept of least squares is used extensively in advanced process control. For example, for problems that can be formulated in a quadratic form, least squares provides the basis for most optimisation techniques. Another example of its use is in identification, the process used to establish dynamic models in which time is the dependent variable.

The principles of the least squares method are covered in this chapter in relation to both simple and multiple regression analysis. The principles of least squares is covered in many texts on modern control and related topics, such as the text on optimisation by Edgar (2001). Regression analysis is obviously covered in most standard texts on statistics.

83.1 Basic Concepts

Consider some phenomenon for which there is an underlying physical relationship of a linear nature:

$$y = \beta_0 + \beta_1.x \qquad (83.1)$$

If the β coefficients are known, for any value of x the corresponding value of y is uniquely determined. Conversely, any two pairs of x and y values will enable the β coefficients to be determined. However, with empirical data, there are measurement errors and other sources of inaccuracy. Provided these errors are small, for a set of values of x the corresponding values of y may be plotted and a straight line graph drawn. There will be little scatter and the β coefficients can be estimated with confidence from the intercept and slope. However, if the errors are significant, for any given value of x there will be an error on the measured value of y according to:

$$y = \beta_0 + \beta_1.x + \varepsilon \qquad (83.2)$$

Often the dependent variable y is referred to as the response or output and the independent variable x as the regressor, predictor or input variable. The error is referred to as the residual.

The objective of regression analysis is to estimate the β coefficients. This is realised by means of the least squares method which, for a set of empirically determined x and y values, minimises the residuals. Thus, for a given value of x the value of y predicted is, at best, an estimate of the true value of y and is often referred to as the fitted value:

$$\hat{y} = \hat{\beta}_0 + \hat{\beta}_1.x \qquad (83.3)$$

where the ˆ denotes an estimated value.

Equation 83.1 is the so called "simple linear regression" model. If the model contains powers of the independent variable, as in the following quadratic example, it is referred to as a polynomial regression model. It is nevertheless still linear with respect to the β coefficients:

$$y = \beta_0 + \beta_1.x + \beta_2.x^2 \qquad (83.4)$$

Indeed, logarithmic and other nonlinear functions may be included in linear regression models, such as:

$$\log(y) = \beta_0 + \beta_1.\log(x) \qquad (83.5)$$

When there are several independent variables involved, the model is referred to as being a multiple linear regression (MLR) model:

$$y = \beta_0 + \beta_1.x_1 + \beta_2.x_2 + \beta_3.x_3 \qquad (83.6)$$

83.2 Pre-Processing of Data

Prior to carrying out a regression analysis, it may be necessary to pre-process the data available. The objective is to identify periods of unrepresentative data and unusual events, with a view to removing suspect data. Pre-processing essentially involves common sense. Plot the data to get a feel for the relationships: trends can be observed much more readily from graphs than from tables. Question any apparent anomalies:

- Points which are inconsistent with the trend, referred to as outliers, are probably either caused by some unusual event or are false readings and may be rejected. However, they should only be rejected if there is strong non-statistical evidence that they are abnormal.
- Sometimes points are known in advance from the nature of the relationship, such as the graph passing through the origin. Check whether the facts are supported by the data.
- Data which is excessively noisy may be filtered.
- Calculate summary statistics, *e.g.* mean, median, standard deviation, correlation coefficients, *etc*.

For MLR analysis:

- Cross-correlation tests using Equation 82.16 on both the input and output variables may be necessary to reveal time delays and dependencies.
- Standardisation of the data so that it has zero mean and unit variance may be necessary. This requires subtraction of the mean and division by the standard deviation. Such standardising makes the data independent of the scale and/or units of measurement and prevents one input overshadowing others.

83.3 Method of Least Squares

Figure 83.1 depicts n sets of measurements $(x_1, y_1), (x_2, y_2), \cdots, (x_n, y_n)$.

The aim is to find the underlying linear relationship of the form of Equation 83.1. Estimates of the regression coefficients β_0 and β_1 are required such that the best fit is obtained.

For each measurement, *i.e.* for each value of x, let the residual (error between the observed value of y and its underlying true value) be:

$$\varepsilon_j = y_j - (\beta_0 + \beta_1.x_j)$$

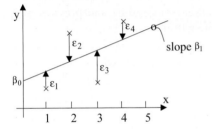

Fig. 83.1 Residuals on n sets of measurements

Dependency upon the sign of the residual is removed by squaring. Thus the sum of the squares of the residuals for all the measurements is given by:

$$Q = \sum_{j=1}^{n} \varepsilon_j^2 = \sum_{j=1}^{n} \left(y_j - (\beta_0 + \beta_1.x_j)\right)^2 \quad (83.7)$$

The straight line best fitting the data corresponds to Q being a minimum. Clearly Q is a function of both β_0 and β_1. Thus, differentiating Q with respect to each of β_0 and β_1 and setting the differentials to zero establishes that minimum.

Differentiating Equation 83.7 with respect to β_0 gives:

$$\frac{\partial Q}{\partial \beta_0} = -2 \sum_{j=1}^{n} \left(y_j - (\beta_0 + \beta_1.x_j)\right)$$

For a minimum:

$$\sum_{j=1}^{n} \left(y_j - (\beta_0 + \beta_1.x_j)\right) = 0$$

Strictly speaking, to prove that this is a minimum a positive second differential should be established: just assume that to be the case. Whence:

$$\sum_{j=1}^{n} y_j = n.\beta_0 + \beta_1 \sum_{j=1}^{n} x_j$$

Dividing throughout by n and rearranging gives:

$$\beta_0 = \bar{y} - \beta_1 \bar{x} \quad (83.8)$$

where $\bar{x}$ and $\bar{y}$ denote the average observed values.

Similarly differentiating Equation 83.7 with respect to β_1 gives:

$$\frac{\partial Q}{\partial \beta_1} = -2 \sum_{j=1}^{n} x_j . \left(y_j - (\beta_0 + \beta_1.x_j)\right)$$

whence, for a minimum:

$$\sum_{j=1}^{n} x_j y_j = \beta_0 \sum_{j=1}^{n} x_j + \beta_1 \sum_{j=1}^{n} x_j^2$$

This, together with Equation 83.8, forms a set of two simultaneous equations which can be solved algebraically for the two unknowns β_0 and β_1.

Extensive manipulation yields:

$$\hat{\beta}_0 = \bar{y} - \hat{\beta}_1 \bar{x}$$

and:

$$\hat{\beta}_1 = \frac{\sum_{j=1}^{n} (x_j - \bar{x}) . (y_j - \bar{y})}{\sum_{j=1}^{n} (x_j - \bar{x})^2} = \frac{(\underline{x} - \bar{x}) . (\underline{y} - \bar{y})^T}{(\underline{x} - \bar{x}) . (\underline{x} - \bar{x})^T}$$

$$(83.9)$$

The two equations at Equation 83.9 enable the regression coefficients to be determined directly from the empirical data available. In essence, the value of β_1 is found first: that value is then used to find the value of β_0. Note that, unless n is large, the coefficients found cannot be anything other than estimates of the regression coefficients, hence they are denoted as estimates in Equation 83.9.

83.4 Model Validation

Validating the model is essentially a question of confirming that the regression analysis has produced sensible results. The most likely problems are due to:

- Insufficient or inadequate empirical data
- The model fitted being inappropriate, *e.g.* a straight line instead of a quadratic
- The effect of variables not included in the model, *e.g.* a simple regression instead of a multiple regression model

Validation is best realised by consideration of the residuals. For any particular input, the residual is the difference between the measured output and its fitted value according to Equation 83.3:

$$\varepsilon_j = y_j - \left(\hat{\beta}_0 + \hat{\beta}_1 . x_j\right) \quad (83.10)$$

If the model is correct, the population of residuals should have zero mean, constant variance and be normally distributed. Any plot of the residuals should demonstrate such characteristics or, if the data is sparse, should at least not contradict them. Otherwise the plot should not exhibit any structure: non-random looking patterns must be considered to be suspect. The most useful diagnostic plot is that of the residuals *vs* the fitted values. The points should have a random distribution along the fitted value axis and a normal distribution along the residuals axis.

An auto-correlation analysis on the residuals is a particularly useful means of validation. The closer the auto correlation function, as determined by Equation 82.18, is to zero the better the regression model fits the data.

83.5 Goodness of Fit

This is a measure of how well a regression equation fits the data from which it was derived. Consider the identity:

$$y - \bar{y} = (y - \hat{y}) + (\hat{y} - \bar{y})$$

Squaring both sides gives:

$$(y - \bar{y})^2 = (y - \hat{y})^2 + 2(y - \hat{y})(\hat{y} - \bar{y}) + (\hat{y} - \bar{y})^2$$

Thus, for a series of n data points:

$$\sum_{j=1}^{n}(y_j - \bar{y})^2 = \sum_{j=1}^{n}(y_j - \hat{y}_j)^2$$

$$+ 2\sum_{j=1}^{n}(y_j - \hat{y}_j)(\hat{y}_j - \bar{y})$$

$$+ \sum_{j=1}^{n}(\hat{y}_j - \bar{y})^2$$

If the regression is a good fit, and provided n is fairly large, then both:

$$\sum_{j=1}^{n}(y_j - \hat{y}_j) \approx 0 \quad \text{and} \quad \sum_{j=1}^{n}(\hat{y}_j - \bar{y}) \approx 0$$

whence:

$$\sum_{j=1}^{n}(y_j - \bar{y})^2 = \sum_{j=1}^{n}(y_j - \hat{y}_j)^2 + \sum_{j=1}^{n}(\hat{y}_j - \bar{y})^2$$

This may be thought of as:

Total variation = unexplained variations
+ explained variations

The coefficient of determination R^2, sometimes referred to as the goodness of fit coefficient, is then articulated as the ratio of explained variations to total variation:

$$R^2 = \frac{\sum_{j=1}^{n}(\hat{y}_j - \bar{y})^2}{\sum_{j=1}^{n}(y_j - \bar{y})^2} = \frac{(\hat{\underline{y}} - \bar{y})^T . (\hat{\underline{y}} - \bar{y})}{(\underline{y} - \bar{y})^T . (\underline{y} - \bar{y})} \quad (83.11)$$

where $\underline{y} = \begin{bmatrix} y_1 & y_2 & \cdots & y_n \end{bmatrix}^T$.

The coefficient of determination lies between 0 and 1. The closer R^2 is to one the nearer the fitted values are to the observed values and the better the regression model fits the data. A value for R^2 of 0.9 is excellent, 0.8 is good, 0.7 is OK, 0.6 is suspect and 0.5 or less is useless.

83.6 Worked Example No 1

An experiment was set up to establish the variation of specific heat of a substance with temperature. Results of measurements taken at each of a series of temperatures are as shown in Table 83.1.

The mean values are $\bar{\theta} = 72.5$ and $\bar{c}_p = 7.039$.

Table 83.1 Specific heat vs temperature data

θ (°C)	50	55	60	65	70	75	80	85	90	95
c_p (kJ/kg °C)	6.72	6.91	6.85	6.97	7.01	7.12	7.14	7.22	7.18	7.27

Noting that the temperature is the input and specific heat is the output, the β coefficients may be calculated from Equation 83.9:

$$\hat{\beta}_1 = \frac{(\underline{\theta} - \bar{\theta}) \cdot (\underline{c}_p - \bar{c}_p)^T}{(\underline{\theta} - \bar{\theta}) \cdot (\underline{\theta} - \bar{\theta})^T} = 0.0113$$

where $\underline{\theta}$ and $\underline{c}_p$ are row vectors of the measured values, and from Equation 83.8:

$$\hat{\beta}_0 = \bar{c}_p - \hat{\beta}_1 \bar{\theta} = 6.221$$

whence:

$$\hat{c}_p = 6.221 + 0.0113 \cdot \theta$$

The coefficient of determination is given by Equation 83.11:

$$R^2 = \frac{(\underline{\hat{c}}_p - \bar{c}_p) \cdot (\underline{\hat{c}}_p - \bar{c}_p)^T}{(\underline{c}_p - \bar{c}_p) \cdot (\underline{c}_p - \bar{c}_p)^T} \approx 0.93$$

where $\underline{\hat{c}}_p$ is the row vector of the fitted values.

The coefficient indicates that some 93% of the variability in the specific heat is explained by the change in temperature.

83.7 Multiple Linear Regression

As seen in Equation 83.6, an MLR model involves several inputs and one output. Due to measurement errors, for any given set of values of x there will be an error on the measured value of y according to:

$$y = \beta_0 + \beta_1 \cdot x_1 + \beta_2 \cdot x_2 + \cdots + \beta_p \cdot x_p + \varepsilon \quad (83.12)$$

Regression analysis involves estimating the various β coefficients for which n sets of empirical data are required. This data may be collated in matrix form as follows:

$$\begin{bmatrix} y_1 \\ y_2 \\ \vdots \\ y_n \end{bmatrix} = \begin{bmatrix} 1 & x_{11} & x_{12} & \cdots & x_{1p} \\ 1 & x_{21} & x_{22} & \cdots & x_{2p} \\ \vdots & \vdots & \vdots & \ddots & \vdots \\ 1 & x_{n1} & x_{n2} & \cdots & x_{np} \end{bmatrix} \begin{bmatrix} \beta_0 \\ \beta_1 \\ \beta_2 \\ \vdots \\ \beta_p \end{bmatrix} + \begin{bmatrix} \varepsilon_1 \\ \varepsilon_2 \\ \vdots \\ \varepsilon_n \end{bmatrix}$$

which is of the general form:

$$\underline{y} = X \cdot \underline{\beta} + \underline{\varepsilon} \quad (83.13)$$

where $\underline{y}$ and $\underline{\varepsilon}$ are (nx1) vectors, X is an (nx(p+1)) matrix and $\underline{\beta}$ is a ((p+1)x1) vector.

The sum of the squares of the residuals may be formulated according to:

$$Q = \sum_{j=1}^{n} \varepsilon_j^2 = \underline{\varepsilon}^T \cdot \underline{\varepsilon}$$

$$= (\underline{y} - X \cdot \underline{\beta})^T \cdot (\underline{y} - X \cdot \underline{\beta})$$

$$= \underline{y}^T \underline{y} - \underline{\beta}^T X^T \underline{y} - \underline{y}^T X \underline{\beta} + \underline{\beta}^T X^T X \underline{\beta}$$

Noting that $(\underline{\beta}^T X^T \underline{y})^T = \underline{y}^T X \underline{\beta}$ and that both $\underline{\beta}^T X^T \underline{y}$ and $\underline{y}^T X \underline{\beta}$ are scalar quantities:

$$Q = \underline{y}^T \underline{y} - 2\underline{\beta}^T X^T \underline{y} + \underline{\beta}^T X^T X \underline{\beta}$$

The regression equation that best fits the data corresponds to the vector $\hat{\underline{\beta}}$ that minimises Q.

Noting that differentiation of a scalar by a vector is covered in Chapter 79, the derivative of Q with respect to $\hat{\underline{\beta}}$ is given by:

$$\frac{\partial Q}{\partial \underline{\beta}} = -2X^T \underline{y} + 2X^T X \underline{\beta}$$

Setting this to zero yields the best estimate of the vector $\hat{\underline{\beta}}$:

$$-X^T \underline{y} + X^T X \hat{\underline{\beta}} = 0$$

whence the so-called batch least squares (BLS) solution:

$$\hat{\underline{\beta}} = (X^T X)^{-1} X^T \underline{y} \quad (83.14)$$

The inverse of $X^T X$ should exist provided that the inputs are linearly independent, i.e. no column of the X matrix is a linear combination of the other columns. The less the "collinearity" the greater the accuracy of the matrix inversion.

The vector of fitted values is thus given by:

$$\underline{\hat{y}} = X.\hat{\underline{\beta}} = X (X^T X)^{-1} X^T \underline{y} = H.\underline{y} \quad (83.15)$$

where $H = X (X^T X)^{-1} X^T$.

Analogous to Equation 83.10, the residuals for MLR are defined to be the difference between the measured outputs and their fitted values:

$$\underline{\varepsilon} = \underline{y} - \underline{\hat{y}} = \underline{y} - X\hat{\underline{\beta}} = \underline{y} - H\underline{y} = (I - H).\underline{y}$$

Note that the formula of Equation 11 used for calculating the coefficient of determination R^2 for simple linear regression also applies to multiple linear regression.

83.8 Variable Selection

The key issue in MLR is the choice of variables. Given a number of possible inputs, how are the most important ones selected? Without doubt, the most important basis of selection is a knowledge of the relationships between the variables gained from an understanding of the underlying process and/or plant. If there is doubt about the significance of possible inputs, then selection may be aided by the use of cross correlation or principal components analysis, as explained in Chapter 101.

Otherwise, a systematic approach has to be adopted in which inputs are added or deleted from a subset of inputs according to the significance of their effect on the regression analysis. The so-called "forward selection" approach starts with the simple linear regression model which contains the input x_1 that has the biggest correlation with the output y. This correlation is in terms of absolute values and does not need to be standardised. The next input x_2 to be added to the model is that which has the second highest sample correlation with the input and/or increases the coefficient of determination value (R^2) by more than any other input. This process of adding inputs continues until all the inputs are included, or the number of inputs is deemed to be sufficient, or the increase in R^2 is no longer significant.

The reverse selection approach is essentially the reverse of forward selection, starting with a model containing all the inputs and then eliminating the least significant. Although these processes have the semblance of being quantitative, it should be recognised that they are essentially subjective.

83.9 Worked Example No 2

A polymerisation is carried out in a reactor batchwise. The end point of the reaction is imprecise, being some function of the mean molecular weight, but normally occurs between 8 and 12 h after the start of the batch. The extent of conversion (fractional) is determined by analysis of samples. The refractive index (dimensionless) and viscosity are measured on line. Data obtained for one batch is given in Table 83.2.

From an understanding of the process the following MLR model is proposed:

$$y = \beta_0 + \beta_1.r + \beta_2.\log(v)$$

There are seven sets of data which can be captured in the form of Equation 83.13:

$$\underline{y} = X.\underline{\beta} + \underline{\varepsilon}$$

$$\begin{bmatrix} 0.781 \\ 0.843 \\ 0.841 \\ 0.840 \\ 0.850 \\ 0.852 \\ 0.855 \end{bmatrix} = \begin{bmatrix} 1.0 & 1.533 & -1.569 \\ 1.0 & 1.428 & -1.181 \\ 1.0 & 1.567 & -0.854 \\ 1.0 & 1.496 & -0.532 \\ 1.0 & 1.560 & -0.267 \\ 1.0 & 1.605 & 0.020 \\ 1.0 & 1.487 & 0.258 \end{bmatrix} \begin{bmatrix} \beta_0 \\ \beta_1 \\ \beta_2 \end{bmatrix} + \begin{bmatrix} \varepsilon_1 \\ \varepsilon_2 \\ \varepsilon_3 \\ \varepsilon_4 \\ \varepsilon_5 \\ \varepsilon_6 \\ \varepsilon_7 \end{bmatrix}$$

The estimates of the β coefficients are given by batch least squares, Equation 83.14:

$$\hat{\underline{\beta}} = \left(X^T X\right)^{-1} X^T \underline{y} = \begin{bmatrix} 1.010 \\ -0.100 \\ 0.0332 \end{bmatrix}$$

whence the regression equation:

$$y = 1.01 - 0.1r + 0.0332 \log(v)$$

The fitted values are given by Equation 83.15:

$$\hat{\underline{y}} = X.\hat{\underline{\beta}} = [\,0.804\ 0.828\ 0.824\ 0.842\ 0.845\ 0.850\ 0.869\,]^T$$

Knowing that $\bar{y} = 0.837$ the coefficient of determination is found from Equation 83.11:

$$R^2 = \frac{(\hat{\underline{y}} - \bar{y})^T \cdot (\hat{\underline{y}} - \bar{y})}{(\underline{y} - \bar{y})^T \cdot (\underline{y} - \bar{y})} \approx 0.67$$

This means that some 67% of the variability in the conversion is explained by the changes in refractive index and viscosity. A relatively low coefficient could have been anticipated because inspection of the data reveals that it is not monotonic: the values of conversion and refractive index do not successively increase or decrease with time.

83.10 Worked Example No 3

A naphtha (C6–C8 hydrocarbons) stream is split in a column as depicted in Figure 83.2, the objective being to operate the column against a constraint on the maximum amount of $\geq$ C7s in the top product stream.

The various measurements and controls are as described in Table 83.3.
Plotting the raw data reveals the following insights:

- The splitter is being forced by the feed rate x_1 that is cyclical, albeit with a small amplitude (period of approximately 25 samples).
- The composition y_3 (of interest) and the top product vapour take off x_6 seem to cycle in response to x_1.
- The column is also forced by the top product liquid take off x_7 which is progressively stepped upwards, and bottom product take offs x_{13} and x_{14} which have large increases.
- The level x_9 decreases to reflect the various increased take offs: the level controller x_9 seems to be detuned because, as the level changes, the take off x_{12} doesn't vary much.
- There is an apparent connection between the level x_9, the feed temp x_2 and the bottom product composition x_{11}.

A total of 145 sets of data for these 17 variables was gathered at 5-min intervals. Summary statistics for pre-processing of the data are as shown in Table 83.4. The standard deviation is calculated from Equation 82.3 and the cross correlation function using Equation 82.16.

Inspection of the column headed "Percent" reveals that for variables x_2, x_8, x_{10} and x_{11} the standard deviation is $\leq$ 1% of the mean which, bearing in mind the accuracy of the instrumentation likely to have been used in their measurement, suggests that the errors in the measurements is likely to be more significant than the trends in the data. These variables are therefore discounted as being statistically suspect. Inspection of the column headed "Xcorr fn" reveals that the cross correlation function for variables x_2, x_5, x_8, x_9 and x_{11} to x_{14} are all $\geq$ 13 samples or 65 min. These are not credible on the basis of a time delay, known *a priori*, of some 30 min for changes in the feed to affect

Table 83.2 Conversion, refractive index and viscosity *vs* time data

Time: t (h)	8.5	9.0	9.5	10.0	10.5	11.0	11.5
Conversion: y	0.781	0.843	0.841	0.840	0.850	0.852	0.855
Refractive index: r	1.533	1.428	1.567	1.496	1.560	1.605	1.487
Viscosity: v (kg/ms)	0.027	0.066	0.140	0.294	0.541	1.048	1.810

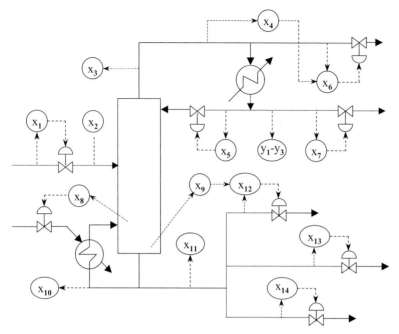

Fig. 83.2 Outline P&I diagram for naphtha column

Table 83.3 Definition of variables for naphtha column

Variable	Units	Description
y_1	%	Amount of benzene in top product (by analysis)
y_2	%	Amount of $\leq C4$s in top product (by analysis)
y_3	%	Amount of $\geq C7$s in top product (by analysis)
x_1	m³/h	Controlled flow rate of feed to splitter
x_2	°C	Inlet temperature of feed stream
x_3	°C	Temperature at top of splitter
x_4	bar (g)	Controlled pressure in overhead system
x_5	m³/h	Controlled reflux stream flow rate
x_6	m³/h	Manipulated flow rate of vapour top product stream (liquid equivalent)
x_7	m³/h	Controlled flow rate of liquid top product stream
x_8	°C	Controlled lower tray temperature
x_9	%	Controlled level in splitter still
x_{10}	°C	Temperature at bottom of splitter
x_{11}	°C	Composition of bottoms (5% cut point)
x_{12}	m³/h	Manipulated flow rate of bottom product stream
x_{13}	m³/h	Controlled flow rate of second bottom product stream
x_{14}	m³/h	Controlled flow rate of third bottom product stream

Table 83.4 Summary statistical data for variables of naphtha column

Variable	Units	Mean	Standard deviation	Percent	Xcorr fn
y_1	%	2.325	0.1735	7.4	–
y_2	%	1.497	0.2771	18.5	–
y_3	%	0.7394	0.2447	33.0	0
x_1	m^3/h	345.5	6.120	1.7	12
x_2	°C	131.0	1.175	0.9	74
x_3	°C	75.05	0.7822	1.0	12
x_4	bar (g)	1.191	0.0396	3.3	13
x_5	m^3/h	95.80	1.744	1.8	113
x_6	m^3/h	102.5	4.520	4.4	10
x_7	m^3/h	11.76	2.570	21.8	3
x_8	°C	126.1	0.3364	0.2	108
x_9	%	61.36	14.52	23.6	87
x_{10}	°C	133.0	0.5096	0.3	11
x_{11}	°C	93.76	0.9124	0.9	117
x_{12}	m^3/h	0.9258	0.4828	52.0	95
x_{13}	m^3/h	118.1	2.386	2.0	63
x_{14}	m^3/h	118.1	2.372	2.0	64

changes in the product streams and so they too are discounted.

Of the remaining variables the biggest delay is of 13 samples in the variable x_4 so, in effect there are 132 complete sets of data representing a period of some 11 h which is an excellent statistical basis. The data for each of these variables is shifted by the appropriate amount, standardised by subtracting the mean and dividing by the standard deviation, and assembled in the form of the X matrix of Equation 83.13. Note that since the data is standardised there is no need for a bias term β_0. The regression coefficients are found by means of batch least squares, Equation 83.14, yielding the MLR model:

$$y_3 = 0.1863.x_1 + 0.3255.x_3 - 0.1492.x_4 + 0.4621.x_6 + 0.1867.x_7$$

The coefficient of determination is found from Equation 83.11 to be 0.824 which indicates that 82% of the variability in the data is accounted for by the model. But this is based upon all five credible inputs. By means of the reverse selection process it is found that a good fit is obtained by eliminating all the variables except for x_3 and x_6 which results in the MLR model:

$$y_3 = 0.5505.x_3 + 0.3988.x_6$$

for which the coefficient of determination is 0.783. Thus the fit is still good but the model much simpler. Given that the reflux rate x_5 is relatively constant, it is to be expected that the top product composition would be highly correlated to the overhead temperature x_3 and to the dominant distillate flow rate x_6. A plot of the residuals versus fitted values is of a random nature which confirms the validity of the model.

83.11 Comments

Regression models are intended for use as interpolation equations and are only as good as the original data from which they were derived. The models are only valid over the range of inputs and outputs used to fit the model and extrapolation beyond these ranges should be treated with suspicion.

And finally, just because a regression model can be fitted to two or more variables, it doesn't necessarily imply a causal relationship. For example, as a reaction approaches completion both the viscosity and refractive index may change. There may well be a regression model between the viscosity and the refractive index, but it is not sensible to say that the change in viscosity is caused by the change in refractive index. In fact, the changes are primarily a function of the extent of reaction.

Section 10

Plant and Process Dynamics

Linearisation

84.1 The Need for Linearisation
84.2 Deviation Variables
84.3 The Process of Linearisation
84.4 Unsteady State Balances
84.5 Transfer Function Model
84.6 Worked Example
84.7 Nomenclature

Usually it is sufficient to have a qualitative feel for the way a plant behaves. For example, decomposing a system into its signals and elements, and being able to articulate their relationships in the form of a block diagram, even if the transfer functions are not known, is often good enough. Sometimes, however, a deeper understanding is required, perhaps because the plant is highly integrated or the process has complex dynamics and is difficult to control. In such circumstances it may be necessary to build a quantitative model as a basis for design. Occasionally, as in model predictive control, a model is essential as it is an integral part of the strategy.

There are various categories of models: the emphasis in this and subsequent chapters is on first principles, deterministic models. Other types of model are covered later on: regression and statistical models, time series models, knowledge based models and so on. First principles models that accurately represent the dynamic behaviour of plants and processes are complex and time consuming to develop. The key to successful modelling, therefore, is knowing how inaccurate a model you can get away with. This is essentially a question of using the right type of model, and deciding what assumptions can be made and what approximations are valid.

There are many excellent texts in which process modelling is covered including those by Coughanowr (1991), Luyben (1990), Marlin (2000), Ogunnaike and Ray (1994), Seborg (2004) and Stephanopoulos (1984) to which the reader is referred for a more comprehensive treatment.

84.1 The Need for Linearisation

Most of the control techniques considered in Section 9 of this Handbook are applicable to linear systems only. Unfortunately, many items of plant and control loop elements have nonlinear characteristics. So, in order to be able to apply these control techniques, the plant models have to be linearised. This involves changing the structure of the model into a linear form. The process of linearisation is arguably the most important modelling technique of all.

Using the characteristic of a control valve and a simple level control system, this chapter introduces a number of basic modelling techniques: the use of deviation variables, the linearisation process, lumped parameter dynamics, transfer function development and integration of plant and control system models. These techniques are further developed in subsequent chapters in which dynamic models for a variety of plant items are established.

84.2 Deviation Variables

The use of deviation variables provides the basis for linearised models. Consider the control valve depicted in Figure 84.1.

Fig. 84.1 Operating variables used to characterise a control valve

As described in Chapter 22, the absolute value of a variable may be expressed in terms of its normal value and some deviation, or perturbation, from that norm:

$$F = \bar{F} + \Delta F$$
$$X = \bar{X} + \Delta X$$
$$\Delta P_V = \overline{\Delta P_V} + \Delta(\Delta P_V)$$

where the bar denotes "normal" conditions. This is a somewhat ambiguous phrase. For example, at the design stage normal could be the specified conditions. For an existing plant normal could mean the average operating conditions. And in the context of control systems, normal would relate to set points or desired values. None of these are necessarily the same thing.

84.3 The Process of Linearisation

The process of linearisation essentially concerns approximating some relationship with its tangent at the point corresponding to normal conditions and considering thereafter only deviations of the variables about that point.

Referring again to the control valve, as explained in Chapter 20, its inherent characteristic is the relationship between flow and stem position assuming a constant pressure drop:

$$F = f(X)|_{\Delta P_V = const} \tag{20.1}$$

If $\Delta P_V = \overline{\Delta P_V}$, then the inherent characteristic inherent characteristic may be as depicted in Figure 84.2.

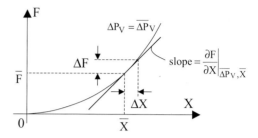

Fig. 84.2 Inherent characteristic of control valve

Clearly, for small changes in operating conditions about the point $\bar{F}, \bar{X}$, the characteristic may be approximated by the tangent to the curve at that point:

$$F = \bar{F} + \Delta F \approx \bar{F} + \left.\frac{\partial F}{\partial X}\right|_{\overline{\Delta P_V}, \bar{X}} . \Delta X \tag{84.1}$$

The extent to which this approximation is valid depends on the curvature of the characteristic at the operating point and on the size of the deviation ΔX. Some judgement is required. Typically, if the curvature is strong, deviations of up to ±10% can be accommodated. For weak curvatures, deviations of ±20% or more can be accommodated. Remember that the feedback nature of a control system inherently minimises the magnitude of the deviation of its signals so, for design purposes, it is not unreasonable to assume a-priori that the deviations will be small if the system functions effectively.

It is common practice, for modelling purposes, to shift the origin to the normal operating point, as depicted in Figure 84.3. Thus the axes become ΔF vs ΔX and the variables are the deviations.

The inherent characteristic thus becomes:

$$\Delta F \approx \left.\frac{\partial F}{\partial X}\right|_{\overline{\Delta P_V}, \bar{X}} . \Delta X$$

It is conventional, at this stage, to drop the Δ notation. This is clearly a potential source of confusion but, with experience, it is easy to distinguish be-

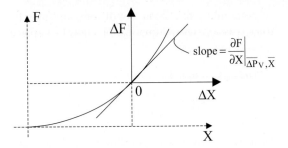

Fig. 84.3 Inherent characteristic with variables in deviation form

tween models whose variables are in absolute and deviation form. Thus:

$$F \approx \left.\frac{\partial F}{\partial X}\right|_{\overline{\Delta P_V}, \overline{X}} . X \qquad (84.2)$$

where it is understood that both F and X represent deviation variables.

Similarly, the relationship between flow and pressure drop, assuming a constant stem position, may be plotted as shown in Figure 84.4.

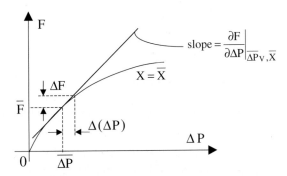

Fig. 84.4 Flow vs pressure drop for fixed valve opening

This too may be linearised about the point corresponding to normal conditions, again assuming deviation variables:

$$F \approx \left.\frac{\partial F}{\partial \Delta P_V}\right|_{\overline{\Delta P_V}, \overline{X}} . \Delta P_V \qquad (84.3)$$

In practice, of course, as the valve opening changes so too does the pressure drop across it resulting in the installed characteristic:

$$F = f(X, \Delta P_V) \qquad (20.2)$$

84.4 Unsteady State Balances

According to the principle of superposition, the net change in flow may be approximated by the sum of the individual affects of changing the valve opening assuming a constant pressure drop, and changing the pressure drop assuming constant valve opening. Thus combining Equations 84.2 and 84.3 gives:

$$F \approx \left.\frac{\partial F}{\partial X}\right|_{\overline{\Delta P_V}, \overline{X}} . X + \left.\frac{\partial F}{\partial \Delta P_V}\right|_{\overline{\Delta P_V}, \overline{X}} . \Delta P_V \qquad (84.4)$$

where the variables are in deviation form. The significance of this becomes apparent when incorporated in the model of a plant.

84.4 Unsteady State Balances

Consider again the simple level control system, as depicted in Figure 84.5, which was analysed from a steady-state point of view in Chapter 22.

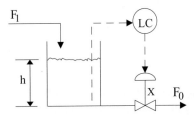

Fig. 84.5 Schematic of simple level control system

Unsteady state balances, whether for mass, energy or whatever, are always of the general form:

Rate of Accumulation = Input − Output

An unsteady state volume balance for the liquid inside the vessel provides the basis for the plant model:

$$\frac{d}{dt}(Ah) = \frac{1}{\rho}(F_1 - F_0) \quad m^3 \ min^{-1}$$

Recognising that A is a constant, this may be rearranged:

$$A\rho \frac{dh}{dt} = F_1 - F_0 \qquad (84.5)$$

The variables may now be put into deviation form:

$$A\rho \frac{d(\bar{h} + \Delta h)}{dt} = \bar{F}_1 + \Delta F_1 - (\bar{F}_0 + \Delta F_0)$$

Since $\bar{h}$ is the normal level which may be assumed to be constant:

$$A\rho \frac{d\Delta h}{dt} = \bar{F}_1 + \Delta F_1 - (\bar{F}_0 + \Delta F_0) \quad (84.6)$$

However, under normal conditions, at steady state, there is no accumulation:

$$0 = \bar{F}_1 - \bar{F}_0 \quad (84.7)$$

Subtracting Equation 84.7 from Equation 84.6 yields:

$$A\rho \frac{d\Delta h}{dt} = \Delta F_1 - \Delta F_0$$

It is this subtraction of the steady state that shifts the origin to the normal operating conditions and establishes the model in deviation form. Dropping the Δ notation yields:

$$A\rho \frac{dh}{dt} = F_1 - F_0 \quad (84.8)$$

Note that Equations 84.5 and 84.8, in which the variables are in absolute and deviation form respectively, appear identical. This happens to be the case for the plant considered but is not usually so, as demonstrated in subsequent chapters.

Substituting for F_0 from Equation 84.4, which is already in deviation form, yields:

$$A\rho \frac{dh}{dt} \approx F_1 - \left(\frac{\partial F_0}{\partial X}\bigg|_{\bar{h},\bar{X}} .X + \frac{\partial F_0}{\partial h}\bigg|_{\bar{h},\bar{X}} .h \right) \quad (84.9)$$

where it is assumed that the pressure drop across the valve is the liquid head in the tank.

84.5 Transfer Function Model

Noting that the two partial differentials are simply coefficients, *i.e.* the slope of two curves at the point corresponding to normal conditions, and may be treated as constants, Equation 84.9 may be Laplace transformed:

$$A\rho (sh(s) - h_0) =$$
$$F_1(s) - \frac{\partial F_0}{\partial X}\bigg|_{\bar{h},\bar{X}} .X(s) - \frac{\partial F_0}{\partial h}\bigg|_{\bar{h},\bar{X}} .h(s)$$

Assuming zero initial conditions, *i.e.* the level started at its set point, such that $h_0 = 0$ for $t \leq 0$, then:

$$A\rho sh(s) = F_1(s) - \frac{\partial F_0}{\partial X}\bigg|_{\bar{h},\bar{X}} .X(s) - \frac{\partial F_0}{\partial h}\bigg|_{\bar{h},\bar{X}} .h(s)$$

which may be rearranged to give:

$$\left(\frac{A\rho}{\frac{\partial F_0}{\partial h}\bigg|_{\bar{h},\bar{X}}} s + 1 \right) h(s) = \quad (84.10)$$

$$\frac{1}{\frac{\partial F_0}{\partial h}\bigg|_{\bar{h},\bar{X}}} F_1(s) - \frac{\frac{\partial F_0}{\partial X}\bigg|_{\bar{h},\bar{X}}}{\frac{\partial F_0}{\partial h}\bigg|_{\bar{h},\bar{X}}} .X(s)$$

which is of the general form:

$$(T_P s + 1) h(s) = K_L.F_1(s) - K_P X(s) \quad (84.11)$$

where T_P, K_L and K_P are defined by Equation 84.10. Clearly the dynamics of the tank are first order with a time constant of T_P, and may be represented by transfer functions, as depicted in Figure 84.6.

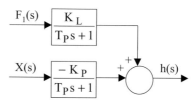

Fig. 84.6 Transfer function model of the process

Note the significance of the signs associated with K_L and K_P. The positive sign of the load gain K_L indicates that following an increase in flow F_1 the

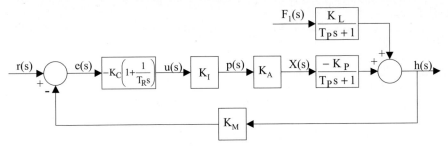

Fig. 84.7 Process model integrated into block diagram of control system

level goes up, whereas the negative sign of the process gain K_P indicates that following an increase in valve opening X the level goes down. These transfer functions may be incorporated in a block diagram for the level control system as a whole, as depicted in Figure 84.7.

Note that a reverse acting PI controller is assumed, and that the dynamics of all the other control loop elements are ignored on the basis that they are fast compared with the process and load.

84.6 Worked Example

The level in a tank is controlled as depicted in Figure 84.5. The tank is 1 m in diameter and 2 m tall and is normally half full. The valve has an equal percentage characteristic and is normally half open.

Thus normal conditions are $\overline{h} = 1$ m and $\overline{X} = 0.5$.

Suppose that the flow through the half open valve is related to the head in the tank by the equation:

$$F_0 = 300\sqrt{h}$$

Substitute for $\overline{h}$ into the tank characteristic gives $\overline{F}_0 = \overline{F}_1 = 300$ kg min^{-1}.

The slope of the tank characteristic is:

$$\left.\frac{\partial F_0}{\partial h}\right|_{\overline{X}} = \frac{300}{2\sqrt{h}}$$

whence:

$$\left.\frac{\partial F_0}{\partial h}\right|_{\overline{h},\overline{X}} = 150.$$

Substitute into Equation 84.10 gives:

$$K_L = \frac{1}{\left.\frac{\partial F_0}{\partial h}\right|_{\overline{h},\overline{X}}} = 6.66 \times 10^{-3} \text{ m min kg}^{-1}.$$

Now suppose that the flow through the valve, assuming a constant head loss of 1 m, is related to its stem position by the equation:

$$F_0 = 60e^{3.22X}$$

Note that substituting $\overline{X} = 0.5$ into this equation also gives $\overline{F}_0 = \overline{F}_1 = 300$ kg min^{-1}.

The slope of the valve characteristic is:

$$\left.\frac{\partial F_0}{\partial X}\right|_{\overline{h}} = 60 \times 3.22 e^{3.22X}$$

whence:

$$\left.\frac{\partial F_0}{\partial X}\right|_{\overline{h},\overline{X}} = 967.$$

Substitute into Equation 84.10 gives:

$$K_P = \frac{\left.\frac{\partial F_0}{\partial X}\right|_{\overline{h},\overline{X}}}{\left.\frac{\partial F_0}{\partial h}\right|_{\overline{h},\overline{X}}} = \frac{967}{150} = 6.66 \text{ m}.$$

The cross sectional area $A = \frac{\pi . 1^2}{4} = 0.785$ m^2 and the density $\rho = 1000$ kg m^{-3}.

Whence the time constant, also from Equation 84.10:

$$T_P = A.\rho.K_L = 0.785 \times 1000 \times 6.66 \times 10^{-3}$$
$$= 5.23 \text{ min}$$

84.7 Nomenclature

A	cross sectional area of the tank	m²
h	level of liquid	m
F	mass flow rate through the valve	kg min⁻¹
K	gain	
ρ	liquid density	kg m⁻³
ΔP	pressure drop	bar
X	fractional opening of the valve	–
T	time constant	min
t	time	min

Subscripts

A	actuator
C	controller
I	I/P converter
L	load
P	process
V	valve or variable resistance
0	outlet stream
1	inlet stream

Lumped Parameter Systems

85.1 Lumped Parameter Models
85.2 Steam Heated Jacketed Vessel
85.3 Water Cooled Jacketed Vessel
85.4 Worked Example
85.5 Nomenclature

A lumped parameter system is one whose parameters are uniform throughout. They are quite common in the process industries, agitated vessels being a classic example. A common characteristic of all lumped parameter systems is their capacity for storage, either of mass or energy. Some simple examples of lumped parameter systems with first-order dynamics were considered in Chapter 69.

85.1 Lumped Parameter Models

Of fundamental importance in the modelling of the dynamics of lumped parameter systems is the concept of an unsteady state balance. This relates the rate of accumulation of some variable to the difference between the system's inputs and outputs. In general:

Rate of Accumulation = Input − Output   (85.1)

In Chapter 84 this equation was used in relation to a level control system. The approach used was to establish the instantaneous volume, to differentiate it with respect to time and to equate this to the difference between inlet and outlet flows. This approach is generic: the central issue is to decide on the nature of the balance and the appropriate variables. The two examples used in this chapter relate to temperature control for which unsteady state heat balances are required. A variety of other examples are used in subsequent chapters. The nature of the more common unsteady state balances, the variables involved and their contexts are summarised in Table 85.1.

85.2 Steam Heated Jacketed Vessel

Consider the agitated vessel depicted in Figure 85.1 through which a process stream flows continuously and is heated up by condensing steam in the vessel's jacket.

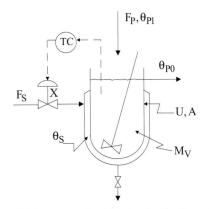

Fig. 85.1 Steam heated jacketed agitated vessel

Table 85.1 The nature of unsteady state balances

Nature of balance	Variables involved	Context
Volume	Height, flow	Level control
Pressure	Pressure, flow	Pressure control
Momentum	Velocity, pressure	Flow control
Heat	Temperature, flow	Temperature control
Mass	Concentration, flow	Composition control

In order to model this it is necessary to make a number of classical assumptions:

1. That the inlet flow displaces an equivalent volume over the overflow such that, to all practical intents and purposes:

$$F_{P1} \approx F_{P0} \approx F_P$$

In practice, following an increase in inlet flow, the level must rise to provide sufficient head to enable the corresponding increase in outlet flow. Clearly there is an associated time lag. This assumption is equivalent to assuming that these hydrodynamics are negligible relative to the dynamics of the thermal processes involved.

2. That the contents of the vessel are well mixed. Thus the physical properties of the liquid, such as its density and specific heat, are uniform throughout the contents of the vessel. Process parameters such as temperature are similarly uniform throughout. It follows that values of such are the same for the outlet stream as for the bulk of the liquid.

3. That the thermal capacity of the vessel wall is negligible relative to that of the mass of the contents of the vessel. In the second example below it is also assumed to be negligible relative to the mass of the contents of the jacket. This depends on the dimensions, thickness and materials involved but, for relatively large vessels, is a reasonable assumption.

4. That the steam pressure is uniform throughout the jacket. For a condensing system such as this, the steam temperature is a function of its pressure only, so the temperature must be uniform throughout the jacket too.

5. That the jacket and vessel form a non-interacting system. This means that the steam temperature is independent of the temperature of the contents of the vessel:

$$\theta_S = f(P_S) \neq g(\theta_{P0})$$

In practice, the temperature in the vessel does affect the steam pressure and hence its temperature, but that is through the effects of feedback.

An unsteady-state heat balance (relative to a datum of 0°C) for the contents of the vessel provides the basis for the plant model:

Rate of Accumulation = Input − Output

$$\frac{d}{dt}(M_V c_P \theta_{P0}) = F_P c_P (\theta_{P1} - \theta_{P0})$$
$$+ UA(\theta_S - \theta_{P0}) \quad \text{kW}$$

Since both M_V and c_p are constants:

$$M_V c_P \frac{d}{dt}\theta_{P0} = F_P c_P (\theta_{P1} - \theta_{P0}) \quad (85.2)$$
$$+ UA(\theta_S - \theta_{P0})$$

Putting the variables in deviation form, assuming that U and A are constants too, and rearranging yields:

$$M_V c_P \frac{d}{dt}\left(\overline{\theta}_{P0} + \Delta\theta_{P0}\right)$$
$$= \left(\overline{F}_P + \Delta F_P\right) c_P \left(\overline{\theta}_{P1} + \Delta\theta_{P1} - \left(\overline{\theta}_{P0} + \Delta\theta_{P0}\right)\right)$$
$$+ UA\left(\overline{\theta}_S + \Delta\theta_S - \left(\overline{\theta}_{P0} + \Delta\theta_{P0}\right)\right)$$

Noting that the differential of a constant is zero, and ignoring the product of two deviations, each of which is relatively small, as being insignificant, this equation may be expanded and rearranged:

$$M_V c_P \frac{d}{dt}\Delta\theta_{P0} = \overline{F}_P c_P \left(\overline{\theta}_{P1} - \overline{\theta}_{P0}\right) + \overline{F}_P c_P \left(\Delta\theta_{P1} - \Delta\theta_{P0}\right)$$
$$+ \left(\overline{\theta}_{P1} - \overline{\theta}_{P0}\right) c_P \Delta F_P$$
$$+ UA \left(\overline{\theta}_S - \overline{\theta}_{P0}\right) + UA \left(\Delta\theta_S - \Delta\theta_{P0}\right)$$

The steady state heat balance, under normal conditions is:

$$0 = \overline{F}_P c_P \left(\overline{\theta}_{P1} - \overline{\theta}_{P0}\right) + UA \left(\overline{\theta}_S - \overline{\theta}_{P0}\right)$$

which, on subtraction, yields:

$$M_V c_P \frac{d}{dt}\Delta\theta_{P0} = \overline{F}_P c_P \left(\Delta\theta_{P1} - \Delta\theta_{P0}\right)$$
$$+ \left(\overline{\theta}_{P1} - \overline{\theta}_{P0}\right) c_P \Delta F_P$$
$$+ UA \left(\Delta\theta_S - \Delta\theta_{P0}\right)$$

Dropping the Δ notation gives the model in deviation form:

$$M_V c_P \frac{d}{dt}\theta_{P0} = \overline{F}_P c_P \left(\theta_{P1} - \theta_{P0}\right) \quad (85.3)$$
$$+ \left(\overline{\theta}_{P1} - \overline{\theta}_{P0}\right) c_P F_P + UA \left(\theta_S - \theta_{P0}\right)$$

Note that Equations 85.2 and 85.3 are different, *i.e.* the absolute and deviation forms of the model are not the same. This is because the first term on the right hand side of the absolute form contained the products of two variables, *i.e.* F and θ, each product resulted in two terms in the deviation form. This concept generalises to terms containing the product of n variables, as summarised in Table 85.2.

Table 85.2 Mapping between absolute and deviation forms of equation

Equation in absolute form No of variables in term	Equation in deviation form No of terms arising
None (*i.e.* constant)	None (it disappears)
One	One (stays the same)
Two	Two
Three (or more)	Three (or more)

With practice it should be possible to write down the deviation form of a model from its absolute form directly, *i.e.* without going through all the intermediate steps. Note that many texts quote models in deviation form: it is presumed that the reader appreciates the distinction between absolute and deviation variables and is capable of generating the model in deviation form from its absolute form in the first place.

Equation 85.3 may be Laplace transformed, assuming zero initial conditions:

$$M_V c_P s \theta_{P0}(s) = \overline{F}_P c_P \left(\theta_{P1}(s) - \theta_{P0}(s)\right)$$
$$+ \left(\overline{\theta}_{P1} - \overline{\theta}_{P0}\right) c_P F_P(s)$$
$$+ UA \left(\theta_S(s) - \theta_{P0}(s)\right)$$

which may be rearranged to give:

$$\left(\frac{M_V c_P}{\overline{F}_P c_P + UA}s + 1\right).\theta_{P0}(s)$$
$$= \frac{\overline{F}_P c_P}{\overline{F}_P c_P + UA}.\theta_{P1}(s) + \frac{\left(\overline{\theta}_{P1} - \overline{\theta}_{P0}\right) c_P}{\overline{F}_P c_P + UA}.F_P(s)$$
$$+ \frac{UA}{\overline{F}_P c_P + UA}.\theta_S(s)$$

which is of the form:

$$(T_P s + 1)\theta_{P0}(s) = K_1 \theta_{P1}(s) + K_2 F_P(s)$$
$$+ K_3 \theta_s(s) \quad (85.4)$$

The corresponding transfer functions may be incorporated in a block diagram for the temperature control system, as depicted in Figure 85.2.

Here θ_{P0} is the controlled variable, θ_S is the manipulated variable, and θ_{P1} and F_P are disturbance variables.

85.3 Water Cooled Jacketed Vessel

Now consider the same stirred vessel, but its contents are cooled by circulating water through the jacket, as depicted in Figure 85.3.

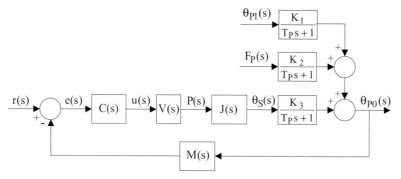

Fig. 85.2 Block diagram of control loop for steam heated vessel

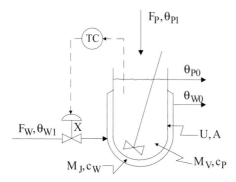

Fig. 85.3 Water cooled jacketed agitated vessel

Assumptions 1 to 3 above apply. In addition it is assumed:

6. That the contents of the jacket are well mixed such that the physical properties of the cooling water are uniform throughout the jacket and the values of parameters are the same for the jacket outlet stream as for its contents. This assumption is valid if the jacket is baffled and the water circulation rate is high. However, it is dubious if there is not significant turbulence in the jacket.
7. That the jacket and vessel form an interacting system. This means that the temperature of the contents of the vessel depends upon that in the jacket, and *vice versa*, i.e.:

$$\theta_{P0} = f(\theta_{W0})$$
$$\theta_{W0} = g(\theta_{P0})$$

Unsteady-state heat balances for the contents of both the vessel and jacket provide the basis for the plant model:

Rate of Accumulation = Input − Output

For the vessel:

$$M_V c_P \frac{d}{dt} \theta_{P0} = F_P c_P (\theta_{P1} - \theta_{P0}) - UA (\theta_{P0} - \theta_{W0}) \text{ kW}$$

For the jacket:

$$M_J c_W \frac{d}{dt} \theta_{W0} = F_W c_W (\theta_{W1} - \theta_{W0}) + UA (\theta_{P0} - \theta_{W0})$$

The interaction between these two equations is clearly established by the common term $UA(\theta_{P0} - \theta_{W0})$.

Putting the equations into deviation form and Laplace transforming assuming zero initial conditions gives:

$$\begin{aligned} M_V c_P s \theta_{P0}(s) &= \bar{F}_P c_P (\theta_{P1}(s) - \theta_{P0}(s)) \\ &\quad + (\bar{\theta}_{P1} - \bar{\theta}_{P0}) c_P F_P(s) \\ &\quad - UA(\theta_{P0}(s) - \theta_{W0}(s)) \end{aligned} \quad (85.5)$$

$$\begin{aligned} M_J c_W s \theta_{W0}(s) &= \bar{F}_W c_W (\theta_{W1}(s) - \theta_{W0}(s)) \\ &\quad + (\bar{\theta}_{W1} - \bar{\theta}_{W0}) c_W F_W(s) \\ &\quad + UA(\theta_{P0}(s) - \theta_{W0}(s)) \end{aligned} \quad (85.6)$$

These two equations have six variables which can be categorised as in Table 85.3. Of these, five variables are of significance from a control point of view. However, the cooling water outlet temperature θ_{W0} is only a state variable and may be eliminated.

85.3 Water Cooled Jacketed Vessel

Table 85.3 Categorisation of variables for water cooled vessel

Variable	Category
θ_{P0}	Controlled
F_W	Manipulated
θ_{P1}, θ_{W1} and F_P	Disturbance
θ_{W0}	State

Elimination from Equations 85.5 and 85.6 is algebraically tedious. It is more convenient, for illustrative purposes, to consider a numerical example. Suppose that:

$$\bar{F}_P = 0.2, \quad \bar{F}_W = 0.1, \quad c_P = 4.2, \quad c_W = 4.2$$
$$M_V = 1000, \quad M_J = 250, \quad U = 0.7, \quad A = 2.4$$
$$\bar{\theta}_{P1} = 90, \quad \bar{\theta}_{P0} = 70, \quad \bar{\theta}_{W1} = 20, \quad \bar{\theta}_{W0} = 60$$

where the units are as defined in the nomenclature. Substitution of these values and re-arrangement yields:

$$(4200s + 2.52)\theta_{P0}(s) = 0.84\theta_{P1}(s) + 84F_P(s) + 1.68\theta_{W0}(s)$$

$$(1050s + 2.1)\theta_{W0}(s) = 0.42\theta_{W1}(s) - 168F_W(s) + 1.68\theta_{P0}(s)$$

Eliminating $\theta_{W0}(s)$, rearranging and factorising gives:

$$(4219s + 1)(423.5s + 1)\theta_{P0}(s)$$
$$= 0.7142(500s + 1)\theta_{P1}(s) + 71.42(500s + 1)F_P(s)$$
$$+ 0.2857\theta_{W1}(s) - 114.3F_W(s) \qquad (85.7)$$

These transfer functions may be incorporated in a block diagram for the temperature control system, as depicted in Figure 85.4.

An important insight into the dynamics of agitated vessels can be gained by consideration of the structure of Equations 85.4 and 85.7. First, consider the steam heated vessel. The lag term in Equation 85.4 indicates that the dynamics are first order. That is to be expected because there is only one capacity, i.e. the contents of the vessel, for energy storage. Note that the dynamics are first order irrespective of the source of disturbance: any change affects the temperature in the vessel directly.

Second, for the water cooled vessel, there are two lag terms in Equation 85.7 indicating that the dynamics are second order. This is because there are two capacities for energy storage, i.e. the vessel and its jacket. Furthermore, the source of disturbance is significant. The transfer functions relating θ_{P0} to changes in θ_{W1} and F_W are truly second order. This is because the affects of any changes in the cooling water stream only indirectly affect the temperature inside the vessel. They are filtered by both the jacket and vessel contents. However, the transfer functions relating θ_{P0} to changes in

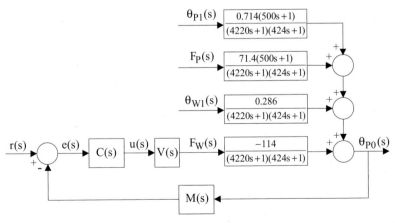

Fig. 85.4 Block diagram of control loop for water cooled vessel

θ_{P1} and F_P each have a lead term. These effectively offset one of the lag terms resulting in quasi first order dynamics. This is because any changes in the process stream directly affect the temperature in the vessel.

85.4 Worked Example

Aqueous streams A and B flow continuously into a stirred tank and react rapidly, releasing 1200 kJ of heat per kg of stream A reacted. The reaction goes substantially to completion, so the heat released is independent of small changes in temperature.

Assume that the cooling water is circulated through an internal coil at a rate such that its rise in temperature is small. This assumption means, in effect, that the thermal capacity of the coil is negligible and that the system can be treated as non-interacting.

The parameters have the following values, their units being as defined in the nomenclature:

$$\overline{F}_A = \overline{F}_B = 0.125, \quad c_A = c_B = 4.2$$
$$M_V = 2300, \quad U = 3.4, \quad A = 2.8$$
$$\overline{\theta}_A = \overline{\theta}_B = 25, \quad \overline{\theta}_{W1} = 20$$

A steady state heat balance for the contents of the reactor yields its temperature:

Input − Output = 0
$$(\overline{F}_A + \overline{F}_B) c_P (\overline{\theta}_A - \overline{\theta}_0) + \overline{F}_A \Delta H - UA(\overline{\theta}_0 - \overline{\theta}_W) = 0$$
$$0.25 \times 4.2(25 - \overline{\theta}_0) + 0.125 \times 1200$$
$$- 3.4 \times 2.8(\overline{\theta}_0 - 20) = 0$$

whence $\overline{\theta}_0 = 34.7°C$.

An unsteady state heat balance for the contents of the reactor is of the form:

Rate of Accumulation = Input − Output

$$Mc_P \frac{d\theta_0}{dt}$$
$$= (F_A + F_B) c_P (\theta_A - \theta_0) + F_A \Delta H - UA(\theta_0 - \theta_W)$$

Assume that it is the response of θ_0 to changes in θ_W that is of interest, in which case F_A, F_B and θ_A may be presumed to be constant:

$$9660 \frac{d\theta_0}{dt} = 1.05(25 - \theta_0) + 150 - 9.52(\theta_0 - \theta_W)$$

Put into deviation form:

$$9660 \frac{d\theta_0}{dt} = -10.57 \cdot \theta_0 + 9.52 \cdot \theta_W$$

Transform and rearrange:

$$\theta_0(s) = \frac{0.9}{(914s + 1)} \cdot \theta_W(s)$$

Suppose a step decrease in cooling water temperature of say 5°C occurs, i.e. $\theta_W(s) = \frac{-5}{s}$.

The reactor's response is:

$$\theta_0(t) = -4.5 \left(1 - e^{-t/914}\right)$$

Thus after say 5 min, i.e. t = 300 s, then $\overline{\theta}_0 = -1.26°C$.

Remember that this is in deviation form. Thus, as an absolute value, the reactor temperature is:

$$\theta_0 = \overline{\theta}_0 + \Delta\theta_0 = 34.7 - 1.26 \approx 33.4°C$$

85.5 Nomenclature

A	effective surface area for heat transfer	m²
C	controller	
F	flow rate	kg s⁻¹
M	mass of contents	kg
c	specific heat	kJ kg⁻¹ K⁻¹
P	pressure	bar
t	time	s
U	overall heat transfer coefficient	kW m⁻² k⁻¹
V	I/P converter/actuator/valve	
θ	temperature	°C

Subscripts

A	reagent stream
B	reagent stream
J	jacket
M	measurement
P	process stream
S	steam
V	vessel
W	cooling water
0	outlet
1	inlet

Zero Capacity Systems

86.1 Steam Injection System
86.2 Worked Example
86.3 Significance of Dynamics
86.4 Dead Time Compensation
86.5 Blending System
86.6 Nomenclature

The systems considered in Chapters 84 and 85 all had the capacity to store mass and/or energy, which gave rise to transfer functions which contained time lags. This is not always so. Some systems have zero capacity which means that their transfer functions are simple steady state gains. In such cases the dynamics of the associated pipework, pumps and valves become dominant. Two classic examples of zero capacity systems are considered in this chapter: steam injection and blending.

86.1 Steam Injection System

One way of heating up an aqueous stream is by direct steam injection through a nozzle, as indicated in Figure 86.1.

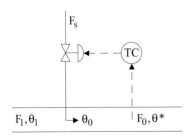

Fig. 86.1 In line steam injection system

The steam condenses virtually instantaneously upon injection, the heat released being dissipated quickly by mixing within the process stream due to extreme turbulence effects. Strictly speaking, the mixing occurs in the section of pipe just downstream of the nozzle. However, to all practical intents and purposes, the heat dissipation can be considered to be immediate.

An unsteady state mass balance at the mixing junction gives:

$$F_1 + F_S = F_0$$

An unsteady state heat balance (relative to a datum of $0°C$) at the mixing junction gives:

$$F_1 c_P \theta_1 + F_S h_g = F_0 c_P \theta_0 = (F_1 + F_S) c_P \theta_0$$

Putting into deviation form and transforming gives:

$$\overline{F}_1 c_P \theta_1(s) + \overline{\theta}_1 c_P F_1(s) + h_g F_S(s)$$
$$= (\overline{F}_1 + \overline{F}_S) c_P \theta_0(s) + \overline{\theta}_0 c_P (F_1(s) + F_S(s))$$

Rearrange gives:

$$\theta_0(s) = \frac{\overline{F}_1}{(\overline{F}_1 + \overline{F}_S)} . \theta_1(s) - \frac{(\overline{\theta}_0 - \overline{\theta}_1)}{(\overline{F}_1 + \overline{F}_S)} . F_1(s)$$
$$+ \frac{(h_g - \overline{\theta}_0 c_P)}{(\overline{F}_1 + \overline{F}_S) c_P} . F_S(s) \qquad (86.1)$$

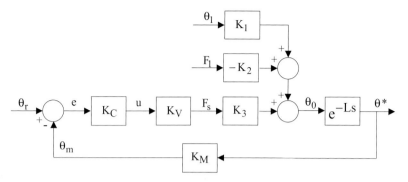

Fig. 86.2 Block diagram of steam injection control system

which is of the general form:

$$\theta_0(s) = K_1.\theta_1(s) - K_2.F_1(s) + K_3.F_S(s) \quad (86.2)$$

These gains may be incorporated in a block diagram of the temperature control system, as depicted in Figure 86.2.

86.2 Worked Example

An aqueous process stream is heated by injecting steam as depicted in Figure 86.1. Normal conditions are as follows:

$$\overline{F}_1 = 1.0\,\text{kg/s}, \quad \overline{\theta}_1 = 20°C, \quad \overline{\theta}_0 = 50°C$$

Assume that the steam is dry saturated and has a total heat h_g of 2700 kJ/kg.

A heat balance under normal conditions (relative to a datum of 0°C) across the mixing junction yields the normal steam flow rate:

$$\overline{F}_1.c_P.\overline{\theta}_1 + \overline{F}_S.h_g = (\overline{F}_1 + \overline{F}_2).c_P.\overline{\theta}_0$$
$$1.0 \times 4.2 \times 20 + \overline{F}_S \times 2700$$
$$= (1.0 + \overline{F}_S) \times 4.2 \times 50$$

whence $\overline{F}_S = 0.0506$ kg/s.

Substituting into Equation 86.1, the unsteady state heat balance:

$$\theta_0(s) = \frac{\overline{F}_1}{(\overline{F}_1 + \overline{F}_s)}.\theta_1(s) - \frac{(\overline{\theta}_0 - \overline{\theta}_1)}{(\overline{F}_1 + \overline{F}_s)}.F_1(s)$$
$$+ \frac{(h_g - \overline{\theta}_0.c_P)}{(\overline{F}_1 + \overline{F}_s)c_P}.F_S(s)$$
$$= 0.952.\theta_1(s) - 28.6.F_1(s) + 564.F_S(s)$$

Comparison with Equation 86.2 reveals that:

$$K_1 = 0.952, \quad K_2 = 28.6°C\,s\,kg^{-1}, \quad K_3 = 564°C\,s\,kg^{-1}.$$

The temperature transmitter has an input range of 25–75°C and an output of 4–20 mA. The controller has a set point of 50°C and a bandwidth of 50%. The I/P converter has an input range of 4–20 mA and an output range of 0.2–1.0 bar. The control valve has an input range of 0.2–1.0 bar and an output range of 0–0.15 kg/s. It may be assumed that the characteristics of all the instrumentation are linear and that their dynamics are negligible. With reference to the block diagram of Figure 86.2, the gains of the instrumentation are as follows:

$$K_M = 0.32\,\text{mA}\,°C^{-1}, \quad K_C = 2.0 \quad \text{and}$$
$$K_V = 0.009375\,\text{kg}\,\text{mA}^{-1}\,s^{-1}.$$

The steady state closed loop response to any disturbance is thus given by:

$$\theta_0 = K_1.\theta_1 - K_2.F_1 + K_3.K_V.K_C.(\theta_r - K_M.\theta_0)$$

Substituting for the various gains and rearranging gives:

$$\theta_0 = 0.217.\theta_1 - 6.52.F_1 + 2.41.\theta_R$$

Thus, for example, assuming a constant set point and a simultaneous increase in F_1 to 1.5 kg/s and decrease in θ_1 to 15°C, the resultant steady state offset would be:

$$\theta_0 = 0.217 \times (-5) - 6.52 \times (+0.5) \approx -4.3°C$$

However, this is in deviation form; the absolute value of the outlet temperature is:

$$\theta_0 = \overline{\theta}_0 + \Delta\theta_0 = 50 - 4.3 = 45.7°C$$

86.3 Significance of Dynamics

Note the time delay L in Figure 86.2 due to distance velocity effects between the mixing junction and the temperature probe:

$$L = \frac{\pi d^2 \ell \rho}{4\,(\overline{F}_1 + \overline{F}_S)}$$

The time delay is clearly dependant upon the flows. Since the steam flow is relatively small, the time delay will not vary much if the process stream flow is roughly constant. This delay, and the time lags associated with the valve and the temperature measurement, are the only dynamics of consequence in a control loop that potentially has a very fast response. This is typical of zero capacity systems.

86.4 Dead Time Compensation

The existence of the time delay within the loop, as depicted in Figure 86.2, is problematic. If the time delay is of the same order of magnitude or greater than the other lags in the loop, as is often the case in zero capacity systems, then tuning the controller is difficult and leads to a response that is more sluggish than would otherwise be the case. Ideally, the time delay needs to be moved outside the loop, as depicted in Figure 86.3, which enables a more effective controller C(s) to be designed.

Dead-time compensation is an approach which enables the controller C(s) to be designed as if the delay was outside the loop. Compensation is realised in practice by the so-called Smith predictor D(s) which utilises the controller output and the measured value, as shown in Figure 86.4.

Open loop analysis of Figures 86.3 yields:

$$u(s) = C(s).(\theta_r(s) - M(s).K_3.V(s).u(s))$$

and of Figure 86.4 yields:

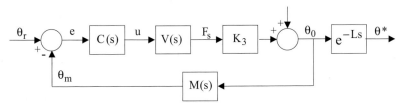

Fig. 86.3 Control loop with delay outside the loop

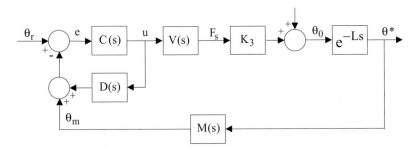

Fig. 86.4 Equivalent control loop with delay inside the loop

$$u(s) = C(s).\left(\theta_r(s) - (D(s).u(s) + M(s).e^{-Ls}.K_3.V(s).u(s))\right)$$

The condition for these to be exactly equivalent is that:

$$M(s).K_3.V(s).u(s) = D(s).u(s) + M(s).e^{-Ls}.K_3.V(s).u(s)$$

Rearranging gives:

$$D(s) = K_3.\left(1 - e^{-Ls}\right).M(s).V(s) \quad (86.3)$$

Dead-time compensation is only possible if the dynamics of the process are known. This means estimating the values of all the parameters in Equation 86.3. Effectiveness is dependant upon the accuracy of these values and is particularly sensitive to the accuracy of the delay L. Realisation of the predictor is only practicable by means of software.

86.5 Blending System

A common operation in the process industries is that of blending two streams together. Often a ratio control scheme is used, as described in Chapter 26. However, in the following example, a multivariable control strategy is used. Consider a concentrated salt solution which is to be diluted with an aqueous process stream, as depicted in Figure 86.5.

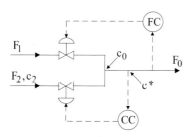

Fig. 86.5 An in line blending system

Dilute product is formed by mixing the two streams. It is assumed that the mixing is complete a short distance downstream of the junction. This is not unreasonable given that flow will be turbulent due to the bends in the pipework and the proximity of the control valves. Otherwise in-line mixers would be necessary.

An overall unsteady state mass balance across the mixing junction gives:

$$F_1 + F_2 = F_0$$

This equation is the same in deviation form. Transforming gives:

$$F_1(s) + F_2(s) = F_0(s) \quad (86.4)$$

Because the streams are liquid and incompressible, there is no delay associated with changes in flow. Any change in inlet flow causes an equal and immediate change in the outlet flow.

An unsteady state mass balance for the salt across the mixing junction gives:

$$F_2 C_2 = F_0 C_0 = (F_1 + F_2) C_0$$

Assume that the concentrated salt solution comes from a reservoir such that its concentration may be assumed to be constant. Putting into deviation form and transforming gives:

$$\overline{C_2} F_2(s) = \overline{F_0} C_0(s) + \overline{C_0} (F_1(s) + F_2(s))$$

Rearrange gives:

$$C_0(s) = \frac{(\overline{C_2} - \overline{C_0})}{\overline{F_0}}.F_2(s) - \frac{\overline{C_0}}{\overline{F_0}}.F_1(s)$$

which is of the general form:

$$C_0(s) = K_2 F_2(s) - K_1 F_1(s) \quad (86.5)$$

Note that there is a time delay due to distance velocity effects associated with the concentration measurement:

$$C^*(s) = e^{-Ls}.C_0(s) \quad (86.6)$$

Equations 86.4–86.6 can be presented in matrix form as follows:

$$\begin{bmatrix} F_0(s) \\ C^*(s) \end{bmatrix} = \begin{bmatrix} 1 & 0 \\ 0 & e^{-Ls} \end{bmatrix} \cdot \begin{bmatrix} F_0(s) \\ C_0(s) \end{bmatrix}$$

$$= \begin{bmatrix} 1 & 0 \\ 0 & e^{-Ls} \end{bmatrix} \cdot \begin{bmatrix} 1 & 1 \\ -K_1 & K_2 \end{bmatrix} \cdot \begin{bmatrix} F_1(s) \\ F_2(s) \end{bmatrix}$$

$$= \begin{bmatrix} 1 & 1 \\ -K_1.e^{-Ls} & K_2.e^{-Ls} \end{bmatrix} \cdot \begin{bmatrix} F_1(s) \\ F_2(s) \end{bmatrix}$$

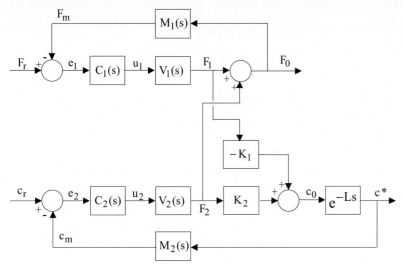

Fig. 86.6 Block diagram of blending system with simple feedback loops

This is a highly interactive system. Its block diagram is depicted in Figure 86.6. Any change in flow will affect the concentration loop, and vice versa.

The design of a multivariable controller for this blending plant is considered in detail in Chapter 81.

86.6 Nomenclature

C	concentration	gm kg^{-1}
c$_P$	specific heat	kJ kg^{-1} K^{-1}
d	diameter	m
F	flow rate	kg s^{-1}
h$_g$	enthalpy	kJ kg^{-1}
ℓ	length	m
L	time delay	s
t	time	s
θ	temperature	°C
ρ	density	kg m^{-3}

Subscripts

S	steam
0	outlet
1	inlet
2	solution

Compressible Flow Systems

Chapter 87

87.1 Resistance to Flow
87.2 Volumetric Capacitance
87.3 Pressure Control
87.4 Worked Example
87.5 Boiler Dynamics
87.6 Nomenclature

The gas law relationships between flow, pressure and, to a lesser extent, temperature are fundamental to the modelling of the dynamics of compressible flow systems. Two typical examples are considered, first the dynamics of pressure control and second the dynamics of a boiler system.

Throughout this chapter pressures are taken to be absolute and volumetric flow rates are assumed to have been measured at 1 bar (abs).

87.1 Resistance to Flow

Consider the flow of gas through a resistance as depicted in Figure 87.1.

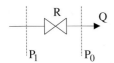

Fig. 87.1 Flow of gas through a fixed resistance

It is assumed that flow is isothermal and that any temperature effects are negligible. In reality, for the purpose of developing a model in deviation form, it is not the absolute conditions up and down stream of the resistance that matter but the variations about those conditions. The assumption of isothermal flow is thus not unreasonable if the pipework is of steel since its thermal capacity acts as a heat sink and tends to average out any variations in temperature.

In general, the flow of gas through any resistance is of the form:

$$Q \propto \sqrt{\frac{(P_1 - P_0)}{\rho}}$$

This has been previously encountered in Equation 12.1, for example, used for orifice plate sizing. For a gas, the density obviously depends upon the pressure. It doesn't particularly matter whether the upstream or downstream pressure is used, or even an average value as *per* Equation 20.10 used for valve sizing. Any discrepancy can be accommodated by the coefficient used as the constant of proportionality. Suppose the downstream pressure is used:

$$Q \propto \sqrt{P_0 \cdot (P_1 - P_0)}$$

For turbulent flow, the requirement is that $P_1 < 2P_0$.

If the downstream pressure is constant and the upstream pressure varied, subject to the constraint, the flow is approximately given by:

$$Q = b\sqrt{(P_1 - P_0)}$$

As depicted in Figure 87.2, this relationship may be linearised about normal conditions.

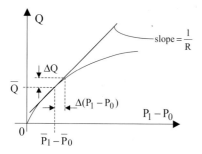

Fig. 87.2 Linearisation of turbulent flow relationship

In deviation form, the relationship becomes:

$$Q = \left. \frac{dQ}{d(P_1 - P_0)} \right|_{\overline{Q}} \cdot (P_1 - P_0)$$

The valve may be compared with an electrical resistance and Ohm's law applied in which flow is analogous to current and pressure drop is analogous to voltage difference:

$$Q = \frac{1}{R} \cdot (P_1 - P_0) \qquad (87.1)$$

where:

$$R = \left. \frac{\Delta(P_1 - P_0)}{\Delta Q} \right|_{\overline{Q}} = \frac{1}{\left. \frac{dQ}{d(P_1 - P_0)} \right|_{\overline{Q}}}$$

However:

$$\frac{dQ}{d(P_1 - P_0)} = \frac{b}{2\sqrt{(P_1 - P_0)}} = \frac{Q}{2(P_1 - P_0)}$$

whence:

$$R = \frac{2(\overline{P}_1 - \overline{P}_0)}{\overline{Q}} \qquad (87.2)$$

If $P_1 > 2P_0$ then the flow is sonic and given approximately by:

$$Q = c.P_1$$

As discussed in Chapter 20, a maximum velocity through the valve is established and the flow rate depends upon the gas density, and hence the upstream pressure, alone. This relationship is depicted in Figure 87.3.

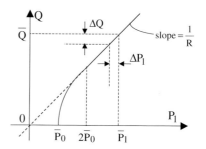

Fig. 87.3 Sonic flow through the fixed resistance

Again, this may be linearised about normal conditions giving, in deviation form:

$$Q = \left. \frac{dQ}{dP_1} \right|_{\overline{Q}} \cdot P_1$$

In this case the electrical analogy is simpler because of the straight line relationship:

$$Q = \frac{1}{R} \cdot P_1 \qquad (87.3)$$

where:

$$R = \left. \frac{\Delta P_1}{\Delta Q} \right|_{\overline{Q}} = \frac{1}{\left. \frac{dQ}{dP_1} \right|_{\overline{Q}}} = \frac{\overline{P}_1}{\overline{Q}} \qquad (87.4)$$

87.2 Volumetric Capacitance

Now consider the effect of capacity by attaching the valve to a pressure vessel as depicted in Figure 87.4.

Fig. 87.4 Capacity with a fixed resistance

The volume of gas in the vessel at $P = V$ m^3. The equivalent volume at 1 bar (abs) = $P.V$ m^{-3}.

Suppose that the pressure in the vessel falls by an amount δP_0 over a period of time δt.

A volume balance gives:

$$-\delta P.V = Q.\delta t$$

In the limit:

$$V \frac{dP}{dt} = -Q$$

Again, by analogy with an electrical capacitance, capacity may be defined by:

$$C\frac{dP}{dt} = -Q \quad (87.5)$$

where $C = V$. Note that although C and V are numerically equal their units are different.

87.3 Pressure Control

Next consider the pressure vessel to have both an inlet and outlet resistance, the outlet discharging into the atmosphere as depicted in Figure 87.5.

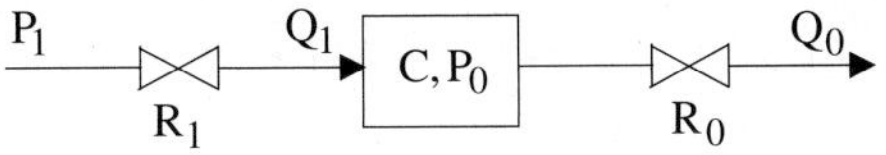

Fig. 87.5 Pressure vessel with inlet and outlet resistances

From Equation 87.5, an unsteady state volume balance for the pressure vessel gives:

Rate of Accumulation = Input − Output

$$C\frac{dP_0}{dt} = Q_1 - Q_0$$

Note that this equation is the same in deviation form.

For sake of argument, assume that the inlet flow is turbulent and the outlet flow is sonic, i.e. $P_1 < 2P_0$ and $P_0 > 2$ bar. Substituting from Equations 87.1 and 87.3 gives:

$$C\frac{dP_0}{dt} = \frac{P_1 - P_0}{R_1} - \frac{P_0}{R_0}$$

Suppose that the valve in the outlet stream is the control valve in a pressure control loop, as shown in Figure 87.6, which is consistent with the pressure vessel being used for anti-surge purposes.

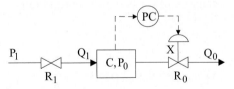

Fig. 87.6 Pressure vessel with variable outlet resistance

The unsteady state balance needs to be modified to take into account the installed characteristic of the valve, as established by Equation 84.4. Remembering that the balance is already in deviation form, this gives:

$$C\frac{dP_0}{dt} = \frac{P_1 - P_0}{R_1} - \frac{P_0}{R_0} - \left.\frac{\partial Q_0}{\partial X}\right|_{\overline{P}_0, \overline{X}} \cdot X$$

This may be transformed, assuming zero initial conditions, to give:

$$CsP_0(s) = \frac{1}{R_1}(P_1(s) - P_0(s))$$
$$- \frac{1}{R_0}P_0(s) - \left.\frac{\partial Q_0}{\partial X}\right|_{\overline{P}_0, \overline{X}} \cdot X(s)$$

Rearranging:

$$\left(\frac{R_0 R_1 C}{R_0 + R_1}s + 1\right) \cdot P_0(s)$$
$$= \frac{R_0}{R_0 + R_1} \cdot P_1(s) - \frac{R_0 R_1}{R_0 + R_1} \cdot \left.\frac{\partial Q_0}{\partial X}\right|_{\overline{P}_0, \overline{X}} \cdot X(s)$$

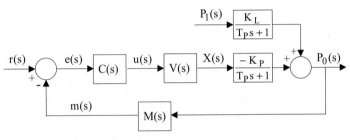

Fig. 87.7 Block diagram of pressure control system

which is of the form:

$$(T_P s + 1) P_0(s) = K_L P_1(s) - K_P X(s) \quad (87.6)$$

These transfer functions may be incorporated in the block diagram of the pressure control loop as shown in Figure 87.7.

87.4 Worked Example

A vacuum control system is as depicted in Figure 87.6. Air is drawn into a process vessel through a manually set needle valve across which the pressure is reduced from atmospheric to P_0. A vacuum of variable pressure P_V is applied to the vessel by means of a control valve. The pressure P_0 is controlled by manipulating the suction flow rate Q_0.

The following data may be assumed:

$$\bar{P}_0 = 0.3 \text{ bar}, \bar{P}_V = 0.2 \text{ bar}, \bar{X} = 0.59, V = 5.0 \text{ m}^3$$

Inspection of the normal conditions indicates sonic flow across R_1 and turbulent flow across R_0. An unsteady state volume balance across the pressure vessel gives:

$$C \frac{dP_0}{dt} = Q_1 - Q_0$$

which, from Equations 87.1 and 87.3, yields in deviation form:

$$C \frac{dP_0}{dt} = 0 - \frac{P_0 - P_V}{R_0} - \left.\frac{\partial Q_0}{\partial X}\right|_{\bar{P}_0, \bar{P}_V, \bar{X}} \cdot X$$

Transform and rearrange gives:

$$(R_0 C s + 1) \cdot P_0(s) = P_V(s) - R_0 \cdot \left.\frac{\partial Q_0}{\partial X}\right|_{\bar{P}_0, \bar{P}_V, \bar{X}} \cdot X(s)$$

which is of the form:

$$(T_P s + 1) \cdot P_0(s) = K_L P_V(s) - K_P X(s) \quad (87.7)$$

By inspection, $K_L = 1.0$.

Assume that the valve's installed characteristic is as follows:

$$Q_0 = 0.014 \sqrt{(P_0 - P_V)} \cdot e^{4.1 X}$$

Substitute the normal conditions gives $\bar{Q}_0 \approx 0.05 \text{ m}^3 \text{ s}^{-1}$. Hence:

$$R_0 = \frac{2(\bar{P}_0 - \bar{P}_V)}{\bar{Q}_0} = 4 \text{ bar s m}^{-3}.$$

However, $C = 5.0 \text{ m}^3 \text{ bar}^{-1}$ which, from Equation 87.7, gives $T_P = R_0 \cdot C = 20$ s.

The process gain comes from the slope of the valve's characteristic:

$$\left.\frac{\partial Q_0}{\partial X}\right|_{\bar{P}_0, \bar{P}_V} = 0.014 \sqrt{(P_0 - P_V)} \cdot 4.1 \cdot e^{4.1 X} = 4.1 Q_0$$

Under normal conditions:

$$\left.\frac{\partial Q_0}{\partial X}\right|_{\bar{P}_0, \bar{P}_V, \bar{X}} = 4.1 \cdot \bar{Q}_0 = 0.205 \text{ m}^3 \text{ s}^{-1}.$$

Whence, from Equation 87.7:

$$K_P = R_0 \cdot \left.\frac{\partial Q_0}{\partial X}\right|_{\bar{P}_0, \bar{P}_V, \bar{X}} = 0.82 \text{ bar}.$$

87.5 Boiler Dynamics

As was seen in Chapter 33, the control of boiler plant can be quite complicated. Their dynamics are complex too. This example shows how, by making a few assumptions, a simple but nevertheless effective model of a boiler's dynamics can be developed.

Consider the boiler system depicted in Figure 87.8 in which a single riser tube is shown for simplicity. The feed water passes through an economiser but there is no superheater. The steam produced is dry saturated but, because the outlet steam pipework is exposed to heat transfer with the flue gases above the drum, it can be presumed that the steam discharged into the steam main is slightly superheated.

The boiler is controlled by simple feedback. Thus, the steam pressure is controlled by manipulating the rate of combustion, the air flow being ratioed to the fuel flow rate, and the drum level is controlled by manipulating the water feed rate. For the purposes of this model any cascade control

87.5 Boiler Dynamics

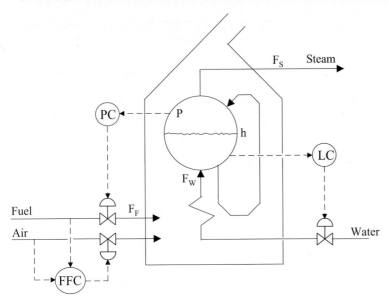

Fig. 87.8 Schematic of boiler featuring drum and single riser

on the drum level, signal selection or ratio trimming are ignored. The objective is to simultaneously control the steam pressure and drum level against a variable steam demand.

Assuming that the drum is always about half full, such that its cross sectional area is approximately constant, an unsteady state mass balance for the water in the drum gives:

$$A\rho \frac{dh}{dt} = F_W - E$$

where E is the rate of evaporation. Noting that this equation is the same in deviation as in absolute form, and assuming zero conditions, it may be transformed:

$$A\rho s h(s) = F_W(s) - E(s) \qquad (87.8)$$

The rate of evaporation in the drum depends on the rate of heat transfer in the riser tubes, which in turn depends on the turbulence in the tubes and on the combustion process. It is probably good enough to assume that there is some second order transfer function that describes this. The parameters of the transfer function would themselves either be determined empirically or by further modelling and simplification. Suppose:

$$E(s) = \frac{K}{(T_1 s + 1)(T_2 s + 1)} F_F(s) \qquad (87.9)$$

An unsteady state mass balance for the steam in the drum and steam main, as far as the first reducing valve in the main, gives:

$$\frac{dM}{dt} = E - F_S$$

which, again, is the same in deviation as in absolute form.

Assuming that some superheating occurs, the steam may be treated as a gas which obeys the ideal gas law rather than as a vapour: this is not unreasonable provided the pipework is well insulated. The gas law may be expressed in the form:

$$PV = kMT$$

Assuming that the temperature is approximately constant:

$$V \frac{dP}{dt} = kT \frac{dM}{dt}$$

whence:

$$\frac{V}{kT}\frac{dP}{dt} = E - F_S$$

Assuming zero initial conditions, this may be transformed:

$$\frac{V}{kT} sP(s) = E(s) - F_S(s) \qquad (87.10)$$

The transfer functions represented by Equations 87.8–87.10 are shown in block diagram form in Figure 87.9.

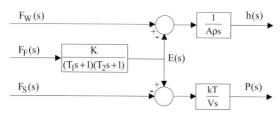

Fig. 87.9 Block diagram of simplified boiler model

Note, in particular, that the dynamics are dominated by two integrator terms. These are classic for boiler dynamics. Typically, all other things being equal, a small step increase in fuel flow will lead to a slow ramp increase in pressure.

There are many refinements that could be made to the model. For example, the effects of swell in the drum, *i.e.* the increase in level due to boil-up following a decrease in pressure, can be significant. It is nevertheless good enough for most purposes.

The interactions between the pressure and level loops are not as severe as in the case of the blending system in Chapter 86. Adequate control can be achieved using independent feedback loops, supplemented with cascade control, signal selection and ratio trimming as in Chapter 33. Multivariable compensation, or even model predictive control, is only necessary for tight control where several boilers are feeding into the same steam main.

87.6 Nomenclature

C	capacity	m³ bar⁻¹
E	rate of evaporation	kg s⁻¹
F	mass flow rate	kg s⁻¹
M	mass	kg
P	pressure	bar (abs)
Q	volumetric flow rate	m³ s⁻¹
R	resistance	bar s m⁻³
t	time	s
T	temperature	K
V	volume	m³
X	fractional valve opening	–
ρ	density	kg/m³

Subscripts

F	fuel
S	steam
W	water
0	outlet
1	inlet

Hydrodynamics

88.1 Nature of the Process
88.2 Energy Considerations
88.3 Energy Balance
88.4 Nomenclature

When the opening of a valve is increased it is normal to assume that the change in flow occurs almost instantaneously. This can only be true if the change in flow is small relative to the capacity of the pump that is producing the flow. If the pipeline is long or of a large diameter, such that the mass of fluid flowing is large, the pump will only be capable of accelerating the fluid at a finite rate depending on its power. The lags associated with such changes are known as hydrodynamics and are potentially of significance in any pumping system. Strictly speaking, hydrodynamics only concerns aqueous flow although the principles apply to liquids in general. The dynamics of gaseous systems has already been covered in Chapter 87.

88.1 Nature of the Process

Consider a flow control loop, as depicted in Figure 88.1, in which the "process" consists of a pump,

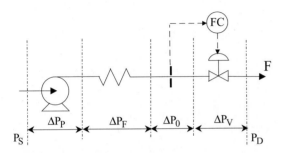

Fig. 88.1 Schematic of pump and pipeline

a pipeline which contains a substantial amount of liquid, an orifice plate and the body of a control valve.

As the valve opens and the flow rate increases, the liquid accelerates. The kinetic energy of the liquid within the pipeline represents the system's capacity for energy storage. Note that, by virtue of the increase in flow rate, both the rate of change of potential and pressure energy across the system vary, as does the rate of energy dissipation within the system. An unsteady state energy balance across the system is thus of the general form:

$$\frac{d}{dt}(\text{kinetic energy})$$
$$= \text{power input from pump}$$
$$\quad - \text{net rate of increase of potential}$$
$$\quad \quad \text{and pressure energy}$$
$$\quad - \text{various rates of energy dissipation}$$

88.2 Energy Considerations

Each of the different energy types is considered in turn.

Kinetic

The mass flow rate is given by:

$$F = \frac{\pi d^2}{4} \cdot v \cdot \rho$$

whence:
$$v^2 = \frac{16}{(\pi d^2 \rho)^2} . F^2$$

The kinetic energy of the mass of liquid in the pipe is thus:
$$E_K = \frac{1}{2} M v^2 = \frac{8M}{(\pi d^2 \rho)^2} . F^2 \qquad (88.1)$$

Potential

Let the vertical difference in height between the inlet and outlet of the pipeline be Δh.

The change in potential energy per unit mass of liquid is thus $\Delta h.g$.

That this is indeed energy per unit mass can be confirmed by considering the units:
$$\frac{J}{kg} = \frac{Nm}{kg} = kg \frac{m}{s^2} . m . \frac{1}{kg} = m . \frac{m}{s^2}$$

Therefore the rate of change of potential energy is:
$$U_H = \Delta h . g . F \qquad (88.2)$$

The rate of change of potential energy is power, also confirmed by considering the units:
$$W = \frac{J}{kg} . \frac{kg}{s}$$

Pressure

Let the difference between the supply and return main pressures be $P_D - P_S$.

The change in pressure energy per unit mass of liquid is thus:
$$\frac{(P_D - P_S)}{\rho}$$

Again, consideration of the units confirms that this is dimensionally correct:
$$\frac{J}{kg} = \frac{Nm}{kg} = \frac{N}{m^2} . \frac{m^3}{kg}$$

Therefore the rate of change of pressure energy is:
$$U_P = \frac{1}{\rho} . (P_D - P_S) . F \qquad (88.3)$$

Frictional

Let the pressure drop due to frictional losses along the pipeline be ΔP_F.

The rate of dissipation of energy is thus given by:
$$U_F = \frac{1}{\rho} \Delta P_F . F$$

However:
$$\Delta P_F = c_F F^2$$

so:
$$U_F = \frac{1}{\rho} c_F . F^3 \qquad (88.4)$$

Other Losses

Similarly, the rate of dissipation of energy across the orifice plate is:
$$U_O = \frac{1}{\rho} c_O . F^3 \qquad (88.5)$$

and the rate of dissipation of energy across the control valve is given by:
$$U_V = \frac{1}{\rho} \Delta P_V . F \qquad (88.6)$$

88.3 Energy Balance

An unsteady state energy balance across the system gives:
$$\frac{d}{dt} E_K = W - U_H - U_P - U_F - U_O - U_V$$

Substituting from Equations 88.1–88.6 gives:
$$\frac{8M}{(\pi d^2 \rho)^2} \frac{d}{dt}(F^2) = W - \Delta h . g . F - \frac{1}{\rho}(P_D - P_S) . F$$
$$- \frac{1}{\rho}(c_F + c_O) . F^3 - \frac{1}{\rho} \Delta P_V . F$$

Putting this into deviation form gives:

$$\frac{16 M \bar{F}}{(\pi d^2 \rho)^2} \frac{dF}{dt} = -\Delta hg.F - \frac{1}{\rho}(\bar{P}_D - P_S).F$$
$$- \frac{1}{\rho}\bar{F}.P_D - \frac{3\bar{F}^2}{\rho}(c_F + c_0).F$$
$$- \frac{1}{\rho}(\overline{\Delta P}_V.F + \bar{F}.\Delta P_V) \quad (88.7)$$

where it is assumed that the supply pressure and power input from the pump are constant.

However, as was seen in Chapter 84, the installed characteristic of a control valve, in deviation form, is:

$$F \approx \left.\frac{\partial F}{\partial X}\right|_{\overline{\Delta P}_V, \bar{X}}.X + \left.\frac{\partial F}{\partial \Delta P_V}\right|_{\overline{\Delta P}_V, \bar{X}}.\Delta P_V \quad (84.3)$$

Substituting into Equation 88.6 for ΔP_V and rearranging gives:

$$U_V = \frac{1}{\rho}.(\overline{\Delta P}_V.F + \bar{F}.\Delta P_V)$$

$$= \frac{1}{\rho}\left(\overline{\Delta P}_V.F + \bar{F}.\frac{\left(F - \left.\frac{\partial F}{\partial X}\right|_{\overline{\Delta P}_V, \bar{X}}.X\right)}{\left.\frac{\partial F}{\partial \Delta P_V}\right|_{\overline{\Delta P}_V, \bar{X}}}\right)$$

$$= \frac{1}{\rho}\left(\overline{\Delta P}_V + \frac{\bar{F}}{\left.\frac{\partial F}{\partial \Delta P_V}\right|_{\overline{\Delta P}_V, \bar{X}}}\right).F$$

$$- \frac{1}{\rho}\left(\frac{\bar{F}.\left.\frac{\partial F}{\partial X}\right|_{\overline{\Delta P}_V, \bar{X}}}{\left.\frac{\partial F}{\partial \Delta P_V}\right|_{\overline{\Delta P}_V, \bar{X}}}\right).X$$

which is of the form:

$$U_V = c_{V1}.F - c_{V2}.X \quad (88.8)$$

Substituting Equation 88.8 into Equation 88.7 gives:

$$\frac{16 M \bar{F}}{(\pi d^2 \rho)^2} \frac{dF}{dt} = -\Delta hg.F - \frac{1}{\rho}(\bar{P}_D - P_S).F$$
$$- \frac{1}{\rho}\bar{F}.P_D - \frac{3\bar{F}^2}{\rho}(c_F + c_0).F$$
$$- c_{V1}.F + c_{V2}.X$$

Rearrange:

$$\frac{16 M \bar{F}}{(\pi d^2 \rho)^2} \frac{dF}{dt}$$
$$+ \left(\Delta hg + \frac{1}{\rho}(\bar{P}_D - P_S) + \frac{3\bar{F}^2}{\rho}(c_F + c_0) + c_{V1}\right).F$$
$$= -\frac{1}{\rho}\bar{F}.P_D + c_{V2}.X$$

Let:

$$\Phi = \left(\Delta hg + \frac{1}{\rho}(\bar{P}_D - P_S) + \frac{3\bar{F}^2}{\rho}(c_F + c_0) + c_{V1}\right)$$

Laplace transform and rearrange:

$$\left(\frac{16 M \bar{F}}{\Phi.(\pi d^2 \rho)^2}s + 1\right).F(s)$$
$$= -\frac{\bar{F}}{\Phi.\rho}.P_D(s) + \frac{c_{V2}}{\Phi}.X(s)$$

which is of the form:

$$(Ts + 1).F(s) = -K_1.P_D(s) + K_2.X(s)$$

These transfer functions are incorporated in the block diagram of a typical flow control loop as depicted in Figure 88.2.

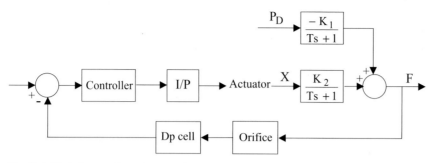

Fig. 88.2 Block diagram of pumping control system

88.4 Nomenclature

c	constant	$m^{-1} kg^{-1}$ (mostly)
d	diameter	m
E	energy	J
F	flow rate	$kg\ s^{-1}$
g	gravitational acceleration	$m\ s^{-2}$
h	height	m
M	mass	kg
P	pressure	$N\ m^{-2}$
t	time	s
U	power	W
v	velocity	$m\ s^{-1}$
W	pump power	W
X	valve opening	–
ρ	density	$kg\ m^{-3}$

Subscripts

D	discharge
F	fixed resistance (pipeline)
H	potential
K	kinetic
O	orifice plate
P	pump
S	supply
V	variable resistance (valve)

Multivariable Systems

Chapter 89

89.1 Semi-Batch Reactor: Temperature Control
89.2 Temperature and Flow Coefficients
89.3 Semi-Batch Reactor: Pressure Control
89.4 Multivariable Control
89.5 Nomenclature

Most process systems involve the simultaneous control of many variables. Usually, if the interaction between the variables is weak, then effective control can be achieved by schemes consisting of multiple independent feedback loops. A good example of this is the evaporator plant of Chapter 31. However, if there are strong interactions as, for example, with the blending system of Chapter 86, then a multivariable controller is required to counter the interactions. The theory of multivariable control system design was introduced in Chapter 81. To apply this theory requires a model of the plant in either matrix or state-space form. For the blending system it was relatively simple to develop the model. However, that is not always the case. This chapter provides an example of the development of a state-space model for a semi-batch reactor which is more complex.

Figure 89.1 depicts a gaseous reagent being bubbled through an agitated semi-batch reactor. Unreacted gas accumulates in the space above the reaction mixture. The reaction is exothermic, heat being removed from the reactor by means of cooling water circulated through a coil. The objective is to simultaneously control the temperature and pressure in the reactor.

89.1 Semi-Batch Reactor: Temperature Control

An unsteady state heat balance on the liquid contents of the reactor yields:

Rate of Accumulation = Input − Output

$$Mc_P \frac{d\theta_0}{dt} = \Phi \Delta H - UA(\theta_0 - \theta_1)$$

Note that any heat transfer between the gas and liquid has been ignored because the thermal capacity of the gas is so small relative to that of the liquid.

Suppose that the rate of reaction is dependant upon the partial pressure of the dissolved gaseous component which, in turn, is proportional to the gas pressure, absorption being a gas film dominated process:

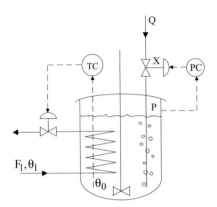

Fig. 89.1 Agitated semi-batch reactor with cooling coil

$$\Phi = k.p = a.k.P$$

Substituting:

$$Mc_P \frac{d\theta_0}{dt} = akP.\Delta H - UA(\theta_0 - \theta_1)$$

Note that the rate constant k is temperature dependant. Also, if the agitator runs at a constant speed, then the overall heat transfer coefficient U is effectively dependant upon the cooling water flow rate only. The balance may thus be put into deviation form:

$$Mc_P \frac{d\theta_0}{dt} = a\overline{k}\Delta H.P + a\overline{P}\Delta H. \left.\frac{dk}{d\theta_0}\right|_{\overline{\theta}_0} .\theta_0$$
$$- \overline{UA}.(\theta_0 - \theta_1)$$
$$- (\overline{\theta}_0 - \overline{\theta}_1) A. \left.\frac{dU}{dF_1}\right|_{\overline{F}_1} .F_1$$

This may be rearranged as follows:

$$\frac{d\theta_0}{dt} = \frac{1}{Mc_P} \left(a\overline{P}\Delta H. \left.\frac{dk}{d\theta_0}\right|_{\overline{\theta}_0} - \overline{UA} \right).\theta_0 + \frac{a\overline{k}\Delta H}{Mc_P}.P$$
$$- \frac{(\overline{\theta}_0 - \overline{\theta}_1) A}{Mc_P} . \left.\frac{dU}{dF_1}\right|_{\overline{F}_1} .F_1 + \frac{\overline{UA}}{Mc_P}.\theta_1 \quad (89.1)$$

which is of the general form:

$$\dot{\theta}_0 = a_{11}.\theta_0 + a_{12}.P + b_{11}.F_1 + b_{12}.\theta_1 \quad (89.2)$$

Assuming zero initial conditions, this may be transformed and rearranged:

$$(s - a_{11}).\theta_0(s) = a_{12}.P(s) + b_{11}.F_1(s)$$
$$+ b_{12}.\theta_1(s) \quad (89.3)$$

Note that for stability the coefficient a_{11} must be negative. This requires that:

$$\overline{UA} > \overline{P}\Delta H. \left.\frac{dk}{d\theta_0}\right|_{\overline{\theta}_0} \quad (89.4)$$

which is consistent with the heat transfer capacity of the coil being greater than the potential rate of heat release by the reaction.

89.2 Temperature and Flow Coefficients

The values of the differential coefficients in Equation 89.1 may be determined from other fundamental relationships as follows. First, the rate constant's temperature dependence is, according to Arrhenius' equation:

$$k = B.e^{\frac{-E_a}{R(\theta_0 + 273)}} \quad (89.5)$$

By differentiation and substitution of normal conditions this yields:

$$\left.\frac{dk}{d\theta_0}\right|_{\overline{\theta}_0} = \frac{\overline{k}E_a}{R(\overline{\theta}_0 + 273)^2} \quad (89.6)$$

And second, the resistance to heat transfer will vary with the thickness of the coil's inside film. Increasing the water flow rate increases the turbulence. This reduces the coil side film's thickness leading to an increase in the film coefficient for heat transfer. The film coefficient is related to the cooling water flow rate by Pratt's adaptation of the dimensionless Dittus Boelter equation:

$$Nu = 0.023 \left(1 + 3.5\frac{d}{d_c}\right) Re^{0.8}.Pr^{0.3} \quad (89.7)$$

Assuming all the physical properties of the cooling water to be approximately constant over the range concerned, this may be simplified to:

$$h_1 = b.F_1^{0.8} \quad (89.8)$$

Even if the coil is thin walled and the reactor strongly agitated, as should be the case, the resistances to heat transfer due to the coil's wall and outside film will be significant but relatively constant. Lumping these resistances together gives the overall heat transfer coefficient:

$$\frac{1}{U} = \frac{1}{h_1} + c \quad (89.9)$$

Substituting and rearranging gives:

$$U = \frac{bF_1^{0.8}}{1 + bcF_1^{0.8}} \quad (89.10)$$

By differentiation and substitution of normal conditions this yields:

$$\left.\frac{dU}{dF_1}\right|_{\bar{F}_1} = \frac{0.8b}{\bar{F}_1^{0.2}\left(1+b c \bar{F}_1^{0.8}\right)}\left(1-\frac{b c \bar{F}_1^{0.8}}{\left(1+b c \bar{F}_1^{0.8}\right)}\right) \quad (89.11)$$

89.3 Semi-Batch Reactor: Pressure Control

Now consider the gas in the space above the liquid inside the reactor. Suppose that the direct affect of temperature changes on the pressure of the gas is negligible, and that there are no significant vaporisation effects. An unsteady state molar balance for the gas gives:

$$C\frac{dP}{dt} = Q - \Phi = Q - akP$$

Assuming that the gas supply is at a constant pressure and that flow through the control valve is turbulent, this balance may be put into deviation form:

$$C\frac{dP}{dt} = \left.\frac{\partial Q}{\partial X}\right|_{\bar{X},\bar{P}}.X - \frac{1}{Z}.P - a\bar{k}.P - a\bar{P}\left.\frac{dk}{d\theta_0}\right|_{\bar{\theta}_0}.\theta_0$$

This may be rearranged:

$$\frac{dP}{dt} = -\frac{a\bar{P}}{C}.\left.\frac{dk}{d\theta_0}\right|_{\bar{\theta}_0}.\theta_0 - \frac{1}{C}.\left(\frac{1}{Z}+a\bar{k}\right).P$$

$$+ \frac{1}{C}.\left.\frac{dQ}{dX}\right|_{\bar{X},\bar{P}}.X \quad (89.12)$$

which is of the general form:

$$\dot{P} = a_{21}.\theta_0 + a_{22}.P + b_{23}.X \quad (89.13)$$

Assuming zero initial conditions, this may be transformed and rearranged:

$$(s - a_{22}).P(s) = a_{21}.\theta_0(s) + b_{23}.X(s) \quad (89.14)$$

Note that the coefficient a_{22} is negative corresponding to a stable system.

89.4 Multivariable Control

Figure 89.2 is a block diagram of the control system which incorporates Equations 89.3 and 89.14.

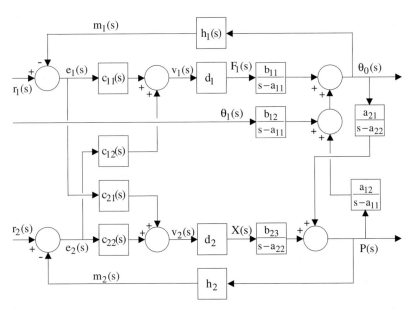

Fig. 89.2 Block diagram of semi batch reactor with multivariable controller

This clearly demonstrates the interactive nature of the reactor. Note that the dynamics of the pressure measurement and of the control valves have been assumed to be negligible.

Equations 89.2 and 89.13 may be assembled into state-space form as follows:

$$\begin{bmatrix} \dot{\theta}_0 \\ \dot{P} \end{bmatrix} = \begin{bmatrix} a_{11} & a_{12} \\ a_{21} & a_{22} \end{bmatrix} \cdot \begin{bmatrix} \theta_0 \\ P \end{bmatrix} \quad (89.15)$$
$$+ \begin{bmatrix} b_{11} & b_{12} & 0 \\ 0 & 0 & b_{23} \end{bmatrix} \cdot \begin{bmatrix} F_1 \\ \theta_1 \\ X \end{bmatrix}$$

which is of the general form

$$\underline{\dot{x}} = A\underline{x} + B\underline{u}$$

The multivariable controller depicted in Figure 89.2 is as described in Chapter 81.

89.5 Nomenclature

a	coefficient	
A	effective heat transfer area	m^2
b	coefficient	
B	frequency factor	$kmol\ bar^{-1}\ s^{-1}$
c	resistance to heat transfer	$m^2\ K\ kW^{-1}$
C	gaseous holdup	$kmol\ bar^{-1}$
c_p	specific heat of liquid	$kJ\ kg^{-1}\ K^{-1}$
d	diameter	m
E_a	activation energy	$kJ\ kmol^{-1}$
F	mass flow rate	$kg\ s^{-1}$
h	film coefficient for heat transfer	$kW\ m^{-2}\ K^{-1}$
ΔH	heat of reaction	$kJ\ kmol^{-1}$
k	rate constant	$kmol\ bar^{-1}\ s^{-1}$
M	mass of liquid	kg
P	gas pressure	bar (abs)
Q	gas feed rate	$kmol\ s^{-1}$
R	universal gas constant	$kJ\ kmol^{-1}\ K^{-1}$
t	time	s
U	overall heat transfer coefficient	$kW\ m^{-2}\ K^{-1}$
X	fractional valve opening	–
Z	normal resistance of control valve	$bar\ s\ kmol^{-1}$
θ	temperature	°C
Φ	rate of reaction	$kmol\ s^{-1}$

Subscripts

C	coil
S	supply
0	reaction mixture
1	cooling water

Multistage Systems

90.1 Vapour Flow Lags
90.2 Liquid Flow Lags
90.3 Concentration Lags
90.4 Dual Composition Control
90.5 Worked Example
90.6 L-V Strategy
90.7 Comments
90.8 Nomenclature

Many items of plant, multi-stage evaporators being a good example, consist of a number of identical units which are operated in series. The general approach to modelling such systems is to decompose the plant into non-interacting units, as far as is practicable, to analyse the dynamics of each unit, and to combine the models of the individual units to obtain the dynamics of the plant as a whole.

The distillation column is used in this chapter as a vehicle for studying multistage dynamics. Distillation is inherently multistage because of the nature of the internal design of the column. Decomposition is done on the basis of individual plates/stages. Analysis and modelling is most effective when the flow and concentration effects are considered separately. There are three types of lag:

1. Vapour flow lags. These are the smallest and may well be negligible. They occur because the vapour holdup of a plate varies with pressure, which in turn depends upon vapour flow rate through the column.
2. Liquid flow lags. These are due to the hydrodynamics of the individual plates. A change in the liquid flow rate onto a plate causes the level on the plate, and hence its liquid holdup, to vary.
3. Concentration lags. These are the largest lags and dominate the column's dynamics. Any change in composition of either of the streams entering a plate will cause a change in the concentration in the liquid holdup on the plate.

The vapour, liquid and concentration lags are essentially independent of each other and may be considered to act in series. Thus, following a step change in external conditions, the response of a plate initially depends mainly on the faster vapour and liquid flows and on the distance of the plate from the source of the disturbance. However, the time required for a 63.2% response depends primarily on the longer concentration lags.

90.1 Vapour Flow Lags

If the capacity for vapour holdup was just the volume of the vapour space, then the vapour lag would be negligible. However, the increase in pressure that is associated with an increase in vapour flow is accompanied by an increased boiling point. This means that vapour must be condensed to heat up the liquid holdup to a higher boiling point. The liquid holdup therefore acts as a capacity for vapour, a holdup which is generally much higher than that due to the capacity of the vapour space.

Consider the nth and n−1th plates of a column, as depicted in Figure 90.1.

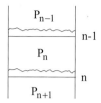

Fig. 90.1 An arbitrary plate of a distillation column

Suppose there is an increase in pressure of dP_n above the plate over a period of time dt. This will cause the boiling point of the liquid holdup on the nth plate to rise by $d\theta$.

Heat required to restore boiling $= H_L c_P d\theta$

Moles of vapour condensed $= \dfrac{H_L c_P d\theta}{\lambda}$

So rate of condensation

$$= \frac{H_L c_P}{\lambda} \cdot \frac{d\theta}{dt} = \frac{H_L c_P}{\lambda} \cdot \left.\frac{d\theta}{dP_n}\right|_{\overline{P_n}} \cdot \frac{dP_n}{dt}$$

Using the analogy of resistance to flow, as described in Chapter 87, an unsteady state molar balance for the vapour across the nth plate thus gives:

Rate of Accumulation = Input − Output

$$H_V \frac{dP_n}{dt} + \frac{H_L c_P}{\lambda} \cdot \left.\frac{d\theta}{dP_n}\right|_{\overline{P_n}} \cdot \frac{dP_n}{dt}$$
$$= \frac{(P_{n+1} - P_n)}{Z} - \frac{(P_n - P_{n-1})}{Z}$$

which is the same in deviation form. Transform, assuming zero initial conditions, and rearrange gives:

$$\left(\frac{Z}{2}\left(H_V + \frac{H_L c_P}{\lambda} \cdot \left.\frac{d\theta}{dP_n}\right|_{\overline{P_n}}\right).s + 1\right).P_n(s)$$
$$= \frac{1}{2} P_{n+1}(s) + \frac{1}{2} P_{n-1}(s)$$

which is of the form:

$$(T_V s + 1) P_n(s) = \frac{1}{2} P_{n+1}(s) + \frac{1}{2} P_{n-1}(s)$$

where the vapour lag T_V is of the order of 2–20 s, depending upon the size of the column.

90.2 Liquid Flow Lags

The lag for liquid flow can be treated as a liquid level problem that is complicated somewhat by the change in level across the plate. Consider the nth plate depicted in Figure 90.2.

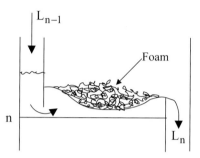

Fig. 90.2 Liquid flow profile across an arbitrary plate

Let h_n be the depth of clear liquid equivalent to the static pressure drop across the centre of the plate. Assume that this represents the average equivalent depth of liquid across the plate. Assuming all the variables to be in deviation form, an unsteady state volume balance for the liquid across the plate gives:

$$A\frac{dh_n}{dt} = \frac{1}{\rho}(L_{n-1} - L_n) = \frac{1}{\rho}.L_{n-1} - \frac{1}{\rho}.\left.\frac{dL}{dh}\right|_{\overline{h}}.h_n$$

Transform, assuming zero initial conditions, and rearrange gives:

$$\left(\frac{A\rho}{\left.\frac{dL}{dh}\right|_{\overline{h}}}.s + 1\right).h_n(s) = \frac{1}{\left.\frac{dL}{dh}\right|_{\overline{h}}}.L_{n-1}(s)$$

which is of the form:

$$(T_L s + 1).h_n(s) = K_L.L_{n-1}(s)$$

where the liquid lag T_L is of the order 10–100 s. As far as liquid flow is concerned, the plates in the column form a series of identical non-interacting first order lags. The transient response several plates from the top to a step increase in reflux rate can be approximated by a time delay plus an exponential response, as described in Chapter 72. It is this time delay that makes control at intermediate plates much more sluggish than control at the top plate.

90.3 Concentration Lags

The composition of the liquid on an isolated plate would show a first order response to any change in either composition or flow rate of either of the two streams entering the plate. However, the plates of a column are not isolated and, because of the counter-current nature of flow within the column, there is considerable dynamic interaction between them. To illustrate this, a model is developed for the concentration lags of the 3-stage column depicted in Figure 90.3.

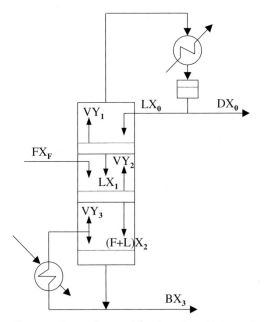

Fig. 90.3 Flows and compositions throughout a 3-stage column

It is recognised that a 3-stage binary separation is not representative of typical industrial distillations: however, it is sufficiently complex to adequately demonstrate all of the issues involved in the modelling of concentration lags. The column consists of two plates and a still. The feed is liquid, at its boiling point, and enters at the second plate.

To prevent the model from becoming too unwieldy, it is necessary to make a number of assumptions and approximations, as follows:

1. That the concentration and flow lags within the column are independent of the dynamics of the overhead condenser and reflux drum.
2. That overhead condensation is complete and there is no sub-cooling of the condensate, *i.e.* the reflux enters as liquid at its boiling point.
3. That there is sufficient turbulence on each plate to assume good mixing such that lumped parameter dynamics apply.
4. That the distillation is of a binary mixture, *i.e.* of two components, which are referred to as the more volatile component (MVC) and the less volatile component.
5. That constant molal overflow occurs, *i.e.* for each mole of vapour that is condensed there is one mole of liquid evaporated. This implies that the molar latent heat of evaporation is the same for both components.
6. That the column is operated at constant overhead pressure. This fixes the vapour-liquid equilibrium data, referred to as the X-Y curve, which relates the composition of the liquid on a plate to that of the vapour leaving the plate.
7. That the efficiency μ of each plate is the same. This is normally articulated in terms of the Murphree efficiency, for which a typical value is 60–70%.

The X–Y curve may be linearised for the nth plate at the point corresponding to its normal condition, as follows:

$$Y_n = m_n X_n + c_n$$

where:

$$m_n = \left. \frac{dY}{dX} \right|_{\overline{X}_n}$$

Thus, for small changes in composition, where the variables are in deviation form:

$$Y_n = m_n . X_n$$

Allowing for the inefficiency of the plate, the change in composition realised is given by:

$$Y_n = \mu m_n . X_n$$

An unsteady state molar balance for the MVC in the liquid on the top plate (n = 1) gives:

$$H_{L1}\frac{dX_1}{dt} = LX_0 + VY_2 - LX_1 - VY_1$$

This may be put into deviation form:

$$H_{L1}\frac{d\overline{X}_1}{dt} = \overline{L}.X_0 + \overline{X}_0.L + \overline{V}.Y_2 + \overline{Y}_2.V$$
$$- \overline{L}.X_1 - \overline{X}_1.L - \overline{V}.Y_1 - \overline{Y}_1.V$$

Since there is total condensation of the overhead vapour, then $X_0 = Y_1$. Hence:

$$H_{L1}\frac{d\overline{X}_1}{dt} = \overline{L}.Y_1 + \overline{X}_0.L + \overline{V}.Y_2 + \overline{Y}_2.V$$
$$- \overline{L}.X_1 - \overline{X}_1.L - \overline{V}.Y_1 - \overline{Y}_1.V$$

Substitute for Y_i in terms of X_i gives:

$$H_{L1}\frac{d\overline{X}_1}{dt} = \overline{L}\mu m_1.X_1 + \overline{X}_0.L + \overline{V}\mu m_2.X_2$$
$$+ \overline{Y}_2.V - \overline{L}.X_1 - \overline{X}_1.L$$
$$- \overline{V}\mu m_1.X_1 - \overline{Y}_1.V$$

This may be rearranged:

$$\frac{d\overline{X}_1}{dt} = -\frac{(\overline{L} + \overline{V}\mu m_1 - \overline{L}\mu m_1)}{H_{L1}}.X_1 + \frac{\overline{V}\mu m_2}{H_{L1}}.X_2$$
$$+ \frac{(\overline{X}_0 - \overline{X}_1)}{H_{L1}}.L - \frac{(\overline{Y}_1 - \overline{Y}_2)}{H_{L1}}.V$$

which is of the general form:

$$\dot{X}_1 = a_{11}.X_1 + a_{12}.X_2 + b_{11}.L + b_{12}.V \quad (90.1)$$

An unsteady state molar balance for the MVC in the liquid on the feed plate (n = 2) gives:

$$H_{L2}\frac{dX_2}{dt} = FX_F + LX_1 + VY_3 - (F + L)X_2 - VY_2$$

This may be put into deviation form:

$$H_{L2}\frac{d\overline{X}_2}{dt} = \overline{F}.X_F + \overline{X}_F.F + \overline{L}.X_1 + \overline{X}_1.L + \overline{V}.Y_3$$
$$+ \overline{Y}_3.V - (\overline{F} + \overline{L}).X_2 - \overline{X}_2.(F + L)$$
$$- \overline{V}.Y_2 - \overline{Y}_2.V$$

Substitute for Y_i in terms of X_i gives:

$$H_{L2}\frac{d\overline{X}_2}{dt} = \overline{F}.X_F + \overline{X}_F.F + \overline{L}.X_1 + \overline{X}_1.L$$
$$+ \overline{V}\mu m_3.X_3 + \overline{Y}_3.V - (\overline{F} + \overline{L}).X_2$$
$$- \overline{X}_2.(F + L) - \overline{V}\mu m_2.X_2 - \overline{Y}_2.V$$

This may be rearranged:

$$\frac{d\overline{X}_2}{dt} = \frac{\overline{L}}{H_{L2}}.X_1 - \frac{(\overline{F} + \overline{L} + \overline{V}\mu m_2)}{H_{L2}}.X_2 + \frac{\overline{V}\mu m_3}{H_{L2}}.X_3$$
$$+ \frac{(\overline{X}_1 - \overline{X}_2)}{H_{L2}}.L - \frac{(\overline{Y}_2 - \overline{Y}_3)}{H_{L2}}.V$$
$$+ \frac{(\overline{X}_F - \overline{X}_2)}{H_{L2}}.F + \frac{\overline{F}}{H_{L2}}.X_F$$

which is of the general form:

$$\dot{X}_2 = a_{21}.X_1 + a_{22}.X_2 + a_{23}.X_3 \quad (90.2)$$
$$+ b_{21}.L + b_{22}.V + b_{23}.F + b_{24}.X_F$$

An unsteady state molar balance for the MVC in the liquid in the still (n = 3) gives:

$$H_{L3}\frac{dX_3}{dt} = (F + L)X_2 - BX_3 - VY_3$$

This may be put into deviation form:

$$H_{L3}\frac{d\overline{X}_3}{dt} = (\overline{F} + \overline{L}).X_2 + \overline{X}_2.(F + L) - \overline{B}.X_3$$
$$- \overline{X}_3.B - \overline{V}.Y_3 - \overline{Y}_3.V$$

Using the overall molar balance for the still to eliminate B gives:

$$H_{L3}\frac{d\overline{X}_3}{dt} = (\overline{F} + \overline{L}).X_2 + \overline{X}_2.(F + L) - \overline{B}.X_3$$
$$- \overline{X}_3.(F + L - V) - \overline{V}.Y_3 - \overline{Y}_3.V$$

Substitute for Y_i in terms of X_i gives :

$$H_{L3}\frac{d\overline{X}_3}{dt} = (\overline{F} + \overline{L}).X_2 + \overline{X}_2.(F + L) - \overline{B}.X_3$$
$$- \overline{X}_3.(F + L - V) - \overline{V}\mu m_3.X_3 - \overline{Y}_3.V$$

Strictly speaking $Y_3 \approx \mu.m_3.X_3$. The vapour of composition Y_3 is in equilibrium with the liquid leaving the reboiler rather than the liquid in the still of composition X_3. However, for high flow rates through the thermosyphon the approximation is good enough.

This may be rearranged:

$$\frac{dX_3}{dt} = \frac{(\overline{F}+\overline{L})}{H_{L3}}.X_2 - \frac{(\overline{B}+\overline{V}\mu m_3)}{H_{L3}}.X_3$$
$$+ \frac{(\overline{X}_2 - \overline{X}_3)}{H_{L3}}.L - \frac{(\overline{Y}_3 - \overline{X}_3)}{H_{L3}}.V$$
$$+ \frac{(\overline{X}_2 - \overline{X}_3)}{H_{L3}}.F$$

which is of the general form:

$$\dot{X}_3 = a_{32}.X_2 + a_{33}.X_3 + b_{31}.L + b_{32}.V + b_{33}F \quad (90.3)$$

Equations 90.1–90.3 may be assembled into state-space form as follows:

$$\begin{bmatrix} \dot{X}_1 \\ \dot{X}_2 \\ \dot{X}_3 \end{bmatrix} = \begin{bmatrix} a_{11} & a_{12} & 0 \\ a_{21} & a_{22} & a_{23} \\ 0 & a_{32} & a_{33} \end{bmatrix} \begin{bmatrix} X_1 \\ X_2 \\ X_3 \end{bmatrix} \quad (90.4)$$
$$+ \begin{bmatrix} b_{11} & b_{12} & 0 & 0 \\ b_{21} & b_{22} & b_{23} & b_{24} \\ b_{31} & b_{32} & b_{33} & 0 \end{bmatrix} \begin{bmatrix} L \\ V \\ F \\ X_F \end{bmatrix}$$

which is of the general form:

$$\underline{\dot{x}} = A\underline{x} + B\underline{u}$$

As explained in Chapter 79, this equation can be solved by Laplace transformation and inversion to yield the composition model:

$$\underline{x}(s) = [sI - A]^{-1} B.\underline{u}(s) = G(s).\underline{u}(s) \quad (90.5)$$

90.4 Dual Composition Control

The model of Equation 90.5 describes the concentration lags of the column in a form that is consistent with the L-V energy balance type of control scheme, as described in Chapter 35 and depicted in Figure 90.4.

Both the feed rate F and its composition X_F vary. Dual control of both top plate X_1 and bottoms X_3 compositions is realised by manipulation of the reflux L and boil-up V rates respectively. Strictly speaking it is the distillate composition X_0 that is of interest, but that is controlled indirectly by controlling X_1 instead. The distillate D and bottoms B streams are manipulated for inventory purposes.

From a control point of view, the transfer functions of particular importance are those relating $X_1(s)$ and $X_3(s)$ to the manipulated variables $L(s)$ and $V(s)$ and to the disturbance variables $F(s)$ and $X_F(s)$. These are the elements $g_{11}(s)$, $g_{13}(s)$, $g_{14}(s)$, $g_{32}(s)$, $g_{33}(s)$ and $g_{34}(s)$ of $G(s)$ respectively.

The denominator of each of these transfer functions is a cubic polynomial in the Laplace operator. It is cubic because the column has three capacities. The smallest of the roots obtained by factorising this polynomial corresponds to the largest time lag which dominates the dynamic response of the column. In general, this lag varies from a minute or so for small columns to several hours for large ones, say 2 min to 4 h, and is approximately equal to the residence time of the column. For this particular 3-stage example:

$$T_{dominant} \approx \frac{H_{L1} + H_{L2} + H_{L3}}{F}$$

The power of the numerator of these transfer functions depends on the position of the source of disturbance relative to the plate in question. In general, the closer the source to the plate the greater the number of lead terms. For example, $g_{11}(s)$ contains two lead terms which, in effect, offset two of the lags resulting in quasi first order dynamics. This is consistent with the reflux stream L entering the top plate and affecting the its composition directly, which has a relatively fast response. By way of a contrast, $g_{13}(s)$ contains one lead term only resulting in quasi second order dynamics: the affect of changes in feed composition, being filtered by two capacities, is more sluggish.

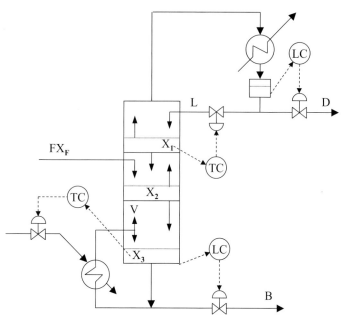

Fig. 90.4 L-V dual composition control of 3-stage column

90.5 Worked Example

The 3-stage column depicted in Figure 90.3 is used to distil a binary mixture. The various assumptions and approximations listed above are valid.

Assuming a relative volatility of 3, the vapour-liquid equilibrium data for the binary mixture is given by the equation:

$$Y_n = \frac{3X_n}{1 + 2X_n}$$

The normal values of the flow rate and composition of the top and bottom product streams are as follows:

$D = 0.4$ kmol min^{-1} $\overline{X}_0 = 0.8$
$\overline{B} = 0.6$ kmol min^{-1} $\overline{X}_3 = 0.433$

This enables the steady state molar balances which yield:

$\overline{F} = 1.0$ kmol min^{-1} $\overline{X}_F = 0.5798$

A reflux ratio L/D of 3 is used such that the normal upper operating line equation is as follows:

$$Y_{n+1} = 0.75X_n + 0.2$$

Hence it can be deduced that the normal values of the flows are:

$\overline{D} = 0.4$, $\overline{B} = 0.6$, $\overline{F} = 1.0$, $\overline{L} = 1.2$
$\overline{V} = 1.6$ kmol min^{-1}

The lower operating line equation is:

$$Y_{n+1} = 1.375X_n - 0.1624$$

from which it can be deduced that the boil-up ratio V/B is 2.67 which is consistent with the values of B and V above.

The Murphree efficiency of each stage is 60%, whence:

$\overline{X}_0 = 0.8$ $\overline{X}_1 = 0.6758$ $\overline{X}_2 = 0.5478$ $\overline{X}_3 = 0.433$
$\overline{X}_F = 0.5798$ $\overline{Y}_1 = 0.8$ $\overline{Y}_2 = 0.7069$ $\overline{Y}_3 = 0.5909$

Differentiating the equation of the X-Y curve gives:

$$\left.\frac{dY}{dX}\right|_{X_n} = \frac{3}{(1 + 2X_n)^2}$$

Substituting the various values of X gives:

$m_1 = 0.5425 \quad m_2 = 0.6831 \quad m_3 = 0.8616$

The liquid holdups H_{L1}, H_{L2} and H_{L3} are 1, 1 and 2 kmols respectively.

Substituting into Equations 90.1–90.3 gives the A and B matrices of Equation 90.4:

$$A = \begin{bmatrix} -1.330 & 0.6558 & 0.0 \\ 1.20 & -2.856 & 0.8271 \\ 0.0 & 1.10 & -0.7136 \end{bmatrix}$$

$$B = \begin{bmatrix} 0.1242 & -0.0931 & 0.0 & 0.0 \\ 0.1280 & -0.1160 & 0.0320 & 1.0 \\ 0.0574 & -0.0789 & 0.0574 & 0.0 \end{bmatrix}$$

Extensive manipulation yields the transfer matrix G(s) of Equation 90.5, of which the key transfer functions are:

$$g_{11}(s) = \frac{X_1(s)}{L(s)}$$
$$= \frac{0.2462\,(2.015s + 1)\,(0.2667s + 1)}{(4.254s + 1)\,(0.8857s + 1)\,(0.2829s + 1)}$$

$$g_{14}(s) = \frac{X_1(s)}{X_F(s)}$$
$$= \frac{0.4987\,(1.401s + 1)}{(4.254s + 1)\,(0.8857s + 1)\,(0.2829s + 1)}$$

$$g_{32}(s) = \frac{X_3(s)}{V(s)}$$
$$= \frac{-0.5650\,(0.6258s + 1)\,(0.2378s + 1)}{(4.254s + 1)\,(0.8857s + 1)\,(0.2829s + 1)}$$

It can be seen that the dominant time constant is 4.254 min which agrees well with the approximation:

$$T_{\text{dominant}} \approx \frac{H_{L1} + H_{L2} + H_{L3}}{F} = 4.0 \text{ min}$$

90.6 L-V Strategy

As can be seen from the concentration model developed, there are very strong interactions between the flows and compositions throughout the column. An L-V energy balance control strategy was depicted in Figure 90.4. Assuming conventional feedback control, the corresponding block diagram is depicted in Figure 90.5.

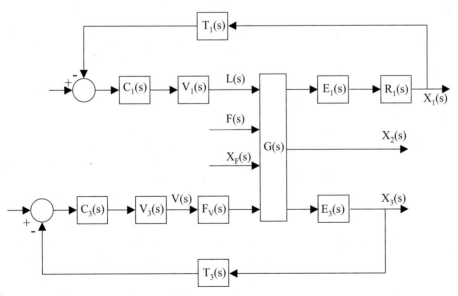

Fig. 90.5 Block diagram of L-V control system

Remember the first assumption in developing the model about the independence of concentration and flow lags. Thus, in Figure 90.5, the flow lags and dynamics of the ancillary plant are superimposed on the composition model. It doesn't particularly matter where the additional dynamics are included, as long as they are appropriate for the loop.

Apart from the controller $C_1(s)$ and valve $V_1(s)$, the top product control loop consists of a the process transfer function $g_{11}(s)$, the overhead condenser $E_1(s)$ and the reflux drum $R_1(s)$:

- There is no liquid flow lag due to the temperature being measured on the top plate onto which the reflux flows directly.
- The top product loop must contain both $E_1(s)$ and $R_1(s)$ because the composition model involved the substitution $X_0 = Y_1$. In practice the reflux composition lags behind that of the vapour due to the dynamics of the condenser and drum.

Apart from the controller $C_3(s)$ and valve $V_3(s)$, the bottom product control loop consists of a vapour flow lag $F_V(s)$, the process transfer function $g_{32}(s)$ and the reboiler $E_3(s)$:

- The single vapour flow lag is due to the effect of changes in boil-up rate on the pressure in the vapour space beneath the feed plate.
- Note that $g_{32}(s)$ incorporates the dynamics of the still.
- The reboiler $E_3(s)$ could have been included before the column, between $V_3(s)$ and $F_V(s)$. In truth, it doesn't matter where the reboiler dynamics are shown, as long as they are included in the loop.

The way that this scheme works is perhaps best explained with reference to the X–Y diagram depicted in Figure 90.6.

Diagrams of this type are used for the steady state design of distillation columns. The design essentially consists of a number of steps between the X-Y curve and the operating lines. Each step corresponds to a stage of the column, i.e. the two plates and the still. The corners of the steps correspond to conditions as follows:

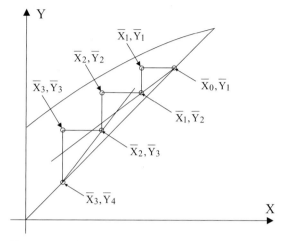

Fig. 90.6 X-Y diagram for 3-stage column

On plates/in still: $(\overline{X}_1, \overline{Y}_1), (\overline{X}_2, \overline{Y}_2)$ and $(\overline{X}_3, \overline{Y}_3)$

Between plates/still: $(\overline{X}_0, \overline{Y}_1), (\overline{X}_1, \overline{Y}_2), (\overline{X}_2, \overline{Y}_3)$ and $(\overline{X}_3, \overline{Y}_4)$.

Note that $\overline{X}_0$ and $\overline{Y}_1$ are numerically equal, as also are $\overline{X}_3$ and $\overline{Y}_4$.

Also note that the steps do not reach the X-Y curve This is because of the Murphree efficiency which has been taken into account at each stage.

The upper operating line, which passes through $\overline{X}_0, \overline{Y}_1$ and $\overline{X}_1, \overline{Y}_2$ has the equation

$$Y_{n+1} = \frac{R}{R+1}.X_n + \frac{1}{R+1}.X_0$$

where the reflux ratio $R = L/D$.

Suppose that a decrease in feed composition X_F occurs. This disturbance will result in a decrease in composition X_1 which in turn will lead to a drop in quality of the overhead product composition X_0. As far as the X-Y diagram is concerned, all the corners would eventually shift left and downwards. The control system will respond to the decrease in X_1 by increasing the reflux rate L which obviously results in a decrease in top product flow rate D. This increase in the reflux ratio may be thought of as the extra recycling necessary to restore the top product quality.

Increasing the reflux ratio R has the effect of increasing the slope $R/(R+1)$ of the upper operating line equation. This means that the size of the steps between the X-Y curve and the upper operating line equation become greater so, for a given number of steps, two in this case, a greater separation can be achieved. Thus the corners of the steps will be driven right and upwards, returning the composition on the top plate to its normal value and restoring the top product quality. When the feed composition returns to its normal value, the control system will maintain top product quality by reducing the reflux ratio.

The lower operating line, which passes through $\overline{X}_2, \overline{Y}_3$ and $\overline{X}_3, \overline{Y}_4$ has the equation

$$Y_{n+1} = \frac{Q+1}{Q}.X_n - \frac{1}{Q}.X_3$$

where the boil-up ratio $Q = V/B$.

A similar analysis can be applied to the bottom product control loop, the controller being reverse acting. An increase in X_3 causes an increase in the boil-up rate V which results in a decrease in bottom product flow rate B. The increase in boil-up ratio has the effect of reducing the slope $(Q+1)/Q$ of the lower operating line equation. This results in a greater separation per stage and drives X_3 down again.

90.7 Comments

The models developed in this chapter may seem complex but they are, nevertheless, realistic and give a clear insight into the dynamics of column behaviour. However, the example is of a binary mixture and for a three stage column only. For more typical industrial columns, the approach is the same but the extent of the equations involved is such that analytical solutions are not practicable and simulation is necessary.

The strength of the interactions between the temperature control loops is considered in Chapter 111 using relative gain analysis. In practice, the interactions are such that conventional feedback is sometimes insufficient and multivariable and/or model predictive controllers, as discussed in Chapters 81 and 117, may have to be used to achieve the necessary decoupling.

90.8 Nomenclature

B	bottoms flow rate	kmol s^{-1}
D	distillate flow rate	kmol s^{-1}
F	feed flow rate	kmol s^{-1}
L	liquid flow rate	kmol s^{-1}
V	vapour flow rate	kmol s^{-1}
A	equivalent cross sectional area	m^2
c_p	specific heat of liquid	kJ kmol^{-1} K^{-1}
h	depth of liquid	m
H_L	liquid holdup of a plate	kmol
H_V	vapour holdup of a plate	kmol bar^{-1}
P	gas pressure	bar (abs)
Q	boil-up ratio (V/B)	–
R	reflux ratio (L/D)	–
t	time	s
X	mol fraction of MVC in liquid phase	–
Y	mol fraction of MVC in vapour phase	–
Z	resistance to flow	bar s kmol^{-1}
λ	latent heat of vaporisation	kJ kmol^{-1}
μ	plate efficiency (fractional)	–
θ	temperature	°C
ρ	density of liquid	kmol m^{-3}

Subscripts

L	liquid
n	nth plate from top of column
V	vapour
0	condenser/reflux drum
1	top plate
2	feed plate
3	still/reboiler

Matrices and Vectors

A	system matrix	3 × 3
B	input matrix	3 × 4
G	transfer matrix	3 × 4
<u>u</u>	input vector	4 × 1
<u>x</u>	state vector	3 × 1

Reacting Systems

91.1 Mass Balance
91.2 Heat Balance
91.3 State Space Model
91.4 Stability Considerations
91.5 Nomenclature

Various schemes for the control of continuous flow stirred tank reactors (CSTR) were covered in Chapter 36. Of particular interest, because of their hazardous nature, is the stability of temperature control systems for reactors in which exothermic reactions are taking place. Figure 91.1 depicts such a system.

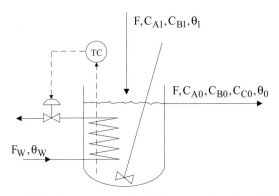

Fig. 91.1 Continuous flow stirred tank reactor with cooling coil

As a basis for designing the control system it is necessary to have an understanding of the kinetics of the reaction as well as the dynamics of the reactor. Suppose that the reaction taking place is of the second order reversible type

$$A + B \Leftrightarrow C$$

where A and B are reagents and C is the product. Although this is only a simple reaction, it will nevertheless demonstrate the essential characteristics of reacting systems. For a treatment of the kinetics of consecutive and parallel reactions, the reader is referred to texts on reactor design, such as those by Coulson (2004) and Metcalfe (1997).

91.1 Mass Balance

Consider unit volume of reaction mixture. The rate at which A and B are consumed by the reaction, which is the same as the rate at which C is produced, is given by:

$$\frac{dC_A}{dt} = \frac{dC_B}{dt} = -\frac{dC_C}{dt} = -k_F C_A C_B + k_R C_C^2 \quad (91.1)$$

Thus, for a volume V of reactants, the rate of reaction is given by:

$$r = V \left(-k_F C_A C_B + k_R C_C^2 \right)$$

An unsteady state molar balance for component A within the reactor gives:

$$V \frac{dC_{A0}}{dt} = F \cdot C_{A1} + V \left(-k_F C_{A0} C_{B0} + k_R C_{C0}^2 \right) - F \cdot C_{A0}$$

Recognising that every term is a variable, apart from V, this equation can be put into deviation form:

$$V\frac{dC_{A0}}{dt} = \overline{F}.C_{A1} + \overline{C}_{A1}.F - V\overline{k}_F\overline{C}_{A0}.C_{B0}$$
$$- V\overline{k}_F\overline{C}_{B0}.C_{A0} - V\overline{C}_{A0}\overline{C}_{B0}.k_F$$
$$+ 2V\overline{k}_R\overline{C}_{C0}.C_{C0} + V\overline{C}_{C0}^2.k_R$$
$$- \overline{F}.C_{A0} - \overline{C}_{A0}.F \quad (91.2)$$

However, as was seen in Chapter 89, k_F is exponentially related to θ_0 by Arrhenius' law:

$$k_F = B_F.e^{\frac{-E_F}{R(\theta_0 + 273)}} \quad (91.3)$$

where B_F, E_F and R are constants. Thus, as with Equation 89.6:

$$\frac{dk_F}{d\theta_0} = B_F.e^{\frac{-E_F}{R(\theta_0 + 273)}} \frac{E_F}{R(\theta_0 + 273)^2}$$
$$= \frac{k_F E_F}{R(\theta_0 + 273)^2}$$

Hence, in deviation form about the normal temperature, k_F is given by:

$$k_F = \left.\frac{dk_F}{d\theta_0}\right|_{\overline{\theta}_0}.\theta_0 = \frac{\overline{k}_F E_F}{R\left(\overline{\theta}_0 + 273\right)^2}.\theta_0$$

Similarly:

$$k_R = \left.\frac{dk_R}{d\theta_0}\right|_{\overline{\theta}_0}.\theta_0 = \frac{\overline{k}_R E_R}{R\left(\overline{\theta}_0 + 273\right)^2}.\theta_0 \quad (91.4)$$

Substituting into 91.2 gives:

$$V\frac{dC_{A0}}{dt} = \overline{F}.C_{A1} + \overline{C}_{A1}.F - V\overline{k}_F\overline{C}_{A0}.C_{B0}$$
$$- V\overline{k}_F\overline{C}_{B0}.C_{A0} - V\overline{C}_{A0}\overline{C}_{B0}.\left.\frac{dk_F}{d\theta_0}\right|_{\overline{\theta}_0}.\theta_0$$
$$+ 2V\overline{k}_R\overline{C}_{C0}.C_{C0} + V\overline{C}_{C0}^2\left.\frac{dk_R}{d\theta_0}\right|_{\overline{\theta}_0}.\theta_0$$
$$- \overline{F}.C_{A0} - \overline{C}_{A0}.F$$

This may be rearranged:

$$\frac{dC_{A0}}{dt} = -\left(\overline{k}_F\overline{C}_{B0} + \frac{\overline{F}}{V}\right).C_{A0} - \overline{k}_F\overline{C}_{A0}.C_{B0}$$
$$+ 2\overline{k}_R\overline{C}_{C0}.C_{C0}$$
$$- \left(\overline{C}_{A0}\overline{C}_{B0}\left.\frac{dk_F}{d\theta_0}\right|_{\overline{\theta}_0} - \overline{C}_{C0}^2\left.\frac{dk_R}{d\theta_0}\right|_{\overline{\theta}_0}\right).\theta_0$$
$$+ \frac{\overline{F}}{V}.C_{A1} + \frac{1}{V}\left(\overline{C}_{A1} - \overline{C}_{A0}\right).F \quad (91.5)$$

which is of the form:

$$\dot{C}_{A0} = -a_{11}.C_{A0} - a_{12}.C_{B0} + a_{13}.C_{C0} - a_{14}.\theta_0$$
$$+ b_{11}.C_{A1} + b_{13}.F \quad (91.6)$$

An unsteady state molar balance for component B within the reactor gives:

$$V\frac{dC_{B0}}{dt} = F.C_{B1} + V.\left(-k_F.C_{A0}.C_{B0} + k_R.C_{C0}^2\right)$$
$$- F.C_{B0}$$

A similar treatment yields:

$$\dot{C}_{B0} = -a_{21}.C_{A0} - a_{22}.C_{B0} + a_{23}.C_{C0} - a_{24}.\theta_0$$
$$+ b_{22}.C_{B1} + b_{23}.F \quad (91.7)$$

An unsteady state molar balance for component C within the reactor gives:

$$V\frac{dC_{C0}}{dt} = V.\left(k_F.C_{A0}.C_{B0} - k_R.C_{C0}^2\right) - F.C_{C0}$$

Again, a similar treatment yields:

$$\dot{C}_{C0} = a_{31}.C_{A0} + a_{32}.C_{B0} - a_{33}.C_{C0} + a_{34}.\theta_0$$
$$- b_{33}.F \quad (91.8)$$

91.2 Heat Balance

If ΔH is the amount of heat released per unit mass of A reacted, then the rate of heat generation by the reaction is given by:

$$Q = V.\left(-k_F.C_{A0}.C_{B0} + k_R.C_{C0}^2\right).\Delta H$$

Note that, by convention, ΔH is negative for exothermic reactions so that Q is a positive quantity for net forward reactions.

An unsteady state heat balance on the contents of the reactor gives:

$$V\rho c_p \frac{d\theta_0}{dt} = F\rho c_p (\theta_1 - \theta_0)$$
$$+ V(-k_F C_{A0} C_{B0} + k_R C_{C0}^2) \Delta H$$
$$- UA(\theta_0 - \theta_W)$$

This may be put into deviation form:

$$V\rho c_p \frac{d\theta_0}{dt} = \overline{F}\rho c_p.(\theta_1 - \theta_0) + (\overline{\theta}_1 - \overline{\theta}_0)\rho c_p.F$$
$$- V\overline{k}_F \overline{C}_{A0} \Delta H.C_{B0} - V\overline{k}_F \overline{C}_{B0} \Delta H.C_{A0}$$
$$- V\overline{C}_{A0}\overline{C}_{B0}\Delta H.k_F + 2V\overline{k}_R \overline{C}_{C0}\Delta H.C_{C0}$$
$$+ V\overline{C}_{C0}^2 \Delta H.k_R - \overline{U}A.(\theta_0 - \theta_W)$$
$$- (\overline{\theta}_0 - \overline{\theta}_W) A.U \qquad (91.9)$$

As was explained in Chapter 89, the coefficient for heat transfer due to the coil's inside film is approximately related to the cooling water flow rate by the equation:

$$h_W = b.F_W^{0.8}$$

Assuming the coil is thin walled and the reactor strongly agitated, as should be the case, the overall heat transfer coefficient is thus given by:

$$\frac{1}{U} = \frac{1}{h_W} + c$$

Substituting and rearranging as with Equation 89.10 gives:

$$U = \frac{bF_W^{0.8}}{1 + bcF_W^{0.8}}$$

Hence, in deviation form about the normal flow, U is given by:

$$U = \left.\frac{dU}{dF_W}\right|_{\overline{F}_W} .F_W \qquad (91.10)$$

$$= \frac{0.8b}{\overline{F}_W^{0.2}(1 + bc\overline{F}_W^{0.8})}\left(1 - \frac{bc\overline{F}_W^{0.8}}{(1 + bc\overline{F}_W^{0.8})}\right).F_W$$

Substituting for $k = f(\theta_0)$ from Equation 91.4 and $U = f(F_W)$ from Equation 91.10 into Equation 91.9 gives:

$$V\rho c_p \frac{d\theta_0}{dt} = \overline{F}\rho c_p.(\theta_1 - \theta_0) + (\overline{\theta}_1 - \overline{\theta}_0)\rho c_p.F$$
$$- V\overline{k}_F \overline{C}_{A0}\Delta H.C_{B0} - V\overline{k}_F \overline{C}_{B0}\Delta H.C_{A0}$$
$$- V\overline{C}_{A0}\overline{C}_{B0}\Delta H \left.\frac{dk_F}{d\theta_0}\right|_{\overline{\theta}_0}.\theta_0$$
$$+ 2V\overline{k}_R \overline{C}_{C0}\Delta H.C_{C0}$$
$$+ V\overline{C}_{C0}^2 \Delta H \left.\frac{dk_R}{d\theta_0}\right|_{\overline{\theta}_0}.\theta_0$$
$$- \overline{U}A.(\theta_0 - \theta_W)$$
$$- (\overline{\theta}_0 - \overline{\theta}_W) A \left.\frac{dU}{dF_W}\right|_{\overline{F}_W}.F_W \qquad (91.11)$$

This may be rearranged:

$$\frac{d\theta_0}{dt} = -\frac{\overline{k}_F \overline{C}_{B0}\Delta H}{\rho c_p}.C_{A0} - \frac{\overline{k}_F \overline{C}_{A0}\Delta H}{\rho c_p}.C_{B0}$$
$$+ \frac{2\overline{k}_R \overline{C}_{C0}\Delta H}{\rho c_p}.C_{C0}$$
$$- \left(\frac{F}{V} + \frac{\overline{U}A}{V\rho c_p} + \frac{\overline{C}_{A0}\overline{C}_{B0}\Delta H}{\rho c_p}\left.\frac{dk_F}{d\theta_0}\right|_{\overline{\theta}_0}\right.$$
$$\left. - \frac{\overline{C}_{C0}^2 \Delta H}{\rho c_p}\left.\frac{dk_R}{d\theta_0}\right|_{\overline{\theta}_0}\right).\theta_0 - \frac{(\overline{\theta}_0 - \overline{\theta}_1)}{V}F$$
$$+ \frac{\overline{F}}{V}.\theta_1 - \frac{(\overline{\theta}_0 - \overline{\theta}_W) A}{V\rho c_p}\left.\frac{dU}{dF_W}\right|_{\overline{F}_W}.F_W$$
$$+ \frac{\overline{U}A}{V\rho c_p}.\theta_W$$

which is of the form:

$$\dot{\theta}_0 = -a_{41}.C_{A0} - a_{42}.C_{B0} + a_{43}.C_{C0} - a_{44}.\theta_0$$
$$- b_{43}.F + b_{44}.\theta_1 - b_{45}.F_W + b_{46}.\theta_W$$
$$(91.12)$$

91.3 State Space Model

Equations 91.6–91.8 and 91.12 can be assembled in state-space form as follows:

$$\begin{bmatrix} \dot{C}_{A0} \\ \dot{C}_{B0} \\ \dot{C}_{C0} \\ \dot{\theta}_0 \end{bmatrix} = \begin{bmatrix} -a_{11} & -a_{12} & a_{13} & -a_{14} \\ -a_{21} & -a_{22} & a_{23} & -a_{24} \\ a_{31} & a_{32} & -a_{33} & a_{34} \\ -a_{41} & -a_{42} & a_{43} & -a_{44} \end{bmatrix} \begin{bmatrix} C_{A0} \\ C_{B0} \\ C_{C0} \\ \theta_0 \end{bmatrix}$$

$$+ \begin{bmatrix} b_{11} & 0 & b_{13} & 0 & 0 & 0 \\ 0 & b_{22} & b_{23} & 0 & 0 & 0 \\ 0 & 0 & -b_{33} & 0 & 0 & 0 \\ 0 & 0 & -b_{43} & b_{44} & -b_{45} & b_{46} \end{bmatrix} \begin{bmatrix} C_{A1} \\ C_{B1} \\ F \\ \theta_1 \\ F_W \\ \theta_W \end{bmatrix}$$

which may be expressed using matrix notation as follows:

$$\underline{\dot{x}} = A.\underline{x} + B.\underline{u} \qquad (91.13)$$

91.4 Stability Considerations

It is evident from the state-space model of the reactor that there are extensive interactions. The extent of these interactions are perhaps best appreciated by consideration of the corresponding block diagram. This requires that Equations 91.6–91.8 and 91.12 are transformed and rearranged in transfer function form.

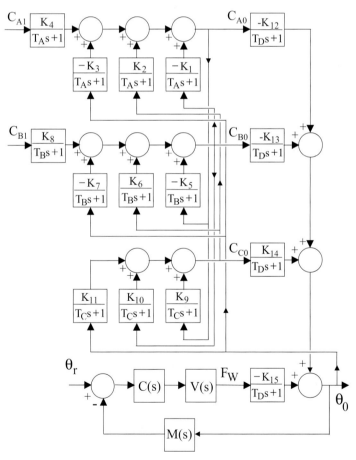

Fig. 91.2 Block diagram of reactor temperature control system

For simplicity, it is assumed that the reactor feed rate and inlet temperature and the cooling water temperature are constant. Thus, in deviation form:

$$F(s) = \theta_1(s) = \theta_W(s) = 0$$

Whence:

$$(T_A s + 1).C_{A0}(s) = -K_1.C_{B0}(s) + K_2.C_{C0}(s)$$
$$- K_3.\theta_0(s) + K_4.C_{A1}.(s)$$

$$(T_B s + 1).C_{B0}(s) = -K_5.C_{A0}(s) + K_6.C_{C0}(s)$$
$$- K_7.\theta_0(s) + K_8.C_{B1}.(s)$$

$$(T_C s + 1).C_{C0}(s) = K_9.C_{A0}(s) + K_{10}.C_{B0}(s)$$
$$+ K_{11}.\theta_0(s)$$

$$(T_D s + 1).\theta_0(s) = K_{12}.C_{A0}(s) + K_{13}.C_{B0}(s)$$
$$- K_{14}.C_{C0}(s) - K_{15}.F_W(s)$$

where:

$$T_A = \frac{V}{V\overline{k_F}\overline{C_{B0}} + \overline{F}} \qquad T_B = \frac{V}{V\overline{k_F}\overline{C_{A0}} + \overline{F}}$$

$$T_C = \frac{V}{2V\overline{k_R}\overline{C_{C0}} + \overline{F}}$$

and:

$$T_D = \frac{V\rho c_p}{\rho c_p \overline{F} + \overline{U}A + V\overline{C_{A0}}\overline{C_{B0}}\Delta H \left.\frac{dk_F}{d\theta_0}\right|_{\overline{\theta_0}} - V\overline{C_{C0}}^2 \Delta H \left.\frac{dk_R}{d\theta_0}\right|_{\overline{\theta_0}}}$$

These transfer functions are incorporated in the block diagram of the temperature control loop as depicted in Figure 91.2.

It is evident from Figure 91.2 that the dominant open loop pole in the temperature control system corresponds to $(T_D s + 1)$. For stability, $T_D > 0$, i.e.:

$$\rho c_p \overline{F} + \overline{U}A > -V\left(\overline{C_{A0}}\overline{C_{B0}}\left.\frac{dk_F}{d\theta_0}\right|_{\overline{\theta_0}} - \overline{C_{C0}}^2 \left.\frac{dk_R}{d\theta_0}\right|_{\overline{\theta_0}}\right)\Delta H$$

Thus, per unit rise in temperature, the increase in rate of heat removal must be greater than the net increase in rate of heat generation. This is entirely consistent with the observation made in Chapter 36, but has been arrived at on a less qualitative basis.

91.5 Nomenclature

A	effective surface area of coil	m^2
b	scaling factor	
B	frequency factor	$m^3 \text{ kmol}^{-1} \text{ s}^{-1}$
c	resistance to heat transfer	$m^2 \text{ K kW}^{-1}$
C	concentration	kmol m^{-3}
c_p	specific heat	$\text{kJ kmol}^{-1} \text{ K}^{-1}$
E	activation energy	kJ kmol^{-1}
F	reactor feed rate	$m^3 \text{ s}^{-1}$
h	film coefficient for heat transfer	$\text{kW m}^{-2} \text{ K}^{-1}$
ΔH	heat of reaction	kJ kmol^{-1}
k	rate constant	$m^3 \text{ kmol}^{-1} \text{ s}^{-1}$
K	steady state gain	
Q	rate of heat generation	kW
r	rate of reaction	kmol s^{-1}
R	universal gas constant (8.314)	$\text{kJ kmol}^{-1} \text{ K}^{-1}$
t	time	s
T	time constant	s
U	overall heat transfer coefficient	$\text{kW m}^{-2} \text{ K}^{-1}$
V	volume of reactor contents	m^3
ρ	density	kmol m^{-3}
θ	temperature	°C

Subscripts

A	reagent
B	reagent
C	product
D	temperature
F	forward reaction
R	reverse reaction
W	cooling water
0	outlet
1	inlet

Distributed Parameter Systems

92.1 Heat Exchanger Dynamics
92.2 Exchanger Process Transfer Function
92.3 Exchanger Load Transfer Function
92.4 Cooling Coil
92.5 Absorption Column Dynamics
92.6 Nomenclature

All the examples of systems modelled so far have been of the lumped parameter type in which a system's parameters vary uniformly, throughout some capacity, with time only. Another important category is that of distributed parameter systems in which the parameters' values vary with both time and distance. The affect of this second dimension is to make the modelling more complex. It is a useful skill to be able to model distributed parameter systems, even though they are often approximated by some lumped parameter equivalent. Two examples are used in this chapter, the first is of a simple heat exchanger and the second is of an absorption column.

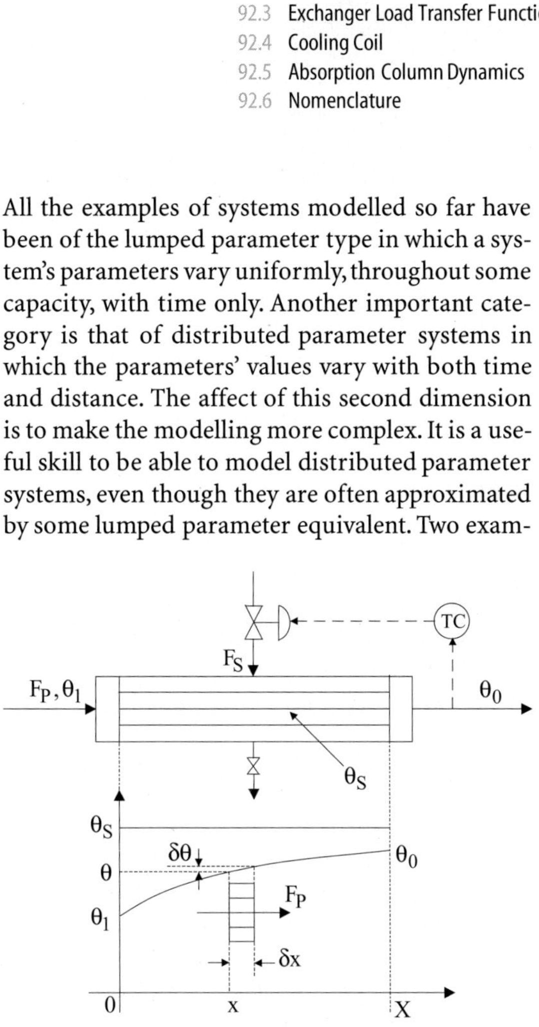

92.1 Heat Exchanger Dynamics

Consider a single pass shell and tube exchanger in which a liquid process stream on the tube side is heated up by condensing steam on the shell side, as depicted in Figure 92.1.

The corresponding block diagram is as shown in Figure 92.2.

Assume that the principle of superposition applies. Each of these transfer functions may be determined by assuming that the others' variables are constant, as indicated in Table 92.1.

Fig. 92.1 Temperature profile in steam heated shell and tube exchanger

Table 92.1 Variables involved in process and load transfer functions

Transfer function	Transfer function relates	Variables assumed constant
$P(s)$	θ_0 to θ_S	F_P and θ_1
$L_1(s)$	θ_0 to θ_1	F_P and θ_S
$L_2(s)$	θ_0 to F_P	θ_S and θ_1

In the following sections, first $P(s)$ and then $L_1(s)$ are established.

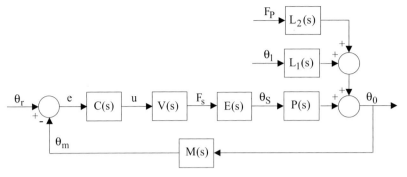

Fig. 92.2 Block diagram of exchanger temperature control loop

92.2 Exchanger Process Transfer Function

As shown in Figure 92.1, the temperature of the process stream varies along the length of the tubes as well as with time. Consider an element of the exchanger of length δx at an arbitrary distance x from the inlet. The process stream enters the element at a temperature of θ and its temperature rises by an amount $\delta\theta$ as it flows through the element. Heat transfer within the element is from the steam through the tube walls and from sensible heat due to flow through the element.

Assuming that F_P is constant and at its normal value, then so too will be U. An unsteady state heat balance on the process stream within the element thus gives:

Rate of Accumulation = Input − Output

$$M\delta x.c_p \frac{\partial \theta}{\partial t} = \overline{U}.A'.\delta x.(\theta_S - \theta) + \overline{F}_P c_p.\theta$$
$$- \overline{F}_P c_p \left(\theta + \frac{\partial \theta}{\partial x}.\delta x\right)$$

This equation is the same in deviation form. It may be simplified and rearranged to give:

$$\frac{Mc_p}{\overline{U}A'}.\frac{\partial \theta}{\partial t} = \theta_S - \theta - \frac{\overline{F}_P c_p}{\overline{U}A'}.\frac{\partial \theta}{\partial x}$$
$$= \theta_S - \theta - \frac{\overline{F}_P}{M}.\frac{Mc_p}{\overline{U}A'}.\frac{\partial \theta}{\partial x}$$

which is of the general form:

$$T_u.\frac{\partial \theta}{\partial t} = \theta_S - \theta - \overline{v}T_u.\frac{\partial \theta}{\partial x} \quad (92.1)$$

This partial differential equation has to be solved to find the required transfer function. It may be Laplace transformed in two stages. First, the transformation is with respect to time and assumes that $x \neq f(t)$. Thus any dependency upon distance is unaffected.

The s operator is defined on the basis that:

$$f(s) = \int_0^\infty f(t)e^{-st}dt.$$

Hence:

$$T_u s.\theta(s) = \theta_S(s) - \theta(s) - \overline{v}T_u.\frac{d}{dx}\theta(s)$$

Note that the partial differential becomes an ordinary differential because transformation has removed the time dimension. Rearranging gives a first order ordinary differential equation:

$$\frac{\overline{v}T_u}{(T_u s + 1)}.\frac{d\theta(s)}{dx} + \theta(s) = \frac{1}{(T_u s + 1)}.\theta_S(s) \quad (92.2)$$

Second, the transformation is with respect to distance and assumes that $\theta_S \neq f(x)$. Thus any dependency upon steam temperature is unaffected.

The p operator is defined on the basis that:

$$f(p) = \int_0^\infty f(x)e^{-px}dx$$

Hence:

$$\frac{\overline{v}T_u}{(T_u s + 1)}p.\theta(p) + \theta(p) = \frac{1}{(T_u s + 1)}.\frac{\theta_S(s)}{p} \quad (92.3)$$

Note that the term in θ_S is, in effect, a constant and is simply treated as a step input.

Solving for $\theta(p)$ gives:

$$\theta(p) = \frac{1}{(T_u s + 1)} \cdot \frac{\theta_S(s)}{p\left(\frac{\overline{v}T_u}{(T_u s + 1)} \cdot p + 1\right)}$$

Inverse transforming, back into the s domain, gives:

$$\theta(s) = \frac{1}{(T_u s + 1)}\left(1 - e^{-\frac{(T_u s + 1)x}{\overline{v}T_u}}\right).\theta_S(s) \quad (92.4)$$

This is the value of θ as a function of θ_S at some arbitrary distance x along the exchanger. What is of interest is the value of θ at the outlet, i.e. when x = X:

$$\theta_0(s) = \frac{1}{(T_u s + 1)}\left(1 - e^{-\frac{(T_u s + 1)X}{\overline{v}T_u}}\right).\theta_S(s)$$

Note that the tube side residence time $\overline{L} = X/\overline{v}$. This may be rearranged to give the required transfer function:

$$\theta_0(s) = \frac{1}{(T_u s + 1)}\left(1 - \frac{e^{-\overline{L}s}}{e^{\overline{L}/T_u}}\right).\theta_S(s) = P(s).\theta_S(s)$$

(92.4)

This is represented in block diagram form in Figure 92.3. Combinations of time lags and delays such as this are typical of the transfer functions of distributed parameter systems.

92.3 Exchanger Load Transfer Function

Next, the transfer function $L_1(s)$ is established. If the steam temperature is assumed constant then, in deviation form, $\theta_S(s) = 0$ and Equation 92.2 may be rearranged:

$$\frac{1}{\theta(s)}.d\theta(s) = -\frac{(T_u s + 1)}{\overline{v}T_u}.dx \quad (92.5)$$

This may be integrated within the limits corresponding to the tubes' inlet and outlet conditions as follows:

$$\int_{\theta_1(s)}^{\theta_0(s)} \frac{d\theta(s)}{\theta(s)} = -\frac{(T_u s + 1)}{\overline{v}T_u} \int_0^X dx$$

Hence:

$$\ln\frac{\theta_0(s)}{\theta_1(s)} = -\frac{(T_u s + 1)}{\overline{v}T_u}.X$$

$$\theta_0(s) = e^{-\frac{(T_u s + 1).X}{\overline{v}T_u}}.\theta_1(s)$$

Again, substituting $\overline{L} = X/\overline{v}$ yields the required transfer function:

$$\theta_0(s) = \frac{e^{-\overline{L}s}}{e^{\overline{L}/T_u}}.\theta_1(s) = L_1(s).\theta_1(s) \quad (92.6)$$

92.4 Cooling Coil

The contents of an agitated vessel are cooled by passing cooling water through an internal coil, as depicted in Figure 92.4.

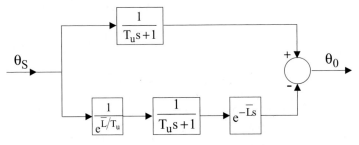

Fig. 92.3 Block diagram representation of the process transfer function

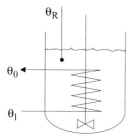

Fig. 92.4 Reaction vessel with internal cooling coil

This system is exactly analogous to the exchanger considered previously: the cooling water temperature being forced in a distributed manner along the length of the coil. Thus, from Equation 92.4:

$$\frac{\theta_0(s)}{\theta_R(s)} = \frac{1}{(T_u s + 1)} \left(1 - \frac{e^{-\bar{L}s}}{e^{\bar{L}/T_u}}\right) \quad (92.7)$$

where, as before: $T_u = \frac{M.c_P}{U.A'}$ and $\bar{L} = \frac{\bar{v}}{X}$.

Also, from Equation 92.6:

$$\frac{\theta_0(s)}{\theta_1(s)} = \frac{e^{-\bar{L}s}}{e^{\bar{L}/T_u}}$$

Hence the transfer function relating the reaction temperature to the coil inlet temperature is of the form:

$$\frac{\theta_R(s)}{\theta_1(s)} = \frac{\theta_R(s)}{\theta_0(s)} \cdot \frac{\theta_0(s)}{\theta_1(s)}$$

$$= \frac{(T_u s + 1)}{\left(1 - \frac{e^{-\bar{L}s}}{e^{\bar{L}/T_u}}\right)} \cdot \frac{e^{-\bar{L}s}}{e^{\bar{L}/T_u}} \quad (92.8)$$

$$= \frac{T_u s + 1}{e^{\bar{L}}(T_u s + 1)/T_u - 1}$$

92.5 Absorption Column Dynamics

This example demonstrates how the same techniques can be applied to a fundamentally different application. Consider the packed bed absorption column as depicted in Figure 92.5.

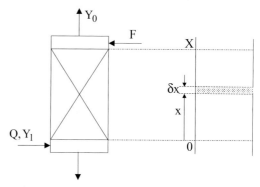

Fig. 92.5 Cross section of a packed bed absorption column

A gaseous stream Q, containing some soluble component, is blown into the bottom of the column. It passes up through the bed, countercurrent to a liquid solvent stream F which is pumped into the top of the column and then trickles down through the bed. The volume fraction (or mol fraction) of the soluble component in the gaseous stream is reduced from Y_1 at the inlet to Y_0 at the outlet.

Assume that the gas flow rate is approximately constant and the concentration of soluble component is low. The flow rate Q may then be assumed to be uniform throughout the column and its pressure P to be constant.

Consider an element of the bed of thickness δx and cross sectional area A. If the surface area of packing per unit volume is a, then

Surface area within the element is $A.\delta x.a$.

The rate of mass transfer within the element is thus

$$N_A = K_G.A\delta xa.(p - p_E)$$

The driving force for mass transfer is the difference in partial pressures at the gas liquid interface. p_E is the partial pressure of the soluble component in the gas stream that would be in equilibrium with soluble component in the liquid stream. Assuming most of the resistance to mass transfer is in the gas film, $p \gg p_E$ and:

$$N_A \approx K_G.A\delta xa.p$$

If the voidage fraction of the packing is e, then the volume of space within the element is $A.\delta x.e$.

Thus the mass of soluble component within that space is $A\delta xe.\rho.Y$.

An unsteady state molar balance for the soluble component in the gas stream within the element thus gives:

Rate of Accumulation = Input − Output

$$A\delta xe.\rho.\frac{\partial Y}{\partial t} = Q\rho.Y - Q\rho.\left(Y + \frac{\partial Y}{\partial x}\delta x\right) - K_G.A\delta xa.p$$

which may be simplified to give:

$$Ae\rho.\frac{\partial Y}{\partial t} = -Q\rho.\frac{\partial Y}{\partial x} - K_G A a.p \quad (92.9)$$

However, from Dalton's law, $p = YP$.

Hence:

$$Ae\rho.\frac{\partial Y}{\partial t} = -Q\rho.\frac{\partial Y}{\partial x} - K_G A a P.Y$$

This equation is already in deviation form since A, e, ρ, P, Q, K_G and a are all approximately constant. It is a partial differential equation and, as before, is firstly transformed with respect to time assuming that $x \neq f(t)$.

The s operator is defined on the basis that $f(s) = \int_0^\infty f(t)e^{-st}dt$.

Hence:

$$Ae\rho.sY(s) = -Q\rho.\frac{d}{dx}.Y(s) - K_G A a P.Y(s)$$

which may be rearranged:

$$\frac{d}{dx}Y(s) = \frac{-A(e\rho.s + K_G aP)}{Q\rho}.Y(s) \quad (92.10)$$

This may be integrated within the limits corresponding to the column's inlet and outlet conditions as follows:

$$\int_{Y_1(s)}^{Y_0(s)} \frac{dY(s)}{Y(s)} = \frac{-A(e\rho.s + K_G aP)}{Q\rho}.\int_0^X dx$$

$$\ln\frac{Y_0(s)}{Y_1(s)} = \frac{-A(e\rho.s + K_G aP)}{Q\rho}.X$$

Substitute for the volume of the column, $V = AX$, and rearrange giving:

$$\ln\frac{Y_0(s)}{Y_1(s)} = -\frac{Ve}{Q}\frac{K_G aP}{e\rho}\left(\frac{e\rho}{K_G aP}s + 1\right) \equiv -\frac{L}{T}(Ts+1)$$

where the time delay $L = \frac{Ve}{Q}$ and the time lag $T = \frac{e\rho}{K_G aP}$. Anti-logging gives:

$$Y_0(s) = e^{-\frac{L}{T}(Ts+1)}.Y_1(s) = \frac{1}{e^{L/T}}.e^{-Ls}.Y_1(s) \quad (92.11)$$

which is the transfer function relating inlet and outlet concentration changes. Note again that the transfer function is a combination of time constants and delays. With practice it is possible to anticipate the structure of the transfer functions for these distributed parameter systems.

92.6 Nomenclature

a	surface area per unit volume	m^{-1}
A	cross sectional area	m^2
A'	surface area per unit length	m
c_p	specific heat	$kJ\ kg^{-1}\ K^{-1}$
e	voidage fraction	
F	mass flow rate	$kg\ s^{-1}$
K_G	mass transfer coefficient of gas film	$kmol\ m^{-2}\ bar^{-1}\ s^{-1}$
L	time delay (residence time)	s
M	mass of liquid per unit length	$kg\ m^{-1}$
N_A	rate of mass transfer	$kmol\ s^{-1}$
p	partial pressure	bar
P	pressure	bar
Q	volumetric flow rate	$m^3\ s^{-1}$
t	time	s
T	time constant	s
U	overall heat transfer coefficient	$kW\ m^{-2}\ K^{-1}$
v	tube side velocity	$m\ s^{-1}$
V	volume	m^3
x	distance from inlet	m
X	overall length	m
Y	volume (or mol) fraction	–
θ	temperature at distance x	°C
ρ	density	$kmol\ m^{-3}$

Subscripts

E	equilibrium
P	process stream
S	steam
u	tube side
0	outlet
1	inlet

Anti-Surge Systems

93.1 Dynamics of Anti-Surge
93.2 Anti-Surge Design
93.3 Worked Example
93.4 Anti-Surge Control: Pressure Context
93.5 Comments
93.6 Nomenclature

There are two main categories of level control system. First, and conventionally, those in which the level is controlled for its own sake. The dynamics of such a level control system were considered in Chapter 84. The second category is where the exact level is unimportant, provided that the tank doesn't overflow or run dry, and surges in flow are damped by allowing the level to rise and fall. The level control system on the reflux drum of a distillation column is a good example, as depicted in Figure 93.1. The reflux rate is controlled by the temperature at the top of the column and a level controller used to average out fluctuations in product rate, as described in more detail in Chapter 35.

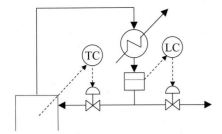

Fig. 93.1 Control of level in a reflux drum

The basic strategy is for the control loop to have P action only with a low gain, such that the closed loop response is overdamped, and to exploit its inherent offset. A tank with a level control system can be designed to provide any degree of damping. It is essentially a question of estimating the size of surge that is likely to occur, specifying the maximum rate of change that is acceptable, determining the necessary tank capacity, and matching the controller gain with the height of the chosen tank.

93.1 Dynamics of Anti-Surge

Consider the simple level control system depicted in Figure 93.2. It is assumed that the dynamics of all the control loop elements other than the tank are negligible.

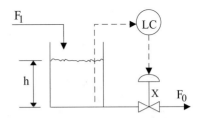

Fig. 93.2 Tank with level control loop

The corresponding block diagram is depicted in Figure 93.3. The process transfer functions are as developed in Chapter 84. Note that all the signals shown are in deviation form.

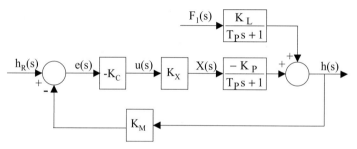

Fig. 93.3 Block diagram of level control system

Using block diagram algebra yields the closed loop response:

$$h(s) = \frac{K_L}{(T_P s + 1)}.F_1(s) + \frac{K_C.K_X.K_P}{(T_P s + 1)}.(h_R(s) - K_M.h(s))$$

Assuming the set point is fixed at the normal level in the tank, which it would be for anti-surge control, then $h_R(s) = 0$.

Whence:

$$h(s) = \frac{\frac{K_L}{(T_P s + 1)}}{1 + \frac{K_M.K_C.K_X.K_P}{(T_P s + 1)}}.F_1(s)$$

$$= \frac{K_L}{(1 + K)}.\frac{1}{(Ts + 1)}.F_1(s)$$

where:

$$K = K_M.K_C.K_X.K_P \quad \text{and} \quad T = \frac{T_P}{1 + K} \quad (93.1)$$

If a step change in feed rate occurs, which is the worst possible surge, i.e. $F_1(s) = \frac{F_1}{s}$:

$$h(s) = \frac{K_L.F_1}{(1 + K)}.\frac{1}{s(Ts + 1)}$$

The response as a function of time is thus:

$$h = \frac{K_L.F_1}{(1 + K)}.\left(1 - e^{-t/T}\right) \quad (93.2)$$

Fig. 93.4 Response to a step change in inlet flow

If the steady state closed loop offset is H, as depicted in Figure 93.4, then:

$$H = \frac{K_L.F_1}{(1 + K)} \quad (93.3)$$

Differentiating Equation 93.2 gives:

$$\frac{dh}{dt} = \frac{K_L.F_1}{(1 + K)}.\frac{e^{-t/T}}{T}$$

$$\frac{d^2h}{dt^2} = -\frac{K_L.F_1}{(1 + K)}.\frac{e^{-t/T}}{T^2}$$

93.2 Anti-Surge Design

Following a step change in feed rate, the important variable for anti-surge control is the rate of change of outlet flowrate, dF_0/dt.

An unsteady state volume balance gives:

$$A\frac{dh}{dt} = F_1 - F_0$$

$$A\frac{d^2h}{dt^2} = -\frac{dF_0}{dt}$$

Thus:
$$\frac{dF_0}{dt} = \frac{A.K_L.F_1}{(1+K)} \cdot \frac{e^{-t/T}}{T^2}$$

Noting that the maximum rate of change of outlet flowrate, the principal constraint, is at $t = 0$, and substituting from Equation 93.1 gives:

$$\left.\frac{dF_0}{dt}\right|_{max} = \frac{A.K_L.F_1}{(1+K).T^2} = \frac{A.K_L.F_1.(1+K)}{T_P^2}$$
(93.4)

Substituting from Equations 84.8 and 84.9, in which it was shown that $T_P = A.K_L$, and also from Equation 93.1 gives:

$$\left.\frac{dF_0}{dt}\right|_{max} = \frac{F_1.(1+K)}{A.K_L} = \frac{F_1^2}{A.H}$$

Whence:
$$A.H = \frac{F_1^2}{\left.\frac{dF_0}{dt}\right|_{max}}$$
(93.5)

This is the change in volume that occurs when the rate of change of outlet flowrate is restricted to some specified maximum value following a step change of magnitude F_1 in inlet flowrate. For any particular anti-surge duty, AH represents the minimum tank volume required. Clearly, in practice, a volume slightly larger than AH is required to prevent flooding, and allowance must be made for the head necessary to produce the outlet flowrate under minimum flow conditions.

The relative values of A and H may be chosen to allow for the shape of tanks available. Having established a value for H, the controller gain K_C to give the required damping may be determined from Equations 93.1 and 93.3:

$$H = \frac{K_L.F_1}{1+K} = \frac{K_L.F_1}{1 + K_M.K_C.K_X.K_P}$$

Hence:
$$K_C = \frac{1}{K_M.K_X.K_P}\left(\frac{K_L.F_1}{H} - 1\right)$$

Note that K_P and K_L are defined explicitly in Chapter 84. K_M and K_X represent the calibrations of the level sensor/transmitter and converter/actuator respectively.

93.3 Worked Example

A reaction is carried out by using three reactors in parallel and the product is sent to a recovery column. The normal flow from each reactor is $10\,m^3/h$. Each reactor is shut down about every 10 days for an 8-h period. To protect the column from changes in feedrate it is proposed to install a buffer vessel between the reactors and the column. What size of storage tank is needed to guarantee a maximum change in column feedrate of 10%/h?

Assume that, because the reactors are shut down every 10 days for an 8-h period, the flow into the buffer vessel will always be either 20 or $30\,m^3/h$. A strategic decision has been made that the buffer vessel is not intended to cope with the situation when two reactors go off stream simultaneously.

The worst case scenario is when the third reactor comes on stream which represents a 50% increase in flow, rather than when the third reactor goes off stream which represents a 33% decrease in flow.

If $F_1 = F_0 = 20\,m^3/h$ and F_0 is to be limited to 10% per hour, then:

$$\left.\frac{dF_0}{dt}\right|_{max} = 2\,m^3\,h^{-2}.$$

Substituting into Equation 93.5 gives the minimum volume necessary:

$$V = \frac{F_1^2}{\left.\frac{dF_0}{dt}\right|_{max}} = \frac{10^2}{2} = 50\,m^3.$$

A vessel of 4 m diameter and 4 m height ($50.3\,m^3$) would be adequate for the job. Remember that allowance must be made for the head required to produce the flowrate of $20\,m^3/h$ when there are only two reactors operating.

93.4 Anti-Surge Control: Pressure Context

Consider the anti-surge control system shown in Figure 87.6 whose block diagram is as depicted in Figure 87.7. Block diagram algebra yields the closed loop response:

$$P_0(s) = \frac{K_L}{(T_Ps + 1)}.P_1(s)$$
$$+ \frac{(-K_C).K_X.(-K_P)}{(T_Ps + 1)}(P_R(s) - K_M.P_0(s))$$

Assuming the set point is fixed at the normal pressure, which it would be for anti-surge control, then $P_R(s) = 0$. Whence:

$$P_0(s) = \frac{K_L}{(1 + K)}.\frac{1}{(Ts + 1)}.P_1(s)$$

where:

$$K = K_M.K_C.K_X.K_P \quad \text{and} \quad T = \frac{T_P}{1 + K}. \quad (93.6)$$

If a step change in supply pressure occurs, which is the worst case scenario, i.e. $P_1(s) = \frac{P_1}{s}$, then

$$P_0(s) = \frac{K_L.P_1}{(1 + K)}.\frac{1}{s(Ts + 1)}$$

The response as a function of time is thus analogous to Equation 93.2:

$$P_0 = \frac{K_L.P_1}{(1 + K)}.\left(1 - e^{-t/T}\right)$$

Differentiating this gives:

$$\frac{d^2P_0}{dt^2} = -\frac{K_L.P_1}{(1 + K)}.\frac{e^{-t/T}}{T^2}$$

An unsteady state volume balance gives:

$$C\frac{dP_0}{dt} = Q_1 - Q_0$$
$$C\frac{d^2P_0}{dt^2} \approx -\frac{dQ_0}{dt}$$

The approximation is in recognition of the fact that the flow Q_1 will not necessarily be constant following the step change in P_1. This depends on whether the flow is sonic or turbulent, and on the relative values of the pressures.

Thus

$$\frac{dQ_0}{dt} = \frac{C.K_L.P_1}{(1 + K)}.\frac{e^{-t/T}}{T^2}$$

Again, analogous to Equation 93.4, the initial rate of change is maximum, i.e. at t = 0:

$$\left.\frac{dQ_0}{dt}\right|_{max} = \frac{C.K_L.P_1}{(1 + K).T^2}$$

Substituting from Equation 93.1 for T and from Equation 87.6 for T_P and K_L:

$$T = \frac{T_P}{1 + K} \quad T_P = \frac{R_0R_1C}{R_0 + R_1} \quad K_L = \frac{R_0}{R_0 + R_1}$$

gives:

$$\left.\frac{dQ_0}{dt}\right|_{max} = \frac{(R_0 + R_1).(1 + K).P_1}{C.R_0.R_1^2} \quad (93.7)$$

which relates the resistances of the valves, the minimum capacity of the anti-surge vessel, the size of the step input and the open loop gain.

93.5 Comments

Some judgement is necessary in deciding upon the set point for the level loop. For example, in a level context, if the tank is to protect against sudden, but reversible, decreases in flow from the normal flow, it would normally be operated full with the set point at the top of the tank. Conversely, for sudden reversible increases, the tank is normally empty with the set point at the bottom. If the tank is required to protect against both sudden increases and decreases about the normal flow, a tank of capacity 2AH would be required with the set point at half full.

Note that these designs are based upon worst case scenarios, i.e. a step change in inlet flow and the maximum slope at time t = 0. In practice, anti-surge systems with smaller capacity will provide adequate protection for most purposes. In this context it is appropriate to introduce a touch of integral action to gently restore the level to the normal value in between disturbances.

And finally, remember that anti-surge control is a capital intensive solution to the problem of averaging out fluctuations. It is much cheaper, if possible, to use feedforward control to manipulate the process down stream to accommodate the surges than to put in an anti-surge vessel.

93.6 Nomenclature

A	cross sectional area	m^2
e	error	m
F	flow rate	m^3 s^{-1}
h	level	m
H	steady state level	m
K	gain	
P	pressure	bar
Q	flow rate (measured at 1 bar)	m^3 s^{-1}
t	time	s
u	controller output	
T	time constant	s
X	fractional opening of the valve	–

Subscripts

C	controller
L	load
M	sensor and transmitter
P	process
R	set point
X	I/P converter and actuator
0	outlet
1	inlet

Psychrometric Systems

94.1 Description of Spray Drier
94.2 Volume Balance
94.3 Mass Balance
94.4 Heat Balance
94.5 Nomenclature

Psychrometric systems are those involving mixtures of air and water vapour. Plant involving air/water mixtures are surprisingly common and include various types of drier and all forms of air conditioning plant. A model is developed in this chapter of a spray drier as a vehicle for considering the issues involved in the modelling of psychrometric systems. Spray driers are undoubtedly the most important category of drier used in bulk drying applications.

94.1 Description of Spray Drier

A spray drier is a large cylindrical vessel in which a slurry is dried by a current of hot and relatively dry air. A typical arrangement is as depicted in Figure 94.1. The slurry, which consists of solids suspended in water, is pumped into an atomiser at the top of the tower and the hot air is blown in at the bottom. There are many designs of atomiser, involving impellers, rotating nozzles, spinning discs, and so on. In essence, the atomiser causes the slurry to form a spray or a conical shaped sheet which breaks up into droplets. As the droplets fall through the air stream, evaporation takes place resulting in dry solid particles which fall to the bottom and are discharged through a rotary valve. The heat for the evaporation process comes from the hot air stream, the moist air stream produced being vented from the top of the drier.

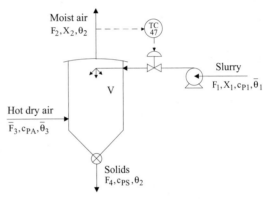

Fig. 94.1 Schematic of spray drier depicting slurry feed and atomiser

Drying is typically controlled by regulating the temperature of the air leaving the drier by means of a simple feedback loop. Clearly the temperature of the moist air stream θ_2 is the controlled variable, the flow rate of the slurry F_1 is the manipulated variable and its water content X_1 is the disturbance variable. In developing the model of the drier it is assumed that the flow rate F_3 and temperature θ_3 of the hot air stream and the temperature θ_1 of the slurry are all constant.

94.2 Volume Balance

The flow rate of the moist air stream is determined by the flow rate of the dry air stream and the amount of water evaporated. Because spray driers are normally operated at atmospheric pressure, there is no capacity for flow changes due to the compressibility of the air stream. Thus, assuming that the volume of solids is negligible relative to that of the air and water vapour, and that all of the water is evaporated, an overall unsteady state volume balance gives:

Rate of Accumulation = Input − Output = 0

$$\frac{1}{\rho_2}.F_2 = \frac{1}{\rho_A}.\overline{F}_3 + \frac{1}{\rho_V}.F_1.X_1 \quad m^3\ s^{-1}$$

Noting that it is presumed that the flow rate of hot dry air is constant, the volume balance may be put into deviation form:

$$F_2 = \frac{\overline{\rho}_2}{\rho_V}.\left(\overline{F}_1.X_1 + \overline{X}_1.F_1\right)$$

This may be Laplace transformed:

$$F_2(s) = \frac{\overline{\rho}_2}{\rho_V}.\left(\overline{F}_1.X_1(s) + \overline{X}_1.F_1(s)\right)$$

which is of the general form:

$$F_2(s) = K_1.X_1(s) + K_2.F_1(s) \quad (94.1)$$

94.3 Mass Balance

It may be presumed that the drier's contents have lumped parameter dynamics because of the turbulence caused by the atomisation. Thus the weight fraction of water inside the drier will be roughly the same as in the moist air stream leaving it. Assuming that the hot air stream entering the drier contains negligible moisture, which is not unreasonable, an unsteady state mass balance for water gives:

Rate of Accumulation = Input − Output

$$V\overline{\rho}_2 \frac{dX_2}{dt} = F_1.X_1 - F_2.X_2 \quad kg\ s^{-1}$$

In deviation form this becomes:

$$V\overline{\rho}_2 \frac{dX_2}{dt} = \overline{F}_1.X_1 + \overline{X}_1.F_1 - \overline{F}_2.X_2 - \overline{X}_2.F_2.$$

Laplace transforming and rearranging gives:

$$\left(\frac{V\overline{\rho}_2}{\overline{F}_2}.s + 1\right).X_2(s) = \frac{\overline{F}_1}{\overline{F}_2}.X_1(s) + \frac{\overline{X}_1}{\overline{F}_2}.F_1(s) - \frac{\overline{X}_2}{\overline{F}_2}.F_2(s)$$

which is of the form:

$$(T_1 s + 1).X_2(s) = K_3.X_1(s) + K_4.F_1(s) - K_5.F_2(s) \quad (94.2)$$

94.4 Heat Balance

The contents of the drier may be considered to be a mixture of air and water vapour only because, to all practical intents and purposes, the evaporation is instantaneous and the residence time of the solids is negligible. Again lumped parameter dynamics are assumed which means that the specific enthalpy of the moist air stream and the drier's contents are roughly equal. Neglecting the thermal capacity of the drier's walls, *etc.*, an unsteady state heat balance (relative to 0°C) gives:

$$V\overline{\rho}_2 \frac{dh_2}{dt} = F_1.h_1 + \overline{F}_3.h_3 - F_2.h_2 - F_4.h_4 \quad kW$$

Substituting from the mass balance for the solids to eliminate F_4 gives:

$$V\overline{\rho}_2 \frac{dh_2}{dt} = F_1.h_1 + \overline{F}_3.h_3 - F_2.h_2 - F_1(1 - X_1).h_4 \quad (94.3)$$

Values for each of these specific enthalpies are next established. Treating the water and solids components of the slurry stream separately:

$$h_1 = c_{P1}.\theta_1 = (X_1.c_{PW} + (1 - X_1).c_{PS}).\theta_1$$
$$= X_1.(c_{PW} - c_{PS}).\theta_1 + c_{PS}.\theta_1 \quad (94.4)$$

Noting that the moist air stream may be considered to be a mixture of dry air and water vapour, the latter having been evaporated at the bulk temperature θ_2:

$$h_2 = (1 - X_2)c_{PA}.\theta_2 + X_2(c_{PW}.\theta_2 + \lambda_W) \quad (94.5)$$

whence:

$$\frac{dh_2}{dt} = (1-\overline{X}_2)c_{PA}\frac{d\theta_2}{dt} - c_{PA}\overline{\theta}_2\frac{dX_2}{dt}$$
$$+ \overline{X}_2.c_{PW}\frac{d\theta_2}{dt} + (c_{PW}.\overline{\theta}_2 + \lambda_W)\frac{dX_2}{dt}$$

which may be rearranged:

$$\frac{dh_2}{dt} = ((1-\overline{X}_2)c_{PA} + \overline{X}_2.c_{PW}) \cdot \frac{d\theta_2}{dt}$$
$$+ ((c_{PW} - c_{PA})\overline{\theta}_2 + \lambda_W)\frac{dX_2}{dt} \quad (94.6)$$

Because the hot air stream is assumed to be dry:

$$h_3 = c_{PA}.\overline{\theta}_3 \quad (94.7)$$

and because evaporation takes place at the bulk temperature:

$$h_4 = c_{PS}.\theta_2 \quad (94.8)$$

Substituting Equations 94.4–94.8 into Equation 94.3 gives:

$$V\overline{\rho}_2 \left\{ ((1-\overline{X}_2)c_{PA} + \overline{X}_2.c_{PW})\frac{d\theta_2}{dt} \right.$$
$$\left. + ((c_{PW} - c_{PA})\overline{\theta}_2 + \lambda_W)\frac{dX_2}{dt} \right\}$$
$$= F_1\left(X_1.(c_{PW} - c_{PS})\overline{\theta}_1 + c_{PS}.\overline{\theta}_1\right)$$
$$+ \overline{F}_3.c_{PA}.\overline{\theta}_3$$
$$- F_2\left((1-X_2)c_{PA}.\theta_2 + X_2(c_{PW}.\theta_2 + \lambda_W)\right)$$
$$- F_1.(1-X_1)c_{PS}.\theta_2$$

Putting this into deviation form gives:

$$V\overline{\rho}_2 \left\{ ((1-\overline{X}_2).c_{PA} + \overline{X}_2 c_{PW})\frac{d\theta_2}{dt} \right.$$
$$\left. + ((c_{PW} - c_{PA})\overline{\theta}_2 + \lambda_W)\frac{dX_2}{dt} \right\}$$
$$= \overline{F}_1(c_{PW} - c_{PS})\overline{\theta}_1.X_1 + \overline{X}_1(c_{PW} - c_{PS})\overline{\theta}_1.F_1$$
$$+ c_{PS}\overline{\theta}_1.F_1 + 0 - \overline{F}_2(1-\overline{X}_2)c_{PA}.\theta_2$$
$$+ \overline{F}_2 c_{PA}\overline{\theta}_2.X_2 - (1-\overline{X}_2)c_{PA}\overline{\theta}_2.F_2$$
$$- \overline{F}_2\overline{X}_2 c_{PW}.\theta_2 - \overline{F}_2(c_{PW}.\overline{\theta}_2 + \lambda_W).X_2$$
$$- \overline{X}_2(c_{PW}.\overline{\theta}_2 + \lambda_W).F_2 - \overline{F}_1(1-\overline{X}_1)c_{PS}.\theta_2$$
$$+ \overline{F}_1 c_{PS}\overline{\theta}_2.X_1 - (1-\overline{X}_1)c_{PS}\overline{\theta}_2.F_1$$

Transforming and rearranging gives:

$$\left\{ V\overline{\rho}_2 \left((1-\overline{X}_2)c_{PA} + \overline{X}_2.c_{PW} \right).s + \overline{F}_2(1-\overline{X}_2)c_{PA} \right.$$
$$\left. + \overline{F}_2\overline{X}_2.c_{PW} + \overline{F}_1(1-\overline{X}_1)c_{PS} \right\}.\theta_2(s)$$
$$+ \left\{ V\overline{\rho}_2 \left((c_{PW} - c_{PA})\overline{\theta}_2 + \lambda_W \right).s \right.$$
$$\left. + \overline{F}_2(c_{PW}.\overline{\theta}_2 + \lambda_W) - \overline{F}_2.c_{PA}\overline{\theta}_2 \right\}.X_2(s)$$
$$= \left(\overline{F}_1(c_{PW} - c_{PS})\overline{\theta}_1 + \overline{F}_1.c_{PS}.\overline{\theta}_2 \right).X_1(s)$$
$$+ \left(\overline{X}_1(c_{PW} - c_{PS})\overline{\theta}_1 + c_{PS}.\overline{\theta}_1 - (1-\overline{X}_1)c_{PS}.\overline{\theta}_2 \right).F_1(s)$$
$$- \left((1-\overline{X}_2)c_{PA}.\overline{\theta}_2 + \overline{X}_2(c_{PW}.\overline{\theta}_2 + \lambda_W) \right).F_2(s)$$

which is of the form:

$$K_6(T_2 s + 1).\theta_2(s) + K_7(T_3 s + 1).X_2(s)$$
$$= K_8.X_1(s) + K_9.F_1(s) - K_{10}.F_2(s) \quad (94.9)$$

where:

$$K_6 = \overline{F}_2(1-\overline{X}_2)c_{PA} + \overline{F}_2\overline{X}_2 c_{PW} + \overline{F}_1(1-\overline{X}_1)c_{PS}$$

$$T_2 = \frac{V\overline{\rho}_2 \left((1-\overline{X}_2)c_{PA} + \overline{X}_2.c_{PW} \right)}{\overline{F}_2(1-\overline{X}_2)c_{PA} + \overline{F}_2\overline{X}_2.c_{PW} + \overline{F}_1(1-\overline{X}_1)c_{PS}}$$

$$K_7 = \overline{F}_2(c_{PW}.\overline{\theta}_2 + \lambda_W) - \overline{F}_2.c_{PA}.\overline{\theta}_2$$

$$T_3 = \frac{V\overline{\rho}_2 \left((c_{PW} - c_{PA})\overline{\theta}_2 + \lambda_W \right)}{\overline{F}_2(c_{PW}.\overline{\theta}_2 + \lambda_W) - \overline{F}_2.c_{PA}.\overline{\theta}_2}$$

$$K_8 = \overline{F}_1(c_{PW} - c_{PS})\overline{\theta}_1 + \overline{F}_1.c_{PS}.\overline{\theta}_2$$

$$K_9 = \overline{X}_1(c_{PW} - c_{PS})\overline{\theta}_1 + c_{PS}.\overline{\theta}_1 - (1-\overline{X}_1)c_{PS}.\overline{\theta}_2$$

$$K_{10} = (1-\overline{X}_2)c_{PA}.\overline{\theta}_2 + \overline{X}_2(c_{PW}.\overline{\theta}_2 + \lambda_W)$$

Equations 94.1, 94.2 and 94.9 are incorporated into a block diagram of the temperature control loop as depicted in Figure 94.2.

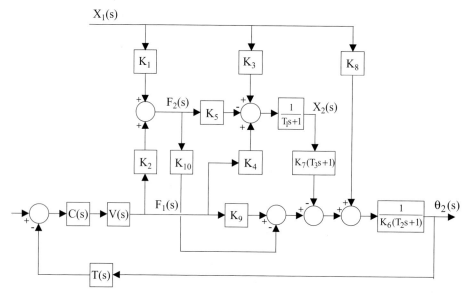

Fig. 94.2 Block diagram of spray drier temperature control system

94.5 Nomenclature

c_P	specific heat	kJ kg K^{-1}
F	mass flow rate	kg s^{-1}
h	specific enthalpy	kJ kg^{-1}
V	volume of spray drier	m^3
X	weight fraction of water	–
λ	latent heat	kJ kg^{-1}
θ	temperature	°C
ρ	density	kg m^{-3}

Subscripts

A	dry air
S	solids
V	water vapour
W	water
1	slurry stream
2	moist air stream
3	hot dry air stream
4	solids stream

Electro-Mechanical Systems

Chapter 95

95.1 Simple Feedback System
95.2 Cascade Control System
95.3 Nomenclature

Electro-mechanical systems are not used much in the process industries. However, they are used extensively throughout other sectors such as manufacturing and aerospace. They are primarily involved in position control, for example in positioning the arm of a robot, the cutting piece on a machine tool or the flap on an aircraft wing. These are all examples of servo-mechanisms which are characterised by a variable set point and small but continuous disturbances, *e.g.* noise. It is inherent in the design of electro-mechanical control systems that there is some power amplification. This is normally realised by means of electric motors, switchgear and gearboxes of appropriate size. However, when massive forces are required, as for example in the positioning of rollers in a steel strip mill, the power amplification is realised through spool valves and hydraulic pistons. It is true to say that the dynamics of electro-mechanical systems are generally much faster than is found in the process industry: time constants are typically measured in milliseconds, seldom in seconds and never in minutes.

This chapter provides an introduction to the design of electro-mechanical position control systems. It is included for completeness rather than of necessity. Most texts on control systems design are written from an electro-mechanical perspective and it is helpful to have some understanding of the concepts involved to be able to relate them to the world of process control.

95.1 Simple Feedback System

The principle elements of a simple electro-mechanical control system are as depicted in Figure 95.1. In essence, the position of the output shaft θ_0 is measured by some transducer and fed back to the power amplifier. A high precision potentiometer is implied but various forms of digital encoding are common practice. The power amplifier is typically thyristor based and, depending upon the difference (error) between the reference (set point) and measured value, produces an output current. This current i is applied to a d.c. motor which causes the drive shaft to rotate, varying its position θ. Switchgear is required to enable the motor to be driven in either direction according to the sign of the error. The motor's drive shaft is connected to a gearbox which enables torque to be applied to the output shaft: it is the output shaft that does the positioning of the cutting piece, flap, or whatever.

The current i from the drive amplifier produces a torque T in the drive motor:

$$T(s) = K_M . i(s)$$

Typically the motor shaft is rigidly connected through to the load. Notwithstanding the gearbox, the motor and load are effectively on a common shaft. Whilst such mechanical components can be modelled separately, it is common practice to develop a combined lumped parameter model for them. If the displacement was linear then, according to Newton's law, its model would be

95 Electro-Mechanical Systems

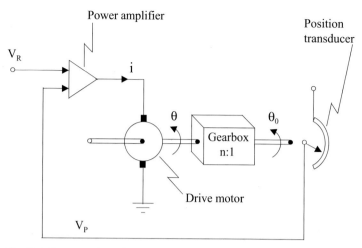

Fig. 95.1 Electro-mechanical system with simple feedback

$$M.\ddot{x} = F - b.\dot{x}$$

In fact the displacement is rotary so, by analogy:

$$J.\ddot{\theta} = T - B.\dot{\theta}$$

where J is a single inertia and B is a common damping coefficient. Thus:

$$\theta(s) = \frac{1}{s.(Js+B)}.T(s)$$

The gear box has a ratio of n:1 so:

$$\theta_0(s) = \frac{1}{n}\theta(s)$$

The feedback signal is proportional to the output position so:

$$V_P(s) = K_P.\theta_0(s)$$

and for the drive amplifier:

$$i(s) = K_A.e(s) = K_A.(V_R(s) - V_P(s))$$

The corresponding block diagram is as depicted in Figure 95.2.

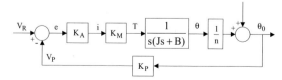

Fig. 95.2 Block diagram of electro-mechanical control system

Note the presence of the integrator which is inherent in the motor/load model. This means that it is not necessary to introduce integral action through the controller to remove offset due to changes in set point or disturbance. Also note that the open loop, and hence the closed loop, transfer function is second order. Position control systems such as this tend to have a fast oscillatory response (high ω_n and low ζ) which is not desirable.

95.2 Cascade Control System

The oscillatory response of the simple system may be dampened by using a tacho generator to provide derivative feedback in a cascade arrangement as depicted in Figure 95.3.

Cascade control was considered in some detail in Chapter 25. The master (outer) loop is controlling the output position θ_0 and the slave (inner) loop is controlling the motor speed $d\theta/dt$. The function of the master loop is to minimise position error by manipulating the set point of the slave loop. Thus, as the error rises, the set point of the slave loop increases causing the motor to speed up and reduce the error faster. Conversely, the smaller the error becomes the slower the motor turns thereby reducing the scope for overshoot.

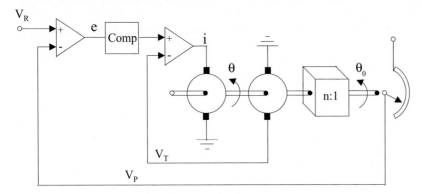

Fig. 95.3 Electro-mechanical system with derivative feedback of motor speed

The transfer functions of most of the elements in the cascade control system are the same as for the simple feedback system but additionally for the tacho generator:

$$V_T = K_T \cdot \dot{\theta}$$
$$V_T(s) = K_T s \cdot \theta(s)$$

Note the inherent derivative action of the tacho generator. Thus the slave loop has both integral action due to the motor/load and derivative action due to the tacho generator. The block diagram of the cascade control system is as depicted in Figure 95.4.

Remember that a fundamental requirement in the design of a cascade control system is that the dynamics of the inner loop must be faster than those of the outer loop. At first sight the system of Figure 95.4 appears to be at odds with this principle. That is because, for modelling purposes, the dynamics of the load and motor were lumped together within the inner loop. In reality, because of the common shaft and rigidity of the system, the dynamics of the motor and load are distributed between both the inner and outer loops. Also, because the tacho generator's dynamics are derivative in nature, they have the effect of speeding up the inner loop relative to the outer loop. Note that it is common practice to use some form of dynamic compensation in the outer loop, rather than simple proportional control, which can be used to make the outer loop more sluggish. The compensator shown in Figure 95.4 is a simple first order lag but lead/lag type compensators are often used to better characterise the overall response for servo control.

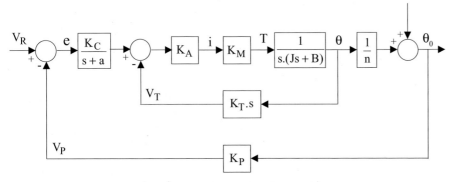

Fig. 95.4 Block diagram of electro-mechanical system with cascade control

95.3 Nomenclature

b	damping coefficient	kg s^{-1}
B	damping coefficient	kN m s rad^{-1}
e	error	V
F	force	N
i	current	A
J	inertia of motor/load	kN m s^2 rad^{-1}
K	gain	various
M	mass	kg
n	gearing ratio	–
T	torque	kN m
V	voltage	V
x	position	m
θ	position	rad

Subscripts

A	amplifier
C	compensator
M	motor
P	position
R	reference (set point)
T	tacho generator
0	output

Section 11

Simulation

Numerical Integration

Chapter 96

96.1 Euler's Explicit Method
96.2 Predictor-Corrector Method
96.3 Worked Example No 1
96.4 Runge Kutta Method
96.5 Euler's Implicit Method
96.6 Worked Example No 2
96.7 Step Length
96.8 Nomenclature

Numerical integration is an approach to solving differential equations, either ordinary or partial, and is at the heart of any dynamic simulation package. In general, numerical integration is used when a solution by classical methods, such as the use of Laplace transforms, is not possible. This may be because the differential equations are:

- nonlinear and/or time dependent,
- mixed with algebraic equations,
- subject to constraints and inequalities,
- too numerous to be handled without difficulty.

A fundamental difference between the solution to a set of differential equations obtained by classical methods and by numerical integration is in the form of the solution. A classical solution is a set of algebraic equations which are explicit relationships between the dependent variables and time. The numerical solution is, in effect, a set of trend diagrams: a graph of each dependent variable as a function of time.

The essential approach to numerical integration is, from a known position at a given point in time, to predict the position at some future point in time. The difference in time is referred to as the period of integration or, more simply, as the step length. The position at the end of the first step is then used to predict the position at the end of the next, and so on. There are various techniques of numerical integration available, the principal differences between them being whether:

- the prediction is explicit or implicit,
- the prediction is corrected or not and, if so, to what extent,
- the step length is fixed or variable.

Each of the main techniques is explained and examples are given. For a more comprehensive treatment the reader is referred to Luyben (1990).

96.1 Euler's Explicit Method

This method was introduced in Chapters 44 and 84 in the context of filtering and linearisation respectively. Consider the ordinary differential equation:

$$\frac{dy}{dt} = f(y, t) \qquad (96.1)$$

The value of y can be predicted at the end of the step length by approximating the curve with a tangent at the current point in time, as depicted in Figure 96.1.

This may be articulated explicitly as follows:

$$y_{j+1} \approx y_j + \left.\frac{dy}{dt}\right|_j . \Delta t = y_j + f(y_j, t_j).\Delta t \qquad (96.2)$$

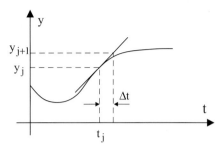

Fig. 96.1 Prediction based on slope at current point

This is the simplest but least accurate approach to numerical integration. As discussed in Chapter 84, its accuracy depends on the curvature of the function and the size of the step length. High levels of accuracy can only be realised by means of very small step lengths.

As seen in Chapter 79, process control systems often consist of sets of first order differential equations which may be described in state space form:

$$\underline{\dot{x}} = A\underline{x} + Bu \qquad (96.3)$$

This may be integrated numerically by Euler's explicit method as follows:

$$\underline{x}_{j+1} \approx \underline{x}_j + \left(A\underline{x}_j + Bu\right).\Delta t \qquad (96.4)$$

96.2 Predictor-Corrector Method

In the predictor-corrector version of Euler's method, an estimate of the value of y at the end of the step is first predicted as above:

$$y^*_{j+1} = y_j + f(y_j, t_j).\Delta t$$

An estimate of the slope at the end of the step is then made:

$$\left.\frac{dy}{dt}\right|^*_{j+1} = f(y^*_{j+1}, t_{j+1})$$

The value of y predicted is then corrected using an average of the values of the slope at the beginning and end of the step:

$$y_{j+1} \approx y_j + \frac{1}{2}\left(f(y_j, t_j) + f(y^*_{j+1}, t_{j+1})\right).\Delta t \qquad (96.5)$$

A much more common problem in process control is to solve sets of equations such as:

$$\frac{dx}{dt} = f(x, y)$$
$$\frac{dy}{dt} = g(x, y) \qquad (96.6)$$

for all values of time in the range $j = 0, 1, 2, \ldots n$ given the initial conditions, i.e. the values of x and y when $j = 0$. The variables x and y could, for example, represent concentrations of reagents in a reaction mixture. Using Euler's predictor-corrector method yields:

$$x^*_{j+1} = x_j + f(x_j, y_j).\Delta t$$
$$y^*_{j+1} = y_j + g(x_j, y_j).\Delta t$$
$$x_{j+1} = x_j + \frac{1}{2}\left(f(x_j, y_j) + f(x^*_{j+1}, y^*_{j+1})\right).\Delta t \qquad (96.7)$$
$$y_{j+1} = y_j + \frac{1}{2}\left(g(x_j, y_j) + g(x^*_{j+1}, y^*_{j+1})\right).\Delta t$$

96.3 Worked Example No 1

Consider the same system as that used in Chapters 74 and 80, whose block diagram is repeated in Figure 96.2.

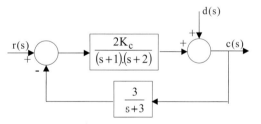

Fig. 96.2 Feedback system of third order with proportional gain

A state space analysis of the system yields:

$$\begin{bmatrix} \dot{x}_1 \\ \dot{x}_2 \\ \dot{x}_3 \end{bmatrix} = \begin{bmatrix} 0 & 1 & 0 \\ -2 & -3 & -2K \\ 3 & 0 & -3 \end{bmatrix} \begin{bmatrix} x_1 \\ x_2 \\ x_3 \end{bmatrix} + \begin{bmatrix} 0 \\ 2K \\ 0 \end{bmatrix} u$$
(96.8)

which is of the general form of Equation 96.3.

Using Euler's predictor-corrector method gives:

$$x^*_{1,j+1} = x_{1,j} + x_{2,j}.\Delta t$$
$$x^*_{2,j+1} = x_{2,j} + \left(-2x_{1,j} - 3x_{2,j} - 2Kx_{3,j} + 2Ku\right).\Delta t$$
$$x^*_{3,j+1} = x_{3,j} + \left(3x_{1,j} - 3x_{3,j}\right).\Delta t$$
$$x_{1,j+1} = x_{1,j} + \frac{1}{2}\left(x_{2,j} + x^*_{2,j+1}\right).\Delta t$$
$$x_{2,j+1} = x_{2,j} + \frac{1}{2}\left(\begin{array}{c}-2x_{1,j} - 3x_{2,j} - 2Kx_{3,j} - 2x^*_{1,j+1} \\ - 3x^*_{2,j+1} - 2Kx^*_{3,j+1} + 4Ku\end{array}\right).\Delta t$$
$$x_{3,j+1} = x_{3,j} + \frac{1}{2}\left(3x_{1,j} - 3x_{3,j} + 3x^*_{1,j+1} - 3x^*_{3,j+1}\right).\Delta t$$

These equations may be articulated in a state space form:

$$\underline{x}^*_{j+1} = \underline{x}_j + \left(A\underline{x}_j + Bu\right).\Delta t$$
$$\underline{x}_{j+1} = \underline{x}_j + \frac{1}{2}\left(A\underline{x}_j + A\underline{x}^*_{j+1} + 2Bu\right).\Delta t \quad (96.9)$$

The predictor corrector method is much more accurate than the explicit Euler method but, obviously, for a given step length involves approximately twice as much computation. There is a trade off between accuracy and step length. In general, for the same accuracy and computing effort, the predictor-corrector method integrates faster.

96.4 Runge Kutta Method

The Runge Kutta methods are even more accurate than the predictor-corrector version of Euler's method because for each prediction there are several corrections. The fourth order Runge Kutta method, which is the most common, involves taking a weighted average of the estimates of the value of the dependent variable based upon four slopes: one at either end of the step and two at intermediate points. Applied to Equation 96.1:

$$\frac{dy}{dt} = f(y, t) \quad (96.1)$$

this involves calculating the following sequence of equations:

$$k_1 = f\left(y_j, t_j\right)$$

$$k_2 = f\left(y_j + \frac{k_1}{2}, t_j + \frac{\Delta t}{2}\right)$$
$$k_3 = f\left(y_j + \frac{k_2}{2}, t_j + \frac{\Delta t}{2}\right)$$
$$k_4 = f\left(y_j + k_3, t_j + \Delta t\right)$$

$$y_{j+1} = y_j + \frac{1}{6}\left(k_1 + 2k_2 + 2k_3 + k_4\right).\Delta t \quad (96.10)$$

Applying Runge Kutta to Equations 96.6:

$$\frac{dx}{dt} = f(x, y)$$
$$\frac{dy}{dt} = g(x, y) \quad (96.6)$$

yields:

$$k_1 = f\left(x_j, y_j\right)$$
$$m_1 = g\left(x_j, y_j\right)$$
$$k_2 = f\left(x_j + \frac{k_1}{2}, y_j + \frac{m_1}{2}\right)$$
$$m_2 = g\left(x_j + \frac{k_1}{2}, y_j + \frac{m_1}{2}\right)$$
$$k_3 = f\left(x_j + \frac{k_2}{2}, y_j + \frac{m_2}{2}\right)$$
$$m_3 = g\left(x_j + \frac{k_2}{2}, y_j + \frac{m_2}{2}\right)$$
$$k_4 = f\left(x_j + k_3, y_j + m_3\right)$$
$$m_4 = g\left(x_j + k_3, y_j + m_3\right)$$

$$x_{j+1} = x_j + \frac{1}{6}\left(k_1 + 2k_2 + 2k_3 + k_4\right).\Delta t$$
$$y_{j+1} = y_j + \frac{1}{6}\left(m_1 + 2m_2 + 2m_3 + m_4\right).\Delta t \quad (96.11)$$

96.5 Euler's Implicit Method

A potential problem, to which all the predictor-corrector methods are prone, is numerical instability. This manifests itself by non-convergence of the solutions. It happens when the step length is too long relative to the curvature of the function being integrated. Instability is exacerbated when the equations being integrated are stiff. A stiff system is one in which there is a wide range of dynamics, characterised by a mixture of large and small time constants. Implicit methods are much more stable.

Consider the previous ordinary differential equation:

$$\frac{dy}{dt} = f(y, t) \qquad (96.1)$$

The value of y can be predicted at the end of the step by approximating the curve with a tangent at the end of the step, as depicted in Figure 96.3.

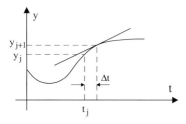

Fig. 96.3 Prediction based on slope at next point

This may be articulated implicitly as follows:

$$\begin{aligned} y_{j+1} &\approx y_j + \left.\frac{dy}{dt}\right|_{j+1} . \Delta t \\ &= y_j + f(y_{j+1}, t_{j+1}) . \Delta t \end{aligned} \qquad (96.12)$$

This is implicit in the sense that the value of y at j+1, which is unknown, appears on both sides of the equation. An explicit solution has to be obtained by isolating the unknown. Consider, for example, the first order response to a step input:

$$T.\frac{dy}{dt} + y = A$$

for which the implicit algorithm is:

$$y_{j+1} \approx y_j + \frac{A - y_{j+1}}{T} . \Delta t$$

Solving for the unknown gives:

$$y_{j+1} \approx \frac{T.y_j + A.\Delta t}{T + \Delta t}$$

Note that as $\Delta t \rightarrow \infty$ then $y_{j+1} \rightarrow A$, and as $T \rightarrow 0$ then $y_{j+1} \rightarrow A$.

Numerical stability depends on the Δt being small enough relative to T. Since the analytical steady state solution to the step input is an output of magnitude A, it is clear that stability is guaranteed irrespective of the values of T and Δt.

96.6 Worked Example No 2

Consider again a system described in state space form:

$$\underline{\dot{x}} = A\underline{x} + Bu \qquad (96.3)$$

The corresponding implicit algorithm is:

$$\underline{x}_{j+1} = \underline{x}_j + \left(A\underline{x}_{j+1} + Bu\right).\Delta t$$

to which the solution is:

$$\underline{x}_{j+1} = (I - A.\Delta t)^{-1} . \left(\underline{x}_j + Bu.\Delta t\right)$$

This may be related to the example of Figure 96.2:

$$\begin{bmatrix} x_{1,j+1} \\ x_{2,j+1} \\ x_{3,j+1} \end{bmatrix} = \left(\begin{bmatrix} 1 & 0 & 0 \\ 0 & 1 & 0 \\ 0 & 0 & 1 \end{bmatrix} - \begin{bmatrix} 0 & 1 & 0 \\ -2 & -3 & -2K \\ 3 & 0 & -3 \end{bmatrix} \Delta t\right)^{-1}$$

$$\times \left(\begin{bmatrix} x_{1,j} \\ x_{2,j} \\ x_{3,j} \end{bmatrix} + \begin{bmatrix} 0 \\ 2K \\ 0 \end{bmatrix} .u\Delta t\right)$$

For values of K = 5 and $\Delta t = 0.1$ the implicit solution is:

$$\begin{bmatrix} x_{1,j+1} \\ x_{2,j+1} \\ x_{3,j+1} \end{bmatrix} =$$

$$\begin{bmatrix} 1.0 & -0.1 & 0 \\ 0.2 & 1.3 & 1.0 \\ -0.3 & 0 & 1.3 \end{bmatrix}^{-1} \left(\begin{bmatrix} x_{1,j} \\ x_{2,j} \\ x_{3,j} \end{bmatrix} + \begin{bmatrix} 0 \\ 1 \\ 0 \end{bmatrix} .u\right) =$$

$$\begin{bmatrix} 0.968 & 0.0745 & -0.0573 \\ -0.321 & 0.745 & -0.573 \\ 0.223 & 0.0172 & 0.756 \end{bmatrix} \left(\begin{bmatrix} x_{1,j} \\ x_{2,j} \\ x_{3,j} \end{bmatrix} + \begin{bmatrix} 0 \\ 1 \\ 0 \end{bmatrix} .u\right)$$

The implicit Euler method suffers from inaccuracy in the same way as the explicit Euler method. However, whereas the explicit Euler and predictor-corrector methods become unstable when the step length is too big, the implicit Euler method is much more stable. Indeed, assuming that an explicit solution can be found, the implicit method can guarantee numerical stability.

96.7 Step Length

Choice of method of numerical integration depends upon factors such as accuracy, numerical stability, speed of computation, number of equations, spread of time constants, *etc*. Often, the explicit Euler method is sufficient. If the step length is short enough then instability should not be a problem. For greater accuracy a predictor-corrector method should be used: fourth order Runge Kutta is the most popular, but runs relatively slowly.

Accuracy is often not critical, the whole point of simulation being to "get a feel" for how a system behaves rather than trying to predict exactly what its behaviour will be. Therefore, in determining the appropriate step length, it is often good enough to start with a small step size for testing purposes, and to keep doubling it until a value is found that gives sufficiently accurate solutions whilst still being numerically stable. Typically this is between a tenth and a hundredth of the largest time constant.

Finding a good step length is particularly important in relation to stiff systems. The step length required for accurate integration of the fast differential equations, *i.e.* those with the small time constants, is much smaller than that required for the slow equations. Yet the overall behaviour of a stiff system is dominated by the slow equations. A sensible strategy is to eliminate any ODE whose time constant is less than, say, one fiftieth of the largest time constant in the system. Replace any such equation with a steady state algebraic relationship. Accuracy on the remaining fast equations may then be sacrificed by using as large a step length as possible, consistent with the fast equations being numerically stable and the slow equations being accurate enough.

There are various algorithms available which vary the step length as the integration proceeds. This approach is attractive when the number of equations involved is large since it can substantially reduce the computation effort involved and speed up the integration. In essence, provided the functions being integrated are relatively smooth, the algorithm relaxes the step length (allows it to increase) subject to some on-going estimate of accuracy and/or stability. However, if the algorithm comes across a sudden increase in curvature in one or more of the functions being integrated, it would reduce the step length to maintain accuracy. The criterion for relaxing (or tightening) the step length would, for example, for the fourth order Runge Kutta algorithm, be based upon the percentage change in the ratio of its k_1 to k_4 values from one iteration to the next.

96.8 Nomenclature

x, y   dependent variables
K   controller gain
t   time
Δt step length

Subscripts

j   denotes current instant
n   range of steps
*   estimated value

Procedural Simulation

97.1 The Matlab User Interface
97.2 Array and Matrix Operations
97.3 Curve Fitting
97.4 Root Finding
97.5 Multiplying and Dividing Polynomials
97.6 Differentiating Polynomials
97.7 Finding Partial Fractions
97.8 Display Functions
97.9 Statistics Functions
97.10 Import and Export of Data
97.11 Script M-Files
97.12 Program Structure
97.13 Control Loop Simulation
97.14 Function M files
97.15 Comments

There are essentially two approaches to dynamic simulation: the procedural approach which is covered in this chapter and the block orientated approach which is covered in Chapter 98. Historically, the procedural approach was realised by programs written in a high level language such as Fortran. Programs would often be massive in terms of the number of lines of code: unless very carefully structured they were generally unintelligible, inflexible and difficult to maintain. This led to the development of the block orientated approach in which functional entities were realised by means of self contained, re-usable, generic, blocks of code. These blocks of code were then strung together as appropriate to create larger programs. Today, they are written in more current languages, such as C# or Java, and are used in a simulation environment with graphical user interfaces (GUI) such that the code itself is hidden from the user.

The purpose of this chapter is to provide some understanding of the functionality of a modern procedural language. It is based upon Matlab which is an interpretive language. All of the Matlab functions are themselves written in C# code and have been pre-compiled ready for execution. The following chapter provides some understanding of the functionality of a block orientated language and is based upon Simulink, the blocks of which are written in Matlab code.

The intention is that the reader should have access to Matlab and Simulink and try out the various functions and programs alongside reading the chapters. It is only by using the languages and exploring their functionality that understanding can be developed and experience gained. Providing a comprehensive coverage of both Matlab and Simulink in a couple of chapters is neither feasible nor desirable.

There are many good texts on Matlab and Simulink. Most introduce the functionality of the languages in a domain context, an excellent example of which is the text by Wilkie (2002). For more detailed information the reader is referred to the Matlab handbooks, in particular to the Student Versions, which are well organised and surprisingly readable.

97.1 The Matlab User Interface

The Matlab user environment is Windows based. The various windows are opened and closed from the Matlab "desktop" menu and all have drop down menus for file handling, editing, viewing, help, and so on. Central to the use of Matlab is the "command window" at which instructions (or statements) are entered and from which programs are controlled. Associated with the command window is the "current directory" window which lists all the files in the Matlab work directory and the "command history" window which provides a log in reverse chronological order of the instructions used.

The variables created in the command window and manipulated by programs reside in the Matlab "workspace". The workspace is virtual but its contents may be viewed in a separate workspace window. The variables may also be listed, or details provided, in the command window by using the functions:

    who
    whos

They can also be deleted individually or collectively from the command window:

    clear

Matlab works in command mode which means that any instruction correctly entered at the prompt, followed by pressing either the "enter" or "return" key, will be executed immediately. It then goes into an idle mode until the next instruction is entered.

There are essentially four different types of instruction that can be entered at the prompt which are used to:

- invoke a valid Matlab operation or function. The results are displayed as appropriate. In effect Matlab is being used as a calculator, albeit an extremely powerful one.
- execute an equation. Variables are assigned and/or their values in the workspace changed according to an equation consisting of valid Matlab operators and functions. Again, results are displayed as appropriate.
- activate a user defined program, referred to as an M-file. The program then runs, interacting with the variables in the workspace, until either it completes according to logic within the program or is aborted.
- launch a Matlab toolbox. Whereas Matlab has a wide range of mathematical functions, they are largely generic. For specific application domains there are a number of toolboxes, each with its own dedicated functions. Simulink is such a toolbox. Other toolboxes are available for handling control systems, signal processing, fuzzy logic, neural nets, and so on.

A particularly useful feature of the command window is use of the ↑ key for entering instructions. Pressing the ↑ key at the prompt will cause the last entry to be repeated. Pressing the ↑ key repeatedly causes previous entries to be repeated in reverse chronological order. If the ↑ key is pressed after the first few characters of an instruction have been entered, the last entry whose first characters match those entered will be repeated. Once the required instruction has been selected, it may be edited if/as appropriate and executed.

Matlab has extensive help facilities. These can be accessed from a menu in the command window. Alternatively, information about specific functions can be listed by typing in "help" followed by the name of the function at the prompt. This provides information about the structure, arguments, format, *etc.*, of the function together with examples. Note the "see also" list of functions at the end of the listing which can give inspiration when uncertain about what function to use.

97.2 Array and Matrix Operations

Fundamental to Matlab's power and flexibility is its array and matrix handling capability. Data is stored either in scalar or array form: arrays may be one dimensional (vectors) or two or more dimensional (matrices). There are multiple instructions for manipulating data within arrays and for carrying out linear and matrix algebra. Some simple examples follow.

Variables are declared at the prompt in the command window as and when required. They then exist within the workspace until deleted. Thus:

x = [1, 2, 3, 4, 5]

creates a row vector x. Note that either commas or spaces are used to separate the elements of the vector. Note also that the vector is created with square brackets: all vector and matrix operations require square brackets. The vector x can be translated into a column vector by using the transpose operator:

x = x′

Alternatively, it could have been entered as a column vector by use of semicolon separators:

x = [1; 2; 3; 4; 5]

The Matlab function for creating vectors of equally spaced values is "linspace". This is useful, for example, in defining axes for data and plots. The arguments of linspace are respectively: first value, last value, number of values.

y = linspace (0, 10, 11)

This produces a row vector of eleven elements of value 0 through to 10. Note that the arguments for linspace are embraced by round brackets, even though a vector is being produced. Arguments of Matlab functions are always contained in round brackets.

Matrices are created in much the same way as vectors. Consider, for example, the A matrix of the numerical example on matrix version of Chapter 79:

A = [1, 2, 0; 3, −1, −2; 1, 0, −3]

Its inverse is found using the "inv" function:

inv(A)

Taking out the determinant of A as a factor yields the adjoint:

d = det(A)
Adj = d ∗ inv(A)

Referring to Worked Example No 2 of Chapter 79, for which the A matrix is as follows:

A = [0, 1, 0; −2, −3, −0.076; 3, 0, −3]

the eigenvalues of A are given by:

E = eig(A)

The eigenvectors and eigenvalues of A are returned if the "eig" function is used with the following format:

[V, D] = eig(A)

where V is the matrix of eigenvectors and D is the trace of eigenvalues.

The eigenvector corresponding to the first eigenvalue is the first column of the matrix V, and so on. That this is the same as the eigenvector r_1 in Worked Example No 2 of Chapter 79 can be demonstrated by the following operation:

V(:, 1) = V(:, 1)/V(1, 1)

Note the use of the colon operator. The above instruction selects the elements in every row of the first column of the matrix V and divides them by the value of the element in the first row of the first column of V (the top left hand corner which happens to have a value of −0.451). The colon operator is a very powerful means of editing values in rows or columns of matrices.

97.3 Curve Fitting

The ability to fit a polynomial to a set of data is a useful empirical technique. Consider the specific heat *vs* temperature data from Worked Example No 1 of Chapter 83:

T = linspace(50, 95, 10)
Cp = [6.72, 6.91, 6.85, 6.97, 7.01, 7.12, 7.14, 7.22, 7.18, 7.27]

Whilst it is known *a priori* that this is essentially a linear relationship, a third order polynomial may nevertheless be fitted to this data using the "polyfit" function:

[P, S] = polyfit(T, Cp, 3)

This returns a row vector P containing the coefficients of the polynomial:

$Cp = 0.T^3 + 0.T^2 + 0.0236T + 5.7299$

Whilst Matlab was asked to fit a third order polynomial to the data, it appears that the best fit that it could find in a least squares sense was linear. However, if the floating point format is changed to provide more significant figures, it can be seen that the higher order coefficients do indeed exist:

format long e

Entering the name of a variable at the prompt causes its value in the workspace to be displayed. Typing in P thus reveals the coefficients to be:

$Cp = -3.4188 \times 10^{-7}T^3 - 4.6853 \times 10^{-5}T^2 + 0.023595T + 5.7299$

Restoring the format to its previous form, the value of Cp predicted by the polynomial at any point, say T = 70, is found to be 7.0347 using the "polyval" function:

format short
polyval (P, 70)

The predicted values of Cp for all the values in T are returned using a row vector:

Cphat = polyval (P, T)

97.4 Root Finding

It is particularly helpful in stability analysis, for example, to be able to find the roots of a polynomial function. Consider the characteristic equation from Worked Example No 1 of Chapter 74:

$s^3 + 6s^2 + 11s + 6(K + 1) = 0$

To find its roots when K = 1.0 requires the polynomial to be declared as a *row* vector as follows:

K = 1.0
f = [1.0, 6.0, 11.0, 6.0 * (K + 1)]
roots(f)

The roots are returned as a column vector corresponding to the factorisation:

$f(s) = (x + 4)(x + 1 + j1.4142)(x + 1 - j1.4142)$

Just as the roots of a polynomial can be found using the "roots" function, so too can the polynomial corresponding to a set of roots declared in a *column* vector be found using the "poly" function:

r = [−4.0; −1.0 − 1.4142j; −1.0 + 1.4142j]
poly(r)

97.5 Multiplying and Dividing Polynomials

Polynomial multiplication is realised by the "conv" function. This is especially useful when working out open and closed loop transfer functions. Consider the characteristic equation of Worked Example No 3 of Chapter 71:

$$\frac{(5s + 1)(1 - 0.5s)}{s.(10s + 1)(0.5s + 1)^2(0.2s + 1)} + 1 = 0$$

The denominator is formed by convolution of den1 and den2, den1 being the quadratic and den2 the product of the other terms. Note the trailing zero in den2 to ensure that it is treated as a third order polynomial:

den1 = [0.25, 1.0, 1.0]
den2 = [2.0, 10.2, 1.0, 0.0]
den = conv (den1, den2)

The characteristic polynomial is thus found by cross multiplication:

num = [0, 0, 0, −2.5, 4.5, 1.0]
fs = num + den

resulting in:

$f(s) = 0.5s^5 + 4.55s^4 + 12.45s^3 + 8.7s^2 + 5.5s + 1$

Note the leading zeros in the numerator necessary for dimensional consistency during the addition.

Polynomial division is realised by the "deconv" function:

[q, r] = deconv(h, g)

This should be interpreted as follows: q(x) is the quotient polynomial obtained by dividing h(x) by g(x), the remainder being r(x)/g(x). Note that g(x) must not have any leading zeros. For example, with:

h = [1, 9, 27, 28]
g = [1, 3]

deconvolution results in

$$\frac{x^3 + 9x^2 + 27x + 28}{x + 3} = x^2 + 6x + 9 + \frac{1}{x + 3}$$

97.6 Differentiating Polynomials

Differentiation of a polynomial is simple. For example:

fsprime = polyder(fs)

gives:

$$\frac{df(s)}{ds} = 2.5s^4 + 18.2s^3 + 37.35s^2 + 17.4s + 5.5$$

So far, all the polynomials considered have been simple. Matlab can also handle rational polynomials, that is ratios of polynomials, for which the numerators and denominators are handled separately. Consider the calculation of the break point in Worked Example No 3 of Chapter 74:

$$K = \frac{-(s^3 + 3s^2 + 2s)}{s + 5}$$

This may be cast in the form of two polynomials:

num = [−1.0, −3.0, −2.0, 0.0]
den = [1.0, 5.0]

The derivative of this rational polynomial with respect to s is also found using the "polyder" function but with different arguments:

[nd, dd] = polyder(num, den)

which results in:

nd = [−2.0, −18.0, −30.0, −10.0]
dd = [1.0, 10.0, 25.0]

and corresponds to:

$$\frac{dK}{ds} = \frac{-(2s^3 + 18s^2 + 30s + 10)}{s^2 + 10s + 25}$$

97.7 Finding Partial Fractions

As seen with inverse Laplace transformation, finding the partial fractions of any rational polynomial can be a non-trivial task. Consider a function of the form considered in Example No 5 of Chapter 70:

$$f(s) = \frac{1}{s^2(5s + 1)}$$

for which:

num = [1.0]
den = [5.0, 1.0, 0.0, 0.0]

Partial fractions are obtained using the "residue" function:

[r, p, k] = residue (num, den)

Note that r and p are column vectors and k is a scalar:

- r contains the residues (coefficients of the partial fractions) which are 5.0, −5.0 and 1.0
- p contains the poles of the partial fractions which are −0.2, 0 and 0 respectively
- k is the integer number of times that den divides into num, 0 in this case

The partial fractions are thus:

$$f(s) = \frac{5}{s + 0.2} - \frac{5}{s} + \frac{1}{s^2} = \frac{25}{5s + 1} - \frac{5}{s} + \frac{1}{s^2}$$

Note that rational polynomials can be constructed from partial fractions by using the residue command in reverse:

[n, d] = residue (r, p, k)

97.8 Display Functions

As stated, whenever an instruction is executed the results are displayed. The standard Matlab display is text in which the values of the variables calculated are displayed in scalar, vector or matrix format. If the instruction is an operation or function, rather than an equation, the result is assigned to the variable ans (short for answer) which can then be used as a variable in subsequent instructions.

The standard displays are rather sparse. To make more effective use of screen space the user is recommended to use the function:

format compact

The original spacing can be restored by:

format loose

Another way of making more effective use of screen space is to put a semicolon at the end of instructions as appropriate. The semicolon suppresses the display of the results but does not stop the instruction from being executed.

Matlab supports a wide range of functions for displaying results as 2D and 3D graphics. An introduction to 2D graphics follows. Consider again the data from Worked Example No 1 of Chapter 83.

T = linspace (50, 95, 10)
Cp = [6.72, 6.91, 6.85, 6.97, 7.01, 7.12, 7.14, 7.22, 7.18, 7.27]

Cp (y axis) can be plotted against T (x axis) using the "plot" function:

plot(T, Cp)

Note that the x axis always comes before the y axis in the arguments of the plot function. Also note that the data for the plot command is held in vectors, and that each vector must have the same number of elements.

The plot is produced in a separate "plot window". Use of the Alt-Tab keys enables the user to toggle between the workspace, the plot and indeed other windows.

For estimating values it is helpful to superimpose grid lines on the plot:

grid

To zoom in on part of the plot:

zoom

and drag the mouse, pressing its LHS button, across the plot to define the zoom area.

Plots can be annotated with axes and a title:

xlabel ('rubbish in')
ylabel ('rubbish out')
title ('a load of rubbish')

Text can be placed on the plot using the co-ordinates of the x and y axes:

text (75.0, 7.1, 'lost the plot')

The default plot uses a solid blue line and no markers. There are many options concerning colour, symbol and line type which are articulated by means of a third argument.

For example, plotting with crosses only at the data points or with red pluses/dotted lines:

plot (T, Cp, 'x')
plot (T, Cp, 'r+:')

Matlab enables multiple plots on the same axes. For each additional plot the x and y axes must be added into the arguments of the plot function. Recall that a third order polynomial was fitted to the Cp T data. The fitted values can be superimposed on the plot as follows:

plot (T, Cp, T, Cphat)

All of the above functions can also be realised from either the insert or the tools menus of the plot window or from icons on the plot window's toolbar.

97.9 Statistics Functions

Matlab supports a wide range of statistical functions. As a basis for demonstrating some of these, the data from Table 83.2 is entered as four vectors:

t = linspace (8.5, 11.5, 7)
y = [0.781, 0.843, 0.841, 0.840, 0.850, 0.852, 0.855]
r = [1.533, 1.428, 1.567, 1.496, 1.560, 1.605, 1.487]
v = [0.027, 0.066, 0.140, 0.294, 0.541, 1.048, 1.810].

Univariate summary statistics, such as average and standard deviation, may be found using the functions:

mean (y)
std (r)

The viscosity v may be linearised, standardised and plotted as follows:

vg = log (v)
vs = (vg − mean (vg))/std (vg)
plot(t, vs)

which confirms that the "log" function is effective for linearising the data which is more or less symmetrical about the mean.

Various forms of cross correlation function are supported by Matlab. It is the unbiased form that is equivalent to the cross correlation function defined by Equation 82.16. For example:

ys = (y − mean (y))/std (y)
xcorr (ys, vs, 'unbiased')

Inspection of the results of the "xcorr" function reveals a maximum at mid range which is not unexpected since there were no delays in the original data.

For finding multivariate summary statistics it is necessary to assemble the data into a matrix in which the columns correspond to the variables y, r and log v and the rows to data sets:

X = [y', r', vg']

Hence the covariance and correlation matrices can be found:

S = cov (X)
P = corrcoef (X)

As explained in Chapter 82, because of the diverse units, it is difficult to interpret the covariance matrix but the correlation matrix reveals a strong relationship between conversion and viscosity.

97.10 Import and Export of Data

Using Matlab often requires that data be imported from a different environment. A typical scenario is that plant data has been gathered on a real-time basis, using some archiving package and/or historian tool, and is to be analysed using Matlab functions. There are several means of importing data into Matlab, but the two most common are as ASCI files and from Excel spreadsheets.

Suppose the data to be imported has been saved in columns in an ASCI file called jumbl with a dot extension (.asc). First the file has to be copied into the working directory. By default, this is designated as the "work" directory but other directories can be used provided they are added to the Matlab search path using the "addpath" function. The file can then be loaded into the workspace as follows:

load jumbl.asc

To check that it has been loaded successfully any column, say the tenth, may be listed using the instruction:

jumbl (:,10)

plotted using the instruction

plot (jumbl(:,10))

Similarly data in an Excel file called jungl with a dot extension (.xls) and saved into the work directory can be loaded into the workspace using the "xlsread" function:

xlsread jungl

Note that there is an "import data" function available under the file menu of the command window which launches an import window. This enables the source directory to be selected (the default is work), the file type to be specified and the datafile to be selected.

It is also often necessary to export data from Matlab to a different environment. Typically results, in the form of a graph, have to be exported into a Word or PowerPoint file. This is easily done by means of cut and paste from the edit menu of the plot window. If pasting into a Word file, it is important to use the paragraph format function beforehand to ensure that there is sufficient spacing before and after the paragraph where the plot is to be pasted. Once pasted, whether in Word or PowerPoint, the size of the plot can be adjusted by means of the handles at the plot corners.

97.11 Script M-Files

Working in command mode is realistic for simple tasks of a once-off nature. However, for more complex tasks, which may be repetitive in nature and involve multiple instructions, command mode is tedious. What is required is a program. A Matlab program is essentially a text file, known as a script file and also referred to as an M-file because of the dot extension (.m). The script file contains Matlab commands which, when the file is initiated, are executed as if they had been typed in command mode.

A script file is created from the command window. Under the file menu, select the new option and then the M-file option. This opens up the "text editor" window in which Matlab commands can be generated and edited. The text editor provides much of the functionality of a conventional word processor with pull down menus to open/close files, cut/copy/paste text, search, help, *etc.*.

Program 97.1 is an example is of a simple script file based on Worked Example No 2 of Chapter 83:

Program 97.1. Determination of multiple linear regression coefficients

```
% X is the matrix of refractive index and
viscosity (input) variables.
% Note that log of raw values of viscosity is
used.
X = [1.0, 1.533, -1.569; 1.0, 1.428, -1.181;
     1.0, 1.567, -0.854; 1.0, 1.496, -0.532;
     1.0, 1.560, -0.267, 1.0, 1.605, 0.020;
     1.0, 1.487, 0.258];
%
% Y is the conversion (output) variable.
Y = [0.781, 0.843, 0.841, 0.840, 0.850,
     0.852, 0.855]';
Ym = mean (Y)
%
% Calculate vector of b coefficients from
Equation 83.14.
B = inv (X'*X)*X'*Y
% Yh is the vector of fitted values from
Equation 83.15.
Yh = X*B;
%
% Validate model by means of a scatter plot.
% Xaxis is fitted values, Yaxis is residuals.
Xaxis = Yh;
Yaxis = [Y-Yh];
% Plot residuals versus fitted values
(scatter plot).
plotmatrix (Xaxis, Yaxis)
%
% Calculate goodness of fit from
Equation 83.11.
R2 = [Yh-Ym]'*[Yh-Ym]/([Y-Ym]'*[Y-Ym])
%
```

Having created and edited the script file, it is saved from the file menu of the text editor in the working directory with a dot extension (.m). Suppose it is saved as jumbl.m. Typing the file name without the dot extension, jumbl in this case, at the prompt in the command window causes the M-file to be executed.

Note the use of comment statements, prefixed by %, to annotate and space out the program.

97.12 Program Structure

Program 97.1 has no structure: starting with the first, each instruction is executed in turn until the last is reached and the program stops. It is much more usual for a program to have structure. Typically there are three sections to a program:

- Initialisation: this consists of instructions for setting up the problem. These include declaration of constants, initialisation of variables, definition of vectors and matrices, input of user defined parameters, *etc.*
- Recursion: the main body of the program, at the heart of which is likely to be one or more routines which are executed recursively. Recursion is characterised by "for" loops and "while" loops. Program flow is driven by the recursion, subject to the constraints of logic which is normally characterised by "if-else" constructs.
- Presentation: this concerns the outcome of the program. In essence, results are generated and data saved during the recursion. Those results, or a subset thereof, are then displayed as text to the user in the command window. Alternatively, the data is manipulated for presentation in graphical format in plot windows.

Program 97.2 illustrates this structure. It has been written to solve the following problem:

$$x = y + \frac{k_1.(a-x)(a-x-y)(2a+2x)^2}{(3x+y)^3.p^2}$$

$$y = \frac{k_2.(x-y)(a-x-y)}{(3x+y)}$$

Program 97.2. Search for equilibrium conditions

```
echo off
% A is initial concentration, 2.5<A<4.0kg/m³.
a = input ('value of A = ');
% P is pressure in system, 4.0<P<8.0bar.
p = input ('value of P = ');
% K1 and K2 are given rate constants.
k1 = 1.022;
k2 = 4.904;
% Initialise error criterion.
emin = inf;
%
% x and y are concentrations in range
0-1.0 kg/m³.
% Search directions have 100 steps so
resolution is 0.01.
%
for j = 1:100
    y = j/100;
    for k = 1:100
        x = k/100;
        X = y+k1*(a-x)*(a-x-y)*(2*a+2*x)^2/
            (3*x+y)^3/p^2;
        E = abs(x-X);
        Y = k2*(x-y)*(a-x-y)/(3*x+y);
        F = abs(y-Y);
        if (E+F) < emin
            emin = E+F;
            kmin = k;
            jmin = j;
        end
    end
end
%
disp (['estimate of x is ', num2str
    (kmin/100)]);
disp (['estimate of y is ', num2str
    (jmin/100)]);
%
```

These equations describe a chemical equilibrium in which x and y are the concentrations at equilibrium of two reagents, a is an initial concentration and p is the reaction pressure. The problem is to find values for x and y that simultaneously satisfy both equations for any specified values of a and p within the ranges of $2.5 < a < 4.0$ kg m^{-3} and $4.0 < p < 8.0$ bar, for a given reaction temperature and hence fixed rate constants of $k_1 = 1.022$ and $k_2 = 4.904$.

There is no obvious analytical solution to this problem so a search approach has been adopted. It is presumed that there is a unique solution. Suppose the correct values of x and y chosen. If these values are substituted into the RHS of the two equations, the values of the RHS should agree with the LHS, *i.e.* the values chosen. Otherwise a discrepancy will exist. The program therefore systematically substitutes values of x and y into the equations and determines discrepancies: the pair of x and y values with the smallest discrepancies correspond to the solution.

Note the use of case: lower case x and y to denote the search values and upper case X and Y to denote evaluations of the RHS of the two equations.

There are many refinements that could be made to this program. For example, the search could be made more efficient by alternating the search direction between x and y axes when the smallest discrepancy along one axis has been found. The accuracy of the solution could be improved by increasing the resolution once the solution has initially been located. However, the point of the program is to illustrate recursion, nested loops in this case, and the use of logic to control program flow rather than to demonstrate perfection.

97.13 Control Loop Simulation

At the heart of any program to simulate a process control system is the numerical integration used to solve the differential equations used to represent the system's dynamics. Program 97.3 illustrates the use of Euler's explicit method of numerical integration, as explained in Chapter 96. It simulates the process control loop described in Worked Example No 3 of Chapter 73 and depicted in Figure 73.10. The variables in the program are named as denoted in Figure 73.10.

Central as the numerical integration may be, it only accounts for a small proportion of Program 97.3. Inspection of the program reveals that, ignoring the comments, there are 68 lines of Matlab code. Notwithstanding the PID controller itself, only three of these lines of code concern the algorithms for numerical integration. The rest concern a variety of features specific to process control, such as the handling of normal values, bias, saturation and integral windup. That is quite typical.

Program 97.3 has a large initialisation section which, amongst other things, includes statements to enable the user to enter the controller's parameters Kc, Tr and Td at the prompt in the command window: it is presumed that other parameters will be changed by editing the M-file itself. The recursion section consists of one big "for loop". In essence, each iteration of the loop progresses the simulation by one step, the values of the variables predicted at the end of one step becoming the starting point for the next step. The presentation section simply generates a plot of the data saved during the main recursion.

Program 97.3. Process control loop simulation

```
% Define system parameters: gains
dimensionless, time in mins.
Kv = 1.0; Tv = 0.2; Kp = 5.0; Tp = 10.0;
L = 1.0; Km = 1.0; Tm = 0.5;
% Note, for numerical stability:
Tv, Tm > 2% of Tp.
%
% Define disturbances: times in mins.
dr = 10; drt = 1; dd = -10; ddt = 10.0;
% Normal values of control signals:
range 0-100%.
r0 = 50; e0 = 0; u0 = 50; f0 = 50; p0 = 50;
d0 = 0; y0 = 50; m0 = 50;
% Initialise simulation variables.
f = f0; ia = 0; m = m0; mp = m0; p = p0;
t = 0; u = u0; y = y0;
%
% Establish step length and run time.
dt = Tp/200; stop = 600;
%
% Initialise vectors for display purposes.
R(1) = r0; T(1) = 0; U(1) = u0; Y(1) = y0;
%
% Establish process delay.
n = round(L/dt);
for i = 1:n
    Q(i) = p0;
end
%
% Enter controller parameters.
echo off;
disp (['classical form of PID with deriv
    feedback']);
Kc = input ('value of prop gain = ');
Tr = input ('value of reset time (mins) = ');
Td = input ('value of rate time (mins) = ');
if Tr == 0
    Tr = 0.01;
end
% Controller gains as per Equation 31.8.
Ki = Kc*dt/Tr; Kd = Kc*Td/dt;
%
% Recursion of loop dynamics.
for j = 1:stop
    t = t+dt;
    T(j+1) = t;
    %
    % Dynamics of controller.
    if t < drt
        r = r0;
    else
        r = r0+dr;
    end
    R(j+1) = r;
    e = r-m;
    % Apply anti-windup to I action.
    if u == 0
        u = u0+Kc*e+ia-Kd*(m-mp);
    elseif u == 100
        u = u0+Kc*e+ia-Kd*(m-mp);
```

```
    else
        ia = ia+Ki*e;
        u = u0+Kc*e+ia-Kd*(m-mp);
    end
    % Force controller saturation.
    if  u < 0
        u = 0;
    end
    if  u > 100
        u =100;
    end
    U(j+1) = u;
    %
    % Dynamics of i/p converter, actuator,
    positioner & valve.
    f = f+(Kv*(u-u0)-(f-f0))/Tv*dt;
    %
    % Dynamics of process.
    p = p+(Kp*(f-f0)-(p-p0))/Tp*dt;
    % Force process saturation.
    if  p < 0
        p = 0;
    end
    if  p > 100
        p = 100;
    end
    %
    % Dynamics of delay.
    if  t < ddt
        d = d0;
    else
        d = d0+dd;
    end
        y = Q(1);
    for i = 1:n-1
        Q(i) = Q(i+1);
        Q(n) = p+d;
    end
    Y(j+1) = y;
    %
    % Dynamics of sensor, transducer &
    transmitter.
    mp = m;
    m = m+(Km*(y-y0)-(m-m0))/Tm*dt;
    %
end
```
```
%
plot(T,R,'g',T,U,'b',T,Y,'r')
zoom
%
```

Using Program 97.3 to apply the continuous cycling test, as explained in Chapter 24, to the system of Figure 73.10, results in a controller gain of K_{CM} = 2.47 for marginal stability. This compares with the true value of K_{CM} = 2.29 derived from frequency response analysis in Worked Example No 3 of Chapter 73. The inaccuracy, some 8%, is due to the numerical integration. Using a better algorithm would clearly improve the accuracy.

The step length is 1/200th of the dominant time constant to allow for a certain amount of stiffness. Assuming the smallest time constant is no less than 1/50th (*i.e.* 2%) of the dominant time constant, the corresponding minimum of four steps per time constant is sufficient to prevent problems with numerical stability. The run time is chosen to be three times the dominant time constant: that is quite sufficient time to observe the dynamic response of the system. For example, in the continuous cycling test, it is long enough for five cycles.

97.14 Function M files

So far, all of the functions used in this chapter have been predefined Matlab functions which, for most purposes, are quite sufficient. However, there is provision to create user defined functions, referred to as function M files. These are created using the text editor in exactly the same way as script M files. It is the syntax of the first instruction of a function M file that distinguishes it from a script M file.

For illustrative purposes it is convenient to use the Matlab function ode45 since it requires the use of a function M file. The function ode45 is a good general purpose ODE solver, the function M file being used to return values of the derivatives for a given time span and set of initial conditions. To demonstrate the use of ode45, the state space description of Worked Example No 1 of Chapter 80 is solved.

The following function M file is created and saved in a file called loop with a dot extension (.m).

```
function xprime = loop (t, x);
K = 4.0;
% K = 10.0 for marginal stability.
U = 1.0;
% Unit step change in set point.
xprime = [x(2); -2*x(1)-3*x(2)-2*K*x(3)
         +2*K*u;3*x(1)-3*x(3)];
```

Note that both xprime and x are 3×1 column vectors.

To use ode45 the time scale and initial conditions have to be specified in the command window:

tspan = [0,10];
x0 = [0, 0, 0];

The first of these sets a time scale of 10 min, the step length being decided by the solver. The solution is obtained by entering, at the prompt, the instruction:

ode45 ('loop', tspan, x0);

The function solves the ODEs and automatically generates a plot. Access to the data generated is provided by means of output arguments:

[t, x] = ode45 ('loop', tspan, x0);

It is worth noting that all of the standard functions in Matlab and its toolboxes are themselves created as function M files.

97.15 Comments

It should be emphasised that Matlab and Simulink have been chosen as vehicles for introducing the functionality of modern simulation languages because they are what the author is familiar with. Whilst they are almost a de-facto standard in academia for dynamic simulation, and are indeed used to a significant extent in industry, it is neither intended to suggest that they are the only languages available for dynamic simulation nor that they are the best.

Copyright

MATLAB and Simulink are registered trademarks of The MathsWorks Inc.

Block Orientated Simulation

98.1 The Simulink Environment
98.2 First Order Step Response
98.3 Control Loop Simulation
98.4 State Space Models in Simulink
98.5 State Feedback Regulator Example
98.6 Dynamic Process Simulators

This chapter focuses on dynamic simulation based upon languages which use configurable function blocks. The main focus of the chapter is Simulink, a simulation environment which supports models developed by configuring functions together as appropriate, the functions being realised by blocks of code written in Matlab. A number of examples are provided, ranging from simple to complex, each of which relates to problems described elsewhere in the Handbook. These examples are used to describe the functionality of Simulink and, in particular, to illustrate the interfaces between Simulink, Matlab and the user. Finally, there is a section, in the interest of balance and completeness, on the use of proprietary dynamic process simulators. These too provide a block orientated simulation environment but are designed around the needs for simulating plant items and process dynamics.

98.1 The Simulink Environment

As with Matlab, Simulink is Windows based. It is accessed either by clicking on the Simulink icon in the Matlab toolbar or else by typing in "Simulink" at the prompt in the Matlab command window. This opens up the "Simulink library browser" window which essentially provides for three types of activity:

- Drop down menus and icons for file handling which enables new Simulink files to be created and existing files to be opened, edited, saved, *etc.* All Simulink files have a dot extension (.mdl).
- Selection of functions which, by means of drag and drop, can be moved into a file being edited. The functions are organised into the following browser categories:

    commonly used blocks
    continuous
    discontinuities
    discrete
    logic and bit operations
    lookup tables
    maths operations
    model verification
    model wide utilities
    ports and sub systems
    signal attributes
    signal routing
    sinks
    sources
    user defined functions
    additional maths and discrete

- Selection of Matlab toolboxes which can themselves be expanded into further categories of functions for use with Simulink. Note that the toolboxes available depend upon the licence but typically could include:

control system toolbox
model predictive control toolbox
neural network blockset
stateflow
system identification toolbox, *etc.*

A Simulink model is created in the "model" window. Clicking on the new file icon in the browser window toolbar opens up a model window with the name of "untitled". It is convenient for model development to have the browser and model windows side by side: the browser window needs to be some 30% of the width of the screen and the model window the other 70%. For large models, the model window can be maximised and the Alt-Tab keys used together to toggle between the browser and model windows.

The model is developed by configuring functions together in the model window. Each function is selected from the panel obtained by opening up the appropriate browser category. Left clicking on the icon enables the function to be dragged from the browser panel and dropped into the model window. The functions are arranged in the model window in a suitable manner: convention is to have the inputs on the LHS and the outputs on the RHS.

Functions are interconnected by clicking (LHS mouse button) on the output handle > of one icon and dragging a line out/across the screen to the input handle > handle on another icon. The same can be achieved, more simply, by clicking on the output icon, holding down the Ctrl key and clicking on the input icon. The connection between any two icons will automatically adjust itself if the position of either function is changed in the model window. Faulty connections *etc.* can be removed by clicking on them and pressing the Del key.

Note that, if necessary, the orientation of a function's icon can be changed by selecting the icon (LHS mouse click) and using the "rotate block" option within the format menu of the model window.

Every function moved into the model window must be parameterised. Double clicking (LHS mouse button) on the function's icon will open up a "block parameters" dialogue box into which the function's parameters have to be entered as appropriate. Specific values may be entered or the names of variables that exist in the Matlab workspace.

A key dialogue box is that for "configuration parameters" which is found under the simulation menu of the model window. This dialogue box has two panels: a selection panel and, for each option selected, a parameterisation panel. The two most commonly used selections are:

- Solver. This enables the run time, that is start/stop times, step size, algorithm for numerical integration, *etc.*, to be specified.
- Data import/export. This provides control over the Matlab/Simulink interface.

A trivial example concerning the step response of a first order system follows to introduce some of the functionality of Simulink.

98.2 First Order Step Response

Open up an untitled model window as explained above. Select the following functions in the browser window and drag their icons into the model window:

transfer function	from	continuous
scope		sinks
out (or outport)		sinks
step		sources

Configure the icons in the model window as depicted in Figure 98.1. Connect the output of the step icon to the input of the transfer function icon, and the output of the transfer function icon to the input of the scope icon. The transfer function icon is also to be connected to the out icon. This is done by back connecting the input of the out icon to the connection between the transfer function and scope icons.

The names of the functions can be changed by clicking on the name and editing as appropriate. For example, change the name of the transfer function to "lag". Also note that text may be written anywhere in the model window: moving the cur-

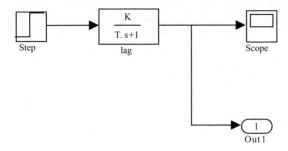

Fig. 98.1 Model of first order lag (Simulink)

sor to the appropriate position and double clicking opens up a text box.

Open up the block parameters dialogue box for the step function and define a unit step after unit time (say, 1 min). Likewise for the transfer function:

numerator [1.0]
denominator [0.5, 1.0]

These correspond to the coefficients of s in the transfer function $1/(0.5 s + 1)$ of a half minute time lag.

To complete model definition, open the configuration parameters dialogue box. Select the solver panel and set the stop time to 10.0 (min) and the maximum step size to 0.01. In the data import/export panel select (tick) both time (tout) and output (yout). Then click OK.

Having built the model it is perhaps wise to save it before going any further. Select "save as" from the "file" menu of the model window and save the model as time_lag *without* any extension. A model file called time_lag.mdl will be saved in the Matlab work directory. The file may be closed from the file menu of the model window. Opening it up again requires use of the file menu or the open icon in the browser window.

The model can be run from the simulation menu. Click on "start" and a beep will be heard when simulation is complete. Because the time_lag model is trivial this is almost instantaneous. The response can be observed by opening up the scope: clicking on the binoculars button automatically adjusts the scaling.

The model can clearly be changed by entering different values for the parameters as appropriate, the simulation repeated and a new response observed. Note that whenever the model definition is changed, the new model must be saved prior to simulation to affect the changes.

The function of the outport is to make values generated within Simulink available within Matlab. Remembering that the Alt-Tab buttons are used to toggle between the Matlab and Simulink environments, typing "who" at the command prompt will reveal that both tout and yout are indeed in the Matlab workspace. The following instruction should produce the same response as was displayed by the scope:

plot (tout, yout)

Similarly, it is important to be able to pass parameters from the Matlab workspace into Simulink models. This is realised by using variables in the model definition. For example, assign values to two variables in the workspace as follows:

K = 1.0
T = 0.5

Then, in the model time_lag, edit the transfer function block's parameters as follows:

numerator [K]
denominator [T, 1.0]

If the model is saved and then run again, the response should be exactly the same as before.

98.3 Control Loop Simulation

An important feature of Simulink is the ability to create subsystems. This is particularly useful when building large models since subsystems:

- are depicted by a single icon, with inputs and outputs as appropriate, which take up much less space in the model window and make for simpler looking models.
- provide a degree of robustness since, once a model is developed, tested and saved as a subsystem, its content cannot inadvertently be changed.

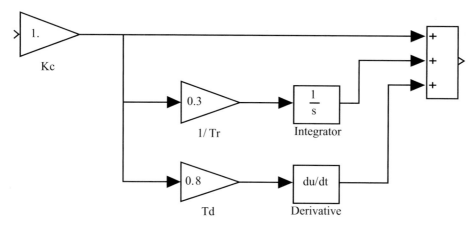

Fig. 98.2 Model of PID controller (Simulink)

- can be nested enabling a layered approach to modelling. For example, subsystems may be created for plant items and combined into a process model which, together with other process models, may become subsystems within a model of a control system.

To illustrate the principle, a Simulink model of a PID controller is developed and used as a subsystem within the Simulink model of a control loop. The controller is of the derivative feedback form of Equation 23.7, apart from not having a bias since the model itself is in deviation form, and is depicted in Figure 98.2.

The controller is created from the following functions:

derivative	from	continuous
integrator		continuous
3 gains		maths operations
sum		maths operations

The values entered for the gain in the dialogue box of the gain function blocks are as follows: $K_C = 1.0$, $T_R = 3.0$ and $T_D = 0.8$. Note that the element-wise (K.*u) method of multiplication must be selected in each case. Otherwise all of the default parameters are sufficient.

Having developed the model of the controller, it is then converted into a subsystem using the "select all" and "create subsystem" options under the edit menu. The subsystem may then be saved as a

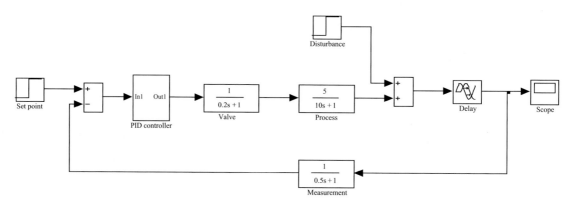

Fig. 98.3 Model of process control loop (Simulink)

file with a dot extension (.mdl) in its own right or else cut and pasted into other Simulink models.

In this case it is incorporated into a model of a control loop, as depicted in Figure 98.3, which is exactly the same loop as that simulated by procedural means in Program 97.3. The rest of the control loop is created from:

3 transfer functions	from	continuous
transport delay		continuous
2 sums		maths operations
scope		sinks
2 steps		sources

with the following parameters: $K_V = 1.0, T_V = 0.2, K_P = 5.0, T_P = 10.0, L = 1.0, K_M = 1.0$ and $T_M = 0.5$. Note that the dialogue box for the transfer function block requires the coefficients of the numerator and denominator polynomials of the transfer function. Thus, for example, for the process transfer function, the numerator would be entered as [5] and the denominator as [10, 1].

To complete model definition, use the configuration parameters dialogue box to set the stop time to 30.0 (min) and select the ode4 (Runge Kutta) solver with a fixed step size of 0.1 (min). Note that the time range in the scope's parameters dialogue box will need to be set either to 30 (min) or else to "auto".

The model of Figure 98.3 may then be simulated. With the above settings an underdamped response to a unit step change in set point is obtained which, to all intents and purposes, reaches steady state with zero offset in 20 min. Applying the continuous cycling test, as explained in Chapter 24, with $T_R = 999$ and $T_D = 0.0$, results in a controller gain of $K_{CM} = 2.31$ for marginal stability. This compares well with the true value of $K_{CM} = 2.29$, derived from frequency response analysis in Worked Example No 3 of Chapter 73. It is much better than the value of $K_{CM} = 2.47$ found using Program 97.3 due to the choice of the ode4 algorithm for numerical integration.

98.4 State Space Models in Simulink

In Chapter 90 a dynamic model of the concentration lags of a 3-stage distillation column was developed in state space form. The parameters of the model were established in the Worked Example of Chapter 90. The L-V strategy for dual composition control as depicted in Figure 90.4 was then considered, its block diagram being as depicted in Figure 90.5. The corresponding Simulink model is shown in Figure 98.4.

Central to the Simulink model is the icon for the state-space function, its functionality being exactly as defined by Equation 80.7. In this particular case it has four inputs: L, V, F and X_F and three outputs: X_1, X_2 and X_3. For configuration purposes, the following functions are required:

demux	from	signal routing
mux		signal routing
state-space		continuous

The number of inputs to the mux (short for multiplex) function needs to be set to 4 in its parameters dialogue box, corresponding to L, V, F and X_F. The number of outputs from the demux (short for demultiplex) function needs to be set to 3 in its parameters dialogue box, corresponding to X_1, X_2 and X_3. It is important that the inputs and outputs are connected in the correct order with respect to the state-space model.

The state space model itself is best set up from within the Matlab workspace. Thus, in the parameters dialogue box for the state-space function, simply enter the values A, B, C and D as appropriate with zero initial conditions and auto tolerance. Then, in the Matlab workspace, the following matrices need to be established, either in command mode or by means of a Matlab M-file:

$$A = \begin{bmatrix} -1.330 & 0.6558 & 0 \\ 1.20 & -2.856 & 0.8271 \\ 0 & 1.10 & -0.7136 \end{bmatrix}$$

$$B = \begin{bmatrix} 0.1242 & -0.0931 & 0.0 & 0.0 \\ 0.1280 & -0.1160 & 0.0320 & 1.0 \\ 0.0574 & -0.0789 & 0.0574 & 0.0 \end{bmatrix}$$

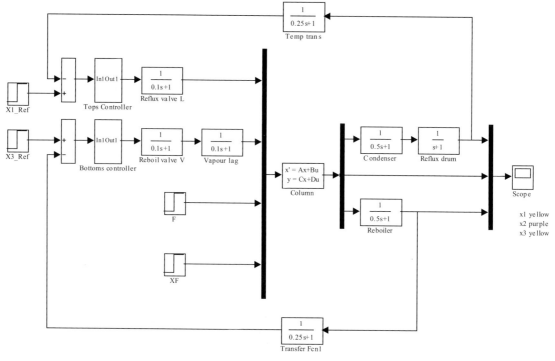

Fig. 98.4 Model of L-V strategy for column control (Simulink)

$$C = \begin{bmatrix} 1.0 & 0.0 & 0.0 \\ 0.0 & 1.0 & 0.0 \\ 0.0 & 0.0 & 1.0 \end{bmatrix}$$

$$D = \begin{bmatrix} 0.0 & 0.0 & 0.0 & 0.0 \\ 0.0 & 0.0 & 0.0 & 0.0 \\ 0.0 & 0.0 & 0.0 & 0.0 \end{bmatrix}$$

These matrices clearly have to be established in the Matlab workspace before the Simulink model can be run.

The rest of the model is realised from transfer function, step, add and scope functions. The reflux drum has a first order lag of 1 min, the condenser and reboiler have lags of 0.5 min, the two temperature measurements have lags of 0.25 min, the reflux and reboil valves have lags of 0.1 min and the vapour flow lag is also presumed to be 0.1 min.

The L-V strategy itself is realised by two PID controllers as articulated in Figure 98.2. The top plate temperature controller has settings of $K_C = 20.0$, $T_R = 1.0$ and $T_D = 1.0$ and manipulates the reflux stream. The bottom plate temperature controller has settings of $K_C = -10.0$, $T_R = 1.0$ and $T_D = 1.0$ and manipulates the reboil stream.

To complete model definition, use the configuration parameters dialogue box to set the stop time to 40.0 (min) and select the variable step type ode45 (Dormand-Prince) solver with the default auto options for the step size. Given that the column has a dominant time constant of some 4.0 min, a run time of 15 min should be sufficient to observe its dynamic response.

Running the model demonstrates that the above settings provide a stable response to a step change of (say) 0.02 in the set point for the tops controller (2% change in X_1). These is much oscillation for several minutes but steady state with zero offset is essentially achieved within 15 minutes. The oscillation is to be expected: the column has only three stages, is subject to severe interactions and dual composition is known to be problematic. It is left as an exercise for the reader to explore the scope for fine tuning the settings to achieve tighter control.

A useful check that the model is behaving as expected is to configure scopes onto the reflux and reboil streams. It can be seen that both achieve an increase of 0.6 kmol min^{-1} which is sensible in relation to the normal flows articulated in the Worked Example of Chapter 90. An increase in the top product purity requires more recycling which, at steady state, requires that both reflux and reboil streams have increased by the same amount.

98.5 State Feedback Regulator Example

This example is also based upon the 3-stage distillation column and uses the state space model developed in the Worked Example of Chapter 90. It concerns the design of a state feedback regulator, as explained in Chapter 112, for controlling the top plate composition X_1 by manipulating the reflux rate L.

Note that the dynamics of the reflux loop are based upon the concentration lags of the column only: clearly first order lags could be introduced before and after the state space function to simulate the dynamics of the reflux drum, overhead condenser, temperature transmitter and control valve.

The design consists of a state feedback regulator with a reference factor for servo operation, integrated with a full order observer, and incorporating integral action as depicted in Figure 112.6. The corresponding Simulink model is as depicted in Figure 98.5. Note that the thick arrows which denote vectors of variables, as opposed to the thin arrows which denote scalar variables, are generated at run time.

A state space model is used to simulate the process, the input being the reflux rate L and the output being the top plate composition X_1. Thus, in the parameters dialogue box of the state-space function block, the values A, B, C and D are entered as appropriate with zero initial conditions and auto tolerance.

For reality, a time delay is included after the state space model to create some plant-model mismatch. In its parameters dialogue box a delay of 0.5 min is entered, together with the default settings of 0 initial input of 0, 1024 buffer size and 0 order Pade approximation.

Whereas the reference factor is established by scalar multiplication, the predictor corrector, observer and state feedback regulator are all realised by matrix algebra. The values entered for the gain in the dialogue box of the gain function blocks are thus as in Table 98.1.

Table 98.1 Gain settings for controller and observer model

Function block	Gain	Multiplication
State matrix	A	K*u
Input matrix	B	K*u
Output matrix	C	K*u
Regulator gain	K	K*u
Observer gain	L	K*u
Reference factor	R	K.*u

The matrix (K*u) method of multiplication must be selected for all of the above gain function blocks, apart from the reference factor for which element wise (K.*u) is appropriate. Otherwise the default parameters are sufficient.

The integrator has a gain of 0.1. This too is realised by means of a gain function block with element wise multiplication.

A variety of step, sinusoidal and noise (band limited white noise) inputs have been configured, all of which come from the sources category in the Simulink library browser. These enable changes in set point and disturbance to be generated for testing purposes. Any sensible combination of parameters will suffice but, for demonstration purposes, those listed in Table 98.2 are used. All others parameters are set to default values.

The rest of the Simulink model consists of various sum function blocks together with a scope

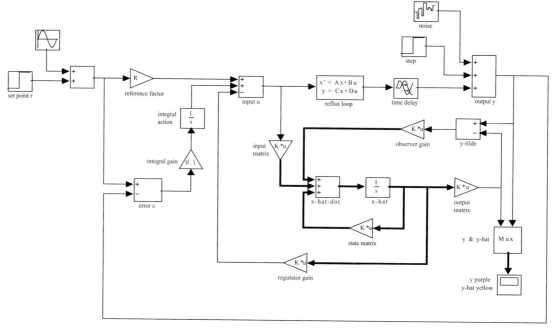

Fig. 98.5 Model of state feedback controller with observer (Simulink)

and mux function block, all of which have been explained in the previous examples.

The Simulink model of Figure 98.5 requires the matrix A, the vectors B, C, K and L and the scalars D and R to be established in the Matlab workspace first. This can be done by running the following Matlab M-file or equivalent:

Program 98.1. Determination of controller and observer parameters

```
% Establish state space model;
G = zpk([-0.4963, -3.7495], [-0.2351,
        -1.1291, -0.5348], 0.1241);
[A, B, C, D] = ssdata (G);
%
% Establish state controller K;
Pk = [-2.5, -0.4+0.3j, -0.4-0.3j];
K = place (A, B, Pk)
%
% Establish reference factor R;
Sys = [A, B; C, D];
Rxu = inv (Sys) * [0, 0, 0, 1]';
Rx = Rxu (1:3);
Ru = Rxu (4);
R = K * Rx + Ru
%
% Establish full order observer L;
Pl = [-10.0, -3.1, -3.0];
L = place (A', C', Pl)'
```

This program needs a little explanation. First, note that the process transfer function, evaluated in the Worked Example of Chapter 90, has been converted into "zero, pole, gain" form as follows:

$$g_{11}(s) = \frac{X_1(s)}{L(s)}$$

$$= \frac{0.2462\,(2.015s + 1)\,(0.2667s + 1)}{(4.254s + 1)\,(0.8857s + 1)\,(0.2829s + 1)}$$

$$= \frac{0.1241\,(s + 0.4963)\,(s + 3.7495)}{(s + 0.2351)\,(s + 1.1291)\,(s + 0.5348)}$$

- The state space model is established using the zpk and ssdata instructions which are functions of the Matlab control systems toolbox.

Table 98.2 Input settings for controller and observer model

Input	Type	Sample time	Parameters
Set point	Step	0	Step time = 5 min Initial value = 0 Final value = 0.02
	Sine wave	0	Amplitude = 0.002 Bias = 0 Frequency = 0.2 rad/min
Disturbance	Step	0	Step time = 20 min Initial value = 0 Final value = 0.01
	Noise	1.0	Noise power = 0.0000001

In essence, the zpk instruction takes the zeros, poles and gain of the transfer function and converts it into a state space model. The A, B, C and D matrices of the state-space model are then abstracted by the ssdata instruction.

- The state feedback controller is determined according to Equation 112.2. The desired closed loop pole positions are first specified as a row vector Pk. The place instruction then determines the parameters of the controller K for the given system matrix A and input matrix B.
- The reference factor R is determined according to Equations 112.9 and 112.10 for a given state space system for which the state controller K is known.
- The observer is determined according to Equation 112.14. The desired pole positions of the observer are specified as a row vector Pl. The place instruction then determines the parameters of the observer L for the given system matrix A and output matrix B. Note that the observer L is a column vector.

To complete model definition, the configuration parameters dialogue box needs values to be completed. In the solver panel, set the stop time to 100.0 (min) and select the variable step type ode45 (Dormand-Prince) solver with the default auto options for the step size.

Running the model enables the design of the state feedback regulator and observer to be validated. There are two aspects to this:

1. Regulator. How well does the regulator control the process given changes in set point and disturbances?
   It can be seen that with a time delay of about 0.5 min the quality of control is good. Notwithstanding the sinusoidal component of the set point and the noise on the disturbance signal, the composition X_1 responds to relatively large changes in set point with a settling time of some 10 min.
   Increasing the time delay to 0.8 min results in a very oscillatory but nevertheless stable response. Increasing it to 0.9 min results in an unstable response. This suggests that the regulator is robust since a delay of say 0.8 min represents a substantial plant model mismatch given that the dominant time constant of the process is 4.25 min.
2. Observer. If the values of the output y (the top tray composition X_1) and its estimate $\hat{y}$ agree, which is when the two traces on the scope coincide, then the observer is working effectively. Inspection of the above responses, even that with the delay of 0.9 min when the system has gone unstable, reveals that the two traces are

virtually indistinguishable. Thus, the observer is very robust across the whole operating envelope of the regulator.

Other parameters, such as the choice of pole positions for both the regulator and the observer as well as the amount of integral action, can be adjusted too. Exploring the effect of changing these on the performance of the system is left as an exercise for the reader.

98.6 Dynamic Process Simulators

As seen, Simulink involves configuring mathematical functions to represent a control system. This requires a dynamic model of the plant items and instrumentation cast in either transfer function or state space form. To develop such models, a deep understanding of both the process dynamics and the control system structures is necessary, together with a good working knowledge of modelling techniques.

The alternative approach is to use a dynamic process simulator which enables generic models of plant items to be configured with pipework connections and control functions to simulate one or more whole plant items and/or processes. Thus models of separation columns, reactors, mixing vessels, heat exchangers, pumps, pipes and branches, valves, controllers, logic function, *etc.*, are selected from appropriate categories of drop down menus and configured on-screen with the same functional layout as the plant itself.

Modern configuration tools work on a "drag and drop" basis and are both flexible and powerful. Once the model icons have been placed on the screen they can be interconnected by simply dragging a connection from one icon to another. The configurator has a number of built in constraints on the types of plant items that can be connected and checks that all connections are complete. It will also enable "rubber banding", a technique that allows the layout to be simplified, or tidied up, by moving the position of one plant icon on the screen relative to others whilst retaining all of its various connections.

Having configured the process, the generic models are then parameterised. This is done by clicking on the icon of the model and opening up its dialogue box. For each stream there are pipework and/or branch models to be parameterised in terms of flow rate, composition, temperature, pressure, initial conditions, *etc*. Instrumentation has to be specified in terms of range and/or calibration. Valves require Cv values for sizing and inherent characteristic types. For items of equipment, such as pumps and compressors, flow vs pressure drop characteristics are entered.

For each plant item, physical and process characteristics are entered as appropriate. For example, for a heat exchanger, the parameters would be as follows:

- connections: to pipes and/or other items, by which means the process stream information is acquired.
- tube information: area, maybe in terms of number of tubes, tube type, tube length, internal tube diameter, tube thickness, tube pitch, no of tube passes, material of construction, *etc*.
- shell information: internal shell diameter, position of internal weirs, number of headers, orientation, *etc*.
- process function: evaporator, condenser, reboiler, thermosyphon, *etc*.
- process conditions: flow regimes, phases, *etc*. in user defined zones.
- model type: choice of model equations, correlations, *etc*.
- simulation: initial conditions, normal conditions, choice of algorithm, *etc*.

Note that dynamic process simulators can be run in both steady state and dynamic modes. The mode of operation determines, in part, the information required during parameterisation. For example, for steady state simulation, the volume of plant items is of no consequence whereas for dynamic simulation it is essential. In steady state mode, it is not necessary to specify the flow and pressure of all streams at the model boundaries: within lim-

Table 98.3 Pros and cons of Simulink *vs* Hysys types of package

Issue	Simulink (or equivalent)	Dynamic Process Simulator
Approach	Configuration of menu based generic components followed by parameterisation.	Configuration of menu based generic components followed by parameterisation.
Components of model	Mathematical functions such as transfer functions, state space functions, comparators, steps, *etc*.	Generic process models such as pumps, pipes, branches, valves, exchangers, vessels, *etc*.
Parameters	Gains, time constants, delays, damping factors, matrices and vectors.	Flows, compositions, temperature, diameters, efficiencies, fouling factors, rate constants, *etc*.
Nature of model	First principles model, linearised, with deviation variables, and converted into transfer function or state-space form.	First principles models in absolute form.
Modelling effort	The equations have to be developed on a bespoke basis to determine the parameters.	The equations for the dynamics are embedded in the generic process models.
Knowledge required	A working knowledge of the plant/process design and a deep understanding of process dynamics, control and model development.	A detailed knowledge of the plant/process design and a working knowledge of process control.
Scope and size of model	Scope tends to be focused on a single section of plant, typically a single unit operation resulting in small to medium sized models which are executed quickly on a laptop.	Scope is more extensive, embracing multiple units, resulting in large to massive models which obviously require much more processor power.
Time take to develop simulation	Given the limited scope, relatively short times (days or maybe weeks) are required to develop the equations and then (days) to build the simulation.	Given the more extensive scope, and the potentially huge volume of data, much longer times (weeks or maybe months) to create the models and do the parameterisation.
Accuracy	Depends on quality of model, but accuracy to within 10% is good enough for most purposes.	Potentially very accurate
Cost of developing model	A function of scope and size of model, cost is relatively low but critically dependant upon availability of expertise.	Noting that dynamic process simulators are usually used for extensive models, the development costs are obviously high, but offset by lower levels of expertise required.
Cost of industrial licence	Modest.	Expensive, but includes library of plant and equipment models, correlations and data for physical property calculations. Cost can be offset by use of model for other purposes, such as validation of plant design, operations simulation and operator training.

its, those that are not specified are determined by the simulation. However, for dynamic simulation, either the flowrate or the pressure of every stream must be specified. In steady state mode, product specifications may be set explicitly but for dynamic mode they are implicit, having to be translated into controller set points.

Also, note that dynamic information such as time constants do not have to be entered. Each of the generic models contains differential equations as appropriate which are automatically solved by numerical integration at run-time. The coefficients of the differential equations consist of combinations of the plant and/or process parameters that have been entered during the parameterisation process.

Once configured and parameterised the simulation can be run. As with Simulink, the step length and run time for numerical integration are required as model independent parameters. In effect, the model of each plant item is executed in turn, the output of one model becoming the input to the next. Once a single iteration has been completed: that is, all of the models have been executed once, the process is repeated. Iterations continue until some specified run time has been completed or some other condition satisfied. During run time, data about key parameters is displayed in the form of trend diagrams to depict the dynamic responses of interest. Simulation can be interrupted to introduce process disturbances as appropriate.

Unlike Simulink type models which use deviation variables and transfer functions, which have themselves been obtained by linearisation and are essentially approximations, dynamic process simulators use absolute variables and are rigorous. Thus, for every stream, for each iteration, the physical properties of every component in that stream are computed at that stream's current temperature and pressure. And for every plant item, for each iteration, the heat and mass transfer between each stream, the change in composition within each stream, and the overall and component mass and energy balances are computed.

It can be seen that there are many differences between packages such as Simulink and dynamic process simulators. Some of the pros and cons are as depicted in Table 98.3.

Copyright

MATLAB and Simulink are registered trademarks of The MathsWorks Inc.

Section 12

Advanced Process Automation

Relational Databases

99.1 Alarm System for RDB Purposes
99.2 Structure and Terminology
99.3 Mappings and Constraints
99.4 Structure and Syntax of SQL
99.5 Other Constructs
99.6 Dependency Theory
99.7 Entity Relationship Modelling
99.8 Database Design

In essence there are three types of database, all of which are of relevance to process automation. The most common, undoubtedly, is the structured type of database, consisting of blocks of data to be used by table driven algorithms for real-time control, as described in Chapter 45. Less common are relational databases (RDB) which are used in management information systems as described in Chapter 100. And occasionally, in the context of knowledge based systems, one comes across object oriented databases, as described in Chapter 107.

With structured databases, the relationships between data values are predefined and not easy to determine without knowledge of the basis on which the database was designed. In an RDB, relationships between values are not defined: they become explicit when the database is queried as opposed to when it is created.

Relational databases are ubiquitous. They are found in personnel departments for storing data about employees and job functions, in universities for keeping student and course records, in banks for accessing information about customers and accounts, in the travel industry for booking accommodation and journeys, and so on. In many instances they have front ends that enable users to interact with them directly over the internet.

The essential characteristic of an RDB, from a user's point of view, is that it enables information to be abstracted from a single database from different perspectives. Consider, for example, the on-line booking system used in the travel industry which enables:

- Individual customers to be provided with the details of their specific bookings: airport, flight number, time of departure, *etc*.
- Travel agents to search for flights going to a particular destination on a given day, to check availability of seats on any particular flight, *etc*.
- An airline to monitor number/class of bookings, check identity of passengers for security purposes, track connecting flights, *etc*.

This chapter provides an overview of the functionality of relational databases. It introduces the structure and terminology of RDB, structured query language (SQL), dependency theory, entity relationship modelling and database design. RDB is an area where there are many high quality proprietary packages available. There is also extensive literature available, both in the form of manuals for particular products and hundreds of textbooks. A comprehensive treatment of RDB is provided, for example, by Elmasri (2003).

99.1 Alarm System for RDB Purposes

An interesting, and not uncommon, application is the development of an off-line RDB during the project build stage of a control system. The RDB is subsequently used to automatically create, using print merge techniques, a conventional block structured database which can be downloaded into the control system. To limit the scope of discussion, the alarm features only of such an RDB are considered as a vehicle for explaining the principles. The use of tag numbers to reference alarms is fundamental: these are the same tag numbers as those used on the P&I diagrams of the plant. Thus, typically:

- Each alarm is uniquely defined in terms of its attributes within a single function block, such as AIN, PID, AOT, *etc.*, as described in Chapter 44.
- Each alarm has an explicit setting (threshold) with a deadband for analogues and a delay for discretes.
- Each alarm may be triggered by change in absolute value, rate of change of value or change of state.
- Each alarm can have different priorities attached to it such as low, urgent, *etc.*
- Each alarm can have different levels attached to it such as warning, action, *etc.*
- Each alarm is implicitly associated with the other function blocks connected to that in which the alarm is defined.
- Each tag is unique (*e.g.* TIC 047) and uniquely identified with a specific alarm within the control system.
- Each tag can be displayed on several mimic diagrams and can be configured into one or more trend diagrams.
- Each tag can reference one or more other tags to form an alarm scheme.
- Each tag can belong to only one alarm unit.
- Each alarm unit is normally associated with a major equipment item (MEI) such as a reactor, filter, *etc.*
- The control system can support a maximum of 100 alarm units.
- Each alarm unit can be assigned to only one area.
- Each area represents a physical area on the plant, such as the boiler plant or a distillation train.
- The control system can support up to 10 areas.
- Each area can have up to 50 trend diagrams.
- *etc., etc.*

99.2 Structure and Terminology

The data of an RDB is essentially organised into relations and tuples. In common practice, a relation is also referred to as a table and a tuple is usually referred to as a record or row within a table. The basic structure and terminology is as depicted in the relation of Table 99.1.

The relation is an instance of an alarm summary called Summary. The intension names the attributes (columns). Attribute names must be unique within a relation. Attributes are values from a particular domain (or range) of legal (or valid) values. For example, the tag numbers must con-

Table 99.1 Basic structure and terminology of a relation

Relation	Summary		Attributes		
Intension	Tagno	Description	Funcblock	Almunit	
Extension	FI046	Reagent flow rate	AIN_FT046	U047	Records
	TIC051	Reflux plate temp	PID_TT051	U101	
	LS122	Filtrate level	null	U120	
	PS123	Bed pressure	DIN_PS123	U120	
	...	...	...	...	
	Prim key		For key	For key	

Table 99.2 Schema for an alarm system

Summary Tagno	Description	For key Funcblock	For key Almunit
FI046	Reagent flow rate	AIN_FT046	U047
TIC047	Reactor temp	PID_TT047	U047
TIC048	Jacket temp	PID_TT048	U047
TI050	Top plate temp	AIN_TT050	U101
TIC051	Reflux plate temp	PID_TT051	U101
PIC052	Column pressure	PID_PT052	U101
FI053	Distillate flow	AIN_FT053	U101
TI120	Slurry temp	AIN_TI120	U120
LS122	Filtrate level	null	U120
PS123	Bed Pressure	DIN_PS123	U120
...	...	...	...

Settings Funcno	Threshold	Units	Trigger	Deadband	Delay	Priority
PID_TT051	110	°C	absolute	1.0	null	low
DIN_PS123	2.5	barabs	off $\to$ on	null	≤ 5.0	urgent
...	...	...	...	...	...	...

Topology Unitno	MEI	Description	Area	Description2
U047	PV047	Reactor#1	A1	Primaries
U101	PV102	Column#2	A4	Separations
U120	PV002	Filter#2	A1	Primaries
...	...	...	...	...

Schemes Scheme	Alarm	Trip	Interlock
S47	TIC047	TZ047	null
S123	PS123	null	PZ123
S52	TI050	PZ052	null
S52	PIC052	PZ052	null
...	...	...	...

Trends Trend	Tag#
T47	TIC047
T47	TIC048
T51	TI050
T51	TIC051
T51	PIC052
T51	FI053
...	...

Mimics Mimic	Mag#
M47	FI046
M47	TIC047
M120	TI120
M120	LS122
M120	PS123
...	...

form with a given convention such as ISA S5.1. The primary key is the attribute, or set of attributes, which uniquely identifies a particular record. It is usually underlined. In the above relation both the tag number and the function block are unique to the record: the tag number is chosen to be the primary key. Sometimes it may be necessary to use a pair of attributes to identify a particular record in which case they are both referred to as being prime and are both underlined.

The semantics are as follows:

- The alarm whose tag number is TIC051 is *associated* with the variable reflux plate temperature and is *defined* in function block PID-TT051 and has been *assigned* to the alarm unit U101.
- The null entry for the level switch LS122 under function block means "not known". That could either be because the level switch is yet to be defined in a function block or because the function block reference was not known when the record for LS122 was created. The null entry can also be taken to mean "not relevant".

There are usually several relations in an RDB, the overall design of which is referred to as a schema. The complete alarm system could consist, for example, of relations for alarm summary, settings, topology, schemes, trends and mimics as depicted in Table 99.2.

Attribute names need not be unique within a schema but must be unique within an individual relation. Relation names must be unique within a schema.

The semantics of a schema, or indeed of a relation, cannot necessarily be deduced from the intensions: some further documentation is usually necessary. For instance, in the Schemes relation, the attribute Trip for Scheme 52 could be interpreted as:

- If TI050 is true then PZ052 is activated.
- If PIC052 is true then PZ052 is activated.
- If both alarms are true then trip is activated.
- Something else entirely.

99.3 Mappings and Constraints

Relations can be considered as mappings from one set of attributes, usually the primary key, to another set of attributes. Consider again the Summary relation.

Funcblock_Of is a mapping from Tagno to Funcblock:

Funcblock_Of (FI046)=AIN_FT046

Record_Of is a mapping from Tagno to the whole record:

Record_Of (PS123)=PS123, "Bed Pressure", DIN_PS123, U120

These are both examples of functional dependencies: the first is a one-to-one mapping and the second an M-to-one mapping. There are various ways of articulating such dependencies:

- If X is known then so too is Y.
- X functionally determines Y.
- Y is functionally determined by X.
- If X matches then Y matches.

Other mappings not involving the primary key are possible. For example:

Almunit_To_Funcblock (U120)={AIN_TI120, null, DIN_PS123}

which may or may not be useful as it results in three answers.

Mappings within a relation or record are explicit. Mappings between relations are implied by use of foreign keys, denoted by "for key" in the schema. The attribute Funcblock in the Summary relation is such a foreign key because it is equivalent to the Funcno attribute of the Settings relation. Likewise Almunit in Summary and Unitno in Topology. Clearly:

Tag PS123 maps through function block DIN_PS123 into Priority urgent.
Tag TIC051 maps through Unit U101 into Area A4.

There are some important rules governing the use of foreign keys, generally referred to as integrity constraints.

1. The foreign key in one relation relates to the primary key in another by means of having the same domain (or range) of values. The domain is the same but, obviously, the number of records in the two relations is likely to be different.
2. The value in the foreign key of a record in the first relation occurs as a value in the primary key of some record in the second.
3. The value in the foreign key of a record of the first relation may be null. This simply means that particular record is not involved in the second relation.
4. No record can have a null entry in its primary key. That is because the primary key is used for identifying records and a null entry would prevent that record from being identified.

In the interest of simplicity, the name of the attribute in the foreign key and that in the primary key to which it relates could be the same. However, it is common practice to use different names as this enables queries to be more specific than would otherwise be the case. Indeed, using different names is essential when a foreign key refers to a primary key within its own relation.

It is possible as a result of using a succession of combinations of foreign and primary keys to establish a daisy chain type of loop between relations within an RDB. Queries based on such recursive relationships cannot be handled effectively by SQL.

Another important concept when considering relationships is totality. Do all members of the domain have to participate in the relationship? If so, the relationship is said to be total. Consider the Schemes relation. Every scheme has an alarm, so that is total, but not every scheme has an interlock, so that is not total.

In addition to the integrity and totality constraints, there are also semantic constraints which can be applied. These are essentially conventional equality and inequality constraints. For example, in the Settings relation, delay ≤ 5.0 s or deadband $= 1.0$ units.

Oracle and SQL Server are the two RDB packages used in the industry although, for smaller applications, Access is common. The former both have a data definition language (DDL) and a database management system (DBMS). The language is used for creating relations, defining attributes, declaring data types, specifying legal domains or valid ranges, and generating records. DDL checks for syntax errors and consistency. It also enables primary and foreign keys to be specified and other constraints to be applied. The database management system (DBMS) executes the RDB, handles the queries and enforces the constraints.

99.4 Structure and Syntax of SQL

The whole point of an RDB is to enable queries formulated in SQL to be processed. This section explains the basic structure and syntax of SQL. It does not address DBMS commands, such as launching an SQL query, as these are specific to the RDB environment.

An SQL query consists of two or more statements of the general form

```
<keyword> <arguments>
```

Structure

The structure of a query is as follows:

```
select     attribute expressions
from       relations
where      conditions
group by   attributes
having     conditions
order by   attributes
```

Only the first two statements are compulsory, the others are optional as appropriate. The order, as listed above, is compulsory too. The use of SQL is illustrated by means of a number of examples.

1. List all tag numbers in the relation Summary:

```
select   Tagno
from     Summary
```

2. Retrieve the alarm unit to which tag number TIC047 belongs:

   ```
   select   Almunit
   from     Summary
   where    Tagno='TIC047'
   ```

3. Find the function blocks of all the alarms whose priority is urgent:

   ```
   select   Funcno
   from     Settings
   where    Priority='urgent'
   ```

4. Display all the information (whole record) about the settings for tag number TIC051:

   ```
   select   *
   from     Settings
   where    Funcno='PID_TT051'
   ```

5. Find the tag numbers of all the alarms in the primaries area.
   This is a more complex query in that it requires both of the Summary and Topology relations and makes use of the Almunit foreign key:

   ```
   select   Tagno
   from     Summary, Topology
   where    Description2='Primaries'
            and Unitno=Almunit
   ```

6. Find all the tag numbers of alarms belonging to alarm units in the primaries area whose priority is low.
   This is an even more complex query. Priorities are defined in the Settings relation and areas in the Topology relation, yet there are no direct mappings between the two relations. However, they can be mapped through the Summary relation as this has foreign keys with both Settings and Topology:

   ```
   select   Tagno
   from     Summary, Settings, Topology
   where    Description2='Primaries'
            and Unitno=Almunit
            and Funcblock=Funcno
            and Priority='low'
   ```

7. List the tag numbers of all the alarms in the separations area, and group them according to major equipment item:

   ```
   select    Tagno
   from      Summary, Topology
   where     Area='A4' and Unitno=Almunit
   group by  MEI
   ```

8. There are built-in aggregate functions for finding the sum, average, maximum, minimum, *etc.* of groups of attributes. These can obviously only be used with attributes that are numerical although the count function can be used with non-numerics.
   For example, identify by number and name all of the major equipment items that have three or more associated alarms, and group them according to the item:

   ```
   select    MEI, Topology.Description
   from      Summary, Topology
   where     Unitno=Almunit
   group by  MEI
   having    count(*)>=3
   ```

   Note the dot extension to Topology in the select field. Since both Summary and Topology have attributes called Description, with different meanings, it is necessary to distinguish between them.
   Note also the inequality operator. SQL supports all standard inequality operators: >, >=, <, <= and !=, the latter meaning not equal to.

9. List all of the tag numbers in the primaries area, grouped according to alarm unit in ascending order:

   ```
   select    Tagno
   from      Summary, Topology
   where     Area='A1' and Unitno=Almunit
   group by  Unitno
   order by  Unitno asc
   ```

99.5 Other Constructs

The basic structure and syntax of SQL have been described. Many other constructs and operations exist, some of the more important ones are described as follows. It is important to appreciate that there are many options on the constructs in terms of syntax and arguments: the user is referred to an SQL manual for the detail.

Updates

These concern maintenance of the RDB: that is, inserting and deleting whole records and changing values of attributes.

10. Insert a new trip into the Schemes relation:

    ```
    insert into    Schemes
    values         (S122, LS122, LZ122, null)
    ```

11. Insert a new record in Settings for tag number PIC052. If only some of the attributes are known the others will be assigned null values:

    ```
    insert into    Settings (Funcno,
                   Threshold, Units)
    values         (PID_PT052, 1.49, barg)
    ```

12. Remove all records relating to trip PZ052 in the Schemes relation:

    ```
    delete from    Schemes
    where          Trip='PZ052'
    ```

13. For tag number TIC051, change its threshold and deadband to 120 and $2.0°C$ respectively:

    ```
    update    Settings
    set       Threshold=120, Deadband=2.0
    where     Funcno='PID_TT051'
    ```

14. Increase the deadband of all temperature measurements to $1.5°C$:

    ```
    update    Settings
    set       Deadband=1.5 * Deadband
    where     Units='°C'
    ```

Nested queries

These are an alternative structure for more complex queries. Consider the so-called flat query of Example 6 above. This query can be broken down into two stages:

- Find the tag numbers of all alarms belonging to alarm units in the primaries area.
- Of those tag numbers identified, find the ones whose priority is low.

15. The corresponding nested query is structured as follows:

    ```
    select  Tagno
    from    Summary, Settings
    where   Funcblock=Funcno
            and Priority='low'
    in      (
            select  Funcblock
            from    Summary, Topology
            where   Description2='Primaries'
                    and Unitno=Almunit
            )
    ```

It is a matter of opinion as to whether the flattened or nested form is clearer to understand for complicated queries. It is tempting to think that the nested structure is computationally more efficient but that is not necessarily so: the inner query has to be evaluated for every record in the outer query.

Joins

A join operation compares two or more relations (or views, see below) by specifying a column from each, comparing the values in those columns, and linking the rows that have matching values. It then displays the results in the form of table which can be saved as a new relation if appropriate.

16. List every alarm scheme in the primaries area together with the alarm unit to which it belongs and the MEI to which it relates:

    ```
    select    MEI, Almunit, Scheme
    from      Summary, Topology, Schemes
    where     Area='A1' and Unitno=Almunit
              and Tagno=Alarm
    group by  MEI
    order by  Scheme desc
    ```

The outcome of this join is a table consisting of three columns: MEI, Almunit and Scheme, in that order, the records of which are grouped according to MEI and, for each MEI, listed in scheme order.

There are several points illustrated by this example of the so-called cross join which are worth noting:

- In every respect the syntax is exactly the same as described previously.
- There are various types of joins: cross, inner, outer, left and right. There is a **join** keyword that can be used with appropriate syntax to make the type of joining being done more explicit.
- The table produced has the same number of columns, in the same order, as the number of attributes named after the select keyword.
- The attributes used in the equalities (or inequalities) must be of the same datatype.
- The attributes used after the select keyword must be unambiguous within the schema or else qualified by means of dot extensions.

Views

A view is a means of looking at the data in one or more relations. Its structure, in the form of a table with rows and columns, is exactly the same as a relation. Views are defined in terms of the relations from which they are derived. The data that is viewed is contained in the underlying relations: there are no separate tables of data associated with the view. From a users perspective, views are used to:

- Save a defined subset of the RDB.
- Focus on the underlying data that is of particular interest to the user.
- Display a perspective of the data that is bespoke to a category of user whilst the RDB is being used by multiple users.
- Define frequently used queries, joins, *etc.*, to save the user from doing so every time the database is accessed.
- Grant or deny others permission to access the user's parts of the RDB, a simple but effective security mechanism.

17. For example, produce a view that lists all alarms, with threshold values, in the separations area whose priority is urgent.

```
create view    Hialarms
as select      Tagno, Threshold
from           Summary, Topology, Settings
where          Area='A4'
               and Unitno=Almunit
               and Funcblock=Funcno
               and Priority="urgent".
```

99.6 Dependency Theory

The concept of functional dependency was introduced earlier in Section 3 on mappings and constraints. Whether a functional dependency is full or not underpins the design of an RDB. Table 99.3 depicts the relation Alternate which is a subset of the Summary and Topology relations considered previously.

There is an explicit functional dependency between Tagno, Description and Almunit:

$$\text{Tagno} \rightarrow \text{Description, Almunit.}$$

There is also an explicit dependency between Almunit, MEI and Area:

$$\text{Almunit} \rightarrow \text{MEI, Area.}$$

However, the dependency between Tagno, MEI and Area is implicit:

$$\text{Tagno} \rightarrow \text{MEI, Area.}$$

Such implicit dependencies are said to be transitive.

An obvious principle in the design of an RDB is to split up complex relations into simpler ones. The original information should, of course, be capable of being reconstructed using a join operation. Recognising that Almunit is a common attribute, the relation Alternate could be decomposed into two separate relations: Left and Right. It could then be reconstructed using the join:

Table 99.3 Alternate: a subset of the summary and topology relations

Alternate Tagno	Description	Almunit	MEI	Area
TIC048	Jacket temp	U047	PV047	A1
TIC051	Top plate temp	U101	PV102	A4

Left Tagno	Description	Almunit
TIC048	Jacket temp	U047
TIC051	Top plate temp	U101

Right Almunit	MEI	Area
U047	PV047	A1
U101	PV102	A4

```
select    Tagno, Description, Almunit, MEI,
          Area
from      Left, Right
where     Left.Almunit=Right.Almunit
```

This join does indeed result in the same data being recreated and the original decomposition into the two simpler tables is said to be lossless. If, however, the recreated data had contained spurious additional records then the decomposition would be said to be non-lossless.

There are three so-called forms of normalisation:

1NF The first normal form of relations is such that only atomic values are held in the attributes, as in all of the relations considered thus far. If, in the design of an RDB, it appears that any record contains a list or set of values for a particular attribute, then that attribute ought to be split into its sub attributes, probably by means of a new relation.

The other normal forms rely upon the functional dependencies which hold between various attributes and the underlying semantics.

2NF The second normal form of relations is such that all non-prime attributes of a relation are fully functionally dependent on the key of the relation. If, for example, a relation has a pair of prime attributes, and other attributes are functionally dependant upon **only** one or other of the prime attributes, then that relation would not be of 2NF form. To be so, the non-prime attributes would have to be functionally dependent on **both** of the prime attributes.

3NF The third normal form of relation is such that it is 2NF and no non-prime attribute is transitively dependant upon the primary key.

These latter normal forms might appear to be somewhat obscure. They are considered to be a poor basis of design for an RDB and should be avoided because, in particular, they:

- result in information being duplicated
- suffer from insert and delete anomalies
- enable inconsistencies in the database

99.7 Entity Relationship Modelling

Entity relationship (ER) models provide the basis for RDB design. An ER model is normally developed in the form of a diagram for which there is a conventional set of symbols, the subset of which used here consists simply of square brackets to denote entities and angular brackets to denote relationships, as shown in Table 99.4.

Table 99.4 Symbols used for entity relationship (ER) models

Symbol	Explanation
[noun]	Entity
(value)	Attribute
⟨verb⟩	Relationship
$[E_1] - \langle R \rangle - [E_2]$	Relationship with 1:1 cardinality.
$[E_1] - {}^M\langle R \rangle^N - [E_2]$	Relationship with M:N cardinality.

There is much more to ER modelling than meets the eye: a deep understanding of the application domain is fundamental. Typically, a set if statements, such as those in Section 1 about the alarm system, is the starting point. The essential approach to ER modelling consists of the following five steps:

1. **Identify entities.** These are objects, normally described by nouns. Note the hierarchy of entities: a useful insight is given by distinguishing between strong and weak ones. A strong entity is one that exists in its own right whereas a weak one depends upon the existence of a stronger one.

   In the alarm system example, all of the entities are strong, notwithstanding that they do in fact all presume the existence of tag numbers in the first place.

2. **Identify relationships.** These are the connections between entities normally described by verbs. A good ER diagram is structured such that connections between the entities clarify, rather than confuse, the relationships.

3. **Identify mapping constraints.** In essence, this means establishing the cardinality for the different relationships and annotating the ER diagram as appropriate. Cardinality is the ratio of the number of entities there are on one side of a relationship per entity on the other side.

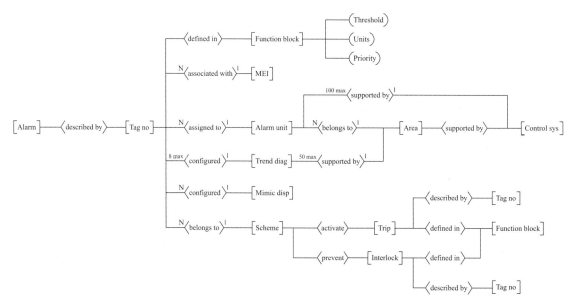

Fig. 99.1 ER model of the alarm system

Maximum and minimum constraints can also be applied.
4. **Identify attributes.** These are the properties (values) specific to the entities which are attached, diagrammatically speaking, to the entities as attributes.
5. **Identify keys.** These are attributes that uniquely identify one instance of an entity from another.

Figure 99.1 is an ER model for the alarm system used in this Chapter. It has been constructed using the above approach. Note that, for clarity, the Tagno entity is repeated: there is clearly only one Tagno entity but it would be confusing to connect all the relations involving Tagno to a single symbol for the entity.

99.8 Database Design

Whilst in this chapter, for illustrative purposes, the RDB was presented first and the ER diagram developed subsequently, it is normal practice to do it the other way round. It is essentially a pragmatic process of mapping. Some informal guidelines are as follows:

- Make the purpose of each relation easy to explain by mapping the semantics into relations.
- Avoid redundant information and hence update anomalies. If redundant information must be introduced, say in the interest of efficiency, then it should be carefully noted.
- Avoid null entries wherever possible. Records containing multiple null entries suggest that either the attributes for the relation or the relation itself is inappropriate.
- Avoid spurious joins: specify relations that can be joined on primary or foreign keys.

It is not the objective of this chapter to go into the detail of the methodologies for and the mechanics of RDB design. There is extensive information available in standard texts to which readers with a particular interest are referred.

Observations

There are many important aspects of the use of RDB that have not been covered in this chapter. For example, protection and security, which are functions of the RDB environment rather than of SQL itself. Another issue is access and the need for prioritisation, to prevent conflicts and contentions, through simultaneous use of the RDB by multiple users. And, of course, there are the special problems of working with real-time data and the need for automatic updating and time stamping of records. Nevertheless, the primary objective of the chapter has been achieved: to develop a feel for SQL: what it can do and how to use it.

Management Information Systems

Chapter 100

100.1 Information Requirements
100.2 Functionality of MIS
100.3 Materials Resource Planning
100.4 Process MRP
100.5 Manufacturing Execution Systems
100.6 Project Planning for the Process Industries
100.7 Integration of Control and Enterprise Systems
100.8 Comments

The concepts of management information systems (MIS) and computer integrated manufacture (CIM) were introduced in Chapter 38. Historically, the demarcation between MIS and CIM was their involvement in manufacturing. MIS were essentially concerned with data processing, *i.e.* the handling of sales and orders, accounts, payroll, *etc.*, and the only management information produced as such were financial summaries of a rudimentary nature. Conversely, the use of CIM in the discrete manufacturing industries largely concerned issues such as stock control, parts ordering and job status: resource planning and production scheduling was fairly primitive.

The scope of both MIS and CIM have evolved and merged to the point where much of their functionality is fairly indistinguishable. Current generation systems offer an integrated environment supporting all aspects of financial planning and presentation, just-in-time manufacturing, work-in-progress monitoring, materials resource planning (MRP), and so on. Of particular importance has been the emergence of process oriented MRP packages, which have begun to penetrate both batch and continuous processing, and manufacturing execution systems (MES). The evolution has been underpinned by the development of client-server technology and the availability of more open systems as outlined in Chapter 49.

This chapter provides an overview of these various aspects of MIS and CIM. For a more comprehensive background the reader is referred to the text by Hicks (1993).

100.1 Information Requirements

An MIS is used for decision making and decision support. The primary objective is to support the decision making process by providing relevant information in the most useful form. Information needs are determined by the management decisions that must be made which, in turn, are determined by the production objectives. Clearly, in designing an MIS, it is necessary to identify the sort of decisions that are to be supported, and to ensure that the data collected is relevant and stored in an appropriate form. Another important aspect of a well designed MIS is that it will accommodate growth in terms of the amount of data being handled and the types of decisions being made.

Three levels of decision making can be identified: strategic, tactical and operational:

- Strategic decisions are made by senior management. They concern establishing objectives and planning goals. These are of a long term nature and are heavily dependent upon financial information and external factors such as commodity prices and market projections. Short term information, such as yesterday's production figures, is of little interest.
- Tactical decisions are made by middle management and are concerned with implementing strategic decisions such as allocating budgets, assigning manning levels and determining overall production schedules. These are medium term in nature and hence more dependent upon real-time data.
- Operational decisions are taken, typically, by plant management and involve implementing the tactical decisions efficiently and effectively. Examples are processing according to an agreed production schedule and the management of planned maintenance. Operational decisions are short term and very dependent on recent data and current performance.

Two types of decision can be identified: programmed and non-programmed:

- In a programmed decision, the rules for making it are explicit. Given a certain set of conditions, the actions are the same and it follows that such decisions can be automated. A good example is automatic stock control in which raw materials are re-ordered according to current inventory, economic order quantity and minimum stock levels. Control systems run almost entirely on programmed decisions with the occasional query, such as the choice of recovery option, being presented to the operator for resolution.
- Non-programmed decisions are related to ill defined problems which require human intervention. A good example of such a decision, for which an MIS is suited, is in defining the objectives for scheduling. In general, the higher the level of decision making the greater the percentage of non-programmed decisions.

In order to make a programmed decision an MIS must have up-to-date, accurate and reliable information. However, to enable management to make non-programmed decisions, an MIS must offer a great deal of flexibility in terms of the range of information available, its manipulation and presentation. Ideally there should be no barriers to information from any part of the organisation: accounts, production, marketing, control, *etc*. This is the real challenge.

The need for flexibility is reinforced by the variability of the human component in the decision making process. The same information can be interpreted quite differently by different people. Furthermore, quite different data may be requested by different people to resolve the same problem. My view of the world is not necessarily the same as yours!

100.2 Functionality of MIS

A modern MIS environment provides access to information, both current and historic, invariably in the form of a relational database, together with an integrated suite of packages (modules) for manipulation of that data. The data normally relates to all aspects of the business activity: sales and marketing, accounts, production, inventory, and so on. Typically, the system will support screen enquiry and analysis facilities. These enable information to be selected and displayed in the form of graphs and spreadsheets. What-if type questions may be asked. Entities may be tracked through the system and audit trails established. In general, the more seamless the interface between the modules, the better the quality of the MIS. The more financially oriented modules of an MIS are as follows:

- Sales order processing
- Sales analysis

These two modules handle all aspects of end product sales. They enable orders to be initiated, tracked, invoiced, costed, documented, and so on.

- Inventory control
- Purchase order processing

These two modules are used for keeping track of stock and ordering raw materials. Inventory embraces raw materials, work in progress (WIP) and finished products. The objective of just-in-time (JIT) inventory systems is to minimise capital tied up in stock subject to not disrupting production through out-of-stock situations. In principle, JIT always delivers raw materials, in the correct quantities, to the right place at the right time, *i.e.* just before it is needed for production purposes. For this to work, the end user has to share production information with its suppliers over electronic data interchange (EDI) links. Ordering is demand driven and takes into account factors such as purchasing policy, supplier history, delivery times, costs, *etc*.

- Accounts payable
- Accounts receivable
- Cash book
- General ledger
- Fixed assets

The functionality of these modules is fairly self-explanatory. They provide an on-going record of all amounts owed, both actual and projected, by an end user to its suppliers (creditors) and to the end-user by its customers (debtors). They also enable the handling of payments *via* the banking automated clearing system (BACS), reconciliation of accounts, generation of forecasts, production of statements, monitoring of cash flow, management of assets and depreciation calculations.

- Payroll
- Human resources

The purpose of these two modules is obvious. They handle payments to employees, both hourly and salaried, and maintain records of deductions and taxes, together with personnel information such as employment history, training, *etc.* for use in career planning and industrial relations.

100.3 Materials Resource Planning

In addition to the above functionality, any sensible MIS provide modules to support materials resource planning (MRP), a concept which evolved from the manufacturing sector. In essence, MRP converts a master schedule for producing some product into detailed schedules of the components required to produce it. Thus, for each product, a "bill of materials" is generated which lists the quantity and type of sub-systems and components. Each sub-system is then further decomposed into sub-sub-systems and components, and so on, until a list of gross needs is generated.

Knowing the gross needs from the bill of materials, MRP would subtract the current inventory for each component and thence establish the various quantities of sub-systems to be produced and/or components to be purchased. By including lead times for purchase, and knowing when components are required from the master schedule, orders can be placed to ensure that they arrive just in time for production. This form of MRP is often referred to as infinite capacity scheduling because it takes into account neither labour requirements nor plant capacity.

MRP II introduced the concept of finite capacity scheduling. A "bill of labour" is defined which establishes the route taken by a product during its manufacture through a factory's resource centres. The bill of labour specifies, for each of the product's resource centres:

- The effort required (hours) for manufacture or assembly.
- The types of machine needed.
- The time required to set-up each machine.
- The milestones for WIP tracking.

This is much more realistic. Conflicts between the master schedule and what is practically realisable are highlighted and resolved, resulting in a control schedule that takes into account delays and equipment availability. Orders are then placed on the basis of achievable manufacture. Milestones enable the loop to be closed. Thus, for each resource centre, WIP is tracked against nominal (target) progress and deviations reported.

In practice, most MRP systems have proved to be generally unsuitable for the process industries because they only cater for scenarios in which many inputs, both materials and labour, combine to produce a single product. Some processes, such as oil refining, use one raw material to produce many products. Others, such as speciality chemicals, use multiple reagents to produce a variety of products, co-products and by-products. This has led to the development of "process MRP" packages which are based upon recipes rather than bills of material and labour. Recipes, as discussed in Chapter 37, contain product formulations, equipment and procedural requirements.

100.4 Process MRP

The more production oriented modules of a process-MRP enable process definition, production scheduling, process monitoring and process costing, as follows:

1. Process definition allows the process and/or plant to be defined using S88 type constructs, the definition providing the basis for production scheduling. Definition is essentially a question of:
   i. Articulating the plant structure in terms of the physical model; that is cells, trains, units and equipment modules, together with their interconnections. Associated with each of these will be information about their capacity and operational constraints.
   ii. Decomposing the processes in terms of the process model: that is streams, stages and process operations as appropriate, and asserting constraints as to which plant items can be used for their processing. Targets for key performance indicators (KPI) such as batch cycle times and yields may be specified.
   iii. Identifying operational requirements for each product type in terms of the procedural model: that is procedures, operations and phases. There will be a nominal processing time for each entity.
   iv. Defining generic recipes for each product type in terms of the recipe model: that is product formulations, equipment requirements, procedural requirements and quality criteria.
2. Production scheduling starts with forecasts for the production of products over medium or long term planning horizons. Clearly these have to be based upon the most up to date sales/order figures available. Forecasts of the raw materials requirements over the same planning horizon are then generated which may be compared with their availability. These forecasts are then converted into a production plan consistent with the S88 activity model. The shorter the forecasting horizon, the less scope there is for change and the less the amount of iteration subsequently required. Note that whilst the production plan generally satisfies the constraints of processing capacity and raw materials availability, there may well be short term conflicts to be resolved.

   The production plan is essentially a queue of generic recipes in the order they are to be run. Each recipe relates a batch of a particular product to the type of unit in which it is to be made and to the procedure to be carried out on that batch. However, to produce a master schedule, the production plan has to be combined with the process definition. Note that the master schedule has been produced off-line and does not take current plant status into account. For implementation, the master schedule is translated into a working schedule, referred to as a control schedule in S88 speak. This is still essentially a list of recipes, but their order is adjusted in the light of current plant status information to allow for unit occupancy, process delays, equipment failures, shortages of raw materials, *etc*. For each recipe, a specific unit is assigned with nominal start and end times.
3. Process monitoring enables tracking of batches of products through the plant in relation to the control schedule and KPIs. It records events as

they occur: analysis of raw materials quality, comparison of lapsed times with nominal, comparison of actual yields with targets, checking of quality metrics against criteria, logging of resources as they are used, *etc*. As production proceeds, discrepancies are reported so that corrective action may be taken.

4. Process costing concerns estimating the true costs of production. Cost categories are established for raw materials, equipment usage, services and so on. Then, as a batch progresses through the plant, from stage to stage, unit to unit, the cumulative processing costs are incremented. This can become quite complex when overheads and indirect costs are factored in.

100.5 Manufacturing Execution Systems

There remains a gap between MIS and CIM systems, with their client server technology and relational databases, and proprietary DCS and PLC systems, despite the trend for the latter to move towards NT based operating systems. That gap has been successfully exploited by SCADA suppliers who have offered manufacturing execution systems (MES) with bridges into a wide cross section of third party systems. Hierarchically speaking, an MES is positioned between the control system and process-MRP, although MES is increasingly taking on the lower level functionality of process-MRP. The essential characteristics of a process oriented MES include:

- Client server architecture
- OLE server supporting OPC as described in Chapter 49
- Standard bridges into all major DCS and PLC systems
- Open interfaces to laboratory information management systems
- SCADA capability
- Real-time operator interface supporting all functionality of Chapter 42
- Relational database
- Process definition tools
- Recipe handling and batch management
- Quality management using SPC and batch history files
- Materials and resource tracking with a lot-history database
- Preventive maintenance and time and attendance modules
- Access to third party document management systems
- Report generation

100.6 Project Planning for the Process Industries

Project planning for the process industries (PP-PI) was developed by a consortium and the SAP PP-PI package is perhaps the best known example of a process oriented MES. It provides a fully integrated process planning system with open interfaces to process control systems and laboratory information management systems (LIMS).

The emphasis of PP-PI is on batch processing. There are eight key modules, the essential characteristics of which are described below in terms of IEC 61512 (S88) constructs as appropriate. It should be recognised that the organisation of PP-PI is not consistent with the structure of the S88 activity model as depicted in Figure 37.11.

Resource Management

This module manages all the resources necessary for the processing of each product, with the exception of raw materials which are managed by the recipe management module. Resources are categorised by the user. Thus, subject to the constraints of the resource network, resources may be categorised as being either primary or secondary, each of which can be sub-categorised into classes:

- Resource networks are routes through the plant, each of which is a potential stream or train. These are determined by the structure of the plant and the flexibility for making change through the use of valve manifolds, flexible pipes and so on. There are always constraints. For example, once a batch has been started in one unit, the choice of unit for the next stage is limited to those units to which the first is physically connected, and to their availability.

- Primary resources are those which are required for the duration of a batch, the obvious example being the units containing the batches. Procedural entities, such as operations, that have the same duration are "committed" to primary resources.
- Secondary resources are those which are not required for the duration of a batch, examples of which are shared equipment modules such as storage vessels and manifolds, labour and laboratory effort. A secondary resource can be allocated a start time relative to the start of any operation or phase in which it is utilised.
- Resource classes are sub-categories of resources that are generically similar. For example, all identically equipped jacketed agitated vessels could be so categorised. Thus, when creating master recipes, equipment requirements may be specified by type or class, the specifics only being of interest when the control recipe is created.

Every resource has attributes in the form of data which is used for planning and scheduling purposes. That data may be about connectivity, say of one unit to another, access to raw materials or about materials of construction. Of particular importance is data about capacity and related formulae:

- Capacity is the ability of a resource to perform a particular task. A reactor, for example, will have a certain volume and heating capacity. An operator will have a certain amount of time available for carrying out tasks. Volume, heating capacity and time available are examples of capacity categories.

  Capacity commitment can be exclusive or non-exclusive. For example, a reactor, even if not filled by its current batch, cannot have its spare capacity filled by another, its commitment is exclusive. Resources shared by a number of batches, such as storage vessels, are non-exclusive. Equipment modules such as manifolds may be either exclusive or otherwise, depending on their usage.
- Formulae can be attached to a resource for use in calculation of, say, processing times, capacity and costs. It is not uncommon, for example, for nominal batch times to be available for standard batch sizes. For larger batches the formula enables the scheduling function to scale up the batch time taking into account non-linear heat transfer effects. Similarly, formulae enable required capacities to be compared with those available in the various resources.

Every resource is allocated to a cost centre so that production costs can be calculated.

Recipe Management

The recipe model of S88 is described detail in Chapter 37. An underlying principle of S88 is that master recipes are relatively generic and control recipes are plant specific. This is reflected in PP-PI in three important ways:

- The master recipe is cast in terms of equipment requirements, *i.e.* equipment types, and is deliberately resource independent. Thus, when the master recipe is converted into a control recipe, *i.e.* when the "process order" is released, resources are assigned. If there are alternative resources available, either primary or secondary, the user is provided with a list to select from.
- Conversion of the master recipe into the control recipe enables its address lists (A lists) to be parameterised. That is because, even though the phases are listed by name only, the physical resources have already been assigned.
- Materials requirements are articulated in the form of a materials list linked to the master recipe. In conversion of the master to the control recipe, standard quantities referred to in the master are operated upon by formulae in relation to the units' capacity to determine the actual quantities required by the process. These quantities can then be combined with other operational requirements into ordered batch parameter lists (B lists).

Note that there are separate master materials records for each raw material which contain not only information about stock levels but also other relevant factors such as purity and yield.

Process Orders

A process order details all the requirements to make a batch of a particular product. It is generated from the master recipe. The order defines in terms of products, batches, recipes, units and so on:

- What and how much is to be manufactured
- Where it is to be manufactured
- When it is to be manufactured
- What information is to be archived
- How much it is expected to cost

Release of an order enables the following to occur:

- Raw materials reservations are confirmed and master records are updated.
- Generic resource classes specified are replaced by equipment specific entities.
- Process instruction sheets are generated.
- Batch inspection records are allocated.

The cost object is a key construct to which the cost of a process order is assigned. This is based upon the planned raw materials consumption and production costs, the latter being based upon in-house charge rates used for use of equipment and utilities. Planned and actual costs may be analysed and the cost object adjusted in the light of such. Costs may be revised, for example, to take reworking into account. When a process order is completed, *i.e.* a batch is sentenced to storage, the cost is transferred to a different cost centre such as to inventory or to the customers account.

Process Planning

Released process orders are used by the process planning module for scheduling purposes. There are two key functions:

- Procedural scheduling. For any order, there will be a specified recipe whose operations and/or phases will each have nominal processing times. Knowing the resources allocated establishes the capacity commitments which enables the true processing times to be estimated. Obviously these times may be aggregated into batch times which enable lead times and completion dates to be predicted. All of this may be presented in Gantt chart form.
- Capacity scheduling. This is carried out once an order is scheduled when the required and available capacity requirements can be compared. The adjustment of any capacity under or overloads is referred to as capacity levelling and is performed interactively by the system user. Capacity information is displayed in Gantt chart form and order data may be manipulated, for example, to enable rescheduling of operations to meet delivery dates.

A campaign is the manufacture of a number of batches of the same product based on the same master recipe, typically utilising a dedicated train of units. This requires long term planning of resources and materials on a campaign basis. The functionality to support campaign planning is not yet available in PP-PI.

Process Management

This module controls the data exchange between PP-PI and LIMS and the control system. The principal functions of the process management module are as follows:

- It enables phase logic to be incorporated in operations as the control recipe is downloaded into the control system. Each phase has a unique name and is stored in a library. Phases, consisting of steps and instructions (or actions), are abstracted according to the phase names listed in the operations, and included within the operation.

The downloading of these operations including their phase logic, together with their address and batch parameter lists, into the target control systems (or nodes of a DCS) is sufficient for execution of the phases and operations, and hence

the complete automatic control of the batch processing.
- Process instruction sheets may be generated for display, or printed out, for those aspects of the processing that are to be carried out manually. This includes the requirements for materials inspection and sampling for laboratory analysis.
- Process messages are used to transfer data from the control system to PP-PI. Typically data is requested *via* process instructions, but unsolicited data such as alarms and events can also be sent. Different categories of data exist, such as analogue measurements, operator confirmations, phase start and stop times, *etc*.

Quality Management

Data in the form of "batch inspection records" are the basis for quality management. These are generated from the master recipe upon release of the process order. The record sets out the quality requirements in terms of inspections and laboratory analyses for each batch, identifying not only what tests are required but when. Inspections may be manual: the reading of analogue signals or results from laboratory analysis. More typically, batch records are compiled from data abstracted from either the control system or LIMS by means of messages. Both in-process and post-process inspections are supported, *i.e.* during a batch and after batch completion.

The capacity and scheduling of laboratory resources can be handled by the process planning module. This is important if production is round the clock but the laboratory is only open during the daytime.

Batch Management

Every material used in production has a generic specification which defines that material in terms of its physical and chemical properties such as state (solid or liquid), colour, viscosity, purity and expiry date. Individual batches (lots) of materials have their own batch inspection records. The characteristics of raw materials in a record come from supplier's information and laboratory analysis, for manufactured products they come through the quality management module. For any new batch, the batch management module checks the batch's record against the material specification to ensure that it conforms.

The batch management module allows materials to be managed in lots. Its essential functionality is as follows:

- Lot number assignment. This can be either automatic or manual, and applies to all lots whether they be raw materials upon receipt or manufactured batches.
- Lot status management. A lot is usable only if it meets its material specification. Usability is indicated by the status types of restricted and unrestricted. Status may be set either automatically by the quality management module or manually.
- Lot tracking. This records the history of a lot of any type of material from procurement through to delivery. For example, all materials and their lot numbers used in the manufacture of a particular batch, or which batches used a particular lot of raw material. This finds its way into the batch records.
- Lot determination. This is used to reserve raw material lots for specific orders. It also enables searches for manufactured lots which meet certain specifications, *e.g.*, those closest to a customer's specification, thereby minimising "giveaway".

Batch Records and Evaluation

FDA and GAMP require companies manufacturing pharmaceuticals to keep documented records of every step in their production, including results of quality related inspections. This data must be archived and can be recalled for inspection and analysis, but cannot be changed. Thus proof of compliance can be demonstrated and the causes of any specific problems identified. Data to be archived is specified through the process order and typically includes the following for every batch:

- Order number
- Process resources utilised
- Planned and actual material quantities
- Operation and phase data: start and stop times
- Time stamped process messages
- Time stamped alarm lists and event logs
- Unscheduled operator interventions, *e.g.* changes to quantities, setpoints, *etc*.

It is in the nature of management execution systems such as PP-PI that the amount of data collected and stored is extensive. Data can be evaluated retrospectively, either by tools provided within the MES or by export to third party packages, such as Excel. Typical evaluations include:

- General planning and cost data
- Equipment utilisation, planned *vs* actual
- Product costs, fixed and variable
- Materials consumption, efficiencies and yields
- Variability, operating conditions *vs* product quality
- Multivariate statistical process control

100.7 Integration of Control and Enterprise Systems

The ISA S95 standard has been published to provide a framework for CIM. Its intent is to improve communications between all parties concerned, thereby reducing the costs, risk and errors involved in implementation.

Part 1 provides standard terminology and a consistent set of concepts and models for integrating control systems with enterprise systems. Its scope includes maintenance management, material and energy control, production scheduling, order processing, product cost accounting, product shipping and administration, product inventory control and quality assurance. The main functions in each area are defined and information flows between them are described. The emphasis throughout is on good integration practice.

Part 2 defines the interface content between manufacturing control functions and other enterprise functions. Its scope is limited to the definition of attributes for the objects of the Part 1 models and does not extend to the attributes that represent the relationships between the objects.

Part 3 on models of manufacturing operations is yet to be published.

100.8 Comments

As will be evident from this chapter, MIS provides a good deal of functionality that, hierarchically speaking, is at a higher level than the control system. Such systems provide a management framework for making decisions about the control of plant and processes. They also enable information to be abstracted from control systems, manipulated and interpreted. It is almost inconceivable these days that an automation project would see a control system being installed without some form of MIS too. Indeed, given the extent of automation that exists within the industry, and the ever reducing scope for realising benefits through control alone, it is inevitable that much more emphasis will be placed upon MIS type functions. Indeed, one can predict that MIS and decision support will merge with process control functions. Pressures from open system developments and network technology will inexorably reinforce this.

Principal Components Analysis

101.1 PCA for Two Variables
101.2 Summary of Bivariate PCA
101.3 PCA for Multiple Variables
101.4 Reduction of Dimensionality
101.5 Worked Example No 1
101.6 Interpretation of Principal Components
101.7 Worked Example No 2
101.8 Loadings Plots
101.9 Worked Example No 3

Multiple linear regression (MLR) was introduced in Chapter 83 as a means of relating an output (response) variable to possible input (regressor or predictor) variables. MLR is undoubtedly the most commonly used means of establishing the relationship between input and output variables, but it does have its limitations. Primarily these are:

- MLR's inability to handle missing data. As will be seen in Chapter 102 on statistical process control there are methods for reconstructing missing or incomplete data, so this problem can be circumvented.
- Collinearity of input variables leads to least squares estimates of coefficients that either do not exist or else are unreliable. Two variables are said to be exactly collinear if there exists an equation such that:

$$c_1.x_1 + c_2.x_2 = 1$$

Knowing the value of x_1 determines precisely the value of x_2. Thus x_2 may be eliminated from the list of inputs for MLR without any loss of information. If collinearity is approximate, then one of the inputs may be eliminated with only nominal loss of information. Clearly this argument may be extended to cases of three or more inputs being collinear.
- Data overload. The amount of data to be processed depends on the number of variables being regressed, the frequency of sampling and the duration of the sampling process. Clearly there is scope for this being massive.

Whilst principal components analysis (PCA) is a linear method, it is nevertheless tolerant of these constraints. It also happens to cope well with data that is noisy, and with situations where the data is sparse, *i.e.* when there is little data relative to the number of inputs being regressed. In essence, PCA reduces the dimensionality of the problem whilst retaining as much of the variability contained in the original data as possible. This is achieved by forming linear transformations of the input variables in such a way that the transformed variables are orthogonal.

This chapter introduces PCA and demonstrates its use. First the underlying principles of PCA are developed in a bivariate context. This is then extended to multivariate problems and followed by an example. For a more detailed treatment reference should be made to texts such as those by Cox (2005) and Manly (1994).

101.1 PCA for Two Variables

Consider two variables x_1 and x_2 for which a number of empirically determined values are plotted against each other as depicted in Figure 101.1. The first principal component y_1 is the best straight line fit to the data. In effect, y_1 explains the maximum possible variance between x_1 and x_2. The second principal component y_2 is another straight line through the same data whose direction is chosen to explain the maximum amount of variance which is unexplained by the first principal component. To satisfy this requirement, y_2 must be at right angles, or orthogonal, to y_1. The fact that the principal components are orthogonal means that they are independent of each other: they are not correlated. Each principal component is said to measure an independent effect. For a typical multivariable process, there may only be a few underlying independent effects.

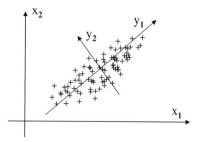

Fig. 101.1 Empirical bivariate data

Suppose the variables x_1 and x_2 are correlated and that they are both in deviation form with zero mean. Let y_1 and y_2 be the first and second principal components respectively. They are defined as linear functions of x_1 and x_2 as follows:

$$\begin{aligned} y_1 &= a_1.x_1 + a_2.x_2 \\ y_2 &= b_1.x_1 + b_2.x_2 \end{aligned} \quad (101.1)$$

First, consider the first principal component. The objective is that it should reflect the variability of the original data so the values of the coefficients a_1 and a_2, referred to as loadings, are such that the variance in y_1 is maximised. Suppose there are n sets of data relating y_1 to x_1 and x_2 such that:

$$\underline{y}_1 = X.\underline{\theta} \quad (101.2)$$

where:

$$\underline{y}_1 = \begin{bmatrix} y_{11} \\ \vdots \\ y_{n1} \end{bmatrix} \quad X = \begin{bmatrix} x_{11} & x_{12} \\ \vdots & \vdots \\ x_{n1} & x_{n2} \end{bmatrix} \quad \underline{\theta} = \begin{bmatrix} a_1 \\ a_2 \end{bmatrix}$$

The variance of $\underline{y}_1$ is given by:

$$\begin{aligned} \text{var}\left(\underline{y}_1\right) &= \frac{y_{11}^2 + y_{21}^2 + \cdots + y_{n1}^2}{n-1} \\ &= \frac{\underline{y}_1^T.\underline{y}_1}{n-1} = \frac{(X.\underline{\theta})^T.X.\underline{\theta}}{n-1} \quad (101.3) \\ &= \frac{\underline{\theta}^T.X^T.X.\underline{\theta}}{n-1} = \underline{\theta}^T.S.\underline{\theta} \end{aligned}$$

where S is the covariance matrix as defined by Equation 82.5. Since there are only two variables, x_1 and x_2, then S must be a 2 × 2 matrix. Note that this variance is of a sample rather than a population so the denominator is (n − 1) as opposed to n. Also note that the variance is positive and a quadratic function of $\underline{\theta}$.

Since both n and S are fixed by the set of data available, the only scope for maximising the variance in $\underline{y}_1$ is by adjusting $\underline{\theta}$.

However, it is necessary to introduce a constraint, otherwise the not very helpful solution of $\underline{\theta} \rightarrow \infty$ is obtained. A sensible constraint may be specified as follows:

$$\underline{\theta}^T.\underline{\theta} = \begin{bmatrix} a_1 & a_2 \end{bmatrix} \begin{bmatrix} a_1 \\ a_2 \end{bmatrix} = a_1^2 + a_2^2 = 1$$

whence:

$$\underline{\theta}^T.\underline{\theta} - 1 = 0. \quad (101.4)$$

In effect, the loadings a_1 and a_2 are limited to the range $-1 \leq a_1, a_2 \leq 1$.

This is quite sufficient for the purposes of formulating the principal component in the first place since, according to Equation 101.1, it enables y_1 to consist of any ratio of x_1 to x_2, ranging from none to all of each. The constraint, which may be thought of as a unit circle in the parameter space, locks both coefficients into a single constraint and, because it is inherently quadratic, is independent of sign. The cost function for maximisation may then be defined to be:

$$Q = \text{var}\left(\underline{y}_1\right) - \lambda \cdot \left(\underline{\theta}^T \cdot \underline{\theta} - 1\right)$$
$$= \underline{\theta}^T \cdot S \cdot \underline{\theta} - \lambda \cdot \left(\underline{\theta}^T \cdot \underline{\theta} - 1\right) \quad (101.5)$$

where λ is known as a Lagrange multiplier. Including the constraint in the cost function in this way guarantees that the constraint is satisfied at the same time as the variance is maximised. The use of Lagrange multipliers is explained in detail in Chapter 105.

Noting that differentiation of a scalar by a vector is covered in Chapter 79, the derivative of Q with respect to the loadings vector $\underline{\theta}$ is given by:

$$\frac{dQ}{d\underline{\theta}} = 2.S.\underline{\theta} - 2.\lambda.\underline{\theta}$$

Let $\hat{\underline{\theta}}$ be the value of $\underline{\theta}$ that maximises the variance. Thus, at the turning point:

$$\frac{dQ}{d\underline{\theta}} = 2.S.\hat{\underline{\theta}} - 2.\lambda.\hat{\underline{\theta}} = 0$$

whence the criterion for maximum variance:

$$S.\hat{\underline{\theta}} = \lambda.\hat{\underline{\theta}} \quad (101.6)$$

It might not be obvious, and certainly requires some interpretation, but this is the solution and identifies the first principal component. Rearrangement gives:

$$[\lambda.I - S].\hat{\underline{\theta}} = 0 \quad (101.7)$$

Inspection of Equation 101.7 reveals that, because S is a 2×2 matrix, there are two pairs of corresponding eigenvalues and eigenvectors that satisfy the maximum variance criterion. Thus there are two maxima, but which pair corresponds to the larger value? Strictly speaking the second differential should be found to confirm that they are maxima rather than minima, but this is taken as read. Let:

λ_1 and λ_2 be the eigenvalues, and
$\hat{\underline{\theta}}_1$ and $\hat{\underline{\theta}}_2$ be the corresponding eigenvectors.

The maximum value of Q corresponding to the pair λ_1 and $\hat{\underline{\theta}}_1$ is given by Equation 101.5:

$$Q_1 = \hat{\underline{\theta}}_1^T.S.\hat{\underline{\theta}}_1 - \lambda_1 \cdot \left(\hat{\underline{\theta}}_1^T.\hat{\underline{\theta}}_1 - 1\right)$$

Substituting from Equation 101.6 gives:

$$Q_1 = \hat{\underline{\theta}}_1^T.\lambda_1.\hat{\underline{\theta}}_1 - \lambda_1 \cdot \left(\hat{\underline{\theta}}_1^T.\hat{\underline{\theta}}_1 - 1\right)$$
$$= \lambda_1.\hat{\underline{\theta}}_1^T.\hat{\underline{\theta}}_1 - \lambda_1 \cdot \left(\hat{\underline{\theta}}_1^T.\hat{\underline{\theta}}_1 - 1\right) = \lambda_1$$

Similarly:

$$Q_2 = \lambda_2$$

The larger of the two possible maximum values for the cost function must correspond to the larger eigenvalue, and hence to the maximum variance in the first principal component.

The fact that the variance of the first principal component has a value of λ_1 can be confirmed by substitution of Equation 101.6 into Equation 101.3:

$$\text{var}\left(\underline{y}_1\right) = \hat{\underline{\theta}}_1^T.S.\hat{\underline{\theta}}_1 = \hat{\underline{\theta}}_1^T.\lambda_1.\hat{\underline{\theta}}_1 = \lambda_1.\hat{\underline{\theta}}_1^T.\hat{\underline{\theta}}_1$$

An important property of an eigenvector is its orthogonality such that $\hat{\underline{\theta}}_1^T.\hat{\underline{\theta}}_1 = 1$. Hence:

$$\text{var}\left(\underline{y}_1\right) = \lambda_1$$

101.2 Summary of Bivariate PCA

The first principal component is of the form:

$$y_1 = a_1.x_1 + a_2.x_2 \quad (101.1a)$$

where the loadings $-1 \leq a_1, a_2 \leq 1$ are given by the eigenvector $\hat{\underline{\theta}}_1 = \begin{bmatrix} a_1 \\ a_2 \end{bmatrix}$ which corresponds to the larger eigenvalue λ_1 of the covariance matrix S.

The variance of the first principal component has a value of λ_1.

Similarly, the second principal component is of the form:

$$y_2 = b_1.x_1 + b_2.x_2 \qquad (101.1b)$$

where the loadings $-1 \leq b_1, b_2 \leq 1$ are given by the eigenvector $\hat{\underline{\theta}}_2 = \begin{bmatrix} b_1 \\ b_2 \end{bmatrix}$ which corresponds to the smaller eigenvalue λ_2 of the same covariance matrix S. The variance of the second principal component has a value of λ_2.

101.3 PCA for Multiple Variables

The principles and formulae developed for bivariate PCA generalise to multivariate PCA. The following is a procedure for finding the principal components where there are multiple variables. Suppose there are p variables and the first principal component is:

$$y_1 = a_1.x_1 + a_2.x_2 + \cdots + a_p x_p \qquad (101.8)$$

where $-1 \leq a_j \leq 1$ and $\sum_{j=1}^{P} a_j^2 = 1$.

Suppose that there are n sets of data:

$$\underline{y}_1 = X.\underline{\theta} \qquad (101.9)$$

where:

$$\underline{y}_1 = \begin{bmatrix} y_{11} \\ \vdots \\ y_{n1} \end{bmatrix} \quad X = \begin{bmatrix} x_{11} & x_{12} & \cdots & x_{1p} \\ \vdots & \vdots & \ddots & \vdots \\ x_{n1} & x_{n2} & \cdots & x_{np} \end{bmatrix} \quad \underline{\theta} = \begin{bmatrix} a_1 \\ a_2 \\ \vdots \\ a_p \end{bmatrix}$$

Remember that it is essential that the data values used in the X matrix are in deviation form, i.e. that they have zero means. The higher the correlation between the columns of X, the fewer the number of principal components required to retain the variability of X.

1. Determine the covariance matrix S, which is of dimension p × p, according to

$$S = \frac{X^T.X}{n-1}$$

2. Given that the units and range of measurement of the variables are likely to be arbitrary, e.g. flow, pressure, temperature, etc., the relative significance of the various elements of S is usually obscured. So it is best to convert the covariance matrix S into the standardised form of the correlation matrix P, as defined in Equations 82.6 and 82.7:

$$\rho_{ij} = \frac{\sigma_{ij}^2}{\sigma_i.\sigma_j} \qquad (82.6)$$

The subsequent procedure is exactly the same, whether it is based upon the correlation matrix P or the covariance matrix S, although the results require different interpretations.

3. Find the p eigenvalues of P and their corresponding eigenvectors. Note that these are different to those of S.

4. Reorder the eigenvalues in descending order such that:

$$\lambda_1 > \lambda_2 > \cdots > \lambda_p.$$

5. Reorder the corresponding eigenvectors:

$$\hat{\underline{\theta}}_1, \hat{\underline{\theta}}_2, \cdots, \hat{\underline{\theta}}_p.$$

6. The first principal component of Equation 101.8 is found from

$$y_1 = \hat{\underline{\theta}}_1^T.\underline{x}$$

where:

$$\underline{x} = \begin{bmatrix} x_1 & x_2 & \cdots & x_p \end{bmatrix}^T.$$

7. The second principal component, which is of the form:

$$y_2 = b_1.x_1 + b_2.x_2 + \cdots + b_p x_p \qquad (101.10)$$

where $-1 \leq b_j \leq 1$ and $\sum_{j=1}^{P} b_j^2 = 1$, is found from:

$$y_2 = \hat{\underline{\theta}}_2^T.\underline{x}$$

8. The jth principal component, where $1 \leq j \leq p$, is found from:

$$y_j = \hat{\underline{\theta}}_j^T \cdot \underline{x}$$

101.4 Reduction of Dimensionality

As explained, an objective of PCA is to reduce the dimensionality of the problem whilst retaining as much of the variability contained in the original data as possible. The central issue is to decide on which of the p principal components to retain. The most commonly used method is based upon the cumulative percentage of the total variance. It is important to appreciate that it is the least significant principal components that are being eliminated and not the original variables: the principal components retained contain all of the original variables.

The total variance is found by summing the variance of each of the original p variables:

$$\sum \sigma^2 = \sigma_1^2 + \sigma_2^2 + \cdots + \sigma_p^2 = \sum_{j=1}^{p} [s_{jj}]$$

These variances are the eigenvalues of the covariance matrix S which happen to lie along its diagonal. Thus the total variance is found by the summation of those diagonal values, often referred to as the trace. Thus the percentage of the total variance attributable to the first principal component is given by:

$$\frac{\lambda_1}{\sum_{j=1}^{p} \lambda_j} \cdot 100$$

The cumulative percentage of the total variance by the first m principal components is thus:

$$\frac{\sum_{j=1}^{m} \lambda_j}{\sum_{j=1}^{p} \lambda_j} \cdot 100$$

The criterion commonly used is to choose the first m principal components such that they contribute some 80% of the total variance. Other criteria involve retaining those components whose variance exceeds the mean variance, the mean being either the arithmetic average or the geometric mean. It should be appreciated that the least significant principal components often relate more to the noise in the data than anything else.

If the correlation matrix P is being used rather than the covariance matrix S, the corresponding trace is of value p and the percentage contribution of each principal component to the total variance is given by:

$$\frac{\lambda_j}{p} \cdot 100 \qquad (101.11)$$

where the λ_j are the eigenvalues of P.

101.5 Worked Example No 1

This example taken from the academic world should be one that every control engineer can relate to. The examinations results for a number of students were recorded for five subjects. The variables are defined as shown in Table 101.1.

Table 101.1 Categorisation or exam types

	Subject	Exam type
x_1	Mechanics	Closed book
x_2	Vectors	Closed book
x_3	Algebra	Open book
x_4	Analysis	Open book
x_5	Statistics	Open book

The correlation coefficients were found to be as follows:

$$P = \begin{bmatrix} 1.000 & 0.557 & 0.545 & 0.411 & 0.400 \\ 0.557 & 1.000 & 0.610 & 0.485 & 0.445 \\ 0.545 & 0.610 & 1.000 & 0.711 & 0.658 \\ 0.411 & 0.485 & 0.711 & 1.000 & 0.603 \\ 0.400 & 0.445 & 0.658 & 0.603 & 1.000 \end{bmatrix}$$

The eigenvalues of P are as shown in the diagonal matrix D and the corresponding eigenvectors are the column vectors of the matrix V:

$$D = \begin{bmatrix} 0.443 & 0.0 & 0.0 & 0.0 & 0.0 \\ 0.0 & 0.394 & 0.0 & 0.0 & 0.0 \\ 0.0 & 0.0 & 0.250 & 0.0 & 0.0 \\ 0.0 & 0.0 & 0.0 & 0.729 & 0.0 \\ 0.0 & 0.0 & 0.0 & 0.0 & 3.185 \end{bmatrix}$$

$$V = \begin{bmatrix} 0.607 & -0.192 & -0.120 & -0.647 & 0.402 \\ -0.686 & 0.338 & -0.184 & -0.441 & 0.433 \\ -0.062 & -0.133 & 0.842 & 0.136 & 0.501 \\ -0.163 & -0.635 & -0.449 & 0.402 & 0.456 \\ 0.362 & 0.655 & -0.204 & 0.455 & 0.439 \end{bmatrix}$$

Inspection of D reveals that the rank order of the eigenvalues is 5th, 4th, 1st, 2nd and 3rd.

The first principal component of P is thus:

$$y_1 = 0.402.x_1 + 0.433.x_2 + 0.501.x_3 + 0.456.x_4 + 0.439.x_5$$

and the second principal component is:

$$y_2 = -0.647.x_1 - 0.441.x_2 + 0.136.x_3 + 0.402.x_4 + 0.455.x_5$$

In fact the first two principal components are sufficient. The trace of P is 5.0 so the percentage contribution of each of the principal components to the total variance is found from Equation 101.11 to be:

$$D\% = \begin{bmatrix} 8.86 & 0.0 & 0.0 & 0.0 & 0.0 \\ 0.0 & 7.87 & 0.0 & 0.0 & 0.0 \\ 0.0 & 0.0 & 5.00 & 0.0 & 0.0 \\ 0.0 & 0.0 & 0.0 & 14.57 & 0.0 \\ 0.0 & 0.0 & 0.0 & 0.0 & 63.70 \end{bmatrix}$$

Thus the combined contribution of the first and second principal components is approximately 78.3% of the total variance, which is good enough.

101.6 Interpretation of Principal Components

Interpretation of the principal components is fairly straightforward. There are two aspects to this, first to distinguish between the different principal components, and second to interpret the relative significance of the variables in each of the principal components.

Because the eigenvectors are orthogonal, the principal components can be thought of as being at right angles to each other. As explained, this means that the patterns of variance which they depict are independent of each other. The first principal component indicates the dominant effects. Considering the second principal component is equivalent to stripping out the dominant effects. This reveals weaker, but nevertheless significant, underlying effects that were previously obscured.

Remember that the loadings of the principal components, which come from the eigenvectors of the correlation matrix P, relate to the correlation coefficients between the variables. Each loading is associated with a variable in a principal component. The value of the loading is a measure of the contribution of that variable to the proportion of the variance in the data accounted for by its principal component. The stronger the loading, *i.e.* the closer it is to unity (+1 or −1), the more dominant is the variable.

In the above example, the first principal component indicates that all the examinations contribute to the variance of the results to a similar extent, algebra being the most significant and mechanics the least. This first principal component is dominant, being some four times stronger than the second principal component. The latter, however, reveals the more interesting effect: a pattern of variance that is associated with the nature of the examination. The negative signs of the loadings for mechanics and vectors indicate that the variances for closed book examinations are in the opposite direction to those for open book. This suggests that candidates who do well in the closed book exams do less well in open book exams, and *vice versa*. This is the power of PCA: it enables weaker patterns of variance to be revealed that would otherwise be obscured by more dominant patterns.

101.7 Worked Example No 2

This is based upon the naphtha splitter used for the MLR analysis in Chapter 83, as depicted in Figure 83.2, and using the same raw data. Again all of the variables whose standard deviation is $\leq 1\%$ of the mean are excluded on the grounds of being not statistically significant. However, to allow lesser principal components associated with inventory to be found, relevant variables for which the cross correlation function is ≥ 13 samples (the threshold used for MLR) are included. This is on the basis that inventory delays could be significantly larger than the known time delay of some 30 min for changes in feed to affect the product streams. Also, note that the variables x_{13} and x_{14} are essentially identical and have been combined into a single variable x_{15}.

Of the variables used the biggest delay is of 95 samples in the variable x_{12} so, in effect, there are 50 complete sets of data representing a period of some 250 min (> 4 h) which is a sufficient statistical basis. The raw data for the variables is shifted in time as appropriate and assembled in the form of the X matrix of Equation 101.9. The correlation coefficients are then found using Equations 82.4 and 82.6 and the correlation matrix established as per Equation 82.7 from which the eigenvalues and eigenvectors are determined.

The first principal component has an eigenvalue of 4.129 which, for a trace of value 9.0, represents 45.9% of the total variance:

$$pc1 = -0.4635.y_3 - 0.4730.x_1 - 0.3358.x_3 \\ - 0.3876.x_4 - 0.4616.x_6 + 0.0972.x_7 \\ - 0.0334.x_9 - 0.2697.x_{12} + 0.0555.x_{15}$$

It can be seen that this is dominated by the variables y_3, x_1, x_3, x_4 and x_6, all of which are related to the top product composition. It might be expected that the top product composition y_3 would be strongly influenced by x_5 and x_7 too. However, both of these flows are controlled directly, as opposed to the top product flow x_6 which is manipulated by a slave loop. Furthermore, the data for the reflux rate x_5 was previously eliminated on the grounds that its cross correlation function was not credible. As for the distillate flow x_7, its flow rate relative to the reflux flow x_5 is not that significant.

The second principal component has an eigenvalue of 2.270 which represents 25.2% of the total variance:

$$pc2 = -0.0268.y_3 - 0.1197.x_1 + 0.2862.x_3 \\ - 0.1391.x_4 - 0.0205.x_6 + 0.5486.x_7 \\ + 0.6160.x_9 + 0.1716.x_{12} - 0.4164.x_{15}$$

It can be seen that this is dominated by the variables x_7, x_9 and x_{15}, all of which are related to the column's liquid inventory. It would normally be expected that the level in the still x_9 be strongly influenced by x_1, x_6 and x_{12} too. However, the feed rate x_1 is relatively constant so its influence is only weak, x_6 is a vapour stream and there is little capacity for vapour inventory and, of the bottom product rates, x_{12} is not significant relative to x_{15}. The strength of x_7 is relatively easy to explain. Closer inspection of the data reveals that, during the period that the data was collected, x_7 monotonically increased as x_9 decreased. This is consistent with the steady state mass balance for the column in which the average feed rate x_1 was 345.5 m^3/h and the sum of the product rates (x_6, x_7 and x_{12} to x_{14}) was 351.4 m^3/h, indicating a slow net depletion of inventory.

The third principal component has an eigenvalue of 1.070 which represents 11.9% of the total variance:

$$pc3 = -0.1973.y_3 + 0.1018.x_1 - 0.1530.x_3 \\ + 0.4717.x_4 + 0.0795.x_6 + 0.0647.x_7 \\ - 0.1245.x_9 - 0.5534.x_{12} - 0.6104.x_{15}$$

It can be seen that this is dominated by the variables x_4, x_{12} and x_{15}, all of which are related to the bottom product composition. It would normally be expected that the bottoms composition be strongly influenced by x_8, x_{10} and x_{11} too. However, these variables were eliminated because their standard deviations are $\leq 1\%$ of their means.

Together, the first three principal components account for 82.9% of the total variance which is good enough.

101.8 Loadings Plots

Further insight into the meaning of principal components can be gained by inspection of their loadings. Suppose that there are two principal components, each with five inputs:

$$y_1 = a_1.x_1 + a_2.x_2 + a_3.x_3 + a_4.x_4 + a_5.x_5$$
$$y_2 = b_1.x_1 + b_2.x_2 + b_3.x_3 + b_4.x_4 + b_5.x_5$$

The bivariate loadings plot for these two principal components can be can be made as depicted in Figure 101.2. Thus A represents the point (a_1, b_1), B is (a_2, b_2), C is (a_3, b_3), and so on. Such a plot enables the meaning of the loadings to be interpreted in relation to the orthogonality of the principal components.

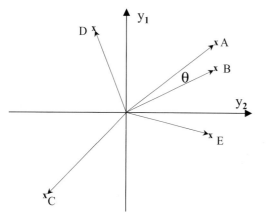

Fig. 101.2 Loadings plot for two principal components

Of particular importance in interpreting loadings plots is the angle θ between the loadings. It is helpful to consider the cosine of that angle, as demonstrated in relation to the following pairs of loadings:

- A and B. Since these vectors are adjacent (θ is approximately 0°), $\cos\theta$ is positive large (approximately +1.0) and the variance of the associated variables is in the same direction. Thus as x_1 increases so does x_2 and *vice versa*.
- A and C. Since these vectors are opposite (θ is approximately 180°), $\cos\theta$ is negative large (approximately −1.0) and the variance of the associated variables is in opposite directions. Thus if x_1 increases then x_3 decreases and *vice versa*.
- A and D. Since these vectors are roughly at right angles (θ is approximately 90°), $\cos\theta$ is small (approximately 0) and the variance of the associated variables is not correlated. Thus if x_1 changes it has no affect on x_4 and *vice versa*.
- A and E. These vectors are adjacent but not close (θ is approximately 45°), $\cos\theta$ is positive medium (approximately +0.7) and the variance of the associated variables is correlated, but not so strongly. So changes in x_1 do have some affect on x_5 and *vice versa*.

101.9 Worked Example No 3

Figure 101.3 depicts the loadings plot of the first two principal components for the naphtha splitter of Worked Example No 2.

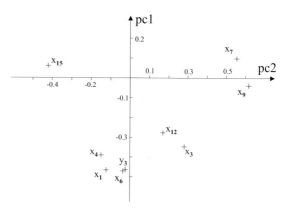

Fig. 101.3 Loadings plot of 1st and 2nd components for naphtha splitter

Great care is necessary in interpreting the loadings plot since the principal components were established from the correlation matrix for which the data was normalised. Thus all the cause and effect relationships are potentially distorted. It is also worth remembering the outcome of the MLR analysis in Worked Example No 3 of Chapter 83 in which a strong relationship was established between x_3, x_6 and y_3:

- x_6 and y_3 are almost exactly aligned indicating that as x_6 increases so does y_3. An increase in the top product rate x_6, for a given reflux rate x_5, would result in a decrease in the reflux ratio and hence to a less pure distillate. So an increase in y_3 which relates to the less volatile components ($\geq$C7s) is to be expected.
- x_3 and y_3 are adjacent but not close which suggests a weak correlation in the same direction. Without doubt an increasing overheads temperature x_3 would be expected to indicate an increasing fraction of $\geq$ C7s. That the correlation is weak is perhaps accounted for by the multicomponent (as opposed to binary) nature of the distillate.
- x_3 and x_4 are not far off being at right angles indicating that the overheads temperature and pressure are not correlated. In fact the overheads temperature x_3 is a function of both pressure x_4 and composition y_3 but, given that the pressure is controlled, the dependence on composition would predominate.
- x_4 and y_3 are adjacent corresponding to a strong correlation. However, an increasing column pressure x_4 would be expected to suppress the amount of $\geq$ C7s in the distillate. Also it is not consistent with the previous comment. These observations cast some doubt on the validity of the PCA model.
- x_{15} and x_9 are opposite each other and so as the bottom product rate x_{15} increases the level in the still x_9 decreases which is to be expected.
- x_1 and x_6 are adjacent indicating that any change in the feed rate x_1 causes the top product flow to increase. Interestingly, x_1 and x_{15} are at right angles indicating that changes in feed have no affect on the bottom product flow. The column is thus operated such that feed flow changes are accommodated by the distillate.
- x_6 is at right angles to x_9 which suggests that the dominant top product flow x_6 does not affect the level in the still. This seems unlikely, although the bottom product flow x_{15} has a more direct and larger influence, and casts some doubt on the validity of the PCA model.
- x_7 and x_9 are adjacent which suggests that as the top product flow x_7 increases then so does the level x_9 which is not credible. However, the flow rate x_7 is so small relative to x_{15} that not too much significance can be attached to this.
- x_1 and x_9 are at right angles which suggests that changes in feed rate x_1 have no affect on still level x_9. That would not normally be so but, as previously observed, the feed rate is relatively constant so its influence is only weak.
- x_4 and x_{15} are at right angles. Since the column pressure x_4 and the dominant bottoms product flow x_{15} are controlled independently they would not be expected to correlate.

Figure 101.4 depicts the loadings plot of the first and third principal components for the naphtha splitter of Worked Example No 2.

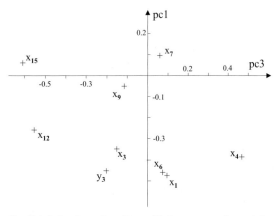

Fig. 101.4 Loadings plot of 1st and 3rd components for naphtha splitter

Comparison of Figures 101.3 and 101.4 reveals that the position of many of the points has changed, some significantly so. However, it is their relative positions that matters given that the underlying relationships between the variables should be the same. If the relative positions are similar on different loadings plots, that reinforces the interpretation and gives confidence in the robustness of the PCA model:

- x_6 and y_3 are no longer aligned, but are nevertheless still adjacent and relatively close.
- x_3 and y_3 are aligned, the stronger correlation being more consistent with expectation.
- x_3 and x_4 are still at right angles.
- x_4 and y_3 are now at right angles which tends to remove the doubt cast earlier.
- x_{15} and x_9 were opposite but are now adjacent. Thus the correlation still exists but in the wrong direction. However, x_9 is close to the origin so its significance is not that great.
- x_1 and x_6 are still adjacent.
- x_6 and x_9 are still at right angles.
- x_7 and x_9 are now opposite which is more sensible, albeit with both close to the origin.
- x_1 and x_9 are still at right angles.
- x_4 and x_{15} are now opposite each other which is not really credible.

The above observations generally provide increased confidence in the PCA model. They are based upon a comparison of the loadings plots of pc1 *vs* pc2 and pc3. These principal components account for 45.9%, 25.2% and 11.9% of the variance, respectively. Clearly the first loadings plot is more meaningful than the second. There is no reason why loadings plots of pc2 *vs* pc3, or even pc1 *vs* pc4 could not have been used although, obviously, the observations become less meaningful as the significance of the principal components decreases.

Statistical Process Control

Chapter 102

102.1 Data Collection
102.2 Data Pre-Screening
102.3 Control Charts
102.4 Average Control Charts
102.5 Moving Average Control Charts
102.6 Spread Control Charts
102.7 Six Sigma
102.8 Capability Indices
102.9 Multivariate SPC
102.10 Geometric Modelling
102.11 Comments

Statistical process control (SPC) concerns the use of stochastics for condition monitoring and for improving the effective operation of processes and plant. Of particular interest are abnormalities and deterioration of performance. Abnormalities are typically due to events, such as failure or malfunction of equipment and unscheduled operator interventions, often characterised by outliers in the data. Deterioration of performance is associated with longer term factors such as fouling of heat transfer surfaces and wear of bearings, more characterised by drifts and trends. The objective of SPC is to detect changes in performance and to isolate the causes of such from the random effects due to the natural variability of the process. This leads to a better understanding of the cause and effect relationships: it provides advance warning of problems and/or enables corrective adjustments to be made. An excellent introduction to SPC is given by Ogunnaike and Ray (1994) and the texts by Shunta (1995) and Oakland (2002) are both comprehensive.

There are two important and related fields: chemometrics and process analytics technology (PAT). Both undoubtedly involve stochastics and techniques such as multiple linear regression (MLR) and principal components analysis (PCA). Both also concern the application of these techniques in a process context. Chemometrics is associated with reaction chemistry and extends to process control. PAT involves analytical measurements and is associated with quality assurance and validation procedures. As to where the boundaries are between SPC, chemometrics and PAT is of semantic interest only: it is sufficient to say that they are all facets of process automation.

102.1 Data Collection

Fundamental to SPC is the choice of variables to be measured. If the set of variables to be monitored is not obvious, then the techniques of correlation, multiple linear regression (MLR) and principal components analysis (PCA), as discussed in Chapters 82, 83 and 101, should be used to aid selection.

SPC is critically dependent on the quality of the data available. The data must be gathered in such a way that subsequent analysis is capable of revealing the underlying facts. This is basically an issue of sampling. It is essential to have some understanding of the dynamics of the plant and/or process and to know what the data is going to be used for. This enables sensible decisions about which measurements are to be taken, how, and the frequency and duration of sampling for each measurement.

Clearly, appropriate instrumentation must be available and the measurements taken must be representative of the variables. The most common problems are with instrumentation that is unreliable, not sensitive enough, incorrectly situated or simply not available. Provision of effort for manual sampling is also a potential problem. What is essential at this stage is willpower, ingenuity and skill. Once measured, there is seldom any difficulty in gathering the data by means of the logging and archiving facilities of DCS and SCADA type systems.

In practice, data is invariably missing or incomplete, and may also be distorted and/or noisy. So, before the stochastics of Chapter 82 may be applied, the data will usually need to be pre-screened to a greater or lesser extent.

102.2 Data Pre-Screening

Aspects of the pre-screening of data have already been commented upon in Chapter 83. Pre-screening is essential prior to analysis. The starting point is to study the data to get a "feel" for it. This includes evaluating summary statistics, making plots and observing trends, noting the pattern of any missing data, locating outliers and identifying the need for transformation.

It goes without saying that the data must be statistically rich enough in the first place. If there is not enough inherent variability then the methods of SPC cannot yield meaningful results. Measurements that are constant are useless. Signals where the noise is more significant than any underlying pattern, as opposed to noisy signals, are also useless. A useful metric here is the magnitude of the standard deviation on a measurement in relation to the accuracy of the instrumentation being used to obtain it.

The most common problem is due to missing and incomplete data. It is invariably the case that there will be gaps in the data collected. Indeed, in large studies, complete sets of data are more the exception than the rule. By inspection, determine which subsets of data and/or variables contain the gaps. There are two approaches which may be used in combination: deletion and invention. Both approaches achieve more or less the same outcome, complete sets of data, although neither is entirely satisfactory.

First, where there are a few missing observations, it is easiest to delete subsets of data and/or variables resulting in smaller, but nevertheless complete sets of data. Subsets containing several missing variables are strong candidates for deletion. Second, missing data can be filled in by inventing plausible values. When inventing values, all known facts and relevant information must be taken into account. Known correlations between variables are particularly helpful for filling in purposes. The effect of filling in missing values on measures such as mean and median should be monitored carefully.

Fundamental to these approaches to handling missing data is the concept of missing at random (MAR). This assumes that the cause of the variables being missing is unrelated to the relationships under study. For example, missing data due to having dropped a test tube or to losing the results counts as MAR and analysis that ignores such data is justified. However, if the missing data is a function of the observation process, MAR does not apply and simple deletion of sub-sets or variables is invalid. A good example of this is where the values of a variable lie outside the range of the instrument used for measurement.

Outliers are observations that appear to be inconsistent with the rest of the data. They may be caused by recording errors or may be genuine values. In some cases outliers may be deleted as being not representative of the process under study: they are simply not plausible. In other cases they may be corrected using *a priori* knowledge. Either way, it is important to respect the data and to only adjust it on the basis of an understanding of the underlying facts.

Data arising from industrial processes is often noisy and if the signal to noise ratio is too low then the measurement is of little use statistically. Note that it is the variation in the signal rather than its absolute value that counts: 3 to 1 is generally considered to be a sensible minimum working ratio. Even above that ratio, the noise may well obscure underlying trends and patterns. The simplest approach is to remove the effects of the noise by filtering the data. There are various ways of doing this:

- Data compression: this involves recording the average value of m observations over consecutive fixed periods of time.
- Moving average, sometimes referred to as running average, in which the mean of the last m observations is recorded.
- First order lag: a numerical method using a lag equivalent to m sampling periods, as described in Chapter 44.

Clearly the value of m must be chosen carefully to remove the high frequency noise whilst retaining the low frequency variations of interest.

If the nature of any non-linearities inherent in the system is known, then transformations may be applied to linearise the data although, as a general rule, use of transformations adds to the complexity of the problem. The more obvious examples of transformations are:

- Logarithmic, for rates of reaction which vary exponentially with temperature
- Square root, for pressure drops which are proportional to the square of flow
- Reciprocal, for time delays which vary in inverse proportion to flow

102.3 Control Charts

The general principle is that continuous processes should run with minimal variation: if variation occurs then it is either due to chance about which nothing can be done, or else it is attributable to some underlying cause which can be detected and eliminated. Having identified the variables of interest, gathered and screened the data, the underlying trends and patterns can be observed. Shewart charts, commonly referred to as control charts, are the most widely used means of depicting and interpreting the data. Three control charts are considered in which average, moving average and spread are plotted as functions of time.

The assumption fundamental to all control charts is that the process from which the data is obtained is under statistical control (said to be well behaved) which means that the variability of the process is both known, or can be determined, and stable. If a process is in statistical control, the distribution of measurements is normal (bell shaped) and symmetrical about the mean, or approximately so. It is also assumed that the samples (measurements) taken are independent of each other, that the frequency of sampling is sufficient to detect changes and that the process of sampling is of sufficient duration to observe trends and patterns.

Irrespective of the type of control chart, the samples are grouped into groups of equal numbers with all groups being weighted equally. The point of grouping is to average out short term variability between samples. The group size m must be such that variations occur between the groups rather than within them. It is essential that grouping be done on a rational basis: for slow processes grouping may have to be on a batch, run, shift or daily basis.

Control charts make use of control limits which have their drawbacks. It is quite possible, through the vagaries of sampling and the inherent variability of the process, that:

- There will be outliers, *i.e.* values beyond the control limits, even though there has been no change in the underlying process.
- Values will lie within the control limits despite there having been a real change in the process.

Choice of the control limits is a compromise between these two scenarios.

102.4 Average Control Charts

These are most useful when there are large group sizes and changes occur slowly. Successive values of the measurement of a single variable are grouped together. For example, the first m measurements form the first group, the next m measurements the second group, and so on until n groups are formed. Thus there are 'm.n' original samples.

The control chart is constructed as follows:

1. Make estimates of the mean $\bar{x}$ and of the variance σ^2 of the measurement. Provided that m.n is large enough the true mean and variance may be estimated from Equations 82.2 and 82.3. Next estimate the standard error of the groups according to:

$$\hat{\sigma}_e = \frac{\sigma}{\sqrt{m}} \qquad (102.1)$$

This is the standard deviation of the distribution of the average value of the n groups. Note that, because of the effect of grouping, the standard error of the groups is less than the standard deviation of the population as a whole.

2. Choose the scale of the chart such that:
   - The mean $\bar{x}$ is fairly central.
   - The scale covers approximately ±4 standard errors, i.e. ±4$\hat{\sigma}_e$, about $\bar{x}$.
3. Mark the control limits as follows:
   - Upper and lower action limit (UAL and LAL) at ±3$\hat{\sigma}_e$ about $\bar{x}$.
   - Upper and lower warning limit (UWL and LWL) at ±2$\hat{\sigma}_e$ about $\bar{x}$.
4. Plot the mean of each group as a function of time. It is normal to continue plotting beyond the first n groups.

An example of an average control chart is as depicted in Figure 102.1.

Interpretation of the control chart is fairly straightforward. If any of the following rules apply, it may be taken as evidence that there is a change in the variability of the data and that some underlying cause must be present:

- One point lies outside either action limit.
- Two successive points lie outside the same warning limit.

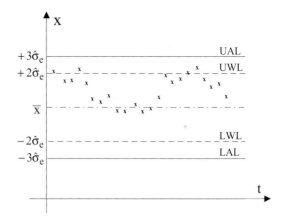

Fig. 102.1 An average control chart

- Seven successive points lie on one side of the mean.
- Seven successive points consistently either increase or decrease.

False alarms may be caused by these seven point rules and, if necessary, the rules may be relaxed or omitted.

102.5 Moving Average Control Charts

These are most useful when there are small group sizes and changes occur quickly. At each sampling instant j, evaluate the mean of the group of the last m values:

$$\bar{x}_j = (x_j + x_{j-1} + x_{j-2} + \cdots + x_{j-m+1})/m \qquad (102.2)$$

The oldest sample is then discarded, the newest included and the average moves on:

$$\bar{x}_{j+1} = (x_{j+1} + x_j + x_{j-1} + \cdots + x_{j-m+2})/m$$

The moving average control chart is constructed and interpreted in exactly the same way as described above for average control charts.

Example

Consider again some of the data used in the example in Chapter 83 on multiple linear regression.

Table 102.1 Moving average for refractive index

Time: t (h)	8.5	9.0	9.5	10.0	10.5	11.0	11.5
Refractive index: r	1.533	1.428	1.567	1.496	1.560	1.605	1.487
Moving average	–	–	1.509	1.497	1.541	1.554	1.551

Assuming groups of size $m = 3$ yields the values shown in Table 102.1.

Noting that there needs to be substantially more data to be meaningful, its summary statistics are nevertheless estimated to be:

$$\bar{r} = 1.525 \qquad \sigma^2 = 0.00352 \qquad \hat{\sigma}_e = 0.0343$$

Hence the action and warning limits are at ± 0.103 and ± 0.069 respectively about 1.525.

102.6 Spread Control Charts

These are constructed as follows:

1. Estimate the mean spread (or range) of the data $\bar{R}$. Again it is assumed there are n groups of m samples. The spread of a group is the difference between the largest and smallest of the m values within that group. The mean spread is the average of the spreads for all n groups.
2. Choose the scale of the chart such that its range goes from zero to twice $\bar{R}$.
3. Mark the control limits as follows:
   - Lower and upper action limits at $L_1.\bar{R}$ and $L_2.\bar{R}$ respectively
   - Lower and upper warning limits at $L_3.\bar{R}$ and $L_4.\bar{R}$ respectively

   where L_1, L_2, L_3 and L_4 are as defined in Table 102.2 according to the group size m.
4. Plot the range of each group as a function of time. It is normal to continue plotting beyond the first n groups.

An example of a spread control chart is as depicted in Figure 102.2.

Interpretation of the control chart for spreads is essentially the same as for averages. Large ranges are indicative of an increase in process spread. A range below the lower action limit probably indicates either that the spread has reduced and the chart needs rescaling, or that there is an instrumentation failure.

Table 102.2 Action and warning limits vs group size

Group size m	LAL L_1	UAL L_2	LWL L_3	UWL L_4
2	0.00	4.12	0.04	2.81
3	0.04	2.99	0.18	2.17
4	0.10	2.58	0.29	1.93
5	0.16	2.36	0.37	1.81
6	0.21	2.22	0.42	1.72
7	0.26	2.12	0.46	1.66
8	0.29	2.04	0.50	1.62
9	0.33	1.99	0.52	1.58
10	0.35	1.94	0.54	1.55

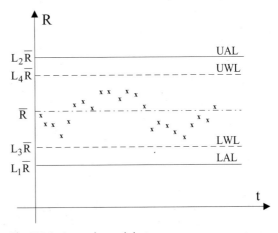

Fig. 102.2 A spread control chart

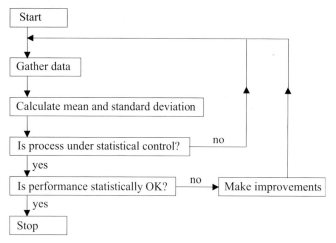

Fig. 102.3 The methodology underlying the six sigma approach

102.7 Six Sigma

Six sigma is a methodology for improving quality, its origins being in the manufacturing industries where the concept of production lines with defective parts is relevant. It consists of a systematic monitoring and review of production, resulting in incremental improvement of quality by eliminating mistakes and/or other causes of defect. The objective is to adjust performance such that the variance of a population of parts falls within a specified tolerance about some target value. For the process industry it is more appropriate to think in terms of some plant with continuous measurements which are sampled. The methodology underlying the six sigma approach is depicted in Figure 102.3 and is self-explanatory.

A simple example illustrates the principle. Suppose that a temperature is used indirectly as a measure of a quality. The specification is a temperature of 120 ±3°C. Presume that the process is under statistical control. The temperature is sampled and recorded over a period of time and the mean and standard deviation evaluated. The mean is found to be at 120°C but with a standard deviation of 1.5°C. Assuming a normal distribution it can be seen that the tolerance of ±3°C corresponds to ±2σ. Thus, from Table 82.2, some 4.5% of the samples lie outside the tolerance. Since the process being sampled is continuous, it is obvious that the quality is off-spec for some 4.5% of the time.

There is clearly scope for improvement. Typically the calibration of the instrumentation would be checked, the control loops retuned and the operating procedures tightened up. More data is gathered and the standard deviation is found to have been reduced to 0.75°C. The tolerance of ±3°C now corresponds to ±4σ. Again referring to Table 82.2, it can be seen that the product is now off-spec for only 0.01% of the time. That is probably good enough for most purposes: the improvement is depicted in Figure 102.4.

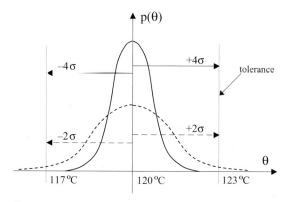

Fig. 102.4 Improvement in quality indicated by temperature measurement

Notwithstanding arguments about the accuracy of the measurements, if the standard deviation could be reduced to 0.5°C, the tolerance would correspond to $\pm 6\sigma$ and the product would be expected to be off-spec for only 2×10^{-7}% of the time. It is important to appreciate that whilst $\pm 6\sigma$ is the target for six sigma methodology, it is only a target: achieving a variance such that $\pm 6\sigma$ lies within the process tolerance may simply be unachievable.

102.8 Capability Indices

There are several so called capability indices used in six sigma defined as follows:

$$Cp = (USL - LSL)/6\sigma$$
$$Cpu = (USL - Mean)/3\sigma$$
$$Cpl = (Mean - LSL)/3\sigma$$
$$Cpk = Min(Cpu, Cpl)$$

where USL and LSL are the upper and lower specification (or service) limits respectively. Put differently, USL-LSL=2T where the tolerance of $\pm T$ on the specification may be thought of as being equivalent to upper and lower control limits. Note that the value of σ used in determining the indices is that of the data gathered and not the value necessary to achieve the desired specification. It can be seen by substitution that:

$$Cp = \frac{(USL - LSL)}{6\sigma}$$
$$= \frac{(USL - mean) + (mean - LSL)}{6\sigma}$$
$$= \frac{3\sigma(Cpu + Cpl)}{6\sigma} = \frac{Cpu + Cpl}{2}$$

Thus if (and only if) the distribution is normal and symmetrical, then:

$$Cp = Cpu = Cpl = Cpk$$

The Cp index is a measure of the spread (variance) of the process data in relation to the tolerance whereas the Cpk index is a measure of both the spread and the mean (setting) of the process data. They are relatively easy to interpret:

- Values for both Cp and Cpk greater than 1 indicate that a spread of $\pm 3\sigma$ lies within the tolerance or specification limits.
- As the value of σ decreases and the spread contained within the tolerance approaches $\pm 6\sigma$, the values of Cp and Cpk tend towards a maximum value of 2.
- Values less than 1 indicate that the process is not capable of achieving the specified tolerances.
- The closer Cpk is to Cp the less the difference between the mean of the process and centre of the distribution.

Few distributions are perfectly normal (Gaussian, bell shaped): there is often a bias or drift on the mean. For six sigma purposes, a positive shift of 1.5σ is factored into the probability statistics. This is depicted in the distribution of Figure 102.5. There is no theoretical basis for this: it is just that a factor of 1.5σ has been found empirically to be appropriate.

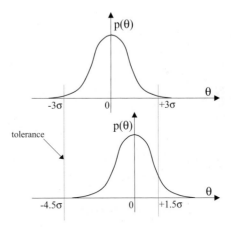

Fig. 102.5 Bias of 1.5σ factored into probability statistics

The basic spread is of $\pm 3\sigma$. If the tolerance of $\pm T$ is fixed but the distribution is shifted to the right by 1.5σ, this results in a left hand tail beyond -4.5σ and a right hand tail beyond $+1.5\sigma$. Referring to Table 82.2, in which the data corresponds to the sum of two identical tails, it can be seen that the percentage of samples in the left hand tail is negligible and in the right hand tail is 13.36/2 = 6.68%. In six sigma speak that represents 66,800 defects per million opportunities (DPMO) as shown in the first row of Table 102.3.

Table 102.3 lists DPMO values and corresponding Cp and Cpk values for various spreads, assuming a 1.5σ shift. Note the oft quoted figure of 3.4 DPMO for $\pm 6\sigma$.

Table 102.3 DPMO and capability indices for six sigma

Cp	Cpk	Number of standard deviations	DPMO
1.00	0.50	$\pm 3.0\sigma$	66,800
1.17	0.67	$\pm 3.5\sigma$	22,760
1.33	0.83	$\pm 4.0\sigma$	6,200
1.50	1.00	$\pm 4.5\sigma$	1,300
1.67	1.17	$\pm 5.0\sigma$	320
1.83	1.33	$\pm 5.5\sigma$	30
2.00	1.50	$\pm 6.0\sigma$	3.4

102.9 Multivariate SPC

The control charts considered thus far depict univariate statistics and are two dimensional. For multivariate systems there is the obvious constraint of only having two or three dimensions to depict patterns. However, this can be obviated by depicting the variables concerned indirectly by means of their principal components, occasionally referred to as latent variables.

For example, consider a system for which there are two principal components, y_1 and y_2. Assume that these are dominant components, accounting for some 80% of the variability of the data, and are defined as follows:

$$y_1 = a_1.x_1 + a_2.x_2 + \cdots + a_p x_p \quad (101.8)$$

$$y_2 = b_1.x_1 + b_2.x_2 + \cdots + b_p x_p \quad (101.10)$$

in which the p variables are all functions of time. First n sets of data are required, each of which is representative of the process under normal operating conditions. From this reference data set the loadings a_j and b_j are determined as explained in Chapter 101. The fitted values, otherwise known as scores, for both of the principal components may then be calculated for each set of data:

$$\underline{y}_1 = [y_{11}\ y_{12}\ \cdots\ y_{1n}]$$
$$\underline{y}_2 = [y_{21}\ y_{22}\ \cdots\ y_{2n}] \quad (102.3)$$

The scores may be plotted against each other in two dimensions resulting in a so-called scores plot as depicted in Figure 102.6.

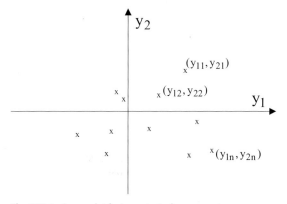

Fig. 102.6 Scores plot for two principal components

Note that because the variables are in deviation form, the mean having been subtracted from each, the distribution is roughly symmetrical about the origin. As time progresses, the principal components may be determined for successive sets of data and plotted:

$$\underline{y}_1 = [y_{11}\ y_{12}\ \cdots\ y_{1n}\ y_{1n+1}\ y_{1n+2}\ \cdots]$$
$$\underline{y}_2 = [y_{21}\ y_{22}\ \cdots\ y_{2n}\ y_{2n+1}\ y_{2n+2}\ \cdots]$$

Plotting the scores of one principal component against another recognises that many of the measurements on a plant are highly correlated and thus different combinations of the variables define the same underlying events. Provided the operating conditions remain relatively constant, the scores should continue to lie within the same region of the plot and indeed are likely to exhibit clustering close to the origin. If some deterioration of performance occurs, the variability of the process data will change and the scores will start to drift and begin to appear elsewhere. However, in the event of a sudden abnormality occurring, or the mode of operation changing, the position of the cluster moves: this is often emphasised by means of colour coding.

Control limits may be applied in multivariate analysis just as in univariate analysis. In general the limits are elliptical as shown in Figure 102.7.

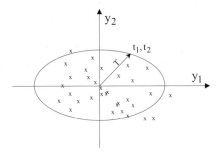

Fig. 102.7 Elliptical control limits for scores plots

Clearly the radius of the ellipse must be based upon the reference data set. The formulae used to determine the radius depend upon whether the data set is big enough to count as a population. If n < 30 the data set counts as a sample, the radius is known as Mahalanobis' distance and a correction has to be introduced to account for the sampling. Otherwise, if n ≥ 30, it counts as a population and the radius is referred to as Hotelling's distance T.

Consider the locus of an arbitrary point on an ellipse with coordinates $\underline{t} = \begin{bmatrix} t_1 & t_2 \end{bmatrix}^T$.

Hotelling's distance is defined as follows:

$$T^2 = \underline{t}^T S^{-1} \underline{t} = \begin{bmatrix} t_1 & t_2 \end{bmatrix} \begin{bmatrix} \sigma_1^2 & \sigma_{12}^2 \\ \sigma_{21}^2 & \sigma_2^2 \end{bmatrix}^{-1} \begin{bmatrix} t_1 \\ t_2 \end{bmatrix} \quad (102.4)$$

where S is the covariance matrix of the vectors of fitted values given by Equation 102.3 for the reference data set. The fact that this is of the general form of the equation of an ellipse can be seen by expansion of Equation 102.4. Noting that σ_{12} and σ_{21} are equal:

$$T^2 = \frac{\sigma_2^2 . t_1^2 - 2\sigma_{12}^2 . t_1 t_2 + \sigma_1^2 . t_2^2}{\sigma_1^2 \sigma_2^2 - \sigma_{12}^4} \quad (102.5)$$

Hotelling's distance is set with a 5% significance (or 95% confidence) level being used for the warning limit. This means, assuming the distribution of the scores is normal, that 5% of the scores are likely to fall outside that limit. Using the statistical χ^2 table for a bivariate distribution (two degrees of freedom) the required setting is 5.991. Whence:

$$5.991 = \underline{t}^T S^{-1} \underline{t} \quad (102.6)$$

The ellipse is thus the locus of points that satisfy this equation. The intercepts with the principal component axes y_1 and y_2 are found by putting t_2 and t_1 to zero respectively. The warning limit is found by solving for t_1 and t_2 between these maxima and is depicted as the inner ellipse in Figure 102.8.

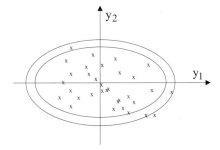

Fig. 102.8 Concentric elliptical action and warning limits for scores plots

Hotellings distance is set with a 1% significance (or 99% confidence) level for the action limit for which the χ^2 setting is 9.210. Whence the locus of the action limit is found by solving:

$$9.210 = \underline{t}^T S^{-1} \underline{t} \quad (102.7)$$

which is depicted as the outer ellipse in Figure 102.8.

Interpretation of the scores plots is much the same as for univariate control charts. If any of the following rules apply, it may be taken as evidence that there is a change in the variability of the data and that some underlying cause must be present:

- One point lies outside the action ellipse.
- Two successive points lie outside the warning ellipse.

- Seven successive points consistently either increase or decrease.

In the event of the reference data set having been standardised (subtraction of the mean *and* division by the standard deviation) prior to establishing the loadings, then the covariance matrix of Equation 102.4 is replaced by the correlation matrix as follows:

$$T^2 = \underline{t}^T P^{-1} \underline{t} = \begin{bmatrix} t_1 & t_2 \end{bmatrix} \begin{bmatrix} 1 & \rho_{12} \\ \rho_{21} & 1 \end{bmatrix}^{-1} \begin{bmatrix} t_1 \\ t_2 \end{bmatrix}$$

$$= \frac{t_1^2 - 2\rho_{12}.t_1 t_2 + t_2^2}{1 - \rho_{12}^2} \qquad (102.8)$$

Otherwise, for standardised data, the ellipses corresponding to action and warning limits are set and interpreted in exactly the same way.

102.10 Geometric Modelling

This is a data driven visual approach to multivariate SPC. It is based upon the use of multiple parallel axes, there being one axis for each variable as depicted in Figure 102.9.

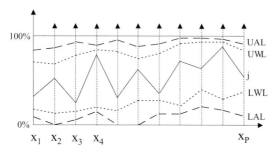

Fig. 102.9 Use of multiple parallel axes for displaying data

Suppose there are p variables of interest, each of which has an ordinate (y axis) equally spaced along the abscissa (x axis) as depicted by the dotted vertical lines. It is not uncommon to have as many as 20 such axes on a single display. Each is scaled, normally on a percentage basis, such that the ordinates are all of the same length. The ordinates are typically annotated with the range in engineering units. Along each ordinate are marked, for example, warning and action limits. These are joined up on a like-for-like basis: the broken lines connecting the upper (UWL) and lower (LWL) warning limits and the dashed lines connecting the upper (UAL) and lower (LAL) action limits respectively. The UWL and LWL polygonal lines form an operating envelope and the UAL and LAL polygonal lines form an action envelope: it can be expected that these envelopes are nested.

At any point in time, the current values of all the variables may be depicted by a single polygonal line, as depicted by the solid line in Figure 102.9. It is not uncommon to have several hundred such lines of data on a single display. The expectation is that they will all lie within the operating envelope. In a real time context, for each new line displayed the oldest line is deleted. As time progresses, a forest of V shaped patterns emerge. Providing that operating conditions do not change, the majority of the polygonal lines will converge on a fairly narrow band on each of the axes as depicted in Figure 102.10.

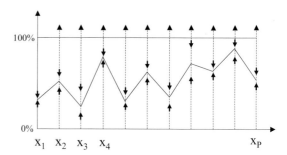

Fig. 102.10 Convergence of polygonal lines within bands on each parallel axis

It is relatively easy to spot trends: as soon as operating conditions change, or a fault occurs, or an abnormal situation develops, several of the V shapes will start to diverge along their axes or converge on a different section as depicted in Figure 102.11.

Thus, in a very compact form, the geometric model provides the operator with an overview of many variables depicting both normal and abnormal conditions. This overview may be enhanced with shading: typically newer polygonal lines are

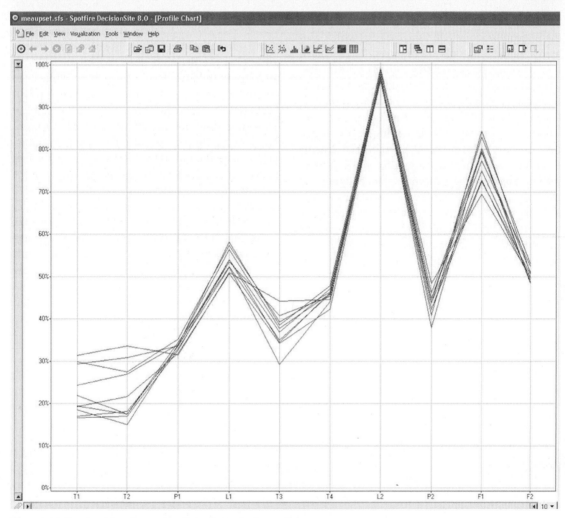

Fig. 102.11 The onset of an abnormal condition depicted using parallel axes

in a brighter shade than older ones. Similarly the overview may be enhanced by colour: polygonal lines which lie outside the operating envelope, usually only partially rather than wholly outside, are identified by a different colour, and those which lie outside the action envelope by yet another colour.

However, it is one thing spotting a trend but quite another thing to interpret the trend and/or decide what to do about it. Because there is no model *per se*, the cause and effect relationships are not necessarily known *a priori*. Bearing in mind that a geometric model is used in a multivariable context, there are likely to be several MVs that will affect each CV. How, by inspection, would an operator know which MV to adjust to correct a CV that was going adrift? Likewise, each MV is likely to affect several CVs. How would the operator know by how much to adjust one MV so as not to adversely affect other CVs? There is much scope for confusion.

One of the more powerful aspects of the use of geometric modelling is in data mining: that is the off-line analysis of historic data, as opposed to the interpretation of real time data, for which geometric modelling tools are required. Thus what-if scenarios may be explored by reusing historic data without compromising current operations. Of particular interest is the effect of tightening and/or relaxing the control limits. It is usually fairly obvious where there is scope for tightening up the warning limits, *i.e.* reducing the UWL and/or increasing the LWL. More difficult to predict, but relatively easy to explore, is the combined effect of tightening some operating limits and relaxing others. A typical scenario concerns key quality measurements: by adjusting the UWL and/or LWL on other process variables, the scope for increasing the number of polygonal lines that satisfy the quality constraints may be established.

102.11 Comments

Fundamental to the effective use of the techniques of SPC is the decision about which variables to monitor in the first place. This can only come from an understanding of the underlying cause and effect relationships, especially from a process point of view, noting that the effect of closed loop control is to strengthen specific relationships. Following this are the non-trivial tasks of data screening and estimating the true mean and variances. In the case of scores plots there may well also be the need to time shift some of the variables to allow for delays in the process dynamics. As will have become apparent, the control charts themselves are surprisingly easy to construct and relatively simple to interpret. For multivariate relationships, the use of principal components and scores plots enables trends in the underlying relationships to be understood and the use of geometric modelling aids their visualisation.

Linear Programming

103.1 Concept of Linear Programming
103.2 Worked Example on LP
103.3 The Simplex Method
103.4 Tableau Representation
103.5 Mixed Integer Linear Programming
103.6 Implicit Enumeration
103.7 Branch and Bound
103.8 Comments

This chapter is the first of several that relate to optimisation. Linear programming (LP) and the handling of constraints is the basis of many optimisation techniques and so provides a sensible starting point. LP is introduced by means of a worked example which is then used as a vehicle for explaining the so-called Simplex method. This naturally leads onto consideration of integer programming (IP) and mixed integer linear programming (MILP). Subsequent chapters cover unconstrained optimisation, quadratic programming and real-time optimisers.

Optimisation is closely allied to process control on the basis that wherever there is the need for automatic control there must be scope for optimisation. Optimisers typically use a control system as a platform because of their dependence upon the control systems' ability to measure and manipulate variables in real-time. Indeed, there is a class of optimiser known as the dynamic optimising controller, covered in Chapter 106, in which a model predictive controller is integrated with a real time optimiser. Because of the scope for realising benefits, even marginal ones, there has been massive investment in optimisation. There has also been much academic interest, drawing upon techniques developed in the field of operations research and resulting in many publications. Two texts to which the reader's attention is drawn are those by Hillier (2004) and, in particular, by Edgar et al. (2001).

103.1 Concept of Linear Programming

Any optimisation problem of a linear nature may generally be cast in the form of an LP problem as follows. The objective is to maximise some objective function $J(\underline{x})$ of the form:

$$J(\underline{x}) = C.\underline{x} = c_1.x_1 + c_2.x_2 + \ldots + c_p.x_p \quad (103.1)$$

subject to the functional constraint $A.\underline{x} \leq \underline{b}$ where

$$\begin{bmatrix} a_{11} & a_{12} & \cdots & a_{1p} \\ a_{21} & a_{22} & \cdots & a_{2p} \\ \vdots & \vdots & \ddots & \vdots \\ a_{n1} & a_{n2} & \cdots & a_{np} \end{bmatrix} \begin{bmatrix} x_1 \\ x_2 \\ \vdots \\ x_p \end{bmatrix} \leq \begin{bmatrix} b_1 \\ b_2 \\ \vdots \\ b_n \end{bmatrix} \quad (103.2)$$

and also subject to the non-negativity constraints that $\underline{x} \geq 0$ where:

$$x_1 \geq 0, x_2 \geq 0, \cdots, x_p \geq 0 \quad (103.3)$$

The variables $x_1, x_2, \ldots, x_p$ are referred to as the decision variables.

The objective function is also referred to as the cost function because it is invariably some profit that has to be maximised. Otherwise it may be some parameter, such as energy consumption, that has to be minimised. The minimisation problem can be cast in LP form by introducing negative signs as appropriate. Functional constraints are largely related to production issues, such as limits on throughput, processing or storage capacity.

Several assumptions are implicit in LP problems cast in the above form:

- Proportionality: that the cost coefficients, *i.e.* the elements of C, are uniform. In practice this means that any start up or shut down costs are the same as the running costs.
- Additivity: that the interactions between activities can be adequately described by the elements of A and B.
- Certainty: that the elements of A, B and C are constant.
- Divisibility: that non-integer values for the decision variables are acceptable.

103.2 Worked Example on LP

The following worked example, based upon that described by Edgar et al. (2001), is a good way to appreciate the formulation of an LP problem. Consider a separation plant in which two crude oil streams are separately distilled into various constituents which are then combined into four distinct product streams, as depicted in Figure 103.1.

The cost of the crude oils as raw materials and their processing costs, together with their separa-

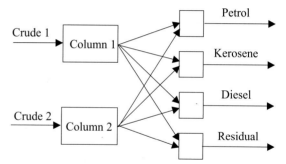

Fig. 103.1 A separation plant with two feed and four product streams

tions, are as listed in Table 103.1. Also indicated is the value and maximum throughput for each of the product streams.

Let x_1 be the flow rate (m^3/h) of Crude 1
x_2 Crude 2
f_3 Petrol
f_4 Kerosene
f_5 Diesel
f_6 Residual

The cost function in this case is the overall operating profit:

Profit (£/h) = Income − Raw materials costs
            − Processing costs

Income = $150.f_3 + 100.f_4 + 75.f_5 + 25.f_6$

Raw materials costs = $100.x_1 + 65.x_2$

Processing costs = $10.x_1 + 15.x_2$

$J(\underline{x}) = 150.f_3 + 100.f_4 + 75.f_5 + 25.f_6$
$\quad\quad - 110.x_1 - 80.x_2$

Table 103.1 Costs and constraints for separation plant

	Separation %v/v Crude 1	Separation %v/v Crude 2	Value £/m³	Maximum m³/h
Petrol	80	45	150	150
Kerosene	5	10	100	13
Diesel	10	35	75	40
Residual	5	10	25	–
Cost of crude (£/m³)	100	65	–	–
Processing costs (£/m³)	10	15	–	–

Mass balances for each of the constituents yield the following equality (functional) constraints:

$$0.8 \cdot x_1 + 0.45 \cdot x_2 = f_3$$
$$0.05 \cdot x_1 + 0.1 \cdot x_2 = f_4$$
$$0.1 \cdot x_1 + 0.35 \cdot x_2 = f_5$$
$$0.05 \cdot x_1 + 0.1 \cdot x_2 = f_6$$

The dimensionality of the problem can be reduced by substituting for f_3 to f_6 in the cost function:

$$J(\underline{x}) = 23.75 \cdot x_1 + 26.25 \cdot x_2 \quad (103.4)$$

There is a set of inequality (functional) constraints imposed by maximum throughput considerations:

$$0.8 \cdot x_1 + 0.45 \cdot x_2 \leq 150 \quad (103.5)$$
$$0.05 \cdot x_1 + 0.1 \cdot x_2 \leq 13 \quad (103.6)$$
$$0.1 \cdot x_1 + 0.35 \cdot x_2 \leq 40 \quad (103.7)$$

And finally, there is set of constraints due to the fact that consumption of raw materials must be non-negative:

$$x_1 \geq 0, \qquad x_2 \geq 0$$

The objective is to maximise $J(x)$ subject to the inequality and non-negativity constraints. The inequality constraints are depicted graphically in Figure 103.2 in which Equations 103.5–103.7 are plotted as if they are equalities. The non-negativity constraints correspond to the positive x_1 and x_2 axes. This yields a feasible solution space: the region bounded by the above equations and axes within which all the constraints are satisfied.

The solution to the problem may be found by rearranging Equation 103.4 into the form:

$$x_2 = -0.905 \cdot x_1 + 0.0381 \cdot J(\underline{x}) \quad (103.4)$$

Noting that for any value of $J(\underline{x})$ this has a slope of -0.905, it may be plotted in the feasible solution space for any arbitrary value of $J(x)$ as depicted in Figure 103.3.

It is evident that moving this plot in such a way that its intercept with the x_2 axis increases, but maintaining a slope of -0.905, causes the value

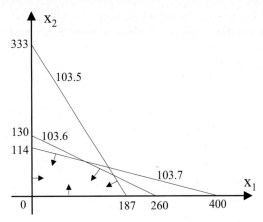

Fig. 103.2 Plot of inequality and non-negativity constraints

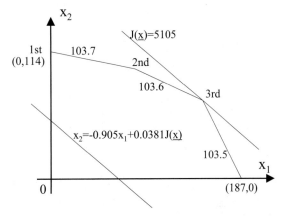

Fig. 103.3 Plot of the cost function in the feasible solution space

of $J(\underline{x})$ to rise. By inspection it can be seen that the maximum profit is where Equations 103.5 and 103.6 intersect, which is when:

$$x_1 = 159.1 \text{ m}^3/\text{h}, \quad x_2 = 50.43 \text{ m}^3/\text{h}$$
$$\text{and} \quad J(\underline{x}) = 5{,}103 \text{ £/h}$$

The values for x_1 and x_2 come from treating Equations 103.5 and 103.6 as equalities and solving them as any two equations in two unknowns. These values exactly satisfy the first two functional constraints and lie within the third.

Notice that the optimum lies at a corner in the feasible solution space. For any well posed problem with two decision variables, x_1 and x_2 in the case of the separation plant, the optimum always occurs at the intersection of two or more functional constraint bounds.

For completeness, there are three other suboptimal corner feasible solutions: at the intersection of the x_1 axis with Equation 103.5, at the intersection of the x_2 axis with Equation 103.7, and at the intersection of Equations 103.6 and 103.7. In the latter case, treating the equations as equalities and solving them yields:

$$x_1 = 73.33, \quad x_2 = 93.33$$
$$\text{and} \quad J(\underline{x}) = 4192 \, £/h$$

There are three categories of ill posed problems:

- Problems in which there are an infinite number of solutions, essentially because the equation for the cost function is parallel to and coincides with one or other of the functional constraint bounds.
- Ones in which the solution is indeterminate, typically infinite in value, because the feasible solution space is open ended, say due to a functional constraint being parallel to one of the axes, or two adjacent constraints diverging.
- Where there is no solution because the constraint bounds are such that a feasible solution space is not formed.

103.3 The Simplex Method

The notion of corner feasible solutions, *i.e.* the optimum occurring at the intersection of two or more constraint bounds, extrapolates to more complex LP problems involving many decision variables and subject to multiple constraints. The Simplex method is an algebraic technique of finding optimum solutions to such problems. The technique is iterative in that it systematically locates the corners of the feasible solution space, evaluating the cost function $J(\underline{x})$ from one corner to the next, until the optimum is found. Whilst the real power of the Simplex method is in handling complex LP problems, it is illustrated here in relation to the separation plant previously considered. Not only is this easy to visualise but the graphical solution already developed provides a basis for explaining the Simplex method.

Step 0

The objective remains, as before, to maximise $J(\underline{x})$ where:

$$J(\underline{x}) = 23.75.x_1 + 26.25.x_2 \quad (103.4)$$

The inequality constraints of Equations 103.5–103.7 are augmented: that is, recast as equality constraints by the introduction of slack variables x_3, x_4 and x_5 as follows:

$$0.8.x_1 + 0.45.x_2 + x_3 = 150 \quad (103.8)$$
$$0.05.x_1 + 0.1.x_2 + x_4 = 13 \quad (103.9)$$
$$0.1.x_1 + 0.35.x_2 + x_5 = 40 \quad (103.10)$$

It is evident that $x_3 \leq 150$, $x_4 \leq 13$ and $x_5 \leq 40$.

The origin is the starting point for the Simplex method, for which the variables are deemed to be either non-basic (independent) or basic (dependant) as follows:

$$\text{Non-basic:} \quad x_1 \text{ and } x_2.$$
$$\text{Basic:} \quad x_3, x_4 \text{ and } x_5.$$

Setting the non-basic variables $x_1 = x_2 = 0$, which locates the origin, yields:

$$x_3 = 150, \quad x_4 = 13, \quad x_5 = 40 \quad \text{and} \quad J(\underline{x}) = 0.$$

Step 1

This step involves finding the first corner feasible solution relative to the origin. The issue is whether to move along the x_1 axis or the x_2 axis. Inspection of Equation 103.4 reveals that $J(\underline{x})$ is more sensitive to (that is, maximised more by) an increase in x_2 rather than in x_1. Thus x_2 becomes a basic variable. The question now is which of x_3, x_4 and x_5 should be deemed to be non-basic? This is decided according to which of Equations 103.8–103.10 is the most limiting constraint with regard to the first corner.

Dividing the RHS of Equations 103.8–103.10 by their respective coefficients in x_2 yields:

$$\frac{150}{0.45} \approx 333, \quad \frac{13}{0.1} = 130 \quad \frac{40}{0.35} \approx 114$$

This essentially determines the intercept of the equations with respect to the x_2 axis. The most limiting constraint is that ratio which has the smallest positive value, 114 in this case, which corresponds to Equation 103.10 for which x_5 is the slack variable. Thus x_5 is deemed to become non-basic resulting in:

$$\text{Non-basic:} \quad x_1 \text{ and } x_5$$
$$\text{Basic:} \quad x_2, x_3 \text{ and } x_4.$$

This is consistent with Figure 103.3, inspection of which shows that the first corner clockwise from the origin is at the intersection of the x_2 axis and the equality constraint of Equation 103.7 for which the slack variable is x_5. The notion of non-basic variables being independent is consistent with the concept of a corner feasible solution. Thus the first corner is at the intersection of the x_2 axis along which x_1 has a value of zero and Equation 103.7 along which the value of x_5 is zero.

The new basic variable x_2 is expressed as a function of the non-basics x_1 and x_5 from Equation 103.10:

$$x_2 = 114.3 - 0.2857.x_1 - 2.857.x_5$$

Eliminate x_2 from Equations 103.4, 103.8 and 103.9 respectively yields:

$$J(\underline{x}) = 16.25.x_1 - 75.x_5 + 3000 \quad (103.11)$$
$$0.6714.x_1 + x_3 - 1.286.x_5 = 98.57 \quad (103.12)$$
$$0.02143.x_1 + x_4 - 0.2857.x_5 = 1.571 \quad (103.13)$$
$$0.2857.x_1 + x_2 + 2.857.x_5 = 114.3 \quad (103.14)$$

Note that these equations are still in the order of the definition of the slack variables. Thus Equation 103.12 is associated with slack variable x_3, 103.13 with x_4 and 103.14 with x_5. It is important that this order be maintained throughout the Simplex method.

By setting $x_1 = x_5 = 0$, these can be solved to yield the value of $J(x)$ at the first corner:

$$x_2 = 114.3, \quad x_3 = 98.57, \quad x_4 = 1.571$$
$$\text{and} \quad J(\underline{x}) = 3000.$$

In moving from the origin to the first corner there has been an increase of £3000/h in the cost function.

Step 2

The aim of this step is to find which of the corner feasible solutions adjacent to the first maximises the cost function. Inspection of Equation 103.11 reveals that $J(\underline{x})$ is more sensitive to a change in x_1 than in x_5. Thus x_1 becomes a basic variable. Dividing the RHS of Equations 103.12–103.14 by their respective coefficients in x_1 yields:

$$\frac{98.57}{0.6714} \approx 147, \quad \frac{1.571}{0.02143} \approx 73, \quad \frac{114.3}{0.2857} \approx 400$$

Clearly the smallest positive ratio is 73 so Equation 103.13 is the most limiting constraint for which x_4 is the slack variable. This is deemed to be non-basic resulting in:

$$\text{Non-basic:} \quad x_4 \text{ and } x_5$$
$$\text{Basic:} \quad x_1, x_2 \text{ and } x_3.$$

This is consistent with Figure 103.3, inspection of which shows that the second corner is at the intersection of Equations 103.6 and 103.7 for which the slack variables are x_4 and x_5.

The new basic variable x_1 is expressed as a function of the non-basics x_4 and x_5 from Equation 103.13:

$$x_1 = 73.33 - 46.67.x_4 + 13.33.x_5$$

Eliminate x_1 from Equations 103.11, 103.12 and 103.14 respectively yields:

$$J(\underline{x}) = -758.3.x_4 + 141.7.x_5 + 4192 \quad (103.15)$$
$$x_3 - 31.33.x_4 + 7.667.x_5 = 49.33 \quad (103.16)$$
$$x_1 + 46.66.x_4 - 13.33.x_5 = 73.31 \quad (103.13)$$
$$x_2 - 13.33.x_4 + 6.666.x_5 = 93.34 \quad (103.17)$$

Setting $x_4 = x_5 = 0$ and solving yields the value of $J(\underline{x})$ at the second corner:

$$x_1 = 73.31, \quad x_2 = 93.33, \quad x_3 = 49.33$$
$$\text{and} \quad J(\underline{x}) = 4192.$$

Moving from the first to the second corner has increased cost function to £4,192/h.

Step 3

This establishes the third corner feasible solution. Inspection of Equation 103.15 reveals that $J(\underline{x})$ is more sensitive to a change in x_5 than in x_4. Thus x_5 becomes a basic variable. Dividing the RHS of Equations 103.16, 103.13 and 103.17 by their respective coefficients in x_5 yields:

$$\frac{49.33}{7.667} \approx 6.4, \quad \frac{73.31}{-13.33} \approx -5.5, \quad \frac{93.34}{6.666} = 14$$

Clearly the smallest positive ratio is 6.4 so Equation 103.14 is the most limiting constraint for which x_3 is the slack variable. This is deemed to be non-basic resulting in:

$$\text{Non-basic:} \quad x_3 \text{ and } x_4$$
$$\text{Basic:} \quad x_1, x_2 \text{ and } x_5.$$

This is consistent with Figure 103.3, inspection of which shows that the third corner is at the intersection of Equations 103.5 and 103.6 for which the slack variables are x_3 and x_4.

The new basic variable x_5 is expressed as a function of the non-basics x_3 and x_4 from Equation 103.16:

$$x_5 = 6.435 - 0.1304.x_3 + 4.087.x_4$$

Eliminate x_5 from Equations 103.15, 103.13 and 103.17 respectively:

$$J(\underline{x}) = -18.48.x_3 - 179.3.x_4 + 5103 \quad (103.18)$$
$$0.1304.x_3 - 4.086.x_4 + x_5 = 6.434 \quad (103.16)$$
$$x_1 + 1.739.x_3 - 7.81.x_4 = 159.1 \quad (103.19)$$
$$x_2 - 0.8694.x_3 + 13.91.x_4 = 50.43 \quad (103.20)$$

Setting $x_3 = x_4 = 0$ and solving yields the value of $J(\underline{x})$ at the third corner:

$$x_1 = 159.1, \quad x_2 = 50.43, \quad x_5 = 6.434$$
$$\text{and} \quad J(\underline{x}) = 5103.$$

This is the optimum corner feasible solution at which the maximum profit is £5103/h as deduced previously. Inspection of Equation 103.18 reveals that there is no scope for identifying a further corner for which there is a larger value of $J(\underline{x})$. Since the signs of x_3 and x_4 are both negative, any increase in either variable will lead to a decrease in the cost function.

For any well posed problem the Simplex method is guaranteed to find the optimum, but it does not necessarily do so in the most efficient way. For example, for the above separation plant, progression from the origin around the corners of the feasible solution space to the optimum took three steps using the Simplex method, whereas inspection of Figure 103.3 reveals that only two steps would have been required using an anti-clockwise progression.

103.4 Tableau Representation

The above algebraic approach enables a relatively simple explanation of the Simplex method. In practice, for complex LP problems, matrix based means of representing the problem are used together with the technique of Gaussian elimination. Progression of the optimisation is depicted by means of tableaux, as below, in which the example of the separation plant is used again as a basis for explanation.

Step 0

As before, the initial corner feasible solution is presumed to be at the origin of the two decision variables x_1 and x_2. The initial tableau, corresponding to Equations 103.4 and 103.8 to 103.10, is laid out as depicted in Table 103.2.

Note the structure of the tableau. The equations are in the form of a weighted sum of variables equal to some constant listed in the RHS column. They are identified by the basic variables x_3 to x_5, as listed in the second column, for which purpose J is also deemed to be basic. The tableau is in the so called proper form for Gaussian elimination. Thus each of the columns of coefficients for the basic variables contains a 1 corresponding to its own ba-

103.4 Tableau Representation

Table 103.2 Tableau for initial corner feasible solution

Step	Basic variable	Equation 103.	J	\multicolumn{5}{c}{Coefficients on LHS}	RHS	Ratio
				x_1	x_2	x_3	x_4	x_5		
0	J	4	1	−23.75	−26.25	0	0	0	0	−
	x_3	8	0	0.8	0.45	1	0	0	150	−
	x_4	9	0	0.05	0.1	0	1	0	13	−
	x_5	10	0	0.1	0.35	0	0	1	40	−

sic variable and a 0 elsewhere. The values of the basic variables at the initial corner feasible solution are given in the RHS column. This is equivalent to setting the two non-basic variables x_1 and x_2 to zero and solving the equations for x_3 to x_5.

Step 1

The new set of basic variables has to be established. The entering basic variable corresponds to the largest negative coefficient in the row corresponding to Equation 103.4 for the cost function. Inspection of that row reveals that the largest negative coefficient is −26.25 so the entering basic variable is x_2. The coefficients in Table 103.3 in the column containing the largest negative coefficient are shaded in light grey: this is known as the pivot column.

The leaving basic variable is established by finding the ratios of the values in the RHS column to the values in the pivot column, the ratios being as shown in the last column of Table 103.3. The leaving basic variable is that which corresponds to the equation with the lowest positive ratio. Equation 103.10 has the lowest positive ratio of 114 so the leaving variable is x_5. The coefficients of that row are shaded in light grey: this is known as the pivot row. The value of the coefficient that is common to both the pivot column and the pivot row is known as the pivot value, 0.35 in the case of Table 103.3.

The tableau corresponding to the first corner feasible solution may now be constructed, as depicted in Table 103.4. The leaving basic variable x_5 is replaced by the entering basic variable x_2 in the second column. Note that the order of the equations is maintained.

The elements of the pivot row are divided by the pivot value such that the coefficient of the entering basic variable in the pivot column becomes 1.0, corresponding to Equation 103.14. The other coefficients in the pivot column are made zero by addition of multiples of Equation 103.14 as appropriate. For example, for the cost function, Equation 103.11 is formed by adding 26.25 times Equation 103.14 to Equation 103.4:

$$(J - 23.75 x_1 - 26.25 x_2)$$
$$+ 26.25 \cdot (0.2857 x_1 + x_2 + 2.857 x_3 - 114.3)$$
$$= J - 16.25 x_1 + 0.0 x_2 + 75 x_5 - 3000$$

Similarly, for the basic variable x_3, Equation 103.12 is formed by subtracting 0.45 times Equation 103.14 from Equation 103.8:

Table 103.3 Pivot value for initial corner feasible solution

Step	Basic variable	Equation 103.	J	\multicolumn{5}{c}{Coefficients on LHS}	RHS	Ratio
				x_1	x_2	x_3	x_4	x_5		
0	J	4	1	−23.75	−26.25	0	0	0	0	−
	x_3	8	0	0.8	0.45	1	0	0	150	333
	x_4	9	0	0.05	0.1	0	1	0	13	130
	x_5	10	0	0.1	0.35	0	0	1	40	114

Table 103.4 Tableau for first corner feasible solution

Step	Basic variable	Equation 103.	J	\multicolumn{5}{c}{Coefficients on LHS}	RHS	Ratio
				x_1	x_2	x_3	x_4	x_5		
1	J	11	1	−16.25	0	0	0	75.0	3000	–
	x_3	12	0	0.6714	0	1	0	−1.2857	98.57	–
	x_4	13	0	0.0214	0	0	1	−0.2857	1.571	–
	x_2	14	0	0.2857	1	0	0	2.857	114.3	–

$(0.8.x_1 + 0.45.x_2 + x_3 − 150)$
$− 0.45.(0.2857.x_1 + x_2 + 2.857.x_5 − 114.3)$
$= 0.6714.x_1 + 0.x_2 + x_3 − 1.2857.x_5 − 98.57$

And for the basic variable x_4, Equation 103.13 is formed by subtracting 0.1 times Equation 103.14 from Equation 103.9:

$(0.05.x_1 + 0.1.x_2 + x_4 − 13)$
$− 0.1.(0.2857.x_1 + x_2 + 2.857.x_5 − 114.3)$
$= 0.0214.x_1 + 0.x_2 + x_4 − 0.2857.x_5 − 1.571$

The RHS column of this tableau gives the values of the cost function and the basic variables at the first corner feasible solution. This is equivalent to setting the two non-basic variables x_1 and x_5 to zero and solving the equations for x_2 to x_4.

Step 2

The methodology for finding the new entering and leaving variables is identical to that for Step 1. Thus the entering basic variable is x_1, the leaving basic variable is x_4 and the pivot value is 0.0214 as depicted in Table 103.5.

Dividing the pivot row by 0.0214 results in a coefficient of 1 in the pivot column for the entering basic variable. The other coefficients in the pivot column are made zero by addition of multiples of Equation 103.13 as appropriate. Hence the tableau depicted in Table 103.6, the RHS column of which contains the values of the cost function and basic variables at the second corner feasible solution.

Table 103.5 Pivot value for first corner feasible solution

Step	Basic variable	Equation 103.	J	\multicolumn{5}{c}{Coefficients on LHS}	RHS	Ratio
				x_1	x_2	x_3	x_4	x_5		
1	J	11	1	−16.25	0	0	0	75.0	3000	–
	x_3	12	0	0.6714	0	1	0	−1.2857	98.57	147
	x_4	13	0	0.0214	0	0	1	−0.2857	1.571	73
	x_2	14	0	0.2857	1	0	0	2.857	114.3	400

Table 103.6 Tableau for second corner feasible solution

Step	Basic variable	Equation 103.	J	\multicolumn{5}{c}{Coefficients on LHS}	RHS	Ratio
				x_1	x_2	x_3	x_4	x_5		
2	J	15	1	0	0	0	758.3	−141.7	4192	–
	x_3	16	0	0	0	1	−31.33	7.667	49.33	–
	x_1	13	0	1	0	0	46.66	−13.33	73.31	–
	x_2	17	0	0	1	0	−13.33	6.666	93.34	–

Table 103.7 Pivot value for second corner feasible solution

Step	Basic variable	Equation 103.	\multicolumn{6}{c	}{Coefficients on LHS}	RHS	Ratio
			J	x_1	x_2	x_3	x_4	x_5		
2	J	15	1	0	0	0	758.3	−141.7	4192	−
	x_3	16	0	0	0	1	−31.33	7.667	49.33	6.4
	x_1	13	0	1	0	0	46.66	−13.33	73.31	−5.5
	x_2	17	0	0	1	0	−13.33	6.666	93.34	14

Table 103.8 Tableau for third corner feasible solution

Step	Basic variable	Equation 103.	J	x_1	x_2	x_3	x_4	x_5	RHS	Ratio
3	J	18	1	0	0	18.48	179.3	0	5103	−
	x_5	16	0	0	0	0.1304	−4.086	1	6.434	−
	x_1	19	0	1	0	1.739	−7.81	0	159.1	−
	x_2	20	0	0	1	−0.8694	13.91	0	50.43	−

Step 3

Likewise for Step 3. The entering basic variable is x_5, the leaving basic variable is x_3 and the pivot value is 7.667, as depicted in Table 103.7.

Manipulation as before results in a coefficient of 1 in the pivot column for the entering basic variable and zeros elsewhere, as depicted in Table 103.8.

The RHS column of the tableau depicted in Table 103.8 contains the values of the cost function and basic variables at the third corner feasible solution:

$$J(\underline{x}) = 5103 \quad \text{when} \quad x_1 = 159.1, \; x_2 = 50.43$$
$$\text{and} \quad x_5 = 6.434.$$

It is evident that this is the optimum, the best corner feasible solution, and no further steps are feasible because the coefficients in the row corresponding to Equation 103.18 for the cost function are both positive.

An obvious question to ask concerns tied entering and leaving variables. In applying the criterion for an entering basic variable, it is possible that two or more coefficients could have the same largest negative coefficients. Similarly, in applying the criterion for a leaving basic variable, two or more ratios could have the same smallest positive ratios. In such circumstances an arbitrary decision can be made. It is theoretically possible that such arbitrary choices will lead to a non-convergent optimisation (degenerate), in which case the iteration becomes a perpetual loop, but the likelihood of such occurring in practice is such that the problem can safely be ignored.

LP is a powerful technique that can be applied to problems with hundreds of decision variables. There are many subtleties and refinements not covered above such as the use of artificial slack variables to get round the problem of initialisation when the origin lies outside the feasible solution space, and the use of shadow prices for exploring the sensitivity of the optimisation.

103.5 Mixed Integer Linear Programming

In the LP type of problems considered so far, it has been assumed that divisibility is acceptable. In other words, the decision variables are allowed to take on non-integer values. However, in certain classes of problem, it is essential that the decision variables are restricted to integer (or binary) values. Depending on whether all or some of the decision variables are so restricted, the problem is said to be either an integer programming (IP) or a mixed integer linear programming (MILP) problem respectively.

A simple and effective way of handling IP and MILP problems is to treat them as if they are LP problems and, having found the optimum values of the decision variables, to round them up or down to the nearest integer values. This obviously results in a slightly sub-optimal solution but, provided the values involved are fairly large, is probably satisfactory from a practical viewpoint. However, if the decision variables are small, this is unsatisfactory. There's a big difference between rounding 99.6 up to 100 compared with 0.6 to 1.0.

Again consider the separation plant. Suppose that for operational reasons the crude oil stream flows are only available in multiples of 10 m³/h:

$$x_1, x_2 = 0, 10, 20, \cdots$$

Knowing that the cost function is, as before:

$$J(\underline{x}) = 23.75.x_1 + 26.25.x_2 \quad (103.4)$$

and that the optimum solution for the continuous problem is:

$$x_1 = 159.1 \text{ m}^3/\text{h}, \quad x_2 = 50.43 \text{ m}^3/\text{h}$$

for which:

$$J(\underline{x}) = 5{,}103 \text{ £/h},$$

sensible combinations of permitted values of the decision variables that are close to the optimum can be tried on a trial and error basis. The obvious combination to go for here is:

$$x_1 = 160 \text{ m}^3/\text{h}, \quad x_2 = 50 \text{ m}^3/\text{h}$$

for which:

$$J(\underline{x}) = 5{,}113 \text{ £/h}.$$

This is probably good enough in practice but does just slightly violate the first constraint, Equation 103.5 on petrol throughput. However, if the constraints are hard, then adjacent combinations of decision variables that satisfy the constraints must be tried, in which case the optimum combination is:

$$x_1 = 150 \text{ m}^3/\text{h}, \quad x_2 = 50 \text{ m}^3/\text{h}$$

for which:

$$J(\underline{x}) = 4{,}875 \text{ £/h}$$

IP and MILP problems are typically concerned with assignment and ordering. Assignment essentially concerns the allocation of resources, such as reactors to batches or utilities to units, whereas ordering concerns the scheduling of resources in a time effective manner. Such problems are often described in the literature according to their generic type: the so called travelling salesman and knapsack problems are typical. Other types include scheduling, location and blending problems. They are normally solved by means of a methodology known as branch and bound for which implicit enumeration is the basis.

103.6 Implicit Enumeration

Consider an assignment problem: the simultaneous manufacture of one batch of each of four products in four different reactors. The objective is to assign batches to reactors in the most cost effective way. Suppose that some simplistic index of operating costs for making the batches in the different reactors may be formulated as depicted in Table 103.9.

Table 103.9 Operating costs for batch assignment problem

		Reactor			
		1	2	3	4
Batch	A	79	19	26	89
	B	18	12	90	26
	C	24	92	12	27
	D	32	46	99	6

For a relatively simple problem such as this it is possible to do an exhaustive search of every possible combination of assignment, of which there are 24, to find the optimum. However, for more

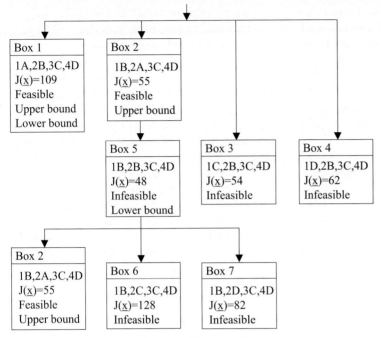

Fig. 103.4 Search strategy based upon implicit enumeration

complex problems, this is somewhat tedious and a strategy is required which searches in the direction of more likely solutions and quickly eliminates others. Such a strategy is depicted in Figure 103.4.

- Four initial assignments are made corresponding to different search directions, or branches. There needs to be a rational basis for such but the initial assignments are essentially arbitrary. Thus reactor 1 has batches A, B, C and D assigned to it, denoted as 1A, 1B, 1C and 1D as indicated in boxes 1 to 4 respectively.
- Now consider the first branch in more detail. A in feasible solution is sought, having already assigned 1A. Assign different batches to each reactor, say in turn, resulting in 1A, 2B, 3C and 4D as indicated in Box 1. Noting the indices of Table 6, the cost function $J(\underline{x})$ for this combination is 109. This is deemed to be the upper bound to the solution on the grounds that it is the best feasible solution found so far. The expectation is that there will be other feasible solutions that are more cost effective and that the upper bound will be reduced.
- Staying with the first branch, the best possible solution, feasible or otherwise, involving all four reactors and starting with 1A is sought. This happens to be the same solution as before, 1A, 2B, 3C and 4D, and is deemed to be the lower bound too.
- Note that it is coincidence that the one solution considered so far happens to be the diagonal of Table 6. It is also coincidence that the upper and lower bounds for Box 1 are the same.
- Now consider the second branch. Starting with 1B, find a better feasible solution by assigning a different batch to each of the reactors on a trial and error basis. Such a solution is 1B, 2A, 3C and 4D for which the cost function is 55 as indicated in Box 2. This is lower than the upper bound for the first branch and so becomes the new upper bound. It is also lower than the lower bound. Thus the first branch can be abandoned: every combination of assignments involving 1A will result in solutions whose cost function is greater than 55.

- Moving onto Box 5, a new lower bound may be established. The best possible solution, feasible or otherwise, involving all four reactors and starting with 1B is sought. Such a solution is 1B, 2B, 3C and 4D for which the cost function is 48. This is an infeasible solution because batch B is being made in both reactors 1 and 2 and there is no batch A. It is nevertheless deemed to be the new lower bound.
- Next consider the third branch which involves the assignment 1C. Proceeding straight to the best possible solution, feasible or otherwise, which is shown in Box 3 to have a cost function of 54. Any feasible solution must cost more than this so cannot be better than the current upper bound of 55 so further searching of the third branch may be abandoned.
- Similarly for the fourth branch involving the assignment 1D. This results in a best infeasible solution in Box 4 with a cost function of 62. This best solution is worse than the current upper bound so the fourth branch can be abandoned too.
- The situation thus far is that the optimum solution cannot involve the assignments 1A, 1C or 1D. For solutions involving the assignment 1B, a feasible upper bound of 55 and an infeasible lower bound of 48 have been established. Further searching of the second branch is now required to reveal whether there are any feasible solutions with a cost function lower than 55, but higher than 48.
- Branch 2 is now split into three sub branches in which assignment 1B is arbitrarily combined with assignments 2A, 2C and 2D. Box 2 is as before. Boxes 6 and 7 indicate the best infeasible solutions for the sub branches involving assignments 2C and 2D. Their cost functions have values of 128 and 82 respectively which are clearly inferior to the branch involving 2A so these two sub branches can be abandoned.
- For completeness, the sub-branch based upon assignments 1B and 2A can be further split and arbitrarily combined with 3C and 3D. This is not shown in Figure 4 but the assignment 1B, 2A, 3D, 4C has a value of 163. Whilst this is feasible it is clearly inferior to that involving 3C. The tree is thus said to have been "fathomed" and Box 2 represents the optimum solution.

This assignment problem shows that by using the bounds as criteria, it is possible to eliminate whole branches of the tree from the need for further searching. In practice about two-thirds of the possible combinations of solutions can be so eliminated. A branch, or node, of the tree is fathomed when further sub branching from that node has been ruled out. Those further sub branches are said to have been "implicitly enumerated". When one branch is fathomed the search backtracks to another that isn't to explore the values of its solutions, feasible or infeasible, against the bound criteria.

103.7 Branch and Bound

In the above problem there were two integer decision variables which had to be assigned, both of which had a very limited range: 1–4 for reactors and A–D for batches. A more realistic scenario would involve many more decision variables, both integer and binary, and maybe continuous ones too. For example, in addition to the integer variables of reactor number and batch type, there could be discrete (binary) variables for unit availability (yes or no), reactor type (glassed or stainless), production priority (high or low), and so on, as well as continuous variables for raw materials inventory and products storage capacity. A much wider range of values can be expected too. It is also likely in practice that enumeration of the cost function would be more complex. For example, cleaning times, idle periods, utilities costs and lateness of delivery would all be explicitly taken into account.

Branch and bound is the methodology normally used for solving MILP problems of this type. It involves the use of LP to solve and evaluate the problem at each node along the branch and implicit enumeration to handle the complexity of the branching. The methodology is as depicted in Figure 103.5.

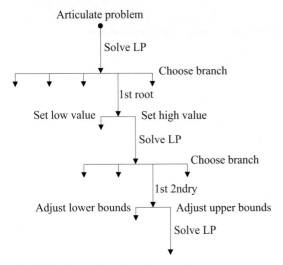

Fig. 103.5 The branch and bound methodology

- The starting point is to articulate the problem in terms of its functional constraints, the cost function and the permitted values of the decision variables. Binary variables will have two values only and integer variables will have a set of values over a given range.
- The MILP problem is then solved as if it was an LP problem in which the decision variables are constrained to their range. For example, a binary variable is allowed to vary between 0 and 1 rather than being set at either 0 or 1. Inspection of the solution will almost certainly reveal that some of the decision variables are at one end or other of their range whilst others will be somewhere in between. Choose a decision variable that is mid range: this becomes the first root branch. Clearly there are potentially as many root branches as there are decision variables.
- Consider the first root branch. Suppose that the decision variable chosen was discrete: this gives rise to two branches. The LP problem is solved twice, once with the decision variable set to its low value and once with it set to its high value. Whichever yields the better value for the cost function is deemed to be the best solution to that node, and is presumed for all subsequent branching down the first root.
- Once the best feasible solution for the first node is established, the most recent solution to the LP problem is examined and a different decision variable that is mid-range selected. This becomes the first of the secondary branches.
- Assume that the decision variable chosen for this secondary branch is integer with a range of seven permissible values: 0, 1, 2, 3, 4, 5 and 6. Suppose that its value in the LP solution was 2.4 which implies a value for the decision variable of either 2 or 3, giving rise to two branches. The LP problem is solved twice with different bounds set on the decision variable: once with the lower range of 0 to 2, and once with the upper range of 3 to 6. Whichever value of the decision variable, 2 or 3, yields the better value for the cost function is deemed to be the best solution to that node, and is presumed for all subsequent branching down the first root.
- This process of choosing a decision variable, creating a branch to force a solution for the node, and then solving the LP problem is repeated, node by node, checking for feasibility at each stage, until all possible combinations of the integer constraints have been applied and/or the branch is fathomed. This occurs when:
  i. The best solution is found which satisfies all the binary and integer constraints
  ii. The best possible value of the cost function is inferior to that found in a different branch
  iii. It is established that no feasible solution exists
- When a branch is fathomed the search routine then systematically backtracks to explore other secondary branches that have been missed out and/or moves onto another root branch in accordance with the process of implicit enumeration.

103.8 Comments

There are many packages available for solving IP and MILP problems, some academic and others commercial, running on a variety of platforms. Some are capable of solving very large and/or complex problems.

Programmes are also available for solving mixed integer non-linear programming (MINLP) problems for which there are many approaches. One such approach involves treating the variables as if they were continuous, solving the continuous LP problem, and then selecting the nearest feasible set of non-linear integer values in the vicinity of the continuous optimum. However, none of the approaches is particularly robust and success tends to be dependant upon problem type. For a given problem one approach may find a better optimum than another, whilst slight changes to the problem definition can lead to very different speeds of convergence.

Unconstrained Optimisation

104.1 One Dimensional Functions
104.2 Two Dimensional Functions
104.3 Newton's Method
104.4 Worked Example No 1
104.5 Search Procedures
104.6 Steepest Descent
104.7 Levenberg Marquardt Algorithm
104.8 Line Searching
104.9 Worked Example No 2
104.10 Comments

In the previous chapter it was demonstrated that for linear problems the optimum is normally found at the intersection of two constraints in the search space, said to be a corner feasible solution. That is not necessarily the case with non-linear problems. The cost function is typically some complex surface and the optimum may be an extremum, that is a maximum or a minimum value, somewhere in the middle of the search space. The basic issues therefore are first in articulating the cost function and second in finding the optimum value. Complications occur when there are several optima in the sense that the search process has to find the global optimum rather than one of the local ones. Optimisation is said to be unconstrained when the search can go in any direction without any equality or inequality constraints being imposed.

This chapter concentrates on the unconstrained optimisation of non-linear functions: constraints are introduced in the next chapter. The basic mathematics required for the study of multivariable optimisation is introduced. On the basis that analytical solutions are not usually available and that ad-hoc trial and error methods are impractical, gradient based search procedures are covered. These include, in particular, Newton's method and the steepest descent approach. For a fuller explanation the reader is referred to the texts by Gill (1982) and Edgar *et al.* (2001).

104.1 One Dimensional Functions

Consider some function $f(x)$ which is assumed to be differentiable. This may be articulated as a power series according to Taylor's theorem, Equation 68.6:

$$f(x+\Delta x) = f(x) + f'(x)\Delta x + \frac{f''(x)}{2}\Delta x^2 + \frac{f'''(x)}{3!}\Delta x^3 + \ldots \quad (104.1)$$

If it is assumed that the point x^* corresponds to an extremum, such that $f'(x^*) = 0$, and third and higher order terms are assumed to be negligible, which is reasonable if Δx is small enough, then:

$$f(x^* + \Delta x) - f(x^*) \approx \frac{f''(x^*)}{2}.\Delta x^2 \quad (104.2)$$

where Δx represents a small move away from the extremum. Clearly if x^* is a minimum then the left hand side of Equation 104.2 is positive. Thus, for a minimum, the first derivative of the function must be zero and the second derivative positive. Conversely, for a maximum the first derivative of the function must be zero and the second derivative negative. A saddle point corresponds to both the first and second derivatives being zero. Such consideration of the first and second derivatives is sufficient to classify the nature of an extremum.

104.2 Two Dimensional Functions

The basic issues are the same as for one dimensional problems except that the mathematics is more complicated. A general non-linear function with two independent variables can also be approximated as a quadratic relationship by taking a second order Taylor series expansion of the function:

$$f(x_1 + \Delta x_1, x_2 + \Delta x_2)$$
$$= f(x_1, x_2) + \frac{\partial f}{\partial x_1}.\Delta x_1 + \frac{\partial f}{\partial x_2}.\Delta x_2 + \frac{\partial^2 f}{\partial x_1^2}.\frac{\Delta x_1^2}{2}$$
$$+ \frac{\partial^2 f}{\partial x_1.\partial x_2}.\frac{\Delta x_1.\Delta x_2}{2} + \frac{\partial^2 f}{\partial x_2.\partial x_1}.\frac{\Delta x_1.\Delta x_2}{2}$$
$$+ \frac{\partial^2 f}{\partial x_2^2}.\frac{\Delta x_2^2}{2} + \ldots \quad (104.3)$$

Define vectors of the independent variables and their increments as:

$$\underline{x} = \begin{bmatrix} x_1 \\ x_2 \end{bmatrix} \qquad \Delta \underline{x} = \begin{bmatrix} \Delta x_1 \\ \Delta x_2 \end{bmatrix}$$

Define a vector of first order partial derivatives known as the Jacobian:

$$J(\underline{x}) = \nabla f(\underline{x}) = \begin{bmatrix} \frac{\partial f(\underline{x})}{\partial x_1} \\ \frac{\partial f(\underline{x})}{\partial x_2} \end{bmatrix} \quad (104.4)$$

Define a matrix of second order partial derivatives known as the Hessian:

$$H(\underline{x}) = \nabla^2 f(\underline{x}) = \begin{bmatrix} \frac{\partial^2 f(\underline{x})}{\partial x_1^2} & \frac{\partial^2 f(\underline{x})}{\partial x_1.\partial x_2} \\ \frac{\partial^2 f(\underline{x})}{\partial x_2.\partial x_1} & \frac{\partial^2 f(\underline{x})}{\partial x_2^2} \end{bmatrix} \quad (104.5)$$

The gradient notation ∇, referred to as grad, is fairly common in the literature on optimisation and is included here for completeness. These definitions enable Equation 104.3 to be written in matrix-vector form:

$$f(\underline{x} + \Delta \underline{x}) = f(\underline{x}) + J(\underline{x})^T.\Delta \underline{x} + \frac{1}{2}.\Delta \underline{x}^T.H(\underline{x}).\Delta \underline{x} + \ldots \quad (104.6)$$

Let the extremum be when $\underline{x}^* = \begin{bmatrix} x_1^* & x_2^* \end{bmatrix}^T$ at which point $J(\underline{x}^*) = 0$. Assuming the Jacobian to be zero gives:

$$f(\underline{x}^* + \Delta \underline{x}) - f(\underline{x}^*) \approx \frac{1}{2}.\Delta \underline{x}^T.H(\underline{x}^*).\Delta \underline{x} \quad (104.7)$$

If the right hand side of Equation 104.7 is positive the extremum is a minimum and if it is negative the extremum is a maximum.

Matrix algebra provides a convenient, albeit more obscure, test through the notion of "definiteness". From matrix algebra it is known that $\Delta \underline{x}^T.H(\underline{x}).\Delta \underline{x}$ will be positive for all $\Delta \underline{x}$ if (and only if) all the eigenvalues of the Hessian $H(\underline{x})$ are positive, this being referred to as positive definite. Similarly, $\Delta \underline{x}^T.H(\underline{x}).\Delta \underline{x}$ will be negative definite if all the eigenvalues are negative. These definiteness properties can be used to test whether an extremum is a maximum or minimum without having to do the calculation explicitly. Note that a mixture of positive and negative eigenvalues indicates a saddle point.

Whilst the above equations have been developed in relation to a two dimensional function, they readily extrapolate to the n-dimensional case.

104.3 Newton's Method

Newton's method is an iterative means of finding extrema. Consider an iterative form of Equation

104.6 in which third and higher order terms are presumed negligible:

$$f(\underline{x}(k+1)) \approx f(\underline{x}(k)) + J(\underline{x}(k))^T . \Delta\underline{x}$$
$$+ \frac{1}{2} . \Delta\underline{x}^T . H(\underline{x}(k)) . \Delta\underline{x} \quad (104.8)$$

where:

$$\Delta\underline{x} = \begin{bmatrix} \Delta x_1 \\ \Delta x_2 \end{bmatrix} = \begin{bmatrix} x_1(k+1) - x_1(k) \\ x_2(k+1) - x_2(k) \end{bmatrix} \quad (104.9)$$

The objective is to find the step $\Delta\underline{x}$, starting from $\underline{x}(k)$, which ensures that $\underline{x}(k+1)$ is an extremum.

Noting that $f(\underline{x}(k))$ is constant for any given $\underline{x}(k)$, Equation 104.8 may be differentiated with respect to $\Delta\underline{x}$. Remember that differentiation by a vector was covered in Chapter 79, and note that the Hessian is symmetrical about its diagonal such that:

$$H(\underline{x}(k))^T = H(\underline{x}(k))$$

Differentiating Equation 104.8 and setting the differential to zero yields the extremum:

$$\frac{\partial f(\underline{x}(k+1))}{\partial \Delta\underline{x}} \approx J(\underline{x}(k)) + H(\underline{x}(k)) . \Delta\underline{x} = 0$$

Solving for $\Delta\underline{x}$ gives:

$$\Delta\underline{x} = -H(\underline{x}(k))^{-1} . J(\underline{x}(k)) \quad (104.10)$$

The latter term is also referred to as the search direction $\underline{s}(k)$:

$$\underline{s}(k) = -H(\underline{x}(k))^{-1} . J(\underline{x}(k)) \quad (104.11)$$

Again the two dimensional case has been considered but the equations readily extrapolate to the n-dimensional case.

104.4 Worked Example No 1

A second order Taylor series was assumed in the derivation of Equation 104.10, so the resulting $\Delta\underline{x}$ will enable the extremum to be found in a single step for any quadratic function $f(\underline{x}(k))$.

Consider the quadratic function:

$$f(\underline{x}) = 3x_1^2 - 4x_1.x_2 + 2x_2^2 - x_1 - x_2.$$

Find the extremum using Newton's method starting from:

$$\underline{x}(0) = \begin{bmatrix} -2 & 2 \end{bmatrix}^T.$$

From Equations 104.9 and 104.10:

$$\Delta\underline{x} = \underline{x}(1) - \underline{x}(0) = -H(\underline{x}(0))^{-1} . J(\underline{x}(0))$$

The Jacobian and Hessian are evaluated from Equations 104.4 and 104.5:

$$J(\underline{x}) = \begin{bmatrix} 6x_1 - 4x_2 - 1 \\ -4x_1 + 4x_2 - 1 \end{bmatrix}$$

so at the start:

$$J(\underline{x}(0)) = \begin{bmatrix} -21 \\ 15 \end{bmatrix}$$

$$H(\underline{x}) = \begin{bmatrix} 6 & -4 \\ -4 & 4 \end{bmatrix}$$

which is constant so:

$$H(\underline{x}(0))^{-1} = \begin{bmatrix} 0.5 & 0.5 \\ 0.5 & 0.75 \end{bmatrix}$$

Substitute:

$$\underline{x}(1) = \begin{bmatrix} -2 \\ 2 \end{bmatrix} - \begin{bmatrix} 0.5 & 0.5 \\ 0.5 & 0.75 \end{bmatrix} \begin{bmatrix} -21 \\ 15 \end{bmatrix} = \begin{bmatrix} 1 \\ 1.25 \end{bmatrix}$$

At this point $J(\underline{x})$ is zero corresponding to an extremum.

The eigenvalues of the Hessian are found by solving the equation:

$$\det[\lambda I - H(\underline{x})] = 0$$

The eigenvalues found are 9.123 and 0.877. Since these are both positive, the Hessian is positive definite so the extremum must be a minimum.

104.5 Search Procedures

The function $f(\underline{x}(k))$ will not necessarily be quadratic so a single iteration of Equation 104.10 is generally insufficient to locate the extremum. Thus recursion is required until convergence is achieved. Multi-dimensional search procedures are normally based upon the following recursion:

$$\underline{x}(k+1) = \underline{x}(k) + \alpha(k).\underline{s}(k) \quad (104.12)$$

where:

k is an iteration index,
$\underline{x}$ is the vector of independent variables,
$\underline{s}(k)$ is the search direction given by Equation 104.11, and
$\alpha(k)$ is a scalar that determines the size of the step in the $\underline{s}(k)$ direction.

The recursion is as depicted in Figure 104.1.

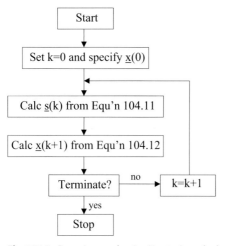

Fig. 104.1 Recursive search using Newton's method

In essence, at each iteration, a search is made along the vector of direction $\underline{s}(k)$ to find the minimum value of the function in that direction. It is largely in the determination of the search direction that the various algorithms differ. There are several limitations:

- When there is more than one extremum in the search field, the algorithm may find local extrema rather than the global one. To increase the chance of finding the global extremum it is advisable to have multiple runs of the algorithm with different starting points.
- The algorithm involves calculating the inverse of the Hessian matrix. This is not always possible and singularity problems are the principal cause of the algorithm breaking down.
- The search is attracted towards extrema as opposed to either maxima or minima. To find a minimum the Hessian must be forced into being positive definite and for a maximum it must be negative definite. This is discussed further under steepest descent.
- If a fixed step length is used then picking a sensible size is key to obtaining a reasonable rate of convergence. This is discussed further under line searching.

Note that there are many potential criteria for ending recursion, such as:

$$|f(\underline{x}(k+1)) - f(\underline{x}(k))| \leq \varepsilon \quad \left|\frac{f(\underline{x}(k+1)) - f(\underline{x}(k))}{f(\underline{x}(k))}\right| \leq \varepsilon$$

$$\|J(\underline{x}(k))\|_2 \leq \varepsilon$$

where ε is some specified very small residual.

Note that the 2-Norm $\|\ \|_2$ of a vector such as $J(\underline{x}(k))$ is as defined in Chapter 111.

104.6 Steepest Descent

In the immediate vicinity of any point $\underline{x}$ on a surface $f(\underline{x})$, the Jacobian (or gradient ∇) of the function $f(\underline{x})$ at that point gives the direction of greatest slope in $f(\underline{x})$ and is orthogonal ("at right angles") to the contour of $f(\underline{x})$ at $\underline{x}$. It follows that if the Hessian matrix in Equation 104.11 is set equal to the identity matrix I the search direction will be that of steepest slope:

$$\underline{s}(k) = -H(\underline{x}(k))^{-1}.J(\underline{x}(k)) = -J(\underline{x}(k)) \quad (104.13)$$

Equation 104.13 corresponds to a search of steepest descent because the identity matrix I is positive definite which ensures that the search proceeds towards a minima. For steepest ascent the sign of the search direction is reversed (made positive).

Substitution into Equation 104.12 yields:

$$\underline{x}(k+1) = \underline{x}(k) - \alpha(k).J(\underline{x}(k))$$

Bearing in mind that in the vicinity of a minimum the elements of $J(\underline{x})$ will all be negative, it can be seen that the search will progress towards the minimum.

A key point to appreciate is that the Jacobian gives the local direction of steepest descent which, for most surfaces, is unlikely to be the most direct route to the minima. As the search proceeds the direction of steepest descent changes. Thus the method is inefficient requiring a relatively large number of iterations before the solution is obtained. However, it is robust and not prone to singularity problems.

104.7 Levenberg Marquardt Algorithm

The problem with the steepest descent approach is that useful information about the curvature of the surface which is contained in the Hessian is discarded when it is replaced by the identity matrix. The Levenberg Marquardt algorithm is a subtle means of using that information to improve the efficiency of searching.

The subtlety is to modify the search direction of Equation 104.11 as follows:

$$\underline{s}(k) = -\left(H(\underline{x}(k)) + \beta.I\right)^{-1}.J(\underline{x}(k)) \quad (104.14)$$

The basic idea is to start with a large value of β which guarantees that the Hessian is dominated by the identity matrix and forces positive definiteness. Thus the first few steps of the search are in the direction of steepest descent. However, as the search proceeds, the value of β is reduced such that the Hessian becomes more dominant and the search is more Newton like in nature. Note that as the search converges on the minimum it may be assumed that the Hessian is positive definite in its own right and the need for forcing such no longer exists. The recursion is depicted in Figure 104.2.

Note that it is assumed that $\|J(\underline{x}(k))\|_2 > \varepsilon$ on the first iteration.

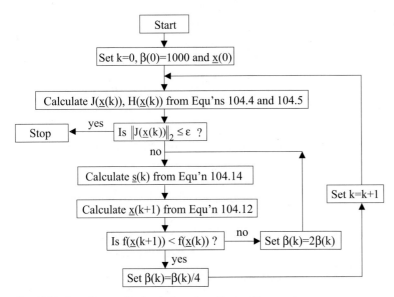

Fig. 104.2 Recursive search using the Levenberg Marquardt approach

104.8 Line Searching

So far it has been implied that the size of the step in the search direction is fixed at a value such as $\alpha = 0.1$. In fact the search algorithms can be significantly improved by conducting line searches to find the optimal value of α. The step size has to be evaluated every time the search direction changes. Two methods are outlined, both of which require the creation of an artificial function $f(\alpha)$ and the use of bracketing.

Consider $f(\alpha)$ to be the variation of $f(\underline{x})$ in the search direction. The minimum in $f(\alpha)$ can be found by "going downhill" in steps of increasing size (say h, 2h, 4h, etc.) until the value of $f(\alpha)$ increases. The initial and final points are deemed to be the positions that "bracket" the minimum.

1. Brent's method assumes that the multi-dimensional function $f(\underline{x})$ can be approximated within the bracket by a uni-dimensional quadratic $f(\alpha)$ where α represents distance in the search direction:

$$f(\alpha) = a.\alpha^2 + b.\alpha + c \quad (104.15)$$

Substitution into this equation of pairs of values of α and $f(\alpha)$ at the two extremes of the bracket and at its mid point yields three simultaneous algebraic equations, the solution of which gives the coefficients a, b and c. Having established the equation for $f(\alpha)$ its minimum is located by differentiation:

$$\frac{df(\alpha)}{d\alpha} = 2.a.\alpha + b = 0$$

If $\alpha(k)$ is the step size required at the kth iteration to roughly locate the minimum then:

$$\alpha(k) = -\frac{b}{2a} \quad (104.16)$$

2. Newton Raphson's method. Suppose that $f(\alpha)$ is approximated by the first three terms of the Taylor series expansion of Equation 104.1:

$$f(\alpha + \Delta\alpha) \approx f(\alpha) + f'(\alpha).\Delta\alpha + \frac{f''(\alpha)}{2}.\Delta\alpha^2$$

For a given α the minimum in $f(\alpha+\Delta\alpha)$ is found by differentiation with respect to $\Delta\alpha$:

$$\frac{df(\alpha + \Delta\alpha)}{d\Delta\alpha} \approx f'(\alpha) + f''(\alpha).\Delta\alpha = 0$$

If, as before, $\alpha(k)$ is the step size required to locate the minimum then:

$$\alpha(k) = -\frac{f'(\alpha)}{f''(\alpha)}$$

As with the multi-dimensional search procedure of Worked Example No 1, the advantage of the Newton Raphson method is that for a quadratic function the minimum is found in a single iteration. The disadvantages of the technique are that the first and second derivatives must be calculated and, for non-quadratic functions, the method can be slow, especially if the second derivative is small.

104.9 Worked Example No 2

Consider again the function of Worked Example No 1:

$$f(\underline{x}) = 3x_1^2 - 4x_1.x_2 + 2x_2^2 - x_1 - x_2.$$

The initial conditions and search direction were:

$$\underline{x}(0) = \begin{bmatrix} -2 \\ 2 \end{bmatrix}$$

$$\underline{s}(0) = -\begin{bmatrix} 0.5 & 0.5 \\ 0.5 & 0.75 \end{bmatrix}\begin{bmatrix} -21 \\ 15 \end{bmatrix} = \begin{bmatrix} 3 \\ -0.75 \end{bmatrix}$$

Let distance in the search direction be articulated according to Equation 104.12 as follows:

$$x_1 = -2 + 3\alpha \qquad x_2 = 2 - 0.75\alpha$$

Substituting into the original function yields the artificial function:

$$f(\alpha) = 3.(-2 + 3\alpha)^2 - 4.(-2 + 3\alpha)(2 - 0.75\alpha) \\ + 2(2 - 0.75\alpha)^2 - (-2 + 3\alpha) - (2 - 0.75\alpha)$$

Expansion gives:

$$f(\alpha) = 37.125.\alpha^2 - 74.25.\alpha + 36$$

Since the original function f($\underline{x}$) was quadratic, this equation for f(α) is already in the form of Brent's equation and the coefficients are already known. Substituting into Equation 104.16 gives

$$\alpha(0) = -\frac{-74.25}{2 \times 37.125} = 1.0$$

Substituting back into Equation 104.12 yields the position of the minimum:

$$\underline{x}(1) = \underline{x}(0) + \alpha(0).\underline{s}(0)$$

$$= \begin{bmatrix} -2 \\ 2 \end{bmatrix} + 1.0 \begin{bmatrix} 3 \\ -0.75 \end{bmatrix} = \begin{bmatrix} 1 \\ 1.25 \end{bmatrix}$$

In this case the solution was relatively trivial with $\alpha = 1.0$ locating the optimum in a single step. That was simply due to the fact that the original function f($\underline{x}$) was quadratic and Brent's method involves a quadratic approximation. However, for non-quadratic functions, bracketing would be required and quadratic approximation would normally yield non-unity values for the step length which would normally require several/many steps to locate the minimum.

104.10 Comments

It is fairly obvious that all of the above techniques for finding an optimum, be it a minimum or a maximum, require that the function being optimised be articulated in the form of an equation. Clearly the more accurate the model the better the prediction of the optimum. If the shape of the surface cannot be articulated then the techniques are of no use.

Constrained Optimisation

105.1 The Lagrangian Function
105.2 Worked Example No 1
105.3 Worked Example No 2
105.4 Generalised Lagrangian Function
105.5 Worked Example No 3
105.6 Sensitivity Analysis
105.7 Kuhn-Tucker Conditions
105.8 Worked Example No 4
105.9 Quadratic Programming
105.10 Worked Example No 5
105.11 Recursive Form of QP
105.12 Sequential Quadratic Programming
105.13 Reduced Gradient Methods
105.14 Penalty Functions
105.15 Nomenclatura

Having dealt with the unconstrained optimisation of non-linear systems in the previous chapter, both equality and inequality constraints are now introduced. These significantly complicate the problem for which more powerful non-linear programming (NLP) techniques are required. First comes the use of Lagrange multipliers and consideration of the Kuhn-Tucker conditions for such. Next comes quadratic programming (QP) and, in particular, sequential quadratic programming (SQP). And finally reduced gradient methods and penalty functions are introduced. The mathematics is non-trivial but is explained carefully as appropriate. For a more comprehensive treatment of what, by any yardstick, is a difficult subject the reader is referred yet again to the texts by Gill (1982) and Edgar (2001).

105.1 The Lagrangian Function

Consider the optimisation of a two-dimensional non-linear problem with a single equality constraint $g(\underline{x})$. Let the cost function be $f(\underline{x})$. The optimisation is cast in the form:

$$\text{minimise} \quad f(\underline{x}) \qquad (105.1)$$
$$\text{subject to} \quad g(\underline{x}) = 0 \qquad (105.2)$$

where

$$\underline{x} = \begin{bmatrix} x_1 & x_2 \end{bmatrix}^T$$

Note that equality constraints are usually set equal to some constant but can always be rearranged into the form of Equation 105.2.

Let there be a local minimum at:

$$\underline{x}^* = \begin{bmatrix} x_1^* & x_2^* \end{bmatrix}^T.$$

This minimum may be an extremum somewhere in the search space. Otherwise it will be at a constraint, either along the edge of the search space or at a corner between constraints. Consider some point at a small distance away from the minimum:

$$f(\underline{x}^* + \Delta \underline{x}) - f(\underline{x}^*) \geq 0$$

Using the Taylor's series expansion of Equation 104.3, and ignoring second and higher order terms:

$$\Delta f(\underline{x}) \approx \left.\frac{\partial f(\underline{x})}{\partial x_1}\right|_{\underline{x}^*} . \Delta x_1 + \left.\frac{\partial f(\underline{x})}{\partial x_2}\right|_{\underline{x}^*} . \Delta x_2 \geq 0 \quad (105.3)$$

The $\geq$ sign needs some explanation. If the minimum is an extremum then the partial derivatives of $f(\underline{x})$ will be zero and $\Delta f(\underline{x})$ will indeed be zero. However, if the minimum is at a constraint it will not necessarily be an extremum, in which case the partial derivatives will be non-zero and $\Delta f(\underline{x})$ will be finite and positive.

As far as the constraint itself is concerned, since all points considered must lie on the constraint:

$$g(\underline{x}^* + \Delta \underline{x}) - g(\underline{x}^*) = 0$$

Using Taylor's series expansion again gives:

$$\Delta g(\underline{x}) \approx \left.\frac{\partial g(\underline{x})}{\partial x_1}\right|_{\underline{x}^*} . \Delta x_1 + \left.\frac{\partial g(\underline{x})}{\partial x_2}\right|_{\underline{x}^*} . \Delta x_2 = 0 \quad (105.4)$$

Equations 105.3 and 105.4 may be combined:

$$\Delta f(\underline{x}) + \lambda . \Delta g(\underline{x})$$
$$\approx \left(\left.\frac{\partial f(\underline{x})}{\partial x_1}\right|_{\underline{x}^*} + \lambda . \left.\frac{\partial g(\underline{x})}{\partial x_1}\right|_{\underline{x}^*}\right) . \Delta x_1 \quad (105.5)$$
$$+ \left(\left.\frac{\partial f(\underline{x})}{\partial x_2}\right|_{\underline{x}^*} + \lambda . \left.\frac{\partial g(\underline{x})}{\partial x_2}\right|_{\underline{x}^*}\right) . \Delta x_2 \geq 0$$

where the algebraic variable λ, known as the Lagrange multiplier, is an extra degree of freedom that has been introduced. Choose λ such that:

$$\left.\frac{\partial f(\underline{x})}{\partial x_1}\right|_{\underline{x}^*} + \lambda . \left.\frac{\partial g(\underline{x})}{\partial x_1}\right|_{\underline{x}^*} = 0 \quad (105.6)$$

whence the term in Δx_1 of Equation 105.5 is zero, irrespective of the value of Δx_1. It can be seen from Equation 105.4 that Δx_1 and Δx_2 are not independent: if Δx_1 (say) is specified then Δx_2 is determined. Since the sign of Δx_1 is arbitrary then so too is that of Δx_2. However, Δx_2 is non-zero, so the only way that the constraint of Equation 105.5 can be satisfied is if:

$$\left.\frac{\partial f(\underline{x})}{\partial x_2}\right|_{\underline{x}^*} + \lambda . \left.\frac{\partial g(\underline{x})}{\partial x_2}\right|_{\underline{x}^*} = 0 \quad (105.7)$$

The above concepts can be generalised in terms of an augmented cost function known as the Lagrangian in which the cost (objective) function and constraint are combined as follows:

$$L(\underline{x}, \lambda) = f(\underline{x}) + \lambda . g(\underline{x}) \quad (105.8)$$

where $f(\underline{x})$ is to be minimised and $g(\underline{x}) = 0$ is the constraint. It can be seen that if the constraint is satisfied then the Lagrangian function reduces to the original cost function, irrespective of the value of the Lagrange multiplier λ, in which case the minimum of the Lagrangian will be the same as the minimum in the cost function. Differentiating Equation 105.8 yields:

$$\frac{\partial L(\underline{x}, \lambda)}{\partial x_1} = \frac{\partial f(\underline{x})}{\partial x_1} + \lambda . \frac{\partial g(\underline{x})}{\partial x_1} \quad (105.9)$$

$$\frac{\partial L(\underline{x}, \lambda)}{\partial x_2} = \frac{\partial f(\underline{x})}{\partial x_2} + \lambda . \frac{\partial g(\underline{x})}{\partial x_2} \quad (105.10)$$

$$\frac{\partial L(\underline{x}, \lambda)}{\partial \lambda} = g(\underline{x}) \quad (105.11)$$

For a minimum:

$$\frac{\partial L(\underline{x}^*, \lambda^*)}{\partial x_1} = \frac{\partial L(\underline{x}^*, \lambda^*)}{\partial x_2} = \frac{\partial L(\underline{x}^*, \lambda^*)}{\partial \lambda} = 0$$

Note that these equations are necessary, but not sufficient, conditions for a minimum.

105.2 Worked Example No 1

Minimise $x + y$ subject to $x^2 + y^2 = 1$, the latter corresponding to the sum of the squares of the co-ordinates of a point on the unit circle. Substituting into Equation 105.8 gives:

$$L(x, y, \lambda) = x + y + \lambda . (x^2 + y^2 - 1)$$

Differentiating:

$$\frac{\partial L}{\partial x} = 1 + 2.\lambda.x = 0 \qquad \frac{\partial L}{\partial y} = 1 + 2.\lambda.y = 0$$

$$\frac{\partial L}{\partial \lambda} = x^2 + y^2 - 1 = 0$$

Solving three equations in three unknowns gives

$$x^* = y^* = -\lambda^* = \pm\frac{1}{\sqrt{2}}.$$

By inspection it can be seen that the minimum value of $x + y$ is $-\sqrt{2}$.

105.3 Worked Example No 2

Minimise $3x_1^2 + 4x_2^2$ subject to $x_1 + 2x_2 = 5$. Substituting into Equation 105.8 gives:

$$L(\underline{x}, \lambda) = 3x_1^2 + 4x_2^2 + \lambda(x_1 + 2x_2 - 5)$$

Differentiating:

$$\frac{\partial L}{\partial x_1} = 6x_1 + \lambda = 0 \qquad \frac{\partial L}{\partial x_2} = 8x_2 + 2\lambda = 0$$

$$\frac{\partial L}{\partial \lambda} = x_1 + 2x_2 - 5 = 0$$

Solving three equations in three unknowns gives $x_1^* = 1.25, x_2^* = 1.875$ and $\lambda^* = -7.5$.

Whence the minimum is at $(1.25, 1.875)$ and has a value of 18.75.

105.4 Generalised Lagrangian Function

In general, if a non-linear constrained optimisation problem contains n decision variables and there are m equality constraints involving the decision variables, and the constraints are independent, and $m < n$, then m Lagrange multipliers are required. The problem will be a function of $n + m$ variables for which $n + m$ necessary conditions must be derived and solved.

The Lagrangian function of Equation 105.8 is thus extended to accommodate multiple equality constraints as follows:

minimise $\qquad f(\underline{x})$ $\qquad$ (105.1)

subject to: $\qquad g_i(\underline{x}) = 0$ $\qquad$ (105.12)

$\qquad\qquad$ where $i = 1, 2, \ldots, m$

where: $\qquad \underline{x} = \begin{bmatrix} x_1 & x_2 & \cdots & x_n \end{bmatrix}^T$

for which the Lagrangian function becomes:

$$L(\underline{x}, \underline{\lambda}) = f(\underline{x}) + \sum_{i=1}^{m} \lambda_i . g_i(\underline{x})$$

Note that it is essential that $m < n$. If $m = n$ then the equality constraints could be solved irrespective of the cost function and if $m > n$ then the optimisation is overdetermined. In practice, it is often the case that the number of decision variables involved in the cost function is less than the number of constraints. At first sight this appears to be inconsistent with the criterion that $m < n$. However, it isn't, because the decision variables involved in the cost function are invariably only a subset of the vector $\underline{x}$. In practice it is most unlikely that there will be as many constraints as there are decision variables (n).

The above generalised form of Lagrangian function may be further extended to accommodate inequality constraints as follows:

minimise $\qquad f(\underline{x})$ $\qquad$ (105.1)

subject to: $\qquad g_i(\underline{x}) = 0$ $\qquad$ (105.12)

$\qquad\qquad$ where $i = 1, 2, \ldots, m$

and: $\qquad h_i(\underline{x}) \leq 0$ $\qquad$ (105.13)

$\qquad\qquad$ where $i = m + 1, \ldots, p$

still with: $\qquad \underline{x} = \begin{bmatrix} x_1 & x_2 & \cdots & x_n \end{bmatrix}^T$

These latter inequality constraints can be cast in the form of equality constraints by the introduction of slack variables as previously considered in Chapter 103.

Thus $h_i(\underline{x}) + \sigma_i^2 = 0$ where $\sigma_i^2 \geq 0$.

Addition of σ_i^2 to $h_i(\underline{x})$ guarantees that the inequality constraint is always satisfied.

Use of a squared terms for the slack variable ensures that the solution approaches the constraint from within the feasible region. The Lagrangian function thus becomes:

$$L(\underline{x}, \underline{\lambda}) = f(\underline{x}) + \sum_{i=1}^{m} \lambda_i \cdot g_i(\underline{x}) \quad (105.14)$$

$$+ \sum_{i=m+1}^{p} \lambda_i \cdot (h_i(\underline{x}) + \sigma_i^2)$$

where the λ_i for $i = 1, 2, \ldots, p$ are independent Lagrange multipliers.

Note that, when there are inequality constraints, finding the minimum requires that $q < n$ where q is the number of equality constraints plus the number of inequalities that are at their constraints. An inequality at its constraint is said to be binding. Thus:

$$i = 1, 2 \ldots m, m + 1 \ldots q \leq p$$

The circumstances under which $L(\underline{x}, \underline{\lambda})$ reduces to $f(\underline{x})$ are:

either $g_i(\underline{x}) = 0$ and $h_i(\underline{x}) = 0$,
in which case the constraints are satisfied and the Lagrange multipliers take on appropriate finite values,
or $g_i(\underline{x}) = 0$ and $h_i(\underline{x}) \leq 0$,
in which case some or all of the inequalities exist and the appropriate Lagrange multipliers must be set to zero. In effect, the inactive inequalities are switched off,
or $g_i(\underline{x}) = 0$ and $h_i(\underline{x}) < 0$,
in which case all of the inequalities exist and all of the Lagrange multipliers must be set to zero.

Thus, provided that every constraint is either satisfied or has its Lagrange multiplier set to zero, the minimum of the Lagrangian function is the same as the minimum of the original cost function. The constrained minimum therefore occurs when:

$$\frac{\partial L(\underline{x}^*, \underline{\lambda}^*)}{\partial x_j} = 0 \quad \text{for } j = 1, 2, \ldots, n \quad (105.15)$$

$$\frac{\partial L(\underline{x}^*, \underline{\lambda}^*)}{\partial \lambda_i} = 0 \quad \text{for } i = 1, 2, \ldots, p \quad (105.16)$$

$$\frac{\partial L(\underline{x}^*, \underline{\lambda}^*)}{\partial \sigma_i} = 2\lambda_i^* \cdot \sigma_i^* = 0 \quad (105.17)$$

$$\text{for } i = m + 1, \ldots, p$$

$$\lambda_i^* \geq 0 \quad \text{for } i = m + 1, \ldots, p \quad (105.18)$$

For a maximum $\lambda_i \geq 0$ is replaced with $\lambda_i \leq 0$.

Again note that these equations are necessary, but not sufficient, conditions for a minimum.

105.5 Worked Example No 3

Optimise $x_1.x_2$ subject to $x_1^2 + x_2^2 \leq 8$. Substituting into 105.1 and 105.13 gives:

$$f(\underline{x}) = x_1.x_2$$
$$h(\underline{x}) = x_1^2 + x_2^2 - 8 \leq 0$$

The Lagrangian function may thus be formulated along the lines of Equation 105.14:

$$L(\underline{x}, \lambda) = x_1.x_2 + \lambda(x_1^2 + x_2^2 - 8 + \sigma^2)$$

The conditions for optima are thus:

$$\frac{\partial L(\underline{x}, \lambda)}{\partial x_1} = x_2 + 2\lambda.x_1 = 0$$

$$\frac{\partial L(\underline{x}, \lambda)}{\partial x_2} = x_1 + 2\lambda.x_2 = 0$$

$$\frac{\partial L(\underline{x}, \lambda)}{\partial \sigma} = 2\lambda.\sigma = 0$$

$$\frac{\partial L(\underline{x}, \lambda)}{\partial \lambda} = x_1^2 + x_2^2 - 8 + \sigma^2 = 0$$

Solving four equations in four unknowns gives the results as shown in Table 105.1.

Table 105.1 Classification of optima for Worked Example No 3

Optima	λ^*	σ^*	x_1^*	x_2^*	$f(\underline{x}^*)$
Saddle	0	$\pm\sqrt{8}$	0	0	0
Minimum	0.5	0	2	-2	-4
Minimum	0.5	0	-2	2	-4
Maximum	-0.5	0	2	2	4
Maximum	-0.5	0	-2	-2	4

That these are the optima can be verified by back substituting into the conditions. Note that the minima occur for positive values of λ and the saddle point when λ is zero.

105.6 Sensitivity Analysis

Consider again the constraint $g(\underline{x}) = 0$ of Equation 105.2 where $\underline{x}$ is of dimension n.

In general such a constraint may be considered to consist of various algebraic terms and a constant, referred to as the excess, of the form

$$\gamma(\underline{x}) - c = 0$$

For the Worked Example No 2, say, these would be $\gamma(\underline{x}) = x_1 + 2x_2$ and $c = 5$.

The Lagrangian of Equation 105.8 may thus be rewritten as:

$$L(\underline{x}, \lambda) = f(\underline{x}) + \lambda.(\gamma(\underline{x}) - c)$$

Differentiating with respect to the constant of the constraint gives:

$$\frac{\partial L(\underline{x}, \lambda)}{\partial c} = -\lambda$$

Provided that the constraint $\gamma(\underline{x}) - c = 0$ is satisfied:

$$\frac{\partial L(\underline{x}, \lambda)}{\partial c} = \frac{\partial f(\underline{x})}{\partial c} = -\lambda$$

whence, at the minimum $\underline{x}^*$:

$$\frac{\partial f(\underline{x}^*)}{\partial c} = -\lambda^*$$

which can be approximated by:

$$\Delta f(\underline{x}^*) = -\lambda^* . \Delta c \qquad (105.19)$$

Thus the Lagrange multiplier can be thought of as the sensitivity, or gain, of the function being optimised with respect to the constant of the constraint. The argument extrapolates to the multiple constraints $g_i(\underline{x}) = 0$ of Equation 105.12 where $i = 1, 2, \ldots, m$.

Sensitivity is helpful when analysing problems with multiple constraints. For example, a process may be optimised to satisfy various constraints on throughput, product quality and energy consumption. Inspection of the magnitude and sign of the Lagrange multipliers will reveal the sensitivity and direction of the cost function to each of the constraints. This provides a practical basis for decisions about which constraints to relax, or not.

105.7 Kuhn-Tucker Conditions

These relate to the generalised Lagrangian function which is reproduced for convenience:

$$L(\underline{x}, \underline{\lambda}) = f(\underline{x}) + \sum_{i=1}^{m} \lambda_i . g_i(\underline{x}) \qquad (105.14)$$
$$+ \sum_{i=m+1}^{p} \lambda_i . (h_i(\underline{x}) + \sigma_i^2)$$

The necessary criteria for $\underline{x}^*$ to be a local minimum of $f(\underline{x})$ are:

1. That $f(\underline{x})$, $g_i(\underline{x})$ and $h_i(\underline{x})$ are all differentiable at $\underline{x}^*$.
2. That the Lagrange multipliers λ_i exist.
3. That the constraints are satisfied at $\underline{x}^*$:

$$g_i(\underline{x}^*) = 0 \quad \text{and} \quad h_i(\underline{x}^*) \leq 0$$

4. That the Lagrange multipliers for the inequality constraints are not negative:

$$\lambda_i^* \geq 0 \quad \text{for} \quad i = m+1, \ldots, p$$

Note that the sign of the multipliers for the equality constraints doesn't matter.

5. That the inequality constraints are:
   i. either active: that is, some of the inequalities are at their constraints for which:

$$h_i(\underline{x}^*) = 0,$$

   that is, they are binding
   ii. or inactive: $h_i(\underline{x}^*) < 0$, in which case the associated:

$$\lambda_i^* = 0 \quad \text{for} \quad i = m+1, \ldots, p$$

such that
$$\lambda_i^* \cdot h_i(\underline{x}^*) = 0.$$

6. That the Lagrangian function is at a stationary point:
$$\frac{\partial L(\underline{x}^*, \underline{\lambda}^*)}{\partial x_j} = \frac{\partial L(\underline{x}^*, \underline{\lambda}^*)}{\partial \lambda_i} = \frac{\partial L(\underline{x}^*, \underline{\lambda}^*)}{\partial \sigma_i} = 0$$

The above criteria are known as the Kuhn-Tucker conditions: they serve as the basis for the design of some optimisation algorithms and as termination criteria for others. Two further necessary conditions are:

7. That the gradients (Jacobians) at $\underline{x}^*$ of all the active constraints (equality constraints and any binding inequality constraints) are linearly independent, the gradients being:

$$\nabla g_i(\underline{x}^*) = \left[\frac{\partial g_i(\underline{x}^*)}{\partial x_1} \quad \frac{\partial g_i(\underline{x}^*)}{\partial x_2} \quad \cdots \quad \frac{\partial g_i(\underline{x}^*)}{\partial x_n} \right]^T$$

for $i = 1, 2, \ldots, m$

$$\nabla h_i(\underline{x}^*) = \left[\frac{\partial h_i(\underline{x}^*)}{\partial x_1} \quad \frac{\partial h_i(\underline{x}^*)}{\partial x_2} \quad \cdots \quad \frac{\partial h_i(\underline{x}^*)}{\partial x_n} \right]^T$$

for $i = m + 1, \ldots, p$

Noting that the gradients are evaluated at $\underline{x}^*$ and each is a column vector of constants, linear independence is essentially a question of checking that no one vector is a linear function of another.

8. That those vectors $\underline{v}$ which, for each active constraint, satisfy:
$$\underline{v}^T \cdot \nabla g_i(\underline{x}^*) = 0 \quad \text{and} \quad \underline{v}^T \cdot \nabla h_i(\underline{x}^*) = 0$$

are such that the Hessian matrix of the Lagrangian function is positive semidefinite, or:
$$\underline{v}^T \cdot \nabla^2 \left(L(\underline{x}^*, \underline{\lambda}) \right) \cdot \underline{v} \geq 0$$

The sufficient conditions for $\underline{x}^*$ to be a local minimum are all of the above necessary criteria subject to a strengthening of the last one:

9. That those vectors $\underline{v}$ which for each active constraint satisfy:
$$\underline{v}^T \cdot \nabla g_i(\underline{x}^*) = 0 \quad \text{and} \quad \underline{v}^T \cdot \nabla h_i(\underline{x}^*) = 0$$

and which for each inactive constraint satisfy:
$$\underline{v}^T \cdot \nabla h_i(\underline{x}^*) > 0$$

are such that the Hessian matrix of the Lagrangian function is positive definite, or:
$$\underline{v}^T \cdot \nabla^2 \left(L(\underline{x}^*, \underline{\lambda}^*) \right) \cdot \underline{v} > 0$$

Note that a local minimum can exist even if the sufficient conditions are not satisfied. Several worked examples on the testing for necessary and sufficient conditions are provided in Edgar (1995).

105.8 Worked Example No 4

Reconsider the Worked Example No 3 to illustrate the use of the Kuhn-Tucker conditions:

$$\text{minimise:} \quad f(\underline{x}) = x_1 \cdot x_2$$
$$\text{subject to:} \quad h(\underline{x}) = x_1^2 + x_2^2 - 8 \leq 0$$

The Lagrangian function is:
$$L(\underline{x}, \lambda) = x_1 \cdot x_2 + \lambda(x_1^2 + x_2^2 - 8 + \sigma^2)$$

for which two minima were found to be when $\lambda^* = 0.5, \sigma^* = 0, x_1* = \pm 2$ and $x_2* = \mp 2$.

To prove that these are indeed true minima it must be demonstrated that the necessary and sufficient conditions are all satisfied:

1. The functions $f(\underline{x})$ and $h(\underline{x})$ are both differentiable:
$$\frac{\partial f(\underline{x})}{\partial x_1} = x_2 \qquad \frac{\partial f(\underline{x})}{\partial x_2} = x_1$$
$$\frac{\partial h(\underline{x})}{\partial x_1} = 2x_1 \qquad \frac{\partial h(\underline{x})}{\partial x_2} = 2x_2$$

These are differentiable for all $\underline{x}$ and are certainly so at $\underline{x}^*$.

2. The Lagrange multiplier does indeed exist: it has a value of $\lambda = 0.5$.

3. The constraint is satisfied at $\underline{x}^*$
If $\underline{x}^* = \pm 2$ or ∓ 2 then $h(\underline{x}^*) = 0$.

4. The Lagrange multiplier for the inequality constraint is positive: $\lambda = 0.5$.
5. The inequality constraint $h(\underline{x}^*) = 0$ is binding.
6. The Lagrangian function is at a stationary point at $\underline{x}^*$:

$$x_2^* + 2\lambda^*.x_1^* = x_1^* + 2\lambda^*.x_2^*$$
$$= x_1^{*2} + x_2^{*2} - 8 + \sigma^{*2}$$
$$= 2\lambda^*\sigma^* = 0$$

7. The gradient (Jacobian) of the only active constraint at $\underline{x}^*$ is:

$$J(\underline{x}) = \begin{bmatrix} \dfrac{\partial h(\underline{x})}{\partial x_1} \\ \dfrac{\partial h(\underline{x})}{\partial x_2} \end{bmatrix} = \begin{bmatrix} 2x_1 \\ 2x_2 \end{bmatrix}$$

whence:

$$J(\underline{x}^*) = \nabla h(\underline{x}^*) = \begin{bmatrix} \pm 4 \\ \mp 4 \end{bmatrix}$$

Since there is only one constraint, linear independence cannot be an issue.

9. The vector $\underline{v}$ which satisfies $\underline{v}^T . \nabla h(\underline{x}^*) = 0$ is found as follows:

$$\begin{bmatrix} v_1 & v_2 \end{bmatrix} \begin{bmatrix} \pm 4 \\ \mp 4 \end{bmatrix} = 0$$

Any value of v such that $v = v_1 = v_2$ will satisfy this criterion.
Choose $\underline{v} = \begin{bmatrix} v & v \end{bmatrix}^T$.
Next evaluate the Hessian of the Lagrangian function at $\underline{x}^*$:

$$H(\underline{x}) = \begin{bmatrix} \dfrac{\partial^2 L(\underline{x}, \lambda)}{\partial x_1^2} & \dfrac{\partial^2 L(\underline{x}, \lambda)}{\partial x_1 \partial x_2} \\ \dfrac{\partial^2 L(\underline{x}, \lambda)}{\partial x_2 \partial x_1} & \dfrac{\partial^2 L(\underline{x}, \lambda)}{\partial x_2^2} \end{bmatrix} = \begin{bmatrix} 2\lambda & 1 \\ 1 & 2\lambda \end{bmatrix}$$

whence:

$$H(\underline{x}^*) = \nabla^2 (L(\underline{x}^*, \lambda^*)) = \begin{bmatrix} 1 & 1 \\ 1 & 1 \end{bmatrix}$$

which is positive definite and hence:

$$\underline{v}^T . \nabla^2 (L(\underline{x}^*, \lambda^*)) . \underline{v} = \begin{bmatrix} v & v \end{bmatrix} \begin{bmatrix} 1 & 1 \\ 1 & 1 \end{bmatrix} \begin{bmatrix} v \\ v \end{bmatrix}$$
$$= 4v^2 > 0$$

Thus all the necessary and sufficient criteria have been satisfied for $\underline{x}^*$ to be local minima.

105.9 Quadratic Programming

Cost functions often contain quadratic functions and quadratic programming (QP) is the name given to the minimisation of a quadratic cost function involving n variables and subject to m linear constraints. Formally, the QP problem can be stated as:

$$\text{minimise:} \quad f(\underline{x}) = \frac{1}{2}.\underline{x}^T.Q.\underline{x} \quad (105.20)$$
$$+ \underline{x}^T.C + c$$

subject to: $A.\underline{x} \leq B$ and $\underline{x} \geq 0$ (105.21)

where the dimensions of $\underline{x}$ and C are $(n \times 1)$, of Q is $(n \times n)$, of A is $(m \times n)$, of B is $(m \times 1)$ and the bias c is a scalar. The matrix A is known as the constraint matrix. Quadratic functions arise naturally in optimisation and, for most purposes in process automation, Equations 105.20 and 105.21 are as complex a scenario as needs to be considered.

Note that because Equation 105.21 is linear, the constraints must be linearised with the vector of decision variables $\underline{x}$ in deviation form.

Introducing slack variables into the inequalities as appropriate:

$$A.\underline{x} - B + \underline{\sigma} = 0 \quad (105.22)$$

and

$$\underline{\rho} - \underline{x} = 0$$

The Lagrangian function may thus be formulated:

$$L(\underline{x}, \underline{\lambda}, \underline{\mu}) = \frac{1}{2}.\underline{x}^T.Q.\underline{x} + \underline{x}^T.C + c \quad (105.23)$$
$$+ \underline{\lambda}^T.(A.\underline{x} - B + \underline{\sigma}) + \underline{\mu}^T.(\underline{\rho} - \underline{x})$$

where:

$$\underline{\lambda} = \begin{bmatrix} \lambda_1 & \lambda_2 & \cdots & \lambda_m \end{bmatrix}^T, \quad \underline{\sigma} = \begin{bmatrix} \sigma_1^2 & \sigma_2^2 & \cdots & \sigma_m^2 \end{bmatrix}^T$$

and

$$\underline{\mu} = \begin{bmatrix} \mu_1 & \mu_2 & \cdots & \mu_n \end{bmatrix}^T, \quad \underline{\rho} = \begin{bmatrix} \rho_1^2 & \rho_2^2 & \cdots & \rho_n^2 \end{bmatrix}^T \quad (105.24)$$

Noting that differentiation by a vector is covered in Chapter 79, the gradient of the Lagrangian function is thus:

$$\frac{dL(\underline{x}, \underline{\lambda}, \underline{\mu})}{d\underline{x}} = Q.\underline{x} + C + A^T.\underline{\lambda} - \underline{\mu}$$

Thus the condition for an extremum is that:

$$Q.\underline{x} + C + A^T.\underline{\lambda} - \underline{\mu} = 0 \quad (105.25)$$

The value of $\underline{x}$ that is the solution to this equation simultaneously minimises the cost function of Equation 105.20 and satisfies the constraints of Equation 105.21.

In addition to the non-negativity constraints on the decision variables, there are also the Kuhn-Tucker conditions for a minimum to be satisfied. Thus:

$$\underline{x} \geq 0, \quad \underline{\lambda} \geq 0, \quad \underline{\mu} \geq 0 \quad \text{and} \quad \underline{\sigma} \geq 0. \quad (105.26)$$

It follows that:

$$\underline{\lambda}^T.\underline{\sigma} = 0 \quad \text{and} \quad \underline{\mu}^T.\underline{x} = 0. \quad (105.27)$$

These latter equations take some explanation: consider the first, although the argument is exactly the same for the second. For any inequality constraint that is:

- active, its value of σ_i^2 is zero
- inactive, its value of λ_i must be zero

Thus for every inequality, irrespective of activity, the product of λ_i and σ_i^2 must be zero.

It is quite common to combine these two equations:

$$\underline{\lambda}^T.\underline{\sigma} + \underline{\mu}^T.\underline{x} = 0$$

Let the set of variables $\underline{x}^*, \underline{\lambda}^*, \underline{\mu}^*$ and $\underline{\sigma}^*$ be the optimal solution to Equation 105.20.

Provided that Q is positive definite, the solution will be a global minimum if the set of variables simultaneously satisfy Equations 105.25–105.27.

105.10 Worked Example No 5

Minimise the function:

$$f(\underline{x}) = \frac{1}{2}x_1^2 + x_2^2 - x_1.x_2 - 3x_1 + 2.x_2 + 5$$

subject to the inequality constraints:

$$x_1 + x_2 \leq 2$$
$$-x_1 + 2.x_2 \leq 2$$
$$2.x_1 + x_2 \leq 3$$

and to the non-negativity constraints:

$$x_1 \geq 0$$
$$x_2 \geq 0$$

The problem may be cast in the form of Equations 105.20 and 105.21 by specifying:

$$A = \begin{bmatrix} 1 & 1 \\ -1 & 2 \\ 2 & 1 \end{bmatrix} \quad B = \begin{bmatrix} 2 \\ 2 \\ 3 \end{bmatrix} \quad C = \begin{bmatrix} -3 \\ 2 \end{bmatrix} \quad c = 5$$

and $Q = \begin{bmatrix} 1 & -1 \\ -1 & 2 \end{bmatrix}$

Substitution into Equation 105.22 and rearrangement gives:

$$x_1 + x_2 + \sigma_1^2 = 2$$
$$-x_1 + 2.x_2 + \sigma_2^2 = 2$$
$$2.x_1 + x_2 + \sigma_3^2 = 3$$

Substitution into Equation 105.25 and rearrangement gives:

$$x_1 - x_2 + \lambda_1 - \lambda_2 + 2.\lambda_3 - \mu_1 = 3$$
$$-x_1 + 2.x_2 + \lambda_1 + 2.\lambda_2 + \lambda_3 - \mu_2 = -2$$

So far there are five equations with ten unknowns. The five extra equations necessary for a solution come from Equation 105.27:

$$\lambda_1.\sigma_1^2 = \lambda_2.\sigma_2^2 = \lambda_3.\sigma_3^2 = \mu_1.x_1 = \mu_2.x_2 = 0$$

Despite the fact that the last five equations are non-linear they do considerably simplify the solution. That is because at least one out of each pair of variables must be zero, that is five zeros in total. Suppose, for sake of argument, that:

$$\sigma_1^2 = \sigma_2^2 = \lambda_3 = \mu_1 = \mu_2 = 0$$

Table 105.2 Feasible solutions for minimisation of Worked Example No 5

x_1	x_2	λ_1	λ_2	λ_3	μ_1	μ_2	σ_1^2	σ_2^2	σ_3^2	$f(\underline{x})$
0.67	1.33	1.11	−2.6	0.0	0.0	0.0	0.0	0.0	0.33	6.778
1.0	1.0	−9.0	0.0	6.0	0.0	0.0	0.0	1.0	0.0	4.50
0.0	1.0	0.0	−2.0	0.0	−2.0	0.0	1.0	0.0	2.0	8.0
−2.0	0.0	0.0	−5.0	0.0	0.0	−6.0	4.0	0.0	7.0	13.0
1.5	0.0	0.0	0.0	0.75	0.0	1.25	0.5	3.5	0.0	1.625
1.69	−0.38	0.0	0.0	0.46	0.0	0.0	0.69	4.46	0.0	1.385
0.0	0.0	0.0	0.0	0.0	−3.0	2.0	2.0	2.0	3.0	5.0
0.0	−1.0	0.0	0.0	0.0	−2.0	0.0	3.0	4.0	4.0	4.0

The original five equations thus reduce to:

$$\begin{bmatrix} 1 & 1 & 0 & 0 & 0 \\ -1 & 2 & 0 & 0 & 0 \\ 2 & 1 & 0 & 0 & 1 \\ 1 & -1 & 1 & -1 & 0 \\ -1 & 2 & 1 & 2 & 0 \end{bmatrix} \begin{bmatrix} x_1 \\ x_2 \\ \lambda_1 \\ \lambda_2 \\ \sigma_3^2 \end{bmatrix} = \begin{bmatrix} 2 \\ 2 \\ 3 \\ 3 \\ -2 \end{bmatrix}$$

the solution to which is found by matrix inversion to be as follows:

$$\begin{bmatrix} 0.667 & 1.333 & 1.111 & -2.556 & 0.333 \end{bmatrix}^T$$

which corresponds to the first row of Table 105.2. In fact there are 2^5 valid combinations of zeros so there are potentially 32 sets of solutions. However, half of the potential solutions are indeterminate and of the remainder many are infeasible in that they do not satisfy the inequality constraints. The feasible solutions are listed in Table 105.2, the value of their cost functions being evaluated from Equation 105.20.

Of these, only the solution at the point (1.5, 0) satisfies the non-negativity constraints and the Kuhn-Tucker conditions of Equation 105.26, corresponding to a global minimum at which the optimum has a value of $f(\underline{x}^*) = 1.625$.

105.11 Recursive Form of QP

Complex constrained optimisation problems are generally solved on a recursive basis for which a very convenient assumption is that any function $f(\underline{x})$ may, on a local basis, be adequately approximated by a quadratic function. The approximation is based upon Newton's method for unconstrained optimisation as previously described in Chapter 104.

Starting with Equation 104.8, Taylor's series expansion of an n dimensional function is:

$$f(\underline{x}(k+1)) - f(\underline{x}(k)) \approx J(\underline{x}(k))^T . \Delta \underline{x} \quad (105.28)$$
$$+ \frac{1}{2} . \Delta \underline{x}^T . H(\underline{x}(k)) . \Delta \underline{x}$$

where:

$$\Delta \underline{x} = \begin{bmatrix} \Delta x_1 \\ \Delta x_2 \\ \vdots \\ \Delta x_n \end{bmatrix} = \begin{bmatrix} x_1(k+1) - x_1(k) \\ x_2(k+1) - x_2(k) \\ \vdots \\ x_n(k+1) - x_n(k) \end{bmatrix} \quad (105.29)$$

Introducing the ∇ notation ($\nabla = d/d\underline{x}$), a merit function $\phi(\Delta \underline{x})$ may be defined:

$$\phi(\Delta \underline{x}) = \frac{1}{2} . \Delta \underline{x}^T . \nabla^2 f(\underline{x}(k)) . \Delta \underline{x} + \Delta \underline{x}^T . \nabla f(\underline{x}(k))$$

Defining the search direction $\underline{s}(k) = \Delta \underline{x}$ gives:

$$\phi(\underline{s}(k)) = \frac{1}{2} . \underline{s}(k)^T . \nabla^2 f(\underline{x}(k)) . \underline{s}(k) \quad (105.30)$$
$$+ \underline{s}(k)^T . \nabla f(\underline{x}(k))$$

This is an iterative form of Equation 105.20 in which $\underline{s}(k)$ is the step in the search direction required to find the minimum of the quadratic. Thus, provided that the approximation holds true, and noting that $\phi(\underline{s}(k))$ relates to the change in $f(\underline{x})$ rather than to its absolute value, the minimisation of $f(\underline{x})$ is equivalent to the following QP problem:

minimise $\phi(\underline{s}(k))$

subject to $g_i(\underline{x}(k)) = 0$
where $i = 1, 2, \ldots, m$ (105.31)

and $h_i(\underline{x}(k)) \leq 0$
where $i = m + 1, \ldots, p$

In principle the only constraints of interest are the equality constraints and any active inequality constraints. Thus for gradient purposes only:

$h_i(\underline{x}(k)) = 0$ where $i = m + 1, \ldots, q$
and $q \leq p$ (105.32)

Note that the h is in italics to distinguish between the active and inactive inequality constraints.

Along with the quadratic approximation, as in QP, it is normal practice to simplify the problem by linearising the constraints. Recognising that the value of the constraint at any point $\underline{x}$ on the constraint is zero, the change in value for a step in the search direction, which has to be along the constraint, must also be zero, whence:

$g_i(\underline{x}(k)) + \underline{s}(k)^T . \nabla g_i(\underline{x}(k)) = 0$ (105.33)
$h_i(\underline{x}(k)) + \underline{s}(k)^T . \nabla h_i(\underline{x}(k)) = 0$ (105.34)

The Lagrangian function may thus be formed from Equations 105.30, 105.33 and 105.34:

$$L(\underline{s}(k), \underline{\lambda}(k)) = \frac{1}{2} . \underline{s}(k)^T . \nabla^2 f(\underline{x}(k)) . \underline{s}(k)$$
$$+ \underline{s}(k)^T . \nabla f(\underline{x}(k))$$
$$+ \sum_{i=1}^{m} \lambda_i \left(g_i(\underline{x}(k)) + \underline{s}(k)^T . \nabla g_i(\underline{x}(k)) \right) \quad (105.35)$$
$$+ \sum_{i=m+1}^{q} \lambda_i \left(h_i(\underline{x}(k)) + \underline{s}(k)^T . \nabla h_i(\underline{x}(k)) \right)$$

Differentiating with respect to $\underline{s}(k)$ gives the gradient of the Lagrangian:

$$\nabla L(\underline{s}(k), \underline{\lambda}(k)) = \nabla^2 f(\underline{x}(k)) . \underline{s}(k) \quad (105.36)$$
$$+ \nabla f(\underline{x}(k)) + G(k) . \underline{\lambda}(k)$$

where:

$$G(k) = \begin{bmatrix} \nabla g_1(\underline{x}(k)) & \nabla g_2(\underline{x}(k)) & \cdots & \nabla g_m(\underline{x}(k)) \\ & \nabla h_{m+1}(\underline{x}(k)) & \cdots & \nabla h_q(\underline{x}(k)) \end{bmatrix} \quad (105.37)$$

Note that there is no need to allow for the active not-negative subset of decision variables $\underline{x}(k)$ in Equation 105.35. If some term involving a vector of Lagrange multipliers $\underline{\mu}(k)$ and $\underline{x}(k)$) was included, as in Equation 105.23, then that term would disappear when the Lagrangian is differentiated. Thus, for a minimum:

$$\nabla^2 f(\underline{x}(k)) . \underline{s}(k) + \nabla f(\underline{x}(k)) + G(k) . \underline{\lambda}(k) = 0 \quad (105.38)$$

This is the gradient criterion. Additionally, all of the constraints of Equations 105.31 and 105.32 must be satisfied, together with any not-negative constraints on the decision variables, and other Kuhn-Tucker conditions as appropriate:

$$\underline{x} \geq 0 \quad \text{and} \quad \underline{\lambda} \geq 0 \quad (105.39)$$

Provided that the Hessian $\nabla^2 f(\underline{x}(k))$ is positive definite, that set of variables $\underline{x}(k)$ and $\underline{\lambda}(k)$ which simultaneously satisfy these criteria corresponds to the global solution of the quadratic approximation.

105.12 Sequential Quadratic Programming

Sequential quadratic programming (SQP) is the most commonly used recursive technique for solving complex optimisation problems. Such problems may be categorised as being more complex than those represented by Equations 105.20 and 105.21: that is when:

- the cost function is cubic or of higher order, and/or
- when the constraints are non-linear and have to be linearised recursively.

The basic SQP recursion is as depicted in Figure 105.1: there are many variations on the theme.

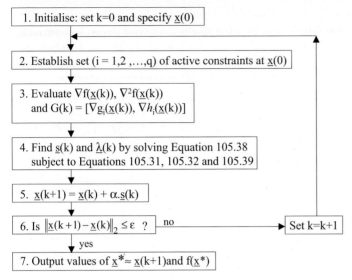

Fig. 105.1 Recursive search using sequential quadratic programming

The objective of SQP is to minimise some cost function f($\underline{x}$). However, rather than minimising f($\underline{x}$) directly some merit function $\phi(\underline{s})$ is minimised. The merit function $\phi(\underline{s})$ is a quadratic approximation of the problem which locally mimics the slope of f($\underline{x}$). Starting from an initial position $\underline{x}(k)$ in the search space, a step is made in the direction $\underline{s}(k)$ of the minimum of the approximation. The position $\underline{x}(k + 1)$ at the end of the step becomes the initial position for the next iteration and the merit function $\phi(\underline{s})$ is redefined. The process is repeated until the global minimum f($\underline{x}^*$) is found.

With reference to the box numbers of Figure 105.1:

1. Since the decision variables are in deviation form, it can be expected that most will have zero initial conditions: that is, $\underline{x}(0) = 0$.
2. Recognising that the constraints which are active will probably vary as $\underline{x}(k)$ progresses through the search space, a set management strategy is required to review and change the set of constraints being applied. Some tolerance is normally applied to each constraint, otherwise those constraints which are almost but not quite active will be excluded resulting in the set changing unnecessarily from one iteration to the next and distorting the optimisation.
3. It is normal practice to use a positive definite approximation for the Hessian $\nabla^2 f(\underline{x}(k))$ to guarantee minimisation. The approximation is updated at each iteration using a so-called secant method such as that due to Broyden, Fletcher, Goldfarb and Shanno (BFGS). This involves first order derivatives only, is computationally less intensive and results in "superlinear convergence".
4. This is the local QP sub problem solved at each iteration. The solution of Equation 105.38 is essentially the same as the Worked Example No 5 above.
5. The size of step taken in the direction of $\underline{s}(k)$ is determined by the factor α. In practice it is normal for the step size to be a variable $\alpha(k)$ as in Equation 104.12. A line searching technique such as described in Section 104.8 may be used.
6. There are many potential end criteria: another obvious one is:

$$\|\phi(\underline{s}(k+1)) - \phi(\underline{s}(k))\|_2 \leq \varepsilon?$$

7. An important characteristic of SQP is that the closer the search gets to the global optimum, the more accurate the quadratic approximation becomes.

The strategy underlying Figure 105.1 is said to be equality constrained because only the equality constraints and the active inequality constraints are considered, as per Equations 105.31 and 105.32. This requires a constraint set management strategy as described. The alternative is an inequality constrained strategy in which all the inequality constraints are included in the Lagrangian function. This avoids the need for a set management strategy but considerably increases the amount of computation.

105.13 Reduced Gradient Methods

Reduced gradient methods are an alternative approach to minimising a cost function. The dimensionality of the search is reduced by using the constraints to eliminate dependent variables. This is done on a local linearised basis. Reconsider the problem:

minimise: $\quad f(\underline{x})$ (105.1)

subject to: $\quad g(\underline{x}) = 0$ (105.2)

where: $\quad \underline{x} = \begin{bmatrix} x_1 & x_2 \end{bmatrix}^T$

The total derivatives are:

$$df(\underline{x}) = \frac{\partial f(\underline{x})}{\partial x_1}.dx_1 + \frac{\partial f(\underline{x})}{\partial x_2}.dx_2 \quad (105.40)$$

$$dg(\underline{x}) = \frac{\partial g(\underline{x})}{\partial x_1}.dx_1 + \frac{\partial g(\underline{x})}{\partial x_2}.dx_2 = 0 \quad (105.41)$$

Suppose x_1 is an independent variable and x_2 is dependent then, from Equation 105.41:

$$dx_2 = -\frac{\partial g(\underline{x})}{\partial x_1}.\frac{\partial x_2}{\partial g(\underline{x})}.dx_1$$

Substituting into Equation 105.40 gives

$$df(\underline{x}) = \left(\frac{\partial f(\underline{x})}{\partial x_1} - \frac{\partial f(\underline{x})}{\partial x_2}.\frac{\partial g(\underline{x})}{\partial x_1}.\frac{\partial x_2}{\partial g(\underline{x})} \right).dx_1$$

The bracketed expression is the reduced gradient: provided the partial derivatives at $\underline{x}$ are known the slope is a scalar quantity. In essence each constraint can be used to eliminate one dependent variable. Suppose $f(\underline{x})$ was a function of five variables and there were two constraints: then the reduced gradient would be in terms of three variables.

Reduced gradient algorithms are recursive. As with SQP the problem is linearised locally. Thus at each step in the search the reduced gradient is re-evaluated in terms of its partial derivatives at $\underline{x}(k)$ which maintains the local feasibility of the constraints. A step of appropriate size and direction is taken to reduce the cost function and the process is repeated until, hopefully, a global minimum is found. However, with the linearisation of the constraints, there is scope for the search to terminate at a non-feasible solution with regard to the original constraints.

105.14 Penalty Functions

Yet another approach to solving constrained optimisation problems involves the use of penalty or barrier functions. These convert the constrained optimisation problem to a systematic series of unconstrained function minimisations of the same form but with different parameters. For example:

minimise $\quad f(\underline{x})$

subject to $\quad h_i(\underline{x}) \leq 0$

where $\quad i = 1, 2, \ldots, m$

This may be recast as a penalty function problem:

minimise $\quad P(\underline{x})$

where $\quad P(\underline{x}) = f(\underline{x}) + \sum_{i=1}^{m} \partial_i.h_i^2(\underline{x})$

with $\quad \partial_i = 0 \quad$ if $\quad h_i(\underline{x}) \leq 0$

or $\quad \partial_i = 1 \quad$ if $\quad h_i(\underline{x}) > 0$

Alternatively it may be recast as a barrier function problem:

minimise $\quad B(\underline{x})$

where $\quad B(\underline{x}) = f(\underline{x}) + \sum_{i=1}^{m} \frac{1}{h_i(\underline{x})}$

The penalty function approach is said to be an exterior method in the sense that it penalises constraint violations. In contrast, the barrier function approach is said to be an interior method as it forces the optimiser to remain inside the constraint boundary by becoming infinite on the boundary. This implies that it is possible to approach a boundary point from the interior of a feasible region. These methods are not the panacea that they may first appear to be as they pose a number of numerical difficulties. Nevertheless they can be used to some effect.

105.15 Nomenclature

Symbol	Description	Dimensions
c	constant	1×1
$f(\underline{x})$	cost (objective) function	1×1
$g(\underline{x})$	equality constraint	1×1
$h(\underline{x})$	inequality constraint	1×1
L	Lagrangian function	1×1
x	decision variable	1×1
$\phi(\underline{s})$	merit function	1×1
σ^2	slack variable	1×1
$\underline{x}$	vector of decision variables	$n \times 1$
$\underline{s}$	search direction	$n \times 1$
$\underline{\lambda}$	vector of Lagrange multipliers (equality)	$m \times 1$ or $q \times 1$
$\underline{\mu}$	vector of Lagrange multipliers (inequality)	$n \times 1$
$\underline{\rho}$	vector of slack variables	$n \times 1$
$\underline{\sigma}$	vector of slack variables	$m \times 1$
A	constraint matrix	$m \times n$
B	vector of excesses	$m \times 1$
C	linear operator	$n \times 1$
G	matrix of gradients	$n \times q$
$H(\underline{x})$	Hessian matrix	$n \times n$
$J(\underline{x})$	Jacobian vector	$n \times 1$
Q	quadratic operator	$n \times n$

Real Time Optimisers

Chapter 106

106.1 Steady State Optimisers
106.2 Models for SS Optimisers
106.3 Methodology for SS Optimisers
106.4 Steady State Detection
106.5 Steady State Model Updating
106.6 Solution of SS Optimisation Problem
106.7 Dynamic Optimising Controllers
106.8 Constraint Handling by DOCs
106.9 QP Solution of DOC Problem
106.10 Formulation of Constraints
106.11 Application of DOCs
106.12 Comments

The purpose of real-time optimisation is to maximise profit, the problem usually being cast in the form of the minimisation of some cost (objective) function. Profit is realised by minimising processing costs, maximising production and/or yield, minimising energy consumption, and so on. However, it is not a universal panacea and it should be recognised that whilst the potential benefits of optimisation are large so too are the costs. Fundamental to the successful implementation of an optimiser is an understanding, not just of the techniques and technology available, but also of the application.

It is evident that there is an important philosophical difference between control and optimisation. With control the objective is to hold the controlled variables at their set points, irrespective of any disturbances or set point changes. There should be no spare degrees of freedom, otherwise the plant is said to be underdetermined. Optimisation implies choice: there are ranges of conditions under which the plant can operate and the best set of conditions is found subject to a variety of constraints. The objective is to hold the process at the optimum, to make changes when any of the constraints are violated or if the process moves away from the optimum, and to manage disturbances allowing for economies.

So, with real-time optimisers (RTO), not only does the optimum have to be found, which is not necessarily easy because the optimum may itself may be changing, but the plant and/or process has to be manipulated dynamically to maintain the optimum set of conditions. There are a variety of approaches and techniques available which, together with the technology to implement them, enable robust applications.

For the purposes of this chapter optimisers are categorised into two generic types: steady state optimisers (SSO) and dynamic optimising controllers (DOC). The essential characteristics are summarised in Table 106.1 and depicted in Figure 106.1. It is worth commenting upon the non-uniqueness of RTOs: they reflect all the variety that one would expect of different suppliers' products. That variety embraces both SSO and DOC types of optimiser, the underlying techniques of LP, QP or SQP, the approaches to constraint handling and

Table 106.1 Steady state *vs* dynamic real-time optimisers

Category	Essential characteristics
Steady state optimiser (SSO) Figures 106.2–106.4	Large, steady state, non-linear models
	Models normally of first principles type
	Parameter estimation to keep model up to date
	Generate set of linear constraints for optimisation
	Optimisation solved by SQP
	Executed infrequently (hours) at steady state
	Operated either off-line or on-line
	If on-line may be either open or closed loop
	SPs downloaded to DCS directly or to MPC
	Expensive to build and maintain
Dynamic optimising controller (DOC) Figures 106.7–106.9	Small, linear models
	Hybrid first principles and/or empirical models
	Model augmented externally if very non-linear
	Model usually biased against process
	Model may be updated by filter or observer
	Optimising capability integrated with MPC
	MPC may be either SISO or MIMO
	Used to predict optimal steady state
	Optimisation solved by QP (older ones by LP)
	Executed frequently (min)
	Constraint management has highest priority
	Cross plant optimisation *via* bridge models

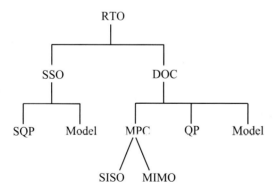

Fig. 106.1 Classification of real-time optimisers

the form of interface between the RTO and plant controllers.

With the theory of LP, QP, *etc.* having been covered previously, the emphasis in this chapter is on the functionality of optimisers, the pros and cons of SSO *vs* DOC, and on implementation issues.

106.1 Steady State Optimisers

The structure of a traditional SSO is as depicted in Figure 106.2. Depending on the type of model, the optimiser is a large LP or, more typically, an SQP program which is executed intermittently, ideally when the plant is at steady state. The SSO receives as inputs the current values of relevant manipulated (independent) variables (MVs) and controlled (dependent) variables (CVs) and evaluates a new set of optimum operating conditions. The output is typically in the form of a number of new set points (SPs) which are downloaded into the

DCS, or equivalent control system. The DCS then moves the plant towards those set points, and holds it there, until they are updated at the next execution of the optimiser. Clearly there is some requirement to filter controller output or to restrict MV changes upon updating the set points to ensure smooth transitions.

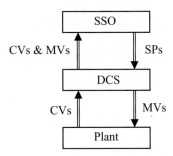

Fig. 106.2 Structure of a traditional steady state optimiser (SSO)

Different scenarios emerge from an operational point of view. Early SSOs were always run in an off-line mode in which case there were no real-time measurements, dynamics were of no concern and disturbances could be ignored. Clearly off-line optimisation was only of any use for case studies and planning purposes. Bringing optimisers on-line introduced all of the above time related issues, together with the decision about whether to operate open or closed loop. In open loop mode the optimiser is essentially advisory: it recommends optimum set points which the operators manually enter into the DCS. There is, however, a danger. The operators may only enter the set points they agree with whereas, of course, they must all be entered to achieve the optimum. Also, it is only feasible if the optimiser is run infrequently, otherwise the operators won't have enough time to make all the changes. In closed loop mode the set points are all downloaded automatically into the DCS as is implied in Figure 106.2. Note that the open loop mode is often used as a means of gaining confidence in an optimiser before going closed loop.

An obvious, but not essential, extension is for the SSO to intermittently determine the required steady state set points and/or targets, and to download them to a model predictive controller (MPC), as depicted in Figure 106.3. Note that MPC is explained in detail in Chapter 117. The MPC, which takes into account the plant's dynamics, regularly adjusts the set points of slave loops which are downloaded to the DCS for realisation. It is not uncommon on large plants for there to be several MPCs as depicted in Figure 106.4, each dedicated to the control of a particular process function or plant unit, with the SSO downloading set points and/or targets to each simultaneously.

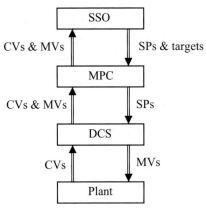

Fig. 106.3 SSO downloading set points to a model predictive controller

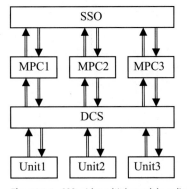

Fig. 106.4 SSO with multiple model predictive controllers (MPC)

106.2 Models for SS Optimisers

There is much variety in the types of model used by SSOs depending upon whether the model is linear or non-linear, of a first principles nature or empirical, and upon the scope of the model in terms of detail, accuracy and size of the plant. It is not unrealistic to expect a rigorous first principles model to have some 10^5 to 10^6 equations for a large and complex plant and/or process. At the other end of the scale, an empirical linear model for a relatively small plant may have just a few hundred equations. Empirical models are typically of a step response nature, although the use of regression and neural net based models are becoming increasingly common. However, in practice, most models used are of a first principles nature, non-linear and large.

These non-linear models are due to the fact that most processes and plant items have non-linear characteristics and are may be operated over a range of conditions for which many different steady states are feasible. The variables are normally in absolute form as opposed to deviation form, and the models are not linearised. Obviously, for a steady state model, the dynamics are ignored. In essence, the model reflects the structure of the plant and the processes being carried out. For each item in which some physical or chemical change takes place, its model will be comprised of a selection of the following as appropriate:

- Physical connections between plant items in terms of pipework and process routes
- Overall mass balance equations that relate the input streams to the output streams
- Mass balance equations for each component in the input and output streams
- Mass transfer equations for each component between streams
- Stoichiometric and equilibria equations for reacting systems
- Formulae or correlations for evaluating physical property data
- Formulae for evaluating coefficients from flow rates, temperatures, *etc.*
- Heat balance equations that relate input and output conditions for each stream
- Heat transfer equations for evaluating heat flow rates between streams
- Formulae for estimating pressure drops across fixed and variable resistances to flow
- Equations for balancing pressure drops in liquid, gaseous or mixed flow systems
- Equations for handling recycle, purge and by-pass streams
- Empirical data for operating characteristics of plant items
- Constraint equations for the physical capacity of the plant/process
- Functional relationships between CVs and MVs for all key control loops
- *etc.*

In addition to the process/plant model there is the production model to be developed which comprises:

- Specification of decision variables: independent (MVs) and dependent (CVs)
- Cost function as a linear or quadratic expression of the decision variables
- Constraints on throughput in relation to decision variables
- Constraints on CVs and MVs for safety operability purposes
- *etc.*

Given the variety of equations, formulae, correlations and constraints that have to be taken into account, and that the process/plant model's equations are largely generated automatically once the relevant parameters are defined, it is easy to see how even a single, relatively simple, plant item can lead to several hundreds of equations.

Models for SSOs are normally generated on a modular basis using proprietary modelling packages of a flowsheeting nature. These have libraries of standardised models for items of plant and process operations which can be selected from menus and configured as appropriate. Modern packages provide a development environment which supports this process with, for example, icons that can be dragged, dropped and interconnected on screen. For each model selected relevant parameters must be specified for which there is a template of database values to be completed. Typically, a context sensitive editor will either deduce database values from the connectivity or prompt for other values, check for consistency and then generate the equations automatically.

Generally speaking, the more rigorous a model the greater its accuracy. However, remember, first principles models are comprised of equations and constraints which describe the underlying chemistry and physics of the process and/or plant. The accuracy of those equations ultimately depends upon the values of the coefficients, constants and physical properties used in them. There is no point in looking for perfection where it doesn't exist or add value. It goes without saying that such models are costly to develop, test and validate in the first place. They are also costly to maintain having to be kept up to date with respect to changes made to the plant, process and operations.

106.3 Methodology for SS Optimisers

The basic methodology for steady state optimisation is as depicted in Figure 106.5.

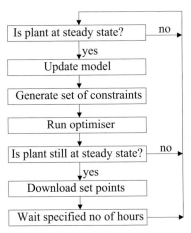

Fig. 106.5 The steady state optimisation process

The methodology is fairly straightforward, even if its implementation isn't. Assuming that the model exists, the methodology may be summarised as follows:

1. Wait for plant to reach steady state. This requires some capability to determine when the plant is at steady state, as described in Section 4 below.
2. Update the model by fitting the model to the process, sometimes referred to as calibrating the model. This is described in Section 5 below.
3. Generate a set of linear equalities and constraints from the non-linear model, and find their slopes, as described in Section 6 below.
4. Solve the steady state optimisation problem to find a new, more profitable, set of operating conditions within the constraints. This too is described in Section 6 below.
5. Check that the plant is still at the same steady state.
   This is an essential step. If the steady state has changed from that established in 1 above, then the solution found in 4 above is unlikely to be optimal.
6. Move to (or towards) the new optimal set of operating conditions. This basically means downloading set points and targets to the DCS and/or MPC as appropriate.
7. Wait an appropriate period to enable new optimum to be established.

106.4 Steady State Detection

Fundamental to this is a steady state detection algorithm, the purpose of which is to detect when the process is probably not at steady state. Typically, values of a number of strategic variables will be collected over a reasonable period of time and their means and standard deviations determined. Referring to Table 82.2, it can be seen that 95% of the measurements for any particular variable should be within two standard deviations of its mean, assuming a normal distribution. So, more than 5% lying beyond the significance limit cannot be accounted for by random process variation, which suggests that the process is not at steady state.

An important point to appreciate is that whilst all strategic variables should be at steady state, inventory variables are permitted to be changing. For example, with a distillation column, the temperature profile and flow rates may all be constant, but the flows in and out don't balance due to changing levels in the still and reflux drum.

Other methods of establishing steady state involve the use of control charts and limits as described in Chapter 102 and the use of principal components as described in Chapter 101. Principal components are often more sensitive to changes in steady state than individual variables and variations in principal components are a convincing indication that the process is not at steady state.

106.5 Steady State Model Updating

Updating of the steady state model makes use of real time plant measurements. Clearly some data validation is required beforehand to confirm that the measurements are in range and that there are no gross errors present. Data pre-screening is covered in detail in Chapter 102. In particular, note the use of average measurements for the filtering of variables.

The presumption made is that the plant/process is correct and that any differences between it and the model are due to either faulty measurements and/or to errors in the model. Model updating involves making adjustments to specific measurements and/or model parameters until the values of the controlled variables predicted by the model agree with, or converge upon, their measurements from the plant:

- Data reconciliation concerns the random errors inherent in process measurements and involves the adjustment of measured values to resolve differences and/or conflicts. For example, the measurements of flow into and out of a unit will never agree, even at steady state, and the difference between them will never agree with the measurement of the change in level. Reconciliation is best done on a unit by unit basis in which all the measurements associated with a particular unit are handled together.
- Parameters in the model can be adjusted until the model fits the process at the current steady state. Clearly only those parameters in the model which could reasonably be expected to change, such as heat transfer coefficients and catalyst activity, should be adjusted. It is unreasonable, for example, to adjust a tank diameter.

Model updating is usually handled as a constrained optimisation problem, the approach being as depicted in Figure 106.6.

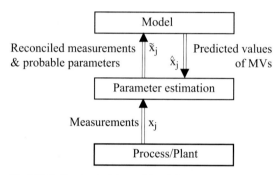

Fig. 106.6 The steady state model updating process

The measurements and/or parameters are adjusted, within bounds, to minimise some objective function:

$$J(\underline{x}) = \sum_{j=1}^{n} \beta_j \cdot \left(\frac{(\hat{x}_j - \tilde{x}_j)}{\sigma_j} \right)^2 \quad (106.1)$$

where it is assumed that there are n variables whose measurements x_j are to be adjusted, and

$\hat{x}$ is the estimate of x produced by the model
$\tilde{x}$ reconciled value of the measurement x
β weighting factor which reflects the variable's importance
σ is the standard deviation of the random error on x

An important characteristic of this approach is that it will detect systematic errors such as drift, faulty calibration and offset in the measurements. Further tests can be then used to confirm those measurements which are most probably in error: these can then be biased if/as appropriate and parameter estimation repeated.

106.6 Solution of SS Optimisation Problem

The steady state optimisation problem is most commonly solved by means of SQP, as explained in Chapter 105, which is a recursive form of a quadratic approximation of the cost function as articulated by Equation 105.28. Fundamental to the implementation of SQP are the constraints. Strictly speaking, it is only the equality and active inequality constraints that are of interest. These are normally cast as a set of linear equations in the form of Equation 105.21:

$$A.\underline{x} = B \qquad (106.2)$$

Generating the constraints as a set of linear equations is a key aspect to the solution. In essence this involves linearising the model about its current steady state condition as explained in Chapter 84:

- Non-linear equations in the model are differentiated and their slopes found about the current steady state, providing they are in a form which is differentiable in the first place.
- The slopes of other non-linear relationships, of an empirical nature, are established by "drawing a tangent" between points that straddle the current steady state conditions.
- Linear constraints and equations, of which there will be many even though the model is inherently non-linear, can all be used directly.

Notwithstanding the fact that all of the inactive inequality constraints can be eliminated, every equation in the non-linear model results in a linear constraint. Recognising that there may be many thousands of equations in the non-linear model, there will be a similar number of linear constraints. Note that the size of the vector $\underline{x}$ is determined by the number of variables involved in the constraints. Typically there are hundreds of such variables: far more than the number of decision variables used in the cost function which, at most, is measured in tens. The resulting sparsity in the formulation of the problem is exploited, in the interests of computational efficiency, by the matrix methods used for solving the problem.

Whereas the non-linear model uses the absolute values of variables, the linearisation process dictates that the constraints, and indeed the cost function, are cast in terms of deviation variables.

At the heart of SQP is Equation 105.38 which, by inspection, can be seen to contain the slopes (first and second derivatives) of the cost function $f(\underline{x})$ with respect to every variable in the cost function. Either $f(\underline{x})$ must be differentiable to establish these slopes, or else empirical values must be available. Also required are the numerous slopes of the equality constraints $g(\underline{x})$ and the active inequality constraints $h(\underline{x})$ with respect to every variable in the vector of decision variables, as defined by Equation 105.37. These are the values of the elements of the A matrix in Equation 106.2.

Finding the solution for a large model is computationally intensive. First, steady state detection and model updating is required. Every time the steady state changes the linearisation process has to be repeated to generate a new set of linear equalities and constraints (the A matrix). Then the QP problem has to be solved recursively, each potential solution being explored at each iteration until convergence on the optimum is achieved. This process takes a long time, 10–20 min not being untypical. For this reason SSOs tend to only be executed infrequently, on an hourly basis at best. Noting that it is necessary to wait for steady state before the optimiser can be run, once *per* shift or once *per* day is more typical.

A fairly fundamental problem with SSOs concerns disturbances: any that occur in between successive optimisations have to be handled by the control system. Clearly any significant disturbance could result in a different optimum and ought to be taken into account by the optimiser. Given the need to wait until a new steady state is reached before another optimisation can take place, it is evident that the effects of disturbances can lead to prolonged periods of sub-optimal operation. Thus, in the event of a major disturbance which is known *a priori* to affect the process adversely, it is not uncommon to run an SSO even if steady state has not been achieved/detected on the grounds that it is better to take some action, albeit imperfect, than no action at all.

It is inevitably the case that an SSO pushes the process over some surface or along a curve corresponding to a constraint in the solution space, or even into some corner feasible solution as described in Chapter 103 on LP. This may well be satisfactory from a steady state point of view but can result in very poor dynamic control. The control system must be able to respond to disturbances by adjusting its MVs: it can't if they are at constraints, and the process becomes uncontrollable. So, in the design of an optimiser, it is essential that some margin is left within the constraints on all the important MVs to enable effective control. This problem is obviated if the set points and/or targets are downloaded by the SSO to MPCs, as depicted in Figures 106.3 and 106.4, since the MPCs provide dynamic constraint management.

106.7 Dynamic Optimising Controllers

Most model predictive controllers (MPC) have an associated dynamic optimising capability: together they are loosely referred to as dynamic optimising controllers (DOC), as depicted in Figure 106.7.

It is appropriate to consider the MPC and its optimiser as being functionally independent of each other such that, for example, the DOC downloads set points to the MPC. In practice that distinction is blurred:

- Deliberately, the DOC and MPC have an integrated user environment.
- The optimisation and control functions are solved using the same QP engine.

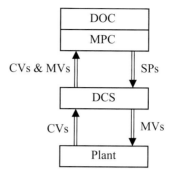

Fig. 106.7 MPC with dynamic optimising capability (DOC)

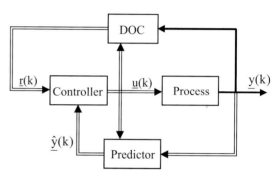

Fig. 106.8 Commonality of I/O signals used by DOC and MPC

- The DOC and MPC have a common set of input and output signals, as depicted in Figure 106.8. In practice the I/O used by an MPC (double arrows) is often a subset of the decision variables used by a DOC (solid arrows). See also Figure 117.3.

Notwithstanding the above comments about integration, the DOC can effectively be switched off resulting in an autonomous MPC.

Consideration of the I/O signals emphasises the fundamental difference between SSOs and DOCs. Because the model used by an SSO is likely to be that of the whole plant, for model updating purposes the number of I/O signals required is typically in the range of hundreds to thousands. In contrast to this the scope of a DOC is limited by the focus of the underlying MPC and is typically restricted to one or more functionally related units in which specific operations are carried out. Given the number of I/O signals required by an MPC, even for a large multivariable MPC, it follows that the number of I/O signals for a DOC is typically counted in tens rather than in hundreds.

Whereas SSOs are run intermittently, on an hourly basis at best, DOCs run at a fixed frequency, typically every few minutes. The frequency at which a DOC is executed relates to that of its underlying MPC. The DOC executes, and downloads setpoints to the MPC, at a frequency consistent with the prediction horizon of the MPC or some five to ten times its control horizon, as described in Chapter 117. For example, if the control horizon was two 10 sec steps ahead, then a DOC executing at a frequency of say once every 2–3 min would be sensible. A higher frequency would simply cause instability of the MPC.

An alternative way of looking at the time scales is to think in terms of the DOC handling the slow dynamics and steady state considerations, whilst the MPC handles the fast dynamics of the plant/process. At each execution of the DOC the QP problem is solved to establish the optimum set of values for the decision variables. This requires that its inputs are filtered to prevent the QP being forced unduly by relatively small changes, and that its outputs are filtered to prevent the underlying MPC being forced unduly by relatively large changes. Thus, despite the fact that the DOC executes frequently and that the solution may not be constant, it can be seen that the DOC can be used to predict the steady state solution.

106.8 Constraint Handling by DOCs

The handling of constraints is fundamental to the QP solution of the optimisation. There are three key scenarios:

- Either the process is operating against or within its constraints, in which case the DOC will attempt to maximise profit by adjusting the SPs downloaded to the MPC subject to keeping the CVs and MVs within their constraints.
  It is invariably the case that the optimum predicted by the DOC involves operating the process against one or more constraints. By downloading those SPs, the DOC is critically dependent upon the underlying MPC for pushing the process against the active constraints and for holding it there.
- Or the process is operating outside one or more of its constraints but, by relaxing (violating) them in the short term, the DOC can continue to maximise profit. Thus, if it appears that a (hard) constraint on an MV will not be satisfied, one or more of the (soft) constraints on the CVs may be relaxed. This is done by adjusting the upper and/or lower limits on CVs as appropriate, typically in some user defined order. The QP problem is then re-evaluated and, subject to the hard and new soft constraints being satisfied, the DOC downloads SPs to the MPC.
- Or else the process is operating outside some of its constraints, which cannot be relaxed, in which case the MPC will be prioritised to bring the process back within the constraints as quickly as possible. During this period the issue is to minimise the sub-optimality as opposed to the DOC maximising profit. This is done by adjusting those SPs which are consistent with moving the process back within the constraints, towards the optimum, albeit in a sub-optimal way.

Note, as discussed in Chapter 105, the significance of the Lagrange multipliers which indicate the sensitivity of the cost function with regard to each of the constraints. Inspection of the sign and magnitude of the multipliers, which will be available within the QP solution of the DOC, will determine which SPs should be changed and in what direction.

On large and/or complex plants there may well be several DOCs optimising the performance of operations being carried out in different units. Clearly there is scope to co-ordinate the activity of the DOCs by passing values between them. This is generally realised by means of bridge models as depicted in Figure 106.9. A bridge is a dynamic model which explains how changes in some variable (say an MV) on one unit affect dependant variables (say CVs) on another downstream unit. This prediction

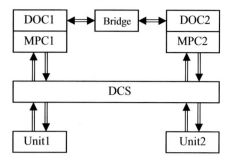

Fig. 106.9 Use of bridge models for co-ordinating multiple DOCs

of the affect of disturbances results in better steady state optimisation of the downstream unit.

106.9 QP Solution of DOC Problem

At the heart of a modern DOC is the QP technique, as described in Chapter 105, fundamental to which is formulation of the cost function and handling of the decision variables.

The cost function $J(\underline{x})$ is quadratic as per Equation 106.3. It is consistent with the general form of Equation 105.20 but with the matrices Q and C being diagonal. Note that there is no bias term since the variables are in deviation form. Deviations are relative to the current QP solution:

$$J(\underline{x}) = \sum_{j=1}^{n} q_j \cdot x_j^2 + \sum_{j=1}^{n} c_j \cdot x_j \qquad (106.3)$$

where $\underline{x}$ is the vector of decision variables (CVs, MVs and perhaps some DVs).

The coefficients c_j are the 1st derivatives (Jacobian) of the cost function with respect to each of the decision variables as appropriate. The coefficients may be either empirically or theoretically determined, or a mixture thereof. Normally they would be constant, consistent with the model being linear. However, if it is known *a priori* that there are significant non-linearities, which would obviously give rise to plant-model mismatch, then the slopes can be adjusted externally from a knowledge of their 2nd derivatives (Hessian) according to the current operating conditions.

It is normal practice to use the coefficients q_j to prevent excessive movement of the plant by penalising deviations of decision variables to which the steady state of the process is known *a priori* to be particularly sensitive.

It is interesting to note that the solution to the QP problem is algebraic: that is, the values of the decision variables that satisfy Equations 105.22, 105.25 and 105.27. The solution is dependant upon the various coefficients, especially the elements of the A matrix of Equation 105.21, and is independent of the decision variables for which no *a priori* values are assumed. So why does the DOC need to know the current value of the decision variables? There are several reasons, all of which relate to the fact that the decision variables are in deviation form:

1. If the constraints are non-linear they must be linearised about the current value of the relevant decision variables, either by differentiation or by "drawing a tangent". Thus the optimum is found relative to the current operating conditions.
2. For display purposes, the optimum values of the decision variables are reported in absolute form. Thus the deviations are added to the current values.
3. Some of the decision variables will correspond to both inputs and outputs. For example, a CV which is an input, may be optimised resulting in an SP being downloaded to the MPC to control that same CV. Thus, for such SPs, the next output x_{0j} is the current input x_{1j} plus the value of the decision variable x_j:

$$x_j = x_{0j} - x_{1j} \qquad (106.4)$$

106.10 Formulation of Constraints

All of the constraints, whether they are equalities or inequalities, of a physical or production nature, associated with process or control variables, are combined into a single matrix equation as appropriate:

$$A.\underline{x} \leq B$$

Remember that every constraint must be cast in a linear form: indeed, that is why the model used for QP is said to be linear. With ingenuity, any constraint, whether it be linear or non-linear, can be cast as an equation whose coefficients are consistent with being elements of the matrices A and B. There are different categories of constraint as follows:

- Process and production constraints, examples of which are as listed in Section 2 above. These define the steady state relationships between key process variables and hence determine the behaviour of the plant.
  Any non-linear constraints must be linearised. This means that they must either be differentiable or else that the values of their slopes about the current steady state are known empirically, as described in Section 6 above.
- Constraints on CVs. These are normally ranges, often determined by plant design considerations such as height of vessel, maximum rate of heat transfer, *etc*. In practice, the effective range of a CV is often determined by the instrumentation being used to measure it.
- Constraints on MVs. These are normally saturation effects, especially the opening of a valve (0, 100%) corresponding to zero and maximum flow. Sometimes there are maximum or minimum limits on flows for safety reasons. Occasionally rate of change constraints are applied to MVs to prevent the plant being forced unduly.

It follows that if any of the constraints change, the relevant elements of the matrices A and B must be updated. That is especially true of the elements associated with non-linear constraints. Typically, the steady state value of an input variable, say a CV associated with a non-linear constraint, will change such that the slope of the constraint about that value will need to be re-evaluated. In extreme cases this can mean updating elements of the matrices at every iteration of the DOC.

106.11 Application of DOCs

Having defined the cost function and articulated the constraints as a QP problem, the optimum is found by solving Equations 105.22, 105.25 and 105.27. The problem is solved by matrix manipulation which, notwithstanding the fact that there are multiple sets of solutions, as seen in Worked Example No 5 of Chapter 105, is relatively straightforward. That, combined with the much smaller size of the problem, means that finding the optimum is nothing like as computationally demanding as solving SQP problems, which explains why DOCs can be executed so frequently relative to SSOs.

The execution of a DOC is essentially as depicted in Figure 106.10.

The inputs to the DOC, a mixture of CVs and MVs, are the values of its input signals at the time of execution. New values for all the decision variables are then established, of which one or more are downloaded as SPs to the underlying MPC. These are held until the next execution of the DOC. When a change in the optimum occurs, the changes to the SPs are filtered resulting in a trajectory of SPs for the MPC over that DOC cycle. Thus the MPC can move the plant/process along a path towards the optimum rather than making a sudden jump.

The solution of the QP optimisation is dynamic in the sense that no attempt is made to wait for steady state conditions. Thus, on a regular basis, irrespective of the state of the plant, the DOC is executed. Subject to filtering to prevent sudden change, its inputs are those at the time of execution, whether they are constant or changing, and its outputs are a prediction of the steady state optimum based upon those inputs. In essence, the DOC does the best it can with the information available. Thus, unlike an SSO, the DOC can get on with pursuing changes in the optimum once a disturbance has occurred.

Most processes and plant items have non-linear characteristics and, strictly speaking, it is necessary to use non-linear models to correctly represent them. Those who have battled long and hard to produce complex, rigorous process models for SSOs often wonder how the simple linear models

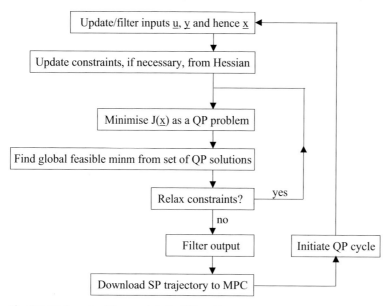

Fig. 106.10 Recursive QP approach to the DOC problem

used in DOCs are able to consistently find solutions that are so close (or even closer) to the true optima. The reason is that the optimisation is continuously supplemented with feedback from the process itself by biasing in real time. To explain this biasing it is helpful to consider the MPC as functionally consisting of two parts, the predictor and the controller proper, as depicted in Figure 106.8.

Consider a single CV. The DOC solves the QP problem as posed and finds the optimum value for the CV. As such this becomes the unbiased SP for the MPC as depicted in Figure 106.11. The predictor then predicts the trajectory of the CV over the prediction horizon in the absence of further control action. This is used to determine the control action required to force the CV towards the SP over the prediction horizon, as indicated by the ideal SP trajectory in Figure 106.11. The initial values of the predicted CV and ideal SP trajectories should be the same. They should also agree with the current measurement of the CV. Any discrepancy at the start of the trajectory is probably due to modelling error and is likely to persist across the prediction horizon such that convergence with the unbiased SP will not occur.

The output of the DOC is therefore biased in proportion to the discrepancy. Subject to filtering, it is this biased output from the DOC which is used as the SP for the controller in determining the values for the MV over the control horizon. Thus the CV is forced towards the unbiased SP. The bias is adjusted at each execution of the DOC. With successive executions of the DOC, the solution to the QP problem converges on the true optimum, and the closer it gets the better able it is to accurately predict that optimum. All this is depicted in Figure 106.12 in which the MPC cycle is depicted with solid arrows and the less frequent QP cycle and biasing with broken arrows. The first box of Figure 106.12, solve the QP problem, essentially corresponds to the whole of Figure 106.10.

Note: a further complication is that the initial value of the CV trajectory may not be used if there is any inherent delay in the model used for the MPC. The first available value is h steps ahead as explained in Chapter 117.

The essential characteristics of some proprietary DOCs are summarised at the end of Chapter 117 in Table 117.2.

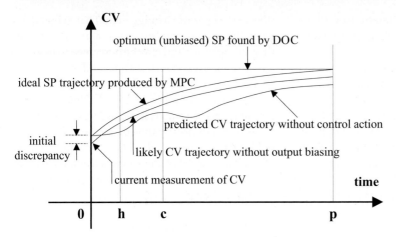

Fig. 106.11 The effect of biasing the output of a DOC in real-time

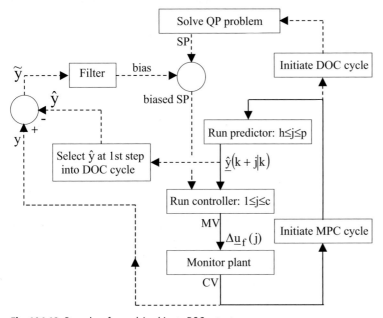

Fig. 106.12 Procedure for applying bias to DOC output

106.12 Comments

For large and complex optimisation problems it is probably relatively easy to justify the cost of building and maintaining large scale, non-linear, rigorous models and SSOs will always have their place. Otherwise, and inevitably, on the grounds of cost effectiveness, DOCs are becoming the preferred option with rigour being sacrificed for speed and robustness. They require only small and linear models which are relatively cheap to build and achieve high utilisation.

DOCs are executed frequently because the model is small and the optimisation can be solved quickly. They do not need to wait for steady state and, through feedback, produce truly optimal solutions. Furthermore, the combination of DOCs and MPCs provide excellent constraint management, both at steady state and dynamically, within a consistent user environment. Using bridges, larger scale optimisation problems can be tackled: there are well established instances of half a dozen or more co-ordinating DOCs. The only obvious drawbacks to the DOC option is the commitment to MPC in the first place and the fact that the DOC models cannot be used for off-line purposes.

A fairly fundamental constraint on the use of optimisers is the difficulty of understanding what is going on. This applies to both SSOs and DOCs but is more problematic with DOCs because of the higher frequency at which they are executed. The basic problem is that the optimiser may decide to change MVs in a direction which is counter intuitive to the operator's judgement. That is inevitable when the optimiser is looking at a broader picture than the operator and may be perceived to have different objectives. Whilst a DOC and its MPC may well have an integrated user environment, there is much scope for improvement in the operator interface in terms of visualisation and user support.

It is worth noting that there are other potential benefits, not directly related to optimisation, of having a model of the process available on-line that is calibrated to match the plant, the more obvious ones being:

- Detection of faulty measurements
- Performance monitoring through the estimation of unmeasured values such as heat transfer coefficients, catalyst activity, tray efficiency, *etc.*

Knowledge Based Systems

107.1 Architecture and Terminology
107.2 Inferencing
107.3 Rule Based Expert Systems
107.4 The Expert System Control Cycle
107.5 Semantic Nets
107.6 Frame Based Systems
107.7 Object Oriented Programming
107.8 Expert System Shells
107.9 Knowledge Elicitation
107.10 Development Life Cycle
107.11 Comments

This is a huge area of activity with its own technology, terminology, methodologies and areas of application. It is also a somewhat misunderstood subject. This chapter, therefore, provides an objective introduction to the basic concepts and essentials of implementation. The emphasis throughout is on generic issues rather than on the various characteristics of specific systems. There are dozens of texts in this area but the classic is that by Jackson (1998) to which the reader is referred for a much more comprehensive treatment.

Some categorisation is perhaps a useful starting point. Artificial intelligence (AI) is a generic term that applies to machines and/or software which emulate the process of reasoning. That process involves search mechanisms and symbolics (names, words, lists, tables, formulae, values, *etc.*) rather than algorithms and numbers as used in conventional programming, a crucial distinction. The objective of AI is to solve problems or to reach independent conclusions, faster and more accurately than a human would given the same information, although the solution to a problem is not guaranteed. The umbrella of AI embraces:

- Knowledge based systems
- Machine learning systems
- Speech recognition and generation
- Robotics

This chapter deals primarily with knowledge based systems (KBS) as these are of the form of AI most relevant to the process sector. KBS are those systems which, by virtue of their architecture, contain both a pre-programmed method of interpreting facts and some means of applying that capability. They provide structured access to large quantities of knowledge. There are many types of KBS which may themselves be categorised:

- Expert systems
- Decision support systems
- Vision systems
- Natural language systems

The main focus of this chapter is on the first two of these, expert systems (ES) and their application in decision support, as they have provided the most promising results and return on investment to date. A widely used definition of an expert system (ES), due to Bramer, is that it "is a computing system which embodies organised human knowledge concerning some specific area of expertise, sufficient to perform as a skilful and cost effective consultant".

An ES is used to emulate the expert's decision making processes. Using the expert's pre-coded knowledge, it reasons (in a non-algorithmic way) with application specific data. Because the information (knowledge and data) may be incomplete or uncertain, an ES will make mistakes. However, the knowledge and data structures of an ES are such that it should always be able to explain its lines of reasoning. An ES is only capable of learning for itself in the sense that its knowledge is developed incrementally, growing with the number of applications from which new relationships or patterns of knowledge may be abstracted.

Within the process industries, ES technology predominantly aims to give the operator advice and decision support (open loop as opposed to closed loop) in relation to control and operability. In principle an ES will provide consistent advice, handle large quantities of data, be available when required and perform quickly in difficult environments. For these reasons ES technology is used for pro-active monitoring purposes with applications in fault diagnosis, testing and troubleshooting, operator training, start-up and emergency shut-down, as well as for scheduling and optimisation.

107.1 Architecture and Terminology

The basic structure of a KBS is as depicted in Figure 107.1.

The structure and the functionality of its elements is designed around the need to support the user and the requirements of the inference engine:

- User interface. Through this the user can build up the knowledge base, enter facts to create models of scenarios (or situations), pose questions, and receive solutions and explanations. In-

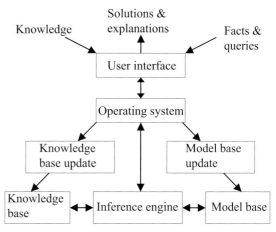

Fig. 107.1 Basic structure of a knowledge based system (KBS)

evitably the user interface is based on templates with menus for selecting features, entering characteristics, and so on.
- Knowledge base, also referred to as the static database. This is where the rules are kept. The rules are usually, but not necessarily, of the form of production rules:

if (some condition)
and/or (some other conditions)
    assert (some outcome)
    else (some other outcomes).

The conditions (or premises) are referred to as antecedents and the outcomes (or actions) as consequents. The knowledge base can itself be subdivided into:
i. The rule base, containing generic rules which are always applicable in the application domain.
ii. The case base, which contains rules specific to certain scenarios, that is model or problem types, referred to as cases.
- Knowledge base update is the means by which the knowledge base is updated, in a systematic and structured way, once the user has entered new knowledge.
- Model base, also referred to as the dynamic database. The model base is a collection of facts and/or data that define the scenario being explored. The facts are structured such that they

may be used by the rules. The model base is domain specific, with different models built for exploring different scenarios.
- Model base update is the means by which the model base is updated once the user has entered new facts.
- The inference engine is a program which interprets the facts of the model by applying the rules of the knowledge base to provide a solution to the query posed. There are different approaches used, referred to as paradigms, as discussed in the next section.
- The operating system is the program which responds to user inputs, synchronises the activities of the various elements of the KBS, co-ordinates the transfer of data between them, manages the inference engine, and so on.

107.2 Inferencing

Inferencing is the ability to reach a conclusion from a set of facts and/or data that does not provide the complete picture. Conclusion, in this context, is the net result of relating a number of cause and effect relationships to the facts of a particular scenario. Thus, for a given condition (cause) or combination of conditions (causes), what is the likely outcome (effect) and *vice versa*? The inference engine is the central part of any KBS or ES and sets them apart from other types of system.

With conventional programming, every possibility has been anticipated and the program/system behaves in a completely predictable and deterministic manner. The use of an inference engine for reasoning removes this predictability. The program's actions and the majority of its conclusions must be deduced or inferred, by pre-programmed paradigms, fresh from the facts and rules as presented to the program. The commonly used means of reasoning are backward and forward chaining, inheritance and adaptive reasoning.

1. Backward chaining was the basis of reasoning used in the early expert systems and is still used extensively. Thus, for a given scenario (effect), the inference engine searches backwards, applying the rules to the data, to ascertain the cause or combination of causes. The inferencing is referred to as being goal (effect) driven. Clearly, for every plausible combination of causes, all the known effects have to have been encoded beforehand to enable the inference engine to chain backwards. This is very efficient for real-time applications: the availability of historic data enables routes back to implausible causes to be eliminated quickly.

This paradigm lends itself to diagnostic tasks in which, typically, the scenario consists of a set of symptoms and the ES works back to find the underlying problems. Applications here range from the diagnosis of medical disease to fault detection on process plant. For example, if the high pressure and low flow alarms associated a process unit occur simultaneously, an ES may search backwards along the following lines:

i. Are the alarms both new or is one of them persistent?
ii. Is the new alarm correct?
iii. Has the cooling water supply pressure dropped?
iv. Is the feed rate higher than normal?
v. Do any product flows appear lower than normal?
vi. Is the heating system steam pressure too high, *etc.*?

Clearly each of these questions potentially gives rise to earlier questions. For example, is the feed rate higher than normal could lead to:

i. Is the product rate from the previous unit higher than normal?
ii. Is the flow measurement correct?
iii. Has the flow controller set point been increased?
iv. Is the controller still in automatic mode?
v. Is the control valve stuck open?
vi. Has a by pass valve been closed, *etc.*?

There will of course be even earlier questions to be asked and answered, and follow up questions. Nevertheless, by a process of elimination, the search will converge upon the potential causes as appropriate. However, if the line of reasoning fails to provide the causes, the method fails. Failure is usually due to not all the of the potential cause and effects having been encoded. If there is the need for previously undetermined reasoning, the application of more complex methods is required.

2. Forward chaining is essentially the opposite of backward chaining. Thus, for a given scenario (cause), the inference engine searches forwards, applying the rules to the data, to ascertain the effect or combination of effects. The inferencing is referred to as being event (cause) driven. The paradigm searches for every possible consequent, or outcome. Since implausible outcomes cannot be eliminated at source, every possible outcome has to be searched. Thus the effort involved in forward chaining is much greater than with backward chaining. To this must be added the possibility of imprecise (or fuzzy) knowledge and the need to make statistical estimates of confidence in the decision making process. Typically, the operator is presented with all outcomes, together with confidence estimates, and left to make a judgement about the most likely as opposed to the ES exclusively predicting it.

3. Inheritance. Whereas forward and backward chaining are explicit means of reasoning, inheritance is more implicit. It relates to the way in which data and/or knowledge is stored. Much of the data stored in KBS can be organised hierarchically into generic classes with the data about individual members of the class being held in the form of "attributes". The distribution of the attributes between the classes and members enables the relationships between them to be inferred.

For example, a pressure filter is a type of filter which is a form of separation unit which is itself a major equipment item (MEI), as depicted in Figure 107.2.

Generic attributes pass down the hierarchy. For example:
i. Every MEI has at least one input and output.
ii. A separation unit has fewer input than output streams.
iii. A filter's input stream is a slurry and its output streams are filtrate and solids.
iv. A pressure filter's slurry and filtrate streams are semi-continuous whereas the solids stream is discontinuous.

Clearly, the further down the hierarchy, the more specific and extensive the attributes become. For example, attributes for the pressure filter would typically embrace its capacity, type/quantity of filter aid required, operating pressure, solids discharge time, and so on. Also, the further down the hierarchy, the more common the attributes. For example, there will be more in common between different types of filter than, say, between evaporators and filters. There are many potential generic classes of a process nature: other obvious ones relate a) units to types of equipment and control module, b) operations to phases and actions, c) transfer of materials to types of manifold, d) control schemes to types of function block, and e) alarms to types of trip and interlock.

The ability to pass attributes down the inheritance path, or to abstract them upwards, enables reasoning about type and properties but is not sufficient in itself to solve many problems. It does, however, when used in conjunction with forward chaining, substantially reduce the extent of searching and reasoning required to yield an understandable outcome. The principal constraint on the exploitation of inheri-

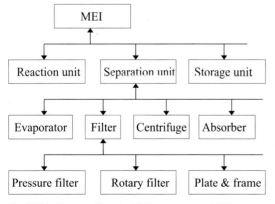

Fig. 107.2 Pressure filter classified as a member of MEI

tance is the difficulty in handling the relationships between members in different classes, say between an alarm and an equipment module, as opposed to the relationships between members of the same class. Inheritance is of particular significance in relation to frames and objects and is discussed further in Sections 6 and 7 below.

4. Adaptive reasoning. When the ability to deduce the required outcome using rules requires more knowledge than is available, adaptive reasoning is an option. This approach is used mostly in case based reasoning which takes an existing set of cause and effect relationships, analogous to the problem in hand, and applies it to the problem using forward chaining and inheritance. It is necessary to provide some indication of confidence as to the level of applicability of the outcome.

In effect, the ES attempts to conclude unknown knowledge given the existence of certain relationships. This is a more complex method of deduction and should not to be confused with learning. A truly cognitive system would deduce the relationship before going on to reason the outcome.

107.3 Rule Based Expert Systems

The facts and/or data of the model base are organised into symbolic structures such as vectors and lists. The construct normally used is the so called object-attribute-value triple. Thus, for example, the current state high (value) of the alarm priority (attribute) associated with temperature transmitter no 47 (object) would be represented:

(TT_47, priority, high).

When an object has several attributes it is normal to gather them together with their values in a single vector such as:

(TT_47 (reading, 99.1), (priority, high), (status, on-scan)).

Nesting enables objects to be explicitly linked to each other. For example, a temperature controller can reference its transmitter as an input attribute which provides access within the controller object to the transmitter's attributes:

(TC_47 (input, TT_47), (set_point, 100), (status, on_scan), (mode, auto)).

Whilst most objects only have a few attributes, tens or less, it is not uncommon for some objects to have hundreds of attributes. In this way the model base is built up with, typically, thousands of such object vectors to describe realistic scenarios. Those objects which are relevant to the scenario in hand are loaded into the working memory of an ES. They are used by the inference engine to activate the rule base in the sense that some object-attribute-value combinations will satisfy (or not) the antecedents of certain rules. The working memory also contains the intermediate results of the reasoning process.

The goal is itself cast as a vector of objects, attributes and values, whether it be the starting point for backward chaining or the end point for forward chaining.

When a production rule is activated, it is said to have been fired and the rule's consequents are asserted. Rules only use the relevant attributes of the appropriate objects. For example, the following rule would ignore the reading and status attributes of the transmitter object and only assert a value for the mode attribute of the controller object:

if (TT_47 (priority, high))
        assert (TC_47 (mode, man)).

The above rule is completely deterministic. In the event of the alarm priority being high the controller is put into manual mode. However, it is often the case with KBS that the rules are being used to reason with facts that are not precise, or indeed with rules that are not certain. In such circumstances a confidence level is attached as an attribute, and the value of the confidence is changed by the rules as the reasoning progresses. For example, suppose that TT_47 has a confidence attribute which, when the transmitter is functioning correctly, has a value of 1.0:

(TT_47 (value, 99.1), (priority, high), (status, on-scan), (confidence, 1.0)).

Suppose the transmitter goes out of range. A rule may then be used to change the confidence to reflect the new situation:

if (TT_47 (status, suspect))
  assert (TT_47 (confidence, 0.8)).

Thus, in any subsequent reasoning, the lack of confidence in the value of TT_47 can be factored in with other confidence values to give a net confidence in the outcome.

The above rules are all relatively simple. In practice a rule base comprises thousands of rules, most of which have multiple antecedents and/or consequents. The order that the rules are stored in the knowledge base doesn't particularly matter. The important thing is that the rules are as complete and as consistent as possible and are organised so that they can be inspected and edited effectively.

The pros and cons of a production rule based ES are as follows.

Advantages:

- The rule base can be developed readily and is easy to modify.
- It is very flexible yet, with effort, can be modularised to enable knowledge to be partitioned.
- The rule base is both readable and understandable.
- It is particularly suitable for changing (dynamic) knowledge.

Disadvantages:

- Knowledge is not integrated into an overall structure and it is difficult to pick out patterns and trends.
- Rules do not lend themselves to knowledge about object and event relationships such as type-subtype, part-whole, before-after.
- Gaps in the rule base are not exposed and interrelated knowledge may not be captured.
- It is easy to create inconsistent and contradictory rules.
- Rule bases are inefficient for handling large amounts of fixed (static) knowledge.

107.4 The Expert System Control Cycle

The order in which the rules of a rule based ES are fired does matter and is determined by the control function of the inference engine. The control cycle, referred to as the recognise-act cycle, is as depicted in Figure 107.3.

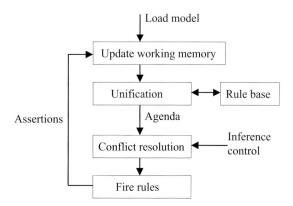

Fig. 107.3 The expert system control cycle

The steps of the control cycle are as follows:

1. The working memory, having been initially loaded, is updated each cycle. Iteration continues until either a goal is found or searching fails to find a solution. Failure occurs when, as a result of firing the rules, there is nothing to be updated.
2. Unification is a question of comparing the antecedents of all the rules with the attributes and values of the objects in the model base to see which ones are satisfied. Those rules whose antecedents are satisfied are said to be the agenda, also referred to as the conflict set. They are conflicting in the sense that a different outcome to the scenario is likely according to which rules

are fired, and indeed to the order in which they are fired.
3. Conflict resolution concerns selecting, from the agenda, which rule or rules are to be fired, and the order of firing. There are two general approaches to this: local control and global control, as explained below.
4. The rules selected are fired and assertions made. These assertions are then used to update the current version of the model in working memory.

In addition to overseeing the conflict resolution, inference control monitors the firing of the rules such that when a goal is found an audit trail of the rules fired can be produced as a basis for explaining the line of reasoning.

The local approach to conflict resolution employs user-defined meta rules (higher level rules) that encode knowledge about how to solve particular problems. The meta rules reason about which rules to fire as opposed to reasoning about the objects and relationships in the application domain. The global approach, which is used by most expert systems, is the opposite in that it is largely domain independent. The mechanisms used are built into the inference engine such that they are not readily accessible to the user. In essence there are four such mechanisms: salience, refractoriness, recency and specificity:

1. Salience is a user-defined property that can be attached to each rule as, for example:

   if (TT_47 (priority, high))
                   assert (TC_47 (mode, man),
                           (salience, 100)).

   Salience is akin to priority. By default, rules have a salience of 0: otherwise the higher the salience the greater the priority. Typically, the agenda is first sorted into classes of equal salience, and then the rules within the classes of equal salience are sorted by the following techniques.
2. Refractoriness relates to rules not being allowed to fire more than once on any object whose data has not changed. In effect, such rules are eliminated from the agenda and priority is given to those rules whose antecedents are newly satisfied.
3. Recency describes how up-to-date the values are of the attributes in the objects referenced by the antecedents of the rules in the agenda. Those rules utilising the most recently updated attributes are fired in preference.
4. Specificity concerns those rules which have the most antecedents and/or the greatest number of attributes. Whilst such rules are more difficult to satisfy than general rules with few conditions, they are likely to lead to quicker convergence on the correct solution. They are given high priority.

The choice of mechanism for conflict resolution has a direct impact on the search strategy. The two extremes are depth-first and breadth-first searches. They are not mutually exclusive: in practice the solution is usually found by some combination of the two with one or other predominating. Figure 107.4 depicts the depth-first approach to searching.

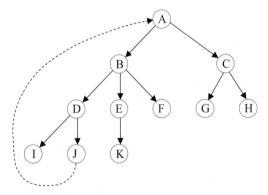

Fig. 107.4 The depth first approach to searching

The search space can be thought of as being hierarchical, starting from the antecedent of a single rule A being satisfied. Firing of that rule results, say, in the antecedents of rules B and C being satisfied. Suppose that the outcome of conflict resolution is that rule B is selected and fired, the result of which is that the antecedents of rules D, E and F are satisfied. Similarly, firing of D results in rules I and J being satisfied. For J to be satisfied, say, the antecedents for A, B, D and J have all had to be satisfied. Such a depth-first search is promoted by both recency and specificity. If, when rule J is fired, there is no change to the model, the search reverts to rule A and refractoriness prevents that particular branch of the tree being searched again. In this way the search space is repeatedly explored, as deeply as possible, until either the goal is achieved or not.

The breadth-first approach to searching is as depicted in Figure 107.5. Rules are fired across the hierarchy rather than through it. This is realised by means of a "broad-brush" approach to salience and giving a low priority to recency and specificity. In this way all possible conclusions are explored at one level before moving onto the next.

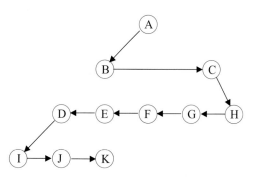

Fig. 107.5 The breadth first approach to searching

Whilst it is evident that the choice of mechanism for conflict resolution determines the search strategy, it is not clear how to decide which search strategy to adopt for any given problem. This is very application dependent although some obvious generalisations can be made:

- Use depth-first search for problems where it is sufficient to find only one solution, even if there are many potential solutions.
- Use depth-first search for problems where it is known that there are only a limited number of potential solutions.
- Use breadth-first search for problems where it is known that there are many potential solutions and they must all be found.

Best-first is an alternative search strategy to depth-first and breadth-first. In essence, at each iteration of the control cycle, a heuristic estimate is made of the difficulty of searching in any particular direction, and the easiest route is pursued. The search strategy has become an optimisation problem and, not surprisingly, is computationally intensive and seldom used.

107.5 Semantic Nets

To a significant extent the use of associative networks can overcome the disadvantages of a production rule based ES, some of which were identified in Section 3 above. An associative network is the generic term used to describe a network which is a graphical representation of knowledge in which concepts or facts are deemed to be nodes and the relationships between them are depicted by interconnecting links. Nodes are also referred to as vertices and links as edges or arcs. The most common type of associative net is the so called semantic net, an example of which for a distillation column is shown in Figure 106.6.

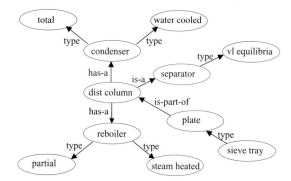

Fig. 107.6 A semantic net for a distillation column

Inspection of Figure 107.6 reveals several key aspects of a semantic net. It is evident that:

- The column is described by a cluster of nodes and there are many different relationships involved.
- The "type" link is used extensively to define one node in terms of others.

- By use of "has-a" and "is-part-of" links, the network is able to distinguish between the condenser and reboiler which are ancillary equipment and plates which are part of the column.
- The column is established as an instance of a separator by means of the "is-a" link.

It is largely through instantiation by means of the "is-a" link that inheritance of properties is established. Thus any other object, such as an absorption column, that is classified as being a separator of the vapour-liquid equilibria type will inherit all the generic properties of that separator node. Similarly, a particular column can be instantiated (a new instance is created) as an instance of the generic distillation column using the "is-a" link and can be presumed to inherit all the properties of the generic node. Inheritance of properties is an extremely widespread technique in KBS and represents some optimum in the trade off between space for storage of data/knowledge and speed of processing. However, this is at the expense of understandability: the meaning of any one node is distributed amongst the type nodes to which it relates.

There are various disadvantages to the use of semantic nets:

1. An obvious problem occurs with exceptions: that is, nodes that do not inherit all the properties of their type. For example, it is not possible to create an instance of the above distillation column with a partial overhead condenser, even if it complied with the generic column in every other respect.
2. Another key problem is with the definition of nodes. For example, it is not clear whether the node "distillation column" refers to the concept of a column, the class of all columns or to a particular column: this potentially gives rise to ambiguity in developing a model base and/or interpreting the results.
3. Semantic nets are logically inadequate, compared with production rules, in the sense that they cannot readily distinguish between a particular column, any column, all columns, no columns, and so on.
4. They are also said to be heuristically inadequate in the sense that there are no meta rules: semantic nets do not contain knowledge about how to go about searching for the data and/or knowledge that leads to the required goal.

Various approaches have been tried to overcome these disadvantages but the resultant systems are unwieldy and semantic nets have largely been superseded by frame based systems and object oriented programming.

107.6 Frame Based Systems

Frames are hierarchical structures for holding knowledge about generic classes which may be either conceptual or physical, such as with operations or equipment. The highest level (most generic) frames in the structure are known as root frames. The knowledge is partly captured by the structure of the individual frames but mostly by the data contained therein, typically held in the form of slots and fillers. Slots may be thought of as spaces in memory reserved for attributes and fillers as the data that is put into those slots. Figure 107.7 depicts a fragment of a frame system used for alarm analysis.

There is a root frame of type alarm with slots for name, reading, setting, status and priority. In practice, there would have to be many more slots for units, limits, message codes, deadband, time and so on, as described in Chapter 43, but the functionality depicted in Figure 107.7 is sufficient for the purpose of explaining frames. Three types of alarm: simple, trip and interlock are depicted, as explained in Chapter 55, each of which has the same structure as the root. Multiple instances of specific alarms of each of these types may be created although only one instance of the simple and trip types are shown.

Inheritance, as described in Section 2 above, is fundamental to frame based KBS. When a new frame is instantiated, the parameters in the slots of the offspring acquire by default the same values as the parameters of its parent. For example, in Fig-

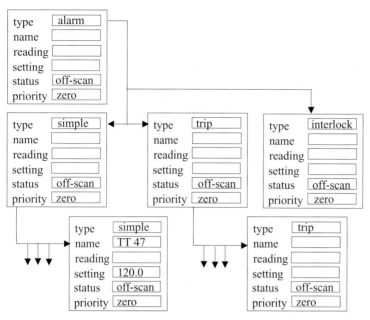

Fig. 107.7 A frame system used for alarm analysis

ure 107.7, the instance of the simple class of alarm, referred to by its tag number TT_47, inherits from its class type the default status of off-scan and the default priority of zero. However, not all the parameters acquire values through inheritance. Thus the values of TT_47 for the frame's name and 120.0 for its setting would be established by editing and, assuming the frame based system is real-time, the current value of TT_47 would be updated automatically when on-scan.

Thus use of a frame base enables the hierarchical nature of knowledge to be captured. It also provides a framework for storing large amounts of data, whole blocks of which can be accessed as single entities. However, there is much more flexibility to frame based KBS than is implied by their structure. Three key aspects are automatic inheritance, deformation and multiple inheritance as follows:

1. Automatic inheritance. If the value of the parameter in a slot is changed, that same change is automatically made to all of its offspring: that is, the parameter in the corresponding slot is changed to the same value in every frame lower down the hierarchy. This is quite different to, say, the configuration of function blocks, as described in Chapter 48, where instances of function blocks are parameterised manually. If the value of a default parameter in the template of a function block is changed, that will only affect new instances of the function block at the point of instantiation. Any changes to that same parameter in prior instances would have to be established by individually editing all of the existing function blocks.

2. Deformation. Frame systems lend themselves to stereotypical representations of knowledge or, in other words, knowledge that can be classified on a generic basis and organised hierarchically. Unfortunately, and inconveniently, whilst much of the knowledge and data encountered in engineering applications is stereotypical, there are always exceptions to the pattern and untoward trends. Thus the knowledge needs to be deformed in some way to capture the reality. This is done by means of procedures, referred to as demons, which are attached to individual slots as appropriate. In

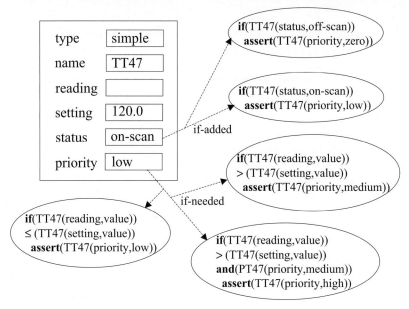

Fig. 107.8 Use of demons for non-standard slot entries

essence, demons are a program control tool which enable non-standard slot entries and allow for situation specific responses, as depicted in Figure 107.8.

There are three main types of demon:

i. If-added. Whenever the value of the parameter in one slot is updated, the demon attached to that slot will work out and update the value of the parameter in another slot.

ii. If-needed. These compute new values for parameters on demand. They are context dependent and, typically, the demon attached to one slot is only triggered when the value in another slot changes. Note that an if-needed demon can be used to override the effects of automatic inheritance by changing the value in its own slot.

iii. If removed. This is similar to the if-added demon, except that it is only triggered when the value of the parameter in the slot to which it is attached is reset to some null value.

3. Multiple inheritance. It is often convenient for one offspring to inherit values of parameters from more than one parent, as depicted in Figure 107.9, in which a trip inherits values from two simple alarms. In this case the priority of the trip defaults to low. However, if the priority of either of the alarms TT_47 or PT_47 changes to medium or high then the priority of the trip changes accordingly, and if the priority of both alarms changes to medium then the that of the trip changes to high.

Multiple inheritance enables more complex, inter-related knowledge to be encoded. This results in frame systems having more of a lattice type of structure than the tree type of structure associated with strictly hierarchical systems. However, there is a danger that ambiguities may be introduced through multiple inheritance. For example, in Figure 107.9, if the priority of the trip is high, it is not apparent whether that is due to just one or both of the parents' priorities being high or whether it is due to both parents' priorities being medium.

In operation the frames are continuously being executed, with demons being exercised and slots up-

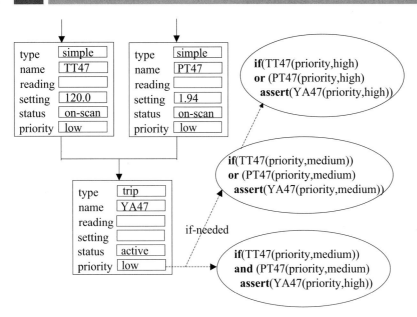

Fig. 107.9 Use of demons to handle multiple inheritance

dated as and when appropriate. Reasoning is an expectation driven paradigm: empty slots, which correspond to unconfirmed expectations, are filled with data when the expectations are confirmed. Although demons are procedural in nature, they only enable a limited amount of reasoning. It is normal for the bulk of the reasoning to be handled by conventional production rules, in a separate rule base, which interact with the frames/slots as appropriate. Thus frame based KBS integrate declarative notions about generic classes, instances and attributes, with procedural notions about how to retrieve information and/or to achieve goals.

Any frame based KBS clearly requires a user interface to enable frames to be instantiated and edited, and some form of interpreter to handle requests and to decide when the goal of the query has been achieved.

107.7 Object Oriented Programming

Object oriented programming (OOP) represents an alternative approach to KBS. Loosely speaking the objects in an OOP environment can be equated to the frames of a frame based KBS in the sense that objects are instantiated (parents and offspring), consist of slots and fillers, and have inheritance properties. However, there are three fundamental differences between objects and frames:

1. The (limited) procedural capability of frames is realised by means of demons attached to slots, whereas with objects whole procedures are attached to the objects themselves. Such a procedure, when applied to its own object, will utilise the values of the parameters in its slots to generate new values for them.
2. Whereas with frames the demons are specific and attached to individual slots as required, the procedures are generic and inherited down the object hierarchy from the root object. Once inherited, a procedure is applied to its object in a localised manner.

3. Unlike frames, special protocols enable messages to be passed between objects. These messages are comprised of data in the form of arguments for the procedures. The messages allow for interaction between the objects in a context dependent way and for control of that interaction. For example, one of the procedures attached to an object can be selected, receive data and be applied to that object by the procedure attached to another object.

Thus the procedural capability is distributed throughout the objects rather than being held in a separate rule base as with frame based systems. In effect, the protocols are the interface between the objects. Some of the terminology used to describe aspects of an OOP environment is explained below:

- Data abstraction. This refers to the level of detail that the knowledge representation encompasses when representing a complicated idea or object. Typically, to define a new data abstraction a new object is defined.
- Encapsulation. This defines the grouping of objects and determines how data flows between and within objects. The protocols enable either wide-open routes in which procedures attached to one object can access data from any other object, or restricted routes in which procedures attached to one object can only access data from other objects within its own predefined group.
- Inheritance. As with frame based systems, objects lower down in the hierarchy inherit slot values from above. However, as a result of execution of an object's procedure, some slot values may change. Such locally derived values always override default values arising from inheritance.
- Message passing. There are two aspects to this: sending and receiving. When the procedure of an object is executed, in addition to changing its own slot values the procedure may generate a message for another object. The relevant procedure of the object receiving the message is then executed using the arguments received as appropriate.
- Polymorphism. This is the capability of different classes of objects to respond to the same set of messages in a unique way that is most appropriate to those receiving objects. Thus, when a procedure attached to an object is invoked, the arguments (referred to as variable bindings) in the message received determine the response of the procedure.
- Modularity. Since all execution is achieved through objects, OOP is inherently modular which makes for easier program maintenance and expansion.

107.8 Expert System Shells

There are several AI languages which can be used for ES development. Java is inherently object oriented and fairly universal, C++ supports object oriented extensions and is used widely in Europe, LISP (*list* *p*rocessing) is favoured in the US, and PROLOG (*pro*gramming for *log*ic) is favoured in Japan. However, for process automation, it is usual to use an ES shell which is itself written in one or other of the AI languages.

An ES shell is a framework, or environment, for development and operation of an expert system which provides a user interface through which the user may interact with the ES without the need to program in an AI language. Typically an ES shell:

- Supports multiple means of knowledge representation, such as frames and objects for the model base and production rules, demons, *etc.* for the rule base.
- Provides a variety of constructs and tools, such as an editor, for ES building and testing.
- Enables inferencing and inference control.
- Offers multiple search strategies.
- Ensures that the ES is able to explain its lines of reasoning through some trace function.
- Accepts data on a real-time basis.

There are various proprietary ES shells available, most of which have been designed for use in a particular domain.

107.9 Knowledge Elicitation

Given that an ES is used to emulate an expert's decision making processes, it follows that a fundamental requirement of the development of an ES is the acquisition of knowledge in a form that can be used for decision making. That knowledge is normally elicited from domain experts who may be considered to be persons who, provided with data about the area of application, are able to use their knowledge to make informed and "accurate" decisions. Clearly the more proven an expert's decision making capability the better the potential of the ES.

The process of knowledge elicitation is enabled by a knowledge engineer whose job is to abstract knowledge from the domain expert and to translate it into some standard and useable form. The emphasis is on translation of the knowledge: it is not a requirement that the knowledge engineer be able to interpret the information, although some understanding of the domain is always helpful. Elicitation is an iterative process, as depicted in Figure 107.10:

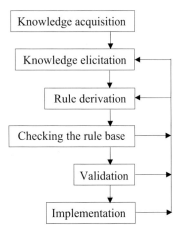

Fig. 107.10 The iterative process of knowledge elicitation

1. Acquisition of knowledge by the domain expert: this is by far the most important step and often involves a lifetime's work.
2. Elicitation of knowledge from the domain expert: the amount of time involved in this step is always grossly underestimated.
3. Rule derivation can be thought of as translating the knowledge into rules.
4. Checking the rule base involves comparing the rule bases from different experts to identify ambiguity and to remove that ambiguity by reasoning.
5. Validation concerns testing the rules of the ES through general usage but in an off-line mode.
6. Implementation is the application of the ES to the domain for which it was developed.

Steps 4 and onwards identify errors and discrepancies resulting in a return to steps 2 or 3 depending upon the fault. To make this iterative process repeatable, and to reduce the cost and improve its effectiveness, several different elicitation methods have been produced. They have been developed by psychologists and are largely geared towards the collection of knowledge rather than towards its usage. In the main they follow similar logic and produce similar results. Five such methods are summarised below, the first two of which are said to be unstructured and the others to be structured:

1. Forward scenario simulation. The knowledge engineer states a series of goals or outcomes to the domain expert and asks what the conditions are that will cause each to happen. This produces a logical response using the expert's own rule structure posing new questions: the process continues until the conditions have been completed. The method provides scrutable rules of a production nature, can be used independently of the domain and does not require the expert to have previous experience of elicitation.

   However, the engineer needs to be knowledgeable about the domain to be able to pose sensible questions. Thus the direction of questioning and the knowledge elicited is inevitably skewed by the engineer, as indeed are the responses of the expert. This is an expensive means of elicitation and the inability of the engineer to maintain the ES independent of the expert has caused it to fall into disrepute.

2. Protocol analysis. This technique is akin to "work shadowing". The engineer observes the

expert at work in the application domain and captures the expert's methods of solving problems. The expert has to give a meaningful commentary on the actions being taken to enable the engineer to translate them into rules, for which reason the engineer has to be familiar with the domain. Again the technique provides scrutable rules of a production nature, can be used independently of the domain and is relatively simple to employ.

Whereas the technique requires little extra time on the expert's behalf, it is quite expensive in terms of engineering effort: the time involved grows exponentially with the extent and complexity of knowledge to be captured. Apart from the ES being difficult to maintain, the major drawback of the technique is the limited number of scenarios that are likely to be revealed: the rare and unusual scenarios, which often justify the use of an ES in the first place, are unlikely to be captured. For these various reasons the so-called structured methods of elicitation evolved.

3. Structured interviews. This is the brainstorming approach to knowledge elicitation and has become the most popular of the various techniques. In essence the expert articulates in an unconstrained manner as many relationships about the domain as possible, whether they be first principles, factual conditions, heuristics (rules of thumb), or even just suspicions and hunches. Once articulated, the engineer attempts to translate them into rules. The rules are then justified in consultation with the expert. Each rule is examined in turn and any that cannot be adequately justified is either amended or set aside. Each has to be rational in itself and not inconsistent with the other rules. Clarification of one rule invariably leads to the creation of other new rules.

Elicitation is an iterative process and can be frustratingly slow and expensive in terms of effort. Success is critically dependent upon the expert's ability to recall and articulate expertise in the form of logical relationships, some are better able to do this than others. Success is also dependent upon the engineers experience of using the technique. With experience the engineer can more readily distinguish between rules which are not viable and should be rejected, and those which are potentially viable but which haven't been fully developed. Research into knowledge acquisition indicates that structured interviews yield significantly better results than any of the other structured methods.

4. Repertory grid analysis. Also known as multidimensional analysis, this technique is based upon a statistical representation of the expert's concepts of the domain and the relationships between them. The expert is asked to identify two objects in the domain that are similar and a third which is different, for example two pumps and a valve. A reason is then given to explain the difference. In a recursive manner every object in the domain is classified and a model base of the entire domain is produced.

The method is domain independent and does not require the expert to be familiar with knowledge elicitation. It provides a sound basis for establishing the structure of a domain, which enables decomposition into frames and/or objects, and for identifying inheritance properties. Because of the emphasis on structure, the output is scrutable and relatively easy to maintain by the engineer without recourse to the expert, although it can be extremely tedious to establish in the first place. Despite the appearance of structure, an incomplete determination of the relationships within the domain is quite possible, even if all the elements have been covered. However, the biggest disadvantage of the approach is that it does not yield a rule base for implementing the ES.

5. Card sort. This technique is similar to repertory grid analysis in most respects. The engineer produces a pack of cards which depict the individual elements within the domain, say exchangers, pumps, valves, *etc*. The expert then uses these cards to produce ordered lists of the elements, according to the "dimension" chosen. The dimension describes some functional rela-

tionship between the elements relative to their domain, examples of such relationships being topology, operational order, reliability, and so on.

Again the method produces a model base but no rule base. What rules that do emerge tend to be unwieldy and incomplete. The approach becomes less successful as the domain increases in size and/or reduces in structure.

107.10 Development Life Cycle

In many respects the development cycle of an ES is not much different from the life cycle of a control engineering project as outlined in Chapters 59–65. The essential phases are as follows:

- Project initialisation. The starting point, as ever, is to define the problem and to articulate the needs from the end-user's point of view. Bearing in mind the cost of ES development there needs to be some justification for going down the ES route: this requires at least some objective comparison of alternative (non-KBS) solutions to the problem. There also has to be meaningful recognition of the amount of management effort and time scales involved in development of the ES.
- Systems analysis and design. This phase concerns the conceptual design of the ES and planning for its development, fundamental to which is identification of the sources of knowledge available and the approach to elicitation. Appraisal of the extent and nature of that knowledge largely determines the best means of representation and the functionality required of the ES shell. That in turn provides a basis for estimating the resources required, both human and physical, and for carrying out some sort of costs and benefits analysis.
- Rapid prototyping. It is normal practice for a prototype ES to be developed first. Thus a small but representative subset of the available knowledge is elicited and translated into model and rule bases and an embryo ES developed. This provides a basis for experimenting with the design of the ES and for demonstrating its feasibility.
- System development. Once the prototype is proven, knowledge elicitation and development of the full ES can progress as parallel activities. As the knowledge becomes available, the model base (say frames or objects) can be built and the rule base (production rules, demons, *etc.*) established. Effective integration is critical: the rules must be systematically applied to the model and the results verified. The development process is iterative, consisting of a combination of testing, evaluation, improvement and expansion. The emphasis on incremental development is deliberate: experience shows that it is far better to provide feedback (mistakes, omissions, *etc.*) from development of the ES into the elicitation process on an on-going basis rather than trying to make many and/or major corrections retrospectively. Note that the bottleneck in ES development is invariably the knowledge elicitation process.
- Acceptance. Just as with any other new system, there is the need to go through some formal acceptance process with the end user in the development environment. This would typically involve demonstrations based on non-trivial problems to which the solutions are known *a priori*.
- Deployment. This essentially concerns installation of the ES in the application domain, integration with other control and management information systems as appropriate, field testing and commissioning. Clearly there is the need for operator training, consideration of access and security arrangements, the provision of documentation, long term support and maintenance.

107.11 Comments

Over the years there have been many unsuccessful attempts to find AI based solutions to problems and much money has been wasted. This was largely due to a combination of ill-conceived projects and unrealistic expectations. However, there have also been many very successful implementations and from these have emerged two cardinal rules about ES development.

First, only build an ES when it is appropriate to do so. There is no point in building an ES if the problem can be solved by common sense: unless cognitive skills are involved and the solution requires heuristics, then alternative solutions will always be more viable. Also, it is not feasible to build an ES unless genuine experts exist and they agree on the solutions. The experts must be willing to have their knowledge elicited and be able to articulate their methods. It follows that the problem must be reasonably well understood and not too big and/or difficult to tackle.

Second, only build an ES when doing so can be justified. Obviously the solutions have to have some practical value to be of benefit to the end-user and there has to be a high enough payback to justify the investment. The lack of availability of expertise is a sensible justification: this may be due, for example, to the scarcity of experts, to the potential loss of expertise or to the need for limited expertise to be deployed in many locations. Another common justification is based upon the deployment of expertise in adverse circumstances or a hostile environment as, for example, in the deployment of an ES for alarm management. It is well known that under emergency conditions operators can become overwhelmed by sudden floods of alarms and are unable to make sound judgements quickly enough to take effective action.

Fuzzy Logic Control

108.1 Controller Structure
108.2 Fuzzification
108.3 Rule Base
108.4 Decision Logic
108.5 Defuzzification
108.6 Worked Example
108.7 Real Time Operation
108.8 Example with Non-Linearity
108.9 Example with Interaction
108.10 Alternative Approaches
108.11 Self Adaptive Fuzzy Control
108.12 Observations

Fuzzy logic is essentially concerned with handling quantities that are qualitative and vague. In the context of process control this usually means variables that are described by imprecise terms such as "nearly empty". These variables are manipulated by linguistic statements to produce a deterministic output. Contrary to what is often supposed, fuzzy logic is completely deterministic. For example, the output of a fuzzy controller, which would typically be the required opening of a valve, must have a specific value: the controller cannot send a signal such as "open alot" to the valve.

This chapter introduces the principal features of a fuzzy logic controller (FLC) and uses the case of a level control loop to illustrate its principle of operation. A typical two-input single-output scenario is considered. A non-linearity is then introduced. This is because one area of application for FLCs, for which significant benefits over PID control are claimed, is in the handling of non-linearities. Next a two-input two-output example is considered to demonstrate the capability of FLC at handling interactions. Having thus covered the basics, self adaptive FLCs are introduced.

Fuzzy logic and the underlying set theory is a large subject to which this chapter is but an introduction. A good introduction is given in McGhee (1990): for an in-depth treatment the reader is referred to the text by Driankov (1996).

108.1 Controller Structure

The basic structure of a Mamdani type of FLC is as depicted in Figure 108.1.

The input and output signals are deterministic, typically analogue signals, and are often referred to as being crisp to distinguish them from the fuzzy signals internal to the controller. It is common practice, but not essential, to scale the crisp signals on a fractional basis on a scale of say 0 to +1. The range of a signal is sometimes referred to as its universe of discourse.

Having been scaled, the input signal is fuzzified and operated upon by the decision logic using the rules in the rule base as appropriate to produce a fuzzy output. This is then defuzzified, and further scaled if necessary, to establish a crisp output signal.

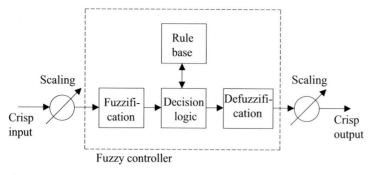

Fig. 108.1 Structure of a Mamdani type of fuzzy logic controller (FLC)

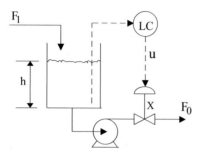

Fig. 108.2 A simple level control system

In practice, something like 80% of all FLCs are of a two-input single-output nature. Typically the inputs are the error e, with respect to the set point, and either the integral or the derivative of the error, as depicted in Figures 108.4 and 108.5 respectively.

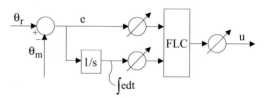

Fig. 108.4 Fuzzy equivalent of a PI controller

Consider the level control system of Chapter 3, as shown in Figure 108.2.

Let h and u be the level measurement and controller output respectively. Suppose that they have been scaled on a fractional basis according to the range of the level transmitter and valve opening. Assuming that the controller is fuzzy in nature, its corresponding block diagram is as depicted in Figure 108.3.

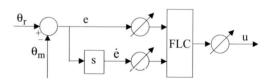

Fig. 108.5 Fuzzy equivalent of a PD controller

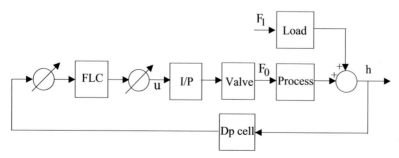

Fig. 108.3 Block diagram of the level control loop with an FLC

These are the fuzzy equivalents of PI and PD controllers, the PD structure being by far the most common.

Not only are FLCs analogous with PID controllers in terms of their structure, there are also analogies with regard to their tuning:

- A separate scaling factor is applied to each of the inputs, and indeed to the output. Adjusting these scaling factors is equivalent to varying the proportional gain of the fuzzy controller.
- The I or D action is inherent and is established by application of the rule base. Tuning the controller in the sense of adjusting its reset or rate time is realised by means of changing the rules.

108.2 Fuzzification

Fuzzification is the process of converting the value of a crisp input signal into memberships of appropriate fuzzy sets.

Consider again the level control system of Figure 108.2 for which the desired value is half full. Suppose that there are two inputs, the error e and its derivative ė, corresponding to the PD type of FLC as depicted in Figure 108.5. Noting that each signal may be either positive or negative, and that some judgement about the likely range of ė is required, they may both be scaled from −0.5 to +0.5. Let the universe of both e and ė be partitioned into seven sub sets, as depicted in Figures 108.6 and 108.7 respectively, using the terms zero ZE, small S, medium M, large L, positive P and negative N.

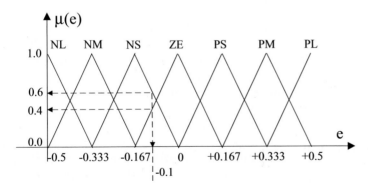

Fig. 108.6 Partitioning of the universe of the error into sub sets

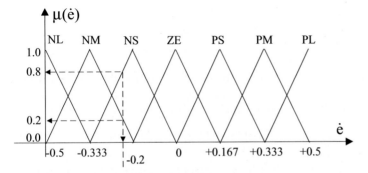

Fig. 108.7 Partitioning of universe of derivative of error into sub sets

Suppose that the level is above its set point and rising, and that the input values are:

$$e = -0.1 \quad \text{and} \quad \dot{e} = -0.2$$

Inspection of Figure 108.6 reveals that the NS subset covers errors in the range of −0.333 to 0 and the ZE subset covers errors in the range of −0.167 to +0.167. Clearly an error of −0.1 lies within both ranges and so the error is said to have a membership of both the NS and ZE subsets. It is also evident that an error of −0.1 is closer to the middle of the range of NS than it is to the middle of ZE, thus it belongs more to NS than it does to ZE. This is articulated in terms of a membership function $\mu(e)$ whose value is a measure of the extent to which the error belongs to a particular subset. Membership varies from 0 (no membership) to 1 (full membership). For symmetrical triangular sub sets of the type used in Figure 108.6 the value of the membership function is established by simple geometry. Thus for $e = -0.1$:

$$\mu_{NS}(e) = 0.6, \qquad \mu_{ZE}(e) = 0.4$$

Similarly, if $\dot{e} = -0.2$ then, from Figure 108.7:

$$\mu_{NS}(\dot{e}) = 0.8 \qquad \mu_{NM}(\dot{e}) = 0.2$$

Note that on this basis of fuzzification, the total membership:

$$\sum \mu_j(e) = 1 \qquad \sum \mu_j(\dot{e}) = 1 \qquad (108.1)$$

The choice of the number of subsets to use in partitioning of the universe, and the extent to which they overlap, is a design decision and is application dependent. Partitioning into seven fuzzy subsets is normal and usually provides sufficient discrimination for control purposes: five subsets are sometimes used for simplicity but more than seven can seldom be justified. Although the triangular basis described thus far is the most common, there are many possible alternatives: skewed triangular as in Figure 108.8, trapezoidal, multiple overlapping as in Figure 108.9, sinusoidal, exponential and Gaussian.

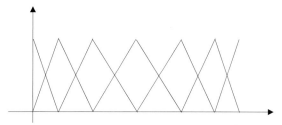

Fig. 108.8 Skewed triangular partitioning of the universe. Partitioning of the universe on a skewed triangular basis

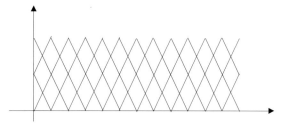

Fig. 108.9 Partitioning of the universe on a multiple overlapping basis

Skewed triangular is used for biasing the input signal. Multiple overlapping gives rise to smoother control action. Gaussian is more natural but makes for more difficult calculation of membership functions. Note that for multiple overlapping:

$$\sum \mu_j \neq 1$$

The universe of the output signal must also be partitioned into a fuzzy set. For the level control problem the output range will be presumed to be partitioned into the same seven subsets with NL corresponding to the valve being fully shut and PL to wide open.

108.3 Rule Base

The rule base of a FLC relates the input fuzzy sets to the output fuzzy sets. It consists of production rules which are of the general form:

if (some condition) **and** (some other condition) **then** (some outcome)

in which the conditional part of the rule is known as its antecedent and the outcome part as its con-

Table 108.1 Standardised rule base for PD type of FLC

		Tank full		e				Tank empty
		NL	NM	NS	ZE	PS	PM	PL
Level falling	PL	ZE	NS	NM	NL	NL	NL	NL
	PM	PS	ZE	NS	NM	NL	NL	NL
	PS	PM	PS	ZE	NS	NM	NL	NL
de/dt	ZE	PL	PM	PS	ZE	NS	NM	NL
	NS	PL	PL	PM	PS	ZE	NS	NM
	NM	PL	PL	PL	PM	PS	ZE	NS
Level rising	NL	PL	PL	PL	PL	PM	PS	ZE

sequent. For a two-input single-output system, the rule base is invariably in the form of a look-up table in which the axes represent the antecedents and the entries in the table represent the consequents. Such a look-up table is referred to as fuzzy associative memory (FAM). A standardised look-up table for the PD type of FLC, annotated in the context of the level control problem, is as depicted in Table 108.1.

The rule base is easy to use. For example, if both the error and its derivative are NS, then the output should be PM.

Note the structure and symmetry of the rule base:

- An equilibrium point is established at mid-table, i.e. if the error and its derivative are both ZE then the output is ZE.
- The leading diagonal consists of ZE values only, and adjacent diagonals consist of adjacent values only.
- The middle column and row correspond to mid range, i.e. if either the error or its derivative is ZE, then the output varies monotonically from NL to PL.
- The middle row explains the rather unexpected ability of the PD type of FLC to handle disturbances and reduce offset. If an offset exists then the level is constant, de/dt = ZE and the output varies from NL to PL according to the size of the error. Note that to eliminate, as opposed to reduce, offset the controller output needs to be in incremental rather than absolute form.
- An operability jacket prevents the tank from flooding or running dry. Thus if the error is NL and its derivative $\leq$ ZE then the output must be PL. Similarly, if the error is PL and its derivative $\geq$ ZE then the output must be NL.

It is normal practice in designing an FLC to utilise a standardised look-up table and to edit individual rules if/as appropriate. This requires the designer to have a good understanding of the application. It should be emphasised that the formulation of a non-standard rule base is a non-trivial task.

108.4 Decision Logic

This concerns applying the rules of the rule base to the inputs to determine the output. Each input signal has membership of two subsets. When there are two inputs there will be four memberships, each combination of which makes a contribution to the output. That contribution is determined by the rules of fuzzy logic which were first established by Zadeh.

Remember that the rules of the rule base are of a production nature:

if (some condition A) **and** (some other condition B)

then (some outcome C).

Such rules are analogous to the Boolean AND operation $C = A \times B$ which gives rise to the truth

Table 108.2 which depicts the logical value of C for inputs A and B:

Table 108.2 Boolean AND operation of $C = A \times B$

A	B	C
0	0	0
1	0	0
0	1	0
1	1	1

The outcome of AND is always the lowest of the two inputs: the equivalent fuzzy operation is to take the minimum membership value. Thus, when the conditions are such that two (or more) subsets overlap, i.e. there is partial membership of both subsets, the resultant membership function is the minimum of the two input membership functions:

$$\mu_C = \min(\mu_A, \mu_B) \qquad (108.2)$$

Consider the level control problem, the scenario when $e = -0.1$ and $\dot{e} = -0.2$, and the combination of memberships in which both the error and its derivative are NS. The value of the membership of the output sub set PM due to that combination of inputs is given by:

$$\mu_{PM}(u) = \min(\mu_{NS}(e) = 0.6, \mu_{NS}(\dot{e}) = 0.8) = 0.6$$

Similarly applying the MIN rule to the other three combinations of input memberships yields the output memberships depicted in Table 108.3.

108.5 Defuzzification

Defuzzification is the process of converting the membership values of the output subsets into a crisp output signal. Each output subset for which there is a membership value contributes to the crisp output. Inevitably there is overlap and over much of the output range there is a choice of membership values for the output function. Choice can be thought of as a production rule of the form:

if (some outcome A) **or** (some other outcome B) **then** (some output C).

Such a choice is analogous to the Boolean OR operation $C = A + B$ which gives rise to the truth Table 108.4 which depicts the logical value of C for inputs A and B:

Table 108.4 Boolean OR operation of $C = A + B$

A	B	C
0	0	0
1	0	1
0	1	1
1	1	1

The outcome of OR is always the highest of the two inputs: the equivalent fuzzy operation is to take the maximum membership value:

$$\mu_C = \max(\mu_A, \mu_B) \qquad (108.3)$$

Table 108.3 Mapping of inputs into output memberships

		\multicolumn{7}{c}{e}						
		NL	NM	NS	ZE	PS	PM	PL
de/dt	PL	0	0	0	0	0	0	0
	PM	0	0	0	0	0	0	0
	PS	0	0	0	0	0	0	0
	ZE	0	0	0	0	0	0	0
	NS	0	0	0.6_{PM}	0.4_{PS}	0	0	0
	NM	0	0	0.2_{PL}	0.2_{PM}	0	0	0
	NL	0	0	0	0	0	0	0

Consider again the level control problem. The membership profile of the output fuzzy set involves weighting the contribution of each subset according to its membership function. Where overlapping of subsets occurs the strongest contribution is chosen. There are two aspects to this:

- when one output subset has two different values for its membership function. This is the situation with PM which has memberships of 0.2 and 0.6, in which case 0.6 is chosen.
- when the memberships of adjacent subsets overlap. Again the larger value is chosen.

This is illustrated in Figure 108.10 which shows the individually weighted subsets and the composite output membership profile.

One method of determining the value of the crisp output involves calculating the centre of gravity (CoG), sometimes referred to as the centroid, of the composite output membership profile about the $\mu(u)$ axis using the formula:

$$u = \frac{\int_{-0.5}^{0.5} \mu(u).u.du}{\int_{-0.5}^{0.5} \mu(u).du} \approx \frac{\sum_{u=-0.5}^{+0.5} \mu(u)_j.u_j}{\sum_{u=-0.5}^{+0.5} \mu(u)_j} \quad (108.4)$$

For $e = -0.1$ and $\dot{e} = -0.2$ this gives $u = 0.2665$ which, allowing for the bias of 0.5, corresponds to a valve opening of some 77%.

Another method of calculating the crisp output is the so called mean of maxima (MoM) which establishes the mid point of the range over which $\mu(u)$ has its maximum value. In the above example it can be seen from Figure 108.10 that $\mu(u)$ has a maximum value of 0.6 over the range of u from 4/15 to 2/5. The MoM is thus 1/3. Allowing for the bias of 0.5 again, this corresponds to a valve opening of some 83%.

The CoG method is the most common means of determining the crisp output. It gives good steady state performance in terms of disturbance rejection and offset reduction. Because of the averaging involved in finding the CoG of the output membership profile, the FLC output varies across its full range. Also, because of the averaging, the method is

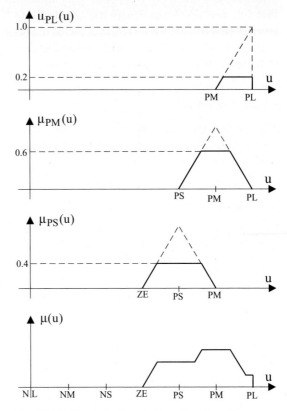

Fig. 108.10 The composite output membership profile

fairly tolerant of errors in the rule base. The MoM method has good dynamic performance. That is because it prejudices the output signal in favour of the subset with the strongest membership function. However, as the dominant subset changes, the output signal jumps from the centre of one subset to another. Thus some form of filtering is required to limit the rate of change of output applied to the control valve. MoM is very sensitive to errors in the rule base.

108.6 Worked Example

The level control system of Figure 108.2 is controlled by the FLC of Figure 108.5. Suppose that the level is above its set point and the inlet flow rate is decreasing such that:

$$e = -0.3 \quad \text{and} \quad \dot{e} = +0.1$$

Presume that the input ranges are divided into seven symmetrical subsets each, as depicted in Figures 108.6 and 108.7. The values of the membership functions of the input signals are as follows:

$$\mu_{NM}(e) = 0.8 \quad \mu_{NS}(e) = 0.2$$
$$\mu_{ZE}(\dot{e}) = 0.4 \quad \mu_{PS}(\dot{e}) = 0.6$$

Assume that the rule base is as depicted in Table 108.1. The values of the membership functions of the output subsets are thus:

$$\mu_{PM}(u) = 0.4 \quad \mu_{PS}(u) = 0.6 \text{ and } 0.2$$
$$\mu_{ZE}(u) = 0.2$$

By inspection, the maximum of the output function is the PS peak truncated at $\mu_{PS}(u) = 0.6$.

The PS peak is centred on u = 0.667.

Hence the value of the output using the mean of maxima criterion, MoM = 0.667 ≡ 67%.

108.7 Real Time Operation

Real time operation is easy to understand. At each sampling instant, the FLC:

- calculates the current values of the membership functions of the input subsets. As e and ė change with time these membership functions vary.
- fires the rule base to determine the output subsets and the values of their membership functions. With time, the group of cells with entries in Table 108.3 will shift about and the values of those entries will vary.
- determines the crisp output from Equation 108.4, or otherwise.

108.8 Example with Non-Linearity

Now consider the level control system of Figure 108.11 in which the lower half of the tank is conical. The variation in cross sectional area, which is a major non-linearity, may be accommodated relatively easily by either an asymmetric triangular basis of fuzzification and/or by changing the rule base. In the latter case the necessary changes to the rules are as shown in Table 108.5.

The number of changes necessary is surprisingly small given the nature of the non-linearity.

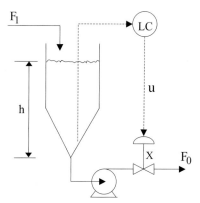

Fig. 108.11 Level control system for conically bottomed vessel

Table 108.5 Editing the rule base to accommodate the non-linearity

		e						
		NL	NM	NS	ZE	PS	PM	PL
de/dt	PL							
	PM							
	PS							
	ZE				NM			
	NS				NS	NM		
	NM				ZE	NS	NM	
	NL			PM	PS	ZE	NS	

The left half of the table, which corresponds to levels in the upper part of the tank, and the operability jacket are unaffected by the non-linearity. It is probably good enough to use the same rules at mid range, *i.e.* when e is ZE. In the top right hand corner of the table which corresponds to the worst scenario, *i.e.* low and falling levels, the rules are all as strong as they can be, *i.e.* the consequence is NL. So the bottom right hand corner is the only area where there is scope for change. The changes proposed represent a modest strengthening of the rules. Otherwise, the implementation of the FLC is exactly the same as for the tank of uniform cross sectional area.

108.9 Example with Interaction

A blending system was considered in Chapter 86. The interactions that are inherent are depicted in the block diagram of Figure 86.6. These interactions can be handled by means of a two-input two-output FLC as depicted in Figure 108.12. Note that this same blending process was used as a vehicle for introducing multivariable controller design in Chapter 81.

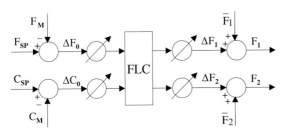

Fig. 108.12

Using the same nomenclature as in Chapter 86, steady state mass balances give:

$$F_1 + F_2 = F_0 \quad \text{and} \quad F_2 C_2 = F_0 C_0$$

which may be rearranged:

$$F_1 = F_0 - F_2 \quad \text{and} \quad F_2 = \frac{F_0 C_0}{C_2}$$

Assuming that the concentrated salt solution comes from a reservoir of constant composition, these may be put into deviation form:

$$\Delta F_2 = \frac{1}{\overline{C_2}} \left(\overline{F_0} . \Delta C_0 + \overline{C_0} . \Delta F_0 \right) = \frac{\overline{F_0}}{\overline{C_2}} \Delta C_0 + \frac{\overline{C_0}}{\overline{C_2}} \Delta F_0$$

$$\Delta F_1 = \Delta F_0 - \Delta F_2$$
$$= \Delta F_0 - \left(\frac{\overline{F_0}}{\overline{C_2}} \Delta C_0 + \frac{\overline{C_0}}{\overline{C_2}} \Delta F_0 \right)$$
$$= \left(1 - \frac{\overline{C_0}}{\overline{C_2}} \right) \Delta F_0 - \frac{\overline{F_0}}{\overline{C_2}} \Delta C_0$$

which may be generalised as follows:

$$\Delta F_1 = K_1 \Delta F_0 - K_2 \Delta C_0$$
$$\Delta F_2 = K_3 \Delta F_0 + K_2 \Delta C_0 \quad (108.5)$$

These two equations provide the necessary understanding of the interactions for designing the rule base for the blending system. Note, in particular, the signs:

- if the concentration C_0 is at its set point and the flow F_0 is too high/low, then both flows F_1 and F_2 need to change in the same direction and proportion to avoid upsetting the composition.
- if the flow F_0 is at its set point and the concentration C_0 is too high/low, then the flows need to change in opposite directions by equal amounts to avoid upsetting the flow.

The rule base is as shown in Table 108.6. As before, the axes represent the antecedents and the entries in the table represent the consequents. However, for each pair of antecedents there are two consequents: the upper value of each entry is the output ΔF_1 and the lower value (in darker shade) is ΔF_2. In effect two look up tables have been combined.

Note that the structure and symmetry of Table 108.6 respects the logic of Equations 108.5. In particular:

- an equilibrium point is established at mid-table, ie if ΔF_0 and ΔC_0 are both ZE, then both outputs ΔF_1 and ΔF_2 are ZE.
- the leading diagonals consists of ZE values only, but the diagonals are in opposite directions for ΔF_1 and ΔF_2.

Table 108.6

Outputs: upper ΔF_1, lower ΔF_2		high flow NL	NM	NS	ΔF_0 ZE	PS	PM	low flow PL
	weak solution PL	NL	NL	NL	NL	NM	NS	ZE
		ZE	PS	PM	PL	PL	PL	PL
	PM	NL	NL	NL	NM	NS	ZE	PS
		NS	ZE	PS	PM	PL	PL	PL
	PS	NL	NL	NM	NS	ZE	PS	PM
		NM	NS	ZE	PS	PM	PL	PL
ΔC_0	ZE	NL	NM	NS	ZE	PS	PM	PL
		NL	NM	NS	ZE	PS	PM	PL
	NS	NM	NS	ZE	PS	PM	PL	PL
		NL	NL	NM	NS	ZE	PS	PM
	NM	NS	ZE	PS	PM	PL	PL	PL
		NL	NL	NL	NM	NS	ZE	PS
	strong solution NL	ZE	PS	PM	PL	PL	PL	PL
		NL	NL	NL	NL	NM	NS	ZE

- the middle column and row correspond to mid range, ie if either ΔF_0 or ΔC_0 is ZE, then the outputs vary monotonically from NL to PL.
- the entries for both ΔF_1 and ΔF_2 go in the same direction along the rows but in opposite directions along the columns.
- an operability jacket prevents the product flow from becoming too high/low and too strong/weak.
- the middle row/column explains the ability of the FLC to handle disturbances and reduce offset. To eliminate offset, the controller output must be in incremental form.

In Figure 108.12 the outputs ΔF_1 and ΔF_2 are added to their normal values to produce outputs in absolute form. However, for incremental outputs, algorithms of the following form would be required:

$$F_{1,j+1} = F_{1,j} + \Delta F_{1,j} \qquad F_{2,j+1} = F_{2,j} + \Delta F_{2,j}$$

Apart from the fact that the FLC for the blending system is 2-input 2-output, and that both outputs have to be evaluated and adjusted at each sampling instant, its operation is otherwise exactly the same as that of the 2-input 1-output FLCs described earlier in the chapter.

108.10 Alternative Approaches

It is the ability of FLC to handle applications with non linearity, interaction and time delay that is of primary interest. The two examples above illustrate the non linear and interaction capability. Regarding time delay, conventional controllers have to be detuned to cope with delay resulting in significant offset. Noting that the middle row of the rule base handles disturbances, it is evident that relaxing rules other than those in the middle row enables the FLC to be detuned whilst maintaining its ability to reduce offset.

Otherwise, it is the flexibility of FLCs that is attractive. All manner of factors can be accommodated within the rule base. For example, an alarm or trip could be activated by one or more specified rules being fired. There is also much flexibility regarding the choice of input and output variables. For example, for the two inputs case of the level control system of Figure 108.2, the inputs chosen were the error e and its derivative ė. Both of these are in deviation form in the sense that their normal values are zero. However, there is no reason why variables in absolute form could not have been chosen, such as the depth of liquid h and the derivative of the inlet flowrate dF_1/dt, albeit with different scaling factors.

Another point to appreciate is that the output of an FLC does not have to be in absolute form as was the case with the level control system. It is fairly common practice for the output to be incremental with the signal being downloaded to (say) a PLC which integrates it. Furthermore, it is not necessary that the output be used to adjust a control valve opening. FLCs may, for example, adjust the set point of a slave loop, manipulate the gain of a controller, as in a gain scheduling strategy, or trim the ratio setting in a ratio control scheme.

As stated, the majority of FLCs have two input signals and a single output. This is largely because the rule base is two dimensional, can be represented in the form of a table, and can be visualised relatively easily. It is possible to have multiple inputs and outputs, but this leads to nesting of the rules. For example, for a fuzzy logic PID controller, each subset of the universe of the integral of the error would require a PD rule base of the form of Table 108.1. This gives rise to a family of rule bases: in effect, the composite rule base is three dimensional.

For developing n dimensional rule bases, adaptive techniques are required which enable the rules to be learned and modified on line. An adaptive approach is also required for handling situations where the process characteristics vary with time.

108.11 Self Adaptive Fuzzy Control

The essential features of a self adaptive FLC are depicted in Figure 108.13.

The performance monitor continuously monitors the performance of the FLC against some independent performance index. The index is typically cast in terms of parameters such as rise time, settling time, overshoot and offset which characterise the desired response of the controller inputs relative to previous changes in output. Performance is deemed to be in need of improvement when the observed response does not fall within some specified tolerance about the performance index. The performance index, monitor and adaptation mechanism may themselves all cast be in terms of production rules.

In the event of performance requiring improvement, the adaptation mechanism is activated. There are three means of adaptation, all of which are typically realised by further production rules:

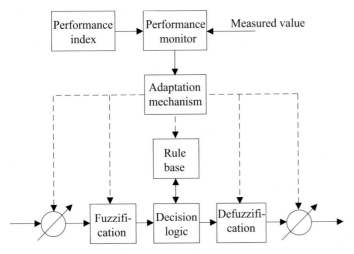

Fig. 108.13 A self adaptive fuzzy logic controller

- Varying the scaling factors. This is the most common means of adaptation and is equivalent to having a variable gain FLC. However, if the I/O signals are not in error or deviation form, care must be taken not to skew the symmetry of the signals relative to the subsets involved.
- Varying the subsets. By varying the position of the centre and the width of the base of the triangular subsets relative to each other, the sensitivity of the fuzzification or defuzzification process can be varied across the universe. This lends itself to adaptation by means of an artificial neural net. The centres and/or widths of the subsets have weights attached which the neural net tunes to optimise performance. Alternatively, the positions and widths may be coded as a chromosome for optimisation by means of a genetic algorithm.
- Changing the rules. By modifying the rules, ie by strengthening or weakening them, by making them more sophisticated or by adding new rules and deleting old ones, the rule base is changed to optimise performance. The principal difficulty is in isolating the rules that are responsible for causing the observed effects prior to making any changes. Clearly a good initial rule base requires little adaptation and will train itself fairly quickly.

If the adaptation mechanism varies the controller's I/O scaling factors or modifies the definition of its subsets, the FLC is said to be self-tuning. If the rule base is modified the FLC is said to be self-organising.

108.12 Observations

There is much evidence that, for the difficult to control problems, FLCs are potentially better than conventional PID controllers. However, the design of a FLC is a non-trivial task which requires expertise and experience. It is an expensive form of control relative to PID because its design and development is of a bespoke nature. FLCs cannot be justified for normal control problems on the grounds of cost.

Implementation is by means of software and is fairly easy to realise and maintain. Some DCS systems support FLC function blocks which can be configured much like any other function block. Proprietary FLC based systems are also available with supporting user interface, engineering tools, *etc.*, which can realise FLC on a MIMO basis: they are not restricted to two-input single output applications. These have been used extensively and successfully in the cement industry, where the non-linearities and delays are extreme. FLC has much to offer the rest of the process industry.

Artificial Neural Networks

109.1 Multi Layer Perceptrons
109.2 Operation of MLP
109.3 Back Propagation Training Algorithms
109.4 Network Size and Generalisation
109.5 Evaluation of Jacobians
109.6 Data Encoding
109.7 Pre-Processing of Data
109.8 Radial Basis Function Networks
109.9 Training of RBF Networks
109.10 Worked Example
109.11 Dynamic Modelling with Neural Nets
109.12 Neural Nets for Inferential Estimation
109.13 Neural Nets for Optimisation
109.14 Comments

Artificial neural networks (ANN), often referred to simply as neural nets, are essentially a means of prediction using computing techniques that are somewhat analogous to the human processes of thought and reasoning. Whereas in conventional computing, algorithms or sets of rules are devised to solve a particular type of problem, and coded as such in software, neuro-computing is the other way round. A general set of software tools is applied to data specific to a particular problem, the objective being to establish the underlying relationships.

It follows that neural nets are most useful when the relationships are unknown. In particular they are good at handling non-linear relationships and dynamics. ANNs should not be used for linear, steady state problems for which the techniques of regression analysis are more appropriate and simpler to use.

The techniques of neuro-computing have been around since the 1950s but, because they are computationally intensive, it was only in the 1990s that they became practicable as an approach for problem solving. The prediction capability of neural nets lends itself to a variety of applications in modelling, estimation, optimisation, classification, *etc.* For further information about applications the reader is referred to the text by Warwick (1992).

109.1 Multi Layer Perceptrons

A simplest form of neural net is as depicted in Figure 109.1.

It can be seen that there are a number of inputs, $x_1, x_2, \ldots x_p$. These are typically the values of process variables, such as flows, temperatures, *etc.* Each of these inputs is connected to a neuron in the input layer. Each of these input neurons is connected to a single neuron in the output layer. These connections are referred to as synapses. Each synapse has an associated weighting, $\beta_1, \beta_2, \ldots \beta_p$. The output y, some derived value, is the weighted sum of the inputs which is no more than a linear regression model as explained in Chapter 83:

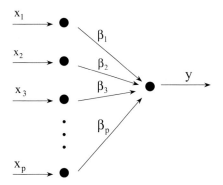

Fig. 109.1 The simplest form of neural net

$$y = \beta_1.x_1 + \beta_2.x_2 + \cdots + \beta_p.x_p$$

The prediction properties of the network can be substantially improved by introducing a hidden layer of neurons between the input and output layers, and by utilising functions other than linear scaling and addition. This structure is the so-called multi layer perceptron (MLP), as depicted in Figure 109.2. Sometimes referred to as a back propagation net or as a feedforward artificial neural net (FANN), it is the most commonly used form of neural network due to its simplicity and effectiveness.

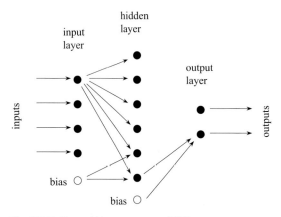

Fig. 109.2 The multi layer perceptron (MLP)

This too has a number of inputs, each of which is connected to one of the neurons in the input layer. The input neurons are passive in the sense that they simply pass on the input values. It is normal practice to have an additional bias neuron in the input layer. Such bias improves the network's approximation capability. Each of the input neurons is connected to every neuron in the hidden layer. So, if there are p inputs and h neurons in the hidden layer then, including the bias, there will be h × (p+1) input synapses each of which has its own weighting. It has been proved that a single hidden layer is sufficient to predict any continuous function but, if h is very large, it may be better to have two or more hidden layers with fewer neurons.

Figure 109.3 depicts the functionality of a single neuron in the hidden layer. All neurons in the hidden layer are identical apart from the weights associated with their inputs and outputs.

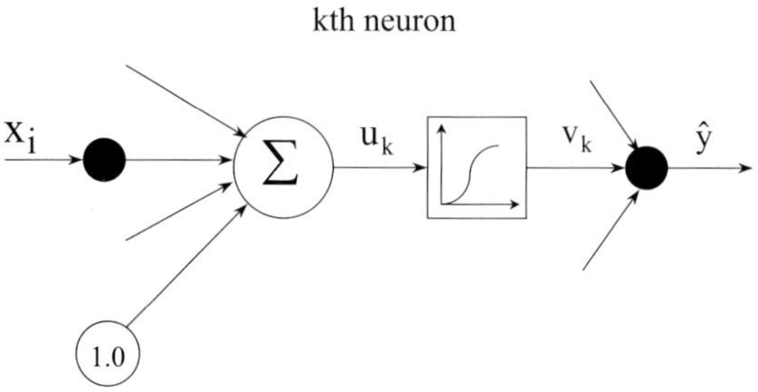

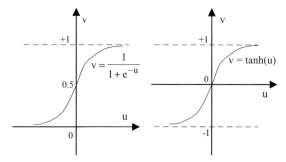

Fig. 109.3 Functionality of a single neuron in the hidden layer

The summation is operated upon by an activation function, sometimes referred to as a squashing function, the prerequisite for which is that it has a bounded derivative. The two most commonly used activation functions are the sigmoid and hyperbolic tangent as depicted in Figure 109.4 and defined as follows:

$$\text{either } v = \frac{1}{1 + e^{-u}} \text{ or } v = \tanh(u) = \frac{e^u - e^{-u}}{e^u + e^{-u}}.$$

Fig. 109.4 Commonly used activation (squashing) functions

where u is the sum of the weighted inputs at that neuron. It is the activation function which provides the network with its ability to handle non-linearities by, in effect, squashing the summation between limits of 0 and 1. Where positive and negative outputs are required then the function can be re-scaled into the interval $(+1, -1)$.

An MLP may have several neurons in its output layer which, in principle, enables several outputs to be calculated simultaneously. However, in practice, it is normal to have just a single neuron in the output layer. The output of each neuron in the hidden layer is connected to the neuron in the output layer. Again it is normal practice to have an additional bias neuron in the hidden layer. Each of these synapses has its own weighting. The output layer neuron normally has a summation function only and no activation function. Thus the output y of the MLP is the weighted sum of the inputs to the output neuron.

109.2 Operation of MLP

Having explained the structure of an MLP, its operation may be articulated. Consider a network comprising an input layer of p neurons ($i = 1 \to p$), a single hidden layer of h neurons ($k = 1 \to h$), and an output layer with a single neuron, as depicted in Figure 109.5. The objective, given any valid combination of inputs, is to predict the output.

The weightings β are identified by subscripts as appropriate. The standard convention uses four subscripts but, when there is only one hidden layer, it is more convenient to use only three subscripts as follows:

1st Denotes whether the weighting is on the net's input (1) or output (0),
2nd Indicates the neuron in the layer from which the synapse originates,
3rd Indicates the neuron in the layer to which the synapse is directed.

The sum of the weighted inputs at the first neuron in the hidden layer is given by:

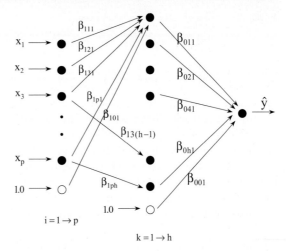

Fig. 109.5 Notation to denote inputs, neurons and synapses

$$u_1 = \beta_{101} + \beta_{111}.x_1 + \beta_{121}.x_2 \\ + \beta_{131}.x_3 + \ldots + \beta_{1p1}.x_p \\ = \beta_{101} + \sum_{i=1}^{p} \beta_{1i1}.x_i \quad (109.1)$$

At the arbitrary kth neuron in the hidden layer:

$$u_k = \beta_{10k} + \sum_{i=1}^{p} \beta_{1ik}.x_i \quad (109.2)$$

Assuming the activation function is sigmoidal, the output from the kth neuron in the hidden layer is given by:

$$v_k = \frac{1}{1+e^{-u_k}} = \frac{1}{1+e^{-\left(\beta_{10k}+\sum_{i=1}^{p}\beta_{1ik}.x_i\right)}} \quad (109.3)$$

The output of the MLP, which is the sum of the weighted outputs from the hidden layer, is thus given by:

$$\hat{y} = \beta_{001} + \sum_{k=1}^{h} \beta_{0k1}.v_k \\ = \beta_{001} + \sum_{k=1}^{h} \frac{\beta_{0k1}}{1+e^{-\left(\beta_{10k}+\sum_{i=1}^{p}\beta_{1ik}.x_i\right)}} \quad (109.4)$$

Note that in the absence of squashing, *i.e.* if the activation function is omitted, the output reduces to that of a multiple linear regression model.

109.3 Back Propagation Training Algorithms

Use of an ANN for prediction is critically dependent on having trained it previously to establish the best values for the weights and biases. It is important to appreciate that prediction (by the net) and training (of the net) are different phases of operation, each of which uses different sets of data.

Also note that, as with any form of extrapolation, the net can only predict outputs with confidence from inputs within the range of the training data used. There is no basis for assuming correct prediction capability when the input and/or output values are outside the range over which the net has been trained.

For training purposes n sets of input data ($j = 1 \rightarrow n$) are required, typically time series of process variables, for each of which the true value of the output y must be known.

During the training phase, a back propagation algorithm adjusts the weights and biases of the network so as to minimise some quadratic error (objective) function. This is normally taken to be the sum of the squared errors for the network, the error being the difference between the true and predicted values of the output, as follows:

$$E_{min} = \frac{1}{2} \cdot \sum_{j=1}^{n} (y_j - \hat{y}_j)^2 \quad (109.5)$$

The adjustment is realised by continually changing the values of the weights and biases in the direction of steepest descent with respect to the error, as previously described in Chapter 104. Thus, for the input weights, the back propagation algorithm is of the form:

$$\beta_{1ik}|_{q+1} = \beta_{1ik}|_q + \alpha_q \cdot \left.\frac{\partial E}{\partial \beta_{1ik}}\right|_q \quad (109.6)$$

The term α_q is known as the learning rate. In principle the learning rate may be varied from one iteration to the next but, in practice, a constant value of less than 0.3 is used. In general the same learning rate is used irrespective of the input data set and synapse. The partial derivative $\partial E/\partial \beta$ is the effect on the error function of varying a particular weighting, in effect a gain. This is referred to as the Jacobian as previously defined by Equation 104.4.

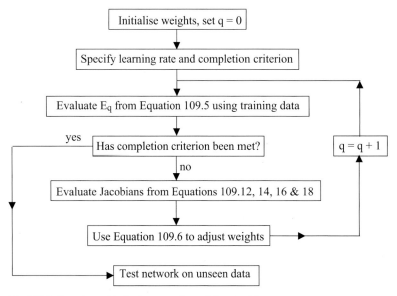

Fig. 109.6 Procedure of the back propagation training algorithm

Thus for each of the input data sets, the output is predicted using the existing weights and biases and the error function is evaluated. New values for the weights and biases are then found using the back propagation algorithm: these values supersede the previous ones. The input data is used again to re-evaluate the error function, and so on. Each iteration of the procedure is referred to as an epoch. The training procedure is depicted in Figure 109.6.

Note that the network has to be initialised: it is not uncommon to set the weights at random between −0.5 and +0.5 unless better values are known *a priori*. Also some completion criterion has to be specified, typically the maximum acceptable error on prediction. Clearly there are trade offs between acceptable error, the number and quality of data sets available for training, and the time and/or processing effort required for training. A typical training profile is depicted in Figure 109.7.

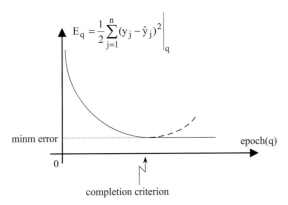

Fig. 109.7 A typical training profile: error function *vs* number of epochs

Whilst steepest descent is the most common approach to network training, it does have some well known disadvantages. These are largely due to the back propagation algorithm using a constant learning rate resulting in the step length (change in weighting) being proportional to the gradient. Thus, if the error surface is shallow in the vicinity of the minimum of the error function, convergence (rate of learning) can become extremely slow. Conversely, with complex error surfaces and steep slopes, the algorithm can jump over minima. More sophisticated algorithms are available for handling such situations. These involve individual learning rates for different parameters, variable learning rates, filtered values of gradients, momentum factors, *etc*. The most popular such algorithm is the Levenberg-Marquardt algorithm for non-linear least-squares optimisation, previously described in Chapter 104.

109.4 Network Size and Generalisation

An important danger to recognise is that too complex a net improves the model fit on the training data at the expense of its prediction ability on unseen data. In general, the training data will be subject to irregularities due to sampling and to noise. During training a complex net is able to improve its fit by capitalising on these peculiarities. However, since the peculiarities do not relate to the underlying process, the net's prediction performance for unseen data is reduced.

For a given network the peculiarities of the training data place a lower limit on the prediction error that can be obtained, as indicated by the solid curve in Figure 109.7. If a net's prediction error is comparable on both the training data and on unseen (validation) data then the model is said to "generalise". However, if a net performs well on the training data but poorly on validation data, as depicted by the broken curve in Figure 109.7, it is said to have "overfitted" the training data.

It follows that a key decision in training concerns the size and complexity of the network required to model a particular system. This depends primarily on the extent and severity of the non-linearities involved. Generalisation is the term used to qualify appropriateness of size. Good generalisation is ensured by adopting the following heuristic approach to training:

1. Start with the minimum complexity of network: typically a single hidden layer with the same number of neurons in the hidden layer as there are inputs to the input layer, and a single neuron in the output layer.

2. If plant data is abundant:
   i. Use validation data sets to check the generalisation performance of a network during training. Training is terminated when the prediction error on the validation data begins to rise relative to the error on the training data.
   ii. Systematically increase the number of neurons in the hidden layer recording the error on the validation data. The optimal structure is the network that minimises the error on the validation data.
3. If there is insufficient data for validation sets as well as for training sets, much caution is required to avoid overfitting.

109.5 Evaluation of Jacobians

Central to the training procedure is the non-trivial task of evaluating the partial derivatives, or Jacobians of Equation 109.6 at the end of each epoch. These are found from analytical expressions for the slope of the error surface with respect to the network parameters. There are four categories of $\partial E/\partial \beta$ that have to be evaluated corresponding, respectively, to the weights and bias from the input layer, the output weights and the bias on the hidden layer as follows:

$$\frac{\partial E}{\partial \beta_{1ik}} \quad \frac{\partial E}{\partial \beta_{10k}} \quad \frac{\partial E}{\partial \beta_{0k1}} \quad \text{and} \quad \frac{\partial E}{\partial \beta_{001}}$$

1. Consider the first category, the input weights, which account for the majority of the synapses.
   i. The error function is given by Equation 109.5:

   $$E = \frac{1}{2} \cdot \sum_{j=1}^{n}(y_j - \hat{y}_j)^2 = \sum_{j=1}^{n} \varepsilon_j \quad (109.5)$$

   where ε is the prediction error for the jth data set of a given epoch:

   $$\varepsilon_j = \frac{1}{2}(y_j - \hat{y}_j)^2$$

   whence:

   $$\frac{\partial \varepsilon_j}{\partial \hat{y}_j} = -(y_j - \hat{y}_j) \quad (109.7)$$

   ii. Summation at the output layer neuron is realised by Equation 109.4: For the jth data set of a given epoch:

   $$\hat{y}_j = \beta_{001} + \sum_{k=1}^{h} \beta_{0k1} \cdot v_{kj}$$

   whence:

   $$\frac{\partial \hat{y}_j}{\partial v_{kj}} = \beta_{0k1} \quad (109.8)$$

   iii. Assuming a sigmoid function, activation at the kth neuron in the hidden layer is realised by Equation 109.3. For the jth data set of a given epoch:

   $$v_{kj} = \frac{1}{1 + e^{-u_{kj}}}$$

   whence:

   $$\frac{\partial v_{kj}}{\partial u_{kj}} = \frac{e^{-u_{kj}}}{(1 + e^{-u_{kj}})^2} = v_{kj} \cdot (1 - v_{kj}) \quad (109.9)$$

   iv. Summation at the kth neuron in the hidden layer is realised by Equation 109.2: For the jth data set of a given epoch:

   $$u_{kj} = \beta_{10k} + \sum_{i=1}^{p} \beta_{1ik} \cdot x_{ij}$$

   whence:

   $$\frac{\partial u_{kj}}{\partial \beta_{1ik}} = x_{ij} \quad (109.10)$$

   Substituting from Equation 109.5 into the Jacobian of interest:

   $$\frac{\partial E}{\partial \beta_{1ik}} = \frac{\partial}{\partial \beta_{1ik}} \sum_{j=1}^{n} \varepsilon_j$$

   $$= \frac{\partial}{\partial \beta_{1ik}} (\varepsilon_1 + \varepsilon_2 + \ldots + \varepsilon_n)$$

   $$= \sum_{j=1}^{n} \frac{\partial \varepsilon_j}{\partial \beta_{1ik}}$$

The partial derivative may be expanded using the chain rule:

$$\frac{\partial E}{\partial \beta_{1ik}} = \sum_{j=1}^{n}\left(\frac{\partial \varepsilon_j}{\partial \hat{y}_j} \cdot \frac{\partial \hat{y}_j}{\partial v_{kj}} \cdot \frac{\partial v_{kj}}{\partial u_{kj}} \cdot \frac{\partial u_{kj}}{\partial \beta_{1ik}}\right) \quad (109.11)$$

Substituting from Equations 109.7 to 109.10 gives:

$$\frac{\partial E}{\partial \beta_{1ik}} = -\sum_{j=1}^{n}\left((y_j - \hat{y}_j).\beta_{0k1}.v_{kj}.(1-v_{kj}).x_{ij}\right) \quad (109.12)$$

The Jacobian is therefore established by summation over an epoch. Note that the training data, that is the values of y_j and x_{ij}, are constant from one epoch to the next, the value of β_{0k1} only changes at the end of an epoch. All the other values change for each data set throughout every epoch.

The notion of back propagation is evident from inspection of Equation 109.11 in relation to Figure 109.3 in that the calculation begins at the output layer and propagates back through the net.

Values for the other three categories of Jacobian are found similarly:

2. The Jacobian for the input bias is found from a slight modification to the chain rule of Equation 109.11 as follows:

$$\frac{\partial E}{\partial \beta_{10k}} = \sum_{j=1}^{n}\left(\frac{\partial \varepsilon_j}{\partial \hat{y}_j} \cdot \frac{\partial \hat{y}_j}{\partial v_{kj}} \cdot \frac{\partial v_{kj}}{\partial u_{kj}} \cdot \frac{\partial u_{kj}}{\partial \beta_{10k}}\right) \quad (109.13)$$

The first three partial derivatives are the same as in Equations 109.7 to 109.9. The fourth comes from Equation 109-2:

$$u_{kj} = \beta_{10k} + \sum_{i=1}^{p} \beta_{1ik}.x_{ij}$$

whence:

$$\frac{\partial u_{kj}}{\partial \beta_{10k}} = 1$$

Substituting into 109.13 gives:

$$\frac{\partial E}{\partial \beta_{10k}} = -\sum_{j=1}^{n}\left((y_j - \hat{y}_j).\beta_{0k1}.v_{kj}.(1-v_{kj})\right) \quad (109.14)$$

3. The calculation of the Jacobian for the output weights is somewhat simpler. As before, summation at the output layer neuron is realised by Equation 109.4: For the jth data set of a given epoch:

$$\hat{y}_j = \beta_{001} + \sum_{k=1}^{h} \beta_{0k1}.v_{kj}$$

whence:

$$\frac{\partial \hat{y}_j}{\partial \beta_{0k1}} = v_{kj} \quad (109.15)$$

A modification to the chain rule expansion of Equation 109.11 yields:

$$\frac{\partial E}{\partial \beta_{0k1}} = \sum_{j=1}^{n}\left(\frac{\partial \varepsilon_j}{\partial \hat{y}_j} \cdot \frac{\partial \hat{y}_j}{\partial \beta_{0k1}}\right)$$

Substitution from Equations 109.7 and 109.15 gives:

$$\frac{\partial E}{\partial \beta_{0k1}} = -\sum_{j=1}^{n}\left((y_j - \hat{y}_j).v_{kj}\right) \quad (109.16)$$

4. And finally, the Jacobian for the bias on the output, due to the bias on the hidden layer. Again, starting with Equation 109.4, for the jth data set of a given epoch:

$$\hat{y}_j = \beta_{001} + \sum_{k=1}^{h} \beta_{0k1}.v_{kj}$$

whence:

$$\frac{\partial \hat{y}_j}{\partial \beta_{001}} = 1 \quad (109.17)$$

A slight modification to the chain rule expansion yields:

$$\frac{\partial E}{\partial \beta_{001}} = \sum_{j=1}^{n}\left(\frac{\partial \varepsilon_j}{\partial \hat{y}_j} \cdot \frac{\partial \hat{y}_j}{\partial \beta_{001}}\right)$$

Substituting from Equation 109.7 and 109.17 gives:

$$\frac{\partial E}{\partial \beta_{001}} = -\sum_{j=1}^{n}(y_j - \hat{y}_j) \quad (109.18)$$

Note that, for back propagation purposes, the evaluation of each partial derivative requires only knowledge of values local to the neuron concerned. That is analogous to the conventional understanding of the working of biological neural systems and goes some way to explaining the concept of neural computing.

Given that the appropriate Jacobian has to be evaluated for every synapse for each iteration of the training algorithm, it is easy to appreciate that the computational load involved in network training is massive. However, inspection of the various expressions for $\partial E/\partial \beta$ reveals that there are several common summations which provides much scope for the recycling of intermediate results.

109.6 Data Encoding

Encoding is essentially concerned with transforming data into a form suitable for presentation to the net. There are various techniques, but by far the most important is the scaling of inputs into similar ranges. The two most common means of scaling are by:

- applying a bias and slope to each input such that values between x_{min} and x_{max} are scaled between -1 and $+1$ proportionately.
- standardising each input by subtracting its mean and dividing by its standard deviation.

Whilst there is no requirement that the weights and bias be constrained by bounds such as $-1.0 < \beta < 1.0$ as, for example, in principle component analysis, they do nevertheless depend upon the range and scaling of the inputs signals. Indeed, wide variations in the ranges of the input signals are likely to lead to a mixture of large and small weights. Such a mixture requires many sets of data for training to satisfy any sensible completion criterion, so the principal advantage of input scaling is in the reduction of training time. Also of benefit is the avoidance of placing undue emphasis on any particular input.

Provided the output layer has a summation function only, the network output $\hat{y}$ is unbounded. In the unlikely event of the input or output layers having activation functions, it is common practice to scale between -0.9 and $+0.9$ because activations close to -1 and $+1$ are difficult to achieve.

Spread encoding linearly distributes, or spreads, the value of an input across several adjacent neurons. This is of particular value for handling relationships that are highly non-linear as it enables different weights to be applied to different parts of the input's range although it does, obviously, increase the network size and training load. Table 109.1 depicts an example in which a single input whose range is 4–20 is linearly spread over five neurons.

Table 109.1 Spread encoding of 4–20 mA signal across five neurons

Neuron	1	2	3	4	5
Input value	4	8	12	16	20
Encoding (8)	0.0	1.0	0.0	0.0	0.0
Encoding (10)	0.0	0.5	0.5	0.0	0.0
Encoding (9)	0.0	0.75	0.25	0.0	0.0

Spread encoding is used for handling cyclic data, such as time of day on a 24-h clock which can be encoded over two neurons as follows:

Neuron 1: $\sin(\theta.\pi/12)$
Neuron 2: $\cos(\theta.\pi/12)$

This avoids a sudden jump at midnight. Spread encoding can also be used for handling non-numerical data such as parameters which can take on 1 of N possible states. For example, days of the week may be encoded using seven neurons, one per day as depicted in Table 109.2.

Another option for non-numerical data is grouping into categories and encoding states onto a single neuron on a fractional basis, as depicted in Table 109.3.

Table 109.2 Spread encoding of non-numerical data

Day	S	M	T	W	T	F	S
Sunday	1	0	0	0	0	0	0
Monday	0	1	0	0	0	0	0

Table 109.3 Spread encoding on a fractional basis

Performance	NBG	Bad	Poor	Avge	OK	Good	Exc
Day		Mon	Tues	Wed	Thurs	Frid	Sat	Sun
Fraction		1/7	2/7	3/7	4/7	5/7	6/7	7/7

However, great care must be taken in ordering and interpreting the categories. For example, encoding performance on the basis of seven states makes sense whereas implying that Sunday is seven times better than Monday doesn't.

109.7 Pre-Processing of Data

The issues are much the same for training as for prediction and, in many respects, are not dissimilar to those explained in Chapter 102 in relation to statistical process control:

- Which variables should be selected? Correlated data causes training problems analogous to the collinearity problem in regression analysis.
- Are there instruments available for measuring the inputs, and are they accurate enough?
- Data must be sampled frequently enough to be useful, and all data must be sampled at the same frequency.
- Inputs must remain within the same range as the training data.
- Outliers may be removed and missing data constructed by means of interpolation.
- Input data may be filtered to prevent spurious noise effects, but the value of the filter constant is not necessarily obvious.
- Where there are time delays between inputs, it is important to time shift the data such that all inputs are on the same time basis. This is relatively easy if the time delays are known and constant. Otherwise it is best handled outside the network.
- Data may be transformed mathematically, e.g. square root for flow, reciprocal for time delay and logarithm for reaction kinetics, to reduce non-linearity. Such *a priori* transformations increase the complexity but do not necessarily improve the model.
- Combinations of inputs may be considered where it is known that such relationships are meaningful. For example, the product of a flow rate and a concentration as suggested by a mass balance.

The importance of understanding the process being modelled, and of interpreting and screening the data prior to modelling cannot be overstated.

109.8 Radial Basis Function Networks

An alternative approach to neural computing is offered by radial basis function (RBF) networks. These too consist of input, hidden and output layers of neurons and, structurally speaking, an RBF is exactly the same as an MLP as depicted in Figure 109.2. Again, consider a network comprising an input layer of p neurons (i = 1 → p), a single hidden layer of h neurons (k = 1 → h), and an output layer with a single neuron.

As with the MLP, the purpose of the input layer neurons is to distribute the scaled inputs to the neurons in the hidden layer. Unlike the MLP though, the synapses between the input and hidden layers each have a weighting of unity. The fundamental difference between MLP and RBF networks is in the hidden layer where each neuron has a different Gaussian function, as depicted in Figure 109.8.

At each neuron in the hidden layer the "distance" between the inputs and the "centre" of that neuron's RBF is calculated using the formula:

$$d_k = \left(\left[\underline{x} - \underline{c}_k\right]^T \left[\underline{x} - \underline{c}_k\right] \right)^{0.5} \quad (109.19)$$

where $\underline{x}$ is the column vector of current inputs, i.e. $\underline{x} = \begin{bmatrix} x_1 & x_2 & \cdots & x_p \end{bmatrix}^T$, and $\underline{c}$ is a column vector,

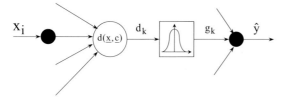

Fig. 109.8 Neuron in hidden layer of a radial basis function (RBF) network

known as the centre vector, of the "co-ordinates" of the centre of the neuron's RBF. The distance is then operated on by the Gaussian function according to:

$$g_k = \exp\left(\frac{-d_k^2}{\sigma_k^2}\right) \qquad (109.20)$$

where σ is a spread parameter (nothing to do with spread encoding). This is depicted in Figure 109.9.

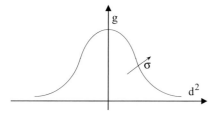

Fig. 109.9 The spread parameter used in the Gaussian function

The network output is found by calculating the weighted sum of the outputs and bias of the hidden neuron layer:

$$\hat{y} = \beta_0 + \sum_{k=1}^{h} \beta_k \cdot g_k \qquad (109.21)$$

Note that, whereas the input vector $\underline{x}$ is common to all the neurons in the hidden layer, the centre vector $\underline{c}$, the spread parameter σ and the output weighting β are unique to each neuron.

109.9 Training of RBF Networks

Training of an RBF network is essentially a question of establishing appropriate values for all these unique parameters.

1. Centres. The most common method of establishing the centre vectors is the so called k-means clustering algorithm, which works as follows:
   i. Gather n sets of input data, *i.e.* n input vectors $\underline{x}_j$ where $j = 1 \rightarrow n$.
   ii. Choose any h of the input vectors as provisional centre vectors $\underline{c}_k$ where $k = 1 \rightarrow h$.
   iii. Calculate the distance of all the other $(n-h)$ input vectors from each of the h centre vectors using Equation 109.19.
   iv. Assign each of the $(n-h)$ input vectors to the nearest centre vector, resulting in h clusters of vectors.
   v. Find the cluster averages, *i.e.* for each cluster, average the centre vector with its various nearest input vectors, and deem these to be the new centre vectors $\underline{c}_k$.
   vi. Return to (iii.) and iterate until no change is observed.

   The above procedure has a degree of randomness associated with it because of the random selection of the provisional centre vectors. This may result in slightly different centre vectors if the procedure is repeated several times. Nevertheless, it is easy to implement and converges rapidly on the centre vectors.

   Each of the centre vectors $\underline{c}_k$ so determined is arbitrarily assigned to one of the h neurons in the hidden layer. It does not matter which centre is assigned to what neuron as all of the synapses between the input and hidden layers have the same weighting of unity.

2. Spread. The so-called nearest neighbour's method is the most common means of establishing the spread parameter σ_k. Each of the hidden layer's neurons has an RBF whose centre vector $\underline{c}_k$ is known. Thus the distance between the centres of any two RBFs may be cal-

culated. For each kth neuron select the m neurons whose centres are nearest. The spread of that neuron's RBF is then evaluated from the equation

$$\sigma_k = \frac{1}{m}\left(\sum_{d=1}^{m}\left([\underline{c}_k - \underline{c}_d]^T[\underline{c}_k - \underline{c}_d]\right)\right)^{0.5} \quad (109.22)$$

where $1 \leq m \leq h - 1$.

The parameter m is adjusted during training to tune the network to give the smallest prediction error. Note that:

- More complex solution spaces, *i.e.* where there is more chance of training the net to find local as opposed to global minima, require smaller spread parameters and hence smaller values of the parameter m.
- If the signal to noise ratio is low then larger spread parameters, and hence larger values of m, may be required to make the network less sensitive to noise.

3. Weights. The weights of the synapses between the hidden and output layers are normally trained by means of multiple linear regression as explained in Chapter 83.

Assuming that the true value of the output y is known, then the error on the network prediction is given by

$$y - \hat{y} = \varepsilon$$

Substituting from Equation 109.21 gives

$$y = \beta_0 + \sum_{k=1}^{h} \beta_k \cdot g_k + \varepsilon$$

Thus, for n sets of input data (j = 1 → n) and h hidden layer neurons (k = 1 → h):

$$\begin{bmatrix} y_1 \\ y_2 \\ \vdots \\ y_n \end{bmatrix} = \begin{bmatrix} 1.0 & g_{11} & g_{12} & \cdots & g_{1h} \\ 1.0 & g_{21} & g_{22} & \cdots & g_{2h} \\ \vdots & \vdots & \vdots & \ddots & \vdots \\ 1.0 & g_{n1} & g_{n2} & \cdots & g_{nh} \end{bmatrix} \begin{bmatrix} \beta_0 \\ \beta_1 \\ \beta_2 \\ \vdots \\ \beta_h \end{bmatrix} + \begin{bmatrix} \varepsilon_1 \\ \varepsilon_2 \\ \vdots \\ \varepsilon_n \end{bmatrix} \quad (109.23)$$

which is of the form

$$\underline{y} = G.\underline{\beta} + \underline{\varepsilon}$$

This is a multiple linear regression model and can be solved by means of batch least squares to find the vector β that minimises $\underline{e}$. Thus, according to Equation 83.14:

$$\hat{\underline{\beta}} = (G^T G)^{-1} G^T \underline{y} \quad (109.24)$$

The pros and cons of MLP *vs* RBF are not clear cut. Both are relatively easy to use, given the availability of ANN toolboxes. Both are good at handling non-linearities and for steady state prediction purposes. For dynamic models, especially if there are a large number of inputs, RBF networks are better. Without doubt RBFs are faster to train than MLPs of a similar size. However, the number of hidden layer neurons required for RBF networks grows exponentially with the number of inputs.

109.10 Worked Example

An RBF network has been developed and trained to predict values of a single output y from values of two inputs x_1 and x_2. The network has the architecture depicted in Figure 109.10.

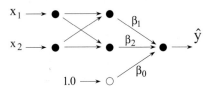

Fig. 109.10 The RBF network used for the Worked Example

The network has two input neurons (p = 2), two hidden layer neurons (h = 2) and a bias on the output. Training of the network has resulted in:

Centres: $\underline{c}_1 = [1.5 \ 1.5]^T \quad \underline{c}_2 = [2.0 \ 2.0]^T$
Spreads: $\sigma_1 = \sigma_2 = 1.0$
Weights: $\beta_0 = 4.3 \quad \beta_1 = -1.2 \quad \beta_2 = 4.2$

What would the prediction of the net be for each of the four sets of data (n = 4) given in Table 109.4, and compare these predictions with the corresponding measured values of the output.

Table 109.4 Data for prediction by the net of the Worked Example

j	x_1	x_2	y
1	1.0	1.0	4.1
2	1.1	1.2	4.4
3	2.0	2.0	7.8
4	1.9	1.8	7.3

Consider the first centre and the second set of input data (j = 2):

$$\underline{x} - \underline{c}_1 = \begin{bmatrix} 1.1 \\ 1.2 \end{bmatrix} - \begin{bmatrix} 1.5 \\ 1.5 \end{bmatrix} = \begin{bmatrix} -0.4 \\ -0.3 \end{bmatrix}$$

Its distance is given by Equation 109.19:

$$d_1 = \left(\begin{bmatrix} -0.4 & -0.3 \end{bmatrix} \begin{bmatrix} -0.4 \\ -0.3 \end{bmatrix} \right)^{0.5} = 0.5$$

Operated upon by the Gaussian function, the output of the first neuron of the hidden layer is given by Equation 109.20:

$$g_1 = \exp\left(\frac{-d_1^2}{\sigma_1^2}\right) = \exp\left(\frac{-0.5^2}{1^2}\right) = 0.7788$$

Similarly for the second centre with the second set of data:

$$\underline{x} - \underline{c}_2 = \begin{bmatrix} 1.1 \\ 1.2 \end{bmatrix} - \begin{bmatrix} 2.0 \\ 2.0 \end{bmatrix} = \begin{bmatrix} -0.9 \\ -0.8 \end{bmatrix}$$

This distance is given by:

$$d_2 = \left(\begin{bmatrix} -0.9 & -0.8 \end{bmatrix} \begin{bmatrix} -0.9 \\ -0.8 \end{bmatrix} \right)^{0.5} = 1.204$$

Operated upon by the Gaussian function, the output of the second hidden neuron becomes:

$$g_2 = \exp\left(\frac{-d_2^2}{\sigma_2^2}\right) = \exp\left(\frac{-1.204^2}{1^2}\right) = 0.2346$$

The predicted output of the network is found from the weighted sum of the outputs and bias of the hidden neuron layer according to Equation 109.21:

$$\hat{y} = \beta_0 + \sum_{k=1}^{h} \beta_k \cdot g_k$$
$$= \beta_0 + \beta_1 \cdot g_1 + \beta_2 \cdot g_2$$
$$= 4.3 - 1.2 \times 0.7788 + 4.2 \times 0.2346 = 4.3508$$

These results are listed in Table 109.5, together with those for the data sets for j = 1, 3 and 4.

The error function is evaluated according to Equation 109.5:

$$E = \frac{1}{2} \cdot \sum_{j=1}^{4} (y_j - \hat{y}_j)^2 = 0.0042$$

It is difficult to pass comment on the results, not knowing the application. Suffice to say that all of the predictions $\hat{y}$ are within 1.2% of the measured value y of the output. This is accurate enough for most purposes and the value of the error function is correspondingly small.

109.11 Dynamic Modelling with Neural Nets

The MLP and RBF nets described thus far are primarily used for steady state prediction purposes.

Table 109.5 Results of predictions by the net for the Worked Example

j	x_1	x_2	d_1	g_1	d_2	g_2	$\hat{y}$	y
1	1.0	1.0	0.7071	0.6065	1.414	0.1353	4.140	4.1
2	1.1	1.2	0.50	0.7788	1.204	0.2346	4.351	4.4
3	2.0	2.0	0.7071	0.6065	0.0	1.0	7.772	7.8
4	1.9	1.8	0.50	0.7788	0.2236	0.9512	7.360	7.3

The non-linearities handled are essentially of a variable gain in nature. However, ANNs can be used to model dynamic systems and to predict future values of variables. There are three structures of importance: conventional nets with time series of inputs, globally recurrent networks and locally recurrent networks:

1. Time series of inputs: this is depicted in Figure 109.11.

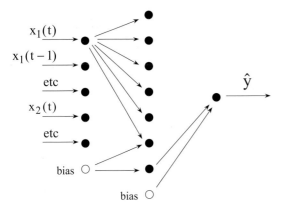

Fig. 109.11 Dynamic modelling with time series of inputs

For each input it is necessary to include some 15 previous measurements if an accurate model is required. This is on the basis of the need to capture approximately three times the dominant time constant's worth of information and a sampling period of about one-fifth of the time constant. Because of the number of previous measurements, ANNs are not practical for time series inputs if there is a large number of inputs.

2. Globally recurrent networks: these are essentially the same as with the time series of inputs, except that there is feedback of the output signal as depicted in Figure 109.12.
Note the analogy here with the SISO sampled data model of Equation 78.6:

$$\frac{y(z)}{x(z)} = \frac{a_0 + a_1 z^{-1} + a_2 z^{-2} + \ldots + a_m z^{-m}}{b_0 + b_1 z^{-1} + b_2 z^{-2} + \ldots + b_n z^{-n}}$$

A globally recurrent network enables process dynamics to be captured more concisely, *i.e.*

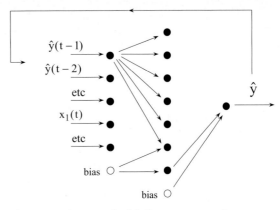

Fig. 109.12 Structure of a globally recurrent network

fewer hidden layer neurons, than by a conventional MLP without feedback. It is therefore more appropriate for models with several inputs. However, the feedback leads to longer training times. Once trained the network will provide one step ahead prediction. Non-standard training procedures are required for multi step ahead prediction involving back propagation from previous predicted values.

3. Locally recurrent networks: the neurons in the hidden layer are provided with memory in the form of a filter, as depicted in Figure 109.13. Otherwise the basic structure is exactly the same as in Figure 109.3.

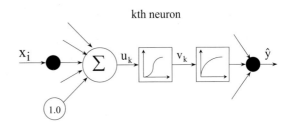

Fig. 109.13 Neuron with memory in the form of a filter

Any filter could be used but a first order lag is usually adequate. Clearly the time constant must be large enough relative to the sampling period of the inputs to avoid numerical instability, but not too large to filter out the underlying variations. Depending on the non-linear

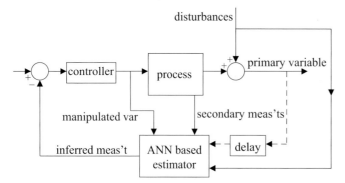

Fig. 109.14 Structure of a neural net for inferential measurement

dynamic effects being modelled, combinations of time lags and time delays may be used.

Locally recurrent networks represent a very compact means of capturing process dynamics since there is no requirement for time series of inputs. They are trained in the same way as a conventional MLP with the filter time constants being trained alongside the network weights. Training is relatively fast, despite having the time constants to train, because of the absence of time series of inputs.

An important point to appreciate in dynamic modelling with neural nets is that there is no attempt to establish the same structure for the net as for the real system. It is the overall dynamics and response that is being modelled, irrespective of whether there are recycle streams, feedforward controls, *etc*, and indeed whether the plant is being operated open loop or otherwise.

109.12 Neural Nets for Inferential Estimation

A not uncommon problem is that the variable of primary interest for control purposes cannot be measured directly and has to be inferred from secondary measurements of temperature, pressure, flow and so on. Examples of primary variables are rate of reaction, amount of biomass produced, composition of product stream, process yield, *etc*.

If the relationship between primary variable and secondary measurements is known *a priori*, the primary can be deduced explicitly. However, that is not usually the case. Neural nets have been successfully used for such inferential measurements, as depicted in Figure 109.14.

The inferential estimator can be considered to be an instrument which is producing a feedback signal to a conventional controller. Clearly the robustness of the ANN used to realise the estimator is critical, fundamental to which is the quality of the training data. Note the use of analytical measurements of the primary variable, indicated by a broken line, used for training purposes only. A characteristic of all such analytical measurements is the delay due to sampling and analysis, on-line or otherwise.

109.13 Neural Nets for Optimisation

If the prediction capability of an ANN can be used to infer or predict some primary process variable, it follows that there is scope for harnessing the ANN for optimising the value of that variable. The philosophy is depicted in Figure 109.15.

The ANN infers the value of the primary variable as described above. The inferred value is then used as an input to a real time optimiser (RTO). The RTO will have many other inputs too: current values of the manipulated variable u, secondary

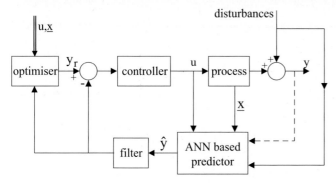

Fig. 109.15 Structure of a neural net for optimisation

measurements (states) $\underline{x}$ and disturbances. The optimisation is solved as a QP problem, or otherwise, as described in Chapter 106, the outcome being a set point downloaded to the primary controller or a target downloaded to a model predictive controller (MPC). Clearly there could be multiple set points and slave loops. Depending on the relative speeds of the optimiser and the predictor, not to mention the dynamics of the process itself, there may well be a need to filter out sudden changes on the predicted value.

109.14 Comments

This chapter has outlined the principal features of neural nets and their principles of operation. It has covered both the structure of ANNs and their training. An insight has been provided into the principal use of networks for prediction in non-linear contexts, together with practical constraints on implementation. Some areas of application have been introduced, especially in relation to dynamic models.

However, the chapter is only an introduction. Not covered are alternative structures such as self organising feature maps, sometimes referred to as Kohonen networks, and alternative techniques such as self adaptive nets, and alternative applications such as in identification, pattern recognition and classification.

Genetic Algorithms

110.1 Chromosomes and Genes
110.2 Cost and Fitness Functions
110.3 Selection
110.4 Crossover
110.5 Mutation
110.6 Reinsertion
110.7 Structure of a GA
110.8 Parallel GAs
110.9 Multi-Objective GAs
110.10 Pareto Ranking
110.11 Pareto Ranking MOGAs
110.12 Visualisation
110.13 Design of a Batch Scheduling MOGA
110.14 Comments

Genetic algorithms (GA) are stochastic search and optimisation methods which are analogous to the reproduction processes found in biological systems. These processes include crossover, mutation, survival of the fittest and evolution. Indeed, GAs have themselves evolved into the field of genetic programming and evolutionary strategies. To refer to a GA as an algorithm is perhaps misleading as this implies the procedural form of a single equation. In fact a GA is best thought of as a procedure, realised by means of a program comprising many algorithms, for which large amounts of data have to be organised and substantive processor power is required.

The chapter starts with an overview of the components of a GA and the processes of selection, crossover, and mutation. These provide the basis for explaining the structure of a GA which is depicted in Figure 110.2. The reader might like to refer to this diagram when reading the earlier parts of the chapter. The concepts of parallel GAs and multi-objective genetic algorithms (MOGA) are then developed. A batch scheduling problem is used as a vehicle for discussion to provide realism. For a good introduction to the subject the reader is referred to the text by Mitchell (1996), for a more comprehensive treatment to the classic text by Goldberg (1989) and for examples of their application in engineering systems to the text by Zalzala (1997).

110.1 Chromosomes and Genes

A chromosome is a potential solution to the optimisation problem in hand. Each chromosome consists of an ordered string of genes, each gene being a bit, an integer number, a letter code or some other entity. Formulation of the chromosome, referred to as encoding, and/or interpretation of the chromosome by decoding in relation to the application are undoubtedly the most difficult aspects of genetic programming.

The most common means of encoding chromosomes is in the form of binary strings. For example, a solution comprised of integer variable x_1

and short integer variable x_2 could be encoded as two bit strings concatenated (joined up) as shown in Table 110.1. It is obvious that without *a priori* knowledge of its structure, and decoding, information about the problem or its solution cannot be obtained by inspection of the chromosome itself.

Table 110.1 Chromosome formed by concatenation

x_1	1001011011010010
x_2	11101101
chromosome	100101101101001011101101

Variables or entities in the solution domain are sometimes referred to as phenotypes. Clearly both x_1 and x_2 have values: these are known as their phenotypic values. Particular instances of chromosomes are referred to as genotypes. There is a clear analogy between a vector of decision variables as, for example, in Equations 105.1 and 105.2, and a chromosome. Whereas the vector contains a number of distinct decision variables, a chromosome consists of a number of phenotypes which are encoded and concatenated.

Binary encoding obscures the nature of the search. To a certain extent this can be avoided by the use of integer and real valued genes which are easier to interpret. For certain classes of problem these can also increase the efficiency of the GA because there is less work involved in converting the chromosomes back into their phenotypes for evaluation of fitness. One such class of problem is batch scheduling.

Imagine that products A, B, C, D and E can be manufactured in a single-stream multi-products batch plant. To keep the problem simple, suppose that all the batches are of the same size, each takes the same time to process, there are no constraints on raw materials or storage, inter-batch cleaning is not required and that the plant is always operational. Presume a scheduling horizon of one week which corresponds to the production of m batches, the optimisation task being to decide what order to make the next m batches in. Suppose that m = 13 and assume that orders have been received to make two batches of A, four batches of B, four batches of C, two batches of D and one batch of E to meet a variety of delivery dates. One possible solution is simply to list the batches in alphabetical order in a chromosome of 13 genes as follows:

A, A, B, B, B, B, C, C, C, C, D, D, E

(2A, 4B, 4C, 2D and 1E)

This implies that two batches of A would be made first, followed by four batches of B, and so on. Given the nature of batch manufacturing it is highly improbable that this schedule would be satisfactory. It is, nevertheless, a possible solution as indeed are the many thousands of others arising from ordering the batches in different combinations. The subset of possible solutions with which a GA is concerned is referred to as the population.

110.2 Cost and Fitness Functions

For the purposes of generating an optimal solution, some measure of the suitability of each possible solution in meeting the delivery schedule is required. Suitability in GA speak is loosely referred to as fitness. It is articulated in terms of a cost or objective function f(t) which may be thought of as a raw value for the fitness of each chromosome. Note that the independent variable is normally, but not necessarily, time. The cost function is then transformed into the chromosome's relative fitness $\phi(t)$, that is relative to the population of chromosomes, by some fitness function:

$$\phi(t) = h(f(t)) \quad (110.1)$$

This is perhaps best illustrated with reference to the batch scheduling problem. Lateness of production, often referred to as tardiness in an optimisation context, is the obvious function to use. The time that each batch (ith gene) is completed, as determined by the schedule (jth chromosome), is compared with its delivery date from which its lateness $g_{ij}(t)$ may be calculated:

$$g_{ij}(t) = t_{ij} - t_{id} \quad \text{when} \quad t_{ij} > t_{id}$$

where

    t_{ij} is the time when batch i of chromosome j is completed, and
    t_{id} is the time that batch i is due for delivery.

Batches produced in advance of their delivery date count nothing towards the total lateness:

$$g_{ij}(t) = 0 \quad \text{when} \quad t_{ij} \leq t_{id}$$

Then, for each schedule, the total number of hours lateness $f_j(t)$ is calculated. This is the cost function:

$$f_j(t) = \sum_{i=1}^{m} g_{ij}(t) \qquad (110.2)$$

The transform most commonly used to establish the relative fitness $\phi_j(t)$ of an individual chromosome is that of "proportional fitness assignment" in which the cost function is expressed as a proportion of that of the population: If there are n chromosomes in the population of possible schedules being considered, then:

$$\phi_j(t) = \frac{f_j(t)}{\sum_{j=1}^{n} f_j(t)} \qquad (110.3)$$

An alternative to proportional fitness assignment is power law scaling. Noting that in many engineering problems it is the relative values of variables that matters rather than their absolute values, power law scaling provides a means of stretching or squashing the range of fitness. The parameter k is problem dependant and may be adjusted during execution of the GA:

$$\phi_j(t) = f_j(t)^k \qquad (110.4)$$

Evaluation of fitness enables the best solution in the population to be identified, ideally that which enables all batches to meet their delivery dates. However, it is seldom the case with scheduling problems that there is a single unique correct solution. Depending on the mix and spread of delivery dates, there may be no solutions at all or several alternatives. If no satisfactory solutions are found, then at least a schedule that minimises the overall lateness will be known. Multiple solutions enable the criterion for optimisation to be strengthened. For example, batches made in advance of their delivery date represent capital tied up in stock and storage costs. They too could contribute towards the total lateness, forcing the solution towards just-in-time manufacture:

$$g_{ij}(t) = |t_{ij} - t_{id}| \quad \text{whether} \quad t_{ij} > \text{or} < t_{id}$$

On the basis that it is more important to meet delivery dates than to save on inventory costs, the constraint on early production could be relaxed by applying a weighting factor of (say) 0.5:

$$g_{ij}(t) = 0.5 \left(t_{id} - t_{ij}\right) \quad \text{when} \quad t_{ij} \leq t_{id}$$

It is evident that there is much more to the chromosome than just the genes themselves: data is associated with and attached to it. In the batch example, for each gene there must be a customer reference, lot number and delivery time for the batch assigned to the gene. Also, for the chromosome itself, values for the overall lateness $f_j(t)$ and fitness $\phi_j(t)$ will be attached. For simple problems the data may be in the form of a look up table associated with the chromosome. For more complex problems the data is associated explicitly with the genes, say in the form of a datablock to which the gene acts as a pointer. Potentially, in extreme cases, the genes could themselves be data objects which are manipulated within an object oriented environment.

110.3 Selection

Selection, sometimes referred to as sampling, is the process of determining the number of times a particular chromosome from the old population is chosen for reproduction. The process is probabilistic, most commonly by means of a simulated roulette wheel. The key issue is allocation of space on the roulette wheel, known as the parent space value (PSV). This is most commonly done in proportion to the relative fitness of the chromosomes as defined by Equation 110.3. In essence, the fitter chromosomes are allocated larger sectors than the weaker ones, thereby increasing the chance of the roulette wheel's pointer coming to rest over their sector.

Returning to the batch example, suppose that the population (unrealistically) comprises (only) four chromosomes α, β, γ and δ whose overall lateness $f_j(t)$ are 10, 20, 30 and 40 h and relative fitness $\phi_j(t)$ are 0.1, 0.2, 0.3 and 0.4 respectively. Clearly the fittest chromosome α is twice as fit as β, three times fitter than γ and four times fitter than δ, so the ratios of expectation of selection are 12, 6 and 4 to 3 respectively. The sectors allocated are thus approximately $173°, 86°, 58°$ and $43°$ as depicted in Figure 110.1.

Fig. 110.1 Selection by simulated roulette wheel

Typically the roulette is spun according to the size of the population, *i.e.* four times in the above example. Several aspects are worth commenting upon:

- Every chromosome from the old population has a sector allocated and thereby has a chance of being selected and propagating its characteristics into the next generation. The fitter chromosomes have a greater chance of selection but the weaker are not precluded.
- Suppose the above roulette wheel was spun 25 times. The expectation would be that chromosomes α, β, γ and δ would be selected 12, 6, 4 and 3 times respectively.
- It is possible, but improbable, that any chromosome could be selected 25 times. Likewise each may not be selected at all.

This leads on to the concepts of bias and spread. Bias is the difference between the expected probability of selection of a particular chromosome and that achieved in practice. Thus, in the above example, if the wheel was spun 25 times and α was selected 13 times the bias would be 1.0. Clearly zero bias is optimal and corresponds to the expected and actual selections being equal. Spread is the range of the number of times an individual chromosome can be selected. The "minimum spread" is the smallest spread that theoretically permits zero bias, 12 for chromosome α in the above example.

The basic method of roulette wheel selection described above is known as stochastic sampling with replacement (SSR). It gives zero bias but the potential spread is limited only by the population size. There are various other methods of selection, of which two are described here.

Stochastic sampling with partial replacement (SSPR) is essentially the same as SSR except that the PSVs are adjusted after each selection. Again consider the roulette wheel of Figure 110.1 being spun 25 times. Suppose the first spin results in selection of chromosome α. The expectation for α is then reduced by 1 such that the ratios of expectation for selection of 12, 6 and 4 to 3 would become 11, 6 and 4 to 3 respectively. The sectors are adjusted in proportion to the new expectations, *i.e.* to $165°, 90°, 60°$ and $45°$, and the process repeated until sampling is complete. SSPR has both upper and lower limits on spread, 12 and 0 for α in the above example, but leads to a higher bias then SSR.

Stochastic universal sampling (SUS) is also essentially the same as SSR except that the roulette wheel has multiple equally-spaced pointers and is only spun once. Yet again consider the roulette wheel and sector allocations of Figure 110.1. Four selections are required so 4 pointers with 90° spacing are required. The wheel is spun and chromosomes are selected according to which sectors are aligned with the pointers. SUS has minimum spread and zero bias, for which reason it is one of the most commonly used selection methods.

Truncation selection is an inherently different approach to roulette wheel based methods of selection. In essence, a pre-specified percentage of the fittest chromosomes are selected from the old population. These are then selected at random and subjected to the processes of reproduction until sufficient offspring are produced for the new population. GAs which use truncation selection are referred to as breeder GAs. Whereas the standard GA models natural selection, the breeder GA models artificial selection.

110.4 Crossover

Crossover is the primary genetic operator. It replicates the natural process of recombination in which the chromosomes of offspring are formed from the genes of their parents. The simplest form of recombination is that of single point crossover. Consider the same schedule as before, for which two possible chromosomes are as follows:

A B B C D D A B B C C C E
        (2A, 4B, 4C, 2D and 1E)
A A D D B B B B C E C C C
        ditto

If an integer position i is selected at random from the range $1 \rightarrow (m-1)$, or $1 \rightarrow 12$ in this case, then two new offspring are created by swapping heads and tails after the crossover point. Supposing that $i = 7$ yields the following offspring:

A B B C D D A B C E C C C
        (2A, 3B, 5C, 2D and 1E)
A A D D B B B B B C C C E
        (2A, 5B, 3C, 2D and 1E)

An important constraint that has to be satisfied for batch scheduling is that of quota balance. There is no point in generating chromosomes that do not meet the production requirements. It is evident that crossover has led to an imbalance in the number of batches of B and C and some adjustment is necessary. One such approach which works satisfactorily is to systematically replace surplus batches with deficit ones. Thus, starting at the crossover point:

A B B C D D A B B E C C C
        (2A, 4B, 4C, 2D and 1E)
A A D D B B B C B C C C E
        ditto

Multi-point crossover is a fairly obvious, but significant, extension to single point crossover. In essence, various crossover points are selected at random from the range $1 \rightarrow (m-1)$, sorted into ascending order, and the genes between alternate pairs of crossover points are swapped between chromosomes. The same scheduling problem as before, and assuming values for i of 2, 5, and 9, yields:

A B D D B D A B B E C C C
        (2A, 4B, 3C, 3D and 1E)
A A B C D B B B C C C C E
        (2A, 4B, 5C, 1D and 1E)

Quota adjustment would be required again to address the imbalance in batches of C and D, also done after the first crossover point:

A B C D B D A B B E C C C
        (2A, 4B, 4C, 2D and 1E)
A A B D D B B B C C C C E
        ditto

With multi-point crossover, the number of points at which crossover is to take place also has to be determined. This may be the same for the whole population or chosen at random for individual chromosomes. It is common practice to place constraints on the locations where crossover may occur, according to the nature of the problem.

The thinking behind multi-point crossover is that those parts of the chromosome that contribute most to its fitness may not necessarily be contained in adjacent genes, or adjacent groups of genes. Furthermore, the disruptive nature of multi-point crossover encourages consideration of a wider range of potential solutions, rather than favouring convergence on highly fit individuals early in the search, thus making the search more robust.

There are other crossover strategies for which various pros and cons are claimed. The reduced surrogate operator constrains crossover to always produce new chromosomes by restricting crossover to those points where the gene values are different. Uniform crossover, which involves a randomly generated mask, is only of practical use for chromosomes that are binary strings. Intermediate recombination is a method of producing new chromosomes whose fitness is between those of their parents.

110.5 Mutation

Mutation is a secondary genetic operator in the sense that it is normally applied after crossover has taken place. Unlike crossover which operates on pairs of chromosomes, mutation operates on single chromosomes. In essence, mutation causes individual genes to be changed according to some probabilistic rule and can take many forms. For chromosomes that are binary strings mutation occurs simply by changing the gene at randomly chosen positions, either from 0 to 1 or *vice versa*.

This approach, however, is inappropriate for the batch scheduling problem. It does not make sense to randomly change E into A, *i.e.* to make three batches of A and none of E, since the new chromosome could never satisfy the production requirements. So, in this case, mutation would be realised by randomly swapping the position of two genes within the chromosome. For example, mutation at genes 5 and 13 yields:

A B B C <u>E</u> D A B B C C C <u>D</u>
(2A, 4B, 4C, 2D and 1E)

As with crossover, constraints may be placed on the locations where mutation may occur according to the nature of the problem.

There are many variations on the mutation operator. One such is to bias the mutation towards the chromosomes with lower fitness values to increase the exploration in the search without losing genetic information from the fitter individuals. Another example is to parameterise the mutation such that the mutation rate decreases with convergence of the population.

The role of mutation should be seen as providing a guarantee that the probability of selection for any given chromosome will never be zero and acting as a safety net to recover good genetic material that may be lost through the selection and crossover operations.

110.6 Reinsertion

Once selection, crossover and mutation are complete there will be two populations: the old and the offspring. Reinsertion is concerned with the means of combining them to produce the new population, and keeping control of the size of the population. There are various possible reinsertion strategies:

- Assuming that selection has produced the same number of offspring as the old population, one such strategy would be to merge the two populations, rank the individuals in order of fitness and select the top half of the list which should create a new population of the same size as the old. The disadvantage of this approach is that it discourages diversity and, within a few generations, eliminates any latent optimal solutions embedded in the weakest chromosomes.
- An alternative strategy which is very effective when a large generation gap exists, *i.e.* when the number of offspring is much less than the old population, is to merge the two populations and reduce it to size by removing the oldest members. This is clearly in keeping with natural reproduction processes as every member of the population will, at some stage, be replaced.
- Another strategy would simply be to discard the old population and use the offspring for the next generation. This is in keeping with the natural principle of survival of the fittest, in that the fitter members of the old population are most likely to have been involved in producing offspring, but it does almost guarantee elimination of the fittest members themselves.
- The most popular method of reinsertion though is the elitist strategy in which a small generation gap, of just one or two individuals, is created at the selection stage. Then, once the offspring have been formed, the population is made up to size by including the fittest members of the old population. This approach has the benefit of guaranteeing survival of the fittest individuals from one generation to the next.

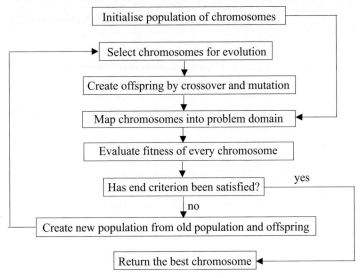

Fig. 110.2 Flow chart of a genetic algorithm (GA) as a procedure

110.7 Structure of a GA

As stated, a GA is basically a procedure. The standard procedure is depicted in flow chart form in Figure 110.2.

Typically a population of between 30 and 100 chromosomes is established. The GA operates on the whole population in the expectation that at least one chromosome will emerge as an optimum solution. The initial population is generated at random. Thus, for example, 40 chromosomes of possible solutions for a scheduling horizon of 15 batches could be generated. If possible, a seed solution is introduced such that the initial population would be comprised of one seed plus 39 randomly generated solutions. The seed is an *a priori* solution that is feasible but almost certainly sub-optimal. Typically the seed is known from previous experience or from attempting to solve the scheduling problem by hand. The use of a feasible seed aids the GA search in terms of speed of convergence.

There then follows the process of selection by means of the roulette wheel, or otherwise, with PSV allocated on the basis of relative fitness. Not only does the method of selection, such as SSR, SSPR or SUS have to be decided, but also the number of new chromosomes to be generated. Typically, the number selected is equal to or slightly smaller than the size of the old population, in which case it is likely that some of the fitter chromosomes will have been replicated, *i.e.* chosen more than once, and some of the weaker ones eliminated, *i.e.* not chosen at all. As stated, if the number selected is different to that of the old population, the difference is referred to as the generation gap.

Having selected the chromosomes for reproduction, offspring are produced by means of crossover and mutation. It is important to appreciate the probabilistic nature of these genetic operators. As discussed, the points at which crossover and mutation occur are determined at random. Furthermore, not all chromosomes in the population are necessarily subject to crossover. Some probability of crossover, typically $P_x = 60\%$, is normally applied to each chromosome. Thus it is likely that a proportion of the chromosomes will not be crossed. Similarly, not all the chromosomes are subject to mutation. Indeed, a low probability of mutation, typically $P_m = 1\%$, is normally applied: higher frequencies cause the GA process to degenerate into random selection. Note that there are always checks to ensure that chromosomes replicated at the selection stage are subject to crossover and/or mutation. There is no point in having two or more identical chromosomes in the new population.

Following decoding of the chromosomes, *i.e.* mapping of the solutions into the problem domain, the fitness of every chromosome in the population is evaluated. Note that this applies to chromosomes surviving from old generations, as well as to their offspring, as their relative fitness values may well have changed. For the batch scheduling problem, evaluation of fitness clearly requires the availability of delivery dates for each batch, calculation of lateness for each chromosome and evaluation of relative fitness values for the whole population.

The end criterion is next tested and if it is not satisfied a new population is created from the fittest members of the old population and the offspring by means of an appropriate reinsertion strategy and the next iteration is commenced.

The end criterion may simply be that an acceptable solution is found, *e.g.* a solution to the batch problem whose overall lateness is zero. However, with GAs being stochastic, convergence can neither be predicted nor guaranteed. It is not unusual for the fitness of a population to remain static for a number of generations (iterations) before a superior solution emerges. A common practice is to terminate the GA after a specified number of iterations, say 50, and then assess the quality of the best solutions against the problem. If no acceptable solution is found iteration may be continued or the population reinitialised and a fresh search started.

It is evident from the above that there is much variety in GAs with much scope for choice in terms of strategy and assignment of probabilistic parameters. Different types of problem lend themselves to different approaches. There is no correct answer: it is simply a question of trying the alternatives and seeing what works best for a given problem.

110.8 Parallel GAs

Parallel GAs operate upon populations that are partitioned into sub-populations, each of which is operated upon by a separate GA. The population itself is often referred to as the global population and the sub-populations as islands. The parallel activities may be carried out as separate tasks sequentially by a single processor or simultaneously on parallel processors. Parallel GAs, even if implemented sequentially on a single processor, usually yield a variety of benefits compared with a standard GA operating upon a global population of the same size as the sum of the islands.

These benefits include not only improvements in speed of execution and efficiency in terms of convergence, but also the ability to tackle larger problems than otherwise and better quality solutions in terms of fitness. The improvement in quality is largely attributable to greater diversity in the search space. In general, increasing the size of the population improves its genetic diversity and reduces the chance of finding sub-optimal solutions. However, there is a trade off between the diversity and speed of execution: the larger the population the slower the execution becomes for a given amount of processor power.

Also, for problems with several objectives, parallel GAs enable multiple, equally satisfactory, solutions to co-exist in the global population simultaneously. A good example of this is in batch scheduling where it may be necessary to simultaneously minimise lateness and maximise plant utilisation.

There are many variations on the theme of parallel GAs but most concern alternative means of partitioning the population and/or selection for reproduction. The three main categories of parallel GAs: global, migration and diffusion, are described in the following sections.

Global

The global GA does not partition the population as such, but exploits parallel aspects of the structure of the GA in its execution. Inspection of the structure of Figure 110.2 reveals that processes such as selection, evaluation of relative fitness, reinsertion and population control, and testing against the end criterion are best carried out on the whole population. However, others such as crossover, mutation

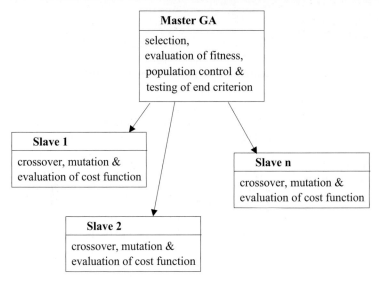

Fig. 110.3 Global GAs: processing by slave routines

and evaluation of the cost functions may be carried out on pairs of chromosomes independent of the rest of the population. These latter operations are candidates for parallel processing by slave routines under the control of a master which manages the GA as depicted in Figure 110.3. Note that in GA speak this master/slave relationship is often referred to as farmer/worker.

Provided that evaluation of the cost function is significantly more demanding computationally than evaluation of fitness and selection, which is usually the case, use of the global GA can lead to near linear speed-up. Otherwise, the slave becomes a bottleneck. Note that if the slave routines are implemented by means of parallel processors operating on a single population in shared memory, care must be taken to ensure that individual chromosomes can only be accessed by one processor at a time.

Migration

In essence, migration GAs partition the population into islands that are each operated upon by an independent standard GA as depicted in Figure 110.4.

To encourage the proliferation of good genetic material throughout the global population, individual chromosomes intermittently migrate between the islands. The migration model is analogous to human reproduction in that the islands replicate races: there is much intra-race breeding but only limited inter-race breeding. The pattern of migration is determined by the number of migration paths between the islands (9 out of 15 possible paths in Figure 110.4). The frequency of migration is determined by the interval (no of generations) between migrations and the rate of transfer (number of chromosomes per migration) between islands.

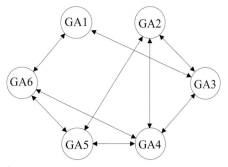

Fig. 110.4 Migration GAs: processing by independent standard GAs

A standard GA can be relatively easily adapted to handle the requirements for migration by the addition of extra routines. Thus parameters are included which enable the migration interval and rate to be specified, subject to constraints. These are probabilistic in nature and applied on a random basis. Note that migration is normally bidirectional.

Typically, chromosomes for migration are selected at random from the subset of the exporting island's chromosomes whose fitness are greater than its average. Likewise, chromosomes selected for replacement may be chosen at random from the subset of the importing island's chromosomes whose fitness are less than its average.

Another approach is to produce more offspring than required at the exporting island. Emigrants are selected at random from the offspring, sufficient to maintain population size, and exported. An equal number of chromosomes on the importing island are eliminated to make room for the immigrants. The philosophy behind this approach is that the most fit individuals are more likely to reproduce and are therefore most likely to migrate.

In terms of diversity, convergence and quality of solution, migration GAs significantly outperform global parallel GAs.

Diffusion

Whereas migration GAs introduce discontinuities into the population through division into islands, diffusion GAs treat the global population as a single geographic distribution of chromosomes. Each is assigned a node on the "surface" and is permitted to breed with chromosomes at neighbouring nodes. This is depicted in the form of a grid in Figure 110.5.

Chromosomes are initialised at random and their cost functions evaluated. Each node then receives a copy of the chromosomes at adjacent nodes. For example, node J_{31} receives copies from J_{21}, J_{32}, J_{35} and J_{41}. The fitness of the members of this subset are evaluated and a candidate for reproduction with J_{31} is selected. Crossover and mutation then takes place and the fittest offspring replaces the chromosome at J_{31}. Rules may be applied to retain the original parent if neither offspring is sufficiently fit to replace it. A single generation of the GA consists of this process being repeated at every node across the surface.

The local selection process leads to the fitter members diffusing through the population. After a few generations, local clusters of chromosomes with similar genetic characteristics and fitness (may) emerge giving rise to virtual islands

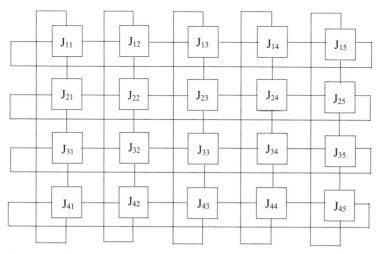

Fig. 110.5 Diffusion GAs: processing by breeding at neighbouring nodes

in the population. Subsequent generations lead to consolidation: the smaller clusters coalesce and the larger ones increase in size. As with all GAs, evolution continues until some end criterion is met.

Diffusion GAs may be thought of as replicating natural selection in the sense that individuals tend to mate in their local environment. The local neighbourhood selection mechanism encourages greater exploration of the search space and helps inhibit the early domination of the population by fitter individuals. Diffusion GAs perform better than migration GAs with comparable population sizes. The search is more robust in that it is exhaustive, and the quality of the solutions in terms of fitness is better.

110.9 Multi-Objective GAs

Consideration of GAs so far has focussed on optimisation problems for which there is a single objective. In practice, many problems require the simultaneous optimisation of multiple objectives. Such problems may be handled by multi objective genetic algorithms (MOGA). Two approaches to MOGAs are considered here: aggregation and pareto ranking. There are other approaches, simulated annealing being quite common. It should be appreciated that MOGAs may be realised either by means of a standard GA or by means of parallel GAs. For the following explanations of MOGAs a standard GA and a single population of chromosomes is assumed.

Apart from evaluation of fitness, the aggregation MOGA is essentially the same as the standard GA as outlined in Figure 110.2. Strictly speaking, it is not fitness as defined by Equation 110.3 that is evaluated but a weighted sum of cost functions. This is evaluated after reproduction, at each generation, for each chromosome as follows. Suppose there are p objectives:

- A separate cost function for each of the p objectives is evaluated as in Equation 110.2:

$$f_{jk}(t) = \sum_{i=1}^{m} g_{ijk}(t) \quad (110.5)$$

where the subscripts i, j and k refer to the gene, chromosome and objective respectively.
- Then the aggregated "fitness" is calculated from the weighted sum of the cost functions for all the objectives:

$$\phi_j(t) = \beta_1.f_{j1} + \beta_2.f_{j2} + \cdots \beta_p.f_{jp}$$
$$= \sum_{k=1}^{P} \beta_k.f_{jk}(t) \quad (110.6)$$

Noting that the essential difference between cost functions and fitness is a scaling factor due to proportional assignment, the fact that cost functions are aggregated rather than true fitness values is offset by appropriate choice of the β coefficients.
- The aggregated fitness is subsequently used as a basis for reinsertion and thence selection.

This process clearly lends itself to matrix manipulation. Thus, the cost functions of each chromosome may be stored as a $(1 \times p)$ row vector. For a population of n chromosomes these may be assembled into a $(n \times p)$ matrix:

$$F(t) = \begin{bmatrix} f_{11}(t) & f_{12}(t) & \cdots & f_{1p}(t) \\ f_{21}(t) & f_{22}(t) & \cdots & f_{2p}(t) \\ \vdots & \vdots & \ddots & \vdots \\ f_{n1}(t) & f_{n2}(t) & \cdots & f_{np}(t) \end{bmatrix} \quad (110.7)$$

Post-multiplication of F(t) by a $(p \times 1)$ column vector of the β coefficients yields the $(n \times 1)$ vector of aggregated fitness values:

$$\phi(t) = F(t).B \quad (110.8)$$

where:

$$B = \begin{bmatrix} \beta_1 & \beta_2 & \cdots & \beta_p \end{bmatrix}^T$$
$$\phi = \begin{bmatrix} \phi_1(t) & \phi_2(t) & \cdots & \phi_n(t) \end{bmatrix}^T$$

The attraction of the aggregated approach is that it yields a single variable that is a simple measure of the optimality of the solution found to a multi-objective problem. The simplicity, however, is misleading because appropriate values for the weighting coefficients are seldom known *a priori* and have to be guessed on the basis of experience. Furthermore, aggregation of the fitness values obscures the relative contribution of each objective to the overall optimisation.

110.10 Pareto Ranking

Within any population of potential solutions, even if there are none which are optimal with regard to every objective, it is likely that there will be several that are sufficiently optimal with regard to most objectives to be acceptable. It is also clear that for any such solution, improvement against one objective can only be achieved at the expense of the others. This subset of solutions is said to be paretooptimal and lies on some surface in the space defined by the objectives' cost-functions. The surface is referred to as the trade-off surface and the space as the objective space. The aim of pareto ranking MOGAs is to establish which members of the population lie on that surface. This enables the user to explore the trade-offs between the various objectives on the basis of experience.

The pareto ranking approach is based upon the concept of dominance, rather than on absolute measures of performance, and is the only method available in the absence of information about the relative importance of the objectives. It is easiest to visualise in two dimensions, so assume that there are only two objectives (p = 2) for each of which there is a cost function $f_1(t)$ and $f_2(t)$. Values of both cost functions are determined for each member of the population, some of which are plotted against each other as depicted in Figure 110.6.

The subset of pareto-optimal solutions are identified on the basis of being non-dominated. They are indicated as being of rank 0. By inspection it can be seen that for each solution (of rank 0) there is no other solution (also of rank 0) that is more optimal, *i.e.* that has a lower value of both $f_1(t)$ *and* $f_2(t)$. Assigning rank to the remaining solutions is more subjective. The approach here is to assign a rank to an individual according to how many other individuals it is dominated by. Figure 110.6 depicts one rank 1 solution: this is

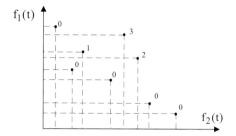

Fig. 110.6 Values of cost functions for each chromosome

clearly dominated by a single rank 0 solution that has a lower value of both $f_1(t)$ and $f_2(t)$. An example of both rank 2 and rank 3 solutions are also depicted.

The above two objectives led to a trade-off curve of rank 0 solutions in the cost function space, albeit with a bulge in the vicinity of the rank 1 solution. Clearly if three or more cost functions are involved, an n-dimensional trade-off surface of the pareto-optimal solutions is established. Note that, because dominance is established on the basis of distance from the origin, it is important to cast the optimisation as a minimisation problem. Thus maxima have to be converted into minima by introducing a minus sign, and finite goals have to be converted into minima by shifting the origin. Also note that the nature of dominance is such that ranking can be based on the cost functions directly: there is no need for the axes to be scaled or normalised.

110.11 Pareto Ranking MOGAs

Operation of a pareto ranking MOGA is as depicted in flow chart form in Figure 110.7.

In many respects operation is essentially the same as explained for the standard GA. Initialisation is by means of creating a random population, perhaps with seeding. Selection is by means of roulette wheel methods such as SSPR or SUS, and reproduction by means of crossover and mutation.

Crossing individuals arbitrarily may lead to large numbers of unfit offspring, referred to as lethals, which could adversely affect the performance of the search. It is far better to encourage

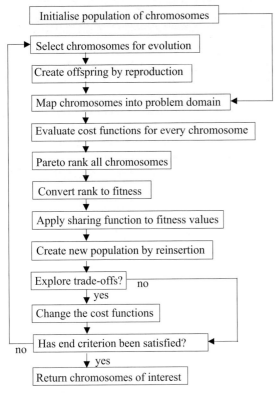

Fig. 110.7 Operation of a pareto ranking multi-objective GA

and error involved in deciding what value of σ_{mate} is most suitable for a given application.

After reproduction comes decoding, *i.e.* mapping of the chromosomes into the application domain, followed by evaluation of the cost functions which obviously involves evaluating multiple cost functions for each chromosome, followed by pareto ranking as described above.

Conversion of rank to relative fitness ϕ is essential as this provides the basis for reinsertion and selection, a basis which discriminates in favour of the fittest individuals. Conversion is typically by means of some exponential rule combined with averaging. Suppose the exponential rule is:

$$\theta_j = 2 \cdot \exp(-(j-1)/n) \quad \text{for} \quad j = 1, 2, \ldots, n. \tag{110.10}$$

Consider the (very) untypical population depicted in Figure 110.6, for which n = 8. Applying the exponential rule yields eight values of θ from 2.0 down to 0.834. The solutions are aligned against these values in rank order to give provisional fitness values. Where several solutions have the same rank, which is normally the case, the mean of those exponential values is used to yield the fitness values ϕ as indicated in Table 110.2.

Table 110.2 Fitness values on basis of common rank

Chromosome j	θ_j	Rank	ϕ_{jav}
1	2.000	0	1.582
2	1.765	0	1.582
3	1.558	0	1.582
4	1.375	0	1.582
5	1.213	0	1.582
6	1.071	1	1.071
7	0.945	2	0.945
8	0.834	3	0.834

individuals on the trade-off surface, or close to it, to mate with each other. Thus mating is restricted to individuals within a given distance of each other. The mating restriction can be thought of as a volume in objective space about each individual, the volume being characterised by a radius σ_{mate}.

Consider two individuals, i and j, in a p-dimensional objective space whose co-ordinates are the values of their cost functions:

$$\underline{f}_i(t) = \begin{bmatrix} f_{i1}(t) & f_{i2}(t) & \cdots & f_{ip}(t) \end{bmatrix}$$
$$\underline{f}_j(t) = \begin{bmatrix} f_{j1}(t) & f_{j2}(t) & \cdots & f_{jp}(t) \end{bmatrix}$$

The Euclidean distance between them is given by:

$$\sigma_{ij} = \sqrt{\left[\underline{f}_i(t) - \underline{f}_j(t)\right]\left[\underline{f}_i(t) - \underline{f}_j(t)\right]^T} \tag{110.9}$$

Individuals are only permitted to mate provided that $\sigma_{ij} \leq \sigma_{mate}$. There is a certain amount of trial

A common phenomenon is genetic drift which means that, as the MOGA evolves, pareto-optimal solutions tend to cluster along the trade-off surface. This is undesirable because the aim of the optimisation is to explore solutions across the whole of the trade-off surface. Genetic drift may be countered by the use of a sharing function which effectively biases the selection process towards individuals that are spread out along the trade-off surface. An important point to appreciate is that to achieve a greater spread along the trade-off surface the whole population has to be spread throughout the objective space. There are two approaches to spreading based upon so-called niche counts or kernel densities.

The niche count approach involves a volume of objective space of radius σ_{share} about each individual. There are no hard and fast rules about the size of σ_{share} and it is not uncommon for it to simply be set equal to σ_{mate}. The niche count Nc of an individual is determined by the number of other individuals within that volume. A sharing function determines the contribution to that niche count of the other individuals according to the distance between them. An obvious example of a sharing function is simple proportion. Suppose there are n_s individuals (i) at a distance σ_{ij} of individual (j). Its niche count is thus:

$$Nc_j = k . \sum_{i=1}^{n_s} (\sigma_{share} - \sigma_{ij}) \quad \text{for all} \quad \sigma_{ij} \leq \sigma_{share} \tag{110.11}$$

This results in individuals that are clustered together having relatively high niche counts and those that are spread out having low niche counts. Each raw fitness value $f_j(t)$ is then weighted according to its niche count. This is done using the inverse of its niche count, normalised by the sum of weights, subject to some constraint on zero division to allow for situations when n_s is zero. Alternatively, weighting may be realised by some inverse exponential function. The total fitness of the population is thus re-distributed throughout the population with the fitness of the more spread out individuals being enhanced relative to the rest.

The basic problem with the niche count approach is that it involves calculating the distance σ_{ij} between large numbers of pairs of individuals throughout the objective space, potentially n factorial calculations, at each generation. This is clearly prohibitive in terms of computational effort so, for large populations, kernel density estimation methods are used. These achieve the same ends by statistical means. So called kernel methods use a smoothing parameter h instead of niche size. Guidelines exist for selection of appropriate values of h for certain kernels such as the standard normal probability density function and the Epanechnikov kernel.

Creation of the new population from the old population and the offspring may be by means of any of the reinsertion strategies outlined earlier but the elitist strategy is common.

Unlike conventional optimisation methods, the use of MOGAs enables a degree of interaction between the user and the optimisation process, thereby enabling user experience to be brought to bear upon the choice of solution from the paretooptimal set of solutions. Thus the MOGA may be stopped at any stage to explore the trade-offs and, if appropriate, the cost functions may be strengthened or relaxed. Alternatively, the MOGA may be left to run for a fixed number of generations, the resultant set of pareto-optimal solutions explored for trade-offs and the MOGA run again. The end criterion itself, as with the standard GA, may be either the evolution of a solution that satisfies some particular combination of cost functions or simply a fixed number of generations.

110.12 Visualisation

Observing the progress and understanding the result of a standard GA with a single objective is fairly straightforward. For example, the value of the cost function of the fittest chromosome in the population can be plotted against number of generations. The cost function will decrease, not necessarily monotonically, and approach some limiting value. As a basis for comparison, and as a measure of the genetic diversity in the population, it is useful to also plot the value of the cost function of that chromosome whose solution is of median fitness. This is typically as depicted in Figure 110.8.

Visualisation of MOGAs is less straightforward. Clearly the same plots as for a standard GA could be generated to observe progress against the cost functions corresponding to each of the separate objectives. However, the focus of interest with

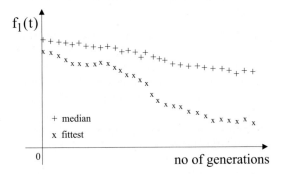

Fig. 110.8 Value of cost function versus number of generations

MOGAs is the set of pareto-optimal solutions and the trade-offs between them in objective space. Since trade-off surfaces cannot be visualised geometrically, other than for 2 or 3 objectives, solutions are best handled as plots using parallel axes, as described in Chapter 102. In the context of GAs, such plots are referred to as trade-off graphs, an example of which is depicted in Figure 110.9.

The x-axis represents the various objectives. Each parallel y-axis represents the cost function corresponding to a particular objective. The cost functions along the y axes are normalised on a common scale, of say 0–100%, to enable comparison. Individuals are depicted by polygonal lines connecting the solution's cost functions. The trade-off graph of Figure 110.9 has five objectives for which four solutions have been plotted. By inspection it can be seen that solution A is dominated by both solutions B and D. However, every other solution is non-dominated, i.e. for each solution there is no other for which all the objectives have a lower valued cost function.

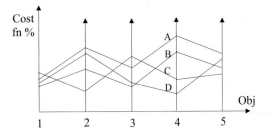

Fig. 110.9 Trade-off graphs using parallel co-ordinates

Trade-off graphs can be used in different ways. Typically, when the MOGA is stopped for interaction at the end of a run or otherwise, all the pareto-optimal solutions will be plotted and the pattern inspected. A bunching together of cost function values, as indicated for the first objective of Figure 110.9, suggests a hard constraint for which the corresponding cost function could be relaxed. A wide spread of cost function values, such as indicated for the fourth objective, suggests there is scope for tightening the cost function. In practice there are usually many solutions on the trade-off

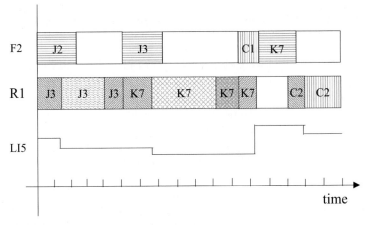

Fig. 110.10 Gantt chart for batch scheduling

graph and the patterns are fairly obvious so identifying the scope for trade-offs is relatively easy.

For batch scheduling purposes, the most effective means of depicting the schedules generated as a function of time is in the form of a Gantt chart, as depicted in Figure 110.10.

Thus for each reactor (R1), filter (F2), *etc.*, batches scheduled such as J2, J3 and K7 and cleaning cycles such as C1 and C2 are as indicated. Use of colour coding and shading can be used to distinguish between transfer operations, reaction, *etc.*, with blanks corresponding to idle periods. Of particular use are predicted inventory levels of raw materials and intermediates, as depicted by the trace LI5, which can be generated fairly easily from the schedule.

110.13 Design of a Batch Scheduling MOGA

Many variations on the implementation of a MOGA are possible and it is common practice for MOGAs to be designed to meet the specific requirements of an application. So, whilst a batch scheduling MOGA is generally consistent with the structure of Figure 110.7, there are inevitably aspects of implementation specific to batch processes. For example, schedules could be generated subject to the simultaneous optimisation of the following objectives:

- Minimise lateness in meeting delivery deadlines for which the cost function is lateness as defined by Equation 110.2.
- Maximise plant utilisation for which the cost function relates to the time spent by units (major equipment items) waiting between operations or standing empty.
- Maximise plant capacity for which the cost function relates to the fractional capacity of units not occupied by a batch during processing.
- Minimise storage requirements for which the cost function relates to the number of intermediates and products being stored, and the volumes thereof.
- Minimise time spent on cleaning for which the cost function relates to the time spent on inter-batch cleaning of units.

The chromosomes will consist of a string of alphabetic codes as discussed. It is important, from the point of view of feasibility, that the total no of batches (length) and the number of batches of each product (genes) be estimated fairly carefully. Note that for a multi-stream, multi-product plant with a scheduling horizon of the order of fortnight to a month, the chromosomes could well be several hundred genes long.

At first sight it may appear that the position of a gene within the chromosome implies a time sequence. That may well be so for a single stream plant with simple, single stage processes. However, with multi-stage processes involving storage of intermediate products and/or with multi-stream plant, the order of the genes does not necessarily define the sequence of batches in a chronological sense. For example:

A B B C D D A B B C C C E
(2A, 4B, 4C, 2D and 1E)

Suppose there are two parallel streams and no shared equipment modules. Manufacture of a batch of product A followed by a batch of C could be assigned to one stream, whilst the first two batches of B are assigned to the other stream. Clearly, depending on the time taken for individual batches, it is possible for the batch of C (fourth gene) to commence before the second batch of B (third gene).

Constraints may be imposed on both crossover and mutation. This may be on a relative or absolute basis. An example of the relative basis would be a number of batches of the same product being forced to be consecutive, irrespective of where the group occurs within the schedule, to enable campaign operation. An example of the absolute basis would be fixing a specified number of batches at the beginning of the schedule, because the schedule is being re-run and the first few batches have already been started.

Of particular importance to the MOGA is the role of the schedule builder which is the program that interprets, or decodes, the chromosomes and translates the solutions back into the application domain. The builder makes use of a range of domain specific data, typically in the form of look up tables. Using the terminology described in Chapter 37, this data includes:

- Production requirements in terms of which products (and grades) are to be made, their throughput (number of tonnes) and delivery dates.
- Topology of the plant in terms of trains, units (with values for maximum holding and minimum running capacities) and equipment modules (especially exclusive use resources).
- Non-availability of units (such as reactors and filters) due to planned shutdowns, breakdowns, staffing shortages, *etc*.
- What recipes to use for which products in terms of procedures and equipment types (especially constraints on the suitability of certain units for particular products).
- Structure of procedures in terms of order of operations and phases, and constraints on such, with time taken for each phase (according to units if necessary).
- Storage facilities (with maximum capacity and initial inventory) for each reagent, intermediate and product (especially constraints on usage) and delivery dates for raw materials.
- Inter-batch cleaning requirements in terms of which procedure (if any) to use between any two consecutive batches (according to units if necessary).

The batch scheduler itself is a fairly complex set of domain specific production rules (of an if-then-else nature) that are consistently applied to every chromosome. The essential tasks of the schedule builder include the following:

- Allocate batches of different products to units according to the order of the batches in the chromosome and subject to the constraints on availability and suitability of units.
- Manage progression of batches from unit to unit (if necessary) according to the procedural requirements and constraints of plant topology.
- Impose inter-batch cleaning operations as appropriate.
- Establish the chronology of events in terms of start and stop times for each operation and/or phase in every unit throughout the schedule, and likewise for periods spent waiting.
- Maintain an inventory of the contents of all units and storage vessels throughout the schedule.
- Manipulate the batch quantities, subject to constraints on unit capacities, to maintain feasible solutions in terms of scheduled throughput compared with production requirements.

It is usually the case that any one batch could be assigned to many units, even after the constraints of availability, topology and suitability are taken into account. Thus, within the allocation process, there is scope for alternative strategies. The two most common approaches are to choose the fastest unit to process the batch even if it means waiting for the unit to become available, and choose the first available unit even if it is the slowest for processing the batch. Other alternatives are to choose the unit at random, to choose that which will complete the batch soonest, or that which will make it just-in-time. In practice, somewhat surprisingly, the choice of allocation strategy does not appear to have much effect on the performance of the MOGA.

The notion of feasibility is important to grasp. Infeasible solutions are those schedules which significantly under- or over-produce in relation to the production requirements. This concerns throughput only (tonnes) and is not directly related to meeting delivery dates. Whilst not encouraged, infeasible solutions are permitted to persist in the population because they preserve genetic diversity. Populations that consistently under-produce have probably got too short a chromosome and require extra genes, the number being found by trial and error. Populations that over-produce may be adjusted by treating the excess as dummy batches. Alternatively, noting that most schedules are re-run before they reach the end of their scheduling horizon, the excess batches may be left in the schedule but not made.

110.14 Comments

It can be seen that GAs differ substantially from conventional search and optimisation methods. In particular, they search a population of parallel solutions rather than for a single solution and invariably come up with several possible solutions to a problem, the final choice being left to the user. For multi-objective optimisations, where there is no unique solution, the MOGA presents a family of pareto-optimal solutions which enables the user to explore the trade-offs between them.

Because of the stochastic nature of the search mechanism, GAs are capable of searching the entire solution space with more likelihood of finding the global optimum than conventional methods. Furthermore, they do not require derivative information or other auxiliary knowledge: it is only the cost functions and relative fitness values that influence the direction of search.

Whilst GAs and the techniques of genetic programming are not new, their use for solving engineering problems is a relatively recent occurrence. Given that they can handle discontinuous and time varying function evaluations, and will tolerate data that is ill-defined, incomplete and uncertain, they have much to offer.

Section 13

Advanced Process Control

Multiloop Systems

111.1 Relative Gain Analysis
111.2 Interpretation of RGA Elements
111.3 Worked Example: L-V Scheme
111.4 Worked Example: L-B Scheme
111.5 Worked Example: D-V Scheme
111.6 Effective Gain
111.7 Singular Values
111.8 Application to Blending System
111.9 Singular Value Decomposition
111.10 Worked Example No 4
111.11 Comments

The essential strategy with multivariable control system design is to establish a diagonal controller such that there is unique stable relationship between each set point and a controlled variable. There are two approaches, leading to the design of either compensators or decouplers. However, as was seen in Chapter 81, compensator and decoupler design is mathematically non-trivial so, although most plants are inherently multivariable in nature, these designs are seldom used.

It is much more common to use multiple single-input single-output (SISO) control loops of a conventional feedback nature. If there are no interactions between the loops then a diagonal structure results. Then it is possible to change the set point in any one loop without affecting the other loops, and each loop may be tuned independently. In practice, it is likely that there will be interactions between the loops, so the design process focuses on pairing off inputs and outputs so as to minimise the extent of interaction. Note that although interaction may be minimised, it is seldom eliminated.

This chapter, therefore, focuses on two techniques, relative gain analysis (RGA) due to Bristol and singular value decomposition (SVD), of which the former is more commonly used, for identifying appropriate input-output pairings and for analysing the consequent extent of interaction. For a more comprehensive overview of the various other techniques, the reader is referred to the texts by Deshpande (1989), Luyben (1990) and Marlin (2000).

111.1 Relative Gain Analysis

Suppose there is some preliminary multiloop control system design, based upon experience, in which n inputs have been paired off with n outputs such that n control loops may be considered to exist. Let these be defined as follows:

n process outputs (CVs): $y_1, y_2, \ldots y_i, \ldots, y_n$, and
n process inputs (MVs): $u_1, u_2, \ldots u_j, \ldots, u_n$.

The process gains define the steady state open-loop relationship between the inputs and outputs:

$$\begin{bmatrix} y_1(s) \\ y_2(s) \\ \vdots \\ y_n(s) \end{bmatrix} = \begin{bmatrix} k_{11} & k_{12} & \cdots & k_{1n} \\ k_{21} & k_{22} & \cdots & k_{2n} \\ \vdots & \vdots & \ddots & \vdots \\ k_{n1} & k_{n2} & \cdots & k_{nn} \end{bmatrix} \begin{bmatrix} u_1(s) \\ u_2(s) \\ \vdots \\ u_n(s) \end{bmatrix}$$

which is of the form:

$$y(s) = K.u(s) \quad (111.1)$$

There may well be some diagonal structure to K, but many of the off-diagonal terms will be non-zero due to interactions. RGA provides a measure of the strength of those interactions which enables judgement about the suitability of the pairings.

Consider the interaction between the jth input and the ith output. Two scenarios are now explored:

1. Put all the controllers into their manual mode with all their outputs at their normal values, and apply a step change to the output of the jth controller. Because of the interactions, it is likely that several of the controlled variables will change: what is of interest is the steady state change in the ith controlled variable. Thus the open loop gain between y_i and u_j when all the other process inputs are held constant is:

$$k_{ij}\big|u = \frac{\Delta y_i}{\Delta u_j}\bigg|_u \quad (111.2)$$

2. Put all the controllers into their automatic mode with all their set points at their normal values, and apply a step change to the ith desired value. Again, because of the interactions, it is likely that several of the manipulated variables will change: what is of interest this time is the steady state change in the jth manipulated variable. Thus the closed loop gain between y_i and u_j when all the other process outputs are held constant is:

$$k_{ij}\big|y = \frac{\Delta y_i}{\Delta u_j}\bigg|_y \quad (111.3)$$

The relative gain between the output y_i and the input u_j is defined to be the ratio:

$$\lambda_{ij} = \frac{k_{ij}\big|u}{k_{ij}\big|y} \quad (111.4)$$

The relative gain array is the array of relative gains for every input and output pair:

$$\Lambda = \begin{bmatrix} \lambda_{11} & \lambda_{12} & \cdots & \lambda_{1n} \\ \lambda_{21} & \lambda_{22} & \cdots & \lambda_{2n} \\ \vdots & \vdots & \ddots & \vdots \\ \lambda_{n1} & \lambda_{n2} & \cdots & \lambda_{nn} \end{bmatrix} \quad (111.5)$$

The elements of the matrix Λ may be determined empirically by means of the step response approach outlined above and using Equation 111.4, or calculated as explained in the following examples. Either way, it is not necessary to determine every value because the sum of the values in any one row or column of Λ is unity. Thus if Λ is n dimensional, then only $(n-1)^2$ elements need be determined: the others may be found by summation.

Alternatively, if the elements of the K matrix of Equation 111.1 are known, say by means of modelling or step testing, then Λ may be found directly from the formula

$$\Lambda = K.* \left(K^T\right)^{-1} \quad (111.6)$$

where the .* denotes an element by element multiplication.

111.2 Interpretation of RGA Elements

Interpretation of the values of the elements of Λ is relatively straightforward and summarised in Table 111.1.

Table 111.1 Interpretation of RGA elements

λ_{ij}	Interpretation
$\lambda_{ij} = 1.0$	This means that the open and closed loop gains are the same: it makes no difference whether the other loops are in manual or auto, or whether the controller's output or set point is changed. Thus there are no interactions between the loop involving the jth input and the ith output and any other loop. The ideal scenario
$\lambda_{ij} = 0.0$	This means either that the open loop gain is zero and/or that the closed loop gain is infinite. If the open loop gain is zero then the jth input has negligible affect on the ith output. If the closed loop gain is infinite then the jth input has a massive affect on the ith output. In both cases there is no sensible basis for control
$0.0 < \lambda_{ij} < 1.0$	This means that there is a high degree of interaction, but the system is nevertheless stable. The maximum interaction occurs when λ is 0.5
$1.0 < \lambda_{ij} < 10$	The open loop gain is relatively large compared with the closed loop gain. This means that there are significant open loop interactions which the closed loop can counter effectively. It is preferable to pair off those inputs and outputs whose RGA elements are closest to 1.0
$\lambda_{ij} > 10$	This means that there are strong interactions. The relative gains are very sensitive to steady state gain error and any decoupling control is likely to become unstable with only a small degree of plant model mismatch (PMM)
$\lambda_{ij} < 0.0$	The jth input and ith output loop pairing is unsuitable for control. This is because the controller changes sign between the open and closed loop scenarios and problems will be encountered with robustness

111.3 Worked Example: L-V Scheme

Consider the distillation column of Chapter 90. A dynamic model for its concentration lags in deviation form was developed:

$$\begin{bmatrix} \dot{X}_1 \\ \dot{X}_2 \\ \dot{X}_3 \end{bmatrix} = \begin{bmatrix} a_{11} & a_{12} & 0 \\ a_{21} & a_{22} & a_{23} \\ 0 & a_{32} & a_{33} \end{bmatrix} \begin{bmatrix} X_1 \\ X_2 \\ X_3 \end{bmatrix} \quad (90.4)$$

$$+ \begin{bmatrix} b_{11} & b_{12} & 0 & 0 \\ b_{21} & b_{22} & b_{23} & b_{24} \\ b_{31} & b_{32} & b_{33} & 0 \end{bmatrix} \begin{bmatrix} L \\ V \\ F \\ X_F \end{bmatrix}$$

or, by notation:

$$\underline{\dot{x}} = A.\underline{x} + B.\underline{u}$$

For RGA purposes a steady state model only is required. Thus, setting $\underline{\dot{x}} = 0$ gives:

$$0 = A.\underline{x} + B.\underline{u}$$
$$\underline{x} = -A^{-1}.B.\underline{u} = G.\underline{u}$$

Assuming the values for the coefficients of the A and B matrices from the Worked Example of Chapter 90 yields:

$$\begin{bmatrix} X_1 \\ X_2 \\ X_3 \end{bmatrix} = \begin{bmatrix} 0.246 & -0.215 & 0.0491 & 0.498 \\ 0.310 & -0.295 & 0.100 & 1.011 \\ 0.558 & -0.565 & 0.234 & 1.558 \end{bmatrix} \begin{bmatrix} L \\ V \\ F \\ X_F \end{bmatrix}$$
(111.7)

Suppose it is proposed to control the column by the L-V scheme of Figure 35.10. In this scheme the concentrations X_1 and X_3 are controlled by manipulating L and V respectively.

Assuming the other inputs to be constant, Equation 111.7 reduces to:

$$\begin{bmatrix} X_1 \\ X_3 \end{bmatrix} = \begin{bmatrix} 0.246 & -0.215 \\ 0.558 & -0.565 \end{bmatrix} \begin{bmatrix} L \\ V \end{bmatrix} \quad (111.8)$$

which is of the form of Equation 111.1 where:

$$K = \begin{bmatrix} 0.246 & -0.215 \\ 0.558 & -0.565 \end{bmatrix}$$

First, find the open loop gain. Keeping the input V constant, differentiate to find the effect of L on X_1:

$$X_1 = 0.246.L - 0.215.V$$

$$k_{11}|V = \left.\frac{\partial X_1}{\partial L}\right|_V = 0.246$$

Second, find the closed loop gain. Keeping the output X_3 constant, manipulate Equation 111.8 to eliminate V and differentiate again to find the effect of L on X_1:

$$X_1 = 0.246.L - 0.215.\left(\frac{0.558.L - X_3}{0.565}\right)$$

$$= 0.0334.L + 0.381.X_3$$

$$k_{11}|x_3 = \left.\frac{\partial X_1}{\partial L}\right|_{X_3} = 0.0334$$

Hence:

$$\lambda_{11} = \frac{k_{11}|V}{k_{11}|x_3} = \frac{0.246}{0.0334} = 7.37$$

Note the same result could have been obtained from:

$$\lambda_{11} = \frac{1}{1 - \left(\dfrac{k_{12}.k_{21}}{k_{11}.k_{22}}\right)}$$

Knowing that the rows and columns of Λ each add up to unity, this yields:

$$\Lambda_{LV} = \begin{bmatrix} 7.37 & -6.37 \\ -6.37 & 7.37 \end{bmatrix}$$

Again, note the same result could have been obtained from Equation 111.6.

This demonstrates that controlling the concentrations X_1 and X_3 by manipulating L and V respectively is a viable strategy but with significant interactions. This should not be surprising because, as depicted in Figure 90.4, the column has only got two plates and one still so there is bound to be much interaction between the top and bottom compositions. The large values of the relative gains suggest that the L-V scheme may be too sensitive.

Clearly controlling X_1 by manipulating V and X_3 by manipulating L will not work.

111.4 Worked Example: L-B Scheme

Now consider the same column for which the control scheme of Figure 35.12 is proposed in which the concentrations X_1 and X_3 are controlled by manipulating L and B respectively.

Noting that $L = B + V$ in deviation form at steady state, substituting for V into Equation 111.8 gives:

$$\begin{bmatrix} X_1 \\ X_3 \end{bmatrix} = \begin{bmatrix} 0.0309 & 0.215 \\ -0.0066 & 0.565 \end{bmatrix} \begin{bmatrix} L \\ B \end{bmatrix} \quad (111.9)$$

First, the open loop gain: keep the input B constant and find the effect of L on X_1:

$$X_1 = 0.0309.L + 0.215.B$$

$$k_{11}|B = \left.\frac{\partial X_1}{\partial L}\right|_B = 0.0309$$

Next the closed loop gain, keeping X_3 constant. Manipulate Equation 111.9 to eliminate B:

$$X_1 = 0.0309.L + 0.215.\left(\frac{0.0066.L + X_3}{0.565}\right)$$

$$= 0.0334.L + 0.381.X_3$$

$$k_{11}|x_3 = \left.\frac{\partial X_1}{\partial L}\right|_{X_3} = 0.0334$$

Hence:

$$\lambda_{11} = \frac{k_{11}|B}{k_{11}|x_3} = \frac{0.0309}{0.0334} = 0.925$$

Again, knowing that the rows and columns of Λ each add up to unity, this yields:

$$\Lambda_{LB} = \begin{bmatrix} 0.925 & 0.075 \\ 0.075 & 0.925 \end{bmatrix}$$

This demonstrates that controlling the concentrations X_1 and X_3 by manipulating L and B is a viable strategy with relatively weak interactions. The closeness of the relative gains to unity means that the L-B strategy is much superior to L-V in this case.

111.5 Worked Example: D-V Scheme

Similarly the D-V scheme of Figure 35.11 may be considered in which the concentrations X_1 and X_3 are controlled by manipulating D and V respectively.

Substituting for L from $V = L+D$ into Equation 111.8 gives:

$$\begin{bmatrix} X_1 \\ X_3 \end{bmatrix} = \begin{bmatrix} -0.246 & 0.0309 \\ -0.558 & -0.0066 \end{bmatrix} \begin{bmatrix} D \\ V \end{bmatrix} \quad (111.10)$$

Keeping the input V constant and finding the effect of D on X_1 yields the open loop gain:

$$k_{11}|V = \left.\frac{\partial X_1}{\partial D}\right|_V = -0.246$$

Keeping X_3 constant and manipulating Equation 111.10 to eliminate V yields:

$$X_1 = -2.86.D - 4.68.X_3$$

whence the closed loop gain:

$$k_{11}|x_3 = \left.\frac{\partial X_1}{\partial D}\right|_{X_3} = -2.86$$

Hence:

$$\lambda_{11} = \frac{k_{11}|V}{k_{11}|x_3} = \frac{-0.246}{-2.86} = 0.0861$$

Again, knowing that the rows and columns of Λ each add up to unity, this yields:

$$\Lambda_{DV} = \begin{bmatrix} 0.086 & 0.914 \\ 0.914 & 0.086 \end{bmatrix}$$

The D-V strategy is only viable if X_1 is controlled by manipulating V and X_3 is controlled by manipulating D. The strategy would be subject to weak interactions. The outcome for the D-V strategy demonstrates the principal weakness of the RGA approach: that it is based upon steady state responses and does not take dynamics into account. Clearly there would be a dynamic penalty if each CV and its MV are at opposite ends of the column.

111.6 Effective Gain

The gain of a single-input single-output system is given by:

$$k = y/u$$

For a multiple-input multiple-output system, the input and output signals are vectors. The strength of a vector is found by summing its elements to find its so-called 2-Norm. Referring to Equation 111.1:

$$\|y\|_2 = \sqrt{y_1^2 + y_2^2 + \cdots + y_n^2}$$

$$\|u\|_2 = \sqrt{u_1^2 + u_2^2 + \cdots + u_n^2}$$

The effective process gain is found from the ratio of the 2-Norms:

$$k = \frac{\|y\|_2}{\|u\|_2}$$

This is a single metric that gives a quantitative feel for the sensitivity of the plant or process. Consider again the L-V system of Equation 111.8:

$$\begin{bmatrix} X_1 \\ X_3 \end{bmatrix} = \begin{bmatrix} 0.246 & -0.215 \\ 0.558 & -0.565 \end{bmatrix} \begin{bmatrix} L \\ V \end{bmatrix} \quad (111.8)$$

The effective gain depends upon the input "direction", as can be seen as follows:

If $\underline{u} = \begin{bmatrix} 1 \\ 0 \end{bmatrix}$ so $\|u\| = 1$ then $\underline{y} = K.\underline{u} = \begin{bmatrix} 0.246 \\ 0.558 \end{bmatrix}$
and $\|y\| = 0.610$ whence $k = 0.610$.

If $\underline{u} = \begin{bmatrix} 0 \\ 1 \end{bmatrix}$ so $\|u\| = 1$ then $\underline{y} = K.\underline{u} = \begin{bmatrix} -0.215 \\ -0.565 \end{bmatrix}$
and $\|y\| = 0.604$ whence $k = 0.604$.

111.7 Singular Values

As the relative values of the inputs change, so the value of k varies between limits. There exist maximum and minimum values of k which are referred to as the singular values of the matrix K. Consider again the L-V system.

Its maximum singular value occurs if
$$\underline{u}_1 = \begin{bmatrix} 1.000 \\ -0.991 \end{bmatrix} \text{ which gives } \underline{y}_1 = K.\underline{u}_1 = \begin{bmatrix} 0.459 \\ 1.118 \end{bmatrix}.$$

Hence $\|u_1\| = 1.408$ and $\|y_1\| = 1.209$ yielding $k_1 = 0.858$.

The minimum singular value occurs if
$$\underline{u}_2 = \begin{bmatrix} -1.000 \\ -1.009 \end{bmatrix} \text{ giving } \underline{y}_2 = K.\underline{u}_2 = \begin{bmatrix} -0.0291 \\ 0.0121 \end{bmatrix}.$$

Hence $\|u_2\| = 1.421$ and $\|y_2\| = 0.0315$ yielding $k_2 = 0.022$.

That these input vectors do indeed correspond to the singular values of the effective gain can be verified by evaluating k with slightly different elements of $\underline{u}_1$ and $\underline{u}_2$.

The condition number is the ratio of the maximum to minimum singular values:

$$C = \frac{S_{MAX}(K)}{S_{MIN}(K)} = \frac{0.858}{0.022} = 39.0$$

The condition number is a measure of controllability, C = 1.0 being ideal. A large condition number, say C > 25, indicates that a system is ill-conditioned. This means that some combinations of inputs have a strong effect on the output whilst other combinations have a weak effect. Also, a large condition number indicates that a control strategy may be overly sensitive to plant-model mismatch.

The fact that the L-V scheme for the worked example is so ill-conditioned should not be surprising as it is known that there are significant interactions between the composition loops. The high condition number indicates that the L-V scheme is very sensitive to plant-model mismatch. The condition numbers for the L-B and D-V schemes are 19.4 and 19.7 respectively, both of which are better than for the L-V scheme but still not well conditioned.

The minimum singular value is often referred to as the Morari resilience index and is itself a useful measure of controllability. The larger the minimum singular value, the more controllable or resilient the process.

111.8 Application to Blending System

Consider the blending system depicted in Figure 81.1, whose process transfer function is:

$$\begin{bmatrix} f_0(s) \\ c^{\#}(s) \end{bmatrix} = \begin{bmatrix} 1 & 1 \\ -e^{Ls} & e^{Ls} \end{bmatrix} \begin{bmatrix} f_1(s) \\ f_2(s) \end{bmatrix}$$

Application of the RGA and SVD techniques yields the following results:

$$\Lambda = \begin{bmatrix} 0.5 & 0.5 \\ 0.5 & 0.5 \end{bmatrix}, \quad C = 1.0$$

The system is clearly subject to very strong interactions but is well conditioned. Thus it is robust to PMM and should be controllable with multivariable compensators or decouplers.

111.9 Singular Value Decomposition

The maximum and minimum singular values, and hence the condition number, can be found directly by a technique known as singular value decomposition. In essence, the singular values are given by the decomposition:

$$K = Y.S.U^T \quad (111.11)$$

where

K   is the steady state gain matrix of the plant, as in Equation 111.1.
S   is a diagonal matrix comprised of the singular values of K in descending order.
U   is a matrix of column vectors that represent the input directions of the plant, arranged in order from maximum to minimum gain.
Y   is a matrix of column vectors that represent the output directions of the plant, arranged in order from maximum to minimum gain.

Note that the column vectors of both U and Y must be orthogonal, *i.e.* at right angles to each other.

Also, for the purposes of comparison, they are orthonormal, *i.e.* of "unit length". Thus, by definition, the matrices U and Y are such that

$$U.U^T = U^T.U = I \quad \text{and} \quad Y.Y^T = Y^T.Y = I \tag{111.12}$$

111.10 Worked Example No 4

Consider again the L-V system of Equation 111.8 for which:

$$K = \begin{bmatrix} 0.246 & -0.215 \\ 0.558 & -0.565 \end{bmatrix}$$

Noting that SVD is a standard function in most matrix manipulation packages, K may be decomposed according to Equation 111.11 as follows:

$$S = \begin{bmatrix} 0.858 & 0 \\ 0 & 0.022 \end{bmatrix}$$

$$U = \begin{bmatrix} 0.710 & -0.704 \\ -0.704 & -0.710 \end{bmatrix}$$

$$Y = \begin{bmatrix} 0.380 & -0.925 \\ 0.925 & 0.380 \end{bmatrix}$$

By inspection it can be seen that:

- The singular values are as found in Section 7 above, giving the same condition number.
- The matrices U and Y are comprised of the column vectors $\underline{u}_1$ and $\underline{u}_2$, $\underline{y}_1$ and $\underline{y}_2$.
- The ratios of the values of the elements of these vectors are the same as those used in Section 7 above, so the vectors have the same directions.

That U and Y are orthonormal can be verified from Equation 111.12.

111.11 Comments

In summary, for the three stage column of Figure 90.3, RGA indicates that the L-W scheme is significantly better than the L-V scheme although SVD indicates that it is somewhat ill conditioned. As for the blending system of Figure 35.10, RGA indicates that there are very strong interactions although SVD indicates that the system is well conditioned.

It is essential to appreciate that the RGA and SVD techniques use steady state criteria to select and/or analyse control loop interactions. The most suitable pairings indicated may be impractical for other reasons.

There is a view, articulated by Luyben (1990), that approaches such as RGA are inherently flawed because they are based on the premise that interaction is undesirable. This may be so for servo control, *i.e.* in response to set point changes, but most process control is of a regulatory nature in which rejection of disturbances is the primary objective. Indeed, limited interaction may assist the controller's task of rejection by absorbing some of the effects of disturbances. Thus RGA should perhaps be viewed as a technique for eliminating the worst pairings rather than identifying the best.

State Feedback Regulators

112.1 The Control Law
112.2 Worked Example No 1
112.3 Set Point for Regulo Control
112.4 Worked Example No 2
112.5 Observer Design
112.6 Full Order Observer
112.7 Worked Example No 3
112.8 Reduced Order Observer
112.9 Integration of Observer and Controller
112.10 Implementation
112.11 Comments

Output feedback is traditional: the variables being controlled are measured and fed back to their respective controllers. However, the outputs do not necessarily describe a system fully. It is the current states of the system that fully describe it which, together with the inputs, determine its future condition. Using the states for feedback therefore offers the potential for improved control. This chapter provides an introduction to the principles of design of state-feedback regulators. There are essentially four stages of design:

- Derive the controller by means of pole placement.
- Introduce a set-point to enable regulo control.
- Specify an observer to provide full state-feedback.
- Integrate the observer with the controller.

The design of state-feedback regulators is covered in detail in most texts on modern control systems design, such as that by Dutton (1997), to which the reader is referred for more detailed information.

112.1 The Control Law

The simplest form of state-feedback is as depicted in Figure 112.1. It is presumed that all of the states can be measured. An important point to appreciate with state-feedback is that the dynamics of the measurements and actuators are either assumed to be negligible or are included in the state-space description of the process and/or plant.

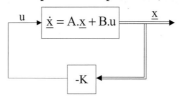

Fig. 112.1 Simple state feedback with a controller matrix

Suppose that the system is single-input single-output (SISO). There are several states and the input u is a scalar, whence the controller matrix K is a row vector of gains:

$$u = -K\underline{x} = -\begin{bmatrix} k_1 & k_2 & \cdots & k_n \end{bmatrix} \begin{bmatrix} x_1 \\ x_2 \\ \vdots \\ x_n \end{bmatrix} \quad (112.1)$$

Each of the states contributes to the control signal, the controller gains being weighting factors as appropriate. Note the minus sign which is necessary to allow for the fact that there is no comparator in this simple form of state-feedback system.

Block diagram algebra thus leads to:

$$\dot{\underline{x}} = A\underline{x} + Bu = A\underline{x} - B.K\underline{x}$$
$$s\underline{x}(s) - \underline{x}(0) = A\underline{x}(s) - BK\underline{x}(s)$$
$$\underline{x}(s) = (sI - (A - BK))^{-1}.\underline{x}(0)$$

The characteristic equation for the closed loop system is thus:

$$\det[sI - (A - BK)] = 0 \quad (112.2)$$

For an nth order system this will be an nth order polynomial with n closed loop poles and n controller gains. The objective is to determine the gains that produce pole positions which correspond to a specified closed loop performance. This is achieved by comparing the characteristic equation of the system with a user specified characteristic polynomial.

112.2 Worked Example No 1

Consider the third order system depicted in Figure 112.2. Note that this is essentially the same system as was considered in Worked Example No 1 of Chapter 74 on root locus and Worked Example No 1 of Chapter 80 on state-space.

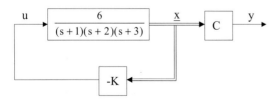

Fig. 112.2 Third order system of Worked Example No 1

$$\begin{bmatrix} \dot{x}_1 \\ \dot{x}_2 \\ \dot{x}_3 \end{bmatrix} = \begin{bmatrix} 0 & 1 & 0 \\ 0 & 0 & 1 \\ -6 & -11 & -6 \end{bmatrix} \begin{bmatrix} x_1 \\ x_2 \\ x_3 \end{bmatrix} + \begin{bmatrix} 0 \\ 0 \\ 6 \end{bmatrix}.u \quad (112.3)$$

$$y = \begin{bmatrix} 1 & 0 & 0 \end{bmatrix} \begin{bmatrix} x_1 \\ x_2 \\ x_3 \end{bmatrix}$$

The characteristic equation for the closed loop system is:

$$\det\left[\begin{bmatrix} s & 0 & 0 \\ 0 & s & 0 \\ 0 & 0 & s \end{bmatrix} - \left(\begin{bmatrix} 0 & 1 & 0 \\ 0 & 0 & 1 \\ -6 & -11 & -6 \end{bmatrix} - \begin{bmatrix} 0 \\ 0 \\ 6 \end{bmatrix}\begin{bmatrix} k_1 & k_2 & k_3 \end{bmatrix}\right)\right] = 0$$

which expands to give:

$$s \begin{vmatrix} s & -1 \\ 11 + 6k_2 & s + 6(1 + k_3) \end{vmatrix} + 6(1 + k_1) \begin{vmatrix} -1 & 0 \\ s & -1 \end{vmatrix} = 0$$

whence:

$$s^3 + 6(1 + k_3)s^2 + (11 + 6k_2)s + 6(1 + k_1) = 0 \quad (112.4)$$

First, suppose that a critically damped response is required, corresponding to repeated real roots. For the purpose of comparison, choose the combination of roots of the characteristic equation given in Table 74.1:

$$(s + 3.155)(s + 1.423)(s + 1.422) = 0$$
$$s^3 + 6.0s^2 + 11.0s + 6.384 = 0$$

Comparing coefficients with Equation 112.4 gives:

$6(1 + k_1) = 6.384$
$11 + 6k_2 = 11$ \quad hence $K = \begin{bmatrix} 0.064 & 0.0 & 0.0 \end{bmatrix}$
$6(1 + k_3) = 6$

Because the roots chosen lie on the root locus of the output feedback system, the design has not surprisingly come up with a single gain element of 0.064. Notice that the gain is applied to the state x_1 which corresponds to the output y. Thus, in every respect, this state-feedback design is equivalent to the output feedback design. However, it doesn't have to be.

Second, suppose that an underdamped response is required. Deliberately choose a characteristic polynomial whose roots do not lie on the root locus of the output feedback design. Specify dominant closed loop pole positions corresponding to a damping factor of 0.8, say, for which the oscillations will decay away quickly:

$$(s + 5.0)(s + 0.4 + j0.3)(s + 0.4 - j0.3) = 0$$

which gives the closed loop characteristic polynomial:

$$s^3 + 5.8s^2 + 4.25s + 1.25 = 0$$

Again, comparing coefficients with Equation 112.4 gives:

$6(1 + k_1) = 1.25$
$11 + 6k_2 = 4.25$ hence $K = -\begin{bmatrix} 0.792 & 1.125 & 0.033 \end{bmatrix}$
$6(1 + k_3) = 5.8$

Three comments are worth making at this stage. First, the system matrix of the state-space model in Equation 112.3 was of the companion form, as described in Chapter 80. This is not essential: the pole placement approach works just as well with other forms of state-space model. However, if the system matrix is not in companion form, comparison of the coefficients of the characteristic and polynomial equations results in a set on n equations in n unknowns which has to be solved for the elements of K.

Second, the magnitude of the elements of the controller matrix K are an indication of the appropriateness of the choice of the closed loop pole positions. In general, very large values imply that a lot of control effort is required to force the system to give the desired response, probably resulting in saturation of the controller output. Conversely, very small values imply that the system is very sensitive and the design will be vulnerable to plant/model mismatch.

Third, a SISO system only has been considered. There is no reason why a multivariable control system cannot be similarly designed although, of course, the interactions must be taken into account.

Thus, for example, for two inputs and n states, Equation 112.1 becomes:

$$\begin{bmatrix} u_1 \\ u_2 \end{bmatrix} = K\underline{x} = \begin{bmatrix} k_{11} & k_{12} & \cdots & k_{1n} \\ k_{21} & k_{22} & \cdots & k_{2n} \end{bmatrix} \begin{bmatrix} x_1 \\ x_2 \\ \vdots \\ x_n \end{bmatrix} \quad (112.5)$$

112.3 Set Point for Regulo Control

The state-feedback system developed thus far is regulatory in nature. However, to be of any practical use, a set-point needs to be incorporated as depicted in Figure 112.3. Note also that a disturbance vector $\underline{d}$ has been introduced, although for design purposes this is assumed to be zero.

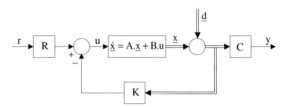

Fig. 112.3 State feedback with a set point

Assuming the controller matrix K has been designed by pole placement such that state-feedback satisfies the dynamic criteria, the design focuses on specifying the reference factor R to satisfy the steady state criterion.

Suppose that a step change in set-point of magnitude r occurs. Once steady state has been reached:

$$u(\infty) = R.r - K.\underline{x}(\infty) \quad (112.6)$$

The reference factor R has to be specified such that at steady state there is no offset, *i.e.*

$$y(\infty) = r \quad (112.7)$$

Let R_X be a column vector defined to relate the steady states of the system to its set-point. And let

R_U be a scalar defined to relate the input, necessary to achieve those steady states, to the set-point:

$$\underline{x}(\infty) = R_x.r$$
$$u(\infty) = R_u.r \quad (112.8)$$

Substitute Equation 112.8 into 112.6 gives:

$$R_u.r = R.r - K.R_x.r$$

hence:

$$R = K.R_x + R_u \quad (112.9)$$

The values of R_X and R_U may themselves be found from steady state considerations. The general state-space equations are:

$$\dot{\underline{x}} = A\underline{x} + Bu$$
$$y = C\underline{x} + Du$$

At steady state:

$$0 = A\underline{x}(\infty) + Bu(\infty)$$
$$y(\infty) = C\underline{x}(\infty) + Du(\infty)$$

Substitute from Equation 112.8 gives:

$$0 = A.R_x r + B.R_u r$$
$$y(\infty) = C.R_x r + D.R_u r$$

These equations may be arranged in matrix form:

$$\begin{bmatrix} A & B \\ C & D \end{bmatrix} \begin{bmatrix} R_x \\ R_u \end{bmatrix}.r = \begin{bmatrix} 0 \\ 1 \end{bmatrix}.y(\infty)$$

However, r and $y(\infty)$ cancel according to Equation 112.7. Matrix inversion yields:

$$\begin{bmatrix} R_x \\ R_u \end{bmatrix} = \begin{bmatrix} A & B \\ C & D \end{bmatrix}^{-1} \begin{bmatrix} 0 \\ 1 \end{bmatrix} \quad (112.10)$$

Note that R_X and R_U are properties of the system and are independent of the size of the set-point change, normal values, *etc*. So, provided that the matrices A, B, C and D are known, R_x and R_u can be found from Equation 112.10; hence R can be found from Equation 112.9.

Also note that by virtue of Equation 112.7, which is the basis of the design, there is no offset for set point changes so, in principle, integral action is not required. In practice integral action is required to cope with offsets due to both process disturbances and plant/model mismatch. This is covered later on.

112.4 Worked Example No 2

Reconsider the system of Worked Example No 1 and depicted in Figure 112.2 with the second state-feedback controller design. From Equation 112.10:

$$\begin{bmatrix} R_x \\ R_u \end{bmatrix} = \begin{bmatrix} A & B \\ C & D \end{bmatrix}^{-1} \begin{bmatrix} 0 \\ 1 \end{bmatrix}$$

$$= \begin{bmatrix} 0 & 1 & 0 & 0 \\ 0 & 0 & 1 & 0 \\ -6 & -11 & -6 & 6 \\ 1 & 0 & 0 & 0 \end{bmatrix}^{-1} \begin{bmatrix} 0 \\ 0 \\ 0 \\ 1 \end{bmatrix} = \begin{bmatrix} 1 \\ 0 \\ 0 \\ 1 \end{bmatrix}$$

whence:

$$R_x = \begin{bmatrix} 1 \\ 0 \\ 0 \end{bmatrix} \quad \text{and} \quad R_u = 1$$

Substituting into Equation 112.9 gives:

$$R = K.R_x + R_u$$
$$= -\begin{bmatrix} 0.792 & 1.125 & 0.033 \end{bmatrix} \begin{bmatrix} 1 \\ 0 \\ 0 \end{bmatrix} + 1 = 0.208$$

112.5 Observer Design

The control law of Equation 112.1 assumed that all the states were available for feedback. This is generally not the case, largely due to the limitations of measurement. Usually the constraint is lack of instrumentation, either it doesn't exist or it's too expensive, but sometimes the signal is implicit and cannot be measured anyway. In such situations it is necessary to estimate the states and to use the estimate of the state vector for feedback purposes. An observer is the name given to an estimator that is used for estimating the states of a system.

From a knowledge of the A and B matrices of the system, a state-space model of the process based upon estimates of its states is of the form:

$$\dot{\hat{\underline{x}}} = A.\hat{\underline{x}} + B.u \quad (112.11)$$

where ˆ denotes an estimate. Knowing the real inputs to the process, and the initial conditions, an estimate of its states could be found directly by solving the model:

$$\hat{\underline{x}}(t) = L^{-1}\left\{(sI - A)^{-1} B.u(s)\right\}$$

In practice, the inputs vary with time and the solution has to be found by numerical integration. Provided that the A matrix is accurate and the system is stable, then the estimates of the states will converge numerically on their true values. This can be demonstrated by considering the errors on the estimates of the states, their residuals. Let:

$$\tilde{\underline{x}} = \underline{x} - \hat{\underline{x}} \quad (112.12)$$

where ~ denotes the residual. Differentiating gives:

$$\begin{aligned}\dot{\tilde{\underline{x}}} &= \dot{\underline{x}} - \dot{\hat{\underline{x}}} \\ &= A.\underline{x} + B.u - \left(A.\hat{\underline{x}} + B.u\right) \\ &= A.\left(\underline{x} - \hat{\underline{x}}\right) \\ &= A.\tilde{\underline{x}}\end{aligned}$$

The solution to this is as stated in Chapter 80:

$$\tilde{\underline{x}} = e^{At}.\tilde{\underline{x}}(0)$$

where the transition matrix:

$$\Phi(t) = e^{At} = L^{-1}\left\{[sI - A]^{-1}\right\}$$

Thus, provided the system's eigenvalues, *i.e.* the roots of:

$$\det[sI - A] = 0$$

are negative real, or have negative real parts, the solution is stable and converges.

112.6 Full Order Observer

The problem with the predictor of Equation 112.11 is that the inputs are seldom known accurately and there is inevitably some plant/model mismatch, *i.e.* errors in the A and B matrices. These always result in offset and can lead to numerical instability. So, to improve upon convergence, the predictor-corrector approach is used. The general structure of a so-called "full order observer" for a SISO system is as depicted in Figure 112.4.

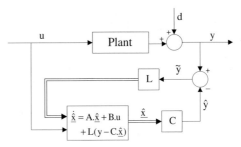

Fig. 112.4 Structure of a full order observer

The predictor corrector is comprised of the state space model of the plant, *i.e.* its system and input matrices A and B respectively, with an extra term for correction purposes. The corrector is in proportion to the size of the prediction error. Remember that the dynamics of the instrumentation, *i.e.* of the actuator and of the measurement must be included in the model:

$$\begin{aligned}\dot{\hat{\underline{x}}} &= A.\hat{\underline{x}} + B.u + L.\tilde{y} \\ &= A.\hat{\underline{x}} + B.u + L.\left(y - \hat{y}\right) \quad (112.13) \\ &= A.\hat{\underline{x}} + B.u + L.\left(y - C.\hat{\underline{x}}\right)\end{aligned}$$

The predictor-corrector is solved by numerical integration and is discussed later on. Thus an estimate of the state vector is obtained. The output matrix C then picks out that state which is measured to provide an estimate of the output. This is compared with the true output and the residual determined. The residual is operated upon by the observer matrix L which corrects the state space model.

The observer matrix L is a column vector of gains, or weighting factors, and is the principal design parameter of the observer. The basic philosophy is to estimate all the states, whether they can be measured or not. Then, by driving the residual on that state which is measured to zero, the estimates of all the states will be driven towards their actual values.

Fundamental to this is convergence (on zero) of the residuals. This is established by consideration of the observer's dynamics. As before:

$$\tilde{\underline{x}} = \underline{x} - \hat{\underline{x}}$$
$$\dot{\tilde{\underline{x}}} = \dot{\underline{x}} - \dot{\hat{\underline{x}}}$$

Substituting from 112.13 gives:

$$\dot{\tilde{\underline{x}}} = A.\underline{x} + B.u - \left(A.\hat{\underline{x}} + B.u + L.\left(y - C.\hat{\underline{x}}\right)\right)$$
$$= A.\left(\underline{x} - \hat{\underline{x}}\right) - L.\left(C.\underline{x} - C.\hat{\underline{x}}\right)$$
$$= (A - L.C).\tilde{\underline{x}}$$

the solution to which is:

$$\tilde{\underline{x}}(s) = (sI - (A - L.C))^{-1}.\tilde{\underline{x}}(0)$$

for which the characteristic equation is:

$$\det[sI - (A - L.C)] = 0 \qquad (112.14)$$

Provided the observer's eigenvalues are negative real, or have negative real parts, stable convergence is assured. The approach is essentially the same as for designing the controller matrix K itself. Thus the desired observer performance is articulated in the form of a characteristic polynomial which, by comparing coefficients with the characteristic equation, yields the observer gains.

By design, the observer's dynamics are much faster than those of the controller considered previously, irrespective of the initial conditions. The predictor-corrector nature of the observer provides robustness: this is particularly important given the likely errors in the matrices A, B and C.

112.7 Worked Example No 3

Again consider the system of Worked Examples No 1 and 2. Since there are three states then the characteristic polynomial must be third order. For the observer's response to be stable choose poles on the negative real axis: this will give exponential decay of the residuals. For the decay to be fast choose pole positions that correspond to lags that are of similar magnitude to the smallest lags of the system itself.

The following characteristic polynomial should be good enough:

$$(s + 4)(s + 3)^2 = 0$$
$$s^3 + 10s^2 + 33s + 36 = 0 \qquad (112.15)$$

The characteristic equation is formed from Equation 112.14:

$$\det\left[\begin{bmatrix} s & 0 & 0 \\ 0 & s & 0 \\ 0 & 0 & s \end{bmatrix} - \left(\begin{bmatrix} 0 & 1 & 0 \\ 0 & 0 & 1 \\ -6 & -11 & -6 \end{bmatrix} - \begin{bmatrix} l_1 \\ l_2 \\ l_3 \end{bmatrix}\begin{bmatrix} 1 & 0 & 0 \end{bmatrix}\right)\right] = 0$$

which yields:

$$s^3 + (6 + l_1).s^2 + (11 + 6.l_1 + l_2).s + (6 + 11.l_1 + 6.l_2 + l_3) = 0$$

Comparing coefficients with Equation 112.15 gives:

$$6 + l_1 = 10$$
$$11 + 6.l_1 + l_2 = 33$$
$$6 + 11.l_1 + 6.l_2 + l_3 = 36$$

which may be solved to give the observer design:

$$L = \begin{bmatrix} 4 \\ -2 \\ -2 \end{bmatrix}$$

112.8 Reduced Order Observer

It is not necessary to estimate all the states of a system if there are good quality, noise free measurements of some of the states available. In such cases estimating a reduced order observer is sufficient.

112.8 Reduced Order Observer

In the following analysis a SISO system is considered in which it is assumed that the single output is measured. The basic approach to designing a reduced order observer is to partition the state vector into the measured state and a vector of unmeasured states as follows:

$$\begin{bmatrix} \dot{x}_1 \\ \underline{\dot{x}}_2 \end{bmatrix} = \begin{bmatrix} A_{11} & A_{12} \\ A_{21} & A_{22} \end{bmatrix} \begin{bmatrix} x_1 \\ \underline{x}_2 \end{bmatrix} + \begin{bmatrix} B_1 \\ B_2 \end{bmatrix}.u \quad (112.16)$$

where subscript 1 denotes the known (measured) state and 2 denotes the unknown states:

$$y = \begin{bmatrix} 1 & 0 \end{bmatrix} \begin{bmatrix} x_1 \\ \underline{x}_2 \end{bmatrix} \quad (112.17)$$

From Equation 112.16 the equation for the measured state is:

$$\dot{x}_1 = A_{11}.x_1 + A_{12}.\underline{x}_2 + B_1.u$$

Substituting from 112.17 and rearranging gives:

$$\dot{y} = A_{11}.y + A_{12}.\underline{x}_2 + B_1.u$$
$$A_{12}.\underline{x}_2 = \dot{y} - A_{11}.y - B_1.u \quad (112.18)$$

Notice that the right hand side of this equation consists of functions of the input and output signals only, and is thus known at all times.

Also from Equation 112.16 comes the equation for the vector of unmeasured states:

$$\underline{\dot{x}}_2 = A_{21}.x_1 + A_{22}.\underline{x}_2 + B_2.u \quad (112.19)$$

A reduced observer may be formulated for this unknown vector in predictor corrector form:

$$\underline{\dot{\hat{x}}}_2 = A_{21}.x_1 + A_{22}.\underline{\hat{x}}_2 + B_2.u + L.A_{12}.\left(\underline{x}_2 - \underline{\hat{x}}_2\right) \quad (112.20)$$

Substituting from Equations 112.17 and 112.18 and rearranging gives:

$$\underline{\dot{\hat{x}}}_2 = A_{21}.y + A_{22}.\underline{\hat{x}}_2 + B_2.u$$
$$+ L.\left(\dot{y} - A_{11}.y - B_1.u - A_{12}.\underline{\hat{x}}_2\right)$$
$$= (A_{22} - L.A_{12}).\underline{\hat{x}}_2 + (A_{21} - L.A_{11}).y$$
$$+ (B_2 - L.B_1).u + L.\dot{y} \quad (112.21)$$

This is the reduced order observer. Note that it includes the derivative of the output. This is problematic if there is any noise and is overcome by introducing a new state. Let:

$$\underline{x}_3 = \underline{\hat{x}}_2 - L.y$$
$$\underline{\dot{x}}_3 = \underline{\dot{\hat{x}}}_2 - L.\dot{y}$$
$$= (A_{22} - L.A_{12}).\underline{\hat{x}}_2$$
$$+ (A_{21} - L.A_{11}).y + (B_2 - L.B_1).u$$

So the estimate of the full state vector is now given by:

$$\underline{\hat{x}} = \begin{bmatrix} x_1 \\ \underline{\hat{x}}_2 \end{bmatrix} = \begin{bmatrix} y \\ \underline{x}_3 + L.y \end{bmatrix}$$

Again the matrix L is a column vector of gains and is the principal design parameter. The observer L is specified by consideration of the dynamics of the state error:

$$\underline{\tilde{x}}_2 = \underline{x}_2 - \underline{\hat{x}}_2$$
$$\underline{\dot{\tilde{x}}}_2 = \underline{\dot{x}}_2 - \underline{\dot{\hat{x}}}_2$$

Substituting from Equations 112.16 and 112.20 gives:

$$\underline{\dot{\tilde{x}}}_2 = A_{21}.x_1 + A_{22}.\underline{x}_2 + B_2.u$$
$$- \left(A_{21}.x_1 + A_{22}.\underline{\hat{x}}_2 + B_2.u \right.$$
$$\left. + L.A_{12}.\left(\underline{x}_2 - \underline{\hat{x}}_2\right)\right)$$
$$= A_{22}.\left(\underline{x}_2 - \underline{\hat{x}}_2\right) - L.A_{12}.\left(\underline{x}_2 - \underline{\hat{x}}_2\right)$$
$$= (A_{22} - L.A_{12}).\underline{\tilde{x}}_2$$

the solution to which is:

$$\underline{\tilde{x}}_2 = (sI - (A_{22} - L.A_{12}))^{-1}.\underline{\tilde{x}}_2(0)$$

for which the characteristic equation is:

$$\det[sI - (A_{22} - L.A_{12})] = 0 \quad (112.22)$$

As with the full order observer, appropriate pole positions are specified for the observer resulting in a characteristic polynomial. Comparison with the characteristic equation yields the gains of the observer L.

Use of a reduced order observer saves on the computation effort involved in the estimation process. However, if there is any doubt about the quality of the measurements, it is probably best to use a full order observer. It is also easier to understand!

112.9 Integration of Observer and Controller

The way in which the controller, set-point and observer are integrated is as depicted in Figure 112.5. Note that the full order observer is used. Also note that the controller input is the state estimate.

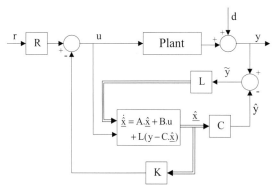

Fig. 112.5 Integration of controller, set point and observer

As stated, it is inevitably the case that plant/model mismatch occurs, *i.e.* there are errors in the A and B matrices. These result in offset. This can be countered by introducing some integral action as depicted in Figure 112.6.

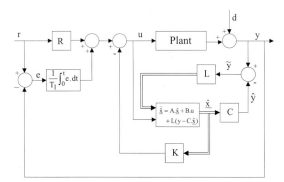

Fig. 112.6 State feedback control with integral action

In general, the amount of integral action required is only slight and may be tuned by adjusting some integral time T_I as in Equation 23.5.

112.10 Implementation

It is very important to appreciate that all the signals referred to in Figures 112.1–112.6 are in deviation form and that the state-space models used presume such. Thus, for realisation, steady state biases have to be added to or subtracted from the various signals as appropriate.

State-feedback control is realised by means of software. In particular, solution of the observer Equation 112.13 is found by numerical integration using, for example, Eulers implicit method as outlined in Chapter 96. Typically, the sampling period of the control system would be the same as the step length of the numerical integration. Thus at each sampling instant, knowing the current values of the plant input and output signals, the numerical integration would be executed once to predict the states. These are then used to calculate the next value of the controller output signal.

It is evident in the design of state-feedback regulators that there are four time scales involved:

- At the level of the predictor-corrector, for stability, numerical integration would need to be some five to ten times faster than the fastest dynamics (smallest lags) of the system.
- At the level of the observer L, dominant pole positions are chosen consistent with the fastest dynamics (smallest lags) of the system. A stable overdamped response is essential. Note that the observer's dynamics must be faster than the controller's, typically some five times faster, since the observer provides a platform for the control action.
- At the level of the controller K, dominant pole positions are chosen that are consistent with the slowest dynamics (largest lags) of the system. A stable underdamped response is typically the requirement.
- And finally, the integral time T_I must be consistent with the dynamics of the integral action being even more sluggish than the largest lags of the system.

112.11 Comments

It has to be said that although the techniques of state-feedback are commonplace in other sectors of industry, such as in aerospace and robotics, they are not used much in process control. This is partly because of the difficulty in establishing accurate enough models and partly because the techniques are not as easy to understand as other methods. They do nevertheless have much to offer in certain circumstances and are covered here to encourage their use through better understanding.

Kalman Filtering

113.1 The Luenberger Observer
113.2 Kalman Filter Design
113.3 Formation of Riccati Equation
113.4 Solution of Riccati Equation
113.5 Realisation of Kalman Filter
113.6 Implementation Issues
113.7 Use of Kalman Filters in Control Systems
113.8 Worked Example
113.9 Nomenclature

Kalman filters are used for estimating the states of a system. They are used, in particular, for estimating the states of dynamic systems whose signals are noisy. Their design draws upon various aspects of observer design which were introduced in the previous Chapter 112. The principal constraints are that they can only be used for linear systems, or for systems whose dynamics are roughly linear about an operating point, for which there exists a model of the system in state-space form. Kalman filtering is a recursive technique with the state estimates being successively refined. Whilst the filters can be operated with fixed gains, it is normal for the gains to be updated continuously. Since Kalman filtering is a real-time technique and is invariably realised digitally, the emphasis in this chapter is on its discrete form.

The origins of Kalman filtering are in electrical engineering and the technique, whilst seldom used in the process industries, is used extensively in other sectors of industry. This chapter, therefore, is included in the hope that, through better understanding, applications will be found in the process industries: it has much to offer. It is also included because some of the concepts are fundamental to the following chapters on identification and model predictive control. Most texts on signal processing and many texts on control theory cover Kalman filtering. The approach taken here follows that by Dutton (1997).

113.1 The Luenberger Observer

The Luenberger observer is a noise free and fixed gain technique for state estimation and is developed as a platform for considering the design of Kalman filters. First, suppose that a SISO system is being considered, with signals in deviation form, and that its state-space description exists in discrete form as follows:

$$\begin{aligned} \underline{x}(k+1) &= A.\underline{x}(k) + B.u(k) \\ y(k) &= C.\underline{x}(k) \end{aligned} \quad (113.1)$$

This is depicted in block diagram form in Figure 113.1.

The basic task is to estimate the values of the states $\underline{x}$ at the current sampling instant k given values for the input u(k) and output y(k) up to the current instant.

The best estimate of the current states that Equation 113.1 can provide is:

$$\underline{\hat{x}}(k|k-1) = A.\underline{\hat{x}}(k-1|k-1) + B.u(k-1) \quad (113.2)$$

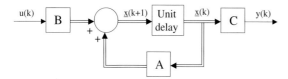

Fig. 113.1 Block diagram representation of state equations in discrete form

Some explanation of the indices is justified:

- The term $\hat{\underline{x}}(k|k)$ means the estimate of $\underline{x}$ at the kth instant based upon data up to and including that instant.
- The term $\hat{\underline{x}}(k|k-1)$ means the estimate of $\underline{x}$ at the kth instant based upon data up to and including the previous $(k-1)$th instant only.

From Equation 113.1 a one step ahead prediction of the output can be made:

$$\hat{y}(k|k-1) = C.\hat{\underline{x}}(k|k-1)$$
$$= C.\left(A.\hat{\underline{x}}(k-1|k-1) + B.u(k-1)\right)$$

However, at instant k the output y is known, so the prediction error may be quantified:

$$\tilde{y}(k) = y(k) - \hat{y}(k|k-1)$$

The state estimates of Equation 113.2 may then be improved by adding a proportion of the prediction error in such a way that the prediction error is normally driven towards zero, whence:

$$\hat{\underline{x}}(k|k) = \hat{\underline{x}}(k|k-1) + K.\tilde{y}(k) \qquad (113.3)$$

The column vector K is referred to as the Kalman gain. It consists of gain terms, one for each state, all of which are less than unity (< 1.0). Equations 113.2 and 113.3 provide a predictor-corrector basis for estimating the system states, known as the Luenberger observer, which is depicted in Figure 113.2. Note that the Kalman gains are fixed in the Luenberger observer.

113.2 Kalman Filter Design

As stated, Kalman filtering is of particular use for estimating the states of dynamic systems whose signals are noisy. Two sources of noise are introduced as depicted in Figure 113.3.

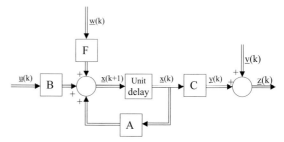

Fig. 113.3 State equations in discrete form with sources of noise

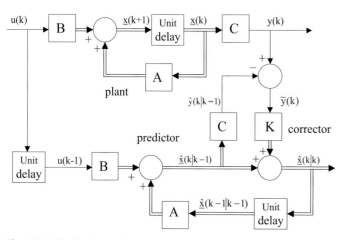

Fig. 113.2 The Luenberger observer

The state equations are:

$$x(k+1) = A.\underline{x}(k) + B.\underline{u}(k) + F.\underline{w}(k)$$
$$\underline{z}(k) = C.\underline{x}(k) + \underline{v}(k) \quad (113.4)$$

Note that the system now being considered is MIMO and that $\underline{z}(k)$ is the vector of measurements of the outputs $\underline{y}(k)$. The vector of random noise signals $\underline{w}(k)$ represents system disturbances, modelling errors, *etc.*, which are coupled into the system by the disturbance matrix F in combinations as appropriate. The vector $\underline{v}(k)$ represents measurement noise, discretisation errors, *etc.*, on the basis of one disturbance in $\underline{v}(k)$ *per* system output.

The covariance matrices of the system and measurement noises are defined as follows:

$$Q = E\left(\underline{w}(k).\underline{w}^T(k)\right) \quad R = E\left(\underline{v}(k).\underline{v}^T(k)\right) \quad (113.5)$$

which are of the form of Equation 82.5 with E being the expectation operator. Note that both $\underline{w}$ and $\underline{v}$ are presumed to be stationary white noise signals, which means that their statistical properties are constant and independent of time, and that they are both of zero mean. Thus Q and R do not vary. Remember that covariance matrices are symmetrical, a property that will be exploited in due course. The estimation error is defined as:

$$\underline{\tilde{x}}(k) = \underline{x}(k) - \underline{\hat{x}}(k|k) \quad (113.6)$$

The covariance matrix of the estimation error is defined as:

$$P(k) = E\left(\underline{\tilde{x}}(k).\underline{\tilde{x}}^T(k)\right) \quad (113.7)$$

Whereas the covariance matrices Q and R are constant, P(k) is a function of time. The objective of filter design is to specify the Kalman gain matrix K such that the covariance matrix P(k) is minimised in real time.

An estimate of the values of the states $\underline{x}$ at instant k, given values for the inputs $\underline{u}(k)$ and measurements $\underline{z}(k)$ up to that instant, and subject to the noise $\underline{w}(k)$ and $\underline{v}(k)$, may be formulated as follows:

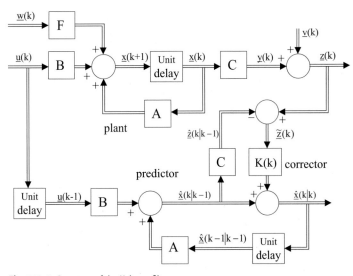

Fig. 113.4 Structure of the Kalman filter

$$\hat{\underline{x}}(k|k) = \hat{\underline{x}}(k|k-1) + K.\underline{\tilde{z}}(k)$$
$$= A.\hat{\underline{x}}(k-1|k-1) + B.\underline{u}(k-1)$$
$$+ K.\left(\underline{z}(k) - \hat{\underline{z}}(k|k-1)\right)$$
$$= A.\hat{\underline{x}}(k-1|k-1) + B.\underline{u}(k-1)$$
$$+ K.\left(\underline{z}(k) - C.\hat{\underline{x}}(k|k-1)\right)$$
$$= A.\hat{\underline{x}}(k-1|k-1) + B.\underline{u}(k-1)$$
$$+ K.\Big(\underline{z}(k) - C.\big(A.\hat{\underline{x}}(k-1|k-1)$$
$$+ B.\underline{u}(k-1)\big)\Big)$$

This may be factorised to provide the basic design equation used to predict the states of the system:

$$\hat{\underline{x}}(k|k) = (I - K.C)\left(A.\hat{\underline{x}}(k-1|k-1) + B.\underline{u}(k-1)\right) + K.\underline{z}(k) \quad (113.8)$$

The structure of a Kalman filter is as depicted in Figure 113.4 which, obviously, is an adaptation of the diagram of the Luenberger observer. Note that the Kalman gain matrix is denoted as K(k) in anticipation of its recursive realisation.

113.3 Formation of Riccati Equation

The estimation error at any point in time may be found by substituting Equations 113.4 and 113.8 into 113.6:

$$\underline{\tilde{x}}(k) = \underline{x}(k) - \hat{\underline{x}}(k|k)$$
$$= A.\underline{x}(k-1) + B.\underline{u}(k-1) + F.\underline{w}(k-1)$$
$$- (I - K.C)\left(A.\hat{\underline{x}}(k-1|k-1) + B.\underline{u}(k-1)\right) - K.\underline{z}(k)$$

Substituting for $\hat{\underline{x}}(k-1|k-1)$ from Equation 113.6 and for $\underline{z}(k)$ from Equation 113.4 gives:

$$\underline{\tilde{x}}(k) = A.\underline{x}(k-1) + B.\underline{u}(k-1) + F.\underline{w}(k-1)$$
$$- (I - K.C)\left(A.\left(\underline{x}(k-1) - \underline{\tilde{x}}(k-1)\right)\right.$$
$$+ B.\underline{u}(k-1)\Big) - K.\Big(C.\big(A.\underline{x}(k-1)$$
$$+ F.\underline{w}(k-1)\big) + \underline{v}(k)\Big)$$

Expansion, cancellation and rearrangement yields the equation used to calculate the errors on the state estimates:

$$\underline{\tilde{x}}(k) = (I - K.C) \quad (113.9)$$
$$\times \left(A.\underline{\tilde{x}}(k-1) + F.\underline{w}(k-1)\right) - K.\underline{v}(k)$$

This provides a recursive basis for minimising the covariance matrix P(k) with respect to K, utilising only previous state estimates and the noise vectors.

Defining $J = (I - K.C)$ to simplify the algebra, substitution of Equation 113.9 into Equation 113.7 yields:

$$P(k) = E\left(\underline{\tilde{x}}(k).\underline{\tilde{x}}^T(k)\right)$$
$$= E\Big(\big(J.A.\underline{\tilde{x}}(k-1) + J.F.\underline{w}(k-1) - K.\underline{v}(k)\big)$$
$$\times \big(J.A.\underline{\tilde{x}}(k-1) + J.F.\underline{w}(k-1) - K.\underline{v}(k)\big)^T\Big)$$
$$= E\Big(\big(J.A.\underline{\tilde{x}}(k-1) + J.F.\underline{w}(k-1) - K.\underline{v}(k)\big)$$
$$\times \big(\underline{\tilde{x}}^T(k-1).A^T J^T + \underline{w}^T(k-1).F^T.J^T - \underline{v}^T(k).K^T\big)\Big)$$

Assume the estimation error vector $\underline{\tilde{x}}$ and the noise vectors $\underline{w}$ and $\underline{v}$ are all uncorrelated.

Thus the expected value of the product of samples of any two of these vectors is zero, so the above equation may be expanded and simplified:

$$P(k) = J.A.E\left(\underline{\tilde{x}}(k-1).\underline{\tilde{x}}^T(k-1)\right).A^T J^T$$
$$+ J.F.E\left(\underline{w}(k-1).\underline{w}^T(k-1)\right).F^T.J^T$$
$$+ K.E\left(\underline{v}(k).\underline{v}^T(k)\right).K^T$$

These three expectations are the covariances defined by Equations 113.5 and 113.7 which, noting that Q and R are constant, give:

$$P(k) = J.A.P(k-1).A^T J^T + J.F.Q.F^T.J^T$$
$$+ K.R.K^T$$
$$= J.\left(A.P(k-1).A^T + F.Q.F^T\right).J^T$$
$$+ K.R.K^T$$
$$= J.P^*(k-1).J^T + K.R.K^T$$

where, by definition:

$$P^*(k-1) = A.P(k-1).A^T + F.Q.F^T \quad (113.10)$$

Reinstating the expansion for J gives:

$$P(k) = (I - K.C).P^*(k-1).(I-K.C)^T + K.R.K^T \quad (113.11)$$

Equation 113.11 is of the form of the so-called matrix Riccati equation. It provides a recursive relationship for predicting the covariance matrix for the error on the state estimates in terms of the covariance matrices for the noise vectors.

113.4 Solution of Riccati Equation

The Kalman gain matrix K that minimises P(k) can now be found. Equation 113.11 may be expanded to give

$$\begin{aligned} P(k) &= K.C.P^*(k-1).C^T K^T + K.R.K^T \\ &\quad - P^*(k-1).C^T K^T - K.C.P^*(k-1) \\ &\quad + P^*(k-1) \\ &= K.\left(C.P^*(k-1).C^T + R\right).K^T \\ &\quad - P^*(k-1).C^T K^T - K.C.P^*(k-1) \\ &\quad + P^*(k-1) \\ &= K.G.K^T - P^*(k-1).C^T K^T \\ &\quad - K.C.P^*(k-1) + P^*(k-1) \quad (113.12) \end{aligned}$$

where, by definition:

$$G = C.P^*(k-1).C^T + R \quad (113.13)$$

Equation 113.12 is solved for K by a technique known as completing the square. Suppose a dummy matrix M exists such that Equation 113.12 can be cast in the following form:

$$\begin{aligned} P(k) &= (K - M).G.(K - M)^T \\ &\quad - M.G.M^T + P^*(k-1) \\ &= K.G.K^T - M.G.K^T \\ &\quad - K.G.M^T + P^*(k-1) \end{aligned} \quad (113.14)$$

It follows that Equations 113.12 and 113.14 are equivalent if the following two identities hold:

$$\begin{aligned} M.G.K^T &= P^*(k-1).C^T.K^T \text{ and} \\ K.G.M^T &= K.C.P^*(k-1) \end{aligned} \quad (113.15)$$

The equivalence of these two identities is now established by means of symmetry. Remember that the covariances P(k), Q and R are all symmetrical. It follows from Equation 113.10 that $P^*(k-1)$ must also be symmetrical, and from Equation 113.13 that G is symmetrical too.

Consider the first of the two identities of Equation 113.15. For any non-zero K matrix:

$$M.G = P^*(k-1).C^T \quad (113.16)$$

Transposing both sides gives:

$$G^T.M^T = C.P^{*T}(k-1)$$

However, G and $P^*(k-1)$ are symmetrical:

$$G.M^T = C.P^*(k-1)$$

Pre-multiplying by K gives:

$$K.G.M^T = K.C.P^*(k-1)$$

This is the same as the second identity of Equation 113.15, so the two identities are mutually consistent and a matrix M must exist such that Equations 113.12 and 113.14 are equivalent. The value of that matrix M is established by isolating M in Equation 113.16:

$$M = P^*(k-1).C^T.G^{-1}$$

Returning to the original problem, the objective was to establish the Kalman gain matrix K that minimises P(k) as given by Equation 113.12. This is clearly the same thing as finding the value of K that minimises P(k) as given by Equation 113.14, inspection of which shows that the minimum occurs when K and M are equal. Noting that the value of K varies from one sample to the next, and denoting K(k) as the value of K that minimises P(k):

$$\begin{aligned} K(k) &= P^*(k-1).C^T.G^{-1} \quad (113.17) \\ &= P^*(k-1).C^T.\left(C.P^*(k-1).C^T + R\right)^{-1} \end{aligned}$$

Substitution of the conditions necessary for the minimum, i.e. (K − M) = 0 and M = K(k), into

Equation 113.14 yields the optimum solution, *i.e.* the minimum of P(k):

$$P(k) = -K(k).G.K^T(k) + P^*(k-1)$$

Substituting for the value of K(k) from Equation 113.17 yields:

$$P(k) = -P^*(k-1).C^T.G^{-1}.G.K^T(k) + P^*(k-1)$$
$$= P^*(k-1).(I - C^T.K^T(k))$$

Transposing both sides gives:

$$P^T(k) = (I - C^T.K^T(k))^T .P^{*T}(k-1)$$

Noting that both P(k) and $P^*(k-1)$ are symmetrical results in:

$$P(k) = (I - K(k).C).P^*(k-1) \qquad (113.18)$$

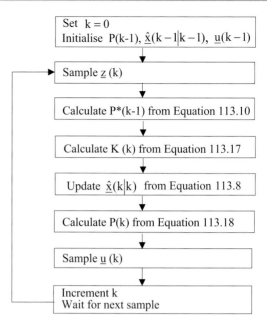

Fig. 113.5 Recursive implementation of Kalman filter in real-time

113.5 Realisation of Kalman Filter

A practical Kalman filter consists of the following set of four equations:

$$P^*(k-1) = A.P(k-1).A^T + F.Q.F^T \qquad (113.10)$$

$$K(k) = P^*(k-1).C^T.(C.P^*(k-1).C^T + R)^{-1} \qquad (113.17)$$

$$\hat{\underline{x}}(k|k) = (I - K(k).C) \qquad (113.8)$$
$$\times (A.\hat{\underline{x}}(k-1|k-1) + B.\underline{u}(k-1))$$
$$+ K(k).\underline{z}(k)$$

$$P(k) = (I - K(k).C).P^*(k-1) \qquad (113.18)$$

Figure 113.5 depicts a routine for the recursive implementation of these equations in real-time.

113.6 Implementation Issues

The set of equations required to realise a Kalman filter is relatively straightforward, even if their derivation is non-trivial. However, as might be expected, there are many implementation issues to be addressed. These are discussed by Dutton (1997) and summarised below.

It is assumed that the state-space model of the system exists, *i.e.* the A, B and C matrices are known. The accuracy of the filter's prediction of the states is fairly tolerant of errors in the A and B matrices, because of the inherent feedback nature of the filter, but is quite sensitive to errors in the C matrix. The effect of errors in the A and B matrices is reflected in a poor dynamic performance of the filter.

For linear systems, the gains of the matrix K(k) converge to constant values. With some loss of dynamic performance, but great savings in computational effort, it is possible to use a stationary Kalman filter with the Kalman gain matrix K(k) of

Figure 113.4 being replaced by a pre-determined and constant matrix K.

Full implementation of the Kalman filter enables it to cope with non-stationary systems, *i.e.* ones in which the A, B and C matrices vary, and with slightly non-linear systems. This is because the Kalman gain matrix K(k) of Equation 113.17 varies from step to step and, each time it is evaluated, will adapt itself to the changes through the inherent feedback.

The Kalman gain matrix is essentially a set of weighting factors used in the correction. The values of the gains in the K(k) matrix are nominally much less than unity. However, inspection of Equation 113.8 reveals that the weightings are applied to the output signals of $\underline{z}(k)$. Given that the outputs are coupled to the states by the C matrix, which may include scaling factors for units conversion, it is sometimes the case that the elements of K(k) converge on values greater than unity.

Information about the characteristics of the noise is contained in the Q and R matrices which must be specified. The F matrix, which establishes how the noise affects the various states, must also be specified. These matrices determine the design of the Kalman gain matrix K(k) and the value of the estimation error covariance matrix P(k). If the characteristics of the noise changes, the Q, R and F matrices must be re-specified.

It is normal for the covariance matrix for the measurement noise R to be assumed diagonal, *i.e.* the various measurements of the noise vector $\underline{v}(k)$ are uncorrelated, with each diagonal element corresponding to the variance of the noise in that particular measurement channel. This can be obtained by squaring the root mean square (RMS) value for the noise or accuracy of the transducer concerned, RMS being the square root of the average squared value of the deviation of the signal from its normal value. RMS values for instrumentation are often quoted in manufacturer's literature.

The covariance matrix for the system disturbances Q is more problematic because little useful information will be readily available. Although it has been assumed that the vector $\underline{w}(k)$ is white noise, in practice it is supposed to cover all disturbances including steps, ramps, spikes, coloured noise (time dependent noise characteristics), modelling errors, and so on. Again, if the various sources of noise are uncorrelated, then the Q matrix will be diagonal. Often the initial values of the elements of Q have to be set more or less at random and then tuned on the basis of experience, either by means of plant trials or simulation studies, taking into account all known disturbances.

Remember that Kalman filtering is inherently a predictor-corrector technique. Suppose that the elements of Q are small and/or the elements of R are large. This scenario is consistent with the system noise being less significant than the measurement noise, in which case the prediction is likely to be more reliable than the correction. The filter adapts by reducing the value of the elements of K(k) accordingly. Conversely, if the elements of Q are large and/or the elements of R are small, the elements of K(k) are increased.

The diagonal elements of Q cannot simply all be set to zero as this implies there is no system noise. The Kalman filter would deduce that its predictions are highly reliable, that corrections are unnecessary and the elements of K(k) would be reduced towards zero. Once the initial transients have decayed away with the gain terms tending to zero, the filter is effectively on open loop and the state estimates deteriorate with time. To prevent this from happening, as far as is practicable, avoid setting the diagonal elements of the covariance matrix Q to zero.

The matrix F can only meaningfully be articulated from an understanding of how the various system disturbances affect the different states. Whilst any one disturbance could directly affect several states, which will be reflected in the values of the corresponding elements of F, the other elements will all be zero. Note also that the combination of matrices specified for A and F must be consistent with the system being controllable.

As indicated in Figure 113.5, it is necessary for the estimation error covariance matrix P(k − 1) to be initialised. It is usual to assume poor initial estimates by assigning large values to the diagonal elements of P(k − 1), and to allow the inherent feedback nature of the filter to drive P(k) downwards. However, in non-stationary and/or non-linear systems, more care is required because the convergence properties of K(k) and P(k) will vary with the initial conditions.

Also indicated by Figure 113.5 is the need for $\hat{x}(k-1|k-1)$ to be initialised. A common practice is to identify any variables in the state vector $\underline{x}$ that appear in the output vector $\underline{z}$, for which the corresponding elements of the output matrix C will be scaling factors. Any such measurements in $\underline{z}$ can be used to initialise the estimate of $\underline{x}$, taking the scaling factors into account, with the other elements being set to zero.

113.7 Use of Kalman Filters in Control Systems

As stated, Kalman filters are used for estimating the states of a system and, for control purposes, there are two principal categories of application:

1. The filter may be used to predict, from input and output signals that can be measured and/or manipulated, values of process variables that cannot be measured. Those predictions can then be used for feedback control in, for example, some multivariable control strategy or state feedback regulator. The Kalman filter is thus within the control loop as depicted in Figure 113.6.

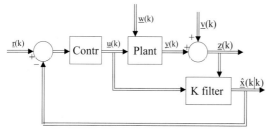

Fig. 113.6 Filter used within a control loop to predict process variables

2. The filter may be used in relation to an existing control system in which certain states are measured and/or manipulated, the filter being used to predict other states. The Kalman filter is thus outside the loop as depicted in Figure 113.7. A worked example of a SISO system in this second category is given in the next section.

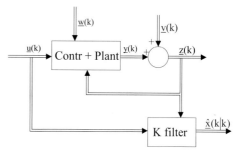

Fig. 113.7 Filter used outside a control loop to predict process variables

It should be noted that feedback is the source of several potential problems in using Kalman filters. The feedback may be due to either the use of control loops and/or to process recycle loops:

- The combination of feedback and the Kalman filter may amplify any noise, leading to non-convergence as can be seen by inspection of Equation 113.8. This can be avoided by having small Kalman gains and a large signal to noise ratio on the output $\underline{z}(k)$.
- Because the output signals become inputs, variance data included in the R matrix may be duplicated in the Q matrix. In effect, it is taken into account twice. In practice, provided the signal to noise ratio is relatively large and the values of the Kalman gains are small, it doesn't seem to matter.
- It was assumed in the formation of the Riccati equation that the estimation error vector and the noise vectors are uncorrelated. That clearly may not be the case if there is feedback. Noise will be propagated around the system and, despite attenuation due to the system's dynamics, may affect the estimation errors. Correlation can be prevented by setting those off-diagonal elements of the Q matrix that relate to the feedback signals to zero.

113.8 Worked Example

Consider the continuous closed loop system depicted in Figure 113.8. It is a slight variation of the system used in the worked examples of Chapters 73, 74, 80 and 112.

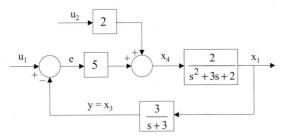

Fig. 113.8 Block diagram of control loop for Worked Example

The system may be thought of as a control loop for controlling the temperature of the contents of an agitated vessel by manipulating the flow rate of steam into its jacket. The time constants have units of minutes, the signals are in deviation form and have all been scaled on a percentage basis. The signals are as follows:

- x_1 Temperature of contents of vessel
- x_2 Temperature of steam in jacket
- x_3 Measured value of contents temperature
- x_4 Steam flow rate
- u_1 Set point for control loop
- u_2 Steam supply pressure

First a state-space model in discrete form must be generated. If the vessel and its jacket are non-interacting, as explained in Chapter 85, the process may be factorised into two transfer functions. The dynamics of the agitated vessel are thus assumed to be:

$$\dot{x}_1 + x_1 = x_2$$

Using Euler's explicit method of numerical integration described in Chapter 96 gives:

$$\frac{x_1(k+1) - x_1(k)}{\Delta t} + x_1(k) = x_2(k)$$

Assuming a sampling period of 0.03 min, which is approximately 2 s and corresponds roughly to 1/30 of the time constant of 1 min, yields:

$$x_1(k+1) = 0.97.x_1(k) + 0.03.x_2(k)$$

The dynamics of the jacket are assumed to be:

$$\dot{x}_2 + 2.x_2 = 2.x_4$$

However, $x_4 = 2.u_2 + 5(u_1 - x_3)$ whence:

$$\dot{x}_2 + 2.x_2 = -10.x_3 + 10.u_1 + 4.u_2$$

Using Eulers explicit method gives:

$$\frac{x_2(k+1) - x_2(k)}{\Delta t} + 2.x_2(k)$$
$$= -10.x_3(k) + 10.u_1(k) + 4.u_2(k)$$

Assume again a sampling period of 0.03 min, which corresponds roughly to 1/15 of the time constant of 0.5 min, yields:

$$x_2(k+1) = 0.94.x_2(k) - 0.3.x_3(k)$$
$$+ 0.3.u_1(k) + 0.12.u_2(k)$$

Similarly for the measurement:

$$\dot{x}_3 + 3.x_3 = 3.x_1$$

giving:

$$x_3(k+1) = 0.09.x_1(k) + 0.91.x_3(k)$$

These equations may now be assembled in the form of Equation 113.1:

$$\begin{bmatrix} x_1(k+1) \\ x_2(k+1) \\ x_3(k+1) \end{bmatrix} = \begin{bmatrix} 0.97 & 0.03 & 0.0 \\ 0.0 & 0.94 & -0.30 \\ 0.09 & 0.0 & 0.91 \end{bmatrix} \begin{bmatrix} x_1(k) \\ x_2(k) \\ x_3(k) \end{bmatrix}$$
$$+ \begin{bmatrix} 0.0 & 0.0 \\ 0.3 & 0.12 \\ 0.0 & 0.0 \end{bmatrix} \begin{bmatrix} u_1 \\ u_2 \end{bmatrix}$$

$$y = \begin{bmatrix} 0.0 & 0.0 & 1.0 \end{bmatrix} \begin{bmatrix} x_1(k) \\ x_2(k) \\ x_3(k) \end{bmatrix}$$

Having established the A, B and C matrices it is now appropriate to think in terms of Figures 113.4 and 113.5 rather than Figure 113.8. What remains is to establish the covariance matrices Q and R, the disturbance matrix F, and to initialise the P(k) matrix and the $\hat{x}$ vector.

Regarding system noise, it is not unreasonable to allow an RMS of 10% for modelling errors on the vessel temperature x_1. In practice it is likely that variations in steam supply pressure will be significant. This will manifest itself as noise on the jacket temperature x_2 for which an RMS of 5% is allowed. Suppose that the temperature measurement x_3 is corrupted by noise with an RMS of 2%. Squaring these RMS values to obtain the variances enables the Q and F matrices to be formulated thus:

$$Q = \begin{bmatrix} 0.01 & 0.0 & 0.0 \\ 0.0 & 0.0025 & 0.0 \\ 0.0 & 0.0 & 0.0004 \end{bmatrix} \quad F = \begin{bmatrix} 1.0 & 0.0 & 0.0 \\ 0.0 & 1.0 & 0.0 \\ 0.0 & 0.0 & 1.0 \end{bmatrix}$$

The measurement x_3 is the same signal as the output y which has already been deemed to have an RMS of 2%. Since there is only one output, R is scalar with a value of 0.0004.

For initialisation purposes, the following may be assumed:

$$P(-1) = \begin{bmatrix} 1,000 & 0 & 0 \\ 0 & 250 & 0 \\ 0 & 0 & 40 \end{bmatrix} \quad \hat{\underline{x}}(-1|-1) = \begin{bmatrix} 0.0 \\ 0.0 \\ 0.0 \end{bmatrix}$$

$$\underline{u}(-1) = \begin{bmatrix} 0.0 \\ 0.0 \end{bmatrix}$$

The solution, or rather the prediction of the output, is in the form of a trace of x_3 *vs* time which, subject to the noise present, should follow any changes in set point u_1 and/or changes in disturbance u_2.

This worked example has shown the approach for establishing the various matrices and initial values. A solution is not meaningful without access to relevant plant data or simulated results. Although the example was realistic, it is nevertheless relatively simple. Kalman filtering can be applied to much more complex problems.

113.9 Nomenclature

Symbol	Description	Dimensions
$\underline{r}$	vector of set points	n × 1
$\underline{u}$	vector of input signals	f × 1
$\underline{v}$	vector of measurement noise	m × 1
$\underline{w}$	vector of system disturbances	h × 1
$\underline{x}$	vector of system states	n × 1
$\underline{y}$	vector of output signals	m × 1
$\underline{z}$	vector of measurements	m × 1
A	system matrix	n × n
B	input matrix	n × f
C	output matrix	m × n
F	system disturbance matrix	n × h
G	dummy matrix used in calculation	m × m
J	dummy matrix used in calculation	n × n
K	Kalman gain (Luenberger)	n × 1
K	Kalman gain matrix	n × m
M	dummy matrix for Riccati solution	n × m
P	covariance matrix of prediction errors	n × n
Q	covariance matrix of system noise	h × h
R	covariance matrix of measurement noise	m × m
k	current sampling instant	

Least Squares Identification

114.1 The Plant Model
114.2 Least Squares Estimation
114.3 Recursive Least Squares
114.4 Least Squares Using Instrumental Variables
114.5 Generalised Least Squares
114.6 Extended Least Squares
114.7 Comparison of Least Squares Estimators
114.8 Extension to Non-Linear Systems
114.9 Extension to Multivariable Systems
114.10 Nomenclatura

Process control relies heavily on simple 3-term feedback control and a variety of strategies such as cascade and ratio control based upon simple feedback, as described in Chapters 22–27. There are two good reasons for this. First, it is relatively simple. Controller design of a PID nature is minimal and tuning is largely realised by empirical means. And second, in something like 95% of all situations, it works remarkably well in terms of accuracy and speed of response. However, PID control does not cope well with:

- time delays, especially when they are variable.
- non-linearities, especially if there are significant variations in set point.
- strong interactions of a multivariable nature.

In these circumstances more advanced design techniques are required, most of which require a deterministic model of the process. There are two approaches to modelling: the classical approach based on first principles and the alternative approach of empirical modelling. They are not necessarily exclusive.

The classical approach of developing models of plant items and processes from consideration of the underlying physical processes and phenomena is described in Chapters 84–92. This gives a good understanding of the structure of the model. Parameters such as gains and time constants are articulated in terms of physical properties, process coefficients and plant characteristics. For accuracy, first principles models usually have to be validated empirically. However, sometimes there is insufficient knowledge of the process or it is too complex to model from first principles. Also, many kinetic and transfer coefficients are too inaccurate for modelling purposes. For such processes, to get the parameters of their models to within 20% is not bad and to within 10% is very good!

The empirical approach embraces statistical (regression and principal components), fuzzy, knowledge based and neural models of the plant and/or process. These types of model are all expensive (labour intensive) to develop and maintain. They are also "bespoke" and used according to their suitability to the application. The alternative time series type of model is generic and software tools are available for generating such models from plant data. This chapter concentrates on the formulation of such time series models, the process of which is usually referred to as identification.

Tests have to be carried out to obtain data about the plant's dynamics in the operating region of interest from which the "best" values of the model's parameters may be found. The various "least squares" methods of finding the best values are considered. These are by far the most common means of parameter estimation. Such identification, carried out in an on-line manner, enables the tracking and hence the control of non-linear as well as time varying systems. Included as an integral part of a self-tuning controller, identification provides the basis for all "self-tuning" control strategies and indeed for many other forms of adaptive control.

Identification and self-tuning are subjects in which there has been extensive publication and dozens of texts written over the last decade or so. A good introduction is given by Dutton (1997): for a more specialised treatment the reader is referred to the text by Wellstead (1991).

114.1 The Plant Model

Consider, for example, the continuous process $G(s)$ as depicted in Figure 114.1 in which the variables are presumed to be in deviation form.

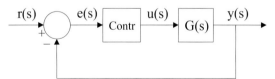

Fig. 114.1 Feedback loop with continuous process $G(s)$

As was demonstrated in Chapter 76, if the input and output of the process are sampled, $G(s)$ is equivalent to a pulse transfer function $G(z)$ which is a time series representation of the system. Thus:

$$G(z) = \frac{y(z)}{u(z)} = \frac{B(z)}{A(z)}$$
$$= \frac{b_0 + b_1.z^{-1} + b_2.z^{-2} + \cdots + b_\beta z^{-\beta}}{1 + a_1.z^{-1} + a_2.z^{-2} + \cdots + a_\alpha.z^{-\alpha}}$$

where u is the process input (MV), typically a self-tuning controller output, and y is the process output (CV) or a measured value. A and B are polynomials of order α and β respectively, A being a monic polynomial, i.e. its leading coefficient is unity.

The above model is referred to as an infinite impulse response (IIR) or, more simply, as a recursive filter since previous values of the output together with current and previous values of the input are used to determine the current output. If B(z) is divided by A(z), assuming it to be either exactly divisible or else the resultant power series in z^{-1} to be truncated after a finite number of terms, the model is said to be a finite impulse response (FIR) or non-recursive filter. That is because the current value of the output depends only on the weighted sum of the current and previous values of the input.

In practice, it is usual for there to be a time delay associated with the plant's dynamics. This is normally accommodated explicitly as follows:

$$y(z) = G(z).u(z) = \frac{z^{-h}.B(z)}{A(z)}.u(z) \quad (114.1)$$
$$= \frac{z^{-h}.\left(b_0 + b_1.z^{-1} + b_2.z^{-2} + \cdots + b_\beta z^{-\beta}\right)}{1 + a_1.z^{-1} + a_2.z^{-2} + \cdots + a_\alpha.z^{-\alpha}}.u(z)$$

where h is the integer number of sample periods equivalent to the time delay. Rearrangement yields:

$$y(z) = -\left(a_1.z^{-1} + a_2.z^{-2} + \cdots + a_\alpha.z^{-\alpha}\right).y(z)$$
$$+ z^{-h}.\left(b_0 + b_1.z^{-1} + b_2.z^{-2} + \cdots + b_\beta.z^{-\beta}\right).u(z)$$

the inverse transform of which is:

$$\begin{aligned}y(k) = & -a_1.y(k-1) - a_2.y(k-2) - \cdots \\ & - a_\alpha.y(k-\alpha) + b_0.u(k-h) \\ & + b_1.u(k-h-1) + b_2.u(k-h-2) \\ & + \cdots + b_\beta.u(k-h-\beta) \quad (114.2)\end{aligned}$$

where k denotes the current sampling instant. This equation is the time series model of the plant $G(s)$ that is to be identified. The parameters to be estimated are the following coefficients:

a_i for all $1 < i < \alpha$ and b_j for all $0 < j < \beta$

There are thus $(\alpha + \beta + 1)$ coefficients to be estimated, together with the delay h. For a linear time invariant (LTI) system operated about a fixed set point, most of these coefficients will be more or less constant. If the system has non-linear characteristics, typically associated with different flow regimes, then the a and b coefficients will vary with the set point. If the system is time dependent as, for example, with the fouling of a heat transfer surface, the coefficients will vary over time. For systems with variable time delay, typically due to distance velocity effects, the coefficient h is likely to vary significantly.

114.2 Least Squares Estimation

For identification purposes, an estimate $\hat{y}(k)$ of the true current value $y(k)$ of the output is made using Equation 114.2 based upon previous values of the outputs and the current and previous values of the inputs:

$$\hat{y}(k) = -a_1.y(k-1) - a_2.y(k-2) - \cdots \\ - a_\alpha.y(k-\alpha) + b_0.u(k-h) \\ + b_1.u(k-h-1) + \cdots \\ + b_\beta.u(k-h-\beta)$$

There will inevitably be an estimation error $\varepsilon(k)$, referred to as the residual:

$$y(k) = \hat{y}(k) + \varepsilon(k) \quad (114.3)$$

which results in:

$$y(k) = -a_1.y(k-1) - a_2.y(k-2) - \cdots \\ - a_\alpha.y(k-\alpha) + b_0.u(k-h) \\ + b_1.u(k-h-1) + \cdots \\ + b_\beta.u(k-h-\beta) + \varepsilon(k)$$

This can be written more compactly in vector form as follows:

$$y(k) = \underline{x}^T(k).\underline{\theta} + \varepsilon(k) \quad (114.4)$$

where y and ε are scalars and $\underline{x}$ and $\underline{\theta}$ are both $(\lambda \times 1)$ vectors with $\lambda = \alpha + \beta + 1$ and:

$$\underline{x}^T(k) = \begin{bmatrix} -y(k-1) & -y(k-2) & \cdots & -y(k-\alpha) \\ u(k-h) & u(k-h-1) & \cdots & u(k-h-\beta) \end{bmatrix}$$

$$\underline{\theta}^T = \begin{bmatrix} a_1 & a_2 & \cdots & a_\alpha & b_0 & b_1 & b_2 & \cdots & b_\beta \end{bmatrix} \quad (114.5)$$

Of fundamental importance to least squares estimation is the premise that the sequence of residuals $\varepsilon(k)$ is random, has zero mean and finite variance, and is not correlated with either the output $y(k)$ or the input $u(k)$. In other words, it is a chance component.

Suppose that pairs of input and output values, u and y, are obtained empirically which enable n equations of the form of Equation 114.4 to be formed, giving $y(k)$ through to $y(k-n+1)$. These n equations may be assembled in matrix form as follows:

$$\begin{bmatrix} y(k) \\ y(k-1) \\ y(k-2) \\ \vdots \\ y(k-n+2) \\ y(k-n+1) \end{bmatrix} = \begin{bmatrix} -y(k-1) & -y(k-2) & \cdots & -y(k-\alpha) & u(k-h) & u(k-h-1) & \cdots & u(k-h-\beta) \\ -y(k-2) & -y(k-3) & \cdots & -y((k-\alpha-1) & u(k-h-1) & u(k-h-2) & \cdots & u(k-h-\beta-1) \\ -y(k-3) & -y(k-4) & \cdots & -y((k-\alpha-2) & u(k-h-2) & u(k-h-3) & \cdots & u(k-h-\beta-2) \\ \vdots & \vdots & \vdots & \vdots & \vdots & \vdots & & \vdots \\ -y(k-n+1) & -y(k-n) & \cdots & -y(k-\alpha-n+2) & u(k-h-n+2) & u(k-h-n+1) & \cdots & u(k-h-\beta-n+2) \\ -y(k-n) & -y(k-n-1) & \cdots & -y(k-\alpha-n+1) & u(k-h-n+1) & u(k-h-n) & \cdots & u(k-h-\beta-n+1) \end{bmatrix} \begin{bmatrix} a_1 \\ a_2 \\ \vdots \\ a_\alpha \\ b_0 \\ b_1 \\ \vdots \\ b_\beta \end{bmatrix} + \begin{bmatrix} \varepsilon_1 \\ \varepsilon_2 \\ \vdots \\ \varepsilon_n \end{bmatrix}$$

which may be simplified by notation:

$$\underline{y}(k) = X(k).\underline{\theta} + \underline{\varepsilon}(k) \quad (114.6)$$

where $\underline{y}$ and $\underline{\varepsilon}$ are $(n \times 1)$ vectors and $X(k)$ is an $(n \times \lambda)$ matrix:

$$X(k) = \begin{bmatrix} \underline{x}^T(k) & \underline{x}^T(k-1) & \cdots & \underline{x}^T(k-n+1) \end{bmatrix}^T$$

The objective is to solve for the elements of the vector $\underline{\theta}$. If $n = \lambda$ there will be n equations with n

unknowns and a unique solution. However, since there is no constraint on n, there will generally be an over-determined set of equations with $n > \lambda$ from which the most probable solution for $\underline{\theta}$ has to be extracted. Define an objective function J as follows:

$$J = \underline{\varepsilon}^T(k).W.\underline{\varepsilon}(k)$$

where $\underline{\varepsilon}(k) = \underline{y}(k) - X(k).\underline{\theta}$ from Equation 114.6 and W is a $(n \times n)$ matrix of constants.

The value of $\underline{\theta}$ which minimises J is the best estimate of the coefficients. This problem has already been solved in Chapter 83 in relation to multiple linear regression, the batch least squares (BLS) solution being:

$$\hat{\underline{\theta}} = \left(X^T(k).W.X(k)\right)^{-1}.X^T(k).W.\underline{y}(k) \quad (114.7)$$

If W is a weighting matrix of positive definite form, the estimation is referred to as weighted least squares (WLS). Choosing the identity matrix, $W = I$, the solution becomes:

$$\hat{\underline{\theta}} = \left(X^T(k).X(k)\right)^{-1}.X^T(k).\underline{y}(k) \quad (114.8)$$

which is identical to the BLS solution of Equation 83.14.

114.3 Recursive Least Squares

The least squares methods described thus far are "batch" or "off-line" methods in that the data, comprising pairs of values u and y, is collected and stored indefinitely prior to estimation. For self-tuning purposes, the estimate of $\underline{\theta}$ has to be made at each sampling instant or some multiple thereof. The obvious approach is to use a fixed data window of the last n samples, discarding old data as new becomes available. This, however, requires a fairly large no of samples to ensure that a sufficient history of process information is taken into account. Unfortunately, because of the matrix inversions involved, the processing effort required is such that it would be unrealistic to estimate $\underline{\theta}$ in real-time. Thus, for an estimator to be practicable, what is required is that it be able to remember the history of the process and that the matrix manipulations be minimised. Both of these objectives can be realised by reformulating the BLS estimator into its recursive least squares (RLS) form.

A recursive solution is sought. Remembering that the index k denotes the current sampling instant, Equation 114.8 may be modified:

$$\begin{aligned}\hat{\underline{\theta}}(k) &= \left(X^T(k).X(k)\right)^{-1} X^T(k).\underline{y}(k) \\ &= R(k).X^T(k).\underline{y}(k)\end{aligned} \quad (114.9)$$

where R(k) is the inverse of the covariance matrix S(k):

$$R(k) = S^{-1}(k) = \left(X^T(k).X(k)\right)^{-1} \quad (114.10)$$

Strictly speaking, S(k) is only of the form of the covariance matrix. For S(k) to be the "true" covariance matrix, as defined by Equations 82.5 and 101.3, a factor of $1/(n-1)$ needs to be introduced. Nevertheless, for RLS purposes, S(k) is usually referred to as the covariance matrix.

The quadratic in X(k) is of particular interest. Recognising that matrix multiplication is simply the sum of a series of vector products, it can be shown that the quadratic may be decomposed as follows:

$$S(k) = X^T(k).X(k) = \sum_{j=k-n+1}^{k-1} \underline{x}(j).\underline{x}^T(j) + \underline{x}(k).\underline{x}^T(k)$$

Thus the current value of the quadratic in X(k) based upon n vectors $\underline{x}$ of sampled values may be thought of as the sum of the quadratics of the $n-1$ previous vectors plus the quadratic of the current vector. The summation may itself be represented as the quadratic based upon the previous n-1 vectors which provides the basis for a recursive relationship:

$$X^T(k).X(k) = X^T(k-1).X(k-1) + \underline{x}(k).\underline{x}^T(k) \quad (114.11)$$

Substituting from Equation 114.10, this may be expressed in terms of R(k) as follows:

$$\begin{aligned}R^{-1}(k) &= R^{-1}(k-1) + \underline{x}(k).\underline{x}^T(k) \\ R(k) &= \left(R^{-1}(k-1) + \underline{x}(k).\underline{x}^T(k)\right)^{-1}\end{aligned} \quad (114.12)$$

This equation is of a recursive nature, *i.e.* a relationship exists between the current and previous values of R(k), and clearly addresses the memory problem. However, two matrix inversions are still required for its evaluation which is unsatisfactory for implementation purposes. Fortunately, this can be overcome by analogy with the "matrix inversion lemma" which is as follows:

$$(A + BCD)^{-1} = A^{-1} - A^{-1}B(C^{-1} + DA^{-1}B)^{-1}DA^{-1}$$

Let $A \equiv R^{-1}(k-1), B \equiv \underline{x}(k), C \equiv I$ and $D \equiv \underline{x}^T(k)$ whence:

$$R(k) = R(k-1) \\ - R(k-1).\underline{x}(k)\left(I + \underline{x}^T(k).R(k-1).\underline{x}(k)\right)^{-1} \\ \times \underline{x}^T(k).R(k-1)$$

This still requires inversion of the expression:

$$I + \underline{x}^T(k).R(k-1).\underline{x}(k)$$

However, inspection of the dimensions reveals that this is a scalar quantity so inversion is trivial, thus:

$$R(k) = R(k-1) - \frac{R(k-1).\underline{x}(k).\underline{x}^T(k).R(k-1)}{1 + \underline{x}^T(k).R(k-1).\underline{x}(k)} \quad (114.13)$$

Substituting back into Equation 114.9 gives:

$$\hat{\underline{\theta}}(k) = \left\{ R(k-1) - \frac{R(k-1).\underline{x}(k).\underline{x}^T(k).R(k-1)}{1 + \underline{x}^T(k).R(k-1).\underline{x}(k)} \right\} \\ \times X^T(k).\underline{y}(k)$$

By analogy with Equation 114.11:

$$X^T(k).\underline{y}(k) = X^T(k-1).\underline{y}(k-1) + \underline{x}(k).y(k)$$

Further substitution yields:

$$\hat{\underline{\theta}}(k) = \left\{ R(k-1) - \frac{R(k-1).\underline{x}(k).\underline{x}^T(k).R(k-1)}{1 + \underline{x}^T(k).R(k-1).\underline{x}(k)} \right\} \\ \times \left(X^T(k-1).\underline{y}(k-1) + \underline{x}(k).y(k) \right) \quad (114.14)$$

This is the basic equation of the RLS estimator. It may look complicated but can be substantially simplified by making the following substitution which comes from applying a time shift of one sample period to Equation 114.9:

$$\hat{\underline{\theta}}(k-1) = R(k-1).X^T(k-1).\underline{y}(k-1)$$

Expansion of Equation 114.14, substitution and rearrangement yields:

$$\hat{\underline{\theta}}(k) = \hat{\underline{\theta}}(k-1) + K(k)\left(y(k) - \underline{x}^T(k).\hat{\underline{\theta}}(k-1)\right)$$

which is inherently recursive in nature. Thus the new estimate of $\underline{\theta}(k)$ consists of the previous estimate plus a correction. The correction consists of a gain vector K(k) applied to the error on the best estimate of the most recent prediction of the output y. Substitution from Equation 114.3 yields:

$$\hat{\underline{\theta}}(k) = \hat{\underline{\theta}}(k-1) + K(k).\varepsilon(k) \quad (114.15)$$

where:

$$\varepsilon(k) = y(k) - \underline{x}^T(k).\hat{\underline{\theta}}(k-1) \quad (114.16)$$

and:

$$K(k) = R(k-1).\underline{x}(k)/d(K) \quad (114.17)$$

where:

$$d(K) = 1 + \underline{x}^T(k).R(k-1).\underline{x}(k) \quad (114.18)$$

Noting that d(K) is scalar and R(k) is symmetrical, Equation 114.13 may be rewritten in terms of K(k):

$$R(k) = R(k-1) - d(K).K(k).K^T(k) \quad (114.19)$$

Equations 114.15–114.19 are the RLS estimator used for identification. Implementation of the RLS estimator is depicted in flow chart form in Figure 114.2.

Through R(k), the inverse of the covariance matrix, the estimator remembers the history of the process: there is no need to use a data window or to store historic values of $\underline{x}$. Note that there are no demanding matrix inversions involved. Also note that both R(k) and $\underline{\theta}(k)$ have to be initialised.

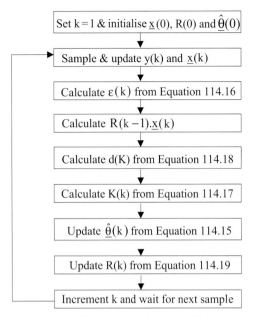

Fig. 114.2 Implementation of the recursive least squares (RLS) estimator

The structure of the RLS estimator is generic. Schemes other than RLS differ only, for example, in the way that the gain vector K(k) is updated and whether the data vector x(k) contains filtered or raw values.

With reference to Equation 114.6, provided that ε is not correlated with X(k), i.e. the stochastic term has a zero mean, the least squares methods will always give unbiased estimates. Unfortunately, especially in the context of control systems, that is not usually the case as is depicted in Figure 114.3.

Any non-deterministic disturbance w(s), often in the form of noise, is propagated through the feedback. This means that the process input is not independent of ε and correlates with X(k), resulting in biased estimates. The basic BLS and RLS methods have to be modified to overcome this limitation. Three modified least squares estimators are considered in the following sections.

114.4 Least Squares Using Instrumental Variables

This approach uses a signal, referred to as the instrumental variable, which yields unbiased estimates of the process parameters when used in place of the original output.

Analysis of Figure 114.3 yields:

$$y(z) = \frac{z^{-h}.B(z)}{A(z)}.u(z) + \frac{1}{A(z)}.w(z)$$

rearrangement of which gives:

$$A(z).y(z) = z^{-h}.B(z).u(z) + w(z) \quad (114.20)$$

Inverse transformation and rearrangement enables this to be cast in the form of Equation 114.4:

$$y(k) = \underline{x}^T(k).\underline{\theta} + w(k) \quad (114.21)$$

where all of the difference between the estimated and true values of the output is attributed to the noise. Suppose, as before, that pairs of input and output values are obtained empirically which enable n such equations to be formed. These may be assembled in matrix form as follows:

$$\underline{y}(k) = X(k).\underline{\theta} + \underline{w}(k) \quad (114.22)$$

where $\underline{y}(k)$ and $X(k)$ are as defined previously and $\underline{w}$ is an $(n \times 1)$ vector.

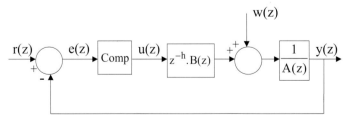

Fig. 114.3 Biased estimates due to noise propagated through feedback

Presume that some matrix V may be specified such that it correlates with the vector of process outputs $\underline{y}(k)$, and by implication with the inputs, but not with the noise $\underline{w}(k)$. By definition, V has the following expectations:

$$E\{V^T.\underline{w}(k)\} = 0 \quad \text{and} \quad E\{V^T.X(k)\} \quad \text{is non-singular.} \tag{114.23}$$

Multiplying Equation 114.22 by V^T gives:

$$\underline{v}(k) = V^T.\underline{y}(k) = V^T.X(k).\underline{\theta}(k) + V^T.\underline{w}(k) \tag{114.24}$$

where the instrumental vector $\underline{v}(k)$ is of dimension ($\lambda \times 1$), and the instrumental matrix V is of dimension ($n \times \lambda$).

The error in the prediction of the instrumental vector, due to the noise, is thus:

$$\underline{\varepsilon}_V(k) = \underline{v}(k) - V^T.X(k).\underline{\theta}(k) = V^T.\underline{w}(k)$$

An objective function J may be defined in terms of the error on the instrumental vector as follows:

$$J = \underline{\varepsilon}_V^T(k).\underline{\varepsilon}_V(k)$$
$$= \left(\underline{v}(k) - V^T.X(k).\underline{\theta}(k)\right)^T \left(\underline{v}(k) - V^T.X(k).\underline{\theta}(k)\right)$$

As before, the objective is to solve for the unknown elements of the vector $\underline{\theta}(k)$ by minimising J. Omitting the (k) notation, in the interest of brevity, the best estimate is found by analogy with Equation 114.8, giving:

$$\hat{\underline{\theta}} = \left((V^T.X)^T.V^T.X\right)^{-1}.(V^T.X)^T.\underline{v}$$
$$= (X^T.V.V^T.X)^{-1}.X^T.V.\underline{v}$$

Substituting from Equation 114.24, it follows that the expected value of the estimate of $\underline{\theta}(k)$ is:

$$E\{\hat{\underline{\theta}}\} = E\{(X^T.V.V^T.X)^{-1}.X^T.V.\underline{v}\}$$
$$= E\{(X^T.V.V^T.X)^{-1}.X^T.V.(V^T.X.\underline{\theta} + V^T.\underline{w})\}$$
$$= E\{(X^T.V.V^T.X)^{-1}.(X^T.V.V^T.X).\underline{\theta}\}$$
$$+ E\{(X^T.V.V^T.X)^{-1}.X^T.V.V^T.\underline{w}\}$$
$$= \underline{\theta} + (X^T.V.V^T.X)^{-1}.X^T.V.E\{V^T.\underline{w}\}$$

Since the latter expectation is of zero mean, through the definition of V by Equation 114.23, the expectation of $\underline{\theta}(k)$ being its true unbiased value is proven.

Least squares using instrumental variables may be implemented on a recursive basis analogous to the method of RLS depicted in Figure 114.2. Remember that with RLS it was sufficient to work with $\underline{x}(k)$, establishing X(k) was not necessary. So too with instrumental variables it is sufficient to work with the $\underline{v}(k)$ rather than to establish the matrix V. The outstanding issue, therefore, is formulation of the instrumental vector $\underline{v}$.

One approach, due to Soderstrom, starts by generating an instrumental variable that relates to the process output y, and by implication to the input u, but not to the noise w, by adapting Equation 114.21 as follows:

$$v(k) = \underline{x}^T(k).\underline{\theta}(k-1)$$

It can be seen that the value of the instrumental variable is a filtered output sequence based upon the most recent estimate of $\underline{\theta}(k)$ in the absence of noise. An instrumental vector is then formed by replacing all the measured outputs in Equation 114.5 with filtered outputs:

$$\underline{v}^T(k) = \begin{bmatrix} -v(k-1) & -v(k-2) & \cdots & -v(k-\alpha) \\ u(k-h) & u(k-h-1) & \cdots & u(k-h-\beta) \end{bmatrix}$$

The recursive formulation is then exactly the same as depicted in Figure 114.2, with the exception that the prediction error of Equation 114.16 is replaced by:

$$\varepsilon(k) = y(k) - \underline{v}^T(k).\hat{\underline{\theta}}(k-1)$$

The estimator gain of Equation 114.17 is replaced by:

$$K(k) = R(k-1).\underline{v}(k)/d(K)$$

for whose denominator Equation 114.18 is replaced by:

$$d(K) = 1 + \underline{v}^T(k).R(k-1).\underline{v}(k)$$

114.5 Generalised Least Squares

With reference to Figure 114.3, suppose that the error on the prediction of the output is some weighted function of the noise according to:

$$\varepsilon(z) = C(z).w(z)$$
$$= \left(1 + c_1.z^{-1} + c_2.z^{-2} + \cdots + c_\gamma.z^{-\gamma}\right) w(z) \quad (114.25)$$

where $C(z)$ is a monic polynomial of order γ. Do not confuse $C(z)$ with the impulse compensator depicted in Figure 114.3 for which the same symbol is commonly used. Substituting into Equation 114.20 gives:

$$A(z).y(z) = z^{-h}.B(z).u(z) + \frac{1}{C(z)}.\varepsilon(z)$$

rearrangement of which yields:

$$A(z).\tilde{y}(z) = z^{-h}.B(z).\tilde{u}(z) + \varepsilon(z) \quad (114.26)$$

where:

$$\tilde{y}(z) = C(z).y(z) \quad \text{and} \quad \tilde{u}(z) = C(z).u(z). \quad (114.27)$$

Expansion and rearrangement yields:

$$\tilde{y}(z) = -\left(a_1.z^{-1} + a_2.z^{-2} + \cdots + a_\alpha.z^{-\alpha}\right).\tilde{y}(z)$$
$$+ z^{-h}.\left(b_0 + b_1.z^{-1} + b_2.z^{-2} + \cdots + b_\beta.z^{-\beta}\right).\tilde{u}(z)$$
$$+ \varepsilon(z)$$

the inverse transform of which is:

$$\tilde{y}(k) = -a_1.\tilde{y}(k-1) - a_2.\tilde{y}(k-2) - \cdots$$
$$- a_\alpha.\tilde{y}(k-\alpha) + b_0.\tilde{u}(k-h)$$
$$+ b_1.\tilde{u}(k-h-1) + b_2.\tilde{u}(k-h-2) + \cdots$$
$$+ b_\beta.\tilde{u}(k-h-\beta) + \varepsilon(k)$$

This can be written more compactly in the familiar vector form of Equation 114.4 as follows:

$$\tilde{y}(k) = \underline{\tilde{x}}^T(k).\underline{\theta} + \varepsilon(k)$$

where the ($\lambda \times 1$) data vector $\underline{\tilde{x}}$ contains $\tilde{y}$ and $\tilde{u}$ terms and $\underline{\theta}$ is as defined in Equation 114.5:

$$\underline{\tilde{x}}^T(k) = \begin{bmatrix} -\tilde{y}(k-1) & -\tilde{y}(k-2) & \cdots & -\tilde{y}(k-\alpha) \\ \tilde{u}(k-h) & \tilde{u}(k-h-1) & \cdots & \tilde{u}(k-h-\beta) \end{bmatrix}$$

Suppose that data is collected that enables n such equations to be assembled in the same matrix form as Equation 114.6 as follows:

$$\underline{\tilde{y}}(k) = \tilde{X}(k).\underline{\theta}(k) + \underline{\varepsilon}(k) \quad (114.28)$$

where $\underline{\tilde{y}}(k)$ and $\underline{\varepsilon}(k)$ are ($n \times 1$) vectors and $\tilde{X}(k)$ is an ($n \times \lambda$) matrix.

This is the generalised least squares (GLS) formulation of the problem. BLS minimisation may be applied to Equation 114.28 to obtain an unbiased estimate of $\underline{\theta}(k)$ of the same form as Equation 114.8:

$$\underline{\hat{\theta}}(k) = \left(\tilde{X}^T(k).\tilde{X}(k)\right)^{-1} \tilde{X}^T(k).\underline{\tilde{y}}(k) \quad (114.29)$$

The problem with this method is that $C(z)$ is unknown so $\underline{\tilde{y}}(k)$ cannot be evaluated.

Clarke's classic approach to solving the problem is iterative in nature. First, a first approximation to $\underline{\theta}(k)$ is obtained from the basic BLS estimate using Equation 114.8:

$$\underline{\hat{\theta}}\bigg|_{1st} = \left(X^T(k).X(k)\right)^{-1} X^T(k).\underline{y}(k)$$

The residuals are then calculated from Equation 114.6:

$$\underline{\varepsilon}(k) = \underline{y}(k) - X(k).\underline{\hat{\theta}}\bigg|_{1st} \quad (114.30)$$

These residuals may then be used as first approximations for the noise in Equation 114.25. By analogy with Equation 114.3:

$$\varepsilon(k) = \hat{\varepsilon}(k) + \Delta(k)$$
$$= -c_1.\varepsilon(k-1) - c_2.\varepsilon(k-2) - \cdots$$
$$- c_\gamma.\varepsilon(k-\gamma) + \Delta(k)$$

where $\Delta(k)$ is the error on the prediction of the residual. Suppose that n such equations are formed at different sampling instants and assembled in matrix form as follows:

$$\underline{\varepsilon}(k) = E(k).\underline{c} + \underline{\Delta}(k)$$

where $\underline{\varepsilon}(k)$ and $\underline{\Delta}(k)$ are $(n \times 1)$ vectors, $\underline{c}$ is a $(\gamma \times 1)$ vector and $E(k)$ is an $(n \times \gamma)$ matrix:

$$\underline{c}^T = \begin{bmatrix} c_1 & c_2 & \cdots & c_\gamma \end{bmatrix}$$

The basic BLS procedure of Equation 114.8 is used to make the first estimate of $\underline{c}$:

$$\hat{\underline{c}} = \left(E^T(k).E(k)\right)^{-1}.E^T(k).\underline{\varepsilon}(k)$$

Having estimated $\underline{c}$, the filtered values $\tilde{y}(k)$ and $\tilde{u}(k)$ are then obtained from Equation 114.27 and subsequently $\underline{\theta}(k)$ from Equation 114.29. This new estimate of $\underline{\theta}$ is used to replace the first estimate in Equation 114.30 and the process is repeated. Iteration is stopped when improvement in $\underline{\theta}(k)$ ceases.

114.6 Extended Least Squares

Also known as the extended matrix method and as the approximate maximum likelihood estimator, the ELS estimator is yet another approach to circumvent the problem of the residuals being correlated with the process data resulting in bias.

In contrast to Equation 114.25, the weighting for ELS operates on the error on the prediction of the output:

$$\begin{aligned} w(z) &= C(z).\varepsilon(z) \\ &= \left(1 + c_1.z^{-1} + c_2.z^{-2} + \cdots + c_\gamma.z^{-\gamma}\right).\varepsilon(z) \end{aligned}$$
(114.31)

Again, take care not to confuse this $C(z)$ with the compensator. From Figure 114.3:

$$A(z).y(z) = z^{-h}.B(z).u(z) + C(z).\varepsilon(z) \quad (114.32)$$

This equation is normally referred to as the auto regressive moving average with exogenous input (ARMAX) model but sometimes as the controlled auto regressive moving average (CARMA) model. Inverse transformation and rearrangement yields:

$$\begin{aligned} y(k) = &-a_1.y(k-1) - a_2.y(k-2) - \cdots \\ &- a_\alpha.y(k-\alpha) + b_0.u(k-h) \\ &+ b_1.u(k-h-1) + \cdots \\ &+ b_\beta.u(k-h-\beta) + \varepsilon(k) + c_1.\varepsilon(k-1) \\ &+ c_2.\varepsilon(k-2) + \cdots + c_\gamma.\varepsilon(k-\gamma) \end{aligned}$$

This can be expressed in vector form as before:

$$y(k) = \underline{x}^T(k).\underline{\theta} + \varepsilon(k) \quad (114.4)$$

where both $\underline{x}(k)$ and $\underline{\theta}(k)$, which are of dimension $((\alpha + \beta + \gamma + 1) \times 1)$, are extended to include the noise terms:

$$\begin{aligned} \underline{x}^T(k) = \big[&-y(k-1) \; -y(k-2) \; \cdots \; -y(k-\alpha) \\ & u(k-h) \; u(k-h-1) \; u(k-h-2) \\ & \cdots \; u(k-h-\beta) \\ & \varepsilon(k-1) \; \varepsilon(k-2) \; \cdots \; \varepsilon(k-\gamma) \big] \end{aligned}$$

$$\begin{aligned} \underline{\theta}^T = \big[&a_1 \; a_2 \; \cdots \; a_\alpha \; b_0 \; b_1 \; b_2 \; \cdots \; b_\beta \\ & c_1 \; c_2 \; \cdots \; c_\gamma \big] \end{aligned}$$
(114.33)

Since the correlation on $\varepsilon(k)$ is taken into account through $C(z)$, the least squares solution will be unbiased. Since $C(z)$ is unknown its coefficients must be found as part of the estimation process. Otherwise, apart from the fact that the dimensions of the vectors and matrices are extended, the ELS estimator is implemented in exactly the same way as RLS as depicted in Figure 114.2.

114.7 Comparison of Least Squares Estimators

The essential problem with the basic BLS and RLS methods, in the context of feedback, is correlation between the residuals $\underline{\varepsilon}(k)$ and the values of the process signals $X(k)$ resulting in biased estimates of $\underline{\theta}(k)$. However, the bias decreases as the order of the model being estimated increases. Clearly if the order of the model, *i.e.* the values of α and β, is high enough the bias becomes insignificant.

This, combined with the efficiency of the RLS algorithm depicted in Figure 114.2, not to mention its fast convergence, means that RLS is the most commonly used method of estimation for self-tuning purposes.

The three other means of estimation, instrumental variables, GLS and ELS, all include measures to overcome the bias problem. Of these ELS is the most effective. However, they all have heavy computational requirements and, in the cases of GLS and ELS, are slower because of the initial uncertainties in estimating the parameters of the noise filter $C(z)$. Furthermore, if the signal to noise ratio is too low, say less than 10 to 1, it is possible that they will converge on false values. RLS may be a lazy choice of estimator, but it works well in practice.

stricted to linear systems. It is possible to identify time series models of non-linear systems such as:

$$\begin{aligned} A(z).y(z) = & z^{-h}.B_1(z).u(z) + z^{-h}.B_2(z).u^2(z) \\ & + z^{-1}.B_3(z).u(z).y(z) \\ & + C(z).\varepsilon(z) \end{aligned} \quad (114.34)$$

Time series models of non-linear systems are referred to as NARMAX (*i.e.* non-linear ARMAX) models because they contain higher order and cross product terms in $u(z)$ and $y(z)$. Provided that the various $A(z)$, $B(z)$ and $C(z)$ are themselves linear series in z^{-1}, *i.e.* the coefficients are all linear, then identification can be realised typically by means of ELS. This makes identification an extremely powerful technique.

114.8 Extension to Non-Linear Systems

Throughout this chapter it has been assumed that the systems being identified are inherently linear and can be represented by time series models. The ARMAX model of Equation 114.32 is such an example. However, identification need not be re-

114.9 Extension to Multivariable Systems

Identification thus far has been concerned with SISO plant but the approach lends itself naturally to processes that are MIMO in nature. Consider, for example, the 2-input 2-output system of Figure 114.4.

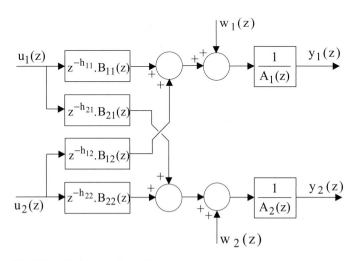

Fig. 114.4 Identification for multivariable systems

Clearly variations in u_1 will affect both y_1 and y_2 so, by applying BLS or RLS between u_1 and y_1 and between u_1 and y_2 separately, the individual models can be established. Likewise, for the affect of variations of u_2 on y_1 and y_2.

However, a more subtle approach enables two or more models to be established simultaneously. The MIMO model corresponding to Figure 114.4 is:

$$\begin{aligned} A_1(z).y_1(z) &= z^{-h_{11}}.B_{11}(z).u_1(z) \\ &\quad + z^{-h_{12}}.B_{12}(z).u_2(z) + w_1(z) \\ A_2(z).y_2(z) &= z^{-h_{21}}.B_{21}(z).u_1(z) \\ &\quad + z^{-h_{22}}.B_{22}(z).u_2(z) + w_2(z) \end{aligned}$$

(114.35)

Consider the first of these two equations, in effect a 2-input 1-output problem. Inverse transformation and rearrangement results in:

$$\begin{aligned} y_1(k) = &-a_{11}.y_1(k-1) - a_{12}.y_1(k-2) - \cdots \\ &- a_{1\alpha_1}.y_1(k-\alpha_1) + b_{110}.u_1(k-h_{11}) \\ &+ b_{111}.u_1(k-h_{11}-1) \\ &+ b_{112}.u_1(k-h_{11}-2) + \cdots \\ &+ b_{11\beta_{11}}.u_1(k-h_{11}-\beta_{11}) \\ &+ b_{120}.u_2(k-h_{12}) \\ &+ b_{121}.u_2(k-h_{12}-1) \\ &+ b_{122}.u_2(k-h_{12}-2) + \cdots \\ &+ b_{12\beta_{12}}.u_2(k-h_{12}-\beta_{12}) \end{aligned}$$

which can be written in the vector form:

$$y_1(k) = \underline{x}_1^T(k).\underline{\theta}_1 + \varepsilon_1(k) \quad (114.36)$$

where:

$$\begin{aligned} \underline{x}_1^T(k) = \big[&-y_1(k-1) - y_1(k-2) \cdots - y_1(k-\alpha_1) \\ &u_1(k-h)u_1(k-h-1)u_1(k-h-2) \\ &\cdots u_1(k-h-\beta_{11}) \\ &u_2(k-h)u_2(k-h-1)u_2(k-h-2) \\ &\cdots u_2(k-h-\beta_{12}) \big] \end{aligned}$$

and:

$$\underline{\theta}_1^T = \big[\begin{matrix} a_{11} & a_{12} & \cdots & a_{1\alpha_1} & b_{110} & b_{111} & b_{112} & \cdots & b_{11\beta_{11}} \\ & & & & b_{120} & b_{121} & b_{122} & \cdots & b_{12\beta_{12}} \end{matrix} \big]$$

Sets of input and output values, u_1, u_2 and y_1, can be obtained empirically and n equations of the form of Equation 114.36 combined into the X matrix of values. The solution, Equation 114.8, can then be found by means of BLS or RLS, extended or otherwise, in exactly the same way as for the SISO scenario. Thus two models are obtained simultaneously. Similarly the solution can be found to the second of Equations 114.35. Clearly this extrapolates to any MISO, multiple inputs (≥ 2) single output, problem.

114.10 Nomenclature

Symbol	Description	Dimensions
e	controller error	1×1
r	controller set point	1×1
u	input variable	1×1
v	instrumental variable	1×1
w	disturbance (noise) variable	1×1
y	output variable	1×1
ε	residual: error in prediction of output y	1×1
Δ	error in prediction of residual ε	1×1
$\underline{c}$	vector of weighting coefficients	$\gamma \times 1$
$\underline{v}$	instrumental vector	$\lambda \times 1$
$\underline{w}$	vector of values of noise	$n \times 1$
$\underline{x}$	vector of input and output values	$\lambda \times 1$
$\underline{y}$	vector of output values	$n \times 1$
$\underline{\varepsilon}$	vector of values of residuals	$n \times 1$
$\underline{\Delta}$	vector of errors on prediction of residuals	$n \times 1$
$\underline{\theta}$	vector of model coefficients	$\lambda \times 1$
E	matrix of values of residuals	$n \times \gamma$
K	gain vector	$\lambda \times 1$
R	inverse of covariance matrix	$\lambda \times \lambda$
S	covariance matrix	$\lambda \times \lambda$
V	instrumental matrix	$n \times \lambda$
W	matrix of weighting factors	$n \times n$
X	matrix of $\underline{x}$ vectors	$n \times \lambda$
h	time delay exponent in process model	
k	current sampling instant	
n	number of vectors of input and output values	
λ	dimension normally but for ELS	$(\alpha + \beta + 1)$ $(\alpha + \beta + \gamma + 1)$

Symbol	Description
A	denominator polynomial of process model (of order α)
B	numerator polynomial of process model (of order β)
C	weighting polynomial on noise (of order γ)

Recursive Estimation

115.1 Setting the Order of the Model
115.2 Initialisation of Parameter Values
115.3 Initialisation of the Covariance Matrix
115.4 Forgetting Factors
115.5 Covariance Resetting
115.6 Numerical Instability
115.7 Covariance Windup
115.8 Variable Forgetting Factors
115.9 Convergence
115.10 Comments

The previous chapter covered the underlying principles of the use of least squares estimators for identification purposes. The emphasis was on the development of the algorithms per-se, with little thought as to their realisation. In practice, of course, there are many prosaic but vital measures that have to be taken to ensure successful implementation. These essentially concern issues such as initialisation, robustness, stability, convergence, and so on. This chapter, therefore, concentrates on these implementation issues with particular reference to the use of estimators for self-tuning. The focus is on the RLS estimator, on the grounds that it is the most common type, but the arguments are generic and largely applicable to all recursive estimators.

115.1 Setting the Order of the Model

This concerns deciding upon the values of α and β in Equation 114.1 and hence the dimension λ of the vector $\underline{\theta}$ containing the coefficients to be estimated. Without doubt, the best way to decide upon the order of the model is from a first principles model of the process. Thus, if the structure of the plant model G(s) in Figure 114.1 is known, even if its parameters aren't, the values of the order of the numerator and denominator of G(s) are assigned to β and α respectively in the first instance. Remember that G(s) includes the actuators and measurements as well as the process. These initial values may be revised according to the estimator's performance characteristics such as bias and convergence. Typically, if G(s) is a high order model there may be scope for reducing the values of α and β, and hence the amount of computation. For low order models the values of α and β may have to be increased.

If a first principles model is not available, it is essentially a question of making a guess at the values of α and β, hopefully from a position of experience, and revising them upwards or downwards depending on the same performance characteristics.

Also, the value of the time delay parameter h must be set. This is typically established by means of an open loop step test and observing the time delay that exists in the response. The delay is always set to the nearest integer number of sample periods.

115.2 Initialisation of Parameter Values

It should be apparent from Figure 114.2 that the vector $\underline{\theta}$ has to be initialised, i.e. initial values assigned to all the parameters of the model being estimated. Provided the estimator is functioning in isolation, all the elements of $\underline{\theta}(0)$ can be set to zero. However, if the estimator is part of a self-tuning controller, this would lead to zero division through singularity, so setting all the elements of $\underline{\theta}(0)$ to zero is not an option.

For the trivial case, which is nevertheless a very common approximation, of the model being a first order lag with a time delay, an estimate of the parameters may be made from the open loop step response using the construction of the reaction curve method outlined in Chapter 24:

$$G(s) = \frac{K.e^{-Ls}}{Ts+1} \equiv \frac{z^{-h}.b_0}{1+a_1.z^{-1}}$$

where $b_0 = K/T$ and, assuming a sampling period of unity, $h = L$ and $a_1 = -\exp(-1/T)$.

Another approach to obtaining $\underline{\theta}(0)$ is to use batch least squares (BLS) on a set of pre-recorded data values, with the bonus that it produces an estimate of R(0). There is clearly a trade-off between accuracy and efficiency: the larger the data set the more accurate the estimates but the greater the computational effort required because of the matrix inversions. This obviously requires that there exists a BLS program for initialisation as well as the RLS program. Alternatively, if the BLS is incorporated in the RLS program, then a large part of the code becomes redundant after the first pass.

Yet another method is to parameterise the model on-line using RLS whilst the process is either on open loop or, more usually, being controlled independently. This would typically be by means of a conventional PID controller. It is important during this initialisation period that the process is repeatedly disturbed by judicious set point changes so that the data used for parameter estimation is sufficiently rich in information content. This method allows $\underline{\theta}(0)$ to be set to the null vector and, once the estimates have converged, the controller can be switched from fixed parameter into self-tuning mode and/or from manual into automatic mode. Provision must be made for bumpless transfer.

115.3 Initialisation of the Covariance Matrix

Also apparent from Figure 114.2 is the need to initialise the (inverse of the) covariance matrix R. As mentioned above, R(0) can be obtained from a BLS estimate prior to switching to RLS. However, a more popular technique is to simply specify that it be diagonal, i.e. $R(0) = \mu.I$ where μ is a scalar quantity. The coefficient μ is usually chosen to be a positive number because, by definition, the covariance matrix S(k) and hence its inverse R(k) have quadratic diagonal elements which must be positive. From the literature it appears that a choice of $100 \leq \mu \leq 10^6$ is common. This range may at first sight seem to be both high and wide but remember, from Equation 82.5, that the individual covariances are formed from raw data values: these may well be in engineering units without any form of standardisation as is the case, for example, with correlation coefficients.

Inspection of Equation 114.17 reveals that a large value of μ causes the gain vector K(k) of the estimator to be high initially which, from Equation 114.15, has the effect of speeding up the initial search for the probable values of $\underline{\theta}(k)$. A consequence is that the initial estimates of the parameters will jump about a lot. If these are then used as a basis for determining the output of a self-tuning controller, closed loop response will be poor initially and may even appear to be unstable. This can be alleviated by specifying a smaller value for μ and compromising on the speed of adaptation.

The value used for μ reflects the confidence in the initial values. If there is little confidence in $\underline{\theta}(0)$ then a value of $\mu = 1000$ works well. If approximate values of $\underline{\theta}(0)$ are available, say through some step response test or BLS, then values in the range $0 \leq \mu \leq 1$ work well. Remember, it doesn't matter exactly what value of μ is chosen, the issue is simply initialisation of R(k) which subsequently adapts itself as appropriate.

Sometimes it is desirable to hold a certain parameter of $\underline{\theta}(k)$ at its initial value which is known *a priori* to be correct. To accomplish this, the corresponding diagonal element of R(0) is set to zero. This will effectively set the corresponding element in the gain vector K(k) to zero which halts adaptation of that parameter.

115.4 Forgetting Factors

The parameters of many real processes are not always constant. They may change slowly with time due to variations in process characteristics such as, for example, when heat exchanger surfaces become fouled. Or they may change relatively quickly due to non-linearities, such as when a set point change forces a different set of operating conditions. Due to the convergent nature of RLS estimators, the elements of the matrix R(k) decrease in magnitude with time. This in turn reduces the size of the gain vector K(k), particularly after a prolonged period of steady operation. Thus, if a parameter change then occurs, the estimator is unable to track it effectively. The estimator is said to have fallen asleep.

One way of tackling this problem is to use a recursive formulation of the weighted least squares (WLS) estimator. The notion of incorporating a weighting matrix W in the objective function was introduced in Chapter 114. Suppose that W is a diagonal matrix defined as follows:

$$J = \sum_{k=1}^{n} w(k).\varepsilon(k)^2 = \underline{\varepsilon}^T(k).W.\underline{\varepsilon}(k)$$

$$= \begin{bmatrix} \varepsilon(n) & \varepsilon(n-1) & \varepsilon(n-2) & \cdots & \varepsilon(1) \end{bmatrix}$$

$$\times \begin{bmatrix} 1 & 0 & 0 & 0 & 0 \\ 0 & \rho & 0 & 0 & 0 \\ 0 & 0 & \rho^2 & 0 & 0 \\ 0 & 0 & 0 & \ddots & 0 \\ 0 & 0 & 0 & 0 & \rho^{n-1} \end{bmatrix} \begin{bmatrix} \varepsilon(n) \\ \varepsilon(n-1) \\ \varepsilon(n-2) \\ \vdots \\ \varepsilon(1) \end{bmatrix}$$

where:

$$w(k) = \rho^{n-k} \quad \text{and} \quad 0 < \rho < 1.0 \quad (115.1)$$

It is evident that the weightings are applied on a reverse time basis, with the strongest weighting of unity being applied to the most recent error. Thus the older the error the less the significance of its contribution to the objective function. The rate at which older information is discounted is determined by the value of the parameter ρ which is generally known as the forgetting factor, but is also referred to as either the data suppression factor or the memory parameter. The rate at which information is discounted is characterised by the asymptotic sample length (ASL) defined as follows:

$$ASL = \frac{1}{1-\rho}$$

ASL is some memory time constant that relates to the length of the data window within which the contribution of the error values is significant. Values of ρ typically lie in the range of 0.95–0.999 corresponding to data windows of 20–1000 samples respectively. Clearly a value of $\rho = 1$ implies infinite memory.

The recursive least squares algorithm incorporating the forgetting factor is exactly the same as depicted in Figure 114.2 except that Equation 114.18 is modified as follows:

$$d(K) = \rho + \underline{x}^T(k).R(k-1).\underline{x}(k) \quad (115.2)$$

and Equation 114.19 is modified as follows:

$$R(k) = \frac{1}{\rho}.\left(R(k-1) - d(K).K(k).K^T(k)\right) \quad (115.3)$$

It can be seen from Equation 115.3, since $\rho < 1$ in the denominator, that R(k) will be prevented from becoming too small, hence keeping the algorithm "alert".

115.5 Covariance Resetting

The use of a forgetting factor is by far the most popular method of improving the tracking of slowly time-varying parameters. However, there exists an alternative approach to preventing the elements of the (inverse of the) covariance matrix R(k) becoming too small. This is known as either covariance resetting or as random walk and involves modifying Equation 114.19 (but not Equation 114.18) of Figure 114.2 as follows:

$$R(k) = R(k-1) - d(K).K(k).K^T(k) + R \quad (115.4)$$

where R is a diagonal, usually time invariant, matrix with positive diagonal elements and has the same dimensions ($\lambda \times \lambda$) as R(k). The attraction of covariance resetting is that R(k) is increased linearly whereas use of the forgetting factor causes R(k) to grow exponentially. A further attraction of covariance resetting is that adjustment of the appropriate diagonal element of R enables the identifier's tracking ability to be varied more or less independently for each parameter.

115.6 Numerical Instability

The recursive estimator is at the heart of most self-tuning control algorithms, so any numerical instability in the estimator will result in the control loop becoming unstable. This can occur due to round off errors in the estimation and/or to use of a forgetting factor.

Inspection of Equation 114.17 reveals a division problem if d(K) is zero or close to zero, and Equation 114.18 shows that this occurs when:

$$\underline{x}^T(k).R(k-1).\underline{x}(k) \approx -1 \quad (115.5)$$

If R(k) was diagonal with positive elements, this condition would never be satisfied. However, R(k) cannot be guaranteed as such since it is the inverse of the covariance matrix S(k) some of whose off diagonal elements are likely to be negative. Also, for R(k) to exist, there is an implicit requirement for S(k) to be non-singular which, for the same reasons, may not be met. Therefore, there will be circumstances in which the RLS algorithm breaks down. These are most likely to occur when some elements of R are very small and round off error becomes significant, typically when the word length used is short and there are large numbers of parameters to estimate.

One way of overcoming the problem is to reduce the RLS estimation to a method known as stochastic approximation. This is achieved by specifying R(k) to be a matrix of constants that is positive definite, the eigenvalues of which are real and positive. This guarantees that Equation 115.5 cannot be satisfied. Hence there is no need for updating R(k) and, as it will always be positive definite, the stability of the estimator cannot be affected by numerical errors. However, least squares is a second order optimisation technique: fixing R(k) reduces it to a first order search algorithm which has a much slower rate of convergence and requires much better estimates of the initial value $\underline{\theta}(0)$.

Another approach involves factorisation. Instead of updating R(k) by Equation 114.19 at each iteration of the RLS algorithm, a factor of R(k) is updated resulting in R(k) itself being guaranteed positive definite. The so-called upper diagonal factored RLS estimator has very good performance characteristics as well as being computationally efficient.

More practical means of realising RLS stability involve testing for the onset of divergence. When this is revealed, execution of Equation 114.19 is bypassed. Alternatively, the onset of divergence can be used to trigger a covariance resetting as described above.

115.7 Covariance Windup

Instability of estimation can also occur when a relatively small forgetting factor is used. If the estimator is quiescent for a long time, corresponding to small errors due to tight control and the absence of noise, the estimator gain will become small and Equation 115.3 reduces to:

$$R(k) = \frac{1}{\rho}.R(k-1) \quad (115.6)$$

Since $\rho < 1$ the elements of R(k) grow exponentially with time leading to covariance windup. The problem can be tackled at source by use of a dither signal which ensures that the system is persistently excited and the input and output signals are rich in data. The dither is a random signal of zero mean and finite variance that is superimposed on either the input signal during open loop identification or on the set point of a closed loop self-tuning controller.

Through Equations 114.15 and 114.17, it can be seen that covariance windup causes the estimator gain K(k) to increase suddenly which makes the estimator very sensitive to the estimation error. A small change in the error caused, say, by a slight change in the process characteristics, can therefore result in large changes in estimates, a phenomenon known as estimator blow-up.

When these estimates are used to calculate the output of a self-tuning controller, the closed loop response becomes unstable. The controller responds with the result that the estimator, and hence the closed loop, becomes stable after a period of oscillation. Whilst the ability to recover from covariance windup is welcome, the intervening period of instability is generally unacceptable.

A common approach to handling estimator blow-up is to include in the RLS algorithm checks on the magnitude of the diagonal elements of R(k). A typical check is to monitor the sum of the diagonal elements, i.e. trace(R(k)). If this becomes too large, R(k) is reset to some acceptable size. One method is to simply replace R(k) with a diagonal matrix whose elements are all of the same specified positive value:

$$R(k) = \alpha.I$$

Alternatively, individual elements along the diagonal of R(k) can be replaced with smaller positive values as and when appropriate.

115.8 Variable Forgetting Factors

A more elegant approach is to use variable forgetting factors to indirectly influence the size of the (inverse of the) covariance matrix. A small forgetting factor enhances the rate of adaptation. This is required whenever there are large estimation errors, typically at start-up, when a disturbance occurs or when a parameter changes. And a large forgetting factor is required when the estimation error is small to prevent covariance windup.

There are many designs of variable forgetting factor. One of the simplest but most effective is as follows:

$$\rho(k) = \rho_{min} + \frac{1 - \rho_{min}}{1 + f(\varepsilon(k))}$$

Here ρ_{min} is the minimum value of the forgetting factor while $f(\varepsilon(k))$ is some positive function of the estimation error, such as $|\varepsilon(k)|$ or $\varepsilon(k)^2$. If the residual is large, then the forgetting factor tends towards ρ_{min} whereas if the residual is small the forgetting factor tends towards unity.

A more sophisticated approach due to Wellstead uses a composite forgetting factor. The requirement for a low value at start-up is handled by the one forgetting factor:

$$\rho_1(k) = \alpha.\rho_1(k-1) + (1-\alpha).\rho_1(\infty)$$

where $\rho_1(\infty) \leq 1$, $\rho_1(0) \leq \rho_1(\infty)$ and α is a constant set to some value between 0.5 and 1.0. To prevent blow-up $\rho_1(\infty)$ is normally set to a value of 1.0. A second forgetting factor is used to handle the tracking of parameter variations:

$$\rho_2(k) = 1 - \frac{\varepsilon^2(k)}{1 + \varepsilon^2(k)}$$

where again $\varepsilon(k)$ is the estimation error. Fulfilment of both criteria is accomplished by combining the two forgetting factors:

$$\rho(k) = \rho_1(k).\rho_2(k)$$

Other approaches include the switching between two forgetting factors depending upon the value of the estimation error, adjustment of the forgetting factor to maintain a constant value of trace(R(k)), and setting the forgetting factor to unity if checks reveal that the data is not exciting.

115.9 Convergence

The error in the value of any particular prediction of the output y is given by Equation 114-16. However, since identification is recursive, the quality of the prediction process is of more interest than any particular predicted values. The two most commonly used measures of quality are consistency and bias.

Consistency is based upon the residuals, *i.e.* on the differences between the estimates of the coefficients of $\underline{\theta}(k)$ and their true values. As time progresses, assuming the identification process is successful, the errors should reduce. Ideally the errors should asymptotically approach zero but seldom do so because of upsets due to changes in value of the coefficients. There are two commonly used means of articulating consistency. The simpler, which is appropriate for use with the BLS estimator, defines consistency in terms of the expected value of the vector product for the errors in $\underline{\theta}$. Thus, as $k \to \infty$:

$$E\left\{(\hat{\underline{\theta}} - \underline{\theta})^T . (\hat{\underline{\theta}} - \underline{\theta})\right\} = E\left\{\sum_{i=1}^{\lambda}(\hat{\theta}_i - \theta_i)^2\right\}$$

$$= \sum_{i=1}^{\lambda} E\left\{(\hat{\theta}_i - \theta_i)^2\right\} \to 0$$

The problem with this measure of consistency is that there is no memory of convergence which makes it inappropriate for use on a recursive basis. For RLS estimators a more sophisticated measure of consistency is required.

From Equations 114.6 and 114.8:

$$\hat{\underline{\theta}} - \underline{\theta} = (X^T.X)^{-1}.X^T.\underline{y} - \underline{\theta}$$
$$= (X^T.X)^{-1}.X^T\left[X.\underline{\theta} + \underline{\varepsilon}\right] - \underline{\theta}$$
$$= (X^T.X)^{-1}.(X^T.X).\underline{\theta}$$
$$\quad + (X^T.X)^{-1}.X^T.\underline{\varepsilon} - \underline{\theta}$$
$$= (X^T.X)^{-1}.X^T.\underline{\varepsilon}$$

Next consider the quadratic in the error on $\underline{\theta}$:

$$(\hat{\underline{\theta}} - \underline{\theta})(\hat{\underline{\theta}} - \underline{\theta})^T$$
$$= ((X^T.X)^{-1}.X^T.\underline{\varepsilon}) . ((X^T.X)^{-1}.X^T.\underline{\varepsilon})^T$$
$$= (X^T.X)^{-1}.X^T.\underline{\varepsilon}.\underline{\varepsilon}^T.X. ((X^T.X)^{-1})^T$$
$$= (X^T.X)^{-1}.X^T.\underline{\varepsilon}.\underline{\varepsilon}^T.X.(X^T.X)^{-1}$$

Of particular interest is the expected value of the embedded quadratic in $\underline{\varepsilon}$. Assuming that the error on each coefficient is random, *i.e.* is not correlated with X, then:

$$E\{\underline{\varepsilon}.\underline{\varepsilon}^T\} = E\begin{Bmatrix} \varepsilon_1^2 & \varepsilon_1\varepsilon_2 & \cdots & \varepsilon_1\varepsilon_n \\ \varepsilon_1\varepsilon_2 & \varepsilon_2^2 & \cdots & \varepsilon_2\varepsilon_n \\ \vdots & \vdots & \ddots & \vdots \\ \varepsilon_1\varepsilon_n & \varepsilon_2\varepsilon_n & \cdots & \varepsilon_n^2 \end{Bmatrix}$$

$$= \begin{bmatrix} \sigma^2 & 0 & \cdots & 0 \\ 0 & \sigma^2 & \cdots & 0 \\ \vdots & \vdots & \ddots & \vdots \\ 0 & 0 & \cdots & \sigma^2 \end{bmatrix} = \sigma^2.I$$

where σ^2 is the variance of the error on the prediction of the output. Since X may be considered as a constant matrix of measured values at any sample instant, it follows that:

$$E\left\{(\hat{\underline{\theta}} - \underline{\theta}) . (\hat{\underline{\theta}} - \underline{\theta})^T\right\} \qquad (115.7)$$
$$= E\left\{(X^T.X)^{-1}.X^T.\sigma^2 I.X.(X^T.X)^{-1}\right\}$$
$$= \sigma^2.(X^T.X)^{-1} = \sigma^2.R$$

This too is inherently diagonal. Clearly, as time progresses and the predictions improve, the diagonal elements must reduce. The sum of the diagonal elements, its trace, should converge on zero. Thus, for consistency, as $k \to \infty$:

$$traceE\left\{(\hat{\underline{\theta}} - \underline{\theta})(\hat{\underline{\theta}} - \underline{\theta})^T\right\} = \sigma^2.trace(R) \to 0$$

This is relatively easy to determine because R(k) is updated at each iteration of the identification process. The true significance of this consistency criterion is that, being based upon R(k), the history of the convergence of the estimates is retained despite the recursive nature of RLS.

Whereas consistency is a measure of convergence, bias is a measure of the magnitude of the residual:

$$bias = E\left\{\hat{\underline{\theta}}\right\} - \underline{\theta}$$

The estimates are said to be unbiased if the following criterion is satisfied as k → ∞:

$$E\left\{\hat{\underline{\theta}}\right\} - \underline{\theta} \to 0$$

Substituting from Equations 114.6 and 114.8 gives:

$$\begin{aligned} E\left\{\hat{\underline{\theta}}\right\} - \underline{\theta} &= E\left\{(X^T.X)^{-1}.X^T\left[X.\underline{\theta} + \underline{\varepsilon}\right]\right\} - \underline{\theta} \\ &= X^{-1}.E\left\{\underline{\varepsilon}\right\} \end{aligned}$$

Provided that $\underline{\varepsilon}$ is not correlated with X, i.e. the stochastic term has a zero mean, the least squares methods will always give unbiased estimates. However, in the context of control systems as depicted in Figure 114.3, the process input is not independent of prediction error $\underline{\varepsilon}$ and correlates with X. Use of RLS within a self-tuning controller results in biased estimates, the bias being given by:

$$E\left\{\hat{\underline{\theta}}\right\} - \underline{\theta} = E\left\{X^{-1}.\underline{\varepsilon}\right\} \qquad (115.8)$$

which will not be equal to zero even if $\underline{\varepsilon}$ has a zero mean. As discussed in Chapter 114, the techniques of instrumental variables, GLS and ELS all include measures to ensure uncorrelated residuals to prevent this bias.

115.10 Comments

It is evident that there are many issues surrounding the use of RLS estimators. However, provided a sensible order model is chosen, the estimator is properly initialised and some form of variable forgetting factor is used to counter the effects of windup, RLS provides a good stable framework for both identification and self tuning applications. It is interesting to note that whilst stability and good convergence of the estimator are essential for its effective operation, the accuracy of the estimates is not in itself critical for self-tuning purposes. Any self-tuning controller should be capable of tolerating moderate residuals. What does matter is that the closed loop error of the regulator is small.

Self Tuning Control

Chapter 116

116.1 Explicit and Implicit STC
116.2 Notation
116.3 Minimum Variance Control
116.4 Implementation of MV Control
116.5 Properties of MV Control
116.6 Generalised Minimum Variance Control
116.7 Implementation of GMV Control
116.8 Properties of GMV Control
116.9 Comments
116.10 Nomenclature

Self tuning control (STC) encompasses any type of controller in which the controller's parameters are automatically adjusted, somehow, whilst controlling a process/plant that is subject to change. Self tuning may be of a tune-on-demand nature or recursive. In the tune-on-demand form, the adjustment mechanism handles the initial tuning of the controller, after which the mechanism is disabled until re-tuning is required. In the recursive form, the adjustment mechanism is executed on a regular basis, up dating the controller's parameters according to the process/plant changes.

Proprietary self-tuners, often referred to as auto-tuners, adopt an empirical approach and are usually used in tune-on-demand mode. Typically a disturbance is introduced, the pattern of response is observed from which the process characteristics are determined, and conventional PID controller settings are adjusted accordingly. There are many variations on the theme:

- The tests may be open loop or closed loop.
- The loop may include a conventional controller and/or a relay.
- The disturbance may be a step change or random, such as a pseudo random binary signal (PRBS).
- The process characteristics may be articulated either in terms of rise time, overshoot, *etc.* or else as parameters such as time constants and gains of some pre-specified model to which the data is best fitted by means of least squares.
- PID settings may be adjusted singly or in combination, using Cohen and Coon type formulae, depending upon whether regulo or servo control is the objective.

This chapter outlines the essential differences between the two principal classes of STC, explicit and implicit. Then, two particular types of implicit STC are considered in detail. First the minimum variance (MV) type is described. This provides a basis for describing the generalised minimum variance (GMV) type which is probably the most common type of STC used in practice.

STC is an area which has attracted, and continues to attract, extensive interest from academia and the literature is littered with references to leading figures such as Astrom, Clarke and Grimble. There are many texts too which range from the difficult to understand to the impossible. For a thorough treatment of the subject the reader is referred in particular to the text by Wellstead (1991).

116.1 Explicit and Implicit STC

Any linear model based controller can be made self-tuning given a suitable on-line parameter estimation technique. Figure 116.1 depicts the structure of an explicit STC which is an intuitive approach to STC design.

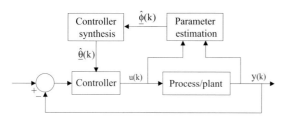

Fig. 116.1 Structure of an explicit self tuning controller

There is a two stage identification and synthesis process. First, the parameters of the model of the process are estimated explicitly from its input and output signals. Second, these are used to calculate the parameters of the controller. The controller output is then evaluated according to some pre-defined control law, typically a pole placement compensator or a state feedback regulator. This structure is also referred to as the indirect approach to STC because the controller's parameters are based upon estimates of the parameters of the process model.

Underlying this approach are the separation (of identification and synthesis) principles and certainty equivalence. The certainty equivalence principle assumes that the estimated parameters of the process are accurate representations of their true values. In other words, the controller parameters are calculated on the basis that the process parameters are both known and correct. This is a strong assumption that is seldom valid due to model inaccuracy such as biased estimates due to correlated noise. So, in practice, the weaker separation principle applies which takes into account the uncertainty of the estimates of the process parameters in calculating the controller parameters.

Figure 116.2 depicts the structure of an implicit STC. The identification and synthesis are integrated. Also known as the direct approach, this type of STC estimates the controller parameters directly from the process' input and output signals. The process parameters are implicitly embedded within the controller parameters.

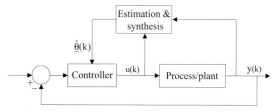

Fig. 116.2 Structure of an implicit self tuning controller

The implicit STC is computationally more efficient than the explicit STC. However, it is less flexible in the sense that the choice of control law is fixed by the STC design: the process parameters are not available to enable a choice of control law.

116.2 Notation

Throughout this text the letter k has been used to denote the current sampling instant. To avoid ambiguity the letter h is used to denote a time advance of h sampling periods in the future. This is the same h as was used to denote process time delay in Equation 114.1 and throughout the rest of Chapter 114. Other texts use the letter k to denote the time advance, hence the common phrase of k-step ahead prediction.

The hat (^) is used to denote an estimate of a future value. Other texts use star (*) notation to denote predicted values but in this text the * denotes the Laplace transform of a sampled data signal, as defined by Equation 75.3. Thus the following expression denotes the h-steps ahead estimate of the value of y on the basis of information available up to (and including) the current sampling instant k:

$$\hat{y}(k+h|k)$$

Also note the non-standard notation used to distinguish between the z transforms of a predicted value:

$$Z\{\hat{y}(k+h)\} = z^h \cdot \hat{y}(z)$$
$$Z\{\hat{y}(k+h|k)\} = z^h \cdot \hat{y}(z)\big|_0$$

The former is the transform of a variable when its values are known h-steps ahead of the current sampling instant whereas the latter implies that information is only available up to the current sampling instant (0 steps ahead).

A fairly common practice in the literature on self-tuning control is to articulate relationships between variables in time series form rather than as z transforms. For convenience these are often expressed as a product, an example of which is as follows:

$$F(k).y(k) = f_0.y(k) + f_1.y(k-1) + \cdots$$
$$+ f_\phi.y(k-\phi)$$
$$\equiv F(z).y(z)$$

where F(k) is some polynomial which operates upon the time series signal y(k), analogous to the equivalent operation in the z domain.

116.3 Minimum Variance Control

The minimum variance (MV) controller is of the implicit type of STC. It generates a control signal u(k) to minimise the following objective function:

$$J = E\left([r(k) - \hat{y}(k+h|k)]^2\right) \quad (116.1)$$

This operand can be thought of as the square of the controller error, in which the error is the difference between the current set point r(k) and the predicted output y(k) some h-steps ahead. The expectation of the square of the error signal corresponds to its variance, the objective being to minimise that variance, hence the term minimum variance control.

To enable minimisation of Equation 116.1 a relationship is required between the output y(k) and the input u(k), both presumed to be in deviation form. Suppose that this is available in the form of an ARMAX model as defined by Equation 114.32:

$$A(z).y(z) = z^{-h}.B(z).u(z) + C(z).\varepsilon(z) \quad (116.2)$$

in which A(z), B(z) and C(z) are polynomials as defined by Equations 114.1 and 114.31:

$$A(z) = 1 + a_1.z^{-1} + a_2.z^{-2} + \ldots + a_\alpha.z^{-\alpha}$$
$$B(z) = b_0 + b_1.z^{-1} + b_2.z^{-2} + \ldots + b_\beta.z^{-\beta}$$
$$C(z) = 1 + c_1.z^{-1} + c_2.z^{-2} + \ldots + c_\gamma.z^{-\gamma}$$
$$(116.3)$$

The random series of residuals $\varepsilon(z)$, of zero mean and finite variance, resulting from fitting the plant data to the model at the current sampling instant, is as defined by Equation 114.3:

$$y(z) = \hat{y}(z) + \varepsilon(z) \quad (116.4)$$

Minimisation of Equation 116.1 also requires an h-step ahead prediction of the output. Let E(z) be some monic polynomial that reconciles the true value of the output h-steps ahead with its current prediction in terms of future residuals:

$$z^h.y(z) = z^h.\hat{y}(z)\big|_0 + z^h.E(z).\varepsilon(z) \quad (116.5)$$

where:

$$E(z) = 1 + e_1 z^{-1} + e_2 z^{-2} + \cdots + e_{h-1} z^{-(h-1)}.$$

Inverse transforming back into the time domain gives:

$$y(k+h) = \hat{y}(k+h|k) + \varepsilon(k+h)$$
$$+ e_1.\varepsilon(k+h-1)$$
$$+ e_2.\varepsilon(k+h-2) + \cdots$$
$$+ e_{h-1}.\varepsilon(k+1)$$

which may be articulated as follows:

$$\hat{y}(k+h|k) = y(k+h) - E(k).\varepsilon(k+h) \quad (116.6)$$

Next presume that there exists another polynomial function F(z) such that the following separation identity is valid:

$$C(z) = A(z)E(z) + z^{-h}.F(z) \quad (116.7)$$

where:

$$F(z) = f_0 + f_1 z^{-1} + f_2 z^{-2} + \cdots + f_{\alpha-1} z^{-(\alpha-1)}.$$

An equation of the following general form is known as a Diophantine identity, the significance of which is that for any given polynomials $A(z)$, $B(z)$ and $C(z)$ there will always be a unique pair of polynomials $X(z)$ and $Y(z)$ that satisfy the identity:

$$A(z).X(z) + B(z).Y(z) = C(z)$$

Thus, in the case of Equation 116.7, for known polynomials $A(z)$ and $C(z)$ and a given delay h, there will exist a pair of polynomials $E(z)$ and $F(z)$ which satisfy the equation. The coefficients of $E(z)$ and $F(z)$ are found by expanding the equation, equating terms in like exponents and solving for the unknowns. Thus multiplying Equation 116.2 throughout by $E(z)$ and substituting from Equation 116.7 gives:

$$\left(C(z) - z^{-h}.F(z)\right).y(z) = z^{-h}.B(z)E(z).u(z) + C(z)E(z).\varepsilon(z)$$

Taking out the time advance as a factor and rearranging gives:

$$C(z).\left(z^h.y(z) - z^h.E(z).\varepsilon(z)\right) = +B(z)E(z).u(z) + F(z).y(z)$$

Substituting from Equation 116.5 gives:

$$C(z).z^h.\left.\hat{y}(z)\right|_0 = B(z)E(z).u(z) + F(z).y(z) \quad (116.8)$$

Notice the structure of this equation: the left hand side is the future prediction of the output whereas the right hand side comprises previous and current values only of the inputs and outputs.

Now define $D(z)$ in such a way that enables $\hat{y}(k+h|k)$ to be isolated:

$$C(z) = 1 - z^{-1}.D(z) \quad (116.9)$$

where:

$$D(z) = d_0 + d_1 z^{-1} + d_2 z^{-2} + \cdots + d_{\gamma-1} z^{-(\gamma-1)}.$$

Substitution into Equation 116.8 and rearrangement gives:

$$z^h.\left.\hat{y}(z)\right|_0 = B(z)E(z).u(z) + F(z).y(z) + z^{h-1}.D(z).\left.\hat{y}(z)\right|_0$$

One final simplification is made. For estimation purposes $B(z), D(z), E(z)$ and $F(z)$ are all unknown. Since $B(z)$ and $E(z)$ occur as a product, they may as well be combined into a single polynomial. Noting that $B(z)$ is not monic:

$$\begin{aligned} G(z) &= B(z).E(z) \\ &= g_0 + g_1 z^{-1} + g_2 z^{-2} + \cdots \\ &\quad + g_{(\beta+h-1)} z^{-(\beta+h-1)} \end{aligned} \quad (116.10)$$

whence:

$$z^h.\left.\hat{y}(z)\right|_0 = G(z).u(z) + F(z).y(z) + z^{h-1}.D(z).\left.\hat{y}(z)\right|_0$$

Inverse transformation yields:

$$\begin{aligned} \hat{y}(k+h|k) &= g_0.u(k) + g_1.u(k-1) \\ &\quad + g_2.u(k-2) + \cdots \\ &\quad + g_{(\beta+h-1)}.u(k-\beta-h+1) \\ &\quad + f_0.y(k) + f_1.y(k-1) \\ &\quad + f_2.y(k-2) + \cdots \\ &\quad + f_{\alpha-1}.y(k-\alpha+1) \\ &\quad + d_0.\hat{y}(k+h-1) \\ &\quad + d_1.\hat{y}(k+h-2) \\ &\quad + d_2.\hat{y}(k+h-3) + \cdots \\ &\quad + d_{\gamma-1}.\hat{y}(k+h-\gamma) \end{aligned}$$

which is equivalent to:

$$\begin{aligned} \hat{y}(k+h|k) &= G(k).u(k) + F(k).y(k) \\ &\quad + D(k).\hat{y}(k+h-1|k) \end{aligned} \quad (116.11)$$

Substituting into the objective function of Equation 116.1 gives:

$$J = E\left(\left[r(k) - \left(G(k).u(k) + F(k).y(k) + D(k).\hat{y}(k+h-1|k)\right)\right]^2\right)$$

Noting that the differential of the time series function $G(k).u(k)$ with respect to $u(k)$ is:

$$\frac{d}{du(k)}\left(g_0.u(k) + g_1.u(k-1) + \cdots + g_{(\beta+h-1)}.u(k-\beta-h+1)\right) = g_0$$

and that the minimum of the expected value of the objective function is the same thing as its true minimum, the control signal u(k) required to minimise the function is found by differentiation:

$$\frac{\partial J}{\partial u(k)} = -2g_0\Big(r(k) - \big(G(k).u(k) + F(k).y(k) + D(k).\hat{y}(k+h-1|k)\big)\Big)$$

Back substituting from Equation 116.11 gives:

$$\frac{\partial J}{\partial u(k)} = -2g_0\left(r(k) - \hat{y}(k+h|k)\right)$$

Clearly the condition that has to be satisfied for a minimum is:

$$r(k) - \hat{y}(k+h|k) = 0$$

which is intuitively obvious by inspection of Equation 116.1. Thus the control signal required to realise the objective function comes from:

$$r(k) - \Big(G(k).u(k) + F(k).y(k) + D(k).\hat{y}(k+h-1|k)\Big) = 0 \quad (116.12)$$

116.4 Implementation of MV Control

The MV type of STC control can be realised by means of the RLS algorithm of Chapter 114. The data and parameter vectors arise from shifting Equation 116.11 backwards by h steps:

$$\hat{y}(k) = G(k).u(k-h) + F(k).y(k-h) + D(k).\hat{y}(k-1)$$

Substituting the estimate of y(k) into Equation 116.4 gives:

$$y(k) = G(k).u(k-h) + F(k).y(k-h) + D(k).\hat{y}(k-1) + \varepsilon(k)$$

which is of the form of Equation 114.4:

$$y(k) = \underline{x}^T(k-h).\underline{\theta} + \varepsilon(k) \quad (116.13)$$

where $\underline{x}(k-h)$ and $\underline{\theta}$ are both $(\sigma \times 1)$ vectors with $\sigma = \alpha + \beta + \gamma + h$ and:

$$\underline{x}^T(k-h) = \begin{bmatrix} u_{k-h} & u_{k-h-1} & \cdots & u_{k-2h-\beta+1} \\ y_{k-h} & y_{k-h-1} & \cdots & y_{k-h-\alpha+1} \\ \hat{y}_{k-1} & \hat{y}_{k-2} & \cdots & \hat{y}_{k-\gamma} \end{bmatrix}$$

$$\underline{\theta}^T = \begin{bmatrix} g_0 & g_1 & \cdots & g_{\beta+h-1} & f_0 & f_1 & \cdots & f_{\alpha-1} \\ d_0 & d_1 & \cdots & d_{\gamma-1} \end{bmatrix} \quad (116.14)$$

Note that the residual in Equation 116.13 is uncorrelated with the data in the regression since it is a future value with respect to the time indices of the input and output signals. The recursive form of estimation is exactly the same as for the RLS algorithm of Figure 114.2 with the exception that the residual of Equation 114.16 is replaced by:

$$\varepsilon(k) = y(k) - \underline{x}^T(k-h).\hat{\underline{\theta}}(k-1) \quad (116.15)$$

Implementation issues are as described in Chapter 115. Once the estimate of the parameters $\hat{\underline{\theta}}(k)$ becomes available, the control signal is evaluated from an expansion and rearrangement of Equation 116.12:

$$u(k) = \frac{1}{g_0}\Bigg(r(k) - \sum_{j=1}^{\beta+h-1} g_j.u(k-j) - \sum_{j=0}^{\alpha-1} f_j.y(k-j) - \sum_{j=0}^{\gamma-1} d_j.\hat{y}(k+h-1-j)\Bigg) \quad (116.16)$$

The combination of estimation by means of RLS and control by means of Equation 116.16, if carried out at each sampling instant, is a self-tuning minimum variance strategy.

116.5 Properties of MV Control

Comparison of Equations 116.11 and 116.12 reveals that:

$$\hat{y}(k+h|k) = r(k) \quad (116.17)$$

Thus the control signal u(k) of Equation 116.16 will drive the controlled variable towards its set point.

Substituting for $\hat{y}(k+h|k)$ from Equation 116.17 into Equation 116.6 gives:

$$r(k) = y(k+h) - E(k).\varepsilon(k+h)$$

Shifting k-steps backwards and rearranging yields:

$$y(k) = r(k-h) + E(k).\varepsilon(k)$$
$$y(z) = z^{-h}.r(z) + E(z).\varepsilon(z) \quad (116.18)$$

This represents the closed loop performance of the MV controller. Its interpretation is that the controlled variable follows the set point exactly subject to a time delay of h steps and a weighted noise signal. MV control inherently eliminates offset due to disturbances, other than the noise component which is of zero mean and finite variance.

However, because the output attempts to follow the input exactly, the response is of a deadbeat nature. This causes excessive control effort and is generally unacceptable for operational reasons.

116.6 Generalised Minimum Variance Control

To get round the deadbeat problem, the objective function of Equation 116.1 is modified:

$$J = E\left([R(k).r(k) - P(k).\hat{y}(k+h|k)]^2 + [Q(k).u(k)]^2 \right) \quad (116.19)$$

where the filters on the process output y(z), input u(z) and set point r(z) are the pulse transfer functions P(z), Q(z) and R(z) respectively: It may be presumed that, in general:

$$P(z) = \frac{n(P(z))}{d(P(z))}$$

where:

$$d(P(z)) = 1 + p_1.z^{-1} + p_2.z^{-2} + \cdots + p_\pi.z^{-\pi}$$
$$Q(z) = q_0 + q_1.z^{-1} + q_2.z^{-2} + \cdots + q_\omega.z^{-\omega}$$
$$R(z) = 1 + r_1.z^{-1} + r_2.z^{-2} + \cdots + r_\rho.z^{-\rho}$$
$$(116.20)$$

Development of the GMV controller then follows along the same lines as for the MV controller. Thus starting with the plant model of Equation 116.2:

$$A(z).y(z) = z^{-h}.B(z).u(z) + C(z).\varepsilon(z) \quad (116.2)$$

and multiplying throughout by E(z) as defined by Equation 116.5 yields:

$$A(z)E(z).y(z) = z^{-h}.B(z)E(z).u(z) + C(z)E(z).\varepsilon(z)$$

A separation identity, different to that of Equation 116.7, is introduced:

$$P(z)C(z) = A(z)E(z) + z^{-h}.\frac{F(z)}{d(P(z))} \quad (116.21)$$

which may be substituted as follows:

$$\left(P(z)C(z) - z^{-h}.\frac{F(z)}{d(P(z))} \right).y(z)$$
$$= z^{-h}.B(z)E(z).u(z) + C(z)E(z).\varepsilon(z)$$

Taking out the time advance as a factor and rearranging gives

$$C(z).\left(z^h.P(z).y(z) - z^h.E(z).\varepsilon(z) \right)$$
$$= B(z)E(z).u(z) + \frac{F(z)}{d(P(z))}.y(z)$$

This may be simplified by introducing similarity transformations:

$$C(z).\left(z^h.\tilde{y}(z) - z^h.E(z).\varepsilon(z) \right)$$
$$= B(z)E(z).u(z) + F(z).\tilde{\tilde{y}}(z) \quad (116.22)$$

where $\tilde{y}(z) = P(z).y(z)$ and $\tilde{\tilde{y}} = \frac{y(z)}{d(P(z))}$ are both filtered values of the output.

A modification to Equation 116.5 is next introduced:

$$z^h . \tilde{y}(z) = z^h . \hat{\tilde{y}}(z)\Big|_0 + z^h . E(z) . \varepsilon(z) \quad (116.23)$$

Substituting into Equation 116.22 gives:

$$C(z) . z^h . \hat{\tilde{y}}(z)\Big|_0 = B(z)E(z).u(z) + F(z).\tilde{y}(z)$$

Again, use Equations 116.9 and 116.10 to isolate $\hat{\tilde{y}}(k+h|k)$ and substitute for $B(z)E(z)$:

$$C(z) = 1 - z^{-1} . D(z) \quad (116.9)$$

$$z^h . \hat{\tilde{y}}(z)\Big|_0 = G(z).u(z) + F(z).\tilde{y}(z) + z^{h-1} . D(z) . \hat{\tilde{y}}(z)\Big|_0$$

the inverse transform of which yields:

$$\hat{\tilde{y}}(k+h|k) = G(k).u(k) + F(k).\tilde{y}(k) + D(k).\hat{\tilde{y}}(k+h-1|k-1) \quad (116.24)$$

Substituting into the objective function of Equation 116.19 gives:

$$J = E\Bigg(\Big[R(k).r(k) - \big(G(k).u(k) + F(k).\tilde{y}(k) + D(k).\hat{\tilde{y}}(k+h-1|k-1)\big)\Big]^2 + [Q(k).u(k)]^2\Bigg)$$

Noting, as before, that the differential of the time series function $G(k).u(k)$ with respect to $u(k)$ is g_0, and remembering that the minimum of the expected value of the objective function is the same thing as its true minimum:

$$\frac{\partial J}{\partial u(k)} = -2g_0 . \Big[R(k).r(k) - \big(G(k).u(k) + F(k).\tilde{y}(k) + D(k).\hat{\tilde{y}}(k+h-1|k-1)\big)\Big] + 2q_0 . (Q(k).u(k))$$

Back substituting from Equation 116.24 yields:

$$\frac{\partial J}{\partial u(k)} = -2.g_0 . \Big(R(k).r(k) - \hat{\tilde{y}}(k+h|k)\Big) + 2q_0 . Q(k).u(k)$$

The condition that has to be satisfied for a minimum is thus:

$$\hat{\tilde{y}}(k+h|k) - R(k).r(k) + \frac{q_0}{g_0} . Q(k).u(k) = 0 \quad (116.25)$$

116.7 Implementation of GMV Control

GMV control can be realised by means of the RLS algorithm of Chapter 114. The data and parameter vectors arise from shifting Equation 116.24 backwards by h steps:

$$\hat{\tilde{y}}(k) = G(k).u(k-h) + F(k).\tilde{y}(k-h) + D(k).\hat{\tilde{y}}(k-1)$$

Substituting the estimate of $\tilde{y}(k)$ into Equation 116.4 gives:

$$\tilde{y}(k) = G(k).u(k-h) + F(k).\tilde{y}(k-h) + D(k).\hat{\tilde{y}}(k-1) + \varepsilon(k)$$

which is of the form of Equation 114.4:

$$\tilde{y}(k) = \underline{x}^T(k-h).\underline{\theta} + \varepsilon(k) \quad (116.26)$$

where $\underline{x}(k-h)$ and $\underline{\theta}$ are both $(\sigma \times 1)$ vectors with $\sigma = \alpha + \beta + \gamma + h$ and:

$$\underline{x}^T(k-h) = \big[\, u_{k-h} \; u_{k-h-1} \; \cdots \; u_{k-2h-\beta+1} \\ \tilde{y}_{k-h} \; \tilde{y}_{k-h-1} \; \cdots \; \tilde{y}_{k-h-\alpha+1} \\ \hat{\tilde{y}}_{k-1} \; \hat{\tilde{y}}_{k-2} \; \cdots \; \hat{\tilde{y}}_{k-\gamma} \,\big]$$

$$\underline{\theta}^T = \big[\, g_0 \; g_1 \; \cdots \; g_{\beta+h-1} \; f_0 \; f_1 \; \cdots \; f_{\alpha-1} \\ d_0 \; d_1 \; \cdots \; d_{\gamma-1} \,\big] \quad (116.27)$$

As with MV control, the residual in Equation 116.26 is uncorrelated with the data in the regression since it is a future value with respect to the time indices of the input and output signals. The recursive form of estimation is exactly the same as for the RLS algorithm of Figure 114.2 with the exception that the residual of Equation 114.16 is replaced by:

$$\varepsilon(k) = \tilde{y}(k) - \underline{x}^T(k-h).\hat{\underline{\theta}}(k-1) \qquad (116.28)$$

Implementation issues are as described in Chapter 115. Once the estimate of the parameters $\hat{\underline{\theta}}(k)$ becomes available, the control signal is evaluated by substituting from Equation 116.24 into Equation 116.25:

$$G(k).u(k) + F(k).\tilde{y}(k)$$
$$+ D(k).\hat{\tilde{y}}(k+h-1|k-1)$$
$$- R(k).r(k) + \frac{q_0}{g_0}.Q(k).u(k) = 0$$

Solving for $u(k)$ gives:

$$u(k) = \frac{g_0}{(g_0^2 + q_0^2)} \left(R(k).r(k) - \sum_{j=1}^{\beta+h-1} g_j.u(k-j) \right.$$
$$- \frac{q_0}{g_0}.\sum_{j=1}^{\omega} q_j.u(k-j) - \sum_{j=0}^{\alpha-1} f_j.\tilde{y}(k-j)$$
$$\left. - \sum_{j=0}^{\gamma-1} d_j.\hat{\tilde{y}}(k+h-1-j) \right) \qquad (116.29)$$

The combination of estimation by means of RLS and control by means of Equation 116.29, if carried out at each sampling instant, is a self-tuning generalised minimum variance strategy.

116.8 Properties of GMV Control

As its name implies, GMV can be applied in almost any circumstances, subject to a sensible choice of $P(z)$, $Q(z)$ and $R(z)$ in Equation 116.19.

Clearly setting $P(z) = 1$, $Q(z) = 0$ and $R(z) = 1$ reduces GMV control to MV control with its deadbeat type of response.

Remember from Equation 116.22 that:

$$\tilde{y}(z) = P(z).y(z).$$

Thus setting $P(z) = 1$, $Q(z) \neq 0$ and $R(z) = 1$ in Equation 116.25 results in:

$$\hat{y}(k+h|k) - r(k) + \frac{q_0}{g_0}.Q(k).u(k) = 0$$
$$z^h.\hat{y}(z)\big|_0 - r(z) + \frac{q_0}{g_0}.Q(z).u(z) = 0 \qquad (116.30)$$

which is depicted in block diagram in Figure 116.3.

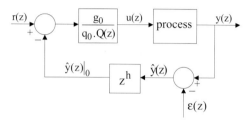

Fig. 116.3 Reduced form of generalised minimum variance (GMV) control

Thus $1/Q(z)$ can be regarded as a controller operating on a predicted feedback signal, somewhat analogous to the Smith predictor strategy of Chapter 86.

Setting $P(z) \neq 1$, $Q(z) = 0$ and $R(z) \neq 0$, and manipulating Equations 116.23 and 116.25, leads to the closed loop response:

$$P(k).y(k) = R(k).r(k-h) + E(k).\varepsilon(k)$$
$$y(z) = z^{-h}.\frac{R(z)}{P(z)}.r(z) + \frac{E(z)}{P(z)}.\varepsilon(z)$$
$$(116.31)$$

Hence the output follows the set point, subject to the time delay, the form of response being determined by the ratio of $R(z)$ to $P(z)$, with the ratio of $E(z)$ to $P(z)$ determining the extent to which the noise is filtered.

For most process control purposes it is normal to make $P(z)$, $Q(z)$ and $R(z)$ as simple as possible, the constraints being that:

- $P(z)$ is chosen such that the ratio of $E(z)$ to $P(z)$ is sluggish enough to filter out noise effects (disturbances).

- Q(z) is normally a simple first order lag consistent with the dynamics of the actuators.
- R(z) is chosen such that the ratio of R(z) to P(z) is realisable and consistent with the closed loop dynamics of the process/plant. Note that R(z) is not necessarily unity even though process control systems are normally of a regulo nature, *i.e.* constant set point.

Appropriate choice of R(z) and P(z) yields the desired closed loop model, for which reason the approach is often referred to as a model following or model reference control strategy.

116.9 Comments

This chapter has concentrated on minimum variance methods. The two methods considered, MV and GMV, are both well understood and accepted. There are other methods not covered here, such as the so-called linear quadratic Gaussian (LQG) method and variations thereof. All of these minimum variance methods are commonly referred to as "optimal" methods of self-tuning in the sense that they are based upon the minimisation of a quadratic performance index. Most of the other methods of self-tuning are based upon pole placement, *i.e.* the control laws are derived by attempting to place the closed loop poles at arbitrary positions. These other methods are commonly referred to as being "sub-optimal" in that they are not designed to minimise any cost function.

116.10 Nomenclature

Symbol	Description
ε	residual: error in prediction of output y
r	controller set point
u	input variable
y	output variable
$\hat{y}$	estimate of y
$\bar{y}$	output y filtered by P
$\tilde{y}$	output y divided by d(P)

Symbol	Description	Dimensions
$\underline{x}$	vector of input and output values	$\sigma \times 1$
$\underline{\theta}$	vector of model coefficients	$\sigma \times 1$
h	time delay exponent in process model	
k	current sampling instant	
σ	dimension of value $(\alpha + \beta + \gamma + h)$	

Symbol	Description
J	objective function

Symbol	Description	Order
A	denominator monic polynomial of process model	α
B	numerator polynomial of process model	β
C	monic polynomial relates residuals to process noise	γ
D	polynomial used for isolating current output	$\gamma - 1$
E	monic polynomial relates residuals to predictions	$h - 1$
F	polynomial used for separation purposes	$\alpha - 1$
G	polynomial product of B and E	$\beta + h - 1$
P	filter on output y in objective function for GMV	
d(P)	monic polynomial denominator of P	π
Q	polynomial filter on output in objective fn for GMV	ω
R	polynomial filter on set point in objective fn for GMV	ρ

Model Predictive Control

117.1 Prediction and Control Horizons
117.2 The Basis of Forward Prediction
117.3 Prediction of the Output
117.4 Controller Output Sequence
117.5 Worked Example
117.6 Recursive Implementation
117.7 QP Solution of MPC Problem
117.8 Extension to Multivariable Systems
117.9 Models for MPC
117.10 Proprietary Packages
117.11 Nomenclature

Model based predictive control, usually referred to simply as model predictive control (MPC), can be thought of as a philosophy with a number of different approaches to its implementation. It is the most sophisticated form of advanced process control that has been realised to any significant extent in practice. That is for three reasons. First, because MPC can handle difficult to control problems such as time delay, non-minimum phase, non-linearity, open loop instability, constraints and combinations thereof. Also, it will handle interactions of a multivariable nature. Second, despite its complexity, implementation of MPC is fairly straightforward because there are good quality robust packages available. The main design parameters have physical meaning and are relatively easy to understand in a process context. Third, because there are large benefits to be derived. According to various surveys, these are potentially some 2–6% of a plant's total operating costs.

It is primarily the proven ability of MPC to reduce variance in controlled variables that realises these benefits. This enables processes to be operated closer to economic and physical constraints than would otherwise be the case. That is particularly so when MPC is used in conjunction with a real-time optimiser in a multivariable context. A typical scenario is that a steady state optimiser (SSO) intermittently downloads set points to one or more MPCs which, in turn, regularly update the set points of conventional feedback, cascade and ratio control loops being implemented by means of a DCS. Alternatively, a dynamic optimisation capability (DOC) is integrated with an MPC. Both scenarios are discussed in Chapter 106.

MPC is advanced in the sense that it is based upon a combination of techniques including state space modelling, identification, time series prediction and least squares optimisation, all of which have been developed in previous chapters. Whilst there are many forms of MPC they all have a common framework. The approach taken here is to develop an appreciation of that framework, together with an in-depth understanding of one particular form of MPC, followed by a survey of the other forms. That will provide a sufficient basis for implementation of any of the other forms of MPC.

The particular form of MPC considered is the generalised predictive control (GPC) form due to Clarke. This form has been chosen as a basis for two reasons. First, GPC is a natural extension to the generalised minimum variance (GMV) form of self-tuning controller considered in the previous chapter. And secondly, because its details are in the public domain. An excellent overview of MPC is provided by Ogunnaike (1994) which is based upon the dynamic matrix control (DMC) form of MPC. The reader is also referred to the texts by Seborg (2004) and Morari (1994) and, for a more comprehensive treatment of GPC, to the text by Camacho (1999).

117.1 Prediction and Control Horizons

For the purposes of MPC it is necessary to predict future values of some process output or controlled variable (y) relative to the current instant (k). Prediction is typically as far ahead as two to three times the dominant time constant of the system. Suppose the process is sampled at say one twentieth of that time constant: the output prediction horizon (p) could then be up to some 60 steps ahead. Note that this includes steps ahead (h) due to any inherent delay in the system. The process input or manipulated variable (u) is adjusted by the MPC. The control horizon is the sequence of input steps ahead (c) that are calculated by the controller. The control horizon is always shorter than the prediction horizon (c < p). For prediction purposes an arbitrary number of steps ahead (j) is considered. These horizons are depicted in Figure 117.1.

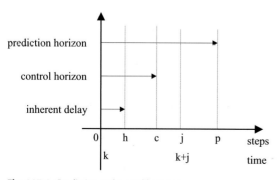

Fig. 117.1 Prediction and control horizons

The basic strategy of MPC is now outlined in the context of a SISO system: its extension to multivariable systems comes later. Given a model of the process, and knowing the past history and current values of both controlled and manipulated variables, future values of the controlled variable (y) may be predicted. On the basis of these predictions and the future set point trajectory, and assuming no corrective action is taken, estimates of the future errors can be made. However, corrective action can be taken, the objective being to eliminate those errors. MPC design, therefore, involves finding the optimal sequence of changes to the manipulated variable (u) that will minimise the predicted errors over the prediction horizon. In practice, it is only the first of that sequence of changes which is implemented: thereafter the whole process is repeated, the so-called receding horizon approach. Figure 117.2 depicts this strategy.

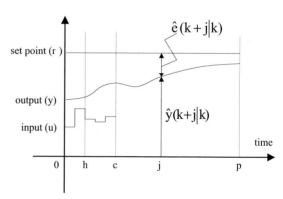

Fig. 117.2 The receding horizon approach to minimising errors

Whilst only the first control step is ever implemented, and the control horizon may be only one step long (c = 1), it is common for a longer control horizon to be used (say c = 2, 3 or 4). Note that the control horizon is not necessarily longer than the inherent delay (c > h) as depicted in Figure 117.2.

The notation used for the input and output signals is summarised in Table 117.1.

Table 117.1 Notation for I/O signals

	Past	Current	Future
Output (CV)	$y(k-1), y(k-2), \ldots$	$y(k)$	$y(k+1), y(k+2), \ldots$
Input (MV)	$u(k-1), u(k-2), \ldots$	$u(k)$	$u(k+1), u(k+2), \ldots$

117.2 The Basis of Forward Prediction

The basis for forward prediction is a model of the process. For MPC purposes it is normal for the model to be in time series form (or z domain) and for the variables to be in deviation form. If the model is not known *a priori* then it would have to be established empirically: step response testing is common but identification by RLS or otherwise, as outlined in Chapter 114, is also used. Some MPC packages provide a neural network function for identification. It doesn't particularly matter what form the model is obtained in, transfer function or state-space, continuous or discrete, because the models can always be converted into time series form. As a basis for design consider the system depicted in Figure 117.3.

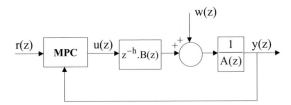

Fig. 117.3 The basis of forward prediction

The pulse transfer function of the process is as given previously by Equation 114.20:

$$A(z).y(z) = z^{-h}.B(z).u(z) + w(z) \quad (117.1)$$

where:

$$A(z) = 1 + a_1.z^{-1} + a_2.z^{-2} + \cdots + a_\alpha.z^{-\alpha}$$
$$B(z) = b_0 + b_1.z^{-1} + b_2.z^{-2} + \cdots + b_\beta z^{-\beta}$$

Suppose that the noise $w(z)$ is related to the residuals $\varepsilon(z)$ on the prediction of the output $y(z)$ by some weighting function as previously defined by Equation 114.31:

$$w(z) = C(z).\varepsilon(z) \quad (117.2)$$

where:

$$C(z) = 1 + c_1.z^{-1} + c_2.z^{-2} + \cdots + c_\gamma.z^{-\gamma}$$

Hence the ARMAX model of Equation 116.2:

$$A(z).y(z) = z^{-h}.B(z).u(z) + C(z).\varepsilon(z) \quad (117.3)$$

For the purposes of GPC consider the simple but special case of:

$$C(z) = \frac{1}{1-z^{-1}} = \frac{1}{\Delta(z)} \quad (117.4)$$

This choice of $C(z)$ introduces an integrator into the model for disturbance rejection: it effectively guarantees zero steady state offset. Substitution for $C(z)$ and rearrangement yields a model of the so called controlled auto regressive integrated moving average (CARIMA) form:

$$A(z)\Delta(z).y(z) = z^{-h}.B(z)\Delta(z).u(z) + \varepsilon(z) \quad (117.5)$$

For GPC purposes it is necessary to separate out current and past values from future ones. This is realised by means of polynomial division. Consider the identity:

$$\frac{1}{A(z)\Delta(z)} = E_j(z) + z^{-j}.\frac{F_j(z)}{A(z)\Delta(z)}$$

In effect, this states that division by a polynomial will result in a numerator (first term) and a remainder (second term). This may be rearranged into the form of the so called Diophantine identity as follows:

$$1 = A(z)E_j(z)\Delta(z) + z^{-j}.F_j(z) \quad (117.6)$$

where:

$$E_j(z) = 1 + e_1 z^{-1} + e_2 z^{-2} + \cdots + e_{j-1} z^{-(j-1)}$$
$$F_j(z) = f_{j,0} + f_{j,1} z^{-1} + f_{j,2} z^{-2} + \cdots + f_{j,\alpha} z^{-\alpha}$$
(117.7)

For any known polynomial A(z) of order α and arbitrary number of steps ahead (j) there will always be a unique pair of polynomials $E_j(z)$ and $F_j(z)$ that satisfy Equation 117.6. Note that the coefficients of $F_j(z)$ vary according to the value of j chosen.

Multiplying the CARIMA model by $E_j(z)$ gives:

$$A(z)E_j(z)\Delta(z).y(z) = z^{-h}.B(z)E_j(z)\Delta(z).u(z) + E_j(z).\varepsilon(z)$$

and substituting from Equation 117.6 gives

$$(1 - z^{-j}.F_j(z)).y(z) = z^{-h}.B(z)E_j(z)\Delta(z).u(z) + E_j(z).\varepsilon(z)$$

Hence:

$$y(z) = z^{-j}.F_j(z).y(z) + z^{-h}.B(z)E_j(z)\Delta(z).u(z) + E_j(z).\varepsilon(z)$$

Consider j steps ahead, i.e. multiply by z^{+j}:

$$z^j.y(z) = F_j(z).y(z) \\ + z^{j-h}.B(z)E_j(z)\Delta(z).u(z) \quad (117.8) \\ + z^j.E_j(z).\varepsilon(z)$$

The polynomials B(z) and E(z) may be combined as previously in Equation 116.10:

$$G_j(z) = B(z).E_j(z) \\ = g_0 + g_1 z^{-1} + g_2 z^{-2} + \cdots + g_{j+\beta-1} z^{-(j+\beta-1)}$$
(117.9)

Substituting Equation 117.9 into Equation 117.8 and inverse transform gives:

$$y(k+j) = F_j(k).y(k) + G_j(k).\Delta u(k+j-h) \\ + E_j(k).\varepsilon(k+j)$$

where the notation is as defined in Chapter 116. Future noise is unknown so future residuals, presumed to be due to prediction error, may be ignored resulting in the estimate:

$$\hat{y}(k+j|k) = F_j(k).y(k) + G_j(k).\Delta u(k+j-h)$$
(117.10)

Thus predictions of future values of the output are separated into two components, the first is based upon known current and previous values of the output and the second upon unknown future values of the input.

The equivalent Z domain representation of Equation 117.10 is:

$$z^j.\hat{y}(z)\big|_0 = F_j(z).y(z) + z^{j-h}.(1-z^{-1})G_j(z).u(z)$$
(117.11)

117.3 Prediction of the Output

Expanding Equation 117.10 gives:

$$\begin{aligned}\hat{y}(k+j|k) &= F_j(k).y(k) + G_j(k).\Delta u(k+j-h) \\ &= F_j(k).y(k) + g_0.\Delta u(k+j-h) \\ &\quad + g_1.\Delta u(k+j-h-1) \\ &\quad + g_2.\Delta u(k+j-h-2) + \ldots \\ &\quad + g_{j-h-1}.\Delta u(k+1) + g_{j-h}.\Delta u(k) \\ &\quad + g_{j-h+1}.\Delta u(k-1) \\ &\quad + g_{j-h+2}.\Delta u(k-2) + \ldots \\ &\quad + g_{j+\beta-1}.\Delta u(k-h-\beta+1)\end{aligned}$$

which may be rearranged as follows:

$$\begin{aligned}\hat{y}(k+j|k) &= \sigma_j + g_{j-h}.\Delta u(k) \\ &\quad + g_{j-h-1}.\Delta u(k+1) \\ &\quad + g_{j-h-2}.\Delta u(k+2) + \ldots \\ &\quad + g_2.\Delta u(k+j-h-2) \\ &\quad + g_1.\Delta u(k+j-h-1) \\ &\quad + g_0.\Delta u(k+j-h)\end{aligned}$$
(117.12)

where:

$$\begin{aligned}\sigma_j &= F_j(k).y(k) + g_{j-h+1}.\Delta u(k-1) \\ &\quad + g_{j-h+2}.\Delta u(k-2) + \ldots \\ &\quad + g_{j+\beta-2}.\Delta u(k-h-\beta+2) \\ &\quad + g_{j+\beta-1}.\Delta u(k-h-\beta+1)\end{aligned}$$
(117.13)

in which:

$$F_j(k).y(k) = f_{j,0}.y(k) + f_{j,1}.y(k-1) + \ldots + f_{j,\alpha}.y(k-\alpha) \quad (117.14)$$

and:

$$\Delta u(k) = u(k) - u(k-1),$$
$$\Delta u(k+1) = u(k+1) - u(k), \quad etc.$$

Inspection of Equation 117.12 reveals that the prediction of the output j steps ahead is comprised of a term σ_j and a series of terms based on unknown current and future values of the input. Equations 117.13 and 117.14 reveal that the term σ_j is itself a series of terms based on known previous values of the input and known current and previous values of the output signal.

Equation 117.12 is the basis for prediction of the output. With reference to Figure 117.1, a prediction of the output can now be made over the prediction horizon. What happens during the inherent delay is not of interest because the control action that affects the output in this period has already taken place so, in effect, the prediction range is $h \leq j \leq p$. The predictions made at each step over that range may be assembled into matrix form as follows:

$$\hat{\underline{y}}(k+j|k) = \begin{bmatrix} \hat{y}(k+h|k) \\ \hat{y}(k+h+1|k) \\ \hat{y}(k+h+2|k) \\ \vdots \\ \hat{y}(k+p|k) \end{bmatrix}$$

$$= \begin{bmatrix} \sigma_h \\ \sigma_{h+1} \\ \sigma_{h+2} \\ \vdots \\ \sigma_p \end{bmatrix} + \begin{bmatrix} g_0 & 0 & 0 & \cdots & 0 \\ g_1 & g_0 & 0 & \cdots & 0 \\ g_2 & g_1 & g_0 & \cdots & 0 \\ \vdots & \vdots & \vdots & \ddots & \vdots \\ g_{p-h} & g_{p-h-1} & g_{p-h-2} & \cdots & g_0 \end{bmatrix}$$

$$\times \begin{bmatrix} \Delta u(k) \\ \Delta u(k+1) \\ \Delta u(k+2) \\ \vdots \\ \Delta u(k+p-h) \end{bmatrix} \quad (117.15)$$

It is perhaps helpful in understanding Equation 117.15 to take note of its dimensions:

$$[(p-h+1) \times 1]$$
$$= [(p-h+1) \times 1]$$
$$+ [(p-h+1) \times (p-h+1)] \times [(p-h+1) \times 1]$$

The upper off diagonal zeros are due to the prediction of the output y at each step ahead j being based upon changes in the input Δu as far as that step only, and not beyond, the earlier steps having fewer prior inputs associated with them.

In practice, the control horizon is only c steps ahead and the values for Δu thereafter are zero, so the above equation reduces to:

$$\hat{\underline{y}}(k+j|k)$$

$$= \begin{bmatrix} \sigma_h \\ \sigma_{h+1} \\ \sigma_{h+2} \\ \vdots \\ \sigma_{h+c-1} \\ \vdots \\ \sigma_p \end{bmatrix} + \begin{bmatrix} g_0 & 0 & 0 & \cdots & 0 \\ g_1 & g_0 & 0 & \cdots & 0 \\ g_2 & g_1 & g_0 & \cdots & 0 \\ \vdots & \vdots & \vdots & \ddots & \vdots \\ g_{c-1} & g_{c-2} & g_{c-3} & \cdots & g_0 \\ \vdots & \vdots & \vdots & \ddots & \vdots \\ g_{p-h} & g_{p-h-1} & g_{p-h-2} & \cdots & g_{p-h-c+1} \end{bmatrix}$$

$$\times \begin{bmatrix} \Delta u(k) \\ \Delta u(k+1) \\ \Delta u(k+2) \\ \vdots \\ \Delta u(k+c-1) \end{bmatrix} \quad (117.16)$$

This may be simplified by notation to:

$$\hat{\underline{y}}(k+j|k) = N(\sigma_j) + M(g_j).\Delta \underline{u}_f(j) \quad (117.17)$$

whose dimensions are:

$$[(p-h+1) \times 1] = [(p-h+1) \times 1]$$
$$+ [(p-h+1) \times c] \times [c \times 1]$$

The vector $N(\sigma_j)$ is comprised of known (backward) values and the matrix $M(g_j)$ operates upon changes in the unknown (forward) inputs $\Delta \underline{u}_f(j)$.

Note that in Equations 117.15 and 117.16 the vectors of the predictions of the output $\underline{y}$ and known values $N(\sigma_j)$ are both of dimension $(p-h+1) \times 1$.

However, whereas in Equation 117.15 the dimension of vector $\Delta \underline{u}_f(j)$ is $(p-h+1) \times 1$, it is reduced in Equation 117.16 to $c \times 1$. Similarly the dimension of matrix $M(g_j)$ is reduced from $(p-h+1) \times (p-h+1)$ to $(p-h+1) \times c$. As explained later, this reduction in size is a major factor in the efficient realisation of MPC.

The vector $N(\sigma_j)$ in Equation 117.17 may be established from Equations 117.13 and 117.14 over the same prediction range of $h \leq j \leq p$ as follows:

$$N(\sigma_j) = \begin{bmatrix} \sigma_h \\ \sigma_{h+1} \\ \sigma_{h+2} \\ \vdots \\ \sigma_p \end{bmatrix}$$

$$= \begin{bmatrix} f_{h,0} & f_{h,1} & \cdots & f_{h,\alpha} \\ f_{h+1,0} & f_{h+1,1} & \cdots & f_{h+1,\alpha} \\ f_{h+2,0} & f_{h+2,1} & \cdots & f_{h+2,\alpha} \\ \vdots & \vdots & \ddots & \vdots \\ f_{p,0} & f_{p,1} & \cdots & f_{p,\alpha} \end{bmatrix} \begin{bmatrix} y(k) \\ y(k-1) \\ y(k-2) \\ \vdots \\ y(k-\alpha) \end{bmatrix}$$

$$+ \begin{bmatrix} g_1 & g_2 & \cdots & g_{h+\beta-2} & g_{h+\beta-1} \\ g_2 & g_3 & \cdots & g_{h+\beta-1} & g_{h+\beta} \\ g_3 & g_4 & \cdots & g_{h+\beta} & g_{h+\beta+1} \\ \vdots & \vdots & \ddots & \vdots & \vdots \\ g_{p-h+1} & g_{p-h+2} & \cdots & g_{p+\beta-2} & g_{p+\beta-1} \end{bmatrix}$$

$$\times \begin{bmatrix} \Delta u(k-1) \\ \Delta u(k-2) \\ \Delta u(k-3) \\ \vdots \\ \Delta u(k-h-\beta+1) \end{bmatrix} \quad (117.18)$$

Equation 117.18 may be simplified by notation to:

$$N(\sigma_j) = N_y(f_j).\underline{y}_b(j) + N_u(g_j).\Delta \underline{u}_b(j)$$

$$= \begin{bmatrix} N_y(f_j) & N_u(g_j) \end{bmatrix} \times \begin{bmatrix} \underline{y}_b(j) \\ \Delta \underline{u}_b(j) \end{bmatrix} \quad (117.19)$$

whose dimensions are:

$[(p-h+1) \times 1]$
$= [(p-h+1) \times (\alpha+1)] \times [(\alpha+1) \times 1]$
$+ [(p-h+1) \times (h+\beta-1)] \times [(h+\beta-1) \times 1]$

where the vector $N(\sigma_j)$ is partitioned into a matrix $N_y(f_j)$ operating upon known (backward) outputs $\underline{y}_b(j)$ and a matrix $N_u(g_j)$ operating upon known (backward) inputs $\Delta \underline{u}_b(j)$.

Note that the order of $G_j(z)$ increases with the number of steps ahead being predicted. This is reflected in the number of terms in the rows of $M(g_j)$ and $N_u(g_j)$. The number of terms in the rows of $M(g_j)$ increases whereas the number of terms in the rows of $N_u(g_j)$ is constant.

117.4 Controller Output Sequence

As stated, the objective of MPC design is to find the optimal sequence of changes to the manipulated variable over the control horizon that minimises the errors over the prediction horizon. With reference to Figure 117.2, the predicted error at any future arbitrary instant j is given by:

$$\hat{e}(k+j|k) = \hat{r}(k+j) - \hat{y}(k+j|k) \quad (117.20)$$

where $r(k)$ is the future set point which may be either constant or variable. A least squares approach to minimisation is taken in which the objective (or cost) function is based upon the sum of the squares of the error over the prediction horizon:

$$J = \sum_{j=h}^{p} \left(\hat{e}(k+j|k) \right)^2 + \omega . \sum_{j=0}^{c-1} \Delta u(k+j)^2$$

$$= \sum_{j=h}^{p} \left(\hat{r}(k+j) - \hat{y}(k+j|k) \right)^2$$

$$+ \omega . \sum_{j=0}^{c-1} \Delta u(k+j)^2 \quad (117.21)$$

Note that the objective function includes the sequence of changes to the manipulated variable. That is because the best minimisation is that which involves least control effort or, in other words, the minimal change in the manipulated variable. The weighting factor ω is user definable: a tuning parameter.

Substituting from Equation 117.17 gives:

$$J = (\underline{r}_f(j) - N(\sigma_j) - M(g_j).\Delta\underline{u}_f(j))^T$$
$$\times (\underline{r}_f(j) - N(\sigma_j) - M(g_j).\Delta\underline{u}_f(j))$$
$$+ \omega.\Delta\underline{u}_f^T(j).\Delta\underline{u}_f(j)$$

expansion of which gives:

$$\begin{aligned}J =\ & \underline{r}_f^T(j).\underline{r}_f(j) - \underline{r}_f^T(j).N(\sigma_j) \\ & - \underline{r}_f^T(j).M(g_j).\Delta\underline{u}_f(j) \\ & - N^T(\sigma_j).\underline{r}_f(j) + N^T(\sigma_j).N(\sigma_j) \\ & + N^T(\sigma_j).M(g_j).\Delta\underline{u}_f(j) \\ & - \Delta\underline{u}_f^T(j).M^T(g_j).\underline{r}_f(j) \\ & + \Delta\underline{u}_f^T(j).M^T(g_j).N(\sigma_j) \\ & + \Delta\underline{u}_f^T(j).M^T(g_j).M(g_j).\Delta\underline{u}_f(j) \\ & + \omega.\Delta\underline{u}_f^T(j).\Delta\underline{u}_f(j)\end{aligned} \quad (117.22)$$

The objective function may be differentiated with respect to the unknown control vector to find a minimum. Noting that differentiation by a vector is covered in Chapter 79:

$$\begin{aligned}\frac{dJ}{d\Delta\underline{u}_f(j)} =\ & -2M^T(g_j).\underline{r}_f(j) + 2M^T(g_j).N(\sigma_j) \\ & + 2.M^T(g_j).M(g_j).\Delta\underline{u}_f(j) \\ & + 2\omega.\Delta\underline{u}_f(j)\end{aligned} \quad (117.23)$$

The minimum occurs when the differential is zero:

$$-M^T(g_j).[\underline{r}_f(j) - N(\sigma_j)]$$
$$+ [M^T(g_j).M(g_j) + \omega.I].\Delta\underline{u}_f(j) = 0$$

which gives upon rearrangement:

$$\Delta\underline{u}_f(j) = [M^T(g_j).M(g_j) + \omega.I]^{-1} \\ \times M^T(g_j).[\underline{r}_f(j) - N(\sigma_j)] \quad (117.24)$$

Equation 117.24 is the basic design equation for the unconstrained GPC form of MPC. For any given matrix of coefficients $M(g_j)$ and matrix of known values $N(\sigma_j)$, it enables the sequence of future changes in the controller output Δu to be calculated which will minimise the error over the prediction horizon. Remember that the variables are in deviation form, so the output signal must be initialised with its normal bias.

117.5 Worked Example

Consider a simple process whose pulse transfer function is as follows:

$$(1 - 0.9.z^{-1}).y(z) = z^{-1}.u(z) + w(z)$$

In fact this process is comprised of a gain of approximately 9.5, a first order lag of approximately 9.5 min and a delay of 1 min with a sampling period of 1 min.

Comparison with Equation 117.1 reveals the process parameters:

$$A(z) = (1 - 0.9.z^{-1}), \quad B(z) = 1,$$
$$\alpha = 1 \quad \beta = 0 \quad h = 1$$

The CARIMA form of the model is given by Equation 117.5 as follows:

$$(1 - 0.9.z^{-1}).\Delta(z).y(z) = z^{-1}.\Delta(z).u(z) + \varepsilon(z)$$

For separation purposes, the Diophantine identity of Equation 117.6 becomes:

$$1 = (1 - 0.9.z^{-1}).E_j(z).(1 - z^{-1}) + z^{-j}.F_j(z)$$

Choose a control horizon of c = 2 and a prediction horizon of p = 3. Thus the steps ahead of interest are j = 1, 2 and 3. It is emphasised that the reduced prediction horizon is used solely to simplify the example.

- Case when j = 1. Equations 117.7 give:

$$E_1(z) = 1$$
$$F_1(z) = f_{10} + f_{11}.z^{-1}$$

whence

$$1 = (1 - 0.9.z^{-1}).1.(1 - z^{-1}) + z^{-1}.(f_{10} + f_{11}.z^{-1})$$

Equating coefficients in z^{-1} yields $f_{10} = 1.9$ and in z^{-2} yields $f_{11} = -0.9$.
Substitute into Equation 117.9 yields:

$$G_1(z) = B(z).E_1(z) = 1.$$

The first forward prediction is thus given by Equations 117.10 and 117.14:

$$\hat{y}(k+1|k) = 1.9.y(k) - 0.9.y(k-1) + \Delta u(k)$$

- Case when j = 2. Equations 117.7 give:

$$E_2(z) = 1 + e_1.z^{-1}$$
$$F_2(z) = f_{20} + f_{21}.z^{-1}$$

whence:

$$1 = (1 - 0.9.z^{-1}).(1 + e_1.z^{-1}).(1 - z^{-1})$$
$$+ z^{-2}.(f_{20} + f_{21}.z^{-1})$$

Equating coefficients yields $e_1 = 1.9$, $f_{20} = 2.71$ and $f_{21} = -1.71$.
From Equation 117.9:

$$G_2(z) = 1 + 1.9.z^{-1}$$

From Equation 117.10 the second forward prediction is:

$$\hat{y}(k+2|k) = 2.71.y(k) - 1.71.y(k-1)$$
$$+ \Delta u(k+1) + 1.9.\Delta u(k)$$

- Case when j = 3. Equations 117.7 give:

$$E_3(z) = 1 + e_1.z^{-1} + e_2.z^{-2}$$
$$F_3(z) = f_{30} + f_{31}.z^{-1}$$

whence:

$$1 = (1 - 0.9.z^{-1}).(1 + e_1.z^{-1} + e_2.z^{-2}).(1 - z^{-1})$$
$$+ z^{-3}.(f_{30} + f_{31}.z^{-1})$$

which yields $e_1 = 1.9$, $e_2 = 2.71$, $f_{30} = 3.439$ and $f_{31} = -2.439$.
From Equation 117.9:

$$G_3(z) = 1 + 1.9.z^{-1} + 2.71.z^{-2}$$

From Equation 117.10 the third forward prediction is:

$$\hat{y}(k+3|k) = 3.439.y(k) - 2.439.y(k-1)$$
$$+ \Delta u(k+2) + 1.9.\Delta u(k+1)$$
$$+ 2.71.\Delta u(k)$$

These three predictions may be combined in the form of Equation 117.15:

$$\begin{bmatrix} \hat{y}(k+1|k) \\ \hat{y}(k+2|k) \\ \hat{y}(k+3|k) \end{bmatrix} = \begin{bmatrix} 1.9 & -0.9 \\ 2.71 & -1.71 \\ 3.44 & -2.44 \end{bmatrix} \begin{bmatrix} y(k) \\ y(k-1) \end{bmatrix}$$
$$+ \begin{bmatrix} 1.0 & 0 & 0 \\ 1.9 & 1.0 & 0 \\ 2.71 & 1.9 & 1.0 \end{bmatrix} \begin{bmatrix} \Delta u(k) \\ \Delta u(k+1) \\ \Delta u(k+2) \end{bmatrix}$$

Noting that the control horizon is only two steps ahead, comparison with Equations 117.17 and 117.19 reveals that:

$$N_y(f_j) = \begin{bmatrix} 1.9 & -0.9 \\ 2.71 & -1.71 \\ 3.44 & -2.44 \end{bmatrix} \quad N_u(g_j) = 0$$

$$M(g_j) = \begin{bmatrix} 1.0 & 0 \\ 1.9 & 1.0 \\ 2.71 & 1.9 \end{bmatrix}$$

Note that the term $N_u(g_j)$ is zero because of the combination of $h = 1$ and $\beta = 0$. Matrix manipulation gives:

$$[M^T(g_j).M(g_j)]^{-1}.M^T(g_j)$$
$$= \begin{bmatrix} 0.8506 & 0.3155 & -0.1661 \\ -1.3006 & -0.2655 & 0.6661 \end{bmatrix}$$

Assume a weighting factor of $\omega = 0$ in which case the controller output sequence is given by Equation 117.24:

$$\begin{bmatrix} \Delta u(k) \\ \Delta u(k+1) \end{bmatrix} = \begin{bmatrix} 0.8506 & 0.3155 & -0.166 \\ -1.301 & -0.266 & 0.6661 \end{bmatrix}$$
$$\times \left\{ \begin{bmatrix} r(k) \\ r(k+1) \\ r(k+2) \end{bmatrix} - \begin{bmatrix} 1.9 & -0.9 \\ 2.71 & -1.71 \\ 3.44 & -2.44 \end{bmatrix} \begin{bmatrix} y(k) \\ y(k-1) \end{bmatrix} \right\}$$

On the basis of the receding horizon approach, this may be simplified:

$$\Delta u(k) = \begin{bmatrix} 0.8506 & 0.3155 & -0.166 \end{bmatrix}$$
$$\times \left\{ \begin{bmatrix} r(k) \\ r(k+1) \\ r(k+2) \end{bmatrix} - \begin{bmatrix} 1.9 & -0.9 \\ 2.71 & -1.71 \\ 3.44 & -2.44 \end{bmatrix} \begin{bmatrix} y(k) \\ y(k-1) \end{bmatrix} \right\}$$

If the set point is constant, this reduces to:

$$\Delta u(k) = r(k) - 1.899.y(k) + 0.899.y(k-1)$$

117.6 Recursive Implementation

To be of any use as a control strategy, MPC has to be capable of being implemented on a recursive basis. Thus, at each sampling instant, the values of the process output (CV) and input (MV) have to be updated and the required change in input calculated. This is depicted in flow chart form in Figure 117.4 and is largely self explanatory.

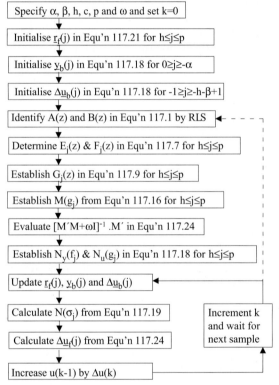

Fig. 117.4 Recursive implementation of model predictive control (MPC)

There are two cases to consider. First, that when the process/plant is stationary, or relatively so, in which case the plant model will be constant and the identification of $A(z)$ and $B(z)$ can be done on a once-off basis by RLS or otherwise. Thus, once $G_j(z)$, $M(g_j)$, $N_y(f_j)$ and $N_u(g_j)$ have been established, the recursion simply entails updating the vectors, calculating $N(\sigma_j)$ and $\Delta \underline{u}_f(j)$ and implementing the first of the changes to the controller output.

Second, in the event of the process/plant varying with time, it is necessary to carry out the identification and re-establish the various matrices recursively, as depicted by the broken line in Figure 117.4. This can be done occasionally or frequently, according to how fast the model changes. Clearly there is a massive increase in the computational effort required, in particular in relation to the separation of $E_j(z)$ and $F_j(z)$ in Equation 117.7, by means of polynomial division or otherwise, and in the matrix inversion in Equation 117.24. This effort can only be justified if the variation of the model with time is significant in the overall context of the control problem.

This increase in effort is partially offset by the use of the control horizon c, rather than the prediction horizon p, in the computation of the $M(g_j)$ matrix in Equation 117.16 and the subsequent matrix inversion in Equation 117.24. Indeed, if it wasn't for this reduction, recursive implementation would not be feasible. To emphasise the point, consider the situation where $h = 2$, $c = 3$ and $p = 10$. The dimensions of $M(g_j)$ are $(p-h+1) \times c$ which leads to

$$M(g_j) = \begin{bmatrix} g_0 & 0 & 0 \\ g_1 & g_0 & 0 \\ g_2 & g_1 & g_0 \\ g_3 & g_2 & g_1 \\ g_4 & g_3 & g_2 \\ \vdots & \vdots & \vdots \\ g_8 & g_7 & g_6 \end{bmatrix}$$

Note the operand in Equation 117.24. An elegant and powerful feature of GPC is that the operand is the difference between the vector of future set point $\underline{r}_f(j)$ values and the vector of previous known values $N(\sigma_j)$. Predictions of future values of the process output are not explicitly involved, they were eliminated in the matrix manipulations that resulted in Equation 117.24. Thus, there are no residuals on the output due to unknown disturbances to be corrected for.

GPC enables the sequence of future changes in the controller output $\Delta \underline{u}$ to be calculated which will minimise the error over the prediction horizon, but only the first change is ever implemented. So why not set the control horizon to unity (c = 1) in the first place? This would considerably simplify the calculations since $M(g_j)$ in Equation 117.24 becomes a vector of dimension $(p - h + 1) \times 1$ and $\Delta \underline{u}_f(j)$ becomes a scalar. The issue here is flexibility regarding the magnitude of change in the controller output. If the minimisation was based upon a single change (c = 1) then a conservative value would be found which satisfies the long term prediction horizon. However, if a control horizon of two steps (c = 2) was used, more aggressive action could be taken. The first change in controller output could be larger than would otherwise be the case, forcing the process to respond faster. This would then be followed by a smaller change in the opposite direction to prevent undue overshoot. In practice, control horizons of of up to five steps ahead (c = 5) are often chosen.

117.7 QP Solution of MPC Problem

As stated, the GPC form of MPC described thus far is of the unconstrained variety in that changes in the controller output are evaluated to minimise the objective function irrespective of any physical constraints. In practice, there are always constraints (limits) on the MV due to saturation effects and rate of change considerations. An alternative approach to the recursive implementation of Equation 117.24 is to treat the MPC problem as a constrained optimisation to be solved by means of quadratic programming. QP is covered in Chapter 105 and of particular interest are Equations 105.20–105.27.

The objective function for GPC was articulated in Equation 117.21 and expanded into Equation 117.22. The expansion may be rearranged as follows:

$$J = \Delta \underline{u}_f^T(j) . \left(M^T(g_j).M(g_j) + \omega I \right) . \Delta \underline{u}_f(j)$$
$$- 2.\Delta \underline{u}_f^T(j).M^T(g_j). \left(\underline{r}_f(j) - N(\sigma_j) \right)$$
$$+ \left(\underline{r}_f(j) - N(\sigma_j) \right)^T \left(\underline{r}_f(j) - N(\sigma_j) \right) \quad (117.25)$$

In doing the above rearrangement it is necessary to substitute the transposes of the following two scalar quantities:

$$\underline{r}_f^T(j).M(g_j).\Delta \underline{u}_f(j) = \left(\Delta \underline{u}_f^T(j).M^T(g_j).\underline{r}_f(j) \right)^T$$
$$N^T(\sigma_j).M(g_j).\Delta \underline{u}_f(j) = \left(\Delta \underline{u}_f^T(j).M^T(g_j).N(\sigma_j) \right)^T$$

By inspection, it can be seen that Equation 117.25 is exactly of the form of the classical QP problem as defined in Chapter 105:

$$f(\underline{x}) = \frac{1}{2}.\underline{x}^T.Q.\underline{x} + \underline{x}^T.C + \text{const} \quad (105.20)$$

The vector of decision variables is $\Delta \underline{u}_f(j)$ and, by analogy:

$$Q \equiv 2 \left(M^T(g_j).M(g_j) + \omega I \right)$$
$$C \equiv -2.M^T(g_j). \left(\underline{r}_f(j) - N(\sigma_j) \right)$$
$$\text{const} \equiv \left(\underline{r}_f(j) - N(\sigma_j) \right)^T \left(\underline{r}_f(j) - N(\sigma_j) \right)$$

where Q is of dimensions $c \times c$ and C is of dimension $c \times 1$.

The constraint matrix for the QP problem is of the general form:

$$A.\underline{x} \leq B \quad \text{and} \quad \underline{x} \geq 0 \quad (105.21)$$

For the QP solution of the MPC problem the constraint matrix is formulated in relation to the vector $\Delta \underline{u}_f(j)$. First consider the saturation effects. In essence there is an upper and lower limit on the value of the MPC's output, the output being either the MV of the plant or the SP for the CV downloaded to a DCS:

$$u_{min} \leq u(j) \leq u_{max}$$

Since the output of the MPC is calculated in incremental form, the value of the output in absolute form has to be established by summation as follows:

$$\begin{bmatrix} \underline{u}(k+1) \\ \underline{u}(k+2) \\ \underline{u}(k+3) \\ \vdots \\ \underline{u}(k+c) \end{bmatrix} = \begin{bmatrix} \underline{u}(k) \\ \underline{u}(k) \\ \underline{u}(k) \\ \vdots \\ \underline{u}(k) \end{bmatrix} + \begin{bmatrix} 1 & 0 & 0 & \cdots & 0 \\ 1 & 1 & 0 & \cdots & 0 \\ 1 & 1 & 1 & \cdots & 0 \\ \vdots & \vdots & \vdots & \ddots & \vdots \\ 1 & 1 & 1 & \cdots & 1 \end{bmatrix} \begin{bmatrix} \Delta\underline{u}(k) \\ \Delta\underline{u}(k+1) \\ \Delta\underline{u}(k+2) \\ \vdots \\ \Delta\underline{u}(k+c-1) \end{bmatrix}$$

This is of the form:

$$\underline{u}(j+1) = \underline{u}(k) + A.\Delta\underline{u}_f(j)$$

The upper saturation limit is thus:

$$\underline{u}(j+1) = \underline{u}(k) + A.\Delta\underline{u}_f(j) \leq \underline{u}_{max}$$

It follows that:

$$A.\Delta\underline{u}_f(j) \leq \underline{u}_{max} - \underline{u}(k) \quad (117.26a)$$

$$\begin{bmatrix} 1 & 0 & 0 & \cdots & 0 \\ 1 & 1 & 0 & \cdots & 0 \\ 1 & 1 & 1 & \cdots & 0 \\ \vdots & \vdots & \vdots & \ddots & \vdots \\ 1 & 1 & 1 & \cdots & 1 \end{bmatrix} \begin{bmatrix} \Delta\underline{u}(k) \\ \Delta\underline{u}(k+1) \\ \Delta\underline{u}(k+2) \\ \vdots \\ \Delta\underline{u}(k+c-1) \end{bmatrix} \leq \begin{bmatrix} \underline{u}_{max} \\ \underline{u}_{max} \\ \underline{u}_{max} \\ \vdots \\ \underline{u}_{max} \end{bmatrix} - \begin{bmatrix} \underline{u}(k) \\ \underline{u}(k) \\ \underline{u}(k) \\ \vdots \\ \underline{u}(k) \end{bmatrix}$$

Similarly, for the lower limit:

$$\underline{u}(j+1) = \underline{u}(k) + A.\Delta\underline{u}_f(j) \geq \underline{u}_{min}$$

whence

$$-A.\Delta\underline{u}_f(j) \leq \underline{u}(k) - \underline{u}_{min} \quad (117.26b)$$

$$\begin{bmatrix} -1 & 0 & 0 & \cdots & 0 \\ -1 & -1 & 0 & \cdots & 0 \\ -1 & -1 & -1 & \cdots & 0 \\ \vdots & \vdots & \vdots & \ddots & \vdots \\ -1 & -1 & -1 & \cdots & -1 \end{bmatrix} \begin{bmatrix} \Delta\underline{u}(k) \\ \Delta\underline{u}(k+1) \\ \Delta\underline{u}(k+2) \\ \vdots \\ \Delta\underline{u}(k+c-1) \end{bmatrix} \leq \begin{bmatrix} \underline{u}(k) \\ \underline{u}(k) \\ \underline{u}(k) \\ \vdots \\ \underline{u}(k) \end{bmatrix} - \begin{bmatrix} \underline{u}_{min} \\ \underline{u}_{min} \\ \underline{u}_{min} \\ \vdots \\ \underline{u}_{min} \end{bmatrix}$$

Second, there may be rate of change constraints. Suppose that there is a limit on the incremental change in output, whether it be an increase or a decrease, as follows:

$$-\Delta\underline{u}_{lim} \leq \Delta\underline{u}(j) \leq \Delta\underline{u}_{lim}$$

Since the MPC output is calculated in incremental form it is relatively easy to apply this constraint. In the case of the upper limit:

$$\begin{bmatrix} 1 & 0 & 0 & \cdots & 0 \\ 0 & 1 & 0 & \cdots & 0 \\ 0 & 0 & 1 & \cdots & 0 \\ \vdots & \vdots & \vdots & \ddots & \vdots \\ 0 & 0 & 0 & \cdots & 1 \end{bmatrix} \begin{bmatrix} \Delta\underline{u}(k) \\ \Delta\underline{u}(k+1) \\ \Delta\underline{u}(k+2) \\ \vdots \\ \Delta\underline{u}(k+c-1) \end{bmatrix} \leq \begin{bmatrix} \Delta\underline{u}_{lim} \\ \Delta\underline{u}_{lim} \\ \Delta\underline{u}_{lim} \\ \vdots \\ \Delta\underline{u}_{lim} \end{bmatrix} \quad (117.27a)$$

Similarly for the lower rate of change limit:

$$\begin{bmatrix} -1 & 0 & 0 & \cdots & 0 \\ 0 & -1 & 0 & \cdots & 0 \\ 0 & 0 & -1 & \cdots & 0 \\ \vdots & \vdots & \vdots & \ddots & \vdots \\ 0 & 0 & 0 & \cdots & -1 \end{bmatrix} \begin{bmatrix} \Delta\underline{u}(k) \\ \Delta\underline{u}(k+1) \\ \Delta\underline{u}(k+2) \\ \vdots \\ \Delta\underline{u}(k+c-1) \end{bmatrix} \leq \begin{bmatrix} \Delta\underline{u}_{lim} \\ \Delta\underline{u}_{lim} \\ \Delta\underline{u}_{lim} \\ \vdots \\ \Delta\underline{u}_{lim} \end{bmatrix} \quad (117.27b)$$

All of these constraints may be combined into a single matrix equation of the form:

$$A . \Delta \underline{u}_f(j) \leq B \qquad (117.28)$$

where the constraint matrix A is of dimension $(m \times c) \times c$, $\Delta \underline{u}_f(j)$ is of dimension $c \times 1$ and the constraint vector B is $m \times 1$.

Note that there are no additional non-negativity constraints: the decision variables are in deviation form and their negativity is handled by Equations 117.26b and 117.27b. Thus the condition for an extremum given by Equation 105.25 and the Kuhn-Tucker conditions of Equation 105.27 reduce respectively to the following:

$$Q . \Delta \underline{u}_f(j) + C + A^T . \underline{\lambda} = 0 \qquad (117.29)$$
$$\underline{\lambda}^T . \underline{\sigma} = 0 \qquad (117.30)$$

where $\underline{\lambda}$ is a vector of Lagrange multipliers and $\underline{\sigma}$ is a vector of slack variables, both of which are of dimension $m \times 1$.

The set of Equations 117.25 and 117.28–117.30 articulate the MPC problem as an optimisation which may be solved as a QP problem by matrix manipulation in exactly the same manner as described in Worked Example No 5 of Chapter 105, and generally as depicted in Figure 106.10.

117.8 Extension to Multivariable Systems

The GPC form of MPC generalises to accommodate systems with multiple inputs and outputs, the basic form of model between each single input-output pair being of the same form as in Equation 117.1. Consider a 2-input 2-output system in which each output is affected by both inputs as depicted in Figure 117.5.

The model is as follows:

$$\begin{aligned} A_1(z).y_1(z) &= z^{-h_{11}}.B_{11}(z).u_1(z) \\ &\quad + z^{-h_{12}}.B_{12}(z).u_2(z) + w_1(z) \\ A_2(z).y_2(z) &= z^{-h_{21}}.B_{21}(z).u_1(z) \\ &\quad + z^{-h_{22}}.B_{22}(z).u_2(z) + w_2(z) \end{aligned} \qquad (117.31)$$

Development of the multivariable form of GPC is exactly the same as before with the various vectors and matrices being partitioned as appropriate. Thus, by extension of Equation 117.17, the multivariable forward prediction is as follows:

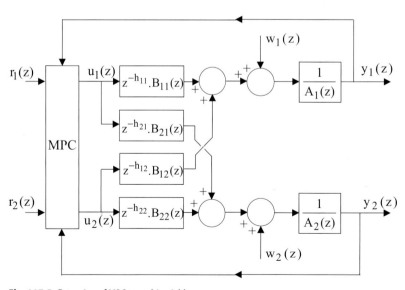

Fig. 117.5 Extension of MPC to multivariable systems

$$\begin{bmatrix} \hat{\underline{y}}_1(k+j|k) \\ \hat{\underline{y}}_2(k+j|k) \end{bmatrix} = \begin{bmatrix} N_1(\sigma_j) \\ N_2(\sigma_j) \end{bmatrix}$$
$$+ \begin{bmatrix} M_{11}(g_j) & M_{12}(g_j) \\ M_{21}(g_j) & M_{22}(g_j) \end{bmatrix} \begin{bmatrix} \Delta \underline{u}_{1f}(j) \\ \Delta \underline{u}_{2f}(j) \end{bmatrix} \quad (117.32)$$

where the known terms are found by extension of Equation 117.19 as follows:

$$\begin{bmatrix} N_1(\sigma_j) \\ N_2(\sigma_j) \end{bmatrix} = \begin{bmatrix} N_{y1}(f_j) & 0 \\ 0 & N_{y2}(f_j) \end{bmatrix} \begin{bmatrix} \underline{y}_{1b}(j) \\ \underline{y}_{2b}(j) \end{bmatrix}$$
$$+ \begin{bmatrix} N_{u11}(g_j) & N_{u12}(g_j) \\ N_{u21}(g_j) & N_{u22}(g_j) \end{bmatrix} \begin{bmatrix} \Delta \underline{u}_{1b}(j) \\ \Delta \underline{u}_{2b}(j) \end{bmatrix}$$
$$(117.33)$$

Noting the equivalence of the vectors and matrices in Equations 117.32 and 117.33 with those in Equations 117.17 and 117.19 respectively, the required control sequence is given by analogy with Equation 117.24 as follows:

$$\begin{bmatrix} \Delta \underline{u}_{1f}(j) \\ \Delta \underline{u}_{2f}(j) \end{bmatrix} = \left[M_{2 \times 2}^T(g_j) . M_{2 \times 2}(g_j) + \underline{\omega} . I \right]^{-1}$$
$$\times M_{2 \times 2}^T(g_j) . \left[\begin{bmatrix} \underline{r}_{1f}(j) \\ \underline{r}_{2f}(j) \end{bmatrix} - \begin{bmatrix} N_1(\sigma_j) \\ N_2(\sigma_j) \end{bmatrix} \right]$$
$$(117.34)$$

where

$$M_{2 \times 2}(g_j) = \begin{bmatrix} M_{11}(g_j) & M_{12}(g_j) \\ M_{21}(g_j) & M_{22}(g_j) \end{bmatrix}$$

Recursive implementation is essentially as depicted in Figure 117.4 for the SISO scenario. The obvious exception is that, having evaluated Equation 117.34, there are two controller outputs to be changed. Thus both MPC outputs u_1 and u_2 are changed by the first elements of $\Delta \underline{u}_{1f}(j)$ and $\Delta \underline{u}_{2f}(j)$ respectively.

117.9 Models for MPC

Critical to the success of MPC is the availability of good quality models. In the context of GPC the models are of a sampled data nature, *i.e.* either in pulse transfer function or time series form as in Equations 117.1 or 117.31. It is implicit, but not essential, that such models are established by means of identification such as RLS as described in Chapter 114. However, in practice, it is quite common to use step or finite impulse response (FIR) tests or otherwise and to convert the model into the required form thereafter.

The MIMO process may be thought of as being comprised of a set of open loop SISO models of the form:

$$\begin{bmatrix} y_1(z) \\ y_2(z) \\ \vdots \\ y_m(z) \end{bmatrix}$$
$$= \begin{bmatrix} G_{11}(z) & G_{12}(z) & \cdots & G_{1n}(z) \\ G_{21}(z) & G_{22}(z) & \cdots & G_{2n}(z) \\ \vdots & \vdots & \ddots & \vdots \\ G_{m1}(z) & G_{m2}(z) & \cdots & G_{mn}(z) \end{bmatrix} \begin{bmatrix} u_1(z) \\ u_2(z) \\ \vdots \\ u_n(z) \end{bmatrix}$$
$$(117.35)$$

By applying disturbances to each of the inputs in turn, and observing the affect on the outputs, the models may be established. For example, a step change in u_1 will enable all the models $G_{11}(z)$, $G_{21}(z)$, *etc.* of the first column of the G matrix to be established. Likewise for each of the other inputs and columns until the G matrix is complete. Knowing each of the SISO models, the time series form can be established and the various matrices for GPC constructed.

Fundamental to the quality of the models are the following points, the importance of which cannot be overemphasised:

- Carefully consider, and carry out tests if necessary, which variables are candidates for inclusion as inputs and outputs of the MPC. The objective is to have as few variables as possible but to include every variable of consequence. Including variables in the model that have little effect or are of marginal interest increases the scope for errors, increases the computation effort and slows down the speed of response.

- Prior to carrying out the RLS or step tests, check that all the instrumentation is properly calibrated, the valves are sensibly sized, and that all slave loops are properly configured, tuned and in automatic mode.
- Whilst carrying out the tests, ensure that the plant and/or process is as well insulated from external disturbances as is reasonably practical.

Robustness, the ability of the MPC to handle plant model mismatch, is a key issue. Lack of robustness arises due to either poor testing, or inappropriate identification or to the fact that the plant's dynamics change with time. This can be addressed by means of intermittently re-identifying the model as depicted in Figure 117.4. Other approaches involve optimal scaling based upon condition numbers and/or singular value thresholding, the concepts of which were covered in Chapter 111.

- Optimal scaling. In essence, the condition number is a measure of the difficulty of control and is evaluated from the system's singular values. The condition number increases with both the extent of interaction and the size of the multivariable problem. The effect is exacerbated by the fact that condition number does not increase linearly with size. Robustness can be improved by optimally scaling the MIMO matrix of Equation 117.35 to minimise the condition number.
- Singular value thresholding. A typical problem with MPC is the decision as to whether or not to include terms in the MIMO matrix with small gains as these can give rise to large changes in controller output. Singular value thresholding, with user defined tolerances, enables such terms to be included in the prediction part of the evaluation but suppresses their effects on the subsequent control actions taken.

117.10 Proprietary Packages

There are many forms of MPC with a host of proprietary DOC packages for their implementation on various platforms in all manner of applications. Table 117.2 is an attempt at summarising some of their functionality.

117.10 Proprietary Packages

Table 117.2 Essential features of proprietary MPC packages

Package	Model form	Principle	Tuning parameters	Platform
Connoisseur (Invensys)	Models in time series form, *i.e.* ARMAX or FIR (finite impulse response). Proprietary tools for RLS identification and converting model types.	Minimises sum of squares of predicted errors. Multiple steps ahead. Constraint handling either by means of long range (LR) or quadratic programming (QP). MVs and CVs declared to be either hard or soft constraints. Uses real-time RBF type neural nets for non-linear prediction.	Control and prediction horizons (c and p). CV, MV move & MV target weights in matrices P, Q and R in objective function J.	Host platforms include UNIX and NT on Sun, DEC and HP machines. Interfaces to all major DCS and PLC systems.
DMC (Aspen)	Step response by proprietary identification package (DMI).	The original MPC system, developed by Shell Oil. Minimises sum of squares of predicted errors. Single step ahead only (c = 1). Needs correction of predicted outputs to avoid offset. Constraints handled by formulating as a QP problem (QDMC).	Prediction horizon (p). Move suppression factor which trades off the change in MVs against the rate of change of CVs. Weighting matrix for relative scaling of CVs.	Host platforms such as VAX with interface to DCS or else runs directly in application nodes of certain DCS.
GPC	Models in ARMAX form identified by RLS or otherwise.	Minimises sum of squares of predicted errors. No need for correction of errors on predicted outputs. Multiple steps ahead.	Control and prediction horizons (c and p). Weighting factor (ω) on MV effort, typically 0.1.	Proprietary package not available, (included for purposes of comparison).
RMPCT (Honeywell)	Impulse response model from step response data, then converted into ARMAX type models by proprietary identifier.	Minimises sum of squares of predicted errors subject to constraints articulated in form of funnels on predicted outputs. If trajectory is within funnel then no change of input occurs, otherwise the input changes subject to minimum control effort. Interface to both LP and QP for integrated MPC and optimisation.	Adjust curvature of funnel by changing performance ratio (0.7 < PR < 1.2) based upon settling time and dominant time constants.	Separate application module (AM) within Honeywell DCS, or on separate NT/DCS node.
SMOC (Shell)	Plant tests with PRBS to establish transfer function type models which are then converted into state space form.	Principle same as for DMC, from which SMOC evolved. Controller design and tuning in simulation mode. Optional optimiser available within objective function J.	Penalties on error and effort (Q & R) in objective function can be adjusted in off-line mode. Limited tuning permitted in real-time environment.	Identification by means of PC. Implemented in real-time on supervisory system (VAX, Dec-Alpha) or in DCS (Fox-IA).

117.11 Nomenclature

Integers

c	control horizon (samples)
h	time delay exponent in process model (samples)
j	arbitrary number of steps ahead
k	current sampling instant
m	number of constraints
p	prediction horizon (samples)

Scalars

J	objective (cost) function
r	controller set point
u	input variable
y	output variable
$\hat{y}$	estimate of y
ε	residual: error in prediction of output y
σ	sum of known (current plus previous) values
ω	weighting factor on output in objective function

Matrices

		Dimensions
A	constraint matrix (from Chapter 105)	$m \times c$
B	constraint vector (from Chapter 105)	$m \times 1$
C	vector of linear coefficients (from Chapter 105)	$c \times 1$
M	matrix operating on future inputs	$(p-h) \times c$ or $(p-h) \times (p-h)$
N	vector of sums of series of known i/o values	$(p-h) \times 1$
N_y	matrix operating on previous outputs	$(p-h) \times \alpha$
N_u	matrix operating on previous inputs	$(p-h) \times (\beta + h - 1)$
Q	matrix of quadratic coefficients (from Chapter 105)	$c \times c$

Vectors

		Dimensions
$\underline{r}$	vector of set point values	$(p-h) \times 1$
$\underline{\hat{y}}$	vector of estimates of output	$(p-h) \times 1$
$\underline{u}$	vector of output values	$c \times 1$ or $(p-h) \times 1$
$\underline{\lambda}$	vector of Lagrange multipliers (from Chapter 105)	$m \times 1$
$\underline{\sigma}$	vector of slack variables (from Chapter 105)	$m \times 1$

Polynomials

		Order
A	denominator monic polynomial of process model	α
B	numerator polynomial of process model	β
C	monic polynomial relates residuals to process noise	γ
E	monic polynomial relates residuals to predictions	$h - 1$
F	polynomial used for separation purposes	$\alpha - 1$
G	polynomial product of B and E	$\beta + h - 1$
Δ	difference operator	1

Subscripts

b	backward values
f	forward values

Non-Linear Control

118.1 L/A Control
118.2 Control Affine Models
118.3 Generic Model Control
118.4 Application of GMC to CSTR
118.5 Worked Example on GMC
118.6 Lie Derivatives and Relative Order
118.7 Globally Linearising Control
118.8 Application of GLC to CSTR
118.9 Comments on GLC
118.10 Nomenclature

Throughout this text there have been many references to non-linearity and the potential problems that it causes. In general, from a control point of view, the more linear a system is the better the quality of control and the easier it is to understand what is going on. There are many sources of non-linearity:

- Processes have discontinuities, such as when a stream's phase changes.
- Reactions have non-linear kinetics, both with regard to rate constants and concentrations.
- Plant characteristics are often non-linear, such as level *vs* volume in conical vessels.
- Flow characteristics, such as flow *vs* head, invariably have a square root characteristic.
- Control loop elements such as =% control valves have, by design, non-linear trims.
- Sensors often introduce non-linearities to measurements, pH being an extreme case.
- Non-linearity also occurs due to non-minimum phase effects and variable time delays.

The best approach is to try to eliminate non-linearity at source, for example by square root extraction in relation to flow measurement, although there is seldom little scope for changing the process itself. Thereafter, non-linear systems are largely treated as if they are linear and tuned to cope with small disturbances within the vicinity of their normal operating conditions. The hope is that they stay there. Alternatively, if it is known a-priori that operating conditions are likely to change significantly, different sets of tuning parameters are established across the range. These are then switched into the controller using gain scheduling. For some systems, and in some cases surprisingly complex ones, this is quite adequate.

Nevertheless, for many non-linear systems, satisfactory control remains a key non-trivial issue. This chapter, therefore, provides an overview of some of the methodologies for handling nonlinearity that have been proposed for use in the chemical and process industries. It covers log antilog (L/A) control, generic model control (GMC) and globally linearising control (GLC).

118.1 L/A Control

The L/A control approach uses non-linear transformations, such as log and anti-log, to map the I/O signals into and out of some linear domain.

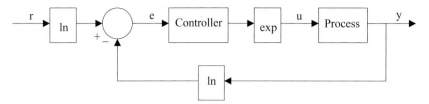

Fig. 118.1 Closed loop structure of L/A strategy

Such transformations can result in more accurate linear-in-the-parameters models of the underlying process and, at the same time, enable conventional control and estimation techniques to be applied. A good example of this is in the development of inferential models for distillation columns where the logarithm of composition measurements is known to be effective for control purposes. The closed loop structure of such an L/A strategy is shown in Figure 118.1.

The error signal is generated from the difference between the logs of the set point r and the feedback of the output y as follows:

$$e = \ln(r) - \ln(y) = \ln\left(\frac{r}{y}\right)$$

The resultant error is operated upon by a linear controller, PID or otherwise, the output of which is anti-logged to produce the input signal to the process. A cited advantage of the L/A approach is that it will always produce a positive output signal u and hence implicitly handles any physical constraints on the plant signals. However, being generic, it will not necessarily produce an optimal solution.

118.2 Control Affine Models

Models for a non-linear SISO system of the following general form are referred to as being control affine:

$$\underline{\dot{x}} = F(\underline{x}) + G(\underline{x}).u$$
$$y = h(\underline{x}) \quad (118.1)$$

where:

$$\frac{dx_1}{dt} = f_1(x_1, x_2, \ldots, x_n) + g_1(x_1, x_2, \ldots, x_n).u$$
$$\vdots$$
$$\frac{dx_n}{dt} = f_n(x_1, x_2, \ldots, x_n) + g_n(x_1, x_2, \ldots, x_n).u$$
$$y = h(x_1, x_2, \ldots, x_n)$$

$F(\underline{x})$ and $G(\underline{x})$ are referred to as vector fields (i.e. vector valued functions of the vector $\underline{x}$), and $h(\underline{x})$ is a scalar field (i.e. a scalar valued function of the vector $\underline{x}$). Note the difference between Equations 118.1 and conventional state space models. In particular, note that f, g and h are functions rather than matrices.

Another important point to appreciate is that control affine models are used for non-linear systems when linearisation and use of deviation variables is inappropriate. That being the case, the variables used in control affine models are in absolute form.

An example of a control affine model can be developed by considering the CSTR depicted in Figure 118.2. Note that control of reactor temperature

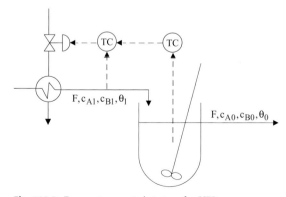

Fig. 118.2 Temperature control strategy for CSTR

θ_0 is to be affected by manipulating the reactor feed temperature θ_1. This feed stream would obviously require a heat exchanger with some slave control loop, the dynamics of which are presumed negligible relative to those of the reactor. The simple but non-linear reversible exothermic reaction of A $\Leftrightarrow$ B is considered. The same conventions for nomenclature and subscripts are used as in Chapters 89 and 91.

Unsteady state mass balance for components A and B across the reactor give:

$$V\frac{dc_{A0}}{dt} = F.(c_{A1} - c_{A0}) + V.(-k_F.c_{A0} + k_R.c_{B0})$$

$$V\frac{dc_{B0}}{dt} = F.(c_{B1} - c_{B0}) + V.(+k_F.c_{A0} - k_R.c_{B0}) \quad (118.2)$$

Noting that ΔH for an exothermic reaction is negative, an unsteady state heat balance gives:

$$V\rho c_p.\frac{d\theta_0}{dt} = F\rho c_p.(\theta_1 - \theta_0) + V.(-k_F.c_{A0} + k_R.c_{B0}).\Delta H \quad (118.3)$$

where the dependence of the rate constants on temperature is given by the Arrhenius equations

$$k_F = B_F.\exp\left(\frac{-E_F}{R(\theta_0 + 273)}\right)$$
$$k_R = B_R.\exp\left(\frac{-E_R}{R(\theta_0 + 273)}\right) \quad (118.4)$$

These equations can be rearranged into the control affine form of Equation 118.1 as follows:

$$\underline{x} = \begin{bmatrix} c_{A0} & c_{B0} & \theta_0 \end{bmatrix}^T, \quad u = \theta_1, \quad (118.5)$$
$$y = h(\underline{x}) = \theta_0$$

$$F(\underline{x}) = \begin{bmatrix} f_1(\underline{x}) & f_2(\underline{x}) & f_3(\underline{x}) \end{bmatrix}^T \quad (118.6)$$

$$f_1(\underline{x}) = \frac{F}{V}.(c_{A1} - c_{A0}) - k_F.c_{A0} + k_R.c_{B0}$$

$$f_2(\underline{x}) = \frac{F}{V}.(c_{B1} - c_{B0}) + k_F.c_{A0} - k_R.c_{B0}$$

$$f_3(\underline{x}) = -\frac{F}{V}.\theta_0 + \frac{\Delta H}{\rho c_p}.(-k_F.c_{A0} + k_R.c_{B0})$$

$$G(\underline{x}) = \begin{bmatrix} g_1(\underline{x}) & g_2(\underline{x}) & g_3(\underline{x}) \end{bmatrix}^T$$
$$= \begin{bmatrix} 0 & 0 & \frac{F}{V} \end{bmatrix}^T \quad (118.7)$$

These vector and scalar fields are important for the design of GMC and GLC laws.

118.3 Generic Model Control

The fundamental difference between the GMC approach, as described by Lee (1998), and other model based control strategies is that the performance objective is cast in terms of the difference between the derivatives of the desired or reference trajectory r(t) and the process output y(t) rather than the difference between their absolute values.

$$J = \int_0^\infty (\dot{r} - \dot{y})^2 dt$$

Consider the same SISO process as described previously:

$$\dot{\underline{x}} = F(\underline{x}) + G(\underline{x}).u \quad (118.1)$$
$$y = h(\underline{x})$$

The derivative of its output may be articulated thus:

$$\frac{dy}{dt} = f(\underline{x}, u)$$

Noting that this involves differentiation of a scalar by a vector, the chain rule may be applied:

$$\frac{dy}{dt} = \frac{dy}{dx_1}.\frac{dx_1}{dt} + \frac{dy}{dx_2}.\frac{dx_2}{dt} + \ldots + \frac{dy}{dx_n}.\frac{dx_n}{dt}$$

$$= \frac{dy}{d\underline{x}}^T.\frac{d\underline{x}}{dt}$$

Substituting from Equation 118.1 gives:

$$\frac{dy}{dt} = \frac{dh(\underline{x})}{d\underline{x}}^T.(F(\underline{x}) + G(\underline{x}).u) \quad (118.8)$$

Now presume that the desired response to an error is to force the output back to its set point at a speed proportional to the error:

$$\left.\frac{dy}{dt}\right|_d = \alpha_1.(r-y)$$

For persistent errors it is desirable to accelerate towards the set point in proportion to the duration of the error, so an integral term can be included:

$$\left.\frac{dy}{dt}\right|_d = \alpha_1.(r-y) + \alpha_2.\int_0^t (r-y)dt \quad (118.9)$$

This clearly has the general form of a PI controller in which the tuning parameters α_1 and α_2 enable the output trajectory to be shaped. The control law comes from forcing the output to follow the desired trajectory:

$$\frac{dy}{dt} = \left.\frac{dy}{dt}\right|_d$$

Substituting from Equations 118.8 and 118.9 gives:

$$\frac{dh(\underline{x})^T}{d\underline{x}} \cdot (F(\underline{x}) + G(\underline{x}).u)$$
$$= \alpha_1.(r-y) + \alpha_2.\int_0^t (r-y)dt \quad (118.10)$$

This control law has two desirable features. First, it incorporates a non-linear model of the process, so any change in the process characteristics due to changes in either the set point or operating conditions are inherently compensated for. Second, because of the integral term, the output is forced towards the set point irrespective of any plant model mismatch.

Equation 118.10 is a non-linear equation with respect to the manipulated variable u which has to be solved at each step for implementation. Sometimes there is an analytical solution which enables u to be found explicitly in which case the implementation is straightforward. Otherwise some root finding method and/or numerical integration has to be used to find the change in u from its current value.

118.4 Application of GMC to CSTR

The use of GMC can be demonstrated by applying it to the CSTR considered previously. Starting from the definition of $\underline{x}$ and y given in Equation 118.5:

$$\frac{dh(\underline{x})^T}{d\underline{x}} = \left[\frac{d\theta_0}{dc_{A0}} \quad \frac{d\theta_0}{dc_{B0}} \quad \frac{d\theta_0}{d\theta_0}\right] = \begin{bmatrix}0 & 0 & 1\end{bmatrix} \quad (118.11)$$

Note that it is the output equation, $y = h(\underline{x})$, of Equation 118.1 that is being differentiated here rather than the state equation. It is the latter which establishes the relationship between the temperature inside the reactor θ_0 and the concentrations of the reagents c_{A0} and c_{B0}.

Substituting from Equations 118.5 to 118.7 into Equation 118.10 gives:

$$f_3(\underline{x}) + g_3(\underline{x}).u = \alpha_1.(r-y) + \alpha_2.\int_0^t (r-y)dt$$

$$-\frac{F}{V}.\theta_0 + \frac{\Delta H}{\rho c_p}.(-k_F.c_{A0} + k_R.c_{B0}) + \frac{F}{V}.\theta_1$$

$$= \alpha_1.(r-\theta_0) + \alpha_2.\int_0^t (r-\theta_0)dt$$

Solving for the manipulated variable gives:

$$\theta_1 = \theta_0 + \frac{V}{F}\left(\alpha_1.(r-\theta_0) + \alpha_2.\int_0^t (r-\theta_0)dt\right.$$
$$\left. -\frac{\Delta H}{\rho c_p}.(-k_F.c_{A0} + k_R.c_{B0})\right) \quad (118.12)$$

The structure of this equation is intuitively correct. The starting point is the temperature in the vessel θ_0. Say this is lower than the set point resulting in a positive error, and assume that both α_1 and α_2 are positive, then both the P and I actions will lead to the manipulated variable θ_1 being increased. Also, suppose that the concentration c_{A0} of reagent in the reactor increases. The net rate of forward reaction will be higher thereby generating more heat and should lead to a lower inlet temperature. Again noting that ΔH for an exothermic reaction is negative, it can be seen that the right hand term is negative which will have the desired effect.

The tuning of a generic model controller is best considered in relation to its ideal closed loop response. Taking the Laplace transform of Equation 118.9 gives:

$$\frac{y(s)}{r(s)} = \frac{\frac{\alpha_1}{\alpha_2}s + 1}{\frac{1}{\alpha_2}s^2 + \frac{\alpha_1}{\alpha_2}s + 1} \quad (118.13)$$

Whilst the closed loop transfer function is not the same as that of a second order system, because of the lag in the numerator, the shape of the response can nevertheless be characterised in terms of second order system parameters as defined in Table 72.1. Thus, for example, a damping factor and a settling time could be specified from which the values of α_1 and α_2 can be derived.

118.5 Worked Example on GMC

Consider a plant whose non linear dynamics are of the form of Equation 118.1:

$$\frac{dx}{dt} = x + 2x^2 u$$

$$y = -\frac{1}{2x}$$

Thus, differentiating:

$$\dot{y} = \frac{dy}{dx} \cdot \frac{dx}{dt} = \frac{1}{2x^2}.(x + 2x^2 u) = \frac{1}{2x} + u = -y + u$$

But GMC requires that $\dot{y} = \dot{y}|_d$ which, from Equation 118.9, results in:

$$u = y + \alpha_1.(r - y) + \alpha_2. \int_0^t (r - y) dt$$

The control action is clearly linear but is dependant upon being able to effectively measure the state. The functionality of the controller is depicted in Figure 118.3.

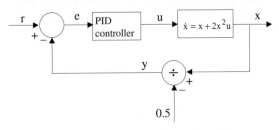

Fig. 118.3 Block diagram for Worked Example on GMC

118.6 Lie Derivatives and Relative Order

The Lie derivative of $h(\underline{x})$ with respect to $f(\underline{x})$ is written as $L_f h(\underline{x})$. It is defined as:

$$L_f h(\underline{x}) = \frac{\partial h(\underline{x})}{\partial x_1}.f_1(\underline{x}) + \frac{\partial h(\underline{x})}{\partial x_2}.f_2(\underline{x}) + \ldots$$

$$+ \frac{\partial h(\underline{x})}{\partial x_n}.f_n(\underline{x}) \quad (118.14)$$

For the CSTR example:

$$\frac{dh(\underline{x})}{dx_1} = \frac{d\theta_0}{dc_{A0}} = 0 \quad \frac{dh(\underline{x})}{dx_2} = \frac{d\theta_0}{dc_{B0}} = 0$$

$$\frac{dh(\underline{x})}{dx_3} = \frac{d\theta_0}{d\theta_0} = 1$$

whence:

$$L_f h(\underline{x}) = f_3(\underline{x}) = -\frac{F}{V}.\theta_0 + \frac{\Delta H}{\rho c_p}.(-k_F.c_{A0} + k_R.c_{B0})$$

Similarly the Lie derivative of $h(\underline{x})$ with respect to $g(\underline{x})$ is written as $L_g h(\underline{x})$ and is defined to be:

$$L_g h(\underline{x}) = \frac{\partial h(\underline{x})}{\partial x_1}.g_1(\underline{x}) + \frac{\partial h(\underline{x})}{\partial x_2}.g_2(\underline{x}) + \ldots$$

$$+ \frac{\partial h(\underline{x})}{\partial x_n}.g_n(\underline{x}) \quad (118.15)$$

and for the CSTR:

$$L_g h(\underline{x}) = g_3(\underline{x}) = \frac{F}{V}$$

Finally, the Lie derivative of $L_f h(\underline{x})$ with respect to $f(\underline{x})$, which is denoted by $L_f^2 h(\underline{x})$, is:

$$L_f^2 h(\underline{x}) = \frac{\partial L_f h(\underline{x})}{\partial x_1} . f_1(\underline{x}) + \frac{\partial L_f h(\underline{x})}{\partial x_2} . f_2(\underline{x}) + \ldots$$
$$+ \frac{\partial L_f h(\underline{x})}{\partial x_n} . f_n(\underline{x}) \qquad (118.16)$$

and likewise:

$$L_g L_f h(\underline{x}) = \frac{\partial L_f h(\underline{x})}{\partial x_1} . g_1(\underline{x}) + \frac{\partial L_f h(\underline{x})}{\partial x_2} . g_2(\underline{x}) + \ldots$$
$$+ \frac{\partial L_f h(\underline{x})}{\partial x_n} . g_n(\underline{x}) \qquad (118.17)$$

The term relative order is a measure of how non-linear a process is. It also gives an insight as to how directly the manipulated variable affects the controlled variable. In essence the greater the relative order the greater the non-linearity but the less the direct affect of u on y.

For the system represented by Equation 118.1, rearrangement of Equation 118.8 gives:

$$\frac{dy}{dt} = \frac{dh(\underline{x})}{d\underline{x}}^T . F(\underline{x}) + \frac{dh(\underline{x})}{d\underline{x}}^T . G(\underline{x}).u$$

Substituting from Equations 118.14 and 118.15, this can be written in terms of lie derivatives as follows:

$$\frac{dy}{dt} = L_f h(\underline{x}) + L_g h(\underline{x}).u \qquad (118.18)$$

If $L_g h(\underline{x}) \neq 0$ then $\dot{y}$ is directly related to u and the relative order of the system is 1.

If $L_g h(\underline{x}) = 0$ then $\dot{y}$ is not explicitly related to u and the order must be greater:

$$\frac{dy}{dt} = L_f h(\underline{x}) \qquad (118.19)$$

To obtain an explicit relationship between the manipulated and controlled variables when the relative order is greater than 1, it is necessary to establish a higher order derivative of the output:

$$\frac{d^2 y}{dt^2} = \frac{d}{dt}\left(\frac{dy}{dt}\right) = \left(\frac{d}{d\underline{x}}\left(\frac{dy}{dt}\right)\right)^T . \frac{d\underline{x}}{dt}$$
$$= \left(\frac{d}{d\underline{x}} L_f h(\underline{x})\right)^T . \dot{\underline{x}}$$

Substituting from Equation 118.1 gives:

$$\frac{d^2 y}{dt^2} = \left(\frac{d}{d\underline{x}} L_f h(\underline{x})\right)^T . (F(\underline{x}) + G(\underline{x}).u)$$
$$= \frac{\partial L_f h(\underline{x})}{\partial x_1} . f_1(\underline{x}) + \frac{\partial L_f h(\underline{x})}{\partial x_2} . f_2(\underline{x}) + \ldots$$
$$+ \frac{\partial L_f h(\underline{x})}{\partial x_n} . f_n(\underline{x})$$
$$+ \left\{ \frac{\partial L_f h(\underline{x})}{\partial x_1} . g_1(\underline{x}) + \frac{\partial L_f h(\underline{x})}{\partial x_2} . g_2(\underline{x}) + \ldots \right.$$
$$\left. + \frac{\partial L_f h(\underline{x})}{\partial x_n} . g_n(\underline{x}) \right\} . u$$

Substituting from Equations 118.16 and 118.17 gives:

$$\frac{d^2 y}{dt^2} = L_f^2 h(\underline{x}) + L_g L_f h(\underline{x}).u$$

If $L_g L_f h(\underline{x}) \neq 0$ then $\ddot{y}$ is a direct function of u and the relative order of the system is 2.

If $L_g L_f h(\underline{x}) = 0$ then $\ddot{y}$ is not explicitly related to u and the order must be greater:

$$\frac{d^2 y}{dt^2} = L_f^2 h(\underline{x}) \qquad (118.20)$$

The output is successively differentiated with respect to time until the resulting derivative is a linear function of the input signal. If the relative order is r, then the result of successive differentiation is:

$$\frac{d^r y}{dt^r} = L_f^r h(\underline{x}) + L_g L_f^{r-1} h(\underline{x}).u \qquad (118.21)$$

118.7 Globally Linearising Control

Figure 118.4 depicts the essential features of a GLC strategy. The objective is to design a linearising transform $\phi(\underline{x}, v)$ so that a linear relationship is obtained between the controlled variable y and the

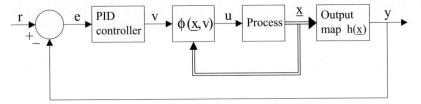

Fig. 118.4 Structure of a globally linearising control (GLC) strategy

controller output v. The transform must be a function of $\underline{x}$ and v, and give a linear relationship between y and v. Furthermore, it must be explicitly solvable for the manipulated variable u.

First the relative order r must be found by successive differentiation. A transform is then chosen of the form:

$$\sum_{k=0}^{r} \beta_k \cdot (d^k y/dt^k) = v$$

which may be expanded thus:

$$v = \beta_0 \cdot y + \beta_1 \cdot \dot{y} + \beta_2 \cdot \ddot{y} + \ldots + \beta_r \cdot y^r \quad (118.22)$$

Substitute in terms of Lie derivatives from Equations 118.19 to 118.21 gives:

$$v = \beta_0 \cdot h(\underline{x}) + \beta_1 \cdot L_f \cdot h(\underline{x}) + \beta_2 \cdot L_f^2 h(\underline{x}) + \ldots$$
$$+ \beta_r \cdot L_f^r h(\underline{x}) + \beta_r \cdot L_g L_f^{r-1} h(\underline{x}) \cdot u$$

This can be solved for the controlled variable:

$$u = \phi(\underline{x}, v)$$
$$= \frac{1}{\beta_r L_g L_f^{r-1} h(\underline{x})} \cdot \left\{ v - \left(\beta_0 h(\underline{x}) \right. \right. \quad (118.23)$$
$$\left. \left. + \beta_1 L_f h(\underline{x}) + \beta_2 L_f^2 h(\underline{x}) + \ldots + \beta_r L_f^r h(\underline{x}) \right) \right\}$$

The characteristic equation of the system of Figure 118.4 provides the basis for design. First, take the Laplace transform of Equation 118.22 and rearrange to yield the open loop transfer function:

$$y(s) = \frac{1}{\beta_0 + \beta_1 s + \beta_2 s^2 + \ldots + \beta_r s^r} \cdot v(s)$$

Next choose a linear control function, such as a PI controller:

$$v(s) = K_C \cdot \left(1 + \frac{1}{T_R s} \right) \cdot e(s)$$

Combining these gives the characteristic equation:

$$\beta_r s^r + \beta_{r-1} s^{r-1} + \ldots + \beta_1 s + (\beta_0 + K_C) + \frac{K_C}{T_R s} = 0$$

The β_k coefficients for $k = 0, 1, \ldots r$ are arbitrarily chosen to ensure the poles of the linearised system are placed in positions that ensure stability of the closed loop. The gain and reset time are then chosen to provide the required form of closed loop response.

118.8 Application of GLC to CSTR

To illustrate globally linearising control consider again the continuous stirred tank reactor of Figure 118.2.

It has been shown that $L_g h(\underline{x}) = \frac{F}{V} \neq 0$ so the relative order of the CSTR must be 1.

Thus, from Equation 118.23:

$$u = \phi(\underline{x}, v) = \frac{v - (\beta_0 \cdot h(\underline{x}) + \beta_1 \cdot L_f \cdot h(\underline{x}))}{\beta_1 \cdot L_g h(\underline{x})}$$

Remember that $y = h(\underline{x}) = \theta_0$ and $u = \theta_1$. Also that $L_f h(\underline{x}) = f_3(\underline{x})$ where:

$$f_3(\underline{x}) = -\frac{F}{V} \cdot \theta_0 + \frac{\Delta H}{\rho c_p} \cdot (-k_F \cdot c_{A0} + k_R \cdot c_{B0})$$

Substitution yields the linearising transformation:

$$\theta_1 = \frac{1}{\beta_1 \frac{F}{V}} \left\{ v - \left(\beta_0 \theta_0 + \beta_1 \left(-\frac{F}{V} \cdot \theta_0 \right. \right. \right.$$
$$\left. \left. \left. + \frac{\Delta H}{\rho c_p} \cdot (-k_F c_{A0} + k_R c_{B0}) \right) \right) \right\} \quad (118.24)$$

118.9 Comments on GLC

In principle, realisation is fairly straight forward. Knowing the controller output, the various coefficients and process constants, and the current value of all the process variables, the output of the transformation is continuously evaluated. This is the manipulated variable which, in effect, becomes the set point of the slave loop.

The major advantage of GLC is that it enables non-linearity to be handled explicitly within the control loop. Note that all the variables used in Equation 118.24 are in absolute form. The use of deviation variables does not make sense in the context of GLC.

There are several drawbacks to GLC. The relative order r must be finite. The method requires accurate non-linear models: otherwise plant model mismatch (PMM) occurs, the effects of which are not well understood. GLC is critically dependent upon being able to continuously measure all of the process variables involved in the linearising transform. Non-linear observers are difficult to design. And finally, the whole concept is based upon analogue signals whereas, in practice, the transform would have to be realised by means of an algorithm using discrete signals subject to sampling errors and dynamics.

There is no pretence here that GLC, or indeed GMC, is used to any significant extent in industry, if at all. The objective has simply been to introduce some of the techniques available for non-linear control system design. Whilst they can be readily applied to simple models, such as that of the ideal CSTR, it is with more complex systems that they probably have most potential benefit. The techniques have promise, but much needs to be done before robust implementations become available. For example, methods for developing accurate reduced-order dynamic non-linear models have yet to emerge. And the consequences of uncertainty in model structure and PMM are still not properly understood. Nevertheless, non-linear control is attracting much research effort and it is techniques such as GLC and GMC that will provide the platform for the next generation of MPC: that is, non-linear model predictive control.

118.10 Nomenclature

B	frequency factor	$m^3\ kmol^{-1}\ s^{-1}$
c	concentration	$kmol\ m^{-3}$
c_p	specific heat	$kJ\ kmol^{-1}\ K^{-1}$
E	activation energy	$kJ\ kmol^{-1}$
F	reactor feed rate	$m^3\ s^{-1}$
ΔH	heat of reaction	$kJ\ kmol^{-1}$
k	rate constant	s^{-1}
R	universal gas constant (8.314)	$kJ\ kmol^{-1}\ K^{-1}$
t	time	s
V	volume of reactor contents	m^3
ρ	density	$kmol\ m^{-3}$
θ	temperature	°C
e	error	
r	reference (set point)	
u	manipulated variable	
v	controller output	
x	state	
y	controlled variable	

Subscripts
A   reagent
B   product
d   desired
F   forward reaction
R   reverse reaction
0   outlet
1   inlet

Suffix
r   relative order

Bibliography

Books

Andrews JD, Moss TR (1993) Reliability and risk assessment. Longman

Astrom K, Hagglund T (1995) PID Controllers: theory, design and tuning, 2nd edn. ISA

Astrom K, Wittenmark B (1997) Computer controlled systems, 4th edn. Prentice Hall

Atkey DA (2005) Cabling installations: user friendly guide, 2nd edn. ERA Technology

Baumann HD (2003) Introduction to control valves, 3rd edn. ISA, Carolina

Bentley JP (2004) Principles of measurement systems, 4th edn. Prentice Hall

Braithwite A, Smith FJ (1995) Chromatographic methods, 5th edn. Blackie

Brosilow C (2002) Techniques of model based control. Prentice Hall

Buckley PS, Luyben WL, Shunta JP (1985) Design of distillation column control schemes. Arnold (for ISA), Carolina

Camacho EF, Bordons C (1999) Model predictive control. Springer, Berlin Heidelberg New York

Caulcutt RA (1991) Statistics in research and development, 2nd edn. Chapman and Hall

Chatfield C (1991) Statistics for technology, 4th edn. Chapman and Hall

Coughanowr DR (1991) Process systems analysis and control, 2nd edn. McGraw Hill

Coulson JM, Richardson JF *et al.* (up to 2004) Chemical engineering, vols 1–6, various editions. Elsevier

Cox TF (2005) An introduction to multivariate data analysis, Hodder, London

Dabney JB, Harman TL (2004) Mastering Simulink. Prentice Hall

Danen GWA (1985) Shell flowmeter engineering handbook, 2nd edn. McGraw Hill

Deshpande PB (1985) Distillation dynamics and control. ISA, Carolina

Deshpande PB (1989) Multivariable process control. ISA, Carolina

Driankov D, Hellendoorn H, Reinfrank M (1996) An introduction to fuzzy control, 2nd edn. Springer, Berlin Heidelberg New York

Driskell LR (1983) Control valve selection and sizing. ISA, Carolina

Dukelow SG (1991) The control of boilers, 2nd edn. ISA, Carolina

Duncan T (1997) Electronics for today and tomorrow, 2nd edn. John Murray

Dutton K, Thompson S, Barraclough B (1997) The art of control engineering. Addison Wesley

Edgar TF, Himmelblau DM, Lasdon LS (2001) Optimisation of chemical processes, 2nd edn. McGraw Hill

Elmasri R, Navathe SB (2003) Fundamentals of database systems. Addison Wesley

Fisher T (1990) Batch control systems: design, application and implementation. ISA, Carolina

Fleming DW, Pillai V (1998) S88 implementation guide. McGraw Hill

Garside R (1991) Electical apparatus and hazardous areas. Hexagon Technology, Bucks

Garside R (1995) Intrinsically safe instrumentation: a guide, 3rd edn. Hexagon Technology, Bucks

Gill EG, Murray W, Wright MH (1986) Practical optimisation. Academic Press

Goble WM (1998) Control system safety evaluation and reliability, 2nd edn. ISA, Carolina

Goldberg DE (1989) Genetic algorithms in search, optimisation and machine learning. Addison Wesley

Hanselman D, Littlefield B (2005) Mastering Matlab. Prentice Hall

Hicks JO (1993) Management information systems: a user perspective, 3rd edn. West

Hillier FS, Lieberman GJ (2004) Introduction to operations research. McGraw-Hill

Horsley D (1998) Process plant commissioning, 2nd edn. IChemE, Rugby

Hughes E (2005) Electrical and electronic technology, 9th edn. Prentice Hall

Jackson P (1998) Introduction to expert systems, 3rd edn. Addison Wesley

Jeffrey A (2002) Advanced engineering mathematics. Harcourt Academic Press

Kleitz W (2003) Digital and microprocessor fundamentals: theory and application, 4th edn. Prentice Hall

Kletz T (1995) Computer control and human error. IChemE, Rugby

Kletz T (1999) HAZOP and HAZAN, 4th edn. IChemE, Rugby

Lee PL, Newell RB, Cameron IT (1998) Process control and management. Blackie (Chapman and Hall)

Lees FP (2005) Loss prevention in the process industries: hazard identification assessment and control, vols 1–3, 3rd edn. Butterworth Heinemann

Leigh JR (1992) Applied digital control, 2nd edn. Prentice Hall

Lewis RW (1998) Programming industrial control systems using IEC 1131-3, 2nd edn. IET, London

Lewis RW (2001) Modelling control systems using IEC 61499. IET, London

Liptak BG (1995) Process control, vol 2. Instrument engineer's handbook, 3rd edn. Chilton

Liptak BG (2002) Process software and digital networks, vol 3. Instrument engineer's handbook, 1st edn. Chilton

Liptak BG (2003) Process measurement analysis, vol 1. Instrument engineer's handbook, 4th edn. Chilton

Luyben WL (1990) Process modelling simulation and control for chemical engineers, 2nd edn McGraw Hill

Luyben WL (1992) Practical distillation control. Van Nostrand

Manly BFJ (1994) Multivariate statistical methods: a primer, 2nd edn. Chapman and Hall

Marlin TE (2000) Process control: designing processes and control systems for dynamic performance, 2nd edn. McGraw Hill

McGhee J, Grimble MJ, Mowforth P (1990) Knowledge based systems for industrial control. IET, London

Metcalfe IS (1999) Chemical reaction engineering: A First Course, Oxford University Press

Mitchell M (1996) An introduction to genetic algorithms. MIT Press, Massachusetts

Montgomerie DC, Runger GC (2006) Applied statistics and probability for engineers, 4th edn. John Wiley & Sons, 2006

Morari M, Garcia CE, Lee JH, Prett DM (1994) Model predictive control. Prentice Hall

Morris A (1993) Principles of measurement and instrumentation, 2nd edn. Prentice Hall

Morris AS (1997) Measurement and calibration requirements: for quality assurance to ISO 9000. Wiley

Morrison DF (1990) Multivariate statistical methods. McGraw Hill

Murrill PW (1988) Application concepts of process control. ISA

Murrill PW (1991) Fundamentals of process control theory. ISA

Oakland JS (2002) Statistical process control, 5th edn. Elsevier

Ogata K (2002) Modern control engineering, 4th edn. Prentice Hall

Ogunnaike BA, Ray WH (1994) Process dynamics, modelling and control. Oxford University Press

Parshall J, Lamb L (2000) Applying S88: batch control from a user's perspective. ISA

Pitt MJ, Preece PE (1990) Instrumentation and automation in process control. Ellis Horwood

Redmill F, Chudleigh M, Catmur J (1999) System safety: HAZOP and software HAZOP. Wiley

Roffel B, Betlem B (2006) Process dynamics and control: Modelling for Control and Prediction, Wiley

Sawyer P (1993) Computer controlled batch processing. IChemE, Rugby

Seborg DE, Edgar TF, Mellichamp DA (2004) Process dynamics and control, 2nd edn Wiley

Shinsky FG (1973) pH and pION control in process and waste streams. Wiley

Shinsky FG (1977) Distillation control. McGraw Hill

Shinsky FG (1996) Process control systems: application, design and tuning, 4th edn. McGraw Hill

Shunta JP (1995) Achieving world class manufacturing through process control. Prentice Hall

Smith DJ (2005) Reliability, maintainability and risk, 7th edn. Butterworth Heineman

Smith CA, Corripio A (2006) Principles and practice of automatic process control, 3rd edn. Wiley

Stephanopoulos G (1984) Chemical process control, 2nd edn. Prentice Hall

Stroud KA (2003) Advanced engineering mathematics, 4th edn. MacMillan (Palgrave)

Svrcek WY, Mahoney DP, Young BR (2000) A real-time approach to process control. Wiley

Thompson LM (1997) Industrial data communications, 2nd edn. ISA

Warnes LA (1994) Electronic and electrical engineering: principles and practice. MacMillan Press

Warnock IG (1988) Programmable controllers: operation and application. Prentice Hall

Warwick K, Irwin GW, Hunt KJ (1992) Neural networks for control and systems. IET, London

Wells G (1996) Hazard identification and risk assessment. IChemE, Rugby

Wellstead PE, Zarrop MB (1991) Self-tuning systems: control and signal processing. Wiley

Wilkie J, Johnson M, Katebi R (2002) Control engineering: an introductory course. McMillan (Palgrave)

Willis N (1993) Computer architecture and communications, 2nd edn. Blackwell Scientific

Wright D (2004) An engineer's guide to the model forms of conditions of contract for process plant, 4th edn. IChemE, Rugby

Wynne RJ, Brook D (1988) Signal processing: principles and applications. Edward Arnold

Zalzala AMS, Fleming PJ (1997) Genetic algorithms in engineering systems. IET, London

Standards and Codes of Practice

AGA Report 12, Cryptographic Protection of SCADA Communications General Recommendations, Draft 3, American Gas Association, 2004

API 1164, Pipeline SCADA Security, 1st edn, American Petroleum Institute, 2004

API RP 14C, Analysis, Design, Installation and Testing of Basic Surface Safety Systems for Offshore Production Platforms, 6th edn, American Petroleum Institute, 1998

ATEX, Equipment Intended for Use in Potentially Explosive Atmospheres, Directive 94/9/EC, 1994

BS 1041, Part 3, Temperature Measurement: Guide to the Selection and Use of Industrial Resistance Thermometers, BSI, London, 1989

BS 1041, Part 4, Temperature Measurement: Guide to the Selection and Use of Thermocouples, BSI, London, 1992

BS 1041, Part 5, Temperature Measurement: Guide to the Selection and Use of Pyrometers, BSI, London, 1989

BS 1042, Measurement of fluid flow in closed circuits, BSI, London, 1991

BS 1646, Symbolic Representation for Process Measurement Control Functions and Instrumentation, BSI, London, 1984

BS 1904, Specification for Industrial Platinum Resistance Thermometer Sensors, BSI, London, 1984

BS 2765, Specification for Dimensions of Temperature Detecting Elements and Corresponding Pockets, BSI, London, 1981

BS 4937, International Thermocouple Reference Tables, BSI, London, 1974

BS 5750, Quality Systems, Parts 1–4, BSI, London, 1987

BS 5792, Measurement of Conductive Liquid Flow in Closed Conduits, BSI, London, 1993

BS 6739, Instrumentation in Process Control Systems: Installation Design and Practice, BSI, London, 1986

BS 7671, The IET Wiring Regulations, Incorporates Amendments 1 and 2, IET, London, 2001

BS 7799, Part 2, Specification for Information Security Management Systems, Revised Edition, London, 1999

DIN 41494, Part 1, Panel Mounting Racks for Electronic Equipment

EN 500XX, Electrical Apparatus for use in Potentially Explosive Atmospheres, BSI, London (see Table 52.5)

IEC 17799, Information Technology - Code of Practice for Information Security Management, 2000

IEC 60079, Electrical Apparatus for Explosive Gas Atmospheres, Various parts, 2004

IEC 60297, Part 1, Dimensions of Mechanical Structures: Panels and Racks, 3rd Edition, 1986

IEC 60534, Parts 1-8, Industrial Process Control Valves, Various editions

IEC 60584, Part 1, Themocouples: Reference Tables, 2nd Edition, 1995

IEC 60584, Part 2, Themocouples: Tolerances, 1982

IEC 61000, Part 5, Electromagnetic Compatibility: Installation and Mitigating Guidelines, Section 2, Earthing and Cabling, 1997

IEC 61131, Part 3, Programmable Controllers: Programming Languages, 1993

IEC 61158, Parts 1-6, Digital Data Communications for Measurement and Control: Fieldbus for use in Industrial Control Systems, 3rd edn. 2004

IEC 61499, Function Blocks for Industrial Process Measurement and Control Systems, Parts 1-4, 2005

IEC 61508, Functional Safety of Electrical, Electronic and Programmable Electronic Safety Related Systems, Parts 1-7, 1998

IEC 61511, Functional Safety of Safety Instrumented Systems for the Process Industry. Draft standard, Parts 1-3, 2003

IEC 61512, Part 1, Batch Control: Models and Teminology, 1997, Formerly known as S88 Part 1, published by ISA, 1995

IEC 61512, Part 2, Batch Control: Data Structures and Guidelines for Languages, 2003, Formerly known as S88 Part 2, published by ISA, 2003

IEC 61713, Guide to Software Dependability through the Software Life Cycle, Draft standard

IEC 61784, Part 1, Digital Data Communications for Measurement and Control: Profile Sets for Continuous and Discrete Manufacturing Relative to Fieldbus Use in Industrial Control Systems, 2003

IEC 61882, Hazard and Operabilty (HAZOP) Studies: A Guideword Approach, 2001

ISA RP31.1, Specification, Installation and Calibration of Turbine Flowmeters, ISA, Carolina, 1977

ISA S5.1, Instrumentation Symbols and Identification, ISA, Carolina, 1984

ISA S5.3, Graphic Symbols for Distributed Control/Shared Display Instrumentation, Logic and Computer Systems, ISA, Carolina, 1982

ISA S84, Application of Safety Instrument Systems for the Process Industries, ISA, Carolina, 1996

ISA S95-1, Enterprise Control System Integration, Part 1: Models and Terminology, ISA, Carolina, 2001

ISA S95-2, Enterprise Control System Integration, Part 2: Data Structures and Attributes, ISA, Carolina, 2001

ISA TR84, Safety Instrumentation Systems (SIS): Safety Integrity Level (SIL): Evaluation Techniques, Parts 1-5, ISA, Carolina, 2002

ISO 5167, Parts 1-4, Measurement of Fluid Flow by Means of Pressure Differential Devices Inserted in Circular Cross-Section Conduits Running Full, 2003

ISO 7498, Part 1, Information Technology – Open Systems Interconnection – Basic Reference Model: The Basic Model, 1995

ISO 8859, Part 1, Eight Bit Single Byte Coded Graphic Character Sets: Specification for Latin Alphabet No 1 (1987)

ISO 9000–9004, Parts 0–4, Quality Management Systems, 1987

ISO 10303, Standard for the Exchange of Product Model Data (STEP), 2002

ISO 12207, Information Technology: Software Lifecycle Processes, 1995

MOD 00-58, Ministry of Defence Standard, HAZOP: Studies on Systems Containing Programmable Electronics, London, 2000

Guides

API: RP 551, Process Measurement Instrumentation, 1993

API: RP 552, Transmission Systems, 1994

API: RP 554, Process Instrumentation and Control, 1995

CIA: A Guide to Hazard and Operability Studies, published by Chemical Industries Association, London, 1992

EEMUA: No 175, Code of Practice for Calibration and Checking Process Analysers, London, 1995

EEMUA: No 178, A Design Guide for the Electrical Safety of Instruments, Instrument/Control Panels and Control Systems, London, 1994

EEMUA: No 187, Analyser Systems: A Guide to Maintenance Management, London, 2000

EEMUA: No 189, A Guide to Fieldbus Application in the Process Industry, London, 1998

EEMUA: No 191, Alarm Systems, A Guide to Design, Management and Procurement, London, 1999

EEMUA: No 201, Process Plant Control Desks Utilising Human-Computer Interfaces: A Guide to Design, Operational and Human Interface Issues, London, 2002

GAMP: Guide for Validation of Automated Systems in Pharmaceutical Manufacture, Version 4, International Society for Pharmaceutical Engineering (ISPE), Florida, 2002

HMSO: Guidance on HAZOP Procedures for Computer Controlled Plants, HSE Contract Research Report No 26, 1991

HMSO (Orange Book), Rules and Guidance for Pharmaceutical Manufacturers and Distributors, London, 1997

HSE: Programmable Electronic Systems in Safety Related Applications, Parts 1 and 2, Health and Safety Executive, HMSO, 1987

IChemE and CIA, HAZOP: Guidelines to Best Practice for the Process and Chemical Industries, Rugby, 1999

IChemE (Green Book): Model Form of Conditions of Contract for Process Plant, Reimbursable Contracts, 3rd Edition, Rugby, 2002

IChemE (Red Book): Model Form of Conditions of Contract for Process Plant, Lump Sum Contracts, 4th Edition, Rugby, 2001

IChemE (Yellow Book): Model Form of Conditions of Contract for Process Plant. Subcontracts, 3rd Edition, Rugby, 2003

IET, Guidelines for the Documentation of Computer Software for Real-Time and Interactive Systems, 2nd Edition, London, 1990

IET, Wiring Regulations (2001), see BS 7671

IMechE/IET, Model Forms of General Conditions of Contract (MF/1), London, 2002

IMechE/IET, Model Forms of General Conditions of Contract (MF/2), London, 1999

InstMC, Instrument Engineers Yearbook, London, Published annually

IoP, Area Classification Code for Petroleum Installations, Part 15 of Model Code of Safe Practice in the Petroleum Industry, Wiley (for the Institute of Petroleum, London), 1990

NISCC, Good Practice Guide to Process Control and SCADA Security, National Infrastructure Security Co-ordination Centre, London, 2005

NISCC, Good Practice Guide on Firewall Deployment for SCADA and Process Control Networks, National Infrastructure Security Co-ordination Centre, London, 2005

OREDA, Offshore Reliability Data Handbook, 4th Edition, DNV Technica, Norway, 2002

STARTS Purchasers Handbook, Procuring Software Based Products, National Computing Centre, Manchester, 1989

TickIT, Guide to Software Quality Management System Construction and Certification Using ISO 9001, DTI, London, 1990

Reports and Papers

Sproston JL, Johnson MW, Pursley WC (1987) Mass flow measurement, flow measurement and instrumentation consortium. National Engineering Laboratory

Thompson A (1964) Operating experience with direct digital control. IFAC Conference, Stockholm

Abbreviations and Acronyms

a.c.	Alternating current	51.2
ACL	Access control list	49.11
A/D	Analogue to digital converter	44.1
AI	Artificial intelligence	107.0
AIN	Analogue input (card or function block)	44.2
ALARP	As low as reasonably practical	56.2
ALU	Arithmetic and logic unit	9.3
AM	Application module	38.5
AMS	Alarm management system	43.6
ANN	Artificial neural network	109.0
AOT	Analogue output (card or function block)	44.5
APC	Advanced process control	60.3
API	American Petroleum Institute	
	Application program interface	49.4
ARK	Archive (package) or history module	41.4
ARMAX	Auto regressive moving average with exogeneous input	114.6
ASL	Asymptotic sample length	115.4
ASM	Abnormal situation management	43.6
ATEX	Explosive atmosphere	52.14
AUTO	Automatic mode of operation	3.4
BACS	Banking automated clearing system	100.2
BASEEFA	British approvals service for electrical equipment in flammable atmospheres	52.7
BLS	Batch least squares	83.7
BPC	Batch process control (package)	41.6
BPCS	Basic process control system	55.1
BS	British Standards Institute	
CAD	Computer aided design	49.5
C&I	Control and instrumentation	59.9

CARIMA	Controlled auto regressive integrated moving average	117.2
CARMA	Controlled auto regressive moving average	114.6
CASE	Computer aided software engineering	63.7
CCF	Cumulative cash flow	59.11
CCITT	International Telegraph and Telephone Consultative Committee	40.4
CD	Compact disc	9.4
CE	Electromagnetic compatibility	52.14
CENELEC	European Electrical Standards Coordination Committee	52.7
CERT	Computer emergency response team	49.9
CF	Cash flow	59.11
CHAZOP	Computer hazard and operability (study)	54.3
CIA	Chemical Industries Association	
CIM	Computer integrated manufacture	100.0
CIP	Common industrial protocol	49.11
CLTF	Closed loop transfer function	71.5
CMS	Code management system	63.7
COMAH	Control of major accident hazards	54.0
COOP	Control and operability study	54.4
COSHH	Control of substances hazardous to health	54.0
CPD	Cumulative probability distribution	82.3
CPU	Central processing unit	9.1
CSMA/CD	Carrier sense multiple access / collision detection	40.2
CSTR	Continuous flow stirred tank reactor	36.0
CV	Controlled variable	3.1
D/A	Digital to analogue converter	44.5
DBMS	Database management system	99.3
d.c.	Direct current	4.2
DC	Diagnostic coverage	53.8
DCF	Discounted cash flow	59.11
DCS	Distributed control system	38.5
DDC	Direct digital control	38.3
DDE	Dynamic data exchange	49.5
DDL	Data definition language	99.3
	Device description language	50.7
DDS	Device description services	50.7
DFS	Detailed functional specification	62.0

DIN	Deutsches Institut für Normung (German standards authority)	
	Discrete input (card or function block)	46.4
DMC	Dynamic matrix control	117
DMZ	Demilitarised zone	49.12
DOC	Dynamic opimising controller	106.7
	Dynamic optimisation capability	106.7
DOT	Discrete output (card or function block)	46.1
DP	cell Differential pressure cell	11.0
DPI	Deep packet inspection	49.11
DPMO	Defects per million opportunities	102.8
DR	Demand rate	53.10
	Discount rate (fractional)	59.11
DV	Disturbance variable	3.1
ECP	Engineers control program	41.8
ECN	Effective carbon number	18.4
EDI	Electronic data interchange	100.2
EEMUA	Engineering Equipment and Materials Users Association	
ELS	Extended least squares	114.6
EMC	Electromagnetic compatibility	51.3
emf	Electro motive force	16.1
ER	Entity relationship	99.7
ERRF	External risk reduction facility	56.2
ES	Expert system	107.3
ESD	Emergency shut down (system)	55.3
FAM	Fuzzy associative memory	108.3
FANN	Feedforward artificial neural network	109.1
FAT	Factory acceptance testing	63.13
F&G	Fire and gas (detection system)	55.3
FBD	Function block diagram	48.5
FDA	United States Food and Drugs Administration	66.3
FDM	Frequency division multiplexing	40.7
FDT	Field device tool	50.7
FF	Foundation Fieldbus	50.3
FID	Flame ionisation detector	18.4
FIL	Filter (routine)	44.3
FIP	Factory instrumentation protocol	50.3

FIR	Finite inpulse response	114.1
FLC	Fuzzy logic controller	108.0
FM	Frequency modulated	40.8
FMEA	Failure mode and effect analysis	54.6
FPMH	Failures per million hours	53.1
FPTF	Forward path transfer function	71.5
FS	File server	38.10
FSK	Frequency shift keying	40.7
FTA	Fault tree analysis	54.7
FTP	Field termination panel	51.1
	File transfer protocol	49.11
GA	Genetic algorithm	110.0
GAMP	Good automated manufacturing practice	66.4
GC	Gas chromatography	18.1
GLC	Globally linearising control	118.5
GLS	Generalised least squares	114.5
GM	Gain margin	73.12
GMC	Generic model control	118.3
GMV	Genaralised minimum variance	116.5
GPC	Generalised predictive control	117.0
GUI	Graphical user interface	97.0
GW	Gateway	38.7
HART	Highway addressable remote transducer	50.1
HAZOP	Hazard and operability	54.1
HDLC	High-level data-link control	40.4
HIPS	High integrity protection system	55.6
HLL	High level language	41.7
HM	History module	38.5
HR	Hazard rate	53.10
HRF	Hazard reduction factor	53.8
HSE	Health and Safety Executive	
HTML	Hypertext markup language	49.7
HTTP	Hypertext transfer protocol	49.11
IC	Integrated circuit	6.0
ICE	Institution of Civil Engineers	

IChemE	Institution of Chemical Engineers	
ICS	Integrated control system	38.4
IEC	International Electrotechnical Commission	
IEE	Institution of Electrical Engineers	
IEEE	Institute of Electrical and Electronic Engineers (US)	
IF	Inflation factor (fractional)	59.11
IIF	Input interface	38.1
IIR	Infinite impulse response	114.1
IMechE	Institution of Mechanical Engineers	
I/O	Input and output signals or channels	2.2
IoP	Institute of Petroleum (constituent of the Energy Institute)	
IP	Integer programming	103.5
	Internet protocol	49.11
	Input (measured) variable	3.1
I/P	Current to pressure converter	44.6
IPSE	Integrated programming support environment	63.7
IR	Instruction register	9.2
IRR	Internal rate of return	59.11
IS	Intrinsically safe	52.0
ISA	International Society for Systems and Automation (formerly Instrument Society of America)	
ISDN	Integrated services digital network	40.7
ISO	International Standards Organisation	
JIT	Just in time	100.2
KBS	Knowledge based systems	107
KPI	Key performance indicator	59.8
LAC	Live activity checking	49.11
LAL	Lower action limit	102.4
LAN	Local area network	40.2
LAS	Link active scheduler	50.5
LCD	Liquid crystal display	23.1
LEC	Local equipment centre	51.1
LED	Light emitting diode	23.1
LEL	Lower explosive limit	52.2
LIMS	Laboratory information management systems	100.6

LISP	List processing	107.8
LLC	Logic link control	40.5
LP	Linear programming	103.1
LSE	Language sensitive editor	63.7
LSL	Lower service limit	102.7
LVDT	Linear variable displacement transducer	13.4
LWL	Lower warning limit	102.4
LTI	Linear time invariant	114.1
MAC	Media access control	40.5
MAP	Manufacturing automation protocol	40.4
MAN	Manual mode of operation	3.4
MAR	Memory address register	9.2
	Missing at random	102.2
MBR	Memory buffer register	9.5
MCC	Motor control centre	51.2
MCS	Minimum cut set	54.8
MDT	Mean down time	53.2
MEI	Major equipment item (or unit)	37.1
MES	Manufacturing execution system	100.5
MILP	Mixed integer linear programming	103.5
MIMO	Multiple input multiple output	80.0
MINLP	Mixed integer non linear programming	103.8
MIS	Management information system	100.0
MLP	Multi-layer perceptron	109.1
MLR	Multiple linear regression	83.1
MMI	Man-machine interface	58.0
MOGA	Multi objective genetic algorithm	110.9
MPC	Model (based) predictive control	117.0
MRP	Materials resource planning	100.3
MTBF	Mean time between failures	53.2
MTTF	Mean time to fail	53.1
MTTR	Mean time to repair	53.2
MV	Manipulated variable	3.1
	Minimum variance	116.3
MVC	More volatile component	90.3

NARMAX	Non linear ARMAX	114.8
NBG	No good	
NISCC	National Infrastructure Security Co-ordination Centre	49.9
NLP	Non linear programming	105
NPV	Net present value	59.11
ODBC	Open database connectivity	49.5
OCP	Operator's control program	41.2
OCS	Operator's control station	38.4
ODE	Ordinary differential equation	70.2
OIF	Output interface	38.2
OLE	Object linking and embedding (see also OPC)	49.5
OLTF	Open loop transfer function	71.5
OOP	Object oriented programming	107.7
OP	Output variable (controller)	3.1
OPC	OLE for Process Control	49.6
Op-amp	Operational amplifier	6.6
Op-code	Instruction or operation code	9.2
OSI	Open systems interconnection	40.4
PAT	Process analytics technology	102
PC	Personal computer	38.7
	Program counter	9.2
PCA	Principal components analysis	101.0
PCB	Printed circuit board	6.0
PCU	Process control unit	38.4
PCM	Pulse code modulation	40.7
PDD	Probability density distribution	82.3
PES	Programmable electronic system	55.5
PFD	Probability of failure on demand	53.8
P&I	Piping and instrumentation (diagram)	2.0
PID	Proportional, integral and derivative control (actions)	23.0
	Proportional, integral and derivative control (routine)	44.4
PIN	Process information network	49.12
	Pulse input (card or function block)	39.3
PISTEP	Process Industries STEP Consortium	60.4
PIU	Plant interface unit	38.4
PLC	Programmable logic controller	38.6

PM	Phase margin	73.12
PMM	Plant model mismatch	111.2
POU	Program organisation unit	8.1
PP-PI	Project planning for the process industries	100.6
PRBS	Pseudo random binary signal	116.0
PROLOG	Programming for logic	107.8
PSD	Process shut-down (system)	55.6
PSV	Parent space value	110.3
PSU	Power supply unit	39.2
PSTN	Public switched telephone network	40.7
PTF	Pulse transfer function	76.2
PTI	Proof test interval	53.3
PTRT	Proof test and repair time	53.3
PV	Present value	59.11
	Process variable	3.1
QA	Quality assurance	66.0
QC	Quality control	66.10
QP	Quadratic programming	105.9
RAM	Random access memory	9.4
RBF	Radial basis function	109.8
RDB	Relational data base	99.0
RGA	Relative gain analysis (or array)	111.1
RIBA	Royal Institute of British Architects	
RISC	Reduced instruction set computing	9.0
RLS	Recursive least squares	114.3
RMS	Root mean square	113.6
ROM	Read only memory	9.4
RTC	Real time clock	39.3
RTD	Resistance temperature device	16.4
RTO	Real time optimiser	106.0
RTOS	Real time operating system	45.5
RTU	Remote terminal unit	40.9
SCADA	Supervisory control and data acquisition	38.7
SIL	Safety integrity level	56.4
SIS	Safety instrumented system	55.6

SISO	Single input single output	71.0
SEX	Sequence executive	41.5
SFC	Sequential function chart (IEC 1131)	29.6
SFD	Sequence flow diagram	29.1
SFF	Safe failure fraction	53.8
SLC	Single loop controller	38.7
SMTP	Simple mail transfer protocol	49.11
SOAP	Simple object access protocol	49.11
SP	Set point (reference)	3.1
SPC	Statistical process control 102.0	
SQL	Structured query language	99.4
SQP	Sequential quadratic programming	105.11
SR	Set-reset	6.5
SRS	Safety related system	55.6
	Safety requirements specification	56.4
SSO	Steady state optimiser	106.1
SSPR	Stochastic sampling with partial replacement	110.3
SSR	Stochastic sampling with replacement	110.3
STC	Self-tuning control, or self-tuning controller	116
STEP	Standards for the exchange of product model data	60.4
SUS	Stochastic universal sampling	110.3
SQL	Structured query language	99.2
SQP	Sequential quadratic programming	105.12
SVD	Singular value decomposition	111.9
TCP/IP	Transmission control protocol / internet protocol	40.5
TDM	Time division multiplexing	40.9
TQM	Total quality management	66.10
TÜV	Technischer Überwachungs-Verein	57.1
UAL	Upper action limit	102.4
UART	Universal asynchronous receiver and transmitter	50.1
UHF	Ultra high frequency	40.8
URS	User requirements specification	60.0
USB	Universal serial bus	49.13
USL	Upper service limit	102.7
UWL	Upper warning limit	102.4

VDU	Visual display unit	38.3
VLSI	Very large scale integrated	9.0
VF	Vulnerability factor	56.4
WAN	Wide area network	40.7
WLAN	Wireless local area network	49.12
WIP	Work in progress	100.2
WLS	Weighted least squares	114.2
WS	Work station	38.10
XML	Extensible markup language	49.7
ZOH	Zero order hold	44.6

Index

Abnormal situation management 43.6
Abscissa 73.3, 102.10
Absolute values/variables 84.2, 84.3, 85.2, 106.2, 106.9, 108.3, 108.9, 108.10, 118.2
Absorption
    column control 30.4
    column dynamics 92.5
    gas film 89.1, 92.5
    rule 54.8
Acceptance 62.4, 64.10, 66.4
Acceptance certificate 67.7, 67.9
Acceptance
    hardware 63.12, 66.5, 66.7, 67.9
    specification 62.4
    test schedule 62.4
Acceptance tests
    factory 62.4, 63.13, 67.9
    works 62.4, 63.13, 67.9
Access
    level of 42.1, 49.8, 64.7
    physical 64.2
Access to system 64.7
Accuracy 10, 13.1, 16.2, 16.4, 17.2, 21.2, 96.7, 102.2
Acids and alkalis/bases 17.2, 17.3
Acoustic noise 19.2
Actions 29.6, 37.5
Active protection 55.3
Active X 49.5
Activity model 100.4
    process management 37.8
    production planning 37.8
    recipe management 37.8
    unit supervision 37.8
A/D converter 44.1
Agitated vessel 69.2
    internal coil 85.4, 89.1, 92.4
    jacketed, steam heated 85.2
    jacketed, water cooled 85.3
Agreement 66.7

Air cooler/separator 51.4
Air dryer/filter 51.4
Air filter/regulator 5.1
Air supply 5.1, 51.4
Air to open/close 19.6, 21.2, 22.2
Alarm
    environment 43, 55.4
    floods 58.5
    handling/management 43.5, 50.9, 58.5
    list 42.8, 43.3
    systems 58.5, 99.1
Amplitude ratio 73.1
Analogue I/O card/channel 4.1, 39.3, 44.1, 44.6
Analogy electrostatic field 74.5
Analyser management 18.10
Angle criterion 74.2, 74.7, 77.5
Anti-
    logging approach 34.3
    systems 93
    virus software 49.14
Anthropometrics 58.3
Application diagnostics 55.4
Application software 41, 54.4, 58.1, 59.10, 62.4
    acceptance 63.13
    breakdown 63.14
    commissioning 64.5
    design 63.3, 63.4
    integration 63.11, 63.12
    management 63.14, 63.15
    manager 62.2, 63.15, 66.8
    metrics 61.5, 61.6
    portability 65.4
    progress 64.5, 66.8
    report 66.8
    specification 66.7
    testing 63.9, 63.10
Application study 62.1
    consultant 62.2
    study group 62.1, 62.2

Arbitration 67.7
Archimede's principle 69.4
Architectural constraints 56.3, 56.8
Archive package 41.4
Argand diagram 68.3, 73.2, 73.4
Argument 68.3, 73.2, 73.4
Arithmetic/logic unit 9.3
Array data type 7.11
Arrhenius' equation 36.1, 89.2, 91.1
Artificial intelligence 107
Asset management 50.11
Asymptotes
    angle of 74.4, 74.6
    frequency 73.3, 73.8
    root locus 74.4
ATEX directive 52.14
Atomiser 94.1
Attenuation 73.1, 73.2, 73.7, 73.9
Attributes 99.2, 99.4
    names of 99.2
    mapping of 99.3
Authorisation 64.6
Auto correlation 82.5, 82.6, 83.4
Auto/manual mode 3.4, 23.1
Auto-tuners 116
Auto regressive moving average
    controlled (CARIMA) 114.6, 117.2, 117.5
    non-linear 114.8
    with exogenous input (ARMAX) 114.6, 117.2
Automation requirements 60.1, 60.3, 62.3
Availability 53.2
Aversion factor 56.1

Back propagation algorithm
    chain rule 109.5
    completion criterion 109.3
    epoch 109.3, 109.5
    Jacobian, evaluation of 109.5
    learning rate 109.3
    Levenberg Marquardt algorithm 109.3
    steepest descent 109.3
Backflushing 18.8
Backplane 39.2
Back-up systems 55.3
Ball valves 19.4
Bandwidth 23.2

Barometric leg 31.1, 31.3
Barrier function 105.14
Barriers 44.1, 44.6, 46.1, 46.3, 52.12
Basic event 54.7
Batch
    cycle times 59.7
    least squares 83.7, 83.9, 83.10, 109.9
    polymeriser 83.9
    process control 37, 41.6
    reactor, semi 89.1
    software 60.4
    status display 42.8
Batch scheduling problem
    allocation strategy 110.13
    chronology 110.13
    constraints, relative/absolute 110.13
    feasibility 110.13
    multi-objective 110.13
    schedule builder 110.13
    single objective 110.2
    tardiness 110.2
Bath tub curve 53.2
Benefits
    estimating 59.8
    intangible 59.4
    sources of 59.4
Bernoulli's equation 12.4
Best endeavours 67.8
Bias 22.1, 22.2, 23.10
Bilinear transform 77.3, 77.6
Bill of labour 100.3
Bill of materials 100.3
Binary mixture/separation 35.1, 90.3
Binary representation 7.1
Binomial expansion 68.1
Bistable 6.5
Blanketing 14.5
Blending system 81.1, 86.5. 108.9, 111.8
Block diagram 3.2, 84.5, 85.2, 85.3, 86.1, 86.5, 87.3, 87.5, 88.3, 90.6, 92.1, 94.4
Block diagram algebra 71.4, 71.7, 76.7, 78.2, 81.4, 81.8, 93.1
Block pointer 45.5
Bluff bodies 13.8
Bode diagram 73.3, 73.6
Bode stability criteria 24.2, 73.9, 73.12

Boil-up rate/ratio 90.4, 90.5, 90.6
Boiler drum level control 25.1, 33.1, 87.5
Boiler dynamics 87.5
Bonded junctions 16.3
Branch and bound 103.7
    binary variables 103.7
    fathomed branches 103.7
    integer variables 103.7
    lower/mid/upper range 103.7
    node 103.7
    root/secondary branches 103.7
    systematic search 103.7
Branches (root locus) 74.4
Break (or breakaway) points 74.4, 74.6, 74.8, 77.6
Brent's method 104.8, 104.9
Bridge circuit 15.2, 15.3, 16.6, 18.3
Broadband transmission 40.6
Bubbles and symbols 2.1
Buffering 17.3
Bumpless transfer 23.5
Butterfly valves 19.3
By-pass control scheme 32.4

Cable rating 52.13
Cable trays 51.2
Cache memory 9.4
Calibration of DP cell 11.4
Calibration of field instrumentation 64.3, 64.4
Canister load cell 15.3
Capability indices 102.8
    interpretation 102.8
    spread 102.8
Capacitance 11.3, 14.6
Capacitance, volumetric 87.2
Capacity for energy storage 85.3
Capillary column 18.2
Card dimensions 39.2
Carrier gas 18.1, 18.3
Cartesian co-ordinates 68.3, 73.2, 77.8
Cascade control 25, 33.1, 36.3, 35.2, 71.8, 95.2
Cascade programs 78.6
CASE tools 63.7
Cash flow 59.11
Categories of failure 56.9
Categories of protection 55.6
Causal relationship 83.11

Cavitation 13.7, 20.8
Centre of gravity 74.4, 74.6
Certification
    barriers 52.13, 52.14
    IS 52.7
    quality 66.1
    safety equipment 57.1
Chain rule 109.5, 118.3
Champion 59.3
Change control 54.5, 62.1, 63.10, 64.6, 65.2, 66.7
Change control policy 64.6
Channel diagram 51.1
Characteristic equation 71.9, 74.2, 74.8, 74.9, 77.2,
    77.6, 80.7, 80.8, 80.10, 112.1, 112.2, 112.6,
    112.7, 118.7
Characteristic equation, roots of 71.9, 74.1, 79.9,
    80.13
Characters 7.3
CHAZOP studies 54.3
Chemometrics 102
Choked flow 20.8
Chromatogram 18.1, 18.8, 18.9
Chromatography 18, 35.5
Circuit breaker 51.2
Client-server technology 49.3, 49.6, 100
Closed loop
    gain 22.3
    operation 3.4
    performance 74.9, 112.1, 116.5
    response 23.3, 23.4, 74.9, 81.4, 116.8
Code management system 63.7
Coefficient of determination 83.5, 83.6, 83.7, 83.9,
    83.10
Coefficient of discharge 12.4
Cohen and Coon formulae 24.4
Collinearity 83.7, 101
Collisions 40.2
Colour coding 42.5
Column feed control 35.9
Column pressure control 35.10
Commissioning 64
    DP cell 11.6
    team 64.8
    time scale 64.1
Common mode failures 53.5, 56.9
Commonality 60.4, 61.6, 62.1

Communications 39.3, 40
Communications channel 54.3
Comparator 3.3, 22.1, 22.2
Compensating cable 16.3
Compensators
    concept 81.2
    design 81.4, 81.8
    lead 71.3
    lead/lag 95.2
    impulse 81.8, 81.10
Compliance 48.7
Compliance commentary 61.2, 62.1
Complementary strips 77.1
Complex conjugate 68.3, 73.2
Complex numbers 68.3
Composition measurement 36.4
Compressible flow systems 87
Compressor 51.4
Computer integrated manufacturing 38.9, 100
Concentration lags 90.3
Conceptual design 59.2
Condition number 111.7, 117.9
Conditional construct 8.6
Confidence limit 82.4, 102.9
Configurable functions 41.3, 60.4, 63.3, 63.14
Configuration 48, 50.8, 60.3, 63.9
Configuration of design 56.10
Configurator 41.8, 47.5, 48.6, 50.10
Conflict resolution (ES)
    global/local resolution 107.4
    meta rules 107.4
    recency 107.4
    refractoriness 107.4
    salience 107.4
    search strategy, impact on 107.4
    specificity 107.4
Conjugate pairs 68.3, 74.1
Consequence 53.10, 56.5
Consequences, categories of 56.6, 56.7, 57.1
Console 39.1
Constant molal overflow 35.1, 90.3
Constrained optimisation 105, 106.5
Constraint handling/management 106, 106.6, 106.8, 106.12
    bridge models 106.8
    disturbances, effect of 106.8
    prioritisation of MPC 106.8
Constraints
    active/inactive 105.4, 105.7, 105.11, 105.12, 106.6, 106.8
    augmented 103.3
    binding 105.4, 105.7
    categories 106.10
    equality 103.2, 105.1, 105.4, 105.11, 106.6
    functional 103.1
    hard/soft 106.8
    inequality 103.2, 105.4, 105.7, 105.9, 105.10, 105.11, 106.6
    intersection of 103.2
    linear/linearised 105.9, 105.11, 105.12, 105.13, 106.6, 106.9, 106.10
    matrix 105.9, 117.7
    most limiting 103.3
    multiple 103.3
    non-negativity 103.1, 103.2, 105.9, 105.10, 117.7
    rate of change limits 117.7
    relaxation of 106.8
    satisfied 105.4, 105.6
    saturation limits 117.7
    sensitivity 105.6
    updating 106.10
    violation of 106.8
Constructs
    array data 7.11
    derived data 7.10
    enumerated data 7.12
    structured data 7.13
Contacts and coils 47.1
Containment 55.2
Continuous cycling method 24.3, 98.3
Continuous operations 60.3
Continuous stirred tank reactor 36, 91, 118.2
Contracts 67
Contract
    back-to-back 67.15
    law of 67.1
    lump sum 67.6
    manager 62.2, 63.15, 66.8, 67.8, 67.10
    model 67.4, 67.5
    reimbursable 67.6, 67.14
    standard 67.4

    sub 67.15
    suspension 67.11
    termination 67.7
    turnkey 67.5
Contractor 59.1, 66.1, 67.3
Control
    affine model 118.2
    horizon 106.11
    loop checklist 64.5
    module 37.3
    philosophy 59.2
    recipe 37.7
    schedule 100.3
    schemes and strategies 60.3
    philosophy 60.3
    unit (microprocessor) 9.2
    valve 19.1, 65.1, 84.3
Control charts 102.3
    average 102.4
    interpretation 102.4, 102.6
    moving average 102.5
    spread 102.6
Control limits 102.3, 102.8
    action limits 102.4, 102.6, 102.9, 102.10
    ellipses 102.9
    polygonal lines 102.10
    warning limits 102.4, 102.6, 102.9, 102.10
Control system
    flow 88.1
    level 22.1, 22.4, 93.1
    position 95
    pressure 87.3, 87.5, 93.4
    temperature 85.2, 85.3, 86.1, 90.4, 91.4, 94.1, 113.8
Controller period 28.3
Convergence/divergence 68.1, 96.5, 115.4, 115.6, 115.9
Convolution 76.1, 76.2, 80.8
COOP studies 54.4, 54.5, 63.5
Corner feasible solution 103.2, 103.3, 103.4
Corner frequency 73.3, 73.6, 73.8
Coriolis effect 13.9
Correlation coefficient 82.2, 82.5
    interpretation of 82.2
Correlation functions
    auto 82.5

    cross 82.5
    properties of 82.5, 82.6
Correlation matrix 82.2, 101.3, 101.7
Cost function 101.1, 103.2, 103.3, 103.4, 103.7, 105.1, 110.2, 110.8, 110.9
    cubic or higher order 105.12
    quadratic 105.9, 106.9
    see also objective function
Costs and benefits analysis 59, 66.9
Costs and benefits audit 59.12
Costs
    categorisation of 59.9
    estimation of 59.10
    field instrumentation 59.10
    running 59.10
Counter 6.6, 47.4
Counterbalanced motion 74.5
Covariance windup 115.7
Critical
    flow 20.8
    frequency 24.2
    item list 54.6
Cross correlation 82.5, 82.6, 83.2, 83.8, 83.10, 83.10
Crossover
    intermediate recombination 110.4
    reduced surrogate operator 110.4
    single/multi point 110.4
    uniform 110.4
Crude oil separation plant 103.2
Crude oil separator control 30.3
Cumulative cash flow 59.11
Curvature 84.3, 96.1, 96.7, 104.7
Cut 35.3
Cyclic patterns 82.5

D/A converter 44.6
Dahlin's method 78.3
Dalton's law of partial pressures 92.5
Damages 67.12
Damped frequency 72.4
Damping 69.2, 74.1, 95.1
    critical 72.3, 112.2
    factor 24.1, 72.1, 74.8, 77.8
    overdamped 72.2, 93
    undamped 72.5
    underdamped 72.4, 112.2

Dangerous failures 53.8, 56.5, 57.2, 57.7
Data
    collection 102.1
    compression 42.7, 102.2
    deletion 102.2
    filtering 102.2
    invention 102.2
    mining 102.10
    missing 102.2
    moving average 102.2
    overload 101
    pre-screening 102.2, 106.5
    transformations 102.2
Data definition language 99.3
Data encoding
    cyclic data 109.6
    input scaling/standardisation 109.6
    spread encoding 109.6
Data flow diagrams, 63.2
Data objects 49.4
Data reconciliation 106.5
Data typing 7
Database
    connectivity 49.5
    management system 99.3
    structure 45.5
Database types
    object oriented 99, 107
    relational 99
    structured 45, 99
Datablocks 43.1, 45, 46
DC coupled logic 57.5
Dead time 71.2
Dead time compensation 86.4
Deadband 43.1
Deadbeat method/response 78.4, 116.5, 116.6
Decay ratio 72.4
Decibel 73.3
Decision
    making 58.2
    support systems 107
    table 29.3
    tree 29.3
    variables 103.1, 103.3, 103.5, 103.7, 105.4, 105.9, 105.11, 106.6, 106.9, 117.7
Decomposition of requirements 37.6

Decoupler design 81.6
Defects 102.7, 102.8
Degrees of freedom 30.1, 30.3, 30.4, 30.5, 31.3, 31.4, 32.4, 35.1, 35.13, 35.14
Delay 71.2
    see also time delay
Deliverables 63.1
Delivery date 67.11
Delta notation 84.3, 85.2
Demand mode of operation 56.4
Demand rate 53.10, 56.5, 57.1
Demand rate evaluation 54.10
Demilitarised zones 49.12
Density measurement 14.3
Dependency factor 53.5
Dependency theory 99.6
Depreciation 59.11
Derivative action 23.4, 78.1, 95.2
Derivative feedback 23.6, 71.3
Derived data types 7.10
Detailed functional specification 54.3, 54.5, 58.6, 66.5, 66.7, 67.6, 67.7
Determinant 79.2
Determination, process of 30, 31.2, 35.1
Determined variable 30.1
Deviation variables 22.2, 81.1, 81.9, 84.2, 84.3, 84.4, 85.2, 86.1, 87.1, 88.3, 89.1, 89.3, 90.3, 91.1, 91.2, 91.4, 94.3, 94.4, 101.1, 101.3, 102.9, 105.9, 106.6, 106.9, 108.9, 112.10, 113.1, 116.3, 117.2
Device description 50.6
Diagnostic coverage 53.8
Diagnostic display 42.10
Diagnostics, self or system 57.2, 57.7, 57.8, 65.1
Diagonalisation 80.11, 81.4
Diagonal structure 111.1
Diaphragm capsule 11.1
Diaphragm valves 19.5
Difference equations 75.5
Differential capacitance 11.3
Differential equations
    first order 69, 80.2
    ordinary 70.2, 92.2, 96.1
    partial 92.2, 92.5
    solving 70.2
    stiff systems 96.5, 96.7

Differential gap 28.3
Digital controller 76.8
Dimensionality 101.4
DIN rails 51.1
Diode, junction 6.2
Diophantine identity 116.3, 117.2, 117.5
Direct acting 22.1
Direct programs 78.5
Dip leg 14.5
Direct digital control 38.3, 41.3
Direct ratio control 26.2
Discount rate 59.11
Discounted cash flow 59.11
Discrepancy checking 46.5
Discrete device 46
Discrete I/O card/channel 4.1, 39.3, 46.1, 46.3
Discretised form of PID 23.9
Display hierarchy 42.4
Display, reserved areas 42.3
Display system 42.4, 43.5, 58.4, 60.4
Dissociation 17.2, 17.7
Distance, Euclidean 109.8, 109.10
Distance velocity lag 71.2, 86.3
Distillation column control 27.2, 35, 90.4
Distillation column, 3-stage 90.3, 98.4, 98.5, 111.3-111.7
Distributed control system 38.5, 57.7
istributed parameter system 92
Disturbance
    external 24.3, 24.4,
    rejection 25.1, 25.2, 25.3, 25.4, 27.1, 108.3, 108.5
    source of 85.4, 90.4
Dittus Boelter equation 32.3, 89.2
Document review 66.7
Documentation 59.10, 64.9, 66.6, 66.7, 67.7, 67.8
Dominant
    pole positions 74.11, 77.7, 112.2
    roots 74.9, 77.5
    time constant 25.4, 90.4, 90.5, 97.13, 117.1
Double seated valve 19.2
DP cell 11.2, 11.3
Dual composition control 35.8, 90.4, 98.4
Dual systems 55.3
Dummy variable 47.1, 74.10, 74.11
Duplex action 21.4, 34.1
Dynamic
    compensation 27.2, 27.3, 95.2
    error 69
    logic 57.6
    packet filtering 49.11
    response 23.2, 25.3, 32.2, 32.4, 36.2, 95.2
    response, MIMO 80.13
    scheduler 37.8
Dynamic modelling (neural)
    globally recurrent networks 109.11
    locally recurrent networks 109.11
    memory 109.11
    time series inputs 109.11
Dynamics
    anti-surge 93.1
    first order 84.5, 85.3
    first order, quasi 85.3, 90.4
    electro-mechanical 95
    hydro 85.2, 88
    lumped parameter 85.1, 90.3, 94.3, 94.4
    multistage 90
    observer 112.6
    second order 85.3
    second order, quasi 90.4
Dynamic optimising controller
    algebraic solution 106.9
    application 106.11
    biasing in real time 106.11
    common I/O signals 106.7
    control horizon 106.11
    cost effectiveness 106.12
    essential characteristics 106
    filtered inputs/outputs 106.7, 106.11
    formulation of constraints 106.10
    frequency of execution 106.7, 106.11
    linear model 106.10, 106.11
    number of I/O signals 106.7
    other benefits 106.12
    plant model mismatch 106.9
    prediction horizon 106.11
    QP solution 106.9, 106.11
    set point trajectory 106.11
    set points, downloaded 106.7, 106.8, 106.11
    slow/fast dynamics 106.7
    steady state solution 106.7
Dynamic process simulator 98.6
    absolute variables 98.6

drag and drop 98.6
differential equations 98.6
generic models 98.6
modes, steady state and dynamic 98.6
parameterisation 98.6
pros and cons 98.6
rubber banding 98.6
run time 98.6

Earthing 13.6, 52.12
Effective carbon number 18.4
Effective gain
    direction, input/output 111.6
    two norm 111.6
Eigenvalues 79.9, 80.10, 80.11, 101.1, 101.3, 104.3, 104.4
    complex 80.12, 80.13
    negative real 80.13, 112.5, 112.6
    repeated 80.13
Eigenvectors 79.9, 80.12, 80.14, 101.1, 101.3
    direction of 79.9
    projections 80.12
Elapsed time 29.6
Electrical filter 69.3
Electrical requirements 64.2
Electrochemical circuit 17.5
Electrolyte 13.6
Electromagnetic
    compatibility 51.3, 52.14
    flowmeter 13.6
    relay logic 57.4
Electro-mechanical systems 95
Electronics 6
Emergency shut-down systems 55.3, 55.6
Encapsulation 52.10
End-user 59.1, 66.1, 66.7, 66.9, 67.3
Energy balance control 35.6, 90.4, 90.6
Energy consumption 59.5
Engineers control program 41.8, 47.5
Entity relationship modelling 99.7
    cardinality 99.7
    repeated entity 99.7
    strong/weak entities 99.7
    symbols for 99.7
Enumerated data type 7.12
Envelope, underdamped response 72.4

Envelope, time constant 72.4
Equal percentage 20.1, 84.6
Equilibrium constant 17.3
Equipment module 37.3
Ethernet 40.2, 49.3
Euler's methods
    explicit 44.3, 96.1, 97.13, 113.8
    implicit 96.5, 96.6, 112.10
    predictor corrector 96.2, 96.3
Evans' rules 74.2, 74.4, 77.5
Evaporation, rate of 31.3, 87.5
Event log 42.8
Excess 105.6
Exchange rate 59.11
Exothermic reaction 36, 36.1, 89, 91.2
Expandability 61.4
Expectation operator 113.2, 115.9, 116.3
Expert systems, 43.4, 107
    appropriateness 107.11
    control cycle 107.4
    development cycle 107.10
    justification 107.10
    prototyping 107.10
    rule based 107.3, 107.4
    shells 107.8
    working memory 107.3
Expert, use of 67.7
Explosion proofing 52.8
Explosive mixture 52.1, 52.5
Exponential decay 70.1, 72.4, 73.1, 75.4, 112.7
Exponential form 68.3
External risk reduction facilities 55.6, 56.2
External variables 8.2
Extrema/extremum 104.1, 104.2, 104.3, 104.4, 105.1, 105.9
    local/global 104.5, 105.7, 105.9, 105.10, 105.11, 105.12

Faceplate 23.1, 42.5, 42.6
Factory acceptance 50.10
Fail-safe design 55.2
Failure
    mode and effect analysis 54.6
    modes 53.8, 57.2
    philosophy 60.3
    rate 53.1

Fathomed branches 103.6, 103.7
Fault tree analysis 54.7
Fault tree evaluation 54.9
Feasible region/solution space 103.2, 105.4
Feedback control 22, 23, 24, 25, 26.3, 27.2, 27.3
Feedback path 3.3
Feedforward
    compensation 27.1
    control 27.3, 35.9
    path 3.3
Fibre optics 40.6
Field termination panel 51.1
Fieldbus 49.6, 50, 51
    segments 50.7, 50.10
    topology 50.4
Filter constant 44.3, 71.2
Filter, first order 44.3, 45.2, 69.3, 71.3
Filtering 69.2
Final certificate 67.12
Final value theorem 71.6, 71.7, 77.8, 78.4
Finite/infinite capacity scheduling 100.3
Fire and gas detection systems 55.3
Firewall 49.8, 49.11
First order reaction 36.1
First order systems 69, 71.1, 76.4, 98.2
    lag 71.2, 90.1, 90.2, 92.2, 92.5
    lead 71.2
Fitness
    aggregated 110.9
    power law scaling 110.2
    proportional fitness assignment 110.2
    relative fitness 110.2, 110.7, 110.11
Fitted values 83.1, 83.7, 102.9
Fixed capital 59.11
Fixed resistance 20.2
Flame ionisation detector 18.4
Flameproof enclosures 52.8
Flapper nozzle 5.3
Flash drum control 30.5
Flashing 20.8
Flexible pipes 15.4
Flip-flop 6.5
Floating variable 30.1
Flow straighteners 12.7, 13.5
Flowmeter types 13.1
Fluid barrier 12.7, 14.5

Force balance 5.5, 11.2, 21.1
Force majeure 67.11
Forcing function 69, 92.4
Foreign key 99.3
Forgetting factors 115.4, 115.8
Forward
    acting 22.1
    difference 75.5
    path gain 22.3
    selection 83.8
Foundation Fieldbus 50.3
Frame (messages) 40.1, 50.1
Frame based systems
    declarative/procedural notions 107.6
    demons and demon types 107.6
    hierarchical structure 107.6
    reasoning, expectation driven 107.6
    root frames 107.6
    slots and fillers 107.6
Frequency
    shift keying 40.7
    damped 72.4
    natural 72.1
Frequency response 73
    compound systems 73.7
    delay 73.5
    integrator 73.5
    lag 73.5
    lead 73.5
    second order 73.6
Friction losses 88.2
Function 47.3, 48.1
    block 41.3, 43.1, 44, 45, 46, 47.3, 48.2, 48.3, 50.8
    block diagram 48.5
Functional dependencies 99.3
    explicit and implicit 99.6
    lossless 99.6
    normal forms 99.6
Functional representation 2.2
Functional specification
    detailed 62
    methodology 62.1
    participants 62.2
Fuzzy logic
    Boolean AND/OR operations 108.4, 108.5
    Zadeh's rules 108.4

Fuzzy logic controller
    block diagram 108.1
    cost effectiveness 108.12
    crisp I/O signals 108.1, 108.5
    fuzzy signals 108.1
    incremental output 108.3, 108.9, 108.10
    multivariable 108.9
    non-linear 108.8, 108.12
    PD structure 108.1, 108.2
    PI structure 108.1
    real-time operation 108.7
    self adaptive 108.11
    tuning 108.1
    user interface 108.12
    variable scaling of signals 108.1, 108.11
Fuzzy signals
    centre of gravity (centroid) 108.5
    defuzzification 108.1, 108.5
    mean of maxima 108.5, 108.6
    fuzzification 108.1, 108.2
Fuzzy sub sets
    asymmetric/skewed triangular 108.2, 108.8

Gain margin 73.12
Gain, see steady state gain
Gain scheduling 34.1
GAMP guide 66.4
Gantt chart/diagram 61.2, 63.14, 67.8, 110.12
Gap meter 13.4
Gas
    law, ideal 87.5
    chromatography 18.1
    group 52.3
Gateway 38.7, 49.3, 49.6
Gauge factor 15.1, 15.2
Gauge glass 14.1
Gaussian 108.2
    number of 108.2
    overlapping 108.2
    triangular 108.2
    variable centres/widths 108.11
    universe of discourse 108.1
Gaussian elimination 103.4
Gearbox 95.1
Generalised minimum variance 116.6
Generalised predictive control 117.4

Generic model control 118.3
Genetic algorithm
    chromosome 110.1, 110.7
    coalescence 110.8
    convergence 110.7
    cost function 110.2, 110.8, 110.9
    crossover 110.4, 110.7, 110.8, 110.11
    encoding/decoding 110.1
    end criterion 110.7, 110.11
    fitness, evaluation of 110.2, 110.7, 110.8, 110.11
    gene 110.1
    generation gap 110.6, 110.7
    multi-objective 110.9
    generations (iterations) 110.7
    mutation 110.5, 110.7, 110.8, 110.11
    parallel 110.8
    phenotype/genotype 110.1
    population 110.1, 110.7
    reinsertion 110.6, 110.11
    seed solution 110.7
    selection 110.3, 110.8, 110.11
    structure 110.7
    visualisation 110.12
Geometric modelling 102.10
    action envelope 102.10
    data mining 102.10
    control limits 102.10
    interpretation 102.10
    operating envelope 102.10
    polygonal lines 102.10
    parallel axes 102.10
Giveaway 59.5
Glass electrode 17.5
Global variables 8.2
Globally linearising control 118.7
Globe valves 19.2
Good automated manufacturing practice 66.4, 100.6
Goodness of fit 83.5
Gradient notation 104.2, 105.7, 105.11
Green book 67.14
Group display 42.6
Guarantees 67.12
Guidewords 54.1, 54.3, 54.5

Hardware design specification 66.7
Hardware estimates 61.4
HART protocol 50.1
Hazard
    analysis 54, 56.3
    rate 53.10, 56.5
    reduction factor 53.8, 56.2
Hazardous area 52.2
Hazardous event severity matrix 56.6
HAZOP studies 54.1, 54.2
Head mounted transmitter 16.8
Heartcutting 18.8
Heat exchanger control 25.2, 27.4, 32
Heat exchanger dynamics 92.1
Heat of reaction 36.1
Heat transfer, film coefficient 89.2, 91.2
Heat transfer, overall coefficient 89.2, 91.2
Hessian 104.2, 104.4, 104.7, 105.7, 105.11, 105.12, 106.9
High level language 41.7
Higher level packages 60.4
Higher order systems 72.6, 78.3
Historian 41.8
Hold device 75.1
Horizons, definition of
    control 117.1
    prediction 117.1
    receding 117.1
Hot standby 53.7
Hotelling's distance 102.9
Human factor in protection system 56.5
Human factors 58, 60.3
Hydraulic valve coefficient 20.6
Hydrodynamics 85.2, 88
Hypertext 49.7

IChemE model conditions 67.5, 67.16
Idempotence rule 54.8
Identification 114
    auto regressive moving average 114.6
    batch least squares 114.2
    biased/unbiased estimates 114.3, 114.4, 114.5, 114.6
    estimation error 114.2
    extended least squares 114.6, 114.7
    generalised least squares 114.5, 114.7
    instrumental variable/vector 114.4, 114.7
    least squares implementation 114.3, 114.4, 114.6
    matrix inversion lemma 114.3
    matrix multiplication 114.3
    multivariable systems 114.9
    noise propagation 114.4
    non-linear systems 114.8
    prediction error 114.5, 114.6
    process history 114.3
    random residuals 114.2
    recursive least squares 114.3, 114.7
    signal noise 114.3, 114.4, 114.5
    weighted least squares 114.2
Identifiers 7.4
Ill defined balance 30.6
Ill/well posed problems (LP) 103.2, 103.3
Implicit enumeration 103.6, 103.7
    assignment 103.6
    fathomed branches 103.6
    feasible/infeasible solution 103.6
    lower/upper bounds 103.6
    optimum solution
    search direction 103.6
Implicit input signal 76.10
Impulse 75.2
    area under 75.2
    compensators 77.7, 77.8, 78, 81.8, 81.10
    Laplace transform of 75.2
    lines 11.5, 12.7, 14.2
    response 76.1, 76.2, 77.5
    response, finite/infinite 114.1
    train of. 76.1
    unit 75.2, 76.1
Incremental form of PID 23.11
Independence 56.11, 59.3
Independent equations 30.3
Indirect addressing 7.11, 29.4
Indirect ratio control 26.3
Industrial Ethernet 40.2, 50.3
Inertia 95.1
Inference engine
    agenda/conflict set 107.4
    audit trail 107.4
    conflict resolution 107.4
    recognise-act cycle 107.4

unification 107.4
Inferencing 107.2
    adaptive reasoning 107.2
    backward chaining 107.2
    best-first search strategy 107.4
    breadth/depth-first searches 107.4
    confidence estimates 107.2
    forward chaining 107.2
    goal/event driven 107.2, 107.3
    inheritance 107.2
    plausibility 107.2
    pre-programmed paradigms 107.2
    real-time applications 107.2
Inferential estimation 109.12
Inflation 59.11
Information security 49.9, 49.10
Information system 49.1
Ingress protection 52.11
Inherent characteristic, valve characteristics
Inherent safety 55.2
Inheritance
    attributes 107.2, 107.6
    automatic inheritance 107.6
    deformation 107.6
    derived vs inherited values 107.7
    generic classes 107.2, 107.6
    instantiation 107.5, 107.6, 107.7
    multiple inheritance 107.6
    parent/offspring 107.6, 107.7
Initial conditions 75.5, 80.1
Initial conditions, zero 69, 70.1, 71.1, 71.4, 72.1, 79.7, 84.3
In-line electrodes 17.6
In-line neutralisation 34.2
Input
    interface 44.1
    scaling 44.2, 45.1
    signal, implicit 76.10
Input/output
    cards 39.3
    channels 63.9
    variables 8.3
Installation 64.2
Installation of DP cell 11.5, 11.6, 12.7
Installed characteristic, see valve characteristics
Insurance 67.12

Integer data types 7.5
Integer programming
    rounding off approach 103.5
Integral action 23.3, 71.7, 71.8, 74.10, 77.7, 78.1, 93.5, 95.1, 112.9
Integral windup 23.7
Integrated programming support environment 63.7
Integrator 71.2, 77.7, 87.5
Integrity constraints 99.3
Intellectual property 67.7
Intension 99.2
Interaction between loops 25.4, 35.2, 35.8, 81.1, 111
Interface position 14.4, 30.3
Internal model control 81
Internal reflux control 35.11
Internet 49.7
Interoperability 49, 50.3, 50.6
Intersection, imaginary axis 74.4
Intersection, negative real axis 73.12
Intrinsic safety 51.3, 52, 55.2
Inventory variable 30.1, 35.2, 35.13, 35.14, 90.4, 101.7, 106.4
I/P converter 3.3, 5.5, 21.1
Isolating valve 19.1, 29.1, 46.2
Isothermal flow 87.1
Iterative construct 8.6

Jacket with refrigerant 36.3
Jacketed stirred tank control 25.3
Jacobian 104.2, 104.4, 105.7, 106.9, 109.3, 109.5
Just in time 100.2

Kalman filters
    covariance matrices 113.2, 113.8
    disturbance matrix 113.2, 113.8
    dynamic performance 113.6
    expectation operator 113.2
    feedback problems 113.7
    initialisation 113.6, 113.8
Kalman gain matrix 113.2, 113.6
    Luenberger observer 113.1, 113.2
    measurement noise 113.2, 113.6
    noise propagation 113.7
    non linearity 113.6
    prediction accuracy 113.6

predictor-corrector 113.6
recursive implementation 113.5
Riccati equation 113.3, 113.4
root mean square values 113.8
square, completing 113.4
stationary signals 113.2, 113.6
system disturbances 113.2, 113.6
use in control systems 113.7
white noise 113.2
Katharometer 18.3
Keyboard 39.1, 42.11
Key components 35.13
Key performance indicator 59.8, 100.4
Kinetic energy 88.2
Knowledge based systems 107
    incomplete information 107
    inference engine 107.1, 107.4
    knowledge base 107.1
    model base 107.1, 107.3
    rule/case base 107.1, 107.3
    scenarios 107.1
    user interface 107.1, 107.8
Knowledge elicitation
    card sort 107.9
    cost effectiveness 107.9
    domain expert 107.9
    forward scenario simulation 107.9
    knowledge expert 107.9
    protocol analysis 107.9
    repertory grid analysis 107.7
    structured interviews 107.9
Kuhn-Tucker conditions 105.7, 105.9, 105.10, 105.11

L/A control 118.1
Laboratory information management 100.6
Ladder execution 47.2
Ladder logic 47
Lag 71.2, 78.3
Lagrange multiplier 101.1, 105.1, 105.4, 105.7, 105.11, 106.8, 117.7
Lagrangian function 105.1, 105.4, 105.5, 105.6, 105.7, 105.9, 105.11
Laplace operator 70.1
Laplace transform 70
    definition 70.1

inverse 70.2, 71.9, 72.1, 73.1, 76.3, 80.6, 92.2
properties of 70.1
table of 70.1
use of 70.2
Laplace transformation 84.5, 85.2, 85.3, 86.1, 86.5, 87.3, 87.5, 88.3, 89.1, 89.3, 90.1, 90.2, 90.3, 92.2, 92.5, 94.3, 94.4
Lateness 67.11
Lead 71.2
Least squares methods
    batch 83.3, 83.7, 83.9, 83.10, 109.9, 114.2, 115.2, 115.9
    extended 114.6, 114.7
    generalised 114.5, 114.7
    instrumental variables 114.4, 114.7
    recursive 114.3, 114.7, 115.4, 115.6, 115.7, 115.9, 116.4, 116.7, 117.2, 117.6
    weighted 114.2, 115.4
Letter code 2.2
Level measurement 14
Levenberg Marquardt algorithm 104.7
Liability 67.12
Licence agreement 67.7
Lie derivative 118.6
Life cycle, product 65.3
Life cycle, project 59.1, 61.7
Likelihood, categories of 56.6, 56.7
Limit cycles 24.3
Linguistic statements 108
Line searching 104.8, 105.12
    artificial function 104.8
    bracketed minimum 104.8
    Brent's method 104.8, 104.9
    Newton Raphson's method 104.8
Linear programming 103.1, 103.2, 103.3, 103.4, 103.7
Linear programming, assumptions 103
    additivity 103.1
    certainty 103.1
    divisibility 103.1, 103.5
    proportionality 103.1
Linear regression analysis 83
Linear regression, definitions of
    predictor 83.1
    regressor 83.1
    residual 83.1

response 83.1
Linearisation 84, 84.3, 87.1
Linearity 4.3, 10, 32.3, 45.1, 118
Link active scheduler 50.5
Liquid holdup 90.1, 90.5
Liquid flow lags 90.2
Liquidated damages 67.8, 67.9, 67.11
Liquidation 67.8
Literals: integer, real, time, string, Boolean 7.5, 7.6, 7.7, 7.8, 7.9
Load
    cells 15.3
    definition of 3.3
    gain 84.5
Loadings 101.1, 101.8
Loadings plots 101.8, 101.9
Local area network 40.2
Local equipment centre 51.1
Logic gates 6.4
Luenberger observer 113.1
    Kalman gain 113.1
    predictor-corrector 113.1
    state estimates 113.1
Lumped approximation 72.6
Lumped parameter systems 69.2, 85
    interacting 85.3
    non interacting 85.2
    uniform properties 85.2, 85.3

Macrocycle 50.5
Magnitude criterion 74.2, 74.11, 77.5
Mahalanobis' distance 102.9
Main memory 9.1
Maintenance 65.1
Maintenance contracts 59.10, 65.1
Malware 49.13
Mamdani structure 108
Man-machine interface 58
Management information systems 38.8, 60.3, 100
    decision support 100.1
    decisions, level and type 100.1
    functionality 100.2
    information requirements 100.1
Manning levels 59.5
Manufacturing execution systems 100.5
Marginal costings 59.11

Marginal stability 24.2, 73.9, 73.10, 73.11, 73.12, 74.10
Marshalling cabinet 51.1
Mass
    balance control 35.6
    flowmeter 13.9
    transfer 19.1, 92.5
Master loop 25.1, 95.2
Master recipe 37.7
Master schedule 100.3, 100.4
Materials 67.7
Materials resource planning 100.3
    process MRP 100.4
Matlab 97
    array and matrix operations 97.2
    colon operator 97.2
    command mode 97.1
    convolution 97.5
    cross correlation function 97.9
    curve fitting 97.3
    display functions 97.8
    dot extensions (.asc, .m, .mdl, .xls) 97.10, 97.11, 97.14
    format 97.3, 97.8
    graphics 97.8
    help facilities 97.1
    import/export of data 97.10
    instruction types 97.1
    leading/trailing zeros 97.5
    partial fractions 97.7
    place instruction 98.5
    polynomials 97.3, 97.4, 97.5
    polynomials, differentiation 97.6
    root finding 97.4
    semicolon separators 97.2
    statistics functions 97.9
    toolboxes 97.1
    variables 97.2
    vectors 97.2
Matlab M files 97.11
    comment statements 97.11
    dominant time constant 97.13
    function M files 97.14
    initialisation 97.12, 97.13
    numerical integration 97.13
    presentation 97.12

program structure 97.12
program flow 97.12
recursion 97.12, 97.13
run time 97.13
script files 97.11
search approach 97.12
step length 97.13
user defined functions 97.14
Matlab user interface 97.1
command history window 97.1
command window 97.1
current directory 97.1
desktop 97.1
plot window 97.8
text editor window 97.11
work directory 97.10
workspace 97.1, 98.1, 98.2, 98.4, 98.5
Matrices
conformable 79.3, 79.7
equal 79.3
use of 79.5
Matrix
adjoint 79.4
cofactors 79.4
companion 80.2
constraint 105.9, 117.7
correlation 82.2, 101.3, 101.7, 102.9
covariance 82.2, 101.1, 101.3, 102.9, 113.2, 115.3
definite 104.2, 104.4, 104.6, 104.7, 105.9, 105.11, 105.12, 115.6
definition of 79.1
diagonal 79.1
dimensions of 79.1
exponential 80.5, 80.13
identity 79.1, 104.7
inverse 79.3, 79.4
inversion lemma 114.3
minors 79.4
singular 79.2, 79.4, 79.7, 104.5, 104.6, 115.6
square 79.1
system 80.1
trace 101.4, 115.9
transfer 80.9, 80.13, 81.4, 81.5, 81.8, 90.5
transition 80.5, 112.5
transpose 79.1

zero 79.1
Matrix laws
inversion 79.3
multiplication 79.3
Matrix operations
addition 79.3
division 79.3
inversion 79.4
multiplication 79.3, 114.3
pre and post multiplication 79.3
subtraction 79.3
Maximum rate of change in flow 93.2, 93.3, 93.4
McClaurin's series 68.1
Mean (average) 82.1, 82.3
Mean square value 82.1, 82.3
Mean time between/to failure 53.1, 53.2
Mean time to repair 53.2
Median 82.1
Mediation 67.7
Membership
functions 108.2, 108.4, 108.5, 108.6
overlapping 108.5
values/profile 108.2, 108.4, 108.5, 108.6
Memory locations 8.4
Memory system 9.4
Merit function 105.11, 105.12
Message frame 40.1
Metrics 59.9, 59.10, 61.5, 63.14
Microprocessor architecture 9.1
Milestones 63.14
Mimic diagram 42.5
Minimal response design 78.4
Minimum
capacity/volume 93.2, 93.3, 93.4
cut sets 54.8
inventory 55.2
variance control 116.3
Missing at random 102.2
Missing data 101, 102.2
Mixed integer linear programming 103.5, 103.7
Mixed integer non linear programming 103.8
Mixing of acids and alkalis 17.4
Model
distributed parameter 92
linearised 84.1
lumped parameter 85.1, 95.1

matrix form 81.3, 81.7, 86.5
order 115.1
state space 89.4, 90.3, 91.3, 112.5, 112.6, 113.1, 113.8
time series 114
transfer function 84.5, 85.2
zero capacity 86.1, 86.5
Model for steady state optimiser
   linear/non-linear nature 106.2
   scope/size of model 106.2
Model predictive control
   benefits 117
   choice of control horizon 117.6
   constrained optimisation 117.7
   constraint matrix 117.7
   controller output sequence 117.4
   constraint matrix 117.7
   decision variables 117.7
   Diophantine identity 117.2, 117.5
   dominant time constant 117.1
   forward/output prediction 117.2, 117.3, 117.5
   generalised 117.4
   horizons 117.1, 117.3, 117.5, 117.6
   implementation, recursive 117.6, 117.8
   inherent delay 117.1
   known (backward) values 117.3
   Lagrange multipliers 117.7
   multivariable systems 117.8
   non-linear 118.9
   objective function 117.4, 117.7
   off-diagonal zeros 117.3
   optimal scaling 117.9
   partitioning of matrix 117.3
   plant model mismatch 117.9
   polynomial division 117.2
   proprietary packages 117.10
   QP solution 117.7
   rate of change limits 117.7
   real-time optimisation 106.1, 106.7, 106.8, 106.11, 109.13
   recursive least squares 117.2, 117.6
   residuals on output 117.6
   robustness. 117.9
   saturation limits 117.7
   separation of predictions 117.2
   step response testing 117.2, 117.9
   unconstrained 117.4
   unknown (forward) values 117.3
   weighting factor 117.4, 117.5
Modes of operation 3.4
Modification control form 64.6, 65.2, 66.7
Modules (software) 63.1, 66.7
Modulus 68.1, 68.3, 73.2, 73.4
Modulus, maximum value 73.6
Monitor 41.8, 47.5
Monotonic values 83.9
Morari resilience index 111.7
Motor control centre 51.2
More volatile component 90.3
Multicolumn separation 35.14
Multicomponent distillation 35.13
Multicores 51.1, 51.4
Multi-disciplinary team 54.1
Multi-layer perceptron
   activation function 109.1, 109.2
   back propagation 109.1, 109.3, 109.5
   non-linearity 109.1
   operation of 109.2
   sigmoid squashing 109.1, 109.5
   training data 109.3, 109.5
Multiloop systems 111
Multi-objective GA
   aggregated fitness 110.9
   distance, Euclidean 110.11
   dominance 110.10
   exponential rule 110.11
   genetic rift 110.11
   kernel density 110.11
   lethal 110.11
   mating restriction 110.11
   niche count 110.11
   objective space 110.10
   operation 110.11
   parallel axes 110.12
   pareto ranking 110.10
   rank 110.10, 110.11
   sharing function 110.11
   trade-off graph 110.12
   trade off surface 110.10
   user interaction 110.11
Multi-products/purpose plant 37.1
Multiple-input multiple-output systems 80.8, 80.9

Multiple linear regression 83.7, 97.11, 109.9
Multiple poles/zeros 74.4
Multistage dynamics 90
Multivariable control 80.9, 81, 86.5, 89
    compensators (s domain) 81.2, 81.5, 89.4
    compensators (z domain) 81.8
    decouplers (s domain) 81.6
    decouplers (z domain) 81.10
Multivariate statistics/SPC 82.2, 102.9, 102.10
Murphree efficiency 90.3, 90.5, 90.6

Naming conventions 66.7
Naphtha column/splitter
    principal components analysis 101.7, 101.9
    regression analysis 83.10
Natural frequency 72.1
Negative feedback 22.1
Nernst equation 17.5
Net present value 59.11
Network access 40.5
Neural networks, artificial 109
    neuron, bias neuron 109.1, 109.5, 109.8
    data encoding 109.6
    dynamic modelling 109.11
    error function 109.3, 109.5, 109.10
    generalisation 109.4
    hidden/input/output layers 109.1, 109.5, 109.8
    identification, use for 117.2
    inferential estimation 109.12
    non-linearity 109.11
    optimisation 109.13
    overfitted data 109.4
    prediction vs training 109.3, 109.10
    pre-processing of data 109.7
    pros and cons, MLP vs RBF 109.9
    subscript convention 109.2
    synapse 109.1, 109.8
    weightings 109.1, 109.5, 109.9
Neuro-computing 109
Neutralisation systems 34.1
Neutrality 17.1
Newton Raphson's method 104.8
Newton's law of motion 69.4
Newton's method 104.3, 104.4, 104.7, 105.11
Noise 23.4
Noise propagation 113.7

Non interacting form of PID 23.3
Non invasive 13.6, 13.7, 14.8
Non linear characteristic 84.1
Non-linear control
    application to CSTR 118.2, 118.4, 118.8
    closed loop transfer function 118.4
    closed loop stability 118.7
    control affine models 118.2
    generic model control 118.3
    globally linearising control 118.7
    integral action 118.3
    L/A control 118.1
    Lie derivative 118.6, 118.7
    linearising transformation 118.1, 118.7
    non linear MPC 118.9
    PI control 118.7
    plant model mismatch 118.3, 118.9
    relative order 118.6, 118.7, 118.9
    scalar field 118.2
    tuning parameters 118.3
    vector field 118.2
Non-linearity 118
Normal conditions 84.2, 84.3, 84.4, 84.5, 84.6, 85.2, 86.2, 87.4, 91.1, 91.2
Normal (Gaussian) distribution 82.4, 83.4, 102.3, 102.7, 102.8
Normal values 22.2, 84.2, 90.5
Nucleonic level measurement 14.8
Null balance 16.6
Null entry 99.2
Numbering convention 2.2
Numerical instability 96.5, 112.6, 115.6
Numerical integration 44.3, 96, 97.13, 112.6, 112.10, 118.3
    see also Euler methods
    see also Runge Kutta method
Nylon tubing 51.4
Nyquist diagram 73.4, 73.6

Object linking 49.5
Object oriented programming
    data abstraction 107.7
    encapsulation 107.7
    message passing 107.7
    modularity 107.7
    polymorphism 107.7

procedural attachments 107.7
procedural inheritance 107.7
root object 107.7
Objective function 103.1, 106.5, 116.3, 116.6, 117.4, 117.7
    maximise profit 103.1, 103.2
    minimise cost 103.1
    see also cost function
Objects (rule based)
    attributes, single/multiple 107.3, 107.4
    nested 107.3
    object-attribute-value triple 107.3
Observer
    characteristic equation 112.6, 112.7
    characteristic polynomial 112.6, 112.7
    convergence 112.5, 112.6
    design 112.5, 112.7
    dynamics 112.6, 112.10
    full order 112.6, 112.7, 112.9
    integration with controller 112.9
    numerical integration/instability 112.6
    observer matrix 112.6
    pole placement 112.7, 112.8
    predictor-corrector 112.6
    reduced order 112.8
    residuals on estimates 112.5, 112.6
    robustness 112.6
    state estimator 112.5, 112.6
    state-space model 112.6
    state vector 112.5
Obsolescence 65.3
Occupancy 56.2, 57.1
Offset 22.3, 27.2, 27.5, 71.7, 71.8, 77.8, 78.4, 86.2, 93.1, 108.3, 108.5, 112.3
Ohm's law analogy 87.1
Oil absorber 51.4
On-line analyser 76.5, 76.10
On-off control 28
One/two dimensional functions 104.1, 104.2
Open
    loop analysis 86.4
    loop gain 22.3
    loop operation 3.4
    process control 49.6, 100.5
    systems 38.10, 49
Operability jacket 34.2

Operating conditions/point,
    see normal conditions
Operating lines, lower and upper 90.5, 90.6
Operation 37.5, 60.4
Operation code (op-code) 9.2, 9.5
Operational amplifier 6.7
Operator
    control program 39.1, 41.2, 42, 58.2
    interface 60.4, 61.5
    involvement 60.3
    role of 58.1
    station 38.4, 39.1, 58.3
Optimisation 103
    constrained 105, 106.5
    methodology 106.3
    non-linear functions 104
    search direction/space 104
    unconstrained 104
    see also constraints, LP and QP
Optimum settings 24.0, 24.3, 24.4
Ordinate 73.3, 102.10
Orifice
    carrier 12.1
    specification 12.6
    tappings 12.7
Orthogonality 101.1, 101.6, 101.8, 104.6
Outliers 83.2, 102.2, 109.7
Output interface 44.6, 46.1
Output scaling 44.5, 45.4
Overdamped response 23.2
Overdetermined system 30.2, 30.3, 30.6, 105.4
Overflow 69.2
Overhead composition control 35.7
Overhead condenser 35.2, 35.10, 90.6
Override control 35.12
Overshoot 72.4
Overt/covert failures 53.8, 57.2, 57.5, 57.6, 57.8
Overtime 61.6, 66.8

Packed sheath 16.3, 16.5
Packing 19.2, 21.2
Pade approximation 71.2, 71.10, 73.11, 74.7
P&I diagram. 2.0, 59.2, 60.2
Parallel axes 102.10
Parallel genetic algorithms
    benefits of 110.8

diffusion GA 110.8
farmer/worker 110.8
global GA 110.8
global/sub populations 110.8
import/export of chromosomes 110.8
islands 110.8
master/slave 110.8
migration GA 110.8
migration paths 110.8
neighbouring nodes 110.8
Parallel operations 37.5
Parallel programs 78.7
Parallelism 29.7, 53.4
Parameter lists (A, B, C and D) 29.4, 37.7, 100.6
Parameter of interest 74.2, 74.4, 74.8, 74.10
Partial batch 37.3, 37.5
Partial fractions 70.2, 70.3, 71.9, 72.1, 73.1, 75.5, 75.6, 78.7
    equating coefficients 70.3
    hidden roots 70.3
    method of residues 70.3
Passive protection 55.2
Payback time 59.3, 59.11
Payments 67.13
Payments, stage 62.4, 64.10
Peak time 72.4
Penalty function 105.14
Performance
    deterioration of 102
    guarantees 67.9
    parameters 67.9
Period of oscillation 72.4, 73.9
Personal computers 39.4
Picture builder 42.5
Phase 37.5
    logic 37.5
    margin 73.12
    plane 80.1
    shift 72.4, 73.1, 73.2, 73.7, 73.9
pH
    control 34.1
    measurement 17, 34.1
    scale 17.1
    scale, quasi 34.3
Physical model 100.4
    control module 37.3

equipment module 37.3
process cell 37.2
unit 37.3
Picture builder 41.8
Pilot valve 5.2, 46.2
Pipe size 13.1, 13.7
PI/PID
    compensator 78.1
    controller 23, 23.9, 24, 25, 26, 73.10, 78.1, 81.6, 81.10, 97.13, 98.3, 98.4
    function block 44.4, 45.3, 48.4
Plant
    interface 60.4
    interface unit 38.4, 51.1
    utilisation 59.7
Plant model mismatch 106.9, 112.2, 112.3, 112.6, 112.9, 117.9, 118.3, 118.9
Platinum resistance thermometer 16.4
Plug and seat assembly 19.2
Pneumatic relay 5.4
Pneumatic actuators 19.6
Pneumatics 5
Pneumercator 14.5
Point count 4.1, 50.10, 59.2
Polar co-ordinates 68.3, 73.2, 73.6, 77.1
Polar plot 73.4
Pole placement 74.11, 112.2, 112.3, 112.7, 112.8
Pole zero cancellation 77.7
Poles, closed loop 74.1, 74.11, 77.5, 112.1, 112.2
Poles, open loop 74.1, 74.8, 77.6
Polygonal lines 102.10
Polynomial, monic 114.1, 114.5, 116.3
Population 82.1
Positioner 21
Positive feedback 71.12
Positive/negative definite 104.2, 104.4, 104.6, 104.7, 105.9, 105.11, 105.12
Potential divider 6.3
Potential energy 88.2
Potentiometer 95.1
Power amplifier 95.1
Power series 68.1, 80.13
    see also time series
Power supply 39.2, 51.2
Pre commissioning, instrumentation 64.4
Pre-processing of data 83.2, 109.7

Prediction horizon 106.11
Preferred supplier 10, 65.1
Present value 59.11
Pressure energy 88.2
Pressurisation 52.9
Preventive action 57.1
Pricing policy 61.4
Primary key 99.2, 99.3
Primary strip 77.1
Prime attributes 99.2
Principal axis 15.2
Principle components 106.4
    analysis. 83.8, 101
    bivariate 101.1, 101.2
    dimensionality 101.4
    dominance 101.6, 101.7, 102.9
    first principal component 101.1, 101.3
    latent variables 102.9
    multivariate 101.3
    scores plot 102.9
Probability 82.3
Probability density function 82.3
Probability distributions 82.3
    cumulative 82.3
    density 59.6, 82.3, 82.4
    joint 82.3
    Maxwell 82.3
    normal (Gaussian) 82.3, 82.4, 83.4
    random 83.4
    Rayleigh 82.3
Probability of failure on demand 53.8
Procedural code 41.5, 61.5, 63.3, 63.14
Procedural model 100.4
    actions 37.5
    operation 37.5
    phase 37.5
    procedure 37.5
    steps 37.5
Procedure 37.5, 60.4, 62.1, 63.3
Process
    cell 37.2
    description 60.1, 62.3, 63.2
    control unit 38.4
    definition of 3.3
    gain 32.2, 32.3, 84.5
    shut-down systems 55.6

Process analytics technology 102
Process model 100.4
    process stages 37.4
    process operations 37.4
    process actions 37.4
Process MRP 100.4
    process costing 100.4
    process definition 100.4
    process monitoring 100.4
    production scheduling 100.4
Processor cycle 9.2, 9.5
Product consistency 59.7
Production plan 100.4
Production planning 37.8
Professional conduct 66.8
Profibus protocol 50.3
Profit 59.10, 61.4, 67.6, 67.10
Program structure 8.6, 97.12
Programmable logic controller 38.6, 47, 57.7
Programming, experience vs efficiency 61.6
Programming, management effort 61.6
Project
    management 59.10
    manager 62.2, 67.7, 67.8, 67.10
    organisation chart 66.7
    plan 61.2, 62.1, 63.14, 66.7
    report 66.8
Project planning
    batch management 100.6
    batch records and evaluation 100.6
    process management 100.6
    process orders 100.6
    process planning 100.6
    quality management 100.6
    recipe management 100.6
    resource management 100.6
Proof testing 53.3, 56.5, 57.4
Proportional action/control 22.1, 22.3, 23.2, 93
Proprietary manifold 12.1
Protection
    IS types 52.6
    systems 53.8, 55.0, 56
    systems design process 56.10
Protocols 40.4
Proximity switch 46.3
Pseudo code 8, 63.4

Psychological model 58.2
Psychrometric systems 94
Pulse 75.1, 75.2
    Laplace transform of 75.2
    train of 75.1, 75.3
    input cards/channels 4.1, 39.3
Pulse transfer function 76.2, 114.1, 116.6, 117.9
    cascaded elements 76.5, 76.6, 78.6
    closed loop 76.7, 76.9, 76.10, 77.2, 77.5, 78.2
    open loop 77.4, 77.5, 77.7, 77.8
    parallel functions 78.7
Pumping system 88.1
Purged flow arrangement 12.7

Quadratic approximation 105.11, 105.12
Quadratic equation/function 68.3, 104.4, 104.5, 104.8, 104.9, 105.9, 105.11
Quadratic programming 105.9, 106.7, 106.8, 106.9, 109.13
    equality constrained 105.12
    feasible/infeasible solutions 105.10
    gradient criterion 105.11
    recursive form of 105.11
    sequential 105.12, 106.6
    set management strategy 105.12
    solution to MPC problem 117.7
    superlinear convergence 105.12
Qualifier 29.6
Quality
    assurance 66, 67.7
    control 66.10
    design 56.10, 66.7
    management system 66.1, 66.6
    manager 66.1
    manual 66.1
    plan 61.2, 66.7
    policy 66.1
    procedures 63.6, 63.8, 66.7
    requirements 66.2
Quasi pH scale 34.3
Query
    aggregate functions 99.4
    dot extensions 99.4
    keywords 99.4
    nested 99.5
    structure 99.4

Quotation generation 61.3

Rack and frame structure 39.2, 51.2
Radial basis function networks
    batch least squares 109.9
    centre 109.8, 109.10
    distance 109.8, 109.10
    Gaussian function 109.8, 109.10
    k-means clustering algorithm 109.9
    multiple linear regression 109.9
    nearest neighbour's method 109.9
    spread parameter 109.8
Radio communications 40.8
Ramp input 69, 70.1, 70.2
Random distribution 83.4
Random failure, components or hardware 53.2, 56.9
Rangeability 10, 34.1
Rare event approximation 54.9
Rate constant 36.1, 89.1
Rate time 23.4
Ratio
    control 26, 33.1, 35.9
    station 26.1
    trimming 33.3
RC network 44.1, 46.3, 69.3
Reaction
    curve method 24.4, 72.6, 78.3, 115.2
    kinetics 91
    rate of 36.1, 89.1, 91.1
Reactor temperature control 36.3
Real data types 7.6
Real-time operating system 41.1
Real-time optimisers 106, 109.13
Realisation 78.1, 78.3, 78.5, 81.4, 81.7, 81.9
Reboiler 35.2, 35.6, 35.7, 35.8, 90.6
Recipe model 100.4
    control recipe 37.7
    general recipe 37.7
    master recipe 37.7
    site recipe 37.7
Recipes 37.7, 41.8
Recovery options 29.9, 55.4, 61.5
Recursive estimation 115.6
    asymptotic sample length 115.4
    bias 115.9

convergence 115.9
convergence/divergence 115.4, 115.6, 115.9
covariance resetting 115.5
covariance windup 115.7
factorisation 115.6
forgetting factors 115.4, 115.8
initialisation 115.2, 115.3
numerical instability 115.6
on-line parameterisation 115.2
order of model 115.1
performance 115.1
prediction error 115.9
round off errors 115.6
stochastic approximation 115.6
Recursive filter 114.1
Red book 67.7, 67.8
Reduced gradient methods 105.13
Reference
    data 102.9
    electrode 17.5
    junction 16.1, 16.3
    mixture 18.9
Reflux drum 35.2, 35.6, 35.9, 35.10, 90.6
Reflux rate/ratio 90.4, 90.5, 90.6
Register, shift 6.6
Regression coefficients 83.3, 83.6, 83.7
Regression models
    simple linear 83.1
    multiple linear 83.1, 83.7, 83.9
    polynomial 83.1
    validation 83.4
    variable selection 83.8
Regulo control 3.4
Relationships, recursive 99.3
Relationships, total 99.3
Relation (table) 99.2
Relational databases 99, 100.2
    design 99.8
    guidelines 99.8
    joins 99.5, 99.6
    updates 99.5
    views 99.5
Relative gain analysis 31.5, 35.2, 111.1
    blending system 111.8
    closed loop gain 111.1
    column, D-V scheme 111.5
    column, L-B scheme 111.4
    column, L-V scheme 111.3, 111.6, 111.7, 111.10
    interpretation 111.2
    open loop gain 111.1
    relative gain/array 111.1
Relative volatility 90.5
Relay 56.5, 57.4
Reliability 53.1, 59.5, 65.1
Reliability of design 56.10
Repairable systems 53.2
Repeatability 10
Replacement 65.4
Reset time 23.3
Residence time 34.1, 69.2, 92.2
Residuals 83.1, 83.3, 83.7
Resistance temperature devices 16.4
Resistance to flow 87.1, 90.1
Resonant frequency 73.6
Resource estimates 61.5, 61.6, 62.3, 63.14, 63.15, 66.8
Response
    closed loop 71.7, 71.8, 71.9, 72.2, 78.3
    second order 72, 73.6, 74.8, 74.9, 74.11, 77.7, 77.8
    steady state 71.6, 71.8, 72.4
Response limit 72.4
Retention time 18.1, 18.7
Reverse acting 22.1, 84.5
Reverse selection 83.8, 83.10
Ribbon cable 39.2
Riccati equation 113.3, 113.4
Ring fencing 59.2, 59.10
Rise time 72.4, 78.4
Risk 53.10, 56.1, 56.5
    acceptable levels 56.2
    commercial 67.1, 67.4, 67.6, 67.8, 67.11
    graph 57.1
Rolling map 42.5
Root locus 74, 77.5, 77.7
    second order systems 74.8
    symmetry 74.4, 74.11
Roots
    complex conjugate 71.9, 71.10, 72.4, 74.1
    imaginary 72.5
    negative real 71.9, 71.10, 72.2

repeated 72.3
stable/unstable 71.9, 71.10
Roots, see also characteristic equation
Rotameter 13.3
Routh array 71.11, 77.4
Routh test 71.11, 77.3
RTD, construction and installation 16.5, 16.6
Rule base (fuzzy)
    antecedents and consequents 108.2, 108.9
    changing rules 108.8, 108.11
    controller, within 108.1
    equilibrium point 108.3
    fuzzy associative memory 108.2
    interactions 108.9
    leading diagonal 108.3
    look-up table 108.2, 108.9
    nesting of rules 108.10
    changing rules 108.8, 108.11
    operability jacket 108.2
    production rules 108.4
    standardised 108.2
    changing rules 108.8, 108.11
Rules (expert)
    antecedents and consequents 107.1, 107.3, 107.4
    assertions 107.3, 107.4
    causes and effects 107.2
    confidence level 107.3
    conditions and outcomes 107.1
    firing of 107.3, 107.4
    meta 107.4
    order 107.4
    premises and actions 107.1
    production rules 107.1, 107.3
    pros and cons 107.3
    salience 107.4
Runge Kutta method 96.4, 98.3
Rungs 47.1
Run time 97.13
Ryskamp scheme 35.8

S-plane 74.4, 74.8, 77.1
Saddle point 104.1, 104.2
Safe
    area 52.2
    failure fraction 53.8, 56.5, 56.8

hold states 29.9, 37.5, 61.5
Safety
    environment 43.4
    equipment approvals 57.1
    integrity levels 53.8, 55.7, 56.3, 56.4, 56.5, 56.6, 56.7, 57.1
    layers of 55
    life cycle 56.3
    related function/signal 54.3, 54.5
    related systems 55.6, 56.2
    requirements specification 56.4
Sample covariance 82.2
Sample size 82.1
Sampled data signal 75.1
Sampled data systems 76, 81.7
Sampler 75.1, 76.5
    imaginary 76.7, 76.9, 81.8
    virtual 76.8
Sampling (stochastic) 102.1
    group size 102.3
    population 82.1
    sample size 82.1, 82.5
    specified band 82.3
Sampling (time series)
    instant 75.1, 76.4, 76.7, 113.1, 114.1, 116.2
    synchronous 76.1, 76.5, 76.8
    period 23.10, 77.7, 78.1, 81.9, 109.7, 112.10, 113.8, 114.1
    system 18.6
Saturation 23.2, 23.6, 23.7
Scaling factor 71.1
Schedule, control/master 100.3
Schema 99.2
Scope of automation 60.3
Scope of project 60.3
Scores plot 102.9
    confidence limit 102.9
    control limits 102.9
    ellipses 102.9
    Hotelling's distance 102.9
    interpretation of 102.9
    Mahalanobis' distance 102.9
Search
    direction/space 104.3, 104.5, 104.8, 105.1, 105.11, 105.12
    efficiency 104.6, 104.7

end criteria 104.5, 105.12
procedure 104.5
steepest descent 104.6, 104.7
Secant method 105.12
Second order systems 72, 73.6, 74.8, 77.7, 77.8, 80.1
see also damping
Secondary strip 77.1
Secondment 66.8
Security 49.8
Seebeck effect 16.1
Segregation policy 51.3, 54.2, 64.4
Selection
parent space value 110.3
simulated roulette wheel 110.3, 110.7
stochastic sampling 110.3
truncation 110.3
universal sampling 110.3
Self adaptive fuzzy control 108.11
performance index 108.11
self-organising 108.11
self-tuning 108.11
Self cleaning 14.5
Self tuning control
certainty equivalence 116.1
closed loop performance 116.5
closed loop response 116.8
control signal 116.3, 116.4, 116.7
deadbeat response 116.5, 116.6
Diophantine identity 116.3
direct/indirect approaches 116.1
explicit and implicit forms 116.1
generalised minimum variance 116.6
h-steps ahead prediction 116.2
implementation 116.4, 116.7
minimum variance control 116.3
notation 116.2
objective function 116.3, 116.6
properties 116.4, 116.8
recursive least squares 116.4, 116.7
separation identity 116.3, 116.6
separation principle 116.1
similarity transformations 116.6
Self-tuning controller 115.2, 115.3, 115.6, 115.7
Semantic nets
associative network 107.5
disadvantages 107.5

links (edges, arcs) 107.5
nodes (vertices) 107.5
Semiconductor 6.1
Sensitivity 10, 26.4
Sensitivity analysis 105.6, 106.8
Separation 35.3
Sequence
check list 64.5
control 29
design 60.3
executive 41.5, 63.9
flow diagram 29.1, 63.4
progression 29.5, 29.6
software 60.4, 63.9
Sequential function chart 29.6, 63.4
Sequential quadratic programming 105.12, 106.6
Series of elements 53.4
Service life 65.3
Servo control 3.4
Servo mechanisms 95
Set point tracking 23.5, 45.3
Sct theory 108
Settling time 72.4, 77.8, 78.4
Shaft, common 95.1, 95.2
Shear beam load cell 15.3
Shut-off 19.2, 21.2
Sidestream 35.13, 35.14
Sight glass 13.2
Signal
types 2.1, 4.0
generic names 3.1
noise 113
selection 33.2
Signal to noise ratio 102.2
Similarity transformation 80.10-80.13, 116.6
Simplex method 103.3
basic variables 103.3
efficiency 103.3
independent variables 103.3
non-basic variables 103.3
order of slack variables 103.3
Simulation, dynami
block oriented approach 97, 98
control loop 97.13
procedural approach 98
process orientated 98.6

see also Matlab, Simulink
Simulink 98
    automatic scaling 98.2
    data import/export 98.1
    dot extensions (.mdl) 98.1, 98.2, 98.3
    element wise multiplication 98.5
    function categories 98.1
    function connections 98.1
    function names 98.2
    function orientation 98.1
    outport 98.2
    polynomial coefficients 98.3
    run time 98.4
    solver 98.1
    state-space function 98.4, 98.5
    subsystems 98.3
    text box 98.2
    toolboxes 98.1
Simulink environment 98.1
    block parameters dialogue box 98.1
    configuration parameters dialogue box 98.1
    library browser 98.1
    model window 98.1
    simulation menu 98.2
Single-input single-output system 71
Single phase flow 13.5
Singular values 111.7
    condition number 111.7
    Morari resilience index 111.7
    singular value decomposition 111.9
    singular value thresholding 117.9
Sinusoidal input 73
Site/works visit 60.6, 66.8
Six sigma 102.7
    capability indices 102.8
    defects 102.7, 102.8
    methodology 102.7
    positive shift 102.8
    service limits 102.8
    specified tolerance 102.7
Shewart charts, see control charts
Slack variables 103.3, 103.4, 105.4, 105.9
Slave loop 25.1, 95.2
Sliding plate valve 18.5
Slots 39.2
Slots and fillers 107.6, 107.7

Smart devices 50, 50.10
Smith predictor 86.4
Soft keys 42.4
Software model, IEC 8.1
Software, see
    application software
    configuration
    function blocks
    procedural code
    resource estimates
    system software
Sonic flow 20.8, 87.1, 87.3
Source of release 52.2
Special purpose PES 57.8
Specific enthalpy 94.4
Speed of response 10, 21.2, 25.1, 69.6, 71.9, 72.3, 95.2
Spray drier 94.1
Spreadsheet, vendor selection 61.7
Spurious failures/trips 53.8, 56.5
Square mean 82.1
Stability
    closed loop. 24.1, 36.1, 36.2, 71.9, 74.1, 77.2, 77.4, 79.9, 89.1, 89.3, 91.4, 118.7
    marginal 71.9, 72.5, 77.2
    numerical 96.5
    Routh test 71.11
Standard
    deviation 82.1, 82.4, 83.10
    error 102.4
    signal ranges 4.2
Standardisation (of data) 82.2, 83.2, 83.8, 83.10
Standby generator 51.2
Standby systems 53.7
Start-up and shut-down 59.5
Starter motor circuit 47.6
State feedback regulators
    characteristic equation 112.1, 112.2
    characteristic polynomial 112.1, 112.2
    closed loop performance 112.1
    closed loop poles 112.1
    controller matrix 112.1, 112.2
    disturbance vector 112.3
    dominant pole positions 112.2, 112.10
    integral action 112.9
    integration with observer 112.9
    multivariable controller 112.2

observer, full/reduced order 112.5, 112.8
output feedback 112.2
pole placement 112.2, 112.3
reference factor 112.3, 112.4
set point for regulo control 112.3
simulation 98.5
time scales 112.10
State space
    feedback control 80.4
    generalised form 79.7, 80.1, 80.3, 90.3, 91.3, 96.1, 96.3, 96.6
    internal variable 80.4
    output equation 80.1, 80.2, 112.6
    solution, MIMO 80.8, 90.3
    solution, SISO 80.7
    state equation 80.1, 80.2
    state vector 112.5, 112.6
    system matrix 80.1
    transfer function conversion 80.3
Static pressure 14.2
Statistical control 102.3, 102.7
Statistical process control 102
Statistics, see summary statistics
Steady state analysis 22.3, 71.6
Steady state gain 22.2, 27.2, 27.3, 71.1, 72.1
Steady state optimiser
    dependent/independent variables 106.1
    detection algorithm 106.4
    disturbances, effect of 106.6
    essential characteristics 106
    filtered outputs 106.1
    frequency of execution 106.6
    methodology 106.3
    model for 106.2
    model updating 106.5
    on/off-line modes 106.1
    open/closed loop modes 106.1
    parameter adjustment 106.5
    set points, downloaded 106.1, 106.6
    scope for adjusting MVs 106.6
    sparsity 106.6
    SQP solution 106.6
    systematic errors 106.5
Steam
    condensing 31.4, 32.1, 85.2, 86.1, 92.1
    ejector 35.10
    injection system 86.1
    superheated 87.5
Steepest descent 104.6, 104.7, 109.3
Step input 69, 70.1, 70.2, 71.2, 71.6, 72.1, 75.4, 77.8, 78.2, 78.3, 78.4, 98.2
Step length/size 78.1, 96, 96.5, 96.7, 97.13, 104.5, 112.10
    variable 96.7, 98.4, 98.5, 104.8, 105.12
    see also sampling period
Step response
    damping factor 72.1
    overdamped 93
    underdamped 72.1
Step response testing
    choice of inputs/outputs 117.9
    external disturbances 117.9
    instrumentation check 117.9
    open loop models 117.9
Steps 29.6, 37.5
Stiff systems 96.5, 96.7
Stirred tank, see agitated vessel
Stochastics 82
Stoichiometric requirements 33.2
Strain gauges 15.2
Strain/resistance effects 15.1
Strategic variable 30.1, 30.6
Stream 37.2
Structure charts 63.2
Structured
    data type 7.13
    programming 63.2
    query language 49.4, 99.4
    text 8, 29.2
Sub contractor 67.8
Sub systems 3.3
Substitution rule 73.2
Summary statistics 82.1
    mean (average) 82.1
    mean square value 82.1
    median 82.1
    square mean 82.1
    standard deviation 82.1
    variance 82.1
Summary statistics (multivariate) 82.2
    correlation coefficient 82.2
    correlation matrix 82.2

covariance matrix 82.2
  sample covariance 82.2
Superposition 3.3, 71.4, 76.1, 84.3, 90.6, 92.1
Supervisory control and data acquisition 38.7
Supplier 59.1, 66.1, 66.7, 66.8, 67.3
Support tools 63.7
Switchgear 95.1
Symbolics 107
System
  architecture 38, 49.3
  cards 39.3
  evolution 65.3
  hardware 39, 60.4, 65.5
  layout 51, 60.3
  modes 80.13
  software 41, 60.4, 61.5, 64.2, 65.5, 66.7
  support 65.2
  upgrades 65.5
System, see also
  control systems
  dynamics
System matrix 80.1
  canonical form 80.11
  Jordan form 80.13
  modal decomposition 80.11, 80.12, 80.13
  modes 80.13
Systematic failures 56.9

Tableau representation 103.4
  entering basic variable 103.4
  leaving basic variable 103.4
  tied variables 103.4
  pivot column/row/value 103.4
Tacho generator 95.2
Tag number 2.0, 42.2, 45.1, 50.8, 99.1
Taxation 59.11
Taylor's series 67.2, 104.1. 104.2, 104.4, 104.8, 105.1, 1 105.11
Telemetry 40.7, 40.9
Temperature
  class 52.4
  measurement 16, 35.5
  profile 18.7, 32.1, 35.5
  resistance effects 15.1, 15.2
Tender analysis 61.7
Tender generation 61.1

Terminal velocity 69.4
Termination rack 39.2
Test
  results sheet 66.7
  signals 64.5
  specifications 63.6, 63.8, 63.10, 63.11, 66.7, 67.8
Testing
  functional 64.5
  physical means 63.9
  simulation means 63.9
Text display 42.9
Thermal capacity 85.2
Thermal system 69
Thermocouple, types and installation 16.2, 16.3
Thermocouple principles 16.1
Thermostat 69.5
Thermosyphon 31.1
Thermowells 16.7
Third party 67.7
Three term controller, see PID controller
Tighter control 59.5
Time
  advance 116.3
  constant 69
  delay 34.1, 71.2, 73.11, 74.7, 77.8, 78.3, 81.1, 86.3, 86.4, 92.2, 92.5, 98.5, 101.7, 106.11, 108.10, 108.12, 114.1, 115.1
  literals (and date) 7.7
  series 75.6, 75.7, 78.1, 78.5, 109.11, 116.3, 117.9
  of the essence 67.11
Timer 47.4
Titration curve 17.2, 17.3, 34.1
Token bus/ring 40.3
Torque 95.1
Top-down approach 63.2, 66.7
Top event 54.7, 54.9, 54.10
Total quality management 66.10
Totality 99.3
Touchscreen 39.1, 42.4
Traceability 66.1, 66.7
Train 37.2
Trajectory 80.1, 80.2
Transfer functions 71
  closed loop 71.2, 71.7, 71.8, 71.9, 79.8, 80.7, 118.3
  delay 71.2

distributed parameter 92.2, 92.3, 92.4, 92.5
first order 71.1
forward path 71.2, 71.8
higher order 72.6
integrator 71.2
lag 71.2
lead 71.2
open loop 71.2, 71.8, 74.1, 118.7
overall 72.6
PID controller 71.3
second order 72.1
state space conversion 80.3
ZOH 75.1
Transient response 69, 90.2
Transistor 6.3
Transit time 13.7
Transportation lag 71.2
Trend diagram 42.7
Trips and interlocks 43.4, 47.1. 55.3
Truncation 54.8, 54.9
Tuple (record) 99.2
Turbine flowmeter 13.5
Turbulence/turbulent flow 87.1, 87.3, 89.2, 89.3
Turnkey project 59.1, 63.16, 67.5
TÜV classes 57.1
Two norm 111.6
Two's complement 7.2

Ultimate period 24.3
Ultrasonic flowmeter 13.7
Ultrasonic level measurement 14.7
Unconstrained optimisation 104
Underdamped response 23.2, 77.8
Unit
    boundary 37.3
    circle (jω domain) 73.12
    circle (z domain) 77.1, 78.1
    physical 37.1, 37.3
Universe of discourse 108.1
Unrepairable systems 53.1
Unsteady state balances 84.4, 85.1
    current 69.3
    energy 88.1, 88.3
    force 69.4
    heat 69.1, 85.2, 85.3, 85.4, 86.1, 86.2, 89.1, 91.2, 92.2, 94.4
    mass 69.2, 86.1, 86.5, 87.5, 94.3
    molar 89.3, 90.1, 90.3, 91.1, 92.5
    volume 84.4, 87.3, 90.2, 93.2, 93.4, 94.2
Upgrade path 65.5
Upthrust 69.4
User requirements
    general 60.5
    particular 60.4
    specification 54.3, 54.5, 60, 61.1, 61.5, 62.1, 66.9, 67.6, 67.7

Vacuum control 31.2
Validation 66.5
Valve characteristics
    equal percentage 20.1
    inherent 20.1, 84.3
    installed 20.2, 20.3, 87.3, 87.4
    linear 20.1
    square root 20.1
Valves 19
    positioners 21
    selection 19.7
    sizing 20.6
    trim 19.2, 20.5
    types 19.1
Vapour flow lags 90.1, 90.6
Variable area meter 13.4
Variable resistance 20.2
Variables
    absolute 22.4, 84.2, 84.3
    deviation 22.4, 68.2, 70.1
    global, external, I/O, memory 8.2, 8.3, 8.4
    normal values 22.4, 68.2, 70.1
Variance 82.1, 101.1
    cumulative percentage 101.4, 101.7
    maximum 101.1
    patterns of 101.6
Variation order 66.7, 67.10
Vectors 79.1
    differentiation by 79.6, 83.7, 101.1, 104.3, 105.9, 117.4
    differentiation of 79.6
    eigen 79.9, 80.12
    initial condition 80.5
    Laplace, transform of 79.7
    orthogonal 101.1, 101.6

Vena contracta 12.2, 12.4
Vendor selection 61.7, 66.9
Venturi 12.2
Verification 63.13
Version control 66.7
Viewing distance 58.2
Virus 49.13
Visual inspection 64.4
Voidage fraction 92.5
Vortex shedding meter 13.8
Voting strategies/systems 53.6, 57.3, 57.8
Vulnerability 56.4, 56.5

Walkthrough, design 54.5, 63.5, 66.7
Waterfall model 63.1, 66.5
Weighted impulses 76.1, 76.4
Wet leg 14.2
Wild variable 30.1
Wiring regulations 51.2
Witnessed tests 63.13
Work in progress 100.2
Working capital 59.11
Working schedule 100.4

X-Y curve/diagram 90.3, 90.6

Yellow book 67.15

Z operator 75.3
Z-plane 77.1
Z transforms 75
    definition 75.3
    inverse 75.6, 78.1, 78.4, 78.5
    prediction 116.2
    properties 75.4
    table 75.4
    use of 75.5
Zeigler and Nichols formulae 24.3, 73.10, 73.11, 74.7
Zener diode 6.2, 46.3, 52.12
Zero capacity systems 86
Zero order hold 44.6, 75.1, 76.8, 78, 81.7
Zeros, open loop 74.1, 77.6
Zone 52.2